Biology of Anaerobic Microorganisms

WILEY SERIES IN
ECOLOGICAL AND APPLIED MICROBIOLOGY

Edited by **Ralph Mitchell**

MICROBIAL LECTINS AND AGGLUTININS: Properties and Biological Activity
David Mirelman Editor

THERMOPHILES: General, Molecular, and Applied Microbiology
Thomas D. Brock Editor

INNOVATIVE APPROACHES TO PLANT DISEASE CONTROL
Ilan Chet Editor

PHAGE ECOLOGY
Sagar M. Goyal, Charles P. Gerba, Gabriel Bitton Editors

BIOLOGY OF ANAEROBIC MICROORGANISMS
Alexander J. B. Zehnder Editor

Biology of Anaerobic Microorganisms

edited by

ALEXANDER J. B. ZEHNDER
Agricultural University
Wageningen, The Netherlands

WILEY

A WILEY-INTERSCIENCE PUBLICATION

JOHN WILEY & SONS

New York • Chichester • Brisbane • Toronto • Singapore

Library of Congress Cataloging-in-Publication Data:

Biology of anaerobic microorganisms.

(Wiley series in ecological and applied microbiology)
''A Wiley-Interscience publication.''
1. Bacteria, Anaerobic. I. Zehnder, Alexander J. B.
II. Series.

QR89.5.B56 1988   589.9   87-28036
ISBN 0-471-88226-7

Printed in the United States of America

10 9 8 7 6 5 4 3 2 1

CONTRIBUTORS

JAN AMESZ Department of Biophysics, Huygens Laboratory, University of Leiden, Wassenaarseweg 78, 2300 RA Leiden, The Netherlands

PATRICIA J. COLBERG Western Research Institute, P.O. Box 3395, University Station, Laramie, WY 82071, USA

JAN DOLFING Swiss Federal Institute for Water Resources and Water Pollution Control (EAWAG), CH-6047, Kastanienbaum, Switzerland

GUY FAUQUE ARBS-Equipe Commune d'Enzymologie, CNRS-CEA, CEN-Cadarache, F-13108-Saint-Paul-Lez-Durance, Cédex, France

WILLIAM C. GHIORSE Department of Microbiology, Cornell University, Ithaca, NY 14853, USA

JAN T. KELTJENS Department of Microbiology, Faculty of Science, University of Nijmegen, Toernooiveld, NL-6525 ED, Nijmegen, The Netherlands

DAVID B. KNAFF Department of Chemistry, Texas Tech University, Lubbock, TX 79409, USA

JEAN LEGALL Biochemistry Department, University of Georgia, Athens, GA 30602, USA

MICHAEL J. McINERNEY Department of Botany and Microbiology, University of Oklahoma, 770 Van Vleet Oval, Norman, OK 73019, USA

MICHAEL T. MADIGAN Department of Microbiology, Southern Illinois University, Carbondale, IL 62901, USA

RONALD S. OREMLAND U.S. Geological Survey, 345 Middlefield Road, Menlo Park, CA 94025, USA

BERNHARD SCHINK Lehrstuhl Mikrobiologie I, Eberhard-Karls-Universität, Auf der Morgenstelle 28, D-7400 Tübingen, Federal Republic of Germany

ADRIAAN H. STOUTHAMER Biological Laboratory, Vrije Universiteit, Postbus 7161, 1007 MC, Amsterdam, The Netherlands

WERNER STUMM Swiss Federal Institute of Technology, Zürich Institute for Water Resources and Water Pollution Control (EAWAG), CH-8600 Dübendorf, Switzerland

JAMES M. TIEDJE Departments of Crop and Soil Sciences and of Microbiology and Public Health, Michigan State University, East Lansing, MI 48824, USA

CHRIS VAN DER DRIFT Department of Microbiology, Faculty of Science, University of Nijmegen, Toernooiveld, NL-6525 ED, Nijmegen, The Netherlands

GODFRIED D. VOGELS Department of Microbiology, Faculty of Science, University of Nijmegen, Toernooiveld, NL-6525 ED, Nijmegen, The Netherlands

FRIEDRICH WIDDEL Mikrobiologie, Fachbereich Biologie, Philipps-Universität, D-3550 Marburg, Federal Republic of Germany

ALEXANDER J. B. ZEHNDER Department of Microbiology, Agricultural University, Hesselink van Suchtelenweg 4, NL-6703 CT Wageningen, The Netherlands

SERIES PREFACE

The Ecological and Applied Microbiology series of monographs and edited volumes is being produced to facilitate the exchange of information relating to the microbiology of specific habitats, biochemical processes of importance in microbial ecology, and evolutionary microbiology. The series will also publish texts in applied microbiology, including biotechnology, medicine, and engineering, and will include such diverse subjects as the biology of anaerobes and thermophiles, paleomicrobiology, and the importance of biofilms in process engineering.

During the past decade we have seen dramatic advances in the study of microbial ecology. It is gratifying that today's microbial ecologists not only cooperate with colleagues in other disciplines but also study the comparative biology of different habitats. Modern microbial ecologists, investigating ecosystems, gain insights into previously unknown biochemical processes, comparative ecology, and evolutionary theory. They also isolate new microorganisms with application to medicine, industry, and agriculture.

Applied microbiology has also undergone a revolution in the past decade. The field of industrial microbiology has been transformed by new techniques in molecular genetics. Because of these advances, we now have the potential to utilize microorganisms for industrial processes in ways microbiologists could not have imagined 20 years ago. At the same time, we face the challenge of determining the consequences of releasing genetically engineered microorganisms into the natural environment.

New concepts and methods to study this extraordinary range of exciting problems in microbiology are now available. Young microbiologists are increasingly being trained in ecological theory, mathematics, biochemistry, and genetics. Barriers between the disciplines essential to the study of modern microbiology are disappearing. It is my hope that this series in Ecological and Applied Microbiology will facilitate the reintegration of microbiology and stimulate research in the tradition of Louis Pasteur.

Anoxic environments are widespread in the biosphere, yet we understand very little about the biological processes occurring in anoxic habitats. In recent years new insights have been gained into the physiology, genetics, and ecological function of anaerobic microorganisms. Alexander Zehnder has been in the forefront of this effort. In this volume he has invited contributions from researchers working in a wide range of disciplines related to the biology of anaerobes. It is my hope that, through the contributions in this volume, readers will gain a better understanding of anaerobic microbial processes and will be stimulated to attempt to answer some of the many unsolved questions about the activities of anaerobes.

RALPH MITCHELL

Cambridge, Massachusetts
October 1987

PREFACE

Anaerobic microorganisms attracted the curiosity of scientists over 100 years ago because it was noted that many of these organisms either improved or spoiled food and drink, or could convert waste into methane. Since the beginning of the twentieth century, an increasing number of anaerobes have been applied to the production of fine and bulk chemicals in industry. In the 1950s and 1960s the interest of most general microbiologists switched from classical fermentation research to the new field of molecular biology, and few remained interested in anaerobic microorganisms and processes. These researchers formed a then small group in which everybody knew everybody. One of the consequences of the energy crisis in 1973 was that many scientists rediscovered the anaerobes and their potential, and research developed around methanogenic and photosynthetic bacteria. At about the same time the growing concern of people about the environment initiated a discussion of the impact of intensive agricultural practice on fertile soil and water quality, thus stimulating research on denitrification. The improvement of cultivation techniques for anaerobes and the discovery of archaebacteria as a new phylogenetic kingdom through study of the 16S rRNA of methanogens led to a burst of research activities with anaerobic microorganisms. Progress in this field is now very rapid and the small group of scientists dealing with anaerobes has become a good-sized community that is steadily growing.

The idea to write a book on fundamental and applied aspects of anaerobes was born in one of many late-night discussions during the First Advanced Course of Microbial Ecology in Kastanienbaum, Switzerland. We mutually agreed that a book on the great diversity of molecular, ecological, and applied aspects of this fascinating group of microorganisms was absolutely needed and we were all convinced that half the work had already been accomplished by the dawn of the new day. But as usually happens with such intentions, the project was put on ice and not pursued further. However, when asked some time later by Trev Leger to edit this book, my initial enthusiasm returned immediately. Luckily, many of the leading scientists (including some from the night-long discussion) who are actively studying various aspects of anaerobic microorganisms became enthusiastic and agreed to contribute chapters to the book. Unfortunately, because of various problems, the book could not be published earlier. I would like to thank all the authors who were on time with their manuscripts for their patience.

It is my hope that the book will not only be useful for those engaged in research with anaerobes and anaerobic processes, but that it will act as a catalyst for students to dedicate themselves to this promising field with its enormous potential for the future.

ALEXANDER J. B. ZEHNDER

Wageningen, The Netherlands
June 1988

CONTENTS

1. GEOCHEMISTRY AND BIOGEOCHEMISTRY OF
    ANAEROBIC HABITATS   1

    Alexander J. B. Zehnder and Werner Stumm

2. MICROBIOLOGY, PHYSIOLOGY, AND ECOLOGY OF
    PHOTOTROPHIC BACTERIA   39

    Michael T. Madigan

3. MOLECULAR MECHANISMS OF BACTERIAL
    PHOTOSYNTHESIS   113

    Jan Amesz and David B. Knaff

4. ECOLOGY OF DENITRIFICATION AND DISSIMILATORY
    NITRATE REDUCTION TO AMMONIUM   179

    James M. Tiedje

5. DISSIMILATORY REDUCTION OF OXIDIZED NITROGEN
    COMPOUNDS   245

    Adriaan H. Stouthamer

6. MICROBIAL REDUCTION OF MANGANESE AND IRON   305

    William C. Ghiorse

7. ANAEROBIC MICROBIAL DEGRADATION OF
    CELLULOSE, LIGNIN, OLIGOLIGNOLS, AND
    MONOAROMATIC LIGNIN DERIVATIVES   333

    Patricia J. Colberg

8. ANAEROBIC HYDROLYSIS AND FERMENTATION OF
    FATS AND PROTEINS   373

    Michael J. McInerney

9. ACETOGENESIS   417

    Jan Dolfing

10. MICROBIOLOGY AND ECOLOGY OF SULFATE- AND
    SULFUR-REDUCING BACTERIA   469

    Friedrich Widdel

11. DISSIMILATORY REDUCTION OF SULFUR COMPOUNDS   587

    Jean LeGall and Guy Fauque

12. BIOGEOCHEMISTRY OF METHANOGENIC BACTERIA   641

    Ronald S. Oremland

13. BIOCHEMISTRY OF METHANE PRODUCTION   707

    Godfried D. Vogels, Jan T. Keltjens, and Chris van der Drift

14. PRINCIPLES AND LIMITS OF ANAEROBIC
    DEGRADATION: ENVIRONMENTAL AND
    TECHNOLOGICAL ASPECTS   771

    Bernhard Schink

    INDEX   847

Biology of Anaerobic
Microorganisms

1

GEOCHEMISTRY AND BIOGEOCHEMISTRY OF ANAEROBIC HABITATS

ALEXANDER J. B. ZEHNDER

Department of Microbiology, Agricultural University, Hesselink van Suchtelenweg 4, NL-6703 CT Wageningen, The Netherlands

WERNER STUMM

Swiss Federal Institute of Technology, Zürich, Institute for Water Resources and Water Pollution Control, CH-8600 Dübendorf, Switzerland

1.1 INTRODUCTION
1.2 REDOX EQUILIBRIA AND ELECTRON ACTIVITY
    1.2.1 Definitions
        1.2.1.1 Electron Activity
    1.2.2 The Electron Free Energy Diagram (EFE) and the Sequence of Redox Reactions
    1.2.3 Microbial Mediation
        1.2.3.1 Nitrogen System
        1.2.3.2 Sulfur System
        1.2.3.3 Iron and Manganese System
        1.2.3.4 Carbon System
    1. 2.4 The Sequence of Microbially Mediated Oxidation and Reduction Reactions
1.3 REDOX REACTION PROGRESS MODELS
1.4 MICROBIOLOGY OF ENVIRONMENTAL REDOX SEQUENCES

1.5   PRINCIPLES OF SYNTROPHIC RELATIONSHIPS
1.6   EVOLUTION OF ANAEROBIC MICROORGANISMS
REFERENCES

1.1   INTRODUCTION

On a global average, the environment is, with regard to a proton and electron balance, in a stationary situation corresponding to a present-day atmosphere of 20.9% O_2, 0.03% CO_2, 79.1% N_2, a world ocean with pH $\approx$ 8 and a redox potential of $E_H = 0.75$ V. This situation is the result of an interplay of different types of reactions, namely geochemical and biochemical processes.

The geochemical processes can best be understood considering the schematic reaction proposed by Sillén (32):

$$\text{igneous rock + volatile substances} \longleftrightarrow \text{air + seawater + sediments} \quad (1)$$

According to this equation, the volatiles (H_2O, CO_2, N_2, HCl, HF, SO_2, CH_4, H_2S, H_2, NH_3) that have leaked from the interior of the earth through different "cracks" (volcaneous, volcanic activities at the oceanic ridges, etc.) have reacted in a giant acid-base and redox reaction with the rocks (silicates, oxides, and carbonates). The result of what is called weathering was a certain kind of atmosphere, seawater, and sediment. Their composition is, over geological time, still changing slightly as a function of volcanic activities. If these activities were to cease, the surface of the earth would reach an equilibrium and all processes would stop. In other words, the earth would be geochemically dead.

However, photosynthesis, quantitatively and qualitatively the most important biochemical process on earth, can be regarded as a local and time-limited reversal in the universal drift toward equilibrium. Photosynthetic reactions capture energy from solar radiation, and photosynthesis may be perceived as a process producing localized centers of highly reduced redox intensities (e.g., organic molecules) and a reservoir of oxygen. The nonphotosynthetic organisms tend to restore the equilibrium through energy-yielding redox reactions (Fig. 1.1). Only the steady shift away from chemical equilibrium as a result of photosynthesis allowed the biosphere to develop and in turn to influence the composition of its environment (i.e., atmosphere, soil, water, and sediments) (20). The distribution and the redox state of the elements in the present-day environment is thus the result of a very sensitive overall balance between a process driven by the chemical heterogeneity of our planet and the impact of solar energy.

On a global average the oxidation states of weathering sources and the oxygen from photosynthesis equal those of soil and sedimentary products. However, the balance might, locally and temporarily, be significantly upset due to the hetero-

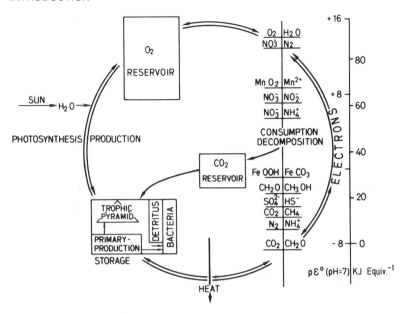

Figure 1.1 Photosynthesis and biochemical cycle. Photosynthesis may be interpreted as a disproportion into an oxygen reservoir and reduced organic matter (biomass containing high-energy bonds made with hydrogen and carbon, nitrogen, sulfur, and phosphorus compounds). The nonphotosynthetic organisms tend to restore equilibrium by catalytically decomposing the unstable products of photosynthesis through energy-yielding redox reactions. The scale of redox couples on the right gives the sequence of the redox reactions observed in an aqueous system. From Stumm and Morgan (35).

geneity of the environment. This heterogeneity can be amplified by various factors, among which the kinetic factors are quantitatively the most important. In well-aerated soils, for example, micro-environments can be found where no oxygen is present and thus anaerobic processes can take place. On a macroscopic scale such a situation should not exist since oxygen abounds. But as a result of the fast rate of oxygen consumption by the biota oxidizing organic material and the slow diffusion of oxygen to the place of its consumption, microscopically and temporarily a situation may occur which is not in equilibrium with the macroscopic (global) environment. Other, larger and more stable anaerobic environments can develop in various sediments. These anaerobic habitats are the result of two processes: (a) the access of oxygen to the sediments is considerably hindered by the water overlaying them because of the low solubility of oxygen in water and the slow diffusion of oxygen within the water body; and (b) part of the photosynthetic organic material produced in the photic zone reaches the sediments far more quickly than does the oxygen. As a consequence of the biological degradation processes, oxygen is depleted within a short time in the surface zones of the sediments. The depth of the oxygenated zone is primarily a function of the input rate of easily degradable organic matter to the sediments. Once oxygen is consumed, a variety of anaerobic oxidation processes will follow in sequence. This sequence is determined by the electron affinity of the electron acceptors. Thus oxygen consumption is followed by the nitrate reduction, sulfate reduction, and methane formation. Environments corresponding to these processes are, according to Berner's (3) classification, oxic, postoxic, sulfidic, and methanic, respectively (Table 1.1). These environments have been shown to succeed one another during early diagenesis.

It is of interest to remember that the processes leading to the formation of anoxic sedimentary environments are partially responsible for the oxygen in the atmosphere. The constant physical separation of the reduced (organics) and oxidized (oxygen) photosynthetic products (e.g., in the sea) leads to a steady-state situation which is, on a global scale, not in equilibrium. This disequilibrium allowed life to proceed on our planet and is the cause of the development of a great variety of organisms and biochemical pathways.

It is the objective of this chapter to discuss the chemical and biological factors that regulate the composition of anoxic habitats. We start by reviewing the equilibrium thermodynamics of redox processes and attempt to illustrate that partial equilibrium models provide insight into the redox chemistry of terrestrial and aquatic systems and into the energetics of microbial processes that mediate these reactions. On the basis of the energetic considerations, we can illustrate what kind of overall microbial redox processes can occur in nature. The sequence of microbially mediated redox reductions from the titration of a local environment with a reductant (e.g., excess organic matter) or with an oxidant is exemplified for lakes, rivers (estuaries), groundwaters, and diagenetic processes in sediments. The microbiology of the different overall redox reactions is discussed. In the light of the various possibilities and limitations of the biochemical pathways, we attempt to

TABLE 1.1 Classification of sedimentary environments, according to Berner (3)

Environment[a]	Characteristic Phases
I. Oxic ($C_{O_2} \geqslant 10^{-6}$)	Hematite, goethite, MnO_2-type minerals; no organic matter
II. Anoxic ($C_{O_2} < 10^{-6}$)	
A. Sulfidic ($C_{H_2S} \geqslant 10^{-6}$)[b]	Pyrite, marcasite, rhodochrosite, alabandite; organic matter
B. Nonsulfidic ($C_{H_2S} < 10^{-6}$)[b]	
1. Post-oxic	Glauconite and other Fe^{2+}–Fe^{3+} silicates (also siderite, vivianite, rhodochrosite); no sulfide minerals; minor organic matter
2. Methanic	Siderite, vivianite, rhodochrosite; earlier formed sulfide minerals; organic matter

[a]C, concentration in mole per liter.
[b]Total sulfide represented by H_2S.

explain some apparent exceptions of the theoretically predicted electron flow in anaerobic communities and the consequences of these deviations for the evolution of living organisms on earth. Finally, we summarize the current view on the evolution of anaerobic microorganisms based on molecular systematics and geological formations.

The various microorganisms and their biochemical pathways are not discussed in detail in this chapter. They are dealt with thoroughly in the succeeding chapters.

1.2   REDOX EQUILIBRIA AND ELECTRON ACTIVITY

Equilibrium considerations can greatly aid our attempts to understand in a general way the microbially mediated redox processes and the redox patterns observed or anticipated in natural soils and waters. In all circumstances equilibrium calculations provide boundary conditions toward which systems must proceed. Most redox processes encountered in aquatic and terrestrial systems are biologically catalyzed or at least indirectly affected by living organisms. This means that the rate with which equilibria is approached depends strongly on the qualitative and quantitative activities of the biota. As a result of high reaction rates and diffusion limitations, oxidation-reduction levels may be established within biotic microenvironments which are quite different from those prevalent in the overall environment. Diffusion and dispersion of products from the microenvironment into the macroenvironment may give an erroneous view of the redox conditions in the latter. Because many redox processes do not couple with one another readily, it is possible to have several different apparent oxidation-reduction levels in the same local. An understanding of the dynamics in natural systems will therefore depend ultimately on the detailed quantitative description of the rates leading to equilibrium, rather than on describing the total or partial equilibrium composition.

1.2.1   Definitions

The "electron level" of elements in various chemical species is given a formal value called the *oxidation state* of the element and is indicated by a roman numeral [e.g., $Fe(+II)$, $S(-II)$, $N(-III)$, etc.]. Table 1.2 gives the formal oxidation states for N, S, and C species of interest in microbial processes.

Biochemical redox processes can conveniently be described by electron transfer; that is, a reductant is a compound that reacts by releasing electrons (an electron donor) and an oxidant is a compound that takes up electrons (an electron acceptor). A "half reaction,"

$$Ox + ne \longleftrightarrow Red \qquad (2)$$

gives together with another half-reaction, $Red_2 \leftrightarrow ne + Ox_2$, the overall redox reaction $Ox + Red_2 \leftrightarrow Red + Ox_2$. For example:

TABLE 1.2 Examples of formal oxidation states, according to Stumm and Morgan (35)

	Nitrogen Compounds		Sulfur Compounds		Carbon Compounds
Substance	Oxidation States	Substance	Oxidation States	Substance	Oxidation States
NH_4^+	$N = -III, H = +I$	H_2S	$S = -II, H = +I$	HCO_3^-	$C = +IV$
N_2	$N = 0$	$S_8(s)$	$S = 0$	$HCOOH$	$C = +II$
NO_2^-	$N = +III, O = -II$	SO_3^{2-}	$S = +IV, O = -II$	$C_6H_{12}O_6$	$C = 0$
NO_3^-	$N = +V, O = -II$	SO_4^{2-}	$S = +VI, O = -II$	CH_3OH	$C = -II$
HCN	$N = -III, C = +II, H = +I$	$S_2O_3^{2-}$	$S = +II, O = -II$	CH_4	$C = -IV$
SCN^-	$S = -I, C = +III, N = -III$	$S_4O_6^{2-}$	$S = +2.5, O = -II$	C_6H_5COOH	$C = -\frac{2}{7}$
		$S_2O_6^{2-}$	$S = +V, O = -II$		

$$4NO_3^- + 24H^+ + 20e^- = 2N_2(g) + 12H_2O \qquad \text{reduction}$$

$$(3)$$

$$5CH_2O + 5H_2O = 5CO_2(g) + 20H^+ + 20e^- \qquad \text{oxidation}$$

$$(4)$$

$$4NO_3^- + 5CH_2O + 4H^+ = 2N_2(g) + 5CO_2(g) + 7H_2O \qquad \text{redox reaction}$$

$$(5)$$

Even if there are no free electrons in solution, we can formulate an equilibrium expression for each half-reaction, for example,

$$\frac{p_{N_2}^{1/2}}{\{NO_3^-\}\{H^+\}^6\{e^-\}^5} = K \qquad \log K = 105.25 \ (25°C) \qquad (6)$$

$$\frac{\{CH_2O\}^{1/4}}{p_{CO_2}^{1/4}\{H^+\}\{e^-\}} = K \qquad \log K = -1.2 \ (25°C) \qquad (7)$$

where K is the equilibrium constant for the reduction process with n electrons.

The half-reaction between the hydrogen ion and hydrogen gas,

$$H^+ + e^- = \tfrac{1}{2}H_2(g) \qquad K = 1, \quad \Delta G° = 0 \qquad (8)$$

has been set by international convention to have a standard free energy change, $\Delta G°$, of zero, which is equivalent to an equilibrium constant of 1. Thus, in a general way, the free energy change of a half-reduction of an oxidant Ox to a reductant Red [equation (3)] is equal to that of a complete redox reaction involving H_2 as a reductant and H^+ as an oxidant.

$$Ox + \frac{n}{2} H_2(g) = Red + nH^+ \qquad (9)$$

$$Ox + ne^- = Red \qquad (2)$$

Both reactions (2) and (9) have thus, per definition, the same $\Delta G°$ and the same K.

The electron can be treated like a ligand in complex formation reactions. Indeed, in *Stability Constants of Metal Ion Complexes* (33) the first inorganic ligand considered is the electron and the equilibrium constants listed there are for reduction reactions corresponding to equations (2) and (9). The free energy changes, equilibrium constants, and standard potentials [see equation (14)] can readily be computed from the free energies of formation of the species involved in the redox reaction, taking into account that the free energy of formation of an electron, as that of a proton, is zero.

1.2.1.1 Electron Activity

The electron can be treated as a redox *component* even if as a species in aqueous solution, it does *not* exist on its own. We can formally and mathematically define an equilibrium *electron activity* for any redox couple as

$$e = \frac{1}{K} \left[\frac{\{\mathrm{Red}\}}{\{\mathrm{Ox}\}} \right]^{1/n} \tag{10}$$

The electron activity is defined in any system where the activities of oxidants [Ox] and reductants [Red] are defined.

$$p\epsilon = -\log \{e\} = p\epsilon^{\circ} + \frac{1}{n} \log \frac{\{\mathrm{Ox}\}}{\{\mathrm{Red}\}} \tag{11}$$

where

$$p\epsilon^{\circ} = \frac{1}{n} \log K \tag{12}$$

Here K is the equilibrium constant for the reduction half-reaction [see equation (10)]. $p\epsilon$ gives the (hypothetical) electron activity at equilibrium and measures the relative tendency of a solution to accept or transfer electrons. In a highly reducing solution the tendency to donate electrons, the hypothetical "electron pressure" or electron activity is relatively large, and $p\epsilon$ is low or negative. On the other hand, a high $p\epsilon$ indicates a relatively high tendency for oxidation.

ΔG of reaction (2) is related to the potential E_H that would be generated at equilibrium at an inert conducting electrode immersed in the solution and coupled with a standard hydrogen electrode, according to the following equation:

$$\Delta G = -nFE_H \tag{13}$$

The redox potentials in published tables refer to reaction (3) with the second half-reaction written as $(n/2) H_2(g) \leftrightarrow nH^+ + ne^-$. Thus $p\epsilon$ is related to the equilibrium redox potential E_H (volts, hydrogen scale) by

$$p\epsilon = E_H / 2.3RTF^{-1} \tag{14}$$

or

$$E_H = \frac{2.3\,RT}{F}\,p\epsilon$$

$$= \frac{2.3\,RT}{F}\frac{1}{n}\left[\log K + \log \frac{\{Ox\}}{\{Red\}}\right] \tag{15}$$

$$= E_H^\circ + \frac{2.3\,RT}{nF}\log \frac{\{Ox\}}{\{Red\}}$$

Equation (15) is known as the Nernst equation. E_H° is the standard redox potential (i.e., the potential that would be obtained if all substances in the reaction were in their standard states of unit activity), R is the gas constant, T the absolute temperature, and F the Faraday constant ($96,490$ C mol^{-1}). At $25°C$, RTF^{-1} is equal to 0.059 V mol^{-1}. For example, reaction (4) can be expressed at equilibrium as

$$p\epsilon = p\epsilon^\circ + \frac{1}{5}\log \frac{\{NO_3^-\}\{H^+\}^6}{p_{N_2}^{1/2}} \tag{16a}$$

or

$$E_H = E_H^\circ + \frac{2.3\,RT}{5F}\log \frac{\{NO_3^-\}\{H^+\}^6}{p_{N_2}^{1/2}} \tag{16b}$$

where $p\epsilon^\circ = \frac{1}{5}\log K = 12.65$ and $E_H^\circ = 0.748$ V.

1.2.2 The Electron Free Energy Diagram (EFE) and the Sequence of Redox Reactions

Only a few elements (i.e., C, H, N, O, S, Fe, and Mn) are predominantly participants of natural redox processes. Table 1.3 presents constants for several couples which are most common in natural environments. The right-hand scale of Fig. 1.1 (enlarged in Fig. 1.2) lists the sequence of redox reactions, calculated for pH 7 (concentration of the protons is at that pH 10^{-7} M), from strong reductants at the bottom to strong oxidants at the top. From such a diagram it can easily be seen that SO_4^{2-} can (thermodynamically speaking) oxidize organic carbon (CH_2O) to CO_2 but also that SO_4^{2-} cannot oxidize NH_4^+ to NO_3^-. Figure 1.2 may be interpreted as an electron free-energy-level diagram. The ordinate measures the energy (expressible in different units, e.g., kcal or kJ mol^{-1} of electrons, eV mol^{-1}, and $p\epsilon$ units) that is required for the transfer of electrons from one free energy level to another.

TABLE 1.3   Equilibrium constants of redox processes pertinent in aquatic conditions (25°C), according to Stumm and Morgan (35)[a]

Reaction	$p\epsilon°$ ($\equiv \log K$)	$p\epsilon°$ (pH)[b]
(1) $\frac{1}{4}O_2(g) + H^+ + e = \frac{1}{2}H_2O$	+20.75	+13.75
(2) $\frac{1}{5}NO_3^- + \frac{6}{5}H^+ + e = \frac{1}{10}N_2(g) + \frac{3}{5}H_2O$	+21.05	+12.65
(3) $\frac{1}{2}MnO_2(s) + \frac{1}{2}HCO_3^-(10^{-3}) + \frac{3}{2}H^+ + e$ $= \frac{1}{2}MnCO_3(s) + H_2O$	—	+8.9[c]
(4) $\frac{1}{2}NO_3^- + H^+ + e = \frac{1}{2}NO_2^- + \frac{1}{2}H_2O$	+14.15	+7.15
(5) $\frac{1}{8}NO_3^- + \frac{5}{4}H^+ + e = \frac{1}{8}NH_4^+ + \frac{3}{8}H_2O$	+14.90	+6.15
(6) $\frac{1}{6}NO_2^- + \frac{4}{3}H^+ + e = \frac{1}{6}NH_4^+ + \frac{1}{3}H_2O$	+15.14	+5.82
(7) $\frac{1}{2}CH_3OH + H^+ + e = \frac{1}{2}CH_4(g) + \frac{1}{2}H_2O$	+9.88	+2.88
(8) $\frac{1}{4}CH_2O + H^+ + e = \frac{1}{4}CH_4(g) + \frac{1}{4}H_2O$	+6.94	−0.06
(9) $FeOOH(s) + HCO_3^-(10^{-3}) + 2H^+ + e$ $= FeCO_3(s) + 2H_2O$	—	−0.8[c]
(10) $\frac{1}{2}CH_2O + H^+ + e = \frac{1}{2}CH_3OH$	+3.99	−3.01
(11) $\frac{1}{6}SO_4^{2-} + \frac{4}{3}H^+ + e = \frac{1}{6}S(s) + \frac{2}{3}H_2O$	+6.03	−3.30
(12) $\frac{1}{8}SO_4^{2-} + \frac{5}{4}H^+ + e = \frac{1}{8}H_2S(g) + \frac{1}{2}H_2O$	+5.25	−3.50
(13) $\frac{1}{8}SO_4^{2-} + \frac{9}{8}H^+ + e = \frac{1}{8}HS^- + \frac{1}{2}H_2O$	+4.25	−3.75
(14) $\frac{1}{2}S(s) + H^+ + e = \frac{1}{2}H_2S(g)$	+2.89	−4.11
(15) $\frac{1}{8}CO_2(g) + H^+ + e = \frac{1}{8}CH_4(g) + \frac{1}{4}H_2O$	+2.87	−4.13
(16) $\frac{1}{6}N_2(g) + \frac{4}{3}H^+ + e = \frac{1}{3}NH_4^+$	+4.68	−4.68
(17) $\frac{1}{2}(NADP^+) + \frac{1}{2}H^+ + e = \frac{1}{2}(NADPH)$	−2.0	−5.5[d]
(18) $H^+ + e = \frac{1}{2}H_2(g)$	0.0	−7.00
(19) Oxidized ferredoxin + e = reduced ferredoxin	−7.1	−7.1[d]
(20) $\frac{1}{4}CO_2(g) + H^+ + e = \frac{1}{24}(glucose) + \frac{1}{4}H_2O$	−0.20	−7.20[d]
(21) $\frac{1}{2}HCOO^- + \frac{3}{2}H^+ + e = \frac{1}{2}CH_2O + \frac{1}{2}H_2O$	+2.82	−7.68
(22) $\frac{1}{4}CO_2(g) + H^+ + e = \frac{1}{4}CH_2O + \frac{1}{4}H_2O$	−1.20	−8.20
(23) $\frac{1}{2}CO_2(g) + \frac{1}{2}H^+ + e = \frac{1}{2}HCOO^-$	−4.83	−8.33

[a] In this table the standard free energy of formation of CH_2O is the same as for dissolved unhydrolyzed formaldehyde.

[b] Values for $p\epsilon°$ (pH) apply to the electron activity for unit activities of oxidant and reductant in neutral water, that is, at pH 7.0 for 25°C.

[c] These data correspond to $(HCO_3^-) = 10^{-3} M$ rather than unity, so are not exactly $p\epsilon°$(pH); they represent typical aquatic conditions more nearly than $p\epsilon°$(pH) values do.

[d] From Thauer et al. (37).

If we add electrons (in the form of reductants, i.e., electron complexes) to a system containing several redox couples, the lowest unoccupied electron levels will be filled up first, followed sequentially by the higher levels, that is, the oxidized species will be reduced in sequence, beginning with the species with the lowest unoccupied electron level. By following this sequence, electron-titration diagrams can be prepared. For example, a reductant such as organic carbon (e.g.,

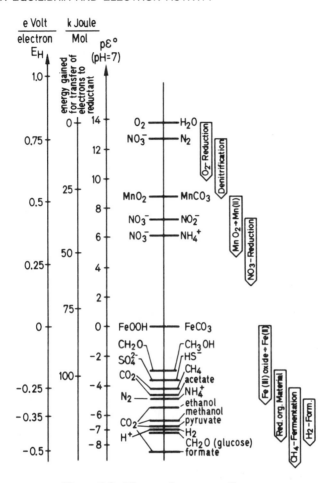

Figure 1.2 Electron-free energy diagram.

CH_2O) will, from a thermodynamic point of view (but not necessarily from a kinetic point of view), first react with O_2, then successively with NO_3^- and MnO_2. The energy gained in such processes per mole of electron transferred can be read from the energy scale: $\Delta G = 2.3RT(p\epsilon_2 - p\epsilon_1)$.

The equilibrium constants given in Table 1.3 or the electron-free energy diagrams can be used to predict what reactions are thermodynamically possible. For example, at pH 7 we see that NO_3^- can oxidize HS^- to SO_4^{2-}, a reaction that is actually catalyzed by *Thiobacillus denitrificans*:

$$\tfrac{1}{5} NO_3^- + \tfrac{6}{5} H^+ + e = \tfrac{1}{10} N_2(g) + \tfrac{3}{5} H_2O$$

$$\log K = 21.05 \tag{17}$$

$$\tfrac{1}{8} HS^- + \tfrac{1}{2} H_2O = \tfrac{1}{8} SO_4^{2-} + \tfrac{9}{8} H^+ + e$$

$$\log K = -4.25 \tag{18}$$

$$\overline{\tfrac{1}{5} NO_3^- + \tfrac{1}{8} HS^- + \tfrac{3}{40} H^+ = \tfrac{1}{8} SO_4^{2-} + \tfrac{1}{10} H_2O}$$

$$\log K = 16.8, \ \Delta G^\circ = -95.87 \text{ kJ mol}^{-1} \ (25^\circ C) \tag{19}$$

or at pH $= 7$,

$$\Delta G^{\circ\prime} (pH \ 7) = -95.79 \text{ kJ mol}^{-1}$$

Thermodynamic considerations can do more than just decide which reactions are thermodynamically possible. From purely energetic considerations, much insight into the catalysis and mediation of known and yet-to-be-discovered redox processes by microorganisms can be gained.

1.2.3   Microbial Mediation

The organisms use the energy of the redox reaction both to synthesize new cells and to maintain the cells already formed (26). The energy exploitation is, of course, not 100% efficient; only a proportion of the free energy released can become available for cell use. Energetic efficiencies of the order of 10 to 50% are typical for many microbially mediated redox processes. It is important to keep in mind that organisms cannot carry out gross reactions that are thermodynamically not possible. From a point of view of overall reactions, these organisms act only as redox catalysts. Since, for example, SO_4^{2-} can be reduced only at a given $p\epsilon$ or redox potential, an equilibrium model characterizes the $p\epsilon$ range in which reduction of sulfate is possible. It has to be remembered that the redox situation inside a microbial cell does not necessarily reflect the $p\epsilon$ of the outside. Therefore, organisms can, for example, reduce sulfate for assimilatory purposes at high outside redox potentials. However, where dissimilatory processes are considered, as in this chapter, $p\epsilon$ is the parameter that characterizes the ecological milieu and thus the possibility of organisms to catalyze particular reactions. The $p\epsilon$ range in which certain oxidation or reduction reactions are possible can be estimated by calculating equilibrium composition as a function of $p\epsilon$. This has been done for nitrogen, manganese, iron, sulfur, and carbon at pH 7. The results are shown in Fig. 1.3.

1.2.3.1   Nitrogen System

Figure 1.3a shows relationships among general oxidation states of nitrogen as a function of $p\epsilon$ for a total atomic concentration of nitrogen-containing species equal

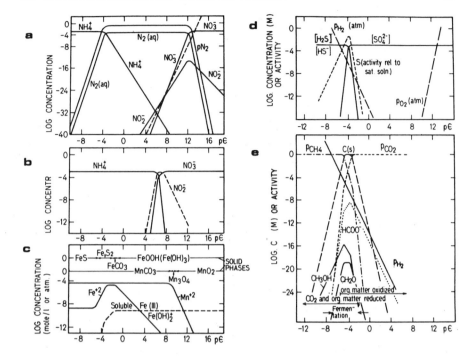

Figure 1.3 Equilibrium concentrations of biochemically important redox components as a function of $p\epsilon$ at pH of 7.0. (*a*) Nitrogen. (*b*) Nitrogen, with elemental nitrogen N_2 ignored. (*c*) Iron and Manganese. (*d*) Sulfur. (*e*) Carbon. These equilibrium diagrams have been constructed from equilibrium constants listed in Table 1.4A for the following concentrations: C_T (total carbonate carbon) $= 10^3 M$; $[H_2S(aq)] + [HS^-] + [SO_4^{2-}] = 10^{-3} M$; $[NO_3^-] + [NO_2^-] + [NH_4^+] = 10^{-3}$; $p_{N_2} = 0.78$ atm and thus $[N_2(aq)] = 0.5 \times 10^{-3} M$. For the construction of (*b*) the species NH_4^+, NO_3^-, and NO_2^- are treated as metastable with regard to N_2. From Stumm and Morgan (35).

to $10^{-3} M$. Maximum $N_2(aq)$ concentration is therefore 5×10^{-4}, corresponding to a p_{N_2} of about 0.77 atm. For most of the aqueous range of $p\epsilon$, N_2 gas is the most stable species, but at quite negative $p\epsilon$ values, ammonia becomes predominant and nitrate dominates for $p\epsilon$ greater than $+12$ at pH 7. The fact that nitrogen gas has not been converted largely into nitrate under prevailing aerobic conditions at the land and water surfaces indicates a lack of efficient biological mediation of this reaction in both directions. It appears, then, that denitrification must occur by an indirect mechanism such as reduction of NO_3^- to NO_2^- followed by a chain of reactions to produce N_2 and N_2O. Because of the nonreversibility in the biological mediation of the $NO_3^- \leftrightarrow N_2$ conversion, the NO_3^-–N_2 couple cannot be used as a reliable redox indicator. For example, NO_3^- may be reduced to N_2 in soil and aquatic systems even *if* the bulk phase contains dissolved oxygen. The reduction may occur in a microenvironment with a $p\epsilon$ value lower than that of the bulk (e.g., in the cell, on the inside of a floc, or within the sediments); the N_2 released to the aerobic bulk phase is not reoxidized, although that is thermodynamically feasible (35).

Several workers have suggested the possibility of aerobic denitrification, largely on the basis that nitrate reductase was present in microorganisms grown aerobically (18, 22). Recently, Robertson and Kuenen (29) not only detected the presence of the appropriate enzyme but also demonstrated the production of nitrogen-containing gases from nitrate by *Thiosphaera pantotropha* at dissolved oxygen concentrations up to 90% of air saturation. This organism was able, concomitantly, to respire oxygen and nitrate. Since many indications are given throughout the literature that denitrification may occur in perfectly well aerated systems, the findings of Robertson and Kuenen is perhaps only the tip of the iceberg.

Besides the dissimilatory production of dinitrogen from nitrate, nitrate can also be reduced to ammonium. The latter reduction is very common for assimilatory purposes; however, some fermentative bacteria are able to gain energy from a dissimilative reduction of nitrate to ammonium (12, 13, 36). Because reduction of N_2 to NH_4^+ (N_2 fixation) at pH 7 can occur to a substantial extent when $p\epsilon$ is less than about -4.5, the level of $p\epsilon$ required is not as negative as for the reduction of CO_2 to CH_2O. It is not surprising, then, that blue-green algae are able to mediate this reduction at the negative $p\epsilon$ levels produced by photosynthetic light energy. What is perhaps surprising is that nitrogen fixation does not occur more widely among photosynthetic organisms and proceeds to a greater extent than CO_2 reduction. Biochemical and evolutionary restrictions must play an important role since nitrogen fixation is limited to prokaryotes (eubacteria and archaebacteria). It is not known why eukaryotes or eukaryotic organelles [which are probably of prokaryotic descendants (45)] have lost or never developed a nitrogen-fixing system. Because of this restriction, it might also be useful to consider a system in which NO_3^-, NO_2^-, and NH_4^+ are treated as components metastable with respect to gaseous N_2. A diagram for such a system (Fig. 1.3*b*) shows the shifts in relative predominance of the three species all within the rather narrow $p\epsilon$ range of 5.8 to 7.2. That each species has a dominant zone within this $p\epsilon$ range would seem to be a contributing factor to the observed highly mobile characteristics of the nitrogen cycle.

1.2.3.2 Sulfur System

The reduction of SO_4^{2-} to H_2S or HS^- provides a good example of the application of equilibrium considerations to natural systems. Figure 1.3*d* shows relative activities of SO_4^{2-} and H_2S at pH 7 and 25°C as a function of $p\epsilon$ when the total concentration of sulfur is 1 mM. It is apparent that significant reduction of SO_4^{2-} to H_2S at this pH requires $p\epsilon < -3$. The biological enzymes that mediate this reduction with the oxidation of organic matter must then operate at or below this $p\epsilon$. Because the system is dynamic rather than static, only an upper bound can be set in this way, for the excess driving force in terms of $p\epsilon$ at the mediation site is not indicated by equilibrium computations. Since, however, many biologically mediated reactions seem to operate with relatively high efficiency for utilizing free energy, it appears likely that the operating $p\epsilon$ value is not greatly different from the equilibrium value. Combining equations (11) and (14) of Table 1.3 gives

$$SO_4^{2-} + 2H^+ \, (pH \, 7) + 3H_2S = 4S(s) + 4H_2O \qquad \log K \, (pH \, 7) = 4.86$$

(20)

This equation indicates a possibility of formation of solid elemental sulfur during the reduction of sulfate at pH 7 and standard concentrations. A concentration of 1 M is unusual. However, if the concentration of SO_4^{2-} is reduced to about 0.018 M (about 1600 mg L^{-1}) and the H_2S activity is taken correspondingly as 0.09 atm (H_2S about half-ionized at pH 7, and the solubility of H_2S at 1 atm is about 0.1 M; thus this condition means about 0.018 M total sulfide), solid sulfur cannot form thermodynamically by this reaction at pH 7.

The solubility of $CaSO_4(s)$ is about 0.016 M at 25°C. According to the foregoing rough calculation, sulfur should form during the reduction of SO_4^{2-} in saturated $CaSO_4(s)$ only if the pH is somewhat below 7. There are some indications that this conclusion agrees with the condition of natural sulfur formation. Elemental sulfur may be formed, however, as an intermediate or as a metastable phase under many natural conditions (15, 17, 34).

1.2.3.3 Iron and Manganese System
In constructing Fig. 1.3c, solid FeOOH ($\Delta G^{\circ\prime} = -410$ kJ mol^{-1}) has been assumed as stable ferric oxide. Although thermodynamically possible, magnetite [$Fe_3O_4(s)$] has been ignored as intermediate in the reduction of ferric oxide to Fe(II). As Fig. 1.3c shows, in the presence of O_2, $p\epsilon < 11$, aqueous iron and manganese are stable only as solid oxidized oxides. Soluble forms are present at concentrations below 10^{-9} M. The concentration of soluble iron and manganese, as Fe^{2+} and Mn^{2+}, increases with decreasing $p\epsilon$, the highest concentrations being controlled by the solubility of $FeCO_3(s)$ and $MnCO_3(s)$, respectively ({ HCO_3^- } $= 10^{-3}$ M has been assumed for the concentration of the diagram).

1.2.3.4 Carbon System
A great number of organic compounds are synthesized, transformed, and decomposed—mostly by microbial catalysis—continually. For the operation of the carbon cycle, degradation is just as important as synthesis. With the exception of CH_4, no organic solutes encountered in natural waters are thermodynamically stable. For example, the disproportionation of acetic acid

$$CH_3COOH = 2H_2O + 2C(s) \qquad \log K = 18 \qquad (21)$$

$$CH_3COOH = CH_4(g) + CO_2(g) \qquad \log K = 9 \qquad (22)$$

is thermodynamically favored, but probably prevented by slow kinetics. Similarly, formaldehyde is unstable with respect to its decomposition into carbon (graphite) and water,

$$CH_2O(aq) = C(s) + H_2O \qquad \log K = 18.7 \qquad (23)$$

but there is no evidence that this reaction occurs.

Even though reversible equilibria cannot be attained at low temperatures, it is of considerable interest to compare the equilibrium constants of the various steps in the oxidation of organic matter. The compounds CH_4, CH_3OH, CH_2O, and $HCOO^-$ given in the diagram (Fig. 1.3e) represent organic material with formal oxidation states of $-IV$, $-II$, 0, and $+II$, respectively. The diagram has been constructed for the condition $p_{CH_4} + p_{CO_2} = 1$ atm. This condition is, however, not fulfilled in the pe range where $[C(s)] = 1$. The major feature of the equilibrium carbon system is simply a conversion of predominant CO_2 to predominant CH_4 with a halfway point at $pe = -4.13$. At this pe value, where the other oxidation states exhibit maximum relative occurrence, formation of graphite is thermodynamically possible.

Methane fermentation may formally be considered a reduction of CO_2 to CH_4; this reduction may be accompanied by oxidation of any one of the intermediate oxidation states. This statement does not imply a mechanism. Methane may be formed directly, for example, from acetic acid: $CH_3COOH \rightarrow CH_4 + CO_2$. This reaction could be classified (thermodynamically) as the sum of the reactions $CH_3COOH + 2H_2O = 2CO_2 + 8H^+ + 8e$; $CO_2 + 8H^+ + 8e = CH_4 + 2H_2O$, which very recently has been shown to occur at temperatures around 60°C, each reaction in a separate bacterium (56). As discussed in Chapters 12 and 13, physiologically different organisms are typically involved in methanogenesis.

Alcohol fermentation may be exemplified by redox disproportionation of CH_2O (or $C_6H_{12}O_6$):

$$CH_2O + CH_2O + H_2O = CH_3OH + HCOOH^- + H^+ \qquad (24)$$

$$CH_2O + 2CH_2O + H_2O = 2CH_3OH + CO_2(g) \qquad (25)$$

$$C_6H_{12}O_6 = 2C_2H_5OH + 2CO_2(g) \qquad \Delta G° = -58.3 \text{ kcal} \qquad (26)$$

The reduction of CH_2O to CH_3OH can occur at $pe < -3$. Because the concomitant oxidation of CH_2O to CO_2 has $pe°(W) = -8.2$, there is no thermodynamic problem.

1.2.4   The Sequence of Microbially Mediated Oxidation and Reduction Reactions

Although, as has been stressed, conclusions regarding chemical dynamics may not generally be drawn from thermodynamic considerations, it appears that all the

TABLE 1.4A Reduction and oxidation reactions that may be combined to result in biologically mediated exergonic processes (pH 7), according to Stumm and Morgan (35)[a]

Reduction	$p\epsilon°$ (pH = 7) log K (pH = 7)	Oxidation	$p\epsilon°$ (pH = 7) − log K (pH = 7)
(A) $\frac{1}{4}O_2(g) + H^+(pH) + e = \frac{1}{2}H_2O$	+13.75	(L) $\frac{1}{4}CH_2O + \frac{1}{4}H_2O = \frac{1}{4}CO_2(g)$ $+ H^+(pH) + e$	−8.20
(B) $\frac{1}{5}NO_3^- + \frac{6}{5}H^+(pH) + e = \frac{1}{10}N_2(g) + \frac{3}{5}H_2O$	+12.65	(L-1) $\frac{1}{2}HCOO^- = \frac{1}{2}CO_2(g) + \frac{1}{2}H^+(pH) + e$	−8.73
(C) $\frac{1}{2}MnO_2(s) + \frac{1}{2}HCO_3^-\,(10^{-3}) + \frac{3}{2}H^+(pH)$ $+ e = \frac{1}{2}MnCO_3(s) + H_2O$	+8.9	(L-2) $\frac{1}{2}CH_2O + \frac{1}{2}H_2O = \frac{1}{2}HCOO^-$ $+ \frac{3}{2}H^+(pH) + e$	−7.68
(D) $\frac{1}{8}NO_3^- + \frac{5}{4}H^+(pH) + e = \frac{1}{8}NH_4^+ + \frac{3}{8}H_2O$	+6.15	(L-3) $\frac{1}{2}CH_3OH = \frac{1}{2}CH_2O + H^+(pH) + e$	−3.01
(E) $FeOOH(s) + HCO_3^-\,(10^{-3}) + 2H^+(pH) + e$ $= FeCO_3(s) + 2H_2O$	−0.8	(L-4) $\frac{1}{2}CH_4(g) + \frac{1}{2}H_2O = \frac{1}{2}CH_3OH$ $+ H^+(pH) + e$	+2.88
(F) $\frac{1}{2}CH_2O + H^+(pH) + e = \frac{1}{2}CH_3OH$	−3.01	(M) $\frac{1}{8}HS^- + \frac{1}{2}H_2O = \frac{1}{8}SO_4^{2-} + \frac{9}{8}H^+(pH) + e$	−3.75
(G) $\frac{1}{8}SO_4^{2-} + \frac{9}{8}H^+(pH) + e = \frac{1}{8}HS^- + \frac{1}{2}H_2O$	−3.75	(N) $FeCO_3(s) + 2H_2O = FeOOH(s)$ $+ HCO_3^-\,(10^{-3}) + 2H^+(pH) + e$	−0.8
(H) $\frac{1}{8}CO_2(g) + H^+(pH) + e = \frac{1}{8}CH_4(g) + \frac{1}{4}H_2O$	−4.13	(O) $\frac{1}{8}NH_4^+ + \frac{3}{8}H_2O = \frac{1}{8}NO_3^- + \frac{5}{4}H^+(pH) + e$	+6.16
(J) $\frac{1}{6}N_2 + \frac{4}{3}H^+(pH) + e = \frac{1}{3}NH_4^+$	−4.68	(P) $\frac{1}{2}MnCO_3(s) + H_2O = \frac{1}{2}MnO_2(s)$ $+ \frac{1}{2}HCO_3^-\,(10^{-3}) + \frac{3}{2}H^+(pH) + e$	+8.9

[a]In this table the standard free energy of formation of CH_2O is the same as for dissolved unhydrolyzed formaldehyde.

17

TABLE 1.4B Combination of appropriate oxidation and reduction processes giving the more important redox processes mediated by bacteria[a]

Examples	Combination	$-\Delta G°$ kJ/equiv
Aerobic respiration	(A) + (L)	125.1
Denitrification	(B) + (L)	118.8
Nitrate reduction	(D) + (L)	81.8
Fermentation	(F) + (L)	29.6
Sulfate reduction	(G) + (L)	25.4
Methane fermentation	(H) + (L)	23.2
N Fixation	(J) + (L)	20.1
Sulfide oxidation	(A) + (M)	99.8
Nitrification	(A) + (O)	43.3
Ferrous oxidation	(A) + (N)	82.9
Mn(II) oxidation	(A) + (P)	27.7

[a]All calculations were done for the exchange of one electron.

reactions discussed in the preceding section, except possibly those involving $N_2(g)$ and $C(s)$, are biologically mediated in the presence of suitable and abundant biota. In Table 1.4A the oxidation and reduction reactions that may be combined to result in exergonic processes are summarized. The combinations listed (Table 1.4B) represent the well-known reactions mediated by hetrotrophic and chemoautotrophic organisms. It appears that in natural habitats, organisms capable of mediating the pertinent redox reactions are nearly always found.

In the absence of light in a closed aqueous system (e.g., a bottle) with organic matter (CH_2O) and nutrients in sufficient quantities, the organisms present will oxidize the organic molecules according to equation (L) in Table 1.4A. This equation represents not only a practical model but describes in general terms the actual reactions taking place in living things. There, organic compounds (e.g., sugars) are first oxidized by a variety of redox reactions to CO_2 (e.g., glycolysis followed by the tricarboxylic acid cycle). The "liberated" electrons are transferred to different electron carriers (coenzymes). The most common carriers are NAD (nicotinamide adenine dinucleotide) and FAD (flavin adenine dinucleotide). Both accept protons as well as electrons. Since these coenzymes are present in only limited amounts in the cell, they have to be regenerated (i.e., reoxidized) in order to keep the process of organic carbon oxidation running. The regeneration takes place by complex redox reactions which are in the cell locally separated from pathways of the organic carbon oxidation (54). Depending on the organisms and the electron carriers, different molecules can act as terminal electron acceptors (Fig. 1.4). In our bottle, oxygen will first be utilized as an electron sink [equation (A), Table 1.4A], since it has the highest affinity for electrons (pe at pH 7 = 13.8). The combination of equations (A) and (L) describes in general terms the aerobic respiration.

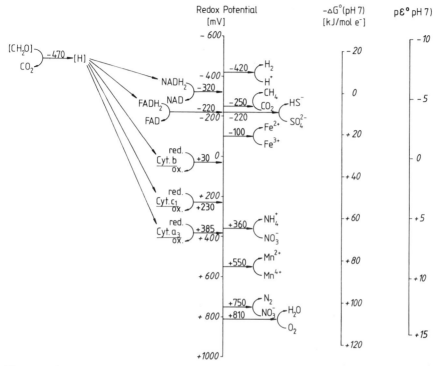

Figure 1.4 Electron free energy diagram for the biologically mediated redox sequence with organic carbon (CH_2O) acting as electron donor. Calculated for standard conditions at pH 7. (CH_2O) represents one-sixth of glucose (i.e. -153 kJ mol^{-1}).

When all the oxygen has been utilized (the bottle is closed!), the system becomes anoxic or anaerobic (i.e., oxygen is no longer available in free form). The population of microorganisms now must look for other electron acceptors. From an energetic standpoint, intrate is the next-best acceptor. It will first be reduced to molecular nitrogen during the process of denitrification [equation (B), Table 1.4A], and probably later, exceptionally, to ammonium [nitrate ammonification, equation (D)]. After denitrification, reduction of MnO_2 should occur. Reduction of $FeOOH(s)$ to Fe^{2+} should follow nitrate ammonification. When sufficiently negative $p\epsilon$ levels have been reached, fermentation reactions and reduction of SO_4^{2-} and CO_2 may occur almost simultaneously. Fermentations [equation (F), not indicated in Fig. 1.4] are processes in which, in general, part of the organic molecule to be oxidized acts as electron acceptor. The result of a typical fermentation are products which are more, and others which are less, oxidized than the original substrate. A good example is the alcohol fermentation: glucose with an overall oxidation state of the carbon of zero is converted to CO_2 and ethanol with an oxidation state of the carbon of $+IV$ and $-II$, respectively (equation 26). The sequence of the redox reactions in our bottle describes in essence the successive filling up of the lowest unoccupied electron level present with electrons from the reductant CH_2O, or in other words, that more-energy-yielding reactions take precedence over processes that are less energy yielding. This sequence of reactions is, of course, not catalyzed by one organism. Rather, it indicates the potential of a mixed microbial population to adapt to different electron acceptors, which means that depending on the acceptor, distinct members of the population become active (see below).

1.3 REDOX REACTION PROGRESS MODELS

In the following we would like to discuss models that can be used in combination with thermodynamic data to predict qualitatively and quantitatively the formation of redox gradients and the relative stability of these gradients. Reaction progress diagrams are useful for showing the effects of initial reactions on the reaction path, on the appearance or disappearance of stable or metastable solid phases, and on the redistribution of aqueous species. Usually, these diagrams are based on the concept of *partial equilibrium*. Partial equilibrium describes a state in which a system is in equilibrium with respect to at least one process (or reaction) but out of equilibrium with respect to others. An irreversible process that involves a series of successive partial equilibrium states may result in a state of *local equilibrium* for the system, that is, a state in which no mutually incompatible phases are in contact, even though the system as a whole is not in equilibrium (39).

The formal equations given in Table 1.4A can be modified to develop stoi-

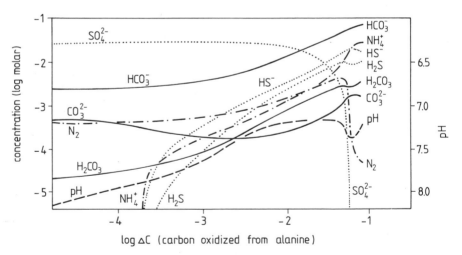

log ΔC (carbon oxidized from alanine)

Figure 1.5 Distribution of predominant dissolved species during the decomposition of alanine ($C_3H_7O_2N$) in seawater at 25°C. The molar concentrations of dissolved species are given as a function of ΔC, the number of moles of organic carbon reacted per liter of solution. Redrawn from Thorstenson (40).

chiometric models that predict how the components of anoxic water will change as a result of the input of organic matter. Such stoichiometric models have been developed for settling plankton by Richards, who corroborated the validity of these models for a large number of anoxic basins and fjords (28).

Thorstenson (40) has predicted in a reaction path calculation the compositional changes in the aqueous phase as a function of the amount of organic matter decomposed. Figure 1.5 gives the distribution of predominant dissolved species at hypothetical equilibrium as a function of increments of organic matter decomposed, represented in this example by $C_3H_7O_2N$ (= alanine). After oxygen dissolved in water has become exhausted, the exodation of organic matter continues, with sulfate serving as oxidant, until the sulfate is entirely used up. The reduced

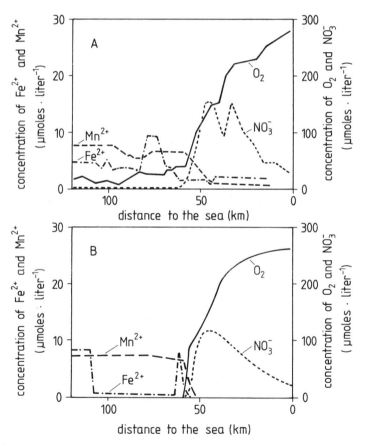

Figure 1.6 Concentration profiles of redox species in the Scheldt estuary (July 1974). (*a*) Experimental. (*b*) According to a computed equilibrium model relating observed organotropic activity and redox intensity. Modified from Billen and Verbeustel (6).

products of the oxidation of organic matter accumulate in the water in addition to the products of this oxidation.

The water quality in highly polluted rivers and estuaries can be described by considering the relations between the balance of the various redox processes, the activity of the bacteria that mediate these redox processes, and the hydrodynamic processes. Billen et al. (5) and Wollast (50) have pioneered the modeling of biological and chemical processes in estuaries and coastal zones. An example of such an application to the Scheldt estuary is briefly reviewed here. This estuary (120 km long) is heavily polluted above Antwerp (km 80) by organic matter. The deterioration of chemical and biological properties is the result of the intense heterotrophic activity, which degrades the organic charge. When the overall rate of oxidant consumption—assimilable to the organotrophic activity (rate of electron transfer)—is higher than the rate of oxidant import, the $p\epsilon$ or redox potential, decreases. If an internal equilibrium exists between the various redox couples involved (i.e., if the chemolithotrophic activities are high enough to restore it), the process resembles a redox titration of the oxidants at the rate of organotrophic activity. This idea has been applied by Billen (4) for calculating the longitudinal

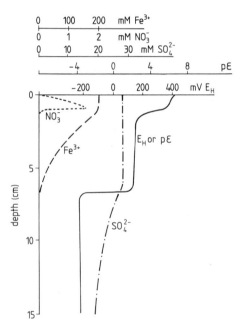

Figure 1.7 Calculated profiles of E_H, or $p\epsilon$, and some redox species as a function of depth in marine sediments. The following assumptions have been made: D (diffusion coefficient) $= 10^5$ cm^2 s^{-1}. The organotrophic activity, rate of electron transfer in the top layer is 10×10^{-11} equivalents per cubic centimeter and second; it decreases exponentially with depth to 0.2×10^{-11} equivalents per cubic centimeter and second. Modified from Billen and Verbeustel (6).

distribution of all redox species in the estuary after heavy discharges of organic matter by the effluent Rupel and the town of Antwerp.

Given the experimentally determined profile of organotrophic activity expressed as a rate of oxidant consumption, the model first calculates the longitudinal profile as a function of F, defined as a weighted sum of all oxidants [$F \equiv i\epsilon\nu_i(\text{Ox})$, where ν_i is the number of electrons susceptible of capture by the oxidant i] and then calculates the distribution of the individual oxidants, resting on the assumption of internal equilibrium, on the one hand, and on conservation equations for each redox couple, on the other hand. One profile obtained for a summer situation is shown in Fig. 1.6 and compared with experimental data.

As has been shown by Berner (2), Billen (4) and others, the same model can also be applied for calculating vertical profiles of redox components in sediments. A calculated profile [Billen and Verbeustel] (6)] for conditions of the North Sea is given in Fig. 1.7.

There are many habitats where the succession of reactions described can be found. The most common are the water column of nutrient-enriched (eutrophied) lakes (Fig. 1.8), oceanic basins, sediments (Fig. 1.9), soils, swamps (43), estu-

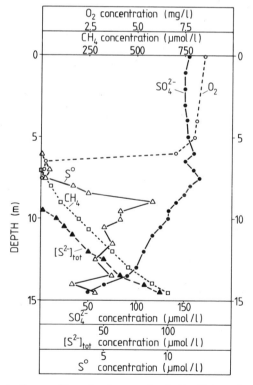

Figure 1.8 Concentration profiles of redox species in the Rotsee, Lucerne, Switzerland from October 1982. Data are partially from Kohler et al. (17).

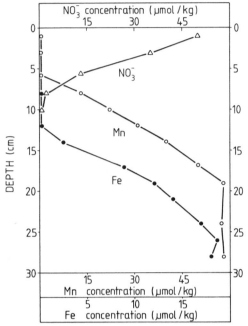

Figure 1.9 Concentrations of NO_3^-, and soluble Mn and Fe in sediment waters from site M, 7BC15BZ in the eastern equatorial Pacific. Data from Klinkhammer (16).

aries (Fig. 1.6), groundwater polluted with organic matter, and closed systems containing excess organic matter, such as a batch digester (anaerobic fermentation unit).

1.4 MICROBIOLOGY OF ENVIRONMENTAL REDOX SEQUENCES

Oxidation-reduction reactions catalyzed by biological systems are the main driving force for diagenic processes. In most of these reactions, molecules of organic carbon represent the electron donor, whereas compounds containing nitrogen, sulfur, manganese, iron, and so on, act in general as electron acceptors. Thus the cycling of the majority of elements at low temperatures ($< 150°C$) is connected directly to the carbon cycle. Depending on the compounds involved, specific microorganisms can be identified which are responsible for the catalysis of a given redox reaction (Table 1.5). No bacterium is yet known that obtains its energy by oxidizing organic carbon molecules and concomitantly reducing manganese and iron oxides (14). Reduction of manganese ($+IV$) and ferric iron is in most cases due to an abiotic process in which, without the aid of enzymes, reduced biogenic compounds (organic, sulfur, etc.) react with manganese and iron oxides. Although

TABLE 1.5 Anaerobic fermentation and respiration processes

Reaction[a]	$-\Delta G^\circ$ per mol of Electrons Exchanged at pH 7 (kJ/mol)	Organisms Catalyzing These Reactions[b]
Transformations of Carbon Compounds[c]		
$3(CH_2O)$[d] $\rightarrow CO_2 + C_2H_6O$	23.4	e.g., yeasts, *Sarcina ventriculi*, *Zymonas*, *Leuconostoc* sp., clostridia, *Thermoanaerobium brockii*, etc.
$n(CH_2O) \rightarrow mCO_2$ and/or fatty acids and/or alcohols and/or hydrogen	5–60	e.g., yeasts, clostridia, enterobacteria, lactobacilli, streptococci, propionibacteria, and many others
$CH_3COOH \rightarrow CH_4 + CO_2$	28[e]	Some methane bacteria (*M. barkeri*, *M. mazei*, *M. söhngenii*)
$CO_2 + 4H_2 \rightarrow CH_4 + 2H_2O$	17.4	Most methane bacteria
$2CO_2 + 4H_2 \rightarrow CH_3COOH + 2H_2O$	14.4	e.g., *Acetobacterium woodii*, *Clostridium aceticum*
$mCO_2 + nH_2 \rightarrow$ fatty acids and/or alcohols	12–20	e.g., *Butyribacterium methylotrophicum*
Reduction of Nitrogen Compounds		
$2NO_3^- + (CH_2O) \rightarrow 2NO_2^- + CO_2 + H_2O$	82.2	e.g., members of the genus *Enterobacter*, *E. coli*, and many others
$\frac{4}{5}NO_3^- + (CH_2O) + \frac{4}{5}H^+ \rightarrow \frac{2}{5}N_2 + CO_2 + \frac{7}{5}H_2O$	112.0	e.g., members of the genus *Pseudomonas*, *Bacillus lichenformis*, *Paracoccus denitrificans*, etc.
$\frac{1}{2}NO_3^- + (CH_2O) + H^+ \rightarrow \frac{1}{2}NH_4^+ + CO_2 + \frac{1}{2}H_2O$	74.0	Members of the genus *Clostridium*
$\frac{6}{5}NO_3^- + S^0 + \frac{2}{5}H_2O \rightarrow \frac{3}{5}N_2 + SO_4^{2-} + \frac{4}{5}H^+$	91.3	Members of the genus *Thiobacillus*
$\frac{8}{5}NO_3^- + HS^- + \frac{3}{5}H^+ \rightarrow \frac{4}{5}N_2 + SO_4^{2-} + \frac{4}{5}H_2O$	93.0	*Thiosphaera pantotropha* and members of the genus *Thiobacillus*[f]

TABLE 1.5 *(Continued)*

Reaction[a]	$-\Delta G°$ per mol of Electrons Exchanged at pH 7 (kJ/mol)	Organisms Catalyzing These Reactions[b]
Reduction of Manganese Compounds[g]		
$2MnO_2 + (CH_2O) + 2H^+ \rightarrow MnCO_3 + Mn^{2+} + 2H_2O$	94.5	Members of the genus *Bacillus*, *Micrococcus*, and *Pseudomonas*
Reduction of Iron Compounds[h]		
$4FeOOH + (CH_2O) + 6H^+ \rightarrow FeCO_3 + 3Fe^{2+} + 6H_2O$	24.3	Members of the genus *Bacillus*
Reduction of Sulfur Compounds		
$\frac{1}{2}SO_4^{2-} + (CH_2O) + \frac{1}{2}H^+ \rightarrow \frac{1}{2}HS^- + CO_2 + H_2O$	18.0	*Desulfobacter* sp., *Desulfovibrio* sp., *Desulfonema* sp., etc.
$S^0 + (CH_2O) + H_2O \rightarrow 2HS^- + CO_2 + 2H^+$	12.0	*Desulfomonas acetoxidans*, *Campylobacter* sp., *Thermoproteus tenax*, *Pyrobaculum islandicum*
$S^0 + H_2 \rightarrow HS^- + H^+$	14.0	*Thermoproteus* sp., *Thermodiscus* sp., *Pyrodictum* sp., various bacteria, etc.

Reduction of Protons[i]

$H_2O + (CH_2O)$ → $CO_2 + 2H_2$ -3.1 Obligate proton reducers such as *Syntrophomonas* sp., *Syntrophobacter* sp., *Syntrophus* sp., etc.[j]

Reductive Dechlorination[k]

$C_2Cl_4 + H_2$ → $C_2HCl_3 + H^+ + Cl^-$ 86.2   Enrichment culture, no pure culture yet known

[a]Some of the more important overall reactions are given.
[b]This list is not complete. Only some of the best known bacteria are indicated.
[c]Only a limited number of reactions are given.
[d]The standard free energy of formation of (CH_2O) was chosen to be one-sixth of glucose (i.e., -153 kJ mol^{-1}) and not that of formaldehyde as in the previous tables.
[e]This is an intramolecular redox reaction and not, as the rest of the listed reactions, an intermolecular redox reaction; 28 kJ mol^{-1} is the energy released by transforming 1 mol of acetate.
[f]HS^- can be replaced by thiosulfate. *Thiomicrospira* sp. can reduce nitrate only with thiosulfate as electron donor.
[g]In the absence of oxygen, MnO_2 can abiotically be reduced at considerable rates.
[h]In the absence of oxygen, FeOOH can also be reduced abiotically at considerable rates.
[i]Only acetate is entirely oxidized to CO_2 and hydrogen. With other fatty acids ($C > 2$), acetate, as well as hydrogen, is an obligate end product.
[j]These bacteria can be grown only in a co-culture with hydrogen-scavenging microorganisms.
[k]Whether this chlorinated compound is reduced with electrons from hydrogen and/or from organic substrate is not yet known.

27

these two metal oxides are reduced chemically, their transformation is strictly dependent on the availability of reduced biogenic compounds and therefore on biological activities. However, it must not be forgotten that manganese and iron oxides can, in certain environments, be reduced by agents (e.g., hydrogen, sulfur compounds) originating from volcanic or anthropogenic activities. Recently, it was shown (31) that synthetic, halogenated compounds can reductively be dechlorinated by biologically mediated reactions and thus act as electron acceptors. The reductive dehalogenation is (fortunately!) not of importance for global cycles but can locally be of considerable interest for the treatment of polluted air, water, and soils.

In relatively stable environments such as lake and ocean sediments, stagnant water bodies, and groundwater systems, a clear, predictable redox sequence will be established (see Section 1.3), which according to the availability of electron acceptors, is colonized by a succession of microorganisms catalyzing the respective redox reaction (51). Some environments are exposed to strong fluctuating conditions: for example, the top of the marine sediments is constantly turned up by large animals (bioturbation), intertidal sediments are subjected to water erosion and alluvial depositions which continuously disturb the creation of a stable environment, shallow water sediments at their water–sediment interface are strongly influenced by diurnal light–dark cycles, and in soil environments the constantly changing water content often limits the access to electron donors and acceptors. Because of the continuously changing conditions, all these unstable environments contain a large diversity of organisms and therefore for microbiologists represent a storehouse for new, often exotic microorganisms.

The effect of bioturbation on biogeochemical processes has been extensively investigated experimentally and theoretically by a variety of researchers (1). Since bioturbation does not result in complete homogenization but rather in random mixing, various redox conditions can exist close together in microniches. The continuous plowing of animals exposes bacteria constantly to new environmental conditions but also brings them in contact with fresh substrates. To survive as a bacterial species in such an ever-changing environment, one must, in addition to escaping from predators, be very versatile (able to use various substrates and/or electron acceptors) or be capable of surviving adverse conditions and/of growing very fast when circumstances improve (9). A very interesting habitat is the water–sediment interface, especially in shallow waters. Advances in microelectrode developments have provided better access to these habitats (27). At the first few millimeters of the marine water–sediment interface which receives sunlight, principally six groups of microorganisms can be found close together: (a) photoautotrophic oxygen producers (e.g., algae), (b) heterotrophic oxygen consumers, (c) "autotrophic" sulfide oxidizing oxygen consumers (e.g., *Beggiatoa*), (d) photoautotrophic sulfide oxidizers (e.g., *Chromatium*), (e) fermenting organisms, and (f) sulfate reducers. A habitat at the water–sediment interface has been studied extensively by Jørgensen and co-workers (15) with special attention given to organisms belonging to groups a, c, and d. During a diurnal cycle, the chemical gradients changed constantly and the microorganisms searched continuously for

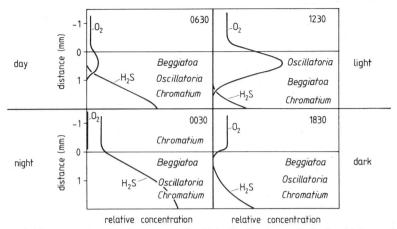

Figure 1.10 Dinural cycle of oxygen and sulfide distribution and of microbial zonation in a marine sulfuretum. Redrawn from Jørgensen (15).

their most optimal place with respect to the physical, chemical, and biological (proximity of other microorganisms) environment (Fig. 1.10).

In soils the aeration status is a continuum from virtually completely aerobic conditions in well-drained sands to permanently anaerobic conditions in marshes and swamps. Within this range there are classes of soils with intermediate drainage and aeration characteristics, which result in temporary periods of anoxia or the presence of anaerobic microsites (mostly in an aggregate) within a generally aerobic matrix. Tiedje et al. (41) postulated that the anaerobic zone in an aggregate expands and contracts in response to the velocity of oxygen consumption. During anaerobiosis in this kind of habitat, denitrification and manganese and iron reduction, especially, will occur. At high inputs of organic carbon compounds into soil, fermentation processes have been observed (21) as well as sulfate reduction and methanogenesis.

Fermenting bacteria (Svensson, unpublished) and methanogens (Zehnder, unpublished) can readily be isolated from topsoil, and some sulfide is formed when soils become flooded for an extended period. It is therefore probable that the strict anaerobic bacteria can survive in anaerobic microsites in soil aggregates within an aerobic environment. However, it is not known if they contribute significantly to the overall mineralization processes in soil. With microelectrodes the chemodynamics within a soil aggregate can now be measured (16), but satisfactory techniques are not yet available that allow us to follow the microbial dynamics at the same time. A further development of Perfil'ev's pedoscope or peloscope (24) might be useful for this purpose in both soil and sediments.

1.5 PRINCIPLES OF SYNTROPHIC RELATIONSHIPS

Methane bacteria have an extremely limited substrate spectrum (53). Their main substrates are acetate or carbon dioxide and hydrogen. Molecules with more than two carbon atoms [except isopropanol (44)] are neither converted to methane nor can their electrons be used to reduce carbon dioxide to methane in methane bacteria. In case only carbon dioxide and protons are present as terminal electron acceptors, other bacteria first have to convert complex organic carbon compounds to acetate or hydrogen and carbon dioxide before methane bacteria can finally produce methane. Fermentation of various compounds leads to the production of these two methanogenic substrates; however, there are other reactions during which fatty acids such as propionate, butyrate, and/or amino acids (alanine, leucine, etc.) are formed (10). The further conversion of these products to hydrogen and carbon dioxide and acetate is, under standard conditions, an endergonic process; thus organisms should not be able to grow in this reaction. In the case of propionate oxidation, the following equation can be written:

$$CH_3CH_2OO^- + 3H_2O \rightleftharpoons CH_3COO^- + HCO_3^- + H^+ + 3H_2$$

$$\Delta G^{\circ\prime} = +74 \text{ kJ mol}^{-1} \quad (27)$$

However, standard conditions do not describe the conditions prevailing in the environment. Therefore, the actual $\Delta G'$ should be calculated rather than the $\Delta G^{\circ\prime}$:

$$\Delta G' = \Delta G^{\circ\prime} + 2.3RT \log \frac{\{CH_3COO^-\}\{HCO_3^-\}\{H_2\}^3}{\{CH_3CH_2COO^-\}} \quad (28)$$

If we assume that the concentrations of acetate and propionate are about equal and the biocarbonate concentration constant, the only factor that is variable and influences ΔG is the hydrogen concentration. In natural systems, steady-state hydrogen concentrations are very low. Values between 3×10^{-6} and 4.4×10^{-8} mol^{-1} have been measured (53), which corresponds to a hydrogen partial pressure of 4×10^{-3} to 6×10^{-5} atm. At a hydrogen partial pressure of about 10^{-4} atm, the oxidation of propionate becomes exergonic (Table 1.6) and organisms can grow at the expense of this reaction. The partial pressure of hydrogen is kept low by methane-producing bacteria or other hydrogen oxidizers (e.g., acetogens, or sulfate reducers). The coupling or syntrophic relationship of hydrogen producers and hydrogen consumers is called *interspecies hydrogen transfer* (49). A unique example of interspecies hydrogen transfer is described by Zinder and Koch (56). They found that acetate can be oxidized in a thermophilic syntrophic association to hydrogen and carbon dioxide, and that these two gases are further converted to methane. This reaction (i.e., first oxidation of acetate and subsequent oxidation of hydrogen) can only proceed within very narrow boundaries, at 60°C between 10^{-4} and 2.5×10^{-3} atm of hydrogen (Fig. 1.11). Although there was doubt 10 years ago that one bacterium could obtain enough energy for growth from the conversion of acetate to methane (55), this example shows that not only one but two bacteria can live from this very small amount of energy. If the energy is distributed equally

TABLE 1.6 Influence of hydrogen partial pressure on the free energy of some typical hydrogen-producing reactions catalyzed by syntrophic microbial associations under methanogenic conditions

Reaction Description	Reactants		Products	Free energy[a] (kJ/reaction)	
				$\Delta G°'$	$\Delta G'$
Fatty acid oxidation	Propionate + $3H_2O$	$\longrightarrow$	Acetate + $HCO_3^- + H^+ + 3H_2$	+74	−1
Alcohol oxidation	Ethanol + H_2O	$\longrightarrow$	Acetate + $H^+ + 2H_2$	+2	−44
Amino acid oxidation	Alanine + $3H_2O$	$\longrightarrow$	Acetate + $HCO_3^- + H^+ + NH_4^+ + 2H_2$	+8	−38
Oxidation of aromatic compounds	Benzoate + $7H_2O$	$\longrightarrow$	3 Acetate + $HCO_3^- + 3H^+ + 3H_2$	+53	−16

[a]$\Delta G°'$, standard conditions, pH 7, 25°C; $\Delta G'$, same as $\Delta G°'$ with the exception that p_{H_2} was 10^{-4} atm.

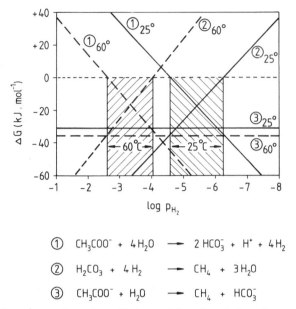

① $CH_3COO^- + 4H_2O \longrightarrow 2HCO_3^- + H^+ + 4H_2$

② $H_2CO_3 + 4H_2 \longrightarrow CH_4 + 3H_2O$

③ $CH_3COO^- + H_2O \longrightarrow CH_4 + HCO_3^-$

Figure 1.11 Free energy from the oxidation and decarboxylation of acetate and methane formation at different hydrogen partial pressures at 60°C and 25°C. The slopes were calculated for pH 7, 20 mM acetate and bicarbonate, and 1 atm of methane. Temperature corrections have been made using the van't Hoff equation (23).

between the two organisms, each will obtain 17 kJ mol^{-1}, which translates into 4.25 kJ or 44.2 mV per pair of electrons [for the calculation, see equation (13)]. For our present apprehension, this is not enough energy to explain chemiosmotic coupling via vectorial protons with both a $\rightarrow H^+/2e^-$ and a $\rightarrow 3H^+/1$ ATP stoichiometry (38). However, the entire 17 kJ mol^{-1} would just be enough to bring one proton across the membrane. It remains to future research to elucidate this intriguing question. Further examples of syntrophic associations and their thermodynamics are discussed in Chapter 9.

Depending on the presence or absence of certain electron acceptors and microorganisms, as well as the nature of the substrate, there are three different possibilities for the oxidation of organic carbon compounds (Fig. 1.12). (a) The first is oxidation of the substrate and concomitant reduction of an inorganic electron acceptor in the same organism (Fig. 1.12A). (b) In the absence of inorganic electron acceptors (except protons and carbon dioxide), the organisms are obliged to use part of the carbon substrate as electron acceptors (Fig. 1.12B). This process is called fermentation. (c) A third possibility is a combination of processes a and b in which the substrate is oxidized in one organism and the electrons liberated are transferred to a second microorganism in the form of hydrogen. There, the hydro-

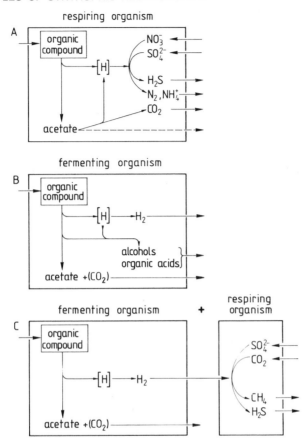

Figure 1.12 Substrate flow in respiring and fermenting organisms and in a combination of both where interspecies hydrogen transfer takes place. Modified from Zehnder and Colberg (52).

gen is oxidized in general with an inorganic electron acceptor. The second organism acts essentially as a terminal electron sink (Fig. 1.12C). Syntrophic associations for catabolic reactions based on interspecies hydrogen transfer are the only ones known so far. One can hypothesize that syntrophic associations which exchange catabolites other than hydrogen may also be important in anaerobic habitats.

In most natural ecosystems, monomeric organic carbon compounds are transformed according to Fig. 1.12A or C or a combination of A and C. Pure fermentations take place only in pure cultures, in habitats where fermentation products are used by a host (e.g., rumen and other gastrointestinal systems), or under conditions in which respiring organisms are either inhibited (e.g., low or high pH) or washed out.

1.6 EVOLUTION OF ANAEROBIC MICROORGANISMS

Since 16S rRNA was chosen for measuring phylogenetic relationships among bacteria (46), discussion of the evolution of life has been given a considerable boost (7). It goes beyond the scope of this chapter to summarize existing knowledge on the evolution of anaerobes. The interested reader should refer to the respective literature. In the following we simply indicate the place of anaerobes in evolution and their traces in ancient rocks and sediments.

From the pioneering work of Woese and co-workers (47), we know that living organisms cluster in three separate kingdoms (Fig. 1.13) and that all three kingdoms contain anaerobic microorganism (48). However, this does not mean that the three branches were essentially developed before the atmosphere became oxygenated. In fact, eukaryotes seem to have evolved about 1500 million years ago when oxygen was already present in the atmosphere (8). The ability of some of them to grow anaerobically might just be a remainder of their ancestors. It is interesting to note that most eukaryotic cells are still able to function for a limited time in the absence of oxygen (54). The other two kingdoms originated in anaerobic periods, and it is therefore not surprising to find anaerobes (which are probably the majority) in both clusters.

Some bacteria produce molecules which are very specific and difficult to degrade. They can therefore serve as markers in geological formations. Archaebacteria, especially methanogens, form some very specific isoprenoid hydrocarbons and ethers (19). Based on the analysis of these compounds, methanogens could be traced back 3 billion years, possibly as far back as 3.8 to 4 billion years (11). In addition to organic markers, stable isotope ratios can be employed to obtain an idea of biological activities in the past. Based on $^{34}S/^{32}S$ ratios from sedimentary

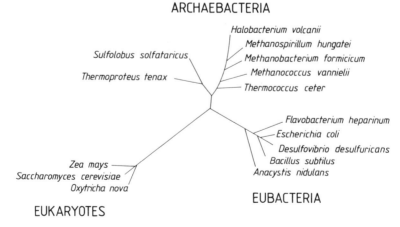

Figure 1.13 Unrooted phylogenetic tree for the three urkingdoms, showing some selected organisms. Modified from Woese and Olsen (48).

pyrites, it can be concluded that sulfate-reducing bacteria existed as early as 3.2 billion years ago (25). The widespread occurrence of barite in Archean sediments suggests that sulfate was present in the ocean 3.5 billion years ago. This first large-scale introduction of sulfate into the environment was probably caused by the activity of photosynthetic sulfur bacteria (25, 42). Carbon isotope ratios from the 3.8-billion-year-old Isua carbonaceous matter shows the signature of autotrophic carbon fixation (30). Thus it can be concluded that life on earth is considerably older than the oldest known rocks (11, 25) and that anaerobes just as billions of years ago still contribute in essential ways to the state of the environment on our planet (20).

REFERENCES

1. Aller, R. C. 1980. Diagenic processes near the sediment-water interface of Long Island Sound. I. Decomposition and nutrient element geochemistry (S, N, P), p. 237–350, in B. Saltzman (ed.), Advances in geophysics, Vol. 22. Academic Press, New York.

2. Berner, R. A. 1980. Early diagenesis. Princeton University Press, Princeton, N.J.

3. Berner, R. A. 1981. A new classification of sedimentary environments. J. Sediment. Petrol. 51:359–366.

4. Billen, G. 1978. A budget of nitrogen recycling in North Sea sediments off the Belgian coast. Estuarine Coastal Mar. Sci. 7:127–246.

5. Billen, G., C. Joiris, J. Wijnant, and G. Gillain. 1980. Concentration and microbial utilization of small organic molecules in the Scheldt estuary, the Belgian coastal zone of the North Sea and the English Channel. Estuarine Coastal Mar. Sci. 11:279–294.

6. Billen, G., and S. Verbeustel. 1980. Distribution of microbial metabolisms in natural environments displaying gradients of oxidation-reduction conditons, p. 291–301, in Biogéochimie de la matière organique à l'interface eau-sédiment marin. Colloq. Int. Cent. Nat. Rech. Sci. no. 293. CNRS, Paris.

7. Carlisle, M. J., J. F. Collins, and B. E. B. Moseley (eds.). 1981. Molecular and cellular aspects of microbial evolution. Cambridge University Press, Cambridge.

8. Ehrlich, H. L. 1981. Geomicrobiology. Marcel Dekker, New York.

9. Gottschal, J. C., S. de Vries, and J. G. Kuenen. 1979. Competition between the facultatively chemolithotrophic *Thiobacillus* A2, an obligately chemolithrophic *Thiobacillus* and a heterotrophic spirillum for inorganic and organic substrates. Arch. Microbiol. 121:241–249.

10. Gottschalk, G. 1986. Bacterial metabolism. Springer-Verlag, New York.

11. Hahn, J., and P. Haug. 1986. Traces of archaebacteria in ancient sediments. Syst. Appl. Microbiol. 7:178–183.

12. Hasan, S. M., and J. B. Hall. 1975. The physiological function of nitrate reduction in *Clostridium perfringens*. J. Gen. Microbiol. 87:120–128.

13. Ishimoto, M. and F. Egami. 1959. Meaning of nitrate and sulfate reduction in the process of metabolic evolution, p. 555–561, in F. Clark and R. L. M. Synge (eds.), Proceedings of the first international symposium on the origin of life on earth. Pergamon Press, Elmsford, N.Y.

14. Jones, J. G. 1986. Iron transformation by freshwater bacteria, p. 149–185, in K. C. Marshall (ed.), Advances in microbial ecology, Vol. 9. Plenum Press, New York.

15. Jørgensen, B. B. 1982. Ecology of the bacteria of the sulfur cycle with special reference to anoxic-oxic interface environments. Philos. Trans. R. Soc. Lond. B298:543–561.

16. Klinkhammer, G. P. 1980. Early diagenesis in sediments from the eastern equatorial pacific. II. Pore water metal results. Earth Planet. Sci. Lett. 49:81–101.

17. Kohler, H. P., B. Ahring, C. Albella, K. Ingvorsen, H. Keweloh, E. Laczkó, E. Stupperich, and F. Tomei. 1984. Bacteriological studies on the sulfur cycle in the anaerobic part of the hypolimnion and the surface sediments of Rotsee in Switzerland. FEMS Microbiol. Lett. 21:279–286.

18. Krul, J. M. 1976. Dissimilatory nitrate and nitrite reduction under aerobic conditions by an aerobically and anaerobically grown *Alcaligenes* sp. and by activated sludge. J. Appl. Bacteriol. 40:245–260.

19. Langworthy, T. A. 1985. Lipids of archaebacteria, p. 459–498, in C. R. Woese and R. S. Wolfe (eds.), The bacteria, Vol. 8, Archaebacteria, Academic Press, New York.

20. Lovelock, J. E. 1979. Gaia. A new look at life on Earth. Oxford University Press, Oxford.

21. Lynch, J. M., K. B. Gunn, and L. M. Panting. 1980. On the concentration of acidic acid in straw and soil. Plant Soil 56:93–98.

22. Meiberg, J. B. M., P. M. Bruinenberg, and W. Harder. 1980. Effect of dissolved oxygen tension on the metabolism of methylated amines in *Hyphomicrobium X* in the absence and presence of nitrate: evidence for 'aerobic' denitrification. J. Gen. Microbiol. 120:453–463.

23. Morel, F. M. M. 1983. Principles of aquatic chemistry. Wiley-Interscience, New York.

24. Perfil'ev, B. V., and D. R. Gabe. 1969. Capillary methods of investigating microorganisms. Oliver & Boyd, Edinburgh.

25. Pflug, H. D. 1986. Morphological and chemical record of the organic particles in precambrian sediments. Syst. Appl. Microbiol. 7:184–189.

26. Pirt, S. J. 1975. Principles of microbe and cell cultivation. Blackwell Scientific, Oxford.

27. Revsbech, N. P., and B. B. Jørgensen. 1986. Microelectrodes: their use in microbial ecology, p. 293–352, in K. C. Marshall (ed.), Advances in microbial ecology, Vol. 9, Plenum Press, New York.

28. Richards, F. A. 1965. Anoxic basins and fjords, p. 611–645, in S. P. Riley and G. Skirrow (eds.), Chemical oceanography, Vol. 1. Academic Press, New York.

29. Robertson, L. A., and J. G. Kuenen. 1984. Aerobic denitrification: a controversy revived. Arch. Microbiol. 139:351–354.

30. Schidlowski, M. 1983. Biologically mediated isotope fractionations: biogeochemistry, geochemical significance and preservation in Earth's oldest sediments, p. 277–322, in C. Ponnamperuma (ed.), Cosmochemistry and the origin of life. D. Reidel, Dordrecht, the Netherlands.

31. Shelton, D. L., and J. M. Tiedje. 1984. Isolation and partial characterization of bacteria in an anaerobic consortium that mineralizes 3-chlorobenzoic acid. Appl. Environ. Microbiol. 48:840–848.

32. Sillén, L. G. 1961. The physical chemistry of seawater, p. 549–581, in M. Sears (ed.), Oceanography (Publ. 67). American Association for the Advancement of Science, Washington, D.C.

33. Sillén, L. G., and A. E. Martell. 1964 and 1971. Stability constants of metal ion complexes. Spec. Publ. 17, and Suppl. 1, Spec. Publ. 25, Chemical Society, London.

34. Smith, R. L., and M. J. Klug. 1981. Reduction of sulfur compounds in the sediments of an eutrophic lake basin. Appl. Environ. Microbiol. 41:1230–1237.

35. Stumm, W., and J. J. Morgan. 1981. Aquatic chemistry, 2nd ed. Wiley-Interscience, New York.

36. Takahashi, H., S. Tanaguchi, and F. Egami. 1983. Inorganic nitrogen compounds distribution and metabolism, p. 91–202, in M. Florkin and H. S. Mason (eds.), Comparative biochemistry, Academic Press, New York.

37. Thauer, R. K., K. Jungermann, and K. Decker, 1977. Energy conservation in chemotrophic anaerobic bacteria. Bacteriol. Rev. 41:100–180.

38. Thauer, R. K., and J. G. Morris. 1984. Metabolism of chemotrophic anaerobes: old view and new aspects, p. 123–168, in D. P. Kelly and N. G. Carr (eds), The microbe 1984, P. II, Prokaryotes and eukaryotes. Cambridge University Press, Cambridge.

39. Thompson, J. B., Jr. 1959. Local equilibrium in metasomatic processes, p. 427–457, in P. H. Abelson (ed.), Researches in geochemistry. Wiley, New York.

40. Thorstenson, D. C. 1970. Equilibrium distribution of small organic molecules in natural waters. Geochim. Cosmochim. Acta 34:745–770.

41. Tiedje, J. M., A. J. Sexstone, T. B. Parkin, N. P. Revsbech, and D. R. Shelton. 1984. Anaerobic processes in soil. Plant Soil 76:197–212.

42. Trüper, H. G. 1982. Microbial processes in the sulfur cycle through time, p. 5–30, in H. D. Holland and M. Schidlowski (eds.), Mineral deposits and the evolution of the biosphere (Dahlem Konferenzen, 1982). Springer-Verlag, Heidelberg.

43. Westermann, P., and B. K. Ahring. 1986. Terminal anarobic carbon mineralization in an alder swamp, p. 305–314, in V. Jensen, A. Kjøller, and L. H. Sørensen (eds.), Microbial communities in soil. Elsevier, Amsterdam.

44. Widdel, F. 1986. Growth of methanogenic bacteria in pure culture with 2-propanol and other alcohols as hydrogen donor. Appl. Environ. Microbiol. 51:1056–1062.

45. Woese, C. R. 1981. Archaebacteria. Sci. Am. 244(6):94–108.

46. Woese, C. R., and G. E. Fox. 1977. The concept of cellular evolution. J. Mol. Evol. 10:1–6.

47. Woese, C. R., L. J. Magrum, and G. E. Fox. 1978. Archaebacteria. J. Mol. Evol. 11:245–252.

48. Woese, C. R., and G. J. Olsen. 1986. Archaebacterial phylogeny: perspectives on the urkingdoms. Syst. Appl. Microbiol. 7:161–177.

49. Wolin, M. J., and T. L. Miller. 1982. Interspecies hydrogen transfer: 15 years later. ASM News 48:561–565.

50. Wollast, R. 1978. Modelling of biological and chemical processes in the Scheldt Estuary, p. 63–77, in J. C. J. Nihoul (ed.), Hydrodynamics in estuaries and fjords. Elsevier, Amsterdam.

51. Zehnder, A. J. B. 1982. The carbon cycle, p. 83–110, in O. Hutzingen (ed.), The handbook of environmental chemistry, vol. 1/B. Springer-Verlag, Heidelberg.

52. Zehnder, A. J. B., and P. J. Colberg. 1986. Anaerobic biotransformation of organic carbon compounds, p. 275–291, in V. Jensen, A. Kjøller, and L. H. Sørensen (eds.), Microbial communities in soil. Elsevier, Amsterdam.

53. Zehnder, A. J. B., K. Ingvorsen, and T. Marti, 1982. Mircrobiology of methane bacteria, p. 45–68, in D. E. Hughes, D. A. Stafford, B. I. Wheatley, W. Baader, G. Lettinga, E. J. Nyns, and W. Verstraeten (eds.), Anaerobic digestion 1981. Elsevier, Amsterdam.

54. Zehnder, A. J. B., and B. H. Svensson. 1986. Life without oxygen: what can and what cannot? Experientia (Basel) 42:1197–1205.

55. Zeikus, J. G., P. J. Weimer, D. R. Nelson, and L. Daniels. 1975. Bacterial methanogenesis: acetate as methane precursor in pure culture. Arch. Microbiol. 104:129–134.

56. Zinder, S. H., and M. Koch. 1984. Non-aceticlastic methanogenesis from acetate: acetate oxidation by a thermophilic synthropic coculture. Arch. Microbiol. 138:263–272.

2

MICROBIOLOGY, PHYSIOLOGY, AND ECOLOGY OF PHOTOTROPHIC BACTERIA

MICHAEL T. MADIGAN

Department of Microbiology, Southern Illinois University, Carbondale, IL 62901

2.1 INTRODUCTION
    2.1.1 Anoxygenic and Anaerobic Nature of Phototrophic Bacteria
    2.1.2 General Taxonomy of Phototrophic Bacteria
        2.1.2.1 Bacteriochlorophylls
        2.1.2.2 Carotenoids
        2.1.2.3 Intracytoplasmic Membranes
    2.1.3 Taxonomy of Purple and Green Sulfur Bacteria
    2.1.4 Taxonomy of Facultative Purple and Green Bacteria
    2.1.5 Phylogenetic Relationships among Phototrophic Bacteria
2.2 PHYSIOLOGY OF PHOTOTROPHIC BACTERIA
    2.2.1 Physiology of Purple Sulfur Bacteria
    2.2.2 Physiology of Green Sulfur Bacteria
    2.2.3 Physiology of Facultative Purple Bacteria
        2.2.3.1 Autotrophic Growth
        2.2.3.2 Photoheterotrophic Growth
        2.2.3.3 Heterotrophic (Aerobic) Growth
        2.2.3.4 Fermentative Growth
        2.2.3.5 Nitrogen Fixation
        2.2.3.6 Genetics of Rhodospirillaceae

2.2.4   Physiology of *Heliobacterium chlorum*

2.2.5   Physiology of Facultative Green Bacteria

2.3   ECOLOGY OF PHOTOTROPHIC BACTERIA

   2.3.1   Stratified Lakes

      2.3.1.1   Wintergreen and Burke Lakes, Michigan

      2.3.1.2   Rotsee, Switzerland

      2.3.1.3   Lake Cisó and Other Spanish Lakes

      2.3.1.4   Colored Wisconsin Lakes

      2.3.1.5   Solar Lake, Sinai

      2.3.1.6   Rhodospirillaceae in Aquatic Habitats

   2.3.2   Sewage and Wastewater as a Habitat for Phototrophic Bacteria

   2.3.3   Hot Springs and Phototrophic Bacteria

      2.3.3.1   *Chloroflexus* in Alkaline Hot Spring Mats

      2.3.3.2   Thermophilic Purple Bacteria

   2.3.4   Marine Microbial Mats

   2.3.5   Chemostat Studies of Phototrophic Bacteria

2.4   ENRICHMENT AND ISOLATION OF PHOTOTROPHIC BACTERIA

   2.4.1   Photosynthetic Sulfur Bacteria

   2.4.2   Facultative Photosynthetic Bacteria

   2.4.3   Preservation of Pure Cultures

ACKNOWLEDGMENTS

REFERENCES

2.1   INTRODUCTION

2.1.1   Anoxygenic and Anaerobic Nature of Phototrophic Bacteria

The phototrophic bacteria are a widespread and metabolically diverse group of photosynthetic organisms which inhabit anaerobic environments exposed to light (151, 156). The terms "green" and "purple" bacteria, used extensively throughout this chapter, are to be considered synonymous with the term "phototrophic" (or "photosynthetic") bacteria. All of these terms are used to differentiate these organisms from other photosynthetic prokaryotes, the cyanobacteria (212), and Prochlorophytes (108).

Although the green and purple bacteria share with the cyanobacteria a prokaryotic cellular architecture and the ability to convert light energy to a chemically usable form, photometabolism in phototrophic bacteria is strictly an *anoxygenic* process; molecular oxygen is *not* evolved as a photosynthetic by-product (155, 212). Pfennig (159) has provided an excellent historical account of the major contributors to the concept of photosynthesis in bacteria. By contrast, in cyanobacteria, molecular oxygen is released as a consequence of the light-driven splitting of water to generate reducing power for CO_2 assimilation (212). The photosynthetic process in cyanobacteria is therefore functionally equivalent to that of green plants.

Anoxygenic phototrophs, on the other hand, require anaerobic conditions for photosynthetic growth and development because synthesis of their pigments, bacteriochlorophylls and carotenoids, is strongly repressed by molecular oxygen (31). This physiological constraint dictates a unique position for phototrophic bacteria in the ecosystem. Their competitive success depends on the previous development of anoxic conditions, which is generally a result of oxygen consumption by heterotrophic organisms during the oxidation of organic materials and the production of H_2S by the obligately anaerobic sulfate- and sulfur-reducing bacteria (144, 156). Once anoxic conditions have been achieved, bacteriochlorophyll synthesis in anoxygenic phototrophs can occur, and in general the physiochemical nature of the habitat (sulfide concentration, pH, light quality and availability, temperature) dictates quantitative and qualitative development of phototrophic bacteria (156).

A fascinating exception to the rule that anaerobic conditions are necessary for the development of phototrophic bacteria has recently been discovered with the isolation from marine environments of a number of strains of gram-negative, aerobic bacteria which produce bacteriochlorophyll *a* (189). Unlike typical anoxygenic phototrophs, these organisms are apparently unable to grow anaerobically and synthesize bacteriochlorophyll only in the presence of molecular oxygen (75). Detailed studies of bacteriochlorophyll synthesis in one strain of aerobic phototrophs have shown pigment synthesis to be positively correlated with oxygen concentration; this indicates that regulatory mechanisms governing pigment synthesis in these organisms function precisely opposite to those of anoxygenic phototrophs (e.g., compare Refs. 31 and 75).

The function of bacteriochlorophyll in aerobic bacteria is not clear, since neither photophosphorylation nor autotrophic growth has been demonstrated (188). Perhaps when the ecology and physiology of these prokaryotes is better understood (the majority of strains were isolated as epiphytes of seaweeds), the function of bacteriochlorophyll *a* in aerobic phototrophic bacteria will become more apparent.

Phototrophic bacteria play an important role in the anaerobic cycling of matter both as primary producers (photoautotrophs) and as light-stimulated consumers of reduced organic compounds (photoheterotrophs). In habitats particularly favorable for their development, purple and green sulfur bacteria have been shown to be more significant as primary producers than oxygenic phototrophs (36–38). In aquatic habitats, in general, however, the role of phototrophic bacteria as sulfide consumers is probably more important than their contribution to primary production per se; hydrogen sulfide is a highly poisonous substance for most forms of aquatic life and for many bacteria. Sulfide oxidation generates nontoxic species of sulfur, allowing the upper layers of a stratified lake to remain oxic and thus suitable for eukaryotic microorganisms and aquatic macroorganisms.

This chapter begins with a consideration of the taxonomy of purple and green bacteria, the main objective being to leave the reader with sufficient background to identify the major groups of anoxygenic phototrophs in natural samples. We will then consider the physiological idiosyncracies of the five families of phototrophic bacteria, discuss examples of the distribution and ecology of photosynthetic bacteria in aquatic ecosystems (drawing from some recent field research in this area), and close with a brief consideration of methods for culturing phototrophic bacteria.

2.1.2 General Taxonomy of Phototrophic Bacteria

Pfennig (156, 158) has proposed that anoxygenic phototrophs be divided into four physiological/ecological subgroups. This view of classification fits nicely with the known metabolic properties of pure culture representatives of each of the various groups (151, 155). The two broad groups, *purple bacteria* and *green bacteria*, consist of three families, with several genera grouped together in each family. Members of the families Chromatiaceae, Ectothiorhodospiraceae, and Chlorobiaceae primarily use reduced sulfur compounds as electron donors for autotrophic CO_2 incorporation; all representatives of these two families grow well in completely inorganic media (some require vitamin B_{12}) containing relatively high levels of sulfide (84, 163, 164, 226). By contrast, the remaining two families, the Rhodospirillaceae and Chloroflexaceae (to be referred to as purple and green nonsulfur bacteria, respectively), contain species that grow best as photoheterotrophs, although most representatives remain capable of autotrophic growth with H_2 or H_2S as electron donor (163, 226).

Over 60 species of phototrophic bacteria are known, comprising most morphological forms found in the prokaryotic world. However, the physiological/ecological traits peculiar to a given taxonomic family are strongly ingrained properties in each species within a group. The currently recognized genera of anoxygenic phototrophic bacteria and their respective morphologies are given in Table 2.1.

2.1.2.1 *Bacteriochlorophylls*

All anoxygenic phototrophs contain bacteriochlorophyll (bchl), a light-sensitive magnesium tetrapyrrole involved in the photochemical reactions that convert light energy into ATP (134; see also Chapter 3 of this volume). Most members of the Chromatiaceae and Rhodospirillaceae contain bchl *a* as their sole bchl (226); intact cells of such species show a strong absorption maximum in the near infrared between 800 and 860 nm (see Table 2.2 and Figure 2.1). A few purple bacteria lack bchl *a* and contain instead bchl *b*, which absorbs in the infrared at slightly over 1000 nm (226). All green bacteria contain small amounts of bchl *a* localized in the cytoplasmic membrane (208), and this pigment plays the central role in cyclic photophosphorylation as in purple bacteria. The sequence of electron carriers and the reduction potential of the primary electron acceptor differs considerably in purple and green bacteria, however, and the reader is referred to Chapter 3 for further details in this connection.

The major bchl in green bacteria is either bchl *c*, *d*, or *e* (61) or, in *Chloroflexus*,

TABLE 2.1 Recognized genera of anoxygenic phototrophic bacteria

Taxonomic Group[a]	Morphology

PURPLE BACTERIA

Purple Sulfur Bacteria (Chromatiaceae and Ectothiorhodospiraceae)

Amoebobacter [2]	Cocci embedded in slime; contain gas vesicles
Chromatium [11]	Large or small rods (see Fig. 2.2*a*)
Lamprocystis [1]	Large cocci or ovoids with gas vesicles
Lamprobacter [1]	Large ovals with gas vesicles
Thiocapsa [2]	Small cocci
Thiocystis [2]	Large cocci or ovoids
Thiodictyon [2]	Large rods with gas vesicles
Thiospirillum [1]	Large spirilla
Thiopedia [1]	Small cocci with gas vesicles, cells arranged in flat sheets
Ectothiorhodospira [4]	Small spirilla, do not store sulfur inside the cell (see Fig. 2.2*c*)

Purple Nonsulfur Bacteria (Rhodospirillaceae)

Rhodocyclus [3]	Half-circle or circle
Rhodomicrobium [1]	Ovoid with stalked budding morphology
Rhodopseudomonas [8]	Rods, dividing by budding
Rhodobacter [5]	Rods and cocci, dividing by binary fission (see Fig. 2.2*b*)
Rhodopila [1]	Cocci, dividing by binary fission
Rhodospirillum [6]	Large or small spirilla

GREEN BACTERIA

Green Sulfur Bacteria (Chlorobiaceae)

Anacalochloris [1][b]	Prosthecate spheres with gas vesicles
Chlorobium [5]	Small rods or vibrios (see Fig. 2.2*d*)
Pelodictyon [3]	Rods or vibrios, some form three-dimensional net; contain gas vesicles
Prosthecochloris [2]	Spheres with prosthecae

Green Gliding Bacteria (Chloroflexaceae)

Chloroflexus [2]	Narrow filaments (multicellular), up to 100 μm long (see Fig. 2.2*e*)
Chloroherpeton [1][c]	Short filaments (unicellular)
Chloronema [1][b]	Large filaments (multicellular), up to 250 μm long; contain gas vesicles
Oscillochloris [1][b]	Very large filaments, up to 2500 μm long; contain gas vesicles

Table 2.1 (*Continued*)

Taxonomic Group[a]	Morphology
GENERA OF UNCERTAIN AFFILIATION	
Heliobacterium [1][d]	Long rods, motile by gliding
Heliospirillum [1]	Large motile spirilla
Heliobacillus [1]	Motile rods
Heliothrix[b][1]	Large filaments

Source: Adapted from Trüper and Pfennig (226) and Pfennig and Trüper (158).

[a] Numbers in brackets indicate the number of species recognized in each genus.

[b] Not in pure culture.

[c] See Gibson et al. (60) for description of this genus.

[d] See Gest and Favinger (56) and Brochmann and Lipinski (17). Due to its unusual bacteriochlorophyll, the relationship of *Heliobacterium* to other phototrophic bacteria is unclear at this time; it is placed in the Rhodospirillaceae in this review because of its strong photoheterotrophic growth properties and inhibition by sulfide.

bchl c_s (62; see Table 2.1). The chlorophylls of green bacteria resemble bchl *a* but differ from it in four ways: (a) the oxidation-reduction level of the tetrapyrrole ring in the bacteriochlorophylls of green bacteria is more oxidized than that of bchl *a*; (b) certain of the substituents on the pyrrole rings vary from that of bchl *a*; (c) the esterifying alcohol is either farnesol or stearyl alcohol instead of phytol or geranylgeraniol, which are present in bchl *a*; and (d) the absorption properties of green

TABLE 2.2   Bacteriochlorophylls of phototrophic bacteria

Bacteriochlorophyll	Absorption Maximum[a]	Distribution
a	825–890	Purple bacteria (small amounts in green bacteria)
b	1020–1040	Purple bacteria (a few species)
c	745–755	Green bacteria (certain species)
c_s	740	*Chloroflexus aurantiacus*[b]
d	705–740	Green bacteria (certain species)
e	719–726	Green bacteria (brown-colored species only)[c]
g	788	*Heliobacterium chlorum*[d] *Heliospirillum* sp.[e] *Heliobacillus* sp.[e]

[a] Major infrared absorption peak as measured in intact cells (see Fig. 2.1).

[b] See Ref. 62.

[c] See Ref. 61.

[d] See Refs. 17 and 56.

[e] Howard Gest, personal communication.

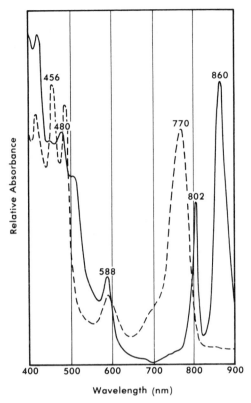

Figure 2.1 Absorption spectra of intact cells (solid line) and acetone : methanol (7 : 2) extracts (dashed line) of *Rhodobacter capsulatus*. Photosynthetically grown cells were suspended in 30% bovine serum albumin for recording of *in vivo* spectra. Note the prominent peaks at 860 and 802 nm in intact cells and the major peak at 770 nm in solvent extracts, due to bacteriochlorophyll *a*.

bacterial chlorophylls differ dramatically from those of bchl *a* [intact cells of green bacteria typically absorb strongly at 710 to 755 nm, depending on the bacteriochlorophyll present (see Table 2.1)].

The recently described *Heliobacterium chlorum* (56) contains a hitherto unknown, structurally unique bacteriochlorophyll, termed bchl *g* (17). Unlike all other bacteriochlorophylls, which contain an alcohol or acetyl group on ring 1 of the tetrapyrrole, bchl *g* has a vinyl group in this position; in this respect bchl *g* resembles chlorophylls *a* and *b* (17). Interestingly, a major physiological property of *Heliobacterium* is its extreme O_2 sensitivity (56). This may be related to the unique photochemistry of *Heliobacterium*, because the primary electron acceptor of this organism has an extremely reducing potential, below -500 mV (141, 173). John Ormerod in Norway has isolated a spirillum containing bchl *g*, and has given the organism the provisional name *Heliospirillum*.

2.1.2.2 Carotenoids

Five major classes of carotenoids are known in anoxygenic phototrophs, and these are presented in Table 2.3. With minor exceptions, carotenoids found in purple bacteria are not found in green bacteria, and vice versa. The distinctive coloration of masses of "purple" bacteria (i.e., red, brown, purple, or violet) or "green" bacteria (i.e, green, brown, or orange) are due to carotenoids (151). The structure, biosynthesis, and occurrence of carotenoids in pure strains of green and purple bacteria have been thoroughly discussed in reviews by Schmidt (184) and Liaaen-Jensen (109).

From studies with pure cultures and natural populations of purple and green bacteria, it has become clear that carotenoids serve not only as protectants against photooxidative damage, but also function as accessory photosynthetic pigments, harvesting wavelengths not absorbed by bacteriochlorophyll and transferring energy to this pigment (63). Hence pure cultures of purple bacteria have been grown in green light [where bchl *a* does not absorb (25, 131, 147)], and brown-colored species of *Chlorobium* have been found in large numbers in the deep layers of stratified lakes where light measurements indicate essentially zero transmission of

TABLE 2.3 Carotenoids of phototrophic bacteria

Carotenoid Group	Major Pigments	Distribution[a]	Color of Mass Cell Suspensions[b]
1. Normal spirilloxanthin series	Lycopene, rhodopin, spirilloxanthin	Purple bacteria	Red, brown, orange
2. Rhodopinal series	Lycopenal, rhodopinal	Purple bacteria	Purple, purple-violet
3. Alternative spirilloxanthin series	Spheroidene, spheroidenone, neurosporene and derivatives	Purple bacteria (small amounts of neurosporene are found in certain green bacteria)	Brown, yellow-brown, yellow-green
4. Okenone series	Okenone, methoxylated keto-carotenoids	Purple bacteria	Purple-red
5. Isorenieratene series	β- and α-carotenes, isorenieratene, chlorobactene	Green bacteria (small amounts of β-carotene are present in *Rhodomicrobium*, a purple bacterium)	Green, brown, orange

Source: Adapted from Schmidt (184) and Trüper and Pfennig (226).

[a] A given species of purple or green bacteria contains various proportions of one or more of the carotenoids listed. For a detailed list of the carotenoid composition of individual species of purple or green bacteria, see Schmidt (184).

[b] Photosynthetically grown cells.

wavelengths utilizable by bacteriochlorophyll (see Section 2.3.1; 138, 225). The major carotenoid of brown-colored *Chlorobium* strains, namely isorenieratene, absorbs strongly at 517 nm (138), and this pigment is undoubtedly responsible for the competitive success of these organisms in nature.

2.1.2.3 *Intracytoplasmic Membranes*

Bacteriochlorophyll(s) and carotenoids of phototrophic bacteria are located in morphologically distinct intracytoplasmic membranes (143). In purple bacteria these membraneous structures presumably originate from the cytoplasmic membrane and are of five major types: (a) vesicles (i.e., "chromatophores") such as those of *Rhodospirillum rubrum* (82) or *Rhodobacter sphaeroides* (42) and *Chromatium* species (176); (b) single irregular membrane invaginations typical of *Rhodocyclus tenuis* and *Rhodocyclus gelatinosus* (249); (c) short lamellar membrane stacks arranged at an angle to the cytoplasmic membrane as found in "brown" *Rhodospirillum* species such as *R. molischianum* (77); (d) paired lamellae running parallel to the cytoplasmic membrane in the budding phototrophic organisms, *Rhodomicrobium vannielii* and in *Rhodopseudomonas* species (33); and (e) bundled tubes, found so far only in the purple sulfur bacterium *Thiocapsa pfennigii* (46). For the interested reader, a consideration of the arrangement of bacteriochlorophyll in photosynthetic membranes can be found in the recent review by Drews (41).

Green bacteria contain a morphologically unique membrane arrangement (170). Bacteriochlorophyll *a* is localized in the cytoplasmic membrane (49, 185, 208), which, unlike purple bacteria, does *not* differentiate within the cell. The majority of the bchl in green bacteria (bchl *c*, *d*, *e* or c_s) is found within a number of oblong vesicles, referred to as chlorosomes,* which line the cell periphery (208, 209). Chlorosomes are bounded by a nonunit lipid membrane 2 to 3 nm thick rather than by a true unit (lipid bilayer) membrane. Chlorosomes are attached to the cytoplasmic membrane via a "base plate" structure, but the ontogeny of the vesicles and related support structures is not yet clear (208). Chlorosomes are also present in *Chloroflexus* (118, 168), and the facultative nature of this organism has stimulated recent studies of chlorosome biogenesis following transfer of cells from heterotrophic dark (areobic) growth conditions [where chlorosomes are not observed (118)] to phototrophic conditions (185, 205, 208, 209).

2.1.3 Taxonomy of Purple and Green Sulfur Bacteria

Purple and green sulfur bacteria have much more in common physiologically and ecologically than either group has to its facultative counterpart (the purple and green "nonsulfur" bacteria, respectively). This is because pigmented sulfur bacteria share a photosynthetic metabolism which is intimately tied to the use of reduced sulfur compounds. Hydrogen sulfide, generated by the activities of the dis-

---

*In the older literature, chlorosomes were referred to as "chlorobium vesicles."

similatory sulfate-reducing bacteria (see Chapter 10), serves as a source of electrons for photoautotrophic metabolism of purple and green sulfur bacteria (154). Depending on several factors, including sulfide concentration, pH, and light intensity, sulfide is anaerobically oxidized by these organisms to elemental sulfur or sulfate (154). If the sulfide concentration is high, elemental sulfur will be formed in the first step of sulfide oxidation and the photosynthetic stoichiometry will approach

$$2(H_2S) + CO_2 \longrightarrow (CH_2O) + H_2O + 2S^0$$

When sulfide concentrations are low, however, elemental sulfur does not accumulate; instead, sulfide is oxidized completely to sulfate during CO_2 reduction as follows:

$$H_2S + 2(CO_2) + 2(H_2O) \longrightarrow 2(CH_2O) + H_2SO_4$$

Experimental conformation of the stoichiometries above were first published over 50 years ago in the monograph of van Niel (241) on the purple and green sulfur bacteria. Thiosulfate ($S_2O_3^{2-}$) can substitute for H_2S as a reductant for photosynthesis in cultures of many purple and green bacteria, but sulfide is considered the most ecologically relevant electron donor, since its biological production is well documented (154, 165). The enzymology of sulfur transformations by phototrophic bacteria is rather complex and is not completely understood; for a good overview of assimilatory and dissimilatory sulfur metabolism in phototrophic bacteria, the reader is referred to the recent reviews of Trüper (222, 223), Trüper and Fischer (224), and Neutzling et al. (140).

Although the nature of the photosynthetic pigments clearly differentiate representatives of the purple and green sulfur bacteria, the process of sulfur production from hydrogen sulfide is also of great taxonomic significance. Figure 2.2a is a photomicrograph of a typical purple sulfur bacterium, *Chromatium vinosum*. Sulfide oxidized by purple sulfur bacteria is converted into globules of elemental sulfur which are temporarily stored *inside* the cells (226); these globules are eventually oxidized to sulfate to generate additional reducing power when sulfide concentrations diminish. Members of the purple sulfur genus *Ectothiorhodospira* (Fig. 2.2c) and all members of the green sulfur bacteria (e.g., *Chlorobium*, see Fig. 2.2d) do *not* store elemental sulfur within the cell, but instead, produce extracellular sulfur which remains dispersed in the medium or occasionally becomes attached to the outside of the cells; in most instances this extracellular sulfur can be used as a source of additional reducing power when sulfide becomes limiting (154, 238).

Morphological diversity is quite apparent within the purple sulfur bacteria. Ten genera are currently recognized (226; see Table 2.1) including spherical, rod, and spiral-shaped cells, all of which divide by binary fission. In the genus *Chromatium* both large- and small-diameter species are known. Depending on environmental conditions, in particular light intensity and sulfide concentration, many Chromatiaceae grow in the form of nonmotile aggregates embedded in slime (156). Many

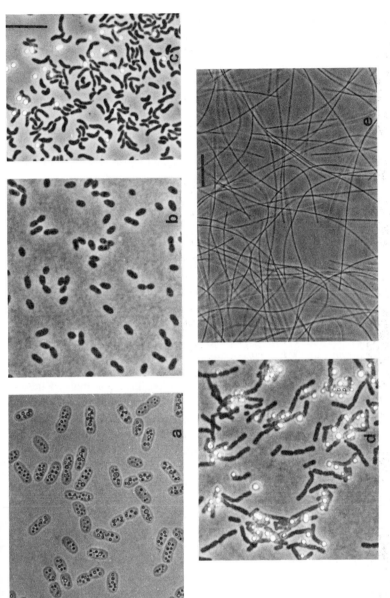

Figure 2.2 Photomicrographs of representative purple and green bacteria. (*a*) Light field micrograph of *Chromatium vinosum*. (*b*) Phase micrograph of *Rhodobacter sphaeroides*. (*c*) Phase micrograph of *Ectothiorhodospira mobilis*. (*d*) Phase micrograph of *Chlorobium limicola*. (*e*) Phase micrograph of *Chloroflexus aurantiacus*. Marker bar for (*a*), (*b*), (*c*), and (*d*) (shown in *c*) equals 10 μm; marker bar in (*e*) equals 20 μm. Micrographs (*a*), (*b*) and (*d*) courtesy of Norbert Pfennig. Micrograph (*a*) from H. G. Trüper and N. Pfennig, 1981, p. 299–312, *in* M. P. Starr, H. Stolp, H. G. Trüper, A. Balows, and H. G. Schlegel (eds.), The prokaryotes, Springer-Verlag, New York. Micrograph (*c*) courtesy of Hans Trüper, from H. G. Trüper, and J. F. Imhoff, 1981, p. 274–278, in Starr et al., The prokaryotes.

purple sulfur bacteria are motile by polar flagella, but a few genera are permanently immotile (226); several of the latter do contain gas vesicles, however, and are therefore capable of movement in the water column.

Only four genera of green sulfur bacteria have been grown in pure culture (163, 226), and the group displays a more limited morphological diversity than the purple sulfur bacteria (Table 2.1). Motility by flagellar means is totally absent in green sulfur bacteria, but *Pelodictyon*, *Ancalochloris*, *Chloronema*, and *Oscillochloris* contain gas vesicles and presumably use them to position themselves in the water column. Also, the newly described green bacterium *Chloroherpeton* (60) is motile, but by gliding motility, not by flagellar means. Most species of green sulfur bacteria divide by the process of binary fission, including the morphologically irregular organisms, *Prosthechochloris aestuarii* and *Ancalochloris perfilievii*. The latter organisms produce cytoplasmic appendages referred to as prosthecae, typical of the heterotrophic prosthecate bacterium *Prosthecomicrobium* (211). *Pelodictyon clathratiforme* is a green bacterium that divides by ternary fission, resulting in a netlike array of cells resembling a branching process (161).

Consortia of green sulfur bacteria and one or more heterotrophic components are frequently observed in natural samples and enrichment cultures (see Sections 2.3.1.1 and 2.3.1.2). The nature of these relationships appears to be syntrophic (symbiotic), with the heterotrophic component producing sulfide which is consumed by the green bacterium (154). Organisms such as the synchronously dividing "*Chlorochromatium*" and "*Pelochromatium*" show such a close association between the phototrophic and heterotrophic components (see Fig. 2.6 for photo and electron micrographs of *Chlorochromatium*) that it is likely that most of the organic matter required for energy generation by the heterotrophic component is excreted by the green bacterium. A rather cozy physiological symbiosis is thus observed in *Chlorochromatium* and *Pelochromatium*, with the phototroph and the heterotroph each producing needed substrates for the other.

Consortia often develop as well between free-living green bacteria and sulfate- or sulfur-reducing bacteria. An instructive example is that of the "*Chloropseudomonas*" syntrophy, a particularly relevant example because this "organism" was used for years in biochemical/physiological studies of green sulfur bacteria before it was shown to consist of a syntrophic partnership between a *Chlorobium* sp. (or a *Prosthechochloris* sp.) and dissimilatory sulfate or sulfur-reducing bacteria (10, 69). "*Chloropseudomonas ethylica*" was used routinely for laboratory studies of green bacteria because it could be grown photoheterotrophically with ethanol and bicarbonate as sole carbon sources in a medium containing little or no added sulfide. The ethanol was not metabolized by the phototroph, but instead was converted to acetate by the heterotroph; the electrons generated from ethanol oxidation were used to reduce sulfate or sulfur to sulfide (10).

Depending on the source of the "*C. ethylica*" culture, either sulfate- or sulfur-reducing heterotrophs were observed (10, 69). The sulfur-reducing species was shown to be a previously undescribed prokaryote, *Desulfuromonas acetoxidans*, which is capable of reducing sulfur but not sulfate. *D. acetoxidans* oxidizes acetate to CO_2 while reducing sulfur to sulfide (160). Sulfide produced by the heterotrophs in "*C. ethylica*" cultures was oxidized back to sulfur or sulfate to complete an anaerobic sulfur cycle as follows:

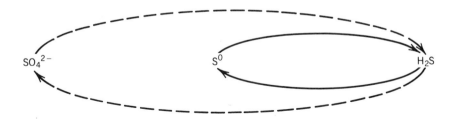

Syntrophic relationships between green bacteria and heterotrophic partners can be established experimentally in the laboratory (10, 256). The value of such relationships to both heterotrophic and phototrophic components is clearly evident by the luxuriant growth observed when the components are co-cultured, in contrast to their growth in pure culture. Syntrophic sulfur cycling does not occur among purple sulfur bacteria because these organisms store elemental sulfur inside the cells, therefore making sulfur unavailable to the heterotroph (238).

2.1.4 Taxonomy of Facultative Purple and Green Bacteria

Members of the families Rhodospirillaceae and Chloroflexaceae are more metabolically versatile than the green and purple sulfur bacteria. Besides growing photoheterotrophically or photoautotrophically, most facultative anoxygenic phototrophs can grow as heterotrophs in darkness at full oxygen tensions (163, 242), although a few species require microaerobic conditions for optimal growth in darkness (163). Dark anaerobic fermentative and anaerobic respiratory growth have also been demonstrated in a variety of Rhodospirillaceae (123, 180, 230, 259). Six genera (with over 20 species) are currently recognized in the Rhodospirillaceae (see Table 2.1): *Rhodopseudomonas*, *Rhodobacter*, *Rhodopila*, *Rhodospirillum*, *Rhodomicrobium*, and *Rhodocyclus* (85). (The species and genera of nonsulfur purple bacteria have recently been rearranged with the creation of two new genera, *Rhodobacter* and *Rhodopila* (85); organisms grouped in these genera were previously recognized as species of *Rhodopseudomonas*.) Morphological heterogeneity is not extensive among nonsulfur purple bacteria; most representatives are either rods (some budding), cocci, or various-sized spirilla (see Table 2.1). A photomicrograph of a typical rod-shaped representative, *Rhodobacter sphaeroides*, is shown in Fig. 2.2*b*.

In contrast to the colored sulfur bacteria, species of Rhodospirillaceae capable of growth below pH 5 are well documented (152, 153). This implies that moderately acidic habitats such as bogs, acid lakes, and streams, and acidic moist soil (habitats usually devoid of significant sulfate reduction and hence pigmented sulfur bacteria), will contain nonsulfur purple bacteria. For example, the author has isolated strains of *Rhodopseudomonas acidophila* from lakes with a pH of 5 (unpublished results), and other acid-tolerant Rhodospirillaceae, such as *Rhodopseudomonas palustris* or *Rhodomicrobium vannielii* from acidic waters or sediments. *Rhodopila globiformis* (originally described as *Rhodopseudomonas globiformis*) has only been isolated from an acidic spring in Yellowstone National Park (153). A second strain was recently isolated by the author from sediment of an acid lake in Yellowstone (unpublished results). *R. globiformis* is unusual in many respects, including its unique carotenoid composition (186), preference for photoheterotrophic growth with sugars or sugar alcohols, and ability to grow at a pH near 4 (153). Cells of *R. globiformis* have been observed by the author in warm sulfur springs in Yellowstone at a pH as low as 3.

Chloroflexus is the only genus in the family Chloroflexaceae that has been grown in pure culture, and contains two species, which differ primarily in their temperature requirements for growth (24). *C. aurantiacus* is restricted to alkaline hot spring effluents at temperatures from 40 to 70°C (3, 22, 168), while mesophilic *Chloroflexus* grow as benthic forms in stratified freshwater lakes (64). Cells of *Chloroflexus* are arranged in filaments (Fig. 2.2e) containing individual cells 0.5 to 1.0 μm in width and 2 to 6 μm in length; filaments can range up to 100 μm in length (168).

2.1.5 Phylogenetic Relationships among Phototrophic Bacteria

Methods are now available for determining phylogenetic (evolutionary) relationships between bacterial species. These methods involve comparative macromolecular sequencing. The most enlightening of these methods is the comparative sequencing of 16S RNA isolated from the 30S subunit of the bacterial ribosome (207, 251a).

Ribosomal RNA sequencing has shown that purple phototrophic bacteria constitute a major group of eubacteria (for a distinction between *eubacteria* and *archaebacteria* one should refer to Ref. 207, 251a). Three broad subdivisions of purple bacteria are recognized. Two of the subdivisions contain purple nonsulfur bacteria, while the third group contains the purple sulfur bacteria; Woese refers to these subdivisions as the alpha, beta, and gamma purple bacteria, respectively (251a, 253–255). The alpha purple bacteria, which contains the majority of nonsulfur purple bacteria, can itself be subdivided into three clusters. The alpha-1 cluster contains most of the species traditionally grouped in the genus *Rhodospirillum*, alpha-2 contains *Rhodomicrobium* and all *Rhodopseudomonas* species (all species in alpha-2 divide by budding), and alpha-3 contains the *Rhodobacter* species (253). The beta group consists exclusively of species of *Rhodocyclus* (255).

Three clusters of purple sulfur bacteria have been defined as well (50). The *Ectothiorhodospira* species forms two clusters, the extremely halophilic species, and the marine and freshwater ectothiorhodospiras, respectively (50, 206). Both clusters of *Ectothiorhodospira* are related (at a fairly distant level) to purple sulfur bacteria, which store sulfur intracellularly, and at a far more distant level to non-sulfur purple bacteria and green bacteria (50). *Chromatium* and other sulfur-storing Chromatiaceae make up the third cluster and appear as a group to be very closely related (50).

Green bacteria contain two major lines of eubacterial descent. Surprisingly, *Chlorobium* and *Chloroflexus* bear no evolutionary relationship to one another (59). *Chloroflexus* is loosely related to a variety of nonphotosynthetic, primarily thermophilic eubacteria, while species of *Chlorobium* cluster tightly among themselves and show a more distant relationship to the gliding green sulfur bacterium, *Chloroherpeton* (59). Considering the major phenotypic similarities between *Chlorobium* and *Chloroflexus*, the presence of bacteriochlorophyll *c* located in chlorosomes (208, 209) and sulfide oxidation with the production of extracellular elemental sulfur (117, 154), the lack of phylogenetic relatedness between these groups is surprising. However, the facultative metabolism of *Chloroflexus* and the arrangement of reaction center bacteriochlorophyll–protein complexes in this organism [which resemble those of purple bacteria more closely than those of green bacteria (172a)] support the phylogenetic evidence.

The recently described strictly anaerobic phototroph *Heliobacterium* shows no affinity (by 16S rRNA sequencing) to any established group of phototrophic bacteria. Instead, *Heliobacterium* shows a specific relationship to certain gram-*positive* bacteria, especially to members of the genus *Bacillus* and *Clostridium* (252). This implies that photosynthesis may indeed have been an ancient phenotypic character widely dispersed among various eubacterial lines.

Phylogenetic studies have shown that phototrophic bacteria have a number of nonphotosynthetic relatives which presumably lost the ability to photosynthesize while evolving to colonize dark habitats successfully. Several nonphotosynthetic relatives of purple sulfur (50) and nonsulfur (253–255) bacteria have been identified. However, nonphotosynthetic relatives of green bacteria are currently unknown (59).

2.2 PHYSIOLOGY OF PHOTOTROPHIC BACTERIA

2.2.1 Physiology of Purple Sulfur Bacteria

The purple and green sulfur bacteria share a phototrophic existence intimately linked to the utilization of sulfide. Sulfide is therefore important for mass development of these organisms, and large numbers of green and purple sulfur bacteria are observed only in aquatic environments where either biogenic or abiogenic sulfide accumulates in significant amounts (154, 156). It is therefore logical to assume that growth of purple and green sulfur bacteria in nature is primarily phototrophic.

The metabolic options available to purple sulfur bacteria, however, are not restricted to phototrophic growth modes. Certain members of the Chromatiaceae, including representatives of the genera *Chromatium*, *Thiocystis*, *Amoebobacter*, and *Thiocapsa*, are capable of heterotrophic and/or lithotrophic growth when the oxygen concentration is significantly reduced ["microaerobic" growth (93)]. *Thiocapsa roseopersicina* and *Thiocystis violacea* appear to be the most oxygen tolerant of the Chromatiaceae (93, 101, 102). In *Chromatium vinosum*, the best substrates for heterotrophic growth are the organic acids malate, fumarate, and propionate; the addition of thiosulfate stimulates growth and the incorporation of other substrates, including acetate and pyruvate (93). The metabolism of organic substrates by *C. vinosum* differs from that of *Amoebobacter* and *Thiocapsa*, in that true heterotrophic growth (i.e., in the absence of thiosulfate or sulfide) occurs only in *C. vinosum*. Mixotrophic conditions (where both organic and inorganic electron donors are present) are required for dark growth of other purple sulfur bacteria (93).

Growth rates of purple sulfur bacteria under dark microaerobic conditions are extremely slow. Although the cell yield per mole of thiosulfate consumed by lithotrophically grown *Thiocapsa roseopersicina* or *Thiocystis violacea* is the same as that of the colorless sulfur bacteria (e.g., thiobacilli), growth rates of phototrophs with thiosulfate as energy source are considerably lower than those of typical lithotrophic bacteria (93). Considering that in nature, heterotrophic growth of purple sulfur bacteria would put them in direct competition with a broad spectrum of heterotrophic bacteria as well as with Rhodospirillaceae, and that growth lithotrophically would put them in competition with lithotrophs that specialize in the use of reduced sulfur compounds as energy sources (104), it is likely that the ecological significance of heterotrophic or lithotrophic metabolism by purple sulfur bacteria is slight. Instead, these metabolic options probably serve as a mechanism for temporary survival in transiently oxygenated environments or as a means of generating ATP at night, rather than as a major means of supporting growth in nature. However, the occasional finding of purple sulfur bacteria in permanently dark habitats (76) suggests that dark metabolism may be of strong survival value to these organisms.

Purple sulfur bacteria probably compete best in nature as phototrophs. For example, migrations of purple sulfur bacteria away from oxic zones have been reported by Jørgensen in studies of a sulfuretum at Kalø Vig near Aarhus, Denmark (89) [a sulfuretum is a community of lithotrophic and phototrophic bacteria which develop in response to a constant supply of sulfide of biological or geochemical origin (see discussion in Refs. 2 and 154)]. The phototrophic component of the sulfuretum (which had developed in the upper layers of marine coastal sediments) was identified as a *Chromatium* sp. (89). The phototroph was observed to move rapidly down into the sediment away from oxygenated regions dominated by cyanobacteria and sulfide-oxidizing lithotrophs during the daylight hours when high levels of oxygenic photosynthesis was occurring; these highly motile swarms of *Chromatium* were presumably retreating to a more sulfide-rich environment in order to continue their phototrophic metabolism (89). Migrations of phototrophic bacteria away from oxic regions probably represents a negative chemotactic response to oxygen. However, because its Calvin cycle enzymes are not repressed by high oxygen concentrations (102), the purple sulfur bacterium *Thiocapsa roseopersicina* may behave differently in a similar situation. Assuming that lithotrophic respiratory processes are not inhibited by light, this organism could stay in one place and switch from photoautotrophic to lithoautotrophic metabolism, depending on oxygen levels in the environment.

Many purple sulfur bacteria can use thiosulfate or H_2 as electron donors for autotrophic growth (224, 260). As is true with sulfide-grown cells, carbon dioxide is assimilated autotrophically by reactions of the reductive pentose (Calvin) cycle (54). Additional carboxylation reactions (catalyzed by phosphoenolpyruvate carboxylase, pyruvate carboxylase, acetyl-CoA carboxylase, and "malic enzyme") can also be demonstrated in extracts of purple sulfur bacteria, but their contribution to net carbon assimilation under autotrophic conditions is minor; these reactions presumably play an ancillary or anaplerotic role (54, 179).

A variety of carbon sources are assimilated by purple sulfur bacteria under photosynthetic growth conditions (139, 202). Organic acids and fatty acids seem to be the preferred substrates, but short-chain alcohols and even carbohydrates are photoassimilated by certain species (202). Highly reduced substrates (e.g., butyrate) require CO_2 for photoassimilation. The function of CO_2 in this case is to act as an electron sink for the assimilation of reduced organic compounds, since the chemical composition of purple bacteria [approximately $C_5H_8O_2N$ (232)] is more oxidized than a long-chain fatty acid such as butyrate ($C_4H_8O_2$). Photohetero-trophic growth of *C. vinosum* and other *Chromatium* species *capable of assimilatory sulfate reduction* is sulfide independent (139, 202).

Several *Chromatium* species are unable to grow in the absence of sulfide and are nutritionally quite restricted. These included the large-celled species such as *C. okenii* and *C. weissei*, as well as the thermophilic species, *C. tepidum* (116). *Thiospirillum* and a few other purple sulfur bacteria also show this pattern. Accordingly, Trüper (223a) has proposed that two nutritional classes of purple sulfur bacteria be recognized, the nutritionally restricted "Chromatiaceae I" and the nutritionally versatile "Chromatiaceae II." The members of group I assimilate only acetate or pyruvate during photoheterotrophic growth, while group II species resemble the nutritional versatility of typical small-celled chromatia (223a).

Anaerobic dark metabolism in *Chromatium vinosum* involves the degradation of photosynthetically produced glycogen with the production of a more reduced polymer, poly-β-hydroxybutyric acid; the excess electrons generated during this transformation are used to reduce stored elemental sulfur to sulfide (233). Sugar units are presumably metabolized via the Embden–Meyerhof–Parnas pathway, but the amount of ATP produced (which is sufficient to power motility and maintain viability anaerobically in darkness for long periods) is insufficient to support growth (233). This is in contrast to Rhodospirillaceae, where several mechanisms supporting anaerobic dark growth are known (see Section 2.2.3.4).

2.2.2 Physiology of Green Sulfur Bacteria

Members of the family Chlorobiaceae are obligate phototrophs, but are capable of assimilating a restricted group of organic compounds as major sources of cell carbon in the presence of sulfide; acetate seems to be a particularly favorable substrate

(178, 196, 202). In addition, *small* amounts of a variety of other organic compounds, including amino acids and glucose, are assimilated by *Chlorobium* (95). In contrast to the photoheterotrophic metabolism of some purple bacteria or of *Chloroflexus*, the assimilation of organic compounds by *Chlorobium* occurs only in the presence of an inorganic electron donor plus CO_2 (79).

Acetate carbon can be distributed throughout all macromolecular cell fractions of *Chlorobium*, although it tends to accumulate in large amounts as storage polysaccharide (196, 200, 219). Poly-β-hydroxybutyrate is not made by *Chlorobium* (79). Under dark anaerobic conditions polyglucose is degraded to succinate, propionate, caproate, and acetate (200). Analogous to the breakdown of glycogen in *Chromatium* (see Section 2.2.1), these transformations presumably serve the dark maintenance energy requirements of *Chlorobium*. Heterotrophic growth has never been observed under any nutritional conditions anaerobically, microaerobically, or aerobically in species of Chlorobiaceae (93).

An intensely studied and highly controversial area of green sulfur bacterial physiology has been the mechanism of autotrophic CO_2 fixation. The controversy has been fueled by repeated reports, from a variety of laboratories, of the absence of Calvin cycle enzymes in *Chlorobium* (18, 48, 195, 199). On the other hand, two reports claiming low levels of Calvin cycle enzymes in *Chlorobium* have been published (201, 215). However, the original contention of Evans et al. (48) that a reversal of reactions of the citric acid cycle involving two novel ferredoxin-dependent carboxylations represents the major route of CO_2 incorporation in *Chlorobium* has gathered increasing experimental support.

The radiolabeling studies of Sirevåg and her colleagues have demonstrated that phosphoglyceric acid, the first labeled product of CO_2 fixation expected in organisms employing the Calvin cycle (54), is not an early labeled product following short-term exposure of whole cells to $^{14}CO_2$ (195, 199). In detailed labeling studies by Fuchs et al. (52, 53), the incorporation of [^{14}C]-pyruvate, propionate, or acetate by cultures of *Chlorobium* resulted in labeling patterns totally inconsistent with those expected from Calvin cycle reactions, yet fully consistent with the distribution of ^{14}C expected from autotrophy based on a reversal of oxidative citric acid cycle reactions.

In a different approach to the same problem, ^{13}C, a stable (nonradioactive) isotope of carbon, has been used to study isotope discrimination of the primary CO_2-fixing enzymes in green and purple bacteria. Interestingly, the usual discrimination against the heavy isotope of CO_2 ($^{13}CO_2$) observed in purple bacteria was *not* observed in cultures of *Chlorobium* grown with $^{13}CO_2$ (174, 197). These results were interpreted to mean that the key CO_2-fixing enzyme of the purple and green bacteria was not the same. Presumably, the main CO_2-fixing enzyme(s) in green bacteria does not discriminate against $^{13}CO_2$ as readily as does ribulose bisphosphate carboxylase, the key enzyme of the Calvin cycle (54).

For many years the absence of detectable citrate lyase activity in extracts of cells of *Chlorobium* was used to argue that green sulfur bacteria could not possibly run a reverse citric acid cycle as a net means of incorporating CO_2 (7). In the absence of citrate lyase, citrate cannot be broken down into acetyl-CoA and oxaloacetate, the latter of which would presumably serve as an acceptor to initiate a new round of the cycle. The recent detection of an ATP-dependent citrate lyase (86) in extracts of *Chlorobium*, however, has now put the reverse citric acid cycle hypothesis on solid ground. Energy-dependent citrate lyases are common in eukaryotes but not in prokaryotes; the only citrate lyase known in any phototrophic bacterium (that of *Rhodocyclus gelatinosus*) is not an ATP-dependent enzyme (182). Citrate lyase activity in *Chlorobium* generates acetyl-CoA (which can be further carboxylated to yield pyruvate and hence cell material) and oxaloacetate, thus completing the cycle.

Evidence against the Calvin cycle and for the Reductive citric acid cycle as the major means of CO_2 assimilation in *Chlorobium* is now quite strong. This poses an interesting evolutionary question: In what ways are the green bacteria related to other autotrophs? The lack of a Calvin cycle in green bacteria implies that a group of autotrophic organisms has survived the rigors of evolutionary selection without the CO_2 reduction mechanism universally distributed in other photoautotrophs. This, of course, implies an ancient origin for the green bacteria. The recent finding that autotrophically grown methanogenic bacteria also lack Calvin cycle enzymes (51) suggests that alternatives to the Calvin cycle in prokaryotes may be more common than previously thought. Ribosomal RNA sequencing has shown the ancient eubacterial nature of green sulfur bacteria (see Section 2.1.5) and thus the reverse citric acid cycle may indeed have predated Calvin cycle mechanisms.

2.2.3 Physiology of Facultative Purple Bacteria

Rhodospirillaceae are a heterogeneous assemblage of photosynthetic bacteria capable of growth either phototrophically or heterotrophically. Particularly well-developed physiological strategies have evolved to support dark growth and metabolism in this group. Most Rhodospirillaceae grow best in mineral media containing an organic acid or fatty acid as carbon source and ammonia as nitrogen source. Many species show one or more vitamin requirements (see Section 2.2.3.1). Photosynthetic growth with organic substances serving as the sole or major carbon source is referred to as *photoheterotrophic* growth. Many of these same organic compounds used photosynthetically can serve as carbon *and* energy sources for heterotrophic respiratory growth of Rhodospirillaceae in the dark. Anaerobically in darkness, fermentative or anaerobic respiratory mechanisms support the growth of a variety of Rhodospirillaceae. Finally, lithotrophic growth of certain Rhodospirillaceae is possible.

2.2.3.1 Autotrophic Growth

The inability of typical Rhodospirillaceae to grow autotrophically at the expense of high concentrations of sulfide delayed investigations of the autotrophy of this group for many years. Sulfide, at levels that support luxuriant growth of purple and green sulfur bacteria (i.e., 2 to 4 mM), is completely growth inhibitory to virtually all Rhodospirillaceae (73, 163, 226). The first demonstration that an inorganic electron donor could support photoautotrophic growth of a nonsulfur purple bacterium, in this case *Rhodospirillum rubrum*, was provided when Ormerod and Gest grew this organism in a synthetic medium with CO_2 as sole carbon source and H_2 as sole electron donor (145). The majority of Rhodospirillaceae are now known to be capable of H_2-supported photoautotrophic growth (163). In darkness, hydrogen can play another role in the metabolism of certain Rhodospirillaceae; lithotrophic growth with H_2 as electron donor, oxygen as electron acceptor, and CO_2 as sole carbon source has been reported in *Rhodobacter capsulatus* and *Rhodopseudomonas acidophila* (124, 193). It is likely that lithotrophic H_2 oxidation is common among Rhodospirillaceae because virtually all of these species possess an uptake hydrogenase (260). Carbon dioxide fixation in autotrophically grown purple bacteria occurs via the Calvin cycle (54).

In 1972, Hansen and van Gemerden showed that the dichotomy between purple sulfur and nonsulfur bacteria, based on the ability of the former, and the inability of the latter, to oxidize sulfide was no longer valid (73). When grown in continuous culture where sulfide levels can be kept constant and very low, many Rhodospirillaceae grow autotrophically at the expense of sulfide (73). Sulfide is oxidized to either sulfur or sulfate, depending on the species, and strains of *R. capsulatus* were shown to be the most sulfide tolerant of all Rhodospirillaceae tested (73). Photoheterotrophic sulfide-containing enrichment cultures yielded a new extremely sulfide-tolerant member of the Rhodospirillaceae, *Rhodobacter sulfidophila*. This organism oxidizes sulfide completely to sulfate without accumulating elemental sulfur (74). Detailed studies of dissimilatory sulfur metabolism by Rhodospirillaceae have been performed by Neutzling et al. (140).

Another inorganic sulfur compound, thiosulfate, is used as a photosynthetic electron donor by *Rhodopseudomonas palustris*, *R. sulfidophila*, and *Rhodopseudomonas sulfoviridis* (224). In addition, several species of Rhodospirillaceae contain rhodanese, the thiosulfate-splitting enzyme (222, 224); however, only the above-mentioned species can actually *grow* photosynthetically at the expense of thiosulfate.

Yeast extract in small amounts has been a common addition to a wide variety of media formulated for the Rhodospirillaceae (11). In general, yeast extract serves as a source of water-soluble B vitamins, one or more of which are required by the majority of recognized species. Requirements for thiamine, nicotinic acid, biotin, and *p*-aminobenzoic acid are most frequently encountered; such requirements are never observed in purple or green sulfur bacteria (243). The latter organisms frequently have a requirement for vitamin B_{12}, and this vitamin is also required by a few Rhodospirillaceae [most notably *Rhodocyclus purpureus* (151) and certain strains of *Rhodocyclus tenuis* and *Rhodopseudomonas palustris* (191, 192)]. The vitamin requirements of the major species of Rhodospirillaceae are discussed by van Niel (243).

2.2.3.2 Photoheterotrophic Growth

A number of organic compounds can support photoheterotrophic/heterotrophic growth of Rhodospirillaceae. Organic acids, amino acids, fatty acids, alcohols, carbohydrates, and even C-1 compounds (in certain species) are metabolized (202). With minor exceptions, the citric acid cycle intermediates malate, succinate, and fumarate are universally used, as are pyruvate and acetate; many species also use ethanol, lactate, propionate, and fumarate (163, 226). Selective utilization of organic compounds by various species has helped to develop a nutritional basis for certain taxonomic divisions within Rhodospirillaceae (226).

The ability or inability to utilize particular organic substances appears to be a rather uniform property of strains of a given species. For instance, in nutritional studies of 33 strains of *Rhodobacter capsulatus*, the inability of the species to utilize ethanol held up in all 33 strains examined (248). A similar picture emerged in the case of fructose, a compound that is not utilized by most Rhodospirillaceae, but serves as an excellent carbon or carbon/energy source for *R. capsulatus*; fructose was utilized by each of the strains examined (248). Certain organic substrates are utilized by only a very restricted group of species. Citrate, for instance, is utilized only by *Rhodopseudomonas acidophila*, *Rhodocyclus gelatinosus*, and *Rhodobacter sphaeroides* (163). One-carbon compounds such as formate or methanol support slow growth of a number of Rhodospirillaceae, but methanol supports particularly good growth only of strains of *Rhodopseudomonas acidophila* (175). Growth of Rhodospirillaceae has not been achieved on methylamine or formaldehyde (175). Enrichment culture studies using methanol–bicarbonate mineral media resulted in the isolation of seven different species of Rhodospirillaceae, with *R. acidophila* being the most frequently enriched species (40).

Bicarbonate is absolutely essential for anaerobic growth of Rhodospirillaceae on methanol (175). The stoichiometry of conversion shows that 1 mol of CO_2 is required for each 2 mol of CH_3OH assimilated:

$$2CH_3OH + CO_2 \longrightarrow 3(CH_2O) + H_2O$$

Apparently, CO_2 is required because of the highly reduced state of methanol; thus CO_2 serves as an electron sink to elevate the organic substrate, methanol, to the approximate oxidation-reduction level of cell material (175).

A small group of Rhodospirillaceae photometabolize aromatic compounds such as benzoate, hydroxy derivatives of benzoate, and cyclohexane carboxylate (45). Enrichment cultures employing benzoate frequently result in isolates of *Rhodospirillum fulvum* (162), or *Rhodopseudomonas* species, in particular *R. palustris* (11, 183, 243). Biochemical studies of benzoate catabolism by *R. palustris* have demonstrated that metabolism of this compound involves reductive, rather than oxidative, ring cleavage (44, 45, 72). Benzoate is reduced and hydroxylated to the seven-carbon saturated dicarboxylic acid, pimelate. The latter compound is degraded to acetate, and further to CO_2 via the citric acid cycle (45).

A variety of hydroxylated benzoate derivatives are metabolized by Rhodospirillaceae only under certain growth conditions. Benzoate itself, for example, is only used as a carbon source by *R. palustris* under photosynthetic conditions, while 3,4-hydroxybenzoate is only degraded heterotrophically (45). These results suggest that certain aromatic compounds can serve as carbon but not energy sources for phototrophic bacteria; this selectivity may be due to the requirement for monooxygenase activity in the catabolism of certain aromatic compounds. Benzene is not attacked photosynthetically, presumably due to the absence of oxygen or a carboxyl group on the ring (however, heterotrophic growth of photosynthetic bacteria on benzene has not been reported either). Cyclohexane carboxylate is utilized particularly well by *Rhodocyclus purpureus* (151). Certain other aromatic compounds, in particular cinnamic acid and its derivatives, are metabolized by *R. palustris* (unpublished observations).

2.2.3.3 Heterotrophic (Aerobic) Growth

Aerobic heterotrophic growth of phototrophic bacteria on pyruvate, succinate, malate, and many other compounds proceeds through the oxidation of these substrates to CO_2 via the citric acid cycle (CAC; 4, 202). Growth on acetate also involves CAC reactions along with the anaplerotic reactions of the glyoxylate cycle (202). The key enzyme of the latter pathway, isocitrate lyase, is present in certain Rhodospirillaceae, yet absent from many others, even if they are grown on acetate (103).

In a detailed study of CAC reactions in *Rhodobacter capsulatus* grown phototrophically or heterotrophically, levels of the enzyme α-ketoglutarate dehydrogenase (KGD) were found to play a prominent role in controlling the overall rate of CAC reactions (4). Heterotrophically, energy conversion in *R. capsulatus* is dependent on operation of the complete CAC, and KGD levels are high; cells grown phototrophically contain significantly lower levels of KGD (4). Molecular oxygen, even at very low concentration, appears to trigger synthesis of KGD in *R. capsulatus*, and in this fashion O_2 serves as an overall regulator of CAC activity. As expected, citrate synthase and isocitrate dehydrogenase also show elevated activities in aerobically grown *R. capsulatus* cells. Anaerobically, the lower levels of key CAC enzymes observed are apparently sufficient to generate biosynthetic intermediates, especially α-ketoglutarate and succinyl ~ CoA.

The picture that emerges from studies of *R. capsulatus* suggests that the CAC is important for heterotrophic growth of all Rhodospirillaceae. Virtually every compound that supports dark aerobic growth can, in some way or another, be converted into a CAC intermediate. Growth on sugars or glycerol can also proceed via CAC reactions once the compound has been converted to pyruvate. In *R. capsulatus* glucose and fructose are converted to pyruvate via two distinctly different catabolic sequences; glucose is metabolized via an inducible Entner–Duodoroff pathway, an enzymatic series commonly found in pseudomonads, while fructose is degraded via the Embden–Meyhof–Parnas scheme (32).

Glycerol is catabolized by a very restricted group of Rhodospirillaceae, but glycerol-utilizing ability can be "induced" in wild-type strains of *R. capsulatus* (which are unable to utilize glycerol) by "gain of function" mutation(s) (107). Spontaneous glycerol-utilizing mutants of *R. capsulatus* arise when wild-type cultures are successively passed for several generations in media containing excess glycerol and growth-limiting amounts of malate (107, 204). The metabolism of glycerol, both heterotrophically and photosynthetically, occurs following the synthesis of two new enzymes, glycerokinase and glycerophosphate dehydrogenase, both of which are present in only trace amounts in wild-type strains (197, 204). The nature of the genetic events leading to the glycerol-utilizing phenotype in *R. capsulatus* is unclear. However, it is likely that the presence of glycerol in the growth medium selects for the growth of spontaneous mutants that have lost the tight regulatory control(s) which for unknown reasons prevent normal transcription of glycerol-degrading genes in the wild type (107, 204).

2.2.3.4 Fermentative Growth

The capacity of purple bacteria to ferment has been studied for some time. The fermentation of pyruvate was noted in early investigations of dark metabolism of *Rhodospirillum rubrum* (100), but anaerobic dark growth of this organism was not achieved until the studies of Uffen and Wolfe (230). In the latter work, growth at the expense of pyruvate of *R. rubrum* and a variety of other Rhodospirillaceae was achieved when the E_h of the medium was adjusted to a very low value, near that of the hydrogen electrode. Subsequent study of the physiology of dark anaerobic growth of *R. rubrum* has implicated the phosphoroclastic cleavage of pyruvate as the energy-yielding reaction (68).

Following the discovery of Yen and Marrs (259) that several Rhodospirillaceae could grow anaerobically in darkness in synthetic media containing glucose as sole carbon and energy source, with dimethylsulfoxide (DMSO) as electron acceptor, detailed studies of the physiology of growth of *R. capsulatus* under anaerobic dark conditions were carried out (35, 120, 123). A related electron acceptor, trimethyl-amine-N-oxide (TMAO), was used in the majority of these studies. During dark anaerobic growth of *R. capsulatus* on fermentable substrates (glucose, fructose, ribose, xylose, pyruvate, and lactate all supported growth, while succinate and malate did not) TMAO was reduced to trimethylamine, and cultures of *R. capsulatus* grew with generation times as short as 6 h (120). Pigmentless mutant strains of *R. capsulatus* grew with a 3-h generation time (121).

After extensive studies of the physiology (120) and biochemistry (35) of energy generation in cells of *R. capsulatus* grown anaerobically in darkness, it has been concluded that growth under these conditions is fermentative, but that fermentation of sugars in Rhodospirillaceae requires an accessory oxidant. Accordingly, the term ''oxidant-dependent fermentation'' was coined to describe growth under these conditions (123). Other interpretations of the need for an oxidant have been proposed, including the possibility that TMAO or DMSO-dependent anaerobic dark growth is really a form of anaerobic respiration (122, 136, 187). However, comparisons of TMAO-dependent growth in *R. capsulatus* and *E. coli*, the latter of which will grow with succinate and TMAO or even H_2 and TMAO (in the presence of yeast extract for cell carbon), support the conclusion that dark growth of *R. capsulatus* with fructose/TMAO is fermentative rather than respiratory in nature (35).

Classical anaerobic respiratory metabolism occurs in a strain of *R. sphaeroides*, where nitrate-dependent growth at the expense of nonfermentable substrates has been shown (180). Nitrate serves as an electron acceptor and is ''denitrified,'' resulting in the formation of N_2 (180). The denitrifying property is apparently quite rare since further reports of denitrification in other Rhodospirillaceae have not appeared.

The description of an unusual carbon monoxide (CO)–oxidizing phototrophic bacterium was published recently by Uffen (228). The organism, which grew anaerobically in darkness with CO as sole energy source, was later classified as a strain of *Rhodocyclus gelatinosus* (39). Of several other strains of *R. gelatinosus* tested for this property, only the neotype strain, ATCC 17011, was capable of dark anaerobic growth on CO (39). The stoichiometry observed was as follows:

$$CO + H_2O \longrightarrow CO_2 + H_2$$

The free energy change associated with this reaction is very low, $-4.8 \, \text{kcal}/\text{mol}^{-1}$, but is obviously sufficient eventually to drive ATP synthesis. With CO as the only carbon compound present, the product of the oxidation, CO_2, must serve as the cell carbon source; reducing power is derived from H_2. It will be interesting to see the extent to which other photosynthetic bacteria participate in CO oxidations, since CO is a common product of incomplete combustion and the microbial breakdown of certain biopolymers. Despite these large CO inputs, biological consumption keeps atmospheric CO levels relatively constant (229). A variety of nonphotosynthetic organisms, both aerobes and anaerobes, also have been shown to consume CO (229).

2.2.3.5 Nitrogen Fixation

Rhodospirillum rubrum was the first photosynthetic bacterium shown to fix nitrogen (67, 92). Shortly after this discovery, a few other species of Rhodospirillaceae, as well as representatives of the purple and green sulfur bacteria, were also shown to be diazotrophic (110, 111). In a recent survey of N_2 fixation in 18 species of Rhodospirillaceae, Madigan et al. (122) concluded that the ability to fix nitrogen was a universal property of species of this family (with one exception, see below), but that the efficacy of the nitrogen-fixing process varied considerably among species. As shown in Fig. 2.3, species such as *Rhodobacter capsulatus*, *R. sphaeroides*, and *Rhodopseudomonas viridis* express high levels of nitrogenase, the enzyme that catalyzes the nitrogen-fixing reaction, $N_2 + 6H \rightarrow 2NH_3$, when they are cultured photosynthetically on N_2. These species are therefore capable of rapid growth in synthetic media containing N_2 as sole nitrogen source. Other species of *Rhodopseudomonas*, and *Rhodobacter* and most *Rhodospirillum* species, are poorer N_2 fixers than is *R. capsulatus*, and a few species are very poor N_2 fixers, growing only very slowly on N_2 (see Fig. 2.3).

The only member of the family Rhodospirillaceae incapable of N_2 fixation is *Rhodocyclus purpureus* (132). A detailed study of nitrogen metabolism in this organism showed that *Rhodocyclus* metabolizes only ammonia or glutamine as nitrogen sources; other nitrogenous compounds are not utilized (132). The nature of the lesion(s) that prevent N_2 fixation in *Rhodocyclus* are not known. Perhaps the unusual habitat (swine waste lagoon) of the single strain of *R. purpureus* avail-

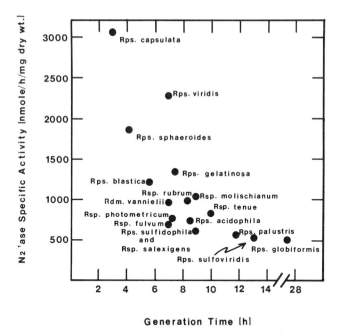

Figure 2.3   Mean *in vivo* nitrogenase (N$_2$ase) contents and growth rates of N$_2$-grown Rhodospirillaceae. These data represent the arithmetic mean of the nitrogenase values and growth rates of several strains of each species in most cases. Rps., *Rhodopseudomonas*; Rsp., *Rhodospirillum*; Rdm., *Rhodomicrobium*. From M. T. Madigan, S. S. Cox, and R. A. Stegeman, 1984, J. Bacteriol. 157:73–78. Note that *Rps. capsulata, Rps. sphaeroides,* and *Rps. sulfidophila* are now recognized as *Rhodobacter* species; *Rps. gelatinosa* and *Rsp. tenue* are now recognized as *Rhodocyclus* species; and that *Rps. globiformis* is now recognized as *Rhodopila globiformis* (see Ref. 85).

able has selected for the nif$^-$ phenotype in this organism, since it would be expected that such a habitat would be rich in amines and ammonia.

It is significant that most Rhodospirillaceae fix N$_2$ in darkness as well as photosynthetically. Photosynthetic growth on N$_2$ is always more rapid, but microaerobic (dark) N$_2$ fixation (194) and dark anaerobic (126) N$_2$ fixation occurs in certain Rhodospirillaceae. Growth under these conditions clearly indicates that nitrogen fixation is *not* a light-dependent process in phototrophic bacteria. Recent studies of the *in vivo* electron donor to nitrogenase in *Rhodospirillum rubrum* indicate that pyruvate, the physiological electron donor to nitrogenase in heterotrophic N$_2$ fixers such as *Clostridium pasteurianum*, probably serves as electron donor for N$_2$ fixation in *R. rubrum* as well (113). Aerobic N$_2$ fixation in phototrophic bacteria occurs only at reduced oxygen tensions (126, 194); oxygen sensitivity is likely to be due to the limited tolerance of nitrogenase for molecular oxygen.

In the absence of NH$_4^+$ and the substrate of nitrogenase, N$_2$, purple nonsulfur bacteria typically evolve large amounts of molecular hydrogen (57, 78, 146). H$_2$

is produced by the nitrogenase enzyme complex, which is capable of reducing protons as well as N_2 (137, 251). The hydrogen evolving capacity of *Rhodobacter capsulatus* is particularly noteworthy, as this organism produces large amounts of H_2 from C_3 or C_4 organic acids (78). A recent report of H_2 production in a related species, *Rhodobacter sphaeroides*, indicates that glucose can be converted by this species into H_2 and CO_2 in near-stoichiometric amounts (114). With yields of this magnitude, it is easy to imagine hydrogen-evolving photosynthetic bacterial systems being employed for the conversion of waste cellulose (after hydrolysis to glucose) to H_2.

Cellulolytic phototrophic bacteria have not been described, but mixed cultures containing cellulolytic fermentative anaerobes and organic acid–consuming, H_2-producing phototrophs have been reported (142). Phototrophic bacteria are ideal candidates for co-culture processes because of their ability to oxidize reduced organic compounds (i.e., fermentation products) completely to $CO_2 + H_2$ anaerobically via the citric acid cycle (58). For recent reviews of hydrogenase/nitrogenase interactions in phototrophic bacteria, the reader is referred to Ref. 137 and 245.

2.2.3.6 *Genetics of Rhodospirillaceae*

Genetic analysis of metabolic processes in photosynthetic bacteria is rapidly maturing following the discovery by Barry Marrs of the first genetic transfer system in phototrophic bacteria, the gene transfer agent (129). The gene transfer agent (GTA) resembles a defective bacteriophage and is produced by cultures of *Rhodobacter capsulatus*; the particles carry a small amount of chromosomal DNA from one strain to another in a fashion analogous to generalized transduction (129–131, 257). Using GTA, a map of the genes coding for bacteriochlorophyll and carotenoid synthesis has been constructed (258), and genetic lesions affecting nitrogenase/hydrogenase in mutant strains of *R. capsulatus* have been genetically corrected (246).

The main limitation in the GTA approach to genetic analysis in *R. capsulatus* rests in the limited amount of genetic information transferrable by one particle. Such constraints do not hinder fine-structure genetic mapping, but certainly prevent linkage analysis of larger pieces of DNA. Conjugative plasmids are the obvious answer to these limitations and considerable progress has been made in recent years in the use of native or foreign plasmids as a means of genetic transfer in Rhodospirillaceae (131, 181). *R. capsulatus* and *R. sphaeroides* appear to be the organisms of choice in this connection, and many of the genetic tricks developed in *E. coli* or *Pseudomonas* sp. seem to work in photosynthetic bacterial systems (131). Therefore, it is now possible to utilize a combination of classical genetics and recombinant DNA techniques to probe the molecular biology of photosynthetic bacteria; rapid progress is being made in areas such as large-scale genetic mapping (217) and the basis of control of pigment synthesis by molecular oxygen (12, 187a) in *R. capsulatus*.

To date, no major gene blocks (e.g., pigment genes, nitrogenase genes, etc.)

have been found on indigenous plasmids of wild-type strains of *R. capsulata*. This also seems to be the case with native plasmids of wild-type strains of enteric or pseudomonad organisms; only helpful genes, not essential genes, seem to be located on plasmids. Interestingly, however, it has been reported that the loss of a large plasmid from several wild-type strains of *Rhodospirillum rubrum* is associated with loss of photosynthetic competence (105). This suggests that the plasmid plays some role in photosynthesis, although the nature of this role is not yet clear. Genetics in *R. rubrum* is still in its infancy compared with the advances in *Rhodobacter* species, while genetic analyses of other families of phototrophic bacteria have not been reported. Good reviews of the genetics of nonsulfur bacteria are given in Refs. 130, 131, 181 and 187a.

2.2.4 Physiology of *Heliobacterium chlorum*

The recently described phototrophic bacterium *Heliobacterium chlorum* resembles members of the Rhodospirillaceae, because its metabolism is primarily photoheterotrophic and sulfide is growth inhibitory (56). Of considerable evolutionary interest is the fact that this organism synthesizes a type of bacteriochlorophyll that shares a key structural feature in common with chlorophyll *a* (17, 56). This pigment, bchl *g*, has not been reported in any other anoxygenic phototroph. Like chlorophyll *a*, and unlike any other bacteriochlorophyll, bchl *g* contains a vinyl ($HC=CH_2$) group on pyrrole ring 1 of the tetrapyrrole molecule (17).

H. chlorum is an obligate anaerobe and thus differs from the majority of Rhodospirillaceae in this regard. It is incapable of autotrophic growth and requires biotin as a growth factor. Systematic studies of carbon nutrition show that *Heliobacterium* is very restricted; pyruvate and lactate are the only carbon sources that support good growth (P. Romero, personal communication). *H. chlorum* utilizes NH_4^+ or N_2 as a nitrogen source (56). The exquisite sensitivity of *H. chlorum* to molecular oxygen is probably one reason this organism has previously eluded culture (enrichments for Rhodospirillaceae are frequently not established using *strictly* anaerobic techniques). *H. chlorum* does not survive exposure to even trace amounts of oxygen, and growth of pure cultures is obtained only when transfer and routine cultivation procedures are performed within the $N_2 + H_2 + CO_2$ atmosphere of an anaerobic glove box (56).

The inoculum for enrichment cultures of *H. chlorum* was obtained from soil; hence this organism may be an important component of the anaerobic zones of soil. Other strains of *Heliobacterium* have been isolated from rice paddy soils (H. Gest, personal communication), and a new genus, provisionally designated *Heliospirillum*, has been isolated from similar habitats by John Ormerod. Perhaps the use of strictly anaerobic techniques in the enrichment and isolation of nonsulfur phototrophic bacteria from a variety of soil habitats will yield new strains of *Heliobacterium* and *Heliospirillum*, or possibly even new species of oxygen-intolerant Rhodospirillaceae.

2.2.5 Physiology of Facultative Green Bacteria

Filamentous gliding prokaryotes containing bacteriochlorophylls were first described by Pierson and Castenholz (167). Further work eventually led to the creation of a new family of phototrophic bacteria, the Chloroflexaceae (221), with *Chloroflexus aurantiacus* (168, 169) serving as type genus. *Chloroflexus aurantiacus* inhabits alkaline hot spring effluents at temperatures from about 40 to 70°C (3, 168). The organism is thus clearly a thermophile, but not an extreme thermophile.

In contrast to typical green sulfur bacteria, *Chloroflexus aurantiacus* is facultative; most strains grow very well either photosynthetically or heterotrophically at full oxygen tensions, the only exception being the dark green obligately phototrophic strains isolated from high sulfide springs (60a, 169). In this regard *Chloroflexus* can be considered the metabolic equivalent of members of the Rhodospirillaceae, in the same way that purple and green sulfur bacteria share a metabolic plan based on the use of reduced sulfur compounds. As is the case with Rhodospirillaceae, *Chloroflexus* is also capable of oxidizing sulfide and can utilize sulfide as sole electron donor to support slow photoautotrophic growth (117). In contrast to most Rhodospirillaceae, however, *Chloroflexus* is very sulfide tolerant, as most strains will grow with up to 6 m*M* sulfide (3). Hydrogen also serves as electron donor for slow photoautotrophic growth of *Chloroflexus* (198). H_2 is assimilated via a sulfide-repressible uptake hydrogenase (43). Photoheterotrophic or heterotrophic growth of *Chloroflexus*, on the other hand, is quite rapid (169), and a variety of organic substrates will serve as sole carbon and energy sources for *Chloroflexus* (125).

Ammonia, complex nitrogen sources (yeast extract, casamino acids), and a variety of amino acids support the growth of *Chloroflexus* (76a, 94, 169). Nitrogen fixation in *Chloroflexus* has not been detected. Inhibitor studies and enzymatic analyses indicate that enzymes of the citric acid and glyoxylate cycles are present in *Chloroflexus* (112). This implies that acetyl-CoA can be completely oxidized to CO_2 by *C. aurantiacus*; such activity is also typical of Rhodospirillaceae (58).

Axenic cultures of a mesophilic strain of *Chloroflexus* isolated from the bottom mud of a stratified lake in the Soviet Union have been reported (64). Except for the difference in temperature optima, it appears as if mesophilic *Chloroflexus* are virtually identical to *C. aurantiacus*. However, detailed studies of mesophilic strains will be required to confirm this suggestion, and such studies have not yet been published.

Other filamentous green bacteria inhabit the metalimnion and hypolimnion of shallow lakes or the surfaces of sulfide-rich muds and sediments. The genera *Oscillochloris* and *Chloronema* have been described, although neither is yet in pure culture (66). These genera differ from *C. aurantiacus* primarily by their large filament diameter (2 to 6 μm) and length (250 to 2500 μm), thus resembling cyanobacteria in overall dimensions. It is thus not surprising that *Oscillochloris* was considered a cyanobacterium for many years (66). The presence of chlorosomes in these two large filamentous prokaryotes, however, indicates a relationship with green bacteria. Whether *Chloronema* or *Oscillochloris* is capable of dark growth is not yet known.

2.3 ECOLOGY OF PHOTOTROPHIC BACTERIA

2.3.1 Stratified Lakes

Large masses of phototrophic sulfur bacteria can develop in anaerobic sulfide-rich ecosystems exposed to light. The majority of such habitats are aquatic. Although blooms of phototrophic sulfur bacteria may occur in shallow lagoons polluted by sewage [which stimulates the activities of sulfate-reducing bacteria (24a)], dense stratified ''plates'' of phototrophic bacteria are most often observed in small deep lakes protected from excessive wind mixing which receive a significant input of organic matter (e.g., leaves, sewage, agricultural and industrial runoff, human pollution inputs, etc.). Depending on its limnological state, a lake is referred to as being either *holomictic*, where the entire body of water turns over one or more times a year, or *meromictic*, where the bottom waters do not mix, regardless of the mixing status of the upper waters (177). The meromictic state results in a stratified water body, the deeper, denser, cooler water (usually referred to as the *hypolimnion*) remaining anaerobic throughout the year. In typical highly productive (eutrophic) lakes, a transitional zone called the *thermocline* exists between upper and lower waters; a rapid decrease in oxygen levels and temperature can be observed through this often very narrow zone (see Fig. 2.10). It is near and below this zone of discontinuity that one frequently finds blooms of phototrophic sulfur bacteria, the extent of which is usually dictated by light penetration and sulfide levels (156).

In the sediments of a highly productive stratified lake, one finds intense microbial activity. Organic material reaching the hypolimnion is attacked by a spectrum of fermentative anaerobes and anaerobic respirers. Cellulose, for example, is hydrolyzed by cellulolytic bacteria, releasing cellobiose and glucose, and these substrates are rapidly fermented to a variety of reduced products, including lactate, ethanol, acetate, H_2, CO_2, and long-chain fatty acids (144). In the light, most of these compounds could be photometabolized by Rhodospirillaceae, but probably because of their sulfide sensitivity, these organisms seldom reach significant numbers in stratified lake ecosystems (see Section 2.3.1.6). Instead, the fermentation products are oxidized by heterotrophic bacteria that reduce nitrate, sulfate (166), or carbonate as terminal electron acceptors. Sulfate, especially in marine waters, appears to be a key electron acceptor (106, 165). In addition, dissimilatory sulfur-reducing bacteria may play an important role in H_2S production (165). Sulfate or sulfur, once reduced to sulfide, can be regenerated anaerobically by activities of the phototrophic sulfur bacteria.

As sulfide accumulates in sediments and is released upward into the water column, the stage is set for development of the purple and green sulfur bacteria. In shallow lakes with relatively low levels of sulfide, mass developments of purple sulfur bacteria have been reported directly on or above the surface of the mud (231). A more common scenario is to have a gradient of sulfide in the hypolimnion of a deep lake, allowing for the vertical migration and layering of phototrophic sulfur bacteria in response to fluctuations in light intensity and sulfide concentrations; numerous examples of the latter can be cited (19, 99, 138, 149, 156, 216).

If cell numbers of phototrophic sulfur bacteria are high enough, the water will become brightly pigmented, and it is possible to get an idea of the types of phototrophs present in such an envrionment by performing absorption spectra on natural samples. For example, in an ecological study by Guerrero and his colleagues of phototrophic bacteria that develop in certain Spanish lakes, absorption spectra of filtered samples gave a rapid indication of whether the predominant *Chlorobium* population contained bacteriochlorophyll *c*, *d*, or *e* (138; see Fig. 2.4). This information is often useful for determining the *predominant* phototrophic species present at various depths in a lake (conclusions based on enrichment and isolation can be misleading if the dominant species is not the most rapid grower under the enrichment conditions chosen). A rapid and sensitive method for measuring absorption spectra of cells on glass fiber filters has been published by Trüper and Yentsch (227).

2.3.1.1 Wintergreen and Burke Lakes, Michigan

A well-illustrated example of the layering of phototrophic bacteria in the hypolimnion of stratified lakes can be found in the work of Caldwell and Tiedje (19, 20).

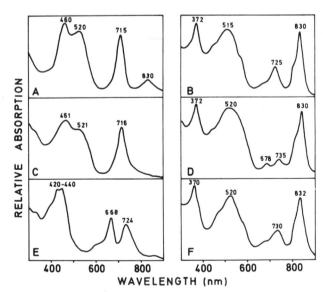

Figure 2.4 *In vivo* absorption spectra of samples of photosynthetic bacteria from Lakes Vilar (*A*), Cisó (*B*), Banyoles III (*C*), Nou (*D*), Negre (*E*), and Estanya (*F*), Spain. Courtesy of R. Guerrero, from E. Montesinos, R. Guerrero, C. Abella, and I. Esteve, 1983, Appl. Envir. Microbiol. 46:1007–1016.

Using phase and electron microscopy of water samples collected at 1-m intervals from two eutrophic lakes, Wintergreen and Burke, located in southwest Michigan, it was possible to follow the predominant species with depth by direct microscopic counts of the morphologically distinct species.

Three bacterial communities, designated A, B, and C, were defined (20). The upper (A) community consisted of sequentially layered suspensions of purple sulfur bacteria of the genera *Thiopedia*, *Thiospirillum*, *Thiocystis*, or *Chromatium*. The average diameter of the phototrophic species in the A community was greater than 2 μm. Phase photomicrographs of the A community from lakes Burke and Wintergreen are shown in Fig. 2.5. In Burke Lake (Fig. 2.5a) small spherical cells (*Thiocystis*) and a few of the large rod-shaped bacteria (*Chromatium* sp.) contained sulfur granules, but the *Thiospirillum* population did not; the latter organism was considered the dominant purple bacterium in Burke Lake (20). In Wintergreen Lake, on the other hand, the composition of the A community was completely different, consisting almost exclusively of *Thiopedia* sp. (Fig. 2.5b).

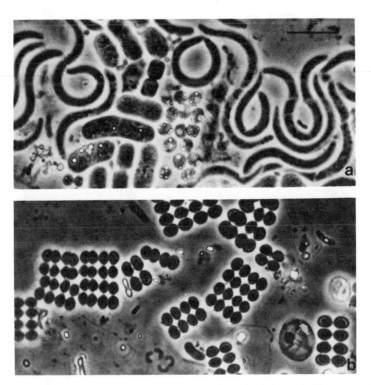

Figure 2.5 Phase photomicrographs of concentrated Burke Lake and Wintergreen Lake (Michigan) hypolimnetic water samples. (*a*) Burke Lake, A community, 8.5 m. (*b*) Wintergreen Lake, A community, 3.0 m. Marker bar equals 10 μm. Micrographs courtesy of Douglas Caldwell, from D. E. Caldwell and J. M. Tiedje, 1975, Can. J. Microbiol. 21:377–385.

The B community was contiguous with the A and C communities and consisted primarily of photosynthetic green bacteria, with an average cell diameter of 2 μm or less (20). In Wintergreen Lake the predominant species were bacteriochloro-phyll *d*–containing *Chlorochromatium*, *Clathrochloris*, and *Prosthechochloris* species (20). Aggregates of *Chlorochromatium*, the synchronously dividing syntrophic association described in Section 2.1.3, were visible by phase microscopy as clumps of green bacteria (Fig. 2.6*a*). The colorless motile central component was resolved only when clumps were spread apart on agar (Fig. 2.6*b*). Electron

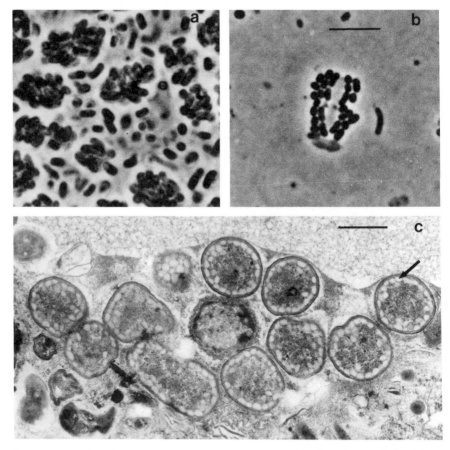

Figure 2.6 *Chlorochromatium aggregatum* from the hypolimnion of Burke Lake, Michigan. (*a*) Phase micrograph of the aggregation as it appears in a wet mount. (*b*) Phase micrograph of the aggregation after mounting on an agar slide. (*c*) Transmission electron micrograph of a transverse section of *C. aggregatum*. Note the central colorless cell and the surrounding green phototrophic bacteria containing peripherally arranged chlorosomes (arrow). Marker bar in (*a*) and (*b*) equals 5 μm; marker bar in (*c*) equals 0.5 μm. Micrographs courtesy of Douglas Caldwell, from D. E. Caldwell and J. M. Tiedje, 1975, Can. J. Microbiol. 21:362–376.

microscopic examination of thin sections of *Chlorochromatium* aggregates from Burke Lake (Fig. 2.6c) showed the intimate relationship between the central cell and the outer shell of green sulfur bacteria, which gives the syntrophic association a "barrel-shaped" configuration. Green bacteria were easily recognized in sections of *Chlorochromatium* because of their peripherally arranged chlorosomes (Fig. 2.6c).

Pelodictyon, a branch-shaped gas vacuolate green bacterium (161) was present in large numbers in Burke Lake and produced truncated prosthecae (small lateral projections) which contained chlorosomes and elongated gas vesicles (Fig. 2.7). In Wintergreen Lake *Pelodictyon* was absent, and the major green bacteria below a depth of 3 m were *Clathrochloris* and *Prosthechochloris* (20). The particular strain of *Prosthechochloris* studied by Caldwell and Tiedje also contained gas vesicles (19).

The C community in the Michigan lakes was the lowermost of the three communities and it extended down to the sediments (6 to 12 m, depending on the lake). A number of morphologically unusual prokaryotes were observed in the C community, but it was concluded that they were neither phototrophic nor lithotrophic species since no intracellular sulfur was observed and organic solvent extracts for bacteriochlorophyll were negative (19). It is likely that community C is a heterotrophic sediment–water interface community which relies on organic matter influx from the upper layers. It is noteworthy, however, that the majority of members of the C community contained gas vesicles; the ability to float is presumably advantageous for the physiological mechanisms that support growth of these unusual prokaryotes.

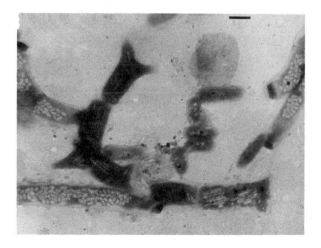

Figure 2.7 Electron micrograph of a shadowed preparation of the phototrophic bacterium *Pelodictyon* from the hypolimnion of Burke Lake, Michigan. Note the long, thin gas vesicles and the shorter, thicker chlorosomes. Marker bar equals 1 μm. Micrograph courtesy of Douglas Caldwell, from D. E. Caldwell and J. M. Tiedje, 1975, Can. J. Microbiol. 21:362–376.

2.3.1.2 Rotsee, Switzerland

In a recent investigation of phototrophic bacteria in the Rotsee (Switzerland), Kohler et al. (99) demonstrated, through solvent extracts of filtered water samples and by microscopic counts, that a bloom of bacteriochlorophyll *e*–containing phototrophs exists at a depth of about 10 m in this water body during late summer stratification (see Fig. 2.8*a*). Bacteriochlorophyll *a*, on the other hand, was maximal at a depth of 8 m. Chlorophyll *a* levels in the Rotsee were considerably lower than that of bacteriochlorophyll *a*, and the greatest abundance of chlorophyll *a* was present from the surface to about 7 m (Fig. 2.8*a*).

Direct microscopic counts of filtered cell material stained with acridine orange showed that the bacteriochlorophyll *a* maximum was due to a bloom of *Thiopedia rosea* and *Lamprocystis roseopersicina*; bacteriochlorophyll *e*, by contrast, was localized in the phototrophic component of the syntrophic consortium, *Pelochromatium roseum* (Fig. 2.8*b*). The latter "organism" is the brown-colored counterpart of the *Chlorochromatium aggregatum* consortium described previously (see Sections 2.1.3 and 2.3.1.1, and Figure 2.6). The phototrophic component of the *Pelochromatium* consortium contains bacteriochlorophyll *e* and carotenoids typical of brown-colored *Chlorobium* sp. (151).

Since all phototrophic components of the water column of the Rotsee contained gas vesicles (or were motile by flagellar means in the case of *Pelochromatium*), it is logical to assume that planktonic phototrophic bacteria benefit by an ability to move vertically in the water column. In the Rotsee study it was demonstrated that diurnal migrations of *Thiopedia* over a distance of about 1 m did occur; the greatest density of *Thiopedia* cells was at a depth of 7 m at 0700 h, 7.7 m at 1200 h, and 8 m at 1900 h (99). It was concluded that the sulfide-consuming activities of the phototrophs forced the population to seek sulfide at lower depths as the day progressed. During the evening hours, however, sulfide accumulates, and the phototrophic community rises in the water column to take advantage of better lighting conditions the next morning.

It also should be pointed out that the appearance of green bacterial consortia in the Rotsee and Michigan lake studies suggests the ecological importance of these syntrophic relationships for vertical migrations. Green sulfur bacteria of the genus *Chlorobium* are immotile. The green sulfur bacteria present in the Michigan and Rotsee examples, however, were either gas vacuolate species (*Clathrochloris*, *Pelodictyon*, or *Prosthechochloris*) or were motile consortia. The consortia thus offer the immotile *Chlorobium* cell an opportunity to migrate in response to sulfide, light or other factors (just as motile purple bacteria do), and this alone may explain why these forms of green bacteria, rather than free *Chlorobium* cells, are more commonly observed in planktonic water samples collected from the hypolimnion.

2.3.1.3 Lake Cisó and Other Spanish Lakes

As light passes down through the water column of a lake, it is attenuated in respect to both intensity and spectral quality (87, 149). The effect of biotic or abiotic light filtering on the species composition of phototrophic bacteria in lakes has recently been studied in some meromictic Spanish lakes, and in some small deep lakes in

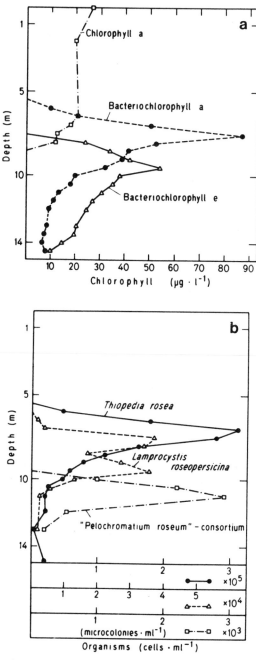

Figure 2.8 Distribution of photosynthetic pigments and major species of phototrophic sulfur bacteria with depth in the Rotsee, Switzerland. (a) Chlorophyll *a* and bacteriochlorophylls *a* and *e*. (b) Purple and green sulfur bacteria. Courtesy of A. J. B. Zehnder, from H. P. Kohler, B. Ahring, C. Abella, K. Ingvorsen, H. Keweloh, E. Laczko, E. Stupperich, and F. Tomei, 1984, FEMS Microbiol. Letts. 21:279–286. Reproduced by permission of FEMS Microbiology Letters.

central Wisconsin (USA). In the Spanish lakes, Guerrero and his colleages found that purple sulfur bacteria, if present, were always positioned in a layer on *top* of the green sulfur bacteria (1; see Fig. 2.9). This was explained by the fact that purple bacteria need higher light intensities to grow at the same rate as green bacteria, due to the better location of light harvesting pigments in the chlorosomes of green bacteria [the ability of cultures of *Chlorobium* to grow at light intensities insufficient to support the growth of cultures of *Chromatium* has been documented (10)]. Due to absorption of light by the *Chromatium* layer, however, light that passes through to the *Chlorobium* zone is of a spectral quality that only supports growth of green-colored *Chlorobium* species; the latter contain carotenoids that absorb maximally at around 450 nm—the "window" of light passed by the *Chromatium* layer (1, 138; see Fig. 2.9).

Several other Spanish lakes contained Chlorobiaceae to the virtual exclusion of Chromatiaceae. In these lakes two distinct populations of *Chlorobium*, one containing green and one containing brown carotenoids, typically developed (138). The brown-colored species were always found in the deeper waters of the lake with green-colored chlorobia layered above them. The explanation for this stratification involves the ability of brown chlorobia to increase their specific carotenoid content far above that of green chlorobia, thus rendering the brown species better able to absorb light at great depths (138). A number of other field studies (sum-

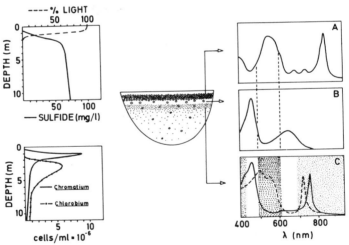

Figure 2.9 Light, sulfide, and photosynthetic bacterial populations in Lake Cisó, Spain. Upper left, limnological profile of light attenuation and sulfide concentration with depth; bottom left, distribution of *Chromatium* and green-colored *Chlorobium* sp. with depth; right, *in vivo* absorption spectra of (*A*) *Chromatium* and (*B*) *Chlorobium* which make up the bulk of the phototrophic population in Lake Cisó. (*C*) Absorption spectra of pure cultures of green (solid line) and brown (dashed line) chlorobia are shown. Note how the spectrum of the green *Chlorobium* complements that of the *Chromatium*. This explains why green-celled species of *Chlorobium* can coexist under a plate of *Chromatium*, whereas brown-celled species cannot. Data courtesy of Ricardo Guerrero.

marized in Ref. 138) strongly support the hypothesis that brown-colored Chloro-biaceae are selected for in the deeper regions of the hypolimnion. In addition, experiments with pure cultures of brown and green *Chlorobium* species suspended *in situ* at various depths confirm the competitive edge of the brown species with increasing depth (133).

In certain lakes brown chlorobia are present at a depth so low in the water column that direct absorption of light by bacteriochlorophyll cannot occur; photosynthesis still occurs, however, due presumably to the efficient transfer of light energy from a large light-harvesting carotenoid pool directly to bacteriochlorophyll (63). Light limitation, even in brown chlorobia, can eventually occur, however, and a study of photosynthesis in a bloom of such organisms in Lake Kinnert clearly emphasized this point (6). Brown Chlorobiaceae also predominate in the upper regions of the hypolimnion when large numbers of cyanobacteria and/or green algae are present in the epilimnion. This is due to the similarity in absorption properties of green chlorobia and oxygenic phototrophs; when the latter are abundant they serve as a biological filter preventing development of green chlorobia. Brown chlorobia, on the other hand, have pigments that complement those of the oxygenic phototrophs, and both groups can therefore coexist.

In more recent studies of the Spanish lakes, van Gemerden, Guerrero, and colleagues have confirmed the general finding that *Chromatium* layers reside above those of green bacteria (71). In studies of Lake Cisó, a small anaerobic, holo-mictic, sulfide-containing lake located in the karstic Banyoles area of northern Spain (71), the metabolism of *Chromatium* (the predominant phototroph in this lake), was followed on a diurnal cycle (240). At a depth of 2 to 3 m sulfide concentrations reached levels of 0.6 mM. During daylight hours sulfide was oxidized rapidly to elemental sulfur by the *Chromatium* layer, and CO_2 was fixed and stored as glycogen. At night, glycogen was used as an energy source and converted to PHB; excess electrons were used to reduce S^0 to H_2S (240). However, since the light gradient in this lake is so sharp (71), cells of *Chromatium*, which rapidly accumulate maximal levels of intracellular sulfur (calculated to be 35 to 40% of total cell volume), increase their densities (sulfur is more dense than cell material) and eventually sink (240). The consequences of this pattern is a rapidly growing upper layer of *Chromatium* cells underlain by layers of cells of decreasing viability; the high levels of sulfide in Lake Cisó prevented any significant oxidation of intracellular sulfur to sulfate (240).

Lake Cisó is highly unusual in that it remains totally anoxic throughout the year (70, 71). It is curious why the lake is not dominated by green sulfur bacteria whose deposition of sulfur (formed from sulfur oxidation) *outside* the cell would not be as suicidal a phenomenon as intracellular storage by purple bacteria. It must be concluded that light and other physiochemical conditions in Lake Cisó favor the growth of purple bacteria over that of green bacteria. Perhaps the fact that the photic zone is limited to the top 3 m in Lake Cisó favors the purple bacteria; it appears that green bacteria are more common in deeper lakes where both light quality and intensity are greatly affected by the size of the water column (see the next section).

2.3.1.4 Colored Wisconsin Lakes

The effects of light quality on photosynthesis and growth of phototrophic bacteria in three Wisconsin lakes was studied by Parkin and Brock (149). Of particular interest in this study was the fact that the lakes differed considerably in their content of dissolved organic material and were either green (Mirror Lake and Fish Lake) or yellow brown (Knaack Lake) in color (Knaack Lake contained large amounts of humic and tannic compounds). All three lakes supported populations of phototrophic bacteria, but only in Fish Lake were significant numbers of purple bacteria observed (Fig. 2.10). The limnological profiles of the three lakes shown in Fig. 2.10 also indicate the differences in sulfide levels and biomass (bacteriochlorophyll) that was observed in each.

The absorption properties of Fish and Mirror Lake were similar and the phototrophic bacteria present, although of differing species composition, were concentrated in a narrow zone at the interface between the joining of the sulfide and oxygen gradients (Fig. 2.10). The Fish Lake population consisted of a mixture of *Pelodictyon* and *Thiopedia*, the Mirror Lake population of large numbers of a brown *Chlorobium* amid small numbers of *Lamprocystis* and *Chromatium*, and the Knaack Lake phototrophs of green-colored species of *Clathrochloris*, *Chlorobium*, and *Pelodictyon* (149).

The Knaack Lake green bacterial community lay suspended at a depth of only 2 to 3 m in the water column; this is in contrast to Fish and Mirror Lakes—both somewhat shallower than Knaack Lake—but whose phototrophic component was localized at a depth of 8 to 11 m (149, 150). The phototrophs in Knaack Lake consisted of green bacteria containing gas vesicles, which presumably allowed the nonmotile chlorobia to remain positioned far up in the water column.

The reason why the phototrophs remained so high in the water column becomes clear when examining the light extinction data shown in Fig. 2.11. Due to the high amounts of dissolved organic material in Knaack Lake, light was rapidly attenuated as it passed through the water column. By a depth of 2 m, less than 0.1% of the surface light intensity remained (149). Due to absorbing materials, the spectral quality of the light that passed to this depth was such that a large proportion of the blue and green wavelengths had already been removed, leaving primarily the red region of the spectrum available for use by phototrophic bacteria (149). This combination of factors, extremely low light intensities and light of a red spectral quality, combined to select first for Chlorobiaceae, and second for green-colored rather than brown-colored species of Chlorobiaceae.

Photosynthesis studies on natural populations of phototrophic bacteria from the Wisconsin Lakes supported the measurements of light quality. Maximal rates of $^{14}CO_2$ incorporation occurred in populations from Knaack Lake exposed to red light, while maximal rates for populations from Fish and Mirror Lakes occurred in green light (149). In enrichment cultures established from lake water samples and incubated under red or green light, red light always led to the development of green-colored species of *Chlorobium*; this occurred even if the predominant green bacterium was a brown-colored *Chlorobium* (149). It can therefore be concluded that when the green region of the spectrum is removed via biological filtration

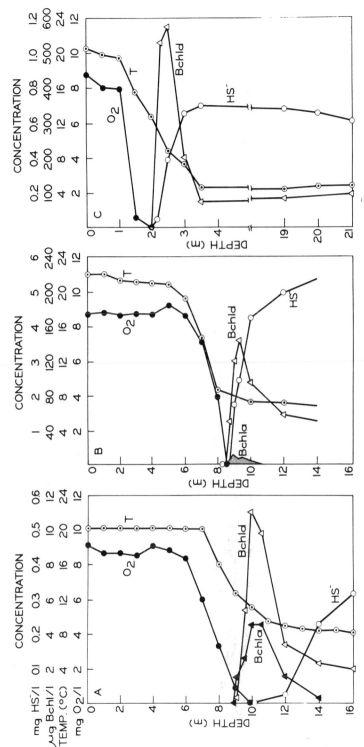

Figure 2.10 Phototrophic bacteria in colored Wisconsin Lakes. Limnological characteristics of (A) Fish Lake (September 22, 1978), (B) Mirror Lake (September 15, 1978), and (C) Knaack Lake (June 10, 1978). Sulfide (○), mg L^{-1}; oxygen (●), mg L^{-1}; temperature (⊙), °C; bchl a (▲), μg L^{-1}; bchl d (△), μg L^{-1}. Courtesy of Tim Parkin, from T. B. Parkin, and T. D. Brock, 1980, Arch. Microbiol. 125:19–27.

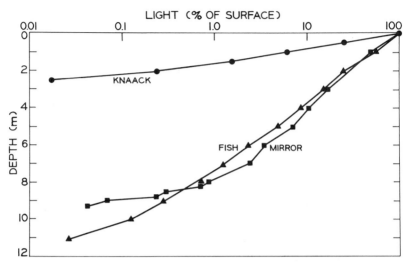

Figure 2.11 Light extinctions in Knaack, Fish, and Mirror Lakes, Wisconsin. Note that less than 0.1% of surface intensities passed below a depth of 2 m in Knaack Lake (see text for discussion). Courtesy of Tim Parkin, from T. B. Parkin and T. D. Brock, 1980, Arch. Microbiol. 125:19–27.

(such as with the purple bacteria in the Spanish lakes study), or when the blue and green regions are removed by abiotic absorption (e.g., by the dissolved organic constituents in the Knaack Lake example), the availability of blue and red light (or red light alone) selects for green-pigmented species of Chlorobiaceae. By contrast, in deep lakes not containing abiotic or biotic filters, brown-colored Chlorobiaceae will predominate because of their superior light-harvesting abilities (149, 150).

2.3.1.5 Solar Lake, Sinai

The last example of layered communities of phototrophic prokaryotes in lakes to be discussed will be those of the Solar Lake in Sinai. This small shallow (4.5 m deep) basin near the edge of the sea, shielded from wind mixing and filled with brine, is inhabited by a variety of photosynthetic prokaryotes (27–29). Both purple and green sulfur bacteria are present in Solar Lake, and during stratification, a *Chromatium* bloom overlays that of *Pelodictyon* (28). What is truly remarkable about the Solar Lake habitat, however, is that cyanobacteria, which are normally found in the epilimnion of a stratifed lake, are instead present at the *bottom* of Solar Lake, underneath the phototrophic bacterial plates (28).

The dominant cyanobacterium in Solar Lake is *Oscillatoria limnetica*, a new species of *Oscillatoria* capable of anoxygenic photosynthesis at high sulfide concentrations. *O. limnetica* stratifies underneath the purple and green layers because of its extreme sulfide tolerance and because it is capable of carrying out anoxygenic (photosystem I only) photosynthesis when sulfide is present as electron donor (26).

The Solar Lake microbial ecosystem therefore represents a unique set of limnological circumstances wherein extremely high sulfide levels and relatively shallow depth stimulate development of an unusual cyanobacterial layer beneath, rather than above, the purple and green bacteria.

Light quality probably has no effect on the positioning of phototrophs in the shallow Solar Lake. Instead, it is likely that the extreme sulfide tolerance of *O. limnetica* dictates its position in the ecosystem; cultures of *O. limnetica* can be grown with 8 m*M* sulfide, which is also near the upper limit for green bacteria (30). The discovery of facultative anoxygenic photosynthesis in *O. limnetica* has led to an investigation of this problem in a variety of cyanobacteria, and it is now known that sulfide-dependent photosynthesis is widespread (but not universal) among cyanobacteria (55). The ecological impact of facultatively anaerobic photoautotrophic metabolism by cyanobacteria has been discussed by Padan (148).

2.3.1.6 Rhodospirillaceae in Aquatic Habitats

Rhodospirillaceae are rarely found in dense blooms in nature, and this is probably a result of their sulfide sensitivity and need for organic compounds as carbon sources; the latter puts them in direct competition with various heterotrophic microbes which use these organic substrates as carbon and energy sources (154, 243). Rhodospirillaceae are unable to attack polymeric substances such as cellulose, chitin, lipids, and proteins, and hence are dependent on heterotrophic organisms for the breakdown of these complex materials (156). Thus Rhodospirillaceae tend to coexist in relatively low numbers [generally, less than 10^5 ml^{-1} (8, 91)] with heterotrophic anaerobes in the hypolimnion, or in shallow aquatic systems such as the mud of eutrophic ponds (156). Rhodospirillaceae can routinely be isolated from aerobic waters of lakes, rivers, and streams (unpublished observations), but their numbers in these habitats are probably not very large.

2.3.2 Sewage and Wastewater as a Habitat for Phototrophic Bacteria

Depending on the source, wastewaters generated from human and industrial sources can contain considerable amounts of organic matter. Purple bacteria, especially Rhodospirillaceae, should therefore thrive in wastewater treatment plants, and the few studies of this problem that have been carried out indicate that significant numbers of purple bacteria are present in sewage (81, 191) and other habitats receiving waste treatment inputs, such as slaughterhouse effluents (34, 157). In Japan, Kobayashi has shown the feasibility of using continuous-flow large-scale processing of wastewaters using photosynthetic bacteria (Rhodospirillaceae) as biological catalysts for the removal of organic matter (i.e., BOD) and volatile sulfur and nitrogen compounds (96, 97). Proposals have also been put forth for the use of phototrophic bacterial cells derived from mass culture on wastewater or animal manure, as animal feed supplements (47, 98). Kobayashi has demonstrated the dramatic positive effects of additions of photosynthetic bacteria to chicken feed (resulting in quantitative and qualitative improvements in eggs) and as a nitrogenous fertilizer for the stimulation of citrus fruit and vegetable production (97, 98).

In various stages of the sewage treatment plant, photosynthetic bacteria, primarily Rhodospirillaceae, abound. In a detailed study of this problem by Siefert et al. (191), it was shown that the mean number of Rhodospirillaceae in the Göttingen, West Germany, sewage plant was highest (10^5 ml^{-1}) in the activated sludge stage of the treatment process; counts flucuated between 10^5 and 10^6 ml^{-1}, but were never greater than 10^6 ml^{-1} [via plate counting techniques (191)]. Purple and green sulfur bacteria, on the other hand, were quantitatively insignificant in sewage and were really detectable only in activated sludge [10^3 cells ml^{-1} of each (191)].

A variety of Rhodospirillaceae were identified in the Göttingen plant, including *Rhodobacter sphaeroides* and *R. capsulatus*, *Rhodopseudomonas palustris* and *R. viridis*, *Rhodocyclus gelatinosus* and *R. tenuis*, and *Rhodospirillum photometricum*. The purple sulfur bacteria present were *Chromatium vinosum* and *Thiocapsa roseopersicina*, and the green sulfur bacterium was *Chlorobium limicola* (191). *R. sphaeroides*, *R. gelatinosus*, *R. palustris*, and *R. capsulata* made up the bulk of the phototrophic nonsulfur species present (191).

Based on the number of Rhodospirillaceae detected in the Göttingen sewage plant, it was concluded that they probably play a minor role in organic matter, transformations in comparison with heterotrophic bacteria, which were present 100- to 1000-fold (191). Since light and oxygen conditions in the activated sludge tank were not ideal for phototrophic growth, it is logical that facultative aerobic species (i.e., *Rhodobacter* and *Rhodopseudomonas*) would predominate over purple sulfur bacteria in the activated sludge environment. Phototrophic bacteria also were present in the strictly anaerobic (and dark) sludge digestor, but it was clear that they represented only transients traveling through the system; digestor sludge incubated anaerobically in darkness showed no increase in phototrophic bacterial numbers (191).

Sewage lagoons offer an ideal habitat for phototrophic bacteria because the lack of mixing characteristic of activated sludge allows for growth of sulfate-reducing bacteria and the subsequent accumulation of sulfide. In the few studies of this problem that have been published, purple sulfur bacteria such as *Chromatium*, *Thiocapsa*, and *Thiopedia* have been implicated as important constituents of the bacterial microflora (34, 81, 135). Rhodospirillaceae were apparently not looked for in these studies; however, one would predict that organic rich sewage lagoons, especially shallow ones where sulfate reduction may not be a major process, would be ideal habitats for growth of facultative purple bacteria.

A bloom of Rhodospirillaceae was reported from the waste lagoon of a corning operation in Minnesota; profilic growth leading to intensely red-colored lagoons occurred, and the bloom was associated with a reduction in offensive odors (88). Extraction of cells and assays for sulfur clearly showed that the bloom did not consist of purple sulfur bacteria, and subsequent culture studies identified *Rhodobacter sphaeroides* and *capsulatus*, *Rhodopseudomonas palustris*, and *Rhodocyclus gelatinosus* as the key components of the bloom (88). It is likely that a combination of the abundant organic load from vegetable processing, coupled with relatively meager sulfate reduction (probably because of low sulfate levels), lead to the development of nonsulfur rather than sulfur phototrophic bacteria.

2.3.3 Hot Springs and Phototrophic Bacteria

2.3.3.1 Chloroflexus in Alkaline Hot Spring Mats

In alkaline hot springs at temperatures below about 73°C, thick mats form consisting primarily of two phototrophic species, the cyanobacterium *Synechococcus lividus* and the facultative green bacterium *Chloroflexus aurantiacus* (3, 14, 21). Certain mats in high sulfide springs lack *Synechococcus*, and consist solely of dark green obligately phototrophic strains of *Chloroflexus* (60a). The mats generally form in the effluent channels or pools of hot springs, the maximum standing crop occurring at a temperature of 50 to 55°C (3).

Chloroflexus is a gliding filamentous green bacterium which shares physiological, biochemical, and ultrastructural relationships with both the green and purple bacterial groups (see Section 2.2.5). In the mat environment it coexists with the unicellular cyanobacterium *Synechococcus* as a 1- to 2-mm-thick upper green layer; beneath this actively photosynthetic layer is a thicker layer, usually brilliant orange in color, composed primarily of *Chloroflexus* (Fig. 2.12). Beneath this layer is a narrow siliceous zone containing dead and lysed *Chloroflexus* and a variety of heterotrophic decomposers (3). Hot spring mats of this type thus represent complete microbial ecosystems where both primary productivity and decompositional processes occur simultaneously (14, 23, 247).

The hot spring environment is one in which physical and chemical conditions can remain unchanged for long periods (14). However, from the standpoint of adaptation, the phototrophic organisms inhabiting these springs must adapt to at least two physical variables which can be considered extreme: (a) temperature, and (b) high light intensities. The temperature of the mat at the point of development is constant but may be as low as 40°C or as high as 73°C, depending on the distance the mat is from the source. Near the top of the mat, light intensities of >80,000 lux are not uncommon, but as the mat continues to grow, overgrown *Chloroflexus* layers experience ever-decreasing light intensities. Thus light and temperature are two important factors governing *in situ* growth of *Chloroflexus*.

Natural populations of *Chloroflexus* taken from mats of differing environmental temperature and tested for photosynthetic competence ($^{14}CO_2$ incorporation) over a range of different temperatures indicate clearly that "temperature strains" of *Chloroflexus* exist (Fig. 2.13). With the exception of mat material taken from the highest temperature, stands of *Chloroflexus* collected at a given environmental temperature were found to be highly adapted to photosynthesize at that particular temperature (3). Thus populations of *Chloroflexus* collected from mats at environmental temperatures of 45°C, 50°C, or 60°C photosynthesized best at 45 to 50°C, 55°C, or 60°C, respectively (3). Similar temperature-adaptation phenomena have been reported for the cyanobacterial population (15).

Adaptation by natural populations of *Chloroflexus* to reduced light intensities in Yellowstone hot springs was studied by Madigan and Brock (119). To study

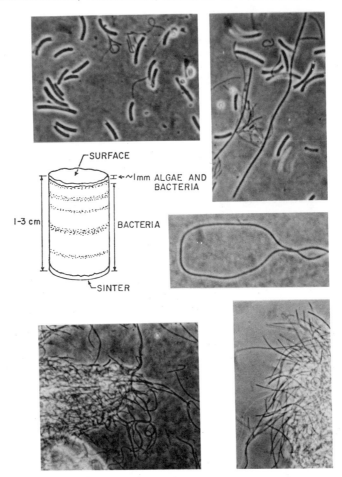

Figure 2.12   Gross anatomy of hot spring algal-bacterial mats and photomicrographs of mat material taken from different levels. Top left, surface layer composed primarily of the cyanobacterium *Synechococcus lividus*; top right, *S. lividus* and *Chloroflexus*; center right, *Chloroflexus* filament from the top bacterial layer about 2 to 3 mm from the surface of the mat; bottom left and right, filamentous bacteria from the lower bacterial layers. Note the absence of *S. lividus*. Courtesy of Thomas Brock, from J. Bauld and T. D. Brock, 1973, Arch. Mikrobiol. 92:267–284.

light-adaptation responses, neutral-density light-reduction filters were installed over a large thick mat adjacent to the source of Octopus Spring. Filters were installed that reduced the intensities by 73, 93, 98, and 100%, respectively; one area was left as an undisturbed control. In photosynthesis experiments on the upper 1 to 1.5 mm of mat which had not been subjected to light-intensity reduction (Fig. 2.14) it was observed that the *Chloroflexus* population was unaffected by high light intensities, and essentially photosynthesized at the maximal rate from 10,000 to

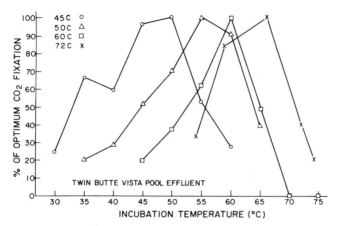

Figure 2.13 Temperature optima for *Chloroflexus* photosynthesis ($^{14}CO_2$ incorporation) from samples at temperatures of 72, 60, 50, and 45°C in the effluent of Twin Butte Vista Pool, Yellowstone National Park. Note the distinct temperature optima for strains growing at different environmental temperatures. Courtesy of Thomas Brock, from J. Bauld and T. D. Brock, 1973, Arch. Mikrobiol. 92:267–284.

80,000 lux. Undisturbed populations of the cyanobacterium *Synechococcus lividus*, on the other hand, always showed peak photosynthetic activity at about 70% of full sunlight intensity (16).

About 2 weeks following installation of the light-reduction filters, similar photosynthesis experiments using material exposed to lower light intensities showed

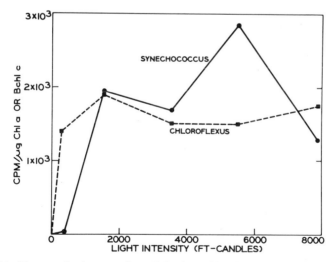

Figure 2.14 Photosynthesis at various light intensities by unadapted populations of *Synechococcus lividus* and *Chloroflexus aurantiacus* in the effluent of Ravine Spring, Yellowstone National Park. From M. T. Madigan and T. D. Brock, 1977, Arch. Microbiol. 113:111–120.

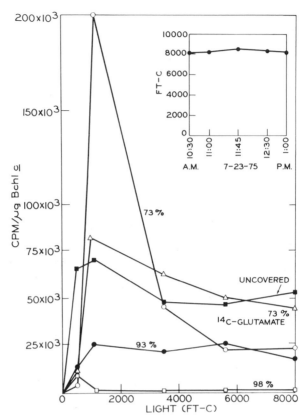

Figure 2.15 Photosynthesis and photoincorporation of glutamate by *Chloroflexus auran-tiacus* from uncovered mat sites and from sites where light intensities were reduced by 73, 93, or 98%, respectively, in Octopus Spring, Yellowstone National Park. Note the absence of high light inhibition in the uncovered (control) population and distinct adaptation to low light by the populations from under the 73% and 98% light reduction filters (see text for details). From M. T. Madigan and T. D. Brock, 1977, Arch. Microbiol. 113:111–120.

completely different results (119). As shown in Fig. 2.15, *Chloroflexus* subjected to a 73% reduction in light intensity had become highly adapted to intensities of less than 25% of full sunlight; similar results were observed if the assay involved photoincorporation of glutamate instead of $^{14}CO_2$ (119). Mat material exposed to a 98% reduction in light intensity underwent a change in species composition. The amount of light passing this filter was insufficient to support growth of the cyanobacterium, and it was observed to wash out after 2 weeks of light reduction, leaving essentially a pure stand of *Chloroflexus* (Fig. 2.16). The latter cells were highly adapted, but only to very low light intensities (< 10% of full incident intensities) and showed complete inhibition of photosynthesis at all higher intensities tested (119).

It is clear from these data that the cyanobacterium has higher light requirements

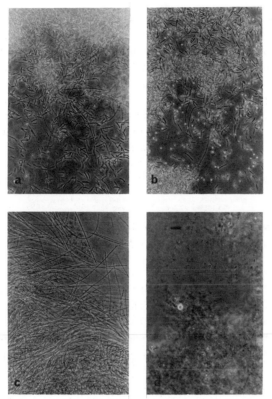

Figure 2.16 Photomicrographs of mat material from the uppermost mat layer of the uncovered (control) site and from under 73% and 98% light reduction filters, and from under a completely darkened site in Octopus Spring, Yellowstone National Park. (*a*) Uncovered site, note presence of *S. lividus* and *Chloroflexus*. (*b*) 73% reduction site. (*c*) 98% reduction site, note absence of *S. lividus*. (*d*) Darkened site, no phototrophs present. Marker bar equals 10 μm. From M. T. Madigan and T. D. Brock, 1977, Arch. Microbiol. 113:111–120.

than *Chloroflexus*, and that at least some light is required for development of *Chloroflexus*; completely darkened mats quickly resulted in washout of both *Synechococcus* and *Chloroflexus* (Fig. 2.16). It is not possible to infer that filaments under the 98% light reduction filter were actually growing at these low intensities, but sufficient light was present to maintain the population and prevent washout [in this regard, van Gemerden (235) has found that the viability of *Chromatium vinosum* can be maintained for long periods at light intensities below that required for growth].

The mechanism(s) governing low light adaptation and high light inhibition in photosynthetic bacteria *in situ* has not been examined. In cultures of purple bacteria, bacteriochlorophyll increases occur in response to low-light challenge (31), while in brown chlorobia carotenoid increases seem to be much more important

(138). Either of these changes could conceivably be at the basis of light-adaptation responses in phototrophic bacteria in nature. In the *Chloroflexus* study, bchl concentrations (per unit protein) remained virtually unchanged in low-light-adapted cells; quantitative or qualitative carotenoid changes were therefore probably responsible for low-light adaptation in *Chloroflexus* (119).

Low-light-adaptation responses are not limited to hot spring phototrophs. Shiokawa *et al.* (190) found that natural populations of *Chromatium* taken at a depth of 5 m from a small reservoir in Japan were optimally adapted to photosynthesize at approximately 10% of full incident illumination; cells taken at a depth of 1 m were optimally adapted to much higher intensities. Similar results were reported by Sorokin (203) in studies of purple bacteria in several meromictic Russian lakes.

2.3.3.2 Thermophilic Purple Bacteria

Enrichment cultures for purple bacteria established by the author using mat material from a small spring in the Mammoth geyser basin of Yellowstone National Park resulted in brightly red pigmented, highly motile rod-shaped bacteria; from the original enrichment, pure cultures of the bacterium were obtained by successive application of the agar shake-culture technique (115). The organism was subsequently designated as a new species of *Chromatium*, *Chromatium tepidum* (116). The phototroph differs from all known purple bacteria in that it grows optimally at a temperature of about 50°C and has a maximal temperature for growth of 57°C (115, 116). *C. tepidum* contains bacteriochlorophyll *a* and grows autotrophically, oxidizing sulfide to elemental sulfur, which is stored as globules inside the cells. The absorption spectrum of *C. tepidum* is unusual. Membranes from *C. tepidum* show peaks at 800, 855, and 920 nm (116); the latter is the longest wavelength absorption band observed for an organism containing only bacteriochlorophyll *a*. The peak is due to an unusual light-harvesting form of bacteriochlorophyll *a* (54a).

The maximal growth temperature of *Chromatium tepidum* agrees well with field observations of *Chromatium*-like organisms in Yellowstone and Oregon thermal springs at temperatures below 60°C (22). *Chromatium tepidum* morphologically resembles *Chromatium vinosum*, but unlike *C. vinosum*, it has an obligate requirement for sulfide (116). The carotenoid composition of *C. tepidum* is not typical of other Chromatiaceae, resembling instead that of certain Rhodospirillaceae [rhodovibrin, or a rhodovibrin-like compound, is the major carotenoid in *C. tepidum* (116)]. Suspensions of *C. tepidum* are much redder than those of *C. vinosum*.

Enrichment cultures established using mat material collected from a New Mexico hot spring (47 to 48°C) have also yielded thermophilic purple sulfur bacteria which strongly resemble the Yellowstone strain. Thermophilic purple bacteria may therefore be ecologically significant primary producers in nonacidic sulfur-containing hot springs, especially in springs where the sulfide concentration prevents the development of cyanobacteria (22). No reports of thermophilic Rhodospirillaceae have been published, although thermotolerant mesophilic nonsulfur purple bacteria are found in some Russian thermal springs (65). More work is needed in this area because of the strong photoheterotrophic potential of *Chloroflexus*; hot spring mats should be ideal habitats for thermophilic Rhodospirillaceae, which are the "premier" photoheterotrophs.

A filamentous, gliding thermophilic phototroph containing bacteriochlorophyll *a* has recently been grown in co-culture with a filamentous heterotrophic bacterium (171). This organism differs from *Chloroflexus* in its absence of bacteriochlorophylls c_s (or any of the other bacteriochlorophylls of green bacteria) and chlorosomes, but resembles *Chloroflexus* in carotenoid complement (171). The organism has been described as a new genus and species of phototrophic bacteria, *Heliothrix oregonensis* (172). *Heliothrix* corresponds to the original F-1 filaments described by Pierson and Castenholz from certain hot spring mats (167).

Heliothrix is moderately thermophilic and shows an upper temperature limit of about 57°C, as does *Chromatium tepidum* (see above). Organic compounds are assimilated in light-dependent fashion either aerobically or anaerobically (under N_2); optimum incorporation is observed around 50°C. Filaments of *Heliothrix* are observed only in certain springs, and then only during those months of the year when light intensities are highest (171). The absence of *Heliothrix* in springs containing moderate or elevated sulfide levels indicates that the physiology of *Heliothrix* may resemble that of the Rhodospirillaceae. If bacteriochlorophyll synthesis in *Heliothrix* is not repressed by molecular oxygen, however, *Heliothrix* may share closer physiological affinities with the aerobic photosynthetic bacteria described in Section 2.1.1.

Through an understanding of the physiological ecology of *Heliothrix*, especially its apparent requirement for high light intensities, it is anticipated that the nutritional barriers that have so far prevented axenic culture of this interesting new phototroph will eventually be overcome. As with *Chloroflexus*, *Chromatium tepidum*, and *Heliobacterium*, *Heliothrix* promises to be an exciting new phototrophic bacterium and possibly one of great evolutionary importance.

2.3.4 Marine Microbial Mats

In a variety of shallow marine basins, microbial mats similar in gross morphology to alkaline hot spring algal-bacterial mats (see Section 2.3.3 and Fig. 2.12) develop containing phototrophic bacteria (210, 213). The mats are dominated by layers of filamentous cyanobacteria, usually *Microcoleus*, *Lyngbya*, or *Oscillatoria* species, and contain a variety of anoxygenic phototrophic bacteria, including *Chromatium*, *Thiocapsa*, and *Chloroflexus*, plus a host of heterotrophic decomposers (210, 213).

In Laguna Figueroa (Baja California) the morphology of thick mats containing a highly diverse phototrophic bacterial component has been studied by electron microscopy (213). In thin-sectioned mat material filamentous bacteria containing chlorosomes and unicellular bacteria containing membrane stacks or membrane vesicles were observed; presumably these represent mesophilic *Chloroflexus*, and *Thiocapsa* and *Chromatium*, respectively. Phototrophic sulfur bacteria exist in such mats at the expense of sulfide generated by sulfate reduction deep in the mat and probably also from the assimilation of excretory products from the cyanobacteria.

Certain mats in Laguna Figuroa contained bands of lithotrophic sulfide oxidizers (*Beggiatoa* sp.) or phototrophic purple bacteria (*Chromatium* sp.). In a microelectrode study of these mats Jørgensen and Des Marais (90) showed that development of the band is dependent on the location of the $H_2S : O_2$ interface in the mat. In mats containing a *Beggiatoa* band, intense photosynthetic activity by dense layers of cyanobacteria in the upper regions of the mat push the $H_2S : O_2$ interface deep enough in the mat that light levels are insufficient to support phototrophic bacteria. By contrast, in mats containing purple bacteria, the cyanobacterial layers are less dense; this establishes the $H_2S : O_2$ interface high enough in the mat for light to penetrate and support growth of phototrophic bacteria.

A different type of microbial mat forms in the upper intertidal zones of the North Sea island of Mellum (210). These mats are initiated by growth of the cyanobacterium *Oscillatoria*, and later by *Microcoleus*. The Mellum mats remain rather thin and occasionally support the development of a thin red layer of *Chromatium* and *Thiocapsa* (210). Unlike the Laguna Figueroa mats, the mats that form at Mellum are usually covered by a fine sand layer. The presence of sand probably prevents the development of a thick gelatinous mat like the Laguna Figueroa mats and is probably also responsible for the relatively high redox potentials measured in even the deepest layers of the mat (210). The latter undoubtedly discourages the formation of thick phototrophic bacterial layers.

No information is available on phototrophic bacterial contributions to N_2 fixation in the Mellum mats, although N_2 fixation, presumably by cyanobacteria, has been implicated as an important process in development and maintenance of the mats (210). This is especially interesting in light of the fact that mature mats primarily contain non N_2-fixing cyanobacterial species (210). Since N_2 fixation is a widespread property of phototrophic bacteria (122, 245), this process could play a major role in fixed nitrogen inputs in mature mats in Mellum.

Numerous other examples of stratified algal bacterial mats containing phototrophic bacteria could be mentioned, but space limits our discussion to the aforementioned examples. In general, however, it can be concluded that marine microbial mats which do contain phototrophic bacteria usually contain only purple bacteria, frequently *Chromatium*, to the exclusion of green sulfur bacteria. This is unlike the situation in stratified lakes, where both purple and green bacteria frequently coexist (see Sections 2.3.1.1 to 2.3.1.5). Although the mats at Laguna Figueroa may contain *Chloroflexus* (213), no cultural evidence for the presence of this organism in marine mats has yet been obtained.

2.3.5 Chemostat Studies of Phototrophic Bacteria

The chemostat has proven to be an excellent tool for the study of microbial ecology (218, 244), especially in examining interactions of two (or more) organisms competing for the same substrate. Van Gemerden has been particularly instrumental in applying chemostat methodology to the study of competition among phototrophic bacteria. In many of his studies he has been able to draw ecologically relevant conclusions from the chemostat and relate them to observations made *in situ*. For example, in attempts to explain the frequently frustrating experience of observing a mass development of large-celled species of *Chromatium* and *Thiospirillum* in nature only to find later that enrichment cultures have been overrun by small-celled *Chromatium* or *Thiocapsa* species, van Gemerden tested the effect of sulfide concentration and light regimen on mixed cultures of *Chromatium weissei* (a large-celled species) and *Chromatium vinosum* (a small-celled species) in the chemostat (234). Although at any sulfide concentration employed in continuous light, *C. vinosum* outgrew *C. weissei*, the introduction of a dark–light cycle began to favor the large *Chromatium*. With a 4-h light/8-h dark cycle, 70% of the population in the chemostat consisted of the large-celled species (234).

Batch culture studies of sulfide oxidation by the two *Chromatium* species showed that *C. weissei* was able to oxidize sulfide (and thus form elemental sulfur) at a rate over twice that of *C. vinosum*. In the chemostat, sulfide levels build up during the dark period, and once illuminated, the large *Chromatium* could rapidly store away large amounts of elemental sulfur for later use as an electron donor. In continuous light, on the other hand, the small *Chromatium*, whose affinity (but not oxidation rate) for sulfide was found to be greater than that of the large *Chromatium*, effectively consumed the majority of the sulfide, because the actual concentration of sulfide in the growth vessel was very low (234). Hence, in natural environments where day–night cycles allow sulfide to accumulate in darkness, the large-celled species should be favored, or at least be able to hold its own, and indeed, these observations are consistent with what one finds in blooms of purple sulfur bacteria in stratified lakes (see Section 2.3.1.1).

Continuous culture investigation of competition between *Chromatium*, which stores elemental sulfur inside the cell, and *Chlorobium*, which does not, has shed light on how these organisms occasionally coexist in stratified lakes (238). At first glance, it would appear that *Chromatium* has a distinct advantage by virtue of the fact that by storing sulfur, all eight electrons derived from the oxidation of sulfide to sulfate ultimately are made available for the reduction of CO_2 to cell material. Elemental sulfur produced by *Chlorobium*, on the other hand, leaves the cell and enters the elemental sulfur "pool" of the environment; this sulfur can be used by other sulfur-oxidizing species, including *Chlorobium*, *Chromatium*, and lithotrophic sulfur oxidizers.

From the standpoint of *Chlorobium*, the solution to this problem of losing electrons by depositing sulfur outside the cell lies in the initial oxidation of sulfide. Although the rate of sulfide oxidation by the *Chlorobium* and *Chromatium* strains studied were quite similar, the *affinity* for sulfide of the green bacterium was found to be far superior to that of the purple bacterium (238). This means that *Chlorobium* can afford to lose some of the sulfur it produces because it will obtain a greater share of any newly generated sulfide. In addition, it was concluded, based on the proportions of the two populations that were obtained at steady state in the chemostat, that *Chlorobium* must also be a better scavenger of extracellular sulfur than *Chromatium*, and probably uses at least some of the sulfur it excretes (238). Relating these observations back to the habitat, it would seem that a balance of *Chlorobium* and *Chromatium* in mixed populations in nature would therefore depend on a steady, although not necessarily large, supply of sulfide being available, and on the ability of green bacteria to utilize some of the sulfur they produce. An excellent survey of the sulfide affinity of various purple and green sulfur bacteria has been published by van Gemerden (237).

The work of Biebl and Pfennig (10) with mixed cultures of sulfur-reducing heterotrophs and sulfide-consuming green bacteria suggest that external deposition of sulfur by green bacteria may not be a disadvantage at all. The rapid conversion of sulfur back to sulfide by dissimilatory sulfur-reducing bacteria, coupled with the high affinity for sulfide of green bacteria (237, 238), probably results in sulfur excretion in *Chlorobium* being an ecologically favorable process. Consortia of green bacteria and sulfur reducers can then be looked at as a means by which green bacteria can successfully compete with purple bacteria in natural habitats.

Other uses of the continuous culture technique in ecological studies of phototrophic bacteria include examination of the effect of the dual substrates sulfide and acetate on the metabolism of *Rhodobacter capsulatus* (250), and *Chlorobium phaeobacteroides* (80), and the kinetics of glycogen production and consumption by *C. vinosum* (5). In a study of acetate assimilation by brown *Chlorobium* species it was found that the sulfide concentration greatly affected the uptake of acetate; the rate of incorporation of acetate increased significantly when sulfide levels diminished (80). Thus it is likely that both photoheterotrophic and photoautotrophic metabolism occurs in natural populations of anoxygenic phototrophs, and that the growth rates of phototrophic bacteria are influenced by the levels of organic compounds as well as by levels of reduced sulfur species. The chemostat has been and will continue to be an excellent tool for modeling these complex nutritional interrelationships. Excellent overviews of the use of continuous culture methods in the study of the physiological ecology of phototrophic bacteria can be found in Refs. 236 and 239.

2.4 ENRICHMENT AND ISOLATION OF PHOTOTROPHIC BACTERIA

2.4.1 Photosynthetic Sulfur Bacteria

A variety of media and techniques have been described for the enrichment and isolation of phototrophic sulfur bacteria. Preparation of ''Pfennig's medium,'' the most successful and widely used of all the recipes that have been published, is outlined clearly in Refs. 164 and 243. Variations for marine strains are listed in (220). Several other variations, mainly from the older literature, are described by Bose (13) and van Niel (242). No attempt will be made here to give a thorough treatment of media and enrichment conditions; only the basic prerequisites for obtaining successful enrichments will be presented. The reader is referred to the paper by Imhoff (83) for details of the mineral requirements of photosynthetic bacteria.

The nutritional needs of photosynthetic sulfur bacteria are actually quite simple.

The medium must contain an assortment of mineral salts, CO_2/bicarbonate, an adequate supply of sulfide to serve as photosynthetic electron donor, and be adjusted to the correct pH. The following is a simplified method for preparing a sulfide-containing medium for dispension into *completely filled* bottles or tubes.

1. *Mineral Salts.* An aspirator bottle (2 to 4 L, depending on the volume of medium desired) containing a rubber hose attached to the exit port (tied off with a hose clamp) and magnetic stir bar is used for final medium assembly. To the end of the hose a glass "bell jar" (Bellco Glass, Vineland, NJ) is attached for aseptic dispensing of medium. For 1 L of medium, the following salts are dissolved in 500 ml of distilled water in the aspirator bottle:

Component	Concentration (per liter)
Ethylenediaminetetraacetic acid	0.01 g
$MgSO_4 \cdot 7H_2O$	0.2 g
$CaCl_2 \cdot 2H_2O$	0.05 g
NaCl	0.4 g
NH_4Cl	0.4 g
KH_2PO_4	0.5 g
Trace elements (see Table 2.4)	1 ml
Vitamin B_{12}	20 μg

The aspirator bottle is plugged with a cotton stopper (covered with aluminum foil) and set aside.

2. *Bicarbonate.* Two grams of $NaHCO_3$ is added to a 1-L screw-capped bottle,

Table 2.4 Composition of trace element solution used in media for phototrophic bacteria

Component	Amount[a]
Distilled water	1000 mL
NaEDTA	5.2 g
$FeCl_2 \cdot 4H_2O$	1.5 g
$ZnCl_2$	70 mg
$MnCl_2 \cdot 4H_2O$	100 mg
H_3BO_3	6 mg
$CoCl_2 \cdot 6H_2O$	190 mg
$CuCl_2 \cdot 2H_2O$	17 mg
$NiCl_2 \cdot 6H_2O$	25 mg
$Na_2MoO_4 \cdot 2H_2O$	188 mg
$VoSO_4 \cdot 2H_2O$	30 mg
$Na_2WO_4 \cdot 2H_2O$	2 mg

[a] Add in the order shown; make sure that the EDTA is fully dissolved before adding remaining components. Store at 4°C.

and the bottle capped and set aside. A second bottle containing 400 ml of distilled water is also prepared and set aside (i.e., do not dissolve the $NaHCO_3$ before autoclaving).

3. *Sulfide*. Depending on the organisms to be enriched (see below) from 0.02% to 0.2% (w/v final concentration) sodium sulfide is added. Sulfide is prepared and autoclaved as a separate solution. Crystals of reagent-grade $Na_2S \cdot 9H_2O$ are removed from the bottle with forceps, dipped in distilled water, dried, and weighed (the wash makes the crystals "crystal clear" and free from metal sulfides and other debris). Washed crystals are immediately dissolved in 100 ml of *freshly boiled* distilled water and the contents transferred to a 250-ml screw-capped bottle.

4. *Sterilization*. All solutions and the dry bicarbonate are autoclaved for 20 min at 121°C (15 psi). Vessels used for dispensing media are autoclaved separately for 30 min along with a bubbling apparatus for sparging the bicarbonate solution with CO_2. The bubbler can be fashioned from a 1-ml pipette placed through a rubber stopper of a size to fit snuggly the 1-L bottle containing the bicarbonate. The pipette, plugged with a cotton filter, is attached to a 1-m length of rubber tubing. An 18-gauge syringe needle is inserted through the stopper to act as a gas exit port. The bubbling apparatus is wrapped in foil before autoclaving. Bottles or tubes used for medium should contain two to five small (~3 mm diameter) glass beads; these are useful for resuspending clumps of cells or settled cell material.

5. *Medium Assembly*

   a. After autoclaving and cooling, the sterile water is aseptically transferred to the bicarbonate powder and the bicarbonate dissolved; then the bubbling apparatus is attached and the solution is bubbled with pure carbon dioxide for 15 to 30 min. The gassed solution is then transferred aseptically to the mineral salt solution (aspirator bottle) and the mixture stirred slowly.

   b. A 5-ml sample of the mixture is removed aseptically and the pH checked. NaOH (1 N) is added to the medium to adjust the pH to 6.7 to 6.8.

   c. When the medium has reached a pH near 6.8, the sulfide solution is added (sodium sulfide is strongly alkaline and it will raise the pH of the bulk solution). Adjust the bulk solution pH to the desired value (usually, pH 6.8 to 7.2) by the addition of 1 N NaOH or 1 N HCl. The medium should be rapidly adjusted to the desired pH without delay.

   d. As soon as the medium is at the proper pH, it is dispensed immediately using the hose clamp to regulate flow through the rubber tubing and the bell jar. Vessels are filled *all the way to the top and capped tightly*. The air space should be no larger than the size of a pea. It is best to check the cap tightness at the end of medium dispensing, tightening them again if necessary. Sulfide media should be left in the dark for at least 24 h before use.

Pfennig and Trüper (164) list in detail the types of habitats in which one would most likely expect to find phototrophic bacteria, and the major ones have been listed in Section 2.3. A small amount of mud, leaf litter, hypolimnetic lake water, activated sludge from a sewage treatment plant, or pigmented material from a

Winogradsky column (see Ref. 243) is added to a bottle of the sulfide medium prepared. In general, 2 mM sulfide (0.05% w/v Na$_2$S · 9H$_2$O) is a good starting point for initial attempts to isolate purple or green sulfur bacteria [slight variations in pH will favor one group over the other (see Ref. 164)]. Yeast extract or other organic supplements are not recommended for initial enrichments; however, subsequent transfers of positive enrichments will be greatly stimulated by the addition of 0.05% ammonium acetate. Vitamin B$_{12}$ is added because many purple and green sulfur bacteria require it; no other vitamin requirements have been reported for typical Chromatiaceae or Chlorobiaceae (151). Inoculated media should always be placed in the dark for a few hours before being placed in the light.

For purple and green sulfur bacteria a temperature of about 25°C and a light intensity of 500 to 1000 lux is recommended for initial enrichments. (*Note:* The light source should be incandescent illumination, i.e., standard light bulbs, not fluorescent illumination. The latter are spectrally very poor in the red and infrared regions.) Successful enrichments are usually evident within 1 to 2 weeks, but it is suggested not to conclude that enrichments are unsuccessful for some time thereafter; large-celled species of *Chromatium*, for example, grow rather slowly (234). If a light box is not available, it is sufficient to incubate enrichments in the light of a north window. Methods for pure culture isolation of purple and green sulfur bacteria are described in (164).

2.4.2 Facultative Photosynthetic Bacteria

A variety of methods for the enrichment and isolation of purple nonsulfur bacteria have been published (11, 13, 243). The major difference between the media used for the sulfur bacteria and those for the Rhodospirillaceae is the fact that sulfide is generally omitted, and this makes for a more straightforward preparation involving a single solution. A medium that has been successful for enrichment of Rhodospirillaceae in the author's laboratory includes the following:

Component	Concentration (per liter)
Ethylenediaminetetraacetic acid	0.010 g
NH$_4$Cl	0.8 g
NaCl	0.4 g
MgSO$_4$ · 7H$_2$O	0.2 g
CaCl$_2$ · 2H$_2$O	0.075 g
K$_2$HPO$_4$	0.45 g
KH$_2$PO$_4$	0.3 g
NaHCO$_3$	1 g
Trace element solution (see Table 2.4)	1 ml
Yeast extract	0.05 g
Vitamin B$_{12}$	20 μg
Organic carbon source	0.1–2 g

The nature of the carbon source employed can be critical to the development of the enrichment (11, 243); fermentable compounds (sugar, pyruvate, lactate) are to be avoided, because they invariably lead to the development of rapidly growing fermentative anaerobes to the exclusion of phototrophic bacteria. Acetate and succinate are excellent carbon sources (usually added at 0.2% w/v to the medium). Malate can also be used, but occasionally leads to problems with overgrowth of heterotrophs. The pH of the medium is adjusted to 6.8 to 7.0 (or lower for certain species, see Ref. 152), dispensed into tubes or bottles, and the medium autoclaved in the usual fashion, cooled, and left tightly capped for storage. Bicarbonate is added after autoclaving from sterile stock solutions.

It is advisable to bubble the medium with N_2 before establishing enrichments in order to reduce the growth of facultative aerobes (nonphototrophs), which can occasionally cause problems in enrichments with, for example, sewage as inoculum (due to the large enteric bacterial load). Alternatively, the medium can be prepared and stored completely anaerobically using the methods described by Malik (127). If the medium is degassed (with N_2 or $N_2 : CO_2$) just before inoculation, the vessel should be filled to the top with medium and sealed tightly or stoppered after inoculation. Incubation conditions are as described for the sulfur phototrophs except that Rhodospirillaceae in general can withstand slightly higher incubation temperatures and higher light intensities (e.g., 28 to 30°C at 1000 to 2000 lux incandescent illumination). Pure cultures can be obtained using agar plates or shake cultures incubated anaerobically in the light (11).

A facile technique for the direct isolation of Rhodospirillaceae on agar plates using membrane filters has been described by Biebl and Drews (8) and Swoager and Lindstrom (214). The author has used the membrane filter technique and found it to work very well; it is particularly valuable for isolating photosynthetic bacteria from lake or pond water samples where they may be present in low numbers and missed by classical liquid enrichment methods. The membrane filter technique is also useful for obtaining cultures of slow-growing Rhodospirillaceae, since at the proper dilution, each viable cell will produce a visible colony; liquid enirchment techniques frequently overlook these species (8, 11).

A rather foolproof method for the isolation of Rhodospirillaceae suitable for use in the classroom to demonstrate the principles of the enrichment culture technique has been described by the author (122). The technique, originally developed by Howard Gest, Indiana University, uses the medium described in this section, modified to contain only one-tenth of the originally specified concentration of NH_4Cl. The medium is dispensed in 40-ml aliquots in 100-ml bottles, inoculated with an appropriate sample, vigorously bubbled with a gas mixture of 99% N_2 + 1% CO_2 for 2 to 3 min and then sealed with a rubber stopper under a stream of anaerobic gas. Enrichments are put at 28 to 30°C and 1000 lux. This protocol takes additional advantage in the enrichment of the ability of most Rhodospirillaceae to fix N_2 (122), and in classroom experiments it has been found virtually 100% successful, even when dried soil samples are used as inocula. One disadvantage of the method is that the number of different enriched species can be rather limited, in that the rapidly growing, good-N_2-fixing species (*R. capsulatus* and *R. sphaeroides*) seem to predominate. However, in a large classroom experiment where each student brings his or her own sample, a reasonable variety of Rhodospirilla-ceae have usually been obtained. Nitrogen-fixing enrichment cultures take about 1 week to develop into brightly pigmented liquid suspensions. Purification is very easy by streaking plates of nitrogen-free media incubated under an N_2/CO_2 gas mixture.

Media and culture conditions for the enrichment and isolation of *Chloroflexus* are described in Ref. 24. Since *Chloroflexus* inhabits hot springs, the medium employed contains a considerably different mixture of inorganic salts than the media described above; tests by the author have shown that cultures of *Chloroflexus* grow poorly in Pfennig's medium. The interested reader should refer to Refs. 23, 125, and 168 for details on how to grow *Chloroflexus*.

2.4.3 Preservation of Pure Cultures

Pure cultures of purple or green sulfur bacteria and purple nonsulfur bacteria can be preserved at liquid nitrogen temperatures ($-195\,°C$) or in ultralow-temperature mechanical freezers (-80 to $-96\,°C$). The usual protocol calls for the use of 5 to 10% dimethylsulfoxide or 10% glycerol (final concentrations) as the protecting agent. Log phase cells are mixed with the protecting agent for 10 to 15 min at $0\,°C$ and then immediately placed in the freezer or in the liquid-nitrogen tank. Five-milliliter sterile polypropylene tubes with tight-fitting "snap cap" tops are ideal for freezing at $-80\,°C$. Frozen cultures can be arranged in the freezer in radioim-munoassay tube supports and cataloged for quick retrieval. Frozen cultures prepared by the author were revivable after 8 years at $-80\,°C$; assuming that the vials do not thaw, they will probably survive much longer periods. Freeze-drying of most Rhodospirillaceae is also possible; sulfur phototrophs, on the other hand, apparently do not survive lyophilization (9). A detailed study of liquid-nitrogen storage of phototrophic bacteria was recently published and should be consulted for additional information on this topic (128).

ACKNOWLEDGMENTS

I thank Thomas Brock, Douglas Caldwell, Tim Parkin, Norbert Pfennig, Hans Trüper, and Alex Zehnder for supplying photomicrographs or copies of published figures, and Ricardo Guerrero for published and unpublished results. I thank Howard Gest and Nancy Spear for reviewing the manuscript, and Sean Andrews and Kita Neal for expert word-processing skills. Work of the author is supported by a grant from the U.S. National Science Foundation (DMB8505492).

REFERENCES

1. Abella, C., E. Montesinos, and R. Guerrero. 1980. Field studies on the competition between purple and green sulfur bacteria for available light (Lake Cisó, Spain), p.

173–181, in M. DoKulil, H. Metz, and D. Jewson (eds.), Shallow lakes. W. Junk Publishers, The Hague.

2. Baas-Becking, L. G. M. 1925. Studies on the sulfur bacteria. Ann. Bot. (Lond.) 39:613–650.

3. Bauld,. J., and T. D. Brock. 1973. Ecological studies of *Chloroflexis*, a gliding photosynthetic bacterium. Arch. Mikrobiol. 92:267–284.

4. Beatty, J. T., and H. Gest. 1981. Biosynthetic and bioenergetic functions of citric acid cycle reactions in *Rhodopseudomonas capsulata*. J. Bacteriol. 148:584–593.

5. Beeftink, H. H., and H. van Gemerden. 1979. Actual and potential rates of substrate oxidation and product formation in continuous cultures of *Chromatium vinosum*. Arch. Microbiol. 121:161–167.

6. Bergstein, T., Y. Hais, and B. Z. Cavari. 1979. Investigation on the photosynthetic sulfur bacterium *Chlorobium phaeobacteroides* causing seasonal blooms in Lake Kinnert. Can. J. Microbiol. 25:999–1007.

7. Beuscher, N., and G. Gottschalk. 1972. Lack of citrate lyase—the key enzyme of the reductive carboxylic acid cycle—in *Chlorobium thiosulfatophilum* and *Rhodospirillum rubrum*. Z. Naturforsch. 276:967–973.

8. Biebl, H., and G. Drews. 1969. Das in-vivo-Spektrum als taxonomisches Merkmal bei Untersuchungen zur Verbreitung von Athiorhodaceae. Zentralbl. Bakteriol. Parasitenkd. Infectionskr. Hyg. 2 Abt. 123:425–452.

9. Biebl, H., and R. A. Malik. 1976. Long term preservation of phototrophic bacteria, p. 31–33, in G. A. Codd and W. D. P. Stewart (eds.), Proceedings of the second international symposium on photosynthetic prokaryotes, Dundee, Scotland.

10. Biebl, H., and N. Pfennig. 1978. Growth yields of green sulfur bacteria in mixed cultures with sulfur and sulfate-reducing bacteria. Arch. Microbiol. 117:9–16.

11. Biebl, H., and N. Pfennig. 1981. Isolation of members of the family Rhodospirillaceae, p. 267–273, in M. P. Starr, H. Stolp, H. G. Trüper, A. Balows, and H. G. Schlegel (eds.), The prokaryotes—a handbook on habitats, isolation and identification of bacteria. Springer-Verlag, New York.

12. Biel, A. J., and B. L. Marrs. 1983. Transcriptional regulation of several genes for bacteriochlorophyll biosynthesis in *Rhodopseudomonas capsulata* in response to oxygen. J. Bacteriol. 156:686–694.

13. Bose, S. K. 1963. Media for anaerobic growth of photosynthetic bacteria, p. 501–510, in H. Gest, A. San Pietro, and L. P. Vernon (eds.), Bacterial photosynthesis. Antioch, Yellow Springs, Ohio.

14. Brock, T. D. 1978. Thermophilic microorganisms and life at high temperature. Springer-Verlag, New York.

15. Brock, T. D., and M. L. Brock. 1966. Temperature optima for algal development in Yellowstone and Icelandic hot springs. Nature (Lond.) 209:733–734.

16. Brock, T. D., and M. L. Brock. 1969. Effect of light intensity on photosynthesis by thermal algae adapted to natural and reduced sunlight. Limnol. Oceanogr. 14:334–341.

17. Brockmann, H. Jr., and A. Lipinski. 1983. Bacteriochlorophyll *g*. A new bacteriochlorophyll from *Heliobacterium chlorum*. Arch. Microbiol. 136:17–19.

18. Buchanan, B. B., and R. Sirevåg. 1976. Ribulose-1,5-diphosphate carboxylase and *Chlorobium thiosulfatophilum*. Arch. Microbiol. 109:15–19.

19. Caldwell, D. E., and J. M. Tiedje. 1975. A morphological study of anaerobic bacteria from the hypolimnia of two Michigan lakes. Can. J. Microbiol. 21:362–376.

20. Caldwell, D. E. and J. M. Tiedje. 1975. The structure of anaerobic bacterial communities in the hypolimnia of several Michigan lakes. Can. J. Microbiol. 21:377–385.

21. Castenholz, R. W. 1973. The possible photosynthetic use of sulfide by the filamentous, phototrophic bacteria of hot springs. Limnol. Oceanogr. 18:863–876.

22. Castenholz, R. W. 1977. The effect of sulfide on the blue-green algae of hot springs. II. Yellowstone National Park. Microbial Ecology 3:79–105.

23. Castenholz, R. W. 1984. Composition of hot spring microbial mats: A summary, p. 101–119, in Y. Cohen, R. W. Castenholz, and H. O. Halvorson (eds.), Microbial Mats: Stromatolites. Alan R. Liss, Inc. New York.

24. Castenholz, R. W., and B. K. Pierson. 1981. Isolation of members of the family Chloroflexaceae, p. 290–298, in M. P. Starr, H. Stolp, H. G. Trüper, A. Balows, and H. G. Schlegel (eds.), The prokaryotes—a handbook on habitats, isolation, and identification of bacteria. Springer-Verlag, New York.

24a. Caumette, P. 1986. Phototrophic sulfur bacteria and sulfate-reducing bacteria causing red waters in a shallow brackish coastal lagoon. FEMS Microbiol. Ecol. 38:113–124.

25. Clayton, R. K. 1953. Studies in the phototaxis of *Rhodospirillum rubrum*. I. Action spectrum, growth in green light, and Weber Law adherence. Arch. Mikrobiol. 19:107–124.

26. Cohen, Y., B. B. Jørgensen, E. Padan, and M. Shilo. 1975. Sulfide-dependent anoxygenic photosynthesis in the cyanobacterium *Oscillatoria limnetica*. Nature (Lond.) 257:486–492.

27. Cohen, Y., W. E. Krumbein, M. Goldberg, and M. Shilo. 1977. Solar Lake (Sinai). 1. Physical and chemical limnology. Limnol. Oceanogr. 22:597–608.

28. Cohen, Y., W. E. Krumbein, and M. Shilo. 1977. Solar Lake (Sinai). 2. Distribution of photosynthetic microorganisms and primary production. Limnol. Oceanogr. 22:609–620.

29. Cohen, Y., W. E. Krumbein, and M. Shilo. 1977. Solar Lake (Sinai). 3. Bacterial distribution and production. Limnol. Oceanogr. 22:621–634.

30. Cohen, Y., E. Padan, and M. Shilo. 1975. Facultative anoxygenic photosynthesis in the cyanobacterium *Oscillatoria limnetica*. J. Bacteriol. 123:855–861.

31. Cohen-Bazire, G., W. R. Sistrom, and R. W. Stanier. 1957. Kinetic studies of pigment synthesis by non-sulfur purple bacteria. J. Cell. Comp. Physiol. 49:25–68.

32. Conrad, R., and H. G. Schlegel. 1977. Different degradation pathways for glucose and fructose in *Rhodopseudomonas capsulata*. Arch. Microbiol. 112:39–48.

33. Conti, S. F., and P. Hirsch. 1965. Biology of budding bacteria. III. Fine structure of *Rhodomicrobium* and *Hyphomicrobium* spp. J. Bacteriol. 89:503–512.

34. Cooper, D. E., M. B. Rands, and C.-P. Woo. 1975. Sulfide reduction in fellmongery effluent by red sulfur bacteria. J. Water Pollution Control Fed. 47:2088–2100.

35. Cox, J. C., M. T. Madigan, J. L. Favinger, and H. Gest. 1980. Redox mechanisms in "oxidant-dependent" hexose fermentation by *Rhodopseudomonas capsulata*. Arch. Biochem. Biophys. 204:10–17.

36. Culver, D. A., and G. J. Brunskill. 1969. Fayetteville Green Lake, New York. V.

Studies of primary production and zooplankton in a meromictic marl lake. Limnol. Oceanogr. 14:862–873.

37. Czeczuga, B. 1968. Primary production of the green hydrosulphuric bacteria *Chlorobium limicola* Nads. (Chlorobiaceae). Photosynthetica (Prague) 2:11–15.

38. Czeczuga, B. 1968. Primary production of the purple sulfuric bacteria, *Thiopedia rosea* Winog. (Thiorhodaceae). Photosynthetica (Prague) 2:161–166.

39. Dashekevicz, M. P., and R. L. Uffen. 1979. Identification of a carbon-monoxide metabolizing bacterium as a strain of *Rhodopseudomonas gelatinosa* (Molisch) van Niel. Int. J. Syst. Bacteriol. 29:145–148.

40. Douhit, H. A., and Pfennig, N. 1976. Isolation and growth rates of methanol utilizing Rhodospirillaceae. Arch. Microbiol. 107:233–234.

41. Drews, G. 1985. Structure and functional organization of light-harvesting complexes and photochemical reaction centers in membranes of phototrophic bacteria. Microbiol. Rev. 49:59–70.

42. Drews, G., and P. Giesbrecht. 1963. Zur Morphogenese der "Chromatophoren" (Thylakoide) und zur Synthese des Bakteriochlorophylls bei *Rhodopseudomonas sphaeroides* and *Rhodospirillum rubrum*. Zentralbl. Bakteriol. Parasitenkd. Infektionskr. Hyg. 1 Abt. Orig. 190:508–536.

43. Drutschmann, M., and J.-H. Klemme. 1985. Sulfide-repressed, membrane-bound hydrogenase in the thermophilic facultative phototroph, *Chloroflexus aurantiacus*. FEMS Microbiol. Lett. 28:231–235.

44. Dutton, P. L., and W. C. Evans. 1969. The metabolism of aromatic compounds by *Rhodopseudomonas palustris:* a new reductive method of aromatic ring metabolism. Biochem. J. 113:525–536.

45. Dutton, P. L., and W. C. Evans. 1978. Metabolism of aromatic compounds by Rhodospirillaceae, p. 719–726, in R. K. Clayton and W. R. Sistrom (eds.), The photosynthetic bacteria. Plenum Press, New York.

46. Eimhjellen, K. E., H. Steensland, and J. Traetteberg. 1967. A *Thiococcus* sp. nov., gen., its pigments and internal membrane system. Arch. Mikrobiol. 59:82–92.

47. Ensign, J. C. 1976. Biomass production from animal wastes by photosynthetic bacteria, p. 455–482, in H. G. Schlegel and J. Barnea (eds.), Microbial energy conversion. E. Goltze KG, Göttingen, West Germany.

48. Evans, M. C. W., B. B. Buchanan, and D. I. Arnon. 1966. A new ferredoxin-dependent carbon reduction cycle in a photosynthetic bacterium. Proc. Natl. Acad. Sci. USA 55:928–934.

49. Feick, R. G., M. Fitzpatrick, and R. C. Fuller. 1982. Isolation and characterization of cytoplasmic membranes and chlorosomes from the green bacterium *Chloroflexus aurantiacus*. J. Bacteriol. 105:905–915.

50. Fowler, V. J., N. Pfennig, W. Schubert, and E. Stackebrandt. 1984. Towards a phylogeny of phototrophic purple sulfur bacteria − 16S rRNA oligonucleotide cataloging of 11 species of Chromatiaceae. Arch. Microbiol. 139:382–387.

51. Fuchs, G., and E. Stupperich. 1982. Autotrophic CO_2 fixation pathway in *Methanobacterium thermoautotrophicum*. Zentralbl. Bakteriol. Mikrobiol. Hyg. 1 Abt. Orig. C3:277–288.

52. Fuchs, G., E. Stupperich, and G. Eden. 1980. Autotrophic CO_2 fixation in *Chloro-*

bium limicola. Evidence for the operation of a reductive tricarboxylic acid cycle in growing cells. Arch. Microbiol. 128:64–71.

53. Fuchs, G., E. Stupperich, and R. Jaenchen. 1980. Autotrophic CO_2 fixation in *Chlorobium limicola.* Evidence against the operation of the Calvin Cycle in growing cells. Arch. Microbiol. 128:56–63.

54. Fuller, R. C. 1978. Photosynthetic carbon metabolism in the green and purple bacteria, p. 691–705, in R. K. Clayton and W. R. Sistrom (eds.), The photosynthetic bacteria. Plenum Press, New York.

54a. Garcia, D., P. Parot, A. Vermeglio, and M. T. Madigan. 1986. The light-harvesting complexes of a thermophilic purple sulfur photosynthetic bacterium *Chromatium tepidum.* Biochim. Biophys. Acta 850:390–395.

55. Garlick, S., A. Oren, and E. Padan. 1977. Occurrence of facultative anoxygenic photosynthesis among filamentous and unicellular cyanobacteria. J. Bacteriol. 129:623–629.

56. Gest, H., and J. L. Favinger. 1983. *Heliobacterium chlorum,* an anoxygenic brownish-green photosynthetic bacterium containing a ''new'' form of bacteriochlorophyll. Arch. Microbiol. 136:11–16.

57. Gest, H., and M. D. Kamen. 1949. Photoproduction of molecular hydrogen by *Rhodospirillum rubrum.* Science 109:558–559.

58. Gest, H., J. G. Ormerod, and K. S. Ormerod. 1962. Photometabolism of *Rhodospirillum rubrum:* light-dependent dissimilation of organic compounds to carbon dioxide and molecular hydrogen by an anaerobic citric acid cycle. Arch. Biochem. Biophys. 97:21–33.

59. Gibson, J., W. Ludwig, E. Stackebrandt, and C. R. Woese. 1985. The phylogeny of the green photosynthetic bacteria: absence of a close relationship between *Chlorobium* and *Chloroflexus.* Syst. Appl. Microbiol. 6:152–156.

60. Gibson, J., N. Pfennig, and J. B. Waterbury. 1984. *Chloroherpeton thalassium* gen. nov. et spec. nov., a nonfilamentous, flexing, and gliding green sulfur bacterium. Arch. Microbiol. 138:96–101.

60a. Giovannoni, S. J., N. P. Revsbech, D. M. Ward, and R. W. Castenholz. 1987. Obligately phototrophic *Chloroflexus:* primary production in anaerobic hot spring microbial mats. Arch. Microbiol. 147:80–87.

61. Gloe, A., N. Pfennig, H. Brockman, Jr., and H. Trowitzsch. 1975. A new bacteriochlorophyll from brown-colored Chlorobiaceae. Arch. Microbiol. 102:103–109.

62. Gloe, A., and N. Risch. 1978. Bacteriochlorophyll c_s, a new bacteriochlorophyll from *Chloroflexus aurantiacus.* Arch. Microbiol. 118:153–156.

63. Goedheer, J. C. 1959. Energy transfer between carotenoids and bacteriochlorophyll in chromatophores of purple bacteria. Biochim. Biophys. Acta 35:1–8.

64. Gorlenko, V. M. 1975. Characteristics of filamentous phototrophic bacteria from freshwater lakes. Microbiology (Engl. Transl. Mikrobiologiya) 44:682–684.

65. Gorlenko, V. M., E. I. Kompantseva, and N. N. Puchkova. 1985. Influence of temperature on phototrophic bacteria in hot springs. Microbiology (Engl. Transl. Mikrobiologiya) 54:681–685.

66. Gorlenko, V. M., and T. A. Pivovarova. 1977. Concerning the membership of blue-

green algae *Oscillatoria coerulescens* Gicklorn, 1921, to a new genus of Chlorobacteria *Oscillatoris* nov. gen. Izv. Akad. Nauk. SSSR Ser. Biol. 3:396–409. (In Russian with an English summary.)

67. Gorlenko, V. M., and T. W. Zhilina. 1968. Study of the ultrastructure of green sulfur bacteria, strain SK 413. Microbiology (Engl. Transl. Mikrobiologiya) 37:892–897.

68. Gorrell, T. E., and R. L. Uffen. 1977. Fermentative metabolism of pyruvate by *Rhodospirillum rubrum* after anaerobic growth in darkness. J. Bacteriol. 131:533–543.

69. Gray, B. H., C. F. Fowler, N. A. Nugent, N. Rigopoulous, and R. C. Fuller. 1973. Reevaluation of *Chloropseudomonas ethylica* strain 2-K. Int. J. Syst. Bacteriol. 23:256–264.

70. Guerreo, R., E. Montesinos, I. Esteve, and C. Abella. 1980. Physiological adaptations and growth of purple and green sulfur bacteria in a meromictic lake as compared to a holomictic lake, p. 161–171, in M. Dokulil, H. Metz, and D. Jewson (eds.), Shallow lakes. W. Junk Publishers, The Hague.

71. Guerrero, R., E. Montesinos, C. Pedros-Alio, I. Esteve, J. Mas, H. van Gemerden, P. A. G. Hoffman, and J. F. Bakker. 1985. Phototrophic sulfur bacteria in two Spanish lakes: vertical distribution and limiting factors. Limnol. Oceanogr. 30:919–931.

72. Guyer, M., and G. D. Hegeman. 1969. Evidence for a reductive pathway for the anaerobic metabolism of benzoate. J. Bacteriol. 99:906–907.

73. Hansen, T. A., and H. van Gemerden. 1972. Sulfide utilization by purple non-sulfur bacteria. Arch. Mikrobiol. 86:49–56.

74. Hansen, T. A., and H. Veldkamp. 1973. *Rhodopseudomonas sulfidophila nov. spec.*, a new species of the purple non-sulfur bacteria. Arch. Mikrobiol. 92:45–58.

75. Harashima, K., J.-I. Hayasaki, T. Ikari, and T. Shiba. 1980. O_2-stimulated synthesis of bacteriochlorophyll and carotenoids in marine bacteria. Plant Cell Physiol. 21:1283–1294.

76. Hashwa, F. A., and H. G. Trüper. 1978. Viable phototrophic sulfur bacteria from the Black Sea bottom. Helgol. Wiss. Meeresunters. 31:249–253.

76a. Heda, G. D., and M. T. Madigan. 1986. Utilization of amino acids and lack of diazotrophy in the thermophilic anoxygenic phototroph *Chloroflexus aurantiacus*. J. Gen. Microbiol. 132:2469–2473.

77. Hickman, D. D., and A. W. Frenkel. 1965. Observation on the structure of *Rhodospirillum molischianum*. J. Cell Biol. 25:261–278.

78. Hillmer, P., and H. Gest. 1977. H_2 metabolism in the photosynthetic bacterium *Rhodopseudomonas capsulata:* H_2 production by growing cultures. J. Bacteriol. 129:724–731.

79. Hoare, D. S., and J. Gibson. 1964. Photoassimilation of acetate and the biosynthesis of amino acids by *Chlorobium thiosulfatophilum*. Biochem. J. 91:546–559.

80. Hofman, P. A. G., M. J. W. Veldhuis, and H. van Gemerden. 1985. Ecological significance of acetate assimilation of *Chlorobium phaeobacterioides*. FEMS Microbiol. Ecol. 31:271–278.

81. Holm, H. W., and J. W. Vennes. 1970. Occurrence of purple sulfur bacteria in a sewage treatment lagoon. Appl. Microbiol. 19:988–996.

82. Holt, S. C., and A. G. Marr. 1965. Location of chlorophyll in *Rhodospirillum rubrum*. J. Bacteriol. 89:1402–1412.

83. Imhoff, J. F. 1982. Response of photosynthetic bacteria to mineral nutrients, p. 135–146, in A. Mitsui and C. C. Black (eds.), Handbook of biosolar resources, Vol. 1, P. 2. CRC Press, Boca Raton, Fla.

84. Imhoff, J. F. 1984. Reassignment of the genus *Ectothiorhodospira* Pelsh 1936 to a new family, *Ectothiorhodospiraceae* fam. nov., and emended description of the *Chromatiaceae* Bavendamm 1924. Int. J. Syst. Bacteriol. 34:338–339.

85. Imhoff, J. F., H. G. Trüper, and N. Pfennig. 1984. Rearrangement of the species and genera of the phototrophic 'purple nonsulfur bacteria.' Int. J. Syst. Bacteriol. 34:340–343.

86. Ivanovsky, R. N., N. V. Sinton, and E. N. Kondratieva. 1980. ATP-linked citrate lyase activity in the green sulfur bacterium *Chlorobium limicola* forma *thiosulfatophilum*. Arch. Microbiol. 128:239–241.

87. James, H. R., and E. A. Birge. 1938. A laboratory study of the absorption of light by lake waters. Trans Wis. Acad. Sci. 31:1–154.

88. Jones, B. R. 1956. Studies of pigmented non-sulfur purple bacteria in relation to cannery waste lagoon odors. Sewage Ind. Wastes 28:883–893.

89. Jørgensen, B. B. 1982. Ecology of the bacteria of the sulfur cycle with special reference to anoxic-oxic interface environments. Philos. Trans. R. Soc. Lond. B298:543–561.

90. Jørgensen, B. B., and D. J. Des Marais. 1986. Competition for sulfide among colorless and purple sulfur bacteria in cyanobacterial mats. FEMS Microbiol. Ecol. 38:179–186.

91. Kaiser, P. 1966. Ecologie des bactéries photosynthetiques. Rev. Ecol. Biol. Sol. 3:409–472.

92. Kamen, M. D., and H. Gest. 1949. Evidence for a nitrogenase system in the photosynthetic bacterium *Rhodospirillum rubrum*. Science 109:560.

93. Kämpf, C., and N. Pfennig. 1980. Capacity of Chromatiaceae for chemotrophic growth. Specific respiration rates of *Thiocystis violacea* and *Chromatium vinosum*. Arch. Microbiol. 127:125–135.

94. Kaulen, J., and J.-H. Klemme. 1983. No evidence of covalent modification of glutamine synthetase in the thermophilic phototrophic bacterium *Chloroflexus aurantiacus*. FEMS Microbiol. Lett. 20:75–79.

95. Kelley, D. P. 1974. Growth and metabolism of the obligate phototroph *Chlorobium thiosulfatophilum* in the presence of added organic nutrients. Arch. Microbiol. 100:163–178.

96. Kobayashi, M. 1975. Role of photosynthetic bacteria in foul water purification. Prog. Water Technol. 7:309–315.

97. Kobayashi, M. 1976. Utilization and disposal of wastes by photosynthetic bacteria, p. 443–453, in H. G. Schlegel and J. Barnea (eds.), Microbial energy conversion. E. Goltze KG, Göttingen, West Germany.

98. Kobayashi, M. 1982. The role of phototrophic bacteria in nature and their utilization, p. 643–666, in N. S. Subba Rao (ed.), Advances in agricultural microbiology. Butterworth, London.

99. Kohler, H.-P., B. Ahring, C. Abella, K. Ingvorsen, H. Keweloh, E. Laczko, E. Stupperich, and F. Tomei. 1984. Bacteriological studies on the sulfur cycle in the

anaerobic part of the hypolimnion and in the surface sediments of Rotsee in Switzerland. FEMS Microbiol. Lett. 21:279–289.

100. Kohlmiller, E. F., Jr., and H. Gest. 1951. A comparative study of the light and dark fermentations of organic acids by *Rhodospirillum rubrum*. J. Bacteriol. 61:269–282.

101. Kondratieva, E. N. 1979. Interrelation between modes of carbon assimilation and energy production in phototrophic purple and green bacteria, p. 117–175, in J. R. Quayle (ed.), Microbial biochemistry (International review of biochemistry, Vol. 21). University Park Press, Baltimore.

102. Kondratieva, E. N., V. G. Zhukov, R. N. Ivanovsky, Y. P. Petuskova, and E. Z. Monosov. 1976. The capacity of phototrophic sulfur bacterium *Thiocapsa roseopersicina* for chemosynthesis. Arch. Microbiol. 108:287–292.

103. Kornber, H. L., and J. Lascelles. 1960. The formation of isocitratase by the Athiorhodaceae. J. Gen. Microbiol. 23:511–517.

104. Kuenen, J. G. 1975. Colourless sulphur bacteria and their role in the sulphur cycle. Plant Soil 43:49–76.

105. Kuhl, S. A., D. W. Nix, and D. C. Yoch. 1983. Characterization of a *Rhodospirillum rubrum* plasmid: loss of photosynthetic growth in plasmidless strains. J. Bacteriol. 156:737–742.

106. Laanbroek, H. J., and N. Pfennig. 1981. Oxidation of short-chain fatty acids by sulfate-reducing bacteria in freshwater and in marine sediments. Arch. Microbiol. 128:330–335.

107. Leuking, D., D. Tokuhisa, and G. Sojka. 1973. Glycerol assimilation by a mutant of *Rhodopseudomonas capsulata*. J. Bacteriol. 115:897–903.

108. Lewin, R. A. 1981. The prochlorophytes, p. 257–266, in M. P. Starr, H. Stolp, H. G. Trüper, A. Balows, and H. G. Schlegel (eds.), The prokaryotes, a handbook on habitats, isolation and identification of bacteria. Springer-Verlag, New York.

109. Liaaen-Jensen, S. 1978. Chemistry of carotenoid pigments, p. 233–247, in R. K. Clayton and W. R. Sistrom (eds.), The photosynthetic bacteria. Plenum Press, New York.

110. Lindstrom, E. S., S. M. Lewis, and M. J. Pinsky. 1951. Nitrogen fixation and hydrogenase in various bacterial species. J. Bacteriol. 61:481–487.

111. Lindstrom, E. S., S. R. Tove, and P. W. Wilson. 1950. Nitrogen fixation by the green and purple sulfur bacteria. Science 112:197–198.

112. Løken, Ø., and R. Sirevåg. 1982. Evidence for the presence of the glyoxalate cycle in *Chloroflexus*. Arch. Microbiol. 132:276–279.

113. Ludden, P. W., and R. H. Burris. 1981. *In vivo* and *in vitro* studies on ATP and electron donors to nitrogenase in *Rhodospirillum rubrum*. Arch. Microbiol. 130:155–158.

114. B. A. Macler, R. A. Pelroy, and J. A. Bassham. 1979. Hydrogen formation in nearly stoichiometric amount from glucose by a *Rhodopseudomonas sphaeroides* mutant. J. Bacteriol. 138:446–452.

115. Madigan, M. T. 1984. A novel photosynthetic purple bacterium isolated from a Yellowstone hot spring. Science 225:313–315.

116. Madigan, M. T. 1986. *Chromatium tepidum* sp. nov., a thermophilic photosynthetic bacterium of the family *Chromatiaceae*. Int. J. Syst. Bacteriol. 36:222–227.

117. Madigan, M. T., and T. D. Brock. 1975. Photosynthetic sulfide oxidation by *Chloroflexus aurantiacus*, a filamentous, photosynthetic, gliding bacterium. J. Bacteriol. 122:782–784.

118. Madigan, M. T., and T. D. Brock. 1977. ''Chlorobium-type'' vesicles of photosynthetically-grown *Chloroflexus aurantiacus* observed using negative staining techniques. J. Gen. Microbiol. 102:279–285.

119. Madigan, M. T., and T. D. Brock. 1977. Adaptation by hot spring phototrophs to reduced light intensities. Arch. Microbiol. 113:111–120.

120. Madigan, M. T., J. C. Cox, and H. Gest. 1980. Physiology of dark fermentative growth of *Rhodopseudomonas capsulata*. J. Bacteriol. 142:908–915.

121. Madigan, M. T., J. C. Cox, and H. Gest. 1982. Photopigments in *Rhodopseudomonas capsulata* cells grown anaerobically in darkness. J. Bacteriol. 150:1422–1429.

122. Madigan, M. T., S. S. Cox, and R. A. Stegeman. 1984. Nitrogen fixation and nitrogenase activities in members of the family Rhodospirillaceae. J. Bacteriol. 157:73–78.

123. Madigan, M. T., and H. Gest. 1978. Growth of a photosynthetic bacterium anaerobically in darkness, supported by oxidant-dependent sugar fermentation. Arch. Microbiol. 117:119–122.

124. Madigan, M. T., and H. Gest. 1979. Growth of the photosynthetic bacterium *Rhodopseudomonas capsulata* chemoautotrophically in darkness with H_2 as the energy source. J. Bacteriol. 137:524–530.

125. Madigan, M. T., S. R. Petersen, and T. D. Brock. 1974. Nutritional studies on *Chloroflexus*, a filamentous, photosynthetic, gliding bacterium. Arch. Microbiol. 100:97–103.

126. Madigan, M. T., J. D. Wall, and H. Gest. 1979. Dark anaerobic dinitrogen fixation by a photosynthetic microorganism. Science 204:1429–1430.

127. Malik, K. A. 1983. A modified method for the cultivation of phototrophic bacteria. J. Microbiol. Methods 1:343–352.

128. Malik, K. A. 1984. A new method for liquid storage of phototrophic bacteria under anaerobic conditions. J. Microbiol. Methods 2:41–47.

129. Marrs, B. L. 1974. Genetic recombination in *Rhodopseudomonas capsulata*. Proc. Natl. Acad. Sci. USA 71:971–973.

130. Marrs, B. L. 1983. Genetics and molecular biology, p. 186–214, in J. G. Ormerod (ed.), The phototrophic bacteria (Studies in microbiology, Vol. 4.) Blackwell Scientific, Oxford.

131. Marrs, B. L. 1978. Mutations and genetic manipulations as probes of bacterial photosynthesis, p. 261–294, in D. R. Sanadi and L. P. Vernon (eds.), Current topics in bioenergetics, Vol. 8, Photosynthesis, Pt. B. Academic Press, New York.

132. Masters, R. A., and M. T. Madigan. 1983. Nitrogen metabolism in the phototrophic bacteria *Rhodocyclus purpureus* and *Rhodospirillum tenue*. J. Bacteriol. 155:222–227.

133. Materon, R., and R. Baulaigue. 1977. Influence de la pénétration de la lumière solaire sur le développement des bactéries phototrophes sulfureuses dans les environnements marins. Can. J. Microbiol. 23:267–270.

134. Mauzerall, D. 1978. Bacteriochlorophyll and photosynthetic evolution, p. 223–231,

in R. K. Clayton and W. R. Sistrom (eds.), The photosynthetic bacteria. Plenum Press, New York.

135. May, D. S., and J. B. Stahl. 1967. The ecology of *Chromatium* in sewage ponds. Bulletin 303, Technical Extension Service, Washington State University, Pullman, Wash.

136. McEwan, A. G., S. J. Ferguson, and J. B. Jackson. 1983. Electron flow to dimethylsulfoxide or trimethylamine-*N*-oxide generates a membrane potential in *Rhodopseudomonas capsulata*. Arch. Microbiol. 136:300–305.

137. Meyer, J., B. C. Kelley, and P. M. Vignais. 1978. Nitrogen fixation and hydrogen metabolism in phototrophic bacteria. Biochemie (Paris) 60:245–260.

138. Montesinos, E., R. Guerrero, C. Abella, and I. Esteve. 1983. Ecology and physiology of the competition for light between *Chlorobium limicola* and *Chlorobium phaeobacteroides* in natural habitats. Appl. Environ. Microbiol. 46:1007–1016.

139. Muller, F. M. 1933. On the metabolism of the purple sulphur bacteria in organic media. Arch. Mikrobiol. 4:131–166.

140. Neutzling, O., C. Pfleiderer, and H. G. Trüper. 1985. Dissimilatory sulphur metabolism in phototrophic 'non-sulphur' bacteria. J. Gen. Microbiol. 131:791–798.

141. Nuijs, A. M., R. J. van Dorssen, L. N. M. Duysens, and J. Amesz. 1985. Excited states and primary photochemical reactions in the photosynthetic bacterium *Heliobacterium chlorum*. Proc. Natl. Acad. Sci. USA 82:6865–6868.

142. Odom, J. M., and J. D. Wall. 1983. Photoproduction of H_2 from cellulose by an anaerobic bacterial coculture. Appl. Environ. Microbiol. 45:1300–1305.

143. Oelze, J., and G. Drews. 1972. Membranes of photosynthetic bacteria. Biochim. Biophys. Acta. 265:209–239.

144. Ormerod, J. G. 1983. The carbon cycle in aquatic ecosystems. Symp. Soc. Gen. Microbiol. 34:463–482.

145. Ormerod, J. G., and H. Gest. 1962. Symposium on metabolism of inorganic compounds. IV. Hydrogen photosynthesis and alternative metabolic pathways in photosynthetic bacteria. Bacteriol. Rev. 26:51–66.

146. Ormerod, J. G., K. S. Ormerod, and H. Gest. 1961. Light dependent utilization of organic compounds and photoproduction of molecular hydrogen by photosynthetic bacteria: Relationships with nitrogen metabolism. Arch. Biochem. Biophys. 94:449–463.

147. Osnitskaya, L. K., and V. I. Chudina. 1977. Photosynthetic growth of purple sulfur bacteria during illumination with green light. Microbiology (Engl. Transl. Mikrobiologiya) 46:44–49.

148. Padan, E. 1979. Impact of facultatively anaerobic photoautotrophic metabolism on ecology of cyanobacteria (blue-green algae). Adv. Microb. Ecol. 3:1–48.

149. Parkin, T. B., and T. D. Brock. 1980. The effects of light quality on the growth of phototrophic bacteria in lakes. Arch. Microbiol. 125:19–27.

150. Parkin, T. B., and T. D. Brock. 1981. The role of phototrophic bacteria in the sulfur cycle of a meromictic lake. Limnol. Oceanogr. 26:880–890.

151. Pfennig, N. 1967. Photosynthetic bacteria. Ann. Rev. Microbiol. 21:285–324.

152. Pfennig, N. 1969. *Rhodopseudomonas acidophila* sp. n., a new species of the budding purple nonsulfur bacteria. J. Bacteriol. 99:597–602.

153. Pfennig, N. 1974. *Rhodopseudomonas globiformis*, sp. n., a new species of the Rhodospirillaceae. Arch. Microbiol. 100:197–206.

154. Pfennig, N. 1975. The phototrophic bacteria and their role in the sulfur cycle. Plant Soil 43:1–16.

155. Pfennig, N. 1977. Phototrophic green and purple bacteria: a comparative systematic survey. Annu. Rev. Microbiol. 31:275–290.

156. Pfennig, N. 1978. General physiology and ecology of photosynthetic bacteria, p. 3–18, in R. K. Clayton and W. R. Sistrom (eds.), The photosynthetic bacteria. Plenum Press, New York.

157. Pfennig, N. 1978. *Rhodocyclus purpureus* gen. nov. and sp. nov., a ring-shape, vitamin B_{12}-requiring member of the family Rhodospirillaceae. Int. J. Syst. Bacteriol. 28:283–288.

158. Pfennig, N., and H. G. Trüper. 1983. Taxonomy of phototrophic green and purple bacteria: a review. Ann. Microbiol. (Paris) B134:9–20.

159. Pfennig, N. 1985. Stages in the recognition of bacteria using light as a source of energy, p. 113–131, in E. R. Leadbetter and J. S. Poindexter (eds.), Bacteria in nature, Vol. 1. Plenum Press, New York.

160. Pfennig, N., and H. Biebl. 1976. *Desulfuromonas acetoxidans*, gen. nov. and sp. nov., a new anaerobic, sulfur-reducing, acetate-oxidizing bacterium. Arch. Microbiol. 110:3–12.

161. Pfennig, N., and G. Cohen-Bazire. 1967. Some properties of the green bacterium *Pelodictyon clathratiforme*. Arch. Microbiol. 59:226–236.

162. Pfennig, N., K. E. Eimhjellen, and S. Liaaen-Jensen. 1965. A new isolate of the *Rhodospirillum fulvum* group and its photosynthetic pigments. Arch. Mikrobiol. 51:258–266.

163. Pfennig, N., and H. G. Trüper. 1977. The Rhodospirillales (phototrophic or photosynthetic bacteria). CRC handbook of microbiology, Vol. 1. CRC Press, Boca Raton, Fla.

164. Pfennig, N., and H. G. Trüper. 1981. Isolation of members of the families Chromatiaceae and Chlorobiaceae, p. 279–289, *in* M. P. Starr, H. Stolp, H. G. Trüper, A. Balows, and H. G. Schlegel (eds.), The prokaryotes, a handbook on habitats, isolation and identification of bacteria. Springer-Verlag, New York.

165. Pfennig, N., and F. Widdell. 1982. The bacteria of the sulfur cycle. Philos. Trans. R. Soc. Lond. B298:433–441.

166. Pfennig, N., F. Widdell, and H. G. Trüper. 1981. The dissimilatory sulfate-reducing bacteria, p. 926–940, in M. P. Starr, H. Stolp, H. G. Trüper, A. Balows, and H. G. Schlegel (eds.), The prokaryotes, a handbook on habitats, isolation and identification of bacteria. Springer-Verlag, New York.

167. Pierson, B. K., and R. W. Castenholz. 1971. Bacteriochlorophylls in gliding filamentous prokaryotes from hot springs. Nature (Lond.) 233:25–27.

168. Pierson, B. K., and R. W. Castenholz. 1974. A phototrophic, gliding filamentous bacterium of hot springs, *Chloroflexus aurantiacus*, gen. and sp. nov. Arch. Microbiol. 100:5–24.

169. Pierson, B. K., and R. W. Castenholz. 1974. Studies of pigments and growth in *Chloroflexus aurantiacus*, a phototrophic, filamentous bacterium. Arch. Microbiol. 100:283–301.

170. Pierson, B. K., and R. W. Castenholz. 1978. Photosynthetic apparatus and cell membranes of green bacteria, p. 179–197, in R. K. Clayton and W. R. Sistrom (eds.), The photosynthetic bacteria. Plenum Press, New York.

171. Pierson, B. K., S. J. Giovannoni, and R. W. Castenholz. 1984. Physiological ecology of a gliding bacterium containing bacteriochlorophyll *a*. Appl. Environ. Microbiol. 47:576–584.

172. Pierson, B. K., S. J. Giovannoni, D. A. Stahl, and R. W. Castenholz. 1985. *Heliothrix oregonensis* gen. nov., spec. nov., a phototrophic filamentous gliding bacterium containing bacteriochlorophyll *a*. Arch. Microbiol. 142:164–167.

172a. Pierson, B. K., and J. P. Thornber. 1983. Isolation and spectral characterization of photochemical reaction centers from the thermophilic green bacterium, *Chloroflexus aurantiacus* strain J-10-fl. Proc. Natl. Acad. Sci. USA 80:80–84.

173. Prince, R. C., H. Gest, and R. E. Blankenship. 1985. Thermodynamic properties of the photochemical reaction center of *Heliobacterium chlorum*. Biochim. Biophys. Acta 810:377–384.

174. Quandt, L., G. Gottschalk, H. Ziegler, and W. Stickler. 1977. Isotope discrimination by photosynthetic bacteria. FEMS Microbiol. Lett. 1:125–128.

175. Quayle, J. R., and N. Pfennig. 1975. Utilization of methanol by Rhodospirillaceae. Arch. Microbiol. 102:193–198.

176. Remsen, C. C. 1978. Comparative subcellular architecture of photosynthetic bacteria, p. 31–60, in R. K. Clayton and W. R. Sistrom (eds.), The photosynthetic bacteria. Plenum Press, New York.

177. Ruttner, F. 1963. Fundamentals of limnology, 3rd ed. University of Toronto Press, Toronto, Canada.

178. Sadler, W. R., and R. Y. Stanier. 1960. The function of acetate in photosynthesis by green bacteria. Proc. Natl. Acad. Sci. USA 46:1328–1334.

179. Sahl, H. G., and H. G. Trüper. 1977. Enzymes of CO_2 fixation in *Chromatiaceae*. FEMS Microbiol. Lett. 2:129–132.

180. Satoh, T., Y. Hoshino, and H. Kitamura. 1976. *Rhodopseudomonas sphaeroides* forma sp. *denitrificans*, a denitrifying strain as a sub-species of *Rhodopseudomonas sphaeroides*. Arch. Microbiol. 108:265–269.

181. Saunders, V. A. 1984. Genetics, metabolic versatility, and differentiation in photosynthetic prokaryotes, p. 241–276, in G. A. Codd (ed.), Aspects of microbial metabolism and ecology. (Society of General Microbiology, special publications, Vol. 11). Academic Press, London.

182. Schaab, C., F. Giffoin, S. Schobert, N. Pfennig, and G. Gottschalk. 1972. Phototrophic growth of *Rhodopseudomonas gelatinosa* on citrate: accumulation and subsequent utilization of cleavage products. Z. Naturforsch. 27:962–967.

183. Scher, S., and M. H. Proctor. 1960. Studies with photosynthetic bacteria: anaerobic oxidation of aromatic compounds, p. 387–393, in M. B. Allen (ed.), Comparative biochemistry of photoreactive pigments. Academic Press, New York.

184. Schmidt, K. 1978. Biosynthesis of carotenoids, p. 729–750, in R. K. Clayton and W. R. Sistrom (eds.), The photosynthetic bacteria. Plenum Press, New York.

185. Schmidt, K. 1980. A comparative study on the composition of chlorosomes (chlorobium vesicles) and cytoplasmic membranes from *Chloroflexus aurantiacus* strain

OK-70-fl and *Chlorobium limicola f. thiosulfatophilum* strain 6230. Arch. Microbiol. 124:21–31.

186. Schmidt, K., and S. Liaaen-Jensen. 1973. Bacterial carotenoids. XLII. New keto-carotenoids from *Rhodopseudomonas globiformis* (Rhodospirillaceae). Acta Chem. Scand. 27:3040–3052.

187. Schultze, J. E., and P. F. Weaver. 1982. Fermentation and anaerobic respiration by *Rhodospirillum rubrum* and *Rhodopseudomonas capsulata*. J. Bacteriol. 149:181–190.

187a. Scolnik, P. A., and B. L. Marrs. 1987. Genetic research with photosynthetic bacteria. Ann. Rev. Microbiol. 41:703–26.

188. Shiba, T., and U. Simidu. 1982. *Erythrobacter longus* gen. nov., sp. nov., an aerobic bacterium which contains bacteriochlorophyll *a*. Int. J. Syst. Bacteriol. 32:211–217.

189. Shiba, T., U. Simidu, and N. Taga. 1979. Distribution of aerobic bacteria which contain bacteriochlorophyll *a*. Appl. Environ. Microbiol. 38:43–45.

190. Shiokawa, K., M. Takahashi, and S. Ichimura. 1973. Physiological adaptation of photosynthetic bacteria to low light and its ecological meaning. Japn. J. Limnol. 34:1–11.

191. Siefert, E., R. L. Irgens, and N. Pfennig. 1978. Phototrophic purple and green bacteria in a sewage treatment plant. Appl. Environ. Microbiol. 35:38–44.

192. Siefert, E., and V. B. Koppenhagen. 1982. Studies on the vitamin B_{12} auxotrophy of *Rhodocyclus purpureus* and two other vitamin B_{12}-requiring purple nonsulfur bacteria. Arch. Microbiol. 132:173–178.

193. Siefert, E., and N. Pfennig. 1979. Chemoautotrophic growth of *Rhodopseudomonas* species with hydrogen and chemotrophic utilization of methanol and formate. Arch. Microbiol. 122:177–182.

194. Siefert, E., and N. Pfennig. 1980. Diazotrophic growth of *Rhodopseudomonas acidophila* and *Rhodopseudomonas capsulata* under microaerobic conditions in the dark. Arch. Microbiol. 125:73–77.

195. Sirevåg, R. 1974. Further studies on carbon dioxide fixation in *Chlorobium*. Arch. Microbiol. 98:3–18.

196. Sirevåg, R. 1975. Photoassimilation of acetate and metabolism of carbohydrate in *Chlorobium thiosulfatophilum*. Arch. Microbiol. 104:105–111.

197. Sirevåg, R., B. B. Buchanan, J. A. Berry, and J. H. Troughton. 1977. Mechanisms of CO_2 fixation in bacterial photosynthesis studied by the carbon isotope fractionation technique. Arch. Microbiol. 122:35–38.

198. Sirevåg, R., and R. W. Castenholz. 1979. Aspects of carbon metabolism in *Chloroflexus*. Arch. Microbiol. 120:151–153.

199. Sirevåg, R., and J. G. Ormerod. 1970. Carbon dioxide fixation in green sulfur bacteria. Biochem. J. 120:399–408.

200. Sirevåg, R., and J. G. Ormerod. 1977. Synthesis, storage, and degradation of polyglucose in *Chlorobium thiosulfatophilum*. Arch. Microbiol. 111:239–244.

201. Smillie, R. M., N. Rigopoulous, and H. Kelly. 1962. Enzymes of the reductive pentose cycle in the purple and in the green photosynthetic bacteria. Biochim. Biophys. Acta. 56:612–614.

202. Sojka, G. A. 1978. Metabolism of nonaromatic organic compounds, p. 707–718, in R. K. Clayton and W. R. Sistrom (eds.), The photosynthetic bacteria. Plenum Press, New York.

203. Sorokin, Yu. I. 1970. Interrelationships between sulfur and carbon turnover in meromictic lakes. Arch. Hydrobiol. 66:391–446.

204. Spear, N., and G. Sojka. 1984. Conversion of two distinct *Rhodopseudomonas capsulata* isolates to the glycerol-utilizing phenotype. FEMS Microbiol. Lett. 22:259–263.

205. Sprague, S. G., L. A. Stahelin, M. J. DiBantolomeis, and R. C. Fuller. 1981. Isolation and development of chlorosomes in the green bacterium *Chloroflexus aurantiacus*. J. Bacteriol. 147:1021–1031.

206. Stackebrandt, E., V. J. Fowler, W. Schubert, and J. F. Imhoff. 1984. Towards a phylogeny of phototrophic purple sulfur bacteria—the genus *Ectothiorhodospira*. Arch. Microbiol. 137:366–370.

207. Stackebrandt, E., and C. R. Woese. 1981. The evolution of prokaryotes, p. 1–31, in M. I. Carlile, I. F. Collins, and B. E. B. Moseley (eds.), Molecular and cellular aspects of microbial evolution. Cambridge University Press, Cambridge.

208. Staehelin, L. A., J. R. Golecki, and G. Drews. 1980. Supramolecular organization of chlorosomes (chlorobium vesicles) and of their membrane attachment sites in *Chlorobium limicola*. Biochim. Biophys. Acta 589:30–45.

209. Staehelin, L. A., R. C. Fuller, and G. Drews. 1978. Visualization of the supramolecular architecture of chlorosomes (chlorobium type vesicles) in freeze-fractured cells of *Chloroflexus aurantiacus*. Arch. Microbiol. 119:269–277.

210. Stal, L. J., H. van Gemerden, and W. E. Krumbein. 1985. Structure and development of a benthic marine microbial mat. FEMS Microbiol. Ecol. 31:111–125.

211. Staley, J. T. 1968. *Prosthecomicrobium* and *Ancalomicrobium*: new prosthecate freshwater bacteria. J. Bacteriol. 95:1921–1942.

212. Stanier, R. Y., and G. Cohen-Bazire. 1977. Phototrophic prokaryotes: the cyanobacteria. Annu. Rev. Microbiol. 31:225–274.

213. Stoltz, J. F. 1983. Fine structure of the stratified microbial community at Laguna Figueroa, Baja California, Mexico. I. Methods of *in situ* study of the laminated sediments. Precambrian Res. 20:479–492.

214. Swoager, W. C., and E. S. Lindstrom. 1971. Isolation and counting of Athiorhodaceae with membrane filters. Appl. Microbiol. 22:683–687.

215. Tabita, F. R., B. A. McFadden, and N. Pfennig. 1974. D-ribulose-1,5-diphosphate carboxylase in *Chlorobium thiosulfatophilum* Tassajara. Biochim. Biophys. Acta 341:187–194.

216. Takahashi, M., and S. E. Ichimura. 1968. Vertical distribution and organic matter production of photosynthetic sulfur bacteria in Japanese lakes. Limnol. Oceanogr. 13:644–655.

217. Taylor, D. P., S. N. Cohen, W. G. Clark, and B. L. Marrs. 1983. Alignment of genetic and restriction maps of the photosynthesis region of the *Rhodopseudomonas capsulata* chromosome by a conjugation-mediated marker rescue technique. J. Bacteriol. 154:580–590.

218. Tempest, D. W., O. M. Neijssel, and W. Zevenboom. 1983. Properties and per-

formance of microorganisms in laboratory culture: their relevance to growth in natural ecosystems. Symp. Soc. Gen. Microbiol. 34:119–152.

219. Thorud, M., and R. Sirevåg. 1982. Kinetics of polyglucose breakdown in *Chlorobium*. Arch. Microbiol. 133:114–117.

220. Trüper, H. G. 1970. Culture and isolation of phototrophic sulfur bacteria from the marine environment. Helgol. Wiss. Meeresunters. 20:6–16.

221. Trüper, H. G. 1976. Higher taxa of the phototrophic bacteria: Chloroflexaceae fam. nov., a family for the gliding filamentous, phototrophic "green" bacteria. Int. J. Syst. Bacteriol. 21:8–10.

222. Trüper, H. G. 1978. Sulfur metabolism, p. 677–690, in R. K. Clayton and W. R. Sistrom (eds.), The photosynthetic bacteria. Plenum Press, New York.

223. Trüper, H. G. 1981. Photolithotrophic sulfur oxidation, p. 199–211, in H. Boethe and A. Tebst (eds.), Biology of inorganic nitrogen and sulfur. Springer-Verlag, New York.

223a. Trüper, H. G. 1981. Versatility of carbon metabolism in the phototrophic bacteria, p. 116–121, in H. Dalton (ed.), Microbial growth on C_1 compounds. Heyden, London.

224. Trüper, H. G., and U. Fischer. 1982. Anaerobic oxidation of sulphur compounds as electron donors for bacterial photosynthesis. Philos. Trans. Soc. Lond. B298:529–542.

225. Trüper, H. G., and S. Genovese. 1968. Characterization of photosynthetic sulfur bacteria causing red water in Lake Faro (Messina, Sicily). Limnol. Oceanogr. 13:225–232.

226. Trüper, H. G., and N. Pfennig. 1981. Characterization and identification of the anoxygenic phototrophic bacteria, p. 299–312, in M. P. Starr, H. Stolp, H. G. Trüper, A. Balows, and H. G. Schlegel (eds.), The prokaryotes, a handbook on habitats, isolation, and identification of bacteria. Springer-Verlag, New York.

227. Trüper, H. G., and C. S. Yentsch. 1967. Use of glass fiber filters for the rapid preparation of *in vivo* absorption spectra of photosynthetic bacteria. J. Bacteriol. 94:1255–1256.

228. Uffen, R. L. 1976. Anaerobic growth of a *Rhodopseudomonas* species in the dark with carbon monoxide as sole carbon and energy source. Proc. Natl. Acad. Sci. USA 73:3298–3302.

229. Uffen, R. L. 1981. Metabolism of carbon monoxide. Enzyme Microb. Technol. 3:197–206.

230. Uffen, R. L., and R. S. Wolfe. 1970. Anaerobic growth of purple nonsulfur bacteria under dark conditions. J. Bacteriol. 104:462–472.

231. Utermohl, H. 1925. Limnologische phytoplankton studien. Arch. Hydrobiol. Suppl. 5:1–527.

232. van Gemerden, H. 1968. Utilization of reducing power in growing cultures of *Chromatium*. Arch. Mikrobiol. 64:111–117.

233. van Gemerden, H. 1968. On the ATP generation by *Chromatium* in darkness. Arch. Mikrobiol. 64:118–124.

234. van Gemerden, H. 1974. Coexistence of organisms competing for the same substrate: an example among the purple sulfur bacteria. Microb. Ecol. 1:104–119.

235. van Gemerden, H. 1980. Survival of *Chromatium vinosum* at low light intensities. Arch. Microbiol. 125:115–121.

236. van Gemerden, H. 1983. Physiological ecology of purple and green bacteria. Ann. Microbiol. (Paris) B134:73–92.

237. van Gemerden, H. 1984. The sulfide affinity of phototrophic bacteria in relation to the location of elemental sulfur. Arch. Microbiol. 139:289–294.

238. van Gemerden, H., and H. H. Beeftink. 1981. Coexistence of *Chlorobium* and *Chromatium* in a sulfide-limited continuous culture. Arch. Microbiol. 129:32–34.

239. van Gemerden, H., and H. H. Beeftink. 1983. Ecology of phototrophic bacteria, p. 146–185, in J. G. Ormerod (ed.), The phototrophic bacteria (Studies in microbiology, Vol. 4). Blackwell Scientific, Oxford.

240. van Gemerden, H., E. Montesinos, J. Mas, and R. Guerrero. 1985. Diel cycle of metabolism of phototrophic sulfur bacteria in Lake Cisó (Spain). Limnol. Oceanogr. 30:932–943.

241. van Niel, C. B. 1931. On the morphology and physiology of the purple and green sulfur bacteria. Arch Mikrobiol. 3:1–112.

242. van Niel, C. B. 1944. The culture, general physiology, morphology, and classification of the non-sulfur purple and brown bacteria. Bacteriol. Rev. 8:1–118.

243. van Niel, C. B. 1971. Techniques for the enrichment, isolation, and maintenance of the photosynthetic bacteria, p. 3–28, in A. San Pietro (ed.), Methods in enzymology, Vol. 23, Pt. A. Academic Press, New York.

244. Veldkamp, H. 1977. Ecological studies with the chemostat. Adv. Microb. Ecol. 1:59–94.

245. Vignais, P. M., A. Colbeau, J. C. Willison, and Y. Jouanneau. 1985. Hydrogenase, nitrogenase, and hydrogen metabolism in the photosynthetic bacteria. Adv. Microb. Physiol. 26:156–234.

246. Wall, J. D., P. F. Weaver, and H. Gest. 1975. Genetic transfer of nitrogenase-hydrogenase activity in *Rhodopseudomonas capsulata*. Nature (Lond.) 258:630–631.

247. Ward, D. M., E. Beck, N. P. Revsbech, K. A. Sandbeck, and M. R. Winfrey. 1984. Decomposition of hot spring mats, p. 191–214, in Y. Cohen, R. W. Castenholz, and H. O. Halvorson (eds.), Microbial mats: stromatolites. Alan R. Liss, New York.

248. Weaver, P. F., J. D. Wall, and H. Gest. 1975. Characterization of *Rhodopseudomonas capsulata*. Arch. Microbiol. 105:207–216.

249. Weckesser, J., G. Drews, and H.-D. Tauschel. 1969. Zur Feinstruktur und Taxonomie von *Rhodopseudomonas gelatinosa*. Arch. Mikrobiol. 65:346–358.

250. Wijbenga, D.-J., and H. van Gemerden. 1981. The influence of acetate on the oxidation of sulfide by *Rhodopseudomonas capsulata*. Arch. Microbiol. 129:115–118.

251. Willison, J. C., Y. Jouanneau, A. Colbeau, and P. M. Vignais. 1983. H$_2$ Metabolism in photosynthetic bacteria and relationship to N$_2$ fixation. Ann. Microbiol. (Paris) 134B:115–135.

251a. Woese, C. R. 1987. Bacterial evolution. Microbiol. Rev. 51:221–271.

252. Woese, C. R., B. A. Debrunner-Vossbrinck, H. Oyaizu, E. Stackebrandt, and W. Ludwig. 1985. Gram positive bacteria: possible photosynthetic ancestry. Science 229:762–765.

253. Woese, C. R., E. Stackebrandt, W. G. Weisburg, B. J. Paster, M. T. Madigan, V. J. Fowler, C. M. Hahn, P. Blanz, R. Gupta, K. H. Nealson, and G. E. Fox. 1984. The phylogeny of purple bacteria: the alpha subdivision. Syst. Appl. Microbiol. 5:315-326.

254. Woese, C. R., W. G. Weisburg, C. M. Hahn, B. J. Paster, L. B. Zablen, B. J. Lewis, T. J. Macke, W. Ludwig, and E. Stackebrandt. 1985. The phylogeny of purple bacteria: the gamma subdivision. Syst. Appl. Microbiol. 6:25-33.

255. Woese, C. R., W. G. Weisburg, B. J. Paster, C. M. Hahn, R. S. Tanner, N. R. Krieg, H.-P. Koops. H. Harms, and E. Stackebrandt. 1984. The phylogeny of purple bacteria: the beta subdivision. Syst. Appl. Microbiol. 5:327-336.

256. Wolfe, R. S., and N. Pfennig. 1977. Reduction of sulfur by *Spirillum* 5175 and syntrophism with *Chlorobium*. Appl. Environ. Microbiol. 33:427-433.

257. Yen, H.-C., N. T. Hu, and B. L. Marrs. 1979. Characterization of the gene transfer agent made by an overproducer mutant of *Rhodopseudomonas capsulata*. J. Mol. Biol. 131:157-168.

258. Yen, H.-C., and B. L. Marrs. 1976. A map of genes for carotenoid and bacterio-chlorophyll biosynthesis in *Rhodopseudomonas capsulata*. J. Bacteriol. 126:619-629.

259. Yen, H.-C., and B. Marrs. 1977. Growth of *Rhodopseudomonas capsulata* under anaerobic dark conditions with dimethyl sulfoxide. Arch. Biochem. Biophys. 181:411-418.

260. Yoch, D. C. 1978. Nitrogen fixation and hydrogen metabolism by photosynthetic bacteria, p. 657-676, in R. K. Clayton and W. R. Sistrom (eds.), The photosynthetic bacteria. Plenum Press, New York.

3

MOLECULAR MECHANISMS OF BACTERIAL PHOTOSYNTHESIS

JAN AMESZ

Department of Biophysics, Huygens Laboratory, University of Leiden, Wassenaarseweg 78, 2300 RA Leiden, The Netherlands

DAVID B. KNAFF

Department of Chemistry, Texas Tech University, Lubbock, TX 79409

3.1 INTRODUCTION

3.2 PIGMENTS AND STRUCTURE

    3.2.1 Morphology and Pigment Composition

    3.2.2 Chlorosomes

    3.2.3 Pigment–Protein Complexes

        3.2.3.1 Organization of the Pigments in the Membrane

        3.2.3.2 The B800-850 Complex

        3.2.3.3 Other Complexes

3.3 PROPERTIES OF THE REACTION CENTER; PRIMARY ELECTRON TRANSPORT AND RELATED PROCESSES

    3.3.1 Purple Bacteria

        3.3.1.1 The Reaction Center

        3.3.1.2 Electron Acceptors

    3.3.2 Green Bacteria

        3.3.2.1 Green Sulfur Bacteria

        3.3.2.2 Gliding Green Bacteria

3.4 SECONDARY ELECTRON TRANSPORT; CYCLIC ELECTRON FLOW

    3.4.1 Introduction

    3.4.2 The Secondary Quinone and the Quinone Pool: Transition from One- to Two-Electron Reactions

3.4.3 c-Type Cytochromes as Donors to P870$^+$ or P985$^+$
    3.4.3.1   Soluble Cytochromes in Purple Bacteria
    3.4.3.2   Membrane-Bound Cytochromes in Purple Bacteria
    3.4.3.3   Green Bacteria
3.4.4   The Cytochrome bc_1 Complex
    3.4.4.1 b-Type Cytochromes
    3.4.4.2   Cytochrome c_1
    3.4.4.3   The Rieske Iron–Sulfur Protein
    3.4.4.4   The Isolated Cytochrome bc_1 Complex
    3.4.4.5   Mechanism of Electron Transport
    3.4.4.6   Respiratory Pathways
3.5   PHOTOPHOSPHORYLATION
  3.5.1   Introduction
  3.5.2   Formation of the Proton Motive Force in Purple Bacteria
3.6   NONCYCLIC ELECTRON FLOW
  3.6.1   NAD(P)$^+$ Reduction
    3.6.1.1   Purple Bacteria
    3.6.1.2   Green Sulfur Bacteria
  3.6.2   Substrate Oxidation
    3.6.2.1   Succinate
    3.6.2.2   Sulfur Compounds
3.7   CONCLUDING REMARKS
REFERENCES

3.1   INTRODUCTION

Traditionally, the term "photosynthetic bacteria" has been applied only to those prokaryotes that contain bacteriochlorophyll as the essential photosynthetic pigment and have an anoxygenic type of photosynthesis. Thus the other important group of prokaryotic photosynthetic organisms, the cyanobacteria (or blue-green algae), are not included. It should be remarked that this distinction is also functionally useful; the photosynthetic mechanism of cyanobacteria is based on two photosystems, like that of all oxygen-evolving organisms, and their major pigment is chlorophyll a. In fact, to the authors' knowledge, systematic differences in the mechanism of photosynthetic electron transport between cyanobacteria and eukaryotic algae and higher plants have never been observed.

The photosynthetic bacteria (Rhodospirillales) are divided into two major groups (suborders): the purple and the green photosynthetic bacteria. Each of these groups is further divided into two families. For the purple bacteria these are the sulfur and the nonsulfur purple bacteria (Chromatiaceae and Rhodospirillaceae, respectively), whereas the green bacteria are divided in the Chlorobiaceae and the Chloroflexaceae (1).* The Chlorobiaceae are the "classical" green or brown sulfur bacteria, discovered by Winogradsky in the nineteenth century; the Chloroflexaceae (gliding green or green filamentous bacteria) were discovered and described only about 15 years ago by Pierson and Castenholz (2, 3). For a long time the green bacteria have been relatively ignored in the study of photosynthesis, but recently important advances have been made in various laboratories.

In terms of morphology and pigment composition the boundary between purple and green bacteria is a well-defined one, but as discussed in later sections, a different line may be drawn when the organization of the photosynthetic membrane and the mechanism of photosynthetic electron transport are considered. On the basis of the latter criteria, the Chloroflexaceae are much more similar to purple bacteria than to the Chlorobiaceae.

In the following sections we try to give a systematic discussion of the various aspects of bacterial photosynthesis. After a brief survey of the morphology of photosynthetic bacteria, the various processes of photosynthesis are discussed according to the approximate order they occupy on the time scale of events. Starting with pigments and antenna complexes, we shall proceed to the reaction center, where the primary charge separation and secondary electron transport take place, and then to a discussion of cyclic electron flow, proton transport, and associated reactions that occur in or near the photosynthetic membrane. The final sections deal with the more terminal electron donors and acceptors of bacterial photosynthesis.

3.2 PIGMENTS AND STRUCTURE

3.2.1 Morphology and Pigment Composition

As noted already, in terms of morphology a clear distinction exists between purple and green bacteria. In purple bacteria the complete photosynthetic apparatus is contained in the intracytoplasmic membrane, which forms invaginations that are continuous with the cytoplasmic membrane (4). In those species that are capable of aerobic and of anaerobic or microaerobic phototrophic growth, the formation of these invaginations and pigmentation is induced by conditions that are favorable for phototrophy (i.e., anaerobiosis and illumination) (4). The shape of these invaginations is different for different species: Fig. 3.1, as an example, shows the form of the invaginations in the nonsulfur purple bacterium *Rhodopseudomonas palustris* as deduced from electron microscopy (5).

Most species of purple bacteria contain bacteriochlorophyll *a* (BChl *a*; Fig.

---

*For recent modifications of the taxonomy of purple bacteria, see Refs. 1a and 1b. New genus names are given in parentheses in the text (Section 3.2.3.1).

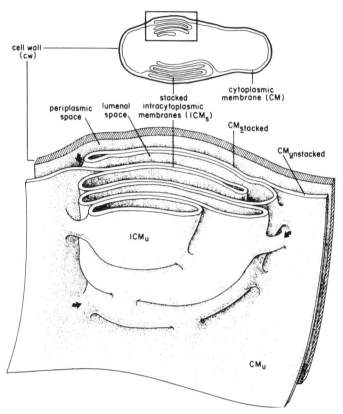

Figure 3.1 Organization of the intracytoplasmic membrane in *Rhodopseudomonas palustris* according to Varga and Staehelin (5). Reproduced by permission.

3.2), but a few species, of which *Rps. viridis* is the best known example, contain the related pigment BChl *b*. Both pigments have a tetrapyrrole structure and are similar to the plant pigment chlorophyll *a*. The main difference with chlorophyll *a* is in ring II, which, in BChl *a* and *b*, is not part of the resonant structure. The long-wavelength band (the Q_y band) of BChl *a in vivo* is situated at 800 to 900 nm; that of BChl *b* is even farther in the infrared, near 1020 nm, which is an extremely long wavelength for an electronic transition. BChl *a* and BChl *b*, together with carotenoid (the nature and composition of which depends on the species), are contained in pigment protein complexes. The so-called light-harvesting or antenna complexes (see Section 3.2.3) contain the bulk of these pigments and are largely responsible for the characteristic absorption spectra of purple bacteria (Fig. 3.3), while only a small fraction is bound to the reaction center where the primary photochemistry takes place (Section 3.3).

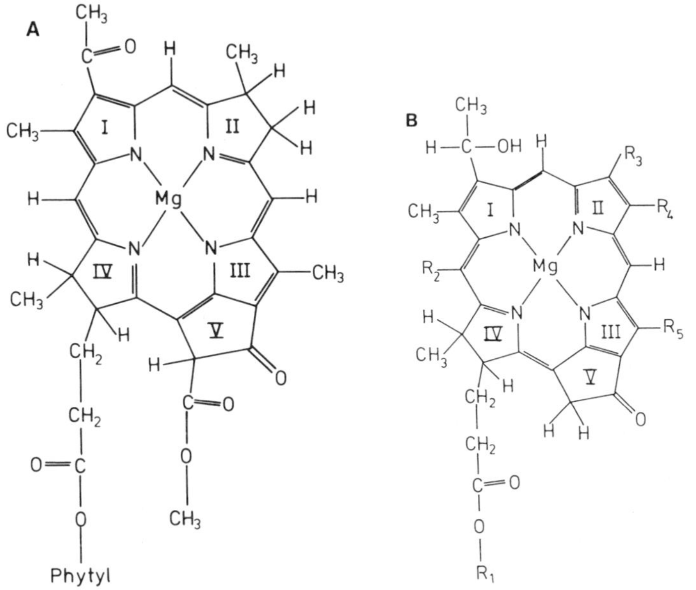

Figure 3.2 Structures of (*A*) bacteriochlorophyll *a* and (*B*) bacteriochlorophylls *c*, *d*, and *e*. For BChl *a*, phytyl may be replaced by geranylgeranyl (6). BChls *c*, *d*, and *e* exist as a number of isomers with different substituents for R_1–R_5 (7, 8, 25).

The absorption spectra of green bacteria (Fig. 3.3) are dominated by strong absorption bands near 460 and 750 nm. These bands belong to the chlorosomes, oblong bodies several hundred angstroms in diameter that are bound to the cytoplasmic membrane. Chlorosomes contain BChl *c*, *d*, or *e* (Fig. 3.2), pigments that are more closely related to chlorophyll than to BChl *a* or *b*. The green sulfur bacteria have a very high antenna-BChl/reaction center ratio, about 1000 to 2000 (9), at least 10 times larger than the ratio observed in purple bacteria. Most of the antenna is accounted for by the chlorosome. The cytoplasmic membrane contains BChl *a* and the reaction center. The few species of green sulfur bacteria that have been investigated in this respect contain 75 to 100 BChl *a* molecules per reaction center (see Section 3.3.2.1), which gives a BChl *c*/BChl *a* ratio of 15 to 20. In the gliding green, filamentous bacterium *Chloroflexus aurantiacus* the ratio of BChl *c* to BChl *a* is lower (10,11).

The newly discovered species *Heliobacterium chlorum* (12) contains a hitherto unknown bacteriochlorophyll, BChl *g* (13), which causes near-infrared absorption bands around 790 nm (12, 13a) and is the major and perhaps the only bacteriochlorophyll present in this bacterium. Electron microscopic examination gave no clear evidence for either intracytoplasmic invaginations or chlorosome-like structures (12).

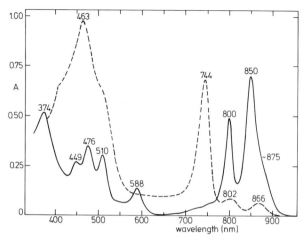

Figure 3.3 Absorption spectra of the purple bacterium *Rhodopseudomonas sphaeroides* (solid line) and of the green bacterium *Chloroflexus aurantiacus* (dashed line). Bands above about 780 nm belong to BChl *a*. Note the strong BChl *c* bands at 463 and 744 nm in the spectrum of *Cfl. aurantiacus*.

3.2.2 Chlorosomes

Chlorosomes are oval-shaped bodies that occur in green bacteria and contain BChl *c*, *d* or *e* (formerly called chlorobium chlorophylls). The chlorosomes constitute about 10 to 20% of the total volume of the bacterial cells and, as noted in the preceding section, they account for the major part of the BChl antenna. They also contain most of the carotenoid present in the bacterium (11). They were first seen by Cohen-Bazire (14) in electron micrographs of green sulfur bacteria and by Madigan and Brock (15) in *Cfl. aurantiacus*. The former name "Chlorobium vesicles" has been replaced by the term "chlorosomes" (16, 17), as it is now clear that they are not vesicles in the normal sense of the word and do not contain an inner aqueous phase.

Chlorosomes can be detached from the cytoplasmic membrane by mechanical disruption (11) or detergent treatment (18). The chemical composition of chlorosomes has been extensively studied by Schmidt (11). Chlorosomes of the Chlorobiacea *Chlorobium limicola* f. *thiosulfatophilum* were found to contain, on a dry weight basis, 55% protein, 10% lipid (mainly glycolipids), 32% BChl *c*, and 1% carotenoid (mainly chlorobactene). One chlorosome may contain approximately 10,000 BChl *c* molecules. The lipid is thought to form a monolayer around the chlorosome (17). Data on different strains and species of *Chlorobium* have been obtained by Cruden and Stanier (19). Chlorosomes of *Cfl. aurantiacus* (Chloroflexaceae), grown at low light intensity, contained 55% protein, 22% BChl *c*, 11% lipid, and 2% carotenoid (11). There is convincing evidence now that chlorosomes of green sulfur bacteria (11, 20, 20a), as well as those of *Cfl. aurantiacus* (11, 21) contain small amounts of BChl *a*, absorbing near 790 to 795 nm (20, 21). The location of the emission band of this BChl *a* (20a, 21) as well as its orientation (20a, 21a) suggest that it may play an important role in energy transfer from the chlorosome to the cytoplasmic membrane.

The amount of chlorosome material and thus the BChl *c*/BChl *a* ratio in *Cfl. aurantiacus* is strongly dependent on light intensity (3, 10), cells grown at low intensity having a higher BChl *c* content. A similar but less pronounced effect has been observed in *Chl. limicola* (22). Electron micrograph studies indicate that chlorosomes of Chlorobiaceae (17), as well as those of Chloroflexaceae (16), contain rod like structures that are parallel to the long axis of the chlorosome and may contain a regular array of BChl *c* protein complexes. A regular arrangement of BChl *c* is also indicated by measurements of linear dichroism (20a, 21, 21a, 23). Optical and electron paramagnetic resonance (EPR) studies suggest an oligomeric structure of BChl *c* (21, 24). For *Cfl. aurantiacus*, the primary structure of the polypeptide to which BChl *c* is bound (24a) has recently been determined (24b). It has a molecular weight of 5.7 kDa; approximately 14 BChl *c* molecules are bound to a dimer of this polypeptide.

Although it is clear that the chlorosomes are the most important antenna constituent of green bacteria, the efficiencies of transfer of absorbed light energy from the chlorosome to the cytoplasmic membrane or to the reaction center are not known precisely (see Ref. 25 for a review). The reason for this lack of knowledge is the experimental difficulty of obtaining reliable action spectra that cover the spectral region of the chlorosome as well as that of the much weaker BChl *a* absorption.

3.2.3 Pigment–Protein Complexes

3.2.3.1 *Organization of the Pigments in the Membrane*

During the last decade, the efforts of various investigators have firmly established that the photosynthetic pigments that are contained in the membrane are bound to specific proteins. This concept thus has replaced the older notion that the pigments are dispersed in the lipid matrix. Pigment–protein complexes have now been isolated and purified from various species of photosynthetic bacteria (see Ref. 25a for a review). In this section we focus on pigment–protein complexes from purple bacteria, where a clear-cut distinction can be made between light-harvesting (or antenna) pigment–protein complexes and reaction center complexes. In green sulfur bacteria a different situation exists (see Section 3.3.2.1).

Most of the bacteriochlorophyll and carotenoid present in the intracytoplasmic membrane of purple bacteria belongs to the light-harvesting complexes. These complexes usually account for 80 to 90% of the total pigment content, the remaining fraction being contained in the reaction center complex. By use of detergents, light-harvesting complexes have now been isolated from various species of purple bacteria (see review in Ref. 26). In *Rps. (Rhodobacter) sphaeroides* and *Rps. capsulata* (*Rb. capsulatus*) two different light-harvesting complexes have been characterized: the B800-850 complex, which is responsible for the BChl *a* absorption bands near 800 and 850 nm of intact cells (Fig. 3.4), and B875 (26–33). *Rhodospirillum rubrum* contains only one complex, which we call B875 (33–36), whereas four different complexes appear to be present in *Rps. acidophila* (37) and in the purple sulfur bacterium *Chromatium vinosum* (38). A light-harvesting complex (B800-1020) was recently isolated from the BChl *b*–containing purple bacterium *Ectothiorhodospira halochloris* with near-infrared absorption maxima near 800, 830, and 1020 nm (39). This complex did not contain carotenoid. In many species, the relative amounts of pigment complexes can be manipulated by varying the culture conditions, such as the light intensity (see, e.g., Refs. 30 and 38).

Excitation spectra of BChl *a* fluorescence and of photochemical activity show that excitation energy is transferred with high efficiency from shortwave bacteriochlorophylls, such as BChl *a* 800 and BChl *a* 850, to BChl *a* 875 and other longwave-absorbing bacteriochlorophylls and hence to the reaction center (see Refs. 25 and 40 for reviews). The efficiency of energy transfer from carotenoids to BChl *a* varies for different complexes and species between about 30 and 90% (25, 40).

The size of the photosynthetic unit (i.e., the number of BChl *a* molecules per reaction center) is between roughly 50 and 200, and studies of the fluorescence

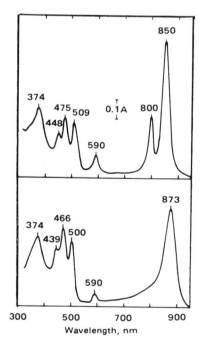

Figure 3.4 Absorption spectra of the B800-850 and the B875 complexes of *Rps. sphaeroides*. Reproduced from Ref. 27 by permission.

yield (41) and of exciton annihilation (42, 42a) have shown that energy transfer occurs over several photosynthetic units. In *Rps. sphaeroides* B800-850 probably occurs in "lakes" that surround an array of B875 complexes which connect reaction centers (42a, 43).

3.2.3.2 The B800-850 Complex

In this section we concentrate on the properties of B800-850 from *Rps. sphaeroides* and *Rps. capsulata*, which to date have been studied more extensively than have the other complexes. The complex contains BChl *a* 850 and BChl *a* 800 in the ratio 2:1 (33, 34, 44, 45). The BChl *a*/carotenoid ratio has been reported to be 3:1 (26, 32, 33), but more recent experiments suggest that 2:1 is a more likely number (46). The type of carotenoid depends on the species of mutant studied. In anaerobically grown *Rps. sphaeroides* the carotenoid is spheroidene, which constitutes about 90% of the carotenoid present in the bacterium.

Two polypeptides, called the α- and β-apoproteins, are present per 3 BChl *a* in the complex (26, 47). The amino acid sequences of these peptides, which have molecular weights of about 6 kDa, and of the peptides of some other light-harvesting complexes have recently been elucidated (48–52, 52a–52f). They all contain a homologous hydrophobic stretch of about 20 amino acids, which probably has an α-helical structures (53, 53a) and is thought to traverse the membrane. On both sides of the hydrophobic section of the chain there are polar regions, the lengths of which vary for different peptides. The α-helices all contain a histidine that is thought to provide a binding site for BChl. In some cases a second histidine that may bind another BChl molecule is present near the transition between the polar and hydrophobic domains. A third polypeptide may be present in the B800-850 from *Rps. capsulata* (31, 54) and some other complexes.

The main fluorescence band of B800-850 is situated at 872 nm (881 nm at 4 K), which fluorescence is emitted by BChl *a* 850 (27, 30, 45). Fluorescence by BChl 800 is much weaker, due to efficient energy transfer to BChl 850 (45). The corresponding distance between BChl 800 and BChl 850 was estimated to be about 20 Å (45). When isolated with lauryl dimethyl amine oxide as detergent, the complex consists of an aggregate of many subunits (42, 55), which form a highly ordered array (29).

BChl *a* 850 and carotenoid show a strong circular dichroism signal of the type referred to as conservative, indicating interaction between pairs of pigment molecules (29, 34, 56, 57). Information about the orientation of the transition dipoles of the pigments was obtained by measurement of linear dichroism and fluorescence polarization (29, 34, 56–60). On the basis of these results, a model was proposed (60) for the pigment organization in the complex, which was recently modified and expanded (29, 30) to accommodate data on energy transfer from carotenoid to bacteriochlorophyll and on polarization of the weak BChl *a* 800 fluorescence. Figure 3.5 gives a schematic presentation of this model.

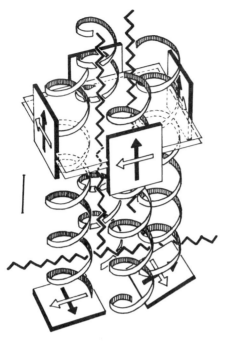

Figure 3.5 Model of the basic unit of the B800-850 antenna complex of *Rps. sphaeroides* (schematic). The plane of the membrane is approximately horizontal in this figure. Squares indicate the porphyrin heads of BChl 850 (top) and BChl 800 (bottom), with open arrows for the Q_y and solid arrows for the Q_x transition dipoles. Carotenoids are indicated by zigzag lines, the helices are those of four apoprotein hydrophobic cores. The bar indicates 0.5 nm. Reproduced from Ref. 29 by permission.

3.2.3.3 Other Complexes

The B875 complexes from *Rps. sphaeroides* and *R. rubrum* appear to be similar to B800-850 with respect to their protein structure. They contain two different peptides (35, 36, 50), which are homologous to the α- and β-apoproteins of B800-850 (50). However, the pigment composition and organization of the two complexes appear to be different. B875 from *R. rubrum* has been reported to contain BChl *a* and carotenoid (spirilloxanthin) in the ratio of 2:1 (33, 35), whereas in the B875 complex from *Rps. sphaeroides*, this ratio appears to be close to 1:1 (27). A different pigment organization is indicated by their circular dichroism spectra (30, 34, 57, 60a).

A light-harvesting pigment with absorption bands at 808 and 866 nm (B808-866) was isolated from the cytoplasmic membrane of the gliding green bacterium *Cfl. aurantiacus* (61). The complex contains only two different polypeptides, which show homology to the α- and β-polypeptides of purple bacteria (61a).

Finally, a pigment–protein complex from green sulfur bacteria should be mentioned here: the so-called soluble light-harvesting BChl *a* protein (62). This complex is water soluble and bound to, rather than embedded in, the cytoplasmic membrane. It does not contain carotenoid. X-ray diffraction studies of the crystalline complex have shown that the complex consists of three subunits, each containing a group of seven BChl *a* molcules. These groups are almost completely surrounded by the protein chain, which consists largely of a β-pleated sheet structure (63,

63a). Electron micrographic studies indicate that the BChl *a* protein functions as a base plate between the chlorosome and the cytoplasmic membrane (17). Energy transfer from the chlorosome to the BChl *a* in the membrane then probably occurs via the BChl *a* in the base plate, but direct evidence for energy transfer between the BChl *a* protein and the pigments in the membrane is lacking (25).

For a discussion of the pigment organization of the membrane of green sulfur bacteria, see Section 3.3.2.1.

3.3 PROPERTIES OF THE REACTION CENTER; PRIMARY ELECTRON TRANSPORT AND RELATED PROCESSES

The primary charge separation in photosynthetic bacteria consists of the transfer of an electron from an excited bacteriochlorophyll to an acceptor molecule. This is true for green as well as for purple bacteria. Normally, the so-called primary electron donor is BChl *a*, but in *Rps. viridis* and other species that contain BChl *b* as light-harvesting pigment, the primary electron donor is likewise BChl *b* (64, 65). There is good evidence now that the primary and secondary electron acceptors are different in purple and green sulfur bacteria, whereas the reaction center in the gliding green bacteria (Chloroflexaceae) is in many respects similar to that of purple bacteria. Therefore, as is also done in other sections, it is useful to discuss these groups separately.

3.3.1 Purple Bacteria

3.3.1.1 The Reaction Center

The primary charge separation and associated reactions have been studied far more extensively in purple bacteria than in the other groups of photosynthetic bacteria. More than 30 years ago, Duysens (66) observed absorption changes upon illumination of *R. rubrum* and *C. vinosum*. The spectrum of these changes showed a bleaching at 870 to 890 nm which was ascribed to photooxidation of BChl *a*. Subsequent research in various laboratories has confirmed this hypothesis, and it is now firmly established that the bleaching in the near-infrared region is due to photooxidation of the primary electron donor, P870, by a reaction that may be written as

$$PI \xrightarrow{h\nu} P*I \longrightarrow P^+I^-$$

where P denotes P870, P* its singlet excited state, and I the primary acceptor molecule.

During the last 15 years most of the work on primary electron transport has been done with isolated reaction centers. Reed and Clayton (67) were the first to obtain a purified reaction center complex by detergent treatment of isolated membranes (chromatophores) of *Rps. sphaeroides*, and the method has been perfected in various laboratories. Reaction centers have now been obtained from many species of purple bacteria, including *Rps. sphaeroides* (68–70), *Rps. capsulata* (71), *Rps. gelatinosa* (72), *R. rubrum* (73), *C. vinosum* (74, 75), and the BChl *b*–containing species *Rps. viridis* (64, 76) and *Thiocapsa pfennigii* (65). A convenient method to obtain purified reaction centers from the carotenoidless mutant *Rps. sphaeroides* R-26, which have been widely used in photosynthesis research, is given by Clayton and Wang (77).

Many of these reaction center preparations have been characterized in terms of their chemical constituents (see reviews in Refs. 78 and 79). They all contain four BChl and two bacteriopheophytin (BPheo) molecules and normally one molecule of carotenoid. Depending on preparation and species, one or two quinone molecules and cytochrome *c* are also present. The chemical nature of the carotenoid depends on the species or strain used. The quinones are either ubiquinone or menaquinone (vitamin K-2): Reaction centers from *Rps. sphaeroides* and *R. rubrum* contain ubiquinone, while in those of *C. vinosum* and *Rps. viridis* one or both ubiquinones are replaced by menaquinone.

Figure 3.6 gives the absorption spectrum of a reaction center preparation from *Rps. sphaeroides*, strain R-26. The spectrum shows bands at 598, 802, and 864 nm due to BChl *a*, whereas BPheo *a* bands are seen at 535 and 756 nm. Illumination or chemical oxidation of the reaction center causes a bleaching of the long-wave band at 864 nm, which can be attributed to the disappearance of the absorption band of the primary electron donor P870 upon oxidation. The midpoint potential of P870 is 450 mV (80). The oxidation of P870 can also be observed by EPR spectroscopy as the formation of the P870$^+$ radical line centered at $g = 2.0026$ (81).

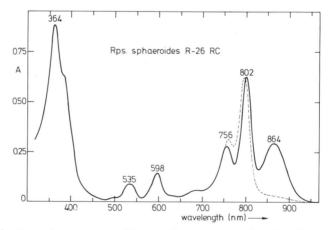

Figure 3.6 Absorption spectrum of the reaction center complex from the carotenoidless mutant R-26 of *Rps. sphaeroides*. Dashed line: after oxidation of the primary electron donor P870.

For most preparations, the protein moiety of the reaction center complex consists of three subunits (in a 1:1:1 ratio), called L, M, and H (69, 78, 79, 82) with apparent molecular weights, as determined by gel electrophoresis, of 21, 24, and 28 kDa, respectively. In some cases only two subunits were obtained (64, 83) and it has been shown that removal of the H polypeptide subunit does not significantly change the characteristics of the primary processes (69, 83, 84). The primary structures of the subunits have been determined for various species (84–86, 86a–86c); the true molecular weights of the L, M, and H subunits of *Rps. capsulata* were found to be 31,565, 34,440, and 28,534, respectively (85), the last one, which is less hydrophobic than the others, in fact turning out to be the smallest. The L and M subunits are homologous to each other and contain five hydrophobic stretches of approximately 20 amino acids, which are situated between polar regions and traverse the membrane in a α-helical configuration (see below). Interestingly, the peptides show a clear homology with the so-called atrazine-binding protein that is associated with system II of green plant photosynthesis (86d). The H subunit has only one membrane-spanning stretch of amino acids.

Techniques have been developed recently to crystallize membrane proteins from aqueous detergent solution (87), and Michel (88) was the first to obtain crystalline reaction centers of *Rps. viridis* in this way. Crystals have now also been obtained of the reaction center of *Rps. sphaeroides* (84, 88a). An obvious advantage of the use of single crystals is that molecular orientations are exactly defined, so that anisotropic optical and spin-state transitions can be studied with much better definition (89, 90). An even more important result is that the crystals turned out to be of sufficient quality to yield a detailed pattern of X-ray scattering, and together with the primary structures of the constituent polypeptides, the results of the X-ray analysis have provided a three-dimensional picture of the organization of the reaction center of *Rps. viridis* that rapidly becomes more detailed (86b, 90a).

The often debated question of whether the primary electron donor P is a (B)Chl dimer (91–96) has been resolved at least for *Rps. viridis* by the X-ray data. The density distribution, at 3-Å resolution, showed the location and orientation of the four BChl *b*, two BPheo *b*, and one menaquinone (Fig. 3.7A) that are contained in the reaction center. Two BChl molecules are close together, at a center-to-center distance of 7 Å, and apparently represent the P dimer. The arrangement of BChl and BPheo shows roughly a twofold rotation symmetry about an axis perpendicular to the membrane (96a) and passing through the middle of P. Electron transfer occurs from P* via the BChl (96b) and BPheo (I) at the right-hand side of the figure to the menaquinone (Q_A). The BChl at the left-hand side can be removed or modified without affecting the quantum efficiency for electron transfer (96c, 96d); no significant electron transfer through the left-hand chain has been observed. The ubiquinone that serves as secondary electron acceptor Q_B (see Section 3.3.1.2) is missing in the crystals. The protein moiety of the reaction center complex contains 11 trans membrane α-helical stretches of amino acid residues that belong to the L, M, and H subunits.

The pigments of the right-hand chain, with the exception of menaquinone, are bound to the L subunit; this chain is therefore commonly referred to as the "L chain." Those on the left-hand side are bound to the M subunit. The second quinone (Q_B) is presumably located symmetrically to Q_A at the end of the M chain. The role of the Fe^{2+} ion halfway between the two quinones will be discussed in Section 3.3.1.2.

In addition to absorbance changes in the long-wave region, the difference spectrum obtained upon oxidation of P870 also shows numerous bands at shorter wavelength. A detailed discussion of the origin of these bands is beyond the scope of this chapter. In general, it may be stated that they reflect changes in the short-wave bands of P870, but also in those of neighboring pigments. In addition, there are small contributions caused by the reduction of the electron acceptor (ubiquinone or menaquinone). In BChl b–containing purple bacteria such as *Rps. viridis*, the main bleaching in the near-infrared is near 985 nm (97) and the primary electron donor in these species is therefore called P985. The absorption and absorption difference spectra of reaction centers have been analyzed by measurements of circular and linear dichroism (for a discussion of these results, see Refs. 59, 96a, 98, and 98a). It is not clear if dimer formation would be responsible, by exciton interaction, for the large red shift (about 100 nm) of the long-wave absorption band

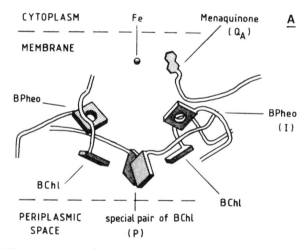

Figure 3.7 (*A*) Arrangement of reaction center components as determined by X-ray diffraction analysis of a crystalline reaction center complex of *Rps. viridis*. Redrawn from Ref. 90a. The figure was kindly provided to us by Dr. H. J. van Gorkom. (*B*) Energy scheme of electron transfer kinetics in the reaction center of the green filamentous bacterium *Cfl. aurantiacus* and the green sulfur bacterium *P. aestuarii*. Excitation of the primary electron donor P is indicated by vertical arrows. Downward pointing solid arrows indicate main electron transport pathways; dashed arrows are "back reactions" that become predominant when main electron transport is blocked. P^T, triplet state of P; FeS, iron–sulfur center; c, cytochrome c; the other symbols are defined in the text. Unless otherwise indicated, the time constants refer to those at room temperature. The scheme for *Cfl. aurantiacus*, with somewhat different time constants and energy levels, also applies to purple bacteria (see text).

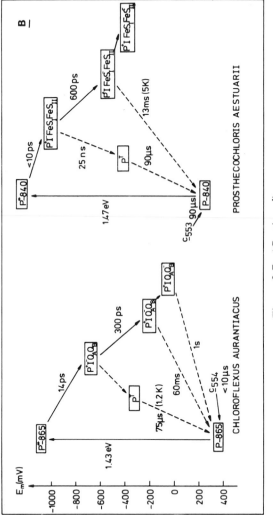

Figure 3.7 (*Continued*)

of P870 as compared to that of BChl a in solution (see Ref. 25). As a matter of fact, the same is true for the similar red shift of the antenna BChl 875.

3.3.1.2 Electron Acceptors

It is well established now that BPheo is at least one of the early electron acceptors in photosynthesis of purple bacteria. Much of the evidence is based on ultrafast spectroscopy of isolated reaction centers of *Rps. sphaeroides*, which has allowed measurement of the kinetics and difference spectra of absorption changes on a picosecond time scale (96b, 99–101, 101a, 101b). These absorbance changes could be attributed to the formation of the charge pair P^+I^- (see Section 3.3.1.1). They are produced in a few picoseconds and normally have a lifetime of about 200 ps (Figs. 3.7B and 3.8). Analysis of the difference spectrum (102) suggested that they are caused by the oxidation of P870 and simultaneous reduction of BPheo a. In about 200 ps the electron is transferred from BPheo a^- to a second acceptor. Identification of ubiquinone as the second acceptor rests in part on the observation that the 200-ps oxidation of BPheo$^-$ is lost on removal of UQ from reaction centers and reappears on reconstitution of the depleted reaction centers with ubiquinone (101; Fig. 3.8).

Analysis of the absorbance changes during the first few picoseconds after the beginning of the flash suggested that one of the other BChl molecules present in the reaction center (to so-called accessory BChl in the L chain) may act as an intermediary electron acceptor in the reduction of BPheo (96b, 103–105). In view of the conditions of the experiments, it seems unlikely that excited states of BChl a may have interfered with the measurements, as thought earlier. Nevertheless, this point is still controversial. Experiments with subpicosecond flashes did not reveal the existence of a short-lived intermediate BChl$^-$ in reaction centers of *Rps. sphaeroides* (101a, 101b) and *Rps. viridis* (105a). Reduction of BPheo was found to occur in 3 to 4 ps (96b, 101a, 101b).

When electron transport to quinone is blocked, as can be done by chemical reduction or extraction of the quinone before illumination (101, 106), the lifetime of the primary radical pair is increased to about 10 ns. Decay of the radical pair is now accompanied by the formation of the triplet state of P870 or P985. This triplet can be observed either optically (95, 99, 107) or by means of EPR (108). For a discussion of the so-called radical-pair mechanism of triplet formation, we refer to Ref. 109.

When conditions are chosen such that electron transfer from BPheo$^-$ to the quinone acceptor is blocked by prior reduction of the quinone, and reduction of P^+ (P870$^+$ or P985$^+$) by another electron donor is fast enough to compete to some extent with reduction by BPheo$^-$, the accumulation of BPheo$^-$ may be observed upon continued illumination. Such an accumulation was observed upon illumination of reaction centers of *C. vinosum* (110, 111) and *Rps. viridis* (112), where reduced cytochrome c served as a fast electron donor. The advantage of this method is that it allows the measurement of the difference spectrum of the reduced acceptor only, without interference by P^+. Nevertheless, the spectra thus obtained were still fairly complicated. In addition to bands of BPheo a (*C. vinosum*) or

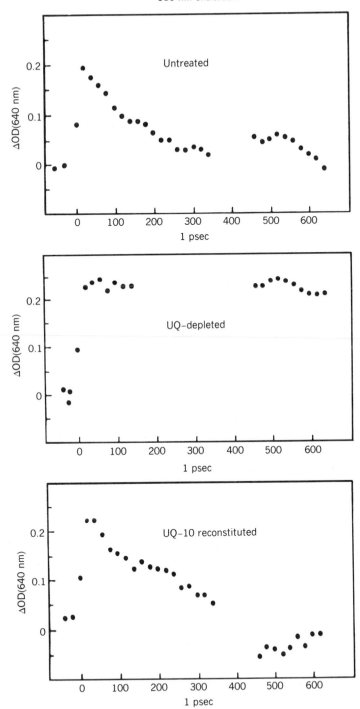

Figure 3.8 BPheo reduction and reoxidation in reaction centers from *Rps. sphaeroides* strain R-26. Reproduced from ref. 101 by permission.

BPheo b (*Rps. viridis*), they showed absorbance changes due to BChl a or b. It is usually assumed that these are due to changes in interaction between BChl and BPheo accompanying the reduction of the latter compound, rather than the reduction of BChl itself.

As mentioned above, the first so-called "stable" electron acceptor is a quinone molecule, which is reduced in about 200 ps. In many species of quinone is ubiquinone, but in *C. vinosum* (75) and *Rps. viridis* (113) menaquinone acts as electron acceptor. Photoreduction of ubiquinone or menaquinone can be observed upon illumination of reaction centers by use of absorption difference spectroscopy in the ultraviolet and blue region (75, 114–116). Although the absorption changes are considerably smaller than those resulting from P870 oxidation, at moderately low redox potentials, the quinone (Q_A) can easily be accumulated in the reduced form (Q_A^-) while P870$^+$ is rereduced by an added electron donor, giving a difference spectrum free from interference by P870$^+$. Spectra obtained by this method, or by kinetic analysis, showed that the quinone is reduced to semiquinone anion radical. Reduction of the quinone is accompanied by small changes in the absorption bands of BPheo and BChl in the near-infrared region (75, 114–116), which may be explained by so-called electrochromic effects due to the negative charge on the quinone molecule. The midpoint potential of the Q_A/Q_A^- couple ranges from -130 to -200 mV, depending on the bacterial species (117).

The quinone radical shows magnetic interaction with the iron normally present in the reaction center. The interaction is not very strong and does not appear to affect significantly the optical properties of the radical, but causes a severe broadening and shift of the EPR radical signal (118, 119). Removal of the iron produces a normal quinone radical signal at $g = 2.005$ for the reduced acceptor (120, 121) and does not interfere strongly with the process of electron transfer, while replacing Fe^{2+} by other divalent cations has no effect on the rate of electron transfer from Q_A to Q_B (122, 123). In fact, recently a strain of *Rps. sphaeroides* was reported to contain manganese instead of iron when grown under appropriate conditions (123a). Mössbauer studies and measurements of magnetic susceptibility have shown that the iron is not directly involved in electron transfer: Irrespective of the redox state of the quinones, the iron remains in the Fe^{2+} valency state (124, 125). The quinones are not directly coordinated to the Fe^{2+} ion (126, 127).

In about 100 μs the electron is transferred from the first to a second quinone molecule (Q_B). In most species this quinone is ubiquinone (116, 128). Difference spectra of the reduction of the second ubiquinone are similar to that of the first (116), again indicating that a semiquinone anion is formed when the first electron is delivered to Q_B. The quinones can be extracted by solvent or detergent treatment. Electron transport can then be reconstituted either by means of the native or by a variety of other quinones (81, 101, 129, 130).

3.3.2 Green Bacteria

3.3.2.1 *Green Sulfur Bacteria*

Until fairly recently, studies of the primary reaction and of electron transport in green sulfur bacteria were confined to intact cells. This implied a low sensitivity and correspondingly low signal-to-noise ratio due to the low reaction center content (Section 3.2.1). A bleaching near 840 nm upon illumination of intact cells was attributed to photooxidation of the primary electron donor (P840), but the difference spectrum was only poorly resolved (131). Better spectra were obtained when methods were found to obtain membrane vesicle preparations from which the chlorosomes had been removed (9, 132). Such preparations contain about 80 BChl a molecules per reaction center. More recently, Swarthoff and Amesz (133), by detergent treatment of such a preparation from *Prosthecochloris aestuarii*, isolated photochemically active pigment–protein complexes which contained about 75 or 35 BChl a molecules per reaction center. So far it has not been possible to obtain isolated reaction centers from green sulfur bacteria similar to those from purple bacteria. The smallest complex obtained, the "core complex," contains about 20 BChl a molecules per reaction center (134, 135).

The optical difference spectrum that accompanies the oxidation of P840 is quite complex. A negative band near 840 nm is probably a direct result of the oxidation of the primary electron donor, which absorbs near 840 nm (136, 137), but in addition a complicated structure of negative and positive bands is visible that is caused mainly by electrical interactions with neighboring pigment molecules (137). The midpoint potential of P840 is 250 mV (138, 139). Like P870 of purple bacteria, P840 is probably a BChl a dimer, as indicated by the width of the EPR line of the P840$^+$ radical (140, 141) and by the line shape of ^{13}C-enriched P840$^+$ (142).

Evidence concerning the electron acceptor chain (see Fig. 3.7) is partly based on photoaccumulation experiments under conditions where P840 formed in the light is rapidly rereduced again, in a manner similar to that discussed in the preceding section for preparations obtained from purple bacteria. Most of these experiments have been done with an isolated pigment–protein complex from *P. aestuarii*. Cooling in the light gave and EPR spectrum at 7 K with g values of 1.94 and 1.85 that indicated the reduction of an iron–sulfur protein with a midpoint potential of about -560 mV (FeS$_I$ in Fig. 3.7B). In agreement with this, the optical difference spectrum obtained at room temperature showed negative bands in the blue region of the spectrum, which were typical for a reduced iron–sulfur protein (143). There is evidence that the electron transport chain contains a second iron–sulfur center (FeS$_{II}$ in Fig. 3.7B), with a midpoint potential more positive than about -400 mV (143). The EPR signal observed on illumination of broken cells of *Chl. limicola* at cryogenic temperatures (139) does not appear to result from either FeS$_I$ or FeS$_{II}$ and may be due to an additional low-potential iron–sulfur center.

By means of laser flash spectroscopy, evidence has recently been obtained concerning the nature of the primary electron acceptor of green sulfur bacteria. Excitation of membranes of *P. aestuarii* with 35-ns flashes produced, in addition to absorbance changes due to formation of P840$^+$, a bleaching at 670 nm, which could be ascribed to reduction of a primary electron acceptor (144, 145). The electron acceptor was identified as a special membrane-bound form of BChl c (145a). Electron transfer to BChl c occurs in less than about 10 ps (145); the electron is subsequently transferred to the next acceptor, possibly FeS$_I$, in 600 to 700 ps (144, 145; see Fig. 3.7B).

The recently discovered, strictly anaerobic photosynthetic bacterium *H. chlorum* (see Section 3.2.1), although it cannot be classified as a green bacterium because it lacks chlorosomes, appears to have an electron acceptor chain that resembles that of green sulfur bacteria. The primary electron donor, P798 (145b, 145c) is probably BChl *g*. The electron acceptor chain, like that of *P. aestuarii*, includes a pigment absorbing at 670 nm which appears to act as primary electron acceptor (145d), together with one or more iron–sulfur centers (145c, 145e, 145f), and possibly a quinone (145e).

3.3.2.2   *Gliding Green Bacteria*

The results discussed in the preceding section clearly demonstrate that the structure of the reaction center of green sulfur bacteria is quite different from that of the purple photosynthetic bacteria, and that the electron acceptor chain shows greater similarity to that of system I of plant photosynthesis than to that of purple bacteria (see Refs. 146, 146a, and 146b).

Research into primary and associated electron transport of Chloroflexaceae has been initiated only quite recently. Nearly all of this work has been done with isolated membranes or reaction centers. The picture that emerged from these studies indicates that *Cfl. aurantiacus*, classified as a green bacterium, but evolutionarily probably quite distant from the green sulfur as well as from the purple bacteria (147), is photochemically very similar to the latter group of organisms.

The evidence is summarized only briefly here. The absorption difference spectrum due to photooxidation of the primary electron donor, P865, is similar to that of purple bacteria (3, 148–150). The absorption spectrum of isolated reaction centers, first obtained by Pierson and Thornber (151) is again similar to that of purple bacteria, and analysis of the absorption and linear and circular dichroism spectra indicates a very similar molecular structure of the reaction center (151a). The most conspicuous difference is that one of the ''accessory'' BChls, presumably that in the M chain (see Fig. 3.7*A*), is replaced by BPheo *a* in *Cfl. aurantiacus*. Measurements of rapid absorption changes by means of flash spectroscopy have shown the reduction of BPheo *a* (152, 153). In about 300 ps the electron is transferred to a secondary electron acceptor (Q_A; see Fig. 3.7*B*), which is menaquinone (149). The second quinone in the electron acceptor chain is again menaquinone, the only quinone present in *Cfl. aurantiacus* (154).

3.4   SECONDARY ELECTRON TRANSPORT; CYCLIC ELECTRON FLOW

3.4.1   Introduction

The major electron transport pathways in purple photosynthetic bacteria are cyclic in nature and produce neither net oxidation nor reduction. The electron flow patterns in these bacteria thus differ from those in respiration and in oxygen-evolving photosynthesis, which involve net electron flow from donor to acceptor. In bacteria, the primary charge separation (discussed in Section 3.3) is followed by a series of energy-releasing reactions in which the electron on the reduced primary acceptor eventually recombines with the positive charge on the photooxidized primary electron donor. This exergonic charge recombination is a stepwise process, channeled through specific secondary electron carriers to ensure that the energy released is available for ATP formation and other energy-requiring processes. Little is known yet concerning the details of secondary electron transfer reactions in green photosynthetic bacteria. Therefore, we concentrate on the more widely studied purple sulfur and purple nonsulfur bacteria.

The vast majority of our knowledge of the details of secondary electron transfer reactions in purple bacteria comes from studies with chromatophores, vesicles derived from the intracytoplasmic membrane by breakage of the cell. These studies are occasionally complemented by investigations with intact cells and, more recently, with isolated electron transfer complexes and proteins. In this discussion we incorporate results obtained with all types of preparations in the main cyclic electron transfer components.

The secondary electron transport chains of all photosynthetic bacteria contain the same types of electron carriers. In general, these include quinones, iron-sulfur proteins, and b and c-type cytochromes. Such carriers are also found in the electron transfer chains of oxygen-evolving photosynthetic membranes and in respiratory chains. Figure 3.9 shows chemical structures for two of these compounds: the Fe_2S_2 cluster of ferredoxin and cytochrome c_2 from *R. rubrum*. Figure 3.10 shows optical absorbance spectra for oxidized and reduced cytochrome c_2. The majority of secondary electron carriers are membrane-bound, intrinsic proteins. However, some carriers such as cytochrome c_2 are soluble and appear to act with considerable mobility along the membrane surface (158, 159). Ubiquinone is present as a pool in considerable excess compared to other secondary electron carriers and the hydrophobic quinone molecules appear to be freely diffusable within the membrane (160, 161). Some quinones are bound at specific sites in the reaction center complex, and in addition, part of the ubiquinone pool may be bound to specific quinone-binding protein(s) (162, 163), similar to those isolated from mitochrondria (164).

3.4.2 The Secondary Quinone and the Quinone Pool: Transition from One- to Two-Electron Reactions

As discussed in section 3.3, the primary reactions in purple photosynthetic bacteria involve the transfer of a single electron from P870 through BPheo to a primary acceptor quinone (Q_A) and on to a secondary acceptor quinone (Q_B). Whereas Q_A appears to function as a strictly one-electron transfer component (operating between the fully oxidized quinone and unprotonated semiquinone anion states), Q_B can function as either a one- or a two-electron carrier. Q_B is reduced to QH_2 (quinol) by Q_A^- in two successive one-electron steps but appears to function in subsequent reactions as a two-electron donor. Under physiological conditions the two-electron oxidant for fully reduced quinol at the Q_B site is probably oxidized

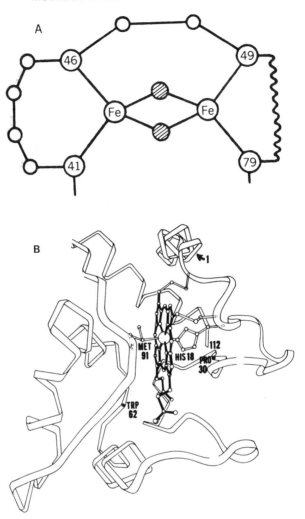

Figure 3.9 (*A*) Schematic representation of the active Fe_2S_2 center from the chloroplast-type ferredoxin of *Spirulina plantensis*. The shaded circles indicate acid-labile sulfide, the numbered circles represent cysteinyl sulfurs and the open circles represent single amino acid residues. Reproduced from Ref. 155 by permission. (*B*) Schematic representation of the structure of *Rhodospirillum rubrum* cytochrome c_2. Reproduced, with permission, from the Annual Review of Biochemistry, Vol. 46.© 1977 by Annual Reviews Inc.

quinone from the quinone pool. It is also possible that quinol of the Q_B site may diffuse to the pool from this site after reduction and be replaced by oxidized quinone diffusing from the pool (161).

The oxidation-reduction properties of Q_B have been studied in detail in *Rps. sphaeroides* (165) and *Rps. viridis* (166, 167). Oxidation-reduction titrations of *Rps. sphaeroides* Q_B show two one-electron transitions: One, associated with the

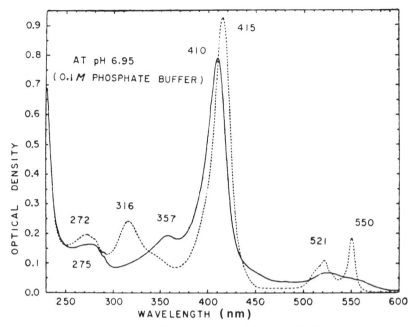

Figure 3.10 Absorption spectra of R. *rubrum* cytochrome c_2. Oxidized form (solid line); reduced form (broken line). Reproduced from Ref. 157 by permission.

one-electron reduction of quinone to semiquinone, has $E_m = +40$ mV (pH 8.0) and the second, associated with the reduction of the paramagnetic semiquinone to the EPR-invisible diamagnetic quinol, has $E_m = -40$ mV (pH 8.0). Both one-electron transitions show pH-dependent E_m values consistent with the existence of a protonated semiquinone as the intermediate redox state. Since optical spectra are more consistent with the presence of the unprotonated semiquinone anion (128, 168), the proton taken up on reduction of Q_B to its semiquinone during these equilibrium measurements may be bound to a protein at the Q_B site rather than by the quinone per se (128, 169). Kinetic studies on proton uptake by reaction center preparations display a periodicity of 2 (128, 170), in agreement with the notion that Q_A^- is not protonated on the time scale of photosynthetic electron transport (see Section 3.3.1.2). If this is the case, both protons necessary for the production of fully reduced quinol at the Q_B site are taken upon accompanying the second of the two electrons. [However, in chromatophores a more complicated situation may exist (see Ref. 161 for a more detailed discussion).]

The quinone pool referred to above is an apparently homogeneous group of quinones (e.q., ubiquinone-10 in photosynthetically grown purple nonsulfur bacteria such as *Rps. sphaeroides*) present in large excess (25 to 40 times P870) over other components (171, 172). However, in *C. vinosum* the pool may contain both ubiquinone and menaquinone (172). The pool in *Rps. sphaeroides* behaves as a two-electron carrier with $E_m = +90$ mV at pH 7.0 (162). The pH dependence of the E_m value for the pool ubiquinone is consistent with the expected reduction to the fully protonated ubiquinol. No semiquinone could be detected during titrations of the pool, suggesting that the equilibrium constant for disproportionation of any pool semiquinone (to fully oxidized and fully reduced quinone) exceeds 10^7 (162). Detection of a stable semiquinone species at the Q_A and Q_B sites thus implies a remarkable stabilization of the semiquinone by protein(s) at these sites. Most (80 to 90%) of the quinone pool can be extracted without altering the rate of electron transfer reactions through the cytochrome chain (162, 173) but the rate of ATP formation coupled to cyclic electron flow was considerably decreased by quinone extraction (174, 175).

3.4.3 c-Type Cytochromes as Donors to P870$^+$ or P985$^+$

As described above, electrons from the reduced primary electron acceptor are eventually transferred to a quinone pool. The other product of the primary photochemical act, P870$^+$ or P985$^+$, must be reduced in order to restore the reaction center to its original state. In all purple bacteria studied to date, the immediate electron donor to P$^+$ is a reduced c-type cytochrome. However, two different patterns for the reduction of P$^+$ are observed. In some cases (the best studied species are *Rps. sphaeroides*, *Rps. capsulata*, and *R. rubrum*), the immediate electron donor to P$^+$ is a soluble c-type cytochrome, cytochrome c_2 (Fig. 3.9*B*). Cytochromes c_2 from purple nonsulfur bacteria have molecular weights ranging from 10.6 to 14.1 kDa and E_m values (pH 7.0) ranging from +290 to +380 mV (176). In other cases (e.g., *C. vinosum*, *Rps. viridis*, *Rps. gelatinosa*, and *Th. pfennigii*) P$^+$ can be reduced by either a high-potential (E_m = +340 mV in *C. vinosum*) or a low-potential (E_m = +10 mV in *C. vinosum*) membrane-bound cytochrome. Rather than describing in detail all the data for each species studied, representative data will be used to illustrate the two different patterns of cytochrome c:P$^+$ interaction.

3.4.3.1 Soluble Cytochromes in Purple Bacteria

In those species where a soluble cytochrome c_2 serves as the immediate electron donor to P$^+$, measurements on the kinetics of the reaction between the reduced cytochrome and P$^+$ have been performed with whole cells, with chromatophores and with a combination of purified cytochrome c_2 and purified reaction centers. Experiments with cells of *R. rubrum* (158) provided convincing evidence for the direct reduction of P$^+$ by cytochrome c_2 in a second-order reaction. A typical time observed for 50% reduction of P870$^+$ in *R. rubrum* cells was 300 μs. These studies also showed mobility of cytochrome c_2 and suggested that a given cytochrome c_2 can diffuse to and reduce any one of approximately four P$^+$ molecules. The reduction of P$^+$ by cytochrome c_2 in chromatophores has been studied most extensively in *Rps. sphaeroides* and *Rps. capsulata*. The cytochrome has been shown to be located in the periplasmic space of these and probably other bacteria (177). During the preparation of chromatophores (which have the opposite membrane sidedness as intact cells) a substantial proportion of the cytochrome c_2 can be trapped inside. The oxidation of cytochrome c_2 in *Rps. sphaeroides* by P^+ occurs with a half-time of approximately 3 μs (178–180).

The oxidation of purified cytochrome c_2 by isolated *Rps. sphaeroides* reaction centers follows second-order kinetics at low cytochrome concentrations. Biphasic kinetics and saturation of the rate of P^+ reduction at higher cytochrome c_2 concentrations have been explained in terms of cytochrome c_2 binding to the reaction center. The binding site has a dissociate constant of approximately $3 \times 10^{-6} M$ at low ionic strength (181–183). No complex formation was observed at higher ionic strength (e.g., [NaCl] $\geq$ 100 mM), strongly suggesting that the forces responsible for complex formation are electrostatic in nature. Immunochemical and chemical cross-linking studies suggest that the cytochrome c_2 binding site involves both the L and M reaction center subunits (181, 184).

The evidence for an electrostatically stabilized complex between cytochrome c_2 and its oxidant in photosynthetic purple nonsulfur bacteria is particularly interesting in the light of the similarities in the three-dimensional structures of cytochrome c_2 and cytochrome c (156), as mitochondrial cytochrome c is known to form similar complexes with its oxidant and reductant (156, 185). In such complexes, a ring of positively charged lysine residues surrounding the single exposed heme edge of mitochondrial cytochrome c contributes the positive charges involved in the electrostatic interaction. These lysine residues have been highly conserved during the evolution of cytochrome c (156). A comparison of the three-dimensional structures indicates that *R. rubrum* cytochrome c_2 contains lysine residues in positions similar to those in mitochondrial cytochrome c (186). It thus appears quite likely that complex formation between cytochrome c_2 and P^+, similar to the electrostatic complexes involving mitochondrial cytochrome c, plays an important role in photosynthetic electron transport (however, see Ref. 186a).

Recently, the techniques of molecular biology have been utilized to construct mutants of *Rps. capsulata* that lack cytochrome c_2 (186b). These mutants are capable of growing photosynthetically (186b), suggesting the presence of an alternate pathway for reducing P870$^+$ that does not involve cytochrome c_2.

3.4.3.2 Membrane-Bound Cytochromes in Purple Bacteria

For species that contain c-type cytochrome as an integral, membrane-bound protein closely associated with the reaction center (e.g., *C. vinosum*, *Rps. viridis*, *Rps. gelatinosa*, and *Th. pfennigii*) there is convincing evidence that these cytochromes serve as the immediate electron donors to P^+. Perhaps the best studied of these bacteria is *C. vinosum*. This species contains two distinct bound c-cytochromes (176, 187): cytochrome c_{555} ($E_m = +340$ mV) and cytochrome c_{552} ($E_m = +10$ mV). Both cytochromes are present in a 2:1, heme/reaction center ratio. When cytochrome c_{552} in chromatophores is reduced prior to flash illumination, it is oxidized by P^+ with $t_{1/2} = 1$ μs, and no oxidation of cytochrome c_{555} is observed. Cytochrome c_{552} (and similar, low-potential membrane-bound cytochromes in other species) appears to be closely linked to the reaction center since oxidation of such cytochromes by P^+ occurs at a high rate even at cryogenic temperatures (187, 188). The oxidation is irreversible at 80 K and it appears that either of the two equivalent cytochrome c_{552} hemes bound to the reaction center can reduce P^+ at low temperature (188). These observations have prompted extensive theoretical studies on electron tunneling in the bacterial reaction center (see Ref. 189 for a review).

The major difficulty in assigning a physiological role to the low-potential cytochrome has been the observation that although the cytochrome is rapidly oxidized, its subsequent reduction at physiological temperatures is extremely slow (187), suggesting that cytochrome c_{552} is not involved in electron flow. In contrast, recent evidence suggests that the higher potential cytochrome c_{555} does participate in cyclic electron flow in *C. vinosum*. At oxidation-reduction potentials where cytochrome c_{555} but not cytochrome c_{552} is reduced prior to flash illumination (+ 70 mV < E_h < +280 mV), cytochrome c_{555} is oxidized directly by P^+ with a half-time of 2 μs (190, 191). Again, either of the two equivalent hemes can be oxidized by the single associated P^+. The oxidized cytochrome is reduced ($t_{1/2}$ = 300 μs) by a soluble c-type cytochrome with a maximum in its difference spectrum at 551 nm (159). No reduction of the low-potential membrane-bound cytochrome c_{552} by the soluble cytochrome c was observed (159), which supports the conclusion that cytochrome c_{552} does not participate in cyclic electron flow in *C. vinosum*. A reduction of the high potential (E_m = +300 mV), membrane-bound cytochrome c_{558} associated with the reaction center of *Rps. viridis* by this bacterium's soluble cytochrome c_2 has also been observed (191a).

The soluble c-type cytochrome from *C. vinosum* has recently been isolated and partially purified. The reduced cytochrome has an α-band absorbance maximum at 550 nm, a molecular weight of 15 kDa, and E_m = +240 mV at pH 7.0 (192, 193). To avoid confusion with another soluble *C. vinosum* cytochrome with α-band maximum at 551 nm (192), we shall refer to the soluble cytochrome of the *C. vinosum* cyclic chain as cytochrome c_{550}.

Measurements with intact *C. vinosum* cells on the kinetics (k = 10 − 15 s^{-1}) and inhibitor sensitivity of reduction of cytochrome c_{550} are consistent with its participation in cyclic electron flow (159). Independent evidence comes from the demonstration that *C. vinosum* spheroplasts (formed by digestion of the cell wall with lysozyme), depleted of the cytochrome, no longer showed light-dependent uptake of alanine (194), a reaction that depends on cyclic electron flow as an energy source (195, 196). Alanine uptake could be restored not only by adding the native cytochrome c_{550} but also with mamalian mitochondrial cytochrome c (194). This indicates that the *C. vinosum* cytochrome is probably related to the family of c-type cytochromes that includes the mitochondrial cytochrome and cytochrome c_2 of the purple nonsulfur bacteria (see above). As cytochrome c_{550} could be removed simply by washing spheroplasts, these results suggest that the cytochrome is located in the periplasmic space of the cells (194).

Molecular weights of 23 to 24 and 40 to 44 kDa have been assigned to *C. vinosum* cytochrome c_{552} and c_{553}, respectively, using heme staining after gel electrophoresis of *C. vinosum* membranes (197, 198). These results and the availability of detergent-solubilized fractions from *C. vinosum* that contain cytochrome c_{552}

but not cytochrome c_{555} (199, 200) indicate that the two species of heme c do not reside on a single peptide. By contrast, in *Rps. viridis* all four hemes (two low potential and two high potential) that are closely associated with the reaction center appear to reside on a single peptide (86b, 201). The 3-Å resolution structure of the *Rps. viridis* reaction center obtained by X-ray analysis shows that the four heme groups are arranged with an approximate two-fold symmetry (86b). The heme groups are positioned roughly along a line that differs by approximately 60° from the axis of symmetry defined by the other reaction center chromophores. The heme closest to P985 is approximately 21 Å away. Heme–heme distances are 14 to 16 Å. EPR investigations of oriented chromatophore preparations have provided evidence that the membrane-bound cytochromes in *C. vinosum* are also specifically oriented with respect to the membrane plane (202).

3.4.3.3 Green Bacteria

As mentioned earlier, considerably less is known about cytochrome reactions in green than in purple bacteria. Nevertheless, it is clear that, as in purple bacteria, a membrane-bound c-type cytochrome is the immediate electron donor to P^+. In the green sulfur bacteria cytochrome c_{553} serves this function (113, 136, 138). Oxidation of cytochrome c by $P840^+$ occurs in the range 10 to 90 μs. Of interest is the fact that E_m for this cytochrome, $+165$ mV (138), is considerably lower than those of the c-type cytochromes serving as immediate donors to P^+ in purple bacteria. This is presumably an evolutionary adjustment to similar differences in the E_m values of the primary electron donors, since, as mentioned above (see Section 3.3.2.1), E_m of P840 is some 200 mV lower than typical values measured for P870. In the gliding green bacterium *Cfl. aurantiacus* cytochrome c_{554}, with $E_m = +260$ mV at pH 8.1, appears to be the electron donor for $P865^+$, with $t_{1/2} < 10$ μs (148).

3.4.4 The Cytochrome bc_1 Complex

3.4.4.1 b-Type Cytochromes

All photosynthetic bacteria examined in detail have been shown to contain b-type cytochromes. The prosthetic group of cytochromes of the b-type is noncovalently bound protoheme. Until fairly recently the presence of membrane-bound b-type cytochromes was thought to be confined to the BChl a–containing purple nonsulfur bacteria. However, b cytochromes have now clearly been demonstrated to be present in the purple sulfur bacteria *Eth. mobilis* and *C. vinosum* (203–205), in the BChl b-containing purple nonsulfur bacterium *Rps. viridis* (205a, 205b), in the green sulfur bacterium *Chl. limicola* f. *thiosulfatophilum* (135, 203, 206), and in the green filamentous bacterium *Cfl. aurantiacus* (205b, 206a, 206b).

As the b-type cytochromes of purple nonsulfur bacteria have been studied in more detail than those of other families of photosynthetic bacteria, in this section we concentrate on the purple nonsulfur bacteria. Representative species such as *Rps. sphaeroides* , *R. rubrum*, and *Rps. capsulata* contain three membrane-bound b cytochromes (80, 187, 207–209). The cytochromes have E_m values (at pH 7.0) of $+155$, $+50$, and -90 mV and are often identified by their E_m values (e.g., cytochrome b_{50}). Cytochromes b_{50} and b_{-90} have been demonstrated to undergo rapid oxidation-reduction reactions upon flash excitation of chromatophores from *Rps. sphaeroides* (see Ref. 161 for a review). The *Rps. sphaeroides* cytochrome b_{155} (α-band absorbance maximum at 561 nm) has been isolated (210) and shown to have cytochrome c_2: oxygen oxidoreductase activity (210, 211). Thus this cytochrome appears to play a role in respiratory rather than photosynthetic electron transport in this facultative aerobe. Evidence also exists for the participation of two distinct *b*-type cytochromes in electron flow in *R. rubrum* chromatophores (212).

The two *b*-type cytochromes that participate in cyclic electron flow in *Rps. sphaeroides* can be distinguished their α-band absorption spectra as well as by their different E_m values. Reduced cytochrome b_{50} has a single symmetric band centered at 560 to 561 nm (80, 209, 213, 214) and thus is often denoted cytochrome b_{560}, while cytochrome b_{-90} has double α-band peaks at 559 and 566 nm (209, 214, 215) and is often called cytochrome b_{566}. The E_m values and α-band maxima of these cytochromes are similar to those measured for the two *b*-cytochromes of the mitochondrial complex III. Reviews that discuss possible relationships between electron transport in photosynthetic bacteria and mitochondria are given in Refs. 161 and 216.

As is the case in mitochondria, the two *b* cytochromes can also be distinguished by the effect of different antibiotic inhibitors. The α-band maximum of cytochrome b_{560} is shifted by 1 to 2 nm to the red on the addition of antimycin A (209, 217–219), but this inhibitor has no effect on the α-band spectrum of cytochrome b_{566}. In contrast, the inhibitor myxothiazol shifts the α-band spectrum of cytochrome b_{566} to the blue and may also increase the E_m value of the cytochrome to -45 mV, without any effect on cytochrome b_{560} (218).

3.4.4.2 Cytochrome c_1

Until quite recently, cytochrome c_2 (see Section 3.4.3.1) was thought to be the only *c*-type cytochrome participating in cyclic electron flow in purple nonsulfur bacteria. However, some years ago, Wood (220, 221) was able to demonstrate the presence of a second *c*-type cytochrome in chromatophores from *Rps. sphaeroides*. The newly discovered cytochrome is membrane bound and differs from the soluble cytochrome c_2 in absorption spectrum (α-band maximum at 552 nm compared to 550 nm for cytochrome c_2) and molecular weight (30 kDa versus 14.1 kDa for cytochrome c_2). Oxidation-reduction titrations in chromatophores and in solubilized complex (see Section 3.4.4.4) gave E_m for the *Rps. sphaeroides* cytochrome ranging from $+260$ to $+295$ mV (220, 222, 223), significantly lower than the E_m value of $+350$ mV measured for cytochrome c_2 from the same bacterium (80, 218, 220). Because of the similarities in spectra, molecular weight, and E_m between the *Rps. sphaeroides* cytochrome and cytochrome c_1 of mitochondrial ubiquinol: cytochrome c oxidoreductase, it was suggested that the membrane-bound *Rps. sphaeroides* cytochrome be named cytochrome c_1 (216, 220). The function of the *Rps. sphaeroides* cytochrome c_1 (see Section 3.4.4.5) also appears to be quite similar to that of mitochondrial cytochrome c_1. The presence of a cytochrome with properties similar to mitochondrial cytochrome c_1 does not appear to be confined to purple nonsulfur bacteria. A similar membrane-bound cytochrome (α-band at 552 to 553 nm, molecular weight = 31 kDa, $E_m \approx +245$ mV) was recently found in *C. vinosum* (198). A membrane-bound *c*-type cytochrome with α-band at 550.5 nm, a molecular weight of 24 kDa, and E_m (pH 6.2) = +220 mV (n = 1) that probably plays a role in electron transport similar to cytochrome c_1 in purple bacteria has recently been identified in *Chl. limicola* (135). Cytochrome c_1 has also been shown to be present in the membranes of *R. rubrum* (205b) and *Rps. viridis* (205a, 205b).

3.4.4.3 The Rieske Iron–Sulfur Protein
Chromatophores from several purple nonsulfur bacteria (224, 225) and from *C. vinosum* (226–228) and cell-free extracts from the green sulfur bacterium *Chl. limicola* f. *thiosulfatophilum* (229) have been shown to contain an iron–sulfur protein that in the reduced state exhibits a characteristic EPR signal with g_z and g_y values of 2.03 and 1.89–1.90, respectively. g_x Features at 1.81 and 1.79 have been observed for the proteins from *Rps. sphaeroides* and *Chl. limicola*, respectively. Recently, a similar protein has also been detected in membranes isolated from phototrophically grown *Cfl. aurantiacus* (205b, 206a). The EPR spectra obtained with these photosynthetic bacteria are very similar (224) to those originally observed in mitochrondria by Rieske and co-workers (230). The molecular weight of the *Rps. sphaeroides* protein has been estimated to be 25 kDa, a value similar to those measured for the proteins from mitochondrial and oxygen-evolving photosynthetic organisms (216). Because of these similarities, the species giving rise to the characteristic EPR signal with g_y = 1.89 to 1.90 is usually referred to as the Rieske iron–sulfur protein. The protein acts as a one-electron carrier and is known to contain two nonheme irons and two inorganic sulfides at its electron-carrying site. However, unlike the Fe_2S_2 group of ferredoxin (see Fig. 3.9), it is likely that the Rieske protein may utilize nitrogen ligands to iron in addition to cysteinyl sulfur and two inorganic sulfides (230a).

E_m values for the Rieske protein in purple bacteria range from +280 to +315 mV (pH 7.0), similar to those measured for the mitochondrial protein (224–228). However, the Rieske proteins from both *Cfl. aurantiacus* [E_m = +100 mV (206a)] and *Chl. limicola* [E_m (pH 6.8) = +165 mV] are considerably less positive than those from the purple bacteria. The much lower value for the Rieske protein in this green sulfur bacterium is probably related to the relatively low E_m values of P840 and cytochrome c_{553} (see Section 3.4.3.3).

3.4.4.4 The Isolated Cytochrome bc_1 Complex
A recent significant development in bioenergetics has been the realization of how similar secondary photosynthetic electron transfer reactions are to mitochondrial electron transfer in the ubiquinol-to-cytochrome *c* region of the respiratory chain and to the cytochrome b_6f region in oxygen-evolving photosynthetic organisms. A detailed discussion of these similarities is beyond the scope of this chapter and

the interested reader should consult Ref. 216 for an excellent review. It has been known for a long time that electron flow from ubiquinol to cytochrome c in mitochondria is catalyzed by a multiprotein complex that can be solubilized from mitochondrial membranes and is referred to as complex III or ubiquinol: cytochrome c oxidoreductase (216, 231, 232). In view of the many similarities between respiratory and bacterial photosynthetic electron transport discussed above, it thus appeared possible that a similar complex, with ubiquinol: cytochrome c_2 oxidoreductase activity, could be solubilized from chromatophore membranes. This has been accomplished with chromatophores form *Rps. sphaeroides* (223, 233, 234) and more recently with *R. rubrum* and *Rps. viridis* (205b). The isolated *Rps. sphaeroides* complex contains three major peptide subunits (40, 34, 25 kDa) and probably one or more additional smaller peptides. Assignments have been made for the three major peptides: cytochrome b, 40 kDa; cytochrome c_1, 34 kDa and Rieske protein, 25 kDa (216). The stoichiometry of cytochrome b (total protoheme): cytochrome c_2:[Fe$_2$S$_2$] cluster is 2:1:1. The complex also contains ubiquinone. Similar electron carrier stoichiometries have been found in the mitochondrial, chloroplast, and cyanobacterial complexes (216). E_m values and optical spectra for the b and c cytochromes of the complex are virtually identical to those determined with chromatophores (209, 216, 223). The isolated *Rps. sphaeroides* complex exhibits ubiquinol: cytochrome c_2 oxidoreductase activity that is sensitive to inhibitors of cyclic electron flow in chromatophores, such as antimycin A, myxothiazol, and 5-n-undecyl-6-hydroxy-4,7-dioxobenzothiazide (UHDBT).

Amino acid sequences, deduced from the nucleotide sequences of the corresponding genes, are available for the cytochrome b, cytochrome c_1, and Rieske peptides of the *Rps. sphaeroides* complex. In this bacterium, the genes coding for the three peptides form an operon, referred to as the *fbc* operon (234a).

3.4.4.5 Mechanism of Electron Transport

Electron transport from ubiquinol to cytochrome c_2 through the cytochrome bc_1 complex in purple nonsulfur bacteria (like the analogous electron transfer from ubiquinol to cytochrome c catalyzed by mitochondrial cytochrome c) exhibits certain anomalies that cannot be explained by a linear, unbranched electron transfer pathway. Perhaps the most striking of these anomalies is the phenomenon referred to as "oxidant-induced reduction," first observed in mitochondria (235, 236) and subsequently observed in *Rps. sphaeroides* chromatophores (237) and in the isolated *Rps. sphaeroides* cytochrome bc_1 complex (209, 223). Addition of an oxidant such as ferricyanide to preparations in which cytochrome c_1 and the Rieske ironsulfur protein were reduced but the b cytochromes were oxidized not only resulted in the expected oxidation of cytochrome c_1 and the Rieske protein but also gave an unexpected reduction of the b cytochromes. The oxidant-induced reduction of the b cytochromes in *Rps. sphaeroides* was enhanced by antimycin A but was inhibited by UHDBT and myxothiazol (209, 223, 237). A detailed discussion of all the various explanations that have been proposed for this phenomenon is beyond the scope of this chapter. Instead, a brief discussion of one model that has been accepted, the Q-cycle (238) will be presented. For a detailed discussion of the relative merits of other models, the reader is referred to recent reviews (161, 169, 216).

The Q-cycle postulates that the oxidized Rieske iron–sulfur protein is the oxidant for a one-electron oxidation of ubiquinol at a specific site in the cytochrome bc_1 complex. The ubisemiquinone anion thus formed in turn acts as the reductant for cytochrome b_{566}. In the mitochondrial bc_1 complex it has been possible to demonstrate the formation of a ubisemiquinone anion under conditions that produce the oxidant-induced reduction of cytochrome b (239). Formation of the ubisemiquinone (239) and the oxidant-induced reduction of cytochrome b (240) both require the presence of the reduced Rieske protein. The demonstration of semiquinone formation, consistent with a Q-cycle mechanism, has not yet been accomplished with preparations from photosynthetic bacteria. It is possible that in photosynthetic bacteria the reductions of cytochrome b and the Rieske protein are so tightly coupled as to constitute a concerted two-electron oxidation of ubiquinol without any appreciable formation of semiquinone.

The enhancement of oxidant-induced cytochrome b reduction caused by antimycin A appears to arise from inhibition of the reoxidation of reduced cytochrome b by this inhibitor (see below) UHDBT, which inhibits the oxidation of the Rieske iron–sulfur protein by cytochrome c_1 (241), and myxothiazol, which appears to inhibit the reduction of the Rieske protein by ubiquinol (218), both inhibit cytochrome b reduction by preventing the formation of the ubisemiquinone anion that serves as the reductant for cytochrome b.

Fig. 3.11 shows a modified Q-cycle scheme proposed by Crofts and co-workers (242) for cyclic electron flow in *Rps. sphaeroides*. The major features of the scheme are:

1. Formation of the fully reduced quinol (QH_2) from quinone (Q) by two

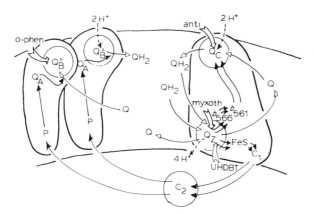

Figure 3.11 A modified Q-cycle mechanism for the cyclic electron transfer chain of *Rps. sphaeroides*. P is P870; c_2, c, b_{566}, and b_{561} (b_{560}) designate the respective cytochromes and FeS the Rieske iron–sulfur protein. Reproduced from Ref. 242 by permission.

turnovers of the reaction center, here shown for illustrative purposes as single turnovers of two separate reaction centers. QH_2 diffuses to a quinol-oxidizing site, labeled Q_z, on the cytochrome bc_1 complex.

2. $P870^+$ oxidizes cytochrome c_2 (see section 3.4.3.1), which in turn oxidizes cytochrome c_1 (221, 222) with $t_{1/2} = 150$ μs (222). Ferricytochrome c_1 then oxidizes the Rieske protein in a UHDBT-sensitive reaction with $t_{1/2} < 200$ μs (241).

3. QH_2 is oxidized in a myxothiazol-sensitive reaction at the Q_z site with one electron donated to the Rieske protein (FeS) and the other to cytochrome b_{566} (b_{-90}). The $t_{1/2}$ for this reaction varies from 0.3 to 7 ms depending on the QH_2 concentration (242). Although this reaction is shown as a concerted two-electron oxidation of QH_2, the possibility certainly exists that the oxidation of QH_2 in *Rps. sphaeroides* proceeds in two one-electron steps, with an initial oxidation by the Rieske protein to yield a semiquinone and a subsequent reduction of cytochrome b_{556} by the semiquinone.

4. Reduction of cytochrome b_{560} (b_{50}) by cytochrome b_{566} with $t_{1/2} < 300$ μs (215). This step has recently been shown to be electrogenic, contributing to the formation of a transmembrane electrical potential (242a).

5. Reduction of Q by cytochrome b in an antimycin A–sensitive reaction at a quinone-reducing site, labeled Q_c, on the cytochrome bc_1 complex with an average $t_{1/2}$ of 4 ms (215). Although this reaction is shown as a concerted two-electron reduction of Q by ferrocytochromes b_{560} plus b_{566} in the scheme, recent evidence suggests that Q reduction at the Q_c site may occur in two one-electron steps. Evidence for an antimycin A–sensitive ubisemiquinone formation in *Rps. sphaeroides* chromatophores (243) supports the involvement of a semiquinone intermediate at this point of the cycle. Evidence for a ubisemiquinone intermediate formed during the oxidation of cytochrome b was obtained earlier with the analogous mitochondrial complex (239, 244).

6. For complete oxidation of one QH_2 to Q, two light quanta are needed, which drive the single turnover of each of two reaction centers. The two P^+ produced in these turnovers each ultimately produce an oxidized Rieske protein, which in turn results in two oxidations of QH_2 at the quinol-oxidizing site (reactions 1 to 4). A single turnover of the quinone-reducing site (reactions 5 and 6) suffices to complete the cycle in which two QH_2 molecules are oxidized to Q and one Q molecule is reduced to QH_2.

7. The asymmetric orientation of the complexes in the membrane results in the obligatory coupling of proton translocation from the outside aqueous phase to the aqueous phase inside the chromatophores with a H^+/e stoichiometry of 2. (Proton translocation and its relation to ATP formation will be discussed in Section 3.5.) In bacterial cells where the sidedness of the membrane is opposite to that of chromatophores, electron transport would result in H^+ efflux rather than in H^+ uptake (196).

Although the details of the scheme of Fig. 3.11 have been derived from experiments with *Rps. sphaeroides* and *Rps. capsulata*, similar Q-cycle schemes have been proposed for other purple nonsulfur bacteria. Recent experiments provide considerable evidence for a Q-cycle in *R. rubrum* (212). In particular, quantitative measurements of electron donation to cytochrome(s) *b* after a flash show that one electron is delivered to high potential acceptors (P870$^+$ and ferricytochrome c_2) for each electron arriving at cytochrome(s) *b*, as predicted by reaction 4 of the scheme in Fig. 3.11. Similar data have also been obtained with *C. vinosum* (244a). Data obtained with *R. rubrum* also support the sequential involvement of two *b* cytochromes in the cycle (212).

Additional support for a Q-cycle model for electron flow through the cytochrome bc_1 complex comes from observations with inhibitors. UHDBT, which bears some structural similarities to quinones, appears to interact directly with the Rieske iron–sulfur protein: It alters the EPR spectrum of the *Rps. sphaeroides* protein and raises its E_m from +280 to +350 mV (241). Another quinone analog, 2,5-dibromomethylisopropyl-*p*-benzoquinone (DBMIB), has been shown to alter both the EPR spectrum and E_m of the *C. vinosum* Rieske protein (228). The fact that these quinone analogs appear to bind to the Rieske protein is consistent with the Q-cycle, which envisages direct interaction between QH$_2$ and the Rieske protein at the Q$_z$ site.

3.4.4.6 *Respiratory Pathways*

Although it may seem out of place to mention aerobic electron transport pathways in a book on anaerobes, it seems useful to dicuss these pathways briefly for the purpose of comparing the photosynthetic and respiratory electron transfer chains in facultative photosynthetic bacteria. It has been known for many years that many purple nonsulfur bacteria are facultative organisms capable of high rates of respiration and of aerobic growth in the dark (245). In the presence of sufficient O$_2$, synthesis of the photosynthetic pigments is repressed and the organisms grow as nonphotosynthetic aerobes. This is also true for the green gliding bacterium *Cfl. aurantiacus* (245a). Details of the alternative energy-generating mechanisms in the purple nonsulfur bacteria are discussed in Chapter 2. Until recently it was thought that the purple sulfur and green sulfur bacteria were obligate anaerobes. Although this still appears to be the case for the green sulfur bacteria, more recent evidence indicates that several species of purple sulfur bacteria can grow and respire in the dark under semi- or microaerobic conditions (246). It has also been shown that *C. vinosum* chromatophores support succinate or NADH-dependent O$_2$ uptake that is sensitive to a number of classical respiratory inhibitors (247). Neither the terminal oxidase nor the other components of the respiratory chain in *C. vinosum* have yet been identified.

The facultative purple nonsulfur bacteria utilize many of the same electron carriers for aerobic electron transport as they do for anaerobic, photosynthetic growth. Reconstitution experiments suggest that the cytochrome bc_1 complex present in aerobically grown cells (where it donates electrons to cytochrome oxidase via cytochrome c_2) is the same as the complex in photosynthetically grown cells, where it serves as an electron donor (via cytochrome c_2) to P870$^+$ (248). It should also be noted that although some purple nonsulfur bacteria use c-type terminal oxidases, the cytochrome oxidase induced in *Rps. sphaeroides* by the presence of O_2, is a heme a–containing oxidase, quite similar to that found in mitochondria (249, 250). The role of this oxidase compared to that of the protoheme-containing oxidase of photosynthetically grown cells (see Section 3.4.4.1) is unclear. Recently, it has also been shown that *Cfl. aurantiacus* can induce the synthesis of a cytochrome aa_3 oxidase during aerobic growth (206b).

3.5 PHOTOPHOSPHORYLATION

3.5.1 Introduction

As discussed previously, the energy released during exergonic cyclic electron flow can be utilized for the phosphorylation of ADP to form ATP. Such ATP formation, coupled to electron flow, is referred to as photophosphorylation and was first observed in chromatophores of *R. rubrum* (251). The detailed mechanism of cyclic photophosphorylation by bacterial chromatophores is still a matter of debate (252). However, most workers accept the basic principles of Mitchell's chemiosmotic hypothesis (253) as a good working hypothesis for understanding the energetics of photophosphorylation. According to this hypothesis, the oxidation-reduction reactions of electron transport are coupled to the stoichiometric translocation of protons across the proton-impermeable membrane containing the electron carriers. This results in the formation of a proton motive force (Δp), which consists of both a concentration component (ΔpH) and an electrical membrane component ($\Delta\psi$). Δp, in turn, provides the energy for ATP synthesis via a proton-translocating enzyme, the ATPase.

ATPases have been characterized for several photosynthetic bacteria (see Ref. 254 for a review). They show strong structural similarity to ATPases from mitochondria, aerobic bacteria, and chloroplasts, which have been reviewed extensively elsewhere (255, 256).

Evidence in favor of proton movements through the ATPase complex in the thermodynamically favorable direction (toward the alkaline, electrically negative compartment) providing the energy for ATP synthesis (as predicted by the chemiosmotic hypothesis) comes from the demonstration that ATP formation in *R. rubrum* chromatophores can be driven by the establishment of a ΔpH across the membrane in the absence of any other apparent source of energy (257). Simultaneous imposition of a $\Delta\Psi$ across the membrane enhanced the amount of ATP formed in such "acid-base" experiments. The stoichiometry of protons translocated per ATP formed is still a matter of debate, values between 2 and 3 being reported (252, 254, 258). Although the chemiosmotic hypothesis seems to be able to account qualitatively for a large body of observations on phosphorylation, it has recently been suggested that more detailed quantitative testing of the hypothesis gives data inconsistent with the hypothesis in its original form (252, 252a).

3.5.2 Formation of the Proton Motive Force in Purple Bacteria

Cells of photosynthetic purple nonsulfur (259) and sulfur bacteria (260) use the energy released during the cyclic electron flow to produce a Δp consisting of two components: $\Delta\Psi$ (outside positive) and a ΔpH (outside acidic). The contribution of ΔpH is important only at acidic pH values, while at pH > 7.0, Δp consists almost entirely of $\Delta\Psi$. The total Δp maintained in the steady state by illuminated cells depends on external pH and the cation composition of the medium, but is generally near 200 mV. (It is common to give the magnitude of Δp in terms of the equivalent electrical potential using a conversion factor of 59 mV for a ΔpH of 1 unit.) Studies on the mechanism of Δp formation have been carried out almost entirely with chromatophores rather than with intact cells. Chromatophores from several species of purple nonsulfur bacteria (see Refs. 252 and 261 for reviews) and from *C. vinosum* (196, 262, 263) maintain a Δp of similar magnitude to that observed with intact cells, but with the opposite polarity (inside positive and acidic).

In chromatophores from purple nonsulfur bacteria, three electrogenic steps that contribute to the development of $\Delta\Psi$ across the membrane have been identified. In these studies advantage has been taken of the fact that electrical potentials across the membrane alter the absorbance spectra of pigments (cartenoid and BChl) contained in the membrane by causing electrochromic band shifts of these pigments (see reviews in Refs. 160 and 264). Kinetic and thermodynamic resolution of the carotenoid band shift in *Rps. sphaeroides* and *Rps. capsulata* chromatophores (178, 217, 265–270) have identified the light-driven charge separation from P870 to Q_A as one of these steps and the reduction of $P870^+$ by cytochrome c_2 as the second. The third, antimycin A–sensitive phase of the carotenoid band shift clearly involves electron flow through the cytochrome bc_1 complex and, as mentioned above, involves (at least in part) electron transfer between the two b cytochromes in the complex (242a). Transport of a single electron through all three electrogenic steps results in a $\Delta\Psi$ of approximately 100 mV. A scheme describing a possible arrangement of electron carriers in the chromatophore membrane that could lead to the generation of $\Delta\Psi$ is given in Fig. 3.12.

The description of the electron flow through Q_B and the cytochrome bc_1 complex presented earlier referred to proton as well as electron movements, and at this point it is necessary to complete that description. In *Rps. sphaeroides* chromatophores, two protons are taken up from the outer aqueous phase per electron that traverses the cyclic electron chain. One of these electrons appears to be taken up during the reduction of Q_B (see Section 3.4.2). The second proton uptake does not occur in the presence of antimycin A and thus can be related to electron flow through the cytochrome bc_1 complex. The Q-cycle scheme of Fig. 3.11 provides for the uptake of the additional proton per electron and the release of both protons to the inner aqueous phase. After the buffering capacity of the inner chromatophore compartment has been consumed, this proton translocation results in the formation of the ΔpH component of Δp. The exact size of the ΔpH that builds up during continuous illumination is subject to some uncertainty (252).

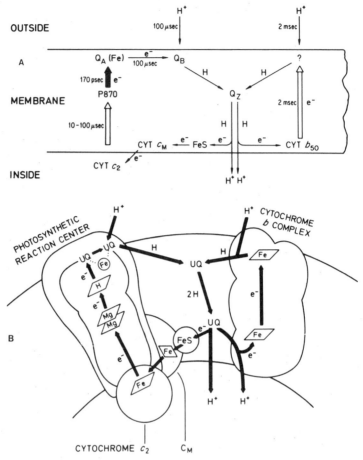

Figure 3.12 Topology and kinetics of electron transfer and proton pumping in chromatophores of purple nonsulfur bacteria. In (*A*), Q_z is the site on the cytochrome bc_1 complex where quinol is oxidized, FeS is the Rieske iron–sulfur protein, and cyt. c_M is cytochrome c_1. In (*B*), the two overlapping squares containing Mg represent the P870 bacteriochlorophyll dimer, the square containing H is the bacteriopheophytin primary acceptor and the squares containing Fe represent the hemes of various cytochromes. Reproduced from Ref. 261 by permission.

3.6 NONCYCLIC ELECTRON FLOW

3.6.1 NAD(P)$^+$ Reduction

3.6.1.1 *Purple Bacteria*

Around 1960 it became clear from spectroscopic measurements with various species of purple bacteria that NAD(P)$^+$ is reduced at fairly high rate upon illumination of intact cells (271). However, the primary quinone (Q_A) acceptors in these bacteria have E_m values that range from 0 to -100 mV at pH 7.0 (117, 272). Thus the first stable reductants generated in the light are considerably weaker reductants than the NADH/NAD$^+$ couple ($E_m = -320$ mV). Even if the unprotonated semiquione anion (Q_A^- or Q_B^-) served as the reductant, reduction of NAD$^+$ could not occur without additional input of energy. The mechanism of such energy-dependent "reverse" electron flow (shown in Fig. 3.13) envisages the use of a high energy state (Δp according to the chemiosmotic hypothesis) generated either by light-driven cyclic electron flow or by ATP hydrolysis via the reversible ATPase (see Section 3.5.2) to "pump" electrons uphill from weak reductants such as succinate to NAD$^+$. The first evidence for such a mechanism came from experiments with *R. rubrum* chromatophores in which it was demonstrated that NAD$^+$ reduction by succinate could occur in the dark if ATP or pyrophosphate hydrolysis was available as source of energy (273). A similar ATP-dependent reduction of NAD$^+$ in the dark was subsequently demonstrated with chromatophores from several other species of purple nonsulfur bacteria (see Ref. 272 for a review).

Further evidence for an energy-dependent photoreduction of NAD$^+$ came from observations that inhibitors of cyclic electron flow also inhibited reduction of NAD$^+$ by succinate in purple nonsulfur (273–276) and sulfur (277) bacteria. Perhaps the most convincing evidence is the observation that uncouplers of phosphorylation (agents that dissipate the high-energy state by making the chromatophore membranes permeable to protons and thus collapsing Δp) completely inhibit NAD$^+$ reduction in both the light and the dark (277 and review in Ref. 272).

Another common feature of both light-dependent and ATP-dependent NAD$^+$ reduction by a variety of electron donors in chromatophores is the inhibition by inhibitors of mitochondrial NADH dehydrogenase such as rotenone (273, 275–278), amytal (273) and piericidin A (275). As these compounds inhibit neither ATP hydrolysis nor cyclic electron flow and do not act as uncouplers, it was concluded that they inhibit the enzyme directly involved in NAD$^+$ reduction (272). This suggests that the bacterial and the mitochondrial enzymes are related, although they function predominantly in opposite directions. Further evidence for similarities between the two enzymes comes from the detection of iron–sulfur proteins in chromatophores (226, 277, 279–281) with EPR spectra and E_m values similar to some of the iron–sulfur centers of mitochondrial NADH dehydrogenase (282). Zannoni and Ingledew (281) tentatively assigned two iron-sulfur centers in *Rps. capsulata* mutants with E_m values of -115 and -370 mV (at pH 7.0) and EPR signals at $g = 1.94$ to NADH hydrogenase. A similar conclusion was obtained for a low-potential iron–sulfur center in *C. vinosum* with an EPR signal at $g = 1.93$ (277).

3.6.1.2 Green Sulfur Bacteria

In terms of thermodynamics, the situation in green sulfur bacteria is quite different from that in purple bacteria. The photoreduced first iron–sulfur center (FeS$_I$, see Section 3.3.2.1), with $E_m < -500$ mV, is able to reduce NAD(P)$^+$ without additional input of energy. In agreement with this, NAD$^+$ photoreduction by cell-free extracts of *Chl. limicola* is not inhibited by uncouplers (203, 283). The re-

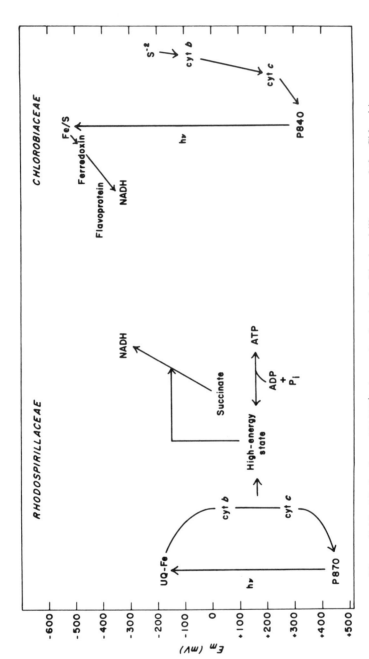

Figure 3.13 Mechanism of NAD$^+$ photoreduction in the Rhodospirillaceae and the Chlorobiaceae. Reproduced from Ref. 272 by permission. A possible role for a b-type cytochrome in the oxidation of sulfide was suggested by the sensitivity of this reaction to antimycin A (203). However, it is unlikely that cytochrome b is the immediate oxidant for sulfide.

action is not affected by rotenone (283), suggesting that no NADH dehydrogenase-like enzyme is involved. Instead, NAD^+ photoreduction in *Chl. limicola* requires soluble ferredoxin and a second soluble protein that is probably a flavin-containing ferredoxin: $NAD(P)^+$ oxidoreductase (203, 283, 284). A possible scheme for NAD^+ photoreduction in green bacteria is shown in Fig. 3.13. It should be pointed out, as was done earlier for the electron acceptor chain (Section 3.3.2.1), that there are many similarities between NAD^+ photoreduction by green sulfur bacteria and $NADP^+$ photoreduction by photosystem I in cyanobacteria and higher plants, strongly suggesting a possible common evolutionary origin for both systems.

3.6.2 Substrate Oxidation

3.6.2.1 *Succinate*

Both purple nonsulfur and purple sulfur bacteria are able to grow with an organic acid as the source of electrons for reductant-requiring biosynthetic reactions. The enzyme responsible for succinate oxidation has been characterized in chromatophores from several species of purple bacteria (279, 280, 281, 285). The bacterial succinic dehydrogenases, like the mitochondrial enzyme, contain two ferredoxin-type iron–sulfur centers with EPR signals at $g = 1.93$ to 1.95 in the reduced form and one iron–sulfur center that is diamagnetic when reduced but paramagnetic when oxidized, exhibiting an EPR signal at $g = 2.01$ to 2.02. The latter, succinate-reducible center, presumably analogous to center S-3 of the mitochondrial complex, has $E_m = +80$ mV in *Rps. sphaeroides* (279), $+60$ mV in *Rps. capsulata* (281), and $E_m = +50$ mV in *C. vinosum* (277). Only the more electropositive of the two other centers ($E_m = +50$ mV in *Rps. sphaeroides*, $E_m = +120$ mV in *Rps. capsulata*, $E_m = -50$ mV in *C. vinosum*) appears to be reducible by succinate, as is the S-1 center of mitochondrial succinic dehydrogenase. The more electronegative center ($E_m = -280$ mV in *Rps. capsulata* and $E_m = -250$ mV in *Rps. sphaeroides*) is similar to the mitochondrial center S-2 in being succinate nonreducible. It has been proposed that S-2 may be a preparative artifact rather than a normal constituent of the enzyme (286). Like the mitochondrial enzyme, the enzymes from *Rps. capsulata* (280) and *C. vinosum* (277) are inhibited by 2-theonyltrifluoroacetone (TTFA). These similarities represent yet another example of close correspondence between mitochondrial respiratory carriers and the electron carriers of purple photosynthetic bacteria.

3.6.2.2 *Sulfur Compounds*

Most species of photosynthetic bacteria are able to use reduced sulfur compounds as electron donors during growth under photosynthetic (anaerobic) conditions. The pathways by which reduced sulfur compounds are oxidized vary considerably from one family to another, are often complex, and in many cases are incompletely understood. In this section we focus on those reactions for which the proteins involved have been reasonably well characterized (for more comprehensive discussions, we refer readers to Refs. 287 to 289).

All species of green sulfur bacteria are capable of oxidizing sulfide to sulfate with elemental sulfur as an intermediate. A flavocytochrome, cytochrome c_{553}, which exhibits sulfide : cytochrome c oxidoreductase activity, has been implicated as the enzyme responsible for the oxidation of sulfide in several species of green sulfur bacteria (290–292). The native electron acceptor for flavocytochrome c_{553} in *Chl. limicola* is a soluble cytochrome c_{555} (290, 291), with $E_m = +150$ mV (293). It is not clear how electrons are transferred from reduced cytochrome c_{555} to P840, nor is the oxidation product known, both thiosulfate and elemental sulfur having been suggested as possible products (287, 294). Yamanaka and Fukumori (294) have presented evidence that the flavocytochrome c_{553} can also catalyze the reduction of elemental sulfur to sulfide with reduced benzyl viologen as a nonphysiological reductant. This reaction may have physiological significance, since it is known that green sulfur bacteria are capable of reducing sulfur to sulfide (295).

Some species of green sulfur bacteria are able to oxidize thiosulfate. Several pathways for this oxidation appear to exist (287). The best characterized pathway involves an enzyme with thiosulfate : cytochrome c oxidoreductase activity that utilizes soluble cytochrome c_{551} as an electron acceptor (291, 296). This cytochrome c_{551} [$E_m = +135$ mV (297)] is absent from non-thiosulfate-utilizing *Chl. limicola*, consistent with its postulated role. Reduced cytochrome c_{551}, in turn, reduces cytochrome c_{555} (291, 296). Thus the latter cytochrome appears to play a control role as an acceptor for electrons from both sulfide and thiosulfate in green sulfur bacteria. It has recently been shown that cytochrome c_{555} forms electrostatically stabilized complexes with both flavocytochrome c_{553} and cytochrome c_{551}, suggesting that these complexes may play a role in the oxidation of sulfide and thiosulfate in *Chl. limicola* (297a).

The purple sulfur bacteria utilize sulfide and, with a few exceptions, thiosulfate as electron donors. In spite of their familiar name, the same is true for several species of the purple nonsulfur bacteria (289, 299). In the presence of reduced sulfur compounds the purple sulfur bacteria produce elemental sulfur during growth, which is usually stored in intracellular, membrane bound, sulfur globules. They are also capable of oxidizing sulfur to sulfate. In the Ectothiorhodospiraceae and Chlorobiaceae the elemental sulfur is excreted (1b). In *Th. roseopersinica*, the oxidation of sulfide to sulfur appears to occur via a nonenzymatic process with a heat-stable cytochrome c serving as the electron acceptor (300, 301). In *C. vinosum* the reaction appears to be catalyzed by flavocytochrome c_{552} (a protein similar to the *Chl. limicola* flavocytochrome c_{553} described above). Fukumori and Yamanaka (302) originally demonstrated that the *C. vinosum* flavocytochrome possesses sulfide : cytochrome c oxidoreductase activity with horse heart cytochrome c as the electron acceptor. This model system now appears more relevant with the subsequent discovery of cytochrome c_{550}, an endogenous *C. vinosum* cytochrome similar to horse heart cytochrome c (see Section 3.4.3.2; 159, 191–193). There is evidence that the active species in the flavocytochrome c_{552}-catalyzed oxidation of sulfide is a complex between flavocytochrome c_{552} and cytochrome c held together by electrostatic forces (303, 303a–303c). As is the case in other cytochrome c–containing complexes, lysine residues on cytochrome c supply the positive charges involved in the complex formation (303b) (see Section 3.4.3.1 for a more detailed discussion of cytochrome c complexes). While the role of flavocytochrome c_{552} in sulfide oxidation seems well established, some uncertainty remains as to whether the oxidized product of this reaction is elemental sulfur (303) or thiosulfate (see the discussion in Ref. 287). Some evidence that sulfur is the product comes from the observation that the *C. vinosum* flavocytochrome (like the similar *Chl. limicola* enzyme) catalyzes the reduction of elemental sulfur to sulfide (294). The reduction of sulfur to sulfide is known to occur in *Chromatium* cells (304).

Photolithoautotrophic cells of *C. vinosum* have been shown to contain a soluble sulfite reductase that possesses siroheme and one or more iron-sulfur clusters as prosthetic groups (305–307). Although similar enzymes in nonphotosynthetic bacteria usually function in the assimilatory direction (i.e., the reduction of sulfite to sulfide), Trüper and co-workers have argued (287, 288) that in *C. vinosum* the enzyme functions in the dissimilatory direction. Consistent with such a role is the observation that the enzyme is absent in cells grown under photoheterotrophic conditions. The identity of the physiological electron acceptor for the sulfite reductase-catalyzed oxidation of sulfite is not known. Thus there appear to be at least two pathways for sulfide oxidation in *C. vinosum*: (a) oxidation to sulfite catalyzed by sulfite reductase, and (b) oxidation to elemental sulfur catalyzed by flavocytochrome c_{552}.

Sulfite formed during the dissimilatory oxidation of sulfide catalyzed by the siroheme-containing sulfite reductase can be oxidized to sulfate in a two-step process catalyzed by a membrane-bound APS (adenylsulfate) reductase (308) and ADP sulfurylase (288). This pathway results in the net formation of one high-energy phosphate bond, since the process utilizes AMP and yields ADP as a product. There also appears to be an alternative pathway for sulfite oxidation (288, 300) involving direct oxidation to sulfate without energy conservation.

The pathway(s) for thiosulfate oxidation in purple sulfur bacteria are less well understood than those for oxidation of sulfide and sulfite. Tetrathionate ($S_4O_6^{-2}$) has been reported as the oxidized product, but the pathway(s) for the conversion of tetrathionate to sulfite are unknown. There is also disagreement as to the identity of the immediate electron acceptor for thiosulfate oxidation. The soluble iron-sulfur protein HiPIP (309) and a membrane-bound cytochrome c (310) have been proposed as possible acceptors. Many purple sulfur bacteria can reductively cleave thiosulfate to sulfide *plus* sulfite (288). Whatever the detailed pathway of thiosulfate oxidation in *C. vinosum* is, it appears that electrons from this donor can reduce the soluble cytochrome c_{550} of the cyclic electron transport chain (159) and can be used (311) for the energy-dependent reduction of $NAD(P)^+$ (see Section 3.6.1.1).

3.7 CONCLUDING REMARKS

Traditionally, the photosynthetic bacteria have played an important role in the study of the mechanism of photosynthesis. Their importance was first realized as a result of the pioneering work of Van Niel and co-workers (312). In the years that followed, studies of bacterial photosynthesis, especially those with purple bacteria, have gained considerable momentum.

These studies have included processes that are as diverse as energy transfer and electronic interaction between pigments, electron transfer and tunneling, heme biochemistry, and sulfur and carbon metabolism. Some of these processes are unique to photosynthetic systems, others are common to a large variety of aerobic and anaerobic assimilatory and dissimilatory systems, but in general it can be said that their study contributed to a broader understanding than that of bacterial photosynthesis alone. In many instances, experiments with preparations from photosynthetic bacteria have yielded a more accurate fundamental description of photosynthetic processes at the molecular level than those with other systems.

Although many aspects had to be omitted or could be treated only in a general way, we hope that this chapter has provided a useful overview of the most important characteristics of these anaerobes, which have captured the fascination of many biochemists, biophysicists, and microbiologists.

REFERENCES

1. Trüper, H. G., and N. Pfennig. 1978. Taxonomy of the Rhodospirillales, p. 19–30, in R. K. Clayton and W. R. Sistrom (eds.), The photosynthetic bacteria. Plenum Press, New York.

1a. Imhoff, J. F., H. G. Trüper, and N. Pfennig. 1984. Rearrangement of the species and genera of the phototrophic 'purple nonsulfur bacteria.' Int. J. Syst. Bacteriol. 34:340–343.

1b. Imhoff, J. F. 1984. Reassignment of the genus *Ectothiorhodospira* Pelsh 1936 to a new family, Ectothiorhodospiraceae fam. nov., and amended description of the Chromatiaceae Bavendamm 1924. Int. J. Syst. Bacteriol. 34:338–339.

2. Pierson, B. K., and R. W. Castenholz. 1974. A phototrophic gliding filamentous bacterium of hot springs, *Chloroflexus aurantiacus* gen. and sp. nov. Arch. Mikrobiol. 100:5–24.

3. Pierson, B. K., and R. W. Castenholz. 1974. Studies of pigments and growth in *Chloroflexus aurantiacus*, a phototrophic filamentous bacterium. Arch. Mikrobiol. 100:283–305.

4. Drews, G. 1978. Structure and development of the membrane system of photosynthetic bacteria, p. 161–207, in D. R. Sanadi and L. P. Vernon (eds.), Current topics in bioenergetics, Vol. 8. Academic Press, New York.

5. Varga, A. R., and L. A. Staehelin. 1983. Spatial differentiation in photosynthetic and non-photosynthetic membranes of *Rhodopseudomonas palustris*. J. Bacteriol. 154:1414–1430.

6. Katz, J. J., H. H. Strain, A. L. Harkness, M. H. Studier, M. A. Svec, T. R. Janson, and B. T. Cope. 1972. Esterifying alcohols in the chlorophylls of purple photosynthetic bacteria. A new chlorophyll, bacteriochlorophyll (*g g*), all-trans geranylgeranyl bacteriochlorophyllide *a*. J. Am. Chem. Soc. 94:7938–7939.

7. Gloe, A., N. Pfennig, H. Brockmann, and W. Trowitsch. 1975. A new bacteriochlorophyll from brown-colored Chlorobiaceae. Arch. Microbiol. 102:103–109.

8. Smith, K. M., G. W. Craig, and L. A. Kehres. 1983. Reversed-phase high-performance liquid chromatography and structural assignments of the bacteriochlorophylls *c*. J. Chromatogr. 281:209–223.

9. Fowler, C. F., N. A. Nugent, and R. C. Fuller. 1971. The isolation and characterization of a photochemically active complex from *Chloropseudomonas ethylica*. Proc. Natl. Acad. Sci. USA 68:2278–2282.

10. Schmidt, K., M. Maarzahl, and F. Mayer. 1980. Development and pigmentation of chlorosomes in *Chloroflexus aurantiacus* strain OK-70-fl. Arch. Microbiol. 127:87–97.

11. Schmidt, K. 1980. A comparative study on the composition of chlorosomes (chlorobium vesicles) and cytoplasmic membranes from *Chloroflexus aurantiacus* strain Ok-70-fl and *Chlorobium limicola* f. *thiosulfatophilum* strain 6230. Arch. Microbiol. 124:21–31.

12. Gest, H., and J. L. Favinger. 1983. *Heliobacterium chlorum*, an anoxygenic brownish-green photosynthetic bacterium containing a 'new' form of bacteriochlorophyll. Arch. Microbiol. 136:11–16.

13. Brockmann, H., and A. Lipinski. 1983. Bacteriochlorophyll *g*. A new bacteriochlorophyll from *Heliobacterium chlorum*. Arch. Microbiol. 136:17–19.

13a. Van Dorssen, R. J., H. Vasmel, and J. Amesz. 1985. Antenna organization and energy transfer in membranes of *Heliobacterium chlorum*. Biochim. Biophys. Acta 809:199–203.

14. Cohen-Bazire, G., N. Pfennig, and R. Kunisaawa. 1964. The fine structure of green bacteria. J. Cell Biol. 22:207–225.

15. Madigan, M. T., and T. D. Brock. 1977. "Chlorobium-type" vesicles of photosynthetically grown *Chloroflexus aurantiacus* observed using negative staining techniques. J. Gen. Microbiol. 102:279–285.

16. Staehelin, L. A., J. R. Golecki, R. C. Fuller, and G. Drews. 1978. Visualization of the supramolecular architecture of chlorosomes (chlorobium type vesicles) in freeze-fractured cells of *Chloroflexus aurantiacus*. Arch. Microbiol. 119:269–277.

17. Staehelin, L. A., J. R. Golecki, and G. Drews. 1980. Supramolecular organization of chlorosomes (chlorobium vesicles) and of their membrane attachment sites in *Chlorobium limicola*. Biochim. Biophys. Acta 589:30–45.

18. Feick, R. G., H. Fitzpatrick, and R. C. Fuller. 1982. Isolation and characterization of cytoplasmic membranes and chlorosomes from the green bacterium *Chloroflexus aurantiacus*. J. Bacteriol. 150:905–915.

19. Cruden, D. L., and R. Y. Stanier. 1970. The characterization of chlorobium vesicles and membranes isolated from green bacteria. Arch. Mikrobiol. 72:115–134.

20. Gerola, P. D., and J. M. Olson. 1986. A new bacteriochlorophyll *a* protein complex associated with chlorosomes of green sulfur bacteria. Biochim. Biophys. Acta 848:69–76.

20a. Van Dorssen, R. J., P. D. Gerola, J. M. Olson, and J. Amesz. 1986. Optical and structural properties of chlorosomes of the photosynthetic green sulfur bacterium *Chlorobium limicola*. Biochim. Biophys. Acta 848:77–82.

21. Betti, J. A., R. E. Blankenship, L. V. Natarajan, L. C. Dickinson, and R. C. Fuller. 1982. Antenna organization and evidence for the function of a new antenna pigment species in the green photosynthetic bacterium *Chloroflexus aurantiacus*. Biochim. Biophys. Acta 680:194–201.

21a. Van Dorssen, R. J., H. Vasmel, and J. Amesz. 1986. Pigment organization and energy transfer in the green photosynthetic bacterium *Chloroflexus aurantiacus*. II. The chlorosome. Photosynth. Res. 9:33–45.

22. Broch-Due, M., J. G. Ormerod, and B. S. Fjerdingen. 1978. Effect of light intensity on vesicle formation in *Chlorobium*. Arch. Microbiol. 116:269–274.

23. Swarthoff, T., B. G. de Grooth, R. F. Meiburg, C. P. Rijgersberg, and J. Amesz. 1980. Orientation of pigments and pigment-protein complexes in the green photosynthetic bacterium *Prosthecochloris aestuarii*. Biochim. Biophys. Acta 593:51–59.

24. Smith, K. M., L. A. Kehres, and J. Fajer. 1983. Aggregation of the bacteriochlorophylls *c*, *d* and *e*. Models for the antenna chlorophyll of green and brown photosynthetic bacteria. J. Am. Chem. Soc. 105:1387–1389.

24a. Feick, R. G., and R. C. Fuller. 1984. Topography of the photosynthetic apparatus of *Chloroflexus aurantiacus*. Biochemistry 23:3693–3700.

24b. Wechsler, T., F. Suter, R. C. Fuller, and H. Zuber. 1984. The complete amino acid sequence of the bacteriochlorophyll *c* binding polypeptide from chlorosomes of the green photosynthetic bacterium *Chloroflexus aurantiacus*. FEBS Lett. 181:173–178.

25. Amesz, J., and H. Vasmel. 1986. Fluorescence properties of photosynthetic bacteria, p. 423–450 in Govindjee, J. Amesz, and D. C. Fork (eds.), Light emission by plants and bacteria. Academic Press, New York.

25a. Drews, G. 1985. Structure and functional organization of light-harvesting complexes and photochemical reaction centers in membranes of phototrophic bacteria. Microbiol. Rev. 49:59–70.

26. Cogdell, R. J., and J. P. Thornber. 1980. Light-harvesting pigment-protein complexes of purple photosynthetic bacteria. FEBS Lett. 122:1–8.

27. Broglie, R. M., C. N. Hunter, P. Delepelaire, R. A. Niederman, N.-H. Chua, and R. K. Clayton. 1980. Isolation and characterization of the pigment-protein complexes of *Rhodopseudomonas sphaeroides* by lithium dodecyl sulfate/polyacrylamide gel electrophoresis. Proc. Natl. Acad. Sci. USA 77:87–91.

28. Clayton, R. K., and B. J. Clayton. 1972. Relations between pigments and proteins in the photosynthetic membranes of *Rhodopseudomonas sphaeroides*. Biochim. Biophys. Acta 283:492–504.

29. Kramer, H. J. M., R. van Grondelle, C. N. Hunter, W. H. J. Westerhuis, and J. Amesz. 1984. Pigment organization of the B800-850 antenna complex of *Rhodopseudomonas sphaeroides*. Biochim. Biophys. Acta 765:156–165.

30. Kramer, H. J. M. 1984. Structural aspects of energy transfer in photosynthesis. Thesis, University of Leiden.

31. Feick, R., and G. Drews. 1978. Isolation and characterization of light harvesting bacteriochlorophyll. Protein complexes from *Rhodopseudomonas capsulata*. Biochim. Biophys. Acta 501:499–513.

32. Cogdell, R. J., and A. R. Crofts. 1978. Analysis of the pigment content of an antenna pigment-protein complex from three strains of *Rhodopseudomonas sphaeroides*. Biochim. Biophys. Acta 502:409–416.

33. Cogdell, R. J. and J. P. Thornber. 1979. The preparation and characterization of different types of light harvesting pigment-protein complexes from some purple bacteria, p. 61–79, in G. Wolstenholme and D. W. Fitzsimons (eds.), Chlorophyll organization and energy transfer in photosynthesis (Ciba Foundation Symposium No. 61, new series). Elsevier/Excerpta Medica, Amsterdam.

34. Sauer, K., and L. A. Austin. 1978. Bacteriochlorophyll-protein complexes from the light-harvesting antenna of photosynthetic bacteria. Biochemistry 17:2011–2019.

35. Cogdell, R. J., J. G. Lindsay, J. Valentine, and I. Durant. 1982. A further characterization of the B890 light-harvesting pigment-protein complex from *Rhodospirillum rubrum* strain S1. FEBS Lett. 150:151–154.

36. Picorel, R., G. Bélanger, and G. Gingras. 1983. Antenna holochrome B880 of *Rhodospirillum rubrum* S1. Pigment, phospholipid and polypeptide composition. Biochemistry 22:2491–2497.

37. Cogdell, R. J., I. Durant, J. Valentine, J. G. Lindsay, and K. Schmidt. 1983. The isolation and partial characterization of the light-harvesting pigment-protein complement of *Rhodopseudomonas acidophila*. Biochim. Biophys. Acta 722:427–435.

38. Hayashi, H., and S. Morita. 1980. Near-infrared absorption spectra of light harvesting bacteriochlorophyll protein complexes from *Chromatium vinosum*. J. Biochem. 88:1251–1258.

39. Steiner, R., and H. Scheer. 1984. Reversible pH induced absorption change in the BChl b-LH-protein of the alkalophile *Ectothiorhodospira halochloris*, p. 221–224, in C. Sybesma (ed.), Advances in photosynthesis research, Vol. 2. Martinus Nijhoff/ W. Junk Publishers, The Hague.

40. Amesz, J. 1978. Fluorescence and energy transfer, p. 333–340, in R. K. Clayton and W. R. Sistrom (eds.), The photosynthetic bacteria. Plenum Press, New York.

41. Vredenberg, W. J., and L. N. M. Duysens. 1963. Transfer of energy from bacteriochlorophyll to a reaction centre during bacterial photosynthesis. Nature (Lond.) 197:355–357.

42. Van Grondelle, R., C. N. Hunter, J. G. C. Bakker, and H. J. M. Kramer. 1983. Size and structure of antenna complexes of photosynthetic bacteria as studied by singlet-singlet quenching of the bacteriochlorophyll fluorescence yield. Biochim. Biophys. Acta 723:30–36.

42a. Vos, M., R. van Grondelle, F. W. van der Kooij, D. van de Poll, J. Amesz, and L. N. M. Duysens. 1986. Singlet-singlet annihilation at low temperatures in the antenna of purple bacteria. Biochim. Biophys. Acta 850:501–512.

43. Monger, T. G., and W. W. Parson. 1977. Singlet-triplet fusion in *Rhodopseudomonas sphaeroides* chromatophores. A probe of the organization of the photosynthetic apparatus. Biochim. Biophys. Acta 460:393–407.

44. Clayton, R. K., and B. J. Clayton. 1981. B850 pigment-protein complex of *Rhodopseudomonas sphaeroides*: extinction coefficients, circular dichroism, and the reversible binding of bacteriochlorophyll. Proc. Natl. Acad. Sci. USA 78:5583–5587.

45. Van Grondelle, R., H. J. M. Kramer, and C. P. Rijgersberg. 1982. Energy transfer in the B800-850-carotenoid light-harvesting complex of various mutants of *Rhodopseudomonas sphaeroides* and of *Rhodopseudomonas capsulata*. Biochim. Biophys. Acta 682:208–215.

46. Radcliffe, C. W., J. D. Pennoyer, R. M. Broglie, and R. A. Niederman. 1984. As-

sociations of pigment-proteins and phospholipids into specific domains in *Rhodopseu-domonas sphaeroides* photosynthetic membranes as determined by lithium dodecyl sulfate/polyacrylamide gel electrophoresis, p. 215–218, in C. Sybesma (ed.), Advances in photosynthesis research, Vol. 2. Martinus Nijhoff/W. Junk Publishers, The Hague.

47. Thornber, J. P., R. J. Cogdell, B. K. Pierson, and R. E. B. Seftor. 1983. Pigment-protein complexes of purple photosynthetic bacteria—an overview. J. Cell Biol. 23:159–169.

48. Cogdell, R. J., and J. Valentine. 1983. Bacterial photosynthesis. Photochem. Photobiol. 38:769–772.

49. Brunisholz, R. A., P. A. Cuendet, R. Theiler, and H. Zuber. 1981. The complete amino acid sequence of the single light harvesting protein from chromatophores of *Rhodospirillum rubrum* G-9+. FEBS Lett. 129:150–154.

50. Gogel, G. E., P. S. Parkes, P. A. Loach, R. A. Brunisholz, and H. Zuber. 1983. The primary structure of a light-harvesting bacteriochlorophyll binding protein of wild-type *Rhodospirillum rubrum*. Biochim. Biophys. Acta 746:32–39.

51. Tadros, M. H., F. Suter, G. Drews, and H. Zuber. 1983. The complete amino acid sequence of the large bacteriochlorophyll-binding polypeptide from the light-harvesting complex II (B800-850) of *Rhodopseudomonas capsulata*. Eur. J. Biochem. 129:533–536.

52. Tadros, M. H., F. Suter, H. H. Seydewitz, I. Witt, H. Zuber, and G. Drews. 1984. Isolation and complete amino-acid sequence of the small polypeptide from light-harvesting pigment-protein complex I (B870) of *Rhodopseudomonas capsulata*. Eur. J. Biochem. 138:209–212.

52a. Nozawa, T., M. Ohta, M. Hatano, H. Hayashi, and K. Shimada. 1985. Sequence homology and structural similarity among B870 (B890) polypeptides of purple photosynthetic bacteria and the mode of bacteriochlorophyll binding. Chem. Lett. 1985, 344–346.

52b. Brunisholz, R. A., F. Jay, F. Suter, and H. Zuber. 1985. The light-harvesting polypeptides of *Rhodopseudomonas viridis*. The complete amino-acid sequences of B1015-α, B1015-β and B1015-γ. Biol. Chem. Hoppe-Seyler 366:87–98.

52c. Theiler, R., F. Suter, V. Wiemken, and H. Zuber. 1984. The light-harvesting polypeptides of *Rhodopseudomonas sphaeroides* R-26.1. Isolation, purification and sequence analysis. Hoppe-Seyler's Z. Physiol. Chem. 365:703–719.

52d. Theiler, R., F. Suter, H. Zuber, and R. J. Cogdell. 1984. Comparison of the primary structures of the two B800-850-apoproteins from wild-type *Rhodopseudomonas sphaeroides* strain 2.4.1 and a carotenoidless mutant strain R 26. FEBS Lett. 175:231–237.

52e. Theiler, R., F. Suter, J. D. Pennoyer, H. Zuber, and R. A. Niederman. 1985. Complete amino acid sequence of the B875 light-harvesting protein of *Rhodopseudomonas sphaeroides* strain 2.4.1. Comparison with R 26.1 carotenoid-less mutant strain. FEBS Lett. 184:231–236.

52f. Youvan, D. C., and S. Ismail. 1985. Light-harvesting II (B800-850 complex) structural genes from *Rhodopseudomonas capsulata*. Proc. Natl. Acad. Sci. USA 82:58–62.

53. Nabedryk, E., and J. Breton. 1981. Orientation of intrinsic protein in photosynthetic

membranes. Polarized infrared spectroscopy of chloroplasts and chromatophores. Biochim. Biophys. Acta 635:515–524.

53a. Cogdell, R. J., and H. Scheer. 1985. Circular dichroism of light-harvesting complexes from purple photosynthetic bacteria. Photochem. Photobiol. 42:669–678.

54. Feick, R., and G. Drews. 1979. Protein subunits of bacteriochlorophylls B802 and B855 of the light-harvesting complex II of *Rhodopseudomonas capsulata*. Z. Naturforsch. C34:196–199.

55. Hunter, C. N., J. D. Pennoyer, and R. A. Niederman. 1982. Assembly and structural organization of pigment-protein complexes in membranes of *Rhodopseudomonas sphaeroides*, p. 257–265, in G. Akoyunoglou, A. E. Evangelopoulos, J. Georgatsos, G. Palaiologos, A. Trakatellis, and C. P. Tsiganos (eds.), Cell function and differentiation, Vol. B. Alan R. Liss, New York.

56. Bolt, J. D., K. Sauer, J. A. Shiozawa, and G. Drews. 1981. Linear and circular dichroism of membranes from *Rhodopseudomonas capsulata*. Biochim. Biophys. Acta 635:535–541.

57. Bolt, J. D., C. N. Hunter, R. A. Niederman, and K. Sauer. 1981. Linear and circular dichroism and fluorescence polarization of the B875 light-harvesting bacteriochlorophyll-protein complex from *Rhodopseudomonas sphaeroides*. Photochem. Photobiol. 34:653–656.

58. Bolt, J., and K. Sauer. 1979. Linear dichroism of light harvesting bacteriochlorophyll proteins from *Rhodopseudomonas sphaeroides* in stretched polyvinyl alcohol films. Biochim. Biophys. Acta 546:54–63.

59. Breton, J., and A. Vermeglio. 1982. Orientation of photosynthetic pigments *in vivo*, p. 153–194, in Govindjee (ed.), Photosynthesis, Vol. 1, Energy conversion by plants and bacteria. Academic Press, New York.

60. Breton, J., A. Vermeglio, M. Garrigos, and G. Paillotin. 1981. Orientation of chromophores in the antenna systems of *Rhodopseudomonas sphaeroides* in photosynthesis, p. 445–459, in G. Akoyunoglou (ed.), Proceedings of the 5th international photosynthesis congress, Vol. 3. Balaban Int. Science Services, Philadelphia.

60a. Kramer, H. J. M., J. D. Pennoyer, R. van Grondelle, W. H. J. Westerhuis, R. A. Niederman, and J. Amesz. 1984. Low-temperature optical properties and pigment organization of the B875 light-harvesting bacteriochlorophyll-protein complex of purple photosynthetic bacteria. Biochim. Biophys. Acta 767:335–344.

61. Fuller, R. C., R. E. Blankenship, and R. G. Feick. 1984. The molecular topography of the photochemical membrane system in the green bacterium *Chloroflexus*, p. 377–380, in C. Sybesma (ed.), Advances in photosynthesis research, Vol. 3. Martinus Nijhoff/W. Junk Publishers, The Hague.

61a. Wechsler, T., R. Brunisholz, F. Suter, R. C. Fuller, and H. Zuber. 1985. The complete amino acid sequence of a bacteriochorophyll *a* binding polypeptide isolated from the cytoplasmic membrane of the green photosynthetic bacterium *Chloroflexus aurantiacus*. FEBS Lett. 191:34–38.

62. Olson, J. M. 1981. Chlorophyll organization in green photosynthetic bacteria. Biochim. Biophys. Acta 549:33–51.

63. Matthews, B. E., R. E. Fenna, M. C. Bolognesi, M. F. Schmid, and J. M. Olson. 1979. Structure of a bacteriochlorophyll *a*-protein from the green photosynthetic bacterium *Prosthecochloris aestuarii*. J. Mol. Biol. 131:259–285.

63a. Tronrud, D. E., M. F. Schmid, and B. W. Matthews. 1986. Structure and X-ray amino acid sequence of a bacteriochlorophyll *a* protein from *Prosthecochloris aestuarii* refined at 1.9 Å resolution. J. Mol. Biol. 188:443–454.

64. Clayton, R. K., and B. J. Clayton. 1978. Molar extinction coefficients and other properties of an improved reaction center preparation from *Rhodopseudomonas viridis*. Biochim. Biophys. Acta 501:478–487.

65. Seftor, R. E. B., and J. P. Thornber. 1984. The photochemical reaction center of the bacteriochlorophyll *b*-containing organism *Thiocapsa pfennigii*. Biochim. Biophys. Acta 764:148–159.

66. Duysens, L. N. M. 1952. Transfer of excitation energy in photosynthesis. Thesis, University of Utrecht.

67. Reed, W., and R. K. Clayton. 1968. Isolation of a reaction center fraction from *Rhodopseudomonas spheroides*. Biochem. Biophys. Res. Commun. 30:471–475.

68. Slooten, L. 1972. Reaction center preparations of *Rhodopseudomonas spheroides*: energy transfer and structure. Biochim. Biophys. Acta 256:452–466.

69. Okamura, M. Y., L. A. Steiner, and G. Feher. 1974. Characterization of reaction centers from photosynthetic bacteria. I. Subunit structure of the protein mediating the primary photochemistry in *Rhodopseudomonas spheroides* R-26. Biochemistry 13:1394–1403.

70. Jolchine, G., and F. Reiss-Husson. 1975. Studies of pigments and lipids in *Rhodopseudomonas spheroides* Y reaction center. FEBS Lett. 52:33–36.

71. Nieth, K. F., G. Drews, and R. Feick. 1975. Photochemical reaction centers from *Rhodopseudomonas capsulata*. Arch. Microbiol. 105:43–45.

72. Clayton, B. J., and R. K. Clayton. 1978. Properties of photochemical reaction centers purified from *Rhodopseudomonas gelatinosa*. Biochim. Biophys. Acta 501:470–477.

73. Noël, H., M. van der Rest, and G. Gingras. 1972. Isolation and partial characterization of P870 reaction center complex from wild type *Rhodospirillum rubrum*. Biochim. Biophys. Acta 275:219–230.

74. Lin, L., and J. P. Thornber. 1975. Isolation and partial characterization of the photochemical reaction center of *Chromatium vinosum* (strain D). Photochem. Photobiol. 22:37–40.

75. Romijn, J. C., and J. Amesz. 1977. Purification and photochemical properties of reaction centers of *Chromatium vinosum*. Evidence for the photoreduction of a naphthoquinone. Biochim. Biophys. Acta 461:327–338.

76. Thornber, J. P., J. M. Olson, D. M. Williams, and M. L. Clayton. 1969. Isolation of the reaction center of *Rhodopseudomonas viridis*. Biochim. Biophys. Acta 172:351–354.

77. Clayton, R. K., and R. T. Wang. 1971. Photochemical reaction centers from *Rhodopseudomonas spheroides*, p. 696–704, in S. P. Colowick and N. O. Kaplan (eds.), Methods of enzymology, Vol. 23. Academic Press, New York.

78. Feher, G., and M. Y. Okamura. 1978. Chemical composition and properties of reaction centers, p. 349–386, in R. K. Clayton and W. R. Sistrom (eds.), The photosynthetic bacteria. Plenum Press, New York.

79. Okamura, M. Y., G. Feher, and N. Nelson. 1982. Reactions centers, p. 195–272, in Govindjee (ed.), Photosynthesis, Vol. 1, Energy conversion by plants and bacteria. Academic Press, New York.

80. Dutton, P. L., and J. B. Jackson. 1972. Thermodynamic and kinetic characterization of electron transport components *in situ* in *Rps. sphaeroides* and *R. rubrum*. Eur. J. Biochem. 30:495–510.

81. McElroy, J. C., G. Feher, and D. Mauzerall. 1972. Characterization of primary reactants in bacterial photosynthesis. I. Comparison of the light-induced EPR signal (g = 2.026) with that of a bacteriochlorophyll radical. Biochim. Biophys. Acta 267:363–374.

82. Clayton, R. K., and R. Haselkorn. 1972. Protein components of bacterial photosynthetic membranes. J. Mol. Biol. 68:97–105.

83. Agalidis, I., and F. Reiss-Husson. 1983. Several properties of the LM unit extracted with sodium dodecyl sulfate from *Rhodopseudomonas sphaeroides* purified reaction centers. Biochim. Biophys. Acta 724:340–351.

84. Feher, G., and M. Y. Okamura. 1984. Structure and function of the reaction center from *Rhodopseudomonas sphaeroides*, p. 155–164, in. C. Sybesma (ed.), Advances in photosynthesis research, Vol. 2. Martinus Nijhoff/W. Junk Publishers, The Hague.

85. Youvan, D. C., E. J. Bylina, M. Alberti, H. Megush, and J. E. Hearst. 1984. Nucleotide and deduced polypeptide sequences of the photosynthetic reaction-center, B870 antenna, and flanking polypeptides from *R. capsulata*. Cell 37:949–957.

86. Williams, J. C., L. A. Steiner, R. C. Ogden, M. I. Simon, and G. Feher. 1983. Primary structure of the M subunit of the reaction center from *Rhodopseudomonas sphaeroides*. Proc. Natl. Acad. Sci. USA 80:6505–6509.

86a. Williams, J. C., L. A. Steiner, G. Feher, and M. I. Simon. 1984. The primary structure of the L subunit of the reaction center from *Rhodospeudomonas sphaeroides*. Proc. Natl. Acad. Sci. USA 81:7303–7307.

86b. Deisenhofer, J., O. Epp, K. Miki, R. Huber, and H. Michel. 1985. Structure of the protein subunits in the photosynthetic reaction centre of *Rhodopseudomonas viridis* at 3 Å resolution. Nature (Lond.) 318:618–624.

86c. Michel, H., K. A. Weyer, H. Gruenberg, and F. Lottspeich. 1985. The 'heavy' subunit of the photosynthetic reaction centre from *Rhodopseudomonas viridis*: isolation of the gene, nucleotide and amino acid sequence. EMBO J. 4:1667–1672.

86d. Kyle, D. J. 1985. The 32 000 Dalton Q_B protein of photosystem II. Photochem. Photobiol. 47:107–116.

87. Michel, H. 1983. Crystallization of membrane proteins. Trends Biochem. Sci. 8:56–59.

88. Michel, H. 1982. Three-dimensional crystals of a membrane protein complex. The photosynthetic reaction centre from *Rhodopseudomonas viridis*. J. Mol. Biol. 158:567–572.

88a. Gast, P., and J. R. Norris. 1984. EPR detected triplet formation in a single crystal of a reaction center protein from the photosynthetic bacterium *Rhodopseudomonas sphaeroides* R-26. FEBS Lett. 177:277–280.

89. Zinth, W., W. Kaiser, and H. Michel. 1983. Efficient photochemical activity and strong dichroism of single crystals of reaction centres from *Rhodopseudomonas viridis*. Biochim. Biophys. Acta 723:128–131.

90. Gast, P., M. A. Wasielewski, M. Schiffer, and J. R. Norris. 1983. Orientation of the primary donor in single crystals of *Rhodopseudomonas viridis* reaction centers. Nature (Lond.) 305:451–452.

90a. Deisenhofer, J., O. Epp, K. Miki, R. Huber, and H. Michel. 1984. X-ray structure analsyis of a membrane protein complex. Electron density map at 3 Å resolution and a model of the chromophores of the photosynthetic reaction center from *Rhodopseudomonas viridis*. J. Mol. Biol. 180:385–398.

91. Norris, J. R., R. A. Uphaus, H. L. Crespi, and J. J. Katz. 1971. Electron spin resonance of chlorophyll and the origin of signal I in photosynthesis. Proc. Natl. Acad. Sci. USA 68:625–628.

92. Feher, G., A. J. Hoff, R. A. Isaacson, and L. C. Ackerson. 1975. ENDOR experiments on chlorophyll and bacteriochlorophyll *in vitro* and in the photosynthetic unit. Ann. N.Y. Acad. Sci. 244:239–259.

93. Norris, J. R., H. Scheer, and J. J. Katz. 1975. Models for antenna and reaction center chlorophylls. Ann. N.Y. Acad. Sci. 244:260–280.

94. Lubitz, W., F. Lendzian, H. Scheer, J. Gottstein, M. Plato, and K. Möbius. 1984. Structural studies of the primary cation radical P^+870 in reaction centers of *Rhodospirillum rubrum* by electron-nuclear double resonance in solution. Proc. Natl. Acad. Sci. USA 81:1401–1405.

95. Den Blanken, H. J., and A. J. Hoff. 1982. High-resolution optical absorption-difference spectra of the triplet state of the primary donor in isolated reaction centers of the photosynthetic bacteria *Rhodopseudomonas sphaeroides* R-26 and *Rhodopseudomonas viridis* measured with optically detected magnetic resonance at 1.2 K. Biochim. Biophys. Acta 681:365–374.

96. O'Malley, P. J., and G. T. Babcock. 1984. The monomeric nature of $P700^+$ as revealed by ENDOR spectroscopy. Proc. Natl. Acad. Sci. USA 81:1098–1101.

96a. Breton, J. 1985. Orientation of the chromophores in the reaction center of *Rhodopseudomonas viridis*. Comparison of low-temperature linear dichroism spectra with a model derived from X-ray crystallography. Biochim. Biophys. Acta 810:235–245.

96b. Shuvalov, V. A., and L. N. M. Duysens. 1986. The primary electron transfer reactions in modified reaction centers from *Rhodopseudomonas sphaeroides*. Proc. Natl. Acad. Sci. USA 83:1690–1694.

96c. Maroti, P., C. Kirmaier, C. Wraight, D. Holten, and R. M. Pearlstein. 1985. Photochemistry and electron transfer in borohydride-treated photosynthetic reaction centers. Biochim. Biophys. Acta 810:132–139.

96d. Shuvalov, V. A., A. Y. Shkuropatov, C. M. Kulakova, M. A. Ismailov, and V. A. Shkuropatova. 1986. Photoreactions of bacteriopheophytins and bacteriochlorophylls in reaction centers of *Rhodopseudomonas sphaeroides* and *Chloroflexus aurantiacus*. Biochim. Biophys. Acta 849:337–346.

97. Holt, A. S., and R. K. Clayton. 1965. Light-induced absorbancy changes in Eimhjellen's *Rhodopseudomonas*. Photochem. Photobiol. 4:829–831.

98. Hoff, A. J. 1982. Photooxidation of the reaction center chlorophylls and structural properties, p. 80–151, in F. K. Fong (ed.), Light reaction path of photosynthesis. Springer-Verlag, Heidelberg.

98a. Knapp, E. W., S. F. Fischer, W. Zinth, M. Sander, W. Kaiser, J. Deisenhofer, and H. Michel. 1985. Analysis of optical spectra from single crystals of *Rhodopseudomonas viridis* reaction centers. Proc. Natl. Acad. Sci. USA 82:8436–8467.

99. Parson, W. W., R. K. Clayton, and R. J. Cogdell. 1975. Excited states of photosynthetic reaction centers at low redox potentials. Biochim. Biophys. Acta 387:265–278.

100. Rockley, M. G., M. W. Windsor, R. J. Cogdell, and W. W. Parson. 1975. Picosecond detection of an intermediate in the photochemical reaction of bacterial photosynthesis. Proc. Natl. Acad. Sci. USA 72:2251–2255.

101. Kaufmann, K. J., K. M. Petty, P. L. Dutton, and P. M. Rentzepis. 1976. Picosecond kinetics of reaction centers of *Rhodopseudomonas sphaeroides* and the effect of ubiquinone extraction and reconstitution. Biochem. Biophys. Res. Commun. 70:839–845.

101a. Martin, J.-L., J. Breton, A. J. Hoff, A. Migus, and A. Antonetti. 1986. Femtosecond spectroscopy of electron transfer in the reaction center of the photosynthetic bacterium *Rhodopseudomonas sphaeroides* R-26: direct electron transfer from the dimeric bacteriochlorophyll primary donor to the bacteriopheophytin acceptor with a time constant of 2.8 ± 0.2 psec. Proc. Natl. Acad. Sci. USA 83:957–981.

101b. Woodbury, N. W., M. Becker, D. Middendorf, and W. W. Parson. 1985. Picosecond kinetics of the initial photochemical electron-transfer reaction in bacterial photosynthetic reaction centers. Biochemistry 24:7516–7521.

102. Fajer, J., D. C. Brune, M. S. Davis, A. Foreman, and L. D. Spaulding. 1975. Primary charge separation in bacterial photosynthesis: oxidized chlorophyll and reduced pheophytin. Proc. Natl. Acad. Sci. USA 72:4956–4960.

103. Shuvalov, V. A., and A. V. Klevanik. 1983. The study of the state $[P870^+B800^-]$ in bacterial reaction centers by selective picosecond and low-temperature spectroscopies. FEBS Lett. 160:51–55.

104. Holten, D., C. Hoganson, M. W. Windsor, C. C. Schenck, W. W. Parson, A. Migus, R. L. Fork, and C. V. Shank. 1980. Subpicosecond and picosecond studies of electron transfer intermediates in *Rhodopseudomonas sphaeroides* reaction centers. Biochim. Biophys. Acta 592:461–477.

105. Shuvalov, V. A., J. Amesz, and L. N. M. Duysens. 1986. Picosecond charge separation upon selective excitation of the primary electron donor in reaction centers of *Rhodopseudomonas viridis*. Biochim. Biophys. Acta 851:327–330.

105a. Wasielewski, M. R., and D. M. Tiede. 1986. Sub-picosecond measurements of primary electron transfer in *Rhodopseudomonas viridis* reaction centers using near-infrared excitation. FEBS Lett. 204:368–372.

106. Parson, W. W., and B. Ke. 1982. Primary photochemical reactions, p. 331–385, in Govindjee (ed.), Photosynthesis, Vol. 1, Energy conversion by plants and bacteria. Academic Press, New York.

107. Monger, T. G., R. J. Cogdell, and W. W. Parson. 1976. Triplet states of bacteriochlorophyll and carotenoids in chromatophores of photosynthetic bacteria. Biochim. Biophys. Acta 449:136–153.

108. Dutton, P. L., J. S. Leigh, and M. Seibert. 1972. Primary processes in photosynthesis: *in situ* ESR studies on the light-induced oxidized and triplet state of reaction center bacteriochlorophyll. Biochem. Biophys. Res. Commun. 46:406–413.

109. Hoff, A. J. 1981. Magnetic field effects on photosynthetic reactions. Q. Rev. Biophys. 14:599–665.

110. Tiede, D. M., R. C. Prince, and P. L. Dutton. 1976. EPR and optical spectroscopic properties of the electron carrier intermediate between the reaction center bacteriochlorophylls and the primary acceptor in *Chromatium vinosum*. Biochim. Biophys. Acta 449:447–467.

111. Van Grondelle, R., J. C. Romijn, and N. G. Holmes. 1976. Photoreduction of the long wavelength bacteriopheophytin in reaction centers and chromatophores of the photosynthetic bacterium *Chromatium vinosum*. FEBS Lett. 72:187–192.

112. Shuvalov, V. A., I. N. Krakhmaleva, and V. V. Klimov. 1976. Photooxidation of P-960 and photoreduction of P-800 (bacteriopheophytin B-800) in reaction centers from *Rhodopseudomonas viridis*. Biochim. Biophys. Acta 449:597–601.

113. Pucheu, N. L., N. L. Kerber, and A. F. Garcia. 1976. Isolation and purification of reaction center of *Rhodopseudomonas viridis* NHTC 133 by means of LDAO. Arch. Microbiol. 109:301–305.

114. Clayton, R. K., and S. C. Straley. 1970. An optical absorption change that could be due to reduction of the primary photochemical electron acceptor in photosynthetic reaction centers. Biochem. Biophys. Res. Commun. 39:1114–1119.

115. Slooten, L. 1972. Electron acceptors in reaction center preparations from photosynthetic bacteria. Biochim. Biophys. Acta 275:208–218.

116. Vermeglio, A., and R. K. Clayton. 1977. Kinetics of electron transfer between the primary and the secondary electron acceptor in reaction centers from *Rhodopseudomonas sphaeroides*. Biochim. Biophys. Acta 461:159–165.

117. Prince, R. C., and P. L. Dutton. 1978. Protonation and the reducing potential of the primary electron acceptor, p. 439–453, in R. K. Clayton and W. R. Sistrom (eds.), The photosynthetic bacteria. Plenum Press, New York.

118. Prince, R. C., J. S. Leigh, and P. L. Dutton. 1976. Thermodynamic properties of the reaction center of *Rhodopseudomonas viridis*. In vivo measurement of the reaction center bacteriochlorophyll-primary acceptor intermediary electron carrier. Biochim. Biophys. Acta 440:623–636.

119. Prince, R. C., and J. P. Thornber. 1977. A novel paramagnetic resonance signal associated with the 'primary' electron acceptor in isolated photochemical reaction centers of *Rhodospirillum rubrum*. FEBS Lett. 81:233–237.

120. Loach, P. A., and R. L. Hall. 1972. The question of the primary electron acceptor in bacterial photosynthesis. Proc. Natl. Acad. Sci. USA 69:786–790.

121. Feher, G., M. Y. Okamura, and J. D. McElroy. 1972. Identification of an electron acceptor in reaction centers of *Rhodopseudomonas spheroides* by EPR spectroscopy. Biochim. Biophys. Acta 267:222–226.

122. Lubitz, W., E. C. Abresch, R. J. Debus, R. A. Isaacson, M. Y. Okamura, and G. Feher. 1985. Electron nuclear double resonance of semiquinones in reaction centers of *Rhodopseudomonas sphaeroides*. Biochim. Biophys. Acta 808:464–469.

123. Debus, R. J., G. Feher, and M. Y. Okamura. 1986. Iron-depleted reaction centers from *Rhodopseudomonas sphaeroides* R-26.1: characterization and reconstitution with Fe^{2+}, Mn^{2+}, Co^{2+}, Ni^{2+}, Cu^{2+} and Zn^{2+}. Biochemistry 25:2276–2287.

123a. Rutherford, A. W., I. Agalidis, and F. Reiss-Husson. 1985. Manganese-quinone interactions in the electron acceptor region of bacterial photosynthetic reaction centres. FEBS Lett. 182:151–157.

124. Boso, B., P. Debrunner, M. Y. Okamura, and G. Feher. 1981. Mössbauer spectroscopy of photosynthetic reaction centers from *Rhodopseudomonas sphaeroides* R-26. Biochim. Biophys. Acta 638:173–177.

125. Butler, W. F., D. J. Johnston, H. B. Shore, D. R. Fredkin, M. Y. Okamura, and G. Feher. 1980. The electronic structure of Fe^{2+} in reaction centers from *Rhodopseu-

domonas sphaeroides. I. Static magnetization measurements. Biophys. J. 32:967–992.

126. Bunker, G., E. A. Stern, R. E. Blankenship, and W. W. Parson. 1982. An X-ray absorption study of the iron site in bacterial photosynthetic reaction centers. Biophys. J. 37:539–551.

127. Eisenberger, P., M. Y. Okamura, and G. Feher. 1982. The electronic structure of Fe^{2+} in reaction centers from *Rhodopseudomonas sphaeroides*. II. Extended X-ray fine structure studies. Biophys. J. 37:523–538.

128. Wraight, C. A. 1979. Electron acceptors of bacterial photosynthetic reaction centers. II. H^+ binding coupled to secondary electron transfer in the quinone acceptor complex. Biochim. Biophys. Acta 548:309–327.

129. Cogdell, R. J., D. C. Brune, and R. K. Clayton. 1974. Effects of extraction and replacement of ubiquinone upon the photochemical activity of reaction centers and chromatophores from *Rhodopseudomonas sphaeroides*. FEBS Lett. 45:344–347.

130. Okamura, M. Y., R. A. Isaacson, and G. Feher. 1975. Primary acceptor in bacterial photosynthesis: obligatory role of ubiquinone in photoactive reaction centers of *Rhodopseudomonas spheroides*. Proc. Natl. Acad. Sci. USA 72:3491–3495.

131. Sybesma, C., and W. J. Vredenberg. 1963. Evidence for a reaction center P840 in the green photosynthetic bacterium *Chloropseudomonas ethylicum*. Biochim. Biophys. Acta 75:439–441.

132. Olson, J. M., K. D. Philipson, and K. Sauer. 1973. Circular dichroism and absorption spectra of bacteriochlorophyll-protein and reaction center complexes from *Chlorobium thiosulfatophilum*. Biochim. Biophys. Acta 292:206–217.

133. Swarthoff, T., and J. Amesz. 1979. Photochemically active pigment-protein complexes from the green photosynthetic bacterium *Prosthecochloris aestaurii*. Biochim. Biophys. Acta. 548:427–432.

134. Vasmel, H., T. Swarthoff, H. J. M. Kramer, and J. Amesz. 1983. Isolation and properties of a pigment-protein complex associated with the reaction center of the green photosynthetic sulfur bacterium *Prosthecochloris aestuarii*. Biochim. Biophys. Acta 723:361–367.

135. Hurt, E. C. and G. Hauska. 1984. Purification of membrane-bound cytochromes and a photoactive P840 protein complex of the green sulfur bacterium *Chlorobium limicola* f. *thiosulfatophilum*. FEBS Lett. 168:149–154.

136. Swarthoff, T., K. M. van der Veek-Horsley, and J. Amesz. 1981. The primary charge separation, cytochrome oxidation and triplet formation in preparations from the green photosynthetic bacterium *Prosthecochloris aestuarii*. Biochim. Biophys. Acta 635:1–12.

137. Swarthoff, T., J. Amesz, H. J. M. Kramer, and C. P. Rijgersberg. 1981. The reaction center and antenna pigments of green photosynthetic bacteria. Isr. J. Chem. 21:332–337.

138. Prince, R. C., and J. M. Olson. 1976. Some thermodynamic and kinetic properties of the primary photochemical reactants in a complex from a green photosynthetic bacterium. Biochim. Biophys. Acta 423:357–362.

139. Jennings, J. V., and M. C. W. Evans. 1977. The irreversible photoreduction of a low potential component at low temperatures in a preparation of the green photosynthetic bacterium *Chlorobium thiosulfatophilum*. FEBS Lett. 75:33–36.

140. Olson, J. M., R. C. Prince, and D. C. Brune. 1977. Reaction center complexes from green bacteria. Brookhaven Symp. Biol. 28:238–245.

141. Swarthoff, T., P. Gast, and A. J. Hoff. 1981. Photooxidation and triplet formation of the primary electron donor of the green photosynthetic bacterium *Prosthecochloris aestuarii*, observed with EPR spectroscopy. FEBS Lett. 127:83–86.

142. Wasielewski, M. R., U. H. Smith, and J. R. Norris. 1982. ESR study of the primary electron donor in highly ^{13}C-enriched *Chlorobium limicola* f. *thiosulfatophilum*. FEBS Lett. 149:138–140.

143. Swarthoff, T., P. Gast, A. J. Hoff, and J. Amesz. 1981. An optical and ESR investigation on the acceptor side of the reaction center of the green photosynthetic bacterium *Prosthecochloris aestuarii*. FEBS Lett. 130:93–98.

144. Nuijs, A. M., H. Vasmel, H. L. P. Joppe, L. N. M. Duysens, and J. Amesz. 1985. Excited states and primary charge separation in the pigment system of the green photosynthetic bacterium *Prosthecochloris aestuarii* as studied by picosecond absorbance difference spectroscopy. Biochim. Biophys. Acta 807:24–34.

145. Shuvalov, V. A., J. Amesz, and L. N. M. Duysens. 1986. Picosecond spectroscopy of isolated membranes of the photosynthetic green sulfur bacterium *Prosthecochloris aestuarii* upon selective excitation of the primary electron donor. Biochim. Biophys. Acta 851:1–5.

145a. Braumman, T., H. Vasmel, L. H. Grimme, and J. Amesz. 1986. Pigment composition of the photosynthetic membrane and reaction center of the green bacterium *Prosthecochloris aestuarii*. Biochim. Biophys. Acta 848:83–91.

145b. Fuller, R. C., S. G. Sprague, H. Gest, and R. E. Blankenship. 1985. Unique photosynthetic reaction center from *Heliobacterium chlorum*. FEBS Lett. 182:345–349.

145c. Prince, R. C., H. Gest, and R. E. Blankenship. 1985. Thermodynamic properties of the photochemical reaction center of *Heliobacterium chlorum*. Biochim. Biophys. Acta 810:377–384.

145d. Nuijs, A. M., R. J. van Dorssen, L. N. M. Duysens, and J. Amesz. 1985. Excited states and primary photochemical reactions in the photosynthetic bacterium *Heliobacterium chlorum*. Proc. Natl. Acad. Sci. USA 82:6865–6868.

145e. Brok, M., H. Vasmel, J. T. G. Horikx, and A. J. Hoff. 1986. Electron transport components of *Heliobacterium chlorum* investigated by EPR spectroscopy at 9 and 35 GHz. FEBS Lett. 194:322–326.

145f. Smit, H. W. J., J. Amesz, M. F. R. van der Hoeven, and L. N. M. Duysens. 1987. Electron transport in *Heliobacterium chlorum*, p. 189–192, in J. Biggins (ed.), Progress in photosynthesis research, Vol. 1. Martinus Nijhoff, Dordrecht.

146. Amesz, J. 1983. Photosynthesis. Photosystems in green plants and green bacteria. Prog. Bot. 45:89–105.

146a. Blankenship, R. E. 1984. Primary photochemistry in green photosynthetic bacteria. Photochem. Photobiol. 40:801–806.

146b. Blankenship, R. E. 1985. Electron transport in green photosynthetic bacteria. Photosynth. Res. 6:317–333.

147. Stackebrandt, E., and C. R. Woese. 1981. The evolution of prokaryotes. Symp. Soc. Gen. Microbiol. 32:1–31.

148. Bruce, B. D., R. C. Fuller, and R. E. Blankenship. 1982. Primary photochemistry

in the facultatively aerobic green photosynthetic bacterium *Chloroflexus aurantiacus*. Proc. Natl. Acad. Sci. USA 79:6532–6536.

149. Vasmel, H., and J. Amesz. 1983. Photoreduction of menaquinone in the reaction center of the green photosynthetic bacterium *Chloroflexus aurantiacus*. Biochim. Biophys. Acta 724:118–122.

150. Vasmel, H., R. F. Meiburg, H. J. M. Kramer, L. J. de Vos, and J. Amesz. 1983. Optical properties of the photosynthetic reaction center of *Chloroflexus aurantiacus* at low temperature. Biochim. Biophys. Acta 724:333–339.

151. Pierson, B. K., and J. P. Thornber. 1983. Isolation and spectral characterization of photochemical reaction centers from the thermophilic green bacterium *Chloroflexus aurantiacus* strain J-10-fl. Proc. Natl. Acad. Sci. USA 80:80–84.

151a. Vasmel, H., J. Amesz, and A. J. Hoff. 1986. Analysis by exciton theory of the optical properties of the reaction center of *Chloroflexus aurantiacus*. Biochim. Biophys. Acta 852:159–168.

152. Kirmaier, C., D. Holten, R. Feick, and R. E. Blankenship. 1983. Picosecond measurements of the primary photochemical events in reaction centers isolated from the facultative green photosynthetic bacterium *Chloroflexus aurantiacus*. Comparison with the purple bacterium *Rhodopseudomonas sphaeroides*. FEBS Lett. 158:73–78.

153. Blankenship, R. E., R. Feick, B. C. Bruce, C. Kirmaier, D. Holten, and R. C. Fuller. 1983. Primary photochemistry in the facultative green photosynthetic bacterium *Chloroflexus aurantiacus*. J. Cell Biol. 22:251–261.

154. Hale, M. B., R. E. Blankenship, and R. C. Fuller. 1983. Menaquinone is the sole quinone in the facultatively aerobic green photosynthetic bacterium *Chloroflexus aurantiacus*. Biochim. Biophys. Acta 723:376–382.

155. Tsukihara, T., K. Fukiyama, H. Tahara, Y. Kotsube, Y. Matsuura, N. Tanaka, M. Kakudo, K. Wada, and H. Matsubara. 1978. X-ray analysis of ferredoxin from *Spirulina platensis*. II. Chelate structure of active center. J. Biochem. 84:1645–1647.

156. Salemme, F. R. 1977. Structure and function of cytochrome *c*. Annu. Rev. Biochem. 46:299–329.

157. Horio, T., and M. D. Kamen. 1961. Preparations and properties of three pure crystalline haem proteins. Biochim. Biophys. Acta 48:266–286.

158. Van Grondelle, R., L. N. M. Duysens, and H. N. van der Wal. 1976. Function of three cytochromes in photosynthesis of whole cells of *Rhodospirillum rubrum* as studied by flash spectroscopy. Biochim. Biophys. Acta 441:169–187.

159. Van Grondelle, R., L. N. M. Duysens, J. A. van der Wel, and H. N. van der Wal. 1977. Function and properties of a soluble *c*-type cytochrome *c*-551 in secondary electron transport in whole cells of *Chromatium vinosum* as studied with flash spectroscopy. Biochim. Biophys. Acta 461:188–201.

160. Cramer, W. A., and A. R. Crofts. 1982. Electron and proton transport, p. 387–467, in Govindjee (ed.), Photosynthesis, Vol. 1, Energy conversion by plants and bacteria. Academic Press, New York.

161. Crofts, A. R., and C. A. Wraight. 1983. The electrochemical domain of photosynthesis. Biochim. Biophys. Acta 726:149–185.

162. Takamiya, K., and P. L. Dutton. 1979. Ubiquinone in *Rhodopseudomonas sphaeroides*. Some thermodynamic properties. Biochim. Biophys. Acta 546:1–16.

163. Takamiya, K.-I., R. C. Prince, and P. L. Dutton. 1979. The recognition of a special ubiquinone functionally central in the ubiquinone-cytochrome b-c_2 oxidoreductase. J. Biol. Chem. 254:11307–11311.

164. Yu, C.-A., and L. Yu. Ubiquinone binding protein in the cytochrome b-c_1 complex, p. 333–350, in B. L. Trumpower (ed.), Function of quinones in energy conserving systems. Academic Press, New York.

165. Rutherford, A. W., and M. C. W. Evans. 1979. Direct measurement of the redox potential of the primary and secondary quinone acceptors in *Rhodopseudomonas sphaeroides* (wild type) by EPR spectrometry. FEBS Lett. 110:257–261.

166. Rutherford, A. W., and M. C. W. Evans. 1979. The high potential semiquinone-iron signal in *Rhodopseudomonas viridis* is the specific quinone secondary electron acceptor in the photosynthetic reaction centre. FEBS Lett. 104:227–230.

167. Rutherford, A. W., P. Heathcote, and M. C. W. Evans. 1979. Electron-paramagnetic-resonance measurements of the electron-transfer components of the reaction centre of *Rhodopseudomonas viridis*. Biochem J. 182:515–523.

168. Vermeglio, A. 1977. Secondary electron transfer in reaction centers of *Rhodopseudomonas sphaeroides*. Out-of-phase periodicity of two for the formation of ubisemiquinone and fully reduced ubiquinone. Biochim. Biophys. Acta 459:516–524.

169. Wraight, C. A. 1982. The involvement of stable semiquinone in the two-electron gates of plant and bacterial photosystems, p. 181–198, in B. L. Trumpower (ed.), Function of quinones in energy conserving systems. Academic Press, New York.

170. Barouch, Y., and R. K. Clayton. 1977. Ubiquinone reduction and proton uptake by chromatophores of *Rhodopseudomonas sphaeroides* R-26. Periodicity of two in consecutive light flashes. Biochem. Biophys. Acta 462:785–788.

171. Parson, W. W. 1978. Quinones as secondary electron acceptors, p. 455–469, in R. K. Clayton and W. R. Sistrom (eds.), The photosynthetic bacteria. Plenum Press, New York.

172. Hauska, G., and Hurt, E. 1982. Pool function and mobility of isoprenoid quinones, p. 87–110, in B. L. Trumpower (ed.), Function of quinones in energy conserving systems. Academic Press, New York.

173. Bowyer, J. F., A. Baccarini-Melandri, B. Melandri, and A. R. Crofts. 1978. The role of ubiquinone-10 in cyclic electron transport in *Rps. capsulata* Ala pho$^+$: Effects of lyophilization and extraction. Z. Naturforsch. C33:704–711.

174. Baccarini-Melandri, A., and B. A. Melandri. 1977. A role for ubiquinone-10 in the b-c_2 segment of the photosynthetic bacterial electron transport chain. FEBS Lett. 80:459–464.

175. Baccarini-Melandri, A., N. Gabellini, and B. A. Melandri. The multifarious role of ubiquinone in bacterial chromatophores, p. 285–298, in B. L. Trumpower (ed.), Function of quinones in energy conserving systems. Academic Press, New York.

176. Bartsch, R. G. 1978. Cytochromes, p. 249–279, in R. K. Clayton and W. R. Sistrom (eds.), The photosynthetic bacteria. Plenum Press, New York.

177. Prince, R. C., A. Baccarini-Melandri, G. A. Hauska, B. A. Melandri, and A. R. Crofts. 1975. Asymmetry of an energy transducing membrane. The location of cytochrome c_2 in *Rps. sphaeroides* and *Rps. capsulata*. Biochim. Biophys. Acta 387:212–227.

178. Dutton, P. L., K. M. Petty, and S. D. Morse. 1975. Cytochrome c_2 and reaction

center of *Rhodopseudomonas spheroides* Ga membranes. Extinction coefficients, content, half-reduction potentials, kinetics and electric field alterations. Biochim. Biophys. Acta 387:536–556.

179. Bowyer, J. R., G. V. Tierney, and A. R. Crofts. 1979. Cytochrome c_2-reaction centre coupling in chromatophores of *Rhodopseudomonas sphaeroides* and *Rhodopseudomonas capsulata*. FEBS Lett. 101:207–212.

180. Overfield, R. E., C. A. Wraight, and D. DeVault. 1979. Microsecond photooxidation kinetics of cytochrome c_2 from *Rhodopseudomonas sphaeroides*: *in vivo* and solution kinetics. FEBS Lett. 105:137–142.

181. Rosen, D., M. Y. Okamura, and G. Feher. 1980. Interaction of cytochrome *c* with reaction centers of *Rhodopseudomonas sphaeroides* R-26: determination of number of binding sites and dissociation constants by equilibrium dialysis. Biochemistry 19:5687–5692.

182. Overfield, R. E., and C. A. Wraight. 1980. Oxidation of cytochromes *c* and c_2 by bacterial photosynthetic reaction centers in phospholipid vesicles. 1. Studies with neutral membranes. Biochemistry 19:3322–3327.

183. Overfield, R. E., and C. A. Wraight. 1980. Oxidation of cytochromes *c* and c_2 by bacterial photosynthetic reaction centers in phospholipid vesicles. 2. Studies with negative membranes. Biochemistry 19:3328–3334.

184. Rosen, D., M. Y. Okamura, E. C. Abresch, G. E. Valkirs, and G. Feher. 1983. Interaction of cytochrome *c* with reaction centers of *Rps. sphaeroides* R-26: localization of the binding site by chemical cross-linking and immunochemical studies. Biochemistry 22:335–341.

185. Margoliash, E., and H. R. Bosshard. 1983. Guided by electrostatics, a textbook protein comes of age. Trends Biol. Sci. 8:316–320.

186. Salemme, F. R., S. T. Freer, N. H. Xuong, R. A. Alden, and J. Kraut. 1973. The structure of oxidized cytochrome c_2 from *R. rubrum*. J. Biol. Chem. 248:3910–3921.

186a. Rieder, R., V. Wiemken, R. Bachofen, and H. R. Bosshard. 1985. Binding of cytochrome c_2 to the isolated reaction center of *Rhodospirillum rubrum* involves the 'backside' of cytochrome c_2. Biochem. Biophys. Res. Commun. 128:120–126.

186b. Daldal, F., S. Cheng, J. Applebaum, E. Davidson, and R. C. Prince. 1986. Cytochrome c_2 is not essential for photosynthetic growth of *Rhodopseudomonas capsulata*. Proc. Natl. Acad. Sci. USA 83:2012–2016.

187. Dutton, P. L., and R. C. Prince. 1978. Reaction-center-driven cytochrome interactions in electron and proton translocation and energy coupling, p. 525–570, in R. K. Clayton and W. R. Sistrom (eds.). The photosynthetic bacteria. Plenum Press, New York.

188. Dutton, P. L. 1971. Oxidation-reduction potential dependence of the interactions of cytochromes, bacteriochlorophyll and carotenoids at 77K in chromatophores of *C. vinosum* and *Rps. gelatinosa*. Biochim. Biophys. Acta 226:63–80.

189. DeVault, D. 1980. Quantum mechanical tunneling in biological systems. Q. Rev. Biophys. 13:387–564.

190. Parson, W. W. 1968. The role of P870 in bacterial photosynthesis. Biochim. Biophys. Acta 153:248–259.

191. Parson, W. W. 1969. Cytochrome photooxidation in *C. vinosum* chromatophores.

Each P870 oxidizes two cytochrome c_{422} hemes. Biochim. Biophys. Acta 189:387–403.

191a. Shill, D. A., and P. M. Wood. 1984. A role for cytochrome c_2 in *Rhodopseudomonas viridis*. Biochim. Biophys. Acta 764:1–7.

192. Gray, G. O., D. F. Gaul, and D. B. Knaff. 1983. Partial purification and characterization of two soluble *c*-type cytochromes from *Chromatium vinosum*. Arch. Biochem. Biophys. 222:78–86.

193. Tomiyama, Y., M. Doi, K.-I. Takamiya, and M. Nishimura. 1983. Isolation, purification, and some properties of cytochrome *c*-551 from *Chromatium vinosum*. Plant Cell Physiol. 24:11–16.

194. Knaff, D. B., R. Whetstone, and J. W. Carr. 1980. The role of soluble cytochrome c_{551} in cyclic-electron flow-driven active transport in *Chromatium vinosum*. Biochim. Biophys. Acta 590:50–58.

195. Knaff, D. B. 1978. Active transport in the photosynthetic bacterium *Chromatium vinosum*. Arch. Biochem. Biophys. 189:225–230.

196. Knaff, D. B., and V. L. Davidson. 1982. Light-dependent active transport in prokaryotes. Photochem. Photobiol. 36:721–724.

197. Doi, M., K.-I. Takamiya, and M. Nishimura. 1983. Isolation and properties of membrane-bound cytochrome *c*-552 from the photosynthetic bacterium *Chromatium vinosum*. Photosynth. Res. 4:49–60.

198. Gaul, D. F., and D. B. Knaff. 1983. The presence of cytochrome c_1 in the purple sulfur bacterium *Chromatium vinosum*. FEBS Lett. 162:69–75.

199. Knaff, D. B., T. M. Worthington, C. C. White, and R. Malkin. 1979. A partial purification of membrane-bound *b* and *c* cytochromes from *Chromatium vinosum*. Arch. Biochem. Biophys. 192:158–163.

200. Knaff, D. B., and S. Kraichoke. 1983. Oxidation-reduction and EPR properties of a cytochrome complex from *Chromatium vinosum*. Photochem. Photobiol. 37:243–246.

201. Thornber, J. P., R. J. Cogdell, R. E. B. Seftor, and G. D. Webster. 1980. Further studies on the composition and spectral properties of the photochemical reaction centers of bacteriochlorophyll *b*-containing bacteria. Biochim. Biophys. Acta 593:60–75.

202. Tiede, D. M., J. S. Leigh, and P. L. Dutton. 1978. Structural organization of the *Chromatium vinosum* reaction center associated *c*-cytochromes. Biochim. Biophys. Acta 503:524–544.

203. Knaff, D. B., and B. B. Buchanan. 1975. Cytochrome *b* and photosynthetic sulfur bacteria. Biochim. Biophys. Acta 376:549–560.

204. Takamiya, K., and H. Hanada. 1980. Cytochrome b_{560} in chromatophores from *Chromatium vinosum*. Plant Cell Physiol. 21:979–988.

205. Bowyer, J. R., and A. R. Crofts. 1980. The photosynthetic transfer chain of *Chromatium vinosum* chromatophores. Biochim. Biophys. Acta 591:298–311.

205a. Wynn, R. M., D. F. Gaul, R. W. Shaw, and D. B. Knaff. 1985. Identification of the components of a putative cytochrome bc_1 complex in *Rhodopseudomonas viridis*. Arch. Biochem. Biophys. 238:373–377.

205b. Wynn, R. M., D. F. Gaul, W. K. Choi, R. W. Shaw, and D. B. Knaff. 1986. Isolation of cytochrome bc_1 complexes from the photosynthetic bacteria *Rhodopseudomonas viridis* and *Rhodospirillum rubrum*. Photosynth. Res. 9:181–195.

206. Fowler, C. F. 1974. Evidence for a cytochrome b in green bacteria. Biochim. Biophys. Acta 357:327–331.

206a. Zannoni, D., and W. J. Ingledew. 1985. A thermodynamic analysis of the plasma membrane electron transport components in photoheterotrophically grown cells of *Chloroflexus aurantiacus*. FEBS Lett. 193:93–98.

206b. Zannoni, D. 1986. The branched respiratory chain of heterotrophically dark-grown *Chloroflexus aurantiacus*. FEBS Lett. 198:119–124.

207. Evans, E. H., and A. R. Crofts. 1974. A thermodynamic characterization of cytochromes of chromatophores of *Rps. capsulata*. Biochim Biophys. Acta 357:78–88.

208. Niederman, R. A., C. N. Hunter, D. E. Mallon, and O. T. G. Jones. 1980. Detection of cytochrome b_{+50} in membranes of *Rhodospirillum rubrum* isolated from aerobically and phototrophically grown cells. Biochem J. 186:453–459.

209. Gabellini, N., and G. Hauska. 1983. Characterization of cytochrome b in the isolated ubiquinol-cytochrome c_2 oxidoreductase from *Rhodopseudomonas sphaeroides* GA. FEBS Lett. 153:146–150.

210. Takamiya, K., and H. Tanake. 1983. Isolation and purification of cytochrome b_{561} from a photosynthetic bacterium, *Rhodopseudomonas sphaeroides*. Plant Cell Physiol. 24:1445–1455.

211. Takamiya, K. 1983. Properties of the cytochrome c oxidase activity of cytochrome b_{561} from photoanaerobically grown *Rhodopseudomonas sphaeroides*. Plant Cell Physiol. 24:1457–1462.

212. Van der Wal, H. N., and R. van Grondelle. 1983. Flash-induced electron transport in b- and c-type cytochromes in *Rhodospirillum rubrum*. Evidence for a Q-cycle. Biochim. Biophys. Acta 725:94–103.

213. O'Keefe, D. P., and P. L. Dutton. 1981. Cytochrome b oxidation and reduction reactions in the ubiquinol-cytochrome b/c_2 oxidoreductase from *Rhodopseudomonas sphaeroides*. Biochim Biophys. Acta 635:149–166.

214. Bowyer, J. R., S. W. Meinhardt, G. V. Tierney, and A. R. Crofts. 1981. Resolved difference spectra of redox centers involved in photosynthetic electron flow in *Rhodopseudomonas capsulata* and *Rhodopseudomonas sphaeroides*. Biochim. Biophys. Acta 635:167–186.

215. Meinhardt, S. W., and A. R. Crofts. 1983. The role of cytochrome b-566 in the electron transport chain of *Rhodopseudomonas sphaeroides*. Biochim. Biophys. Acta 723:219–230.

216. Hauska, G., E. Hurt, N. Gabellini, and W. Lockau. 1983. Comparative aspects of quinol-cytochrome c/plastocyanin oxidoreductase. Biochim. Biophys. Acta 726:97–133.

217. Van der Berg, W. H., R. C. Prince, C. L. Bashford, K.-I. Takamiya, W. D. Bonner, and P. L. Dutton. 1979. Electron and proton transport in the ubiquinone cytochrome bc_2 oxidoreductase of *Rhodopseudomonas sphaeroides*. J. Biol. Chem. 254:8594–8604.

218. Meinhardt, S. W., and A. R. Crofts. 1982. The site and mechanism of action of myxothiazol as an inhibitor of electron transfer in *Rhodopseudomonas sphaeroides*. FEBS Lett. 149:217–222.

219. Takamiya, K. 1980. Interaction between antimycin and cytochrome b_{560} in *Chromatium* chromatophores. Plant Cell Physiol. 21:1551–1557.

220. Wood, P. M. 1980. Do photosynthetic bacteria contain cytochrome c_1? Biochem. J. 189:385–391.

221. Wood, P. M. 1980. The interrelation of the two *c*-type cytochromes in *Rps. sphaeroides* photosynthesis. Biochem. J. 192:761–764.

222. Meinhardt, S. W., and A. R. Crofts. 1982. Kinetic and thermodynamic resolution of cytochrome c_1 and cytochrome c_2 from *Rhodopseudomonas sphaeroides*. FEBS Lett. 149:223–227.

223. Gabellini, N., J. R. Bowyer, E. Hurt, B. A. Melandri, and G. Hauska. 1982. A cytochrome b/c_2 complex with ubiquinol-cytochrome c_2 oxidoreductase activity from *Rhodopseudomonas sphaeroides*. Eur. J. Biochem. 126:105–111.

224. Prince, R. C., J. G. Lindsay, and P. L. Dutton. 1975. The Rieske iron–sulfur center in mitochondrial and photosynthetic systems. E_m/pH relationships. FEBS Lett. 51:108–111.

225. Prince, R. C., and P. L. Dutton. 1976. Further studies on the Rieske iron–sulfur center in mitochondrial and photosynthetic systems: a pK on the oxidized form. FEBS Lett. 65:117–119.

226. Dutton, P. L., and J. S. Leigh. 1973. Electron spin resonance characterization of *Chromatium* D hemes. Non-heme irons and the components involved in primary photochemistry. Biochim. Biophys. Acta 314:178–190.

227. Evans, M. C. W., A. V. Lord, and S. G. Reeves. 1974. The detection and characterization by electron-paramagnetic-resonance spectroscopy of iron–sulfur proteins and other electron-transport components in chromatophores from the purple bacterium *Chromatium*. Biochem J. 138:177–183.

228. Malkin, R. 1981. Interaction of the quinone analogue, DBMIB, with the photosynthetic Rieske iron–sulfur center. Isr. J. Chem. 21:301–305.

229. Knaff, D. B., and R. Malkin. 1976. Iron–sulfur proteins of the green photosynthetic bacterium *Chlorobium*. Biochim. Biophys. Acta 430:244–252.

230. Rieske, J. S., R. E. Hanssen, and W. S. Zaugg. 1964. Studies on the electron transfer system. Properties of a new oxidation reduction component of the respiratory chain as studied by electron paramagnetic resonance spectroscopy. J. Biol. Chem. 239:3017–3022.

230a. Cline, J. F., B. M. Hoffman, W. B. Mims, E. Lattaie, D. P. Ballov, and J. A. Fee. 1985. Evidence for N coordination to Fe in the [2Fe-2S] clusters of *Thermus* Rieske protein and phtalate dehydrogenase from *Pseudomonas*. J. Biol. Chem. 260:3251–3254.

231. Rieske, J. S. 1976. Composition, structure, and function of complex III of the respiratory chain. Biochim. Biophys. Acta 456:195–247.

232. Trumpower, B. L., and A. G. Katki. 1979. Succinate-cytochrome *c* reductase complex of the mitochondrial electron transport chain, p. 89–200, in R. A. Capaldi (ed.), Membrane proteins in energy transduction. Marcel Dekker, New York.

233. Takamiya, K.-I., M. Doi, and H. Okimatsu. 1982. Isolation and purification of a ubiquinone-cytochrome b-c_1 complex from a photosynthetic bacterium, *Rhodopseudomonas sphaeroides*. Plant Cell Physiol. 23:987–997.

234. Yu, L., and C.-A. Yu. 1982. Isolation and properties of the cytochrome b-c_1 complex from *Rhodopseudomonas sphaeroides*. Biochem. Biophys. Res. Commun. 108:1285–1292.

234a. Gabellini, N., and W. Sebald. 1986. Nucleotide sequence and transcription of the *fbc* operon from *Rhodopseudomonas sphaeroides*. Eur. J. Biochem. 154:569–579.

235. Baum, H., J. S. Rieske, H. I. Silman, and S. H. Lipton. 1967. On the mechanism of electron transfer in complex III of the electron transfer chain. Proc. Natl. Acad. Sci. USA 57:798–805.

236. Wikström, M. and J. Berden. 1972. Oxidoreduction of cytochrome b in the presence of antimycin. Biochim. Biophys. Acta 283:403–420.

237. Dutton, P. L., and R. C. Prince. 1978. Equilibrium and disequilibrium in the ubiquinone-cytochrome b-c_2 oxidoreductase of *Rhodopseudomonas sphaeroides*. FEBS Lett. 91:15–20.

238. Mitchell, P. 1976. Possible molecular mechanism of the protonmotive function of cytochrome systems. J. Theor. Biol. 62:327–367.

239. De Vries, S., S. P. J. Albracht, J. A. Berden, and E. C. Slater. 1982. The pathway of electrons through QH_2 : cytochrome c oxidoreductase studied by pre-steady-state kinetics. Biochim. Biophys. Acta 681:41–53.

240. Trumpower, B. L., and C. A. Edwards. 1979. Purification of a reconstitutively active iron–sulfur protein (oxidation factor) from succinate-cytochrome c reductase complex of bovine heart mitochondria. J. Biol. Chem. 254:8697–8706.

241. Bowyer, J. R., P. L. Dutton, R. C. Prince, and A. R. Crofts. 1980. The role of the Rieske iron–sulfur center as the electron donor to ferricytochrome c_2 in *Rhodopseudomonas sphaeroides*. Biochim. Biophys. Acta 592:445–460.

242. Crofts, A. R., S. W. Meinhardt, K. R. Jones, and M. Snozzi. 1983. The role of the quinone pool in the cyclic electron transfer chain of *Rhodopseudomonas sphaeroides*. Biochim. Biophys. Acta 723:202–218.

242a. Glaser, E. G., and A. R. Crofts. 1984. A new electrogenic step in the ubiquinol : cytochrome c_2 oxidoreductase complex of *Rhodopseudomonas sphaeroides*. Biochim. Biophys. Acta 766:322–333.

243. Robertson, D. E., R. C. Prince, J. R. Bowyer, K. Matsuura, P. L. Dutton, and T. Ohnishi. 1984. Thermodynamic properties of the semiquinone and its binding site the ubiquinol : cytochrome c (c_2) oxidoreductase of respiratory and photosynthetic systems. J. Biol. Chem. 259:1758–1763.

244. Konstantinov, A. A., and E. K. Ruuge. 1977. Semiquinone Q in the respiratory chain of electron transport particles. Electron spin resonance studies. FEBS Lett. 81:137–141.

244a. Coremans, J. M. C. C., H. N. van der Wal, R. van Grondelle, J. Amesz, and D. B. Knaff. 1985. The pathway of cyclic transport in chromatophores of *Chromatium vinosum*. Evidence for a Q-cycle mechanism. Biochim. Biophys. Acta 807:134–142.

245. Smith, L., and P. B. Pinder. 1978. Oxygen-linked electron transport and energy conservation, p. 641–659, in R. K. Clayton and W. R. Sistrom (eds.), The photosynthetic bacteria. Plenum Press, New York.

245a. Fuller, R. C., and T. E. Redlinger. 1985. Light and oxygen regulation of the development of the photosynthetic apparatus in *Chloroflexus*, p. 155–162, in K. E.

Steinback, S. Bonitz, C. J. Arntzen, and L. Bogorad (eds.), Molecular biology of the photosynthetic apparatus. Cold Spring Harbor Laboratory, Cold Spring Harbor, N.Y.

246. Kämpf, C., and N. Pfennig. 1980. Capacity of Chromatiaceae for chemotrophic growth. Specific respiration rates of *Thiocystis violacea* and *Chromatium vinosum*. Arch Microbiol. 127:125–135.

247. Takamiya, K.-I., K. Kimura, M. Doi, and M. Nishimura. 1980. $NADH_2$- and succinate-dependent O_2 uptake in chromatophores from *Chromatium vinosum*. Plant Cell Physiol. 21:405–411.

248. Jones, O. T. G., and K. M. Plewis. 1974. Reconstitution of light-dependent electron transport in membranes from a bacteriochlorophyll-less mutant of *R. spheroides*. Biochim. Biophys. Acta 357:204–214.

249. Connelly, J. L., O. T. G. Jones, V. A. Saunders, and D. W. Yates. 1973. Kinetic and thermodynamic properties of membrane-bound cytochromes of aerobically and photosynthetically grown *R. spheroides*. Biochim Biophys. Acta 292:644–653.

250. Saunders, V. A., and O. T. G. Jones. 1974. Properties of the cytochrome *a*-like material developed in the photosynthetic bacterium *R. spheroides* when grown aerobically. Biochim. Biophys. Acta 333:439–445.

251. Frenkel, A. W. 1954. Light induced phosphorylation by cell-free preparations of photosynthetic bacteria. J. Am. Chem. Soc. 76:5568–5569.

252. Ort, D. R., and B. A. Melandri. 1982. Mechanism of ATP synthesis, p. 537–587, in Govindjee (ed.), Photosynthesis, Vol. 1, Energy conversion by plants and bacteria. Academic Press, New York.

252a. Elferink, M. G. L., K. J. Hellingwerf, and W. N. Konings. 1986. The relation between electron transfer, proton-motive force and energy-consuming processes in cells of *Rhodopseudomonas sphaeroides*. Biochim. Biophys. Acta 848:58–68.

253. Mitchell, P. 1972. Chemiosmotic coupling in energy transduction: a logical development of biological knowledge. J. Bioenerg. 3:5–24.

254. McCarty, R. E., and C. Carmeli. 1982. Proton translocating ATPases of photosynthetic membranes, p. 647–655, in Govindjee (ed.), Photosynthesis, Vol. 1, Energy conversion by plants and bacteria. Academic Press, New York.

255. Senior, A. L. 1979. The mitochondrial ATPase, p. 234–278, in R. A. Capaldi (ed.), Membrane proteins in energy transduction. Marcel Dekker, New York.

256. Kagawa, Y. 1978. Reconstitution of the energy transformer, gate and channel subunit reassembly, crystalline ATPase and ATP synthesis. Biochim. Biophys. Acta 505:45–93.

257. Gromet-Elhanan, Z., and M. Leiser, 1975. Postillumination adenosine triphosphate synthesis in *Rhodospirillum rubrum* chromatophores. II. Stimulation by a K^+ diffusion potential. J. Biol. Chem. 250:90–93.

258. Clark, A., N. P. J. Cotton, and J. B. Jackson. 1983. The relation between membrane ionic current and ATP synthesis in chromatophores from *Rhodopseudomonas capsulata*. Biochim. Biophys. Acta 723:440–453.

259. Michels, P. A. M., and W. N. Konings. 1978. The electrochemical proton gradient generated by light in membrane vesicles and chromatophores from *Rhodopseudomonas sphaeroides*. Eur. J. Biochem. 85:147–155.

260. Davidson, V. L., and D. B. Knaff. 1982. The electrochemical proton gradient in the

photosynthetic purple sulfur bacterium *Chromatium vinosum*. Photochem. Photobiol. 36:551–558.

261. Junge, W., and J. B. Jackson. 1982. The development of electrochemical potential gradients across the photosynthetic membrane, p. 589–646, in Govindjee (ed.), Photosynthesis, Vol. 1, Energy conversion by plants and bacteria. Academic Press, New York.

262. Davidson, V. L., and D. B. Knaff. 1981. Calcium-proton antiports in photosynthetic purple bacteria. Biochim. Biophys. Acta 673:53–60.

263. Davidson, V. L., and D. B. Knaff. 1981. Properties of a potassium/proton antiport in the photosynthetic bacterium *Chromatium vinosum*. Photobiochem. Photobiophys. 3:167–174.

264. Amesz, J. 1977. Photosynthesis: biophysical aspects. Prog. Bot. 39:48–61.

265. Jackson, J. B., and A. R. Crofts. 1971. The kinetics of light induced carotenoid changes in *Rhodopseudomonas spheroides* and their relation to electrical field generation across the chromatophore membrane. Eur. J. Biochem. 18:120–130.

266. Jackson, J. B., and P. L. Dutton. 1973. The kinetic and redox potentiometric resolution of the carotenoid shifts in *Rhodopseudomonas spheroides* chromatophores: their relationship to electric field alterations in electron transport and energy coupling. Biochim. Biophys. Acta 325:102–113.

267. De Grooth, B. G., and J. Amesz. 1977. Electrochromic absorbance changes of photosynthetic pigments in *Rhodopseudomonas sphaeroides*. I. Stimulation by secondary electron transport at low temperature. Biochim. Biophys. Acta 462:237–246.

268. Zannoni, D., R. C. Prince, P. L. Dutton, and B. L. Marrs. 1980. Isolation and characterization of a cytochrome c_2-deficient mutant of *Rhodopseudomonas capsulata*. FEBS Lett. 113:289–293.

269. Bashford, C. L., C. Prince, K.-I. Takamiya, and P. L. Dutton. 1979. Electrogenic events in the ubiquinone-cytochrome b/c_2 oxidoreductase of *Rhodopseudomonas sphaeroides*. Biochim. Biophys. Acta 545:223–235.

270. Matsuura, K., K. Masamoto, S. Itoh, and M. Nishimura. 1979. Effect of surface potential on the intramembrane electrical field measured with carotenoid spectral shifts in chromatophores from *Rhodopseudomonas sphaeroides*. Biochim. Biophys. Acta 547:91–102.

271. Amesz, J. 1963. Kinetics, quantum requirement and action spectrum of light-induced phosphopyridine nucleotide reduction in *Rhodospirillum rubrum* and *Rhodopseudomonas spheroides*. Biochim. Biophys. Acta 66:22–36.

272. Knaff, D. B. 1978. Reducing potentials and the pathway of NAD^+ reduction, p. 629–640, in R. K. Clayton and W. R. Sistrom (eds.), The photosynthetic bacteria. Plenum Press, New York.

273. Keister, D., and J. J. Yike. 1967. Energy-linked reactions in photosynthetic bacteria. I. Succinate-linked ATP-driven NAD^+ reduction by *R. rubrum* chromatophores. Arch. Biochem. Biophys. 121:415–422.

274. Jones, C. W., and L. P. Vernon. 1969. NAD photoreduction in *R. rubrum* chromatophores. Biochim. Biophys. Acta 180:149–161.

275. Klemme, J. H. 1969. Studies on the mechanism of NAD-photoreduction by chromatophores of the facultative phototroph, *Rhodopseudomonas capsulata*. Z. Naturforsch. B24:67–76.

276. Jones, O. T. G., and V. A. Saunders. 1972. Energy-linked electron transfer reactions in *Rps. viridis*. Biochim. Biophys. Acta 275:427–436.

277. Malkin, R., R. K. Chain, S. Kraichoke, and D. B. Knaff. 1981. Studies of the function of the membrane-bound iron–sulfur centers of the photosynthetic bacterium *Chromatium vinosum*. Biochim. Biophys. Acta 637:88–95.

278. Gromet-Elhanan, Z. 1969. Inhibitors of photophosphorylation and photoreduction by chromatophores from *R. rubrum*. Arch. Biochem. Biophys. 131:299–315.

279. Ingledew, W. J., and R. C. Prince. 1977. Thermodynamical resolution of the iron-sulfur centers of the succinic dehydrogenase of *Rhodopseudomonas sphaeroides*. Arch. Biochem. Biophys. 178:303–307.

280. Zannoni, D., and W. J. Ingledew. 1983. A functional characterization of the membrane and iron sulphur centres of *Rhodopseudomonas capsulata*. Arch. Microbiol. 135:176–181.

281. Zannoni, D., and W. J. Ingledew. 1983. *Rhodopseudomonas capsulata* respiratory dehydrogenase mutants: an EPR study. FEMS Microbiol. Lett. 17:331–334.

282. Ohnishi, T. 1973. Mechanism of electron transport and energy conservation on the site I region of the respiratory chain. Biochim. Biophys. Acta 301:105–128.

283. Evans, M. C. W. 1969. Ferredoxin: NAD reductases and the photoreduction of NAD by *Chlorobium thiosulfatophilum*, p. 1474–1475, in H. Metzner (ed.), Progress in photosynthesis research. Laupp, Tübingen, West Germany.

284. Evans, M. C. W., and B. B. Buchanan. 1965. Photoreduction of ferredoxin and its use in CO_2 fixation by a subcellular system from a photosynthetic bacterium. Proc. Natl. Acad. Sci. USA 53:1420–1425.

285. Carrithers, R. P., D. C. Yoch, and D. I. Arnon. 1977. Isolation and characterization of bound iron sulfur proteins from bacterial photosynthetic membranes. II. Succinate dehydrogenase from *R. rubrum* chromatophores. J. Biol. Chem. 252:7461–7467.

286. Albracht, S. P. J. 1980. The prosthetic groups in succinate dehydrogenase. Number and stoichiometry. Biochim. Biophys. Acta 612:11–28.

287. Trüper, H. G., and U. Fischer. 1982. Anaerobic oxidation of sulphur compounds as electron donors for bacterial photosynthesis. Philos. Trans. R. Soc. Lond. B298:529–542.

288. Trüper, H. G. 1981. Photolithotrophic sulfur oxidation, p. 199–211, in H. Bothe and A. Trebst (eds.), Biology of inorganic nitrogen and sulfur. Springer-Verlag, Heidelberg.

289. Trüper, H. G. 1978. Sulfur metabolism, p. 677–690, in R. K. Clayton and W. R. Sistrom (eds.), The photosynthetic bacteria. Plenum Press, New York.

290. Kusai, A., and T. Yamanaka. 1973. Cytochrome c_{553} (*Chlorobium thiosulfatophilum*) is a sulfide-cytochrome *c* oxidoreductase. FEBS Lett. 34:235–237.

291. Kusai, T., and T. Yamanaka. 1973. The oxidation mechanism of thiosulfate and sulfide in *Chlorobium thiosulfatophilum*: roles of cytochrome c-551 and cytochrome c-553. Biochim. Biophys. Acta 325:304–314.

292. Steinmetz, M. A., and U. Fischer. 1981. Cytochromes of the non-thiosulfate-utilizing green sulfur bacterium *Chlorobium limicola*. Arch. Microbiol. 130:31–37.

293. Gibson, J. 1961. Cytochrome pigments from the green photosynthetic bacterium *Chlorobium thiosulfatophilum*. Biochem. J. 79:151–158.

294. Yamanaka, T., and Y. Fukumori. 1980. A biochemical comparison between *Chlorobium* and *Chromatium* flavocytochromes *c*, p. 631–639, in K. Yagi and T. Yamano (eds.), Flavins and flavoproteins. Japan Scientific Societies Press, Tokyo.

295. Paschinger, H., J. Paschinger, and H. Gaffron. 1974. Photochemical disproportion of sulfur into sulfide and sulfate by *Chlorobium limocola* f. *thiosulfatophilum*. Arch. Mikrobiol. 96:341–351.

296. Kusai, A., and T. Yamanaka. 1973. A novel function of cytochrome c_{555} (*Chlorobium thiosulfatophilum*) in oxidation of thiosulfate. Biochem. Biophys. Res. Commun. 51:107–112.

297. Meyer, T. E., R. G. Bartsch, M. A. Cusanovich, and J. H. Mathewson. 1968. The cytochromes of *Chlorobium thiosulfatophilum*. Biochim. Biophys. Acta 153:854–861.

297a. Davidson, M. W., T. E. Meyer, M. A. Cusanovich, and D. B. Knaff. 1986. Complex formation between *Chlorobium limicola* f. *thiosulfatophilum* *c*-type cytochromes. Biochim. Biophys. Acta 850:396–401.

298. Van Niel, C. B. 1944. The culture, general physiology, morphology and classification of the non-sulfur purple and brown bacteria. Bacteriol. Rev. 8:1–118.

299. Hansen, T. A., and H. van Gemerden. 1972. Sulfide utilization by purple nonsulfur bacteria. Arch. Mikrobiol. 86:49–56.

300. Petushkova, Y. P., and R. N. Ivanovsky. 1976. Respiration of *Thiocapsa roseopersinica*. Mikrobiologiya 45:592–597.

301. Fischer, U., and H. G. Trüper. 1977. Cytochrome *c*-550 of *Thiocapsa roseopersinica*, properties and reduction by sulfide. FEMS Microbiol. Lett. 1:87–90.

302. Fukumori, Y., and T. Yamanaka. 1979. Flavocytochrome *c* of *Chromatium vinosum*. Some enzymatic properties and subunit structure. J. Biochem. 85:1405–1419.

303. Gray, G. O., and D. B. Knaff. 1982. The role of a cytochrome c_{552} : cytochrome *c* complex in the oxidation of sulfide in *Chromatium vinosum*. Biochim. Biophys. Acta 680:290–296.

303a. Davidson, M. W., G. O. Gray, and D. B. Knaff. 1985. Interaction of *Chromatium vinosum* flavocytochrome *c*-552 with cytochromes *c* studied by affinity chromatography. FEBS Lett. 187:155–159.

303b. Bosshard, H. R. M. W. Davidson, D. B. Knaff, and F. Millett. 1986. Complex formation and electron transfer between mitochondrial cytochrome *c* and flavocytochrome c_{552} from *Chromatium vinosum*. J. Biol. Chem. 261:190–193.

303c. Vieira, B., M. Davidson, D. Knaff, and F. Millett. 1986. The use of a water-soluble carbodiimide to study the interaction between *Chromatium vinosum* flavocytochrome c_{552} and cytochrome *c*. Biochim. Biophys. Acta 848:131–136.

304. Van Gemerden, H. 1968. On ATP generation by *Chromatium* in darkness. Arch. Mikrobiol. 64:118–124.

305. Kobayashi, K., E. Kastura, T. Kondo, and M. Ishimoto. 1978. *Chromatium* sulfite reductase I. Characterization of thiosulfate forming activity at the cell extract level. J. Biochem. 84:1205–1215.

306. Schedel, M., M. Vanselow, and H. G. Trüper. 1979. Siroheme sulfite reductase isolated from *Chromatium vinosum*. Arch. Microbiol. 121:29–36.

307. Seki, Y. N. Sogawa, and M. Ishimoto. 1981. Siroheme as an active catalyst in sulfite reduction. J. Biochem. 90:1487–1492.

308. Schwenn, J. D., and M. Biere, 1979. APS reductase activity in the chromatophores of *Chromatium vinosum* strain D. FEMS Microbiol. Lett. 6:19–22.

309. Fukumori, Y., and T. Yamanaka. 1979. A HiPIP-linked thiosulfate oxidizing enzyme derived from *Chromatium vinosum*. Curr. Microbiol. 3:119–120.

310. Schmitt, W., G. Schliefer, and K. Knobloch. 1981. The enzymatic system thiosulfate:cytochrome *c* oxidoreductase from photolithoautotrophically grown *Chromatium vinosum*. Arch. Microbiol. 130:328–333.

311. Doi, M., K. Takamiya, and M. Nishimura. 1980. Light- and thiosulfate-dependent reduction of nictotinamide adenine dinucleotides in whole cells of *Chromatium vinosum*. Plant Cell Physiol. 21:1015–1022.

312. Van Niel, C. B. 1941. The bacterial photosynthesis and their importance for the general problem of photosynthesis. Adv. Enzymol. 1:263–328.

4

ECOLOGY OF DENITRIFICATION AND DISSIMILATORY NITRATE REDUCTION TO AMMONIUM

JAMES M. TIEDJE

Departments of Crop and Soil Sciences and of Microbiology and Public Health, Michigan State University, East Lansing, MI 48824

4.1   INTRODUCTION
4.2   DEFINITIONS
    4.2.1   Chemodenitrification
    4.2.2   Respiratory Denitrification
    4.2.3   Nonrespiratory N_2O Production
    4.2.4   Dissimilatory Nitrate Reduction to Ammonium
4.3   IMPORTANCE OF DISSIMILATORY NITRATE REDUCING PROCESSES
4.4   DISSIMILATORY NITRATE REDUCING ORGANISMS AND THEIR PHYSIOLOGY
    4.4.1   Denitrification (Respiratory)
        4.4.1.1   Growth
        4.4.1.2   Organisms
        4.4.1.3   Special Oxygen Physiology
        4.4.1.4   Pollutant Metabolism
        4.4.1.5   Regulation of Intermediate Production
    4.4.2   Dissimilatory Nitrate Reduction to Ammonium
        4.4.2.1   General Features
        4.4.2.2   Organisms
        4.4.2.3   Physiology

4.5   POPULATION ECOLOGY OF DISSIMILATORY NITRATE REDUCTION
      4.5.1   Denitrification
              4.5.1.1   What Denitrifier Populations Are Found?
              4.5.1.2   What Selects Denitrifier Populations?
              4.5.1.3   Adaptive Traits of Denitrifiers
              4.5.1.4   Survival of Denitrifying Activity
      4.5.2   Dissimilatory Nitrate Reduction to Ammonium (DNRA)
              4.5.2.1   What DNRA Populations Are Found
              4.5.2.2   What Selects DNRA Populations?
4.6   REGULATION OF DENITRIFICATION IN THE ENVIRONMENT
      4.6.1   Characteristics of Oxygen Regulation
              4.6.1.1   Oxygen as a Dominant Regulator
              4.6.1.2   Oxygen Threshold for Denitrification
              4.6.1.3   Synthesis of Denitrifying Enzymes
              4.6.1.4   Effect of Oxygen on the Different Steps in the
                        Denitrification Sequence
      4.6.2   Conceptual Model of Environmental Regulation of Denitrification
      4.6.3   Regulatory Factors
              4.6.3.1   Oxygen
              4.6.3.2   Nitrate
              4.6.3.3   Carbon as Electron Donor
              4.6.3.4   Effect of Macrofeatures: Burrows and Roots
4.7   SUMMARY OF METHODS USED TO MEASURE DENITRIFICATION
REFERENCES

4.1   INTRODUCTION

Reduction of nitrate in anaerobic environments is dominated by two dissimilatory processes: respiratory denitrification and dissimilatory nitrate reduction to ammonium. Assimilatory nitrate reduction could also occur, but most anaerobic environments have large concentrations of ammonium and organic nitrogen, which repress this process or make it quantitatively insignificant. Dissimilatory processes are distinguished from the assimilatory process by the fact that the nitrogen reduced is not used by the cell. Rather, the nitrogen oxide serves as an electron acceptor for the cell's metabolism. Since the dissimilatory processes are inhibited by oxygen, they occur only in anaerobic environments. The main features of these three processes are summarized in Table 4.1.

4.2   DEFINITIONS

Denitrification is the process by which nitrogenous oxides, principally nitrate and nitrite, are reduced to dinitrogen gases, N_2O and N_2. Historically, this process was

TABLE 4.1 Biological nitrate reduction mechanisms

Process	Pathway of Free Intermediates	Regulated by:	Groups of Organisms Possessing Process
Assimilatory mechanism			
Assimilatory nitrate reduction	$NO_3^- \rightarrow NO_2^- \rightarrow NH_4^+$	NH_4^+, organic N	Plants, fungi, algae, bacteria
Dissimilatory mechanisms			
Denitrification	$NO_3^- \rightarrow NO_2^- \rightarrow N_2O \rightarrow N_2$	O_2	Aerobic bacteria also capable of anaerobic growth with NO_3^- or NO_2^-
Dissimilatory nitrate reduction to ammonium	$NO_3^- \rightarrow NO_2^- \rightarrow NH_4^+$	O_2	Anaerobic and facultatively anaerobic bacteria

not associated with any particular mechanism, and from the perspective of global nitrogen budgets, this remains a relevant definition. However, there are several types of mechanisms that convert nitrogen oxides to dinitrogen gases.

4.2.1 Chemodenitrification

Some nitrogen-gas-generating reactions are catalyzed by abiologic agents and are known as chemodenitrification (99). The most significant of these is the acid-catalyzed destruction of nitrite, which can become significant at pH values < 5.0. Although N_2, N_2O, and NO have all been reported as products of chemodenitrification, the predominant product is usually NO (26, 205). Since NO is a very minor product of biological denitrification (90, 141), its occurrence can be used as a preliminary indication of chemodenitrification. The second most likely chemodenitrification mechanism involves the oxidation of organic N by NO_2^-, forming N_2 gas (a van Slyke type of reaction). This mechanism has been shown to be of potential significance in frozen soils, where NO_2^- is concentrated by the salting-out effect (36). Chemodenitrification is not thought to be a major process on a global scale, but there are occasional situations, such as nitrite-containing acid or frozen soils, in which it could occur.

4.2.2 Respiratory Denitrification

The majority of denitrification is carried out by biological denitrifying processes, particularly by the bacterial respiratory process that microbiologists typically refer to as denitrification. To differentiate among the biological processes, it is important to define the characteristics of typical respiratory denitrification. The distinctive feature of this process is that the nitrogen oxide reduction is coupled to electron transport phosphorylation (ETP; 103, 104). Therefore, the major criterion for a respiratory denitrifier is that the growth yield should be enhanced proportional to the amount of N oxide present, and that the increase should be greater than that provided if the N oxide simply served as an electron sink. Other measurable features that seem to be uniformly characteristic of respiratory denitrification are: (a) the conversion of 80% or more of NO_3^- or NO_2^- nitrogen to $N_2O + N_2$ (I. Mahne and J. M. Tiedje, unpublished data), (b) a rapid rate of nitrate reduction (e.g., >10 nmol N gas mg^{-1} protein min^{-1}) (I. Mahne and J. M. Tiedje, unpublished data), and (c) the presence of one of the two denitrifying-type nitrite reductases (i.e., the cytochrome cd or the Cu type) (141). A convenient preliminary identification of most respiratory denitrifiers is to observe, in response to acetylene addition to the culture, a nearly stoichiometric accumulation of N_2O from NO_3^- or NO_2^-. Acetylene is now well established as an effective inhibitor of nitrous oxide reductase (9, 56, 222) and it does not alter N_2O production by nonrespiratory denitrifiers (200) or by chemodenitrification. However, this method will not distinguish denitrifiers lacking the $N_2O \rightarrow N_2$ step.

4.2.3 Nonrespiratory N₂O Production

The nonrespiratory denitrifiers typically produce N_2O and not N_2 as the product of N-oxide reduction. Several groups of organisms produce N_2O, including representatives of the following: bacteria that assimilate nitrate, bacteria that dissimilate nitrate to ammonium, yeasts, filamentous fungi, algae, and (presumably) microorganisms living in association with animals. The percentage of nitrate-N or nitrite-N converted to N_2O-N by some of these organisms is summarized in Table 4.2. The mechanism(s) for this N_2O production is not yet known, but in the case of the nitrate respirers it may be due to nitrate reductase acting on nitrite since mutants lacking nitrate reductase did not produce N_2O, but mutants lacking nitrite reductase did (181). Since the type of nitrate reductase in nitrate respirers is restricted to the Enterobacteriaceae, this enzyme cannot be responsible for N_2O production in the other groups of organisms. Nitrite is a much better source of N_2O than is NO_3^- for all groups except the Enterobacteriaceae (Table 4.2).

It has not yet been demonstrated that the nondenitrifying sources of N_2O are of environmental significance. The difficulty is that experimental methods do not exist to separate these sources of N_2O unequivocally from the respiratory denitrifier sources. If the nondenitrifying sources are important, it would seem that the most likely environments would be those in which respiratory denitrifiers do not predominate but other N_2O producers do, for example well-aerated, acid forest soils (154) or the animal alimentary tract. We recently obtained presumptive evidence that nonrespiratory and nonnitrifying sources of N_2O were significant in forest soils since in some cores as much as one-half of the N_2O production was not inhibited by 100% O_2 (to inhibit denitrification), only 10 Pa acetylene (to inhibit nitrification), but was inhibited by sterilization (to rule out chemodenitrification) (155).

4.2.4 Dissimilatory Nitrate Reduction to Ammonium

Dissimilatory nitrate reduction to ammonium (DNRA) has recently been more widely recognized as a process distant from assimilatory nitrate reduction. Both the assimilatory and dissimilatory processes result in ammonium production, but the regulation of the two pathways is different (Table 4.1). The dissimilatory pathway is regulated by oxygen and is unaffected by ammonium, while the opposite is true of assimilatory reduction. Therefore, the dissimilatory pathway is well suited to anaerobic environments. Furthermore, in anaerobic environments where the limitation in electron acceptors often restricts metabolism, the accommodation of eight electrons per nitrogen in the nitrate to ammonium step makes this one of the most favorable electron acceptors available to anaerobic environments (200).

The major criterion that identifies dissimilatory nitrate reduction to ammonium is the production of ammonium from nitrate in excess of the reduced nitrogen needed for growth. The easiest way to distinguish this process from assimilatory reduction experimentally is to measure $^{15}NH_4^+$ plus organic ^{15}N production from $^{15}NO_3^-$ in the presence of sufficient ammonium (e.g., 1 mM) to repress nitrate-assimilating pathways.

TABLE 4.2 Extent of N_2O production by nonrespiratory denitrifiers

Group, Organism	Typical values for % of NO_x-N recovered as N_2O-N[a]		References
	From NO_3^-	From NO_2^-	
Enterobacteriaceae			
Escherichia coli			
Klebsiella pneumoniae			
Enterobacter aerogenes			
Erwinia carotovora	6–35%	6–35%	180, 17, 179, 168
Serratia marcescens	(100–1000)	(100–1000)	
Citrobacter sp.			
Salmonella typhimurium			
Other bacteria[b]			
Bacillus spp.	3–13%	—	220, 17, 179
Azotobacter vinelandii	0.3%	—	17
Propionibacterium spp.	—	90–100%	91
	(0.2)	(8–42)	
Lactobacillus lactis		77–98%	50
		(400–4000)	
Yeasts			
Hansenula	0.06–0.18%	—	17
Rhodotorula	(40)	—	
Fungi			
Fusarium	0.03	6.5–13	17, 19, 28
Alternaria			
Aspergillus			
Geotrichum	0–0.03%	>5%	28, 17, 220
Penicillum			
Acremonium			
Algae			
Chlorella species			
Scenedesmus obliquus	0%	5–14%	213
Coelastrum spp.			
Chlorosoccum vacuolatum			
Plants	None documented on noninjured, noninfected plants		17
Animals			
Humans	Elevated concentrations in breath		17
Bovine			

[a]Values in parentheses are typical rates of N_2O production in pmol N_2O/mg cells · min.
[b]Nitrifiers are not listed in this table since their N_2O production appears to result from respiratory denitrification and is discussed later.

4.3 IMPORTANCE OF DISSIMILATORY NITRATE-REDUCING PROCESSES

The microorganisms capable of dissimilatory nitrate reduction are widely distributed in nature, including habitats such as soil, fresh water, marine waters, sediments, waste treatment systems, and animal gastrointestinal tracts. Thus, any restriction in nitrate-reducing activity in natural habitats is probably not caused by the lack of organisms but by the environmental conditions (principally oxygen) that regulate these processes.

Denitrification is of interest for several reasons. First, it has long been a concern in agriculture since nitrogen is the nutrient most limiting to crop production. Fertilizer losses to denitrification range from virtually none to over 70 percent but losses are more commonly in the range 20 to 30% (59). The principal importance of dissimilatory nitrate reduction to ammonium is that it conserves combined nitrogen in contrast to denitrification and thus would be beneficial in noneutrophic environments. Second, denitrification is of great potential in waste treatment, since it can permanently remove excess combined nitrogen from local environments by completing the natural nitrogen cycle. Considerable effort by engineers has been devoted to improving designs for the efficient and economical removal of nitrate from waste water by denitrification (102). Third, in the last decade the role of N_2O in contributing to destruction of the protective ozone layer and in warming of the planet was recognized (43, 116, 211). Denitrification can act both as a source and a sink of atmospheric N_2O. Fourth, the denitrification pathway produces toxic intermediates, NO_2^- and NO, which could produce local hazards; nitrite also reacts with secondary amines to produce nitrosamines, some of which are strongly carcinogenic (111). Fifth, denitrification completes the global nitrogen cycle. It is presumed that denitrification roughly balances total nitrogen fixation on the planet.

The importance of denitrification to the global nitrogen budget is depicted in Fig. 4.1. Denitrification recycles from 52 to 110% of the total nitrogen inputs. The major difficulty, particularly with the denitrification estimates in these budgets, is that they are not based on direct measurements of the process. The lack of methodology and the extreme temporal and spatial variability of denitrification have made this an impossible task. Nonetheless, there is no evidence to suggest that the global nitrogen cycle is grossly out of balance, and denitrification, as the only known major loss mechanism, must be responsible for keeping the cycle nearly in balance.

One shift that is occurring in global budgets is the site of denitrification. With increased fertilizer use, more nitrogen has been leached from lands to surface waters. Furthermore, the increased harvested nitrogen has been more heavily concentrated by urbanization, where it is directly routed to surface waters. Thus the lakes, bays, and coastal areas that have received this increased nitrogen load would be expected to have accelerated denitrification. Since the increased nitrogen load in these aquatic environments is usually accompanied by an increase in respiratory oxygen consumption, the conditions for denitrification are also favored (160). Thus,

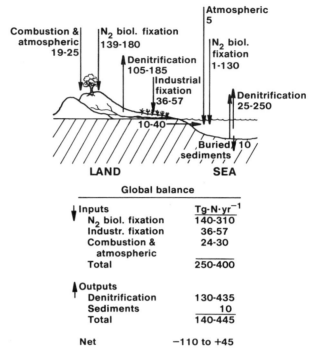

Figure 4.1 Global fluxes affecting the available nitrogen budget. The figure separately represents the terrestrial and marine components. Drawn from the summaries of flux data provided by Knowles (100, 102).

rather than a balanced terrestrial cycle, which may have been more typical in historical times, there has been an increased export to the aquatic system, where the excess nitrogen is thought to be denitrified.

4.4   THE ORGANISMS AND THEIR PHYSIOLOGY

4.4.1   Denitrification (Respiratory)

4.4.1.1   Growth
Denitrifiers are basically aerobic bacteria which have the alternative capacity to reduce nitrogen oxides when oxygen becomes limiting. Hence they do not require strict anaerobic media or procedures; in fact, growth is more reliable and rapid when initially aerobic medium is inoculated with the denitrifying culture. The aerobic respiratory growth of the culture will then consume the oxygen and gradually allow the culture to shift to denitrifying metabolism. If an aerobic inoculum is transferred to a strictly anaerobic medium, growth will be delayed or prevented because of the inability of the inoculum to generate energy to synthesize the required denitrifying enzymes (5; N. V. Caskey and J. M. Tiedje, unpublished data).

The most favorable general medium for the most common heterotrophic denitrifiers is tryptic soy broth (TSB) supplemented with 0.1% (9.9 mM) KNO_3. The growth rate of many heterotrophic denitrifiers may be up to two times faster in TSB than in nutrient broth plus nitrate, another commonly used medium. Defined media are often satisfactory for many denitrifiers, but complex media are routinely used for simplicity and acceptability for growth of many strains. Since denitrification consumes NO_3^-, one must guard against an excessive pH increase caused by the residual cation.

The other electron acceptors in the denitrifying pathway, NO_2^- and N_2O, can be substituted for NO_3^- in growth media. However, if NO_2^- is used, the concentration may need to be reduced since some denitrifier strains are inhibited by nitrite concentrations as low as 2 to 3 mM. Nitrous oxide is not toxic when the medium is in equilibrium with 100% N_2O in the headspace. Most commercial nitrous oxide contains a few percent O_2 as a contaminant, and this cannot be selectively removed by chemical catalysts. Thus controls using obligate respiring aerobes may be necessary (e.g., growth on plates) to ensure that the growth observed under a N_2O atmosphere is N_2O dependent. Acetylene used as 10% of the headspace gas is almost always effective in completely inhibiting N_2O reduction and in not causing toxicity to the growing culture (199).

Most denitrifiers freshly isolated from nature have the entire pathway, $NO_3^- \rightarrow N_2$, but strains that lack the ability to reduce $NO_3^- \rightarrow NO_2^-$ and $N_2O \rightarrow N_2$ are common. Only one organism, *Wolinella succinogenes*, is known to lack the $NO_2^- \rightarrow N_2O$ step (224). New isolates from nature often lose the ability to reduce N_2O after cultivation on laboratory media (1, 62).

4.4.1.2 Denitrifying Organisms

The denitrification capacity is spread among a wide variety of physiologic and taxonomic groups. The genera with strains that have respiratory denitrification capacity are shown in Table 4.3. They are grouped according to physiological features to illustrate the diversity and to aid in identifying which genera would be expected in a particular type of environment. In the past some of the organisms were termed denitrifiers but without adequate evidence (e.g., simply production of N_2O or disappearance of NO_3^- and NO_2^-, both of which can be due to other processes). We have recently examined some of these unconfirmed denitrifiers using the criteria described above (Section 4.2.2) and have included this information in Table 4.3 (I. Mahne and J. M. Tiedje, Abstr. 4th Int. Symp. Microbiol. Ecol., 1986, p. 58). The genera shown in brackets in the table are not yet confirmed as containing denitrifier strains.

The energy sources of denitrifiers include all three classes known to be used by microorganisms: organic (organotrophs), inorganic (lithotrophs), and light (phototrophs). Organic substrates are the most common energy source, both in the number of genera as well as in the dominance of populations in nature. The most common denitrifiers in nature are species of *Pseudomonas* followed by the closely related *Alcaligenes* (Section 4.5.1.1). Relative to these two the other genera are much more rare or prevalent only in specialized environments. Among the pseu-

TABLE 4.3   Genera with denitrifying species grouped according to distinctive physiological features[a,b]

	References		References
ORGANOTROPHIC		**N₂ fixing**	
General aerobic		*Rhizobium*	45, 225
Pseudomonas	107	*Bradyrhizobium*[j]	45, 225
Alcaligenes	107	*Azospirillum*[g]	54, 132, 193
Flavobacterium[c]	14, 145	*Pseudomonas*	33
(Achromobacter)[d]	85	*Rhodopseudomonas*	167
Paracoccus	107	*Agrobacterium*[k]	146, 162
(Corynebacterium)[e]	151	**Animal or pathogenic association**	
[*Acinetobacter*]	60	*Neisseria*	72
Cytophaga	2, 3	*Kingella*	72
[*Gluconobacter*][f]	60	*(Moraxella)*[l]	215
[*Xanthomonas*]	117	*Wolinella*[i]	224
Oligocarbophilic		**PHOTOTROPHIC**	
Hyphomicrobium	118, 186	*Rhodopseudomonas*	167
Aquaspirillum[g]	106	**LITHOTROPHIC**	
Fermentative		**H₂ use**	
Azospirillum[g]	54, 132, 193	*Paracoccus*	107
(Chromobacterium)[h]	13, 72	*Alcaligenes*	107
Bacillus	144, 147	*Bradyrhizobium*	128, 129
Wolinella[i]	224	*Pseudomonas*	33
Halophilic		**S use**	
Halobacterium	203, 214	*Thiobacillus*	107
Paracoccus	80	*Thiomicrospira*	201
Thermophilic		*Thiospaera*	156
Bacillus	65	[*Thermothrix*]	23
[*Thermothrix*]	23	**NH₄⁺ use**	
Sporeformer		*Nitrosomonas*	148, 149
Bacillus	144, 147		
Magnetotactic			
Aquaspirillum	12		

[a] Genera in parentheses are of uncertain or discontinued taxonomic status or the denitrifying species have been transferred to another genus.

[b] Genera in brackets indicate that characterization as having respiratory denitrification is incomplete.

[c] It has been suggested that denitrifying *Flavobacteria* are actually strains of *Pseudomonas* (J. DeLey, personal communication).

[d] Most strains now considered to be *Alcaligenes* (77).

[e] It has been suggested that the main strain studied, *C. nephridii*, is an *Alcaligenes* sp. (75).

[f] Described as *Acetomonas* (60) by the authors, but now designated as *Gluconobacter* (27, 107).

[g] *Aquaspirillum* and *Azospirillum* were derived from the genus *Spirillum*; the remaining *Spirillum* species does not denitrify (107).

[h] *C. violaceum*, the species *C. lividum* now termed *Janthinobacterium* (47, 183).

[i] Originally, *Vibrio succinogenes*; cannot reduce NO_2^- to N_2O (224).

[j] Slow-growing rhizobium classes (e.g., *Rhizobium japonicum*) are now designated as *Bradyrhizobium*.

[k] *Agrobacterium* is not N₂ fixing but is closely related to *Rhizobium* and is usually associated with plants, due to its plant pathogenic properties.

[l] All denitrifying strains of *Moraxella* are now considered *Kingella denitrificans* (107, 141) except for one poorly characterized soil isolate (215).

domonads, *Ps. fluorescens*, especially biotype II, seems to be the most prevalent in soils (62). It is unfortunate that most of the biochemical and biophysical work has been done on species other than *Ps. fluorescens*. The predominance of the pseudomonads is likely due to their versatility and competitiveness for carbon substrates in natural soil and water environments.

In aquatic habitats low in dissolved organic carbon, the oligocarbophiles (Table 4.3) would be expected to be prevalent along with the pseudomonads. Hyphomicrobium, also a specialist in C-1 metabolism (118), is known to predominate in methanol fed waste treatment systems designed to remove nitrate by denitrification (202).

Organisms that ferment as well as denitrify are rare. *Wolinella succinogenes*, which lacks the key enzyme nitrite reductase, is restricted to the rumen habitat (224). The other three fermentative organisms listed in Table 4.3—*Azospirillum*, *Chromobacterium*, and *Bacillus*—are all weak fermentors, with fermentative growth yields an order of magnitude less than for *Escherichia coli* (12, 64; I. Mahne and J. M. Tiedje, unpublished data). Furthermore, none are obligate anaerobes. Previously, *Propionibacterium acidi-propionici* was listed as a denitrifying fermentative anaerobe (141). However, Kaspar showed that this strain as well as four other species of propionibacteria did not have respiratory denitrifying ability, nor did they produce N_2O at a rate typical of respiratory denitrifiers (Section 4.2.2), but they did reduce nitrite to nitrous oxide in nearly stoichiometric amounts (91). Bazylinski et al. reported similar behavior by *Chromobacterium violaceum* isolates which did not conserve energy in association with the NO_2^--to-N_2O step, but they did convert >85% of the NO_2^--N to N_2O-N (13). In contrast, however, the Chromobacteria formed N_2O at a rate more typical of denitrifiers. This could be an NO_2^--detoxifying reaction like the one Kaspar suggested for the propionibacteria, but it is more likely that they are variants of respiratory denitrifiers in which a lesion has caused a loss in the ability to conserve energy in association with this reduction. The apparent absence of respiratory denitrification in organisms that are effective fermenters is puzzling, especially in view of the high quantity of denitrifying enzymes found in deep sediments and anaerobic digester sludge, both habitats devoid of O_2 and NO_3^- (200).

Certain denitrifier strains are specialized to withstand particular stresses or to possess taxis responses (Table 4.3). Both spore formation and thermophily are exhibited by certain denitrifiers. Adaption to salt is shown by organisms with an obligate high NaCl requirement, *Halobacterium*, by a moderately halophilic bacterium, *Paracoccus halodenitrificans*, and by a well-studied organism adapted to the low salt concentration in the marine environment (*Pseudomonas perfectomarinus*) (141). The latter is probably a marine-adapted version of *Ps. stutzeri*. Denitrification has also been found in the magnetotactic bacterium *Aquaspirillum magnetotaticum* strain MS-1.

Somewhat more fastidious organisms with the denitrification property are found in association with plants and animals. The *Rhizobium* and *Agrobacterium* species survive well in the absence of host plants but proliferate more extensively in association with their specific hosts. *Azospirillum* grows well in the rhizosphere of certain grasses. *Agrobacterium* spp. and *Ps. solanacearum* are plant pathogens that denitrify (107). Some strains of the plant-associated nitrogen fixers *Rhizobium*, *Bradyrhizobium*, and *Azospirillum* also denitrify (Table 4.3). The value of denitrification to rhizobia is only speculative (134). Since nitrate inhibits nodulation, reduction of the nitrate concentration in the rhizosphere would favor nodulation and therefore be a selective advantage for *Rhizobium*. Also, respiratory denitrification does provide ATP that could drive N_2 fixation. Thus under oxygen stress the nitrogen-fixation system could be maintained, although this would be at greater energy cost to the plant than would direct assimilation of the nitrate.

The denitrification capacity among *Rhizobiaceae* has recently been surveyed for a number of strains (Table 4.4). This is particularly interesting because large numbers of strains isolated without regard to denitrification capacity have been tested for the presence of this process. Hence this gives some indication of the prevalence of this capacity among a taxonomic group and for which the selective process for retaining the denitrification capacity is not readily apparent. Respiratory denitrification is very common in both fast (*R. fredii*) and slow (*B. japonicum*)-growing soybean rhizobia, and in the cowpea, *lupini*, and *meliloti* groups; 82% of these strains denitrify. However, in the closely related fast-growing *Rhizobium trifiollii*, *R. leguminosarum*, and *R. phaseoli*, only one of 140 strains denitrify. It was suggested previously that fast-growing rhizobia did not denitrify, but this conclusion was premature, as many fast-growing soybean, cowpea, *lupini*, and *meliloti* strains denitrify.

The most fastidious denitrifiers are the strains of *Neisseria* and the related *Kingella*, which are usually isolated from animal specimens (Table 4.3). Environmental denitrifier isolates of this type have been reported as *Moraxella* species;

TABLE 4.4   Frequency of denitrification capacity in isolates of the *Rhizobaceae*

Taxon	Number Positive for Denitrification/Number Strains Tested	References
Slow growing		
Bradyrhizobium japonicum	251/321	204
Cowpea rhizobia	5/14	225
Rhizobium lupini[a]	125/135	134, 187
Fast growing		
Rhizobium fredii	11/11	83, 204
Rhizobium meliloti	5/5	46
Rhizobium phaseoli	0/5	46, 204
Rhizobium leguminosarum	1/3	46, 204
Rhizobium trifolii[b]	0/132	134, 46, 204
Agrobacterium species	Present	162, 146

[a] Includes lotus rhizobia (*R. loti*). Both fast- and slow-growing rhizobia are in this group.
[b] Some strains produce N_2O (29), but it is not clear that they meet the criteria for respiratory denitrification. If not, they should be added to Table 4.2.

however, most of the currently accepted *Moraxella* species are parasitic on mucous membranes of warm-blooded animals (107). Three of the four denitrifying species of *Neisseria* studied by Grant and Payne were lacking the $NO_3^- \rightarrow NO_2^-$ portion of the pathway (72).

The only phototroph that has been shown to denitrify is *Rhodopseudomonas sphaeroides*. Betlach, in his analysis of the phylogeny and evolution of denitrification, suggested that denitrification may have arisen from an ancestral prototroph denitrifier, perhaps similar to *Rhodopseudomonas* (15). If this hypothesis is true, one might expect other photosynthetic relatives to have retained this capacity. It is unlikely that an extensive search for denitrification among the *Rhodospirillaceae* has been done.

The denitrification capacity is well distributed among organisms with chemolithotrophic metabolism. Facultative chemolithotrops include all the H_2-oxidizers listed in Table 4.3 and *Thiomicrospira* and *Thiospaera*. The obligate chemolithotrophs are *Thiobacillus denitrificans* and *Nitrosomonas europeae*. Ammonium-oxidizing bacteria have been known to produce N_2O for some time, but it was assumed to originate from the oxidation of ammonium rather than the reduction of the oxidized product, NO_2^-. However, *N. europeae* contains a denitrifying type of nitrite reductase (119), and recently Poth and Focht presented kinetic evidence showing that the N_2O was produced from NO_2^- reduction and not from NH_4^+ oxidation (148). Poth has also isolated a strain of *Nitrosomonas* that reduced N_2O to N_2 (149). Furthermore, N_2O production is stimulated under low O_2 concentrations, a relationship found for most heterotrophic denitrifiers (16, 71).

Among the biogeochemical cycles on earth, there are no inorganic biotransformations that are carried out by a wider distribution and diversity of organisms than is the case for denitrification. Denitrification is present in strains contained in 10 different procaryotic families (107): Rhodospirillaceae, Cytophagaceae, "budding or appendaged" bacteria, Spirileaceaea, Pseudomonaceae, Rhizobiaceae, Halobacteriaceae, Neisseriaceae, Nitrobacteraceae, and Bacillaceae. Given this widespread distribution, it may be more instructive to know what groups are missing representatives that denitrify. Three are notable: (a) obligate anaerobes, (b) gram-positive organisms other than *Bacillus*, and (c) the Enterobacteriaceae.

4.4.1.3 *Special Oxygen Physiology*

As is well known and discussed later (Section 4.6.1), denitrification is inhibited by oxygen. Two organisms, *Aquaspirillum magnetotacticum* and *Nitrosomonas europeae*, will not grow without O_2 and thus are obligately microaerophilic denitrifiers. For *Nitrosomonas*, oxygen is required by the ammonium oxygenase which initiates the ammonium oxidation pathway. For *A. magnetotacticum* the basis of the oxygen requirement is not known but is presumed to be required for an essential oxygenase reaction (12). Neither the use of complex media nor the addition of heme to media alleviated the oxygen requirement, however.

Two other denitrifiers that grow anaerobically on nitrate show enhanced growth with O_2. One of these, an unnamed gram-positive rod, will not grow with more than 1% O_2 in the headspace, nor will it grow well as a fermenter (N. J. Toltschin and M. K. Firestone, Abstr. Annu. Meet. Am. Soc. Microbiol. 1984, p. 187). Therefore, it seems to be specialized as a microaerophilic denitrifier. The other bacterium, *Thiosphaera pantropha*, appears to have a restricted electron flow to oxygen such that nitrate respiration allows for a more rapid growth (157). All of the organisms noted above must have a relaxed oxygen control of denitrification since they carry out denitrification in the presence of oxygen.

4.4.1.4 Metabolism of Pollutants

Besides the variety of natural products that are typically used as denitrifier carbon substrates, certain aromatic and anthropogenic compounds have also been shown to be degraded by denitrifiers. The use of denitrifiers in pollutant destruction is attractive for future research because (a) denitrifiers have the highest growth yield and are the easiest to grow of any bacteria capable of anoxic growth; (b) genetic manipulations and understanding is advanced in the most prevalent, competitive denitrifiers, the pseudomonads; and (c) because of its solubility and low cost, NO_3^- would be among the most feasible additions that could be made to polluted sites to enhance a desired anaerobic process. Only a few classes of chemicals have been studied for their metabolism by denitrifiers, but they illustrate that these organisms do possess potential for pollutant degradation.

Aromatic compounds, more widely recognized as being metabolized by O_2-requiring oxygenase pathways, can also be metabolized anaerobically. Anaerobic benzoate metabolism has been demonstrated in denitrifiers (66, 194), a photosynthetic anaerobe (*Rhodopseudomonas palustris*) (55), a sulfate reducer (142), and methanogenic consortia, especially by *Syntrophus* species (122). All seem to use a similar pathway. The proposed pathway of degradation of benzoates by denitrifiers is shown in Fig. 4.2. The aromatic ring is first reduced to cyclohexane carboxylate after which beta oxidation is thought to proceed via CoA-activated intermediates, eventually yielding adipate (215). As shown in Fig. 4.2, several substituted benzoates are first converted to benzoate and, therefore, presumably support denitrifier growth by the same pathway. *Pseudomonas* strain PN-1 released F^- form *o*- and *p*-F-benzoate but did not grow in the presence of either of *o*-F- or *o*-Cl-benzoates (195). The halogenated analog apparently inhibits benzoate oxidation. Other *Pseudomonas* strains subsequently isolated on *o*-aminobenzoate both grew and defluorinated *o*-F-benzoate, although these strains were also slowed in their benzoate metabolism by the presence of F- and Cl-benzoates (170). Since the defluorination seems to require the same enzymes as benzoate metabolism, the dehalogenation is assumed to be fortuitous. These dehalogenations are restricted to the smallest halogen, F. This is in contrast to the reductive dehalogenation by methanogenic consortia in which Cl, Br, and I groups were removed, but F groups were not (190). Hence there must be a basic difference in the enzymology responsible for these two different classes of dehalogenation.

As shown in Fig. 4.2, phthalate, a base unit used in plastics (4), and *o*-aminobenzoate (24), a product of some azo dyes and nitro compounds, can also be substrates for denitrifiers. All the isolates that grow on the substrates shown in Fig. 4.2 are *Pseudomonas* species except for the phthalate using organism, which

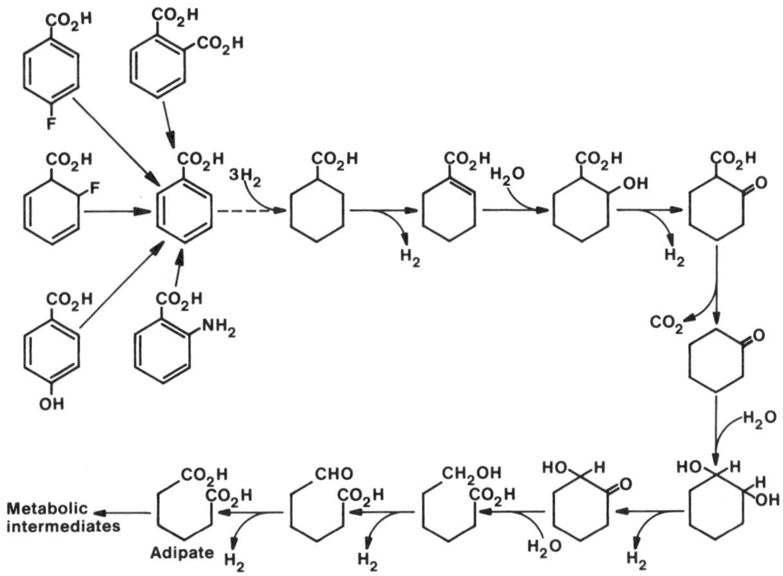

Figure 4.2 Pathways of benzoate metabolism by denitrifiers. The sequence for the conversion of benzoate to metabolic intermediates is proposed by Evans (55), but not all steps in this reaction have been proved.

is a *Bacillus* species (4). Thus the aromatic degradation capacity must also exist in gram-positive denitrifiers. It is notable that the substitutions in the ortho position apparently more readily yield denitrifying isolates. Hydroxy substitutions, regardless of position, are also known to be oxidized by denitrifiers. Apparently, denitrifiers can use other aromatic compounds as well. Williams and Evans reported that their benzoate-degrading *Moraxella* species would grow on the following substrates: benzaldehyde, benzyl alcohol, benzylglycine, protocatechuic acid, phenylacetic acid, cinnamic acid, *p*-hydroxycinnamic acid, caffeic acid, and phloroglucinol (215).

No denitrifiers have yet been isolated on phenol, but denitrifiers with this capacity apparently exist since phenol wastes have served as the carbon source in a denitrifying sludge reactor (8). None of the benzoate-utilizing denitrifier strains will grow on or oxidize either phenol or catechol (24, 194, 215).

Denitrifiers also degrade nonionic detergents. Nine isolates were enriched from river water using SDS (sodium dodecyl sulfate) as the sole carbon source under denitrifying conditions (51). The isolates studied grew both aerobically and anaerobically. They grew on alkyl sulfates of chain length C-6 to C-12 but not on shorter analogs. The alkyl sulfatase present in both aerobic and anaerobic grown cells converted the detergents to their corresponding fatty acid, which is presumably metabolized by beta oxidation.

Chlorinated solvents have been degraded under denitrifying conditions with primary sludge effluent as the inoculum (22). Four halomethanes at concentrations typical of polluted surface and groundwaters (42 to 115 $\mu g/L$) were degraded within 8 weeks; these were carbon tetrachloride, bromodichloromethane, dibromodichloromethane, and bromoform. Substrates not degraded were several chlorobenzenes, chloroform and trichloroethane. Using $[^{14}C]Cl_4$ as substrate, the following distribution of products was observed; 25% principally as $CHCl_3$, 43%; CO_2, 20% in cells; and 12% hydrophilic metabolites. Thus CCl_4 was both hydrolyzed and reduced. In this enrichment the concurrent degradation and denitrification stoichiometry were not studied, so that the degree of association of this solvent-degrading activity with denitrification is not well established.

If the bioconversion reaction and pollutant treatment system does not have an obligate O_2 requirement, denitrifiers could offer several advantages. Chemicals sensitive to oxidation could be produced, construction costs would generally be less since lower hydraulic volumes are possible in nonaerated units, odorous waste products are eliminated by the more oxidizing denitrification system, and because of the high growth yield, reaction rates should be higher than for other anaerobic bioprocessing systems. The potential of denitrifiers, especially with the genetic engineering now possible in pseudomonads, is just beginning to be investigated.

4.4.1.5 Regulation of Intermediate Production

There has been a large volume of work reporting the accumulation of intermediates in the denitrification pathway. This has been described for pure cultures, for cultures under stress conditions, and for natural samples. However, without unifying principles to explain such patterns, individual descriptions contribute little toward achieving a general understanding of the process and its control mechanisms. In an attempt to establish a general explanation for these observations, Betlach and Tiedje (14) proposed that the pattern of intermediate production could be explained by the reaction sequence $NO_3^- \xrightarrow{1} NO_2^- \xrightarrow{2} NO \xrightarrow{3} N_2O \xrightarrow{4} N_2$, with each step being described by the Michaelis–Menten model of enzyme kinetics. Furthermore, a general inhibition of any step could be described by the addition of a term, P, which could be scaled from 0 to 100% as an empirical description of the degree of inhibition:

$$v_i = P\,\frac{V_{\mathrm{max}_i}S_i}{K_{m_i} + S_i}$$

where i refers to the specific step (1 to 4) in the reaction sequence. Thus P, V_{max}, K_m, and S all interact to control the observed pattern of denitrification intermediates.

The classes of factors that affect denitrification kinetics and thus the terms described above are outlined below:

1. Quantity of enzyme (V_{max})
   a. Induction (if present): by NO_3^-, NO_2^-, and N_2O.
   b. Repression: by O_2.

2. Inhibition of activity (P)

    a. O_2: by competition for electrons or by enzyme inactivation.

    b. Limited electron donor: by shortage of available carbon.

    c. Other inhibition (e.g., pH, temperature, toxic chemicals).

3. Half-saturation constant (K_m) should remain a constant.

4. Substrate concentration (S)

    a. May show time-dependent changes.

    b. Diffusional or phase transfer limitation may mean that concentration at cell surface is not what is found by measurement of the bulk solution.

When denitrification intermediates accumulate, it can be due to the total lack of an enzyme in the sequence or to the kinetic parameters of the steps in the sequence as described by the model above. In the latter case the major products found in a culture or an environmental sample should be dependent on the conditions of the incubation since a sequential reaction should eventually go to completion unless an essential component is exhaused. The most common limiting factors are: (a) too little electron donor (carbon source) relative to electron acceptor (N oxides) (thus the reaction slows or stops at NO_2^- or N_2O), (b) too short an incubation time relative to the reaction rate (hence the characterization is time dependent and probably difficult to repeat); and (c) production of toxic products in a confined incubation vessel such that metabolism comes to a halt before completion of the conversion (e.g., stationary phase, pH shift).

Since the sequential Michaelis–Menten model can readily be implemented on personal computers, one can examine how changes in the various parameters would be expected to affect the pattern of intermediates and their concentration. Furthermore, with verified parameters for the particular organism or system, one may be able to use this model to predict denitrification intermediates and their relative concentration. Figure 4.3 shows examples of experimental data on the top panel and the model results generated by computer on the bottom panel. Dimensionless parameters for time (t) and S were used to illustrate the pattern of intermediates.

Typically, after NO_3^- is added to a denitrifying system, NO_2^- accumulates, as seen in Fig. 4.3A. In this particular case, NO_2^- accumulation is due entirely to a two-fold-greater rate of NO_3^- to NO_2^- reduction. The linear nature of the pattern of NO_2^- accumulation and consumption shows that there were no time-dependent changes of any inhibitory factor in the incubation. The bottom panel (Fig. 4.3A) shows the pattern of NO_2^- accumulation when the NO_3^- reduction rate was fivefold greater than the NO_2^- reduction rate. Thus the proportion of NO_3^--N that accumulates as NO_2^- should be greater, and the peak should be earlier than for the experimental results shown in the top panel. Figure 4.3B shows a very different NO_2^- pattern. This was caused by, and could be modeled by, a NO_2^- reduction rate that is fivefold greater than the NO_3^- reduction rate. This pattern is much more rare; it was found in a *Flavobacterium* strain (14). These two examples show

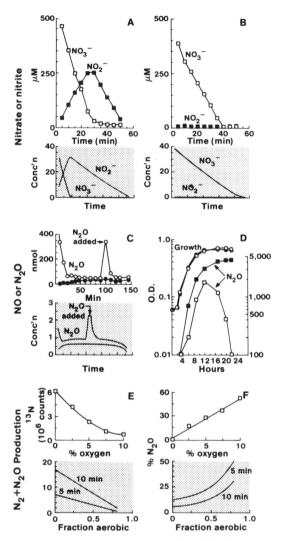

Figure 4.3 Kinetic explanation for the pattern of denitrification intermediate production and consumption. The top panel in each set is the experimental data from a denitrifying culture and the bottom panel (shaded) is the pattern produced by a mathematical model based on a Michaelis–Menten expression for each step in the denitrification sequence. Units are not given for the model version since they are illustrative. In (*A*) temporary nitrite accumulation is seen if the velocity of the nitrate reduction is greater than nitrite reduction. In (*B*) no nitrate accumulates when the nitrite reduction rate is greater than the nitrate reduction rate. (*C*) shows the behavior of NO (lower line) and N_2O under steady-state conditions, nitrate-reducing conditions, and after an addition of exogenous N_2O. (*D*) shows experimental data of a *Rhizobium* strain that temporarily accumulates N_2O during batch growth (solid symbols were plus 1% acetylene). (*E*) shows how a general inhibitor (e.g., O_2) reduces the total denitrification rate, which results in an increase in the proportion of N_2O (*F*). (*D*) redrawn from El Hassan et al. (53); (*A*)–(*C*) and (*E*) from Betlach and Tiedje (14).

196

that the relative enzyme activity for the steps in the denitrifying sequence can differ widely among bacterial strains.

The pattern of NO and N_2O accumulation often shows a steady-state concentration (Fig. 4.3C). In this case additional N_2O was added initially and at 100 min. As expected from the kinetic model and the appropriate parameters for this organism (*Ps. fluorescens*), the additional N_2O was rapidly consumed until a concentration range was reached where substrate concentration became the major parameter limiting the rate of N_2O consumption. In this experimental system (Fig. 4.3C), the electron donor was in excess and the competition for electrons by NO_3^- or NO_2^- was not significant. This is not always the pattern observed as shown in Fig. 4.3D, where N_2O accumulates in a temporary manner much as shown for NO_2^- in Fig. 4.3A. It is also important to note in Fig. 4.3D that the presence of acetylene had no effect on the growth rate of this fast-growing cowpea *Rhizobium*.

NO, which may not be a free intermediate in denitrification (7, 63), does accumulate in low concentrations in a pattern typical of a free intermediate. However, since labeling studies show that NO can be in rapid equilibrium with an enzyme-bound intermediate, the pattern of NO would be expected to behave in a manner typical of an intermediate (58) even if it was not a true intermediate. NO is commonly measured in any denitrifying culture or system if the conditions for its analysis are adequate (14, 90, 93). Since NO is extremely reactive, it can easily be lost before analysis is complete.

A particularly important prediction from the sequential Michaelis–Menten model is that anything that generally slows the rate metabolism should increase the proportion of N_2O. Experimental results consistent with this explanation are shown in Fig. 4.3E and F, with O_2 as the inhibitor. However, other examples in the literature show an increased proportion of N_2O at lower temperatures (94, 100) or with high concentrations of inhibitory pesticide (20), both examples where a general reduced rate of metabolism is expected. This model also predicts that any reduction in available carbon (and therefore electron donor) should also increase the proportion of N_2O, even if NO_2^- and NO_3^- were not preferentially competing for electrons.

4.4.2 Dissimilatory Nitrate Reduction to Ammonium

4.4.2.1 General Features

Dissimilatory nitrate reduction to ammonium (DNRA) is a process that can be recognized on physiological and ecological grounds, but because of the diversity of enzymes and organisms that carry out this process, it has not been well characterized on a biochemical level. This is in contrast to denitrification, which is characterized by two similar and identifiable nitrite reductases that carry out the key step forming the N-N bond. The process has also not been systematically studied at the microbiological level. Nonetheless, it is a process well recognized as important in anaerobic habitats such as the rumen (89, 110), anaerobic sludge digestors (90) and anoxic sediments (21, 92, 93, 104, 185), and in a number of obligately anaerobic and facultatively anerobic bacteria.

4.4.2.2 Organisms

DNRA has so far been found in bacteria that have fermentative rather than oxidative metabolism, which is the opposite of denitrification. Organisms in which this process has been well characterized are listed in Table 4.5. DNRA was recognized long ago in some *Clostridia* and rumen bacteria (e.g., *Wolinella* and *Selenomonas*) but has been studied primarily in the last decade. This process has now been documented in *Veillonella* and *Desulfovibrio*. Of the other obligate anaerobes, only *Bacteroides* has species that reduce nitrate (107). Whether some of them can also dissimilate nitrite to ammonium is not clear.

Among the facultative anaerobes, the capacity to dissimilate nitrate to ammonium is rather common, with it being reported in seven of the most common genera

TABLE 4.5 Organisms that have been reported to dissimilate nitrate or nitrite to ammonium (DNRA)

	Typical Habitat	References
Obligate anaerobes		
Clostridium (several species)	Soil, sediments	31, 30, 76, 217
Veillonella alcalescens	Intestinal tract	84, 219
Wolinella (Vibrio)[a] succinogenes	Rumen	18, 113, 216
Desulfovibrio desulfuricans	Sediment	113, 188
Desulfovibrio gigas	Sediment	11
Desulfovibrio species	Sediment	120
Selenomonas ruminantium	Rumen	48
Facultative anaerobes		
Escherichia coli	Soil, wastewater	17, 40
Citrobacter sp.	Soil, wastewater	180
Salmonella typhimurium	Sewage	168
Klebsiella species	Soil, wastewater	17, 41, 52, 168
Enterobacter (Aerobacter)[a] aerogenes	Soil, wastewater	17
Serratia marcescens		17
Erwinia carotovora	Soil	17
Photobacterium (Achromobacter)[a] fischeri	Sea	150
Vibrio (several species)	Sediment	114
Microaerophile		
Campylobacter sputorum	Oral cavity	49
Aerobes		
Pseudomonas (several strains)[b]	Soil, water	166
Neisseria subflava[b]	Mucous membranes	72
Bacillus (several strains)	Soil, food	30, 179, 207

[a] Genera of original description, but former genera have been discarded or corrected in later accepted taxonomy.
[b] Process not well characterized in these organisms.

of the Enterobacteriaceae, in *Photobacterium*, and in *Bacillus* species. Since members of virtually all of the Enterobacteriaceae and Vibrionaceae reduce nitrate, the DNRA capacity may be even more widespread than recognized. *Aeromonas* and *Vibrio* appear to be the most prevalent anaerobic nitrate-reducing organisms in estuarine sediments (41, 114).

The occurrence of this process in organisms with only an aerobic respiratory metabolism is not well documented except for *Campylobacter*, which is a microaerophile that is also capable of anaerobic growth with fumarate or nitrate as an electron acceptor. This organism is more like a denitrifier in that it has nonfermentative electron sinks to supplement its basically aerobic metabolism (49). There is some preliminary information that some other aerobes—*Pseudomonas* and *Neisseria subflava*—can also dissimilate nitrate to ammonium under certain conditions (72, 166).

4.4.2.3 Physiology

The first reaction in the pathway is nitrate reduction to nitrite, termed nitrate respiration, a process well known in *E. coli* and other enteric organisms. This step is coupled to energy production in most organisms except the *Clostridia* (15). This reaction is found in many more organisms than those listed in Table 4.5 and is the basis of the nitrate reduction test (really nitrite accumulation) often used in bacterial identification. This step, although necessary, is not the distinctive step in DNRA since it is usually not the rate-limiting step. Furthermore, nitrite accumulated by nitrate respirers could readily be converted to gas by denitrifiers coexisting in the community. Hence the critical step in DNRA is the nitrite-to-ammonium conversion.

The benefit to the cell of DNRA is thought to be (a) detoxification of accumulated nitrite, (b) an electron sink which allows the reoxidation of NADH, and (c) the production of energy through electron transport phosphorylation (ETP), such as occurs for the denitrifying nitrite reduction. Of these, the most commonly postulated function is that of an electron sink.

The high capacity for accepting electrons per nitrate reduction—eight—makes this reduction one of the most favorable to anaerobes faced with a shortage of electron acceptors (200). In an organism such as *Clostridium*, which has no ETP coupled with nitrate reduction to ammonium, the molar growth yield on glucose was increased 15.7% by use of this dissimilatory pathway (31). The increased growth yield is thought to be due to the increased substrate-level phosphorylation allowed by use of the alternative electron acceptor. There is little evidence to show how important the detoxification function is, but for *Clostridium* strain KDHS2, nitrite concentrations greater than 2mM inhibit growth. Detoxification is most likely to be a consequence of the electron sink function rather than the main benefit of the reduction.

Energy conservation coupled to nitrate reduction to ammonium had not been reported until 1982, but it has now been demonstrated in three genera: *Campylobacter sputorum* (48), *Desulfovibrio gigas* (11), and *Wolinella succinogenes* (18). All three organisms can produce ATP by coupling the oxidation of H_2 or formate to NO_2^- reduction to NH_4^+. *E. coli* has also been shown to couple energy production to nitrite reduction to ammonium but only with a formate-coupled nitrite reductase (121). This pathway, however, accounts for only 20% of the nitrate reduction to ammonium in *E. coli*. The organisms above all use a membrane-bound nitrite reductase. In the other DNRA organisms where the enzymology has been studied, the nitrite reductase is soluble and therefore is not expected to conserve energy.

The lack of energy conservation in the nitrite-to-ammonium step for most DNRA organisms probably explains the tendency of these organisms to accumulate nitrite under carbon-limited conditions. Since ATP is produced from the nitrate-to-nitrite step but not from nitrite reduction, it is more advantageous for the organism to divert its limited electron flow to the energy-producing step. However, under nitrate-limited conditions the need for a high-capacity electron sink is more important, so the reduction to ammonium is expected. This shift under carbon versus nitrate limiting conditions is illustrated in Table 4.6 for two strains of *Klebsiella* grown in continuous culture. This table also shows the shift of the nitrate products from ammonium to nitrite for most of the 70 environmental isolates when grown in a carbon-rich versus a carbon-poor medium. Thus the behavior of DNRA under well-defined continuous culture conditions is consistent with the behavior of the predominant anaerobic DNRA organisms from soil. The non-energy-linked nitrite-to-ammonium producers are thought to be the predominant type in nature.

The non-energy-linked nitrite reducers seem to carry out this reaction by a variety of soluble nitrite reductases (42). Some of these are NADH linked (e.g., *E. coli*), which is consistent with its assumed role of NADH reoxidation. One of these enzymes is a sulfite reductase; since DNRA is often high in sulfate-contain-

TABLE 4.6 Effect of carbon versus NO_3^- limitation on products of NO_3^- dissimilation in strains that dissimilate NO_3^- to NH_4^+

Organism/Source	Limitation	Fate of NO_3^- in medium		References
		NO_2^-	NH_4^+	
Products in continuous culture		(% of NO_3^-)		
Klebsiella K312	NO_3^- limited	<0.1	63	41, 52
	C limited	11	1.6	41, 52
Klebsiella aerogenes				
(NCIB 418)	NO_3^- limited	<0.1	51	41
	C limited	48	<0.1	41
Major products of 70 isolates		(% of isolates)		
Soil	Tryptic soy	79	81	179
	broth[a]	90	10	179
	Nutrient broth			

[a] Tryptic soy broth (TSB) has more available carbon than nutrient broth, usually making TSB NO_3^- limited.

ing marine sediments, nitrite reduction to ammonium may be enhanced because it is an alternative substrate for the sulfite reduction pathway. Some of the nitrite reductases may have primarily an assimilatory function. For example, the synthesis of nitrite reductase in *Veillonella* requires nitrite or nitrate and is repressed by ammonium and tryptone, yet the organism excretes ammonium from nitrite reduction and benefits from the electron sink provided (219). It is reasonable that a number of organisms with the assimilatory nitrite-reducing pathway may use this pathway during periods of oxygen stress to accept electrons and thereby aid survival. The high number of DNRA organisms in soil (179, 200)—higher than expected from the generally aerobic nature of soil—would suggest that this activity may be due to other than the obligate anaerobes and the *Enterobacteriaceae*. The pseudomonads mentioned in Table 4.5 could be examples.

4.5 POPULATION ECOLOGY OF DISSIMILATORY NITRATE REDUCTION

4.5.1 Denitrification

The population ecology of denitrifiers has been neglected. Many data exist on denitrifier enumeration of natural samples, but much of this information was taken in studies where the major goal was to make conclusions on denitrification activity. This approach is unsound because factors other than population density dominate the control of denitrification (see Section 4.6). Information on denitrifier populations, however, is important to answer a number of other questions about denitrification.

Autecological studies, which are numerous for *Rhizobium* and nitrifiers, for example, are virtually nonexistent for denitrifiers. One study using antibiotic-resistant strains (178) and one using fluorescent antibodies (197) have demonstrated that these autecological techniques can be used. The difficulty, however, is that with so many diverse denitrifiers, it is hard to generalize from the behavior of one marked strain. Nonetheless, identification of specific features important to competition of denitrifiers can best be answered by this more specific approach.

4.5.1.1 *What Denitrifier Populations Are Found?*
Although it is recognized that denitrifiers are widely distributed in nature, the similarity in denitrifier populations between soil and aquatic environments is quite high (Table 4.7). In studies by Gamble et al. on the numerically dominant denitrifiers in soils from eight countries, the majority of the isolates were *Pseudomonas* species with *Alcaligenes* species as the second most prominent (62). Since that paper appeared, some of the *Alcaligenes* strains have been found to be *Ps. pseudoalcaligenes* (14), which further expands the proportion of pseudomonads in the study to at least 75% of the 146 N_2-producing strains. In Terai's study (196) of denitrifier isolates in Lake Kizaki, Japan, 77% were characterized as *Pseudomonas*, with the remainder *Alcaligenes* or *Achromobacter* (the genus *Achromo*-

TABLE 4.7 A comparison of some denitrifier population characteristics between soil and water habitats[a]

Characteristic	Soil (62, 208)	Lake Water (197, 196, 127)	Freshwater Sediment (62)	Marine Water (214)
Most prevalent genera	*Pseudomonas* >> *Alcaligenes* *Flavobacterium* > *Bacillus*	*Pseudomonas* >> *Alcaligenes*	*Pseudomonas* > *Alcaligenes*	*Pseudomonas* >>> of *Bacillus, Alcaligenes*
Ratio of fluorescent *Pseudomonas*: achromogenic *Pseudomonas*	50:50	50:50	50:50	
Percent of total population[b] that denitrifiers comprise	0.1–5%	0.01–5%	—	0.01–9% $\bar{x} = 1.4\%$
Denitrifier density[c] (number of organisms)	10^4–10^6 g^{-1}	5–10^3 mL^{-1}	10^6 g^{-1}	10^{-1}–10^2 mL^{-1}

[a] References shown in parentheses.

[b] Total population that will grow in complex media.

[c] In some studies, methodology limitations have probably caused an underestimation of the denitrifier population density.

bacter has been discarded and most of these strains are now *Alcaligenes*). Suga-hara et al. found that 98.1% of the 264 denitrifying isolates from marine waters were *Pseudomonas* species (191). Taxonomic studies of denitrifiers in sediments also show *Pseudomonas* to dominate, with *Alcaligenes* next in importance (62). The media used in these studies may be selective for these two genera; however, most other denitrifiers are more specialized (Table 4.3) and would not be expected to be found in such high densities (Table 4.7). Furthermore, general complex me-dia rather than selective media were used. One case where this generalization might not be true is with *Rhizobium*, since they do not grow on these media. In some New Zealand pasture soils, *R. lupini* numbers were equivalent or greater than other denitrifier numbers; 90% of *R. lupini* strains are claimed to be denitrifiers (133, 187).

It is also striking that the proportion of fluorescent versus nonchromogenic pseu-domonads in the population is the same in soil, water, and sediment habitats (Table 4.7). Furthermore, the proportion of the total viable population comprised of de-nitrifiers is also nearly similar for all habitats. The denitrifier population density is higher for soils and sediments, but this is thought to be a function of the higher carbon content and permanent support found in these habitats.

Terai and Yoshioka were the first to use fluorescent antibodies to study denitri-fiers in nature (197). Using fluorescent antibodies to six different taxonomic groups, they were able to identify by agglutination and fluorescent staining 25% of the 60 new isolates and the species in 13% of denitrification positive MPN tubes. Given the diversity of denitrifiers, this degree of serological similarity is encouraging for use of this method.

4.5.1.2 What Selects Denitrifier Populations?

The similarities in denitrifier species and proportion of the total heterotrophic pop-ulation between soil and water habitats suggest that the major factor controlling denitrifier populations is not the denitrification property but rather their general ability to compete for natural carbon substrates irrespective of habitat. Since growth must occur to establish and maintain populations, one can examine this thesis by looking at conditions in which competitive growth has occurred.

There are several lines of direct evidence which show that competition for car-bon under aerobic conditions is the major factor responsible for the population distribution of denitrifiers. Myrold and Tiedje (125) showed that when ground alfalfa was added to soil, the denitrifier biomass increased in proportion to the total population under conditions which should not favor organisms with the denitrifi-cation capacity (less than 0.1% of the soil could be calculated as having an an-aerobic volume). Thus the denitrifier growth response must have been based on the ability of this group to compete for added carbon as aerobic heterotrophs. Smith and Tiedje, using rifampicin-resistant denitrifier strains, showed that at least one very common soil denitrifier, *Ps. fluorescens*, actually grew better as an aerobe in the face of natural competition than under denitrifying conditions (178). The suc-cess of denitrifiers under aerobic condition is also illustrated by the fluorescent antibody studies of Terai and Yoshioka, who found that the two most common serogroups (a *Ps. fluorescens* and a nonfluorescent *Pseudomonas*) were found only in lake waters with oxygen concentrations >10 and >5 mg L^{-1}, respectively (197). Since both of these common denitrifiers were detected only in the oxic waters, they must have been growing as aerobic heterotrophs and with no benefit from their denitrification capacity. Similar data exist for soils, although in this habitat it is more difficult to discount the presence of anaerobic microsites. Nonetheless, it is striking that denitrifiers are found in nonaggregated, course sandy soils including desert soils (209), conditions where it is difficult to imagine that denitrification ever plays a role in their success.

Although the examples above show specific growth responses, there are also considerable data that show a correlation between organic carbon on the habitat and denitrifier population density. For example, in soil both denitrifier density and organic carbon decrease with depth in soil (208). Sediments also show a correlation between the organic carbon content and denitrification capacity and not with NO_3^- availability (102, 104). Nakajima's study of lakes and river waters shows that the numbers of denitrifiers increased with increased BOD and with increased particulate nitrogen for attached streambed populations (127).

Thus my general impression is that most of the denitrifiers in nature exist because they are effective aerobic competitors for carbon and never use their denitrification capacity. One would, of course, expect selective enrichment of denitrifiers when nitrate is available and oxygen is limiting, but these situations are rare, and when they occur it is only for brief periods. Jacobson and Alexander showed that it takes 2 to 8 μg of $NO_3^- - N$ to grow 10^6 denitrifiers (86). In most natural habitats (except fertilized fields), there would usually not be enough nitrate to support the large extant population of denitrifiers if their growth was predominantly due to denitrifying respiration. Although I believe that competitiveness for carbon is the dominant factor responsible for denitrifier populations in nature, I do not want to imply that it is the only factor.

4.5.1.3 Adaptive Traits of Denitrifiers

Denitrifiers have recently been shown by Kennedy and Lawless to exhibit positive chemotaxis for nitrate and nitrite beginning at about 10^{-5} M nitrate and with a maximum response at 10^{-3} M (95). The taxis response occurs for organisms grown both aerobically and anaerobically and with and without nitrate. In the three soils surveyed, 62% of the bacteria that moved to the nitrate source were *Pseudomonas* species. In soil, taxis responses are generally not thought to be useful for survival since movement is so restricted by the extensive surface area, the attraction of bacteria to these surfaces, and limited water. Furthermore, NO_3^- is very soluble in soil, and it would seem to be a very rare circumstance where NO_3^- flux would not be greater than bacterial movement. But in microbial films and in aquatic habitats, chemotaxis to NO_3^- could be beneficial. Perhaps this is further evidence that these dominant denitrifiers are not specialized for soil versus water habitats.

The adaption by denitrifiers to environmental stress (e.g., extremes of pH, temperature, salt, pressure, light) are not well studied. One would expect such an adaptation if the selective conditions existed, and indeed, thermophilic and halophilic denitrifiers are known (Table 4.3). One historical and problematic question is whether denitrifiers are adapted to the low pH that is often found in soils. It is now well established that some soils with pH values as low as 4.0 denitrify, although the denitrification rate is much reduced (69, 105, 123, 137). Parkin et al. (137) recently showed an apparent population adaptation to low-pH condition, since the population in a soil of pH 4.08 had an optimum denitrification rate of pH 3.9, while a similar and adjacent soil had a pH of 6.05 and denitrification optimum of 6.5 (Fig. 4.4.). However, we have not been able to isolate an acid-tolerant denitrifier from this site. Initial enrichments seem to lose their acid tolerance when repeatedly subcultured in the laboratory.

4.5.1.4 Survival of Denitrifying Activity

Recently, Smith and Parsons reported the surprising observation that denitrification enzyme activity persisted extremely well in air-dried soils and that this activity was not due to newly grown denitrifier cells (182). Although this phenomenon reflects enzyme activity more than denitrifier populations, it is important to interpretation of denitrifier population data and may be suggestive of some unique denitrifier populations. Representative data illustrating the key features of this study are summarized in Table 4.8.

The major feature to be noted is that when soils were air dried, only a minor portion (16 to 29%) of the denitrifying enzyme activity was lost (Table 4.8, column 2), but when cells newly grown in soil were subjected to the same treatment, virtually all of the enzymatic activity was lost (Table 4.8, columns 3 to 5). This was shown by three treatments. In the first treatment (column 3) the soil was preincubated anaerobically for 3 days, which increased the denitrifying population, but when the soil was air dried, the activity that remained (8.8 ng of N_2O-N/g soil $\cdot$ min^{-1}) was similar to the residual activity in the original air-dried soil. In the second treatment (column 4), when the soil was first heat shocked to kill the indigenous denitrifying organisms and treated as above, all the denitrifier activity

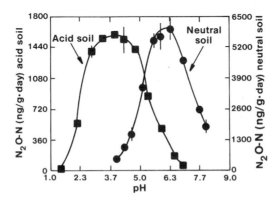

Figure 4.4 pH optima of denitrification for two soils; the pH of the acid soil was 4.08 and the neutral soils was 6.05. Redrawn from Parkin et al. (137).

TABLE 4.8 **Differences between indigenous soil denitrifier activity and freshly grown soil denitrifier activity as revealed by the effect of air drying and the accompanying aeration on soil denitrifying enzymes, populations, and activity**

Soil Condition	Denitrification enzyme activity (ng N_2O-N/g soil · min)				Denitrifier Counts ($\log_{10} g^{-1}$)	N_2O flux from soil cores after irrigation (ng N_2O-N/g · min)	
	Fresh Soil	3-Day Anaerobic Preincubation	Heat Shocked[a] Then 3-Day Anaerobic Preincubation	*Ps. fluorescens* Added to Sterile Soil		10 h	37 h
Fresh soil, 27% moisture	9.0	26.0	30.8	28.0	6.9	<0.01	0.15
Air dried for 7 days, 2–4% moisture	7.6	8.8	0.6	0	5.0	0.76	0.28

Source: Data on Donerail soil summarized from Smith and Parsons (182).

[a] Heat-shocked fresh soil (78°C for 15 min) which destroyed most initial denitrifying activity (<0.01 ng N_2O-N/g soil · min); 3-day anaerobic preincubation allowed regrowth of new denitrifier cells.

was lost. In the third treatment (column 5), when a denitrifying *Pseudomonas* strain was added to the same soil that was previously sterilized, allowed to grow for 3 days, and then air dried, all the denitrifying activity was again lost. Thus only the indigenous activity, not that of newly grown denitrifier populations, withstood the stress of air drying.

Two other features of this air-drying effect are the reduction in the recoverable denitrifier population (Table 4.8, column 6) and the much faster initiation of denitrifying activity by the air-dried soil compared to the moist soil (Table 4.8, column 7 and 8). The declining population was probably due to the reduction in the ability to recover these stressed cells since the per cell denitrifying activity under these conditions is higher than reported for any denitrifier in culture under optimum denitrifying conditions. The more rapid denitrification response of the air-dried soil is surprising because of the oxygen sensitivity of these enzymes. However, this could be explained by the more rapid wetting of the dried soil and the pulse in respiration provided by the released carbon, thereby more rapidly creating anaerobic microsites. This explanation has not been proved, but the observation of the faster denitrification response in the air-dried soils has been seen by several authors (25, 115, 139), and thus the phenomenon seems to be a general one. This rapid response also emphasizes how fast denitrification can be initiated which is certainly faster than can be explained by either population growth or even enzyme synthesis.

The explanations offered by Smith and Parsons for the air-dried, resistant, denitrifying enzyme activity are (a) the physical association of the indigenous microbes with the soil matrix is protective (but the mechanism is not protective to the newly grown cells); (b) a second population is present that is slow growing and not easily isolated but is persistent on drying; and (c) the denitrifying enzymes of the aged indigenous organisms or their carcasses are stabilized. There are no data yet to support or refute any of these hypotheses.

A second environment in which denitrifying enzyme activity persists when it is not expected is in the constantly anaerobic habitats, those not receiving either oxygen or nitrate. The high-denitrifying enzyme activity in these environments is surprising because, as discussed earlier, there are not denitrifiers yet reported that can live in the absence of these electron acceptors (Section 4.4.1.2). There are two denitrifiers (*Bacillus* and *Azospirillum*) that are weak fermenters, and there are two organisms that can grow anaerobically and reduce nitrite to N_2O but with no energy gain (*Propionibacterium* and *Chromobacterium*). Perhaps these organisms or ones like them can explain this observation, but it seems unlikely since they either grow very poorly under these conditions, are not known to be sediment organisms, and/or carry out a rather feeble reduction of nitrite to N_2O. The amount of denitrifying enzyme activity in two of these anerobic habitats is shown in Table 4.9, where these rates are compared to the activity measured in the same way for a number of aerobic soils. The fact that enzyme activity in the anaerobic habitats is 20 to 1000 times higher than in soils suggests that this is not a feeble activity due to decaying enzymes or cells or to a secondary metabolic activity. The most likely explanations of this paradox are that there is a population of yet undiscov-

TABLE 4.9   Comparison of denitrifying enzyme concentration in aerobic habitats versus anaerobic habitats that should not allow respiratory growth

Source	Denitrification Activity (nmol N_2O/g dry wt · h)
Aerobic	
Forest soils	3–14
Agricultural soils	0.2–10
Anaerobic habitats free of O_2 and NO_3^-	
Eutrophic lake sediments, depth 9–12 cm	230
Sludge from anaerobic digester	1090

Source: Summarized from Ref. 200.

ered obligate anaerobes that are capable of respiratory denitrification, or that known respiratory denitrifiers can form syntrophic associations with other populations that serve as their electron acceptors. It is perhaps also feasible that the stabilized enzyme explanation suggested for the air-dried soils can be offered for the sediment observation.

These two examples show the value of having a process assay in addition to a population assay since the lack of parallel results may lead to uncovering novel features about the biology of the process. In both cases described above, the enzyme activity assay showed that the present understanding of known populations and their physiology is not sufficient to explain the process as it is expressed in nature.

4.5.2   Dissimilatory Nitrate Reduction to Ammonium

4.5.2.1   *What DNRA Populations Are Found?*

Data on the population ecology of DNRA organisms are virtually nonexistent. One study has identified a few of the many DNRA isolates obtained from soil (179). Of the 14 strains identified, the order of prevalence was *Bacillus* > *Enterobacter* > *Citrobacter*. Another study showed *Clostridia* capable of DNRA to be very numerous in soil (30). Cole and Brown, studying DNRA populations in marine sediments, found *Aeromonas* and *Vibrio* species to be the most prevalent nitrate reducers, but their particular isolates did not reduce nitrite to ammonium (41). The only isolates which they obtained that produced ammonium were enterobacteria, especially *Klebsiella*. Later MacFarlane and Herbert also found *Aeromonas/Vibrio* species to be the most prevalent nitrate-respiring bacteria in estuarine sediments, followed by Enterobacteriaceae and then by *Pseudomonads* and gram-positive bacteria (114). However, their selected *Vibrio* isolates accumulated ammonium in the medium when grown under NO_3^--limited conditions (114). Strains of *Clostridia*, *Desulfovibrio*, and *Photobacterium* are capable of DNRA (Table 4.5) and would be expected in marine sediments, but their importance in DNRA in nature has not yet been shown. From this limited data, it is clear that the question of which DNRA populations are most important in soils or sediments has not been answered. Two population studies done with soils using MPN methodology have both shown that DNRA populations are slightly more numerous then denitrifier populations (179, 200). Although the enterobacteria are the most commonly encountered, they are not yet thought to be sufficiently successful environmental inhabitants to account for the expected high density of organisms with this activity.

4.5.2.2 What Selects DNRA Populations?

Nitrate use does not appear to be an important selector of DNRA populations since these populations seem successful when nitrate is very low or absent (sediment) or when anoxic conditions are limited (soils). Thus, like denitrifiers, the populations of DNRA organisms are probably a result of their successful competition for carbon coupled with either respiratory metabolism (soil) or fermentative metabolism (sediment).

We have also noted that DNRA activity is highest in carbon-rich, electron-acceptor-poor environments. This condition is suggested to contribute to selection for DNRA organisms. We previously postulated (Fig. 4.5) that the ratio of available carbon-to-electron acceptor controls whether nitrate partitions to DNRA or to denitrification (200). This hypothesis is based on the fact that under a high ratio the greatest need in metabolism is for maximum electron acceptor capacity. When the ratio is low, the greatest advantage is to the organism that gains the most energy per nitrate; this is denitrification (200). The projected outcome of nitrate competition between the two processes could be a result of populations selected by this ratio, which is probably the case in the rumen and the sludge digestor, or to different kinetic parameters of nitrate uptake by the existing populations. The latter has been supported by King and Nedwell, who found that in sediment slurry the proportion of nitrate converted to ammonium increased as the nitrate concentration decreased (96).

In summary, habitats that are continuously anoxic would be expected to select for vigorous fermenters and obligate anaerobes, features of some DNRAs but not

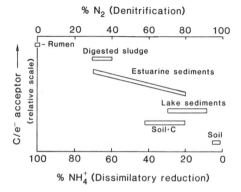

Figure 4.5 Partitioning of nitrate between denitrification and DNRA as a function of available carbon/electron acceptor ratio. Reproduced from J. M. Tiedje, A. J. Sexstone, D. D. Myrold, and J. A. Robinson, Denitrification: ecological niches, competition and survival, Antonie van Leeuwenhoek, 48:569–583, 1982.

of denitrifiers. Habitats that are predominantly aerobic would be expected to select populations that are most efficient and competitive at using the available carbon under respiratory conditions. The use of nitrate is a secondary feature but would favor DNRA organisms under continuous anoxia, and denitrifiers when anoxia is only temporary.

4.6 REGULATION OF DENITRIFICATION IN THE ENVIRONMENT

4.6.1 Characteristics of Oxygen Regulation

4.6.1.1 Oxygen as a Dominant Regulator

Oxygen inhibition of denitrification has been known since Gayon and Dupetit first reported the process in 1882 (68). In the past century we have come to recognize the generality and magnitude of importance of oxygen to control this process in nature. But the biochemical mechanism(s) of inhibition is still not understood nor is the relationship between oxygen concentration and denitrification well quantified. The impact of oxygen on *in situ* denitrification activity is illustrated by studies in which denitrification rates in soil cores with their natural aerobic atmospheres were compared to the rates in the same cores with the atmosphere replaced with an oxygen-free gas. These studies show that the typical aerobic denitrification rate is only 0.3 to 3% of the anaerobic rate (135) and thus directly implicate oxygen as the major regulator of denitrification activity. The magnitude of the oxygen inhibition and the response of denitrification rate to oxygen concentration are illustrated by the two sets of data in Fig. 4.6. The pattern is similar even though the data are from two very different experimental systems: a soil core (135) and a wheat–*Azospirillum* rhizosphere association grown in soft agar (131). Note the dramatic drop in denitrification rate with a slight increase in oxygen concentration. Although these curves illustrate the dramatic effect of oxygen, they do not directly quantify the effect of oxygen on the process since the position and shape of these curves are controlled by oxygen diffusion to the active cells and by the strength of the respiratory sink for oxygen.

4.6.1.2 Oxygen Threshold for Denitrification

The concept of a threshold oxygen concentration below which denitrification is allowed has arisen because of the dramatic responses to oxygen, such as those shown in Fig. 4.6. A threshold implies a critical value for a breakpoint rather than a linear or gradual inhibitory response. Further evidence of a sharp breakpoint is shown by the relationship between N_2O concentration and oxygen concentration for lake, sea, and ocean data (70, 101, 159, 221). As shown for the Baltic Sea data (159), the N_2O production increased as oxygen decreased (due to nitrification) until a critical point of approximately 9 μmol O_2/L^{-1} where N_2O consumption dramatically dominated the profile (Fig. 4.7). The sharpness of the breakpoint for denitrification is easily observed when compared to the gradual change noted for the oxygen effect on N_2O production. Even though the breakpoint is sharp, it is

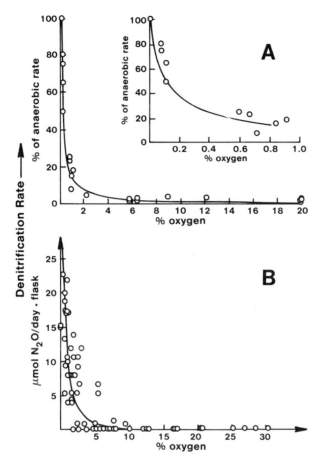

Figure 4.6 Effect of O_2 concentration on denitrification in (*A*) a soil core and (*B*) a wheat-*Azospirillum* rhizosphere association. (Redrawn from Refs. 135 and 131, respectively).

unlikely that there is a single critical oxygen concentration. It is more reasonable to presume a narrow range of values that describes the oxygen threshold.

Threshold values have been quoted in the literature for some time (e.g., 0.2 mL L^{-1} (39) and 0.2 mg L^{-1}) (32, 102), but the actual data in the papers that have been used to derive those values are more of a qualitative than quantitative character. The quoted and other data that provide information on the oxygen threshold are summarized in Table 4.10 along with the type of measurement used. I have excluded data in which the oxygen concentration near the denitrifier cell surface could not be assured, which is often the case because of the high respiratory activity of typical cultures and the transport limitations on oxygen flux to the cells. The two situations that minimize these confounding effects are low-density cultures in oxygen-controlled chemostats, and nitrogen profiles from the relatively

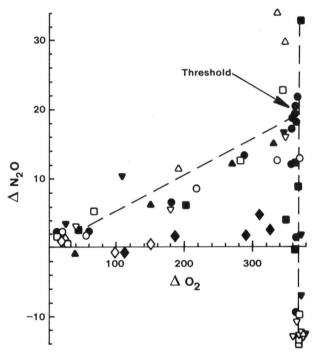

Figure 4.7 A threshold for N_2O reduction is suggested by the relationship between N_2O and O_2 concentrations in the deep oxygen-deficient waters of the Baltic Sea. The data are expressed as the deviations (Δ) from saturation of N_2O (nM, obs.-sat) and of O_2 (μM, sat.-obs.). The points are from different sampling stations and dates. Redrawn from Rönner (159).

oligotrophic marine waters; these two types of data dominate Table 4.10. The most definitive data are those of Rönner and Sorensson (161), who measured $^{15}NO_3^-$ conversion to $^{15}N_2$ in bottles of Baltic Sea water along with measuring the oxygen concentration at the time of analysis, and several chemostat studies in which oxygen was monitored by an electrode (80, 87, 130). Unfortunately in the latter cases, no denitrification was measured at any oxygen concentration other than zero; the threshold for denitrification appears to be below the oxygen detection limit of the Clark electrode. Nonetheless, from Table 4.10 one can safely generalize that the threshold for denitrification is at least as low as 10 μmol L^{-1} and is probably lower, but this has not been verified because of the inability to reliably measure oxygen at the necessary concentrations. It is also notable that the oxygen concentrations reported in the denitrification literature are in at least 10 different units: μg-atoms L^{-1}, μmol L^{-1}, mL L^{-1}, mg L^{-1}, % O_2 saturation, % air saturation, atmospheres, pascals, ppm, and mmHg; this situation does not foster understanding. Although some of these units are easily derived from another, others are not because of the effect of temperature, salt, and pressure. I have used μmol L^{-1} as

TABLE 4.10 Summary of the threshold value of O_2 concentrations that permits denitrification

Threshold O_2 concentrations				
Reported Units	Converted to Equivalent Unit (μmol L^{-1})	System Measured [a]	Type of Measurement	Reference
~0.2 mL L^{-1}	8.9	Eastern tropical North Pacific	$^{15}N_2$ produced from $^{15}NO_3^-$ amended samples taken from water columns with measured O_2 profile	70
0.2 mg L^{-1}	6.2	Lake 227, Manitoba	$^{15}N_2$ produced from $^{15}NO_3^-$ amended samples taken from water columns with measured O_2 profile	32
<10 μmol L^{-1}	<10	Saanich Inlet, British Columbia	Profile of rapid N_2O depletion	38
9 μmol L^{-1}		Baltic Sea	Profiles of rapid N_2O depletion	159
0.2–0.25 mL L^{-1}	8.9–11.1	Baltic Sea	$^{15}N_2$ produced from $^{15}NO_3^-$ amended samples taken from water column and with O_2 measured in bottles at each time point	161
17 μmol L^{-1}	17	Eastern tropical North Pacific	N_2O profiles: breakpoint between production and consumption	39
1.5% O_2	$(21)^b$	*Pseudomonas denitrificans*	Rapid increase in NO_2^- consumption	165
0.2 ppm[c]	6.2	*Pseudomonas denitrificans*	Nitrate reduction in stirred cell suspension with solution O_2 measured by a dropping mercury electrode	174

TABLE 4.10 (*Continued*)

Reported Units	Threshold O_2 concentrations Converted to Equivalent Unit (μmol L^{-1})	System Measured [a]	Type of Measurement	Reference
0.25–10 μmol L^{-1}	0.25–10	3 marine denitrifiers	Culture study by Ozretich quoted by Cohen and Gordon	39
<12 μmol L^{-1d}	<12	*Thiobacillus denitrificans*	Chemostat-grown cultures with solution O_2 concentration measured by Clark electrode	87
<0.0–0.0035 atm^d	<4.9	*Azospirillum brasilense*	Chemostat-grown cultures with solution O_2 concentration measured by Clark electrode	130
<3 nmol mL^{-1d}	<3	*Paracoccus halodenitrificans*	Chemostat-grown cultures with solution O_2 concentration measured by Clark electrode	80

[a] Most marine and freshwater O_2 profiles are determined by the Winkler method, which gives O_2 values too high by ~0.1 mL L^{-1} (~4.5 μmol L^{-1}) (37).

[b] Threshold likely too high due to O_2 phase transfer limitation.

[c] Also detection limit of method.

[d] Denitrification did not occur at the listed value, but occurred at the next lower value studied, which was reported as zero O_2 concentration because the detection limits of the electrode had been exceeded.

the standard reference unit, and I encourage others to do likewise because this is the basic unit used to describe the substrate and inhibitor concentrations for cell metabolism.

From the data above and other studies on oxygen concentration effects on denitrification, it can be concluded that the active denitrifying flora of marine, freshwater, and soil environments have very low thresholds for oxygen. Since the process cannot be measured in nature at higher oxygen concentrations, the overwhelming majority of strains in nature do not have less sensitive oxygen control. This may not be true for all denitrifiers in waste processing units, however, since two very interesting isolates with much more relaxed oxygen control have been reported. Krul and Veeningen (109) studied denitrifiers from activated sludge and found that they showed more dissimilatory nitrate reductase activity in the presence of oxygen than did any of the isolates from drinking water. One of their isolates, *Alcaligenes* strain 15, showed the most tolerance to oxygen with an aerobic nitrate reduction rate 20% of that under anaerobic conditions (108). The other even more remarkable strain is *Thiosphaera pantropha*, described by Robertson and Kuenen (157). This isolate denitrifies in the presence of air under conditions in which the oxygen solution concentration was verified always to be greater than 80% of air saturation. It may be that in wastewater environments the excess of electron donor and the sometimes fluctuating status of nitrate and oxygen would select for organisms that could make simultaneous use of both electron acceptors.

4.6.1.3 Synthesis of Denitrifying Enzymes

The discussion above on the threshold for denitrification is without respect to whether the inhibition is on synthesis of denitrifying enzymes or on the activity of these enzymes, nor does it distinguish which steps in the denitrifying pathway are most affected by oxygen. It has long been known that oxygen has two modes of regulation: repression of enzyme synthesis and inhibition of enzyme activity. However, differentiation of the critical oxygen concentrations for each mechanism is poorly characterized. Our limited state of knowledge is summarized in Fig. 4.8

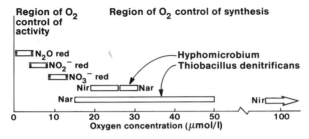

Figure 4.8 Approximate regions of O_2 concentration which inhibit the enzyme activity and synthesis for the three steps in the denitrification sequence. Nir, nitrite reductase; Nar, nitrate reductase. Data for *Hyphomicrobium* from Ref. 118 and for *Thiobacillus* from Ref. 87.

(excluding the two unusual organisms mentioned above). For the two cases in which denitrifying enzyme synthesis has been examined [*Hyphomicrobium* X (118) and *Thiobacillus denitrificans* (87)], denitrifying enzyme synthesis is permitted at higher oxygen concentrations than the threshold for activity of these enzymes. In one case nitrite reductase synthesis is allowed at higher oxygen concentrations than nitrate reductase synthesis; in the other organism the order is reversed. No data have been reported on the oxygen concentrations that permit nitrous oxide reductase synthesis. It is reasonable that oxygen concentrations permitting synthesis should be higher than the concentrations permitting activity since the cell would then be prepared for immediate use of nitrate under the common condition of declining oxygen concentration.

4.6.1.4 Effect of Oxygen on the Different Steps in the Denitrification Sequence

Different thresholds of activity are also sketched in Fig. 4.8 for the three major denitrification steps. The position is not based on direct threshold measurements for each step since those numbers do not exist or are not comparable (Table 4.10). They are based on the general environmental observations, which show a tendency for nitrite to accumulate and for the N_2O/N_2 ratio to increase with increased oxygen concentrations. Athough Fig. 4.8 suggests some trends, it should not be viewed as established principle until experiments are done under well-defined conditions.

The sequence of denitrification intermediates that appear in response to oxygen is illustrated by the data of Hochstein et al. for chemostat-grown cultures (Fig. 4.9) and for marine nitrogen profiles (Fig. 4.10). In the nitrate-limited chemostat the order of the steps according to decreasing oxygen sensitivity is: nitrous oxide reduction > nitrate reduction > nitrate reduction. The Baltic Sea profile shows typical features of marine denitrification, as well as the typical appearance of an N_2O peak at higher rather than lower oxygen concentrations. This N_2O peak is ascribed to nitrification since it parallels nitrite production and occurs well above the threshold for denitrification (70, 159). The lower nitrite peak, known as the secondary nitrite maximum, is thought to be due to denitrification (i.e., nitrate reduction to nitrite) (37, 39, 70, 159). Nitrite reduction presumably is restricted by the O_2 in this zone, while nitrate reduction proceeds. Nitrite then declines in the lower oxygen zone below. N_2O accumulation from denitrification is rarely seen in fresh and marine water columns, but when it occurs it is in the zone of nitrite accumulation or below and can be found in very large concentrations. It is the latter feature—the extremely high concentrations (e.g., $> 1 \mu M N_2O$) (101, 159)— that almost certainly identifies this N_2O as coming from denitrification. The O_2 concentration–nitrogen oxide relationships that do not conform to those illustrated above are for the Eastern Tropical North Pacific, in which the secondary nitrite maximum corresponds to nitrate and N_2O minima, and this nitrite peak occurs only at extremely low oxygen concentrations ($< 1 \mu mol\ O_2\ L^{-1}$) (37). Since both minima must be due to denitrification, it can only be interpreted that the nitrite reduction step is more limited than the other two, but probably not by oxygen.

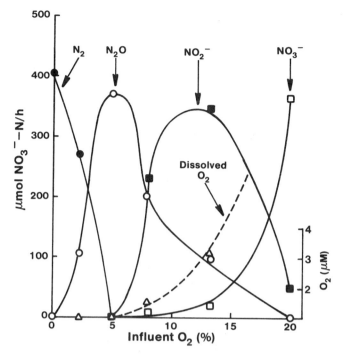

Figure 4.9 Effect of O_2 concentration on denitrification intermediates during nitrate-limited, steady-state (chemostat) growth of *Paracoccus halodenitrificans*. The influent O_2 concentration is shown on the x-axis and the measured dissolved O_2 concentrations is shown by a dashed line. Redrawn from Hochstein et al. (80).

4.6.2 Conceptual Model of Environment Regulation of Denitrification

Of microbial processes in nature, denitrification is the process that is most sensitive to environmental regulation. This is because many factors participate in the regulation and because some of these factors, especially oxygen, cause such a rapid and dramatic response. To simplify this complexity, a two-dimensional conceptual model for environmental regulation of denitrification is given in Fig. 4.11. It should be noted that several of the regulatory factors—water, plants, organic matter—can have multiple effects on the status of the proximate regulators. These effects are additive in the case of water and carbon, but they may be antagonistic as in the case of plants (e.g., plants remove nitrate and water but add carbon and reduce the rhizosphere oxygen concentration by respiration). The understanding of denitrification can be greatly enhanced with more effort spent on dissecting the denitrification response into its component mechanisms. This will also help avoid what sometimes appear to become conflicting results if the response to the parameter is viewed only at the gross level. An example of misunderstanding caused by failure to understand the underlying mechanisms is in the role of carbon in soil denitrifi-

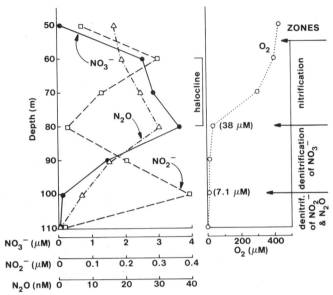

Figure 4.10 Relationship among denitrification N species and O_2 illustrated by vertical profiles from the Baltic Sea. Redrawn from Figure 2c in Rönner (159).

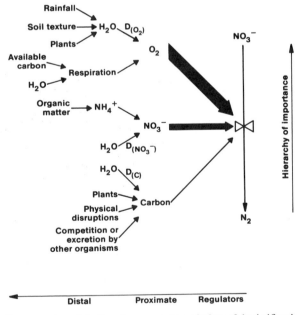

Figure 4.11 Conceptual model of environmental regulation of denitrification. The vertical dimension shows the hierarchy of importance of the three major regulators of denitrification. The horizontal dimension illustrates the proximity of the regulatory factors or features to the enzymatic process. D refers to an effect on diffusion rates of the indicated regulator.

218

cation. As shown in the model, carbon (organic matter) has three potential roles: source of electron donor, source of nitrate, and to drive respiration which reduces oxygen concentration. It has often been assumed that the major role of carbon is to supply electron donor, but it now appears that a much greater role is in creating anaerobic microsites, thereby influencing the major regulator of denitrification.

The importance of the three proximate regulators in limiting denitrification varies with habitat (Table 4.11). In habitats that are exposed to the atmosphere, oxygen is the principal factor limiting denitrification; this together with the great sensitivity to oxygen discussed above is the reason for ranking this as the most important regulator of denitrification. The habitats that are primarily anaerobic—sediments, anaerobic sludge and at least some intestinal communities—the lack of nitrate limits denitrification because nitrate is quickly denitrified and its resupply (nitrification) is blocked by the absence of O_2. Carbon as the electron donor almost never prevents denitrification, although it is often not present in amounts to saturate the reaction.

The following discussion of regulation is based on the organization provided by this conceptual model. The soil environment is more commonly discussed because of the greater diversity of regulatory factors experienced in this habitat.

4.6.3 Regulatory Factors

4.6.3.1 Oxygen

The oxygen status of a habitat microsite is controlled by the rate of oxygen supply to that site and the rate of oxygen consumption (respiration). The interplay of these two processes is more important than any others to understanding where denitrification occurs. Oxygen transport is heavily influenced by impermeable particles (e.g., rocks, sand, and clay). Restriction of oxygen transport also occurs due to biological surfaces (e.g., cell membranes, plant roots, and organic debris). Water

TABLE 4.11 Ranking of importance of the factors that limit denitrification in various habitats[a]

| Habitat | Regulator | | |
	Oxygen	Nitrate	Carbon (Electron Donor)
Fertilized soil	1	3	2
Nonfertilized soil	1	2	3
Sediments	2	1	
Lake water column	1	2	
Ocean waters	1	3	2
Intestinal tract	2	1	
Anaerobic sludge		1	
Secondary waste treatment	1	2	

[a] A ranking of 1 indicates that this factor most commonly limits denitrification.

is an important regulator of oxygen transport in nonsaturated habitats because of its dynamic temporal and variable spatial character. Water greatly restricts diffusion in porous media; for example, in sediments the oxygen diffusion coefficient is in the range 10^{-7} cm^{-2} s^{-1} versus still particle-free water, where it is in the range 10^{-5} cm^{-2} s^{-1} (152). In soil, water restricts diffusion by increasing the distance to air channels and by ensuring that oxygen transport is through water rather than through the gas phase, where the diffusion coefficient is much greater, in the range 10^{-3} cm^{-2} s^{-1}. Because of the large influence water status can have on the oxygen diffusion coefficient, it is easy to see how changes in water can readily influence denitrification. The major parameters that control water status, other than rainfall, are plants, which can rapidly remove the water due to evapotranspiration, and soil texture, because of its influence on water-holding capacity.

The respiratory activity of the habitat is the major mechanism that removes oxygen. Respiration, in turn, is driven by the supply of available carbon and, in droughty sites, by water, which can stimulate metabolic activity and foster transport of carbon to respiring organisms. The importance of respiratory activity to denitrification is too often underestimated, and in aerobic environments, it is sometimes the most important factor in controlling denitrification. As an illustration of this point, Parkin recently advanced the concept that denitrification in soils is often found in "hot spots" which are created by decaying organic matter that generates the anaerobic microsite (T. P. Parkin, Abstr. 4th Int. Symp. Microbiol. Ecol., 1986, p. 118). This is a major advancement in understanding soil denitrification, since it helps explain the very large spatial variability characteristic for soil denitrification. Stimulation of denitrification resulting from organic waste addition to soils and marsh sediments is another example of increased anaerobiosis caused by enhanced respiration.

The interplay of oxygen diffusion and oxygen consumption is exhibited in the form of oxygen gradients in habitats. The gradient profiles of oxygen, nitrate, and available carbon usually define the denitrification zone within various subhabitats, as illustrated in Fig. 4.12. Distance from oxygen-saturated water to the denitrifying zone ranges from only 20 μm in biofilms (189) to over 100 m in the Eastern Tropical North Pacific Ocean waters (39). This distance to anaerobiosis can change readily depending on the rate of carbon supply for respiration and to any change in the oxygen diffusion coefficient. An example of a rapid change is in the pulse of soil denitrification (Fig. 4.13) often seen after rainfall or irrigation (74, 158, 164, 172). In fact, in the study illustrated, about one-half of the total denitrification of these sites occurred within the 48-h periods after rainfall (172). The position of oxygen gradients in sediments, algal mats, and soil aggregates has now been measured by microoxygen electrodes (152, 153, 171). Denitrification could only be measured in soil aggregates in which anaerobic centers could be measured. (Fig. 4.14). If the aggregates were gently rolled, the gradients became steeper, illustrating oxygen-restricted diffusion due to compaction (171). This aggregate model could be expected to occur in fecal pellets (169) and other particulate organic matter where denitrification is measured.

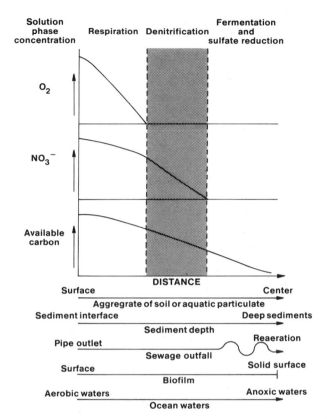

Figure 4.12 Gradient profiles of O_2, NO_3^-, and available carbon which define the zone of denitrification. This principle can be applied to any habitat and the distances can range from μm to meters.

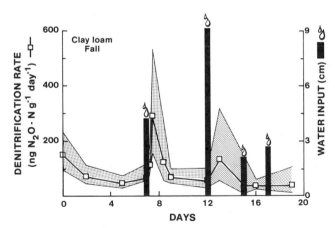

Figure 4.13 Pulse in soil denitrification seen after rainfall. Redrawn from Sexstone et al. (172). Mean rates and 95% confidence intervals are shown.

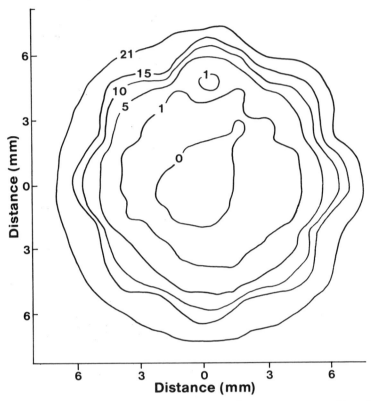

Figure 4.14 Map of O_2 concentration (%) within a soil aggregate measured by microoxygen electrodes. An anaerobic center (0-% contour) is observed in this aggregate. Reproduced from Sexstone et al. (171).

4.6.3.2 Nitrate

The response of denitrification to nitrate follows a Michaelis–Menten relationship, as would be expected for a membrane ion transport system (14). However, in soil and sediment environments the Michaelis–Menten response is not always seen, and if it is, the K_m for nitrate is often much higher (13 to 1300 times) than the values of 5 to 10 μM found for denitrifying cultures (124). One explanation for the differing kinetic behavior of natural samples relative to pure cultures is that nitrate diffusion (143) in the natural sample influences the pattern and inflates the K_m value. This is the explanation for some of the high values in the literature, but not for all. The other explanation is that denitrifiers have more than one nitrate affinity system, and it is usually the higher-affinity system that is measured in cultures (124). However, there is not yet any direct experimental evidence for a lower-affinity system. With a low-affinity system the nitrate concentration in natural habitats will often be rate limiting, while with a high-affinity system, nitrate concentration is much less likely to be rate limiting. In an experimental and theoretical study with soil, we found that a formal nitrate diffusion limitation would be likely in many soils, but in practice it would not be realized in fertilized soils because the nitrate concentration would be > 100 times the denitrifier K_m for nitrate (124). In contrast, in nonfertilized soils, nitrate diffusion would often be rate limiting because of the much lower nitrate concentration. This interpretation is borne out by our experience, in which NO_3^- more commonly limited denitrification in forest (nonfertilized soils) than in agricultural (fertilized) soils (124).

As illustrated in the regulation model of denitrification, the nitrate concentration is controlled by its rate of supply to the denitrifying site, which in turn is controlled by the water content in soils and by the rate of nitrate production. Nitrate is derived principally from nitrification and the ammonium in turn derived from mineralization. Constantly anaerobic habitats have limited denitrification because nitrification is blocked. In other habitats, nitrification is usually not limiting. In more oligotrophic habitats, the low mineralization rate may limit nitrate availability (154).

There are several competing fates for nitrate: nitrate assimilation, DNRA, plant competition, and soluble losses. The fate that most commonly influences denitrification is plant uptake of nitrate. In agricultural soils, and especially in nonfertilized soils, plant removal of nitrate substantially reduces denitrification (177).

The competition for nitrate between two competing organisms with Michaelis–Menten uptake kinetics has been modeled (200). The essential results are summarized as a function of V_{max} (organism density), K_m, and nitrate concentration (Fig. 4.15). The lines in this figure show the conditions in which nitrate is partitioned 50:50 to the two competing populations. At low nitrate concentrations the conditions for equal partitioning is illustrated by curve A, and for high (saturating)

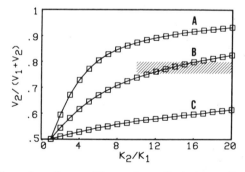

Figure 4.15 The lines show the conditions where nitrate is equally partitioned between two competing nitrate-using organisms, illustrated as organisms 1 and 2. The parameters are V (V_{max}, which reflects organism density), K (K_m of NO_3^- uptake), and three nitrate concentrations (lines A, B, C equal, respectively, low, intermediate, and high nitrate concentrations). The hatched area represents the range of parameters for competition between denitrifiers and DNRA organisms. Reproduced from J. M. Tiedje, A. J. Sexstone, D. D. Myrold, and J. A. Robinson, Denitrification: ecological niches, competition and survival, Antonie van Leeuwenhoek, 48:569–583. 1982, by permission.

nitrate concentrations the partition is illustrated by curve C. At low concentrations the K_m plays a major role in predicting the outcome of competition, but at high concentrations the outcome is dominated by V_{max}. This theoretically derived figure can be used to analyze competition by any two populations (e.g., roots versus denitrifiers or DNRA versus denitrifiers). In the latter case the high nitrate K_m of the DNRA organisms suggests that DNRA organisms must have a tenfold greater V_{max} (or an approximately tenfold greater population) than denitrifiers to process 50% of the substrate. This explains why DNRA does not compete with denitrification, except in habitats that greatly favor the population of DNRA organisms over denitrifiers (Section 4.5.2.2).

4.6.3.3 Carbon as Electron Donor

The role of carbon is to provide the electron donor for nitrate reduction. As discussed earlier, denitrifiers can be considered similar to other heterotrophic bacteria in their competitiveness for carbon. Since denitrifiers comprise only a few percent of the heterotrophic population, they would be expected to process only a few percent of the carbon, except under anoxic conditions where the obligate aerobes can no longer compete. Thus under denitrifying conditions a greater percent of the carbon should be available to the denitrifying population. Also, under anoxic conditions more fermentation products are available and oxygen-stressed cells excrete more available carbon. For these three reasons the carbon limitation to denitrification is less likely under anoxic conditions. The situation in which carbon would be rate limiting is when the available carbon used to drive the respiratory removal of oxygen is exhausted at the same time or location (in gradients) that carbon is exhausted (in Fig. 4.12 if the available carbon gradient was identical to the oxygen gradient). This situation is probably uncommon, but one example of where it might have occurred is in Fig. 4.13, where increased denitrification did not occur in response to the third and fourth rain events in a series. Since nitrate was present in this soil, the interpretation is that the available carbon was exhausted following the earlier two rainfalls. The situation is rare, however, because it occurred only twice (those shown) of 11 rain events studied over 2 years in two different soils (172).

In the regulation model (Fig. 4.11) the major controllers of available carbon are shown to be (a) water, which stimulates metabolism in dry soil and transports available carbon; (b) plants, which excrete and deposit carbon; (c) physical disruption of protected carbon in the habitat by freezing and thawing, wetting and drying cycles, cultivation or natural turbation; and (d) competition or excretion by the other organism. The importance of these mechanisms and, in fact, the importance of carbon per se as an electron donor has rarely been isolated from its other effects. There are two experiments, however, which reveal that extra electron donor (carbon) does enhance denitrification (Table 4.12). Under aerobic conditions disruption of soil structure stimulated denitrification threefold probably because more carbon is made available, although the major mechanism is probably the enhancement of respiration. But under anaerobic conditions, the role of respiration is removed and the effect of added carbon as an electron donor can be directly

TABLE 4.12   The effect on denitrification rates of added carbon from disturbance of soil structure and from soluble carbon additions

| | Denitrification rate (μg N$_2$O-N $\cdot$ kg^{-1} $\cdot$ h^{-1}) | | |
Treatment	Natural Core	Mixed Soil	Reference
Aerobic	2.7	7.8 (repacked core)	138
Anaerobic	67	508 (slurry)	124
Anaerobic + succinate[a]	554		124

[a] Succinate allowed to diffuse throughout soil core at 4°C prior to warming the core and making denitrification measurements.

observed. To test this without disturbing soil structure, a diffusable carbon source (succinate) was added to cores held at 4°C to allow diffusion to distribute the available carbon and yet suppress biological activity (124). When the cores were warmed to field temperature, the denitrification rate was enhanced eightfold over the presuccinate rate. When the soil structure was destroyed by making the core into a soil slurry, the rate was also enhanced eightfold, suggesting that disruption provides a major influx of carbon, probably saturating the denitrification capacity of the soil. These studies show that under aerobic soil conditions, denitrifiers, like other heterotrophic respirers, are carbon limited. But in naturally aerated soils, this limitation is masked by the much greater effect of oxygen on denitrification.

4.6.3.4 Effect of Macrofeatures: Burrows and Roots

The foregoing regulators are affected by macrofeatures such as small animal turbation, plant roots, and sites of carbon debris. Several reports establish that benthic organisms enhanced denitrification in freshwater and marine sediments (34, 78, 169). The postulated mechanisms are enhanced nitrification rates associated with the burrow walls, greater transport of nitrate into the sediments due to the burrow channels, and increased denitrification associated with fecal pellets, burrow walls, and benthic organisms themselves. The increased transport of both oxygen for nitrification and nitrate into the sediments is probably the main reason why sediment denitrification is stimulated by bioturbation (34, 169).

The rhizosphere has long been studied as a site of potential denitrification (see Refs. 59 and 102). Most of the work has been done with agricultural crops and soils. About half of the reports show that denitrification is enhanced by plants, and the other half show that roots have no effect on denitrification. The former response is usually explained by the increased respiration associated with roots, and the latter is explained by plants removing the nitrate, making it unavailable to denitrifiers. Roots have both positive and negative effects on denitrification, which could account for the variability in these results. Roots can (a) reduce the oxygen tension of soil by both root respiration and bacterial respiration stimulated by root-produced carbon; (b) utilize water drying the soil, creating a more aerated rhizosphere; (c) consume nitrate, making it unavailable to denitrifiers; and (d) in wetland plants, transport oxygen down the stem and root to the rhizosphere, which can stimulate nitrification in a nitrate-limited anoxic sediment.

Denitrification has been shown to be higher in fallow soils (plant-free) than in the same soil planted to crops. This was shown in both a fertile Chernozem soil (6) and a flooded rice soil (57, 67). Haider et al. showed by ^{14}C labeling studies that rhizo-deposited carbon was not immediately available for mineralization or denitrification but entered an organic matter fraction that turned over more slowly (73). Christensen and Sørensen, studying macrophytes in littoral sediments, found that the root zone accounted for 50 to 70% of the annual denitrification of these lake sediments (35). Denitrifying activity was low in the winter and in sediments without plants, but by midsummer sediment supporting plants had high denitrification rates which were found throughout the rhizosphere zone. Nitrification was enhanced in the root zone, probably due to oxygen release from the roots. Two zones of activity were noted: surface sediments in which denitrification was reduced during light periods and the underlying sediment (root zone), which was not affected by light. In this environment the plants clearly stimulated denitrification, with the key effect probably being the plant's supply of oxygen through the stem to overcome the lack of nitrification, which usually prevents much denitrification in sediments. This same mechanism exists in rice, but there are agricultural management strategies to reduce fertilizer losses in paddy rice soils to 5 to 10% or less (57, 67, 175).

4.7 SUMMARY OF METHODS USED TO MEASURE DENITRIFICATION

No discussion of the ecology of denitrification could be complete without mention of methodology to measure this process since this has been a major obstacle to the study of denitrification in natural habitats. The direct approach to measuring a process is to measure the product of the reaction. In this case the product is principally N_2, but since our atmosphere is 80% N_2, this approach is impossible. Therefore, indirect methods are used, but each has its assumptions and limitations. No one method has been the obvious method of choice for denitrification studies, but after a decade of use the acetylene inhibition method has become the method of first consideration because of its convenience and its proven reliability in many habitats.

The purpose of this section is to summarize briefly the major denitrification methods used, to give their advantages and limitations, and to provide the key features to be considered in choosing a method. For reviews of the extensive literature on use of denitrification methodology and more supportive documentation of the discussion that follows, see the following references (61, 99, 102, 141, 198, 199). The basic methods for measuring natural denitrification are summarized in Table 4.13. These methods are grouped according to approach. In the first group the attempt is to measure an absolute change in products or substrates. Because of the limitation our atmosphere puts on measurement of N_2 and the multiple fates of

TABLE 4.13 Summary of methods used to measure denitrification activity

Method	Analysis	Where Used	Comments
Change in substrates or products			
Acetylene inhibition–N_2O production	Gas chromatography (electron capture)	Natural samples of all types; microbial cultures	The most popular method because of its sensitivity, capacity for large numbers of samples, simplicity, low cost, and does not require any addition to the natural NO_3^- pool. Its major limitation is that acetylene also inhibits nitrification (81, 82, 210) and thus will reduce the denitrification rates in samples where NO_3^- is very low.
N_2 production	Gas chromatography (thermal conductivity)	Microbial cultures; rarely used successfully for natural samples	Because of the large N_2 background in the atmosphere, this method works only when the sample atmosphere has been exchanged for another gas (e.g., He). Even then the sample must be very active for detection of denitrification over leaks, outgassing, and the relative insensitivity of the thermal conductivity detector.
Consumption of NO_3^- and NO_2^-	Colorimetric methods, nitrate electrode, steam distillation, UV adsorption (direct or after HPLC)	Microbial cultures; aquatic habitats	Not considered definitive for denitrification since NO_3^- and NO_2^- can be reduced to NH_4^+. Can be used together with total NH_4^+ + organic-N (mass balance), in which unrecovered N is equated to denitrification. The method is relatively insensitive and works best when total N in the sample is very low.

227

TABLE 4.13 (*Continued*)

Method	Analysis	Where Used	Comments
Change Relative to Conservative Property			
NO_3^-/Cl^- (or Br^-) ratio	Colorimetric methods, specific ion electrodes	Soil	The method is based on the principle that both anions have the same mobility in natural systems, but only NO_3^- is consumed (61, 79). It will not work where other fates of NO_3^- (e.g., plant uptake) are significant. The method is qualitative rather than quantitative, but it has the advantage of requiring only a single measurement, and it reflects the process as it occurred naturally.
N_2/Ar ratio	Gas chromatography (thermal conductivity)	Marine water column	The method is based on the principle that N_2 production can be distinguished from the atmospheric N_2 profile since the latter profile should be the same as the atmospheric Ar profile (10). N_2 consumption by N_2 fixation can confound this measurement, although this is usually not significant in the deeper water column where this method is usually used. The method is not sensitive, but it directly reflects the process over a large scale in nature.
Isotopic Methods The isotopes. ^{15}N	Mass spectrometry magnetic sector quadrupole UV emission		Mass spectrometry is the preferred analysis method. Isotope ratio instruments (magnetic sector) are designed for high precision, measurements but most require a large amount of N_2. The quadrupole units are coupled to a gas chromatograph and have the ability to separate gases as well as detect much smaller amounts. The UV emission instrument is less expensive and requires only a small N_2 sample, but is much less precise.

Isotopic Methods

^{13}N (production by cyclotron or Van de Graaff accelerator)	β^+ liquid scintillation, proportional counter; γ NaI (T1) detectors		^{13}N is the longest-lived radioactive N isotope; half-life 9.96 min. It can be produced by a variety of nuclear reactions. Because of the short half-life, it provides the greatest sensitivity of any denitrification method, but it can be used for only short periods, must be used near an accelerator, and requires expensive equipment (198). It is the only direct method to measure denitrification in the natural atmosphere.

Use of the isotopes

Labeled N-gas production $^{13}N_2$, $^{13}N_2O$, $^{28}N_2$, $^{29}N_2$, $^{30}N_2$	Proportional counter, mass spectrometry	Natural sample, culture	Isotopes can be used to improve the sensitivity and specificity of measurement of denitrification. A limitation is that the isotope must be uniformly mixed with the natural NO_3^- pool over space and time and in amounts that will not increase the natural nitrate concentration (136). Because ^{15}N is only a moderately sensitive method, the ^{15}N required often enhances the denitrification rate in habitats that have low NO_3^- pools. The detection of the $^{30}N_2$ mass is the most sensitive because of the low $^{30}N_2$ in the atmosphere (173).
Mass balance [loss of enriched (^{15}N) or depleted (^{14}N) nitrogen]	Mass spectrometry	Natural samples, cultures	
Isotope dilution (using ^{15}N)	Mass spectrometry	Natural samples	Isotope dilution to measure denitrification can be used by observing dilution of ^{15}N-enriched atmosphere (112) or by the use of modeling in conjunction with the measurement of other N-cycle processes (126).

nitrate, two of the three approaches listed are usually not satisfactory. The second group of methods rely on measuring a change in substrate or product relative to a naturally occurring, conserved molecule with similar behavior. These methods have the advantage of measuring what has actually occurred naturally—no additions or incubations—but they are not particularly sensitive or quantitative. The N_2/Ar method is the only method that directly and unambiguously reflects that denitrification has naturally occurred. The third group of methods—the use of isotopes— is a subset of the first group in that they help improve the sensitivity and specificity of these methods for denitrification. A limitation of the isotope methods is that the isotope must be added to the sample, which will enhance the denitrification rate if the process is not saturated with nitrate, and in nonaqueous habitats, it is often difficult to achieve uniform distribution over time and space of the labeled nitrate with the natural nitrate pool. The isotopes are also costly and require expensive equipment for analysis. Nonetheless, much of the more definitive knowledge on denitrification in nature has come from the use of isotopes.

The acetylene inhibition method warrants more discussion because it has proved to be the most practical for ecological studies. It is based on the principle that acetylene ($HC\equiv CH$), being structurally similar to nitrous oxide ($N\!=\!N\!=\!O$), blocks the reduction of nitrous oxide and, therefore, nitrous oxide accumulates stoichiometrically from nitrate or nitrite. This inhibition was first noted by Fedorova in 1973 (56), and its use for denitrification studies was demonstrated in pure cultures in 1976 independently by Balderston et al. (9) and by Yoshinari and Knowles (222). Immediately thereafter its use was extended to samples from natural habitats (97, 184, 223). The validity of the acetylene method to measure denitrification in natural samples was established since the results by the acetylene method compared favorably to those using ^{15}N and ^{13}N methods (136, 140, 163, 176).

To use the acetylene method successfully, one must distribute an adequate concentration of the inhibitor—acetylene gas—throughout the sample and recover the product gas—N_2O—without affecting the principal regulators of the process (i.e., oxygen, nitrate, and available carbon). Following are the potential problems to be aware of for successful use of the method:

1. The sample may be disrupted such that the denitrification regulators, oxygen and available carbon, are altered.
2. Acetylene inhibits nitrification, which may alter the nitrate concentration.
3. Gas diffusion may be restricted, which can limit acetylene dispersal and quantitative N_2O recovery.
4. The inhibitor may not completely block N_2O reduction.
5. Contaminants in commercial acetylene may alter the denitrification rate.

Each of these possible problems can be assessed and the assay modified to eliminate or minimize any error. The only problem that is difficult to overcome is number 2, the inhibition of nitrification, but it is only a problem for those samples in which the nitrate concentration is rate limiting. The assay can be useful even in this situation if the nitrate concentration is sufficient to allow definition of the pattern of N_2O production with time so that the effect of the declining nitrate concentration can be corrected. It is only in anaerobic habitats, where nitrate is the primary limiting regulator (Table 4.11), that the method will not work; the nitrate concentrations are so low that any denitrification is directly dependent on nitrification. It is important to note that isotopic methods have also not been successfully used in these habitats, since they also alter (in this case increase) the nitrate concentration. Acetylene blocks nitrification by irreversibly reacting with the active site of ammonium monooxygenase, the ammonia-oxidizing enzyme of nitrifiers (81). This suicidal inactivation of the nitrifying enzyme means that the activity of this enzyme is not restored when acetylene is removed. Only synthesis of new enzymes will restore nitrification.

The other four potential problems of the acetylene method can usually be dealt with by proper attention to the methodology used. Physical disruption of samples, especially soils and sediments, changes the oxygen and carbon availability. A change in gas composition (e.g., large acetylene introductions, which reduce the oxygen concentration) would change oxygen gradients in denitrifying microsites. Maintaining natural physical structure and avoiding large changes in the oxygen composition of the atmosphere can overcome both limitations. The problem of achieving acetylene dispersal and N_2O recovery is a problem when the gas diffusion coefficient is low. Small sample sizes, good spatial distribution of acetylene, and observing a linear production of N_2O with time are used to minimize the effect of severe diffusion limitations. The failure of acetylene to block N_2O reduction has not proven to be a major problem. The situations where it can occur are when there is a combination of high biological activity and very low nitrate concentrations [e.g., sludge (90), sediments (98, 206), and carbon-enriched soils (218)] or when sulfide is present (192). Acetylene is biodegraded under both aerobic and anaerobic conditions (44, 88, 212), but the natural population of acetylene degraders is so low that they cannot remove acetylene in a short-term incubation (e.g., <4 days). However, longer-term incubations result in an enrichment of acetylene degraders which will rapidly remove the inhibitor. Commercial acetylene tanks contain acetone and CO plus other contaminants; they may affect the measurement if not removed. They are easily removed in a gas scrubbing train (210), or can be avoided by generating acetylene by reacting carbide rock with water. The latter reaction is also more convenient for generating acetylene on site for field studies.

REFERENCES

1. Abd-el-Malek, Y., I. Hosny, and N. F. Emam. 1974. Evaluation of media, used for enumeration of denitrifying bacteria. Zentralbl. Bakteriol. Parasitenkd. Infektionski. Hyg. 2 Abt.:415–421.

2. Adkins, A. M., and R. Knowles. 1984. Reduction of nitrous oxide by a soil *Cytophaga* in the presence of acetylene and sulfide. FEMS Microbiol. Lett. 232:171–174.

3. Adkins, A. M., and R. Knowles. 1986. Denitrification by *Cytophage johnsonae* strains and by a gliding bacterium able to reduce nitrous oxide in the presence of acetylene and sulfide. Can. J. Microbiol. 32:421–424.

4. Aftring, R. P., and B. F. Taylor. 1981. Aerobic and anaerobic catabolism of phthalic acid by a nitrate-respiring bacterium. Arch. Microbiol. 130:101–104.

5. Aida, T., S. Hata, and H. Kusunoki. 1986. Temporary low oxygen conditions for the formation of nitrate reductase and nitrous oxide reductase by denitrifying *Pseudomonas* sp. G59. Can. J. Microbiol. 32:543–547.

6. Aulakh, M. S., D. A. Rennie, and E. A. Paul. 1983. Field studies on gaseous nitrogen losses under continuous wheat versus a wheat-fallow rotation. Plant Soil 75:15–27.

7. Averill, B. A., and J. M. Tiedje. 1982. The chemical mechanism of microbial denitrification. FEBS Lett. 138:1–12.

8. Bakker, G. 1977. Anaerobic degradation of aromatic compounds in the presence of nitrate. FEMS Microbiol. Lett. 1:103–108.

9. Balderston, W. L., B. Sherr, and W. J. Payne. 1976. Blockage by acetylene of nitrous oxide reduction in *Pseudomonas perfectomarinus*. Appl. Environ. Microbiol. 31:504–508.

10. Barnes, R. O., K. K. Bertine, and E. D. Goldberg. 1975. N_2:Ar, nitrification and denitrification in southern California borderland basin sediments. Limnol. Oceanogr. 20:962–970.

11. Barton, L. L., J. LeGall, J. M. Odom, and H. D. Peck, Jr. 1983. Energy coupling to nitrite respiration in the sulfate-reducing bacterium *Desulfovibrio gigas*. J. Bacteriol. 153:867–871.

12. Bazylinski, D. A., and R. P. Blakemore. 1983. Denitrification and assimilatory nitrate reduction in *Aquaspirillum magnetotactium*. Appl. Environ. Microbiol. 46:1118–1124.

13. Bazylinski, D. A., E. Palome, N. A. Blakemore, and R. P. Blakemore. 1986. Denitrification by *Chromobacterium violaceum*. Appl. Environ. Microbiol. 52:696–699.

14. Betlach, M. R., and J. M. Tiedje. 1981. Kinetic explanation for accumulation of nitrite, nitric oxide, and nitrous oxide during bacterial denitrification. Appl. Environ. Microibol. 42:1074–1084.

15. Betlach, M. R. 1982. Evolution of bacterial denitrification and denitrifier diversity. Antonie Leeuwenhoek J. Microbiol. 48:585–607.

16. Blackmer, A. M., J. M. Bremner, and E. L. Schmidt. 1980. Production of nitrous oxide by ammonia-oxidizing chemoautotrophic microorganisms in soil. Appl. Environ. Microbiol. 40:1060–1066.

17. Bleakley, B. H., and J. M. Tiedje. 1982. Nitrous oxide production by organisms other than nitrifiers or denitrifiers. Appl. Environ. Microbiol. 44:1342–1348.

18. Bokranz, M., J. Katz, I. Schroder, A. M. Roberton, and A. Kroger. 1983. Energy metabolism and biosynthesis of *Vibrio succinogenes* growing with nitrate or nitrite as terminal electron acceptor. Arch. Microbiol. 135:36–41.

19. Bollag, J.-M., and G. Tung. 1972. Nitrous oxide release by soil fungi. Soil Biol. Biochem. 4:271–276.

20. Bollag, J.-M., and E. J. Kurek. 1980. Nitrite and nitrous oxide accumulation during denitrification in the presence of pesticide derivatives. Appl. Environ. Microbiol. 39:845–849.

21. Boon, P. I., D. J. W. Moriarty, and P. G. Saffigna. 1986. Nitrate metabolism in sediments from seagrass (*Zostera capricorni*) beds of Moreton Bay, Australia. Mar. Biol. (Berl.) 91:269–275.

22. Bouwer, E. J., and P. L. McCarty. 1983. Transformations of halogenated organic compounds under denitrification conditions. Appl. Environ. Microbiol. 45:1295–1299.

23. Brannan, D. K., and D. E. Caldwell. 1980. *Thermothrix thiopara*: growth and metabolism of a newly isolated thermophile capable of oxidizing sulfur and sulfur compounds. Appl. Environ. Microbiol. 40:211–216.

24. Braun, K., and D. T. Gibson. 1984. Anaerobic degradation of 2-aminobenzoate (anthranilic acid) by denitrifying bacteria. Appl. Environ. Microbiol. 48:102–107.

25. Bremner, J. M., and K. Shaw. 1958. Denitrification in soil. I. Methods of investigation. J. Agric. Sci. 51:22–39.

26. Broadbent, F. E., and F. G. Clark. 1965. Denitrification, p. 344–359, in W. V. Bartholomew, and F. E. Clark (eds.), Soil nitrogen. American Society of Agronomy, Madison, Wis.

27. Buchanan, R. E., and N. E. Gibbons. 1974. Bergey's manual of determinative bacteriology, 8th ed. Williams & Wilkins, Baltimore.

28. Burth, I., G. Benckiser, and J. C. G. Ottow. 1982. N_2O-Freisetzung aus Nitrit (Denitrifikation) durch ubiquitare Pilze unter aeroben Bedingungen. Naturwissenschaften 69 S 598.

29. Casella, S., C. Leporini, and M. P. Nuti. 1984. Nitrous oxide production by nitrogen-fixing, fast-growing rhizoba. Microbiol. Ecol. 10:107–114.

30. Caskey, W. H., and J. M. Tiedje. 1979. Evidence for clostridia as agents of dissimilatory reduction of nitrate to ammonium in soils. Soil Sci. Soc. Am. J. 43:931–935.

31. Caskey, W. H., and J. M. Tiedje. 1980. The reduction of nitrate to ammonium by a *Clostridium* sp. isolated from soil. J. Gen. Microbiol. 119:217–223.

32. Chan, Y. K., and N. E. R. Campbell. 1980. Denitrification in Lake 227 during summer stratification. Can. J. Fish. Aquat. Sci. 37:506–512.

33. Chan, Y.-K. 1985. Denitrification by a diazotrophic *Pseudomonas* species. Can. J. Microbiol. 31:1136–1141.

34. Chatarpaul, L., J. B. Robinson, and N. K. Kaushik. 1980. Effects of tubificid worms on denitrification and nitrification in stream sediment. Can. J. Fish. Aquat. Sci. 37:656–663.

35. Christensen, P. B., and J. Sørensen. 1986. Temporal variation of denitrification activity in plant-covered, littoral sediment from Lake Hampen, Denmark. Appl. Environ. Microbiol. 52:1174–1179.

36. Christianson, C. B., and C. M. Cho. 1983. Chemical denitrification of nitrite in frozen soils. Soil Sci. Soc. Am. J. 47:38–42.

37. Cline, J. D., and F. A. Richards. 1972. Oxygen deficient conditions and nitrate reduction in the eastern tropical north Pacific Ocean. Limnol. Oceanogr. 17:885–900.

38. Cohen, Y. 1978. Consumption of dissolved nitrous oxide in an anoxic basin, Saanich Inlet, British Columbia. Nature (Lond.) 272:235–237.

39. Cohen, Y., and L. I. Gordon. 1978. Nitrous oxide in the oxygen minimum of the eastern tropical North Pacific: evidence for its consumption during denitrification and possible mechanisms for its production. Deep Sea Res. 25:509–524.

40. Cole, J. A. 1978. The rapid accumulation of large quantities of ammonia during nitrite reduction by *Escherichia coli*. FEMS Microbiol. Lett. 4:327–329.

41. Cole, J. A., and C. M. Brown. 1980. Nitrite reduction to ammonia by fermentative bacteria: a short circuit in the biological nitrogen cycle. FEMS Microbiol. Lett. 7:65–72.

42. Coleman, K. J., A. Cornish-Bowden, and J. A. Cole. 1978. Purification and properties of nitrite reductase from *Escherichia coli* K12. Biochem. J. 175:483–493.

43. Crutzen, P. J. 1976. Upper limits on atmospheric ozone reductions following increased application of fixed nitrogen to the soil. Geophys. Res. Lett. 3:169–172.

44. Culbertson, C. W., A. J. B. Zehnder, and R. S. Oremland. 1981. Anaerobic oxidation of acetylene by estuarine sediments and enrichment cultures. Appl. Environ. Microbiol. 41:396–403.

45. Daniel, R. M., I. M. Smith, J. A. D. Phillip, H. D. Ratcliffe, J. W. Drozd, and A. T. Bull. 1980. Anaerobic growth and denitrification by *Rhizobium japonicum* and other rhizobia. J. Gen. Microbiol. 120:517–521.

46. Daniel, R. M., A. W. Limmer, K. W. Steele, and I. M. Smith. 1982. Anaerobic growth, nitrate reduction and denitrification in 46 *Rhizobium* strains. J. Gen. Microbiol. 128:1811–1815.

47. DeLey, J., P. Segers, and M. Gillis. 1978. Intra- and intergenic similarities of *Chromobacterium* and *Janthinobacterium* ribosomal ribonucleic acid cistrons. Int. J. Syst. Bacteriol. 28:154–168.

48. deVries, W., W. M. C. van Wyck-Kapteyn, and S. K. H. Oosterhuis. 1974. The presence and function of cytochromes in *Selenomonas ruminantium*, *Anaerovibrio lipolytica*, and *Veillonella alcalescens*. J. Gen. Microbiol. 81:69–78.

49. deVries, W., H. G. D. Niekus, M. Boellaard, and A. J. Stouthamer. 1980. Growth yields and energy generation by *Campylobacter sputorum* subspecies *bubulus* during regrowth in continuous culture with different hydrogen acceptors. Arch. Microbiol. 124:221–227.

50. Dodds, K. L., and D. L. Collins-Thompson. 1985. Production of N_2O and CO_2 during the reduction of NO_2^- by *Lactobacillus lactis*. Appl. Environ. Microbiol. 50:1550–1552.

51. Dodgson, K. S., G. F. White, J. A. Massey, J. Shapleigh, and W. J. Payne, 1984. Utilization of sodium dodecyl sulphate by denitrifying bacteria under anaerobic conditions. FEMS Microbiol. Lett. 24:53–56.

52. Dunn, G. M., R. A. Herbert, and C. M. Brown. 1979. Influence of oxygen tension on nitrate reduction by a *Klebsiella* sp. growing in chemostat culture. J. Gen. Microbiol. 112:379–383.

53. El-Hassan, G. A., R. M. Zablotowicz, and D. D. Focht. 1985. Kinetics of denitrifying growth by fast-growing cowpea *Rhizobia*. Appl. Environ. Microbiol. 49:517–521.

54. Eskew, D. L., D. D. Focht, and I. P. Ting. 1977. Nitrogen fixation, denitrification, and pleomorphic growth in a highly pigmented *Spirillum lipoferum*. Appl. Environ. Microbiol. 34:582–585.

55. Evans, W. C. 1977. Biochemistry of the bacterial catabolism of aromatic compounds in anaerobic environments. Nature (Lond.) 270:17–22.

56. Fedorova, R. I., E. I. Milekhina, and N. I. Il'yukhina. 1973. Evaluation of the method of "gas metabolism" for detecting extraterrestrial life. Identification of nitrogen-fixing microorganisms. Izv. Akad. Nauk SSSR Ser. Biol. 6:797–806.

57. Fillery, I. R. P. and P. L. G. Vlek. 1982. The significance of denitrification of applied nitrogen in fallow and cropped rice soils under different flooding regimes. I. Greenhouse experiments. Plant Soil 65:153–169.

58. Firestone, M. K., R. B. Firestone, and J. M. Tiedje. 1979. Nitric oxide as an intermediate in denitrification: evidence from nitrogen-13 isotope exchange. Biochem. Biophys. Res. Commun. 91:10–16.

59. Firestone, M. K. 1982. Biological denitrification, p. 289–326, in F. J. Stevenson (ed.), Nitrogen in agricultural soils. Agronomy Monograph. No. 22. American Society of Agronomy, Madison, Wis.

60. Focht, D. D., and H. Joseph. 1974. Degradation of 1,1-diphenyl-ethylene by mixed culture. Can. J. Microbiol. 20:631–635.

61. Focht, D. D. 1978. Methods for analysis of denitrification in soils, p. 433–490, *in* D. R. Nielsen and J. G. MacDonald (eds.), Nitrogen in the environment, Vol. 2, Soil-plant-nitrogen relationships. Academic Press, New York.

62. Gamble, T. N., M. R. Betlach, and J. M. Tiedje. 1977. Numerically dominant denitrifying bacteria from world soils. Appl. Environ. Microbiol. 33:926–939.

63. Garber, E. A. E., and T. C. Hollocher. 1981. [15]N Tracer studies on the role of NO in denitrification. J. Biol. Chem. 256:5459–5465.

64. Garcia, J.-L. 1977. Analyses de différents groupes composant la microflore dénitrifiante des sols de rizière du Sénégal. Ann. Microbiol. (Paris) A128:433–446.

65. Garcia, J.-L. 1977. Étude la dénitrification chez une bactérie thermophile sporulée. Ann. Microbiol. (Paris) A128:447–458.

66. Garcia, J.-L., S. Roussos, and M. Bensonssan. 1981. Études taxonomique de bactéries dénitrifiantes isolées sur benzoate dans des sols de rizières du Sénégal. Cah. ORSTOM Ser. Biol. 43:13–25.

67. Garcia, J.-L., and J. M. Tiedje. 1982. Denitrification in rice soils, p. 187–208, in Y. R. Dommergues and H. G. Diem (eds.), Microbiology of tropical soils and plant productivity. Martinus Nijhoff, The Hague.

68. Gayon, U., and G. Dupetit. 1882. Sur la fermentation des nitrates. C. R. Acad. Sci. 95:644–646.

69. Gilliam, J. W., and R. P. Gambrell. 1978. Temperature and pH as limiting factors in loss of nitrate from saturated Atlantic Coastal plain soils. J. Environ. Qual. 7:526–532.

70. Goering, J. J. 1968. Denitrification in the oxygen minimum layer of the eastern tropical Pacific Ocean. Deep Sea Res. 15:157–164.

71. Goreau, T. J., W. A. Kaplan, S. C. Wofsy, M. B. McElroy, F. W. Valois, and S. W. Watson. 1980. Production of NO_2^- and N_2O by nitrifying bacteria at reduced concentrations of oxygen. Appl. Environ. Microbiol. 40:526–532.

72. Grant, M. A., and W. J. Payne. 1981. Denitrification by strains of *Neisseria, Kingella*, and *Chromobacterium*. Int. J. Syst. Bacteriol. 31:276–279.

73. Haider, K., A. Mosier, and O. Heinemeyer. 1985. Pytotron experiments to evaluate the effect of growing plants on denitrification. Soil Sci. Soc. Am. J. 49:636–641.

74. Hallmark, S. L., and R. E. Terry. 1985. Field measurement of denitrification in irrigated soils. Soil Sci. 140:35–44.

75. Hart, L. T., A. D. Larson, and C. S. McCleskey. 1965. Denitrification by *Corynebacterium nephridii*. J. Bacteriol. 89:1104–1108.

76. Hasan, S. M., and J. B. Hall. 1975. The physiological function of nitrate reduction in *Clostridium perfringens*. J. Gen. Microbiol. 87:120–128.

77. Hendrie, M. S., A. J. Holding, and J. M. Shewan. 1974. Emended descriptions of the genus *Alcaligenes* and of *Alcaligenes faecalis* and proposal that the generic name *Achromobacter* be rejected: status of the named species of *Alcaligenes* and *Achromobacter*. Int. J. Syst. Bacteriol. 24:534–550.

78. Henriksen, K., J. I. Hansen, and T. H. Blackburn. 1980. The influence of benthic in fauna on exchange rates of inorganic nitrogen between sediment and water. Ophelia (Suppl 1):249–256.

79. Hill, A. R. 1986. Nitrate and chloride distribution and balance under continuous potatoe cropping. Agric. Ecosyst. & Environ. 15:267–280.

80. Hochstein, L. I., M. Betlach, and G. Kritikos. 1984. The effect of oxygen on denitrification during steady-state growth of *Paracoccus halodenitrificans*. Arch. Microbiol. 137:74–78.

81. Hyman, M. R., and P. M. Wood. 1985. Suicidal inactivation and labelling of ammonia mono-oxygenase by acetylene. Biochem. J. 227:719–725.

82. Hynes, R. K., and R. Knowles. 1978. Inhibition by acetylene of ammonia oxidation in *Nitrosomonas europaea*. FEMS Microbiol. Lett. 4:319–321.

83. Hynes, R. K., A.-L. Ding, and L. M. Nelson. 1985. Denitrification by *Rhizobium fredii*. FEMS Microbiol. Lett. 30:183–186.

84. Inderlied, C. B., and E. A. Delwiche. 1973. Nitrate reduction and the growth of *Veillonella alcalescens*. J. Bacteriol. 114:1206–1212.

85. Iwasaki, H., and T. Matsubara. 1972. A nitrite reductase from *Achromobacter cycloclastes*. J. Biochem. (Tokyo) 71:645–652.

86. Jacobson, S. N., and M. Alexander. 1980. Nitrate losses from soil in relation to temperature, carbon source and denitrifier populations. Soil. Biol. Biochem. 12:501–505.

87. Justin, P., and D. P. Kelly. 1978. Metabolic changes in *Thiobacillus denitrificans* accompanying the transition from aerobic to anaerobic growth in continuous culture. J. Gen. Microbiol. 107:131–137.

88. Kanner, D., and R. Bartha. 1979. Growth of *Nocardia rhodochrous* on acetylene gas. J. Bacteriol. 139:225–230.

89. Kaspar, H. F., and J. M. Tiedje. 1981. Dissimilatory reduction of nitrate and nitrite in the bovine rumen: nitrous oxide production and effect of acetylene. Appl. Environ. Microbiol. 41:705–709.

90. Kaspar, H. F., J. M. Tiedje, and R. B. Firestone. 1981. Denitrification and dissimilatory nitrate reduction to ammonium in digested sludge. Can. J. Microbiol. 27:878–885.

91. Kaspar, H. F. 1982. Nitrite reduction to nitrous oxide by propionibacteria: detoxication mechanism. Arch. Microbiol. 133:126–130.

92. Kaspar, H. F. 1983. Denitrification, nitrate reduction to ammonium, and inorganic nitrogen pools in intertidal sediments. Mar. Biol. 74:133–140.

93. Kaspar, H. F. 1985. The denitrification capacity of sediment from a hypereutrophic lake. Freshwater Biol. 15:449–453.

94. Keeney, D. R., I. R. Fillery, and G. P. Marx. 1979. Effect of temperature on the gaseous nitrogen products of denitrification in a silt loam soil. Soil Sci. Soc. Am. J. 43:1124–1128.

95. Kennedy, M. J., and J. G. Lawless. 1985. Role of chemotaxis in the ecology of denitrifiers. Appl. Environ. Microbiol. 49:109–114.

96. King, D., and D. B. Nedwell. 1985. The influence of nitrate concentration upon the end-products of nitrate dissimilation by bacteria in anaerobic salt marsh sediments. FEMS Microbiol. Ecol. 31:23–28.

97. Klemedtsson, L., B. H. Svensson, T. Lindberg, and T. Rosswall. 1977. The use of acetylene inhibition of nitrous oxide reductase in quantifying denitrification in soils. Swed. J. Agric. Res. 7:179–185.

98. Knowles, R. 1979. Denitrification, acetylene reduction, and methane metabolism in lake sediment exposed to acetylene. Appl. Environ. Microbiol. 38:486–493.

99. Knowles, R. 1981. Denitrification, p. 323–369, in E. A. Paul and J. Ladd (eds.), Soil biochemistry, Vol. 5. Marcel Dekker, New York.

100. Knowles, R. 1981. Denitrification, p. 315–329, in F. E. Clark and T. Rosswall (eds.), Terrestrial nitrogen cycles. Ecol. Bull. (Stockholm) No. 33. Swedish Natural Science Research Council, Stockholm.

101. Knowles, R., D. R. S. Lean, and Y. K. Chan. 1981. Nitrous oxide concentrations in lakes: variations with depth and time. Limnol. Oceanogr. 26:855–866.

102. Knowles, R. 1982. Denitrification. Microbiol. Rev. 46:43–70.

103. Koike, I., and A. Hattori. 1975. Energy yield of denitrification: an estimate for growth yield in continuous culture of Pseudomonas denitrificans under nitrate-, nitrite- and nitrous oxide-limited conditions. J. Gen. Microbiol. 88:11–19.

104. Koike, I., and A. Hattori. 1978. Denitrification and ammonia formation in anaerobic coastal sediments. Appl. Environ. Microbiol. 35:278–282.

105. Koskinen, W. C., and D. R. Keeney. 1982. Effect of pH on the rate of gaseous products of denitrification in a silt loam soil. Soil Sci. Soc. Am. J. 46:1165–1167.

106. Krieg, N. R. 1976. Biology of the chemoheterotrophic spirilla. Bacteriol. Rev. 40:55–115.

107. Krieg, N. R. and J. G. Holt (eds.). 1984. Bergey's manual of systematic bacteriology, Vol. 1. Williams & Wilkins, Baltimore, p. 964.

108. Krul, J. M. 1976. Dissimilatory nitrate and nitrite reduction under aerobic conditions by an aerobically and anaerobically grown Alcaligenes sp. and by activated sludge. J. Appl. Bacteriol. 4:245–260.

109. Krul, J. M., and R. Veeningen. 1977. The synthesis of the dissimilatory nitrate reductase under aerobic conditions in a number of denitrifying bacteria, isolated from activated-sludge and drinking water. Water Res. 11:39–43.

110. Lewis, D. 1951. The metabolism of nitrate and nitrite in sheep. I. The reduction of nitrate in the rumen of the sheep. Biochem. J. 48:175–180.

111. Lijinsky, W. 1977. How nitrosamines cause cancer. New Sci. 27:216–217.

112. Limmer, A. W., K. W. Steele, and A. T. Wilson. 1982. Direct measurement of N_2 and N_2O evolution from soil. J. Soil. Sci. 33:499–507.

113. Liu, M. C., and H. D. Peck, Jr. 1981. The isolation of a hexaheme cytochrome from *Desulfovibrio desulfuricans* and its identification as a new type of nitrite reductase. J. Biol. Chem. 256:13159–13164.

114. MacFarlane, G. T., and R. A. Herbert. 1982. Nitrate dissimilation by *Vibrio* spp. isolated from estuarine sediments. J. Gen. Microbiol. 128:2463–2468.

115. MacGregor, A. N. 1972. Gaseous losses of nitrogen from freshly wetted desert soil. Soil Sci. Soc. Am. Proc. 36:594–596.

116. McElroy, M. B., S. C. Wofsy, and Y. L. Yung. 1977. Nitrogen cycle perturbations due to man and their impact on atmospheric N_2O and O_3. Philos. Trans. R. Soc. Lond. B277:159–181.

117. Mechsner, R., and K. Wuhrmann. 1963. Beitrag zur Kenntnis der mikrobiellen Denitrifikation. Pathol. Microbiol. 26:579–591.

118. Meiberg, J. B. M., P. M. Bruinenberg, and W. Harder. 1980. Effect of dissolved oxygen tension on the metabolism of methylated amines in *Hyphomicrobium* X in the absence and presence of nitrate: evidence for "*aerobic*" denitrification. J. Gen. Microbiol. 120:453–463.

119. Miller, D. J., and P. Wood. 1983. The soluble cytochrome oxidase of *Nitrosomonas europaea*. J. Gen. Microbiol. 129:1645–1650.

120. Mitchell, G. J., J. G. Jones, and J. A. Cole. 1986. Distribution and regulation of nitrate and nitrite reduction by *Desulfovibrio* and *Desulfotomaculum* species. Arch. Microbiol. 144:35–40.

121. Motteram, P. A. S., J. E. G. McCarthy, S. J. Ferguson, J. B. Jackson, and J. A. Cole. 1981. Energy conservation during the formate dependent reduction of nitrite by *Escherichia coli*. FEMS Microbiol. Lett. 12:317–320.

122. Mountfort, D. O., and M. P. Bryant. 1982. Isolation and characterization of an anaerobic syntrophic benzoate-degrading bacterium from sewage sludge. Arch. Microbiol. 133:249–256.

123. Muller, M. M., V. Sundman, and J. Skujins. 1980. Denitrification in low pH spodsols and peats determined with the acetylene inhibition method. Appl. Environ. Microbiol. 40:235–239.

124. Myrold, D. D., and J. M. Tiedje. 1985. Diffusional constraints on denitrification in soil. Soil Sci. Soc. Am. J. 49:651–657.

125. Myrold, D. D., and J. M. Tiedje. 1985. Establishment of denitrification capacity in soil: effects of carbon, nitrate and moisture. Soil. Biol. Biochem. 17:819–822.

126. Myrold, D. D., and J. M. Tiedje. 1986. Simultaneous estimation of several nitrogen cycle rates using ^{15}N: theory and application. Soil. Biol. Biochem. 18:559–568.

127. Nakajima, T. 1982. Distribution of denitrifying bacteria and its controlling factors in freshwater environments. Jpn. J. Limnol. 43:17–26.

128. Neal, J. L., G. C. Allen, R. D. Morse, and D. D. Wolf. 1983. Anaerobic nitrate-dependent chemolithotrophic growth by *Rhizobium japonicum*. Can. J. Microbiol. 29:316–320.

129. Neal, J. L., G. C. Allen, R. D. Morse, and D. D. Wolf. 1983. Nitrate, nitrite, nitrous oxide and oxygen-dependent hydrogen uptake by *Rhizobium*. FEMS Microbiol. Lett. 17:335–338.

130. Nelson, L. M., and R. Knowles. 1978. Effect of oxygen and nitrate on nitrogen fixation and denitrification by *Azospirillum brasilense* grown in continuous culture. Can. J. Microbiol. 24:1395–1403.

131. Neuer, G., A. Kronenberg, and H. Bothe. 1985. Denitrification and nitrogen fixation by *Azospirillum* IIII. Properties of a wheat-*Azospirillum* association. Arch. Microbiol. 141:364–370.

132. Neyra, C. A., J. Dobereiner, R. Lalande, and R. Knowles. 1977. Denitrification by N_2-fixing *Spirillum lipoferum*. Can. J. Microbiol. 23:300–305.

133. O'Hara, G. W., R. M. Daniel, K. W. Steele, and P. M. Bonish. 1984. Nitrogen losses from soils caused by *Rhizobium*-dependent denitrification. Soil Biol. Biochem. 4:429–431.

134. O'Hara, G. W., and R. M. Daniel. 1985. *Rhizobium* denitrification: a review. Soil Biol. Biochem. 17:1–9.

135. Parkin, T. B., and J. M. Tiedje. 1984. Application of a soil core method to investigate the effect of oxygen concentration on denitrification. Soil. Biol. Biochem. 16:331–334.

136. Parkin, T. B., A. J. Sexstone, and J. M. Tiedje. 1985. Comparison of field denitrification rates by acetylene-based soil core and nitrogen methods. Soil Sci. Soc. Am. J. 49:94–99.

137. Parkin, T. B., A. J. Sexstone, and J. M. Tiedje. 1985. Adaptation of denitrifying populations to low soil pH. Appl. Environ. Microbiol. 49:1053–1056.

138. Parkin, T. B., H. F. Kaspar, A. J. Sexstone, and J. M. Tiedje. 1985. A gas-flow soil core method to measure field denitrification rates. Soil Biol. Biochem. 16:323–330.

139. Patrick, W. H., Jr., and R. Wyatt. 1964. Soil nitrogen loss as a result of alternate submergence and drying. Soil Sci. Soc. Am. Proc. 28:647–653.

140. Paul, E. A., and R. L. Victoria. 1978. Nitrogen transfer between the soil and atmosphere, p. 525–541, in W. E. Krumbein (ed.), Environmental biogeochemistry and geomicrobiology, Vol. 2, The terrestrial environment. Ann Arbor Science Publishers, Ann Arbor, Mich.

141. Payne, W. J. 1981. Denitrification. Wiley, New York, p. 241.

142. Pfennig, N., F. Widdel, and H. G. Trüper. 1981. The dissimilatory sulfate-reducing bacteria, p. 926–940, in M. P. Starr, H. Stolp, H. G. Trüper, A. Balows, and H. G. Schlegel (eds.), The prokaryotes, a handbook on habitats, isolation and identification of bacteria. Springer-Verlag, Heidelberg.

143. Phillips, R. E., K. R. Reddy, and W. H. Patrick. 1978. The role of nitrate diffusion in determining the order and rate of denitrification in flooded soil. II. Theoretical analysis and interpretation. Soil Sci. Soc. Am. J. 42:272–278.

144. Pichinoty, F., H. de Barjac, M. Mandel, B. Greenway, and J.-L. Garcia. 1976. Une nouvelle bactérie sporulée denitrifiante, mésophile: *Bacillus azotoformans* n. sp. Ann. Microbiol. (Paris) B127:351–361.

145. Pichinoty, F., J. Bagliardi-Rouvier, M. Mandel, B. Greenway, G. Metanier, and J.-L. Garcia. 1976. The isolation and properties of a denitrifying bacterium of the genus *Flavobacterium*. Antonie Leeuwenhoek J. Microbiol. Serol. 42:349–354.

146. Pichinoty, F., M. Mandel, and J.-L. Garcia. 1977. Étude de six souches de *Agrobacterium tumefaciens* et *A. radiobacter*. Ann. Microbiol. (Paris) A128:303–310.

147. Pichinoty, F., M. Mandel, and J.-L. Garcia. 1979. The properties of novel meso-

philic denitrifying *Bacillus* cultures found in tropical soils. J. Gen. Microbiol. 115:419–430.

148. Poth, M., and D. D. Focht. 1985. [15]N kinetic analysis of N_2O production by *Nitrosomonas europaea*: an examination of nitrifier denitrification. Appl. Environ. Microbiol. 49:1134–1141.

149. Poth, M. 1986. Dinitrogen production from nitrite by a *Nitrosomonas* isolate. Appl. Environ. Microbiol. 52:957–959.

150. Prakash, O., and J. C. Sadana. 1973. Metabolism of nitrate by *Achromobacter fischeri*. Can. J. Microbiol. 19:15–25.

151. Renner, E. D., and G. E. Becker. 1970. Production of nitric oxide and nitrous oxide during denitrification by *Corynebacterium nephridii*. J. Bacteriol. 101:821–826.

152. Revsbech, N. P., J. Sørensen, and T. H. Blackburn. 1980. Distribution of oxygen in marine sediments measured with microelectrodes. Limnol. Oceanogr. 25:403–411.

153. Revsbech, N. P., and D. M. Ward. 1983. Oxygen microelectrode that is insensitive to medium chemical composition: use in an acid microbial mat dominated by *Cyanidium caldarium*. Appl. Environ. Microbiol. 45:755–759.

154. Robertson, G. P., and J. M. Tiedje. 1984. Denitrification and nitrous oxide production in successional and old-growth Michigan forests. Soil. Sci. Soc. Am. J. 48:383–389.

155. Robertson, G. P., and J. M. Tiedje. 1987. Nitrous oxide sources in aerobic soils: nitrification, denitrification, and other biological processes. Soil Biol. Biochem. 19:187–193.

156. Robertson, L. A., and J. G. Kuenen. 1983. *Thiosphaera pantotropha* gen. nov. sp. nov., a facultatively anaerobic, facultatively autotrophic sulphur bacterium. J. Gen. Microbiol. 129:2847–2855.

157. Robertson, L. A., and J. G. Kuenen. 1984. Aerobic denitrification: a controversy revived. Arch. Microbiol. 139:351–354.

158. Rolston, D. E., A. N. Sharpley, D. W. Troy, and F. E. Broadbent. 1982. Field measurement of denitrification. III. Rates during irrigation cycles. Soil Sci. Soc. Am. J. 46:289–296.

159. Rönner, U. 1983. Distribution, production and consumption of nitrous oxide in the Baltic Sea. Geochim. Cosmochim. Acta 47:2179–2188.

160. Rönner, U. 1985. Nitrogen transformations in the Baltic Proper: denitrification counteracts eutrophication. Ambio 14:134–138.

161. Rönner, U., and F. Sorensson. 1985. Denitrification rates in the low-oxygen waters of the stratified Baltic Proper. Appl. Environ. Microbiol. 50:801–806.

162. Reuger, H. J., T. L. Tau, G. Hentzschel, and M. Naguib. 1983. New denitrifying *Agrobacterium* and *Alcaligenes* strains from Weser Estuary, West Germany sediments. Taxonomy and physiology. Veroeff Inst. Meeresforsch. Bremerhaven 19:229–244.

163. Ryden, J. C., L. J. Lund, and D. D. Focht. 1979. Direct measurement of denitrification loss from soils. I. Laboratory evaluation of acetylene inhibition of nitrous oxide reduction. Soil Sci. Soc. Am. J. 43:104–110.

164. Ryden, J. C. 1983. Denitrification loss from a grassland soil in the field receiving different rates of nitrogen as ammonium nitrate. J. Soil Sci. 34:355–365.

165. Sacks, L. E., and H. A. Barker. 1949. The influence of oxygen on nitrate and nitrite reduction. J. Bacteriol. 58:11–22.

166. Samuelsson, M.-O. 1985. Dissimilatory nitrate reduction to nitrite, nitrous oxide, and ammonium by *Pseudomonas putrefaciens*. Appl. Environ. Microbiol. 50:812–815.

167. Satoh, T., Y. Hoshimo, and H. Kitamura. 1976. Rhodopseudomonas sphaeroides *forma* sp. *denitrificans*, a denitrifying strain as a subspecies of *Rhodopseudomonas sphaeroides*. Arch. Microbiol. 108:265–269.

168. Satoh, T. H., S. S. M. Hom, and K. T. Shanmugam. 1981. Production of nitrous oxide as a product of nitrite metabolism by enteric bacteria, p. 481–497, in J. M. Lyons, R. C. Valentine, D. A. Phillips, D. W. Rains, and R. C. Huffaker (eds.), Genetic engineering of symbiotic nitrogen fixation and conservation of fixed nitrogen. Plenum Press, New York.

169. Sayama, M., and Y. Kurihara. 1983. Relationship between burrowing activity of the polychaetous annelid, *Neanthes japonica* (Izuka) and nitrification-denitrification processes in the sediments. J. Exp. Mar. Biol. Ecol. 72:233–241.

170. Schenneu, U., K. Braun, and H. J. Knackmuss. 1985. Anaerobic degradation of 2-fluorobenzoate by benzoate degrading, denitrifying bacteria. J. Bacteriol. 161:321–325.

171. Sexstone, A. J., N. P. Revsbech, T. B. Parkin, and J. M. Tiedje. 1985. Direct measurement of oxygen profiles and denitrification rates in soil aggregates. Soil Sci. Soc. Am. J. 49:645–651.

172. Sexstone, A. J., T. B. Parkin, and J. M. Tiedje. 1985. Temporal response of soil denitrification rates to rainfall and irrigation. Soil Sci. Soc. Am. J. 49:99–103.

173. Siegel, R. S., R. D. Hauck, and L. T. Kurtz. 1982. Determination of $^{30}N_2$ and application to measurement of N_2 evolution during denitrification. Soil Sci. Soc. Am. J. 46:68–74.

174. Skerman, V. B. D., and I. C. MacRae. 1957. The influence of oxygen availability on the degree of nitrate reduction by *Pseudomonas denitrificans*. Can. J. Microbiol. 3:505–530.

175. Smith, C. J., and R. D. Delaune. 1984. Effect of rice plants on nitrification-denitrification loss of nitrogen under greenhouse conditions. Plant Soil 79:287–290.

176. Smith, M. S., M. K. Firestone, and J. M. Tiedje. 1978. The acetylene inhibition method for short-term measurement of soil denitrification and its evaluation using nitrogen-13. Soil Sci. Soc. Am. J. 42:611–615.

177. Smith, M. S., and J. M. Tiedje. 1979. The effect of roots on soil denitrification. Soil Sci. Soc. Am. J. 43:951–955.

178. Smith, M. S., and J. M. Tiedje. 1980. Growth and survival of antibiotic resistant denitrifier strains in soil. Can. J. Microbiol. 26:854–856.

179. Smith, M. S., and K. Zimmerman. 1981. Nitrous oxide production by nondenitrifying soil nitrate reducers. Soil Sci. Soc. Am. J. 45:865–871.

180. Smith, M. S. 1982. Dissimilatory reduction of nitrite to ammonium and nitrous oxide by a soil *Citrobacter* sp. Appl. Environ. Microbiol. 43:854–860.

181. Smith, M. S. 1983. Nitrous oxide production by *Escherichia coli* is correlated with nitrate reductase activity. Appl. Environ. Microbiol. 45:1545–1547.

182. Smith, M. S. and L. L. Parsons. 1985. Persistence of denitrifying enzyme activity in dried soils. Appl. Environ. Microbiol. 49:316–320.

183. Sneath, P. H. A. 1984. Genus *Chromobacterium* Bergonzini, 1881, p. 580–582, N. R. Krieg and J. G. Holt (eds.), Bergey's manual of systematic bacteriology, Vol. 1. Williams & Wilkins, Baltimore.

184. Sørensen, J. 1978. Denitrification rates in a marine sediment as measured by the acetylene inhibition technique. Appl. Environ. Microbiol. 36:139–143.

185. Sørensen, J. 1978. Capacity for denitrification and reduction of nitrate to ammonia in a coastal marine sediment. Appl. Environ. Microbiol. 35:301–305.

186. Sperl, G. T., and D. S. Hoare. 1971. Denitrification with methanol: selective enrichment for *Hyphomicrobium* species. J. Bacteriol. 108:733–736.

187. Steele, K. W., P. M. Bonish, and S. U. Saratchandra. 1984. Denitrification potentials and microbiological characteristics of some northern North Island New Zealand soils. N.Z. J. Agric. Res. 27:525–530.

188. Steenkamp, D. J., and H. D. Peck, Jr. 1981. Proton translocation associated with nitrite respiration in *Desulfovibrio desulfuricans*. J. Biol. Chem. 256:5450–5458.

189. Strand, S. E., and A. J. McDonnell. 1985. Mathematical analysis of oxygen and nitrate consumption in deep microbial films. Water Res. 19:345–352.

190. Suflita, J. M., A. Horowitz, D. R. Shelton, and J. M. Tiedje. 1982. Dehalogenation: a novel pathway for the anaerobic biodegradation of haloaromatic compounds. Science 218:1115–1117.

191. Sugahara, I., K. Hayashi, and T. Kimura. 1986. Distribution and generic composition of denitrifying bacteria in coastal and oceanic waters. Bull. Jpn. Soc. Sci. Fish. 52:497–503.

192. Tam, T.-Y., and R. Knowles. 1979. Effects of sulfide and acetylene on nitrous oxide reduction by soil and by *Pseudomonas aeruginosa*. Can. J. Microbiol. 25:1133–1138.

193. Tarrand, J. J., N. R. Krieg, and J. Dobereiner. 1978. A taxonomic study of the *spirillum lipoferum* group with descriptions of a new genus, *Azospirillum* gen. nov. and two new species, *Azospirillum lipoferum* (Beijerinck) comb. nov. and *Azospirillum brasilense* sp. nov. Can. J. Microbiol. 24:967–980.

194. Taylor, B. F., W. L. Campbell, and I. Chinoy. 1970. Anaerobic degradation of the benzene nucleus by a facultatively anaerobic microorganism. J. Bacteriol. 102:430–437.

195. Taylor, B. F., W. L. Hearn, and S. Pincus. 1979. Metabolism of monofluoro and monochlorobenzoates by a denitrifying bacterium. Arch. Microbiol. 122:301–306.

196. Terai, H. 1979. Taxonomic study and distribution of denitrifying bacteria in Lake Kizaki. Jpn. J. Limnol. 40:81–92.

197. Terai, H., and T. Yoshioka. 1983. Serological study on seasonal and vertical distribution of specific denitrifying bacteria in Lake Kizaki. Jpn. J. Limnol. 44:81–92.

198. Tiedje, J. M., R. B. Firestone, M. K. Firestone, M. R. Betlach, H. F. Kaspar, and J. Sørensen. 1981. Use of ^{13}N in studies of denitrification, p. 295–317, in J. W. Root and K. A. Krohn (eds.), Short-lived radionuclides in chemistry and biology (Advances in Chemistry Series No. 197). American Chemical Society, Washington, D.C.

199. Tiedje, J. M. 1982. Denitrification, p. 1011–1026, in A. L. Page, R. H. Miller, and D. R. Keeney (eds.), Methods of soil analysis, Pt. 2, Agronomy Monograph No. 9, American Society of Agronomy, Madison, Wis.

200. Tiedje, J. M., A. J. Sexstone, D. D. Myrold, and J. A. Robinson. 1982. Denitrifi-

cation: ecological niches, competition and survival. Antonie Leeuwenhoek J. Microbiol. 48:569–583.

201. Timmer ten Hoor, A. 1975. A new type of thiosulfate oxidizing, nitrate-reducing microorganism: *Thiomicrospira denitrificans* sp. nov. Neth. J. Sea Res. 9:344–350.

202. Timmermans, P., and A. Van Haute. 1983. Denitrification with methanol: fundamental study of the growth and denitrification capacity of *Hyphomicrobium* sp. Water Res. 17:1249–1255.

203. Tomlinson, G. A., L. L. Jahnke, and L. I. Hockstein. 1986. *Halobacterium denitrificans* new species an extremely halophilic denitrifying bacterium. Int. J. Syst. Bacteriol. 36:66–70.

204. van Berkum, P., and H. H. Keyser. 1985. Anaerobic growth and denitrification among different serogroups of soybean rhizobia. Appl. Environ. Microbiol. 49:772–777.

205. Van Cleemput, O., W. H. Patrick, and R. C. McIlhenny. 1976. Nitrite decomposition in flooded soil under different pH and redox potential conditions. Soil. Sci. Soc. Am. J. 40:55–60.

206. Van Raalte, C. D., and D. G. Patriquin. 1979. Use of the "acetylene blockage" technique for assaying denitrification in a salt marsh. Mar. Biol. 52:315–320.

207. Verhoeven, W. 1950. On a spore-forming bacterium causing the swelling of cans containing cured ham. Antonie Leeuwenhoek J. Microbiol. Serol. 16:269–281.

208. Vinther, F. P., H. G. Memon, and V. Jensen. 1982. Populations of denitrifying bacteria in agricultural soils under continuous barley cultivation. Pedobiologia 24:319–328.

209. Virginia, R. A., W. M. Jarrell, and E. Franco-Vizcaino. 1982. Direct measurements of denitrification in a *Prosopis* (mesquite) dominated Sonoran desert ecosystem. Oecologia (Berl.) 53:120–122.

210. Walter, H. M., and D. R. Keeney, and I. R. Fillery. 1979. Inhibition of nitrification by acetylene. Soil Sci. Soc. Am. J. 43:195–196.

211. Wang, W. C., Y. L. Yung, A. L. Lacis, T. Mo., and J. E. Hanson. 1976. Greenhouse effects due to man-made perturbations of trace gases. Science 194:685–689.

212. Watanabe, I., and M. R. de Guzman. 1980. Effect of nitrate on acetylene disappearance from anaerobic soil. Soil Biol. Biochem. 12:193–194.

213. Weathers, P. J., 1984. N_2O evolution by green algae. Appl. Environ. Microbiol. 48:1251–1253.

214. Werber, M. M., and M. Mevarech. 1978. Induction of a dissimilatory reduction pathway of nitrate in *Halobacterium* of the Dead Sea. Arch. Biochem. Biophys. 186:60–65.

215. Williams, R. J., and W. C. Evans. 1975. The metabolism of benzoate by *Moraxella* species through anaerobic nitrate respiration. Biochem. J. 148:1–10.

216. Wolin, M. J., E. A. Wolin, and N. J. Jacobs. 1961. Cytochrome-producing anaerobic vibrio, *Vibrio succinogenes* sp. n. J. Bacteriol. 81:911–917.

217. Woods, D. D. 1938. The reduction of nitrate to ammonia by *Clostridium welchii*. Biochem. J. 32:2000–2012.

218. Yeomans, J. C., and E. G. Beauchamp. 1978. Limited inhibition of nitrous oxide reduction in soil in the presence of acetylene. Soil. Biol. Biochem. 10:517–519.

219. Yordy, D. M., and E. A. Delwiche. 1979. Nitrite reduction in *Veillonella alcalescens*. J. Bacteriol. 137:905–911.

220. Yoshida, T., and M. Alexander. 1970. Nitrous oxide formation by *Nitromonas europaea* and heterotrophic microorganisms. Soil Sci. Soc. Am. Proc. 34:880–882.

221. Yoshinari, T. 1975. Nitrous oxide in the sea. Mar. Chem. 4:189–202.

222. Yoshinari, T., and R. Knowles. 1976. Acetylene inhibition of nitrous oxide reduction by denitrifying bacteria. Biochem. Biophys. Res. Commun. 69:705–710.

223. Yoshinari, T., R. Hynes, and R. Knowles. 1977. Acetylene inhibition of nitrous oxide reduction and measurement of denitrification and nitrogen fixation in soil. Soil Biol. Biochem. 9:177–183.

224. Yoshinari, T. 1980. N_2O reduction by *Vibrio succinogenes*. Appl. Environ. Microbiol. 39:81–84.

225. Zablotowicz, R. M., D. L. Eskew, and D. D. Focht. 1978. Denitrification in *Rhizobium*. Can. J. Microbiol. 24:757–760.

5

DISSIMILATORY REDUCTION OF OXIDIZED NITROGEN COMPOUNDS

ADRIAAN H. STOUTHAMER

Biological Laboratory, Vrije Universiteit, Postbus 7161, 1007 MC, Amsterdam, The Netherlands

5.1 INTRODUCTION
5.2 PROPERTIES OF THE ENZYMES INVOLVED IN NITRATE RESPIRATION
    5.2.1 Properties of Nitrate Reductase
    5.2.2 Involvement of Molybdenum and the Molybdenum Cofactor in Nitrate Reduction
    5.2.3 Properties of Nitrite Reductase
        5.2.3.1 Nitrite Reductases Involved in Denitrification
        5.2.3.2 Denitrification Pathway between Nitrite and N_2O
        5.2.3.3 Nitrite Reductase Involved in Dissimilatory Nitrate Reduction to Ammonia
    5.2.4 Properties of Nitrous Oxide Reductase
    5.2.5 Production of N_2O by Nondenitrifiers
5.3 BIOENERGETICS OF NITRATE REDUCTION
    5.3.1 Respiratory Chain to Nitrogenous Oxides
        5.3.1.1 *Escherichia coli*
        5.3.1.2 *Paracoccus denitrificans*
        5.3.1.3 *Thiobacillus denitrificans*
        5.3.1.4 Organisms with Dissimilatory Reduction of Nitrate to Ammonia

5.3.2   Determination of the Proton-Consuming Site in the Reduction of
        Nitrogenous Oxides

5.3.3   Nitrate Transport

5.3.4   Stoichiometry of Proton Translocation and Energy Production during
        Nitrate Reduction

    5.3.4.1   *Escherichia coli*, the Proton Motive Q Cycle and
        Cytochrome *b* Cycle

    5.3.4.2   *Paracoccus denitrificans*

    5.3.4.3   *Campylobacter sputorum* subsp. *bubulus*

    5.3.4.4   *Desulfovibrio* sp.

5.4   REGULATORY ASPECTS

5.4.1   Regulation of the Formation of Nitrate Reductase

5.4.2   Control of Electron Flow to Nitrogenous Oxides and Oxygen

5.5   CONCLUSIONS

REFERENCES

5.1   INTRODUCTION

Many bacteria can utilize nitrate instead of oxygen as a terminal electron acceptor. This process, called nitrate respiration or dissimilatory nitrate reduction, occurs generally only under anaerobic conditions. The first step in nitrate reduction is a reduction of nitrate to nitrite. In a number of organisms nitrite can also be used as a terminal hydrogen acceptor under anaerobic conditions. It has become clear during the last four years that two different methods of dissimilatory nitrite reduction can be distinguished. In the first, nitrite can be converted to gaseous products, such as nitrogen or nitrous oxide, a process called denitrification. In addition, nitrite can be converted into ammonia in a dissimilatory process. The standard free energy changes of the reactions $NO_3^- + 2H^+ + 4H_2 \rightarrow NH_4^+ + 3H_2O$ and $2NO_3^- + 2H^+ + 5H_2 \rightarrow N_2 + 6H_2O$ are -599.6 and -1120.5 kJ per reaction, respectively (240). On the basis of these data it would be expected that the energy gain during denitrification and during dissimilatory reduction of nitrate to ammonia would be about the same. However, there is a large difference in the amount of ATP gained during denitrification and during dissimilatory reduction of nitrate to ammonia. The distribution of the capacity of denitrification has recently been reviewed (14, 127, 193). The properties of nitrate reductase, the composition of the electron transport chain toward nitrogenous oxides, and energy production during nitrate respiration have also been treated in a number of recent reviews (130, 231, 236, 240). Since these reviews were written, considerable progress has been made on various aspects of nitrate respiration.

5.2 PROPERTIES OF THE ENZYMES INVOLVED IN NITRATE RESPIRATION

5.2.1 Properties of Nitrate Reductase

Nitrate reductase is a membrane-bound enzyme. Therefore, the first step in the purification of nitrate reductase must be its solubilization from the cytoplasmic membrane. The methods which have been used for that purpose are treatment of cytoplasmic membranes by detergents, organic solvents, or heat (236). The different procedures used for the solubilization of nitrate reductase have led to the isolation of different nitrate reductase complexes. Nitrate reductase from *Escherichia coli* can be solubilized while still complexed with formate dehydrogenase and cytochrome *b* (101, 155). Nitrate reductase has been purified from *Escherichia coli* (43, 63, 72, 154, 160, 238), *Klebsiella aerogenes* (246), *Proteus mirabilis* (186), *Bacillus licheniformis* (249), *Paracoccus denitrificans* and *halodenitrificans* (70, 144, 204), and *Pseudomonas aeruginosa* (30). It is remarkable that nitrate reductases from these various organisms are very similar in subunit structure, molecular weight of the subunits, and other properties. Purified enzyme from all these species, which still reduces NO_3^- to NO_2^- with artificial electron donors like reduced viologens, consists of only two polypeptides. These A and B subunits have molecular weights of about 150,000 and 60,000, and are present in equimolar amounts. *E. coli* nitrate reductase contains about 12 Fe-S groups and one atom of molybdenum for every 200,000 Da. (71, 154, 160). The enzymes of *B. licheniformis* (249), *Pa. denitrificans* (70), and *K. aerogenes* (23, 248) contain about eight Fe-S groups and one atom of molybdenum. These results indicate the presence of two or three 4Fe-S iron–sulfur clusters in nitrate reductase. Electron paramagnetic resonance studies clearly show the participation of Fe-S centers and the molybdenum in the nitrate-reducing activity of the enzyme (reviewed in Ref. 236). It has recently been demonstrated that the iron–sulfur clusters and the molybdenum are associated with subunit A (33), which is therefore the catalytic subunit. Subunit B has been implicated in membrane binding (53, 154, 155). It has been noted that subunit B is always isolated in two different forms, which have been referred to as B and B^1 (43, 55, 63, 154, 160). Subunit B^1 has a slightly higher electrophoretic mobility than subunit B on sodium dodecyl sulfate–polyacrylamide gels. An enzymatic activity that modifies nitrate reductase has been identified in the cytoplasmic membrane from *E. coli* (32). It was concluded that the modification of B to B^1 is a reversible process and is due to the removal of one or more nonprotein molecules. Probably the conversion of B^1 to B involves the attachment of fatty acid molecules to the protein. The conversion of B^1 to B is accompanied by binding of subunits A and B to the membrane, which occur by their binding to subunit C. By various purification procedures a purified nitrate reductase is obtained which contains three subunits: A, B, and C. Subunit C is a cytochrome *b* with a molecular weight of about 20,000 (43, 63, 156). This cytochrome is essential for the functional association of the enzyme with physiological electron donor systems, such as formate dehydrogenase (63). It is also required for association of nitrate reductase with the cell membrane (156, 158). Studies with a strain unable to synthesize heme (*hem* A) demonstrated that the complete cytochrome must be available in order for subunits A and B to be properly integrated into the membrane (158). A new purification procedure for isolation of a nitrate reductase complex containing subunits A, B, and C in the ratio 2:2:4 has recently been described (34). This dimeric form is supposed to be the native form of the enzyme. The concentration of nitrate reductase in the cytoplasmic membrane may be as high as about 25% of the total membrane protein (23, 64, 160). The localization of nitrate reductase in the membrane has been studied by various methods (24, 80, 159): labeling of tyrosine residues by treatment with [^{125}I]lactoperoxidase, labeling of glutamine residues by treatment with transglutaminase, and indirect immunofluorescence with antibodies specific for the A and B subunits. A schematic representation of the localization of nitrate reductase in the membrane of *E. coli* is shown in Fig. 5.1. The nitrate reductase complex is a transmembrane protein. The C subunit can be labeled from the periplasmic aspect of the membrane, whereas the A subunit can be labeled from the cytoplasmic aspect of the membrane. The B subunit was not labeled from either side of the membrane. However, experiments with the non-permeant reagent, diazotized [^{125}I]diidosulfanilic acid or diazobenezene [^{35}S]sulfonate, show that this subunit is located solely on the cytoplasmic face of the membrane (81). Nitrate is reduced at the cytoplasmic side of the membrane. This aspect is discussed in more detail in Section 5.3.2. The formation of nitrate reductase is affected by mutations at several distinct genetic loci. These are designated *chl* (from resistance against chlorate) or *nar*. Only one of these *chl* C has been implicated as a structural gene (54, 157). More recent work has demonstrated that in the *chl* C region a multicistronic operon is present, which contains the genetic information for the A, B, and C subunits of the nitrate reductase (17, 61, 227).

5.2.2 Involvement of Molybdenum and the Molybdenum Cofactor in Nitrate Reduction

As stated earlier, purified nitrate reductase contains 1 atom of molybdenum per monomer. A model for the molecular mechanism of molybdenum in enzymes has

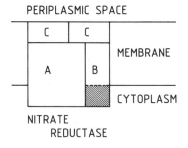

Figure 5.1 Schematic representation of the localization of nitrate reductase in the cytoplasmic membrane of *Escherichia coli*. The hatched area in the B subunit indicates that this subunit reacts only with some reagents. Adapted from Ref. 236.

Figure 5.2 Mechanism of coupled proton-electron transfer during nitrate reduction. Adapted from Ref. 228.

been proposed by Stiefel (228). This mechanism, shown in Fig. 5.2, involves coupled proton and electron transfer. According to this model molybdenum must be in oxidation state IV to interact with nitrate. A fully protonated ligand is supposed to be attached to the Mo atom. As the Mo(IV) reduces NO_3^- by two electrons, it becomes Mo(VI) and the donor atom (shown here as XH) becomes acidic and transfers its proton to nitrate, which then splits to nitrite and hydroxide. The Mo in the VI state is then reduced again by electron transfer, most probably from the Fe-S clusters in the enzyme. By electron paramagnetic resonance studies it has been demonstrated that upon reduction by dithionite, Mo was present in oxidation state IV in nitrate reductase (25). It has been proposed on the basis of such studies that molybdenum shuttles between oxidation states IV and VI in nitrate reductase (23, 95). However, there is also a possibility of a Mo(IV)–Mo(V) couple, with two one-electron transfer steps occurring sequentially (236). In samples containing purified resting nitrate reductase, 25% of the molybdenum was found to be in oxidation state V (259). Alternatively, a two-electron transfer followed by a rapid intermolecular equilibration of molybdenum and Fe-S cluster may account for the observed Mo(V) signal in oxidized molybdoenzymes (26). It is generally assumed that sulfur and oxygen acts as ligands of molybdenum in all molybdoenzymes (25, 49).

It has been known for years that a supply of tungstate in high concentrations with respect to molybdate during growth generally causes the loss of the activities of various molybdoenzymes. As examples, we may mention the loss of nitrate reductase and formate dehydrogenase activity in *E. coli* (62, 211), *P. mirabilis* (185), and *Pa. denitrificans* (28). For nitrate reductase in *Neurospora crassa*, the same feature has been reported (148). During growth in the presence of tungstate, de-molybdoenzyme is formed, and in addition, enzyme in which tungsten has replaced molybdenum. In many papers, *in vivo* activation of inactive demolybdoenzymes was reported when tungstate-grown cells were incubated in the presence of molybdate (185, 212). This even occurred in the presence of chloramphenicol (or cycloheximide), indicating that reactivation was not a consequence of *de novo* protein synthesis but rather must be ascribed to a conversion of a de-molybdoprotein into an active molybdenum-containing enzyme. Attempts to achieve reactivation of dissimilatory de-molybdonitrate reductase *in vitro* have failed so far, although successful reactivation studies have been reported for several other molybdoenzymes.

In 1964, mutants of *Aspergillus nidulans* were described, which were unable to grow on nitrate or hypoxanthine as sole nitrogen source (192). Since both enzymes involved were molybdoproteins, the authors suggested that they might share a common molybdenum-containing cofactor. Now it is clear that five or six *cnx* genes are concerned with the expression of the molybdenum cofactor (47). Similar mutants have been isolated from *N. crassa nit* mutants (194, 242), *Ps. aeruginosa* (243, 244), and *Rhizobium meliloti* (126). Mutants of a large number of microorganisms selected for resistance against chlorate are deficient in various molybdoenzymes (87, 231). Based on the hypothesis of a common cofactor in molybdoenzymes, Nason and co-workers started a series of complementation experiments with the *N. crassa nit-1* mutant, which is also blocked in nitrate reductase and xanthine dehydrogenase activities. In wild-type *N. crassa* grown under appropriate conditions, an enzyme complex showing cytochrome *c* reductase and nitrate reductase activities is present (188). The *nit-1* mutant grown under identical conditions shows cytochrome *c* reductase activity only since the molybdenum-containing moiety, present in the wild-type, cannot be synthesized (178). Nitrate reductase activity could not be restored by addition of molybdate or other inorganic molybdo-complexes. However, during incubation of extracts of *nit-1* with extracts of several molybdoenzymes (e.g., intestinal or bovine milk xanthine oxidase, rabbit liver aldehyde oxidase, chicken liver xanthine dehydrogenase, *E. coli* nitrate reductase, and rat liver sulfite oxidase), the nitrate reductase activity in the *nit-1* mutant could be restored (125, 179) The extracts of enzymes were prepared by acidification, and it was reported that this was a rather critical procedure since the complementing factor was acid labile. The complementation was stimulated by the presence of physiologically unusually high amounts of molybdate, indicating that the cofactor was partially devoid of molybdenum (149). With this assay it has indeed been demonstrated that some *chl* mutants of *E. coli* (173, 227) and of *P. mirabilis* (41) are devoid of molybdenum cofactor. The complementation assay utilizing the *N. crassa nit-1* is also the basis for the purification of the molybdenum cofactor. This purification is severely hampered by the great sensitivity toward oxygen and by the presence of the cofactor in very small amounts. The purification was made easier by new extraction procedures: heat treatment (9) or heat treatment in the presence of sodium dodecyl sulfate (40). In this way the yield of cofactor was increased, and it was obtained in a higher concentration and free of protein. Further steps used in purification procedures are gel filtration (112), ultrafiltration, and high-performance liquid chromatography (42). Procedures to obtain purified molybdenum cofactor in sufficient amounts for structure analysis have not been reported so far. It has been demonstrated that the loss of molybdenum cofactor activity is accompanied by the development of fluorescence, especially when cofactor-containing solutions were incubated aerobically (112). From the recorded excitation and emission spectra of the fluorescent species it was concluded that the molybdenum cofactor contains a pteridin structure (112). It was shown that upon treatment of molybdenum cofactor with alkaline permanganate pterin, 6-carboxylic acid is formed. Based on these observations, the name "molybdopterin" has been suggested for the molybdenum cofactor. Later a structural relationship was suggested between urothione, a pigment in urine whose metabolic function is un-

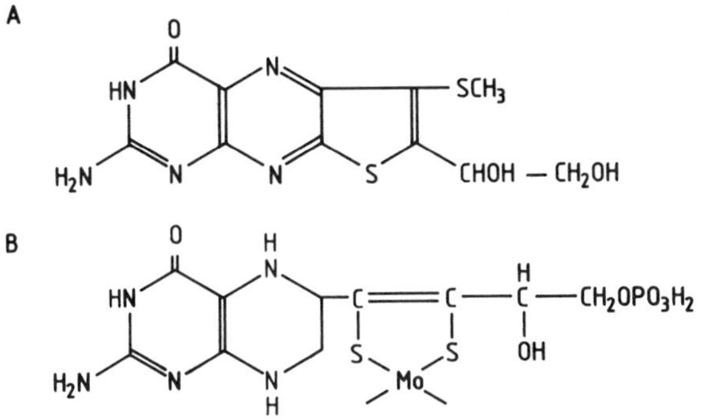

Figure 5.3 Molecular structure of (*A*) urothione (79) and (*B*) molybdenum cofactor as proposed by Johnson and Rajagopolan (113).

known, and the molybdenum cofactor (113). The structure of urothione (79) and the proposed structure of molybdopterin are shown in Fig. 5.3. A metabolic link between the two molecules was demonstrated by the finding that patients deficient in the molybdenum cofactor do not excrete urothione in urine (113). Molybdopterin contains a phosphate group, and the biological activity is lost upon removal of this phosphate group by treatment with alkaline phosphatase (113). However, it seems that the structure proposed in Fig. 5.3*B* cannot be correct on the basis of the following observations

1. The fluorescent oxidation products of molybdopterin do not penetrate a YM-2 ultrafiltration membrane filter, which retains molecules with a molecular weight larger than about 1000 (40). Similarly, molybdopterin elutes in the void volume upon gelfitration on Sephadex G10 (38).
2. The molecular weights of two oxidation products of molybdopterin have been determined by field-desorption mass spectrometric analysis as 1113 and 886 (39). Both compounds contain at most 24 C atoms and lack S and Mo (which is lost completely from the molybdenum cofactor upon purification).
3. It could be demonstrated that the extraction procedure used by Johnson and Rajagopolan (112, 113), which involves boiling for 20 min at pH 2.5, leads to fragmentation of molybdopterin (38).

Based on these observations, it seems that the structure proposed in Fig. 5.3*B* is too simple. More work seems necessary to elucidate the structure of this very important compound.

5.2.3 Properties of Nitrite Reductase

Various nitrite reductase are known in bacteria. There are nitrite reductases involved in denitrification, of which several forms are known to exist. In addition, there are nitrite reductases involved in dissimilatory nitrite reduction to ammonia. These enzymes will be discussed in turn.

5.2.3.1 Nitrite Reductases Involved in Denitrification

Nitrite reductase has been purified from several bacteria and appears to be of two main types: a multiheme enzyme and a copper-containing enzyme. The first type contains two types of prosthetic group, heme c, which is covalently linked to the protein, and a noncovalently bound chlorin type, heme $d1$ (196). Therefore, this type of nitrite reductase is often referred to as cytochrome cd. The enzyme has been purified from various organisms: *Ps. aeruginosa* (83, 99, 141), *Pa. denitrificans* (145, 241), *Alcaligenes faecalis* (102), and *Thiobacillus denitrificans* (200). The enzyme of *Ps. perfectomarinus* is also of this type (270). The molecular weight of cytochrome cd is about 120,000, consisting of two identical subunits, each of which contains one c-type and one $d1$-type heme (83, 141, 142). Electrons enter cytochrome cd at heme c and are then transferred to heme d. The end product of nitrite reduction by purified enzymes is a mixture of NO and N_2O (209, 260). In these cases the products of nitrite reduction were identified by using a combination of gas chromatography and mass spectrometry. Furthermore, it has been demonstrated that the purified enzyme of *A. faecalis* has NO reductase activity (162). However, in a number of cases only NO was detected as end product of nitrite reduction (150, 269). This raises the important question of whether NO is an obligatory intermediate in denitrification and whether a separate NO-reducing activity plays a role in the main line of the denitrification pathway. This point is discussed in Section 5.2.3.2. Cytochrome cd has cytochrome oxidase activity. However, the main function of cytochrome cd is the reduction of nitrite. This is evident from the K_m values of the enzyme of *Pa. denitrificans* for O_2 and nitrite which were estimated as 80 and 6-μM, respectively (241). The K_m of cytochrome cd for O_2 is much higher than the K_m of other cytochrome oxidases (196), which are also present in denitrifying cells. The side of the membrane at which nitrite is reduced is discussed in Section 5.3.2. A copper-containing nitrite reductase has been purified from *Achromobacter cycloclastes* (103, 104) and *Alcaligenes* sp. [(106); previously, this strain was incorrectly described as *Ps. denitrificans*] and denitrifying strains of *Rhodopseudomonas sphaeroides* (207). The enzyme of *Corynebacterium nephridii* is also of this type (201). The molecular weight of the enzyme estimated by gelfiltration was 80,000. The enzyme contained two identical subunits and two copper atoms per molecule. The product of nitrite reduction was in all cases identified as NO (103, 104, 106, 207). The enzyme also catalyzed nitrous oxide production from a mixture of hydroxylamine and nitrite. In that reaction the nitrous oxide evolved contained nitrogen atoms from both nitrite and hydroxylamine (103). The copper-containing nitrite reductase also has cytochrome oxidase activity.

5.2.3.2 Denitrification Pathway between Nitrite and N_2O

The role of NO and the existence of a unique nitric oxide reductase remain debatable. Recent experiments suggest that NO is not a free obligatory intermediate in

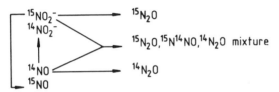

Figure 5.4 Possible conversions during concomittant denitrification of ^{15}N nitrite and ^{14}NO.

denitrification. The evidence has come from several kinds of isotope experiments, which were reviewed recently (97). An example of such an experiment is to allow [^{15}N]nitrite and normal NO to undergo denitrification concomitantly. Instead of the stable isotope ^{15}N, the radioactive isotope ^{13}N was used in such experiments. The questions that can be studied in such an experiments are outlined in Fig. 5.4. In the first place, one can look for trapping of ^{15}NO from ^{15}N nitrite in a pool of normal NO. The results differ with various bacteria. In *Ps. denitrificans*, *Ps. aeruginosa*, *Ps. stutzeri*, and an *Alcaligenes* sp., the release of ^{15}NO is very small and the rate of reduction of NO is much smaller than that of nitrite (73, 97, 229). These data suggest that NO cannot be a free intermediate in denitrification in these bacteria. Furthermore, it could be shown that with *Pa. denitrificans*, *Ps. aeruginosa*, and the *Alcaligenes* sp., the production of isotopically mixed N_2O is very small, indicating that the reduction of nitrite and NO involve entirely separate pathways. However, with *Ps. denitrificans* and *Ps. aureofaciens*, production of ^{15}NO is observed in these experiments. Furthermore, in these organisms a rapid isotope exchange reaction was found to take place by which ^{14}N from NO entered the nitrite pool (68, 73). This exchange reaction serves to explain the rather extensive scrambling of isotopes in the product N_2O. *Ps. stutzeri* is an exceptional bacterium. On the one hand, NO trapping experiments were negative and ^{14}N was not transferred from NO to nitrite. On the other hand, the N_2O produced from ^{15}N nitrite and NO showed extensive isotopic scrambling. This might indicate the occurrence of a free intermediate in the reduction of both nitrite and NO (97). An important observation was that the nitrogen atoms in the isotopically mixed N_2O were positionally equivalent, which means that the isotopically mixed N_2O contained equal amounts of $^{14}N^{15}NO$ and $^{15}N^{14}NO$ (74, 97). This suggests that the isotopically mixed N_2O arose by the random dimerization of identical monomeric units or, if not that, by way of an effectively symmetric dinitrogen compound. It has been suggested on the basis of these results that nitroxyl (HNO) is a free intermediate in the reduction of both nitrite and NO (74, 97). This means that isotopic scrambling must occur at the 1 + redox level of nitrogen.

These very complex results are illustrated in Fig. 5.5. The results with *Ps. denitrificans* and *Ps. aureofaciens* can be explained in two ways. In scheme *B* NO is supposed to be a free intermediate in the pathway of nitrite to N_2O. In scheme *C* the NO is bound to an enzyme complex. Experiments of this kind can also be performed with a mixture of ^{15}N nitrite and $^{14}N_2O$. In these cases the results in-

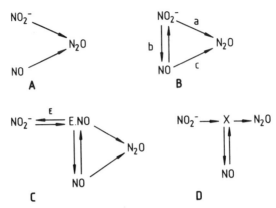

Figure 5.5   Schemes proposed for the role of NO in the reduction of nitrite to N_2O. Scheme *A* is valid for *Pa. denitrificans*, *Ps. aeruginosa*, and an *Alcaligenes* sp. Scheme *B* is valid for *Ps. denitrificans* (path *a* mainly) and *Ps. aureofaciens* (path *bc* mainly). Scheme *C* is an alternative for scheme *B* in which an enzyme-bound NO is an intermediate denitrification pathway. Scheme *D* is valid for *Ps. stutzeri*. Compound X is a free intermediate in the reduction of both nitrite and NO and is probably identical with nitroxyl [HNO (74, 97)].

dicate that N_2O is an obligatory free intermediate of denitrification (97) and that the formation of N_2O from nitrite is an irreversible reaction.

These conflicting data can be reconciled by some recent schemes for denitrification (10, 75, 76, 97). The scheme shown in Fig. 5.6 is a modification of these schemes. The scheme is drawn for an enzyme utilizing a ferrous heme prosthetic group. However, an analogous scheme can be drawn for the copper-containing enzymes. In the first step, nitrite binds to a ferrous center (I), followed by protonation and dehydration to a ferrous-nitrosyl complex (III). It is proposed that complex III can lose NO, forming the ferric porphyrin VI and providing a facile route for exchange of isotope between NO and nitrite. In the scheme of Averill and Tiedje (10), complex III undergoes attack by a second free nitrite molecule to form N_2O_3 coordinated to the iron. The resulting compound is then reduced by two subsequent two-electron steps. In the first step coordinated oxyhyponitrite would be produced. The essential point is that in this scheme the first N—N bond is formed at the 3+ redox level. However, the demonstration of nitroxyl as intermediate in *Pa. denitrificans* and *Ps. stutzeri* by ^{15}N-tracer and NMR studies has provided indirect evidence for N—N-bound formation at the +1 redox level (74, 76, 97). The exclusion of oxyhyponitrite (HN_2O_3) as intermediate in four denitrifying bacteria (77) also argues against the scheme of Averill and Tiedje (10), who postulated such an intermediate although in an enzyme-bound form. The related hyponitrite (HN_2O_2) has also been excluded as an intermediate in denitrification in *Pa. denitrificans* (98). In the scheme in Fig. 5.6 the idea of Averill and Tiedje (10) that the reduction of nitrite proceeds in two-electron steps is main-

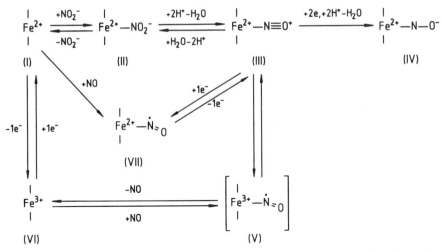

Figure 5.6 Scheme for the mechanism of reduction of nitrite and NO. The scheme is a modification of earlier ones (10, 75, 77, 97).

tained, since all other known biological oxidation-reduction reactions coupled to phosphorylation proceed in two-electron steps. The reduction of nitrite to NO would be a one-electron step. The final stages in the formation of N_2O are not known. It is possible that on one enzyme molecule two nitroxyl residues are attached from which N_2O is formed. The lower part of the scheme describes how the evidence previously interpreted in favor of NO as an intermediate can be accounted for. The most important step in the formation of NO is considered to be the decomposition of the ferrous nitrosyl complex (III) to ferric porphyrin (VI), presumably via internal electron transfer. The ferrous NO complex (VII), which has been observed frequently by EPR spectroscopy in nitrite reductase systems of various bacteria (48, 114, 150, 151, 176), is considered not to be involved in either the reduction of nitrite to N_2O or in isotope exchange between nitrite and NO. The amount of ferrous NO (VII) is dependent on the relative rate constants for III → IV and III → VI. The relative rates of the various steps in this scheme can be expected to be sensitive to minor changes in the environment of the enzyme. This can explain why the end product of the purified enzyme in many cases is NO (or a mixture of NO and N_2O), whereas the enzyme in the intact cell yields N_2O (103, 104, 106, 150, 207, 209, 210, 260, 269). The scheme presented above can give an explanation for many conflicting observations. However, the molecular mechanism of N_2O formation has not yet been completely elucidated. The fact that reduction of NO in a number of bacteria occurs by a separate pathway (Fig. 5.5A) indicates that there should exist a unique nitric oxide reductase. As an alternative, NO might be oxidized by a second reductive function of another reductase for a nitrogenous oxide. More work on these points is necessary.

5.2.3.3 Nitrite Reductase Involved in Dissimilatory Nitrate Reduction to Ammonia

Dissimilatory nitrate reduction to ammonia takes place in many more bacteria than thought previously. This reaction has been proposed as a short circuit in the nitrogen cycle (44). The enzyme involved in the six-electron reduction of nitrite to ammonia has been purified from *E. coli* (46, 108), *Desulfovibrio desulfuricans* (152), and *A. fischeri* (100, 198). In *E. coli* there are two systems for reduction of nitrite to ammonia. One utilizes NADH as electron donor (46, 108), the other formate (1, 2). Only the enzyme involved in the first reaction has been purified. It is a soluble enzyme with a molecular weight of 190,000. The native enzyme is considered to be a dimer with two identical subunits or similar-sized subunits (46). The enzyme contains one noncovalently bound FAD molecule and about five Fe and four acid-labile S atoms per subunit. In addition, the enzyme contains siroheme. Consequently, one Fe is thought to be associated with the heme and the others with a 4Fe-S cluster (108). The occurrence of siroheme in this nitrite reductase explains the identity of the *nir* B and *cys* G mutations (45). Sulfite reductase, which also performs a six-electron reduction step of sulfite to sulfide, also contains siroheme (177). Probably in *nir* B and *cys* G mutants the formation of siroheme is blocked (45). The enzyme is identical to assimilatory nitrite reductase from a large number of eucaryotic organisms such as *N. crassa* (143, 258) with respect to the oxidation-reduction centers in the enzyme. A nitrite reductase containing six *c*-type heme groups per molecule has been purified from *D. desulfuricans* (152). This enzyme is a membrane-bound nitrite reductase and the first step in the purification procedure involves solubilization from the membrane. It has a minimal molecular weight of 66,000. NADH and NADPH do not function as direct electron donor for this enzyme, which is in contrast with the purified enzyme of *E. coli*. Reduced FAD, however, can act as electron donor for the enzyme of *D. desulfuricans*. The enzyme of *A. fischeri* contains only two *c*-type heme groups and can also utilize reduced flavins as electron donor (198). The molecular weight is about 80,000 (100).

5.2.4 Properties of Nitrous Oxide Reductase

Nitrous oxide reductase is a very labile enzyme, and this has made purification very difficult. Two approaches have been used to obtain more information about the properties of this enzyme. Kristjansson and Hollocher (130, 131) could demonstrate nitrous oxide reductase activity in osmotically shocked spheroplasts of *Pa. denitrificans* with dithionite-reduced viologens as electron donor. The enzyme was found to be rapidly inactivated by oxygen, and therefore anaerobic conditions were required during lysis and purification. A partial purification was achieved. The molecular weight was estimated by gel filtration to be about 85,000 and by sodium dodecyl sulfate electrophoresis, proteins with apparent molecular weight of 58,000 and 25,000 were reported to be correlated with the enzyme activity. However, only 30% of the protein was considered to be nitrous oxide reductase, and therefore these estimations must be regarded with caution. It has recently been shown that cells of an *Alcaligenes* sp. (105) and of *Ps. perfectomarinus* (163) were devoid of nitrous oxide reductase after growth in copper-deficient medium. Consequently, cells grown in that medium converted nitrate and nitrite only to nitrous oxide. On fractionation by gel filtration on Sephadex G-200 of a cell-free extract from anaerobically nitrate-grown and copper-sufficient cells of *Ps. perfectomarinus*, three copper-containing components were observed (164). N_2O respiration was consistently associated with the presence of the component with the highest molecular weight. This copper-containing protein was also detected in denitrifying cells of *Ps. stutzeri* and *Ps. fluorescens*. However, it was absent from strains that are unable to reduce N_2O as *Ps. chlororaphis*, *Ps. aureofaciens*, and some other strains of *Ps. fluorescens* (164). Under growth conditions by which the specific activity of nitrous oxide reductase in *Ps. perfectomarinus* was reduced, the content of the high-molecular-weight copper protein was also reduced. In subsequent work the molecular weight was estimated to be about 120,000 (268). The subunit structure was dimeric, with two peptides of equal size. In a recent publication evidence was presented that nitrous oxide reductase of *R. sphaeroides* f.sp. *denitrificans* also consisted of subunits with a molecular weight of 60,000 (172). The enzyme of *Ps. perfectomarinus* contained about eight copper atoms per molecule (268). N_2O was reduced by the purified enzyme with reduced methylviologen as electron donor. The reported properties of this copper protein are at variance with the properties reported for the enzyme preparation from *Pa. denitrificans*. Recently, it was reported, however, that the copper protein and nitrous oxide reductase have different molecular weights and exhibit different behavior upon anoin-exchange chromatography (222a). Furthermore, the specific activity of purified nitrous oxide reductase was much higher than those reported for the copper protein. In these studies similar results were reported for *Pa. denitrificans*, *Ps. denitrificans*, and *Ps. perfectomarinus*. The molecular weight estimates for nitrous oxide reductase from these organisms was 87 to 89 $\times$ 10^3. It is evident that more work is necessary to find a solution for these contradictory results on nitrous oxide reductase.

Nitrous oxide reduction by whole cells is inhibited by acetylene (12, 65, 267). Therefore, nitrate and nitrite are quantitatively converted into N_2O during denitrification in the presence of acetylene. This provides the basis for a reliable method for the determination of the rate of denitrification in soil (127). Another inhibitor of nitrous oxide reduction is thiocyanate (20, 97). This compound is normally included in reaction mixtures which are used in studies on proton translocation coupled to electron transport. Thiocyanate is used in such studies to dissipate the membrane potential. This effect has complicated the determination of the stoichiometry of proton translocation (the $\rightarrow H^+/2e^-$ ratio) for electron transport to nitrogenous oxides (20, 132, 168). Furthermore, the effect was of importance in the determination of the location of the proton-consuming site on the cytoplasmic membrane of the reduction of nitrite and nitrous oxide. These aspects are discussed in Sections 5.3.2 and 5.3.4. Nitrous oxide reductase is also inhibited by sulfide (223, 237). This may lead to N_2O as an end product of denitrification in natural environments in which sulfide is abundant.

In a number of denitrifying bacteria, nitrous oxide reductase is absent. In these organisms N_2O is the end product of denitrification. This has been reported for *Co. nephridii* (90, 201), some rhizobia (50), *Ps. clororaphis*, *Ps. aureofaciens*, some *Ps. fluorescens* strains (82), and propionibacteria (121). In the latter organism the reduction of nitrite to N_2O is considered to be a detoxication mechanism rather than part of an energy transformation system.

In *Vibrio succinogenes* [recently, a proposal was made to rename this organism *Wolinella succenogenes* (239)] and *Campylobacter sputorum*, nitrite is reduced to ammonia (56, 133, 181, 235, 266). However, both organisms are able to reduce N_2O (266; W. de Vries, unpublished data) and are able to grow anaerobically with formate and N_2O as hydrogen acceptor.

5.2.5   Production of N_2O by Nondenitrifiers

By sensitive detection methods it has recently been demonstrated that small amounts of N_2O can be produced by nondenitrifying organisms which are able to reduce nitrate to ammonia. This has been found for *K. pmeumoniae* (203), a *Citrobacter* sp. (220), and *E. coli* (221). Furthermore, it has been found that virtually all organisms capable of dissimilatory nitrate reduction in soil produce N_2O, even though many of them are nondenitrifiers (222). The production of N_2O in the rumen is also supposed to be associated with NO_2^- reduction to ammonia and not denitrification (122). It has been demonstrated that mutants defective in nitrate respiration are also defective in the production of N_2O from nitrite (206, 221). On the other hand, mutants defective in the reduction of nitrite to ammonia still produce N_2O from nitrite (206). Therefore, it has been suggested that N_2O is produced by the action of nitrate reductase on nitrite (221). In these organisms the production of N_2O from nitrite is not supposed to have a physiological function.

5.3   BIOENERGETICS OF NITRATE REDUCTION

5.3.1   Respiratory Chain to Nitrogenous Oxides

The composition of the respiratory chain to nitrogenous oxides has been reviewed several times recently (88, 231, 236, 240). The discussion in this chapter is restricted to only a few organisms and to some general phenomena. The composition of the respiratory chain to nitrogenous oxides in a number of organisms is shown in Fig. 5.7.

5.3.1.1   *Escherichia coli*

It has been concluded that in whole cells NADH is the preferential electron donor for nitrate reduction (236). The respiratory chain to oxygen and nitrate, shown in

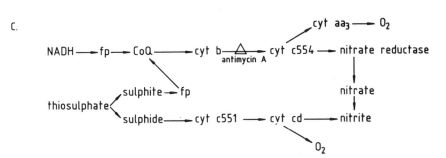

Figure 5.7 Respiratory chains of (*A*) *Escherichia coli*, (*B*) *Paraococcus denitrificans*, and (*C*) *Thiobacillus denitrificans* to oxygen, nitrate, nitrite, and nitrous oxide. Phosphorylation sites are indicated by 1, 2, and 3, respectively. For *Pa. denitrificans* and the *T. denitrificans* the site of inhibition by antimycin A is indicated.

Fig. 5.7*A*, which is a modification of that in earlier reviews (88, 240), has been published before. The methods for the characterization of bacterial cytochromes have been strongly refined during the last years. The methods used were fourth-order difference analysis of absorption spectra (213, 215), a combination of spectrum deconvolution and potentiometric analysis (255), and the utilization of a variety of techniques, including difference spectroscopy, potentiometric titration, and solubilization of cytochrome complexes (85). These studies have shown that the picture in Fig. 5.7 is a simplification. In cells grown aerobically, four (255) or even six (85) *b*-type cytochromes were detected. These *b*-type cytochromes are characterized by their absorbance maximum and their midpoint potential. On the basis of the midpoint potentials, the following sequence of electron transfer to oxygen has been proposed: cytochrome *b*556 → cytochrome *b*555 → cytochrome *b*562 → O_2 (85). This chain should replace the sequence cytochrome *b*556 → cytochrome *o* in Fig. 5.7*A*. In Section 5.3.4.1 the sequence of the cytochromes in the related bacterium *P. mirabilis* is described. However, more work is necessary to elucidate the function and the order of these cytochromes in detail. The large number of *b*-type cytochromes may be explained in the following way: (a) Several *b*-type cytochromes are involved in the proton motive Q cycle, or cytochrome *b* cycle, which have been proposed to explain electron transport and proton translocation in the QH_2 : cytochrome oxidase segment of the respiratory chain. This point is discussed in Section 5.3.4.1. (b) Some *b*-type cytochromes are involved in specific pathways: the *b*-type cytochromes associated with formate dehydrogenase and nitrate reductase are examples of this kind. The *b*-type cytochrome associated with nitrate reductase has been identified as a cytochrome *b*556 with a midpoint potential of +120 mV (85) or +149 mV (64, 257). The *b*-type cytochrome associated with formate dehydrogenase is also a cytochrome *b*556 with a midpoint potential of −100 mV (85, 86, 257). With these methods small amounts of *c*-type cytochromes can be detected (257). A low-potential cytochrome *c*552 plays a role in the soluble NADH–nitrite reductase activity (153). The formate nitrite reaction is inhibited by 2-*n*-heptyl-4-hydroxyquinoline-*N*-oxide (HQNO) (100), which inhibits after cytochrome *b*. These results indicate that in contrast to the soluble NADH–nitrite reductase system, membrane-bound cytochromes are involved in the reduction of nitrite by formate. Ascorbate-$N, N^1, {}^1N$-tetramethyl-*p*-phenylenediamine (TMPD) cannot act as an electron donor for nitrite reduction in *E. coli*.

5.3.1.2 *Paracoccus denitrificans*

The respiratory chain of *Pa. denitrificans* to oxygen and nitrogenous oxides is shown in Fig. 5.7*B*. The chain has a branched structure. There are two cytochrome oxidases: cytochrome aa_3 and most probably cytochrome *o* (233, 234, 254). The phosphorylation sites are indicated 1, 2, and 3 in Fig. 5.7*B*. The functioning of these phosphorylation sites is strongly dependent on the growth conditions, which can be varied by continuous cultivation in different media. In cells grown with succinate or mannitol as the carbon and energy source, respiration occurs mainly through cytochrome *o*, and consequently, during the oxidation of NADH, two phosphorylation sites are passed. Electrons from methanol enter the respiratory at the level of cytochrome c (233). In cells grown with methanol as the carbon and energy source, respiration occurs mainly through cytochrome aa_3, and consequently during the oxidation of NADH, three phosphorylation sites are passed. Recently, mutants that lack cytochrome c were isolated (252, 263). These mutants grow normally under aerobic conditions, but they cannot grow aerobically using methanol as carbon and energy source or anaerobically using nitrate as hydrogen acceptor (Fig. 5.7B). After anaerobic growth, *Pa. denitrificans* cells lack cytochrome aa_3. Electron transport to nitrate involves cytochrome b (144). However, electron transport to nitrite involves cytochrome c (111). A similar conclusion has been reached for electron transport to N_2O (19, 161). An important observation was that in cells in which electron transport from NADH and succinate to cytochrome c was inhibited by antimycin A, nitrite and nitrous oxide could be reduced with methanol (Fig. 5.7B).

In Fig. 5.7B the respiratory chain is presented as a linear array of electron carriers. This picture suggests that energy generation with oxygen, nitrite, and nitrous oxide will be the same, since in all these cases the same number of phosphorylation sites is passed. However, it must be realized that the picture in Fig. 5.7B is misleading in this aspect, since it does not take into account the side of the membrane where the reductions take place. The place (cytoplasmic or periplasmic aspect of the membrane) where the reduction takes place has a very great influence on the stoichiometry of ATP generation. This point is discussed in Section 5.3.4.2.

5.3.1.3 *Thiobacillus denitrificans*

The respiratory chain of this organism to oxygen, nitrate, and nitrite has been studied intensively by Nicholas and associates (7, 8, 208, 209). This organism is an obligate chemolithotroph which can derive energy needed for growth by performing the reaction $5S_2O_3^{2-} + 8NO_3^- + H_2O \rightarrow 10SO_4^{2-} + 4N_2 + 2H^+$. An unusual point in the respiratory chain of this organism (Fig. 5.7C) is that cytochrome c is involved in nitrate reduction. Normally, nitrate reductase accepts electrons from cytochrome b instead of cytochrome c (231). However, in *T. denitrificans* nitrate reduction is strongly inhibited by antimycin A, which inhibits electron transfer between cytochrome b and cytochrome c (208). It is evident from Fig. 5.7C that separate chains for electron transport exist for sulfide and sulfite, which both arise from thiosulfate. Electrons from sulfite are transferred to nitrate, whereas electrons from sulfide are transferred to nitrite (209). Different c-type cytochromes are involved in these two branches of the respiratory chain.

5.3.1.4 *Organisms with Dissimilatory Reduction of Nitrate to Ammonia*

Although the bioenergetics of dissimilatory nitrate reduction to ammonia has been studied intensively in a number of organisms, it is remarkable that no studies were devoted to the elucidation of the respiratory chain to nitrate. However, more work was done to identify the electron donor for nitrite reduction, which in general proved to be a cytochrome c. In *C. sputorum* the formate–nitrite system was sensitive to inhibition by (HQNO) and antimycin A (57). Also from experiments in which the reduction level of cytochrome b and c was measured, it was concluded that cytochrome c was the immediate electron donor for nitrite reduction (57). Also, in *A. fischeri* (198) and *D. desulfuricans* (224) cytochrome c was suggested to be involved in electron transport to nitrite. In *Clostridium perfringens* (91) and *Clostridium tertium* (92), however, cytochromes are not involved in the reduction of nitrate and nitrite. In these organisms reduced ferredoxin is the electron donor for the reduction of nitrogenous oxides. These *Clostridia* are regarded as agents for dissimilatory reduction of nitrate to ammonia in soils (31).

5.3.2 Determination of the Proton-Consuming Site in the Reduction of Nitrogenous Oxides

It has been mentioned before that with regard to the stoichiometry of energy generation, whether the reduction of nitrogenous oxides takes place at the cytoplasmic or at the periplasmic side of the membrane is of great importance. In most of these studies use is made of the low permeability of the cytoplasmic membrane to protons. Consequently, if the reduction of a nitrogenous oxide takes place at the periplasmic side of the membrane, the utilization of protons can be detected directly. However, this is not the case if the reaction takes place at the cytoplasmic side of the membrane. In the latter case the utilization of protons for the reduction of nitrogenous oxides can be detected only when membrane has been made permeable to protons by the addition of uncoupling agents. In *E. coli* the localization of the proton-consuming site of nitrate reduction was determined by studies on the reaction between reduced benzylviologen (BV) and nitrate. The reduced BV (the radical BV^+) is able to cross the cytoplasmic membrane. The reduction of nitrate proceeds by the equation $2BV^+ + 2H^+ + NO_3^- \rightarrow 2BV^{2+} + NO_2^- + H_2O$. The consumption of the protons could not be detected by a pH electrode in the extracellular bulk phase of a suspension of spheroplasts unless the cytoplasmic membrane was made permeable to protons (115). This indicated that nitrate is reduced at the cytoplasmic side of the membrane. The same conclusion was reached for *Pa. denitrificans* on the basis of the following observations.

1. In cell-free extracts nitrate reductase of *Pa. denitrificans* reduces both nitrate and chlorate readily (70). However, intact cells reduce chlorate at a very low rate, which is strongly increased by addition of the detergent Triton-X-100 (109). These observations can be explained by reduction of nitrate at the cytoplasmic side of the membrane and a low permeability of the cells for chlorate.

2. Purified preparations of nitrate reductase are very sensitive to inhibition by thiocyanate; the K_i value was estimated to be about 2 mM (144, 199, 245). However, in intact cells, 60 mM thiocyanate gave only a small inhibition

of the rate of nitrate reduction after the occurrence of a lag phase (18). The latter is due to interference of thiocyanate with nitrate uptake (see Section 5.3.3). Due to the existence of the proton motive force, the concentrations of the anion thiocyanate in whole cells will be much lower than in the outer bulk phase. This explains the difference in sensitivity of nitrate reduction to thiocyanate in whole cells and in cell-free preparations.

The location of the proton-consuming sites of the reduction of nitrite and nitrous oxide has been studied with ascorbate + TMPD as electron donor. At neutral pH ascorbate is a two-electron and one-proton donor, which donates its electrons via TMPD to cytochrome c (Fig. 5.7B; 252). The overall reactions of ascorbate (AH^-) with nitrite and nitrous oxide are

$$1.5AH^- + NO_2^- + 2.5H^+ \longrightarrow 1.5A + 0.5N_2 + 2H_2O$$
$$AH^- + NO_2^- + 2.0H^+ \longrightarrow A + 0.5N_2O + 1.5H_2O$$
$$AH^- + N_2O + 1.0H^+ \longrightarrow A + N_2 + 1.0H_2O$$

The values found for H^+/NO_2^- for reduction to N_2, H^+/NO_2^- for reduction to N_2O (thiocyanate present to inhibit nitrous oxide reductase), and H^+/N_2O for reduction to N_2 were -2.33, -1.90, and -0.84, respectively. These results indicate that nitrite and nitrous oxide are reduced at the periplasmic side of the membrane. Studies on the stoichiometry of proton translocation during electron transfer to nitrogenous oxides are in accordance with these conclusions (see Section 5.3.4.2). In *D. desulfuricans* the location of hydrogenase, formate dehydrogenase, and nitrite reductase was also studied by utilization of benzylviologen (225). The reactions catalyzed by these enzymes are, respectively,

$$H_2 + 2BV^{2+} \longrightarrow 2H^+ + 2BV^+$$
$$HCOO^- + 2BV^{2+} + H_2O \longrightarrow 0.82HCO_3^- + O18H_2CO_3$$
$$+ 1.82H^+ + 2BV^+$$
$$NO_2^- + 6BV^+ + 8H^+ \longrightarrow NH_4^+ + 6BV^{2+} + 2H_2O$$

It was observed that the production and consumption of protons by these reactions was immediately apparent in the suspending medium, indicating that these reactions occur at the periplasmic side of the membrane. These conclusions were in accordance with the results of studies on the stoichiometry of proton translocation during the reduction of nitrite with formate or hydrogen as electron donor. The location of the proton consuming site of nitrate reductase could not be established by the approaches outlined above. By other experiments it could be demonstrated that lactate donates electrons at the cytoplasmic side of the membrane (225). In *C. sputorum* the localization of enzymes was determined by measurements of the stoichiometry of proton translocation with various hydrogen acceptors and donors (56–58). The results indicate that formate dehydrogenase and hydrogenase donate electrons at the periplasmic side of the membrane, whereas lactate dehydrogenase donates electrons at the cytoplasmic side of the membrane. The proton-consuming site of nitrate reductase and nitrite reductase was at the cytoplasmic and periplasmic side of the membrane, respectively (57, 58). The evidence is discussed later (Section 5.3.4.3) in more detail. The results discussed in this section indicate that in general, nitrate is reduced at the cytoplasmic side of the membrane and nitrite, irrespective of whether the end product of the reduction is nitrous oxide or ammonia at the periplasmic side of the membrane. In Section 5.3.4 these results are incorporated in schemes for the organization of the respiratory chain for a number of organisms. Recently, an exception to the rule that nitrate is reduced at the cytoplasmic side of the membrane has been reported (166a, 207a) for *R. sphaeroides* f. sp. *denitrificans* and *R. capsulata*. Nitrate reductase in these organisms is located in the periplasmic space. The evidence was based on cell fractionation procedures.

5.3.3 Nitrate Transport

The results discussed in the preceding section have demonstrated that nitrate is reduced at the cytoplasmic side of the membrane. Therefore, nitrate has to be imported inside the cell. The ability of the cell to discriminate between nitrate and chlorate (109) suggests the presence of a specific carrier system in the cytoplasmic membrane. Studies on the transport of nitrate have been performed with *Pa. denitrificans* cells grown on succinate and nitrate or H_2, CO_2, and nitrate (21, 234). The kinetics and energetics of NO_3^- uptake in whole cells was studied by measuring H_2 consumption or N_2O production after addition of nitrate to a cell suspension. When the membrane potential was dissipated by addition of thiocyanate, carbonyl cyanide *m*-chlorophenylhydrazone (an uncoupler), or triphenylmethylphosphonium bromide (a permeant cation), a lag phase in H_2 consumption or N_2O production was observed. However, these lag phases were not observed when nitrite was present at the moment of introduction of nitrate. When the reduction of nitrate by H_2 was studied in the presence of antimycin A nitrate was only reduced to nitrite (Fig. 5.7B). Under these circumstances in the presence of thiocyanate a lag in the H_2 consumption was observed upon the first addition of nitrate. No lag was observed when a second addition of nitrate was made. In that case the lag was prevented by the presence of nitrite formed after the first addition of nitrite. If antimycin A was omitted, which makes possible the further reduction of nitrite, a lag in the reduction of nitrate is also observed at the second, third, and fourth addition of nitrate. When the reduction of nitrous oxide is blocked by a high concentration of thiocyanate, a long lag phase in N_2O production is observed after the addition of nitrate. When nitrate is added during the lag phase, nitrite and part of the nitrate are converted into N_2O. The reduction of nitrate comes at an immediate stop after complete conversion of the nitrate added. Upon a second addition of nitrite, nitrite and part of the nitrate are again reduced to N_2O (21). These observations can be explained by the presence of two systems for the entry of nitrate in the cell: an active uptake process driven by the proton motive force and

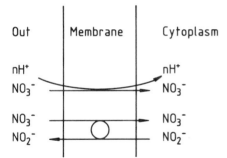

Figure 5.8 Transport systems for nitrate in *Paracoccus denitrificans*.

a nitrate-nitrite antiport system (Fig. 5.8). The existence of the latter system has the advantage that after the reduction of nitrate has started, its further uptake does not require energy. The observations indicate that the initial uptake of nitrate is electrogenic, which implies that nitrate is taken up together with two or more protons. During nitrate assimilation nitrite is further reduced inside the cell to ammonia. In that case the nitrate-nitrite antiport system cannot function. It has been demonstrated that in *Ps. fluorescens* (15) and *R. capsulata* (107) the uptake of nitrate is driven by the proton motive force. The consequence is that nitrate uptake for assimilatory nitrate reduction is always an energy-requiring process. The presence of the two uptake systems for nitrate has not been studied in other denitrifying bacteria, but it seems a logical assumption that they will be present in all denitrifying bacteria.

5.3.4 The Stoichiometry of Proton Translocation and Energy Production During Nitrate Reduction

According to the chemiosmotic hypothesis of the mechanism of oxidative phosphorylation, oxidation of substrates must be accompanied by the extrusion of protons. It is implicit in this hypothesis that $\rightarrow H^+/2e^-$ ratios reflect the efficiency of oxidative phosphorylation. In microorganisms there is a great diversity in the composition of the respiratory chain and therefore also in the stoichiometry of ATP formation during oxidation. The comparative study of the mechanism of energy generation can therefore yield data which are of general importance. Data on the stoichiometry of proton translocation are discussed subsequently for various organisms, and its implication for the mechanism of proton translocation is given.

5.3.4.1 *Escherichia coli, the Proton Motive Q Cycle and the* *Cytochrome b Cycle*

For the oxidation of endogenous substrates (yielding mostly NADH) $\rightarrow H^+/O$ and $\rightarrow H^+/NO_3^-$ ratios of 4.0 and 4.2, respectively, have been measured (27). For the related *K. pneumoniae*, the $\rightarrow H^+/O$ and $\rightarrow H^+/NO_3^-$ ratios were 3.92 and 3.85, respectively (27). These values are about the same, indicating that the efficiency of oxidative phosphorylation with oxygen and nitrate is about the same, which had been concluded earlier for the related *K. aerogenes* from the measurement of molar growth yields (89, 230). It has been demonstrated that the oxidation of flavin-linked substrates such as succinate, D-lactate, and glycerol-3-phosphate yielded a $\rightarrow$ H$^+$/O ratio of about 2, whereas the oxidation of a NAD$^+$-linked substrate such as L-malate produced a $\rightarrow$ H$^+$/O ratio close to 4 (77, 147). Oxidation of ubiquinol-1 gave $\rightarrow$ H$^+$/2e$^-$ ratios of 1.49 when nitrate was the acceptor and 2.28 when oxygen was the acceptor (115). Release of protons by the oxidation of added ubiquinol-1 was also observed in cells from a double quinone-deficient mutant (*ubi* A *men* A). In these cases the protons appearing in the medium must be derived from the reductant. On the basis of these results it can be concluded that in *E. coli* two sites of energy generation are present. It is very remarkable that energy generation during the reduction of nitrate to nitrite is about the same in *E. coli* and *Pa. denitrificans* (see Section 5.3.4.2) whereas energy generation during the reduction of oxygen is very different. The data available on *E. coli* can be represented as in Fig. 5.9, which is a revison of earlier schemes (88, 236). In this scheme the electron transport chain is organized in redox loops. A redox loop is

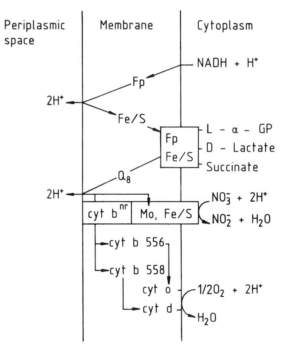

Figure 5.9 Proposed functional organization of the redox carriers with oxygen and nitrate as terminal hydrogen acceptors in *Escherichia coli*. The scheme is a revision of earlier ones (88, 236). Fp, flavoprotein; Fe/S, iron–sulfur center; Q, ubiquinone, L-α-GP, L-α-glycerolphosphate.

an alternating sequence of hydrogen and electron carriers. However, it has recently become evident that in mitochondria (262) and in *Pa. denitrificans* (233), the stoichiometry of proton translation cannot be explained by the arrangement of the respiratory chain in redox loops as suggested by the chemiosmotic theory of Mitchell. Furthermore, there are indications that this organization in redox loops is not correct for *E. coli* either. It was observed that cytochrome *b* functioned between two ubiquinone sites (60). Similarly, in the related *P. mirabilis*, two HQNO inhibition sites were recently observed, before and after the *b*-type cytochromes (256). Both facts can be explained better by a mechanism for cyclic electron flow. For this purpose a modification of the proton motive Q cycle of Mitchell can be proposed (174). Two versions of such modifications are demonstrated in Fig. 5.10. In both models three ubiquinone species are distinguished: Q, ubiquinone; QH_2, ubiquinol; and $Q\cdot^-$, ubisemiquinone radical. In Fig. 5.10*a* a diagram illustrating the Q-cycle model for the oxidation of ubiquinol (QH_2) by cytochrome *c* in mitochondria (or *Pa. denitrificans*) is shown (219). The net result is $QH_2 + 2$ cyt $c^{3+} + 4H_{in}^+ \rightarrow Q + 2$ cyt $c^{2+} + 4H_{out}^+ + 2H^+$. In this version the $4H^+$ trans-

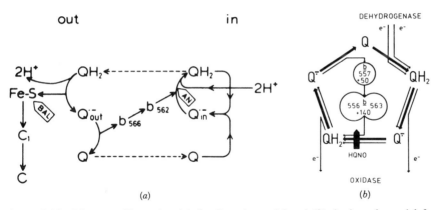

Figure 5.10 Diagrams illustrating (*a*) the Q-cycle model and (*b*) the b-cycle model for the oxidation of ubiquinol (QH_2). In diagram (*a*) curved arrows indicate chemical reactions, straight dashed lines represent diffusion across the membrane and straight solid arrows show the direction of electron transfer. The cycle, shown here for mitochondria, shows the net transfer of one electron from one-half molecule of QH_2 to one molecule of cytochrome *c*, coupled with the translocation of one proton across the membrane. $Q\cdot^-_{out}$ and $Q\cdot^-_{in}$ represent ubisemiquinone radical at the outer and at the inner side of the membrane, respectively. The sites of inhibition of antimycin A and BAL (British Anti Lewisite, 2,3-dimercaptopropanol) are indicated. The scheme is from Ref. 219. In the b-cycle model the destiny of Q is depicted by thick arrows; the flow of electrons is depicted by thin arrows. The scheme is for the aerobic respiratory chain of *P. mirabilis* (25). The wavelength of the absorbance maxima and the midpoint potentials of the *b*-type cytochromes are given. The circled areas are proportional to the intensities of the α-bands of the cytochromes. Furthermore, the site of inhibition by HQNO is indicated. (*b*) Reproduced from Ref. 256 by permission.

located that are translocated in the site II region of mitochondria originate completely from QH_2. In Fig. 5.10b a modification of the b cycle model of Wikström and Krab (262) is given, which was used to give a scheme for the aerobic respiratory chain of *P. mirabilis* (256). In the b-cycle model the translocation of four protons in the site II region of the respiratory chain of mitochondria may be explained by the release of two protons from QH_2 and the translocation of two protons by a cytochrome b, which acts as a proton pump. The stoichiometry of proton translocation in the respiratory chain of bacteria can be more easily explained by the second model since it can give an explanation for differences in the number of protons translocated in the site II region of various bacteria. In this scheme proton translocation in *E. coli* can be explained by a redox loop or a proton pump at the site I region of the respiratory chain and the involvement of ubiquinone in the translocation of two other protons. In subsequent sections schemes for proton translocation and electron transport for *Pa. denitrificans* and *C. sputorum* are presented which indicate a large difference in the stoichiometry between these organisms. Also, this difference can be better explained by the b-cycle model. It has been demonstrated that the NADH–nitrite reaction is not associated with energy generation (44). However a membrane potential can be generated by the formate–nitrite reaction, since this reaction led to the accumulation of the lipophilic cation butyltriphenylphosphonium (175, 197).

5.3.4.2 *Paracoccus denitrificans*

It has been demonstrated that 2–3 protons are extruded at site I and 4 protons at site II (252). In methanol-grown cells, which contain cytochrome aa_3, it was shown that this cytochrome oxidase acts as a proton pump (252). Electron transport between cytochrome c and oxygen was found to be associated with the extrusion of two protons. In this aspect the cytochrome oxidase of *Pa. denitrificans* was similar to that of mitochondria (196, 261). In earlier studies it had been demonstrated that iron–sulfur center 2 was essential for proton translocation at site I (171); possibly, this compound acts as a proton pump. Consequently, in *Pa. denitrificans* there is a difference in the stoichiometry of proton translocation by the three sites of energy conservation (252).

As mentioned before, normally thiocyanate is included in reaction mixtures for the study of proton translocation. However, thiocyanate inhibits nitrous oxide reductase (Section 5.2.4) and has an effect on nitrate transport (Section 5.3.3). Therefore, thiocyanate had to be replaced by another agent, which is able to collapse the membrane potential. Triphenylmethylphosphonium was used for that purpose (20, 21). The stoichiometries of proton translocation associated with electron transfer from various donors to oxygen and nitrogenous oxides are shown in Table 5.1. A scheme for proton translocation and electron transport is given in Fig. 5.11. In this scheme protons for the reduction of nitrite and nitrous oxide are utilized from the periplasmic side of the membrane (Section 5.3.2). In this scheme two protons arise from ubiquinol (site IIa) and two protons are pumped by cytochrome b (site IIb). Consequently, this segment is a simplification of a b-cycle

TABLE 5.1 Respiration-driven proton translocation with endogenous substrate (ES) or H_2 as electron donor and O_2 or nitrogenous oxides as electron acceptor in *Pa. denitrificans*

	$\rightarrow$ H$^+$/oxidant ratio[a]			Theoretical $\rightarrow$ H$^+$/oxidant ratio[b]	
Acceptor Couple	ES	ES	H_2	Es	H_2
O_2/H_2O	6.32	7.5	3.55, 3.92	6.0–7.0	4.0
NO_3^-/N_2		10.75	4.03	9.0–11.5	4.0
NO_2^-/N_2	5.79	5.55	2.87	5–6.5	2.0
NO_2^-/N_2O	3.37	3.3	1.68	3–4	1.0
NO_2/N_2	4.02	4.5	2.06	4–5	2.0
Reference	20	132	21		

[a]For the $\rightarrow$ H$^+$/O ratio with H_2, two values are given; the first one was obtained with triphenylmethylphonium cation as permeant ion, the second with thiocyanate.
[b]The theoretical $\rightarrow$ H$^+$/oxidant ratios are calculated according to Fig. 5.11. For endogenous substrate (NADH) the calculation is made for a translocation of two or three protons at site I.

model. Electrons for nitrate reductase do not pass site IIb, since electron transfer to nitrate is not sensitive to inhibition by antimycin A (21, 94). From this scheme it is possible to calculate the theoretical $\rightarrow$ H$^+$/oxidant ratios, which are included in Table 5.1. In general, there is a good correspondence between the experimental values and those calculated from the scheme in Fig. 5.11. The experimental $\rightarrow$ H$^+$/ oxidant ratios can be converted into $\rightarrow$ H$^+$/2e$^-$ ratios by correcting for scalar proton consumption in the case of reduction of nitrite and nitrous oxide. The $\rightarrow$ H$^+$/2e$^-$ values for the reduction of nitrite and nitrous oxide with endogenous substrate or H_2 are between 6 and 7 or about 4, respectively. The $\rightarrow$ H$^+$/2e$^-$ ratio for the reduction of nitrate is lower than that for the reduction of nitrite and nitrous oxide. The $\rightarrow$ H$^+$/2e$^-$ ratios calculated from the experimental $\rightarrow$ H$^+$/oxidant ratios correspond very well with those calculated from the scheme in Fig. 5.11. The efficiency of oxidative phosphorylation is related to the $\rightarrow$ q$^+$/2e$^-$ ratio (number of charges translocated across the membrane during flow of 2e$^-$ to an acceptor), the charge separation (261). For aerobic electron transport the $\rightarrow$ q$^+$/2e$^-$ for sites I, II, and III of oxidative phosphorylation are 2 to 3, 2, and 4, respectively (252). Based on Fig. 5.11, the q$^+$/2e$^-$ values for electron transport from NADH, succinate, and H_2 to oxygen and nitrogenous oxides can be calculated. The results are given in Table 5.2. It is remarkable that although the electron-transport chain to nitrate is different from that to nitrite and nitrous oxide, the charge separation is the same, which is due to the difference in location of the reductases. The efficiency of oxidative phosphorylation during electron transport from NADH to nitrogenous oxides is 67 to 71% of that during electron transport to oxygen.

Recently, it has been demonstrated that the reduction of NO is associated with

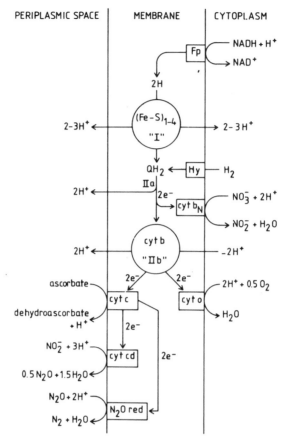

Figure 5.11 Scheme for proton translocation and electron transport to oxygen and nitrogenous oxides in denitrifying cells of *Paracoccus denitrificans*. Fp, flavoprotein; Fe-S, iron–sulfur center; QH$_2$, ubiquinol; Hy, hydrogenase; N$_2$O-R, nitrous oxide reductase. I, IIa, and IIb are the traditional sites of energy conservation (21, 234).

proton translocation in *Pa. denitrificans* (72). The H$^+$/2e$^-$ ratio for both the NO → $\frac{1}{2}$N$_2$O and NO → $\frac{1}{2}$N$_2$ reactions were about 3.7. Active transport of proline was supported by these two reactions. These experiments demonstrate that reduction of endogenous NO is energy yielding.

The considerations given above are in accordance with measurements of the efficiency of oxidative phosphorylation in cell-free preparations and of molar growth yields. Measurements of oxidative phosphorylation coupled to reduction of nitrate with NADH and succinate yielded P/2e$^-$ ratios of 0.9 and 0.06, respectively (110). P/2e$^-$ ratios as high as 1.8 and 0.6 were measured for electron transport from NADH and succinate, respectively, to oxygen (110). The observation that almost no ATP synthesis could be detected to accompany electron flow

TABLE 5.2 **Number of charges translocated across the cytoplasmic membrane during flow of 2e⁻ ($\rightarrow q^+/2e^-$ Ratio) from NADH, Succinate, or H_2 to an electron acceptor in** *Paracoccus denitrificans*

Electron Acceptor Reaction	$\rightarrow q^+/2e^-$	
	NADH	H_2 or Succinate
$O_2 \rightarrow H_2O$ (cyt o)	6–7	4
$O_2 \rightarrow H_2O$ (cyt aa_3)[a]	8–9	6
$NO_3^- \rightarrow NO_2^-$	4–5	2
$NO_3^- \rightarrow 0.5\ N_2O$	4–5	2
$NO_3^- \rightarrow 0.5 N_2$	4–5	2
$NO_2^- \rightarrow 0.5 N_2O$	4–5	2
$NO_2^- \rightarrow 0.5 N_2$	4–5	2
$N_2O \rightarrow N_2$	4–5	2

Source: Data from Refs. 20, 21, and 234.

[a] Cyt aa_3 is not present in cells grown anaerobically.

from succinate to nitrate is surprising. However, in recent experiments it was demonstrated that in accordance with Fig. 5.11 and Table 5.2, electron flow from succinate to nitrate generates a proton electrochemical gradient (189). In addition, a $P/2e^-$ ratio of 0.16 was measured. A number of reasons were discussed why measurement of a $P/2e^-$ ratio in membrane vesicles can be an unreliable indicator of whether a given segment of an electron transfer chain generates a proton electrochemical gradient (189). The scheme in Fig. 5.10 and $q^+/2e^-$ ratios in Table 5.2 can be considered of general validity for denitrifying organisms. Therefore, these data can be compared with molar growth yields which have been measured for *Ps. denitrificans* (128, 129), *Pa. denitrificans* (22, 170, 234, 250, 251, 253), and *T. denitrificans* (116), and *Thiobacillus* A2 (265). Molar growth yields of *Ps. denitrificans* for chemostat cultures with glutamate as carbon and energy source in the presence of nitrate, nitrite, and nitrous oxide were found to be 28.6 g mol⁻¹ nitrite, and 8.8 g mol⁻¹ nitrous oxide after correction for the requirement of maintenance energy (129). The energy yield was proportional to the oxidation number of the nitrogen in the hydrogen acceptor. It was therefore concluded that oxidative phosphorylations coupled to reduction of nitrate, nitrite, and nitrous oxide have similar efficiencies that are all lower than the efficiency of oxidative phosphorylation with oxygen. The molar growth yields of *Thiobacillus* A2 on glucose were 66.4 and 68.6 g mol⁻¹ for growth with nitrate or nitrous oxide as hydrogen acceptor, respectively (265). This indicates that the efficiency of oxidative phosphorylation is the same for electron transport to nitrate and nitrous oxide. In *Ps. denitrificans* the Y_{el}^{max} values for growth with glutamate (molar growth yield per electron transferred to a hydrogen acceptor after correction for maintenance energy) were 7.7 and 4.5 for electron transport to oxygen and nitrate, respectively (128). This indicates that the efficiency of oxidative phosphorylation of electron transport to nitrate is about 60% of that to oxygen. From the chemostat studies with *T.*

denitrificans it has been concluded that 4 to 5 mol ATP and 6 to 7 mol of ATP is produced per mole of thiosulfate with nitrate and oxygen as hydrogen acceptor, respectively (116). From the experiments with *T. denitrificans* (116) and *Thiobacillus* A2 (265) it was concluded that energy conservation anaerobically is about two-thirds of the aerobic value. These conclusions are in accordance with the data on the $\rightarrow q^+/2e^-$ values in Table 5.2. The most extensive studies on molar growth yields with various hydrogen acceptor have been performed with *Pa. denitrificans* (22, 170, 234, 250, 251, 253). For these data it was even possible to give an exact calculation of the ATP production during growth (22). In this case the molar growth yields were related to the $\rightarrow q^+/2e^-$ values (Table 5.2). Previously, it was assumed that the H^+/ATP ratio is 2 (88). However, in a recent review it has been stated that in bacteria the H^+/ATP ratio is generally greater than 2 (66). In particular, comparison of the phosphorylation potential with the proton motive force in membrane vesicles and whole cells of *Pa. denitrificans* indicated an H^+/ATP ratio of 3.0 or even 4.0 (124, 165, 166). Furthermore, in cells of *E. coli* growing aerobically (119) or anaerobically (120), the H^+/ATP ratio was determined to be 3.0. Therefore, it seems reasonable to use the assumption that synthesis of one ATP needs three translocated charges. Hence the oxidation of NADH and succinate by oxygen will yield 2.33 and 1.33 mol ATP, respectively, whereas oxidation of these compounds by nitrogenous oxides yields 1.67 and 0.67 mol ATP, respectively. For the calculation of the ATP production from the molar growth yields shown in Table 5.3 use is made of assimilation and dissimilation reactions (232, 250). As an example, the calculations will be given for growth with gluconate ($C_6H_{12}O_7$). The reactions for assimilation [equation (1)] and dissimilation [equation (2)] are

$$C_6H_{12}O_7 + 1.5NH_3 + 1.25(NADH + H^+) \tag{1}$$
$$\longrightarrow C_6H_{10.8}N_{1.5}O_{2.9} \text{ (cell material)} + 4.1H_2O + 1.25NAD^+$$

$$C_6H_{12}O_7 + 9NAD^+ + 2FAD + 5H_2O \tag{2}$$
$$\longrightarrow 6CO_2 + 9(NADH + H^+) + 2FADH_2$$

The molecular weight of cell material is thus 150.2. The parts of gluconate that are assimilated and dissimilated are, respectively, $Y_{gluc}/150.2$ and $(1 - Y_{gluc}/150.2)$. The amounts of reduction equivalents and ATP produced by oxidative phosphorylation are given by

$$v_{H_2} = (1 - Y_{gluc}/150.2)11 - 1.25Y_{gluc}/150.2 \tag{3}$$

$$v_{ATP} = 2(1 - Y_{gluc}/150.2)P/2e^-(FADH_2)$$
$$+ [9(1 - Y_{gluc}/150.2) - 1.25Y_{gluc}/150.2]P/2e^-(NADH) \tag{4}$$

The overall $P/2e^-$ ratio can easily be calculated with

$$P/2e^- = v_{ATP}/v_{H_2} \tag{5}$$

TABLE 5.3 Growth parameters of *Paracoccus denitrificans* for different growth conditions with various substrates as carbon and energy source[a,b]

Growth Condition	H-ac	Y_{sub}	q_{sub}	Y_{2e^-}	q_{2e^-}	v_{H_2}	v_{ATP}	$P/2e^-$ Theor.	$P/2e^-$ Exp.	Ref.
Gluconate	O_2	73.6	2.04	18.18	8.25	5.00	10.62	2.12	—	170
Gluconate, sulfate	O_2	57.0	2.63	12.70	11.81	6.35	8.45	1.33	1.33	170
Gluconate	NO_3^-	69.4	2.16	13.00	11.54	5.34	7.84	1.47	1.48	253
Mannitol	O_2	91.2	1.64	16.26	9.23	5.56	12.17	2.19	—	250
Mannitol	NO_3^-	74.8	2.00	13.46	11.14	6.90	10.52	1.53	1.72	22
Mannitol	NO_2^-	76.1	1.97	12.70	11.81	6.79	10.36	1.53	1.63	22
Succinate	O_2	35.1	4.27	12.70	11.81	4.14	8.17	1.97	—	170
Succinate, sulfate	O_2	29.8	5.03	8.54	17.56	4.57	6.08	1.33	1.28	170
Succinate	NO_2^-	37.9	3.96	9.30	16.13	3.91	5.10	1.30	1.46	253
Succinate	O_2	47.7	3.14	15.28	9.81	3.11	5.96	1.92	—	251
Succinate	NO_3^-	38.1	3.94	10.30	13.27	3.89	5.07	1.30	1.29	22
Succinate	NO_2^-	38.8	3.87	10.70	14.02	3.84	4.99	1.30	1.36	22

[a]The specific growth rate is in all cases 0.15 h^{-1}; the growth-limiting factor during growth in the chemostat is underlined.

[b]H-acc, hydrogen acceptor; Y_{sub}, g dry weight/mol substrate; q_{sub}, specific rate of substrate consumption (mmol substrate g^{-1} h^{-1}); Y_{2e^-}, growth yield per electron pair (g dry weight/mol "electron pair"); q_{2e^-}, specific rate of electron transport to the hydrogen acceptor (mmol "electron pair" g^{-1} h^{-1}); v_{H_2}, number of reduction equivalents/mol substrate; v_{ATP}, number of ATP molecules formed by oxidative phosphorylation/mol substrate; theoretical $P/2e^-$, overall $P/2e^-$ ratio calculated with the equation v_{ATP}/v_{H_2}; experimental $P/2e^-$ ratio, overall $P/2e^-$ ratio calculated from the q_{ATP} for the preceding conditions for which no experimental $P/2e^-$ ratio is given.

It must be emphasized that in these calculations ATP is used in the meaning of "ATP equivalent," for the proton motive force is not only indirectly employed in energy-consuming reactions as ATP, but is also used directly (e.g., for transport processes). In these equations the $P/2e^-$ ratios calculated from the charge separations must be substituted. During sulfate-limited growth under aerobic conditions, phosphorylation site I is lost (170, 171), and therefore in this case $P/2e^-(NADH) = P/2e^-(FADH_2)$. In Table 5.3 the experimental Y_{sub} and Y_{2e^-} values are given. Since all data are for a growth rate of 0.15 h$^-$, the corresponding q values can be calculated with the equations $q_{sub} = 150/Y_{sub}$ and $q_{2e^-} = 150/Y_{2e^-}$, respectively. The v_{H_2}, v_{ATP}, and theoretical $P/2e^-$ ratio are calculated from the Y_{sub} values with equations (3), (4), and (5) or similar equations for the other substrates. Now the specific rate of ATP production (q_{ATP}) can be calculated from q_{sub}, q_{2e^-}, and the theoretical $P/2e^-$ ratio with the aid of the relation

$$q_{ATP} = q_{gluc}(1 - Y_{gluc}/150.2)3 + q_{2e^-} \cdot P/2e^- \tag{6}$$

For aerobic growth with gluconate a q_{ATP} of 20.61 mmol ATP g^{-1} h^{-1} can be calculated in this way. Consequently, Y_{ATP} for growth with gluconate at $u = 0.15$ h$^-$ is 7.3. Now it is possible to postulate that the q_{ATP} for the other growth conditions with gluconate is equal to the q_{ATP} for aerobic gluconate-limited growth, since under these conditions biomass is formed from the same carbon source, gluconate, and at the same rate. From the q_{ATP} for aerobic gluconate-limited growth, the q_{sub} and q_{2e^-} for the other growth conditions, it is possible to calculate the P/2e$^-$ (called experimental P/2e$^-$ ratio) for those conditions. Then the experimental and the theoretical P/2e$^-$ ratios can be compared. It is evident from the results in Table 5.3 that in all cases there is a very good correspondence between the experimental and the theoretical P/2e$^-$ ratios. This gives strong support for the suppositions made and for the validity of the scheme in Fig. 5.11. It also indicates that the uptake of nitrate under physiological conditions takes place completely by the nitrate/nitrite antiport system and does not require energy. By the methods outlined above, it is also possible to see if a good material and redox balance during growth has been achieved (232). This is a point which is seriously neglected in many yield studies. If a good material and redox balance has been obtained, the data must fit.

$$Y_{2e^-} = Y_{sub}/v_{H_2} \tag{7}$$

From the data in Table 5.3 it is evident that in some cases the material and redox balance do not fit very well, which is due to the observation that in some cases 10 to 15% of the carbon source was used for the production of siderochromes (250). This is also the reason that two values for aerobic succinate-limited growth are given for succinate. Accumulation of siderochromes does not interfere with the calculations given above and results collected in Table 5.3. In the same way, aerobic growth with mannitol and gluconate with nitrate as nitrogen source has been studied (22). In these cases it is not allowed to calculate an experimental $P/2e^-$ ratio from the q_{ATP} for aerobic growth with these substrates and ammonia. On the assumption that the theoretical $P/2e^-$ ratio is the true value, the q_{ATP} for growth under these circumstances can be calculated. The difference between the q_{ATP} for growth with nitrate and ammonia as nitrogen source is then the energy cost for the assimilation of nitrate. For growth with gluconate and mannitol, respectively, 2.99 and 2.84 ATP equivalents were calculated to be necessary for the assimilation of 1 mol of nitrate (22). This energy may be necessary for uptake of nitrate, the possible formation of NADPH from NADH by transhydrogenation, and additional costs in the pathway of incorporation of ammonia into organic material. Before the differences in charge separation between electron transport to oxygen and nitrogenous oxides were known, the difference in molar growth yields between aerobic growth and anaerobic growth with nitrate or nitrite were explained by the assumption that nitrite had an uncoupling effect on oxidative phosphorylation (169). To explain the toxic effect which nitrite in general has on bacteria, it was assumed that nitrite acts as an uncoupler and thus dissipates the proton motive force. However, this hypothesis has been shown to be incorrect. It was demonstrated that nitrite inhibited active transport, oxygen uptake, and oxidative phosphorylation in a number of bacteria (205). Evidence was presented that nitrite exerted its effect by oxidizing ferrous iron of electron carriers, such as cytochrome oxidase, to ferric iron. Subsequent work has indeed demonstrated that nitrite inhibited oxidase activity in *Pa. denitrificans* (137, 140). An uncoupling effect of nitrite could not be demonstrated on a number of bacteria (200). In accordance with this view, evidence has been presented which indicates that nitrite enters the cell as anion (137, 140).

5.3.4.3 *Campylobacter sputorum* subsp. *bubulus*

C. sputorum is a microaerophilic organism which grows optimally at reduced oxygen tension (235). However, it can also grow anaerobically with various hydrogen acceptors (235). The stoichiometries of proton translocation associated with electron transfer from various donors to oxygen and nitrogenous oxides are shown in Table 5.4. A scheme for proton translocation and electron transport is given in Fig. 5.12. It is evident that the $\rightarrow H^+/O$ ratios for this organism are much lower than those for *Pa. denitrificans*. Consequently, it must be concluded that the respiratory chain in this organism is relatively simple, since the $\rightarrow H^+/2e^-$ ratio is only about 2. It has been suggested that a quinone is an intermediate in electron transfer. This quinone is supposed to be the only place in the respiratory chain of *C. sputorum* where two protons are translocated (57). Consequently, proton pumps are supposed not to be present in *C. sputorum* (57).

There is a good correspondence between the theoretical $\rightarrow H^+/O$ and $\rightarrow H^+/NO_2^-$ ratios calculated from the scheme in Fig. 5.12 and the experimental values, which can be regarded as strong evidence for the validity of the scheme. However, the theoretical $\rightarrow H^+/NO_3^-$ ratios are much higher than the experimental values. On the basis of this observation it might be questioned whether

TABLE 5.4 Respiration-driven proton translocation with various substrates and O_2 or nitrogenous oxides as electron acceptor in *Campylobacter sputorum*.

Electron Donor	Experimental $\rightarrow H^+/$Oxidant ratio			Theoretical $\rightarrow$ $H^+/$Oxidant ratio[a]		
	H^+/O	H^+/NO_3^-	H^+/NO_2^-	H^+/O	H^+/NO_3^-	H^+/NO_2^-
Formate	3.51	4.03	2.79	3.8	7.2	3.4
Hydrogen[b,c]	3.74 (2.33)	4.36 (1.31)	2.61 (−0.86)	4.0	8.0 (2.0)	4.0
Lactate[c]	2.15	(1.12)	−1.64	2.0	2.0 (2.0)	−2.0

Source: Data from Refs. 56 to 58.

[a] The theoretical values are calculated from Fig. 5.12.

[b] For hydrogen the values for cells with a low hydrogenase activity are given in parentheses.

[c] For cells with low hydrogenase activity with hydrogen and for other cells with lactate, the reduction of nitrate to nitrite must be separated from the reduction of nitrite to ammonia. The H^+/NO_3^- ratios in parentheses represent the proton translocation for the conversion of nitrate to nitrite.

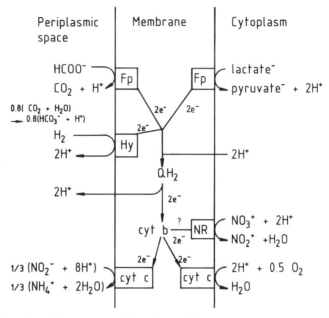

Figure 5.12 Scheme for proton translocation and electron transport to oxygen and nitrogenous oxides in *Campylobacter sputorum* subsp. *bubulus*. Fp, flavoprotein; QH_2, ubiquinol; Hy, hydrogenase; NR, nitrate reductase. Revised from Ref. 57.

nitrate is indeed reduced at the cytoplasmic side of the membrane. However, in the experiments with nitrate a rapid proton backflow was usually found, and this may lead to an underestimation of the $\rightarrow H^+/NO_3^-$ ratios (58). With cells with a low hydrogenase activity, the $\rightarrow H^+/O$ values with H_2 was lower than the value obtained in cells with high hydrogenase activity. The $\rightarrow H^+/NO_2^-$ value (-0.86) was even negative. These results may be explained by the functioning of reduced

TABLE 5.5 **Number of charges translocated across the cytoplasmic membrane during the flow of $2e^-$ ($\rightarrow q^+/2e^-$ ratio) from various substrates to an electron acceptor in *Campylobacter sputorum* subsp. *bubulus***

| | $\rightarrow q^+/2e^-$ | | | |
| | Electron acceptor reaction | | | |
Substrate	O_2/H_2O	NO_3^-/NO_2^-	NO_3^-/NH_4^+	NO_2^-/NH_4^+
Formate	4	4	2.5	2
Hydrogen	4	4	2.5	2
Lactate	2	2	0.5	0

Source: Data from Refs. 56 to 58.

components of the respiratory chain, other than hydrogen, as the electron donor during the measurement of the $\rightarrow H^+$/oxidant values. The negative $\rightarrow H^+/NO_2^-$ ratio is in accordance with the periplasmic orientation of nitrite reductase. The $\rightarrow H^+/NO_3^-$ value in cells with low hydrogenase activity was 1.31 (Table 5.5).

Furthermore, during the measurement of the $\rightarrow H^+/NO_3^-$ value, a transient accumulation of nitrite was observed. In *C. sputorum* the lactate dehydrogenase activity is low; therefore, experiments with lactate as an electron donor should be similar to those obtained with hydrogen in cells with a low hydrogenase activity. This is indeed the case. The positive $\rightarrow H^+/NO_3^-$ values are strong evidence for the location of nitrate reductase at the cytoplasmic face of the membrane, as shown in Fig. 5.12. It is not known, however, whether cytochrome *b* or cytochrome *c* is the immediate electron donor for the reduction of nitrate. Oxygen is reduced at the cytoplasmic face of the membrane. Cytochromes of the *a*-type and cytochrome *o* were not detected, and it is assumed that a carbon monoxide–binding cytochrome *c* acts as a cytochrome oxidase (182, 183).

The location of the electron-donating site of formate dehydrogenase and hydrogenase at the periplasmic side of the membrane can be regarded as a clever trick to increase the charge separation during electron transport from formate and hydrogen to oxygen. The charge separation ($\rightarrow q^+/2e^-$) values for electron transport from various substrates to oxygen and nitrogenous oxides are shown in Table 5.5. The $\rightarrow q^+/2e^-$ value for the oxidation of formate and hydrogen by oxygen is 4, and notwithstanding the simple respiratory chain of this organism, the $\rightarrow q^+/2e^-$ value is the same as that for hydrogen oxidation in *Pa. denitrificans* (Table 5.2). The electron-donating site of formate dehydrogenase and hydrogenase is also at the periplasmic side of the membrane in *V. succinogenes* (134, 135), *D. desulfuricans*, and in *D. gigas* (184, 225) and *C. fetus* subsp. *jejuni* (96). Therefore, this seems a general phenomenon in this type of bacteria. For electron transport from lactate to nitrate and nitrite, the $\rightarrow q^+/2e^-$ values are 2 and 0, respectively. Thus although the electron transport chains from formate and lactate to nitrite are similar, only the first is energy yielding, and this is merely due to the fact that formate and lactate donate their electrons to different sides of the membrane.

These conclusions are in accordance with measurements of molar growth yields (Table 5.6). During anaerobic growth with lactate and fumarate no oxidative phosphorylation occurs and energy generation occurs completely by substrate phosphorylation associated with acetate formation (217). Consequently, Y_{ATP} is about 16.5. From the growth yields a P/O ratio of 0.63 was calculated and a $P/2e^-$ ratio for the reaction nitrate $\rightarrow$ nitrite of 0.30. In accordance with the scheme in Fig. 5.12 and the results in Table 5.5, the reduction of nitrite by lactate was concluded not to be associated with oxidative phosphorylation. The observation that with formate the growth yields with nitrate are higher than with nitrite are also in accordance with the scheme in Fig. 5.12 and the results in Table 5.5.

An important conclusion is that the $\rightarrow q^+/2e^-$ values for electron transport to nitrogenous oxides in *C. sputorum* are much lower than those in *Pa. denitrificans*. This is due to a difference in the complexity of the respiratory chain. Thus although

TABLE 5.6   Molar growth yields of *Campylobacter sputorum* subsp. *bubulus* growing with various substrates in continuous culture with growth-limiting amounts of hydrogen acceptor

Hydrogen Acceptor	Molar growth yield (g/g mol)		
	Lactate	Formate	Hydrogen[a]
O_2	42.2	5.9	n.d.
Fumarate	16.2	3.8	n.d.
Nitrate	20.2	3.9–4.3	5.6
Nitrite	18.0	3.7	n.d.

Source: Data from Refs. 56 and 58.
[a] n.d., not determined.

the standard free energy changes of reduction of nitrate to nitrogen and reduction of nitrate to ammonia are about the same, the ATP yield for the first process is much higher. This is a general phenomenon which is due to the fact that the respiratory systems associated with the reduction of nitrate to ammonia are less complicated and have no or fewer proton pumps than those in denitrifying organisms.

In recent experiments it was demonstrated that in cell-free extracts of *V. succinogenes* the reduction of nitrite by H_2 was associated with ATP formation (16). The observed $P/2e^-$ ratio was 0.11. Nitrite reduction by formate was not associated with ATP formation, however. From growth yields it was concluded that

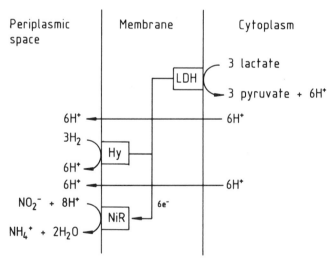

Figure 5.13   Scheme for proton translocation during the reduction of nitrite by lactate or H_2 in *Desulfovibrio desulfuricans*. LDH, lactate dehydrogenase; Hy, hydrogenase, NiR, nitrite reductase. Adapted from Ref. 225.

the ATP gain with nitrate or nitrite is about 0.5 mol ATP per mole of formate or H_2.

5.3.4.4 Desulfovibrio sp.

Desulfovibrio sp. are anaerobic, sulfate-reducing bacteria. It is very remarkable that some strains can reduce nitrate. This has been known for a long time (195, 214). Recently, similar strains were isolated and it was demonstrated that nitrate and sulfate could be used simultaneously as a hydrogen acceptor (123). Based on the results described in Section 5.3.2, a scheme for proton translocation during the reduction of nitrite by lactate or H_2 in *D. desulfuricans* is presented in Fig. 5.13. In the electron transport chain from lactate and hydrogen two and one proton-translocating site, respectively, are assumed to be present (225). For *D. desulfuricans* (225) as for *D. gigas* (13), $H^+/2e^-$ ratios of about 2 were reported for electron transport from hydrogen to nitrite. Membrane fractions of *D. gigas* formed ATP coupled to the reduction of nitrite by H_2 with a $P/2e^-$ ratio of 0.32 (13).

5.4 REGULATORY ASPECTS

In general, nitrate reduction is an inducible process, which occurs only in the absence of oxygen and in the presence of nitrate (193, 231, 236). This is very logical since the standard free energy changes of the reactions

$$NADH + H^+ + NO_3 \rightarrow NAD^+ + NO_2^- + H_2O$$

and

$$NADH + H^+ + 1/2O_2 \rightarrow NAD^+ + H_2O$$

are -142.5 and -216.5 kJ per reaction, respectively. Thus the largest amount of energy can be gained by the utilization of oxygen. In a number of bacteria, anaerobic conditions are sufficient to obtain formation of nitrate reductase [e.g., *Bacillus licheniformis* (210) and *Haemophilus influenzae* (218)], but even then, the amount of enzyme formed is increased when nitrate is present. Recently, also, a number of organisms have been described, which are able to denitrify under aerobic conditions. This was found for a number of denitrifying organisms isolated from activated sludge (136). In a *Hyphomicrobium* sp. growing with dimethylamine as carbon and energy source, denitrification occurred under partly aerobic conditions (167). The dissolved oxygen tension at which nitrate reductase was formed in the culture was inversely proportional to the growth rate. At low growth rates (0.01 h^{-1}) nitrate reductase was synthesized up to a dissolved oxygen tension (DOT) of 50 mmHg, at intermediate growth rates (0.1 h^{-1}) the enzyme was formed up to a DOT 20 mmHg, but at the highest growth rate (0.15 h^{-1}) the enzyme was formed only up to a DOT of 8 mmHg (167). No explanation is available for these interesting observations. Recently it was demonstrated that *Thiosphaera pantotropha*, a bacterium isolated from a desulfurizing, denitrifying effluent treatment system (202), denitrifies under fully aerobic conditions (203). In this organism oxygen and nitrate were used simultaneously at a DOT above 80% of air saturation. Oxygen uptake and denitrification both contributed to the energy budget of the cells. Further studies on these interesting organisms will be rewarding.

In general, it can be stated that the effect of oxygen on the synthesis of nitrate reductase occurs in parallel with an effect of oxygen on the activity of the enzyme. When a culture of *K. aerogenes*, growing anaerobically with nitrate as the hydrogen acceptor, is shifted to aerobic conditions, three effects are observed: (a) an immediate cessation of the formation of nitrate reductase (repression by oxygen); (b) an immediate cessation of nitrite production, which means that electron transfer to nitrate reductase stops immediately; and (c) a partial inactivation of the nitrate reductase already present (247). We will now discuss the regulation of the formation of nitrate reductase, and subsequently, the control of nitrate respiration by oxygen.

5.4.1 The Regulation of the Formation of Nitrate Reductase

A model for the regulation of nitrate reductase was presented in a previous review (236). Since then a number of new observations have been made. However, these have not yet led to a good understanding of the regulation of the formation of nitrate reductase. Even less is known about the regulation of the formation of nitrite reductase and nitrous oxide reductase. For the moment it will be supposed that the formation of these other reductases is coregulated with that of nitrate reductase. We must realize, however, that this will not always be the case, for example, in *Pa. denitrificans*, nitrite reductase and nitrous oxide reductase but not nitrate reductase are derepressed in oxygen-limited chemostat cultures (251). Similarly, in a denitrifying *Flavobacterium* sp. shifted to anaerobic growth, it was found that the synthesis of nitrous oxide reductase lagged behind that of nitrate reductase and nitrite reductase (69). Consequently, after the shift to anaerobic conditions, N_2O was initially the main end product of denitrification. Similar observations were made for soil (69).

Detailed knowledge of the regulation of the formation of the molybdenum cofactor is also lacking, although it has been suggested that in *Proteus mirabilis* the synthesis of nitrate reductase and the molybdenum cofactor are coregulated under anaerobic conditions (41). In *E. coli* aerobic cultures have extremely low levels of nitrate reductase activity, whereas anaerobic cultures have a significant activity. Addition of nitrate to an anaerobic culture further induces nitrate reductase activity some 20-fold. Nitrate has no effect on the activity in an aerobic culture. On the basis of these observations, the existence of two separate regulatory signals was suggested (216). The effect of oxygen has been explained in several ways: (a) Oxygen itself is the corepressor (118). (b) Electron transport to oxygen rather than oxygen itself regulates enzyme formation, by blocking its assembly into the membrane-bound active form (11, 51, 52, 187, 231, 236). (c) The repressor that mediates regulation is sensitive to the intracellular redox potential (216, 218, 264). In recent experiments the appearance of nitrate reductase in *E. coli* after induction by nitrate was analyzed under different conditions, and the inhibitory effects of oxygen, chloramphenicol, and rifampin were compared (117). It was concluded that there are two sites of oxygen control. In the first place oxygen appeared to inhibit the synthesis of nitrate reductase at the level of transcription. In addition, the translation or some later steps (perhaps assembly) were found to be blocked. This suggests that nitrate reductase cannot be complexed in an active membrane-bound form with other electron transfer components when electron flow to these enzymes is impeded (11, 51, 52, 187, 231, 236). Recently, it has been observed that soluble nitrate reductase cannot be inserted into the membrane in the presence of oxygen (84). It was mentioned before (Section 5.2.1) that cytochrome b^{NR} is required for the association of nitrate reductase with the cell membrane (156, 158). Thus it has been assumed that oxygen prevents the association of cytochrome b^{NR} with the membrane (156, 158, 236), and in this way also, the association of nitrate reductase with the membrane. That oxygen prevents the transcription of nitrate reductase has also been demonstrated by the study of *chl* C-*lac* operon fusions (56, 67, 220). In such strains the *lac* operon is under the control of the regulatory elements of the *chl* C operon. Under anaerobic conditions β-galactosidase is induced by nitrate (35, 67, 226). Under aerobic conditions β-galactosidase remains repressed.

Although this indeed proves that oxygen prevents transcription of the nitrate reductase operon, it must be realized that these experiments do not give any indication about the mechanism by which oxygen prevents transcription. In this aspect it is worthwhile to make a comparison with the expression of the 17 nitrogen fixation (*nif*) genes in *K. pneumoniae*, which are regulated in response to both the presence of combined nitrogen and the oxygen tension (59). Nitrogen control of *nif* transcription is maintained at two levels: first by the *ntr* system, which exerts general control on nitrogen metabolism, and second, by regulatory proteins, encoded by two *nif* genes, *nif* A and *nif* L. The *nif* A product is required for the transcriptional activation of the *nif* operons, whereas the *nif* L product has been implicated in regression of *nif* transcription in response to combined nitrogen and oxygen. Hence a specific repression of nitrogenase synthesis can be maintained by *nif* L without repressing other operons under *ntr* control. It has been suggested that the *nif* L product exerts repression by inactivating the *nif* A gene product. The promotor of the *nif* LA operon is subject to positive control by the *ntr* system and is also regulated autogenously by the *nif* L and *nif* A products. Furthermore, the complete amino acid sequence of the components in nitrogen fixation has been determined from the nucleotide sequences in the genes of several organisms (93). Also, the nucleotide sequences of several promoter regions has been determined (58, 93). These data show that our knowledge of the regulation of the *nif* genes is much greater than that of the regulation of the nitrate reductase operon. However, further discussion will indicate that there are a number of parallels in the regulation of both systems.

Several groups have isolated mutants termed *fnr*, *nir* A, or *nir* R which result in very low levels of anaerobic respiratory enzymes (36, 146, 180, 227), as formate hydrogenlyase, and reductases for nitrate, nitrite, and fumarate. It has been suggested that this locus encodes a protein that activates the transcription of anaerobic respiratory enzyme structural genes. In strains with a *chl* C-*lac* fusion and a *fnr* mutation there is a complete lack of β-galactosidase induction under anaerobic conditions in the presence of nitrate (226), which supports this suggestion. By transposon mutagenesis a mutant strain designated *nar* L. was obtained (227) in which a nitrate-specific positive regulatory element is lacking (226). There are a number of observations, which cannot be explained by the mechanisms described so far:

1. Nitrate reductase is induced by azide and nitrite (28a, 37, 51). It is a remarkable observation that after anaerobic growth of *P. mirabilis* and azide the specific activity is about fourfold higher than after anaerobic growth with nitrate (51). In *Pa. denitrificans* the specific activity of nitrate reductase after incubation with azide under conditions of low aeration was tenfold higher than after growth with nitrate (28b). The concentration of cytochrome *b* was also considerably increased by growth in the presence of azide under conditions of low aeration.

2. When an aerobic culture of *E. coli* or *P. mirabilis* is shifted to anaerobic conditions, nitrate reductase synthesis is temporarily derepressed (51, 216).

3. Analysis of cytoplasmic membrane proteins of a chlorate-resistant mutant of *P. mirabilis*, which is defective in the formation of the molybdenum cofactor (40), suggested the presence of inactive nitrate reductase after anaerobic growth in the absence of nitrate (187). A similar observation was made for a chlorate-resistant mutant of *Pa. denitrificans* (28).

4. During anaerobic growth of *P. mirabilis* in the presence of tungstate a relatively high concentration of inactive nitrate reductase was formed, although nitrate was not present in the medium. The inactive enzyme could be reactivated in a few minutes upon addition of molybdate (185).

5. In *chl* C-*lac* operon fusions in *E. coli* the introduction of *chl* A, *chl* B, or *chl* E mutations, which are supposed to be defective in the formation of the molybdenum cofactor, leads to partial or total constitutive expression of β-galactosidase (190). It was suggested that the molybdenum cofactor is a corepressor of the nitrate reductase operon (190, 191).

However, it seems advisable to try to explain the five observations mentioned above in one way. It seems that under none of these conditions, is a repressor sensitive to nitrate formed, or alternatively, that such a repressor becomes inactivated. These observations can be explained in various ways. It has been remarked that under each of the conditions mentioned above in which derepression of nitrate reductase occurs in the absence of nitrate, the activity of the formate hydrogen lyase system is impaired (51, 231). The possibility has therefore been considered that a component of that system is involved in the regulation of the formulation of nitrate reductase (51, 231). As an alternative it has been suggested that the regulation of the formation of nitrate reductase occurs by a mechanism known as autogenous regulation of gene expression (78). The supposition is that nitrate reductase or a soluble precursor of the enzyme functions as an autogenous repressor of its own biosynthesis (187, 236). Interaction of nitrate, nitrite, or azide or absence of an active molybdenum cofactor may result in conformational shifts in the molecule, rendering this molecule unfit for the repressor function. This regulation model also includes a feedback control of the amount of nitrate reductase and cytochrome b^{NR} which will be synthesized during derepression by nitrate. When the intracellular nitrate concentration is lowered sufficiently by the operative nitrate reductase system, the repressor function of the freshly synthesized nitrate reductase will not be completely eliminated by interaction with nitrate and further synthesis will be reduced (187, 236). During growth with azide and tungstate, this effect will not occur and an unbalanced formation of large amounts of nitrate reductase will take place. The formation of the assimilatory nitrate reductase of *N. crassa* has also been suggested to be subject to autogenous control (242). Although this hypothetic model can explain the induction and the level of induction of nitrate reductase under many experimental conditions, more work is necessary to confirm the model.

The results described above indicate that the regulation of the formation of nitrate reductase is very complex. A number of mechanisms which are involved in this regulation have been demonstrated beyond doubt. However, we are far from a complete scheme for the regulation of the nitrate reductase operon. Much work will be necessary to unravel the mechanism of the regulation of nitrate reductase.

5.4.2 Control of Electron Flow to Nitrogenous Oxides and Oxygen

It has been mentioned before that when a culture of *K. aerogenes* growing anaerobically with nitrate as the hydrogen acceptor is shifted to aerobic conditions, there is an immediate cessation of nitrate reduction (247). The same phenomenon occurs very generally (e.g., also in *Pa. denitrificans*) (109). In electron transport chains (Fig. 5.7) electron transfer to nitrate branches from the aerobic chain in the region between ubiquinone and the *b*-type cytochromes (Fig. 5.11). It is an interesting question by what mechanism electron flow in branched respiratory chains is regulated and how nitrate reduction is blocked in the presence of oxygen. The first possibility is that oxygen might itself inhibit via an interaction with a component of the electron transfer chain to nitrate. This possibility has been excluded, however, on the basis of many experimental observations:

1. A mutant of *Ps. aeruginosa* has been described which is affected in aerobic growth (245). The mutant has a defect in heme biosynthesis and accumulated coproporhyrin under aerobic conditions. In this mutant nitrite accumulation proceeded under aerobic conditions.

2. The artificial electron acceptor ferricyanide, which accepts electrons from cytochrome *c* (20, 252), mimics oxygen in blocking nitrate reduction in *Pa. denitrificans* (6, 139). Inhibition of electron flow to ferricyamide by antimycin A substantially reversed this blockade.

3. Interference with respiration by inhibiting terminal oxidases by hydroxyl-amine abolished the effect of oxygen on nitrate reduction in *Pa. denitrificans* (139).

4. Also, nitrous oxide and sometimes nitrite behave like oxygen by inhibiting nitrate reductase (5, 138). Again nitrate reduction starts when electron flow to these nitrogenous oxides is prevented by antimycin A or acetylene (5, 138).

The observations mentioned above can better be explained by the hypothesis that the redox state of a component of the electron transfer chain is involved in the control of nitrate reduction by oxygen. Several intermediates of the respiratory chain have been considered as possible controlling redox carriers. The demonstration of control of nitrate reduction by oxygen in cytochrome *c*–deficient mutants (5, 263) indicates that the redox-controlled switch is before cytochrome *c*. On the basis of theoretical arguments, a specific role for *b*-type cytochromes in regulating nitrate reduction has been rejected (5). In view of the fact that ubiquinone is the last common component of the respiratory chain before pathways of electron flow to oxygen and nitrate diverge, this redox carrier is a plausible candidate for a role in the regulation of nitrate reduction. In fact, some evidence, albeit indirect, has been presented in favor of such a proposal in which relatively small changes in the redox state of ubiquinone critically control the functioning of nitrate reductase (5). In this context it is conceivable that the degree of reduction of ubiquinone determines whether nitrate is reduced. Since nitrate reductase is located at the cytoplasmic side of the membrane, the control of nitrate reductase activity could in principle be achieved preventing movement of nitrate to nitrate reductase. It has been demonstrated with inverted membrane vesicles of *Pa. denitrificans* that nitrate reduction occurred under aerobic conditions (3, 5, 109, 140). In addition, three different detergents that allowed chlorate to be reduced in cells also permitted aerobic nitrate reduction (3, 5). On the basis of these observations it has been suggested that the redox state of ubiquinone controls the movement of nitrate to its reductase (5). However, an alternative explanation for the observation that in permeabilized cells and in inverted membrane vesicles aerobic nitrate reduction is possible has been presented (140). It was pointed out that upon permeabilization of the cytoplasmic membrane, the rate of reduction of terminal acceptors and the degree of cytochrome *c* reduction strongly increased (140). The increase in the rate of reduction of terminal acceptors after permeabilization was supposed to be due to an enhancement of the influx of redox equivalents to the respiratory chain due to stimulation of the succinate dehydrogenase reaction. The latter effect could be due to the fact that in whole cells the succinate transport into the cells is limiting for its utilization. Another effect that is of importance for the explanation of aerobic nitrate reduction in permeabilized cells and in inverted membrane vesicles is in the inhibition of oxidase activity by endogenously formed nitrite (140). After permeabilization there was a marked increase in the rate of aerobic nitrite reduction (140), whereas in whole cells the rate of nitrite reduction is severely inhibited by oxygen (109). This effect can never be explained by a control of the access of nitrite to its reductase by the redox state of ubiquinone, since nitrite is reduced at the periplasmic side of the membrane (3, 20, 168). In whole cells the preferential utilization of oxygen was abolished in the presence of the uncoupler CCCP, which brought about a switch to the reduction of nitrite (137). Due to the dissipation of the proton motive force by CCCP, nitrite can enter the cell to a higher concentration than in the absence of the uncoupler. The intracellular accumulation of nitrite to a high concentration leads to an inhibition of oxidases. In the presence of oxygen and nitrite, the degree of reductase of cytochrome c was increased by the presence of CCCP. This enables the flow of electrons to nitrite (137).

The most likely explanation for the effects of oxygen on the reduction of nitrate and nitrite is that the redox state of the components in the respiratory chain in the branching sites is the controlling factor in electron flow to nitrate and nitrite. A competition between reductases for nitrate and nitrite and oxidases for a limited supply of electrons from primary dehydrogens seems to play an important role (5, 6, 138–140).

Oxygen has also been reported to inhibit electron transfer to nitrous oxide in whole cells (4, 137). This inhibition is possibly explained by a reversible inactivation of nitrous oxide reductase, since the partial purified reductase is rapidly inactivated by oxygen, which inactivation can be reversed to a large extent (131).

Some remarks may be made on the control of nitrate reduction under anaerobic conditions. In general, the reduction of nitrogenous oxides proceeds simultaneously without accumulation of intermediates in resting (5, 73, 138) and growing cells (18). So it seems that the reduction of nitrate to nitrite is finely accommodated to the concentration of endogenously formed nitrite and nitrous oxide to a rate at which it is accompanied by simultaneous reduction of nitrite and nitrous oxide (5, 138).

5.5 CONCLUSIONS

During the last few years great progress has been made in our knowledge of all aspects of the reduction of oxidized nitrogen compounds. However, it is very peculiar that our knowledge is restricted to only a few organisms. The understanding of the regulatory aspects of the formation of reductases is fully restricted to *E. coli*, which is easily explained by the availability of well-characterized mutants and sophisticated methods for genetic analysis. However, even for *E. coli* the picture of the regulation of the formation of nitrate reductase is far from complete. In this aspect our knowledge on nitrate reduction lags behind that of some other complex systems. For instance, knowledge of the regulation of the system for nitrogen fixation is much more detailed than that for nitrate reduction. Much work has thus still to be done to advance our knowledge of nitrate reduction to the same level as that for nitrogen fixation. In denitrifying organisms a detailed analysis of the regulation of the formation of reductase for oxidized nitrogen compounds is not yet possible. The number of available mutants is very restricted, and methods for genetic analysis are only available for *Pseudomonas* sp. (29, 243, 244). Even in those organisms methods for genetic analysis as operon fusions are not yet available. Our knowledge of the bioenergetics of denitrification and the control of electron flow to nitrogenous oxides and oxygen is most complete in *Pa. denitrificans*. It is regrettable that methods for genetic analysis in this organism are not available. Therefore, there is an urgent need for the development of methods for genetic analysis in those organisms and for their application to the study of the genetic control of denitrification.

It would be very fruitful if detailed knowledge on all aspects of nitrate reduction would be available for a larger variety of microorganisms. This will need a large effort, however. As an example, the study of denitrification in organisms capable of aerobic denitrification may yield new insights in the regulation of the synthesis of reductases for nitrogenous oxides and for the regulation of electron flow to oxygen and nitrogenous oxides.

Although our knowledge on all aspects of the reduction of oxidized nitrogen compounds has increased considerably during the last few years, a number of areas can be mentioned in which our knowledge still shows great gaps. In this aspect it is remarkable that the structure of the molybdenum cofactor for nitrate reductase is still not known. This cofactor, molybdopterin, plays an important role in the metabolism of bacteria. However, it may be mentioned that this compound also plays an essential role in the metabolism of humans (113). Another area in which much further work has to be done is the mechanism of nitrite reduction by nitrite reductase, in particular the way in which the first $-N-N-$ bound is formed. The knowledge of organisms performing dissimilatory reduction of nitrate to ammonia is still scanty, and study of these organisms also deserves more attention. Much additional work is necessary to clarify the many points on which uncertainties still exist and to broaden our present knowledge to a larger number of organisms.

REFERENCES

1. Abou-Jaoudé, A., M. Chippaux, M.-C. Pascal, and F. Casse, 1977. Formate: a new electron donor for nitrite reduction in *Escherichia coli* K12. Biochem. Biophys. Res. Commun. 78:579–583.

2. Abou-Jaoudé, A., M.-C. Pascal, and M. Chippaux. 1979. Formate-nitrite reduction in *Escherichia coli* K12.2. Identification of components involved in electron transfer. Eur. J. Biochem. 95:315–321.

3. Alefounder, P. R., and S. J. Ferguson. 1980. The location of dissimilatory nitrite reductase and the control of dissimilatory nitrate reductase by oxygen. Biochem. J. 192:231–240.

4. Alefounder, P. R., and S. J. Ferguson. 1982. Electron-transport linked nitrous oxide synthesis and reduction monitored with an electrode. Biochem. Biophys. Res. Commun. 104:1149–1155.

5. Alefounder, R., A. J. Greenfield, J. E. G. McCarthy, and S. J. Ferguson. 1983.

Selection and organisation of denitrifying electron-transfer pathways in *Paracoccus denitrificans*. Biochem. Biophys. Acta 724:20–39.

6. Alefounder, P. R., J. E. G. McCarthy, and S. J. Ferguson. 1981. The basis of control of nitrate reduction by oxygen in *Paracoccus denitrificans*. FEMS Microbiol. Lett. 12:321–326.

7. Aminuddin, M., and D. J. D. Nicholas. 1973. Sulphide oxidation linked to the reduction of nitrate and nitrite in *Thiobacillus denitrificans*. Biochim. Biophys. Acta 325:81–93.

8. Aminuddin, M., and D. J. D. Nicholas. 1974. Electron transfer during sulphide and sulphite oxidation in *Thiobacillus denitrificans*. J. Gen. Microbiol. 28:115–123.

9. Amy, N. K. 1981. Identification of the molybdenum cofactor in chlorate resistant mutants of *Escherichia coli*. J. Bacteriol. 148:274–282.

10. Averill, B. A., and J. M. Tiedje. 1982. The chemical mechanism of microbial denitrification. FEBS Lett. 138:8–12.

11. Azoulay, E., J. Puig, and P. Couchoud-Beaumont. 1969. Étude des mutants chlorate-résistants chez *Escherichia coli* K12. I. Reconstitution in vitro de l'activité nitrate réductase particulaire chez *Escherichia coli* K12. Biochim. Biophys. Acta 171:238–252.

12. Balderston, W. L., B. Sherr, and W. J. Payne. 1976. Blockage by acetylene of nitrous oxide reduction in *Pseudomonas perfectomarinus*. Appl. Environ. Microbiol. 31:504–508.

13. Barton, L. L., J. Le Gall, J. M. Odom, and H. D. Peck, Jr. 1983. Energy coupling to nitrite respiration in the sulfate-reducing bacterium *Desulfovibrio gigas*. J. Bacteriol. 153:867–871.

14. Betlach, M. R. 1982. Evolution of bacterial denitrification and denitrifier diversity. Antonie Leeuwenhoek J. Microbiol. 48:585–607.

15. Betlach, M. R., J. M. Tiedje, and R. B. Firestone. 1981. Assimilatory nitrate uptake in *Pseudomonas fluorescens* studied using nitrogen-13. Arch. Microbiol. 129:135–140.

16. Bokranz, M., J. Katz, I. Schröder, A. M. Robertson, and A. Kröger. 1983. Energy metabolism and biosynthesis of *Vibrio succinogenes* growing with nitrate or nitrite as terminal electron acceptor. Arch. Microbiol. 135:36–41.

17. Bonnefoy-Orth, V., M. Lepelletier, M.-C. Pascal, and M. Chippaux. 1981. Nitrate reductase and cytochrome b-nitrate reductase structural genes as parts of the nitrate reductase operon. Mol. & Gen. Genet. 181:535–540.

18. Boogerd, F. C., K. J. Appeldoorn, and A. H. Stouthamer. 1983. Effects of electron transport inhibitors and uncouplers on denitrification in *Paracoccus denitrificans*. FEMS Microbiol. Lett. 20:455–460.

19. Boogerd, F. C., H. W. van Verseveld, and A. H. Stouthamer. 1980. Electron transport to nitrous oxide in *Paracoccus denitrificans* FEBS Lett. 113:279–284.

20. Boogerd, F. C., H. W. van Verseveld, and A. H. Stouthamer. 1981. Respiration-driven proton translocation with nitrite and nitrous oxide in *Paracoccus denitrificans*. Biochim. Biophys. Acta 638:181–191.

21. Boogerd, F. C., H. W. van Verseveld, and A. H. Stouthamer. 1983. Dissimilatory nitrate uptake in *Paracoccus denitrificans* via a Δu_{H^+}-dependent system and a nitrate-nitrite antiport system. Biochim. Biophys. Acta 723:415–427.

22. Boogerd, F. C., H. W. van Verseveld, D. Torenvliet, M. Braster, and A. H. Stouthamer. 1984. Reconsideration of the efficiency of energy transduction in *Paracoccus denitrificans* during growth under a variety of culture conditions. A new approach for the calculation of $P/2e^-$ ratios. Arch. Microbiol. 139:344–350.

23. Bosma, J. H., R. Wever, and J. A. van't Riet. 1978. Electron paramagnetic resonance studies on membrane-bound respiratory nitrate reductase of *Klebsiella aerogenes*. FEBS Lett. 90:107–111.

24. Boxer, D. H., and R. A. Clegg. 1975. A transmembrane location for the proton-translocating reduced ubiquinone-nitrate reductase segment of the respiratory chain of *Escherichia coli*. FEBS Lett. 60:54–57.

25. Bray, R. C., G. N. George, S. Gutteridge, L. Norlander, J. G. P. Stell, and C. Stubley. 1982. Studies by EPR spectroscopy of the molybdenum centre of aldehyde oxidase. Biochem J. 203:263–268.

26. Bray, R. C., S. Gutteridge, D. A. Stotter, and S. I. Tanner. 1979. The mechanism of action of xanthineoxidase. The relationship between the rapid and the very rapid molybdenum electron paramagnetic resonance signals. Biochem J. 177:357–360.

27. Brice, J. M., J. F. Law, D. J. Meyer, and C. W. Jones. 1974. Energy conservation in *Escherichia coli* and *Klebsiella pneumoniae*. Biochem. Soc. Trans. 2:523–526.

28. Burke, K. A., K. Calder, and J. Lascelles, 1980. Effects of molybdenum and tungsten on induction of nitrate reductase and formate dehydrogenase in wild type and mutant *Paracoccus denitrificans*. Arch. Microbiol. 126:155–159.

28a. Calder, K., K. A. Burke and J. A. Lascelles. 1980. Induction of nitrate reductase and cytochromes in wild type and chlorate-resistant *Paracoccus denitrificans*. Arch. Microbiol. 126:149–153.

28b. Calder, K. M., and J. Lascelles. 1984. Analysis of cytoplasmic membrane from wild type and mutant *Paracoccus denitrificans* containing excess nitrate reductase and cytochrome b. Arch. Microbiol. 137:226–230.

29. Carlson, C. A. 1982. The physiological genetics of denitrifying bacteria. Antonie Leeuwenhoek J. Microbiol. 48:555–567.

30. Carlson, C. A., L. P. Ferguson, and J. L. Ingraham. 1982. Properties of dissimilatory nitrate reductase purified from the denitrifier *Pseudomonas aeruginosa*. J. Bacteriol. 151:162–171.

31. Caskey, W. H., and J. M. Tiedje. 1979. Evidence for clostridia as agents of dissimilatory reduction of nitrate to ammonium in soils. Soil Sci. Soc. Am. J. 43:931–936.

32. Chaudhry, G. R., I. M. Chaiken, and C. H. MacGregor. 1983. An activity from *Escherichia coli* membranes responsible for the modification of nitrate reductase to its precursor form. J. Biol. Chem. 258:5828–5833.

33. Chaudhry, G. R., and C. H. MacGregor. 1983. *Escherichia coli* nitrate reductase subunit A. Its role as the catalytic site and evidence for its modification. J. Bacteriol. 154:387–394.

34. Chaudhry, G. R., and C. H. MacGregor. 1983. Cytochrome b from *Escherichia coli* nitrate reductase. Its properties and association with the enzyme complex. J. Biol. Chem. 258:5819–5827.

35. Chippaux, M., V. Bonnefoy-Orth, J. Ratouchniak, and M.-C. Pascal. 1981. Operon

fusions in the nitrate reductase operon and study of the control gene *nir* R in *Escherichia coli*. Mol. & Gen. Genet. 182:477–479.

36. Chippaux, M., D. Giudici, A. Abou-Jaoudé, F. Casse, and M.-C. Pascal. 1978. A mutation leading to total lack of nitrite reductase activity in *Escherichia coli* K12. Mol. & Gen. Genet. 160:225–229.

37. Chippaux, M., and F. Pichinoty. 1970. Les nitrate-reductases bacteriennes V. Introduction de la biosynthèse de l'enzyme A par l'azoture. Arch. Microbiol. 71:361–366.

38. Claassen V. P. 1983. Studies on the molybdenum cofactor common to several enzymes. Vrije Universiteit, Amsterdam.

39. Claassen, V. P., L. F. Oltmann, C. W. Bettenhaussen, A. H. Stouthamer, J. van't Riet, F. A. Pinkse, R. H. Fokkens, and N. M. M. Nibbering. 1984. Mass spectrometric analysis of the fluorescent products originating from the molybdenum cofactor. Biochem. Int. 8:127–134.

40. Claassen, V. P., L. F. Oltmann, S. Bus, J. van't Riet, and A. H. Stouthamer. 1981. The influence of growth conditions on the synthesis of molybdenum cofactor in *Proteus mirabilis*. Arch. Microbiol. 130:44–49.

41. Claassen, V. P., L. F. Oltmann, C. E. M. Vader, J. van't Riet, and A. H. Stouthamer. 1982. Molybdenum cofactor from the cytoplasmic membrane of *Proteus mirabilis*. Arch. Microbiol. 133:283–288.

42. Claassen, V. P., L. F. Oltmann, J. van't Riet, U. A. Th. Brinkman, and A. H. Stouthamer. 1982. Purification of molybdenum cofactor and its fluorescent oxidation products. FEBS Lett. 142:133–137.

43. Clegg, R. A. 1976. Purification and some properties of nitrate reductase (EC 1.7.99.4) from *Escherichia coli* K12. Biochem. J. 153:533–541.

44. Cole, J. A., and C. M. Brown. 1980. Nitrite reduction to ammonia by fermentative bacteria: a short circuit in the biological nitrogen cycle. FEMS Microbiol. Lett. 7:65–72.

45. Cole, J. A., B. M. Newman, and P. White. 1980. Biochemical and genetic characterization of *nir* B mutants of *Escherichia coli* pleiotropically defective in nitrite and sulphite reduction. J. Gen. Microbiol. 120:475–483.

46. Coleman, K. J., A. Cornish-Bowden, and J. A. Cole. 1978. Purification and properties of nitrite reductase from *Escherichia coli* K12. Biochem J. 175:483–493.

47. Cove, D. J., 1979. Genetic studies of nitrate assimilation in *Aspergillus nidulans*. Biol. Rev. 54:291–327.

48. Cox, C. D., W. J. Payne, and D. DerVartanian. 1971. Electron paramagnetic resonance studies on the nature of hemoproteins in nitrite and nitric oxide reduction. Biochim. Biophys. Acta 253:290–294.

49. Cramer, S. P., R. Wahl, and K. V. Rajagopolan. 1981. Molybdenum sites of sulfite oxidase and xanthine dehydrogenase. A comparison by extended X-ray absorption fine structure. J. Am. Chem. Soc. 103:7721–7727.

50. Daniel, R. M., J. M. Smith, J. A. D. Phillip, H. D. Ratcliffe, J. W. Drozd, and A. T. Bull. 1980. Anaerobic growth and denitrification by *Rhizobium japonicum* and other rhizobia. J. Gen. Microbiol. 120:517–521.

51. De Groot, G. N., and A. H. Stouthamer. 1970. Regulation of reductase formation in

Proteus mirabilis. II. Influence of growth with azide and of haem deficiency on nitrate reductase formation. Biochim. Biophys. Acta 208:414–427.

52. De Groot, G. N., and A. H. Stouthamer. 1970. Regulation of reductase formation in *Proteus mirabilis*. III. Influence of oxygen, nitrate and azide on thiosulfate reductase and tetrathionate reductase formation. Arch. Microbiol. 74:326–339.

53. DeMoss, J. A. 1977. Limited proteolysis of nitrate reductase purified from membranes of *Escherichia coli*. J. Biol. Chem. 252:1696–1701.

54. DeMoss, J. A. 1978. Role of *chl* C gene in formation of the formate-nitrate reductase pathway in *Escherichia coli*. J. Bacteriol. 133:626–630.

55. DeMoss, J. A., T. Y. Fan, and R. Scott. 1981. Characterization of subunit structural alterations which occur during purification of nitrate reductase from *Escherichia coli*. Arch. Biochem. Biophys. 206:54–64.

56. De Vries, W., H. G. D. Niekus, M. Boellaard, and A. H. Stouthamer. 1980. Growth yields and energy generation by *Campylobacter sputorum* subspecies *bubulus* during growth in continuous culture with different hydrogen acceptors Arch. Microbiol. 124:221–227.

57. De Vries, W., H. G. D. Niekus, H. van Berchum, and A. H. Stouthamer. 1982. Electron-transport linked proton translocation at nitrite reduction in *Campylobacter sputorum* subspecies *bubulus*. Arch. Microbiol. 131:132–139.

58. De Vries, W., H. van Berchum, and A. H. Stouthamer. 1984. Localization of hydrogenase and nitrate reductase in *Campylobacter sputorum* subspecies *bubulus*. Antonie Leeuwenhoek J. Microbiol. 50:63–73.

59. Dixon, R. A., A. Alvarez-Morales, J. Clements, M. Drummond, M. Merrick, and J. R. Postgate. 1984. Transcriptional control of the nif regulon in *Klebsiella pneumoniae*, p. 635–642, in C. Veeger and W. E. Newton, (eds.), Advances in nitrogen fixation research. Martinus Nijhoff/W. Junk Publishers, The Hague.

60. Downie, J. A., and G. B. Cox. 1978. Sequence of b cytochrome relative to ubiquinone in the electron transport chain of *Escherichia coli*. J. Bacteriol. 133:477–484.

61. Edwards, E. S., S. S. Rondeau, and J. A. DeMoss. 1983. Chl c (nar) Operon of *Escherichia coli* includes structural genes for α and β subunits of nitrate reductase. J. Bacteriol. 153:1513–1520.

62. Enoch, H. G., and R. L. Lester. 1972. Effects of molybdate, tungstate and selenium compounds on formate dehydrogenase and other enzyme systems in *Escherichia coli*. J. Bacteriol. 110:1032–1040.

63. Enoch, H. G. and R. L. Lester. 1974. The role of a novel cytochrome b-containing nitrate reductase and quinone in the in vitro reconstruction of formate-nitrate reductase activity of *E. coli*. Biochem. Biophys. Res. Commun. 61:1234–1241.

64. Enoch, H. G., and R. L. Lester. 1975. The purification and properties of formate dehydrogenase and nitrate reductase from *Escherichia coli*. J. Biol. Chem. 250:6693–6705.

65. Federova R. I., E. I. Milekhina, and N. I. Il'yukhina. 1973. Evaluation of the method of ''gas metabolism'' for detecting extraterrestrial life. Identification of nitrogen-fixing micro-organisms. Izv. Akad. Nauk SSSR Ser. Biol. 6:797–806.

66. Ferguson, S. J., and M. C. Sorgato. 1982. Proton electrochemical gradients and energy-transduction processes. Annu. Rev. Biochem. 51:185–217.

67. Fimmel, A. L., and B. A. Haddock. 1979. Use of *chl* C-*lac* fusions to determine regulation of *chl*-C in *Escherichia coli* K12. J. Bacteriol. 138:726–730.

68. Firestone, M. K., R. B. Firestone, and J. M. Tiedje. 1979. Nitric oxide as an intermediate in denitrification: evidence from nitrogen-13 isotope exchange. Biochem. Biophys. Res. Commun. 91:10–16.

69. Firestone, M. K., and J. M. Tiedje. 1979. Temporal changes in nitrous oxide and dinitrogen from denitrification following onset of anaerobiosis. Appl. Environ. Microbiol. 38:673–679.

70. Forget, P. 1971. Les nitrate-réductases bactériennes. Solubilisation, purification et propriétés de l'enzyme A de *Micrococcus denitrificans*. Eur. J. Biochem. 18:442–458.

71. Forget, P. 1974. The bacterial nitrate reductases. Solubilization, purification and properties of the enzyme A of *Escherichia coli* K12. Eur. J. Biochem. 42:325–332.

72. Garber, E. A. E., D. Castignetti, and T. C. Hollocher. 1982. Proton translocation and proline uptake associated with reduction of nitric oxide by denitrifying *Paracoccus denitrificans*. Biochem. Biophys. Res. Commun. 107:1504–1507.

73. Garber, E. A. E., and T. C. Hollocher. 1981. [15]N tracer studies on the role of NO in denitrification. J. Biol. Chem. 256:5459–5465.

74. Garber, E. A. E., and T. C. Hollocher. 1982. Positional isotopic equivalence of nitrogen in N_2O produced by the denitrifying bacterium *Pseudomonas stutzeri*. Indirect evidence for a nitroxyl pathway. J. Biol. Chem. 257:4705–4708.

75. Garber, E. A. E., and T. C. Hollocher. 1982. [15]N ^{18}O-tracer studies on the activation of nitrite by denitrifying bacteria. Nitrite/water-oxygen exchange and nitrosation reactions as indicators of electrophilic catalysis. J. Biol. Chem. 257:8091–8097.

76. Garber, E. A. E., S. Wehrli, and T. C. Hollocher. 1983. ^{15}N-tracer and NMR studies on the pathway of denitrification. Evidence against trioxodinitrate but for nitroxyl as an intermediate. J. Biol. Chem. 258:3587–3591.

77. Garland, P. B., J. A. Downie, and B. A. Haddock. 1975. Proton translocation and the respiratory nitrate reductase of *Escherichia coli*. Biochem. J. 152:547–559.

78. Goldberger, R. F. 1974. Autogenous regulation of gene expression. Science 183:810–816.

79. Goto, M., A. Sakurai, K. Ohta, and H. Yamakani. 1969. Die Struktur des Urothions. J. Biochem. (Tokyo) 65:611–620.

80. Graham, A., and D. H. Boxer. 1978. Immunochemical localisation of nitrate reductase in *Escherichia coli*. Biochem. Soc. Trans. 6:1210–1211.

81. Graham, A., and D. H. Boxer. 1980. Arrangement of respiratory nitrate reductase in the cytoplasmic membrane of *Escherichia coli*: location of β subunit. FEBS Lett. 113:15–20.

82. Greenberg, E. P., and G. E. Becker. 1977. Nitrous oxide as end product of denitrification by strains of fluorescent pseudomonads. Can. J. Microbiol. 23:903–907.

83. Gudat, J. C., J. Singh, and D. C. Wharton. 1973. Cytochrome oxidase from *Pseudomonas aeruginosa*. I. Purification and some properties. Biochim. Biophys. Acta 292:376–390.

84. Hackett, C. S., and C. H. MacGregor. 1981. Synthesis and degradation of nitrate reductase in *Escherichia coli*. J. Bacteriol. 146:352–359.

85. Hackett, N. R., and P. D. Bragg. 1983. Membrane cytochromes of *Escherichia coli* grown aerobically and anaerobically with nitrate. J. Bacteriol. 154:708–718.

86. Hackett, N. R., and P. D. Bragg. 1983. Membrane cytochromes of *Escherichia coli* *chl* mutants. J. Bacteriol. 154:719–727.

87. Haddock, B. A. 1977. The isolation of phenotypic and genotypic variants for functional characterisation of bacterial oxidative phosphorylation. Symp. Soc. Gen. Microbiol. 27:95–120.

88. Haddock, B. A., and C. W. Jones. 1977. Bacterial respiration. Bacteriol. Rev. 41:47–99.

89. Hadjipetrou, L. P., and A. H. Stouthamer. 1965. Energy production during nitrate respiration by *Aerobacter aerogenes*. J. Gen. Microbiol. 38:29–34.

90. Hart, L. T., A. D. Larson, and C. S. McCleskey. 1965. Denitrification by *Corynebacterium nephridii*. J. Bacteriol. 89:1104–1108.

91. Hasan, S. M., and J. B. Hall. 1975. The physiological function of nitrate reduction in *Clostridium perfringens*. J. Gen. Microbiol. 87:120–128.

92. Hasan, S. M., and J. B. Hall. 1977. Dissimilatory nitrate reduction in *Clostridium tertium*. Z. Allg. Mikrobiol. 17:501–506.

93. Haselkorn, R., P. J. Lammers, D. Rice, and S. J. Robinson. 1984. Organization and transcription of nitrogenase genes in the cyanobacterium *Anabaena*, p. 653–659, in C. Veeger and W. E. Newton (eds.), Advances in nitrogen fixation research. Martinus Nijhoff/W. Junk Publishers, The Hague.

94. Henry, M.-F., and P. M. Vignias. 1983. Electron pathways from H_2 to nitrate in *Paracoccus denitrificans*. Effects of inhibitors of UQ-cytochrome b-region. Arch. Microbiol. 136:64–68.

95. Hewitt, E. J., B. A. Notton, and C. D. Garner. 1979. Nitrate reductases: Properties and possible mechanism. Biochem. Soc. Trans. 7:629–633.

96. Hoffman, P. S., and T. G. Goodman. 1982. Respiratory physiology and energy conservation efficiency of *Campylobacter jejuni*. J. Bacteriol. 150:319–326.

97. Hollocher, T. C. 1982. The pathway of nitrogen and reduction enzymes of denitrification. Antonie Leeuwenhoek J. Microbiol. 48:531–544.

98. Hollocher, T. C., E. Garber, A. J. L. Cooper, and R. E. Reiman. 1980. ^{13}N, ^{15}N isotope and kinetic evidence against hyponitrite as an intermediate in denitrification. J. Biol. Chem. 255:5027–5030.

99. Horio, T. T., T. Higashi, T. Yamanaka, H. Matsubara, and K. Okunuki. 1981. Purification and properties of cytochrome oxidase from *Pseudomonas aeruginosa*. J. Biol. Chem. 236:944–951.

100. Husain, M., and J. Sadana. 1974. Nitrite reductase from *Achromobacter fischeri*: molecular weight and subunit structure. Eur. J. Biochem. 42:283–289.

101. Iida, K., and S. Taniguchi. 1959. Studies on nitrate redictase system of *Escherichia coli*. I. Particulate electron transport system to nitrate and its solubilization. J. Biochem. (Tokyo) 46:1041–1055.

102. Iwasaki, H., and T. Matsubara. 1971. Cytochrome c-557 (551) and cytochrome cd. from *Alcaligenes faecalis*. J. Biochem. (Tokyo) 69:847–857.

103. Iwasaki, H., and T. Matsubara. 1972. A nitrate reductase from *Achromobacter cycloclastes*. *J. Biochem.* (Tokyo) 71:645–652.

104. Iwasaki, H., S. Noji, and S. Shidara. 1975. *Achromobacter cycloclastes* nitrite reductase. The function of copper, amino acid composition and ESR spectra. J. Biochem. (Tokyo) 78:355–361.

105. Iwasaki, H., T. Saigo, and T. Matsubara. 1981. Copper as a controlling factor of anaerobic growth under N_2O and biosynthesis of N_2O reductase in denitrifying bacteria. Plant Cell Physiol. 21:1573–1584.

106. Iwasaki, H., S. Shidara, H. Suzuki, and T. Mori. 1963. Studies on denitrification. VIII. Further purification and properties of denitrifying enzyme. J. Biochem. (Tokyo) 53:299–303.

107. Jackson, M. A., J. B. Jackson, and S. J. Ferguson. 1981. Direct observation with an electrode of uncoupler-sensitive assimilatory nitrate uptake by *Rhodopseudomonas capsulata*. FEBS Lett. 136:275–278.

108. Jackson, R. H., A. Cornish-Bowden, and J. A. Cole. 1981. Prosthetic groups of the NADH-dependent nitrite reductase from *Escherichia coli* K12. Biochem. J. 193:861–867.

109. John, P. 1977. Aerobic and anaerobic bacterial respiration monitored by electrodes. J. Gen. Microbiol. 98:231–238.

110. John, P., and F. R. Whatley. 1970. Oxidative phosphorylation coupled to oxygen uptake and nitrate reduction in *Micrococcus denitrificans*. Biochim. Biophys. Acta 216:342–352.

111. John, P., and F. R. Whatley. 1975. *Paracoccus denitrificans* and the evolutionary origin of the mothochondrion. Nature (Lond.) 254:495–498.

112. Johnson, J. L., B. E. Hainline, and K. V. Rajagopalan. 1980. Characterization of the molybdenum cofactor of sulfite oxidase, xanthine oxidase and nitrate reductase. Identification of a pteridine as a structural component. J. Biol. Chem. 255:1783–1786.

113. Johnson, J. L., and K. V. Rajagopalan. 1982. Structural and metabolic relationship between the molybdenum cofactor and urothione. Proc. Natl. Acad. Sci. USA 79:6856–6860.

114. Johnson, M. K., A. J. Thomson, T. A. Walsh, D. Barber, and C. Greenwood. 1980. Electron paramagnetic resonance studies on *Pseudomonas* nitrosyl nitrite reductase. Evidence for multiple species in the electron paramagnetic spectra of nitrosyl haemoproteins. Biochem J. 189:285–294.

115. Jones, R. W., A. Lamont, and P. B. Garland. 1980. The mechanism of proton translocation driven by the respiratory nitrate reductase complex of *Escherichia coli*. Biochim. J. 190:79–94.

116. Justin, P., and D. P. Kelly. 1978. Growth kinetics of *Thiobacillus denitrificans* in anaerobic and aerobic chemostat culture. J. Gen. Microbiol. 107:123–130.

117. Kapralek, F., E. Jechova, and M. Otavova. 1982. Two sites of oxygen control in induced synthesis of respiratory nitrate reductase in *Escherichia coli*. J. Bacteriol. 149:1142–1145.

118. Kapralek, F., and F. Pichinoty. 1970. The effects of oxygen on tetrathionate reductase activity and biosynthesis. J. Gen. Microbiol. 62:95–105.

119. Kashket, E. R. 1982. Stoichiometry of the H^+-ATPase of growing and resting, aerobic *Escherichia coli*. Biochemistry 21:5534–5538.

120. Kashket, E. R., 1983. Stoichiometry of the H⁺-ATPase of *Escherichia coli* cells during anaerobic growth. FEBS Lett. 154:343–346.

121. Kaspar, H. F. 1982. Nitrite reduction to nitrous oxide by propionibacteria. Detoxification mechanism. Arch. Microbiol. 133:126–130.

122. Kaspar, H. F., and J. M. Tiedje. 1981. Dissimilatory reduction of nitrate and nitrite in the bovine rumen: nitrous oxide production and effect of acetylene. Appl. Environ. Microbiol. 41:705–709.

123. Keith, S. M., and R. A. Herbert. 1983. Dissimilatory nitrate reduction by a strain of *Desulfovibrio desulfuricans*. FEMS Microbiol. Lett. 18:55–59.

124. Kell, D. B., P. John, and S. J. Ferguson 1978. The protonmotive force in phosphorylating membrane vesicles from *Paracoccus denitrificans*. Biochem. J. 174:257–266.

125. Ketchum, P. A., H. A. Cambier, W. A. Frazier III, H. A. Madansky, and A. Nason. 1970. In vitro assembly of a *Neurospora* assimilatory nitrate reductase from protein subunits of a *Neurospora* mutant and the xanthine oxidizing or aldehyde oxidase systems of higher animals. Proc. Natl. Acad. Sci. USA 66:1016–1023.

126. Kiss, G. B., E. Vincze, Z. Kalman, T. Forrai, and A. Kondorosi. 1979. Genetic and biochemical analysis of mutants affected in nitrate reduction of *Rhizobium meliloti*. J. Gen Microbiol. 113:105–118.

127. Knowles, R. 1982. Denitrification. Microbiol. Rev. 46:43–70.

128. Koike, I., and A. Hattori. 1975. Growth yield of a denitrifying bacterium *Pseudomonas denitrificans* under aerobic and denitrifying conditions. J. Gen. Microbiol. 88:1–10.

129. Koike, I., and A. Hattori. 1975. Energy yield of denitrification: an estimate from growth yield in continuous cultures of *Pseudomonas denitrificans* under nitrate-, nitrite- and nitrous oxide-limited conditions. J. Gen Microbiol. 88:11–19.

130. Kristjansson, J. K., and T. C. Hollocher. 1980. First practical assay for soluble nitrous oxide reductase of denitrifying bacteria and a partial kinetic characterization. J. Biol. Chem. 255:704–707.

131. Kristjansson, J. K., and T. C. Hollocher. 1981. Partial purification and characterization of nitrous oxide reductase from *Paracoccus denitrificans*. Curr. Microbiol. 6:247–251.

132. Kristjansson, J. K., B. Walter, and T. C. Hollocher. 1978. Respiration-dependent proton translocation and the transport of nitrate and nitrite in *Paracoccus denitrificans* and other denitrifying bacteria. Biochemistry 17:5014–5019.

133. Kröger, A. 1977. Phosphorylative electron transport with fumarate and nitrate as terminal hydrogen acceptors. Symp. Soc. Gen. Microbiol. 27:61–93.

134. Kröger, A., E. Dorrer, and E. Winkler. 1980. The orientation of the substrate sites of formate dehydrogenase and fumarate reductase in the membrane of *Vibrio succinogenes*. Biochim. Biophys. Acta 589:118–136.

135. Kröger, A., and E. Winkler. 1981. Phosphorylative fumarate reduction in *Vibrio succinogenes*: stoichiometry of ATP synthesis. Arch. Microbiol. 129:100–104.

136. Krul, J. M., and R. Veeningen. 1977. The synthesis of the dissimilatory nitrate reductase under aerobic conditions in a number of denitrifying bacteria isolated from activated sludge and drinking water. Water Res. 11:39–43.

137. Kucera, J., and V. Dadak. 1983. The effect of uncoupler on the distribution of the

electron flow between the terminal acceptors oxygen and nitrite in the cells of *Paracoccus denitrificans*. Biochem. Biophys. Res. Commun. 117:252–258.

138. Kucera, J., V. Dadak, and T. Dobry. 1983. The distribution of redox equivalents in the anaerobic respiratory chain of *Paracoccus denitrificans*. Eur. J. Biochem. 130:359–364.

139. Kucera, J., P. Karlovsky, and V. Dadak. 1981. Control of nitrate respiration in *Paracoccus denitrificans*. FEMS Microbiol. Lett. 12:391–394.

140. Kucera, J., J. Lancik, and V. Dadak. 1983. The function of cytoplasmic membrane of *Paracoccus denitrificans* in controlling the rate of reduction of terminal acceptors. Eur. J. Biochem. 136:135–140.

141. Kuronen, T., and N. Effolk. 1972. A new purification procedure and molecular properties of *Pseudomonas* cytochrome oxidase. Biochim. Biophys. Acta 275:308–318.

142. Kuronen, T., M. Saraste, and N. Ellfolk. 1975. The subunit structure of *Pseudomonas* cytochrome oxidase. Biochim. Biophys. Acta 393:48–54.

143. Lafferty, M. A., and R. H. Garrett. 1974. Purification and properties of the *Neurospora crassa* assimilatory nitrite reductase. J. Biol. Chem. 249:7555–7567.

144. Lam, Y., and D. J. D. Nicholas. 1969. A nitrite reductase from *Micrococcus denitrificans*. Biochim. Biophys. Acta 178:225–234.

145. Lam, Y., and D. J. D. Nicholas. 1969. A nitrite reductase with cytochrome oxidase activity from *Micrococcus denitrificans*. Biochim. Biophys. Acta 180:459–472.

146. Lambden, P. R., and J. R. Guest. 1976. Mutants of *Escherichia coli* K12 unable to use fumarate as an anaerobic electron acceptor. J. Gen. Microbiol. 97:145–160.

147. Lawford, H. G., and B. A. Haddock. 1973. Respiration-driven proton translocation in *Escherichia coli*. Biochem. J. 136:217–220.

148. Lee, K. Y., R. Erickson, S. S. Pan, G. Jones, F. May, and A. Nason. 1974. Effect of tungsten and vanadium on the *in vitro* assembly of assimilatory nitrate reductase utilizing *Neurospora* mutant *nit*-1. J. Biol. Chem. 249:3941–3952.

149. Lee, K. Y., S. S. Pan, R. Erickson, and A. Nason. 1974. Involvement of molybdenum and iron in the *in vitro* assembly of assimilatory nitrate reductase utilizing *Neurospora* mutant *nit*-1. J. Biol. Chem. 249:3941–3952.

150. LeGall, J., W. J. Payne, T. V. Morgan, and D. DerVertanian. 1979. On the purification of nitrate reductase from *Thiobacillus denitrificans* and its reaction with nitrite under reducing conditions. Biochem. Biophys. Res. Commun. 87:355–362.

151. Liu, M. C., D. DerVertanian, and H. D. Peck, Jr. 1980. On the nature of the oxidation-reduction properties of nitrite reductase from *Desulfovibrio desulfuricans*. Biochem. Biophys. Res. Commun. 96:278–285.

152. Liu, M. C., and H. D. Peck, Jr. 1981. The isolation of a hexaheme cytochrome from *Desulfovibrio desulfuricans* and its identification of a new type of nitrite reductase. J. Biol. Chem. 256:13159–13164.

153. Liu, M. C., H. D. Peck, Jr., A. Abou-Jaoudé, M. Chippaux, and J. LeGall. 1981. A reappraisal of the role of the low potential c-type cytochrome (cytochrome *c*-552) in NADH-dependent nitrite reduction and its relationship with co-purified NADH oxidase in *Escherichia coli* K12. FEMS Microbiol. Lett. 10:333–337.

154. Lund, K., and J. A. DeMoss. 1976. Association-dissociation behavior and sub-unit structure of heat-released nitrate reductase from *E. coli*. J. Biol. Chem. 251:2207–2216.

155. MacGregor, C. H. 1975. Solubilization of *Escherichia coli* nitrate reductase by a membrane-bound protease. J. Bacteriol. 121:1102–1110.

156. MacGregor, C. H. 1975. Anaerobic cytochrome b, in *Escherichia coli*: association with and regulation of nitrate reductase. J. Bacteriol. 121:1111–1116.

157. MacGregor, C. H. 1975. Synthesis of nitrate reductase components in chlorate-resistant mutants of *Escherichia coli*. J. Bacteriol. 121:1117–1121.

158. MacGregor, C. H. 1976. Biosynthesis of membrane-bound nitrate reductase in *Escherichia coli*: evidence for a soluble precursor. J. Bacteriol. 126:122–131.

159. MacGregor, C. H., and A. R. Christopher. 1978. A symmetric distribution of nitrate reductase subunits in the cytoplasmic membrane of *Escherichia coli*. Evidence derived from surface labeling studies with transglutaminase. Arch. Biochem. Biophys. 185:204–213.

160. MacGregor, C. H., A. Schnaitman, D. E. Normansell, and M. G. Hodgins. 1974. Purification and properties of nitrate reductase from *Escherichia coli* K12. J. Biol. Chem. 249:5321–5327.

161. Matsubara, T. 1975. The participation of cytochromes in the reduction of N_2O to N_2 by a denitrifying bacterium. J. Biochem. (Tokyo) 77:627–632.

162. Matsubara, T., and Iwasaki, H. 1972. Nitrix oxide-reducing activity of *Alcaligenes faecalis* cytochrome cd. J. Biochem. (Tokyo) 72:57–64.

163. Matsubara, T., K. Frunzke, and W. G. Zumft. 1982. Modulation by copper of the products of nitrite respiration in *Pseudomonas perfectomarinus*. J. Bacteriol. 149:816–823.

164. Matsubara, T., and W. G. Zumft. 1982. Identification of a copper protein as part of the nitrous oxide-reducing system in nitrite-respiring (denitrifying) Pseudomonad. Arch. Microbiol. 132:322–328.

165. McCarthny, J. E. G., and S. J. Ferguson. 1983. Characterisation of membrane vesicles from *Paracoccus denitrificans* and measurements of the effect of partial uncoupling on their thermodynamics of oxidative phosphorylation. Eur. J. Biochem. 132:417–424.

166. McCarthny, J. E. G., S. J. Ferguson, and D. B. Kell. 1981. Estimation with an ion-selective in cells of *Paracoccus denitrificans* from the uptake of butyltriphenylphosphonium cation during aerobic and anaerobic respiration. Biochem. J. 196:311–321.

166a. McEwan, A. G., J. B. Jackson, and S. J. Ferguson (1984). Rationalization of properties of nitrate reductases in *Rhodopseudomonas capsulata*. Arch. Microbiol. 137:344–349.

167. Meiberg, J. B. M., P. M. Bruinenberg, and W. Harder. 1980. Effect of dissolved oxygen tension on the metabolism of methylated amines in *Hyphomicrobium* X in the absence and presence of nitrate: evidence for aerobic denitrification. J. Gen. Microbiol. 120:453–463.

168. Meijer, E. M., J. W. van der Zwaan, and A. H. Stouthamer. 1979. Location of the proton consuming site in nitrite reduction and stoichiometries for proton pumping in aerobically grown *Paracoccus denitrificans*. FEMS Microbiol. Lett. 5:369–372.

169. Meijer, E. M., J. W. van der Zwaan, R. Wever, and A. H. Stouthamer. 1979. Anaerobic respiration and energy conservation in *Paracoccus denitrificans*. Functioning of iron–sulfur centers and the uncoupling effect of nitrite. Eur. J. Biochem. 96:69–76.

170. Meijer, E. M., H. W. van Verseveld, E. G. van der Beek, and A. H. Stouthamer. 1977. Energy conservation during aerobic growth in *Paracoccus denitrificans*. Arch. Microbiol. 112:25–34.

171. Meijer, E. M., R. Wever, and A. H. Stouthamer. 1977. The role of iron–sulfur center 2 in electron transport and energy conservation in the NADH-ubiquinone segment of the respiratory chain of *Paracoccus denitrificans*. Eur. J. Biochem. 81:267–275.

172. Michalski, W. P., and D. J. D. Nicholas. 1984. The adaptation of *Rhodopseudomonas sphaeroides* f. sp. *denitrificans* for growth under denitrifying conditions. J. Gen. Microbiol. 130:155–165.

173. Miller, J. B., and N. K. Amy. 1983. Molybdenum cofactor chlorate-resistant and nitrate reductase-deficient insertion mutants of *Escherichia coli*. J. Bacteriol. 155:793–801.

174. Mitschell, P. 1976. Possible mechanisms of the protonmotion function of cytochrome systems. J. Theor. Biol. 62:327–367.

175. Motteram, P. A. S., J. E. G. McCarthny, S. J. Ferguson, J. B. Jackson, and J. A. Cole. 1981. Energy conservation during the formate-dependent reduction of nitrite by *Escherichia coli*. FEMS Microbiol. Lett. 12:317–320.

176. Muhoberac, B. M., and D. C. Wharton. 1980. EPR study of heme. NO complexes of ascorbic acid-reduced *Pseudomonas* cytochrome oxidase and corresponding model complexes. J. Biol. Chem. 255:8437–8442.

177. Murphy, M. J., L. M. Siegel, S. R. Tove, and H. Kamin. 1974. Siroheme: a new prosthetic group participating in six-electron reduction reactions catalyzed by both sulfite and nitrite reductases. Proc. Natl. Acad. Sci. USA 71:612–616.

178. Nason, A., A. D. Antonine, P. A. Ketchum, W. A. Frazier III, and D. K. Lee. 1970. Formation of assimilatory nitrate reductase by *in vitro* intercistronic complementation in *Neurospora crassa*. Proc. Natl. Acad. Sci. USA 65:137–144.

179. Nason, A., D. K. Lee, S. S. Pan, P. A. Ketchum, A. Lamberti, and J. De Vries. 1971. *In vitro* formation of assimilatory reduced nicotinamide adenine dinucleotide phosphate: nitrate reductase from a *Neurospora* mutant and a component of molybdenum enzymes. Proc. Natl. Acad. Sci. USA 68:3242–3246.

180. Newman, B. M., and J. A. Cole. 1978. The chromosomal location and pleiotropic effects of mutations of the *nir* A$^+$ gene of *Escherichia coli*: the essential role of *nir* A$^+$ in nitrite reduction and in other anaerobic redox reactions. J. Gen. Microbiol. 106:1–12.

181. Niedermann, R. A., and M. J. Wolin. 1972. Requirement of succinate for growth of *Vibrio succinogenes*. J. Bacteriol. 109:546–549.

182. Niekus, H. G. D., W. de Vries, and A. H. Stouthamer. 1977. The effect of different dissolved oxygen tensions on growth and enzyme activities of *Campylobacter sputorum* subspecies *bubulus*. J. Gen. Microbiol. 103:215–222.

183. Niekus, H. G. D., E. van Doorn, W. de Vries, and A. H. Stouthamer. 1980. Aerobic growth of *Campylobacter sputorum* subspecies *bubulus* with formate. J. Gen. Microbiol. 118:419–428.

184. Odom, J. M., and H. D. Peck, Jr. 1981. Localization of dehydrogenes, reductases and electron transfer components in the sulfate-reducing bacterium *Desulfovibrio gigas*. J. Bacteriol. 147:161–169.

185. Oltmann, L. F., V. P. Claassen, P. Kastelein, W. N. M. Reijnder, and A. H.

Stouthamer. 1979. Influence of tungstate on the formation and activities of four reductases in *Proteus mirabilis*. Identification of two new molybo-enzymes: chlorate reductase and tetrathionate reductase. FEBS Lett. 106:43–46.

186. Oltmann, L. F., W. N. M. Reijnders, and A. H. Stouthamer. 1976. Characterization of purified nitrate reductase A and chlorate reductase C from *Proteus mirabilis*. Arch. Microbiol. 111:25–35.

187. Oltmann, L. F., W. N. M. Reijnders, and A. H. Stouthamer. 1976. The correlation between the protein composition of cytoplasmic membranes and the formation of nitrate reductase A, chlorate reductase C and tetrathionate reductase in *Proteus mirabilis* wild type and some chlorate-resistant mutants. Arch. Microbiol. 111:37–43.

188. Pan, S. S., and A. Nason. 1978. Purification and characterization of homogeneous assimilatory NADPH-dependent nitrate reductase from *Neurosporora crassa*. Biochim. Biophys. Acta 523:297–313.

189. Parsonage, D., and S. J. Ferguson. 1983. Reassessment of pathways of electron flow to nitrate reductase that are coupled to energy conservation in *Paracoccus denitrificans*. FEBS Lett. 153:108–112.

190. Pascal, M.-C., J.-F. Burini, J. Ratouchniak, and M. Chippaux. 1982. Regulation of the nitrate reductase operon: effect of mutations in *chl* A, B, D, and E genes. Mol. & Gen. Genet. 188:103–106.

191. Pascal, M.-C., and M. Chippaux. 1982. Involvement of a gene of the *chl* E locus in the regulation of the nitrate reductase operon. Mol. & Gen. Genet. 185:334–338.

192. Pateman, J. A., D. J. Cove, B. M. Rever, and D. B. Roberts. 1964. A common cofactor for nitrate reductase and xanthine dehydrogenase which also regulates the synthesis of nitrate reductase. Nature (Lond.) 201:58–60.

193. Payne, W. J. 1981. Denitrification. Wiley, New York.

194. Perkings, D. D., A. Radford, D. Newmeijer, and M. Björkman. 1982. Chromosomal loci of *Neurospora crassa*. Microbiol. Rev. 46:426–470.

195. Pichinoty, F., and J. C. Senez. 1956. Réactivation par le cytochrome c_3 de l'hydroxylamine- et de la nitrite-réductase des bactéries sulfato-réductrices. C. R. Soc. Biol. 150:744–745.

196. Poole, R. K. 1983. Bacterial cytochrome oxidases. A structurally and functionally diverse group of electron-transfer proteins. Biochim. Biophys. Acta 726:205–243.

197. Pope, N. R., and J. A. Cole. 1982. Generation of a membrane potential by one or two independent pathways for nitrite reduction by *Escherichia coli*. J. Gen. Microbiol. 128:219–222.

198. Prakash, D., and J. C. Sadana. 1972. Purification, characterisation and properties of nitrite reductase of *Achromobacter fischeri*. Arch. Biochem. Biophys. 148:614–632.

199. Radcliffe, B. C., and D. J. D. Nicholas. 1970. Some properties of a nitrate reductase from *Pseudomonas denitrificans*. Biochim. Biophys. Acta 205:273–287.

200. Rake, J. B., and R. G. Eagon. 1980. Inhibition, but not uncoupling of respiratory energy coupling of three bacterial species by nitrite. J. Bacteriol. 144:975–982.

201. Reuner, E. D., and G. E. Becker. 1970. Production of nitric oxide and nitrous oxide during denitrification by *Corynebacterium nephridii*. J. Bacteriol. 101:821–826.

202. Robertson, L. A., and J. G. Kuenen. 1983. *Thiosphaera pantotropha* gen. nov. sp. nov., a facultatively anaerobic, facultatively autotrophic sulphur bacterium. J. Gen. Microbiol. 129:2847–2855.

203. Robertson, L. A., and J. G. Kuenen. 1984. Aerobic denitrification: a controversy revived. Arch. Microbiol. 139:351–354.

204. Rosso, J. P., P. Forget, and F. Pichinoty. 1973. Les nitrate-réductases bactériennes. Solubilisation, purification et propriétés de l'enzyme A de *Micrococcus halodenitrificans*. Biochim. Biophys. Acta. 321:443–455.

205. Rowe, J. J., J. M. Yarbrough, J. B. Rake, and R. G. Eagon. 1979. Nitrite inhibition of aerobic bacteria. Curr. Microbiol. 2:51–54.

206. Satoh, T., S. S. M. Hom, and K. T. Shanmugam. 1983. Production of nitrous oxide from nitrite in *Klebsiella pneumoniae*: mutants altered in nitrogen metabolism. J. Bacteriol. 155:454–458.

207. Sawada, E., T. Satoh, and H. Kitamura. 1978. Purification and properties of a dissimilatory nitrite reductase of a denitrifying phototrophic bacterium. Plant. Cell. Physiol. 19:1339–1351.

207a. Sawada, E., and T. Satoh. 1980. Periplasmic location of dissimilatory nitrate and nitrite reductases in a denitrifying phototrophic bacterium, *Rhodopseudomonas sphaeroides* forma sp. *denitrificans*. Plant Cell Physiol. 24:501–508.

208. Sawnhey, V., and D. J. D. Nicholas. 1977. Sulphite- and NADH-dependent nitrate reductase from *Thiobacillus denitrificans*. J. Gen. Microbiol. 100:49–58.

209. Sawnhey, V., and D. J. D. Nicholas. 1978. Sulphide-linked nitrite reductase from *Thiobacillus denitrificans* with cytochrome oxidase activity: purification and properties. J. Gen. Microbiol. 106:119–128.

210. Schulp, J. A., and A. H. Stouthamer. 1970. The influence of oxygen, glucose and nitrate upon the formation of nitrate reductase and the respiratory system in *Bacillus licheniformis*. J. Gen. Microbiol. 64:195–203.

211. Scott, R. H., and J. A. DeMoss. 1976. Formation of the formate nitrate electron transport pathway from inactive components in *Escherichia coli*. J. Bacteriol. 126:478–486.

212. Scott, R. H., G. T. Sperl, and J. A. DeMoss. 1979. *In vitro* incorporation of molybdate into demolybdo-proteins in *Escherichia coli*. J. Bacteriol. 137:719–726.

213. Scott, R. I., and R. K. Poole. 1982. A reexamination of the cytochromes of *Escherichia coli* using fourth-order finite difference analysis: their characterization under different growth conditions and accumulation during the cell cycles. J. Gen Microbiol. 128:1685–1696.

214. Senez, J. C., and F. Pichinoty. 1985. Sur la réduction du nitrite aux dépens de l'hydrogène moléculaire par *Desulfovibrio desulfuricans* et d'autres espèces bactériennes. Bull. Soc. Chim. Biol. 40:2099–2117.

215. Shipp, W. S. 1972. Cytochromes of *Escherichia coli*. Arch. Biochem. Biophys. 150:459–472.

216. Showe, M. K., and J. A. DeMoss. 1968. Localization and regulation of synthesis of nitrate reductase in *Escherichia coli*. J. Bacteriol. 95:1305–1313.

217. Siegel, L. M., M. J. Murphy, and H. Kamin. 1973. Reduced nicotinamide adenine dinucleotide phosphate-sulfite reductase of enterobacteria. I. The *Escherichia coli* hemoflavoprotein: molecular parameters and prosthetic groups. J. Biol. Chem. 248:251–264.

218. Sinclair, P. R., and D. C. White. 1970. Effect of nitrate, fumarate and oxygen on

the formation of membrane-bound electron transport system of *Haemophilus influenzae*. J. Bacteriol. 101:365–372.

219. Slater, E. C. 1983. The Q cycle, a ubiquitous mechanism of electron transfer. Tr. Biochem. Sci. 8:239–242.

220. Smith, M. S. 1982. Dissimilatory reduction of NO_2^- to NH_4^+ and N_2O by a soil *Citrobacter* sp. Appl. Environ. Microbiol. 43:854–860.

221. Smith, M. S. 1983. Nitrous oxide production by *Escherichia coli* is correlated with nitrate reductase activity. Appl. Environ. Microbiol. 45:1545–1547.

222. Smith, M. S., and K. Zimmerman. 1981. Nitrous oxide production by nondenitrifying soil nitrate reducers. Soil Sci. Soc. Am. J. 45:865–871.

222a. Snijder, S. W., and T. C. Hollocher. 1984. Nitrous oxide reductase and the 120.000 MW copper protein of N_2-producing denitrifying bacteria are different entities. Biochem. Biophys. Res. Commun. 119:588–592.

223. Sörensen, J., J. M. Tiedje, and R. B. Firestone. 1980. Inhibition by sulfide of nitric and nitrous oxide reduction by denitrifying *Pseudomonas fluorences*. Appl. Environ. Microbiol. 39:105–108.

224. Steenkamp, D. J., and H. D. Peck, Jr. 1980. The association of hydrogenase and dithionite reductase activities with the nitrite reductase of *Desulfovibrio desulfuricans*. Biochem. Biophys. Res. Commun. 94:41–48.

225. Steenkamp, D. J., and H. D. Peck, Jr. 1981. Proton translocation associated with nitrite respiration in *Desulfovibrio desulfuricans*. J. Biol. Chem. 256:5450–5458.

226. Stewart, V. 1982. Requirement of *fnr* and *nar* L functions for nitrate reductase expression in *Escherichia coli* K12. J. Bacteriol. 151:1320–1325.

227. Stewart, V., and C. H. MacGregor. 1982. Nitrate reductase in *Escherichia coli* K12: involvement of *chl* C, *chl* E, and *chl* G loci. J. Bacteriol. 151:788–799.

228. Stiefel, E. I. 1973. Proposed molecular mechanism for the action of molybdenum in enzymes: coupled proton and electron transfer. Proc. Natl. Acad. Sci. USA 70:988–992.

229. St. John, R. T., and T. C. Hollocher. 1977. Nitrogen 15 tracer studies on the pathway of denitrification in *Pseudomonas aeruginosa*. J. Biol. Chem. 252:212–218.

230. Stouthamer, A. H. 1967. Mutant strains of *Aerobacter aerogenes* which require both methionine and lysine for aerobic growth. J. Gen. Microbiol. 46:389–398.

231. Stouthamer, A. H. 1976. Biochemistry and genetics of nitrate reduction in bacteria. Adv. Microb. Physiol. 14:315–375.

232. Stouthamer, A. H. 1977. Theoretical calculations on the influence of the inorganic nitrogen source on parameters for aerobic growth of microorganisms. Antonie Leeuwenhoek J. Microbiol. Serol. 43:351–367.

233. Stouthamer, A. H. 1980. Bioenergetic studies on *Paracoccus denitrificans*. Tr. Biochem. Sci. 5:164–166.

234. Stouthamer, A. H., F. C. Boogerd, and H. W. van Verseveld. 1982. The bioenergetics of denitrification. Antonie Leeuwenhoeck J. Microbiol. 48:545–553.

235. Stouthamer, A. H., W. de Vries, and H. G. D. Niekus. 1979. Microaerophily. Antonie Leeuwenhoek J. Microbiol. Serol. 45:5–12.

236. Stouthamer, A. H., J. A. van't Riet, and L. F. Oltmann. 1980. Respiration with

nitrate as acceptor, p. 19–48, in C. J. Knowles (ed.), Diversity in bacterial respiratory systems, Vol. 2. CRC Press, Boca Raton, Fl.

237. Tam, T.-Y., and R. Knowles. 1979. Effects of sulfide and acetylene on nitrous oxide reduction by soil and by *Pseudomonas aeruginosa*. Can. J. Microbiol. 25:1133–1138.

238. Taniguchi, S., and E. Itagaki. 1960. Nitrate reductase of nitrate respiration type from *Escherichia coli*. I. Solubilization and purification from the particulate system with molecular characterization as a metalloprotein. Biochim. Biophys. Acta 44:263–279.

239. Taumer, A. C. R., S. Badger, C. H. Lai, M. A. Listgarten, R. A. Visconti, and S. S. Socranski. 1981. *Wolinella* gen. nov., *Wolinella recta* sp. nov., *Campylobacter concisus* sp. nov. and *Eikenella corrodens* from humans with peridontal disease. Int. J. Syst. Bacteriol. 31:432–445.

240. Thauer, R. K., R. K. Jungermann, and K. Decker. 1977. Energy conservation in the chemotrophic anaerobic bacteria. Bacteriol. Rev. 41:100–180.

241. Timkovich, R., R. Dhesi, K. J. Martinkus, M. K. Robinson, and T. M. Rea. 1982. Isolation of *Paracoccus denitrificans* cytochrome cd 1: comparative kinetics with other nitrite reductases. Arch. Biochem. Biophys. 215:47–58.

242. Tomsett, A. B., and R. H. Garrett. 1981. Biochemical analysis of mutants defective in nitrate assimilation in *Neurospora crassa*. Evidence for autogenous control by nitrate reductase. Mol. & Gen. Genet. 184:183–190.

243. Van Hartingsveldt, J., M. G. Marinus, and A. H. Stouthamer. 1971. Mutants of *Pseudomonas aeruginosa* blocked in nitrate or nitrite dissimilation. Genetics 67:469–482.

244. Van Hartingsveldt, J., and A. H. Stouthamer. 1973. Mapping and characterization of mutants of *Pseudomonas aeruginosa* affected in nitrate respiration in aerobic or anaerobic growth. J. Gen. Microbiol. 74:97–106.

245. Van Hartingsveldt, J., and A. H. Stouthamer. 1974. Properties of a mutant of *Pseudomonas aeruginosa* affected in aerobic growth. J. Gen. Microbiol. 83:303–310.

246. van't Riet, J. A., and R. J. Planta. 1975. Purification, structure and properties of the respiratory nitrate reductase of *Klebsiella aerogenes*. Biochim. Biophys. Acta 379:81–94.

247. van't Riet, J. A., A. H. Stouthamer, and R. J. Planta. 1968. Regulation of nitrate assimilation and nitrate respiration in *Aerobacter aerogenes*. J. Bacteriol. 96:1455–1464.

248. van't Riet, J. A., J. H. van Ee, R. Wever, B. F. van Gelder, and R. J. Planta. 1975. Characterization of the respiratory nitrate reductase of *Klebsiella aerogenes* as a molybdenum-containing iron–sulfur enzyme. Biochim. Biophys. Acta 405:306–317.

249. van't Riet, J. A., F. B. Wientjes, J. van Doorn, and R. J. Planta. 1979. Purification and characterization of the respiratory nitrate reductase of *Bacillus lichenformis*. Biochim. Biophys. Acta 576:347–360.

250. van Verseveld, H. W., J. P. Boon, and A. H. Stouthamer. 1979. Growth yields and the efficiency of oxidative phosphorylation of *Paracoccus denitrificans* during two (carbon) substrate-limited growth. Arch. Microbiol. 121:213–223.

251. van Verseveld, H. W., M. Braster, F. C. Boogerd, B. Chance, and A. H. Stouthamer. 1983. Energetic aspects of growth of *Paracoccus denitrificans*: oxygen-limitation and

shift from anaerobic nitrate-limitation to aerobic succinate-limitation. Evidence for a new alternative oxidase, cytochrome a 1. Arch. Microbiol. 135:229–236.

252. van Verseveld, H. W., K. Krab, and A. H. Stouthamer. 1981. Proton pump coupled to cytochrome *c* oxidase in *Paracoccus denitrificans*. Biochim. Biophys. Acta 635:525–534.

253. van Verseveld, H. W., E. M. Meijer, and A. H. Stouthamer. 1977. Energy conservation during nitrate respiration in *Paracoccus denitrificans*. Arch. Microbiol. 112:17–23.

254. van Verseveld, H. W., and H. Stouthamer. 1978. Electron transport chain and coupled oxidative phosphorylation in methanol-grown *Paracoccus denitrificans*. Arch. Microbiol. 118:13–20.

255. Van Wielink, J. E., L. F. Oltmann, J. Leeuwerik, J. A. de Hollander, and A. H. Stouthamer. 1982. A method for in situ characterization of *b*- and *c*-type cytochromes in *Escherichia coli* and in complex III from beef heart mitochondria by combined spectrum deconvolution and potentiometric analysis. Biochim. Biophys. Acta 681:177–190.

256. Van Wielink, J. E., W. N. M. Reijnders, L. F. Oltmann, and A. H. Stouthamer. 1983. Electron transport and cytochromes in aerobically grown *Proteus mirabilis*. Arch. Microbiol. 136:152–157.

257. Van Wielink, J. E., W. N. M. Reijnders, L. F. Oltmann, and A. H. Stouthamer. 1983. The characterization of the membrane-bound- *b*- and *c*-type cytochromes of differently grown *Escherichia coli* cells by means of coupled potentiometric analysis and spectrum deconvolution. FEMS Microbiol. Lett. 18:167–172.

258. Vega, J. M., R. H. Garrett, and L. M. Siegel. 1975. Siroheme: a prosthetic group of the *Neurospora crassa* assimilatory nitrite reductase. J. Biol. Chem. 250:7980–7989.

259. Vincent, S. P., and R. C. Bray. 1978. Electron paramagnetic resonance studies on nitrate reductase from *Escherichia coli* K12. Biochem. J. 171:639–647.

260. Wharton, D. C., and S. T. Weintraub. 1980. Identification of nitric oxide and nitrous oxide as products of nitrite reduction by *Pseudomonas* cytochrome oxidase (nitrite reductase). Biochem. Biophys. Res. Commun. 97:236–242.

261. Wikström, M., and K. Krab. 1979. Proton-pumping cytochrome *c* oxidase Biochim. Biophys. Acta 549:177–222.

262. Wikström, M., and K. Krab. 1980. Respiration-linked H^+ translocation in mitochondria: stoichiometry and mechanism. Curr. Top. Bioenerg. 10:51–101.

263. Willison, J. C., and P. John. 1979. Mutants of *Paracoccus denitrificans* deficient in *c*-type cytochromes. J. Gen. Microbiol. 115:443–450.

264. Wimpenny, J. W. T., J. A. Cole. 1967. The regulation of metabolism in facultative bacteria. III. The effect of nitrate. Biochim. Biophys. Acta 148:233–242.

265. Wood, A. P., and D. P. Kelly. 1983. Autotropic, mixotrophic and heterotrophic growth with denitrification by *Thiobacillus* A2 under anaerobic conditions. FEMS Microbiol. Lett. 16:363–370.

266. Yoshinari, T. 1980. N_2O reduction by *Vibrio succinogenes*. Appl. Environ. Microbiol. 39:81–84.

267. Yoshinari, T., and R. Knowles. 1976. Acetylene inhibition of nitrous oxide reduction by denitrifying bacteria. Biochem. Biophys. Res. Commun. 69:705–710.

268. Zumft, W. G., and T. Matsubara. 1982. A novel kind of multi-copper protein as terminal oxidoreductase of nitrous oxide respiration in *Pseudomonas perfectomarinus*. FEBS Lett. 148:107–112.

269. Zumft, W. G., B. F. Sherr, and W. J. Payne. 1979. A reappraisal of the nitric-oxide-binding protein of denitrifying *Pseudomonas*. Biochem. Biophys. Res. Commun. 88:1230–1236.

270. Zumft, W. G., and J. M. Vega. 1979. Reduction of nitrite to nitrous oxide by cytoplasmic membrane fraction from the marine denitrifier *Pseudomonas perfectomarinus*. Biochim. Biophys. Acta 548:484–499.

6

MICROBIAL REDUCTION OF MANGANESE AND IRON

WILLIAM C. GHIORSE

Department of Microbiology, Cornell University, Ithaca, NY 14853

6.1 INTRODUCTION
6.2 NATURAL OCCURRENCE OF Mn- AND Fe-REDUCING
     MICROORGANISMS
6.3 ISOLATION, ENUMERATION, AND CULTIVATION
     6.3.1 Mn Reducers
     6.3.2 Fe Reducers
6.4 MECHANISMS OF MICROBIAL Mn AND Fe REDUCTION
     6.4.1 Overview
     6.4.2 Enzymatic Reduction of Mn in Cell-Free Systems
     6.4.3 Enzymatic Reduction of Fe in Cell-Free Systems
     6.4.4 Reduction of Mn and Fe by End Products of Metabolism
6.5 IMPORTANCE OF CELLULAR BINDING
6.6 MICROBIAL Mn AND Fe REDUCTION IN NATURAL ENVIRONMENTS
     6.6.1 Examples from Anaerobic Sediments
6.7 SUMMARY AND CONCLUSIONS
REFERENCES

6.1 INTRODUCTION

The study of microbial Mn and Fe reduction presents some complex problems to students and researchers interested in the microbiology of anaerobic environments.

305

As discussed in Chapter 1 (see also Refs. 45, 52, and 64), thermodynamic considerations predict that in closed aqueous systems containing living organisms, after O_2 is removed during the oxidation of organic matter, biological reduction of the alternative electron acceptors, NO_3^-, $MnO_2(s)$, $Fe(OH)_3(s)$, and SO_4^{2-}, should proceed in that order. In neutral-pH waters, the solid phases of Mn and Fe will dominate. In fact, when profiles of stable anaerobic systems such as marine sediments are examined, first NO_3^- disappears, and then Mn^{2+}, Fe^{2+}, and SH^- appear sequentially with depth bearing out the thermodynamic predictions (45). It has been argued (64) that because each of the reactions involved in this sequence may be microbially mediated, the sequence should be paralleled by a succession of microorganisms that specialize in each reduction to gain energy from the redox reactions involved. This argument is supported in the cases of denitrifying and SO_4^{2-}-reducing bacteria by strong evidence of their activity in natural environments; however, evidence for the existence of Mn- and Fe-reducing specialists is scanty.

Numerous microbiological studies reviewed in Sections 6.2 and 6.3 show that a variety of microorganisms with the capacity to reduce solid Mn and Fe oxides (mostly bacteria and fungi) are found in both aerobic and anaerobic habitats; however, neither their exact role in the natural processes of Mn and Fe reduction nor, except in one case, their special dependence on the processes for energy gain have been clearly established. Furthermore, examination of isolated Mn- and Fe-reducing microorganisms in axenic cultures has revealed some apparent anomalies in the predicted trend toward the sequential use of alternative electron acceptors. For example, a facultatively anaerobic marine *Bacillus* sp. studied by Ehrlich (21) was found to reduce Mn oxides in the presence and absence of O_2, apparently showing no preference for O_2, the energetically more favorable electron acceptor. Another marine *Bacillus* sp. studied by deVrind et al. (15a) also reduced MnO_2 in the presence of O_2. Similarly, Ottow and Munch (52) found that an anaerobic soil isolate, *Clostridium butyricum*, reduced Mn and Fe oxides in the presence and absence of NO_3^-. Because growth was not measured in the presence and absence of oxide in any of these cases, it cannot be ascertained whether the bacteria derived any useful energy during the reduction. However, because respiratory enzymes were involved, linkage to energy-conserving metabolic reactions certainly was possible.

Indeed, many of the papers reviewed in Section 6.4 are laboratory studies indicating that enzymatic mechanisms, usually electron transport chains, are involved in microbial reduction of Mn and Fe. In some cases indirect mechanisms, such as reduction by metabolic endproducts, have also been demonstrated. However, because most of the results have been obtained under laboratory conditions, they are not interpreted easily in relation to more complex natural systems.

An important, but somewhat neglected aspect of microbial Fe and Mn reduction, emphasized in Section 6.5, is the effect of cellular binding on the rate of reduction. Because Mn and Fe oxides are practically insoluble in waters at neutral pH [$< 10^{-15}$ M at pH 7 (46, 62)], any proposed enzymatic mechanism for their reduction in neutral pH environments must include contact of enzymes with the oxide surface. Furthermore, recent evidence (62, 63) shows that nonbiological surface reactions of MnO_2 with a variety of organic compounds, including common metabolic end products, also may be very important in the reduction of Mn oxides in natural waters. Because end-product concentrations will be higher at the cell surface, it is likely that indirect reduction by metabolic end products also will be enhanced by cellular contact with the oxides. Enhancement of Mn and Fe reduction by cellular binding has been demonstrated by several workers; however, in most cases the enhancement has been attributed to improved enzyme–oxide interaction. Generally, the possibility that surface reactions of Mn and Fe oxides with metabolic end products may also be enhanced by cellular binding has not been fully appreciated.

Finally, Section 6.6 provides a review of experimental evidence indicating that microorganisms actually do participate in the processes of Mn and Fe reduction in nature. A large portion of this evidence is based on the application of poisons and other antibiological treatments to natural samples to establish that microorganisms are indeed involved in the natural Mn- and Fe-reducing processes. Surprisingly little evidence has been obtained that reveals the exact biological and chemical mechanisms; however, several recent studies of Mn and Fe reduction in natural sediments (5a, 12a, 36, 37, 39a, 39b, 39c, 39d, 60) indicate some promising experimental approaches that may help to elucidate the mechanisms of microbial Mn and Fe reduction in nature.

6.2 NATURAL OCCURRENCE OF Mn- AND Fe-REDUCING MICROORGANISMS

The existence of microorganisms that can reduce Mn oxides has been known for nearly a century. The first published report of such activity appears to be that of Adeny in 1894 (1), who demonstrated that reduction of Mn oxide to form $MnCO_3$ was mediated by microorganisms in sewage. Three decades later, Starkey and Halvorson (61) explained Fe reduction in nature as an indirect microbial process in which the microorganisms lowered the pH and O_2 concentration in their environment, creating conditions favorable for reduction of $Fe(OH)_3$. Since these early investigations, subsequent work (3, 20, 21, 43–45, 51) has shown that Mn- and Fe-reducing bacteria and fungi occur widely in aerated and waterlogged surface soil, gleyed subsurface soil, fresh and marine waters, and sediments, including ferromanganese concretions. Recently, Mn-reducing microorganisms have also been studied in groundwater-bearing subsurface soil (31).

Table 6.1 illustrates typical results of a study of Mn-reducing bacteria in which the proportion of Mn reducers in a heterotrophic population was estimated. These results show that 15 to 36% of the total heterotrophic bacteria in ferromanganese concretions and sediment from the western Baltic Sea were capable of reducing Mn oxides. During this investigation, detailed electron microscopic examination of the same samples (23) showed that a variety of morphological types of bacteria and fungi were present in the surface coatings of the concretions. Furthermore,

TABLE 6.1 Proportion of Mn-reducing bacteria in Baltic Sea ferromanganese concretions and sediments[a]

| | CFU (S.D.) $\times$ 10^4 $\cdot$ g dry wt^{-1} | | |
Sample	Total Heterotrophs[b]	Mn Reducers[c]	% Mn Reducers
Concretion			
B10	3.9 (1.6)	0.6 (0.6)	15
B11	8.2 (2.9)	1.7 (1.4)	21
B12	1.5 (1.0)	0.4 (0.3)	27
B13	6.0(1.7)	1.1(0.7)	18
Sediment			
B10s	2.4 (0.7)	0.6 (0.5)	25
B11s	7.0 (5.8)	2.5 (1.8)	36

[a] See Ref. 23 and 29 for sampling details.
[b] Plate counts on Mn oxide-containing nutrient agar (Difco) made up in artificial seawater (28). 2% $KMnO_4$ (10 ml $\cdot$ L^{-1}) was added after autoclaving to form a fine suspension of Mn oxide in the agar (for details of preparation of other similar media, see Ref. 31).
[c] Colonies surrounded by a clear zone after flooding plates with acidified leucoberbelin blue spot-test solution (28).

when oxalic acid or acidified leucoberbelin blue was employed to dissolve ferromanganese oxides, rod-shaped bacteria enmeshed in a residual organic matrix were observed by light microscopy (curved arrows, Fig. 6.1). Many of these bacteria were sporeforming bacilli (straight arrows, Fig. 6.1), a group that has frequently been identified as carrying out microbial Fe and Mn reduction (Table 6.2).

Table 6.2 lists many of the bacterial and fungal genera capable of reducing Mn and Fe oxides in axenic cultures. Although this list may not be complete, it shows that microorganisms capable of Mn and Fe reduction are found in many different groups of microorganisms, ranging from highly aerobic bacteria and fungi to strictly anaerobic bacteria. Such a list also shows that Mn- and Fe-reducing microorganisms may be more ubiquitous in the biosphere than those capable of Mn and Fe oxidation (deposition) because Mn and Fe depositors are found only among genera containing aerobes and facultative anaerobes (25).

6.3 ISOLATION, ENUMERATION, AND CULTIVATION

6.3.1 Mn Reducers

Mn-reducing microorganisms can be isolated employing various culture media containing Mn oxide and mixtures of inorganic salts (see references in Ref. 21). If an organic nitrogen source is employed, NH_4^+ is preferred over NO_3^- because NO_3^- may inhibit reduction (Table 6.2). For enumeration of Mn reducers, either the most probable number (MPN) or the plate-count method may be used. For the MPN method, synthetic or natural Mn oxides are usually added to liquid media in a suspension of fine particles. Mn reduction is detected by assaying for Mn^{2+} or

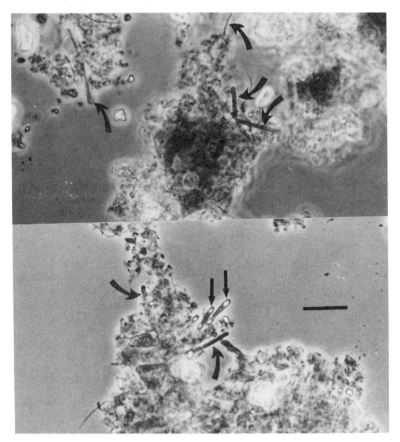

Figure 6.1 Phase-contrast photomicrographs showing rod-shaped bacteria (curved arrows in surface material of Baltic Sea ferromanganese concretions (see Table 6.1). Freshly collected concretions were fixed with glutaraldehyde (23, 29) and small pieces of materia from their surface were treated with acidified leucoberbelin blue to dissolve Mn oxides (28). Note refractile bodies (straight arrows) resembling endospores in some of the bacteria. Bar = 10 μm.

by visually observing Mn oxide dissolution. For the plate-count method, Mn oxides can be formed in agar-containing nutrient media by adding $KMnO_4$ after autoclaving (e.g., see Table 6.1). The $KMnO_4$ forms Mn oxide when it is reduced by organic constituents in the agar. Apparently, the Mn oxide formed in this way is very susceptible to non-biological reduction; therefore, to avoid this problem, Ehrlich (21) recommends that Mn oxide be synthesized separately and then mixed with the test sample in a layer of capping agar overlaying a basal agar nutrient medium. In both types of media, colonies of Mn-reducing microorganisms produce clear zones around them as Mn oxides are dissolved. According to Ehrlich (21), the $KMnO_4$–agar method will favor detection of microorganisms that reduce

TABLE 6.2 Genera of bacteria and fungi capable of Mn and Fe oxide reduction and its inhibition by O_2 and NO_3^- [a]

Genus	Reduction of:		Inhibition by:		References
	Mn Oxide	Fe Oxide	O_2	NO_3^-	
Aerobic and facultatively anaerobic bacteria					
Bacillus	+	+	+ or −	+ or −	20, 21, 43, 44
Pseudomonas	+	+	+	+ or −	20, 21, 43, 44
Leptothrix	+	NT	NT	NTU	16, 17
Paracoccus	NT	+	+	−	35
Micrococcus	+	+	NT	NTU	5, 67
Corynebacterium	NT	+	−[b]	NT	22
Arthrobacter	+	NT	−[b]	NTU	10
Streptomyces	+	NT	NT	NTU	30
Escherichia	NT	+	NTL	NTL	8, 47
Enterobacter	NT	+	+ or −	NTL	8, 47
Citrobacter	NT	+	+	+	47, 48
Serratia	NT	+	+	+	47, 48
Proteus	+	+	NTL	NTL	67
Alcaligenes	NT	+	+	+	48
Vibrio	+	+	+	+ or −	36
Anaerobic bacteria					
Clostridium	+	+	+[c]	+ or −	35, 49
Bacteroides	NT	+	+[c]	−	35
Desulfovibrio	+	+	+[c]	+ or −	35, 49
Fungi					
Aspergillus	+	NT	NT	NT	30
Penicillium	+	NT	NT	NT	30
Fusarium	NT	+	−	−	53
Actinomucor	NT	+	−	−	53
Alternaria	NT	+	−	−	53
Pichia	+	NT	NT	NT	5

[a] NT, not tested; NTU, not tested, but unlikely, because bacteria in these genera employ only O_2 as a terminal electron acceptor (11); NTL, not tested, but likely, based on comparison with other enteric bacteria.
[b] Microaerobic conditions employed.
[c] Do not grow in presence of O_2.

Mn indirectly by excreting reducing end products, whereas the Mn oxide-capping agar will favor detection of those that reduce Mn by direct enzymatic mechanisms, especially if an artificial electron carrier such as ferricyanide is included in the capping agar.

A clever variation of the Mn oxide–agar technique was employed by Troshanov (67) to screen for Mn- and Fe-reducing bacteria in sediments from Karelian lakes in the USSR. First, he mixed finely ground ferromanganese lake ore with sediment samples in molten agar and then sucked the mixture into 4-to-5-mm-diameter glass tubes and small capillaries that could be examined microscopically. After 20 to 30 days of incubation, clear zones developed around bacterial colonies and he followed the progression of colony development and particle dissolution in the capillaries microscopically. Finally, the bacteria in the colonies that dissolved ferromanganese were isolated in liquid medium without oxide, and these were later tested for their ability to reduce Mn and Fe oxides. Ehrlich (18, 19) has also studied several Mn-reducing bacteria derived from enrichment cultures of deep-sea ferromanganese nodules by first isolating bacteria from the enrichments and then testing them individually for their ability to reduce Mn oxides.

Many of the enrichment, isolation, and enumeration procedures for Mn reducers, unlike those for Fe reducers (see Section 6.3.2), have been carried out under aerobic conditions without special precautions to control the O_2 concentration or redox potential of the media. One reason for this may be that Mn^{2+}, unlike Fe^{2+}, is stable chemically in the presence of O_2; Mn^{2+} produced during Mn oxide reduction will accumulate in aerobic culture media, whereas Fe^{2+} will be reoxidized and will not accumulate. Thus detection of Mn-reducing microorganisms is possible under aerobic conditions; anaerobic incubation techniques are not required. However, as for Fe reducers (see Section 6.3.2), if aerobic conditions are employed, dissimilatory Mn-reducing bacteria may not be detected.

Recently, Wollast et al. (69) and Burdige and Nealson (12,12a) employed enrichment culture procedures on marine sediments to find bacteria that can couple dissimilatory Mn oxide reduction to the oxidation of organic matter. The enrichment strategy (12,12a) was to provide MnO_2 as the only electron acceptor with a nonfermentable carbon substrate. Enrichment cultures were obtained in which an appreciable amount of synthetic Mn oxide was solubilized within 30 days. Furthermore, Buridge and Nealson (12,12a) demonstrated that a mixed population containing at least two different types of bacteria was responsible for the Mn oxide reduction and that the respiratory inhibitor, azide, inhibited the process, but molybdate, an inhibitor of SO_4^{2-} reduction, did not. Thus it appears that dissimilatory Mn oxide reducers probably were present in their cultures, but they were not isolated or further characterized in axenic cultures.

Another possible enrichment strategy is microaerobic incubation in the presence of MnO_2. This is suggested by the recent work of deVrind et al. (15a), who showed that electron transport–linked Mn reduction by a Mn-oxidizing marine Bacillus sp. occurred maximally at low O_2 concentration, suggesting that spore germination under low-O_2 conditions may have required MnO_2 as an alternate electron acceptor. Incubation under low-O_2 conditions may select for microaerobic bacteria that can dissimilate MnO_2.

6.3.2 Fe Reducers

Fe-reducing microorganisms can also be isolated employing media similar to those employed for Mn-reducing bacteria, except that Fe oxides are added instead of Mn oxides (8). In the past, enrichment cultures for Fe reducers generally have been incubated either aerobically or anaerobically (9, 15), but recent work (4, 34,

35, 39a, 58a) shows that microaerobic or strictly anaerobic incubation may be necessary to demonstrate dissimilatory Fe reduction. For example, anaerobic techniques were necessary to isolate an Fe-reducing, H_2-oxidizing pseudomonad from swampy soil (4). The organism grew in a mineral medium containing $Fe(OH_3)$ or ferrihydrite under a gas mixture consisting of H_2 and CO_2. Because it could also grow microaerobically, strict anaerobic methods were employed to demonstrate that this bacterium could couple Fe oxide reduction to H_2 consumption promoting cellular growth. Strictly anaerobic conditions also were employed for MPN counts and isolation of Fe-reducing bacteria from lake water and sediment by Jones and co-workers (34, 36). In these studies semisolid (0.25%) agar media were found to be critical for the development of the maximum number of Fe-reducing bacteria.

In studies of bacterial Fe-reducing activity, Fe reduction is frequently detected by the production of Fe^{2+}, which can be trapped by a complexing reagent such as 2,2'-dipyridyl (8) or 2,4,6-tripyridyl-1,3,5-triazine (36). The colored complexes that form as Fe^{2+} is trapped can be measured spectrophotometrically. It has been noted during some of these studies that chelated forms of Fe^{3+} are more readily reduced by bacteria than are crystalline Fe oxides. In addition, it has been observed that the less crystalline oxides, such as amorphous ferrihydrite or ferric oxyhydroxide, are dissolved more readily by microorganisms than are highly crystalline oxides (15, 22, 34, 39a, 41; see Section 6.5 for further discussion of this point).

Microorganisms that reduce Fe oxides, like those that reduce Mn oxides, are not difficult to isolate and study in the laboratory. However, in nature Mn and Fe reducers may function optimally in mixed populations. This is apparent from the recent findings of Jones and co-workers (36), which show that mixed cultures of Fe reducers solubilized more Fe than the pure cultures derived from them. As noted above, similar situations may exist for Mn-reducing microorganisms (12, 12a), however, this possibility has not been tested rigorously. If syntrophic consortia of bacteria are important for Mn and Fe reduction in natural environments, isolation and study of the individual microorganisms responsible for the reduction may require special cultural techniques designed to support the consortia.

6.4 MECHANISMS OF MICROBIAL Mn AND Fe REDUCTION

6.4.1 Overview

The capacity of bacteria and fungi to reduce Mn and Fe oxides in axenic culture frequently is inhibited by O_2 or NO_3^- (Table 6.2). This inhibition might be predicted based on the thermodynamics of the chemical reactions involved (see Section 6.1). Apparently, some bacterial NO_3^- reductases are capable of employing chelated Fe^{3+} or Fe oxide and alternative electron acceptors (42, 58a). However, neither possession of dissimilatory NO_3^- reductase nor anaerobic conditions are prerequisites for microbial Mn or Fe reduction (15, 21, 50, 58a; Table 6.2). Indeed, some of the Mn- and Fe-reducing bacteria listed in Table 6.2 (*Arthrobacter*,

Leptothrix, Micrococcus) are strict aerobes that do not employ any other terminal electron acceptor than O_2 to support their growth (11). It may be concluded, therefore, that at least the strictly aerobic bacteria must employ enzyme systems other than NO_3^- reductase for Mn and Fe reduction. As discussed below, other enzymatic mechanisms responsible for Mn and Fe reduction may include non-energy-yielding branches of electron transport systems in which Mn or Fe oxides may serve as sinks for excess reductants. Nonenzymatic reactions with metabolic end products are also possible.

It should be mentioned here that in addition to the bacteria listed in Table 6.2, the chemolithotrophic acidophiles *Thiobacillus thiooxidans* and *Sulfolobus acidocaldarius* are also capable of reducing soluble Fe^{3+} (6). These bacteria were not included in Table 6.2 because they live in highly acidic environments where Fe^{3+} solubility is high. Interestingly, *T. ferrooxidans* has been shown to reduce Fe^{3+} only under anaerobic conditions because it was rapidly reoxidized to Fe^{2+} in the presence of O_2 (6). In addition, *T. denitrificans* apparently cannot reduce Fe oxide anaerobically (35), indicating that its NO_3^- reductase system does not reduce Fe^{3+}.

Also omitted from Table 6.2 are the facultatively chemolithotrophic, H_2-oxidizing bacteria mentioned in Section 6.3.2 and discussed more fully in Section 6.6.1. These bacteria have not been fully characterized as yet (36), but they probably are pseudomonads (4) that can couple Fe oxide reduction to H_2 oxidation in the complete absence of O_2. As discussed in Section 6.6.1, they may be the only clear examples yet described of bacteria that have been observed to grow by coupling anaerobic reduction of Fe oxide to energy-conserving metabolic reactions (iron respiration).

6.4.2 Enzymatic Reduction of Mn in Cell-Free Systems

The biochemical investigations of Hochster and Quastel in 1952 (32) showed that under anaerobic conditions, amorphous Mn oxide could serve as an electron acceptor during enzymatic oxidation of ethanol, lactate, succinate, and reduced cytochrome *c* in cell-free extracts of yeast and animal tissue. Ten years later Woolfolk and Whiteley (70) showed that crude extracts of *Veillonella alcalescens* could reduce a large number of inorganic compounds enzymatically including Mn and Fe oxides employing H_2 as reductant, presumably via a hydrogenase system. Later, Bautista and Alexander (5) demonstrated that cell-free extracts of two soil isolates, a *Micrococcus* sp. and a yeast, *Pichia guillermondii*, could also reduce many inorganic compounds enzymatically, including MnO_2. In each case the reductions were observed in crude cell-free extracts and were presumed to be enzymatic; however, no enzymes were isolated and the systems were not well characterized. Because reducing substances may have been present in the extracts, unknown nonenzymatic reactions cannot be ruled out in these cases.

By far the most extensively characterized enzymatic Mn-reducing system is that of the marine bacterium *Bacillus* 29, studied by Ehrlich and his students (26, 27, 65, 66). In the *Bacillus* 29 system, involvement of enzymes in the MnO_2-reducing system has been established; however, many details of the system are not clear. *Bacillus* 29 whole cells and cell-free extracts can reduce MnO_2 in a reaction mixture containing diluted seawater or an equivalent salts solution, glucose, and reagent-grade MnO_2. At least one component enzyme of the system is synthesized (induced) when cells are grown in Mn^{2+}-containing media. The activity of cells and extracts is optimal at concentrations of cations found in fivefold diluted seawater and at various temperatures from 25 to 40°C, depending on whether induced or uninduced cells or cell extracts are employed. If the cells are not induced, an artificial electron carrier, potassium ferricyanide, is necessary for Mn reduction. Cells and extracts reduce MnO_2 in the presence and absence of O_2; however, the inducible enzyme component is synthesized only in the presence of O_2. Metabolic inhibitor studies show that the Mn-reducing activity of induced cells is stimulated by 1.0 to 10 mM azide, but this activity of uninduced cells is not. Azide concentrations above 10 mM, 1 mM cyanide, and dicumarol inhibit Mn reduction in both types of cells. Atebrine inhibits only induced cells. Other respiratory inhibitors, such as CO, 2-n-nonyl-4-hydroxy-quinoline-N-oxide (NOQNO), and antimycin A, do not inhibit MnO_2-reducing activity in either induced or uninduced cells. The terminal portion of the system is inhibited by cyanide and azide but not by dicumarol or atebrine. This portion appears to be loosely associated with the cell membranes. Much of the terminal reductase activity is found in the soluble fraction, which contains flavoprotein but no cytochromes.

These data were consistent with a model of the *Bacillus* 29 Mn-reducing system (27) which has been modified recently based on additional experiments done in my laboratory (see Section 6.4.4 and Fig. 6.2). In the new model, MnO_2 is reduced via a truncated electron transport system that contains a dicumarol-sensitive quinone, a Mn^{2+}-inducible, atebrine-sensitive flavoprotein, and a cyanide- and azide-sensitive metalloenzyme, the terminal "MnO_2 reductase." Stimulation of

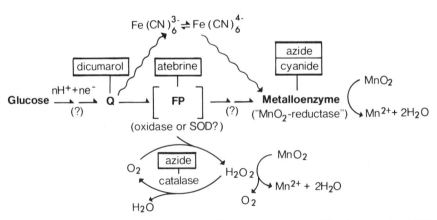

Figure 6.2 MnO_2-reducing system of *Bacillus* 29. Q, quinone; FP, flavoprotein; SOD, superoxide dismutase; [induced by Mn^{2+}], inhibitor ; $\rightsquigarrow$, electron flow via artificial carrier; (?), unknown enzymes. Modified after Ref. 49.

TABLE 6.3 H$_2$O$_2$ production by resting cells of *Bacillus* 29[a]

	nmol H$_2$O$_2$ produced · mg dry wt^{-1}		
Time (minutes)	Induced Cells	Induced Cells + NaN$_3$	Uninduced Cells
0	0	0	0
15	7.3	9.0	4.2
30	6.5	8.1	3.6
60	7.6	15.4	0

[a] Cells from a 24-hour culture of *Bacillus* 29 were induced for MnO$_2$-reducing activity by incubating them at 37°C for 12 to 15 h in an artificial seawater (ASW)-nutrient medium containing 0.01 M MnSO$_4$ · H$_2$O (66). The induced cells were washed to remove Mn^{2+} and resuspended in fivefold-diluted ASW to a density of 450 mg dry wt · mL^{-1}. A similar cell suspension was prepared from uninduced cells. 1.0 mL of the induced suspension was added to eight 50-mL Erlenmeyer flasks containing 5.0 mL of fivefold-diluted ASW and 1.0 mL of 1% glucose, giving a final cell density of 64.3 mg dry wt · mL^{-1}. 1.0 mM NaN$_3$ was added to four of these flasks. 1.0 mL of the uninduced cell suspension was added to four flasks without NaN$_3$. The flasks were incubated in a reciprocating shaker incubator at 37°C. At each time interval, one flask of each condition was removed and 5.0 mL of the suspension was filtered (0.2 μm) to remove cells. The filtrate was assayed immediately for H$_2$O$_2$ with titanium oxysulfate reagent prepared in 2 N H$_2$SO$_4$ (39). Absorbance at 400 nm was read in a spectrophotometer after removing a precipitate that formed when the acidic reagent was added. The sensitivity of the assay was ±1.2 nmol H$_2$O$_2$ · mg cell dry wt^{-1}.

the induced system by 1.0 to 10 mM azide originally suggested (27) that azide inhibited another enzyme as well as the terminal reductase in induced cells. It was proposed (27) that inhibition of the second azide-sensitive enzyme allowed more electrons to flow to MnO$_2$ when azide concentrations were low. Our recent experiments (Tables 6.3 and 6.4) suggest that the second azide-sensitive enzyme in *Bacillus* 29 may be catalase. As shown in Fig. 6.2 and discussed in Section 6.4.4, inhibition of catalase by azide would allow H$_2$O$_2$ to accumulate and reduce MnO$_2$ nonenzymatically.

TABLE 6.4 Effect of azide on MnO$_2$-reducing and catalase activities of induced and uninduced *Bacillus* 29[a,b]

	MnO$_2$-reduced activity[c]		Catalase activity[d]	
Azide Concentration	Induced Cells	Uninduced Cells[e]	Induced Cells	Uninduced Cells
0	38.4 (6.0)	21.9 (0.9)	432 (10.0)	698 (6.7)
1 mM	65.3 (10.2)	17.5 (0.7)	0	0

[a] Induced and uninduced cell suspensions were prepared as described in Table 6.3. Aliquots of the cell suspensions were assayed for MnO$_2$-reducing activity (26, 27) and catalase activity (39) after 3 h of incubation at 37°C.
[b] Numbers in parentheses are standard deviations of duplicate flasks.
[c] nmol Mn^{2+} released · h^{-1} · mg dry wt^{-1} assuming linear rate of MnO$_2$ reduction during 3 h of incubation.
[d] nmoles H$_2$O$_2$ destroyed · min^{-1} · mg dry wt^{-1}. Measured at end of 3 h of incubation.
[e] 0.1 mL of 0.01 M K$_4$Fe(CN)$_6$ was added to flasks as an artificial electron carrier (48).

Still other more recent work shows that the MnO_2-reducing system of *Bacillus* 29 may be quite different from the systems of other MnO_2-reducing bacteria. Ehrlich (20) compared the inhibition patterns of two gram-negative MnO_2-reducing marine isolates with that of *Bacillus* 29. The gram-negative isolates, like *Bacillus* 29 and all other marine MnO_2-reducing bacteria so far examined by Ehrlich, possess a Mn^{2+}-inducible, MnO_2-reducing activity in which ferricyanide can replace the inducible component. Unlike *Bacillus* 29, however, the MnO_2-reducing systems of Ehrlich's gram-negative isolates (20) were not inhibited by dicumarol and only one was inhibited by atebrine. In further contrast to *Bacillus* 29, Ehrlich's gram-negative isolates were not stimulated by azide and the activity in both isolates was inhibited by antimycin A and NOQNO, suggesting that cytochromes *b* and *c* were part of their Mn-reducing enzyme systems. Adding more diversity to the picture is the Mn-reducing system of the marine *Bacillus* sp. studied by de Vrind et al. (15a), which like the gram-negative isolates, was inhibited by antimycin A and NOQNO, and like *Bacillus* 29, was stimulated by azide when O_2 concentrations were high.

The results of these laboratory studies clearly indicate that respiratory enzyme activity can be involved in bacterial Mn reduction; however, clear evidence of manganese respiration is lacking. It is yet to be established experimentally that bacteria can couple MnO_2 reduction to proton translocation or ATP synthesis.

6.4.3 Enzymatic Reduction of Fe in Cell-Free Systems

Bacteria possess a number of Fe-reducing enzyme systems, "iron reductases" (14), that probably are employed for the purpose of "assimilating" (21) Fe^{2+} needed for biosynthesis of heme or other Fe-containing cofactors and enzymes. The assimilatory iron reductase systems employ soluble, chelated Fe^{3+} rather than insoluble Fe oxides as substrates. To study these enzyme systems, chelated Fe^{3+} is added to buffered reaction mixtures containing bacterial cell suspensions or crude cell extracts and the Fe^{2+} produced is trapped for analysis employing a highly specific Fe^{2+} chelator such as ferrozine (14). Alternatively, anaerobic conditions can be employed to keep Fe^{2+} reduced. An advantage of the ferrozine-trapping procedure is that it can be employed in vessels exposed to air, so that anaerobic conditions are not required (38).

The iron reductase systems that have been studied in cell-free preparations of *Aquaspirillum itersonii* (14, 58a), *Staphylococcus aureus* (38), *Pseudomonas aeruginosa* (13), and several other bacteria (14) all appear to contain electron transport enzymes that occur before cytochromes *b* or *c*. In *A. itersonii* and *S. aureus*, Fe^{3+} apparently is reduced by side reactions of the NO_3^- or O_2-terminated respiratory chains of these bacteria. Although these Fe-reducing systems obviously can provide Fe^{2+} for heme synthesis, they have not been linked to any energy-yielding reactions in the cell. On the other hand, Short and Blakemore (58a) showed that proton translocation was linked to iron respiration in *A. magnetotacticum*, *Bacillus subtilus*, *and Escherichia coli*, indicating that dissimilatory Fe reduction is possible in these bacteria.

Another bacterial Fe oxide-reducing enzyme system was studied in cell-free extracts of a *Vibrio* sp. isolated from an English lake (36). Anaerobic conditions and a buffered reaction mixture containing ADP, NADH, and $FeCl_3$ were employed to study the system in crude cell-free extracts. Presumably, the $FeCl_3$ hydrolyzed immediately in the reaction mixture, forming colloidal $Fe(OH)_3$. The Fe oxide was reduced by cell extracts, forming Fe^{2+} which was trapped with 2,4,6-tripyridyl-1,3,5-triazine. The Fe-reducing activity was inhibited by NOQNO, azide, rotenone, antimycin A, and chlorate, suggesting that a respiratory chain was involved. However, because no increased growth yield was noted in anaerobic cultures provided with Fe oxide as the sole electron acceptor, it was assumed that the organism used its Fe-terminated electron transport system to dissipate excess reductant without synthesizing ATP.

Yet another Fe-reducing enzyme system has been investigated in a *Pseudomonas* sp., strain 200, isolated from crude oil (3a, 46a, 46b). Arnold et al. (3a) showed by inhibitor and spectrophotometric studies of nongrowing whole cells that cells grown at high O_2 concentrations reduced chelated Fe^{3+} by an abbreviated electron transport chain. Cells grown at low O_2 concentration increased their Fe-reducing activity six- to eight-fold, but the increased activity apparently was uncoupled from oxidative phosphorylation.

The results of these studies on Fe-reducing enzyme systems, like those of studies on Mn-reducing enzyme systems, indicate that respiratory pathways can be involved in Fe reduction and respiratory energy gain is possible, but except for the work of Short and Blakemore (58a), direct experimental evidence for coupling of Fe reduction to ATP synthesizing systems of the cell has not been obtained. In many instances it appears that iron reduction occurs as a side reaction of electron transport to other electron acceptors.

6.4.4 Reduction of Mn and Fe by End Products of Metabolism

Although details of the enzymatic mechanisms of microbial Mn and Fe oxide reduction are not always clear, it is clear that microorganisms can also mediate reduction of the Mn and Fe oxides by excreting a variety of end products that reduce Mn and Fe oxides (8, 20, 21, 36, 43–45, 68). Examples of end products that can reduce the oxides include H_2S, formate, oxalate, and other organic acids (e.g., malate, oxaloacetate, and pyruvate) that have been shown to reduce Fe and Mn oxides under slightly acidic conditions (33). With the possible exception of H_2S, the metabolic end products listed above may be expected to reduce Mn and Fe oxides in natural environments under either aerobic or anaerobic conditions.

Mn oxides appear to be more readily reduced than Fe oxides by organic substances. Troshanov (68) found that glucose, xylose, acetate, and butyrate caused a release of Mn^{2+}, but not Fe^{2+}, from ferromanganese ore in neutral or weakly acidic media. Furthermore, Jauregui and Reisenauer (33) showed that malate and pyruvate reduced Mn oxides more readily than they reduced Fe oxides at pH values between 3.0 and 7.0. In addition, these workers showed that low levels of Mn oxides enhanced the reduction of Fe oxides by stimulating the production of pyruvate from malate, which promoted reduction of Fe oxides (33). On the other hand, elevated levels of Mn oxide depressed Fe reduction because Fe^{2+} was absorbed and reoxidized by the Mn oxide.

Stone and Morgan (62, 63) demonstrated clearly that a variety of naturally occurring organic compounds, including some microbial metabolites, can reduce Mn oxides chemically. Of the microbial metabolites tested, only oxalate and pyruvate reduced Mn oxide significantly during 3 h of reaction at pH 7.2. Other microbial metabolites tested (e.g. formate, fumarate, glycerol, lactate, malonate, and propionate) did not reduce Mn oxides under the neutral conditions employed. By far the most reactive compounds tested in these studies were phenols (e.g., catechol), phenolic acids (e.g., hydrobenzoic acid) and quinones (e.g., hydroquinone), which form the core structures of humic acids. These studies show conclusively that the humic residues which are ubiquitous in aquatic environments, soil, and sediment can greatly influence Mn oxide reduction. These studies also support a proposal (58) that Mn oxides can promote the abiotic oxidation of humic substances in natural environments.

H_2O_2 is another end product of microbial metabolism that can readily reduce Mn oxide. Because H_2O_2 may be produced only during aerobic metabolism, Mn oxide reduction by H_2O_2 would be expected to occur only under aerobic or microaerobic conditions. Most bacteria and fungi dispose of their excess H_2O_2 rapidly and efficiently employing the enzymes catalase and peroxidase. However, as was shown by Dubinina (16, 17), when organic substrates are oxidized by certain bacteria, such as Mn-depositing *Leptothrix* spp., excess H_2O_2 may accumulate in the culture medium. If Mn oxide is also present in the medium, especially under slightly acidic conditions, it will be reduced chemically to Mn^{2+} and O_2 will be evolved:

$$H_2O_2 + MnO_2 + 2H^+ \longrightarrow Mn^{2+} + 2H_2O + O_2$$

This reaction has been found to mimic catalase activity in cultures of Mn-depositing bacteria (29).

Further, the reduction of Mn oxides by H_2O_2 has been demonstrated by Dubinina in cultures of *Leptothrix pseudoochraceae*, *Arthrobacter siderocapsulatus* (*Siderocapsa*), and "*Metallogenium*" (16, 17). Similarly, Mn oxide reduction was observed in resting cell suspensions of another *Arthrobacter* sp. (formerly called *Corynebacterium* strain B) (10). An earlier study by Bromfield (9) showed that Mn oxidation by this strain was inhibited by H_2O_2 and the inhibition was overcome by adding purified catalase or by employing a culture of strain B that showed elevated catalase activity. Both Dubinina (16, 17) and Bromfield (9) suggested that H_2O_2 and catalase may be involved in bacterial Mn oxidation, but only Dubinina demonstrated that metabolically produced H_2O_2 can also be responsible for bacterial Mn reduction.

Recent experiments in my laboratory suggest that excess H_2O_2 produced during glucose oxidation by Mn^{2+}-induced cells of *Bacillus* 29 can reduce MnO_2. Table 6.3 shows that resting *Bacillus* 29 cells, induced by prior exposure to Mn^{2+}, accumulated nearly twice as much H_2O_2 in the medium than did uninduced cells. The induced cells accumulated even greater amounts of H_2O_2 when 1 mM azide was added to the reaction mixture (Table 6.3). Additional experiments (Table 6.4) showed that the addition of 1 mM azide to the reaction mixture strongly inhibited the catalase activity of both induced and uninduced cells; but as was demonstrated previously (27), the addition of azide also stimulated MnO_2-reducing activity of induced cells nearly two-fold over untreated cells. Uninduced cells were not affected by the addition of azide (Table 6.4). These results strongly suggest that the production of excess H_2O_2 was responsible for the elevated Mn-reducing activity of induced cells and that azide stimulated this activity by inhibiting catalase. Further support for this hypothesis is provided by the recent work of de Vrind et al. (15a), who showed that azide stimulated the Mn-reducing activity in cell suspensions of a marine *Bacillus* only at nonlimiting O_2 concentrations.

These results demand that the previous model for the MnO_2-reducing system of *Bacillus* 29 (27; see also Section 6.4.2) be modified. In the modified model (Fig. 6.2) it is proposed that MnO_2 induces the synthesis of H_2O_2-producing enzymes in *Bacillus* 29 as part of or in addition to the inducible component(s) of the MnO_2 reductase system. As indicated in Fig. 6.2, when glucose is oxidized by Mn^{2+}-induced *Bacillus* 29 cells, H_2O_2 is produced in excess, which promotes increased MnO_2 reduction. Furthermore, the model predicts that the situation by azide of the MnO_2-reducing activity of induced cells actually resulted from inhibition of catalase by azide, which would allow for the accumulation of even more H_2O_2 and consequently more MnO_2 reduction. The work of de Vrind et al. (15a) suggests that such a model may apply to other *Bacillus* sp. as well.

It may be fruitful to speculate on the possible identity of the Mn^{2+}-inducible components(s) of the MnO_2-reducing system of *Bacillus* 29. Because NADH oxidases frequently contain flavoproteins, and because the Mn^{2+}-inducible system is inhibited by atebrine, a flavoprotein inhibitor, it is possible that the inducible component is a flavoprotein oxidase that generates H_2O_2. Such a flavoprotein may also convey electrons to the terminal metalloenzyme of the MnO_2 reductase system, as proposed previously (27). Another possibility is that the Mn^{2+} induces the synthesis of a superoxide dismutase (SOD), another H_2O_2-generating enzyme. Recently, a Mn^{2+}-inducible MnSOD has been described in *E. coli* (55). Significantly, the SOD activity of *E. coli*, like the *Bacillus* 29 MnO_2 reductase activity (see Section 6.4.2), was induced by Mn^{2+} only in the presence of O_2. Both the oxidase and SOD are indicated in parentheses in the new model (Fig. 6.2) to emphasize that the involvement of either component is strictly speculative. To proceed further with the identification of the inducible component(s) of this system, further characterization of the components will be necessary.

6.5 IMPORTANCE OF CELLULAR BINDING

Several workers have reported that the crystalline form of Mn and Fe oxide affects its reduction. As indicated in Section 6.3.2, amorphous or poorly crystallized oxides are more readily reduced by microorganisms than are highly crystalline minerals (15, 21, 22, 34, 39a, 41, 48, 68). Although this phenomenon is probably best explained by the less-ordered minerals being more easily reduced chemically, it might also be explained as an effect of enzymes that recognize specific binding sites at the surface of the different mineral types. In either case, direct contact of microbial cells to the oxide surface (cellular binding) should enhance the rate and extent of reduction. If enzymatic reactions such as those discussed in Sections 6.4.2 and 6.4.3 are involved, tight binding would be essential for the proper operation of membrane-associated terminal reductases, which must contact oxide in order to reduce it. If metabolic end products (see Section 6.4.4) are involved in the reduction mechanism, looser binding could be tolerated. In the latter case, the tighter the binding of cells, the higher the concentration of metabolite would be at the oxide surface and therefore the greater would be the rate of reduction. In support of this argument is recent experimental evidence (33, 62) showing that the mechanisms of nonbiological Fe and Mn oxide reduction by organic compounds may depend to a large extent on adsorption of the organic reductant to the oxide surfaces.

In some cases, Fe oxide reduction by metabolic end products has been tested by separating active cells from Fe oxide, either by a dialysis membrane (36, 42) or by centrifuging the active cells and subsequently testing the cell-free supernatant for activity (7, 15). Generally, little or no oxide is reduced during these tests, leading to the questionable conclusion that extracellular products do not contribute significantly to reduction by the organism involved. In one case, however, 30% of the Fe reduced by a *Vibrio* sp. was attributed to indirect reduction by extracellular components (36). Furthermore, a close examination of the original data in the papers cited above shows that a small amount of Fe^{2+} above background usually was released by the cell-free medium. Based on the data available, it can be concluded that Fe reduction by metabolic end products is not important if cells are widely separated from the oxide. However, in most natural systems bacterial cells probably are bound to oxide-containing particles, which would enhance their ability to cause reduction. Thus it can also be concluded that in natural systems, cellular binding should greatly enhance Fe reduction caused by end products of metabolism.

Using the same arguments, cellular binding should also enhance both enzymatic and end-product reduction of Mn oxides. Although few experiments have been done to test this hypothesis directly, several observations on *Bacillus* 29 suggest that cellular binding may be important in MnO_2 reduction. First, increasing concentrations of Mg^{2+} and Ca^{2+} enhanced the MnO_2-reducing activity of *Bacillus* 29 biphasically (26), suggesting that increasing the divalent cation concentration may have affected cellular binding as well as enzyme activity. In contrast, when concentrations of Na^+ and K^+ were increased, optimum concentrations of the cations were observed above which MnO_2 reduction was inhibited. Second, whole cells were very tightly bound to synthetic Mn-Fe oxide when viewed by electron microscopy (27; Fig. 6.3). From these observations it was argued (27) that if cellular binding is an important factor in MnO_2 reduction, whole cells should be

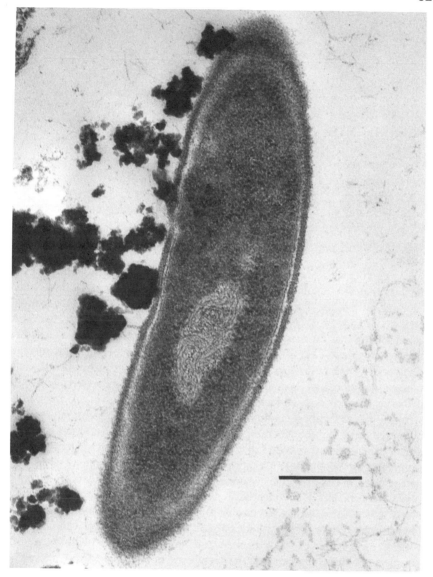

Figure 6.3 Transmission electron photomicrograph of a *Bacillus* 29 cell attached to synthetic Mn-Fe oxide (see Ref. 27 for details of EM preparation). Note that the oxide particles are very tightly bound to the cell wall fabric. Bar = 0.25 μm.

more active than cell-free extracts because of the possibility of strong binding of the cell wall to Mn oxide. In fact, when the activity of *Bacillus* 29 cells was compared with the activities of extracts, invariably the cells had the greater specific activity (26, 27). These observations indicate that the binding of *Bacillus* 29 wall polymers to the oxide surface, as suggested by tight binding of oxide particles to the wall fabric (Fig. 6.3), could bring Mn oxide-reducing enzymes in the membrane or a continuous supply of H_2O_2 produced at the cell surface, in very close proximity to the oxide surface, resulting in the higher rate of reduction observed for whole cells.

6.6 MICROBIAL Mn AND Fe REDUCTION IN NATURAL ENVIRONMENTS

The previous sections of this chapter show clearly that Mn- and Fe-reducing microorganisms can be isolated from a wide variety of natural environments and that their mechanisms of Mn and Fe reduction in laboratory cultures may be quite varied, including both respiratory enzyme systems and nonezymatic reactions that may be enhanced by cellular binding. The intent of this final section is to assess the actual significance of microorganisms in the processes of Mn and Fe reduction in nature.

A large volume of indirect evidence on this subject has been produced by soil scientists and soil microbiologists, particularly those interested in the effects of waterlogging and the availability of Fe and Mn in soil and on the formation of gley soils (2, 54, 56, 57). Geochemists and aquatic microbiologists interested in cycling of Mn and Fe and mineralization of organic matter in marine and freshwater environments (21, 36, 37, 39a, 45, 48, 59, 60, 69, 71) have also contributed. Usually, the primary goal was to show that microorganisms do participate in the natural processes. Specific mechanisms of microbial reduction under natural conditions have rarely been identified or studied in detail.

By and large, the most common approach has been to bring natural samples of soil or sediment into the laboratory for experiments under controlled conditions. In such experiments it is hoped that the natural populations of microorganisms in the samples are maintained in a close-to-natural state. This strategy offers an advantage over studies of axenic cultures, in which natural conditions are difficult or impossible to maintain. Frequently, metabolic poisons or heat treatments that adversely affect biological activity are employed as controls for biological activity. However, the results obtained with poisons must be interpreted with caution because, in most instances, the precise interactions of the poison with the microorganisms and other components in the sample are not known. Despite these drawbacks, the approach has been employed successfully by numerous workers. The results demonstrate that a large portion of the Mn and Fe reduction that occurs in nature may be mediated by microorganisms. Furthermore, in a few cases, especially where Fe reduction in sediments have been investigated (36, 37, 39a, 60), biological mechanisms of the reduction processes can be inferred.

6.6.1 Examples from Anaerobic Sediments

Microbial Fe reduction was studied by Sørensen (60) in anaerobic slurries of marine surface sediment obtained from three different coastal locations in Denmark.

Measurements of Fe^{2+} released from the sediment (Fe reduction), NO_3^- depletion (NO_3^- reduction), and $^{35}SO_4^{2-}$ conversion to HCl-volatile ^{35}S-sulfide (SO_4^{2-} reduction) showed that Fe reduction occurred immediately after most of the endogenous NO_3^- was reduced, but before the onset of SO_4^{2-} reduction. When NO_3^- and NO_2^- were added, Fe and SO_4^{2-} reduction was prevented until the added NO_3^- or NO_2^- was depleted. Pasteurization blocked all activity. The possibility that chemical oxidation of Fe^{2+} by NO_2^- (40) had caused the inhibition by NO_3^- and NO_2^- was ruled out, because NO_2^- consumption was not observed in pasteurized slurries. Overnight aeration of the sediment samples before anaerobic conditions were imposed did not affect the capacity of the sediments to reduce Fe or NO_3^-, but it did lower their capacity for SO_4^{2-} reduction. Pretreatment with NO_3^- increased the capacity for Fe reduction but did not affect the capacity for SO_4^{2-} reduction. Addition of MoO_4^{2-} completely inhibited SO_4^{2-} reduction without affecting the rate of Fe reduction, indicating that SO_4^{2-}-reducing bacteria, or the sulfide they produced, were not involved in the reduction of Fe. Sørensen (60) concluded from these results that facultatively anaerobic, NO_3^--reducing microorganisms were actively involved in Fe reduction. Furthermore, the results suggested that NO_3^--reducing bacteria similar to those studied in axenic cultures by Ottow and co-workers (42, 47, 48, 50, 52) may have been responsible. Such bacteria may have employed their NO_3^- reductase enzyme systems to reduce Fe in the sediment. Indeed, the observed inhibition of Fe reduction by NO_3^- and NO_2^- and the stimulation by pretreatment with NO_3^- would be expected if NO_3^- reductase systems were involved.

Although an enzymatic mechanism (i.e., NO_3^- reductase) is implied by these results, indirect mechanisms cannot be ruled out completely. Reduction by H_2S produced by SO_4^{2-} reducers probably was not involved; however, reduction by organic metabolites such as formate or other organic acids that may have been produced during anaerobic incubation of the sediments (see below) is a possibility that was not investigated.

Jones et al. (36, 37) conducted somewhat more complete experiments to investigate the involvement of bacteria in Fe^{3+} reduction in sediments of an English lake. In their initial work (36), the rate of Fe^{2+} released (Fe reduction) was measured in sealed N_2-filled tubes in which sediment cores were overlayed with a defined salts solution based on the composition of the lake water. Fe was not reduced appreciably until O_2 and NO_3^- were depleted in the overlying water. The Fe-reducing activity was optimal at a temperature of 30°C and was inhibited equally by $HgCl_2$ and heating at 90°C. These experiments suggest that Fe reduction was mediated, at least partly, by NO_3^--reducing bacteria similar to those implicated in Sørensen's studies (60).

To study further the Fe-reducing microorganisms in the sediments, enrichment cultures were prepared from the top 1 cm of sediment. A carbon and energy source was added and the cultures were incubated in serum vials under anaerobic conditions. Fe-reducing activity of bacteria in the cultures was suppressed by NO_3^-, ClO_3^-, MnO_2, Mn_2O_3, and O_2. A glucose-fermenting *Vibrio* sp. was isolated that required contact with Fe_2O_3 to reduce it maximally. However, when cells were separated from the oxide by a dialysis membrane or when cell-free supernatants were tested, it was found that at least 30% of the Fe reduced by a cell suspension could be accounted for by reactions with extracellular components (probably the end products of glucose metabolism). Fe-reducing activity in whole cells and cell-free extracts was inhibited by rotenone, HOQNO, antimycin A, and azide, suggesting that respiratory enzymes were involved; however, because molar growth yields of the cultures were not affected by Fe oxide, the authors concluded that the *Vibrio* sp. probably used Fe as a sink for excess reductant rather than as a terminal electron acceptor of an energy-yielding pathway.

In addition to the *Vibrio* sp., a gram-negative, coccoid, facultatively anaerobic, facultatively chemolithotrophic, H_2-oxidizing bacterium was isolated in this study. This bacterium was similar in its physiological properties to the H_2-oxidizing, Fe-reducing pseudomonad studied by Balashova and Zavarzin (4), both organisms being capable of using H_2 as a major source of reducing power for Fe reduction and of increasing their biomass yield anaerobically by coupling H_2 oxidation to Fe reduction. These bacteria clearly are examples of bacteria that can couple anaerobic iron respiration to energy-yielding systems of the cell using Fe oxide as the alternative electron acceptor. The biochemical properties of their iron reductase enzyme system remain to be elucidated.

In further studies by Jones et al. (37), it was found that the addition of $FeCl_3$ to sediments decreased the amount of volatile fatty acids which accumulated and that the addition of glucose, ethanol, and malate stimulated the microbial reduction of naturally occurring Fe compounds in the sediment. These results can be compared with those of Lovley and Phillips (39a), who showed that glucose, acetate, H_2, proprionate, butyrate, ethanol, and methanol stimulated Fe reduction in anaerobic sediment enrichments ammended with iron oxides. These authors concluded that Fe^{3+} reduction may out-compete methanogenic food chains for sediment organic matter, especially when amorphous ferric oxyhydroxides are available.

During the study by Jones et al. (37), a malate-fermenting *Vibrio* sp. was isolated from the enrichment cultures. Like many other facultatively anaerobic Fe-reducing bacteria (Table 6.2), the malate-fermenting *Vibrio* sp. was also capable of reducing MnO_2 and NO_3^-, both of which inhibited Fe reduction. Therefore, it appeared that this *Vibrio* sp., like a strain of *Bacillus polymyxa* studied by Munch and Ottow (42), preferentially reduced NO_3^- and MnO_2 before $Fe(OH)_3$. Interestingly, the addition of MnO_2 to anaerobic cultures increased the molar growth yield of cells by 20%. These results suggested that this *Vibrio* sp. may be able to employ both MnO_2 and $Fe(OH)_3$ as alternative electron acceptors for anaerobic respiration.

A few studies have been reported on microbial Mn reduction in sediments that can be compared to the work on Fe reduction. For example, the work of Wollast *et al.* (69) and Burdige and Nealson (12, 12a) shows that an anaerobic enrichment cultures derived from marine sediment, mixed populations of bacteria were involved in Mn oxide reduction. However, no isolates were obtained for enzymatic characterization. On the other hand, Burdige (12) has demonstrated that SO_4^{2-}-

reducing bacteria probably participate in Mn oxide reduction in stratified marine waters by producing H_2S, which can reduce Mn abiotically.

It should also be recognized that biologically mediated Mn reduction can occur under aerobic conditions. For example, in wetland environments, where Mn oxidation is catalyzed by *Leptothrix* and other Fe- Mn-depositing bacteria in highly active surface films (24), Mn reduction may also occur by reaction with metabolic end products, especially in the root zone of *Lemna* sp., where organic acids and hydrogen peroxide may be produced as a result of photosynthetic activity.

The work discussed above reveals the variety and complexity of processes that may be involved in Mn and Fe reduction anaerobic environments. It can be concluded from this work that both enzymatic and abiotic mechanisms are important and that both must be understood before a complete understanding of the role of Mn and Fe in anaerobic processes is achieved.

6.7 SUMMARY AND CONCLUSIONS

This chapter began with a discussion of problems facing students and researchers interested in elucidating the role of microorganisms in the natural processes of Mn and Fe reduction. The widespread occurrence of Mn- and Fe-reducing bacteria and fungi in natural environments was noted and the literature on their enumeration, isolation, cultivation, and study in laboratory cultures was surveyed. Studies on the mechanisms of microbial Mn and Fe reduction under laboratory conditions show that both Mn and Fe oxides can be reduced enzymatically by bacterial respiratory enzyme systems that may or may not yield ATP when the oxides are employed as alternative electron acceptors.

For Fe reduction, many bacteria appear to use NO_3^- reductase systems that can employ either chelated Fe^{3+} or Fe oxide as alternative electron acceptors; however, it is unclear whether these systems provide energy to the cell during Fe reduction. Non-energy-yielding branches or side reactions of the NO_3^- reductase systems or O_2-terminated respiratory chains may also reduce Fe. These systems may operate to some extent in the presence of O_2, NO_3^-, and Mn oxides, but most are inhibited by them. Energy-yielding systems that may be involved in Fe reduction are not well characterized.

In some respects, bacterial Mn-reducing enzyme systems appear to be similar to Fe-reducing systems. For example, bacteria appear to employ Mn oxides as alternative electron acceptors in branches or side reactions of common respiratory systems. On the other hand, it is unclear whether NO_3^- reductase systems are employed for Mn oxide reduction or if energy is derived during bacterial electron transport to Mn oxide. In general, Mn-reducing enzyme systems are not inhibited by Fe oxide and they are less sensitive to NO_3^- and O_2 inhibition than are Fe-reducing systems.

Microorganisms can also mediate Mn and Fe reduction indirectly via abiotic reactions with their metabolic end products. It has been demonstrated that common microbial metabolites such as H_2S, formate, oxalate, and other organic acids can reduce Mn and Fe chemically, and therefore production of these metabolites may be involved in the microbial reduction processes. Acidic environmental conditions generated by microbial metabolism generally enhance the reduction of Mn and Fe oxides by end products of metabolism. Such reactions would be favored in anaerobic environments. Under aerobic conditions when organic substances are being oxidized, enough H_2O_2 may be produced to reduce Mn oxide. This reaction would be favored in aerobic environments containing excess organic matter.

Cellular binding will favor both enzymatic and metabolic end-product reactions. The rates of both types of reactions will be enhanced by cellular contact with oxide surfaces. It was argued that Mn- or Fe-reducing enzymes in microbial cells must contact their insoluble substrate in order to affect reduction. Similarly, reduction by metabolic end products should proceed at a higher rate when cells are bound to the oxide surface, because a constant supply and higher concentration of the end product will be available at reactive sites. It can be concluded that cellular binding is a very important factor that must not be overlooked in any proposed mechanism of microbial Mn and Fe reduction.

Finally, the question of the actual contribution of microorganisms to Fe reduction in natural environments was considered. Recent studies on microbial Fe reduction in anaerobic sediments reveal that microorganisms play a major role in the natural processes of Fe reduction. As might be predicted from axenic culture studies, facultatively anaerobic NO_3^--reducing bacteria appeared to be involved in Fe reduction in both marine and freshwater sediments. Detailed investigations of bacteria isolated from freshwater sediments show that enzymatic respiratory pathways and reactions with metabolic end products are involved in microbially mediated Fe reduction. These studies support the conclusion that microbial Fe reduction in anaerobic environments is a complex process in which enzymatic and nonenzymatic mechanisms may operate simultaneously.

Comparable studies on microbial Mn reduction in anaerobic sediments are fewer in number, but they indicate that similar enzymatic and nonenzymatic mechanisms may be involved in natural Mn reduction processes. However, because of the limited evidence available, it is not yet possible to form definite conclusions on Mn reduction. Furthermore, it is necessary to consider additional mechanisms to account for Mn reduction in aerobic environments.

Note added in proof

Several papers on microbial iron reduction in freshwater sediments have been published since this chapter was completed, (5a, 39a, 39b, 39c, 39d). These papers emphasize the key role that iron reduction may play in carbon mineralization in anaerobic sediments. In one case (39c) a dissimilatory iron-reducing bacterium was isolated that, apparently, couples anaerobic oxidation of organic matter to iron oxide reduction, producing magnetite extracellularly. Magnetite was also found to be the product of dissimilatory iron reduction carried out by consortia of anaerobic bacteria, which simultaneously reduced iron oxide and altered pH and Eh conditions to favor magnetite formation (5a). The existence of dissimilatory iron-reducing bacteria in modern anaerobic sediments lends support to the recent hypothesis that oxidized iron may have been a major electron acceptor for reduced organic compounds in the anaerobic Archaean biosphere (68a).

ACKNOWLEDGMENTS

Some of the research reported in this chapter was done while I was supported by a fellowship from the Alexander von Humboldt Foundation in the laboratory of Peter Hirsch in Kiel, West Germany. Subsequently, support was provided by a National Science Foundation grant DAR-7924494. Presently, research in my laboratory is funded by the U.S. Environmental Protection Agency through Cooperative Agreement CR81148-02.

The experiments reported in Tables 6.3 and 6.4 were part of an undergraduate honors thesis project by Lynn M. Kozma. The electron micrograph in Fig. 6.3 was made while I worked in the laboratory of Mercedes R. Edwards at the New York State Department of Health, Albany.

The skillful secretarial assistance of Patti L. Lisk is gratefully acknowledged.

References

1. Adeny, W. F. 1894. On the reduction of manganese peroxide in sewage. Sci. Proc. R. Soc. Edinb. Nat. Sci. 3:247–251.

2. Alexander, M. 1977. Introduction to soil microbiology, 2nd ed. Wiley, New York.

3. Aristovskaya, T. V., and G. A. Zarvarzin. 1971. Biochemistry of iron in soil, p. 385–408, in A. D. McLaren and J. Skujins (eds.), Soil biochemistry, vol. 2. Marcel Dekker, New York.

3a. Arnold, R. G., T. J. DiChristina, and M. R. Hoffmann. 1986. Inhibitor studies of dissimilative Fe(III) reduction by Pseudomonas sp. Strain 200 ("Pseudomonas ferri-reductans"). Appl. Environ. Microbiol. 52:281–289.

4. Balashova, V. V., and G. A. Zavarzin. 1979. Anaerobic reduction of ferric iron by hydrogen bacteria. Microbiology (Engl. Transl. Mikrobiologiya) 48:635–639.

5. Bautista, E. M., and M. A. Alexander. 1972. Reduction of inorganic compounds by soil microorganisms. Soil Sci. Soc. Am. Proc. 36:918–920.

5a. Bell, P. E., A. L. Mills, and J. S. Herman. 1987. Biogeochemical conditions favoring magnetite formation during anaerobic iron reduction. Appl. Environ. Microbiol. 53: 2610–2616.

6. Brock, T. D., and J. Gustafson. 1976. Ferric iron reduction by sulfur- and iron-oxidizing bacteria. Appl. Environ. Microbiol. 32:567–571.

7. Bromfield, S. M. 1954. Reduction of ferric compounds by soil bacteria. J. Gen. Microbiol. 11:1–6.

8. Bromfield, S. M. 1954. The reduction of iron oxide by bacteria. J. Soil Sci. 5:129–139.

9. Bromfield, S. M. 1956. Oxidation of manganese by soil microorganisms. Aust. J. Biol. Sci. 9:238–252.

10. Bromfield, S. M., and D. J. David. 1976. Sorption and oxidation of manganous ions and reduction of manganese oxide by cell suspensions of a manganese oxidizing bacterium. Soil Biol. Biochem. 8:37–43.

11. Buchanan, R. E., and N. E. Gibbons (eds.). 1974. Bergey's manual of determinative bacteriology, 8th ed. Williams & Wilkins, Baltimore.

12. Burdige, D. J. 1983. The biogeochemistry of manganese redox reactions: rates and mechanisms. Ph.D. dissertation, University of California, San Diego.

12a. Burdige, D. J., and K. H. Nealson. 1985. Microbial manganaese reduction by enrichment cultures from coastal marine sediments. Appl. Environ. Microbiol. 50:491–497.

13. Cox, C. D. 1980. Iron reductases from *Pseudomonas aeruginosa*. J. Bacteriol. 141:199–204.

14. Dailey, H. A., and J. Lascelles. 1977. Reduction of iron and synthesis of protoheme by *Spirillum itersonii* and other organisms. J. Bacteriol. 129:815–820.

15. DeCastro, A. F., and H. L. Ehrlich. 1970. Reduction of iron oxide minerals by a marine *Bacillus*. Antonie Leeuwenhoek J. Microbiol. Serol. 36:317–327.

15a. deVrind, J. P. M., F. C. Boogerd, and E. W. deVrind-deJong. 1986. Manganese reduction by a marine *Bacillus* species. J. Bacteriol. 167:30–34.

16. Dubinina, G. A. 1979. Mechanism of the oxidation of divalent iron and manganese by iron bacteria growing at neutral pH of the medium. Microbiology (Engl. Transl. Mikrobiologiya) 47:471–478.

17. Dubinina, G. A. 1979. Functional role of bivalent iron and manganese oxidation in *Leptothrix pseudoochraceae*. Microbiology (Engl. Transl. Mikrobiologiya) 47:631–636.

18. Ehrlich, H. L. 1963. Bacteriology of managese nodules. I. Bacterial action on manganese in nodule enrichments. Appl. Microbiol. 11:15–19.

19. Ehrlich, H. L. 1966. Reactions with managanese by bacteria from marine ferromanganese nodules. Dev. Ind. Microbiol. 7:279–286.

20. Ehrlich, H. L. 1980. Bacterial leaching of manganese ores, p. 609–614, in P. A. Trudinger, M. R. Walter, and B. J. Ralph (eds.), Biogeochemistry of ancient and modern environments. Springer-Verlag, New York.

21. Ehrlich, H. L. 1981. Geomicrobiology. Marcel Dekker, New York,

22. Fischer, W. R., and T. Pfanneberg. 1984. An improved method for testing the rate of iron (III) oxide reduction by bacteria. Zentralbl. Mikrobiol. 139:163–166.

23. Ghiorse, W. C. 1980. Electron microscopic analysis of metal-depositing microorganisms in surface layers of Baltic Sea ferromanganese concretions, p. 345–354, in P. A. Trudinger, M. R. Walter, and B. J. Ralph (eds.), Biogeochemistry of ancient and modern environments. Springer-Verlag, New York.

24. Ghiorse, W. C. 1984. Bacterial transformations of manganese in wetland environments, p. 615–622, in M. J. Klug and C. A. Reddy (eds.), Current perspectives in microbial ecology. American Society for Microbiology, Washington, D.C.

25. Ghiorse, W. C. 1984. Biology of iron- and manganese-depositing bacteria. Annu. Rev. Microbiol. 38:515–520.

26. Ghiorse, W. C., and H. L. Ehrlich. 1974. Effects of seawater cations and temperature

on manganese dioxide-reductase activity in a marine *Bacillus*. Appl. Microbiol. 28:785–792.

27. Ghiorse, W. C., and H. L. Ehrlich. 1976. Electron transport components of the MnO_2 reductase system and location of the terminal reductase in a marine *Bacillus*. Appl. Environ. Microbiol. 31:977–985.

28. Ghiorse, W. C., and P. Hirsch. 1978. Iron and manganese deposition by budding bacteria, p. 897–909, in W. E. Krumbein (eds.), Environmental biogeochemistry and geomicrobiology, Vol. 3. Ann Arbor Science Publishers, Ann Arbor, Mich.

29. Ghiorse, W. C., and P. Hirsch. 1982. Isolation and properties of ferromanganese-depositing budding bacteria from Baltic Sea ferromanganese concretions. Appl. Environ. Microbiol. 43:1464–1472.

30. Gomah, A. M., M. S. Awad Soliman, and A. S. Abdel-Ghaffar. 1980. Manganese mobility in Egyptian soils as affected by inoculation with manganese reducing organisms. Z. Pflanzenernaehr. Bodenkd. 143:274–281.

31. Gottfreund, J., and R. Schweisfurth. 1983. Mikrobiologische Oxidation und Reduktion von Manganspecies. Fresenius Z. Anal. Chem. 316:634–638.

32. Hochster, R. M., and J. H. Quastel. 1952. Manganese dioxide as a terminal hydrogen acceptor in the study of respiratory systems. Arch. Biochem. Biophys. 36:132–146.

33. Jauregui, M. A., and H. M. Reisenauer. 1982. Dissolution of oxides of manganese and iron by root exudate components. Soil. Sic. Soc. Am. J. 46:314–317.

34. Jones, J. G. 1983. A note on the isolation and enumeration of bacteria which deposit and reduce ferric iron. J. Appl. Bacteriol. 54:305–310.

35. Jones, J. G., W. Davidson, and S. Gardener. 1984. Iron reduction by bacteria: range of organisms involved and metals reduced. FEMS Microbiol. Lett. 21:133–136.

36. Jones, J. G., S. Gardener, and B. M. Simon. 1983. Bacterial reduction of ferric iron in a stratified eutrophic lake. J. Gen. Microbiol. 129:131–139.

37. Jones, J. G., S. Gardener, and B. M. Simon. 1984. Reduction of ferric iron by heterotrophic bacteria in lake sediments. J. Gen. Microbiol. 130:45–51.

38. Lascelles, J., and K. A. Burke. 1978. Reduction of ferric iron by L-lactate and DL-glycerol-3-phosphate in membrane preparations from *Staphylococcus aureus* and interactions with the nitrate reductase system. J. Bacteriol. 134:585–589.

39. Leighton, F., B. Poole, H. Beaufay, P. Baudhuin, J. W. Coffey, S. Fowler, and C. de Duve. 1968. The large-scale separation of peroxisomes, mitochondria, and lysosomes from the livers of rats injected with Triton WR01339. Improved isolation procedures, automated analysis, biochemical and morphological properties of fractions. J. Cell Biol. 37:482–513.

39a. Lovley, D. R., and E. J. P. Phillips. 1986. Organic matter mineralization with reduction of ferric iron in anaerobic sediments. Appl. Environ. Microbiol. 51:683–689.

39b. Lovley, D. R., and E. J. P. Phillips. 1987. Rapid assay for microbially reducible ferric iron in aquatic sediments. Appl. Environ. Miorobiol. 53:1536–1540.

39c. Lovley, D. R., and E. J. P. Phillips. 1987. Competitive mechanisms for inhibition of sulfate reduction and methane production in zones of ferric iron reduction in sediments. Appl. Env. Microbiol. 53:2636–2641.

39d. Lovley, D. R., J. F. Stolz, G. L. Nord, Jr., and E. J. P. Phillips. 1987. Anaerobic production of magnetite by a dissimilatory iron-reducing microorganism. Nature 330:252–254.

40. Moraghan, J. T., and R. J. Buresh. 1977. Chemical reduction of nitrite and nitrous oxide by ferrous iron. Soil Sci. Soc. Am. J. 41:47–50.

41. Munch, J. C., and J. C. G. Ottow. 1980. Preferential reduction of amorphous crystalline iron oxides by bacterial activity. Soil Sci. 129:15–21.

42. Munch, J. C., and J. C. G. Ottow. 1983. Reductive transformation mechanism of ferric oxides in hydromorphic soils. p. 383–394, in R. Hallberg (ed.), Environmental biogeochemistry. Swedish Natural Scientific Research Council, Stockholm, Ecol. Bull. (Stockholm) No. 35.

43. Nealson, K. H. 1983. The microbial iron cycle, p. 159–190, in W. E. Krumbein (ed.), Microbial geochemistry. Blackwell Scientific, Boston.

44. Nealson, K. H. 1983. The microbial manganese cycle, p. 191–222, in W. E. Krumbein (ed.), Microbial geochemistry. Blackwell Scientific, Boston.

45. Nealson, K. H. 1983. Microbial oxidation and reduction of manganese and iron, p. 459–479, in P. Westbroek and E. W. de Jong (eds.), biomineralization and biological metal accumulation. D. Reidel, Boston.

46. Neilands, J. B. 1984. Siderophores of bacteria and fungi. Microbiol. Sci. 1:9–14.

46a. Obuekwe, C. O., and D. W. S. Westlake. 1982. Effects of medium composition on cell pigmentation, cytochrome content, and ferric iron reduction in a *Pseudomonas* sp. isolated from crude oil. Can. J. Microbiol. 28:989–992.

46b. Obuekwe, C. O., D. W. S. Westlake, and F. D. Cook. 1981. Effect of nitrate ion on reduction of ferric iron by a bacterium isolated from crude oil. Can. J. Microbiol. 27:692–697.

47. Ottow, J. C. G. 1968. Evaluation of iron-reducing bacteria in soil and the physiological mechanism of iron-reduction in *Aerobacter aerogenes*. Z. Allg. Mikrobiol. 8:441–443.

48. Ottow, J. C. G. 1969. Der Einfluss von Nitrat, Chlorat, Sulfat, Eisenoxidform und Wachstumsbedingungen auf das Ausmass der bakteriellen Eisenreduction. Z. Pflanzenernaehr. Bodenkd. 124:238–253.

49. Ottow, J. C. G. 1969. The distribution and differentiation of iron-reducing bacteria in gley soils. Zentralbl. Bakteriol. Parasitenkd. Infektionskr. Hyg. 2 Abt. 123:600–615.

50. Ottow, J. C. G. 1970. Selection, characterization and iron-reducing capacity of nitrate reductaseless (nit⁻) mutants of iron-reducing bacteria. Z. Allg. Mikrobiol. 10:55–62.

51. Ottow, J. C. G., and H. Glathe. 1973. Pedochemie und Pedomikrobiologie hydromorpher Böden. Markmale, Voraussetzungen und Ursache der Eisenreduction. Chem. Erde 32:1–44.

52. Ottow, J. C. G., and J. C. Munch. 1978. Mechanisms of reductive transformations in the anaerobic microenvironment of hydromorphic soils, p. 483–491, in W. E. Krumbein (ed.), Environmental biogeochemistry and geomicrobiology, Vol. 2. Ann Arbor Science Publishers, Ann Arbor, Mich.

53. Ottow, J. C. G., and A. von Klopotek. 1969. Enzymatic reduction of iron oxide by fungi. Appl. Microbiol. 18:41–43.

54. Patrick, W. H., Jr., and R. E. Henderson. 1981. Reduction and reoxidation cycles of manganese and iron in flooded soil and water solution. Soil Sci. Soc. Am. J. 45:855–859.

55. Pugh, S. Y. R., J. L. DiGuiseppi, and I. Fridovich. 1984. Induction of superoxide dismutases in *Escherichia coli* by manganese and iron. J. Bacteriol. 160:137–142.

56. Schwab, A. P., and W. L. Lindsay. 1983. Effect of redox on the solubility and availability of iron. Soil Sci. Soc. Am. J. 47:201–220.

57. Schwab, A. P., and W. L. Lindsay. 1983. The effect of redox on the solubility and availability of manganese in a calcareous soil. Soil Sci. Soc. Am. J. 47:217–220.

58. Shindo, H., and P. M. Huang. 1984. Significance of Mn(IV) oxide in abiotic formation of organic nitrogen complexes in natural environments. Nature (Lond.) 308:57–58.

58a. Short, K. A., and R. P. Blakemore. 1986. Iron respiration-driven proton translocation in aerobic bacteria. J. Bacteriol. 167:729–731.

59. Sokolova-Dubinina, G. A., and Z. P. Deryugina. 1967. On the role of microorganisms in the formation of rhodochrosite in Punnus-Yarvi Lake. Microbiology (Engl. Transl. Mikrobiologiya) 36:445–451.

60. Sørensen, J. 1982. Reduction of ferric iron in anaerobic, marine sediment and interaction with reduction of nitrate and sulfate. Appl. Environ. Microbiol. 43:319–324.

61. Starkey, R. L., and H. O. Halvorson. 1927. Studies on the transformations of iron in nature. II. Concerning the importance of microorganisms in the solution and precipitation of iron. Soil Sci. 24:381–402.

62. Stone, A. T., and J. J. Morgan. 1984. Reduction and dissolution of manganese(III) and manganese(IV) oxides by organics. 1. Reaction with hydroquinone. Environ. Sci. Technol. 18:450–456.

63. Stone, A. T., and J. J. Morgan. 1984. Reduction and dissolution of manganese(III) and manganese(IV) oxides by organics. 2. Survey of the reactivity of organics. Environ. Sci. Technol. 18:617–624.

64. Stumm, W., and J. J. Morgan. 1981. Aquatic chemistry, 2nd ed. Wiley, New York.

65. Trimble, R. B., and H. L. Ehrlich. 1968. Bacteriology of manganese nodules. III. Reduction of MnO_2 by two strains of nodule bacteria. Appl. Microbiol. 16:695–702.

66. Trimble, R. B., and H. L. Ehrlich. 1968. Bacteriology of manganese nodules. IV. Induction of an MnO_2-reductase system in a marine *Bacillus*. Appl. Microbiol. 19:966–972.

67. Troshanov, E. P. 1968. Iron- and manganese-reducing microorganisms in ore-containing lakes of the Karelian Isthmus. Microbiology (Engl. Transl. Mikrobiologiya) 37:786–791.

68. Troshanov, E. P. 1969. Conditions affecting the reduction of iron and manganese by bacteria in the ore-bearing lakes of the Karelian Isthmus. Microbiology (Engl. Transl. Mikrobiologiya) 38:528–535.

68a. Walker, J. C. G. 1987. Was the Archaean biosphere upside down? Nature 329:710–712.

69. Wollast, R., G. Billen, and J. C. Duinker. 1979. Behavior of manganese in the Rhine and Scheldt estuaries. I. Physico-chemical aspects. Estuarine Coastal Mar. Sci. 9:161–169.

70. Woolfolk, C. A., and H. R. Whiteley. 1962. Reduction of inorganic compounds with molecular hydrogen by *Micrococcus lactilyticus*. I. Stoichiometry with compounds of arsenic, selenium, tellurium, transition and other elements. J. Bacteriol. 84:647–685.

71. Zehnder, A. J. B., and T. D. Brock. 1980. Anaerobic methane oxidation: occurrence and ecology. Appl. Environ. Microbiol. 39:194–204.

7

ANAEROBIC MICROBIAL DEGRADATION OF CELLULOSE, LIGNIN, OLIGOLIGNOLS, AND MONOAROMATIC LIGNIN DERIVATIVES

PATRICIA J. COLBERG*

Swiss Federal Institute for Water Resources and Water Pollution Control (EAWAG), CH-8600 Dübendorf, Switzerland

7.1 INTRODUCTION
7.2 LIGNOCELLULOSE DEFINED
7.3 HYDROLYSIS AND FERMENTATION OF CELLULOSE BY ANAEROBIC MICROORGANISMS
    7.3.1 Cellulose Degradation in the Rumen
    7.3.2 Cellulose Digestion in Termites
    7.3.3 Anaerobic Cellulose Degradation in Aquatic Ecosystems
7.4 ANAEROBIC DEGRADATION OF MONOAROMATIC LIGNIN DERIVATIVES
    7.4.1 Monoaromatic Cleavage during Anaerobic Photometabolism, Nitrate Reduction, and Sulfate Respiration
    7.4.2 Degradation of Lignin-Derived Monomers under Methanogenic Conditions
        7.4.2.1 Mixed Culture Studies with Benzoate

*Present Address: Western Research Institute, P.O. Box 3395, University Station, Laramie, WY 82071.

7.4.2.2   Mixed Culture Studies with other Aromatic Compounds

7.4.2.3   Removal of Aromatic Ring Substituent Groups

7.4.3   Aromatic Ring Fission by Pure Cultures of Anaerobic Bacteria

7.5   ANAEROBIC DEGRADATION OF LIGNIN DERIVATIVES: OLIGOLIGNOLS

7.5.1   Degradation Studies using [^{14}C]Oligolignols Derived from Lignin

7.5.2   Identification of Metabolic Intermediates

7.5.3   Oligolignols Produced by Aerobic Lignolytic Microorganisms

7.6   ANAEROBIC DEGRADATION OF POLYMERIC LIGNIN

7.6.1   Degradation in the Rumen

7.6.2   Degradation in Termites

7.6.3   Degradation in Aquatic Ecosystems

7.7   SUMMARY AND CONCLUSIONS

REFERENCES

7.1   INTRODUCTION

Almost half of the global carbon fixed annually via photosynthesis is incorporated into the lignocellulosic cell walls of arborescent plants (12). However, lignin-derived carbon is highly recalcitrant and, due to its association with cellulose as lignocellulose, probably impedes the turnover of large quantities of photosynthate carbon (37). Lignin decomposition is therefore considered to be the rate-limiting step in the biospheric carbon cycle.

Despite recent progress made in understanding aerobic mechanisms of lignocellulose catabolism (37, 91), relatively little attention has been given to its microbial degradation under anaerobic conditions. The results of one study, in which no degradation of synthetic lignin was observed in a variety of anoxic sediments and soils (63), have led to the general belief that the lignin component of lignocellulose is inert in anaerobic habitats (162, 163). However, recent long-term studies suggest that at least in some aquatic ecosystems, polymeric lignin is degraded slowly and at environmentally significant rates (17). In addition, two recent reports indicate that cleavage of the most common intermonomeric bond in natural lignin (133) is possible in the absence of molecular oxygen (33, 164). Although aerobic degradation of lignin is known to require highly oxidative conditions (53, 59), it is now well established that oxygen is not needed for the biodegradation of aromatic monomers (52), including those derived from lignin (67, 68, 86).

Aerobic microbial communities are the primary lignocellulose decomposers in most natural systems; however, results from several recent studies collectively suggest that the contribution of strictly anaerobic microorganisms to the turnover of lignocellulose-derived carbon has probably been underestimated and merits reevaluation.

The objective of this chapter is to summarize briefly what is currently known of the environmental fate of lignocellulose and its derivatives—cellulose, lignin, oligolignols, and aromatic monomers—under anaerobic conditions. Natural ecosystems, rather than wastewater or biotechnological applications, are emphasized.

7.2 LIGNOCELLULOSE DEFINED

Lignocellulose is a collective term for the three major components of plant vascular tissue: cellulose, hemicellulose, and lignin (see Fig. 7.1). Lignocellulose may constitute as much as 89 to 98% of the dry weight of wood (37). The holocellulose fraction, which includes both cellulose and hemicellulose, is the largest component (63 to 78%), while estimates of 15 to 38% lignin have been made for various hardwood and softwood trees (133).

Cellulose is the most abundant natural organic compound on earth (37) and is located in the primary and secondary cell walls of plants. A piece of one cellulose chain is represented in Fig. 7.1. It is an unbranched polymer of several thousand D-glucose units joined via β-1,4-glycosidic linkages and is insoluble in water. Cellobiose is a disaccharide produced upon partial hydrolysis of cellulose and contains the β-1,4-bond, while complete hydrolysis results in release of glucose.

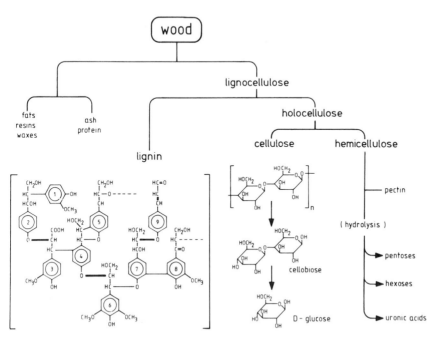

Figure 7.1 Major components of wood, including the chemical structures of lignin (model), cellulose, and its hydrolysis products, cellobiose and D-glucose.

Hemicellulose is a rather ill-defined group of other structural poly saccharides in plants. They are not considered cellulose derivatives, but rather, are linear polymers of D-glucose linked β-1,4 (34); however, hydrolysis products also include L-arabinose, uronic acids, and hexoses such as D-glucose, D-mannose, and D-galactose (21). Pectin is sometimes considered to be a type of hemicellulose (44, 72) and, for simplicity, is depicted as such in Fig. 7.1. It consists mainly of rhamnose and α-1,4-linked D-galacturonic acid residues whose carboxylic acid groups may be methoxylated (132, 134). While young plants have relatively high pectin contents (12 to 30%; 84), it is only a minor component of mature plants and wood (1 to 5%; 35). The microbial degradation of pectin has recently been reviewed by Schink (134).

Lignin, which ranks second only to cellulose as the most abundant naturally occurring polymer in the biosphere (37), is located primarily in the cell walls of vascular plants. It is a highly branched, constitutionally undefined aromatic polymer composed of phenylpropane subunits which are randomly linked by a variety of carbon–carbon and ether bonds. A portion of an "average lignin" is illustrated in Fig. 7.1. The arylglycerol-β-aryl ether bond, located between aromatic rings 2–3, 4–6, and 7–9, is the most common intermonomeric bridge in lignin (2) and is discussed futher in Section 7.5.

One unresolved question concerning the structure of lignocellulose is whether or not lignin and holocellulose are covalently bound. Although lignin physically surrounds but is probably not covalently linked to cellulose (162), available data do suggest the existence of some types of chemical bonds between other plant polysaccharides and lignin (71, 92, 101, 157). Adler (2) suggests that galactose and arabinose may act as links between lignin and hemicellulose. The significance of such interactions and knowledge as to how they affect the recalcitrance of lignocellulose remain unknown.

7.3 HYDROLYSIS AND FERMENTATION OF CELLULOSE BY ANAEROBIC MICROORGANISMS

Hydrolysis is a prerequisite to microbial utilization of complex biopolymers such as cellulose. In anaerobic environments, the initial enzymatic attack on cellulose is dependent on the activity of a relatively select number of microorganisms. The primary hydrolysis products, cellobiose and glucose (see Fig. 7.1), may then be fermented to organic acids, CO_2, H_2, and in the presence of methanogenic bacteria, to CH_4.

Until the recent introduction of techniques for the specific isotopic labeling of the cellulose component of lignocellulose (38), however, the unequivocable study of cellulose degradation in its natural association with lignin and other plant polysaccharides was not possible. Consequently, much of our knowledge of anaerobic cellulose decomposition comes from investigations that used pure cellulose as substrates and isolated organisms from environments in which anaerobic cellulose digestion was known or suspected to occur. One such habitat is the rumen, a strictly anaerobic ecosystem in which the carbohydrate components of lignocellulose are degraded and fermented through the combined efforts of a diverse microbial community. Although the total number of cellulose degraders is probably small (78), several strictly anaerobic, cellulolytic microorganisms have been isolated from the rumen and are listed in Table 7.1. All are strict anaerobes and ferment cellulose to organic acids, which in the ruminant pass to the bloodstream, where they are oxidized by the animal as its main source of energy.

7.3.1 Cellulose Degradation in the Rumen

Bacteria are perhaps the most studied of the rumen inhabitants. Ruminococci (e.g., *Ruminococcus albus*, *R. flavefaciens*) are the most numerous of the bacteria isolated from the bovine rumen (78, 80, 95). Other species of anaerobic, cellulose-degrading bacteria indigenous to the rumen include *Bacteroides succinogenes*, *Butyrivibrio fibrisolvens*, and *Eubacterium cellulosolvens*. Products of cellulose fermentation include succinate, propionate, acetate, butyrate, formate, lactate, H_2, and CO_2 (see Table 7.1).

Cellulolytic enzymes in anaerobic bacteria examined thus far appear to be somewhat different from one another, as well as from those of aerobic organisms (161). In general, however, microbial cellulases have three parts: (a) an exoglucanase that acts on the nonreducing end of the cellulose chain; (b) an endoglucanase that randomly depolymerizes internal units; and (c) a cellobiase that hydrolyzes cellobiose to yield D-glucose units. When purified, each component exhibits characteristic activities (65, 129, 130). *R. albus* produces two extracellular enzyme components, termed the affinity and hydrolytic factors, which when combined hydrolyze cellulose. Even though the hydrolytic factor exhibits enzymatic activity on soluble cellulose derivatives, the role of the affinity factor is to bind the hydrolytic factor to the cellulose polymer (98). In contrast, the cellulase complex of *B. succinogenes* is cell bound and is therefore not released into the surrounding medium (65). Among anaerobic bacteria, the thermophile *Clostridium thermocellum* exhibits the greatest known cellulase activity. Ng and his colleagues (109) recently described the cellulase complex of this organism. It was concluded to be nonconstitutive, nonoxygen labile, heat stable at 70°C for 45 min, and strongly adsorbing to cellulose. The highest enzyme activities were obtained on native cellulose and α-cellulose, where endoglucanase activity predominated. The enzymology of other anaerobic cellulolytic organisms is not yet well elucidated (45).

In addition to bacteria, some protozoa are also known to be actively cellulolytic in the rumen (1, 19, 78, 80, 104, 127, 150). Orpin (114, 115, 117) has demonstrated that some of the previously described flagellated protozoa in the rumen of sheep are, in fact, zoospores of anaerobic phycomycetous fungi. Cellulose digestion among the fungi is not an unusual characteristic. They are, for example, believed chiefly responsible for the decomposition of cellulose-containing leaf litter in forests (37). However, until Orpin's rather surprising discovery of strictly anaerobic types, virtually all fungi were believed to require oxygen for growth. To date, three such species of strictly anaerobic fungi have been described and are

TABLE 7.1 Some known anaerobic cellulolytic microorganisms

Organism	Fermentation Product	Habitat	References
Bacteroides succinogenes	Succinate, propionate, acetate, formate	Rumen	22, 23, 77
Butyrivibrio fibrisolvens	Lactate, butyrate, acetate, formate, H_2, CO_2	Rumen	24, 28, 93
Ruminococcus albus	Succinate, acetate, formate, H_2	Rumen	105, 141
Ruminococcus flavefaciens	Succinate, lactate, acetate, ethanol, formate, H_2, CO_2	Rumen	94, 96, 139
Eubacterium cellulosolvens	Lactate, butyrate, acetate, formate	Rumen	24
Clostridium thermocellum	Succinate, lactate, acetate, ethanol, formate, H_2, CO_2	Soil, marine mud, sewage/dairy waste	48, 99, 109, 149, 151
Clostridium cellobioparum	Lactate, acetate, ethanol, formate, H_2, CO_2	Soil, rumen, lake sediments	75
Acetivibrio cellulolyticus	Acetate, ethanol, H_2, CO_2	Sewage sludge	88, 89, 97, 122
Micromonospora propionici	Propionate, acetate	Termites, rumen protozoa	76
Anaerobic fungi (e.g., *Neocallimastix frontalis*, *Piromonas communis*, *Sphaeromonas communis*)	Lactate, acetate, ethanol, formate, H_2 (*in vitro*), short-chain fatty acids (*in vivo*), CO_2	Rumen, horse cecum	6, 13, 14, 15, 107, 116, 117, 118, 119
Ciliated protozoans (e.g., *Diplodinium* spp.)	Organic acids, H_2, CO_2	Rumen, termites	1, 19, 78, 104, 127, 150

included in Table 7.1. Morphologically similar anaerobic fungi have also been found in the cecum of a horse (118), where they are also believed to degrade cellulose.

Bauchop (13, 14) has since clearly demonstrated that plant fragments in the rumen of cattle and sheep are rapidly colonized by fungal zoospores. The zoospores attach and produce hyphae that grow and penetrate plant tissues. Anaerobic cellulase activity has been demonstrated *in vitro* (14). His findings strongly suggest that these organisms may be responsible, in large part, for the primary hydrolysis of lignocellulose-containing plant fibers in some ruminants.

7.3.2 Cellulose Digestion in Termites

Another apparently anaerobic environment where lignocellulose degradation is important is in the hindgut of wood-eating termites (19, 25, 138). Termites are known to harbor anaerobic cellulolytic protozoa (19, 104); however, the role of bacteria in cellulose digestion in the termite remains unresolved. Hungate (76) isolated an anaerobic, cellulolytic actinomycete (*Micromonospora propionici*) from the termite gut, but concluded that it was of limited importance *in situ*. Other attempts to isolate anaerobic, cellulolytic procaryotes from termites have been unsuccessful, suggesting that protozoa are the main cellulose-degrading community in these insects.

7.3.3 Anaerobic Cellulose Degradation in Aquatic Ecosystems

As already mentioned, until the recent development of methods to independently radiolabel the cellulose and lignin components of lignocellulose (38, 90), it was not possible to conduct quantitatively reliable biodegradation studies in natural terrestrial and aquatic environments. Much progress has since been made using ^{14}C -labeled plant preparations in aerobic degradation studies (see Ref. 37). However, to date, there are only two reports concerning the microbial degradation of cellulose in lignocellulose under anaerobic conditions (17, 54).

The first such study was made by Federle and Vestal (54), who examined the mineralization of lignocellulose in the sediments of an extremely oligotrophic arctic lake. They prepared [^{14}C-CELLULOSE]lignocellulose (38) from white pine by feeding live plant cuttings a ^{14}C-labeled cellulose precursor (i.e., D-[U-^{14}C]glucose). In an effort to simulate oxygen-depleted conditions in the sediments, they flushed some test bottles with nitrogen gas. After 50 days of incubation (13.5°C) under anoxic conditions, a range of 16 to 23% of [^{14}C-CELLULOSE]white pine was mineralized to $^{14}CO_2$ and $^{14}CH_4$. The aerobic controls released only slightly more or 29% of original ^{14}C as $^{14}CO_2$. Subsequent additions of ammonia and nitrate had no effect on cellulose utilization, although phosphorous additions significantly enhanced mineralization, resulting in as much cellulose degradation without oxygen as with it. The authors postulated that perhaps, as is the case in poorly drained soils, anaerobic cellulolytic bacteria grew preferentially with the decline in activity of the aerobic cellulolytic population (7). They also suggested that nutrient availability, more than any other factor, controls cellulose turnover in this arctic lake.

Lignocellulose-containing detritus is a significant source of particulate organic carbon in aquatic ecosystems, especially in swamps and salt marshes (20, 73, 103). These environments are characterized by both waterlogged sediments and anoxic conditions within a few millimeters of the sediment–water interface (17). Roughly 50 to 80% of aquatic plant biomass is lignocellulose, with polysaccharide/lignin ratios ranging from 2:1 to 7:1 (73, 102). In addition, below-ground production of some aquatic plants, such as *Spartina alterniflora* and *Juncus roemerianus*, is often considered greater than that above ground (74, 126).

In an effort to assess the contribution of anaerobic microbial processes to the carbon budget of such ecosystems, Benner et al. (17) conducted long-term studies in three predominantly anaerobic wetlands: a salt marsh, an acidic freshwater marsh, and a mangrove swamp. [^{14}C- POLYSACCHARIDE] lignocelluloses were prepared from indigenous plants (16) and incubated under strictly anaerobic conditions with anoxic sediments collected from the habitat sites. The samples were periodically analyzed for the release of ^{14}C as ^{14}CO$_2$ and ^{14}CH$_4$ over a period of 40 weeks. The results of these incubations are summarized in Fig. 7.2. The radiolabeled polysaccharide fraction of *S. alterniflora*, the ubiquitous salt marsh cord grass, was the most amenable to degradation, with almost 30% of the original ^{14}C recovered as labeled gases. Twenty-four percent of the labeled polysaccharide in the needlerush, *J. roemerianus*, was mineralized to ^{14}CO$_2$ and ^{14}CH$_4$ after 280 days, while a surprising 20% of *Carex walteriana* radiolabel (a freshwater macrophyte) was mineralized in the highly acidic (pH 3.8) anoxic peat sediments of Okefenokee Swamp. These results indicate that slow but significant turnover of native cellulose, present in the lignocellulose complex, may take place in anoxic sediments.

Although the number of anaerobic studies of cellulose degradation in aquatic environments is limited to the two briefly described here, together they suggest that at least in the sediments examined, the potential anaerobic mineralization of

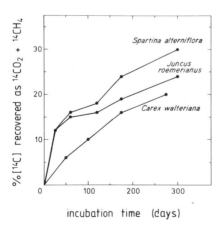

Figure 7.2 Anaerobic mineralization of [^{14}C-POLYSACCHARIDE] lignocelluloses in anoxic sediments. *S. alterniflora* (salt marsh cord grass) and *J. roemerianus* (needlerush) were incubated in salt marsh sediments at 30°C. *C. walteriana* (freshwater sedge) was incubated in acidic peat sediments (pH 3.8) at 30°C. Data excerpted from Benner et al. (17).

cellulose in the natural lignocellulose complex is significant. These results have important implications for the biospheric cycling of lignocellulose-derived carbon (17), as discussed in Section 7.7.

7.4 ANAEROBIC DEGRADATION OF MONOAROMATIC LIGNIN DERIVATIVES

Since polymeric lignin is composed of aromatic subunits, any discussion of its microbial degradation requires an understanding of the mechanisms by which aromatic rings are metabolized. Presumably, elucidation of pathways whereby simple lignin-derived aromatic compounds are degraded may contribute to an understanding of the microbial degradation of polymeric lignin as well (37). This approach has been used extensively by investigators interested in aerobic lignin catabolism, where molecular oxygen is required both as the terminal electron acceptor in respiration and for hydroxylation and ring fission reactions. [Aerobic metabolism of aromatics has been reviewed elsewhere (41–43, 60, 142).]

Aromatic monomers are also known to be released by aerobic lignolytic organisms (i.e., white rot fungi; 8, 83). Their fate in natural systems has yet to be determined. However, it has been postulated that if further biodegradation does not occur in aerobic environments, such components may eventually enter anaerobic sediments, where they would then be subjected to anaerobic transformations (86, 158). Aromatic monomers derived from lignin are also believed to contribute to the formation of soil humus (55, 64, 82, 112, 137). Consequently, apart from their consideration as lignin model compounds, aromatic monomers probably have an environmental fate as such, once released from the lignocellulose matrix.

Until recently, catabolism of aromatic compounds was thought to be a strictly aerobic process (40, 51, 113). The first evidence of anaerobic degradation of aromatic compounds was provided in 1934 by Boruff and Buswell (18), who reported that 54% of the lignin in cornstalks was converted to CO_2 and CH_4 after 600 days of incubation. More direct evidence was provided later that year by Tarvin and Buswell (143). They demonstrated that four aromatic compounds (benzoic, phenylacetic, phenylpropionic, and cinnamic acids) were completely mineralized to CO_2 and CH_4 by an inoculum of sewage sludge. However, in the absence of molecular oxygen, the degradation of aromatic compounds was necessarily postulated to require a novel means of ring fission. Based on their studies using benzoate, Dutton and Evans (47) subsequently proposed a reductive ring cleavage mechanism whereby hydrogenation of the aromatic nucleus results in the corresponding cyclohexane derivative, the ring of which is then opened.

Although elucidation of detailed pathways is still lacking, anaerobic metabolism of aromatic monomers in the absence of molecular oxygen is now known to occur during anaerobic photometabolism (47, 123, 124, 152), under nitrate-reducing conditions (120, 144–146, 155), under sulfate-reducing conditions (153), in microbial consortia where fermentation is often coupled with methanogenesis (10,

26, 31, 56–58, 61c, 62, 67–69, 86, 87, 110, 136, 140), and by pure cultures of fermentative bacteria (61b, 86, 121, 135, 147, 148.)

The purpose of this brief discussion is to consider selected aspects of the biochemistry of anaerobic aromatic degradation and to merge some of the proposed reactions into a common pathway. Discussion will focus on monoaromatic derivatives of lignin. For a more detailed treatment of the anaerobic fate of aromatic compounds, including those of synthetic origin, the reader is referred to an excellent review by Young (158).

7.4.1 Monoaromatic Cleavage during Anaerobic Photometabolism, Nitrate Reduction, and Sulfate Respiration

Although identical pathways often operate during aerobic and anaerobic respiration by some organisms (52), this is clearly not the case for aromatic substrates. Several species of purple, nonsulfur bacteria are able to use simple aromatic compounds as sole sources of carbon, both anaerobically in the light during photosynthesis and aerobically in the dark during respiration [*Rhodopseudomonas palustris* (47), *R. gelatinosa* (152), *Rhodocyclus purpureus* (123), and *Rhodospirillum fulvum* (124)]. Proctor and Scher (128) proposed an oxidative pathway for the observed degradation of benzoate during photosynthetic growth of *R. palustris*. They suggested that an unknown oxidant, believed to be generated during photosynthesis, was biochemically equivalent to molecular oxygen. However, Dutton and Evans (46, 47) showed that this organism could not grow on benzoate if incubated in air, nor could any enzymes of the aerobic pathway be detected during anaerobic growth. Based on these results, they proposed an anaerobic pathway whereby the aromatic ring is reduced prior to cleavage. This reduction was thought to be catalyzed by reductive enzymes coupled to a low-redox-potential component of the light-induced electron transport system (i.e., ferredoxin).

As was the case in the early work with photosynthetic bacteria, the first report of aromatic cleavage during nitrate respiration resulted in the proposal of a degradation scheme not unlike that found in aerobic organisms. Based on mixed culture studies, Oshima (120) surmised that during ring cleavage of both hydroxybenzoate and protocatechuate, the oxygen atoms of nitrate (NO_3^-) behaved as if they were O_2. Later investigations with a *Pseudomonas* sp. (145, 146), however, suggested that the aromatic metabolism of benzoate in the obligatory presence of nitrate must be different from that of aerobic pathways since no oxygenase enzymes could be detected. Definitive studies by Williams and Evans (154, 155), using an aromatic-degrading *Moraxella* sp. that reduces nitrate to N_2, resulted in identification of the same metabolic intermediates found during reductive ring cleavage of benzoate by *R. palustris*. Additional reports of aromatic cleavage during nitrate respiration have recently been published (3, 4, 144).

Use of sulfate as an electron acceptor in the metabolism of aromatic substrates has not been so readily demonstrated. Attempts by Williams and Evans to grow either their ring-cleaving *Moraxella* (154, 155) or *Desulfovibrio* (52) on benzoate under sulfate-reducing conditions have failed. However, Widdel (153) recently observed cleavage of the aromatic rings of benzoate, phenylacetate, and phenyl-propionate by pure cultures of sulfate-respiring bacteria.

7.4.2 Degradation of Lignin-Derived Monomers under Methanogenic Conditions

Methanogenic bacteria are known to form methane from only five substrates: (a) from methanol, (b) from formate, (c) from methylamines, (d) by decarboxylating acetate, and (e) by utilizing hydrogen as an electron donor during CO_2 reduction (156). The production of methane from more complex substrates therefore depends on the activity of nonmethanogenic bacteria in association with methanogens. Consequently, studies of the dissimilation of aromatic compounds under methanogenic conditions have necessarily relied on mixed cultures, so-called microbial consortia.

Figure 7.3 merges some of the reported reactions involving benzene ring–containing lignin derivatives into a common pathway. With some exceptions (see Section 7.4.3), all were mixed culture studies in which an aromatic compound was provided as a sole carbon source, with bicarbonate (HCO_3^-) present as electron acceptor. The figure legend for the pathway contains reference citations to investigations in which the compound was used as a substrate or was identified as an intermediate during the degradation of another aromatic compound.

7.4.2.1 Mixed Culture Studies with Benzoate

Benzoate was used as a model substrate in most early studies concerned with aromatic ring catabolism under methanogenic conditions. Tarvin and Buswell (143) and Clark and Fina (31) were the first to report the methanogenic mixed culture fermentation of benzoate (compound 11 in Fig. 7.3). Fina and Fiskin (58) later showed that $^{14}CH_4$ was produced both in anaerobic digesters and by rumen fluid enrichment cultures that had been fed $[1-^{14}C]$- and $[7-^{14}C]$ benzoic acid. Using mixed cultures obtained from sewage sludge with no previous exposure to benzoate, Nottingham and Hungate (110) obtained radiolabeled methane and CO_2 from uniformly labeled benzoate.

Several years later, Ferry and Wolfe (56) identified acetate as a key intermediate in the conversion of benzoate to methane. Methanogens were isolated from their mixed cultures but were not able to degrade benzoate. The three predominant organisms observed in their stabilized consortia are shown in Fig. 7.4. They were unable to isolate organism 1 but believed it to be an acetate-cleaving methanogen. This was soon accomplished by Zehnder et al. (160), who described *Methanothrix söhngenii*, a non-hydrogen-oxidizing, acetate-decarboxylating methanogen. Organism 2 also was not isolated but was believed to be involved in initial cleavage of the aromatic ring.

Based on thermodynamic considerations, Ferry and Wolfe (56) suggested that a microbial food chain was needed for complete mineralization of benzoate. The methanogens were postulated to remove metabolic intermediates produced by other members of the consortium (i.e., hydrogen), thereby providing thermodynamically

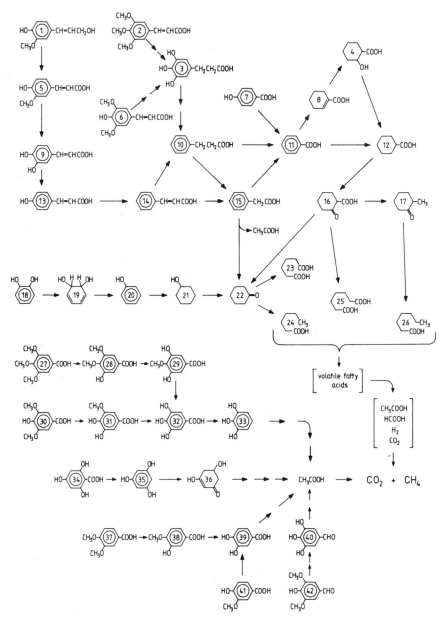

Figure 7.3 Summary pathway merging some reported reactions involving the anaerobic metabolism of lignin-derived monoaromatic compounds under methanogenic conditions (HCO$_3^-$ as electron acceptor). 1 = coniferyl alcohol (61a); 2 = 3,4,5-trimethoxycinnamic acid (9); 3 = 3,4,5-trihydroxyl-3-phenylpropionic acid (9); 4 = 2-hydroxycyclohexane-carboxylic acid (56, 136); 5 = ferulic acid (9, 61a, 69); 6 = sinapic acid (9); 7 = p-hydroxybenzoic acid (68); 8 = 1-cyclohexene-1-carboxylic acid (10, 87, 136); 9 = caffeic acid (61c); 10 = 3-phenylpropionic acid (61a, 61c, 69); 11 = benzoic acid (52, 56, 61c, 68, 69, 87, 108, 136); 12 = cyclohexane carboxylic acid (52, 56, 61c, 69, 87); 13 = p-

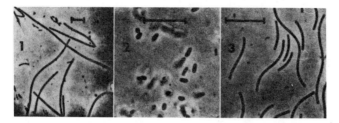

Figure 7.4 Phase-contrast micrographs of the three predominant organisms in a stabilized consortium that anaerobically degrades benzoate to CO_2 and CH_4 (56). Photographs reproduced from Ref. 56 with permission.

favorable conditions for degradation of the aromatic substrate. This implies that ring fission is tightly coupled to product removal and is supported by the work of Mountfort and Bryant (108). They were able to isolate a benzoate-utilizing bacterium only in co-culture with a *Desulfovibrio* sp. that scavenged the hydrogen. However, Grbić-Galić and Young (61a) have shown that it is possible to uncouple methanogenesis and benzoate degradation. 2-Bromoethanesulfonic acid (BESA), a structural analog of coenzyme M (found uniquely in methanogens), acts as a metabolic inhibitor of methane formation by specifically blocking the methyl transfer reaction in the last step of methanogenesis. Addition of BESA to their active benzoate-degrading consortium reduced methane production to 5% of normal levels, but did not affect ring cleavage of benzoate. Their results suggest that some other, as yet unknown, compound(s) must act as a hydrogen sink in these consortia (158).

Using benzoate-enrichment cultures obtained from sheep rumen fluid and sewage sludge, Balba and Evans (10) detected several potential pathway intermediates, including 1-cyclohexene-1-carboxylic acid (compound 8 in Fig. 7.3), cyclohexane carboxylic acid (compound 12), adipic acid (compound 23), caproic acid (compound 24), heptanoic acid (compound 26), propionate, and acetate. Similar results were obtained by Keith et al. (87), who studied mixed consortia that had been

---

hydroxycinnamic acid (61c); 14 = cinnamic acid (61c, 68, 69); 15 = phenylacetic acid (61c, 69); 16 = 2-oxocyclohexanecarboxylic acid (136); 17 = methylcyclohexanone (61c, proposed by 52); 18 = catechol (67, 68, 85, 86); 19 = *cis*-1,2-dihydroxyl-1,2-dihydrobenzene (see 158); 20 = phenol (9, 26, 52, 67, 158a); 21 = cyclohexanol (52); 22 = cyclohexanone (10, 52, 61c); 23 = adipic acid (52); 24 = caproic acid (52, 136); 25 = pimelic acid (52, 56, 61c, 136); 26 = heptanoic acid (52, 61c, 87); 27 = 3,4,5-trimethoxybenzoic acid (9, 85, 86); 28 = 3-hydroxy-4,5-dimethoxybenzoic acid (85, 86); 29 = 3,5-dihydroxy-4-methoxybenzoic acid (85, 86); 30 = syringic acid (9, 68, 85, 86); 31 = 3,4-dihydroxy-5-methoxybenzoic acid (85, 86); 32 = gallic acid (9, 86, 135); 33 = pyrogallol (86, 135); 34 = 2,4,6-trihydroxybenzoic acid (135); 35 = phloroglucinol (121, 135, 147, 158a); 36 = dihydrophloroglucinol (121, 135); 37 = veratric acid (85, 86); 38 = 3-hydroxy-4-methoxybenzoic acid (86); 39 = protocatechuic acid (9, 68); 40 = 3,4,5-trihydroxybenzaldehyde (9); 41 = vanillic acid (9, 68, 85, 86); 42 = syringaldehyde (9, 68).

maintained on benzoate for 10 years. In cultures seeded with anaerobic mud from a polluted river, Schlomi et al. (136) did not find cyclohexane carboxylic acid (compound 12), but did identify two new compounds in the pathway, 2-hydroxycyclohexanecarboxylic acid (compound 4) and 2-oxocyclohexanecarboxylic acid (compound 16). Considering the number of different sources used to obtain benzoate-degrading cultures, it is not surprising that such a variety of intermediates has been proposed. The sequence of reactions in the anaerobic dissimilation of benzoate under methanogenic conditions, however, remains the same; that is, reduction of the aromatic ring with the formation of a cyclohexane derivative, reductive cleavage of the cyclohexane ring yielding aliphatic acids, which, in turn, are substrates for members of the consortia that release methane precursors (159).

7.4.2.2 Mixed Culture Studies with other Aromatic Compounds

None of the substituted monoaromatic lignin derivatives examined thus far have proven to be resistant to anaerobic biotransformation, although ring cleavage has not always been observed (9, 85, 86). Healy and Young (68) extensively surveyed the potential for methanogenic degradation of monoaromatics other than benzoate, including vanillic acid (compound 41), ferulic acid (compound 5), phenol (compound 20), catechol (compound 18), cinnamic acid (compound 14), protocatechuic acid (compound 39), p-hydroxybenzoic acid (compound 7), syringic acid (compound 30), and syringaldehyde (compound 42). Ring fission was observed in all cases with the resultant production of CO_2 and CH_4. Healy et al. (69) later focused on the methanogenic degradation of ferulic acid (compound 5). Based on chromatographic analyses, they proposed cinnamic acid (compound 14) and acetate as major intermediates. Upon inhibition of methane formation by addition of BESA to their mixed cultures, they were able to detect several other compounds, including phenylpropionic acid (compound 10), phenylacetic acid (compound 15), benzoate (compound 11), cyclohexane carboxylic acid (compound 12), butyric acid, isovalerate, and propionate. The presence of benzoate and compound 12, in particular, suggested that the ferulic acid pathway might overlap the one already proposed by Evans (152) for benzoate. Additional studies have since been completed by Grbić-Galić and Young (61a) with stablilized methanogenic consortia which degrade both ferulic acid and benzoate. They have identified most of the intermediates detected by other groups (10, 87, 136) as well as two previously unreported compounds, caffeic acid (compound 9) and p-hydroxycinnamic acid (compound 13). They have also found methylcyclohexanone (compound 17), which Evans (52) hypothesized but never detected. The proposed pathways for benzoate and ferulic acid are included in Fig. 7.3.

Phenol degradation to methane was first reported by Chemielowski et al. (26). Healy and Young (67) later showed that both catechol (compound 18) and phenol (compound 20) could be fermented to CO_2 and CH_4 by mixed consortia after a short acclimation period. Balba and Evans (see Ref. 158) recently worked out a degradation scheme for catechol that includes phenol as an intermediate. Catechol is first dehydroxylated to phenol through cis-benzenediol (compound 19). Phenol is then reduced to cyclohexanol (compound 21), with subsequent formation of cyclohexanone (compound 22). Ring fission yields adipic acid (compound 23), caproic acid (compound 24), acetate, succinate, and propionate.

Using freshwater sediments as inoculum, Kaiser and Hanselmann (86) demonstrated that syringic acid (compound 30) and related hydroxy- and methoxy-substituted compounds are mineralized to CO_2 and CH_4 by a microbial community that was at first believed to be composed of four or five different organisms. An electron micrograph of their syringic acid–degrading consortium is shown in Fig. 7.5. Recently, three different populations have been isolated (K. W. Hanselmann, personal communication) and maintained in pure culture on methanol (organism I), on gallic acid (organism II), and on acetate (organism III). Methoxylated aromatic rings cannot be cleaved by organisms I and II; however, organism II can attack such compounds after the methyl groups have been removed. Organism I

Figure 7.5 Diversity of microbial populations in a community that degrades syringic acid fermentatively (85). Photograph provided by K. W. Hanselmann from Ref. 85 and reproduced with permission.

has been characterized as a sporeforming sulfate reducer that grows with methanol (also ferments methanol in the presence of CO_2), suggesting it to be a *Desulfotomaculum*-type organism. The pathways proposed for both syringic acid and 3,4,5-trimethoxybenzoic acid (compound 27) intersect with the formation of gallic acid (compound 32), as evident in Fig. 7.3. Decarboxylation of gallic acid to yield pyrogallol (compound 33) has been postulated by both Kaiser and Hanselmann (86) and Schink and Pfenning (135).

7.4.2.3 Removal of Aromatic Ring Substituent Groups

Since monoaromatic compounds such as benzoate (52), ferulic acid (69), and syringic acid (86) may be stoichiometrically fermented to CO_2 and CH_4, it follows that demethoxylation, decarboxylation, and dehydrogenation reactions are all possible under strictly anaerobic conditions. There are several reports of mixed methanogenic culture systems in which only the substituent groups are removed, leaving the aromatic ring intact. For example, Kasier and Hanselmann (85) found that when vanillic acid and its analogs were fed to syringic acid–adapted cultures, they were converted to catechol with only small amounts of acetate and methane formed, presumably by demethoxylation reactions. Bache and Pfenning (9) identified *Acetobacterium woodii* as the organism responsible for demethoxylation of lignin monomers such as 3,4,5-trimethoxycinnamic acid (compound 2), 3,4,5-trihydroxy-3-phenylpropionic acid (compound 3), and sinapic acid (compound 6). The aromatic rings were not cleaved, and acetate was the only fermentation product. They were also able to correlate organism growth yields with the number of methoxyl groups per substrate molecule, suggesting that this organism can use methoxyl groups as a sole source of carbon and energy. These reactions have been included in the pathway shown in Fig. 7.3 since, in the presence of other microorganisms, cleavage of their aromatic rings may be possible.

Although not included in Fig. 7.3, anaerobic demethylation reactions may also be important. Young and Rivera (158a) recently demonstrated that *p*-cresol is demethylated to phenol (compound 20 in Fig. 7.3) by mixed anaerobic cultures obtained from a municipal sludge digester. Since acclimation did not significantly increase the rate of *p*-cresol metabolism, they suggest that demethylation is the rate-limiting step.

7.4.3 Aromatic Ring Fission by Pure Cultures of Anaerobic Bacteria

Several strictly anaerobic bacteria have been described that are able to cleave the rings of aromatic monomers. Phloroglucinol (compound 35), for example, is fermented by two rumen isolates, *Coprococcus* sp. and *Streptococcus bovis* (121, 147, 148). As already mentioned, Mountfort and Bryant (108) isolated an organism that degrades benzoate only when grown in co-culture with a *Desulfovibrio* sp. More recently, Barik and Bryant (Abstr. Annu. Meet. Am. Soc. Microbiol. 1984, I72, p. 133) described two anaerobic syntrophic bacteria that degrade benzoate, phenylacetate, and phenol in co-culture with *Wolinella succinogenes*. A new genus and species *Pelobacter acidigallici* has been described by Schink and

Pfennig (135). This organism grows on gallic acid (compound 32), phloroglucinol (compound 35), pyrogallol (compound 33), and 2,4,6-trihydroxybenzoic acid (compound 34) with the production of acetate and CO_2. Kasier and Hanselmann (86) have isolated an anaerobic coccus from their syringic acid–degrading consortia that degrades gallic acid (compound 32). Finally, Grbić-Galić (61b) has isolated a facultatively anaerobic bacterium from a ferulic acid–degrading community that transforms ring substituents (i.e., demethoxylates, dehydroxylates, saturates the double bond in the side chain); however, it cleaves the ring only when grown aerobically.

Despite the suggestion that anaerobic ring fission is tightly coupled to product removal (56), it is apparently possible to isolate important members of anaerobic consortia. Further characterization of their physiology should provide valuable insight into both their population dynamics and the importance of these anaerobic communities in natural systems.

7.5 ANAEROBIC DEGRADATION OF LIGNIN DERIVATIVES: OLIGOLIGNOLS

The pathways by which polymeric lignin is degraded are not well understood (37). In fact, it has been only recently that investigators studying aerobic lignin-degrading organisms have reported the formation of metabolic products more complex than carbon dioxide (27, 29, 30, 39, 125, 131). These studies have provided new evidence that an appreciable fraction of polymeric lignin may be incompletely oxidized by lignolytic bacteria and fungi, and subsequently released as soluble lignin fragments of reduced molecular size. These lignin-derived complexes, which are larger than the monoaromatic compounds discussed in Section 7.4, are generally termed oligomers (30), lignin fragments (131), or oligolignols (33a, 33b). Their potential contribution to the global carbon budget has not been assessed, although one study of land-derived organic matter in offshore sediments had results which suggested that most of the lignin-related compounds were not in the form of humic acids or simple phenols, but rather, as plant fragments (70).

7.5.1 Degradation Studies Using [^{14}C]Oligolignols Derived from Lignin

The first report of anaerobic microbial degradation of oligolignols was that of Colberg and Young (33). They prepared [^{14}C-LIGNIN]lignocellulose from Douglas fir (38). To produce oligolignols which could be used as substrates in methanogenic enrichment cultures, the radiolabeled wood was heat-treated under alkaline conditions, which resulted in the release of water-soluble, lignin-derived fragments (MW < 1400). The carbohydrate-free oligolignol mixture was then neutralized and preparatively separated into its component molecular sizes by gel permeation chromatography. After a series of such separations, similar fractions were pooled together, dried, and reconstituted in a defined medium for use as sole sources of carbon in biodegradation studies (33, 33a, 33b).

Using sewage sludge as inoculum, enrichment cultures were prepared under strictly anaerobic conditions (67, 68, 106) and periodically analyzed for the formation of radiolabeled gases. As may be seen in Fig. 7.6, lignin fragments corresponding to molecular weights of 1000 to 1400 were only minimally degraded; however, greater than 25% of the original ^{14}C of the smaller oligolignol mixture (MW 400 to 1000 with an average MW 600) was recovered as $^{14}CO_2$ and $^{14}CH_4$ after less than 20 days of incubation.

On the basis of these preliminary incubations, as well as corresponding changes observed in the molecular size profiles when the substrates were reeluted on gel columns (not shown), Colberg and Young (33) postulated that cleavage of intermonomeric bonds must have occurred during anaerobic degradation of the oligolignols. Their hypothesis was subsequently confirmed by Zeikus et al. (164) who incubated a ^{14}C-labeled synthetic dilignol compound with anaerobic lake sediments. The structure of this compound and results of their study are shown together in Fig. 7.6. The position of the arylglycerol-β-aryl ether bond, the most common type in polymeric lignin (133), is denoted by the heavy line and joins the two aromatic rings. Since the synthetic dilignol was readily degraded to $^{14}CO_2$ and $^{14}CH_4$, the authors suggested that the β-aryl ether bond per se does not limit anaerobic mineralization, but is readily cleaved in the absence of molecular oxygen.

7 5.2 Identification of Metabolic Intermediates

One of the major difficulties in elucidating lignin pathways is the inability to sufficiently characterize high-molecular-weight catabolic intermediates (39). The ad-

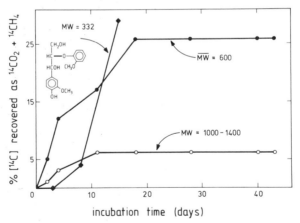

Figure 7.6 Anaerobic biodegradation of two [^{14}C]oligolignol mixtures derived from [^{14}C-LIGNIN] Douglas fir and a [^{14}C]synthetic dilignol model compound containing the aryl-glycerol-β-aryl ether bond (structure shown). Data for oligolignols taken from Ref. 33a, data for dilignol model compound excerpted from Ref. 164.

vantage of studying oligolignols is that the metabolic intermediates expected to be released upon cleavage of intermonomeric linkages are monoaromatic compounds, which are readily identified by capillary gas chromatography. It was for this reason that Colberg and Young (33b) pursued more detailed characterization of intermediate products formed during the anaerobic degradation of [^{14}C] oligolignols (33a).

Methanogenic mixed cultures that had been acclimated for 2 years to use oligolignols (average MW 600) as a sole source of carbon, were analyzed for the presence of lignin monomers and volatile fatty acids, which were postulated to be released during anaerobic dissimilation. Accumulation of potential intermediates was enhanced by inhibition of methane formation with BESA, as described in a previous study (see Section 7.4.2.1). The results of their analyses are summarized in the series of chromatographic profiles shown in Fig. 7.7. Figure 7.7a is a chromatogram of the oligolignol substrate prior to its anaerobic degradation. The peak that eluted just before the internal standard was never identified and did not correspond with any monoaromatic standard compounds. The other smaller peaks were solvent contaminants and are evident in subsequent profiles. As expected, no aromatic monomers were present in the oligolignol mixture prior to its degradation. Fig. 7.7b is a typical chromatogram obtained from an uninhibited (no BESA added), methane-producing culture that had been routinely provided oligolignols as a sole carbon source. After 30 days of incubation with a new spike of substrate (350 mg/L carbon), low concentrations of four aromatic monomers were identified: benzoate, 3-phenylpropionic acid, cinnamic acid, and vanillin. Since it was possible that other aromatic intermediates were also formed, but not at detectable levels, BESA was added to some culture in order to block methane formation in an effort to enhance their accumulation. With methanogenesis completely inhibited, six additional intermediates were identified (see Fig. 7.7c): catechol, phenylacetic acid, syringic acid, vanillic acid, ferulic acid, and caffeic acid. Addition of BESA caused monoaromatic compound accumulation to increase fivefold, while volatile acids concentrations (not shown) were almost 10 times greater than in the uninhibited cultures.

These results clearly indicated that anaerobic microorganisms are able to transform oligolignols derived from natural wood and release lignin monomers which are known to be degraded to CO_2 and CH_4 under strictly anaerobic conditions (68). Detection of aromatic intermediates, particularly ferulic, syringic, and vanillic acids, is suggestive of β-aryl ether cleavage. The presence of heptanoic acid, detected in both inhibited and uninhibited systems, indicated that aromatic rings were cleaved.

7.5.3 Oligolignols Produced by Aerobic Lignolytic Microorganisms

Based on the results of several recent investigations (27, 29, 30, 39, 125, 131), it appears that the ability of aerobic lignolytic microorganisms to solubilize polymeric lignin to fragments of reduced molecular size (i.e., oligolignols) may not be uncommon. Futhermore, it is interesting to note that the lignin-derived oligolignols used by Colberg and Young (33) were found to have a molecular size

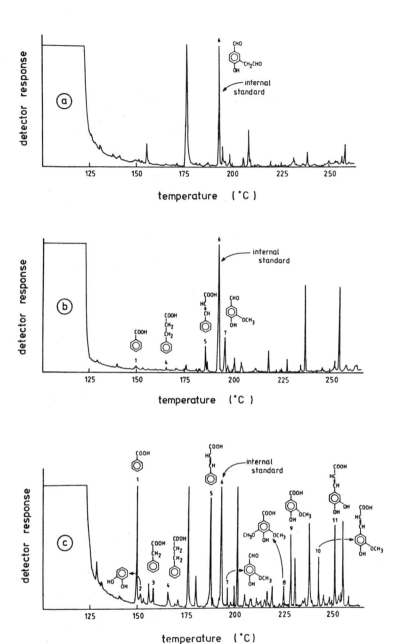

Figure 7.7 (*a*) Capillary gas chromatogram of oligolignol substrate (MW 400 to 1000) prior to its anaerobic degradation. 6 = ethyl vanillin (internal standard). (*b*) Capillary gas chromatogram obtained from an active, methane-producing culture (no BESA added) provided [^{14}C]oligolignols (MW 400 to 1000) as a sole carbon source. 1 = benzoic acid; 4 = 3-phenylpropionic acid; 5 = cinnamic acid; 6 = internal standard; 7 = vanillin. (*c*) Capillary gas chromatogram obtained from an inhibited (BESA added/no methane produced) culture provided [^{14}C]oligolignols (MW 400 to 1000) as a sole carbon source. 1 = benzoic azid; 2 = catechol; 3 = phenylacetic acid; 4 = 3-phenylpropionic acid; 5 = cinnamic acid; 6 = internal standard; 7 = vanillin; 8 = syringic acid; 9 = vanillic acid; 10 = ferulic acid; 11 = caffeic acid.

352

TABLE 7.2 Percentage of original ^{14}C Recovered as $^{14}CO_2$ and radiolabeled water-soluble products ($[^{14}C]_{ws}$) from various $[^{14}C\text{-LIGNIN}]$lignocelluloses degraded by aerobic lignolytic microorganisms

$[^{14}C]$Wood	Inoculum	% radiolabel recovered as:		Total: $^{14}CO_2 + [^{14}C]_{ws}$	% $[^{14}C]_{ws}$ of Total: $[^{14}C]_{ws}/^{14}CO_2 + [^{14}C]_{ws}$	Reference
		$^{14}CO_2$	$[^{14}C]_{ws}$			
Douglas fir	*Streptomyces badius*	9	16	25	64	125
Aspen	*Phanerochaete chrysosporium*	39.6	26.3	65.9	40	131
Douglas fir	*Streptomyces* spp.	2.8	5.3	8.1	65	125
		3.5	12.4	15.9	78	
		6.9	3.1	10	31	
		8.2	12.9	21.1	61	
		6.5	13.9	20.4	68	
White oak	Freshwater stream	9.8	3.1	12.9	24	K. H. Baker, personal communication

distribution similar to that of lignin fragments released during lignin degradation by the white rot fungus *Phanerochaete chrysosporium* (32). In the laboratory cultures examined thus far, on the average almost half of the total carbon conversion is to such water-soluble intermediates. Some of these data are summarized in Table 7.2. For example, Phelan et al. (125) found that during growth of several lignolytic *Streptomyces* isolates on [^{14}C-LIGNIN] Douglas fir, a range of 3 to 16% of the original ^{14}C label could be recovered as ^{14}C water-soluble products. As may be seen in Table 7.2, this corresponds to 31 to 78% of the total radiolabeled carbon conversion, with the balance recovered in the gas phase as $^{14}CO_2$. Similarly, when [^{14}C-LIGNIN] aspen was degraded by *P. chrysosporium*, Reid et al. (131) reported that 26% of the original ^{14}C was recovered as ^{14}C water-soluble products, which translates to 65% of the total ^{14}C-labeled carbon conversion observed. In addition, lignin solubilization has also been observed in field incubation studies (K. H. Baker, personal communication).

The fate of oligolignols in natural ecosystems is unknown. However, the results of biodegradation studies reported to date (32, 33, 164) suggest that if they are of sufficiently small molecular size (i.e., < 1400), and become available in anoxic zones, anaerobic microbial communities may be able to degrade them partially to CO_2 and CH_4, thereby limiting their accumulation in anaerobic sinks.

7.6 ANAEROBIC DEGRADATION OF POLYMERIC LIGNIN

There have been very few studies concerning lignin degradation under anaerobic conditions (17, 63, 111, 164). The first such report was that of Boruff and Buswell (18), who reported a 54% reduction in the lignin content of cornstalks after almost 2 years of anaerobic incubation. However, since they relied on gravimetric analyses to determine lignin weight loss, it is difficult to assess the validity of the data.

With the recent introduction of radioisotopic techniques, both the sensitivity and accuracy of lignin biodegradation assays have dramatically increased. Several types of natural ([^{14}C- LIGNIN] lignocelluloses; 16, 36, 38) and synthetic (DHPs = dehydropolmerizates; 90) radiolabeled lignin preparations are now available. In addition, the development of novel techniques for the cultivation of strictly anaerobic bacteria (11, 79, 105) has made the study of these organisms possible in a variety of anoxic habitats.

7.6.1 Degradation in the Rumen

Hungate and Stack (81) have noted that the amount of phenylpropionic acid in rumen fluid is significantly greater than may be accounted for by amino acid utilization of the microflora. However, since fermentation of lignin in ruminants has never been established, they were hesitant to postulate that some of the phenylpropionate may have come from the anaerobic breakdown of lignin. One recent investigation, however, suggests that anaerobic microorganisms may be involved in some transformations of lignified tissues in forage grasses. In 1980, Akin (5) described a facultatively anaerobic bacterium isolated from rumen fluid that preferentially attacked lignin-containing cell walls of Bermuda grass. The organism, designated 7-1, was not cellulolytic and did not grow on pectin; however, it did ferment several lignin-derived monomers, including sinapic acid, ferulic acid, and *p*-coumaric acid.

Transmission and scanning electron microscopy (SEM) revealed that lignified grass cell walls were apparently digested by the bacterium when incubated under strictly anaerobic conditions. Some of Akin's documentation is based on the photomicrographs shown in Fig. 7.8. The SEM photos labeled (*a*) and (*b*) are cross-sections of the blade and stem of Bermuda grass, respectively. Following 8 days of anaerobic incubation in uninoculated culture medium, the grass parts are apparently intact and unaltered. Figure 7.8*c* is a section of blade examined after 8 days of anaerobic incubation with organism 7-1. The inner bundle sheath is notably deformed, while the sclerenchyma is separated into individual cells. A large portion of predominantly unlignified tissue remains. Figure 7.8*d* is an SEM of blade

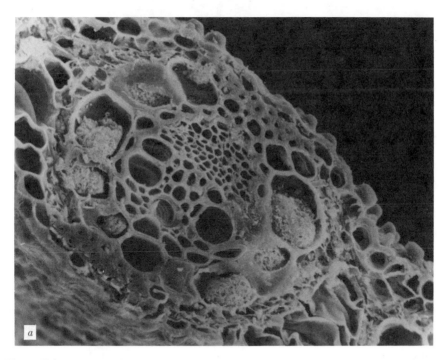

Figure 7.8 Scanning electron micrographs (SEMs) of Bermuda grass in the presence and absence of organism 7-1 (5). (*a*) Cross section of Bermuda grass blade after 8 days of anaerobic incubation in uninoculated culture medium. (*b*) Cross section of Bermuda grass stem after 8 days of anaerobic incubation in uninoculated culture medium. (*c*) SEM of blade section after 8 days of anaerobic incubation with organism 7-1. (*d*) SEM of blade sclerenchyma degraded by organism 7-1 under aerobic conditions. (*e*) SEM of blade segment incubated anaerobically with organism 7-1 for 28 days. (Micrographs provided by D. E. Akin from Ref. 5 and reproduced with permission.)

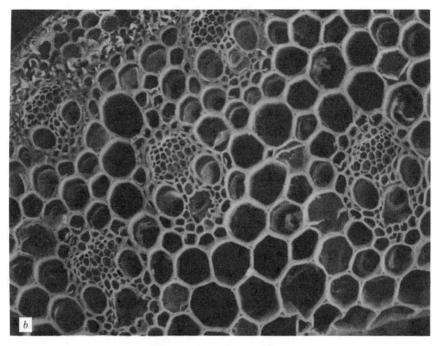

Figure 7.8 (*Continued*)

sclerenchyma incubated with 7-1 under aerobic conditions. Rod-shaped bacteria and short filaments, both characteristic morphologies of the organism, are associated with degraded zones in the cell walls. Finally, Fig. 7.8*e* shows a blade segment that was incubated with organism 7-1 for 28 days in anaerobic culture medium. The sclerenchyma is extensively degraded and many rods and filaments are located within the disrupted cells. The inner bundle sheath is also partially destroyed.

Interpretation of these observations is complicated by two factors: (a) although organism 7-1 could use lignin monomers as sole carbon sources, utilization of lignin was inferred only on the basis that the cell walls that were degraded had a high lignin content; and (b) the histological stains that were used are only qualitative (D. E. Akin, personal communication). Nevertheless, Akin's observations (5) suggest that anaerobic microorganisms in the rumen may be able to alter, if not partially degrade, portions of lignified plant cells. The significance of such activity in the utilization of lignin by ruminants, if confirmed, has not yet been established.

7.6.2 Degradation in Termites

Work by Butler and Buckerfield (25) implicates the microflora of the termite hindgut in the anaerobic degradation of lignin. In their novel experiments with an Aus-

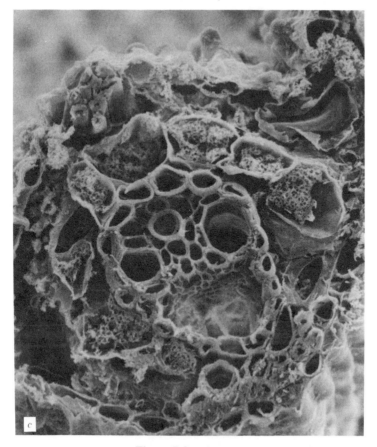

Figure 7.8   (*Continued*)

tralian wood-eating species, *Nasutitermes exitiosus*, termites were fed side-chain-, methoxy-, and ring-labeled [^{14}C-LIGNIN] lignocellulose and DHPs (synthetic lignins). In incubation periods ranging from 49 to 69 days, 13.5 to 32.4% of the [^{14}C] DHPs were respired by the termites as $^{14}CO_2$. In the case of [^{14}C-LIGNIN] maize, approximately 15% of the ring-labeled lignocellulose was released as $^{14}CO_2$, while 32.4% and 63.1% of side-chain- and methoxy-labeled lignins, respectively, were released in the same period. Their results suggest that the termite microflora is able to demethylate, depolymerize, and degrade both natural and synthetic lignins. Although there is some debate concerning the anaerobiosis of the termite hindgut (49, 50), Butler and Buckerfield (25) maintain on the basis of work by Hungate (76) and others (100) that there is too much similarity in the mechanisms and end products of cellulose degradation between the termite and ruminants for one to be aerobic and the other to be anaerobic.

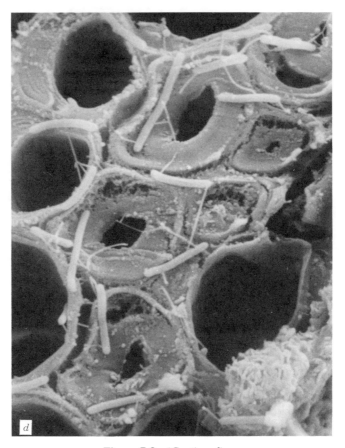

Figure 7.8   (*Continued*)

7.6.3   Degradation in Aquatic Ecosystems

Using both ring- and side-chain-labeled [^{14}C] DHPs that were synthesized in the laboratory, Hackett and his group (63) reported no conversion of ^{14}C to ^{14}CO$_2$ and ^{14}CH$_4$ when incubated with a variety of anaerobic inocula, including rumen fluid, several soil samples, and lake sediments (see also Ref. 164). Despite the relatively short incubation times allowed (60 to 78 days) and the lack of any associated polysaccharides (such as found in natural lignocellulose), they concluded that lignin is, in essence, inert in anaerobic environments (162). Recent long-term studies by Benner et al. (17), however, indicate that polymeric lignin is slowly degraded in some aquatic wetlands and at environmentally significant rates.

As already mentioned in Section 7.3.3, swamps and salt marshes are characterized by both waterlogged sediments and anoxic conditions immediately below the sediment–water interface. Since primary production of higher plants in wet-

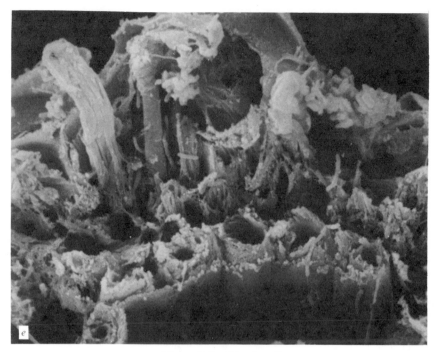

Figure 7.8   (*Continued*)

lands is among the highest known for natural ecosystems (17, 126), turnover of lignocellulose is crucial to the continued operation of these predominantly anaerobic, detritus-based habitats.

Benner et al. (17) prepared [^{14}C-LIGNIN] lignocelluloses from indigenous plants (16), which were then incubated under strictly anaerobic conditions with anoxic sediments from a salt marsh, an acidic freshwater marsh, and a mangrove swamp. The samples were analayzed for the release of radiolabeled gases over a period of 280 to 300 days. In addition [^{14}C] DHPs were tested for comparison with the natural lignins. The results of these incubations are plotted together in Fig. 7.9. The lignin fraction of *S. alterniflora* (salt marsh cord grass) was the most readily degraded with 17% of the original ^{14}C converted to $^{14}CO_2$ and $^{14}CH_4$. Almost 10% of the lignin component of the freshwater macrophyte *C. walteriana* was recovered in the gas phase, although incubated in sediments with a pH of 3.8. Synthetic DHPs and *J. roemerianus* responded similarly when incubated in anoxic salt marsh sediments with less than 5% of the original label recovered as labeled gases. These results clearly demonstrate that slow but significant mineralization of polymeric lignin is possible in anoxic sediments.

It is important to note, however, that anaerobic rates of both lignin and polysaccharide turnover were, in general, significantly less than the corresponding

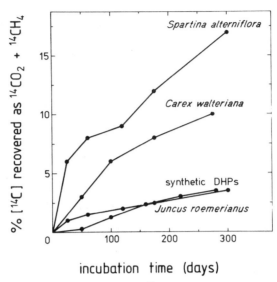

incubation time (days)

Figure 7.9 Anaerobic mineralization of [^{14}C-LIGNIN]lignocelluloses and synthetic [^{14}C]DHPs in anoxic sediments. *S. alterniflora* (salt marsh cord grass) and *J. roemerianus* (needlerush) were incubated in salt marsh sediments at 30°C. *C. walteriana* (freshwater sedge) was incubated in acidic peat sediments (pH 3.8) at 30°C. Data excerpted from Benner et al. (17).

aerobic rates determined in the same sediments. A comparison of the observed anaerobic mineralization rates, expressed as percentages of the aerobic rates, is made in Table 7.3.

Despite the limited number of investigations that have attempted to assess the potential anaerobic transformation of polymeric lignin (17, 63, 111, 164), results

TABLE 7.3 Rates of anaerobic biodegradation of specifically radiolabeled lignocelluloses expressed as percentages of observed aerobic rates (17)

	% of aerobic mineralization rates	
Radiolabeled Substrate	[^{14}C-LIGNIN] Lignocellulose	[^{14}C-POLYSACCHARIDE] Lignocellulose
Spartina alterniflora (salt marsh cord grass)	7.5	8.2
Juncus roemerianus (needlerush)	6.0	3.8
Carex walteriana (freshwater sedge)	33.0	40.2
Rhizophora mangle (red mangrove, leaves)	2.9	4.9
R. mangle (red mangrove, wood)	3.7	12.6

of future studies, employing more sensitive biodegradation assays combined with long incubation periods, should increase our understanding of the role of anaerobic microbial communities in the turnover of complex lignaceous biomass in the environment.

7.7 SUMMARY AND CONCLUSIONS

The biosphere is composed of a complex mixture of organic compounds in a continuous state of formation, transformation, and decomposition. Through photosynthesis, autotrophic organisms trap solar energy, the cycle's driving force, thereby converting carbon dioxide to complex cellular carbon. With the death of photosynthetic organisms and higher members of the food chain, biologically sequestered organic carbon is eventually mineralized to its original constituents: carbon

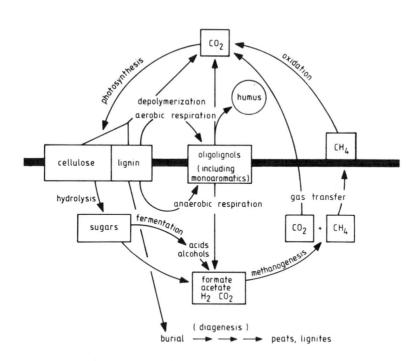

Figure 7.10 Biological cycling of lignocellulose-derived carbon. Based on an original drawing by Healy (66).

dioxide and dihydrogen oxide. Since most organic carbon on earth is in the form of lignocellulose (and its derivatives), its degradation is essential for the continued operation of the biological global carbon cycle (see Fig. 7.10).

Traditionally, little attention has been given to the study of potential anaerobic mechanisms of lignocellulose turnover, particularly in aquatic ecosystems. However, several recent studies collectively suggest that anaerobic microbial communities are not only capable of transforming lignocellulose and its derivatives, but that their contribution to the turnover of lignin-derived carbon has perhaps been underestimated. Figure 7.10 depicts microbially mediated transformations, both aerobic and anaerobic, involved in the global cycling of lignocellulose-derived organic carbon. Although the significance of the various anaerobic mechanisms to turnover and/or preservation are unknown at this time, the following statements may be made in summary:

1. Significant quantities of cellulose, present in the lignocellulose complex, may be degraded anaerobically by microbes found in the rumen, in termites, and in aquatic sediments.

2. Microbial metabolism of monoaromatic lignin derivatives occurs in the absence of molecular oxygen during anaerobic photometabolism, under nitrate-reducing conditions, during sulfate respiration, in microbial consortia where aromatic cleavage is often coupled to methanogensis, and by individual organisms that are able to either cleave the aromatic ring or demethoxylate substitute groups on the ring.

3. Oligolignols (MW < 1400) may be released as intermediate products during lignin degradation by aerobic fungi and bacteria and may be partially degraded to CO_2 and CH_4 by anaerobic microorganisms.

4. Polymeric lignin is mineralized to CO_2 and CH_4 in anoxic sediments at slow but environmentally significant rates.

ACKNOWLEDGMENTS

I am grateful to Lily Y. Young and Dunja Grbić-Galić for their advice in preparation of the consolidated aromatic pathway as well as for allowing me access to their data prior to publication. I thank Danny Akin and Kurt Hanselmann for providing photomicrographs from their respective studies. I also thank Kurt Hanselmann and Dunja Grbić-Galić for reviewing the final draft. My greatest appreciation is reserved for Norbert Swoboda-Colberg, who made all the drawings, assisted with the literature search, triple-checked all reference citations and chemical structures, and cooked gourmet meals at all hours. Financial support from the Swiss National Science Foundation is gratefully acknowledged.

REFERENCES

1. Abou-Akkada, A. R., and B. H. Howard. 1960. The biochemistry of rumen protozoa. 3. The carbohydrate metabolism of *Entodinium*. Biochem. J. 76:445–451.

2. Adler, E. 1977. Lignin chemistry—past, present, and future. Wood Sci. Technol. 11:169–218.

3. Aftring, R. P., B. E. Chalker, and B. F. Taylor. 1981. Degradation of phthalic acids by denitrifying, mixed cultures of bacteria. Appl. Environ. Microbiol. 41:1177–1183.

4. Aftring, R. P., and B. F. Taylor. 1981. Aerobic and anaerobic catabolism of phthalic acid by a nitrate-respiring bacterium. Arch. Microbiol. 130:101–104.

5. Akin, D. E. 1980. Attack on lignified cell walls by a facultatively anaerobic bacterium. Appl. Environ. Microbiol. 40:809–820.

6. Akin, D. E., G. L. R. Gordon, and J. P. Hogan. 1983. Rumen bacterial and fungal degradation of *Digitaria pentzii* grown with or without sulfur. Appl. Environ. Microbiol. 46:738–748.

7. Alexander, M. 1977. Introduction to soil microbiology, 2nd ed. Wiley, New York.

8. Ander, P., and K. E. Eriksson. 1978. Lignin degradation and utilization by microorganisms. Prog. Ind. Microbiol. 14:1–58.

9. Bache, R., and N. Pfennig. 1981. Selective isolation of *Acetobacterium woodii* on methoxylated aromatic acids and determination of growth yields. Arch. Microbiol. 130:255–261.

10. Balba, M. T., and W. C. Evans. 1977. The methanogenic fermentation of aromatic substrates. Biochem. Soc. Trans. 5:302–304.

11. Balch, W. E., and R. S. Wolfe. 1976. New approach to the cultivation of methanogenic bacteria: 2-mercaptoethanesulfonic acid (HS-CoM)-dependent growth of *Methanobacterium ruminantium* in a pressurized atmosphere. Appl. Environ. Microbiol. 32:781–791.

12. Bassham, J. A. 1975. General considerations, in C. R. Wilke (ed.), p. 9–19, Cellulose as a chemical and energy source. Wiley, New York.

13. Bauchop, T. 1979. Rumen anaerobic fungi of cattle and sheep. Appl. Environ. Microbiol. 38:148–158.

14. Bauchop, T. 1981. The anaerobic fungi in rumen fibre digestion. Agric. Environ. 6:339–348.

15. Bauchop, T., and D. O. Mountfort. 1981. Cellulose fermentation by a rumen anaerobic fungus in both the absence and the presence of rumen methanogens. Appl. Environ. Microbiol. 42:1103–1110.

16. Benner, R., A. E. Maccubbin, and R. E. Hodson. 1984. Preparation, characterization, and microbial degradation of specifically radiolabeled [^{14}C] lignocelluloses from marine and freshwater macrophytes. Appl. Environ. Microbiol. 47:381–389.

17. Benner, R., A. E. Maccubbin, and R. E. Hodson. 1984. Anaerobic biodegradation of the lignin and polysaccharide components of lignocellulose and synthetic lignin by sediment microflora. Appl. Environ. Microbiol. 47:998–1004.

18. Boruff, C. S., and A. M. Buswell. 1934. The anaerobic fermentation of lignin. J. Am. Chem. Soc. 56:886–887.

19. Breznak, J. A. 1975. Symbiotic relationships between termites and their intestinal microbiota, p. 559–580, in D. H. Jennings and D. L. Lee (eds.), Symbiosis (Society for Experimental Biology Symposium series No. 29). Cambridge University Press, Cambridge.

20. Brinson, M. M., A. E. Lugo, and S. Brown. 1981. Primary productivity, decomposition and consumer activity in freshwater wetlands. Annu. Rev. Ecol. Syst. 12:123–161.

21. Browning, B. L. 1963. The chemistry of wood. R. E. Krieger, Melbourne, Fla.

22. Bryant, M. P., and R. N. Doetsch. 1954. A study of actively cellulolytic rod-shaped bacteria of the bovine rumen. J. Dairy. Sci. 37:1176–1183.

23. Bryant, M. P., I. M. Robinson, and H. Chu. 1959. Observations on the nutrition of *Bacteroides succinogens*—a ruminal cellulolytic bacterium. J. Dairy Sci. 42:1831–1847.

24. Bryant, M. P., N. Small, C. Bouma, and H. Chu. 1958. *Bacteroides ruminococcus* n. sp. and the new genus and species *Succinomonas amylolytica*. Species of succinic acid-producing anaerobic bacteria of the bovine rumen. J. Bacteriol. 76:15–23.

25. Butler, J. H. A., and J. C. Buckerfield. 1979. Digestion of lignin by termites. Soil Biol. Biochem. 11:507–513.

26. Chemielowski, J., A. Grossman, and T. Wegrzynowska. 1964. The anaerobic decomposition of phenol during methane fermentation. Zesz. Nauk. Politech. Slask. Inz. Sanit. 8:97–122.

27. Chan, C. L., H.-m. Chang, and T. K. Kirk. 1982. Aromatic acids produced during degradation of lignin in spruce wood by *Phanerochaete chrysosporium*. *Holzforschung 36:3–9*.

28. Cheng, K.-J., and J. W. Costerton. 1977. Ultrastructure of *Butyrivibrio fibrisolvens*: a gram-positive bacterium? J. Bacteriol. 129:1506–1512.

29. Chua, M. G. S., C. L. Chen, and H.-m. Chang. 1982. [13]NMR spectroscopic study of lignin degraded by *Phanerochaete chrysosporium*. *I. New structures. Holzforschung 36:165–172*.

30. Chua, M. G. S., S. Choi, and T. K. Kirk. 1983. Mycelium binding and depolymerization of synthetic ^{14}C-labeled lignin during decomposition by *Phanerochaete chrysosporium*. Holzforschung 37:55–61.

31. Clark, F. M., and L. R. Fina. 1952. The anaerobic decomposition of benzoic acid during methane fermentation. Arch. Biochem. 36:26–32.

32. Colberg, P. J. 1982. Microbial degradation of lignin-derived compounds under anaerobic conditions. Ph.D. thesis, Stanford University.

33. Colberg, P. J., and L. Y. Young. 1982. Biodegradation of lignin-derived molecules under anaerobic conditions. Can. J. Microbiol. 28:886–889.

33a. Colberg, P. J., and L. Y. Young. 1985. Anaerobic degradation of soluble fractions of [^{14}C-LIGNIN] lignocellulose. Appl. Environ. Microbiol. 49:345–349.

33b. Colberg, P. J., and L. Y. Young. 1985. Aromatic and volatile acid intermediates observed during anaerobic metabolism of lignin-derived oligomers. Appl. Environ. Microbiol. 49:350–358.

34. Conn, E. E., and P. K. Stumpf. 1972. Outlines of biochemistry, 3rd ed. Wiley, New York.

35. Côté, W. A. 1977. Wood ultrastructure in relation to chemical composition, p. 1–44, in F. A. Loewus, and V. C. Runeckles (eds.), The structure, biosynthesis and degradation of wood. Plenum Press, New York.

36. Crawford, D. L. 1981. Microbial conversion of lignin to useful chemicals using a lignin-degrading *Streptomyces*. Biotechnol. Bioeng. Symp. 11:275–291.

37. Crawford, R. L. 1981. Lignin biodegradation and transformation. Wiley-Interscience, New York.

38. Crawford, D. L., and R. L. Crawford. 1976. Microbial degradation of lignocellulose: the lignin component. Appl. Environ. Microbiol. 31:714–717.

39. Crawford, D. L., A. L. Pometto III, and R. L. Crawford. 1983. Lignin degradation by *Streptomyces viridosporus*: isolation and characterization of new polymeric degradation intermediate. Appl. Environ. Microbiol. 45:898–904.

40. Dagley, S. 1967. The microbial metabolism of phenolics, p. 287–317, in A. D. McLaren, and G. H. Peterson (eds.), Soil biochemistry. Marcel-Dekker, New York.

41. Dagley, S. 1971. Catabolism of aromatic compounds by microorganisms. Adv. Microb. Physiol. 6:1–46.

42. Dagley, S. 1975. A biochemical approach to some problems of environmental pollution. Essays Biochem. 11:81–138.

43. Dagley, S. 1977. Microbial degradation of organic compounds in the biosphere. Surv. Prog. Chem. 8:121–170.

44. Delmer, D. P. 1977. The biosynthesis of cellulose and other plant cell wall polysaccharides, p. 45–77, in F. A. Loewus and V. C. Runeckles (eds.), The structure, biosynthesis and degradation of wood. Plenum Press, New York.

45. Dinsdale, D., E. J. Morris, and J. S. D. Bacon. 1978. Electron microscopy of the microbial populations present and their modes of attack on various cellulosic substrates undergoing digestion in the sheep rumen. Appl. Environ. Microbiol. 36:160–168.

46. Dutton, P. L., and W. C. Evans. 1968. The photometabolism of benzoic acid by *Rhodopseudomonas palustris*: a new pathway of aromatic ring metabolism. Biochem. J. 109:5P.

47. Dutton, P. L., and W. C. Evans. 1969. The metabolism of aromatic compounds by *Rhodopseudomonas palustris*. Biochem. J. 113:525–536.

48. Enebo, L., and H. Lundin. 1951. On three bacteria connected with thermophilic cellulose fermentation. Physiol. Plant. 4:652–660.

49. Eutick, M. L., R. W. O'Brien, and M. Slaytor. 1976. Aerobic state of the gut of *Nasutitermes excitiosus* and *Coptotermes lacteus*, high and low caste termites. J. Insect Physiol. 22:1377–1380.

50. Eutick, M. L., P. Veivers, R. W. O'Brien, and M. Slaytor. 1978. Dependence of the higher termite *Nasutitermes exitiosus* and the lower termite *Coptotermes lacteus* on their gut flora. J. Insect Physiol. 24:363–368.

51. Evans, W. C. 1963. The microbiological degradation of aromatic compounds. J. Gen. Microbiol. 32:177–184.

52. Evans, W. C. 1977. Biochemistry of the bacterial catabolism of aromatic compounds in anaerobic environments. Nature (Lond.) 270:17–22.

53. Falson, B. D., and T. K. Kirk. 1983. Relationship between lignin degradation and production of reduced oxygen species by *Phanerochaete chrysosporium*. Appl. Environ. Microbiol. 46:1140–1145.

54. Federle, T. W., and J. R. Vestal. 1980. Lignocellulose mineralization by arctic lake sediments in response to nutrient manipulation. Appl. Environ. Microbiol. 40:32–39.

55. Felbeck, G. T. 1965. Structural chemistry of soil humic substances. Adv. Agron. 17:327–368.

56. Ferry, J. G., and R. S. Wolfe. 1976. Anaerobic degradation of benzoate to methane by a microbial consortium. Arch. Microbiol. 107:33–40.

57. Fina, L. R., R. L. Bridges, T. L. Coblentz, and F. F. Roberts. 1978. The anaerobic decomposition of benzoic acid during methane fermentation. III. The fate of carbon four and the identification of propanoic acid. Arch. Microbiol. 118:169–172.

58. Fina, L. R., and A. M. Fiskin. 1960. The anaerobic decompositon of benzoic acid during methane fermentation. II. Fate of carbons one and seven. Arch. Biochem. Biophys. 91:163–165.

59. Forney, L. J., C. A. Reddy, M. Tien, and S. D. Aust. 1982. The involvement of hydroxyl radical derived from hydrogen peroxide in lignin degradation by the white rot fungus *Phanerochaete chrysosporium*. J. Biol. Chem. 257:11455–11462.

60. Gibson, D. T. 1984. Microbial degradation of organic compounds. Marcel-Dekker, New York.

61a. Grbić-Galić, D. 1983. Anaerobic degradation of coniferyl alcohol by methanogenic consortia. Appl. Environ. Microbiol. 46:1442–1446.

61b. Grbić-Galić, D. 1985. Fermentative and oxidative transformations of ferulate by a facultatively anaerobic bacterium isolated from sewage sludge. Appl. Environ. Microbiol. 50:1052–1057.

61c. Grbić-Galić, D., and L. Y. Young. 1985. Methane fermentation of ferulate and benzoate: anaerobic degradation pathways. Appl. Environ. Microbiol. 50:292–297.

62. Guyer, M., and G. Hegeman. 1969. Evidence for a reductive pathway for the anaerobic metabolism of benzoate. J. Bacteriol. 99:906–907.

63. Hackett, W. F., W. J. Connors, T. K. Kirk, and J. G. Zeikus. 1977. Microbial decomposition of synthetic ^{14}C-labeled lignins in nature: lignin biodegradation in a variety of natural materials. Appl. Environ. Microbiol. 33:43–51.

64. Haider, K., J. P. Martin, and Z. Filip. 1975. Humus biochemistry, p. 195, in E. A. Paul and A. D. McLaren (eds.), Soil biochemistry. Marcel Dekker, New York.

65. Halliwell, G. 1979. Microbial β-glucanases. Prog. Ind. Microbiol. 15:1–86.

66. Healy, J. B., Jr. 1979. Biodegradation of simple aromatic compounds under methanogenic conditions. Ph.D. thesis, Stanford University.

67. Healy, J. B., Jr., and L. Y. Young. 1978. Catechol and phenol degradation by a methanogenic population of bacteria. Appl. Environ. Microbiol. 35:216–218.

68. Healy, J. B., Jr., and L. Y. Young. 1979. Anaerobic biodegradation of eleven aromatic compounds to methane. Appl. Environ. Microbiol. 38:84–89.

69. Healy, J. B., Jr., L. Y. Young, and M. Reinhard. 1980. Methanogenic decomposition of ferulic acid, a model lignin derivative. Appl. Environ. Microbiol. 39:436–444.

70. Hedges, J. I., and P. L. Parker. 1976. Land-derived organic matter in surface sediments from the Gulf of Mexico. Geochim. Cosmochim. Acta 40:1019–1029.

71. Hemmingson, J. A. 1979. A new way of forming lignin-carbohydrate bonds. Etherification of model benzyl alcohols in alcohol/water mixtures. Aust. J. Chem. 32:225–229.

72. Herrick, R. W., and H. L. Hergert. 1977. Utilization of chemicals from wood: retrospect and prospect, p. 443–515, in F. A. Loewus and V. C. Runeckles (eds.), The structure, biosynthesis and degradation of wood. Plenum Press, New York.

73. Hodson, R. E., R. Benner, and A. E. Maccubbin. 1983. Transformations and fate of lignocellulosic detritus in marine environments, p. 185–195, in T. A. Oxley and S. Barry (eds.), Biodeterioration 5. Wiley, New York.

74. Howarth, R. W., and J. E. Hobbie. 1982. The regulation of decomposition and heterotrophic microbial activity in salt marsh soils: a review, p. 183–207, in V. S. Kennedy (ed.), Estuarine comparisons. Academic Press, New York.

75. Hungate, R. E. 1944. Cellulose-digesting bacteria in *Armitermes*. J. Bacteriol. 48:380–381.

76. Hungate, R. E. 1946. Studies on cellulose fermentation. II. An anaerobic cellulose-decomposing actinomycete, *Micromonospora propionici* n. sp. J. Bacteriol. 51:51–56.

77. Hungate, R. E. 1950. The anaerobic mesophilic cellulolytic bacteria. Bacteriol. Rev. 14:1–49.

78. Hungate, R. E. 1966. The rumen and its microbes. Academic Press, New York.

79. Hungate, R. E. 1969. A roll tube method for cultivation of strict anaerobes, p. 117–132, in J. R. Norris and D. W. Ribbons (eds.), Methods in microbiology, Vol. 3B. Academic Press, London.

80. Hungate, R. E. 1975. The rumen microbial ecosystem. Annu. Rev. Syst. Ecol. 6:39–66.

81. Hungate, R. E., and R. J. Stack. 1982. Phenylpropanoic acid: growth factor for *Ruminococcus albus*. Appl. Environ. Microbiol. 44:79–83.

82. Hurst, H. M., and N. A. Burges. 1967. Lignin and humic acids, p. 260–286, in A. D. McLaren and G. H. Peterson (eds.), Soil biochemistry. Marcel Dekker, New York.

83. Ishihara, T., and M. Miyazaki. 1972. Oxidation of milled wood lignin by fungal laccase. J. Japn. Wood Res. Soc. 18:371–375.

84. Jarvis, M. C. 1982. The proportion of calcium-bound pectin in plant cell walls. Planta (Berl.) 154:344–346.

85. Kaiser, J.-P., and K. W. Hanselmann. 1982. Aromatic chemicals through microbial conversion of lignin monomers. Experientia (Basel) 38:167–176.

86. Kaiser, J.-P., and K. W. Hanselmann. 1982. Fermentative metabolism of substituted monoaromatic compounds by a bacterial community from aerobic sediments. Arch. Microbiol. 133:185–194.

87. Keith, C. L., R. L. Bridges, L. R. Fina, K. L. Iverson, and J. A. Cloran. 1978. The anaerobic decomposition of benzoic acid during methane fermentation. IV. Dearomatization of the ring and volatile fatty acids formed on ring rupture. Arch. Microbiol. 118:173–176.

88. Khan, A. W. 1980. Degradation of cellulose to methane by a coculture of *Acetivibrio cellulolyticus* and *Methanosarcina barkeri*. FEMS Microbiol. Lett. 9:233–235.

89. Khan, A. W., J. N. Saddler, G. B. Patel, J. R. Colvin, and S. M. Martin. 1980.

Degradation of cellulose by a newly isolated mesophilic anaerobe. Bacteroidaceae family. FEMS Microbiol. Lett. 7:47–50.

90. Kirk, T. K., W. J. Connors, R. D. Bleam, W. F. Hackett, and J. G. Zeikus. 1975. Preparation and microbial decomposition of synthetic [^{14}C] lignins. Proc. Natl. Acad. Sci. USA 72:2515–2519.

91. Kirk, T. K., W. J. Connors, and J. G. Zeikus. 1977. Advances in understanding the microbiological degradation of lignin. Rec. Adv. Phytopath. 11:369–394.

92. Kosikova, B., D. Joniak, and L. Kosakova. 1979. On the properties of benzyl ether bonds in the lignin-saccharidic complex isolated from spruce. Holzforschung 33:11–14.

93. Krishnamurty, H. G., K.-J. Cheng, G. A. Jones, F. J. Simpson, and J. E. Watkin. 1970. Identification of products produced by the anaerobic degradation of rutin and related flavonoids by *Butyrivibrio* sp. c$_3$. Can. J. Microbiol. 16:759–767.

94. Latham, M. J., B. E. Brooker, G. L. Pettipher, and P. J. Harris. 1978. *Ruminococcus flavefaciens* cell coat and adhesion to cotton cellulose and to cell walls in leaves of perennial ryegrass (*Lolium perenne*). Appl. Environ. Microbiol. 35:156–165.

95. Latham, M. J., M. E. Sharpe, and J. D. Sutton. 1971. The microbial flora of the rumen of cows fed hay and high cereal rations and its relationship to the rumen fermentation. J. Appl. Bacteriol. 34:425–434.

96. Latham, M. J., and M. J. Wolin. 1977. Fermentation of cellulose by *Ruminococcus flavefaciens* in the presence and absence of *Methanobacterium ruminantium*. Appl. Environ. Microbiol. 34:297–301.

97. Laube, V. W., and S. M. Martin. 1981. Conversion of cellulose to methane and carbon dioxide by triculture of *Acetivibrio cellulolyticus*, *Desulfovibrio* sp. and *Methanosarcina barkeri*. Appl. Environ. Microbiol. 42:413–420.

98. Leatherwood, J. M. 1973. Cellulose degradation by *Ruminococcus*. Fed. Proc. 32:1814–1818.

99. Lee, B. H., and T. H. Blackburn. 1975. Cellulase production by a thermophilic *Clostridium* species. Appl. Microbiol. 30:346–353.

100. Lee, K. E., and T. H. Blackburn. 1975. Cellulase production by role of *Nasusitermes exitiosus* (Hill) in the cycling of organic matter in a yellow podzolic soil under dry schlerophyll forest in South Australia. Trans. 9th Int. Cong. Soil Sci., Adelaide. 2:11–18.

101. Lindgren, B. O. 1958. The lignin-carbohydrate linkage. Acta Chem. Scand. 12:447–452.

102. Maccubbin, A. E., and R. E. Hodson. 1980. Mineralization of detrital lignocelluloses by salt marsh sediment microflora. Appl. Environ. Microbiol. 40:735–740.

103. Mann, K. H. 1972. Macrophyte production and detritus food chains in coastal waters. Mem. Ist. Ital. Idrobiol. Dott. Marco Marchi 29 (Suppl):353–383.

104. Mauldin, J. K. 1977. Cellulose catabolism and lipid synthesis by normally and abnormally faunated termites *Reticulitermes flavipes*. Insect Biochem. 7:27–31.

105. Miller, T. L., and M. J. Wolin. 1973. Formation of hydrogen and formate by *Ruminococcus albus*. J. Bacteriol. 118:836–846.

106. Miller, T. L., and M. J. Wolin. 1974. A serum bottle modification of the Hungate technique for cultivating obligate anaerobes. Appl. Microbiol. 27:985–987.

107. Mountfort, D. O., R. A. Asher, and T. Bauchop. 1982. Fermentation of cellulose to

methane and carbon dioxide by a rumen anaerobic fungus in a triculture with *Methanobrevibacter* sp. Strain RA1 and *Methanosarcina barkeri*. Appl. Environ. Microbiol. 44:128–134.

108. Mountfort, D. O., and M. P. Bryant. 1982. Isolation and characterization of an anaerobic syntrophic benzoate-degrading bacterium from sewage sludge. Arch. Microbiol. 133:249–256.

109. Ng, T. K., P. J. Weimer, and J. G. Zeikus. 1977. Cellulolytic and physiological properties of *Clostridium thermocellum*. Arch. Microbiol. 114:1–7.

110. Nottingham, P. M., and R. E. Hungate. 1969. Methanogenic fermentation of benzoate. J. Bacteriol. 98:1170–1172.

111. Odier, E., and B. Monties. 1983. Absence of microbial mineralization of lignin in anaerobic enrichment cultures. Appl. Environ. Microbiol. 46:661–665.

112. Oglesby, R. T., R. F. Christman, and C. H. Driver. 1967. The biotransformation of lignin to humus—facts and postulates. Adv. Appl. Microbiol. 9:111–184.

113. Ornston, L. N., and R. Y. Stanier. 1964. Mechanism of β-ketoadipate formation by bacteria. Nature (Lond.) 204:1279–1283.

114. Orpin, C. G. 1975. Studies on the rumen flagellate, *Neocallimastix frontalis*. J. Gen. Microbiol. 91:249–262.

115. Orpin, C. G. 1976. Studies on the rumen flagellate, *Sphaeromonas communis*. J. Gen. Microbiol. 94:270–280.

116. Orpin, C. G. 1977. Invasion of plant tissue in the rumen by the flagellate *Neocallimastix frontalis*. J. Gen. Microbiol. 98:423–430.

117. Orpin, C. G. 1977. The rumen flagellate *Piromonas communis*: its life-history and invasion of plant materials in the rumen. J. Gen. Microbiol. 99:107–117.

118. Orpin, C. G. 1981. Isolation of cellulolytic phycomycete fungi from the caecum of the horse. J. Gen. Microbiol. 123:287–296.

119. Orpin, C. G., and A. J. Letcher. 1979. Utilization of cellulose, starch, xylan and other hemicelluloses for growth by the rumen phycomycete *Neocallimastix frontalis*. Curr. Microbiol. 3:121–124.

120. Oshima, T. 1965. On the anaerobic metabolism of aromatic compounds in presence of nitrate by soil microorganisms. Z. Allg. Mikrobiol. 5:386–394.

121. Patel, T. R., K. G. Jure, and G. A. Jones. 1981. Catabolism of phloroglucinol by the rumen anaerobe *Coprococcus*. Appl. Environ. Microbiol. 42:1010–1017.

122. Patel, G. B., A. W. Khan, B. J. Agnew, and J. R. Colvin. 1980. Isolation and characterization of an anaerobic microorganism, *Acetivibrio cellulolyticus* gen. nov. sp. nov. Int. J. Syst. Bacteriol. 30:179–185.

123. Pfennig, N. 1978. *Rhodocyclus purpureus* gen. nov. sp. nov. a ring-shaped, vitamin B_{12}-requiring member of the family Rhodospirillaceae. Int. J. Syst. Bacteriol. 28:283–288.

124. Pfenning, N., K. E. Eimhjellen, and S. Liaaen-Jensen. 1965. A new isolate of the *Rhodospirillum fulvum* group and its photosynthetic pigments. Arch. Mikrobiol. 83:165–171.

125. Phelan, M. B., D. L. Crawford, and A. L. Pometto III. 1979. Isolation of lignocellulose-decomposing actinomycetes and degradation of specifically ^{14}C-labeled lignocelluloses by six *Streptomyces* strains. Can. J. Microbiol. 25:1270–1276.

126. Pomeroy, L. R., W. M. Darley, E. L. Dunn, J. L. Gallagher, E. B. Haines, and D. M. Whitney. 1981. Primary production, p. 39–67, in L. R. Pomeroy and R. G. Weigert (eds.), The ecology of a salt marsh. Springer-Verlag, New York.

127. Prins, R. A., and E. R. Prast. 1973. Oxidation of NADH in a coupled oxidaseperoxidase reaction and its significance for the fermentation of rumen protozoa of the genus *Isotricha*. J. Protozool. 20:471–477.

128. Proctor, M. H., and S. Scher. 1960. Decomposition of benzoate by a photosynthetic bacterium. Biochem. J. 76:33P.

129. Reese, E. T. 1975. Summary statement on the enzyme system. Biotechnol. Bioeng. Symp. 5:77–80.

130. Reese, E. T. 1977. Degradation of polymeric carbohydrates by microbial enzymes, p. 311–367, in F. A. Loewus and V. C. Runeckles (eds.), The structure, biosynthesis, and degradation of wood. Plenum Press, New York.

131. Reid, I. D., G. D. Abrams, and J. M. Pepper. 1982. Water-soluble products from the degradation of aspen lignin by *Phanerochaete chrysosporium*. Can. J. Bot. 60:2357–2364.

132. Rexová-Benková, L., and O. Marković. 1976. Pectic enzymes. Adv. Carbohydr. Chem. Biochem. 33:323–385.

133. Sarkanen, K. V., and C. H. Ludwig. 1971. Lignins: occurrence, formation, structure, and reactions. Wiley-Interscience, New York.

134. Schink, B. 1984. Microbial degradation of pectin in plants and aquatic environments, p. 580–587, in M. J. Klug and C. Reddy (eds.), Current perspectives in microbial ecology. American Society for Microbiology, Washington, D.C.

135. Schink, B., and N. Pfennig. 1982. Fermentation of trihydroxybenzenes by *Pellobacter acidigallici* gen. nov. sp. nov., a new strictly anaerobic, non-sporeforming bacterium. Arch. Microbiol. 133:195–201.

136. Schlomi, E. R., A. Lankhorst, and R. A. Prins. 1978. Methanogenic fermentation of benzoate in an enrichment culture. Microb. Ecol. 4:249–261.

137. Schubert, W. J. 1965. Lignin biochemistry. Academic Press, New York.

138. Schultz, J. E., and J. A. Breznak. 1978. Heterotrophic bacteria present in hindguts of wood-eating termites [*Reticulitermes flavipes* (Kollar)]. Appl. Environ. Microbiol. 35:930–936.

139. Sijpesteijn, A. K. 1951. On *Ruminococcus flavefaciens*, a cellulose-decomposing bacterium from the rumen of sheep and cattle. J. Gen. Microbiol. 3:289–311.

140. Simpson, F. J., G. A. Jones, and E. A. Wolin. 1969. Anaerobic degradation of some bioflavonoids by microflora of the rumen. Can. J. Microbiol. 15:972–974.

141. Smith, W. R., I. Yu, and R. E. Hungate. 1973. Factors affecting cellulolysis of *Ruminococcus albus*. J. Bacteriol. 114:729–737.

142. Sugumaran, M., and C. S. Vaidyanathan. 1978. Metabolism of aromatic compounds. J. Ind. Inst. Sci. 60:57–123.

143. Tarvin, D., and A. M. Buswell. 1934. The methane fermentation of organic acids and carbohydrates. J. Am. Chem. Soc. 56:1751–1755.

144. Taylor, B. F. 1983. Aerobic and anaerobic catabolism of vanillic acid and some other ethoxy-aromatic compounds by *Pseudomonas* sp. Strain PN-1. Appl. Environ. Microbiol. 46:1286–1292.

145. Taylor, B. F., W. L. Campbell, and I. Chinoy. 1970. Anaerobic degradation of the benzene nucleus by a facultatively anaerobic microorganism. J. Bacteriol. 102:430–437.

146. Taylor, B. F., and M. J. Heeb. 1972. The anaerobic degradation of aromatic compounds by a denitrifying bacterium. Arch. Mikrobiol. 83:165–171.

147. Tsai, C. G., D. M. Gates, W. M. Ingledew, and G. A. Jones. 1976. Products of anaerobic phloroglucinol degradation by *Coprococcus* sp. Pe₁5. Can. J. Microbiol. 22:159–164.

148. Tsai, C. G., and G. A. Jones. 1975. Isolation and identification of rumen bacteria capable of anaerobic phloroglucinol degradation. Can. J. Microbiol. 21:794–801.

149. Viljoen, J. A., E. B. Fred, and W. H. Peterson. 1926. The fermentation of cellulose by thermophilic bacteria. J. Agric. Sci. 16:1–17.

150. Vogels, G. D., W. F. Hoppe, and C. K. Stumm. 1980. Association of methanogenic bacteria with rumen ciliates. Appl. Environ. Microbiol. 40:608–612.

151. Weimer, P. J., and J. G. Zeikus. 1977. Fermentation of cellulose and cellobiose by *Clostridium thermocellum* in the absence of *Methanobacterium thermoautotrophicum*. Appl. Environ. Microbiol. 33:289–297.

152. Whittle, P. J., D. O. Lunt, and W. C. Evans. 1976. Anaerobic photometabolism of aromatic compounds by *Rhodopseudomonas* sp. Biochem. Soc. Trans. 4:490–491.

153. Widdel, F. 1980. Anaerober Abbau von Fettsäuren und Benzoesäure durch neu isolierte Arten Sulfat-reduzierender Bakterien. Ph.D. thesis, University of Göttingen.

154. Williams, R. J., and W. C. Evans. 1973. Anaerobic metabolism of aromatic substrates by certain microorganisms. Biochem. Soc. Trans. 1:186–187.

155. Williams, R. J., and W. C. Evans. 1975. The metabolism of benzoate by *Moraxella* species through anaerobic nitrite respiration. Evidence for a reductive pathway. Biochem. J. 148:1–10.

156. Wuhrmann, K. 1982. Ecology of methanogenic systems in nature. Experientia (Basel) 38:193–198.

157. Yaku, F., S. Tsuji, and T. Koshijima. 1979. Lignin carbohydrate complex. Part III. Formation of micelles in the aqueous solution of acidic lignin carbohydrate complex. Holzforschung 33:54–59.

158. Young, L. Y. 1984. Anaerobic degradation of aromatic compounds, p. 487–523, in D. T. Gibson (ed.), Microbial degradation of organic compounds. Marcel Dekker, New York.

158a. Young, L. Y., and M. D. Rivera. 1985. Methanogenic degradation of four phenolic compounds. Water Res. 19:1325–1332.

159. Zehnder, A. J. B. 1978. Ecology of methane formation, p. 349–376, in R. Mitchell (ed.), Water pollution microbiology, Vol. 2. Wiley, New York.

160. Zehnder, A. J. B., B. A. Huser, T. D. Brock, and K. Wuhrmann. 1980. Characterization of an acetate-decarboxylating, non-hydrogen-oxidizing methane bacterium. Arch. Microbiol. 124:1–11.

161. Zeikus, J. G. 1980. Chemical and fuel production by anaerobic bacteria. Annu. Rev. Microbiol. 34:423–464.

162. Zeikus, J. G. 1980. Fate of lignin and related aromatic substrates in anaerobic environments, p. 101–109, in T. K. Kirk, T. Higuchi, and H.-m. Chang (eds.), Lignin

biodegradation: microbiology chemistry and potential applications. CRC Press, Boca Raton, Fla.

163. Zeikus, J. G. 1981. Lignin metabolism and the carbon cycle: polymer biosynthesis, biodegradation and environmental recalcitrance. Adv. Microb. Ecol. 5:211–243.

164. Zeikus, J. G., A. L. Wellstein, and T. K. Kirk. 1982. Molecular basis for the biodegradative recalcitrance of lignin in anaerobic environments. FEMS Microbiol. Lett. 15:193–197.

8

ANAEROBIC HYDROLYSIS AND FERMENTATION OF FATS AND PROTEINS

MICHAEL J. McINERNEY

Department of Botany and Microbiology, University of Oklahoma, 770 Van Vleet Oval, Norman, OK 73019

8.1 INTRODUCTION
8.2 DEGRADATION OF LIPIDS
    8.2.1   Hydrolysis of Glycerol Esters
    8.2.2   Hydrogenation of Unsaturated Fatty Acids
    8.2.3   Utilization of Exogeneous Fatty Acids
8.3 DEGRADATION OF PROTEINS
    8.3.1   Proteolytic Bacteria
    8.3.2   Proteases
    8.3.3   Factors Affecting Protein Degradation
8.4 AMINO ACID FERMENTATION
    8.4.1   Stickland Reactions
    8.4.2   Oxidation of Branched-Chain and Other Amino Acids
    8.4.3   Aromatic Amino Acid Degradation
    8.4.4   Arginine
    8.4.5   Threonine
    8.4.6   Glycine
    8.4.7   Glutamate
    8.4.8   Lysine

8.4.9   Other Amino Acids
8.4.10   Importance of Interspecies H_2 Transfer
8.5   SUMMARY
REFERENCES

8.1   INTRODUCTION

The input of proteins and lipids into anaerobic environments is large. As an example, Table 8.1 shows the composition of various materials used as substrates for methanogenic digestors. Crude protein is always a significant component of these materials. The amount of lipid (ether extract) is usually much less than that of proteins or polysaccharides, but in some cases, such as domestic sewage sludge, it is a major component. Although proteins and lipids are quantitatively important substrates for anaerobic degradation and the production of H_2S and NH_4^+ from proteins is important in S and N cycles, the degradation of proteins and lipids has not been studied as extensively as carbohydrate fermentations. In some environments, the major proteolytic or lipolytic bacteria are not known and the biochemical pathways for the degradation of certain amino acids have only recently been elucidated (14, 56). Although much information has been obtained since Barker's early studies on anaerobic fermentations of nitrogenous compounds (12), many unanswered questions remain.

The degradation of proteins and lipids is a complex process involving many different kinds of anaerobic microorganisms. The kinds of organisms involved in these reactions can vary depending on the particular ecosystem. Although we emphasize the fermentations as carried out by bacteria, it should be noted that anaerobic protozoa and possibly fungi are also important (27, 106, 150, 181b). The great diversity of species involved in these reactions is also reflected in the number of biochemical pathways involved (10, 76). In general proteins are hydrolyzed to peptides and amino acids which are fermented to volatile fatty acids, CO_2, H_2, NH_4^+, and S^{2-} by fermentative bacteria such as the clostridia (10–12, 26, 132, 134). Lipids are hydrolyzed with the release of free fatty acids and the galactose and glycerol moieties are fermented to similar products without NH_4^+ or S^{2-} production. The fermentative bacteria do not attack the aromatic nucleus of amino acids, nor do they oxidize fatty acids to any appreciable extent. The latter reactions are carried out by other metabolic groups of bacteria such as the sulfate-reducing bacteria (F. Widdel, Chapter 10) or H_2-producing syntrophs (132, 133, 138), which are often associated with the fermentative bacteria in many anaerobic environments.

In this chapter we attempt to summarize the work done on the biochemistry, microbiology and ecology of protein and lipid degradation. Several extensive reviews on these subjects are available (1, 3, 10–12, 27, 35, 50, 67, 99a, 100, 150,

TABLE 8.1 Major components of various substrates used for anaerobic digestion and methane production[a]

Component	Cattle Manure	Chicken Manure	Pig Manure	Domestic Sewage	Sludge	Coastal Bermuda Grass	Giant Brown Kelp
Volatile solids	72.00	76.0	82.4	75.9	60.8	—	—
Ether extract (lipid)	3.5	1.5	7.7	34.4	25.2	—	—
Cellulose	17.0	28.3	33.0[b]	3.8	1.4	31.7	8.9
Hemicellulose	19.0	11.9	53.8[b]	3.2	—	40.2	—
Lignin	6.8	9.2	10.1	5.8	—	4.1	—
Crude protein	19.0	28.8	20.9	27.1	19.4	12.3	29.3
Ash	28.0	24.0	17.6	24.1	—	—	—
References	(177)	(105)	(102)	(96, 129)	(35)	(118)	(118)

[a] —, not reported.
[b] Cellulose and hemicellulose were originally reported as acid detergent fiber and neutral detergent fiber, respectively (102).

156, 166) and have served as important sources for the material present herein. Reviews by Barker (11, 12) provide an excellent treatment of the earlier work on protein and amino acid fermentations.

8.2 DEGRADATION OF LIPIDS

Different anaerobic environments receive different inputs of lipid material. Ruminant diets contain mainly galactoglycerides as well as sterols, waxes, phospholipids, and some free fatty acids (12, 97, 102). A large proportion of the fatty acids are unsaturated with linolenic, linoleic, and oleic acids being the most common. Domestic sewage sludge usually has a larger amount of lipid on a percent dry weight basis than ruminant feeds (97, 102, 174; Table 8.1). Wastes from meatpacking and slaughterhouses are especially high in lipids. Sewage sludge may also contain large amounts of oils and greases. The amount of unsaturated fatty acids present in sewage sludge is usually lower than that found in the plant materials fed ruminants, and the unsaturated fatty acids are usually monoenoic rather than polyunsaturated. The concentration of free fatty acids in sewage sludge is usually much higher than that found in ruminant diets.

The degradation of lipids is a major source of acetate in sludge (42). However, very little is known about anaerobic lipid degradation, and much of this work has focus on just one ecosystem, the rumen. Several reviews on lipid degradation in the rumen and other parts of the gastrointestinal tract are available (27, 50, 71, 100, 150). The biotransformations of bile acids, cholesterol, and steroids by intestinal microorganisms have recently been reviewed (20a, 109a) and are not discussed here.

8.2.1 Hydrolysis of Glycerol Esters

In anaerobic environments, glycerol esters are first hydrolyzed with the release of free fatty acids (51, 71–73). Glycerol, galactose, choline, and other non-fatty acid components released by hydrolysis are fermented mainly to volatile fatty acids by the fermentative bacteria (26, 27, 132). The fermentative bacteria do not further degrade the free fatty acids, although some species may hydrogenate unsaturated fatty acids (50). Fatty acids are not degraded in the rumen or other parts of the gastrointestinal tract. However, in environments such as sludge digestors or aquatic sediments where the retention time of the material in the system is long, both long-chain and short-chain fatty acids are β-oxidized to acetate by H_2-producing syntrophic bacteria (132, 133). In sulfate-rich environments such as marine sediments, fatty acids are degraded by sulfate-reducing bacteria (147).

Several lipolytic bacteria have been isolated from the rumen. *Anaerovibrio lipolytica* is a gram-negative curved rod that ferments only glycerol, ribose, and fructose (93, 101). Products produced from these compounds include acetate, propionate, and succinate. *A. lipolytica* hydrolyzes triglycerides and phospholipids to glycerol and free fatty acids (152). However, it does not attack galactolipids, which are the major plant lipid present in the feed unless the galactose residue has been removed (71). Galactolipids as well as phospholipids and sulpholipids are deacylated by *Butyrivibrio* strain S2 that requires long-chain fatty acids for growth (51, 89). The long-chain fatty acids released by hydrolysis are incorporated into membrane lipids of the plasmalogen type, partially as diabolic acids, a new type of hydrophobic group derived from the condensation of two molecules of the fatty acid. Strain S2 ferments galactose with the production of butyrate and some pyruvate (89). A noncellulolytic *Butyrivibrio fibrisolvens* (strain LM8/1B) rapidly hydrolyzes phospholipids when grown with a fermentable sugar as the carbon source (88). Butyrate, ethanol, and hydrogen are the main products produced by strain LM8/1B from galactose. In addition to the ruminal species above, several types of *Clostridium botulinum* produce lypase and lecithinase activities (103). *Eubacterium* species, which comprise about 4% of the culturable bacteria in a eutrophic lake sediment, are able to hydrolyze tributyrin, indicating the presence of lipase activity (135a).

    A. lipolytica produces an extracellular lipase during the exponential phase of growth (94, 95). The lipase activity has been partially purified and characterized (94, 95). The lipase activity is excluded from a Sephadex G-200 column and the purest preparations contain two components and a large amount of nucleic acid. The lipolytic activity is enhanced by $CaCl_2$ or $BaCl_2$, while $ZnCl_2$, $HgCl_2$, and high concentrations of NaCl are inhibitory. Trilaurin is the most rapidly hydrolyzed triglyceride, but diglycerides are more rapidly degraded than triglycerides. Phospholipids and galactolipids are not hydrolyzed.

    Figure 8.1 shows the reactions catalyzed by various phospholipases. *B. fibrisolvens* strain LM8/1B contains a phospholipase active against both phosphatidyl-

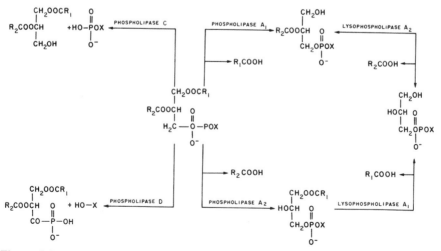

Figure 8.1 Reactions catalyzed by various phospholipases. Reproduced from Finnerty (67) with permission from Academic Press, Inc.

choline and phosphatidylethanolamine (88). Lysophospholipid and a free fatty acid are released by phospholipase activity. The rapid appearance of a glycerylphosphorylated derivative and the minimal accumulation of lysophospholipid indicates that an active lysophospholipase (phospholipase B) is also present. The phospholipase activity in strain LM8/1B has many unusual characteristics (88). It is activated by cysteine, suggesting that a low redox potential is necessary to prevent oxygen poisoning of the enzyme. The addition of calcium does not stimulate phospholipase A activity nor is ethylenediaminetetraacetate inhibitory. This is unusual since most bacterial phospholipase A's have an absolute requirement for Ca^{2+} ions (88). Anionic detergents stimulate activity when a zwitterionic phospholipid is the substrate, suggesting that the enzyme prefers a substrate–water interface that has an excess of negative charges. Dawson (49) has shown that many phospholipid-metabolizing enzymes have a very precise requirement for the zeta potential on the surface of the substrate molecule. *Butyrivibrio* strain S2 contains phospholipase A and B activities as well as phospholipase C activity (89). The latter activity has only limited distribution among bacterial genera (67). The α-toxin of *Clostridium perfringens* type A was identified by McFarlane and Knight (127) as phospholipase C. Phospholipase C hydrolyzes phosphatidylcholine to phosphorylcholine and diglyceride. The clostridial enzyme is a zinc metalloprotein which requires calcium as an activator (39, 172). The clostridial enzyme also hydrolyzes sphingomyelin.

Some members of the genera *Treponema* and *Borrelia* are lipolytic (148, 175). Most of the lipolytic spirochetes require long-chain fatty acids for growth which are obtained by hydrolysis of serum components. *Borrelia* species contains a phospholipase B activity but not a lipase activity (148). Five species of *Treponema*, *T. phagedenis*, *T. denticola*, *T. refringins*, *T. minutum*, and *T. vincentii*, contain phospholipase B and glycerophosphorylcholine diesterase activities (175). *T. phagedenis* and *T. denticola* have phosphatase activity and can catabolize lysophosphatidylcholine to glycerol. *T. refringens*, *T. minutum*, and *T. vincentii* catabolize lysophosphatidylcholine to glycerol phosphate and choline.

Choline released from phosphatidylcholine during lipid metabolism is fermented by several strains of *Desulfovibrio desulfuricans* and by a species of *Clostridium* as shown in reaction (1) (22, 65, 85, 141):

$$2\ choline^+ + H_2O \rightarrow 2\ trimethylamine^+ + acetate^- + ethanol + H^+ \quad (1)$$

Trimethylamine is further degraded to methane and ammonia by *Methanosarcina barkeri* (65, 141).

8.2.2 Hydrogenation of Unsaturated Fatty Acids

The hydrogenation of unsaturated fatty acids by mixed ruminal bacteria is well documented (50, 114). Hydrogenation occurs only after the fatty acid has been liberated from its ester linkages (51). α-Linolenic acid (*cis-cis-cis*-octadec-9,12,15-enoic acid), γ-linolenic acid (*cis-cis-cis*-octadec-6,9,12-enoic), and linoleic acid (*cis-cis*-octadec-9,12-enoic) can be reduced to stearic acid, although a small amount of *trans*-octadec-11-enoic acid escapes reduction (50, 114–116). The bacteria that perform these reactions may be placed into two groups (114–116). The first group consists of those species that cannot hydrogenate octadecenoic acid but hydrogenate linoleic acid and α-linolenic acid to *trans*-octadec-11-enoic acid and related isomers (89, 114–116). This group contains most of the isolates, including one strain of *Ruminococcus albus* (F2/6) (116), two strains of *Eubacterium* (114), and several strains of *Butyrivibrio* (88, 89, 114). The second group has only three known bacteria, which are two strains of *Fusocillus* and an unidentified gram-negative rod (115, 115a, 116). These bacteria can hydrogenate many octadecenoic acids, including oleic acid, linoleic acid, and *trans*-octadec-11-enoic acid to stearic acid. α-Linolenic acid is hydrogenated mainly to *cis*- and *trans*-octadec-15-enoic acids which are not further hydrogenated even by mixed ruminal bacteria (50, 114, 115). Ruminal protozoa may also be active in hydrogenation reactions (75, 181b).

Thirty-three strains of *Eubacterium lentum* isolated from intestinal contents of animals and humans transform linoleic, α-linolenic, and γ-linolenic acids by an isomerization of the 12-cis double bond into an 11-trans double bond and a reduction of the 9-cis double bond to yield *trans*-octadec-11-enoic acid (Fig. 8.2; 63a, 178a). Twenty-nine of these strains also isomerize cis double bonds located at C-11 and C-14 as found in homo-γ-linolenic acid and arachidonic acid to the Δ^{14}cis, Δ^{13} trans configuration or those at C-10 and C-13 (*cis-cis*-nonadeca-10,13-enoic acid) to the Δ^{13}cis, Δ^{12}trans configuration. Many of these *E. lentum* strains also biotransform steroids (20b, 63a). Interestingly, patients with recently diagnosed, untreated colorectal cancer have less than 1% of the normal levels of *E. lentum*

Hydrogenation of γ-Linolenic Acid

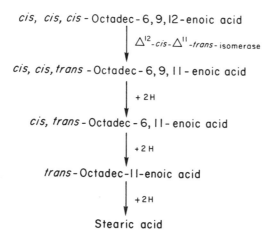

cis, cis, cis - Octadec - 6,9,12 - enoic acid

↓ Δ^{12}-*cis*-Δ^{11}-*trans*- isomerase

cis, cis, trans - Octadec - 6,9,11 - enoic acid

↓ + 2 H

cis, trans - Octadec - 6,11 - enoic acid

↓ + 2 H

trans - Octadec - 11 - enoic acid

↓ + 2 H

Stearic acid

Figure 8.2 Sequence of reactions in the hydrogenation of γ-linolenic acid by pure cultures of ruminal bacteria. Reproduced by permission from Biochemical Journal, Vol. 216, pp. 519–522, copyright © 1983, The Biochemical Society, London.

and phenotypically similar organisms in their feces, suggesting that there is some relationship between these organisms and colorectal cancer (20a). *E. lentum* is known to convert tetrahydrodeoxycorticosterone, which is carcinogenic for normal hamster embryonic cells, to the noncarcinogen, pregnanolone (20a). These findings need further study before definitive conclusions can be drawn about the role of these organisms in colorectal cancer. The hydrogenation of polyunsaturated fatty acids may have a survival value for these organisms since polyunsaturated fatty acids are inhibitory to the growth of ruminal and intestinal bacteria (115a,115b).

The pathway for the conversion of linolenic acid is shown in Fig. 8.2 (115, 116). The first reaction involves the action of linoleic acid Δ^{12}-*cis*, Δ^{11}-*trans*-isomerase which produces a conjugated diene or triene by the stereospecific addition of hydrogen (108, 117). A free C-1 carboxyl group and a *cis*-9, *cis*-12-diene system are absolutely required (117). The fact that arachidonic acid, homo-γ-linolenic acid, and *cis-cis*-nonadec-10,13-enoic acid are isomerized by many *E. lentum* strains suggests that the two double bonds do not have to be strictly located at C-9 and C-12 (178a). The reduction of the conjugated intermediate proceeds by the stereospecific cis-addition of hydrogen atoms that are in rapid equilibrium with water. The enzyme responsible for the reductive steps has not been isolated, nor is the nature of the electron donor(s) unequivocally known. Yokoyama and Davis (186) showed that hydrogenation of the conjugated intermediate in a *Borrelia* sp. occurred only in the presence of viologen dyes or dithionite, suggesting that ferredoxin or a similar molecule is involved. Yamazaki and Tove (184) found that a fraction obtained after Sepharose 6B chromatography of *B. fibrisolvens* cell-free extracts served as the electron donor for the hydrogenation reaction. After its oxidation, this fraction could be reduced by dithionite, which was present in large amounts in extracts of *B. fibrisolvens*.

8.2.3 Utilization of Exogenous Fatty Acids

The *de novo* synthesis of lipids, particularly phospholipids by ruminal bacteria is significant (27, 100). Maczulak et al. (128) did not find that the addition of low concentrations of long-chain fatty acids to the growth medium had any significant energy-sparing effect in the seven ruminal bacterial studied. However, Demeyer et al. (52) concluded that the amount of long-chain fatty acids directly incorporated into microbial lipids in the rumen was higher than that synthesized *de novo*.

Elsden et al. (59) studied the fatty acid composition of 23 species of proteolytic clostridia and found that these species could be separated into two groups based on their fatty acid composition. One group contained those species that had n, iso, or anteiso fatty acids. All of the organisms that had iso and anteiso fatty acids also oxidized valine, leucine, and isoleucine to the corresponding branched-chain fatty acid. The iso fatty acids were both even-numbered and odd-numbered and were probably derived from valine and leucine, respectively. All the anteiso fatty acids were odd-numbered and were probably derived from isoleucine and 2-methylbutyrate. The other group of organisms contained only *n*-fatty acids and these bacteria produced acetic or acetic and *n*-butyric acids from the amino acid substrate.

Many ruminal bacteria are unable to synthesize various long-chain fatty acids from carbohydrate or amino acid carbon. These strains require acids such as *n*-valeric, isovaleric, 2-methylbutyric, and isobutyric for growth (4, 27–29, 62). *Bacteroides succinogenes* requires *n*-valeric and isobutyric acids for growth and uses these acids to synthesize *n*-pentadecanoic acid and isotetradecanoic acid, respectively. Some species such as *Butyrivibrio* strain S2 require C_{13} to C_{18} straight-chain saturated or monoenoic fatty acids for growth (89).

Butyrivibrio strain S2 contains a newly discovered lipid component, a long-chain vicinyl dimethyl–substituted dicarboxylic acid where two saturated fatty acids appear to be linked together by their penultimate carbon atoms (43, 119). These diabolic acids are linked at each carboxyl group to the C-2 hydroxyl of an alkenyl glycerol (43, 119). To date, diabolic acids have been found only in the rumen and hindgut of termites (86). Since *Butyrivibrio* strain S2 rapidly hydrogenates unsaturated fatty acids, it was unclear how membrane fluidity was maintained in this bacterium. The membranes of strain S2 become fluid at 34.5°C and the presence of diabolic acids and butyryl groups in glycolipids and phospholipids contribute to the fluidity (84).

Selenomonas ruminantium produces long-chain even-numbered fatty acids from odd-numbered volatile fatty acids (61, 113), suggesting that α-oxidation occurs in this bacterium. Emmanuel (61, 113) showed that $^{14}CO_2$ was released when [1-^{14}C]palmitate was incubated in growing cultures of several ruminal bacterial, including *S. ruminantium*, confirming that α-oxidation does occur. Odd-numbered, long-chain fatty acids can also be synthesized by the α-oxidation of even-numbered, long-chain fatty acids (61).

8.3 DEGRADATION OF PROTEINS

The anaerobic degradation of proteins is important for the cycling of N and S as well as C in nature. In the ruminant, the microbial protein synthesized in the rumen is the major source of proteins and amino acids for the animal. For this reason, the degradation of protein in the rumen has been studied intensively (3, 27, 98, 99a, 100, 106). In the rumen, proteins are hydrolyzed to peptides and amino acids (27, 100, 182; Fig. 8.3). The amino acids are rapidly fermented and only very low concentrations of amino acids are found in the rumen. The major products produced from amino acid fermentation are short-chain or branched-chain fatty acids, ammonia, and CO_2. A substantial amount of the ammonia pool in the rumen is derived from these proteolytic processes. Ammonia serves as the main nitrogen source for microbial growth in the rumen. The degradation of protein in digestor sludge as well as other anaerobic environments is probably very similar to that found in the rumen (96, 99a, 102, 132, 174). The major difference seems to be that in the rumen, proteins are degraded by carbohydrate-fermenting species and

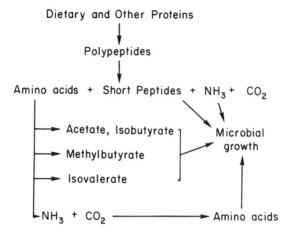

Figure 8.3 Protein degradation and nitrogen metabolism in the rumen. Reproduced from Hespell (98) by permission of the Society for Industrial Microbiology.

the fermentation of peptides and/or amino acids alone does not provide sufficient energy for growth (98, 99a, 182). In digestors and other anaerobic environments, a specialized group of anaerobic bacteria such as the proteolytic clostridia are responsible for protein degradation and these processes are energy yeilding (10–12, 96, 99, 100, 102, 174).

8.3.1 Proteolytic Bacteria

Most of the proteolytic activity in the rumen resides with the bacterial fraction (100, 179b). However, ruminal protozoa are important for recycling microbial protein in faunated animals (27, 106, 181b). The main proteolytic activity found in the protozoal fraction was a amino acid arylamidase (leucine aminopeptidase) (69). The number of proteolytic bacteria in the rumen is large. Up to 43% of the viable bacteria isolated from cows fed grass and concentrates grew on medium with casein as the major substrate (70, 151). Proteolytic species include *Bacteroides ruminicola* (87, 90, 92), *Bacteroides amylophilus* (19, 70), *Butyrivibrio* sp. (97), *Selenomonas* sp. (97), *Streptococcus bovis*, *Lachnospira multiparus*, and isolates of *Eubacterium ruminantium* (19, 99a, 153). The population attached to the wall of the rumen is also proteolytic (40). These bacteria are mainly facultative anaerobes, such as staphylococci (40). In general, proteins, peptides, or amino acids are not used as sole energy sources by these bacteria.

Analysis of human gut contents shows the presence of substantial amounts of soluble protein, branched-chain volatile fatty acids, and ammonia throughout the large intestine (126a). The distribution of proteolytic activity suggests that a substantial portion of the fecal proteolytic activity is bound to the bacterial cell fraction. The predominant proteolytic bacteria in fecal samples are *Bacteroides* species (about 1 to 10 × 10^{11} per gram dry weight of feces) and *Propionibacterium* species (about 1 to 100 × 10^8 per gram dry weight of feces) (126a).

Digestor sludge also contains large numbers of proteolytic bacteria, around 6.5 × 10^7 bacteria per milliliter (161). Most of the isolates are clostridia, but *Peptococcus anaerobicus*, *Bifidobacterium* sp., *Staphylococcus* sp., and some gram-negative, nonsporeforming bacilli are also proteolytic. Most studies show that the main proteolytic bacteria in sludge are gram-positive bacteria, principally clostridia (96, 102, 174). This is also the case in a eutrophic lake sediment, where 66% of the culturable bacteria are the proteolytic species *C. bifermentens* and *C. sporogenes* (135a). This is in contrast to the rumen, where the principal proteolytic bacteria are gram-negative and of the same species as those that ferment carbohydrates.

Both *Clostridium* and *Eubacterium* species are the active proteolytic species in a high-strength acid whey digestor (181a). Jain and Zeikus (109b) have recently isolated a new species, *Clostridium proteolyticum*, from sewage sludge, which hydrolyzes various proteins and ferments the amino acids mainly to acetate and CO_2 with small amounts of H_2, isovalerate, and isobutyrate. This organism consumes H_2/CO_2 in stationary phase, suggesting that it may be an acetogen (109b). *Thermobacteroides proteolyticus* is a thermophilic, gram-negative nonsporeforming, nonmotile rod isolated from an enrichment derived from tannery wastes and cattle manure (144a). Various protein substrates are used by *T. proteolyticus*, but sugars are poorly used. The extreme thermophilic archaebacteria, *Thermofilum pendens*, *Thermococcus celer*, and *Desulfurococcus* sp., use casein or the peptides in yeast extract or tryptone as energy sources (189–191). In the latter two organisms, the use of these compounds is coupled to the reduction of elemental sulfur.

Bacteroides gingivalis has recently been shown to be the only species of the black-pigmented *Bacteroides* group which possesses collagenolytic activity (126a). Other species in this group such as *Bacteroides melaninogenicus* have other proteolytic activities, but do not possess collagenolytic activity as once thought (74, 125, 176a). *B. gingivalis* is the most virulent of the black-pigmented *Bacteroides* species in experimental animals, causing spreading types of infection. The high proteolyic activity of this species may account for its virulence (66). Various proteolytic clostridia are frequently found in wounds and other traumatic injuries and are important etiologic factors in the development of gas gangrene or tetanus (162).

Proteolytic clostridia such as *Clostridium prefringens*, *C. bifermentans*, *C. histolyticum*, and *C. sporogenes* are important in the spoilage of meats and canned foods (9, 109). *Clostridium botulinum* is also actively proteolytic and causes food spoilage (9). The type A and B botulinal toxins are synthesized as prototoxins, which are activated by clostridial proteases or trypsin (137). The caseinase activity present in *C. perfringens* ATCC3626B is plasmid encoded (20).

8.3.2 Proteases

Proteases are generally classified by identification of the essential catalytic group through the use of inhibitors (Table 8.2). This approach has been used to identify

TABLE 8.2 Characteristics and classification of proteases

Enzyme	pH Optima	Metal Requirement	Inhibitors	Examples
Acid or carboxyl proteases	1-5	—	Diazoketones	Pepsin
Thiolproteases	4-8	—	Iodine, iodoacetate, organic mercury, H_2O_2	Papain, ficin, fimelin, cathepsin
Metalloproteases	7-8	Zn, Mg, Co, Fe, Mn, Ni	EDTA, o-phenanthroline	Carboxypeptidase A, thermolysin
Alkaline or serine proteases	9-11	—	Diisopropyl phosphorofluoridate, phenylmethanesulphonylfluoride, lima bean or soybean trypsin inhibitor	Trypsin, chymotrysin, subtilisin

Source: Data from Loll and Bollag (126).

the types of proteases present in ruminal fluid. Most of the protease activity found in the microbial fraction of ruminal fluid is removed by gentle physical means, indicating that this activity is associated with the coat and capsular material (121). These coat proteases have a broad substrate specificity, which includes several kinds of exo and endopeptidase activities (121, 180). Elastin congo red is not attacked by coat proteases, indicating that similar fibrous proteins in plant material might resist proteolytic attack in the rumen (121, 180). Inhibitor studies show that mixed ruminal bacteria have primarily serine, cysteine, metalloproteinases (24). Prins et al. (151) found that diet affects the kind of protease present in rumen fluid. The main bacterial protease present in a hay-fed cow is a trypsin-like, cysteine protease. In a cow fed grass and concentrates, this activity plus a serine-protease activity are found.

Proteases present in pure cultures of several ruminal bacteria have been studied. *Bacteroides amylophilus* H18 has two forms of proteases, one that is liberated into the growth medium during the exponential phase of growth and another which is cell bound and presumably located at the cell surface (17). Both forms of proteases are sensitive to serine protease inhibitors and show trypsin-like specificity (18). Growth of *B. amylophilus* H18 in medium with tryptone, proteose peptone, or casamino acids does not stimulate protease activity nor is protease activity repressed by the addition of ammonia (17). Another strain of *B. ruminicola* (R8/4) was isolated in medium containing leaf fraction 1 (ribulosebisphosphate carboxylase) as the sole nitrogen source (90–92). This strain has cell-associated protease activity and casein is the most rapidly degraded substrate. However, the affinity for leaf fraction 1 is higher than that for casein. The proteolytic activity of strain R8/4 is composed of a mixture of serine, cysteine, and aspartic acid proteases with the possibility that at least some of the activity is dependent on the presence of metal ions (87, 92). The substrate specificities and some inhibition characteristics indicate that the *B. ruminicola* R8/4 proteolytic activity is similar to that found in ruminal contents (92). Both *B. ruminicola* (90, 149) and *B. melaninogenicus* (124) require peptides for growth. Oligopeptide nitrogen but not small peptides or amino acid nitrogen serves as the sole source of cellular nitrogen in *B. ruminicola* (149). In the presence of large amounts of ammonia, oligopeptides are neither required nor stimulatory (149).

The proteolytic activity of *B. fibrisolvens* is sensitive to serine protease inhibitors but does not show trypsin or chymotrypsin-like specificity (47a). The activity is modulated by changes in the types or concentrations of nitrogen sources in the medium contrary to that found with *B. amylophilus*.

C. histolyticum produces a wide variety of proteolytic enzymes (135), including collagenase, a cysteine-activated protease, and clostripain (clostridiopeptidase B). The properties of clostripain (135) and the collagenases of *C. histolyticum* and *B. melaninogenicus* (135) have been reviewed.

8.3.3 Factors Affecting Protein Degradation

Solubility of the protein has often been cited as the main determinant controlling protein degradation in the rumen, with the least soluble proteins being more slowly degraded than those that are highly soluble (3, 100). However, bovine albumin or ovalbumin, which are highly soluble proteins, are degraded more slowly than casein or other kinds of soluble proteins (3, 100). Treatment of bovine albumin with dithiothreitol increases its rate of degradation in the rumen (144), suggesting that the number of disulfide bridges and the tertiary structure of the protein are also important factors. The kind of end group on the protein may also be important in determining the rate of degradation.

Modification of the protein by heating or chemical treatment has been the only way to protect the protein from degradation in the rumen (100). Increasing the amount of dietary urea does not spare protein from degradation (146). The degradation of maize protein is unaffected by ruminal ammonia concentrations (142). The pH of the system is important for the optimal growth of proteolytic bacteria. In a continuous culture study of whole ruminal contents, the proteolytic bacteria are washed out of the system as the pH dropped below 5.5 (63). Factors affecting protein degradation in soils have been reviewed (126).

8.4 AMINO ACID FERMENTATION

The degradation of amino acids derived from protein hydrolysis is a complex process. A large number of amino acids and other nitrogenous compounds can serve as energy sources for anaerobic bacteria. These bacteria often require complex media for growth, making it difficult to follow the degradation of individual amino acids in such media. Recently, these problems have been overcome and the kinds of amino acids used and the products produced from amino acids have been studied in a variety of anaerobic bacteria (57–60, 134). The anaerobic degradation of amino acids involves oxidation-reduction reactions between one or more amino acids or nonnitrogenous compounds derived from amino acids (10–12). The oxidation reactions are often similar to if not identical to those found in aerobic bacteria and include oxidative deamination, transamination, and α-keto oxidations. However, reactions involving molecular oxygen or other high-potential oxidants are not found in anaerobic bacteria. The reductive reactions used by anaerobes are much more distinct. Amino acids, α- and β-keto acids, α, β-unsaturated acids or their coenzyme A (CoA) derivatives, and protons are the major electron acceptors used. The major end products produced by these reactions include short-chain fatty acids, succinate, δ-aminovalerate, and hydrogen.

Mead (134) divided the amino acid–fermenting clostridia into four major groups and several subgroups, based on the patterns of amino acids used or formed (Table 8.3). Group I consists of species that carry out the Strickland reaction between pairs of amino acids. These bacteria reduce proline or arginine to δ-aminovalerate by oxidizing serine, phenylalanine, and other amino acids. Groups II, III, and IV do not produce δ-aminovalerate, α-aminobutyrate, or γ-aminobutyrate from casein hydrolysate. All group II bacteria use arginine and/or glycine. Group III bacteria are glutamate fermentors such as *Clostridium tetanomorphum*. Group III bacteria also use serine and histidine. Group IV contains only *Clostridium putrefaciens*, which readily uses serine and threonine.

TABLE 8.3 Amino acid utilization patterns of clostridial species[a]

Group Species	Characteristics[b]

I. *Clostridium bifermentans*
  C. sordellii
  C. botulinum types A, B, F
  C. caloritolerans
  C. sporogenes
  C. cochlearium (one
  strain)
  C. difficile
  C. putrificum
  C. sticklandii
  C. ghoni
  C. mangenoti
  C. scatologenes[c]
  C. lituseburense[c]
	Organisms that carry out Stickland reaction; proline reduced and δ-aminovalerate produced; serine, phenylalanine, and other amino acids oxidized; α-aminobutyrate produced from threonine or methionine and γ-aminobutyrate produced from glutamate by most species

II. *C. botulinum* types C, G
  C. histolyticum
  C. cochlearium (one strain)
  C. subterminale
	δ-Aminovalerate not produced; arginine and/or glycine used by all species

III. *C. cochlearium* (one strain)
  C. tetani
  C. tetanomorphum
  C. microsporum
  C. lentoputrescens
  C. limosum
  C. malenomenatum
	δ-Aminovalerate not produced; glutamate, serine, and histidine utilized

IV. *C. putrefaciens*	δ-Aminovalerate not produced; serine and threonine utilized

[a] Classification of Mead (134) with additional species added (57–60).

[b] Major characteristics of each group are given. Other amino acids may be used by various species (see Refs. 57 to 60 and 134).

[c] α-Aminobutyrate and γ-aminobutyrate but not δ-aminovalerate are formed (134).

Elsden and co-workers (57–60) extended the studies of Mead to include additional species as well as including the end products produced by these bacteria. In general, their work confirms that of Mead and shows the importance of these kinds of analyses in clostridial taxonomy. As an example, Elsden and Hilton (57) found *C. botulinum* types A, B, and F closely resemble *Clostridium sporogenes* in the pattern of amino acid utilization. *C. botulinum* type G could not be differentiated from *Clostridium subterminale* (57).

The pathways for amino acid utilization have been reviewed (1, 10–12, 130, 143, 156, 166). The degradation of amino acids and other nitrogenous compounds in the rumen and other paths of the gastrointestinal tract has also been reviewed (3, 27, 100, 150, 182).

8.4.1 Stickland Reactions

Group I contains the majority of amino acid–fermenting clostridia (Table 8.3). These bacteria carry out coupled oxidation-reduction reactions between pairs of amino acids called Strickland reactions (11, 143, 156). An example of a Strickland reaction is shown in reaction (2) for the oxidation of alanine with glycine serving as the electron acceptor

$$CH_3CHNH_2COOH + 2H_2O \longrightarrow CH_3COOH + CO_2 + NH_3 + 4H$$
$$2CH_2NH_2COOH + 4H \longrightarrow 2CH_3COOH + 2NH_3$$

$$CH_3CHNH_2COOH + 2CH_2NH_2COOH + 2H_2O \longrightarrow 3CH_3COOH + 3NH_3 + CO_2$$

$$(2)$$

A partial list of amino acids that serve as electron donors include alanine, leucine, isoleucine, valine, and histidine. Electron acceptors include glycine, proline, hydroxyproline, ornithine, arginine, and tryptophan (143, 156). Not all these amino acids are used by the same species and the rates of utilization of different amino acids by a particular species differ.

The oxidation of amino acids such as alanine, leucine, isoleucine, and valine proceeds via an α-keto acid [reaction (3)] (11, 143):

$$R-\underset{\underset{NH_2}{|}}{CH}-COOH \xrightarrow[2H]{+H_2O} R-\overset{\overset{O}{\|}}{C}-COOH + NH_3 \xrightarrow[2H]{+H_2O} R-COOH + CO_2 \quad (3)$$

These oxidative reactions are discussed in more detail in Section 8.4.2.

The reduction of glycine [reaction (4)] and proline [reaction (5)] has been studied in some detail.

$$\text{glycine} + R(SH)_2 + ADP + P_i \longrightarrow \text{acetate} + NH_3 + ATP + R\overset{S}{\underset{S}{\diagup|}} \quad (4)$$

$$\text{proline} + R(SH_2) \longrightarrow \delta\text{-aminovalerate} + R\overset{S}{\underset{S}{\diagup|}} \quad (5)$$

The reduction of 1 mole of glycine but not proline is stoichiometrically coupled to the synthesis of 1 mol of ATP (171). Either NADH or a dimercaptan can serve as the electron donor in equations (3) and (4) (156, 168). Presumably, NADH is the physiological reductant. With NADH, a flavoprotein, ferredoxin, and an uncharacterized labile protein are required for glycine reduction (167, 168). The dimercaptan interacts directly with glycine reductase complex and the electron carriers above are not required.

The glycine reductase complex of *C. sporogenes* is composed of at least three different components, designated A, B, and C (44, 45, 167, 173, 176). Proteins A and B have been extensively purified, whereas component C is a mixture of proteins (173). Protein A is a selenoprotein of about 12,000 molecular weight with 1 gram-atom of Se per mole of protein (44, 165, 176). Selenium is present in the protein as Se cysteine (45). *C. sporogenes* has an absolute requirement for selenium when grown with glycine as the electron acceptor (46). When *C. sporogenes* is grown in a selenium-deficient medium with proline as the electron acceptor, protein A is not synthesized (178). Protein A is readily oxidized by molecular oxygen, which results in a change in the absorption spectrum between 240 and 280 nm. Absorbance in this region is thought to be due to the selanide anion (44). The oxidized form of protein A is readily reduced by KBH_4. These observations have led to the hypothesis that protein A functions as an electron carrier in glycine reduction.

Protein B and the proteins found in component C have molecular weights in excess of 200,000 and are membrane bound (173). Protein B has a carboxyl group that is essential for activity and the electronic absorption spectrum of the pure protein indicates that this carboxyl group is not pyridoxal phosphate (173). Little is known about component C except that it contains iron. All three components (A, B, C) are required for glycine reduction (173). The mechanism of ATP synthesis is not known.

The reductive ring cleavage of proline to δ-aminovalerate is catalyzed by proline reductase (156, 170). A FAD-containing NADH dehydrogenase and a metalloprotein are required to transfer electrons from NADH to proline reductase (158). Dithiols interact directly with proline reductase. Proline reductase has a molecular weight of 300,000 and is composed of 10 identical subunits (159). Each subunit has a pyruvate residue that is essential for activity (157). Proline reductase exhibits no absorbance in the visible region, but absorbance due to phenylalanine and tyrosine is discernible in the ultraviolet region (159).

The use of ornithine as an oxidant in the Stickland reaction first involves its conversion to proline (10, 47). By using [δ-^{15}N]ornithine, it was found that the α-aminonitrogen is converted to ammonia and the δ-aminonitrogen is retained in proline (139). Thus the conversion of ornithine to proline apparently involves an oxidative deamination at the α-position, forming 2-keto-5-aminopentanoic acid and Δ'-pyrroline-2-carboxylic acid as intermediates. The latter compound is then reduced to proline (47).

Britz and Wilkinson (23) found that leucine can serve either as an electron donor or electron acceptor in the Strickland reaction, depending on the nature of the other amino acid supplied to the medium. When electron donors were present, leucine was reduced to isocaproate, whereas in the presence of electron acceptors, leucine was oxidized to isovalerate. Tyrosine and tryptophan also serve either as electron donors or electron acceptors. Leucine, however, is unique in this regard, in that it can be reduced and oxidized simultaneously (23). Bader et al. (8) demonstrated the reductive steps involving the use of phenylalanine, leucine, and isoleucine as electron acceptors. Betaine has recently been shown to serve as an oxidant in the Strickland reaction in *C. sporogenes*. (140b).

8.4.2 Oxidation of Branched-Chain and Other Amino Acids

Many amino acids are oxidatively deaminated and decarboxylated to the corresponding carboxylic acid:

$$RCH_2NH_2COOH + 2H_2O \longrightarrow RCOOH + CO_2 + NH_3 + 4H \qquad (6)$$

The amino acids that are probably catabolized in this way and the corresponding carboxylic acid are alanine (acetate), valine (isobutyrate), leucine (isovalerate), isoleucine (2-methylbutyrate), phenylalanine (phenylacetate), tryptophan (indolyacetate), and tyrosine (*p*-hydroxyphenylacetate) (2, 3, 27). In group I clostridia, these oxidative reactions are coupled to the reduction of glycine, proline, or another amino acid in the Stickland reaction (see Section 8.4.1; 134, 143, 156). In the rumen, reaction (4) is the major route for amino acid degradation (3, 27, 100). The catabolism of amino acids in the rumen is much faster than the use of amino acids for biosynthetic purposes (27, 99). However, it is not clear whether ruminal bacteria derive energy from these processes or whether amino acids serve as growth substrates for ruminal bacteria (27, 99, 179a). Increased cell yields of ruminal bacteria have been observed when amino acids are present (47b), but this may be due to decreased energetic uncoupling (99a). Recently, *Spirochaeta isovalerica* has been shown to degrade branched-chain amino acids (valine, leucine, and isoleucine) to the corresponding branched-chain fatty acid according to reaction (4) (80–83). These amino acids cannot be used as the sole energy source for growth by *S. isovalerica*, but ATP generated from the degradation of these amino acids can be used to prolong survival during periods of starvation (82, 83).

The oxidation of amino acids usually proceeds via an α-keto acid as shown in reaction (3). The amino acid is oxidatively deaminated by one of two routes: an enzyme analogous to glutamate dehydrogenase or by transamination with α-ketoglutarate to form the corresponding α-keto acid and glutamate (11, 83, 130, 145). The regeneration of α-ketoglutarate is accomplished by glutamate dehydrogenase. The α-keto acid is oxidatively decarboxylated by enzymes analogous to pyruvate-ferredoxin oxidoreductase, which requires thiamin pyrophosphate and CoA. The CoA-thioester generated by this reaction is further metabolized by the action of phosphotransacetylase and acetate kinase to the corresponding fatty acid and ATP (83).

The pathway for branched-chain amino acid degradation in *S. isovalerica* is shown in Fig. 8.4. Leucine, isoleucine, and valine are converted to 2-ketoisocaproate, 2-keto-3-methylvalerate, and 2-ketovalerate, respectively, by α-ketoglu-

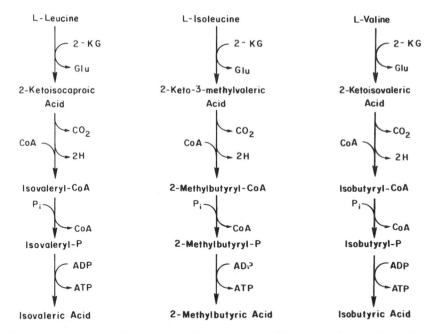

Figure 8.4 Pathways for the anaerobic degradation of branched-chain amino acids by *Spirochaeta isovalerica*. 2-KG, 2-ketoglutarate; Glu, glutamate; CoA, coenzyme A; P_i, inorganic phosphate; ADP, adenosine 5'diphosphate; ATP, adenosine 5'-triphosphate; 2H, reducing equivalents. Reproduced from Harwood and Canale-Parola (83) by permission of the American Society for Microbiology.

tarate-dependent aminotransferase activity. The branched-chain keto acids are oxidatively decarboxylated by crude extracts to isovaleryl-coenzyme A (CoA), 2-methylbutyryl-CoA, and isobutyry-CoA, respectively, in the presence of CoA and benzyl viologen. The acyl-CoA compounds are converted to acyl phosphates by phosphate branched-chain acyltransferase activity. Branched-chain fatty acid kinase activity catalyzes the formation of ATP and of isovalerate, 2-methylbutyrate, and isobutyrate from the respective acyl phosphate. The branched-chain fatty acid kinase activity is a different enzyme than acetate kinase (81).

C. *sporogenes* and other clostridia produce isobutyrate, acetate, and isocaproate from leucine (10). Leucine is oxidized to β-leucine by leucine-2,3-aminomutase. β-Leucine is further oxidized to isobutyrate and acetate. The production of isocaproate from leucine was discussed in Section 8.4.1.

Many anaerobic bacteria require one or more monocarboxylic acids for growth (27–29). The monocarboxylic acids produced by oxidative decarboxylation of amino acids [reaction (5)] by some bacteria are used by other bacteria as carbon skeletons for the synthesis of amino acids and membrane lipids (3, 4, 27–29, 99a). Amino acids are synthesized from monocarboxylic acids by reductive carboxylation reactions:

$$R-COOH + CO_2 + NH_3 + 4H \longrightarrow RCHNH_2COOH + 2H_2O \quad (7)$$

Amino acids synthesized in this manner include leucine, isoleucine, valine, tryptophan, and phenylalanine from isovalerate, 2-methylbutyrate, isobutyrate, indolyacetate, and phenylacetate, respectively (7, 107, 164). Reductive carboxylation reactions are used to synthesize α-ketoglutarate from propionate or succinate in many anaerobes (5, 6, 136). *Bacteroides ruminicola* and *Bacteroides fragilis*, which can synthesize leucine *de novo* from glucose, preferentially use the reductive carboxylation reactions rather than the *de novo* pathway (7a).

8.4.3   Aromatic Amino Acid Degradation

The aromatic products produced by 23 clostridial species include phenylacetate, phenylpropionate, and phenylacetate from phenylalanine, hydroxyphenylacetate, hydroxyphenyllactate, hydroxyphenylpropionate, phenol, and ρ-cresol from tyrosine, and indole, indolyacetate, and indolypropionate from tryptophan (10–12, 21, 27–29, 64, 131). In pig feces, the end products from tyrosine were mainly ρ-cresol and 3-phenylpropionate while in stored piggery wastes, mainly phenol and ρ-cresol with small amounts of phenylpropionate were formed from tyrosine (163). The degradation of tryptophan in the rumen involves its conversion to indolyacetate. Indolyacetate can be decarboxylated to skatole by a *Lactobacillus* species which cannot use tryptophan directly (187). This bacterium also decarboxylates ρ-hydroxyphenylacetate to ρ-cresol (185), as does *C. difficile* (48b).

Phenol is produced from tyrosine by tyrosine-phenol lyase in *Clostridium tetanomorphum* (25). This enzyme converts tyrosine to phenol, pyruvate, and ammonia and requires pyridoxal phosphate and Mg^{2+} as cofactors (25). Indole is probably produced from tryptophan by the tryptophase reaction (10). The enzymes responsible for skatole or cresol formation in obligate anaerobes have not been characterized (10).

The reduction of aromatic amino acids to the substituted aromatic acids presumably involves the formation of α-keto acids by transamination and either a direct oxidative decarboxylation of the keto acid [reaction (2)] or a decarboxylation followed by the oxidation of the resulting aldehyde (34, 145). The reduction of phenylalanine, tyrosine, and tryptophan to phenylpropionate, ρ-hydroxyphenylpropionate, and indopropionate, respectively, by clostridia probably goes via the α-keto, α-hydroxy, and α,β-unsaturated acids (10, 34, 110). The transaminases and the α-hydroxyacid dehydrogenases have been purified or partially purified, but the conversion of the α-hydroxyaromatic acid to the α,β-unsaturated aromatic acid has only been demonstrated in crude extracts or cell suspensions.

The aromatic ring is not altered by fermentative bacteria such as the clostridia. Other bacteria, such as *Syntrophus buswellii* (138), *Pelobacter acidigallii* (155), and certain sulfate-reducing bacteria (147), are responsible for ring cleavage.

8.4.4 Arginine

Arginine is degraded to ornithine, carbon dioxide, and ammonia with the formation of 1 mol of ATP per mole of arginine via the arginine deiminase pathway (formerly referred to as the arginine dehydrolase pathway) (1). Three enzymes are involved in this pathway, arginine deiminase [reaction (8)], ornithine transcarbonylase [reaction (9)], and carbamate kinase [reaction (10)].

$$\text{L-arginine} + H_2O \longrightarrow \text{L-citrulline} + NH_3 \tag{8}$$
$$\text{L-citrulline} + PO_4^{2-} \rightleftharpoons \text{L-ornithine} + \text{carbamyl phosphate} \tag{9}$$
$$\text{carbamyl phosphate} + ADP \rightleftharpoons ATP + CO_2 + NH_3 \tag{10}$$

Although all three enzymes have been clearly demonstrated only in *Streptococcus faecalis*, *S. lactis*, *Pseudomonas putida*, and *Mycoplasma arthritidis* (1, 48, 160), the presence of arginine deiminase has been demonstrated in many bacteria, including the cyanobacterium *Aphanocapsa* (1).

8.4.5 Threonine

Analysis of end products produced from threonine by several clostridia show three distinct patterns (58). Most species produce much more propionate and some *n*-butyrate when the medium is supplemented with threonine. A few species form only acetate from threonine, and a number of clostridia produce substantial amounts of 2-aminobutyrate from threonine.

The biochemistry of anaerobic threonine degradation is not well understood. Barker (11, 13) postulated that the formation of only acetate from threonine could involve the direct cleavage of threonine to acetaldehyde and glycine by threonine aldolase [reaction (11)] followed by the oxidation of acetaldehyde to acetate and the conversion of glycine to acetate [reaction (12)].

$$CH_3CHOHCHNH_2COOH \longrightarrow CH_3CHO + CH_2NH_2COOH \tag{11}$$
$$CH_3CHO + CH_2NH_2COOH + H_2O \longrightarrow 2CH_3COOH + NH_3 \tag{12}$$

Threonine aldolase activity has been found in *Clostridium pasteurianum* (48a). Cardon and Barker (36) showed that *C. propionicum* degraded threonine to propionate and butyrate:

$$3CH_3CHOHCHNH_2COOH + H_2O \longrightarrow CH_3CH_2CH_2COOH$$
$$+ 2CH_3CH_2COOH + 2CO_2 + 3NH_3 \tag{11}$$

The overall process represents the oxidation of 2 mol of threonine (or α-ketobutyrate) to propionate and carbon dioxide coupled with the reduction of 1 mol of threonine to butyrate. Tracer experiments showed that butyrate is not formed by the condensation of two acetyl CoA molecules, suggesting that the carbon skeleton of threonine remains intact (13, 36, 58).

8.4.6 Glycine

Cardon and Barker (36, 37) isolated *Peptococcus (Diplococcus) glycinophilus*, which ferments glycine to acetate, carbon dioxide, and ammonia:

$$4CH_2NH_2COOH + 2H_2O \longrightarrow 4NH_3 + 2CO_2 + 3CH_3COOH \qquad (14)$$

Hydrogen is produced in variable amounts depending on the partial pressure of hydrogen (36). Strains of *Peptococcus (Micrococcus) anaerobius*, *Peptococcus (Micrococcus) magnus* (formerly *P. variabilis*), *C. histolyticum*, and *Clostridium purinolyticum* (53, 54, 77) ferment glycine according to reaction (14).

Tracer experiments (36) showed that most of the ^{14}C is found in carbon dioxide when $[1\text{-}^{14}C]$glycine is the substrate, but radioactivity is also found in lesser amounts in both carbons of acetate. With $[2\text{-}^{14}C]$glycine, both carbon atoms of acetate are strongly labeled, whereas carbon dioxide is weakly labeled. With $^{14}CO_2$, both carbon atoms of acetate are labeled. From these data, the complete oxidation of glycine to carbon dioxide and significant reduction of glycine to acetate were excluded. The tracer experiments plus enzymatic studies (120, 154) lead to the scheme shown in Fig. 8.5*A*. The main reactions include the splitting of glycine into methylenetetrahydrofolate (THFA), CO_2, NH_3, and subsequent condensation of methylene-THFA with another glycine yielding serine. Serine is converted to pyruvate from which acetate is formed. However, more recent studies (54–56) strongly indicate that glycine is degraded according to Fig. 8.5*B*. One mole of glycine is oxidized to CO_2 via methylene-THFA, methenyl-THFA, formyl-THFA, and formate. The reducing equivalents generated in the oxidation of 1 mol of glycine are used to reduce 3 mol of glycine to 3 mol each of acetate and ammonia [reaction (4)]. The latter reaction is catalyzed by the glycine reductase system discussed previously (Section 8.4.1). The discrepancy between the work of Dürre and Andreesen (54–56) and the earlier studies (36, 120, 154) was the result of the use of a selenium-deficient medium in the earlier studies, so glycine reductase activity was low. Also, the complex labeling pattern observed by earlier workers were the result of exchange reactions (54–56).

8.4.7 Glutamate

Glutamate is degraded to acetate, butyrate, CO_2, NH_3, and H_2 by a variety of anaerobic bacteria (10–12, 134). Although the end products are the same, two different pathways exist for glutamate fermentation (Figure 8.6). In the methylaspartate pathway found in *Clostridium tetanomorphum* (Fig. 8.6*A*), glutamate is converted to $2S,3S$-β-methylaspartate, mesaconate, *S*-citramalate, and pyruvate

A.

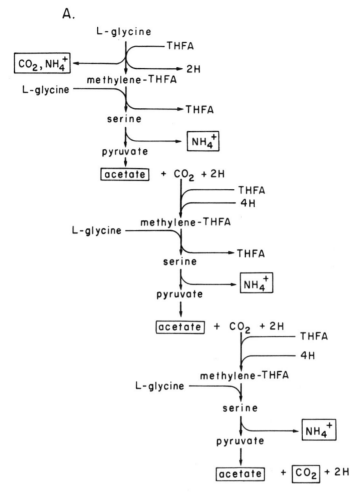

Figure 8.5 Pathway for the degradation of glycine by *Peptococcus glycinophilus*. (*A*) Glycine conversion to acetate via serine and pyruvate (36, 120, 154). (*B*) Glycine conversion to acetate involving glycine decarboxylase and glycine reductase (54–56) where THFA is tetrahydrofolate. Boxes indicate end products. Reproduced from Ref. 55 by permission of the Society for General Microbiology.

(179). Pyruvate is then converted to the final products via oxidative decarboxylation of pyruvate to acetyl-CoA, CO_2, and reduced ferredoxin, the condensation of acetyl-CoA to acetoacetyl-CoA, which is then reduced to butyryl-CoA (76). Glutamate mutase contains deoxyadenosylcobalamin as cofactor (15, 16). Citramalate lyase, which converts L-citramalate to pyruvate and acetate, contains pantothenate and cysteamine residues (32). This enzyme has three subunits, α, β, γ. The pantothenate and cysteamine residues are found exclusively on the γ-subunit, which

B.

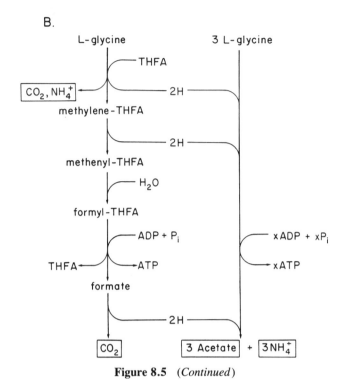

Figure 8.5   (*Continued*)

serves as the acyl-carrier protein of citramalate lyase (10, 32). The enzyme is inactivated in crude extracts by the deacetylation of the cysteamide residue.

Peptococcus aerogenes and *Acidaminococcus fermentans* use the hydroxyglutarate pathway for glutamate degradation (Fig. 8.6*B*; 123, 181). Glutamate dehydrogenase catalyzes the conversion of glutamate to α-ketoglutarate, which is converted to (R)-2-hydroxyglutarate by α-hydroxyglutarate dehydrogenase (104, 123). Both enzymes are pyridine nucleotide dependent. The conversion of (R)-2-hydroxyglutarate to glutaconate by extracts of *A. fermentans* requires catalytic amounts of coenzyme A and acetyl phosphate (30). CoA transferase reversibly transfers CoA between acetate and (R)-2-hydroxyglutarate or (E)-glutaconate (33). (R)-2-hydroxyglutaryl-CoA is dehydrated to (E)-glutaconyl-CoA by a membrane-bound dehydratase that requires CoA, acetylphosphate, NADH, $FeSO_4$, $MgCl_2$, dithioerythritol, and anaerobic conditions for full activity. The decarboxylation of glutaconyl-CoA to crotonyl-CoA is catalyzed by a membrane-bound, biotin-dependent enzyme that requires Na^+ for activity (31). The decarboxylase functions as an electrogenic sodium ion pump where the free energy of the decarboxylation reaction is conserved in an electrochemical gradient of Na^+ (31).

In the hydroxyglutarate pathway, the linear carbon chain of glutamate remains unchanged in butyrate, but in the methylaspartate pathway the carbon chain is rearranged (104). Nine of the 15 clostridial species studied have the methylaspar-

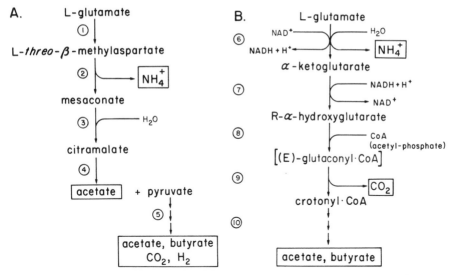

Figure 8.6 Pathways for the anaerobic degradation of glutamate. (*A*) The methylaspartate pathway (10). 1. glutamate mutase (deoxyadenosyl-cobalamin dependent); 2. β-methylaspartase; 3. citramalate dehydratase; 4. citramalate lyase; 5. reactions as found in saccharolytic clostridia (76). Modified from Gottschalk (76), reproduced by permission of Springer-Verlag. (*B*) The hydroxyglutarate pathway (10). 6. glutamate dehydrogenase (NAD dependent); 7. 2-hydroxyglutarate dehydrogenase (NAD dependent); 8. coenzyme A : transferase; 9. membrane-bound dehydratase; 10. glutaconyl-CoA decarboxylase (membrane bound, biotin dependent). The conversion of (R)-2-hydroxyglutarate to glutaconate requires coenzyme A and acetyl phosphate. Glutaconyl-CoA decarboxylase functions as a Na$^+$ pump. Boxes indicate end products.

tate pathway, whereas all four of the species in the genera *Peptococcus*, *Acidaminococcus*, and *Fusobacterium* use the hydroxyglutarate pathway (10).

8.4.8 Lysine

Lysine is fermented by *C. sticklandii* and *C. subterminale* to acetate, butyrate, and ammonia (10–12, 166):

$$\underset{\underset{NH_2}{|}}{CH_2}CH_2CH_2CH_2\underset{\underset{NH_2}{|}}{CH}COOH + 2H_2O \longrightarrow$$

$$CH_3CH_2CH_2COOH + CH_3COOH + 2NH_3 \quad (15)$$

Tracer experiments indicated that lysine is degraded by three different routes, one where butyrate is derived from carbons 3 to 6 of lysine, another where butyrate is derived from carbons 1 to 4 of lysine, and the third involves butyrate formation from two acetate or acetyl moieties (166). Evidence suggests that in extracts lysine is degraded almost exclusively by the first pathway, where acetate is formed from carbons 1 and 2 of lysine and butyrate is formed from carbons 3 to 6 of lysine (166). The pathway for lysine formation is shown in Fig. 8.7. The first intermediates in the pathway are β-lysine (3,6-diaminohexanoate), 3,5-diaminohexanoate, and 3-keto-5-aminohexanoate (41, 166, 169). L-Lysine is converted to β-lysine by L-lysine-2,3-amino mutase, which contains pyridoxal phosphate, is Fe^{2+}-dependent, and is stimulated by S-adenosylmethionine (41). L-β-Lysine mutase catalyzes

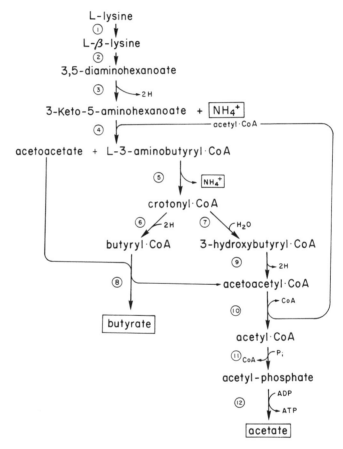

Figure 8.7 The clostridial pathway for lysine fermentation (10, 166). 1. L-Lysine-2,3-amino mutase (pyridoxal phosphate and Fe^{2+}-dependent, activated by S-adenosylmethionine); 2. β-lysine mutase (deoxyadenosylcobalamin dependent); 3. 3,5-deaminohexanoate dehydrogenase (NAD or NADP dependent); 4. 3-keto-5-aminohexanoate cleavage enzyme; 5. 3-aminobutyryl-CoA deaminase; 6. acyl-CoA dehydrogenase; 7. crotonase; 8. butyryl-CoA: acetoacetate-CoA transferase; 9. 3-hydroxyacyl-CoA dehydrogenase; 10. β-ketoacetyl-CoA thiolase; 11. phosphotransacetylase; 12. acetate kinase. Boxes indicate end products.

the coenzyme B_{12}-dependent migration of the terminal amino group of β-lysine from carbon 6 to 5 carbon to form 3,5-diaminohexanoate (169). The latter compound is oxidized to 3-keto-5-aminohexanoate via NAD or NADP-dependent 3,5-diaminohexanoate dehydrogenase (166).

The 3-keto-5-aminohexanoate cleavage enzyme degrades 3-keto-5-aminohexanoate by reacting with acetyl-CoA to form L-3-aminobutyryl-CoA and acetoacetate (188). L-3-Aminobutyryl-CoA is deaminated to crotonyl-CoA (111). The latter compound is presumably reduced to butyryl-CoA. Butyryl-CoA can react with acetoacetate to form butyrate and acetoacetyl-CoA via a coenzyme A transferase (14). Acetoacetyl-CoA is converted by a 3-ketoacyl-CoA thiolase to 2 mol of acetyl-CoA, one that participates in 3-keto-5-aminohexanoate cleavage reaction; the other is converted to acetate with ATP formation (14).

In *Selenomonas ruminantium*, lysine is decarboxylated to cadaverine and CO_2 via a constitutive lysine decarboxylase (112).

8.4.9 Other Amino Acids

Alanine is fermented to propionate, acetate, and CO_2 [reaction (16)] via the acrylate pathway (Fig. 8.8) in *Clostridium propionicum* (121a, 155a).

$$3CH_3CHNH_2COOH + 2H_2O \longrightarrow 3NH_3$$
$$+ 2CH_3CH_2COOH + CH_3COOH + CO_2 \quad (16)$$

The presence of alanine : α-ketoglutarate aminotransferase and L-glutamate dehydrogenase in *C. propionicum* (155a) suggests that the conversion of alanine to pyruvate is catalyzed by these two enzymes rather than by a direct dehydrogenation as found in other clostridia and certain *Bacillus* species. One-third of the pyruvate is oxidized to acetate and CO_2, resulting in ATP synthesis via an acetyl phosphate intermediate. The other two thirds of the pyruvate are reduced to (R)-lactate. The dehydration of (R)-lactate in crude extracts requires CoA and acetyl phosphate, indicating that lactate activation is required prior to its dehydration. A propionate CoA-transferase has been purified homogeneity, which catalyzes the transfer of CoA from propionyl-CoA to lactate (155a). The (R) isomer of lactate, which is the product of pyruvate reduction, is the preferred substrate. Isotopic labeling studies show that the abstraction of a β-hydrogen from lactate during dehydration is partially rate limiting, which is consistent with acrylyl-CoA as an intermediate (121a). The lactyl-CoA dehydratase has been purified to greater than 90% of homogeneity and one of the subunits contains equal amounts of riboflavin and flavin mononucleotide (121a). The use of riboflavin by an enzyme has not been previously reported.

Aspartate is fermented by *Campylobacter* species and *Bacteroides melaninogenicus* (10, 122, 183). Part of the aspartate is oxidized to CO_2, acetate, succinate, and ammonia, while the remainder is reduced to succinate and ammonia by the fumarate reductase system.

Cysteine is degraded to pyruvate, ammonia, and H_2S via cysteine desulhydrase

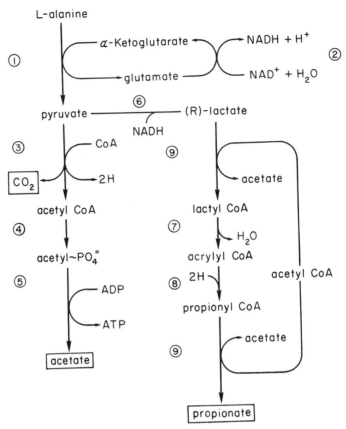

Figure 8.8 Acrylate pathway for alanine fermentation in *Clostridium propionicum*. 1. alanine:ketoglutarate aminotransferase; 2. L-glutamate dehydrogenase; 3. pyruvate:ferredoxin oxidoreductase; 4. phosphotransacetylase; 5. acetate kinase; 6. (R)-lactyl-CoA dehydratase; 7. 8. propionyl-CoA transferase; 9. Boxes indicate end products. Adapted from Ref. 155a.

(68). Pyruvate is further metabolized depending on the enzymatic machinary present in the cell.

Histidine is degraded by *Clostridium* and *Fusobacterium* species by a pathway similar to that found in aerobic organisms involving urocanate and formimino-glutamate as intermediates (11). Formiminoglutamate is degraded to formamide and glutamate, which is degraded by the methylaspartate pathway (11).

Serine is converted to pyruvate and ammonia via serine dehydratase, which is a pyridoxal phosphate-containing enzyme (38).

A clostridial species degrades δ-aminovalerate to valerate, propionate, acetate, and ammonia (79). Another species converts γ-aminobutyrate to butyrate, acetate, and ammonia (78).

8.4.10 Importance of Interspecies H₂ Transfer

The importance of interspecies H_2-transfer reactions in amino acid degradation is now beginning to be understood. Carbon monoxide, an inhibitor of bacterial hydrogenase, inhibits the degradation of branched-chain amino acids in the rumen, suggesting that interspecies H_2-transfer is required for the degradation of these compounds (99b, 152a). Nagase and Matsuo (140) found that the degradation of alanine, valine, and leucine is inhibited by chloroform, an inhibitor of methanogenesis. Their data suggest that H_2 is produced during the degradation of these compounds. The addition of H_2-using sulfate reducers to sewage sludge slurries stimulates the degradation of alanine, serine, valine, and leucine (140a). Little degradation of valine and leucine occurs in these samples unless the partial pressure of H_2 is low (140a).

The studies above strongly suggest that interspecies H_2 transfer is important in the degradation of certain amino acids. The oxidative deamination/decarboxylation with H_2 production of reduced amino acids such as alanine and leucine is energetically unfavorable at standard conditions (Table 8.4). Thus H_2 production from these compounds will occur only if the partial pressure of H_2 is maintained at a low level by H_2-using species. Hydrogen production from other, more oxidized amino acids, such as glutamate, is favorable at standard conditions (Table 8.4), so the degradation of these compounds may not depend on interspecies H_2 transfer.

Direct isolation of anaerobic glutamate-fermenting bacteria from estuary sediments shows that *Acidaminobacter hydrogenoformans* is the most numerous glutamate-using anaerobe (171a). In pure culture, *A. hydrogenoformans* produces acetate, CO_2, formate, and H_2 from glutamate, and the addition of H_2 to the gas phases inhibits growth. When grown in co-culture with H_2-using species, a shift in the fermentation products is observed, with propionate being produced in large amounts in addition to acetate and CO_2. Propionate is also produced in large amounts by the co-cultures when histidine and α-ketoglutarate are the substrates.

The formation of acetate rather than propionate from glutamate is favored at standard conditions (Table 8.4). However, as the partial pressure of H_2 is decreased, the formation of propionate from glutamate becomes more favorable

TABLE 8.4 Energetics of H₂ production from amino acids

Reactions	$\Delta G^{\circ\prime}$ (kJ/reaction)
a. Glutamate⁻ + 3H₂O → 2 acetate⁻ + HCO₃⁻ + H⁺ + NH₄⁺ + H₂	−33.9
b. Glutamate⁻ → propionate⁻ + NH₄⁺, 2HCO₃⁻, + 2H₂ + H⁺ + 4H₂O	−5.8
c. Glutamate⁻ + H₂O → α-ketoglutarate²⁻ + NH₄⁺ + H₂	+59.9
d. Alanine + 3H₂O → acetate⁻ + HCO₃⁻ + NH₄⁺ + 2H₂ + H⁺	+7.5
e. Leucine + 3H₂O → isovalerate⁻ + HCO₃⁻ + H⁺ + NH₄⁺ + 2H₂	+4.2

Source: Reaction stoichiometries and free energy changes from Stams and Hansen (171a) and Thauer et al. (173a).

(171a). The reason why propionate is not made in large amounts by pure cultures of *A. hydrogenoformans* is that the first step, the oxidative deamination of glutamate to α-ketoglutarate, is strongly dependent on the partial pressure of H_2 and is quite unfavorable at standard conditions (Table 8.4). This is probably the reason why the degradation of branched-chain amino acids is highly dependent on H_2 partial pressure in the rumen and in sewage sludge (99b, 140, 140a, 152a). *A. hydrogenoformans* degrades alanine, *valine*, *leucine*, and *isoleucine* only in co-culture with H_2-using bacteria (171a). Other amino acid–degrading syntrophic bacteria include an anaerobic gram-negative rod which grows on alanine, asparate, or malate only in the presence of H_2-using bacteria (Widdel and Schnell, Abstr. 3rd Int. Symp. Microb. Ecol., 1983 A2 p. 21). The benzenoid degrader, PA-1, which degrades aspartate only in co-culture with H_2-using bacteria (9a). The existence of amino acid–degrading syntrophic bacteria, particularly as the most numerous species in certain environments (171a), suggests that interspecies H_2-transfer reactions may be quite important in the initial degradation of amino acids. *T. proteolyticus* produces mainly acetate, H_2, and isovalerate from gelatin in pure culture. When co-cultured with a H_2-using methanogen, large amounts of propionate and isobutyrate are made in addition to acetate and isovalerate (144b).

8.5 Summary

The degradation of proteins and lipids in anaerobic environments is a complex process involving many different kinds of anaerobic microorganisms. The kinds of organisms involved in these reactions can vary depending on the particular eco-system. The major proteolytic or lipolytic species in some environments have not been determined. In general, proteins are hydrolyzed to peptides and amino acids which are fermented to volatile fatty acids, CO_2, H_2, NH_4^+, and S^{2-}. Lipids are hydrolyzed with the release of free fatty acids, and the galactose and glycerol moieties are fermented to similar products without NH_4^+ or S^{2-} production.

There is a great diversity of species which degrade into amino acids, and diversity is also reflected in the number of biochemical pathways involved. Although much is known about the biochemical aspects of amino acid degradation in individual species, little is known about the degradative routes used in anaerobic bacterial communities.

Recent studies indicate that interactions based on H_2 production and use are important for the degradation of certain amino acids. The energetics of anaerobic amino acid metabolism contains interesting, but as yet unanswered questions, such as why the most numerous proteolytic species in the rumen are organisms which cannot use proteins as sole energy sources.

REFERENCES

1. Abdelal, A. 1979. Arginine catabolism by microorganisms. Annu. Rev. Microbiol. 33:139–168.

2. Allison, M. J. 1978. Production of branched-chain volatile fatty acids by certain anaerobic bacteria. Appl. Environ. Microbiol. 35:872–877.

3. Allison, M. J. 1970. Nitrogen metabolism in rumen microorganisms, p. 456–473, in A. T. Phillipson (ed.), Physiology and digestion and metabolism in the ruminant. Oriel Press, Stocksfield, Northumberland, England.

4. Allison, M. J. 1969. Biosynthesis of amino acids by ruminal microorganisms. J. Anim. Sci. 29:797–807.

5. Allison, M. J., and I. M. Robinson. 1970. Biosynthesis of α-ketoglutarate by the reductive carboxylation of succinate in *Bacteroides ruminicola*. J. Bacteriol. 104:50–56.

6. Allison, M. J., I. M. Robinson, and A. L. Baetz. 1979. Synthesis of α-ketoglutarate by reductive carboxylation of succinate in *Veillonella, Selenomonas*, and *Bacteroides*. J. Bacteriol. 140:980–986.

7. Allison, M. J., J. A. Bucklen, and I. M. Robinson. 1966. Importance of the isovalerate carboxylation pathway of leucine biosynthesis in the rumen. Appl. Microbiol. 14:807–814.

7a. Allison, M. J., A. L. Baetz, and J. Wiegel. 1984. Alternative pathways for biosynthesis of leucine and other amino acids in *Bacteroides ruminicola* and *Bacteroides fragilis*. Appl. Environ. Microbiol. 48:1111–1117.

8. Bader, J., P. Rauschenbach, and H. Simon. 1982. On a hitherto unknown fermentation path of several amino acids by proteolytic clostridia. FEBS Lett. 140:67–72.

9. Banwart, G. J. 1979. Basic food microbiology. AVI, Westport, Conn.

9a. Barik, S., W. J. Brulla, and M. P. Bryant. 1985. PA-1, a versatile anaerobe obtained in pure culture, catabolizes benzenoids and other compounds in syntrophy with hydrogenotrophs, and P-2 plus *Wolinella* sp. degrades benzenoids. Appl. Environ. Microbiol. 50:304–310.

10. Barker, H. A. 1981. Amino acid degradation by anaerobic bacteria. Annu. Rev. Biochem. 50:23–40.

11. Barker, H. A. 1961. Fermentations of nitrogenous organic compounds, p. 151–207, in I. C. Gunsalus and R. Y. Stanier (eds.), The bacteria, Vol. 2. Academic Press, New York.

12. Barker, H. A. 1956. Bacterial fermentation. Wiley, New York.

13. Barker, H. A., and T. Wiken. 1948. The origin of butyric acid in the fermentation of threonine by *Clostridium propionicum*. Arch. Biochem. 17:149–151.

14. Barker, H. A., I.-M. Jeng, N. Neff, J. M. Robertson, F. K. Tam, and S. Hosaka. 1978. Butyryl-CoA: acetoacetate CoA-transferase from a lysine-fermenting *Clostridium*. J. Biol. Chem. 253:1219–1225.

15. Barker, H. A., R. D. Smyth, R. M. Wilson, and H. Weissback. 1959. The purification and properties of β-methylaspartase, J. Biol. Chem. 234:320–328.

16. Barker, H. A., H. Weissback, and R. D. Smyth. 1958. A coenzyme containing pseudovitamin B_{12}. Proc. Natl. Acad. Sci. USA 44:1093–1097.

17. Blackburn, T. H. 1968. Protease production by *Bacteroides amylophilus* strain H18. J. Gen. Microbiol. 53:27–36.

18. Blackburn, T. H. 1968. The protease liberated from *Bacteroides amylophilus* strain H18 by mechanical disintegration. J. Gen. Microbiol. 53:37–51.

19. Blackburn, T. H., and P. N. Hobson. 1962. Further studies on the isolation of proteolytic bacteria from the sheep rumen. J. Gen. Microbiol. 29:69–81.

20. Blaschek, H. P., and M. Solberg. 1981. Isolation of a plasmid responsible for caseinase activity in *Clostridium perfringens*. ATCC 3626B. J. Bacteriol. 147:262–266.

20a. Bokkenheuser, V. D., and J. Winter. 1983. Biotransformation of steroids, p. 215–239, in D. J. Hentges (ed.), Human intestinal microflora in health disease. Academic Press, New York.

20b. Bokkenheuser, V. D., J. Winter, S. M. Finegold, V. L. Sutter, A. E. Ritchie, W. E. C. Moore, and L. V. Holdeman. 1979. New markers for *Eubacterium lentum*. Appl. Environ. Microbiol. 37:1001–1006.

21. Bourgeau, G., and D. Mayrand. 1983. Phenylacetic acid production by *Bacteroides gingivalis* from phenylalanine and phenylalanine-containing peptides. Can. J. Microbiol. 29:1184—1189.

22. Bradbeer, C. 1965. The clostridia fermentations of choline and ethanolamine. J. Biol. Chem. 240:4669–4674.

23. Britz, M. L., and R. G. Wilkinson. 1982. Leucine dissimilation to isovaleric and isocaproic acids by cell suspensions of amino acid-fermenting anaerobes: the Stickland reaction revisited. Can. J. Microbiol. 28:291–300.

24. Brock, F. M., C. W. Forsberg, and J. G. Buchanan-Smith. 1982. Proteolytic activity of rumen microorganisms and effects of proteinase inhibitors. Appl. Environ. Microbiol. 44:561–569.

25. Brot, N., Z. Smit, and H. Weissbach. 1965. Conversion of L-tyrosine to phenol by *Clostridium tetanomorphum*. Arch. Biochem. Biophys. 112:1–6.

26. Bryant, M. P. 1979. Microbial methane production—theoretical aspects. J. Anim. Sci. 48:193–201.

27. Bryant, M. P. 1977. Microbiology of the rumen, p. 287–304, in M. J. Stevenson (ed.), Duke's physiology of domestic animals, 9th ed. Cornell University Press, Ithaca, N.Y.

28. Bryant, M. P. 1974. Nutritional features and ecology of predominant anaerobic bacteria of the intestinal tract. Am. J. Clin. Nutr. 27:1313–1319.

29. Bryant, M. P. 1973. Nutritional requirements of the predominant rumen cellulolytic bacteria. Fed. Proc. 32:1809–1813.

30. Buckel, W. 1980. The reversible dehydration of (R)-2-hydroxyglutarate to (E)-gluconate. Eur. J. Biochem. 106:439–447.

31. Buckel, W., and R. Semmler. 1982. A biotin-dependent sodium pump: glutaconyl-CoA decarboxylase from *Acidaminococcus fermentans*. FEBS Lett. 148:35–38.

32. Buckel, W., and A. Bobi. 1976. The enzyme complex citramalate lyase from *Clostridium tetanomorphum*. Eur. J. Biochem. 64:255–262.

33. Buckel, W., U. Dorn, and R. Semmler. 1981. Glutaconate CoA-transferase from *Acidaminococcus fermentans*. Eur. J. Biochem. 118:315–321.

34. Bühler, M., H. Gresel, W. Tischer, and H. Simon. 1980. Occurrence and the possible physiological role of 2-enoate reductases. FEMS. Lett. 109:244–246.

35. Buswell, A. M., and S. L. Neave. 1930. Laboratory studies on sludge digestion. III. State Water Surv. Bull. 30:1–84.

36. Cardon, B. P., and H. A. Barker. 1947. Amino acid fermentations by *Clostridium propionicum* and *Diplococcus glycinophilus*. Arch. Biochem. 12:165–180.

37. Cardon, B. P., and H. A. Barker. 1946. Two amino acid-fermenting bacteria. *Clostridium propionicum* and *Diplococcus glycinophilus*. J. Bacteriol. 52:629–634.

38. Carter, J. E., and R. D. Sagers. 1972. Ferrous ion-dependent L-serine dehydratase from *Clostridium acidiurici*. J. Bacteriol. 109:757–763.

39. Casu, A., V. Pala, R. Monacelli, and G. Nanni. 1971. Phospholipase C from *Clostridium perfringens*: purification by electrophoresis on acrylamide-agarose gel. Italian J. Biochem. (Engl. Ed.) 20:166–178.

40. Cheng, K.-J., R. P. McCowan, and J. W. Costerton. 1979. Adherent epithelial bacteria in ruminants and their roles in digestive tract function. Am. J. Clin. Nutr. 32:139–148.

41. Chirpich, T. P., V. Zappia, R. N. Costilow, and H. A. Barker. 1970. Lysine 2,3-aminomutase. Purification and properties of a pyridoxal phosphate and S-adenosylmethionine activated enzyme. J. Biol. Chem. 245:1778–1789.

42. Chynoweth, D. P., and R. A. Mah. 1971. Volatile acid formation in sludge digestion. Adv. Chem. Ser. 105:41–54.

43. Clark, N. G., G. P. Hazlewood, and R. M. C. Dawson. 1980. Structure of diabolic acid-containing phospholipids isolated from *Butyrivibrio* sp. Biochem. J. 191:561–569.

44. Cone, J. E., R. Martin del Rio, J. N. Davis, and T. C. Stadtman. 1976. Chemical characterization of the selenoprotein component of clostridial glycine reductase: identification of selenocysteine as the organoselenium activity. Proc. Natl. Acad. Sci. USA 73:2659–2663.

45. Cone, J. F., R. Martin del Rio, and T. C. Stadtman. 1977. Clostridial glycine reductase complex: purification and characterization of the selenoprotein component. J. Biol. Chem. 252:5337–5344.

46. Costilow, R. N. 1977. Selenium requirement for the growth of *Clostridium sporogenes* with glycine as the oxidant in the Stickland systems. J. Bacteriol. 131:366–368.

47. Costilow, R. N., and D. Cooper. 1978. Identity of proline dehydrogenase and Δ^1-pyrroline-5-carboxylic acid reductase in *Clostridium sporogenes*. J. Bacteriol. 134:139–146.

47a. Cotta, M. A., and R. B. Hespell. 1986. Proteolytic activity of the ruminal bacterium *Butyrivibrio fibrisolvens*. Appl. Environ. Microbiol. 52:51–58.

47b. Cotta, M. A., and J. B. Russell. 1982. Effects of peptides and amino acids on the efficiency of rumen bacterial protein synthesis in continuous culture. J. Dairy Sci. 65:226–234.

48. Crow, V. L., and T. D. Thomas. 1982. Arginine metabolism in lactic streptococci. J. Bacteriol. 150:1024–1032.

48a. Dainty, R. H. 1970. Purification of threonine aldolase from *Clostridium pasteurianum*. Biochem. J. 117:585–592.

48b. D'Ari, L., and H. A. Barker. 1985. p-Cresol formation by cell extracts of *Clostridium difficile*. Arch. Microbiol. 143:311–312.

49. Dawson, R. M. C. 1973. Specificity of enzymes involved in the metabolism of phospholipids, p. 97–116, in G. B. Ansell, J. N. Hawthorne and R. M. C. Dawson (eds.), Form and function of phospholipids, Vol. 3, Elsevier, Amsterdam.

50. Dawson, R. M. C., and P. Kemp. 1970. Biohydrogenation of dietary fats in ruminants, p. 504–518, in A. T. Phillipson (ed.), Physiology of digestion and metabolism in the ruminant. Oriel Press, Stocksfield, Northumberland, England.

51. Dawson, R. M. C., N. Hemington, D. Grime, D. Lander, and P. Kemp. 1974. Lipolysis and hydrogenation of galactolipids and the accumulation of phytanic acid in the rumen. Biochem. J. 144:169–171.

52. Demeyer, D. I., C. Henderson, and R. A. Prins. 1978. Relative significance of exogenous and *de novo* fatty acid synthesis in formation of rumen microbial lipids *in vitro*. Appl. Environ. Microbiol. 35:24–31.

53. Douglas, H. C. 1951. Glycine fermentation by nongas-forming anaerobic micrococci. J. Bacteriol. 62:517–518.

54. Dürre, P., and J. R. Andreesen. 1983. Purine and glycine metabolism by purinolytic clostridia. J. Bacteriol. 154:192–199.

55. Dürre, P., and J. R. Andreesen. 1982. Selenium-dependent growth and glycine fermentation by *Clostridium purinolyticium*. J. Gen. Microbiol. 128:1457–1466.

56. Dürre, P., R. Spahr, and J. R. Andreesen. 1983. Glycine fermentation via a glycine reductase in *Peptococcus glycinophilus* and *Peptococcus magnus*. Arch. Microbiol. 134:127–135.

57. Elsden, S. R., and M. G. Hilton. 1979. Amino acid utilization patterns in clostridial taxonomy. Arch Microbiol. 123:137–141.

58. Elsden, S. R., and M. G. Hilton. 1978. Volatile acid production from threonine, valine, leucine and isoleucine by clostridia. Arch. Microbiol. 117:165–172.

59. Elsden, S. R., M. G. Hilton, K. R. Parsley, and R. Self. 1980. The lipid fatty acids of proteolytic clostridia. J. Gen. Microbiol. 118:115–123.

60. Elsden, S. R., M. G. Hilton, and J. M. Waller. 1976. The end products of the metabolism of aromatic amino acids by clostridia. Arch. Microbiol. 107:283–288.

61. Emmanuel, B. 1978. The relative contribution of propionate and long-chain even-numbered fatty acids to the production of long-chain odd-numbered fatty acids in rumen bacteria. Biochem. Biophys. Acta 528:239–246.

62. Emmanuel, B. 1974. On the origin of rumen protozoan fatty acids. Biochem. Biophys. Acta 337:404–413.

63. Erfle, J. P., R. J. Boila, R. M. Teather, S. Mabadevan, and F. D. Sauer. 1982. Effect of pH on fermentation characteristics and protein degradation by rumen microorganisms in vitro. J. Dairy Sci. 65:1457–1464.

63a. Eyssen, H., and A. Verhulst. 1984. Biotransformation of linoleic acid and bile acids by *Eubacterium lentum*. Appl. Environ. Microbiol. 47:39–43.

64. Ferguson, K. P., W. R. Mayberry, and D. W. Lambe, Jr. 1983. Production of phenylacetic acid by *Bacteroides melaninogenicus* ssp. *intermedius* serogroup C-1. Can. J. Microbiol. 29:276–279.

65. Fiebig, K., and G. Gottschalk. 1983. Methanogenesis from choline by a culture of *Desulfovibrio* sp. and *Methanosarcina barkeri*. Appl. Environ. Microbiol. 45:161–168.

66. Finegold, S. M. 1980. Survey of human diseases due to anaerobes and isolation of anaerobes from clinical specimens, p. 45–51, in G. Gottschalk, N. Pfennig, and H. Werner (eds.), Anaerobes and anaerobic infections. Gustav Fischer, Stuttgart, West Germany.

67. Finnerty, W. R. 1978. Physiology and biochemistry of bacterial phospholipid metabolism. Adv. Microbiol. Physiol. 18:177–233.

68. Forsberg, C. W. 1980. Sulfide production from cysteine by *Desulfovibrio desulfuricans*. Appl. Environ. Microbiol. 39:453–455.

69. Forsberg, C. W., L. K. Lovelock, L. Krumholz, and J. G. Buchanan-Smith. 1984. Protease activities of rumen protozoa. Appl. Environ. Microbiol. 47:101–110.

70. Fudghum, R. S., and W. E. C. Moore. 1963. Isolation, enumeration and characterization of proteolytic ruminal bacteria. J. Bacteriol. 85:808–815.

71. Garton, G. A. 1977. Fatty acid metabolism in ruminants, p. 337–370, in T. W. Goodwin (ed.), International review of biochemistry of lipids II, Vol. 14. University Park Press, Baltimore.

72. Garton, G. A., A. K. Lough, and E. Vioque. 1961. Glyceride hydrolysis and glycerol fermentation by sheep rumen contents. J. Gen. Microbiol. 25:215–225.

73. Garton, G. A., P. N. Hobson, and A. K. Lough. 1958. Lipolysis in the rumen. Nature (Lond.) 182:1511–1512.

74. Gibbons, R. J., and J. B. MacDonald. 1961. Degradation of collagenous substrates by *Bacteroides melaninogenicus*. J. Bacteriol. 81:614–621.

75. Girard, V., and J. C. Hawke. 1978. The role of holotrichs in the metabolism of dietary linoleic acid in the rumen. Biochem. Biophys. Acta 528:17–27.

76. Gottschalk, G. 1979. Bacterial metabolism. Springer-Verlag, New York.

77. Guillaume, J., H. Beerens, and H. Osteux. 1956. Production de gaz carbonique et fermentation du glycocolle par *Clostridium histolyticum*. Ann. Inst. Pasteur (Paris) 91:721–726.

78. Hardman, J. K., and T. C. Stadtman. 1960. Metabolism of ω-amino acids. I. Fermentation of γ-aminobutyric acid by *Clostridium aminobutyricum* n. sp. J. Bacteriol. 79:544–548.

79. Hardman, J. K., and T. C. Stadtman. 1960. Metabolism of ω-amino acids. II. Fermentation of Δ-aminovaleric acid by *Clostridium aminovalericum* n. sp. J. Bacteriol. 79:549–552.

80. Harwood, C. S. and E. Canale-Parola. 1983. *Spirochaeta isovalerica* sp. nov., a marine anaerobe that forms branched-chain fatty acids as fermentation products. Int. J. Syst. Bacteriol. 33:573–579.

81. Harwood, C. S., and E. Canale-Parola. 1982. Properties of acetate kinase isoenzymes and branched-chain fatty acid kinase from a spirochete. J. Bacteriol. 152:246–254.

82. Harwood, C. S., and E. Canale-Parola. 1981. Branched-chain amino acid fermentation by a marine spirochete: strategy for starvation survival. J. Bacteriol. 148:109–116.

83. Harwood, C. S., and E. Canale-Parola. 1981. Adenosine 5′-triphosphate-yielding pathways of branched-chain amino acid fermentation by a marine spirochete. J. Bacteriol. 148:117–123.

84. Hauser, H., G. P. Hazlewood, and R. M. C. Dawson. 1979. Membrane fluidity of a fatty acid auxotroph grown with palmitic acid. Nature (Lond.) 279:536–538.

85. Hayward, H. R., and T. C. Stadtman. 1959. Anaerobic degradation of choline. I. Fermentation of choline by an anaerobic, cytochrome-producing bacterium, *Vibrio cholinicus*, n. sp. J. Bacteriol. 78:557–561.

86. Hazlewood, G. P., A. J. Northrop, and R. M. C. Dawson. 1981. Diabolic acids: occurrence and identification in natural products and their metabolism by simply stomached and ruminant animals. Br. J. Nutr. 45:159–166.

87. Hazlewood, G. P., and R. Edwards. 1981. Proteolytic activities of a rumen bacterium, *Bacteroides ruminicola* R8/4. J. Gen. Microbiol. 125:11–15.

88. Hazlewood, G. P., and R. M. C. Dawson. 1975. Isolation and properties of a phospholipid-hydrolyzing bacterium from ovine rumen fluid. J. Gen. Microbiol. 89:163–174.

89. Hazlewood, G., and R. M. C. Dawson. 1979. Characteristics of a lipolytic and fatty acid-requiring *Butyrivibrio* sp. isolated from the ovine rumen. J. Gen. Microbiol. 112:15–27.

90. Hazlewood, G. P., and J. H. A. Nugent. 1978. Leaf fraction 1 protein as a nitrogen source for the growth of a proteolytic rumen bacterium. J. Gen. Microbiol. 106:369–371.

91. Hazlewood, G. P., and C. G. Orpin, Y. Greenwood, and M. E. Black. 1983. Isolation of proteolytic rumen bacteria by use of selective medium containing leaf fraction 1 protein (ribulose-bisphosphate carboxylase). Appl. Environ. Microbiol. 45:1780–1784.

92. Hazlewood, G. P., G. A. Jones, and J. L. Mangan. 1981. Hydrolysis of leaf fraction protein by proteolytic rumen bacterium *Bacteroides ruminicola* R8/4. J. Gen. Microbiol. 123:223–232.

93. Henderson, C. 1975. The isolation and chracterization of strains of lipolytic bacteria from the ovine rumen. J. Appl. Bacteriol. 39:101–109.

94. Henderson, C. 1971. A study of the lipase produced by *Anaerovibrio lipolytica*, a rumen bacterium. J. Gen. Microbiol. 65:81–89.

95. Henderson, C. 1968. A study of the lipase of *Anaerovibrio lipolytica*, a rumen bacterium. Ph.D. thesis, University of Aberdeen.

96. Henkelekian, H. 1958. Basic principles of sludge digestion, p. 25–43, in J. McCabe and W. W. Eckenfelder, Jr. (eds.), Biological treatment of sewage and industrial wastes. Reinhold, New York.

97. Hespell, R. B., and M. P. Bryant. 1980. The genera *Butyrivibrio*, *Succinovibrio*, and *Selenomonas*, p. 1439–1494, in M. P. Starr, H. Stolp, H. G. Trüper, A. Balows, and H. G. Schlegel (eds.), The procaryotes—A handbook on habitats, isolation and identification of bacteria, Vol. 1. Springer-Verlag, New York.

98. Hespell, R. B. 1981. Ruminal microorganisms—their significance and nutritional value. Dev. Ind. Microbiol. 22:261–275.

99. Hespell, R. B., and E. Canale-Parola. 1971. Amino acid and glucose fermentation by *Treponema denticola*. Arch. Microbiol. 78:234–251.

99a. Hespell, R. B., and C. J. Smith. 1983. Utilization of nitrogen sources by gastrointestinal tract bacteria, p. 167–187, in D. J. Hentges (ed.), Human intestinal microflora in health and disease. Academic Press, New York.

99b. Hino, T., and J. B. Russell. 1985. Effect of reducing-equivalent disposal and NADH/NAD on deamination of amino acids by intact rumen microorganisms and their cell extracts. Appl. Environ. Microbiol. 50:1368–1374.

100. Hobson, P. N., and R. J. Wallace. 1982. Microbial ecology and activities in the rumen. Crit. Rev. Microbiol. 9:253–320.

101. Hobson, P. N., and S. O. Mann. 1961. The isolation of glycerol-fermenting and lipolytic bacteria from the rumen of sheep. J. Gen. Microbiol. 25:227–240.

102. Hobson, P. N., S. Bousfield, and R. Summers. 1974. Anaerobic digestion organic matter. CRC Crit. Rev. Environ. Control 4:131–191.

103. Holdeman, L. V., and W. E. Moore. 1972. Anaerobe laboratory manual, 2nd, ed. Anaerobe Laboratory, Virginia Polytechnic Institute and State University, Blacksburg, VA.

104. Horler, D. F., W. B. McConnell, and D. W. S. Westlake, 1966. Glutaconic acid, a product of glutamic acid by *Peptococcus aerogenes*. Can. J. Microbiol. 12:1247–1252.

105. Huang, J. J. H., and J. C. H. Shih. 1981. The potential of biological methane generation from chicken manure. Biotechnol. Bioeng. 23:2307–2314.

106. Hungate, R. E. 1966. The rumen and its microbes. Academic Press, New York.

107. Hungate, R. E., and R. J. Stack. 1982. Phenylpropionic acid: growth factor for *Ruminococcus albus*. Appl. Environ. Microbiol. 44:79–83.

108. Hunter, W. J., F. C. Baker, I. S. Rosenfeld, J. B. Keyser, and S. B. Tove. 1976. Biohydrogenation of unsaturated fatty acids. Hydrogenation by cell-free preparations of *Butyrivibrio fibrisolvens*. J. Biol. Chem. 251:2241–2247.

109. Hurst, A., and D. L. Collins-Thompson. 1979. Food as a bacterial habitat. Adv. Microb. Ecol. 3:79–134.

109a. Hylemon, P. B., and T. L. Glass. 1983. Biotransformation of bile acids and cholesterol by the intestinal microflora, p. 189–213, in D. J. Hentges (ed.), Human intestinal microflora in health and disease. Academic Press, New York.

109b. Jain, M. K., and J. G. Ziekus. 1985. A defined starter culture for biomethanation of proteinaceous wastes, p. 125–139, in A. A. Antonopoulos (ed.), Biotechnological advances in processing municipal wastes for fuels and chemicals. Argonne National Laboratory, Argonne, Ill.

110. Jean, M., and R. D. DeMoss. 1968. Indolelactate dehydrogenase from *Clostridium sporogenes*. Can. J. Microbiol. 14:429–435.

111. Jeng, I.-M., and H. A. Barker. 1974. Purification and properties of L-3-aminobutyryl coenzyme A deaminase from a lysine-fermenting clostridium. J. Biol. Chem. 249:6578–6584.

112. Kamio, Y., and Y. Terawaki. 1983. Purification and properties of *Selenomonas ruminantium* lysine decarboxylase. J. Bacteriol. 153:658–654.

113. Kamio, Y., H. Inagaki, and H. Takahashi. 1970. Possible occurrence of α-oxidation in phospholipid biosynthesis from *S. ruminantium*. J. Gen. Appl. Microbiol. 16:463–478.

114. Kemp, P., R. W. White, and D. L. Lander. 1975. The hydrogenation of unsaturated fatty acids by five bacterial isolates from the sheep rumen, including a new species. J. Gen. Microbiol. 90:100–114.

115. Kemp, P., and D. J. Lander. 1984. Hydrogenation *in vitro* of α-linolenic acid to stearic acid by mixed cultures of pure strains of rumen bacteria. J. Gen. Microbiol. 130:527–533.

115a. Kemp, P., and D. J. Lander. 1984. The hydrogenation of some *cis*- and *trans*-octadenenoic acids to stearic acid by a rumen *Fusocillus* sp. Br. J. Nutr. 52:165–170.

115b. Kemp, P., and D. J. Lander. 1984. The hydrogenation of the series of methylene-interrupted *cis,cis*-octadecadienoic acids by pure cultures of six rumen bacteria. Br. J. Nutr. 52:171–177.

116. Kemp, P., and D. J. Lander. 1983. The hydrogenation of γ-linolenic acid by pure cultures of two rumen bacteria. Biochem. J. 216:519–522.

117. Kepler, C. R., W. P. Tucker, and G. B. Tove. 1970. Biohydrogenation of unsaturated fatty acids. IV. Substrate specificity and inhibition of linoleate Δ^{12}-*cis*, Δ^{11}-*trans*-isomerase from *Butyrivibrio fibrisolvens*. J. Biol. Chem. 245:3612–3620.

118. Klass, 1984. Methane from anaerobic fermentation. Science 223:1021–1028.

119. Klein, R. A., G. P. Hazlewood, P. Kemp, and R. M. C. Dawson. 1979. A new series of long-chain dicarboxylic acids with vicinyl dimethyl branching found as major components of the lipids of *Butyrivibrio* spp. Biochem. J. 183:691–700.

120. Klein, S. M., and R. D. Sagers. 1962. Intermediary metabolism of *Diplococcus glycinophilus*. II. Enzymes of the acetate-generating system. J. Bacteriol. 83:121–126.

121. Kopecny, J., and R. J. Wallace. 1982. Cellular location and some properties of proteolytic enzymes of rumen bacteria. Appl. Environ. Microbiol. 43:1026–1033.

121a. Kuchta, R. D., and R. H. Abeles. 1985. Lactate reduction in *Clostridium propionicum*. Purification and properties of lactyl-CoA dehydratase. J. Biol. Chem. 260:13181–13189.

122. Laanbroek, H. J., L. J. Stal, and H. Veldkamp. 1978. Utilization of hydrogen and formate by *Campylobacter* spec. under aerobic and anaerobic conditions. Arch. Microbiol. 119:99–102.

123. Lerud, R. F., and H. R. Whitely. 1971. Purification and properties of α-ketoglutarate reductase from *Mirococcus aerogenes*. J. Bacteriol. 106:571–577.

124. Lev, M. 1980. Glutamate-stimulated amino acid and peptide incorporation in *Bacteroides melaninogenicus*. J. Bacteriol. 143:753–760.

125. Loesche, W. J. 1980. The role of anaerobes in peridontal disease, p. 61–71, in G. Gottschalk, N. Pfennig, and H. Werner (eds.), Anaerobes and anaerobic infections. Gustave Fischer, Stuttgart, West Germany.

126. Loll, M. S., and J.-M. Bollag. 1984. Protein transformation in soil. Adv. Agron. 36:351–382.

126a. MacFarlane, G. T., J. H. Cummings, and C. Allison. 1986. Protein degradation by human intestinal bacteria. J. Gen. Microbiol. 132:1647–1656.

127. MacFarlane, M. G., and B. C. J. G. Knight. 1941. The biochemistry of bacterial toxins. I. The lecithinase activity of *Cl. welchii* toxins. Biochem. J. 35:884–902.

128. Maczulak, A. E., B. A. Dehority, and D. L. Palmquist. 1981. Effects of long-chain fatty acids on growth of rumen bacteria. Appl. Environ. Microbiol. 42:856–862.

129. Maki, L. R. 1954. Experiments on the microbiology of cellulose decomposition in a municipal sewage plant. Antonie Leeuwenhoek, J. Microbiol. Serol. 20:185–200.

130. Massey, L. K., J. R. Sokatch, and R. S. Conrad. 1976. Branched-chain amino acid catabolism in bacteria. Bacteriol. Rev. 40:42–54.

131. Mayrand, D. 1979. Identification of clinical isolates of selected species of *Bacteroides*: production of phenylacetic acid. Can. J. Microbiol. 25:927–928.

132. McInerney, M. J., and M. P. Bryant. 1981. Review of methane fermentation fundamentals, p. 19–47, in D. L. Wise (ed.), Fuel gas production from biomass, Vol. 1. CRC Press, Boca Raton, Fla.

133. McInerney, M. P. Bryant, and N. Pfennig. 1979. An anaerobic bacterium that degrades fatty acids in syntrophic association with methanogens. Arch. Microbiol. 122:129–135.

134. Mead, G. C. 1971. The amino acid-fermenting clostridia. J. Gen. Microbiol. 67:47–56.

135. Mitchell, W. M., and W. F. Harrington. 1971. Clostripain, p. 699–719, in P. D. Boyer (ed.), The enzymes, Vol. 3. Academic Press, New York.

135a. Molongoski, J. J., and M. J. Klug. 1976. Characterization of anaerobic heterotrophic bacteria isolated from fresh-water lake sediments. Appl. Environ. Microbiol. 31:83–90.

136. Monticello, D. J., and R. N. Costilow. 1982. Interconversion of valine and leucine by *Clostridium sporogenes*. J. Bacteriol. 152:946–949.

137. Montville, T. J. 1983. Dependence of *Clostridium botulinum* gas and protease production on culture conditions. Appl. Environ. Microbiol. 45:571–575.

138. Mountfort, D. O., and M. P. Bryant. 1982. Isolation and characterization of an anaerobic syntrophic benzoate-degrading bacterium from sewage sludge. Arch. Microbiol. 133:249–256.

139. Muth, W. L., and R. N. Costilow. 1974. Ornithine cyclase (deaminating). II. Properties of the homogeneous enzyme. J. Biol. Chem. 249:7457–7462.

140. Nagase, M., and T. Matsuo. 1982. Interactions between amino acid-degrading bacteria and methanogenic bacteria in anaerobic digestion. Biotechnol. Bioeng. 24:2227–2239.

140a. Nanninga, H. J., and J. C. Gottschal. 1985. Amino acid fermentation and hydrogen transfer in mixed cultures. FEMS Microbiol. Lett. 31:261–269.

140b. Naumann, E. L., H. Hippe, and G. Gottschalk. 1983. Betaine: new oxidant in the Stickland reaction and methanogenesis from betaine and L-alanine by a *Clostridium sporogenes-Methanosarcina barkeri* coculture. Appl. Environ. Microbiol. 45:474–483.

141. Neill, A. R., D. W. Grime, and R. M. C. Dawson. 1978. Conversion of choline methyl groups through trimethylamine into methane in the rumen. Biochem. J. 170:529–535.

142. Nikolic, J. A., and R. Filipovic. 1981. Degradation of maize protein in rumen contents. Influence of ammonia concentration. Br. J. Nutr. 45:111–116.

143. Nisman, B. 1954. The Stickland reaction. Bacteriol. Rev. 18:16–42.

144. Nugent, J. H. A., and J. L. Mangan. 1978. Rumen proteolysis of fraction 1 leaf protein, casein, and bovine serum albumin. Proc. Nutr. Soc. 37:48A.

144a. Ollivier, B. M., R. A. Mah, T. J. Ferguson, D. R. Boone, J. L. Garcia, and R. Robinson. 1985. Emendation of the genus *Thermobacteroides*: *Thermobacteroides proteolyticus* sp. nov., a proteolytic acetogen from a methanogenic enrichment. Int. J. Syst. Bacteriol. 35:425–428.

144b. Ollivier, B., N. Smiti, R. A. Mah, and J. L. Garcia. 1986. Thermophilic methanogenesis from gelatin by a mixed defined bacterial culture. Appl. Microbiol. Biotechnol. 24:79–83.

145. O'Neil, S. R., and R. D. DeMoss. 1968. Tryptophan transaminase from *Clostridium sporogenes*. Arch. Biochem. Biosphys. 127:361–369.

146. Orskov, E. R., C. Fraser, I. McDonald, and R. I. Smart. 1974. Digestion of sheep concentrates. 5. The effect of adding fishmeal and urea together on protein digestion and utilization by young sheep. Br. J. Nutr. 31:89–98.

147. Pfennig, N., and F. Widdel. 1981. Ecology and physiology of some anaerobes from the microbial sulfur cycle, p. 169–177, in H. Bothe and A. Trebest (eds.), Biology of inorganic nitrogen and sulfur. Springer-Verlag, Heidelberg.

148. Pickett, J., and R. Kelly. 1975. Lipid catabolism of relapsing fever borreliae. Infect. Immun. 9:279–285.

149. Pittman, K. A., and M. P. Bryant. 1964. Peptides and other nitrogen sources for growth of *Bacteroides ruminicola*. J. Bacteriol. 88:401–410.

150. Prins, R. A. 1977. Biochemical activities of gut microorganisms, p. 73–183, in R. T. J. Clarke and T. Bauchop (eds.), Microbial ecology of the gut. Academic Press, London.

151. Prins, R. A., D. L. van Rheenen, and T. van't Klooster. 1983. Characterization of microbiol proteolytic enzymes in the rumen. Antonie Leeuwenhoek, J. Microbiol. Serol. 49:585–595.

152. Prins, R. A., A. Lankhorst, P. Van der Meer, and C. J. Van Nevel. 1975. Some characteristics of *Anaerovibrio lipolytica*, a rumen lipolytic organism. Antonie Leeuwenhoek, J. Microbiol. Serol. 41:1–11.

152a. Russell, J. B., and J. L. Jeraci. 1984. Effect of carbon monoxide on fermentation of fiber, starch, and amino acids by mixed rumen microorganisms in vitro. Appl. Environ. Microbiol. 48:211–217.

153. Russell, J. B., W. G. Bottje, and M. A. Cotta. 1981. Degradation of protein by mixed cultures of rumen bacteria: identification of *Streptococcus bovis* as an actively proteolytic bacterium. J. Anim. Sci. 53:242–252.

154. Sagers, R. D., and I. C. Gunsalus. 1961. Intermediary metabolisms of *Diplococcus glycinophilus*. I. Glycine cleavage and one-carbon interconversions. J. Bacteriol. 81:541–549.

155. Schink, B., and N. Pfennig. 1982. Fermentation of trihydroxybenzenes by *Pelobacter acidigallici* gen. nov., sp. nov., a new strictly anaerobic, non-spore-forming bacterium. Arch. Microbiol. 133:195–201.

155a. Schweiger, G., and W. Buckel. 1984. On the degradation of (R)-lactate in the fermentation of alanine to propionate by *Clostridium propionicum*. FEBS Lett. 171:79–84.

156. Seto, B. 1980. The Stickland reaction, p. 50–64, in C. J. Knowles (ed.), Diversity of bacterial respiratory systems, Vol. 2. CRC Press, Boca Raton, Fla.

157. Seto, B. 1978. A pyruvate-containing peptide of proline reductase from *Clostridium sticklandii*. J. Biol. Chem. 253:4525–4529.

158. Seto, B. 1978. Electron transport proteins associated with proline fermentation in *Clostridium sticklandii*. Fed. Proc. 37:1521.

159. Seto, B., and T. C. Stadtman. 1976. Purification and properties of proline reductase from *Clostridium sticklandii*. J. Biol. Chem. 251:2435–2439.

160. Shoesmith, J. G., and J. C. Sherris. 1960. Studies on the mechanism of arginine-activated motility in a *Pseudomonas* strain. J. Gen. Microbiol. 22:10–24.

161. Siebert, M. L., and D. F. Toerien. 1969. The proteolytic bacteria present in anaerobic digestion of raw sewage sludge. Water Res. 3:241–250.

162. Smith, D. T. 1972. Anaerobic bacilli and wound infections, p. 589–598, in W. K. Jollin and D. T. Smith (eds.), Zinsser's microbiology, 15th ed. Appleton-Century-Crofts, New York.

163. Spoelstra, S. F. 1978. Degradation of tyrosine in anaerobically stored piggery wastes and in pig feces. Appl. Environ. Microbiol. 36:631–638.

164. Stack, R. J., R. E. Hungate, and W. P. Opsahl. 1983. Phenylacetic acid stimulation of cellulose digestion by *Ruminococcus albus* 8. Appl. Environ. Microbiol. 46:539–544.

165. Stadtman, T. C. 1980. Selenium-dependent enzymes. Annu. Rev. Biochem. 49:93–110.

166. Stadtman, T. C. 1973. Lysine metabolism by clostridia. Adv. Enzymol. 38:418–448.

167. Stadtman, T. C. 1966. Glycine reduction to acetate and ammonia: Identification of ferredoxin and another low molecular weight acidic protein as components of the reductase systems. Arch. Biochem. Biophys. 113:9–19.

168. Stadtman, T. C. 1962. Coupling of a DPNH-generating system to glycine reduction. Arch. Biochem. Biophys. 99:36–44.

169. Stadtman, T. C., and P. Renz. 1968. Anaerobic degradation of lysine. V. Some properties of the cobamide coenzyme-dependent β-lysine mutase of *Clostridium sticklandii*. Arch. Biochem. Biophys. 125:226–239.

170. Stadtman, T. C., and P. Elliot. 1957. Studies on the enzymatic reduction of amino acids. II. Purification and properties of D-proline reductase and a proline racemase from *Clostridium sticklandii*. J. Biol. Chem. 228:983–997.

171. Stadtman, T. C., P. Elliot, and L. Tiemann. 1958. Phosphate esterification coupled with glycine reduction. J. Biol. Chem. 231:961–973.

171a. Stams, A. J. M., and T. A. Hansen. 1984. Fermentation of glutamate and other compounds by *Acidaminobacter hydrogenoformans* gen. nov. sp. nov., an obligate anaerobe isolated from black mud. Studies with pure cultures and mixed cultures with sulfate-reducing and methanogenic bacteria. Arch. Microbiol. 137:329–337.

172. Takahashi, T., T. Sugahara, and A. Oksaka. 1974. Purification of *Clostridium perfringens* phospholipase C (α-toxin) by affinity chromatography on agarose-linked egg-yolk lipoprotein. Biochem. Biophys. Acta 351:155–171.

173. Tanaka, H., and T. C. Stadtman. 1979. Selenium-dependent clostridial glycine reductase: purification and characterization of the two membrane-associated protein components. J. Biol. Chem. 254:447–452.

173a. Thauer, R. K., K. Jungermann, and K. Decker. 1977. Energy conservation in chemotrophic anaerobic bacteria. Bacteriol. Rev. 41:100–180.

174. Toerien, D. F., and W. H. J. Haltingh. 1969. Anaerobic digestion. I. The microbiology of anaerobic digestion. Water Res. 3:385–416.

175. Trevathan, C. A., R. M. Smibert, and H. A. George. 1982. Lipid catabolism of cultivated treponemes. Can. J. Microbiol. 28:672–678.

176. Turner, D. C., and T. C. Stadtman. 1973. Purification of protein components of the

clostridial glycine reductase system and characterization of protein A as a selenoprotein. Arch. Biochem. Biophys. 154:366–381.

176a. van Steenbergen, T. J. M., and J. de Graff. 1986. Proteolytic activity of black-pigmented *Bacteroides* strains. FEMS Microbiol. Lett. 33:219–222.

177. Varel, V. H., H. R. Isaacson, and M. P. Bryant. 1977. Thermophilic methane production from cattle waste. Appl. Environ. Microbiol. 33:298–307.

178. Venugopslan, V. 1980. Influence of growth conditions on glycine reductase of *Clostridium sporogenes*. J. Bacteriol. 141:386–388.

178a. Verhulst, A., G. Parmentier, G. Janssen, S. Asselberghs, and H. Eyssen. 1986. Biotransmation of unsaturated long-chain fatty acids by *Eubacterium lentum*. Appl. Environ. Microbiol. 51:532–538.

179. Wachsman, J. T., and H. A. Barker. 1955. Tracer experiments on glutamate fermentation by *Clostridium tetanomorphum*. J. Biol. Chem. 217:695–702.

179a. Wallace, R. J. 1986. Catabolism of amino acids by *Megasphaera elsdenii* LC1. Appl. Environ. Microbiol. 51:1141–1143.

179b. Wallace, R. J., and M. L. Brammall. 1985. The role of different species of bacteria in the hydrolysis of protein in the rumen. J. Gen. Microbiol. 131:821–832.

180. Wallace, R. J., and J. Kopecny. 1983. Breakdown of diazotized proteins and synthetic substrates by rumen bacterial proteases. Appl. Environ. Microbiol. 45:212–217.

181. Whitely, H. R. 1957. Fermentation of amino acids by *Micrococcus aerogenes*. J. Bacteriol. 74:324–330.

181a. Wildenauer, F. X., and J. Winter. 1985. Anaerobic digestion of high-strength acidic whey in a pH-controlled up-flow fixed film loop reactor. Arch. Microbiol. 22:367–372.

181b. Williams, A. G. 1986. Rumen holotrich ciliate protozoa. Microbiol. Rev. 50:25–49.

182. Wolin, M. J. 1979. The rumen fermentation: a model for microbial interactions in anaerobic ecosystems. Adv. Microbiol. Ecol. 3:49–78.

183. Wong, J. C., J. K. Dyer, and J. L. Tribble. 1977. Fermentation of L-aspartate by a saccharolytic strain of *Bacteroides melaninogenecus*. Appl. Environ. Microbiol. 33:69–73.

184. Yamazaki, S., and S. B. Tove. 1979. Biohydrogenation of unsaturated fatty acids. Presence of dithionite and an endogenous electron donor in *Butyrivibrio fibrisolvens*. J. Biol. Chem. 254:3812–3817.

185. Yokoyama, M. T., and J. R. Carlson. 1981. Production of skatole and para-cresol by a rumen *Lactobacillus* sp. Appl. Environ. Microbiol. 41:71–76.

186. Yokoyama, M. T., and C. L. Davis. 1971. Stimulation of hydrogenation of linoleate in *Treponema* (*Borrelia*) sp. strain $B_2 5$ by reduced methyl viologen and by reduced benzyl viologen. Biochem. J. 125:913–915.

187. Yokoyama, M. T., J. R. Carlson, and L. V. Holdeman. 1977. Isolation and characterization of a skatole-producing *Lactobacillus* sp. from the bovine rumen. Appl. Environ. Microbiol. 34:837–842.

188. Yorifuji, T., I.-M. Jeng, and H. A. Barker. 1977. Purification and properties of 3-

keto-5-aminohexanoate cleavage enzyme from a lysine-fermenting clostridium. J. Biol. Chem. 252:20–31.

189. Zillig, W., A. Grerl, G. Schreiver, S. Wunderl, D. Janekovic, K. O. Stetter, and H. P. Klink. 1983. The archaebacterium *Thermofilum pendens* represents, a novel genus of the thermophilic, anaerobic, sulfur-respiring *Thermoproteales*. Syst. Appl. Microbiol. 4:79–87.

190. Zillig, W., I. Holz, D. Janekovec, W. D. Recter. 1983. The archaebacterium *Thermococcus celer* represents, a novel genus within the thermophilic branch of the Archaebacteria. Syst. Appl. Microbiol. 4:88–94.

191. Zillig, W., K. O. Stetter, D. Prangishvilli, W. Schäfer, S. Wunderi, D. Janekovic, I. Holz, and P. Palm. 1982. *Desulfurococcaceae*, the second family of the extremely thermophilic, anaerobic, sulfur-respiring *Thermoproteales*. Zentralbl. Bakteriol. Hyg. Microbiol. 1 Abt. Orig. C3:304–317.

9

ACETOGENESIS

Jan Dolfing

Department of Microbiology, Agricultural University, 6703 CT Wageningen, The Netherlands

9.1 INTRODUCTION
    9.1.1 Definitions
        9.1.1.1 Acetogenesis
        9.1.1.2 Interspecies Hydrogen Transfer
    9.1.2 Topography and Scope
9.2 ACETOGENIC DEHYDROGENATIONS
    9.2.1 Obligate Proton Reducing Bacteria
    9.2.2 Facultative Proton Reducing Bacteria
    9.2.3 Energy Conservation in Acetogenic Dehydrogenations
    9.2.4 Ecology of Acetogenic Dehydrogenations
9.3 ACETOGENIC HYDROGENATIONS
    9.3.1 Homoacetogens on Multicarbons Compounds
    9.3.2 Homoacetogens on One Carbon Compounds
    9.3.3 Energy Conservation in Acetogenic Hydrogenations
    9.3.4 Ecology of Acetogenic Hydrogenations
9.4 FORMATION OF HIGHER FATTY ACIDS FROM C_2 COMPOUNDS
    9.4.1 Organisms Involved
    9.4.2 Biotechnological Applications
    9.4.3 Ecological Implications
9.5 CONCLUDING REMARKS
REFERENCES

9.1 INTRODUCTION

In the catabolic metabolism of methanogenic ecosystems, energy is obtained from a series of decarboxylations and redox reactions in which electrons eventually end up in methane. Our understanding of the flow of carbon and electrons in anaerobic environments is reflected in the schemes developed to explain the inevitable formation of methane in such systems; these schemes generally show the degradation of organic compounds as a stepwise process in which carbon–carbon bonds are step by step broken up through the combined activities of various metabolic groups of microorganisms.

In the current concepts of methanogenic conversion processes, methanogens pull these reactions by removing acetate and hydrogen (electrons) in the form of methane from the system. The success of this concept in explaining the ultimate fate of electrons, still enhanced by its correct prediction of the effects of various disturbances of the methanogenic environment, suggests that a subdivision of the anaerobic microflora in three metabolic groups of organisms is adequate to describe the carbon and electron flow in methanogenic environments. However, the influence of the ability of microorganisms to use more than one energy substrate at a time, and to switch their catabolism from breaking carbon–carbon bonds to forming new carbon–carbon bonds has only been studied to a small extent and is generally not taken into account in the current schemes. These physiological properties may well have an ecological basis and exert influence on the effective flow of carbon and electrons. These flows may not always take the shortest route: A part of the electrons may occasionally be used to reduce organic compounds instead of carbon dioxide or to reduce carbon dioxide to other compounds than methane.

There are indications that "reversed electron flow" occurs in anaerobic ecosystems, particularly through the activities of acetogenic microorganisms. In this chapter we review the literature on these acetogenic bacteria (including the obligate proton-reducing bacteria).

9.1.1 Definitions

9.1.1.1 Acetogenesis

Different kinds of fermentations have frequently been identified by the major types of end products formed, a nomenclature recently advocated by Zeikus (140). Consequently, within the diverse group of anaerobic acidogens, acetate-forming organisms have conveniently been addressed as acetogens. Usually, homoacetogens (i.e., organisms that form acetyl-CoA as a consequence of C_1 catabolism) are implicitly treated as obligate acetogens. By means of interspecies hydrogen transfer (51), however, homoacetogens can be persuaded to produce hydrogen and CO_2 as well as acetate (128), so homoacetogens are not obligately homoacetogenic, and not even obligately acetogenic, as "butyrogenic homoacetogens" also exist (140). Since culture conditions may greatly influence microbial catabolism, a nomenclature along these lines is more appropriate for naming reactions than for specifying organisms. Two different types of acetogenic mechanisms can be distinguished: acetogenic hydrogenations and acetogenic dehydrogenations, which occasionally can be coupled intracellular in so-called homoacetogenic bacteria (65).

9.1.1.2 *Interspecies Hydrogen Transfer*

Hungate (49, 50) was the first to express the idea that hydrogen production and utilization can profoundly influence the course of fermentations in anaerobic ecosystems. The regulatory role of hydrogen in the course of anaerobic fermentations is effectuated by interspecies hydrogen transfer, (i.e., stimulation of hydrogenogenic organisms), induced by hydrogenotrophic organisms, to produce more hydrogen than would have been produced in the absence of hydrogenotrophs (131). The special feature of interspecies hydrogen transfer is that hydrogenotrophs, by using hydrogen, can create conditions for obligate hydrogenogens to perform catabolic oxidations which would not have been energy-yielding in the absence of hydrogenotrophs. The thermodynamical rationale behind these syntrophic relationships has been introduced in Chapter 1.

9.1.2 Topography and Scope

Under anaerobic conditions in natural habitats, microbial decomposition of organic material to inorganic products is predominantly accomplished by microbial oxidations, with protons, sulfur, or carbon atoms as the exclusive electron sinks. In the absence of light and inorganic hydrogen acceptors, biodegradable organic material can be completely fermented to methane and carbon dioxide (20). Substrate conversion in such environments is thought to occur in a multistep process (Fig. 9.1; 74, 139). The number of subdivisions that can be made in these processes depends on the concept with which the overall process is approached. The reaction sequence can be divided into various "distinct" subprocesses (44), or it can be dealt with at the level of stages (74), or metabolic groups of organisms involved (142) or of individual microorganisms (127). The purpose of this chapter is to examine the ecophysiology of acetogenic organisms. This implies a discussion of acetogenesis on various levels, aimed at an integration of what is known about the ecology and the physiology of acetogenic processes.

9.2 ACETOGENIC DEHYDROGENATIONS

Acetogenic dehydrogenations are those reactions in which the oxidation of the substrate is coupled to a reduction of protons, with as a result the formation of hydrogen plus acetate (or a longer-chain fatty acid) as end products. Note that acetogenic sulfate reduction is not included in this definition. Acetogenic dehydrogenations are growth supporting for two groups of microorganisms (a) the obligate proton-reducing bacteria and (b) the fermentative bacteria. The latter group

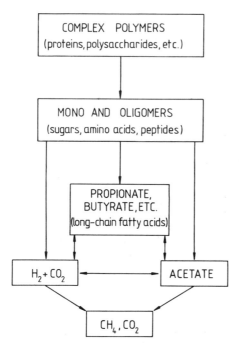

Figure 9.1 Multistep scheme for the flow of carbon in the complete anaerobic conversion of organic matter of methane. Adapted from Ref. 142.

(see also Chapter 8) has the possibility of using various electron acceptors for the disposal of electrons. The obligate proton reducers, however, can utilize only protons as electron acceptors. The idea underlying this distinction is that the two groups of organisms use different substrates with different "energy contents" (i.e., the difference between the amount of energy present in the substrate and the amount of energy present in the oxidized product). The oxidation of the substrates used by fermentative organisms yields a considerable amount of energy. This amount of energy is maximal if hydrogen can be formed as electron sink product, but as increased hydrogen concentrations inhibit the formation of hydrogen, these organisms have the possibility to dispose of their electrons by forming other more reduced organic end products, and still gain energy of such a fermentation. The oxidations catalyzed by obligate proton reducers, however, yield only small amounts of energy, provided that the concentration of the produced hydrogen is kept low. At elevated concentrations of hydrogen, these oxidations become endergonic, and the relatively small "energy content" of the substrates used by obligate proton reducers does not allow that the electrons are disposed of in a less efficient way by forming less reduced oxidation products (130).

Figure 9.2 shows the linear relationship between the free energy change and the logarithm of the equilibrium constant for various reactions in which hydrogen is produced or consumed. This graph illustrates that hydrogen formation from pyruvate is an energy-yielding reaction under standard conditions, while the formation of hydrogen from $NAD(P)H + H^+$ under the same conditions is an energy-

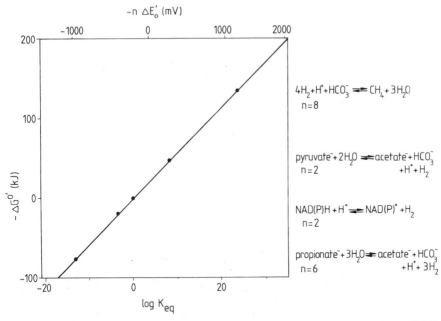

Figure 9.2 Thermodynamic relationship between free energy change at pH 7.0 and 25 °C. $\Delta G°'$, oxidation-reduction potential ($n\,\Delta E_0'$), and equilibrium constant (K_{eq}) for various reactions in which hydrogen participates (n, number of electrons involved).

consuming reaction; the natural equilibrium for these hydrogen-producing reactions is to the right for the former and to the left for the latter, respectively. As a result, these reactions are differently influenced by the presence of hydrogen. While hydrogen affects the dehydrogenation of pyruvate only to a limited extent, hydrogen formation via NAD(P)H is thermodynamically already unfavorable in the presence of low hydrogen concentrations. The equilibrium constant is of prime importance for a hydrogen-yielding reaction: It is an indicator for the thermodynamically determined sensitivity of a reaction for hydrogen. A second important factor is the molar ratio between amount of substrate consumed and amount of hydrogen formed. The relationship between the aforementioned parameters for a hypothetical reaction $aA + bB \rightarrow cC + dD$ can be formally expressed as

$$\Delta G' = -RT \ln K_{eq}' + RT \ln \frac{[C]^c\,[D]^d}{[A]^a\,[B]^b}$$

(see also Chapter 1).

It is generally assumed that obligate proton-reducing bacteria form hydrogen via NAD(P)H, while fermentative organisms have the option to direct their electron flow via either NAD(P)H or pyruvate (130). The implications of these characteristics are discussed in later sections.

9.2.1   Obligate Proton-Reducing Bacteria

The history of the obligate proton-reducing (or obligate hydrogen-producing) bacteria goes back to the classical studies of Bryant, Wolin, and Wolfe who reported in 1967 that *"Methanobacillus omelianskii"* was actually a symbiotic association of two different species (22). Microscopic observations suggested that *M. omelianskii* consisted of two organisms of different shape, and it was found that the "organism" failed to ferment ethanol after it had been grown in media with H_2–CO_2 rather than ethanol–CO_2 as the energy source. Both organisms were isolated and it was shown that both are necessary to convert ethanol–CO_2 to acetate and methane. The production of hydrogen and acetate from ethanol by the so-called S-organism occurred only at lower hydrogen concentrations. The significance of this paper (22) is their speculation that alcohols other than methanol and fatty acids other than formate and acetate are not catabolized by methanogens per se, but by another group of nonmethanogenic bacteria. Furthermore, Bryant et al. cited a paper of Johns and Barker (52), who had calculated that the free energy change for the anaerobic oxidation of alcohol to acetate and hydrogen was unfavorable unless the partial pressure of hydrogen was kept low.

Thereby the foundations for the discovery of obligate proton-reducing bacteria were laid. It took until 1979, however, before McInerney and Bryant were able to isolate an obligate proton-reducing bacterium in co-culture with a hydrogen-consuming bacterium (73): They obtained *Syntrophomonas wolfei* (75), an anaerobic butyrate-oxidizing bacterium in coculture with either a hydrogenotrophic sulfate reducer or a hydrogenotrophic methanogen. A year later Boone and Bryant reported on a syntrophic association of an anaerobic propionate oxidizer in co-culture with a hydrogenotrophic organism (13). The currently known mesophilic obligate proton-reducing bacteria are listed in Table 9.1. The description of *Syntrophus buswellii* has its roots in the observation by Tarvin and Buswell that benzoate can be converted anaerobically to methane (110). Nottingham and Hungate (81) subsequently demonstrated that uniformly ring-labeled [^{14}C]benzoate was fermented exclusively to $^{14}CO_2$ and $^{14}CH_4$. Various attempts to isolate the benzoate-degrading organisms failed, however, and after the schism of *"M. omelianskii,"* Ferry and Wolfe identified acetate, formate, H_2, and CO_2 as intermediates during benzoate degradation and suggested that methanogenic benzoate degradation occurs only via interspecies hydrogen transfer between a benzoate-degrading organism and a hydrogenotrophic methanogen (39). This hypothesis was substantiated by Mountfort and Bryant (80), who obtained *S. buswellii* (79) in co-culture with a hydrogenotrophic sulfate reducer.

There are several reasons why it took such a long time before these syntrophic associations were isolated. As shown in Table 9.2, the growth rates of these fastidious anaerobes are extremely low, so much patience is needed. Furthermore, to date chemical methods to remove hydrogen from growing cultures have not been effective enough to allow growth of the obligate hydrogen producers in the absence of hydrogen consumers. The most successful approach was the addition of an ac-

TABLE 9.1 Currently known mesophilic obligate proton-reducing bacteria[a]

Organisms	In Co-culture with:	Source/Habitat	Substrate	References
Syntrophobacter wolinii	*Desulfovibrio* G11, or *M. hungatei* JF1 plus *D.* G11	Digestor sludge	Propionic acid	13
Syntrophomonas wolfei	*Desulfovibrio* G11, *M. hungatei* JF1	Digestor sludge, aquatic sediments, rumen	Monocarboxylic saturated fatty acids (14- to 8-carbon)	73, 75
Syntrophus buswellii	*Desulfovibrio* G11, *M. hungatei* JF1 plus *D.* G11	Digestor sludge, aquatic sediments	Benzoic acid	79, 80
Clostridium bryantii	*Desulfovibrio* E70, *M. hungatei* M1h	Marine and freshwater sediments, digestor sludge	Monocarboxylic fatty acids (to 11-carbon)	106a
Strain Gra I val	*Desulfovibrio* E70	Marine sediment	Isovaleric acid	106b
Strain Gö I Val	*Desulfovibrio* (Marburg)	Digestor sludge	Isovaleric acid	106b
Strains SF-1 and NSF-2	*M. hungatei* or *Desulfovibrio* sp.	Digestor sludge	Monocarboxylic saturated fatty acids (4- to 6-carbon)	102a
Strain BZ-2	*Desulfovibrio* PS-1 or *Methanospirillum* PM-1 plus *Desulfovibrio* PS-1	Digestor sludge	Benzoic acid	102a

[a]The prefix "syntroph" has been used for the generic names of these bacteria to stress their obligate syntrophic dependency on hydrogen-consuming partners. Recently, however, this restriction has been relaxed by assigning the name *Syntrophococcus sucromutans* to an organism that uses carbohydrates as electron donors, and formate, methoxymonobenzenoids, or methanogens as electron acceptors (58a).

TABLE 9.2   Growth rates of obligate proton-reducing bacteria

Organism	Substrate	$+SO_4^{2-}$	$-SO_4^{2-}$	Reference
		$\mu(day^{-1})$		
Syntrophobacter wolinii	Propionate	0.192	0.103	13
Syntrophomonas wolfei[a]	Butyrate	0.307	0.185	73
Syntrophus buswellii	Benzoate	0.127	0.101	80

[a] The growth rates on valerate and caproate with *M. hungatei* as hydrogen consumer were 0.185 and 0.092 day^{-1}, respectively.

tively growing culture of a hydrogen utilizing desulfovibrio to serial dilutions in roll tubes of enrichments of the desired hydrogen producer. This method resulted in colonies of the hydrogenic organism and the sulfate reducer, and allowed the "isolation" of the obligate syntrophic organisms of Table 9.1. In the case of the butyrate catabolizer, the *Desulfovibrio* could be purified away by using the method described above in sulfate-free medium and *Methanospirillum hungatei* as the hydrogen consumer (73), but for the propionate degrader and the benzoate degrader, this approach never resulted in a complete removal of the sulfate reducer.

Figure 9.3 illustrates that the partial pressure of hydrogen must be very low for the oxidation of ethanol, propionate, butyrate, and benzoate to become exergonic. Under these conditions the hydrogenotrophic organisms obtain less energy per mole hydrogen oxidized than under standard conditions. Table 9.3 lists the free energy changes for the reactions catalyzed by the obligate proton reducers and hydrogen consumers mentioned above. It illustrates that the reactions catalyzed by obligate proton reducers per se are endergonic, while hydrogenotrophic methane formation and sulfate reduction are exergonic under standard conditions. It also illustrates that the coupled reactions as performed by the cocultures listed in Table 9.1 are exergonic for both methanogenic and sulfidogenic couples. The free energy change per mole HS$^-$ formed is more negative than the free energy change per mole methane formed as a result of the fact that the free energy change for the formation of sulfide from four hydrogen is more negative than the formation of methane from four hydrogen. Note that it makes a difference whether the free energy change is evaluated per mole substrate oxidized or per mole methane or sulfate formed.

The values given in Table 9.3 are valid under standard conditions but may be significantly different under nonstandard conditions. This is illustrated by Fig. 9.3, which shows the effect of the partial pressure of hydrogen (p_{H_2}) on the change in free energy ($\Delta G'$) for the reactions listed in Table 9.3. The concentrations of the other compounds are representative of those encountered in methanogenic environments. Figure 9.3 shows that $\Delta G'$ for these reactions depends strongly on the p_{H_2}, especially the oxidations of propionate and benzoate, which both yield 3 mol of hydrogen per mole of substrate oxidized. Table 9.4 illustrates that the influences on $\Delta G'$ of the concentrations of the substrate and of acetate, the other major prod-

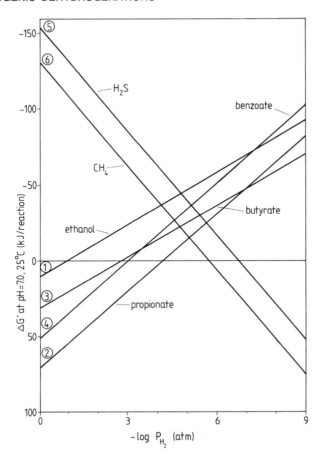

Figure 9.3 Effect of the partial pressure of H_2 (p_{H_2}) on the change in free energy ($\Delta G'$) for the reactions involving ethanol, propionate, butyrate, and benzoate degradation with hydrogen and acetate production, and methane formation from H_2 plus CO_2, and sulfate reduction with hydrogen. The calculations were based on the assumption that the concentrations of ethanol, acetate, propionate, butyrate, and benzoate were 1 mM each and that of bicarbonate was 100 mM and that SO_4^{2-} and HS^- were present in equimolar concentrations and that the partial pressure of methane was 0.5 atm. using the equation: $\Delta G' = \Delta G^{\circ\prime} + 5.71 \log (C^c \cdot D^d / A^a \cdot B^b)$, for a hypothetical reaction according to aA + bB → cC + dD. The stoichiometry of the reactions is given in Table 9.3.

uct in these oxidations, is relatively small compared to the influence of hydrogen on $\Delta G'$. Only in the oxidation of benzoate, which yields 3 mol of acetate per mole of benzoate, the acetate concentration significantly influences the change in free energy.

The concurrent drop in $\Delta G'$ for hydrogen-consuming organisms and rise in $\Delta G'$ for hydrogen-producing organisms with decreasing partial pressures of hydrogen

TABLE 9.3 Equations and free energy changes for reactions involving anaerobic oxidation of ethanol, propionate, butyrate, and benzoate to acetate or acetate and HCO_3^-, with protons, HCO_3^-, or SO_4^{2-} serving as electron acceptor, by ethanol-, propionate-, butyrate-, or benzoate-catabolizing bacteria in pure culture or in co-culture with H_2-utilizing methanogens or sulfate reducers under standard conditions[a]

Equation	$\Delta G°'$ (kJ/reaction)
I. Proton-reducing (H_2-producing) acetogenic bacteria	
(1) $CH_3CH_2OH + H_2O \rightarrow CH_3COO^- + H^+ + 2H_2$	+9.6
(2) $CH_3CH_2COO^- + 3H_2O \rightarrow CH_3COO^- + H^+ + 3H_2 + HCO_3^-$	+76.1
(3) $CH_3CH_2CH_2COO^- + 2H_2O \rightarrow 2CH_3COO^- + H^+ + 2H_2$	+48.1
(4) $C_7H_5O_2^- + 7H_2O \rightarrow 3CH_3COO^- + 3H^+ + 3H_2 + HCO_3^-$	+53
II. H_2 using methanogens and sulfate reducers	
(5) $4H_2 + HCO_3^- + H^+ \rightarrow CH_4 + 3H_2O$	−135.6
(6) $4H_2 + SO_4^{2-} + H^+ \rightarrow HS^- + 4H_2O$	−151.9

$\Delta G°'$ of Co-cultures with

Substrate	H_2-utilizing methanogen		H_2-utilizing sulfate reducer	
	kJ/mol Substrate	kJ/mol CH_4	kJ/mol Substrate	kJ/mol HS^-
Ethanol	−58.2	−116.4	−66.4	−132.7
Propionate	−25.6	−34.1	−37.8	−50.4
Butyrate	−19.7	−39.4	−27.9	−55.7
Benzoate	−10.7	−14.3	−22.9	−30.6

Source: Data from Thauer et al. (112) or calculated from data therein.
[a] H_2 and CH_4 in the gaseous state; all other compounds in aqueous solution at 1 mol per kilogram activity; 25°C.

TABLE 9.4 Free energy changes for reactions involving anaerobic oxidation of ethanol, propionate, butyrate, and benzoate to acetate or acetate and HCO_3^-, with protons, HCO_3^- or SO_4^{2-} serving as electron acceptor, by ethanol-, prionate-, butyrate-, or benzoate-catabolizing bacteria in co-culture with H_2-utilizing methanogens or sulfate reducers under nonstandard conditions

	$\Delta G'$ of Co-cultures with							
	H_2-utilizing methanogen				H_2-utilizing sulfate reducer			
	kJ/mol Substrate		kJ/mol CH_4		kJ/mol Substrate		kJ/mol HS^-	
Substrate Acetate[a] (mM):	10	1	10	1	10	1	10	1
Ethanol	−58.2	−58.2	−116.4	−116.4	−66.4	−66.4	−132.7	−132.7
Propionate	−25.6	−25.6	−34.1	−34.1	−37.8	−37.8	−50.4	−50.4
Butyrate	−31.0	−36.7	−62.0	−73.4	−39.3	−45.0	−78.6	−90.0
Benzoate	−33.3	−44.7	−44.4	−59.6	−45.7	−57.1	−60.9	−76.1

[a]Acetate and substrate concentrations as indicated. Other conditions as defined in Table 9.3.

(Fig. 9.3) implies that hydrogen concentrations in syntrophic associations of hydrogen-producing and hydrogen-consuming organisms must be within certain limits. Outside a certain well-defined traject of hydrogen concentrations, either the hydrogen-producing or the hydrogen-consuming reactions become endergonic. As the coupling of hydrogen oxidation to sulfate reduction yields more energy than the coupling of hydrogen oxidation to carbonate reduction, this traject is wider when hydrogen is consumed by a sulfate reducer. This opens the possibility that the hydrogen concentration in a co-culture with a hydrogenotrophic sulfate-reducing bacterium is lower than when a methanogen acts as hydrogen scavenger. In theory this will make it possible that hydrogenotrophic sulfate reducers make more energy available for the syntroph, thereby allowing a higher growth rate and a larger growth yield. Indeed, hydrogen-producing syntrophs in co-culture with a hydrogen-consuming sulfate reducer have higher growth rates than in co-culture with a methanogen as hydrogen scavenger. (Table 9.2). In this respect an interesting observation was reported recently that *Syntrophus buswellii* has a higher growth rate when co-cultured with *Wolinella succinogenes* and fumarate as the electron acceptor than when the oxidation of benzoate is coupled to sulfate reduction (79). The reduction of fumarate to succinate yields 86 kJ per mole of hydrogen, while the reduction of sulfate to sulfide yields 38 kJ per mole of hydrogen under standard conditions (112).

The V_{max} for butyrate conversion by co-cultures of *Syntrophomonas wolfei* with *Methanospirillum hungatei* was 350 μmol $\cdot$ g^{-1} $\cdot$ min^{-1} (128a; McInerney personal communication, 1986), corresponding to a rate of methane production of 175 μmol $\cdot$ g^{-1} $\cdot$ min^{-1}. Direct cell counts indicated that *Syntrophomonas wolfei* made up about 30% of the population versus *Methanospirillum hungatei* 70%. Thus at about 1 mmol $\cdot$ butyrate $\cdot$ g^{-1} $\cdot$ min^{-1} the apparent V_{max} of the butyrate degrader was significantly higher than the apparent V_{max} of the methanogen (122 μmol CH$_4$ $\cdot$ g^{-1} $\cdot$ min^{-1}). The term apparent V_{max} is used here to indicate that their values are indicative only for the test conditions, especially the hydrogen concentrations under which they were obtained. As hydrogen inhibits butyrate degradation, it is highly likely that the apparent V_{max} of the butyrate degrader would have been higher at lower hydrogen concentrations, and conversely, that the V_{max} of the methanogen would have been higher under hydrogen-saturated conditions.

The first thermophilic syntrophic co-culture dependent on interspecies hydrogen transfer was reported recently by Zinder and Koch (144). It metabolizes acetate via a mechanism in which acetate is first oxidized to carbon dioxide and hydrogen by one organism, while hydrogen is subsequently used by a second organism to reduce carbon dioxide to methane. This mechanism was originally proposed for a single organism by Van Niel in 1936 (7). Such a acetotrophic co-culture does not yet have a mesophilic counterpart. It is also unique in that it was directly isolated as a methanogenic co-culture, without a detour via hydrogenotrophic sulfate reduction. Preliminary results indicated that acetate is the mandatory substrate for the acetate oxidizer. Thermodynamical calculations corroborate that this organism is highly dependent on a low partial pressure of hydrogen if acetate oxidation is to be energy yielding. The effects of co-culturing this organism with a thermophilic hydrogen-consuming sulfate reducer have not yet been studied. The growth yield of the hydrogenotrophic methanogen (that could not catabolize acetate and resembled *Methanobacterium thermoautotrophicum*) on H$_2$–CO$_2$ was 2.0 g dry weight per mole of methane formed, and the yield of the co-culture grown on acetate was 2.7 g dry weight per mole of methane formed. This suggests that 25% of the co-culture consisted of the acetate oxidizer, although it is not excluded that the yield of the hydrogenotrophs decreases at the lower hydrogen concentrations which must prevail in the co-culture to make the acetate oxidation exergonic. The addition of excess hydrogenotrophic methanogens did not affect the rate of methanogenesis, suggesting that acetate oxidation was the rate-limiting step in methanogenesis from acetate in this co-culture. The growth rate of the co-culture at 60°C (the optimal temperature for growth and methanogenesis, which were coupled) was about 0.42 to 0.55 day^{-1} on acetate, while the growth rate of the methanogen on H$_2$–CO$_2$ was 12.8 day^{-1}. Interestingly, the acetate oxidizer could not grow as a hydrogen-oxidizing acetogen, suggesting that this organism is not simply an acetogenic organism which catabolizes acetate backwards (see Section 9.3 on acetogenic hydrogenations).

Hardly any information is presently available on the biochemical pathways in obligate proton-reducing bacteria, as pure cultures are not yet available. Tracer studies with enrichment cultures that converted propionate to methane and acetate (58) showed that the ^{14}C-labeled carboxyl group exclusively appeared in carbon dioxide, whereas the accumulated acetate was practically free of labeled carbon. Both [2-^{14}C]- and [3-^{14}C]propionate lead to the production of radioactive acetate. The methyl group and the carboxyl group of the acetate produced were equally labeled, regardless of whether [2-^{14}C]- or [3-^{14}C]propionate had been used. These observations suggest that propionate was degraded through a randomizing pathway.

The anaerobic oxidation of butyrate in *S. wolfei* has been postulated to go via β-oxidation (75), but detailed labeling studies are lacking, so this assumption remains to be proved. There is, however, some evidence available: The observation that the conversion of normal monocarboxylic, saturated four- to eight-carbon fatty acids yields H_2 and acetate, or H_2 and acetate plus propionate, and the observation that isoheptanoate was converted to hydrogen and acetate plus isovalerate, are consistent with a β-oxidation mechanism for the degradation of fatty acids by *Syntrophomonas wolfei*. Selective lysing of cells of *Syntrophomonas wolfei*, mass cultured in the presence of *Methanospirillum hungatei*, has recently allowed McInerney and co-workers (128a) to demonstrate the presence of enzyme activities consistent with a β-oxidation pathway for butyrate degradation. This paper (128a) also reports that *S. wolfei* can be grown in pure culture on crotonate, a breakthrough in the study of this hitherto "obligate syntrophe."

Older tracer experiments by Stadtman and Barker (105) with an enrichment culture called "*Methanobacterium suboxydans*" already suggested that β-oxidation was involved in the conversion of valeric acid to acetate and propionate and the oxidation of butyrate to acetate plus methane. The major thermodynamic barrier to hydrogen production in anaerobic β-oxidation-like mechanisms is the oxidation of the saturated to the unsaturated acid, as can be seen by comparing the free energy changes for the oxidation of butyric, crotonic, and β-hydroxybutyric acids to acetate and hydrogen, which are -48.1, -26.9, and -35.1 kJ·mol^{-1}, respectively (132). Comparative biochemistry suggests that this oxidation step is catalyzed by a flavin-dependent enzyme (132).

For the conversion of benzoate by *Syntrophus buswellii*, a reductive cleavage of the benzene ring followed by a mechanism similar to β-oxidation as proposed for the catabolism of fatty acids by *Syntrophomonas wolfei* has been suggested (80). The absence of mono- and dicarboxylic acids (chain length C_3–C_7) in the co-cultures of *Syntrophus buswellii* together with the absence of growth on these compounds is consistent with the existence of such a mechanism.

Results with various enrichment cultures indicate that different pathways may be operative in different benzoate-degrading consortia. Shlomi et al. (103) obtained considerable evidence that the aromatic ring is cleaved after the conversion of benzoate to 2-oxo-cyclohexanecarboxylate (Fig. 9.4*A*), resulting in a pathway via pimelate, caproate, and butyrate, which were all regular components of the culture fluid. Fina et al. (40, 55), on the other hand, suggested the pathway of Fig. 9.4*B* for their enrichment. These authors identified cyclohexanecarboxylate and 1-cyclohexene-1-carboxylate as intermediates, and obtained evidence that the conversion of benzoate goes via propionate. Interestingly, the consortium of Ferry and Wolfe (39) would not utilize 1-cyclohexene-1-carboxylate to produce methane, while Fina's consortium converted this compound without a lag. This discrepancy is probably due to different populations and suggests that novel benzoate degraders are waiting to be isolated. The anaerobic metabolism of aromatic compounds de-

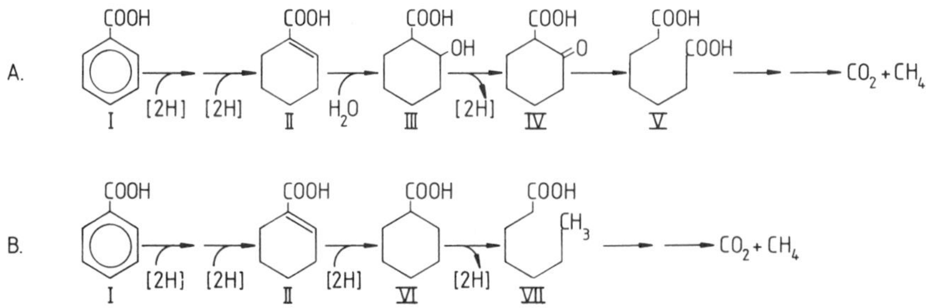

Figure 9.4 Proposed pathways for the methanogenic conversion of benzoic acid by various enrichment cultures. Compounds involved are benzoic acid (I), 1-cyclohexene-1-carboxylic acid (II), 2-hydroxycyclohexane carboxylic acid (III), 2-oxocyclohexane carboxylic acid (IV), pimelic acid (V), cyclohexane carboxylic acid (VI), and heptanoic acid (VII). (*A*) After Shlomi et al. (103); (*B*) after Keith et al. (55).

serves special interest because aromatic molecules are widespread in nature as constituents of phenolic plant products and aromatic amino acids, and appear in large amounts in lignin, from which they may be released to the terrestrial and aquatic environment by the activities of wood-rotting fungi.

9.2.2 Facultative Proton-Reducing Bacteria

Fermentative organisms may be induced to act as proton reducers by virtue of the mechanism of interspecies hydrogen transfer (129). In the presence of hydrogenotrophs, fermentative organisms shift their metabolism to the formation of more oxidized end products plus hydrogen. Therefore, these organisms can be labeled as facultative proton reducers. The first experiments to illustrate this principle were reported by the group of Wolin (51), in a paper by Ianotti et al., where these authors also introduced the term "interspecies hydrogen transfer." In their experiments, *Ruminococcus albus*, an organism that ferments glucose to acetate, ethanol, CO_2, and H_2 (Fig. 9.5*A*), was coupled to *Vibrio* (*Wolinella*) *succinogenes*, an organism that can only reduce fumarate to succinate. Interspecies hydrogen transfer occurred when this co-culture was grown on glucose in the presence of fumarate: The electron flow was shifted toward hydrogen formation and fermentation of glucose toward ethanol stopped (Fig. 9.5*B*).

Analogously, co-culture experiments of a variety of fermentative organisms with hydrogenotrophic methanogens have been shown to display interspecies hydrogen transfer. Experiments with *Selenomonas ruminantium* showed that an organism that does not produce (significant amounts of) hydrogen in monoculture, from glucose, but only lactate, acetate, and propionate may shift its metabolism toward hydrogen production in the presence of a hydrogen consumer (24, 95). As various fermentative organisms did not produce (significant amounts of) hydrogen in monoculture, these organisms can be regarded as facultative proton reducers. It is, however, likely that such organisms, even in monoculture, produce some hy-

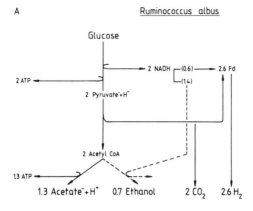

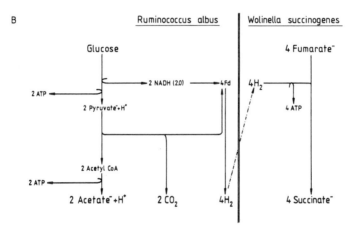

Figure 9.5 Effect of interspecies hydrogen transfer on electron flow and energy gain for glucose catabolism by *Ruminococcus albus*. (*A*) Part of the acetylcoenzyme A is used as an electron sink at the expense of its ATP yielding conversion to acetate. (*B*) The presence of *Wolinella succinogenes* plus a suitable hydrogen acceptor specific for *W. succinogenes* results in the diversion of the electron flow away from acetylcoenzyme A reduction, thereby saving acetylcoenzyme A for its ATP yielding conversion to acetate.

drogen, as the hydrogen concentration seems the most likely trigger for the regulation of hydrogen formation.

In theory, the energy that becomes available to the fermentative organism per mole substrate converted increases when electrons can be disposed of by the reduction of protons. Production of hydrogen removes, for example, the necessity to use acetyl-CoA for electron disposal in fermentations of glucose or ethanol (Fig. 9.5), but it has yet to be shown to what extent fermentative organisms have the ability to translate these advantages into measurable gains in growth yield. In the literature on interspecies H_2 transfer generally, only information is given on the extent of the metabolic shift and not on the effects of this shift on growth. This is probably due to the fact that the experiments generally were performed in batch culture, where the growth rates of the hydrogenotrophic and the hydrogenogenic organisms differed greatly. Chung (25) was the first to give numerical data on the effect of interspecies hydrogen transfer on the growth yield. From his results an increase of 30% can be calculated for the growth yield of *Clostridium cellobioparum* on glucose, as a result of interspecies hydrogen transfer.

More recently, Traore et al. (115) reported that *Desulfovibrio vulgaris* growth yields on pyruvate were 3.5 times higher in the presence of *M. barkeri* as hydrogen acceptor than when hydrogen accumulated, so by removing hydrogen from the medium, hydrogen utilizers may increase biosynthetic efficiency in their partners. Interspecies hydrogen transfer also resulted in an increase of the growth efficiency of *D. vulgaris* on lactate as compared with growth yields of *D. vulgaris* with sulfate as hydrogen acceptor. The experiments with *D. vulgaris* were performed in a chemostat, an approach that allows a meaningful evaluation of the growth parameters thus obtained.

Interspecies hydrogen transfer alters the conversion of hexoses to acetate and hydrogen in such a way that the hexose converting organism can maximalize its energy gain by shifting its metabolism to the formation of hydrogen due to the activity of the hydrogen-consuming organism (130). No bacterium is known, however, that can mediate the degradation of glucose completely to CO_2 and H_2, although theoretically it is possible to construct a food chain that operates via this route by including the recently described syntrophic acetate (144) degrading coculture.

Figure 9.6 illustrates that below partial pressures of hydrogen of about 10^{-4} atm, the energy yield of a glucose-converting organism is favorably influenced when it shifts its catabolism to the (exclusive) formation of hydrogen plus carbon dioxide. Theoretically, it would thus be possible, and at low hydrogen concentrations it would even be advantageous for a hexose degrader to convert hexoses completely to hydrogen plus carbon dioxide.

In summary, interspecies hydrogen transfer causes (a) a shift in the flow of electrons in fermentative organisms (and thereby also in the flow of carbon and electrons through anaerobic ecosystems) toward the formation of H_2 plus more oxidized products, (b) a theoretical energy gain for facultative proton reducing fermentative organisms, and (c) a distinct increase in biosynthetic efficiency in hydrogen-producing organisms.

9.2.3 Energy Conservation in Acetogenic Dehydrogenations

As all living cells, acetogenic bacteria are endowed with a system that transforms chemical energy into biologically useful energy and that utilizes the latter to perform work. The universal molecular carrier for biological energy is the ATP system, which stoichiometrically couples catabolism (112). Dependent on the intra-

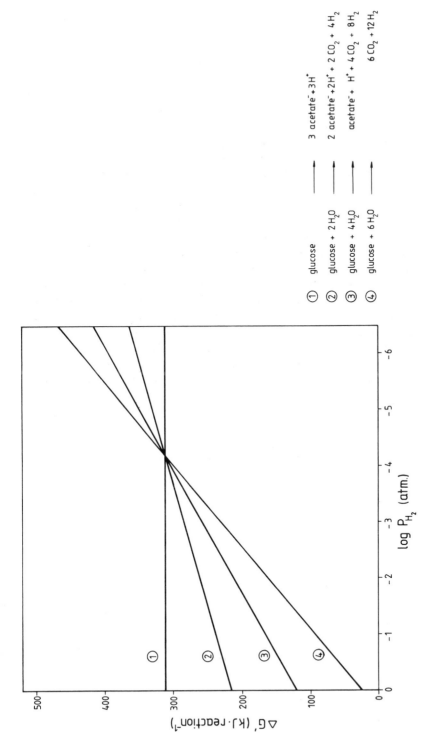

Figure 9.6 The influence of hydrogen and carbon dioxide formation on the change in free energy of glucose conversion at various partial pressures of hydrogen.

① glucose → 3 acetate⁻ + 3H⁺

② glucose + 2 H₂O → 2 acetate⁻ + 2H⁺ + 2 CO₂ + 4 H₂

③ glucose + 4 H₂O → acetate⁻ + H⁺ + 4 CO₂ + 8 H₂

④ glucose + 6 H₂O → 6 CO₂ + 12 H₂

cellular pH, between -42 and -50 kJ is required for the synthesis of 1 mol of ATP from ADP and P_i in anaerobic bacteria (92). A catabolic process under physiological conditions must therefore be associated with a free energy change of -42 to -50 kJ $\cdot$ mol^{-1} if it is to be coupled directly with phosphorylation. Via electron transport phosphorylation, catabolic reactions can be coupled with the formation of various amounts of ATP, depending on the amount of protons required per mole of ATP formed and on the free energy change of the catabolic reaction under physiological conditions. A fixed number of protons required for the formation of 1 ATP leads to a quantization of the energy transfer. Variability in the number of protons required to form 1 ATP opens up the possibility of a highly efficient energy conservation but might impede the possibility to regulate such an energy metabolism, as reactions at equilibrium cannot be regulated.

The energy metabolism of organisms is either linear or branched (Fig. 9.7). A linear catabolism generally results in a constant energy (ATP) gain per mole substrate converted. This is sometimes denominated as constant thermodynamic efficiency (112), defined as 100 times the ratio between amount of energy (ATP) conserved and amount of free energy available from catabolism. Since the free energy from catabolism changes with varying substrate and product concentrations, thermodynamic efficiency may, indeed, vary. In a branched catabolism, each branch may lead to a different ATP gain, so the ATP gained per mole of substrate converted may vary with the relative rates of the two partial processes.

Catabolism in obligate proton-reducing bacteria such as *S. wolinii* and *S. buswellii* on propionate and benzoate, respectively, resembles the scheme for linear catabolism presented in Fig. 9.8 to a certain extent. The amount of ATP gained per mole substrate converted cannot be varied by directing the flow of electrons

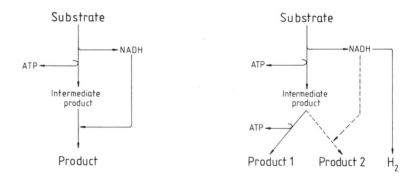

Linear Catabolism   Branched Catabolism

Figure 9.7 Comparison of a linear catabolism having a constant ATP gain per mole of substrate converted with a branched catabolism having a variable ATP gain. Modified from Thauer et al. (112).

Substrate

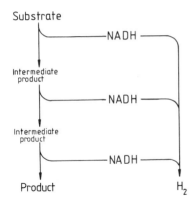

Figure 9.8 Hypothetical scheme for the linear catabolism of propionate and benzoate in obligate proton-reducing bacteria, illustrating that the amount of ATP gained per mole of substrate converted cannot be varied by directing the flow of electrons from one acceptor to another.

from one acceptor to another. These bacteria dump their electrons per definition in the form of H_2. Whether the amount of ATP gained per mole substrate converted in these bacteria varies according to the free energy change of the reaction remains to be elucidated. Table 9.4 shows that these reactions are generally not exergonic enough under physiological conditions to be coupled with ATP formation via a substrate-level phosphorylation at the expense of -42 to -50 kJ $\cdot$ mol^{-1}. Therefore, energy conservation probably proceeds via electron transport phosphorylation. The free energy changes listed in Table 9.4 must be partitioned among the hydrogen-producing and hydrogen-consuming microorganisms: The actual free energy change available to either species is a fraction of these ΔG values. For propionate-oxidizing cultures, assuming similar thermodynamic efficiencies and a ratio between hydrogen producer and hydrogen consumer of $1:10$ to $1:20$ (Koch, personal communication), this would imply that propionate oxidizers can conserve at most between 1 and 3 kJ per mole of propionate converted. Assuming a cost of -42 to -50 kJ $\cdot$ mol^{-1} ATP, this suggests that 14 to 50 molecules of propionate must be oxidized to gain 1 ATP. This suggests that energy can be conserved in rather small "quants."

Thus it is evident that electron transport can in principle effect the net formation of fractional numbers of moles of ATP per 2e$^-$, but it is not clear how chemotrophic anaerobes can couple redox processes with the net synthesis of less than 1 mol of ATP per mole of substrate oxidized. Thauer and Morris (113) have speculated that acetate formation from butyrate in *S. wolfei* proceeds via butyryl-CoA, crotonyl-CoA, β-hydroxybutyryl-CoA, acetoacetyl-CoA, and two acetyl-CoA, and hence results in the net synthesis by means of substrate-level phosphorylation of 1 mol of ATP per mole of butyrate consumed under *in vivo* conditions. In this pathway the hydrolysis of $\frac{2}{3}$ mol of ATP is required to drive the oxidation of butyryl-CoA to crotonyl-CoA with protons as the electron acceptor, resulting in a net gain of $\frac{1}{3}$ mol of ATP per mole of butyrate. Recent work (128a) demonstrates that *S. wolfei* has indeed many of the enzyme activities of the proposed pathway. The mechanism of energy conservation via substrate-level phosphorylation is still prominently present in this scheme, and it is not clear how the energy liberated with the hydrolysis of $\frac{2}{3}$ mol of ATP is coupled to the oxidation of butyryl-CoA. A straightforward coupling between electron transport and energy conservation is more appealing, and should not be dismissed before an alternative has been proved. A nice mechanism for energy generation in these fastidious anaerobes would be a mechanism similar to the sodium-gradient-driven energy conservation in *Propionigenium modestum* (47). *P. modestum* grows rapidly ($t_d = 2.5$ to 4.5 h) on the decarboxylation of succinate to propionate (98). This reaction yields only 22 kJ per reaction, and it was shown that the sodium gradient was the sole mechanism to conserve energy in this exciting organism (47).

Another, indirect argument against the postulated involvement of substrate-level phosphorylation in anaerobic butyrate oxidation is the inability of Reddy et al. (90, 91) to show ATP or acetyl phosphate formation from ethanol by the S organism. These authors proposed (90, 91) that the oxidation of ethanol to acetate proceeds via acetaldehyde, where the energy-yielding step is the oxidation of acetaldehyde to acetate, this by analogy to known reactions in *Clostridium kluyveri* (23). However, the oxidation of ethanol to acetate yields only 36 kJ $\cdot$ mol^{-1} at an hydrogen partial pressure as low as 10^{-4} atm. (112), thus a straightforward coupling between substrate-level phosphorylation and ethanol oxidation was precluded, and an indirect coupling was not detected.

Before starting with a discussion of the ecological aspects of acetogenic dehydrogenations, it should be stressed that thermodynamics determine the amount of free energy that is available from a reaction, and that the amount of available free energy is independent of the route. Thus microorganisms cannot beat thermodynamics by transferring part of the reducing equivalents via, for example, formate rather than via hydrogen. The syntrophic partnerships are deadlocked in a zero-sum situation with respect to free available energy. There may, however, be kinetic advantages to modes of interspecies reductant transfer other than via the renowned but largely unproven route through hydrogen.

9.2.4 Ecology of Acetogenic Dehydrogenations

Acetogenic dehydrogenating organisms are the vital link between hydrolysis/acidogenesis and methanogenesis is anaerobic ecosystems. Acetogenesis provides the two main substrates for the terminal step in the methanogenic conversion of organic material. Both hydrogen and acetate can be converted directly into membrane by methanogenic organisms. Interspecies hydrogen transfer with methanogens as hydrogen consumers makes acetogenic dehydrogenations possible for both obligate and facultative proton-reducing bacteria. The latter group consists of fermentative organisms, and also of those sulfate-reducing bacteria that can act as hydrogen producers when confronted with potential hydrogenogenic substrates (21, 72, 115) under sulfate-limited conditions in the presence of a hydrogen-consuming organism, and further contains organisms that are able to perform acetogenic dehydrogenations with substrates like ethanol (i.e., obligate proton reductions) but are, as well, able to grow on other substrates. Examples of the latter subgroup are

Thermoanaerobium brockii (12), *Desulfovibrio desulfuricans* (21), *Desulfovibrio vulgaris* (21), and the S-organism (22).

Acetogenic dehydrogenations have been assessed in various ecosystems. Kaspar and Wuhrmann (54) showed that about 60% of the flow of hydrogen in digesting sewage sludge was influenced by increasing the partial pressure of hydrogen 0.5 atm and concluded that about 50% of the flow of electrons in sewage sludge goes via the oxidation of H_2 via pyridinenucleotide/ferredoxin. A part of this flow obviously went via fermentative organisms that made use of their ability to form hydrogen via this route at reduced hydrogen concentrations. Estimates from turnover rates showed that 15% of the methane formed was from propionate via acetate and hydrogen. This value for sewage sludge agrees well with direct estimates from turnover rates in cattle waste digestors, where about 23% of the methane formed comes from propionate (15%) and butyrate (8%) (70).

The specificity of the propionate oxidizing and the benzoate oxidizing syntrophs now available in co-culture for, respectively, propionate (13) and benzoate (80) indicates that the hydrogenogenic conversions of these compounds in ecosystems such as anaerobic digestor sludge and anaerobic aquatic sediments, where they have their habitat, is significant. Thus interspecies hydrogen transfer plays a dual role in these habitats both by preventing the formation of propionate and butyrate, and by making syntrophic oxidation of these acids possible. In the presence of sulfate these compounds can be metabolized by sulfate-reducing bacteria (see Chapter 10). Sulfate-reducing bacteria which may utilize these compounds in the presence of sulfate have not been successfully grown (yet?) by substituting a hydrogen-oxidizing methanogen for sulfate (122) and hence may not compete for this niche. Chemostat experiments with propionate-converting methanogenic enrichments indicate that various propionate-degrading subpopulations can be distinguished. Depending on the dilution rate imposed, a population with a high μ_{max} and a high K_s for propionate or a low μ_{max} and a low K_s for propionate developed (46). It is interesting that the growth rate of the "high-K_s population" was significantly higher than the growth rate of *S. wolinii*, even though this organism was enriched in batch culture previously to its "isolation," a technique which generally results in the dominancy of organisms that growth rapidly at high substrate concentrations.

Methanospirillum hungatei is the most frequently observed hydrogenotrophic methanogen in propionate- and butyrate-degrading consortia, enriched from sewage sludge and sediments (13, 76). This has been ascribed to its high affinity for H_2 at low hydrogen concentrations (92). In butyrate-converting enrichments from bovine rumen fluid *Methanosarcina* rather than *Methanobrevibacter ruminantium*, the dominant hydrogenotrophic methanogen in the rumen, was the dominant hydrogenotrophic methanogen (76), possibly because of its ability to use H_2 and acetate simultaneously for methanogenesis at low growth rates. The hydrogen-scavenging ability of *Methanosarcina* in the absence of acetate is lower than this ability of *Methanospirillum* (92). It should be pointed out that *Methanospirillum* and *Methanothrix* were absent in the bovine rumen fluid from which these enrichments were started (76). Degradation of longer-chain fatty acids and acetate does not normally occur to a significant extent in the rumen, probably because the mean residence time in this system is too short to allow the development of the slow-growing bacteria that convert these compounds. In addition to *Methanospirillum* spp., other hydrogenotrophic methanogens are also able to keep the partial pressure of H_2 low enough to allow growth of propionate oxidizers (Fig. 9.9). The enrichment called "*Methanobacterium propionicum*" (105) contained no *Methanospirillum*-like organisms, and in our own methanogenic propionate-degrading enrichments a small fluorescent coc was the hydrogenotrophic methanogen. These observations stress that it is important to realize that the source of the original inoculum in combination with the enrichment technique employed determines which organism(s) will become dominant, and that it is very tricky to make general statements on growth rates, kinetics, and energetics of this new group of organisms. The interpretation of such data in an ecological context is even more hazardous. At present it seems justified to conclude only that the obligate proton-reducing process play an important role in various ecosystems. The processes are partially catalyzed by obligate proton-reducing bacteria, but another part of these processes may be catalyzed by "facultative" proton-reducing bacteria (i.e., bacteria that perform reactions in which proton reduction is obligate, but that are also able to catalyze reactions in which proton reduction is not obligate). These organisms are obligate syntrophic only for the catalyzation of proton-reducing reactions. To what extent these organisms act as proton reducers in (their) natural environments is not known. They may be important in environments where the supply of substrates for proton-reducing reactions is fluctuating or where they can oxidize their compounds mixotrophically.

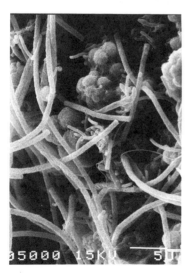

Figure 9.9 Scanning electron micrograph of an enrichment of a propionate-degrading methanogenic consortium. Courtesy M. Koch and A. J. B. Zehnder.

The novel thermophilic acetotrophic co-culture isolated by Zinder and Koch (144) indicates that Van Niels' proposed pathway of methanogenic acetate conversion via carbon dioxide and hydrogen is indeed possible in anaerobic ecosystems, with the annotation that more than one organism is involved in such a route. Methanogenic monocultures have been shown to carry out methanogenesis from acetate by an acetoclastic reaction (136; see Chapters 12 and 13), and for methanogenic environments such as sewage sludge digestors it was shown that 70% of the methane formed was derived from the methyl group of acetate (104). Therefore, it is doubtful if methane formation from acetate via H_2–CO_2 is of importance in a mesophilic system, although there is some evidence for methanogenic acetate oxidation in anoxic marine sediments (26, 94, 121).

The existence of an obligate acetotrophic syntrophic co-culture and pure cultures of organisms catalyzing the identical overall reaction (143) opens up the possibility of comparing the physiology and energetics of these different bacterial strategies to convert compounds with a low energy content. Figure 9.10 illustrates the fact that the overall free energy change is not influenced by the strategy chosen. Therefore, it is tempting to speculate that the acetate oxidizer optimizes the overall reaction by specializing in the rate-limiting step, without spending biosynthetic energy for the catalyzation of subsequent steps. A sufficiently low partial pressure of hydrogen would preclude a significant loss of available free energy per mole of acetate converted. Thus the co-culture might be able to achieve a higher specific activity than that of the monoculture. In mixed systems the hydrogenotrophic methanogens would, furthermore, benefit from hydrogen produced by other nonacetotrophic organisms. Note that in this context methanogens are virtually "mixotrophic" by using hydrogen liberated from a range of substrates they cannot convert on their own.

Interspecies hydrogen transfer is probably also the physiological background of the symbiosis of methanogens with anaerobic ciliates (Fig. 9.11). Episymbiontic as well as endosymbiontic interactions of methanogens with ciliates have been described (107, 116, 117). In the rumen of sheep, the best studied environment in this respect, hydrogen is released by both ciliates and bacteria. The relative strength of these hydrogen sources may vary during the process of digestion of plant material, and the hydrogen-consuming methanogens change place in response to such alterations. Consequently, the extent of the episymbiontic association varies as a function of the feeding regime and the nature of the food source of the sheep (107). Endosymbiontic interactions between methanogens and ciliates have been observed for sapropelic systems such as mud and paddy fields. In such systems large spatial variations may occur in the availability of substrate, and therefore endosymbiosis may be the optimal solution for both organisms: the protozoa graze for food and carry their electron scavengers, which show some resemblance with mitochondria, in the form of methanogens with them, an alliance that is beneficial for both. The frequent observation of dividing bacteria in close association with a microbody-like organelle indicates that the methanogens (identified by fluorescence microscopy and subsequent isolation as *Methanobacterium formicicum*) were

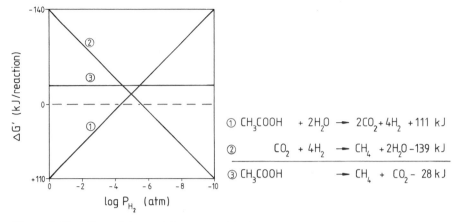

Figure 9.10 Effect of the partial pressure of hydrogen (p_{H_2}) on the change in free energy ($\Delta G'$) for the reactions involved in the syntrophic oxidation of acetate to methane and carbon dioxide. Calculated with $\Delta G' = \Delta G^{\circ\prime} + 6.5 \log$ (products/substrates).

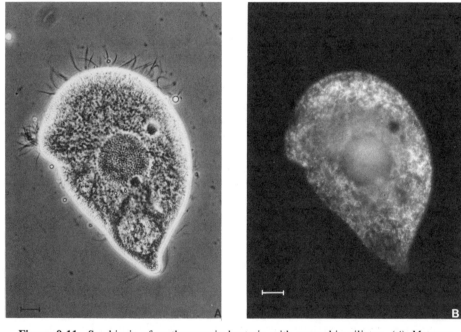

Figure 9.11 Symbiosis of methanogenic bacteria with anaerobic ciliates. (*A*) *Metopus striatus* from anoxic freshwater mud (sapropel). The round bodies near the middle of the cell are the macro- and micronucleus and a food vacuole. Phase contrast. (*B*) Same cell; by means of epifluorescence microscopy the endosymbiotic *Methanobacterium formicicum* is visible due to the specific fluorescent cell constituents coenzymes F_{420} and F_{342}. Because of the small depth of field, only a small part of the bacterial population is in focus. Bar represents 10 μm. (*C*) Long chains of episymbiotic bacteria are attached to the cell surface of *Eudiplodinium maggii* from the rumen of a sheep; bright field. (*D*) Same cell. Fluorescence microscopy identifies the bacteria as methanogens (probably *Methanobrevibacter ruminantium*). The foggy background fluorescence results from plant fibers that were engulfed by the ciliate. Bar represents 10 μm. All photomicrographs courtesy of C. K. Stumm.

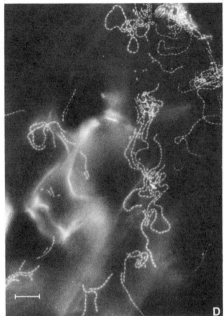

Figure 9.11 (*Continued*)

actively functioning as hydrogen scavengers in the proximity of organelles with a hydrogenosome-like function in one of these protozoa: *Metopus striatus* (117). The absence of mitochondria in *Metopus* cells suggests that these hydrogenosomes play the role of energetic organelles, which convert pyruvate to hydrogen, carbon dioxide, and acetate under anaerobic conditions (119). These endosymbiontic associations may be the source of a significant portion of the total amount of methane produced in sediments, as a rough estimation of the numbers of endosymbiontic and free-living methanogens suggests that most methanogens occur as endosymbionts in such environments (116).

These comparisons were made for hydrogenotrophic methanogens. Acetotrophic methanogens are rarely encountered in nonmarine sediments, so it has been speculated that such organisms could be identical to endosymbiotic bacteria in protozoa (116). In fact, the phase-contrast micrograph of endosymbiontic bacteria presented in the paper of Van Bruggen et al. (116) shows a nonfluorescing bacterium that closely resembles the acetoclastic methanogen *Methanothrix soehngenii* (136). This suggests that ciliates can be included in the physiological group of acetogenic organisms.

9.3 ACETOGENIC HYDROGENATIONS

Apart from the acetogenic hydrogenations described in the preceding section, acetogenesis can also proceed via acetate-yielding hydrogenations. In the homoacetate fermentation of hexoses, these two processes are coupled:

$$C_6H_{12}O_6 + H_2O \longrightarrow CH_3COCOOH + CH_3COOH + CO_2 + 6H$$
$$CH_3COCOOH + CO_2 + 6H \longrightarrow 2CH_3COOH + H_2O$$

$$C_6H_{12}O_6 \longrightarrow 3CH_3COOH$$

Thus homoacetate fermentations are characterized by the formation of acetate as the sole end product.

Acetogenic hydrogenation reactions occur not only with endogenous substrates but also in "fermentations" where carbon dioxide and hydrogen as exogenous substrates are coupled to form acetate in an energy-yielding (catabolic) reaction:

$$2CO_2 + 4H_2 \rightarrow CH_3COOH + 2H_2O$$

Both types of acetogenesis are discussed in the following sections.

9.3.1 Homoacetogens on Multicarbon Compounds

Homoacetogenic metabolism was discovered in 1942 by Fontaine (41), who found that *Clostridium thermoaceticum* ferments glucose with the formation of ~ 3 mol of acetate per mole of glucose, and that one of the acetates is formed by reduction of CO_2. It thus was postulated that an autotrophic type of synthesis accompanies the homoacetate fermentation of glucose, and this idea provoked extensive research into the mechanism behind this pathway. Barker and Kamen (10) showed with the use of labeled CO_2 that the homoacetogenic fermentation of glucose can be viewed as a partial oxidation of glucose to acetic acid and CO_2 accompanied by the reduction of 2 mol of carbon dioxide to acetic acid. They presumed that the oxidation of glucose goes via pyruvate as an intermediate. Their results excluded a mechanism in which glucose was first completely oxidized to carbon dioxide. From their results it was not clear whether carbon dioxide was an obligate intermediate for both the methyl and the carboxyl group of the third acetate. Data obtained from fermentations of a variety of labeled glucose by *C. thermoaceticum* revealed that the labeling of the acetate was in accord with the Embden–Meyerhof pathway; that is, [1- or 6-^{14}C]glucose gave methyl-labeled acetate and [2-^{14}C]glucose gave carboxyl-labeled acetate (133). Synthesis of acetate totally from carbon dioxide was suggested by Wood (134), who showed that part of the acetate formed from glucose in the presence of [^{13}C]carbon dioxide was two mass units heavier than unlabeled acetate. He argued that actual discrepancies with straightforward theoretical predictions on the distribution of label over the carbon atoms of acetate could be rationalized by exchange reactions between gaseous carbon dioxide and the carboxyl group of acetate.

Schulman et al. (100), however, presented evidence that the carboxylation of the methyl group occurred via transcarboxylation with pyruvate, and not by direct fixation of carbon dioxide. The requirement for pyruvate or other α-keto acids in the carboxylation of the methyl group could not be removed by ATP and an electron-generating system. The first step in the formation of this methyl group of acetate from carbon dioxide is the reduction of carbon dioxide to formate. Lentz and Wood (62) found that [^{14}C] formate was a better precursor of the methyl group of acetate than $^{14}CO_2$ and that [^{14}C] formate is selectively incorporated into the methyl group of acetate. The production of the NADP-linked formate dehydrogenase enzyme in *C. thermoaceticum* depends on the presence of selenite, tungstate, and molybdate in the growth medium. The role of these elements in formate dehydrogenases has recently been reviewed (66).

An important observation was reported in 1964 as Poston et al. (86) showed that the methyl group of methyl B$_{12}$ is converted to the methyl group of acetate in the presence of pyruvate. In analogy to the involvement of methyltetrahydrofolate in the methylation of homocysteine to methionine, it was shown that the methyl group of methyltetrahydrofolate is a good precursor of the methyl group of acetate (43). Formate was subsequently shown to be converted to the methyl group of methyltetrahydrofolate (4). Ljungdahl et al. have purified and characterized the involved enzymes, which all required folate intermediates as electron carriers (4).

Thus the outlines of the pathways involved in homoacetate fermentation of hexoses seemed to have been elucidated, although many details of the mechanism of the synthesis of acetate were still unclear. Authoritative reviews have been presented recently (65, 135). In one of them, Ljungdahl and Wood have proposed a tentative scheme for the conversion of glucose to acetate in *C. thermoaceticum* (65). Two acetates are formed by an Embden–Meyerhoff type of cleavage of glucose, while the third acetate originates from the two pyruvate–carboxyl moieties. The methyl group of acetate is formed via carbon dioxide and its subsequent reduction through free formate and tetrahydrofolate derivatives, and is finally carried over the corrinoid one-carbon carriers. The carboxyl group of acetate arises from the carboxyl group of the second pyruvate "directly" without the formation of free CO_2.

A modification of this scheme was proposed subsequently, as Wood et al. (135) presented a scheme where the reduction of CO_2 to formate (by formate dehydrogenase) and the conversion of formate, tetrahydrofolate, and ATP to formyl tetrahydrofolate and ADP (by formyl tetrahydrofolate synthetase) were not indicated. Although it has been demonstrated that *C. thermoaceticum* contains the enzymes necessary to convert carbon dioxide to formate, it was considered likely that carbon dioxide is converted to a formyl intermediate [H—COOH] by carbon monoxide dehydrogenase, and that formyl tetrahydrofolate is formed directly from it. This conversion would be attractive since it might occur without involvement of ATP and thus would conserve energy. Hu et al. (48) furthermore proposed that carbon monoxide dehydrogenase *in vivo* may serve to incorporate a bound formyl derivative formed from carbon dioxide or C$_1$ of pyruvate into acetate. They have purified a carbon monoxide dehydrogenase containing enzyme system from cells of *C. thermoautotrophicum*, which was shown to convert ^{14}CO, [$^{12}CH_3$] tetrahydrofolate, and CoA to acetyl-CoA labeled in the C$_1$ position of the acetate (note that in this system carbon monoxide could serve as the carboxyl donor in place of pyruvate). Diekert et al. (28), however, showed that *C. thermoaceticum* growing on glucose produced significant amounts of carbon monoxide and obtained strong evidence in inhibition experiments with cyanide that carbon monoxide, probably in a bound form, is an intermediate in carbon dioxide reduction to acetate, as incorporation of carbon monoxide into C_1 of acetate was not affected by cyanide, while cyanide inactivates CO dehydrogenase of acetogenic bacteria (31). Furthermore, this is evidence that carbon monoxide dehydrogenase served to reduce CO_2 to CO rather than to incorporate the carbonyl into the C_1 of acetate. Simultaneously, Pezacka and Wood (85) showed that in the presence of hydrogenase, an enzyme that had been demonstrated to be present in *C. thermoaceticum* (32), H_2 was capable of replacing pyruvate as the electron donor for generation of reduced ferredoxin and for conversion of CO_2 to CO by CO dehydrogenase. Prior to this finding, Kerby and Zeikus (57) had already shown that *C. thermoaceticum* has the ability to grow with CO or CO_2–H_2 as energy sources, thereby demonstrating that *C. thermoaceticum* is a facultative autotroph. Based on these data the present concepts of the mechanism of synthesis of acetate by *C. thermoaceticum* are outlined in Fig. 9.12.

The current research into the pathways of the homoacetate fermentation has received a strong impetus from the discovery that carbon monoxide was oxidized and served as an electron donor during growth of various *Clostridia* (31, 42), and the subsequent establishment of growth of acetogens on CO (69, 101). All ace-

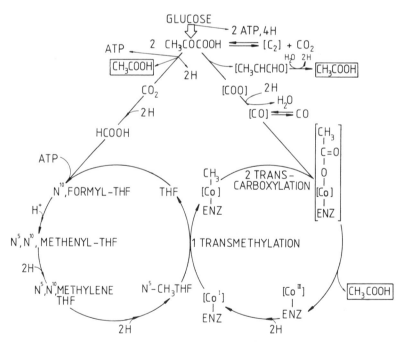

Figure 9.12 Outline of the present concepts of the homoacetogenic pathway in *Clostridium thermoautotrophicum* growing on glucose. Co, corrinoid; enz, enzyme; THF, tetrahydrofolate. Modified from Ljungdahl and Wood (65).

togenic bacteria tested contain carbon monoxide dehydrogenase with high specific activity, but its role is not completely understood. Nickel is a component of carbon monoxide dehydrogenase in *C. thermoaceticum* (88). The nickel component, as well as the (4Fe-4S) iron-sulfur clusters, is redox active (87). Molybdenum, present in CO dehydrogenase of *Pseudomonas* (77), is absent from the CO dehydrogenase of acetogenic bacteria (89).

In their original paper describing the isolation of *Clostridium thermoaceticum* and the discovery that at least 85% of the carbon from glucose goes to acetic acid (about 5% goes into cells), Fontaine et al. (41) did put forward the question of whether glucose is cleaved such as to yield a 1-carbon compound that is produced and then reabsorbed. After more than 40 years of extensive research, this question still cannot be answered unambiguously. Especially, (isotopic) exchange reactions between different carbon-containing compounds (intermediates?) have prevented clear-cut answers until now. The present availability of microcomputers should make it possible, however, to model these exchange reactions in such a way that the carbon flow during homoacetate fermentation of hexoses becomes fully understood. Furthermore, nuclear magnetic resonance studies of whole cells and the enzyme systems involved might prove to be a useful tool in this area of biological research.

The question of Fontaine et al. can however be answered to a certain extent: The recent finding of Diekert et al. (28) that *C. thermoaceticum* excretes carbon monoxide during growth on glucose answers their question at least partially. Furthermore, the findings of Kerby and Zeikus that *C. thermoaceticum* can be grown on H_2–CO_2 (57) suggested that during growth on glucose these compounds may be excreted, too (in minor quantities), as most cellular processes can be viewed as equilibrium processes. Indeed, Martin et al. (71) subsequently reported evolution of hydrogen by *C. thermoaceticum* during growth on dextrose. Obviously, this capacity may be of importance to the cell in its native environment, where it may benefit from interspecies hydrogen transfer in a manner similar to the way that growth of *Acetobacterium woodii* is affected by interspecies hydrogen transfer when cultivated in the presence of hydrogenotrophic methanogens (128). Martin et al. (71) also noticed that the culture conditions may greatly affect the evolution of hydrogen. Therefore, it seems of importance to give more attention to this aspect and to perform future studies with cells cultivated under well-defined conditions in, for example, chemostat cultures.

9.3.2 Homoacetogens on One-Carbon Compounds

Wieringa isolated *Clostridium aceticum* and demonstrated for the first time that bacteria can grow on H_2–CO_2 with concomitant acetate production (125, 126). After the nutritional requirements of *C. aceticum* had been studied in more detail (53), the organism was thought to be lost since 1948. Efforts in various laboratories to reisolate it were not successful (3, 37, 38) until Adamse, in the same laboratory where Wieringa worked in 1936, isolated a new strain (of this species) in 1980 (1). Simultaneously a spore preparation of the original strain was found in the culture collection of H. A. Barker. These spores could be revived, and the original (?) *C. aceticum* has been redescribed by Braun et al. (18). Meanwhile a new non-spore forming species, *Acetobacterium woodii*, was described (6), which was at that time the only organism in pure culture that was able to grow at the expense of acetate formation from H_2 and CO_2. Various chemolithotrophic acetate-producing species have been isolated since then (Table 9.5). Note that most of these organisms were enriched from their natural environment (usually, anaerobic mud) prior to their actual isolation. Only *Acetobacterium wieringae* was isolated directly from its natural environment by streaking out dilutions in flat agar bottles (17). It is not clear if these organisms function as acetate-forming hydrogen consumers in their natural environment, especially as none of these organisms is specialized in the conversion of hydrogen and carbon dioxide to acetate. All bacteria listed in Table 9.5 are able to ferment sugars to acetic acid. Thus it did not come as a surprise that *C. thermoaceticum*, too, possesses the capacity to grow on H_2–CO_2 with concomitant acetate production (57).

Currently, only one hydrogenotrophic acetogen is known which does not possess the capacity to grow with hexoses, but this *Acetobacterium* possesses the capacity to grow with lactate as the energy source (60).

In this context it is also of interest that *Acetobacterium woodii* in co-culture with a hydrogen-consuming methanogen converted fructose to 2 mol of acetate and 1 mol of methane and 1 mol of carbon dioxide (128), via interspecies hydrogen transfer, instead of to 3 mol of acetate as it does in pure culture. The growth yield of *A. woodii* in co-culture increased, showing that more energy could be gained by disposal of electrons via H_2 at low hydrogen concentrations than via the formation of acetate. This finding suggests that by forming methane from hydrogen plus carbon dioxide, methanogens will be able to outcompete acetate formers for hydrogen. Unfortunately, the growth yields for *A. woodii* in these (128) experiments were unusually low, suggesting a nutrient limitation that may have influenced the outcome of the experiments.

Hydrogenase was shown to be constitutive in *A. woodii* (16). In media low in bicarbonate (less than 0.04%) hydrogen was evolved during growth on hexoses, and a hydrogen atmosphere completely inhibited growth. Under a H_2–CO_2 atmo-

TABLE 9.5 Acetogenic bacteria isolated with H_2–CO_2

Organism	Method	Source	Reference
Acetobacterium wieringae	Flat agar bottles	Sludge digestor	17
Acetobacterium woodii	Enrichment on H_2–CO_2	Mud	6
Acetogenium kivui	Enrichment on H_2–CO_2	Mud	61
Clostridium aceticum	Enrichment on H_2–CO_2	Mud	126
Clostridium aceticum	Enrichment on H_2–CO_2	Mud	1
Clostridium thermoaceticum	Enrichment on H_2–CO_2 plus an inhibitor of methane formation	Mud	124

sphere, rapid growth occurred and $H_2 + CO_2$ and the organic substrate were used simultaneously. *Clostridium aceticum*, however uses only the sugar in the presence of H_2-CO_2 (16). This illustrates that it is dangerous to base general conclusions for the physiology of metabolic groups on results obtained with only one organism, and underlines the contention that processes in ecosystems cannot always be extrapolated from results obtained with only one microorganism.

The following ranges for the kinetic parameters of mesophilic acetogens growing on H_2-CO_2 have been obtained (2, 68): $\mu = 0.4$ to 1.9 day^{-1}; $Y = 0.7$ to 1.7 g dry weight $\cdot$ mol^{-1} H_2; $q = 700$ μmol $H_2 \cdot$ g$^{-1} \cdot$ min^{-1}. These values are in the same range as those reported for methanogens growing on H_2-CO_2 (137), probably their main competitors for hydrogen in sulfate-depleted environments. Thermodynamics predict that the amount of free energy that becomes available per mole of H_2 converted is about equal for these two groups under standard conditions:

$$4H_2 + HCO_3^- + H^+ \longrightarrow CH_4 + 3H_2O$$
$$\Delta G°' = -135.6 \text{ kJ/reaction}$$
$$4H_2 + 2HCO_3^- + H^+ \longrightarrow CH_3COO^- + 4H_2O$$
$$\Delta G°' = -104.6 \text{ kJ/reaction}$$

At a partial pressure for hydrogen of 10^{-4} atm and a partial pressure of 0.5 atm for methane, a bicarbonate concentration of 100 mM, and an acetate concentration of 10 mM, the Gibbs free energy drops to -40 kJ per reaction for the methanogenic conversion and to -13 kJ per reaction for the acetogenic consumption of H_2, respectively (i.e., a ratio of about $3:1$).

The kinetic parameters given above have been obtained in batch culture. A meaningful evaluation of the energy conservation and the selection mechanism that govern the outcome of competition between methanogens and acetogens has to wait until more data are available on μ, K_s, K_m, Y, and q of these organisms, preferably obtained in chemostat experiments.

Apart from acetogenesis from H_2-CO_2, CO, and the homoacetate fermentation of hexoses, acetogenesis can also occur from methanol. Under specific conditions in artificial anaerobic environments, this process may play an important role in the total carbon flow. For an anaerobic reactor treating a methanolic wastewater, a homoacetogenic *Clostridium*, strain CV-AAI, was isolated using methanol as the substrate (2). This organism is the first acetogen isolated directly on methanol. Bacteria that are able to convert methanol to acetate include *Acetobacterium woodii* (5), although in this organism the utilization of methanol as an energy source appears to be either strain specific or requires cultural adaptation (140), *Clostridium thermoautotrophicum* (124), *Eubacterium limosum* (102), and *Butyribacterium methylotrophicum* (141). None of these bacteria was isolated directly with methanol as the substrate. *B. methylotrophicum* was isolated after previous enrichment with methanol plus acetate (141). Methanol as the only carbon and energy source resulted in this case in enrichment of *Methanosarcina*, a methanogen able to convert methanol to methane plus carbon dioxide. Interestingly, the conversion of methanol to acetate in the reactor from which *Clostridium* CV-AAI (2) was isolated resulted in a medium that consisted of methanol plus acetate. The prominent acetogenesis in this methanogenic treatment system was related to the concentration of bicarbonate and the availability of trace elements (63, 64).

Studies into the biochemical pathways of acetate formation from methanol have been initiated only recently. The first results indicate that there may be analogies with the primary events in the formation of methane from methanol (118). The first step in methanol conversion in *Eubacterium limosum* is catalyzed by methanol: cobalamin methyltransferase.

Due to the absence of a model organism since the disappearance of *Clostridium aceticum*, studies on the autotrophic reduction of CO_2 with hydrogen to acetate were only resumed as *A. woodii* was shown to grow on this reaction (6). High levels of enzymes of the tetrahydrofolate pathway and a pyruvate-dependent formation of acetate from methyl-B_{12} and methyl tetrahydrofolate (109) indicated that acetate synthesis from carbon dioxide by *A. woodii* may occur via mechanisms similar to those in homoacetate fermenting *Clostridia*. Therefore, the demonstration that CO was oxidized and served as an electron donor during the growth of various *Clostridia* (42) prompted research into the role of carbon monoxide in hydrogen-consuming acetogenic bacteria as well. Diekert and Ritter (30) showed that carbon monoxide is fixed into the carboxyl group of acetate during growth of *A. woodii* on H_2–CO_2. The low rate of carbon monoxide incorporation into acetate suggested that a bound carbonyl rather than free CO is formed as an intermediate from carbon dioxide. With the use of cyanide, a selective inhibitor of CO-dehydrogenase supporting evidence was collected for the hypothesis that acetate is synthesized from $2CO_2$ according to the scheme of Fig. 9.13 (28). This tentative scheme assumes that carbon monoxide in a bound form is an intermediate formed from carbon dioxide in a reaction catalyzed by carbon monoxide dehydrogenase. Although ^{14}CO and $^{14}CO_2$ incorporation by cell suspensions of *A. woodii* was strictly dependent on pyruvate, $^{14}CO_2$ and ^{14}CO were not incorporated into pyruvate. These findings argue against a role of pyruvate as the intermediate carboxyl donor for acetate formation in *A. woodii*. The role of pyruvate in total synthesis of acetate from CO_2 or CO is not yet understood (28).

Autotrophic acetogenic bacteria may use the total synthesis of acetyl-CoA from CO_2 via methyl tetrahydrofolate not only for energy metabolism but also for cell carbon fixation (36).

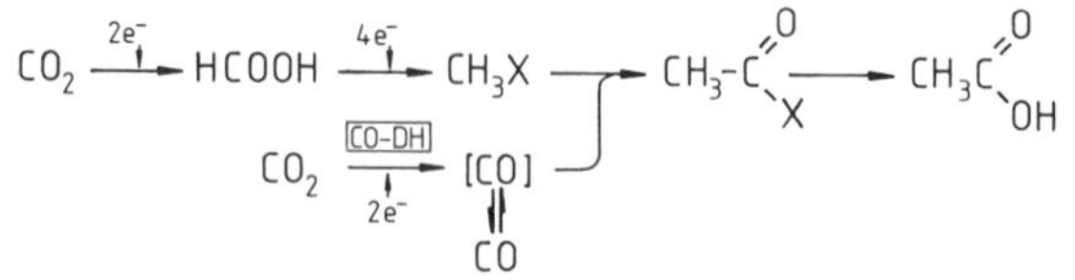

Figure 9.13 Tentative scheme for the total synthesis of acetate from carbon dioxide in *Acetobacterium woodii*. CO-DH, CO dehydrogenase. After Diekert et al. (28).

9.3.3 Energy Conservation in Acetogenic Hydrogenations

The finding that acetogenic bacteria can grow on hydrogen plus carbon dioxide demonstrates conclusively that a net formation of ATP coupled to the oxidation of hydrogen to acetate in homoacetogenic fermentations is possible, and explains the otherwise inexplicable high growth yields of *C. thermoaceticum* on glucose (4). This contention contradicts the conclusion of Winter and Wolfe that "*de novo*" acetate synthesis in *A. woodii* would be an energy-consuming process (128).

Apparent similarities between the oxidation of hydrogen with either sulfate or carbon dioxide have tempted Odom and Peck (84) to speculate that the generation of a proton gradient in acetogenic bacteria may occur by means of vectorial electron transfer in these organisms and that ATP formation would involve a hydrogen-cycling mechanism similar to the mechanism occurring in *Desulfovibrio gigas* (83). In the postulated scheme (Fig. 9.14), fructose is fermented to two acetate, two carbon dioxide, and four hydrogen molecules in the cytoplasma. The hydrogen diffuses across the cytoplasmic membrane and may continue its way by diffusion through the cell wall out of the cell where it may be consumed by other organisms. In the hydrogen-cycling model, hydrogen is oxidized by a periplasmatic hydrogenase to yield eight electrons and eight protons. The electrons are transferred back across the cytoplasmatic membrane, where they are used to reduce carbon dioxide to acetate with the consumption of eight protons. This leaves eight protons at the external surface of the membrane, thereby establishing the proton gradient and the potential energy that can be used for the synthesis of ATP.

This mechanism requires that there is no production or consumption of protons

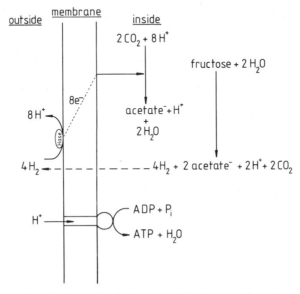

Figure 9.14 Postulated scheme for hydrogen cycling in *Acetobacterium woodii*. Hase, hydrogenase. After Odom and Peck (83, 84).

on the same side of the membrane and that there is no direct electron transfer bypassing the hydrogenase-like systems. The uptake hydrogenase in *A. woodii* might be a nickel-containing enzyme, as in the absence of nickel hydrogen formation from protons rather than acetate formation from carbon dioxide was used as the electron sink by this bacterium: Cultures of *A. woodii* growing heterotrophically and deprived of nickel evolved hydrogen according to (29)

$$1 \text{ fructose} + 2H_2O \longrightarrow 2 \text{ acetate}^- + 1 \text{ formate}^- + 3H_2 + 1CO_2 + 3H^+$$

However, alternative explanations for the forced hydrogen evolution by nickel-deprived *A. woodii* are possible, such as a lack of carbon monoxide dehydrogenase by nickel-deprived cells which may result in the prevention of the formation of the carboxyl group of "*de novo* acetate." Furthermore, the hydrogen-cycle model is not compatible with observations made with *Clostridium aceticum* that hydrogen does not affect heterotrophic growth of this organism on fructose (16). There is, however, no reason to assume that the energy conservation in all acetogenic bacteria goes via the same mechanism. Tanner et al. recently showed that the acetogens are a phylogenetically diverse group with sometimes only a distant relatedness between the different species (108). And, more convincing, Waber and Wood (120) have shown that there is a pathway for the synthesis of acetate from CO_2 that does not involve corrinoids. In this so-called glycine decarboxylase pathway, methylene tetrahydrofolate, ammonia and CO_2 form glycine in a reaction that is catalyzed by glycine decarboxylase. Glycine is then converted to serine by accepting a second methylene tetrahydrofolate; the serine is converted to pyruvate by serine dehydrase and subsequently metabolized in *Clostridium acidiurici* to acetate. However, recent results indicate that this pathway is operative only under selenium-limited conditions. Under non-selenium-limited conditions, this organism [like *Clostridium purinolyticum* and probably *Clostridium cylindrosporum* (33, 34) as well as *Peptococcus glycinophilus* and *Peptococcus magnus* (35)] form acetate from glycine via glycine reductase. The initial steps in this pathway in *Clostridium acidiurici* (120), which give rise to the formation of methylene tetrahydrofolate, are identical to those in *Clostridium thermoaceticum* (65). The occurrence of at least two different pathways for the synthesis of acetate from carbon dioxide suggests that energy may be conserved via different mechanisms in the various species.

9.3.4 Ecology of Acetogenic Hydrogenations

Acetogenic bacteria are typical facultative chemolithotrophs capable of growing with $H_2 + CO_2$ but also with sugars and acids such as glycerate and lactate. The simultaneous utilization of organic and inorganic substrates for energy conservation by facultative chemolithotrophs, as observed for *A. woodii*, may be of importance under conditions of energy limitation, while the versatility of other strains, which do not possess this capacity to grow mixotrophically, may help these organisms in their competition with methanogens and other hydrogen scavengers for hydrogen. As already pointed out, methanogens gain more energy per mole of hydrogen converted than acetogens, so mixotrophy and versatility are probably the factors that secure the presence of acetogens in various ecosystems.

Cell counts for acetogens with H$_2$–CO$_2$ as the substrate in mud samples and sludge ranged, however, in the order of 10^5 per milliliter (i.e., a factor 100 lower than the cell counts for methanogens in these environments) (19). Thus their influence as hydrogen consumers on the flow of carbon and electrons is probably not large in these environments. This assumption is corroborated for sediments of a eutrophic lake by results obtained in studies with ^{14}C tracer techniques by Lovley and Klug (67), who found that hydrogen consumption by hydrogen-oxidizing acetogens was 5% or less of the total hydrogen uptake by acetogens and methanogens. In their studies hydrogen utilization by hydrogen-oxidizing acetogens was probably overestimated since bacteria that do not use hydrogen as an electron donor can also give rise to the production of [^{14}C] acetate from [^{14}C]CO$_2$.

Similarly, acetate formation from CO$_2$ in the mesophilic methanogenic conversion of cattle waste accounted for less than 2% of the total amount of acetate synthesized (70), indicating that acetate formation from carbon dioxide was not in the main path of electron flow in this environment.

Occasionally, however, acetogenesis from one-carbon compounds may be significant. The sediment of a hypereutrophic, mildly acidic (pH 6.2) lake contained 100 times more hydrogenotrophic acetogens than hydrogenotrophic methanogens. In this particular environment methane was produced predominantly from acetate and only 4% was derived from CO$_2$, so these anaerobic conversion processes were dominanted by acetogens. One of the explanations of their apparent ability to maintain low partial pressures of hydrogen may be the ability of acetogens not to evolve hydrogen during the consumption of multicarbon substrates (85a). Another environment where acetate formation plays an important role is in the hindgut of termites, where protozoa dominate acetogenesis from wood polysaccharides. Careful balance studies indicate that acetogenic bacteria constitute important "electron sink" organisms in the hindgut foodweb and are able to outcompete methanogens for H$_2$–CO$_2$ (82). Experiments to isolate the responsible acetogens are currently being carried out (J. A. Breznak, personal communication, 1984).

These examples show that it is not possible to make generalizations as to the importance of acetogens in anaerobic conversion processes. More information on their physiological abilities (mixotrophy, versatility, hydrogen-scavenging abilities), and competition experiments are needed to understand their dominance in some environments and their relative absence in others. It is thereby necessary to include factors other than the energy source, such as pH, temperature, and the presence or absence of trace elements and growth factors.

9.4 FORMATION OF HIGHER FATTY ACIDS FROM C$_2$ COMPOUNDS

9.4.1 Organisms Involved

Besides the homoacetate fermentations discussed in the preceding sections, some organisms have been discovered recently that can perform homoacetate fermen-

tations but are not "true homoacetogens." Under specific conditions, these acetogens are able to ferment saccharides or single-carbon compounds to butyrate.
Butyribacterum methylotrophicum, for example, utilizes acetyl-CoA/acetate as an
electron acceptor when growing on hexoses or methanol. However, this organism
produces essentially no butyrate when grown on carbon monoxide (69) or on H_2–
CO_2 (68). Its close relative *Eubacterium limosum* (*Butyribacterium rettgeri*) (80)
has similar physiological characteristics and is in addition also able to form caproate from C_1 and C_2 compounds (102). It has been claimed that *B. methylotrophicum* in co-culture with *Methanosarcina barkeri* would be able to convert
butyrate to methane via interspecies hydrogen transfer (139). This would imply
the availability in monoculture of a bacterium that can function as an obligate
proton-reducing acetogen. Unfortunately, this claim has never been substantiated.
It remains intriguing why *B. methylotrophicum* is not able to catalyze the oxidation
of butyrate to hydrogen and acetate.

It has been speculated that butyrate in *B. methylotrophicum* is formed from 2
acetyl-CoA (140). An energetically more attractive pathway would involve butyrate formation from acetate plus acetyl-CoA in a cyclic pathway, as shown in Fig.
9.15. Butyrate synthesis may occur through an initial condensation of two molecules of acetyl-CoA to form acetoacetyl-CoA, which may be subsequently reduced
to butyryl-CoA. Butyrate would then be produced by CoA transfer to acetate.
Kerby et al. (56) have presented data that support this pathway. In experiments
with ^{13}C-labeled methanol fed to *B. methylotrophicum* no double ^{13}C-labeled butyrate was found, as would be expected when butyrate was to be reformed exclu-

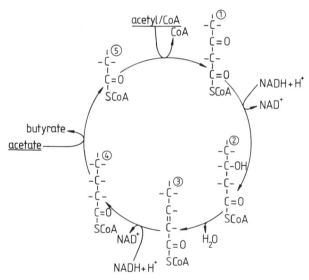

Figure 9.15 Proposed cyclic pathway for the formation of butyrate from acetyl-CoA plus
acetate in *Butyribacterium methylotrophicum*. 1 = acetoacetyl-CoA; 2 = β-OH butyryl-
CoA; 3 = crotonyl-CoA; 4 = butyryl-CoA; 5 = acetyl-CoA.

sively from 2 acetyl-CoA. Dumping electrons by forming butyrate instead of 2 acetate costs 1 ATP per mole of glucose fermented, and therefore this pathway will be operative only at higher acetate concentrations.

B. *methylotrophicum* fermentations display the typical substrate/cell carbon synthesis conversion ratios for an anaerobe growing on hexoses, and cells are a significant product. In elegant calculations by Lynd and Zeikus (68) it was shown that the thermodynamic efficiency of cell synthesis of this organism ranged from 58 to 74% for its various energy sources: 58% on CO, 63% on H_2–CO_2, 74% on glucose, and 76% on methanol. For these calculations the following equation was used:

$$\text{thermodynamic efficiency} = \frac{(\text{mol cells})\,(\Delta G_{ox}^{\circ\prime}\ \text{cells})\times 100}{\Sigma\,(\text{mol substrate})\,(\Delta G_{ox}^{\circ\prime}\ \text{substrate}) - \Sigma\,(\text{mol product})\,(\Delta G_{ox}^{\circ\prime}\ \text{product})}$$

The actual free energy change after 50% substrate utilization ($\Delta G'$) as associated with batch culture and the standard free energy change at pH 7 ($\Delta G^{\circ\prime}$) for the fermentations studied were shown to be about equal; thus this approximation seems a valid approach. The similarity in the thermodynamic efficiency on methanol and on glucose suggests that the number of carbon–carbon bounds in the substrate has a minor effect on the efficiency of cell synthesis in this organism. Due to the different energy contents of the various substrates, a cell synthesis efficiency of 63% on H_2–CO_2 results in a substrate/cell synthesis ratio per carbon of 12%, while the slightly higher thermodynamic efficiency on methanol results in a much higher ratio of 24%. As growth of B. *methylotrophicum* on glucose ($Y = 43$ g of cells/mol glucose) is not more efficient than growth of *Clostridium thermoaceticum* on glucose ($Y = 50$ g of cells/mol glucose) (4), it appears that high thermodynamic efficiencies are prevalent among anaerobic microorganisms. Whether the high yields on glucose are achieved solely by ATP gain via substrate-level phosphorylation, which is theoretically within the realm of possibility (68), remains to be elucidated.

The ability to form butyrate and caproate from C_1 and C_2 compounds calls in mind the well-known *Clostridium kluyveri*. This microorganism has been isolated from an enrichment of *Methanobacillus omelianskii* and forms butyric and caproic acid from ethanol plus acetate (11). The underlying metabolic processes and their regulation have been discussed in detail (112). Other, more recently discovered organisms that ferment ethanol to higher fatty acids are *Desulfobulbus propionicus* (59), *Pelobacter propionicus* (98), and an as yet unnamed organism isolated by Samain et al. (93). D. *propionicus* is one of the ''new'' sulfate reducers described by Widdel and Pfennig (123). It appeared to be an extremely versatile organism that is able to exploit at least 11 different catabolic reactions. In seven of these reactions sulfate is the final electron acceptor, but D. *propionicus* in the absence of sulfate is also able to act as a ''fermentative'' organism (59). Then D. *propionicus* ferments ethanol to propionate and acetate, or produces propionate from ethanol + H_2 + HCO_3^-, or from acetate + H_2 + HCO_3^-, thus in part turning back the propionate oxidation reaction on which it was first isolated. Until these observations on propionate formation, chain elongations of fatty acids had only been described to proceed in steps of C_2 units by reductive condensation of acetate residues, as in the case of *Clostridium kluyveri* (14, 15) or *Butyribacterium* (8, 9, 141). The presence of sulfate is essential for the growth of *D. propionicus* on propionate, as *D. propionicus* is not able to grow on this compound by the mechanism of interspecies hydrogen transfer (122).

The formation of propionate from acetate and other C_2 units deserves interest as a new means of utilizing carbon dioxide as an electron acceptor which is different from the homoacetate fermentation and the methane formation process. It is also of major ecological interest, although its ecological significance *in situ* remains to be studied.

The utilization of bicarbonate as electron acceptor allows a fermentative organism to obtain higher growth yields, as illustrated with various fermentations of butanediol in Table 9.6. *Pelobacter carbinolicus* carries out a disproportionation reaction with 2,3-butanediol, leading to ethanol and acetate as products, whereas *Pelobacter propionicus* ferments butanediol to acetate and propionate with concomitant reduction of 1 mol of bicarbonate per mole of butanediol (96). *Clostridium magnum* ferments butanediol in a homoacetate fermentation (97). This example shows that bacteria can effectively conserve the extra free energy change that comes available when bicarbonate is used as electron acceptor for the formation of fatty acids in either a homoacetate fermentation or a propionate-yielding fermentation.

Fermentation of glucose to propionate is a characteristic of *Propionispira arboris*, a nitrogen-fixing anaerobe isolated from ''wetwood'' trees (99). In the absence of exogenous hydrogen, this organism ferments 1.5 mol of glucose to 2 mol of propionate, 1 mol of acetate, and 1 mol of carbon dioxide. But the addition of exogenous hydrogen during growth of glucose results in an increased propionate production at the expense of acetate formation (114). Thus *P. arboris* features a hydrogen-dependent homopropionate fermentation of glucose. The hydrogenase activity of *P. arboris* is reversible, suggesting that hydrogenase controls the flow of carbon and electrons during growth and propionate production. Under fermentation conditions hydrogenase prevents overreduction of electron carriers, and hydrogen is produced as electron sink, while under conditions of excess exogenous hydrogen, hydrogenase functions as an uptake hydrogenase and donates the electrons for a homopropionate fermentation. The use of exogenous hydrogen by propionate-forming bacteria may be of importance in anaerobic ecosystems, especially under mild acidic conditions or in environments low in sulfate and fixed nitrogen (114). Several other fermentative bacteria, especially from genera found in the rumen, such as *Anaerovibrio*, *Bacteroides*, and *Selenomonas*, are able to consume exogenous hydrogen, but in these organisms hydrogen is generally oxidized with fumarate to succinate (45). Homopropionate fermentations may, however, be of importance in the rumen, as addition of hydrogen to hexoses fed to rumen populations resulted in the formation of propionate as sole fermentation

TABLE 9.6 Equations, free energy changes, and growth yields for reactions involving anaerobic conversion of butanediol

Organism	Reaction	$\Delta G^{o\prime}$ (kJ·mol^{-1})	$\Delta G^{\prime a}$ (kJ·mol^{-1})	Y^b (g·mol^{-1})
Pelobacter carbinolicus	Butanediol + $\frac{1}{2}$H$_2$O → $\frac{1}{2}$ acetate + $1\frac{1}{2}$ ethanol	−38.2	−50.7	3.60
Pelobacter propionicus	Butanediol + bicarbonate → acetate + propionate	−100.4	−107.8	6.25
Clostridium magnum	Butanediol + $1\frac{1}{2}$ bicarbonate → $2\frac{3}{4}$ acetate	−101.7	−105.2	7.05

Source: Data from Refs. 96 and 97.

[a] Calculated for conditions where 50% of the substrate had been depleted.

[b] Y; gram of cells per mole of butanediol converted.

product (132). Nothing has been mentioned (114) on the possible oxidative catabolism of propionate by *Propionispira arboris* in syntrophism with a hydrogen-consuming methanogen, but it is most likely that this possibility has been tested with a negative result.

9.4.2 Biotechnological Applications

It has been proposed (138) that acetogenic bacteria may be useful as agents of chemical and fuel production, for example in the conversion of pyrolysis products (mainly CO), or of methanol to acetate, butyrate, and other longer-chain fatty acids. Product formation can be directed when more fundamental knowledge is gained on such conversions. A promising start was reported by Datta and Ogeltree (27), who found that methanol/bicarbonate ratios during methanol conversion by *Butyribacterium methylotrophicum* determine the formation of butyrate.

9.4.3 Ecological Implications

The formation of higher fatty acids from C_1 and C_2 compounds is a part of anaerobic microbiology that attracts more and more attention. Apart from biotechnological applications (see also Chapter 14), attention should also be focussed on the ecological implications of such conversions. The substrate flow in anaerobic ecosystems is generally depicted in multistep top-down schemes, where carbon–carbon bonds are gradually disrupted, but never newly formed, with the formation of acetate from carbon dioxide as the only exception on this trend in the current schemes. The conversion of ethanol by syntrophic associations such as "*M. omelianskii*" is implicitly assumed to go via acetate plus hydrogen directly to methane in those schemes. The total number of ethanol-degrading organisms in freshwater sediments, however, is on the same order as the number of propionate-forming ethanol degraders, which suggests that propionate formation from ethanol may be important in ethanol conversion in those sediments.

The conversion of ethanol to acetate plus propionate yields 37.3 kJ/mole of ethanol (Table 9.7) and is independent of the partial pressure of hydrogen. The

TABLE 9.7 Reactions involved in methane formation from ethanol

Reaction	Substrates	Products	$\Delta G^{\circ\prime}$ (kJ/reaction)
1	Ethanol + H_2O	Acetate$^-$ + $2H_2$ + H^+	+9.6
2	Ethanol + H_2 + HCO_3^-	Propionate$^-$ + $2H_2O$	−66.5
3	Ethanol + $\frac{2}{3} HCO_3^-$	$\frac{1}{3}$ Acetate$^-$ + $\frac{2}{3}$ propionate$^-$ + $\frac{1}{3} H^+$ + H_2O	−41.1
4	H_2 + $\frac{1}{4} HCO_3^-$	$\frac{1}{4} CH_4$ + $\frac{3}{4} H_2O$	−33.9
5	Propionate$^-$ + $3H_2O$	Acetate$^-$ + $3H_2$ + H^+ + HCO_3^-	+76.1
6	Acetate$^-$ + H_2O	CH_4 + HCO_3^-	−31.0

Source: Data are from Thauer et al. (112) or calculated from data therein.

conversion of ethanol to H_2 plus acetate, and the conversion of ethanol to propionate with the use of H_2, are both dependent on the partial pressure of H_2. The free energy change increases with increasing partial pressures of hydrogen for the formation of propionate, but decreases for the formation of acetate. For $p_{H_2} = 2 \times 10^{-4}$ atm the free energy change per mole of ethanol converted is higher for propionate formation, and below this value the free energy change is higher for acetate formation. Figure 9.16 shows that both processes are exergonic over a wide range of hydrogen concentrations. Thus they may proceed simultaneously, especially as the outcome of competition for ethanol is determined by more factors than the availability of free energy from ethanol alone.

Figure 9.16 also shows that the amount of free energy available from the conversion of H_2 to methane is smaller per mole of hydrogen oxidized to methane than per mole of hydrogen oxidized with ethanol to propionate. This suggests that propionate-forming ethanol converters may be able to scavenge hydrogen very efficiently at extremely low hydrogen partial pressures, where methane formation is endergonic. All known bacteria that are able to form propionate from ethanol are also able to carry out other reactions (i.e., they are not specialists but are versatile and/or able to grow mixotrophic). Such organisms have an advantage over specialist organisms in environments with fluctuating substrate fluxes. Therefore, it is very likely that in such environments, biodegradable organic material is not simply converted along the "fastest" top-down route to volatile end products

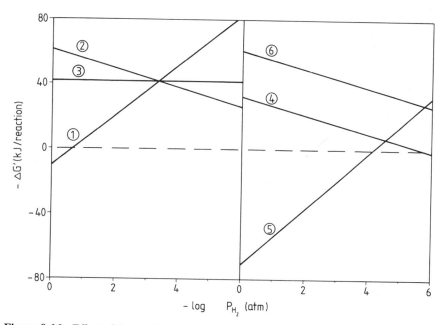

Figure 9.16 Effect of the partial pressure of hydrogen (p_{H_2}) on the change in free energy ($\Delta G'$) for reactions involved in the conversion of ethanol into methane via propionate. (Reaction designations as in Table 9.7).

such as CH_4, CO_2, and H_2S, with the involvement of a minimal number of different metabolic groups of bacteria, but that such conversions go via a larger number of different bacteria than should theoretically be necessary, as nonspecialist bacteria happen to be present and use (part of) their enzymatic machinery to gain energy by forming new carbon–carbon bounds. Such conditions can be expected in various diverse habitats, from industrial waste treatment plants to anaerobic sediments. It is of special importance to note that in such systems bacteria are frequently immobilized or attached, so that once present, they can maintain themselves for a long time if they are able to exploit the environmental conditions.

Formation of propionate instead of acetate also occurs from glutamate by means of interspecies hydrogen transfer between *Acidaminobacter hydrogenoformans* and hydrogenotrophic organisms (106). This is an intriguing example where interspecies hydrogen transfer induces the formation of a more reduced product (propionate) by the hydrogen-donating organism than is formed in the absence of a proton-reducing bacterium (acetate). It is of ecological importance to stress that *Acidaminobacter* was directly isolated from mud, not via enrichments on glutamate in batch cultures, where butyrate-forming bacteria usually become predominant (106).

9.5 CONCLUDING REMARKS

The last decade has brought some significant steps forward in our understanding of acetogenic processes. Ten years ago the existence of obligate proton-reducing acetogens was still only hypothetical, and progress in understanding the pathways involved in homoacetogenic processes was seriously hampered by lack of acetogens able to grow with one-carbon compounds as energy source.

Since then tremendous progress has been made. Various obligate proton-reducing acetogenic bacteria have been obtained in co-culture with hydrogen scavengers, and a number of lithotrophic acetogens has been isolated. While our knowledge of the biochemical processes involved in acetogenic dehydrogenations is still minimal, mainly because of the dogma that biochemical studies must be carried out with pure cultures, our understanding of the physiology of homoacetogenic organisms is increasing rapidly. Various pathways involved in acetogenic processes have been identified and described in much detail. The potential of most, if not all, acetogens to grow on carbon monoxide has given a new impetus to research in this field, and it is now becoming clear that many of the individual steps in lithotrophic acetate formation closely resemble or are even similar to those occurring in the catabolism of methane-forming bacteria, where carbon monoxide (dehydrogenase) is also heavily involved.

As the pathways occurring in homoacetogenic bacteria are now becoming well defined, the time seems ripe to attack questions regarding energy conservation in these organisms. Surprisingly, little is still known on, for example, the occurrence of compounds involved in energy generation via electron transport phosphorylation, let alone on the mechanism of energy conservation in these bacteria. Speculations on this mechanism are frequently based on Y_{ATP} calculations, a parameter that should be used with utmost prudence (111). Going through the literature it is manifest that many authors seem unwilling to accept that energy can be generated via mechanisms other than substrate-level phosphorylation. The influence of product concentrations on the growth yield of lactate-producing organisms suggests that energy conservation is fine tuned to the thermodynamically determined free energy change that accompanies the catalyzation of a biochemical reaction. Microorganisms probably have optimized their ability to convert these, sometimes small changes in free energy into biologically useful energy, and the employment of gradients seems a more sophisticated way to achieve this than the rather crude mechanism of substrate-level phosphorylation. A striking example of this principle is *Propionigenium modestum* (98), which conserves energy via a sodium gradient (47) and thus is able to exploit a very small free energy change of substrate conversion.

Analogous mechanisms may play a role in obligate proton-reducing bacteria, and it is tempting to speculate that the hydrogen concentration in syntrophic associations influences the amount of conserved energy, as reflected in growth yields, of both syntrophic partners.

Interspecies hydrogen transfer has generally been accepted as a major regulator in the flow of carbon and electrons through anaerobic ecosystems. The phenomenon itself has been made clear in many co-culture experiments, but its physiological and bioenergetic implications are still poorly understood. Recent discoveries of the ability of hitherto unknown types of hydrogen-consuming anaerobes to remove hydrogen via the formation of higher fatty acids indicate that the flow of carbon and electron in the environment may not always be as straightforward as was previously assumed. By virtue of their metabolic flexibility and their ability to grow mixotrophically, these new organisms create new ecological niches that would not exist when there was a straightforward selection between two species, based on their growth rate as a function of the concentration of one limiting substrate. Substrate supply in natural ecosystems is subjected to large temporal variations, thus favoring the occurrence of versatile and mixotrophic organisms. Once present these organisms may survive for a long time, even during periods when they have no apparent advantage of their ''special'' capacities, and continue to influence the flow of carbon and electrons in their habitat. This is especially true for sessile (immobilized) bacteria in renowned anaerobic environments such as high-rate wastewater treatment systems and sediments, where ''reversed electron flow'' may occur through the activities of acetogenic bacteria.

ACKNOWLEDGMENTS

I thank Alexander J. B. Zehnder for suggestions and discussions and Claudius K. Stumm and Markus Koch for their willingness to provide micrographs. The artwork of Nees Slotboom and typing skills of Bep Scheffers are gratefully acknowledged.

REFERENCES

1. Adamse, A. D. 1980. New isolation of *Clostridium aceticum* (Wieringa). Antonie Leeuwenhoek J. Microbiol. 46:523–531.

2. Adamse, A. D., and C. T. M. Velzeboer. 1982. Features of a *Clostridium*, strain CV-AA1, an obligatory anaerobic bacterium producing acetic acid from methanol. Antonie Leeuwenhoek J. Microbiol. 48:305–313.

3. Andreesen, J. R., G. Gottschalk, and H. G. Schelgel. 1970. *Clostridium formicoaceticum* nov. spec. Isolation, description, and distinction from *C.aceticum* and *C.thermoaceticum*. Arch. Mikrobiol. 72:154–174.

4. Andreesen, J. R., A. Schaupp, C. Neurauter, A. Brown, and L. G. Ljungdahl. 1973. Fermentation of glucose, fructose and xylose by *Clostridium thermoaceticum*: effect of metals on growth yield, enzymes, and the synthesis of acetate from CO_2. J. Bacteriol. 114:742–751.

5. Bache, R., and N. Pfenning. 1981. Selective isolation of *Acetobacterium woodii* on methoxylated aromatic acids and determination of growth yields. Arch. Microbiol. 130:255–261.

6. Balch, W. E., S. Schoberth, R. S. Tanner, and R. S. Wolfe. 1977. *Acetobacterium*, a new genus of hydrogen oxidizing, carbon-dioxide-reducing, anaerobic bacteria. Int. J. Syst. Bacteriol. 27:355–361.

7. Barker, H. A. 1936. On the biochemistry of the methane fermentation. Arch. Mikrobiol. 7:404–419.

8. Barker, H. A. 1937. The production of caproic and butyric acids by the methane fermentation of ethyl alcohol. Arch. Mikrobiol. 8:416–421.

9. Barker, H. A., and V. Haas. 1944. *Butyribacterium*, a new genus of Gram-positive, non sporulating anaerobic bacteria of intestinal origin. J. Bacteriol. 47:301–305.

10. Barker, H. A., and M. D. Kamen. 1945. Carbon dioxide utilization in the synthesis of acetate by *Clostridium thermoaceticum*. Proc. Natl. Acad. Sci. USA 31:219–225

11. Barker, H. A., and S. M. Taha. 1942. *Clostridium kluyverii*, an organism concerned in the formation of caproic acid from ethyl alcohol. J. Bacteriol. 43:347–363.

12. Ben-Bassat, A., R. Lamed, and J. G. Zeikus. 1981. Ethanol production by thermophilic bacteria: metabolic control of end product formation in *Thermoanaerobium brockii* J. Bacteriol. 146:192–199.

13. Boone, D. R., and M. P. Bryant. Propionate-degrading bacterium, *Syntrophobacter wolinii* sp. nov. gen. nov., from methanogenic ecosystems. Appl. Environ. Microbiol. 40:626–632.

14. Bornstein, B. T., and H. A. Barker. 1948. The nutrition of *Clostridium kluyveri*. J. Bacteriol. 55:223–230.

15. Bornstein, B. T., and H. A. Barker. 1948. The energy metabolism of *Clostridium kluyveri* and the synthesis of fatty acids. J. Biol. Chem. 172:659–669.

16. Braun, K., and G. Gottschalk. 1981. Effect of molecular hydrogen and carbon dioxide on chemo-organotrophic growth of *Acetobacterium woodii* and *Clostridium aceticum*. Arch. Microbiol. 128:294–298.

17. Braun, M., and G. Gottschalk. 1982. *Acetobacterium wieringae* sp. nov., a new species producing acetic acid from molecular hydrogen and carbon dioxide. Zentralbl. Bakteriol. Microbiol. Hyg. 1 Abt. Orig. C3:368–376.

18. Braun, M., F. Mayer, and G. Gottschalk. 1981. *Clostridium aceticum* (Wieringa), a microorganism producing acetic acid from molecular hydrogen and carbon dioxide. Arch. Microbiol. 128:288–293.

19. Braun, M., S. Schoberth, and G. Gottschalk. 1979. Enumeration of bacteria forming acetate from H_2 and CO_2 in anaerobic habitats. Arch. Microbiol. 120:201–204.

20. Bryant, M. P. 1979. Microbial methane production-theoretical aspects. J. Anim. Sci. 48:193–201.

21. Bryant, M. P., L. L. Campbell, C. A. Reddy, and M. R. Crabill. 1977. Growth of *Desulfovibrio* in lactate or ethanol media low in sulfate in association with H_2-utilizing methanogenic bacteria. Appl. Environ. Microbiol. 33:1162–1169.

22. Bryant, M. P., E. A. Wolin, M. J. Wolin, and R. S. Wolfe. 1967. *Methanobacillus omelianskii*, a symbiotic association of two species of bacteria. Arch. Microbiol. 59:20–31.

23. Burton, R. M., and E. R. Stadtman. 1953. The oxidation of acetaldehyde to acetyl coenzyme A. J. Biol. Chem. 202:873–890.

24. Chen, M., and M. J. Wolin. 1977. Influence of CH_4 production by *Methanobacterium ruminantium* on the fermentation of glucose and lactate by *Selenomonos ruminantium*. Appl. Environ. Microbiol. 34:756–759.

25. Chung, K. T. 1976. Inhibitory effects of H_2 on growth of *Clostridium cellobioparum*. Appl. Environ. Microbiol. 31:342–348.

26. Claypool, G. E., and I. R. Kaplan. 1974. The origin and distribution of methane in marine sediments, p. 99–140, in I. R. Kaplan (ed.), Natural gases in marine sediments. Plenum Press, New York.

27. Datta, R., and J. Ogeltree. 1983. Methanol bioconversion by *Butyribacterium methylotrophicum*—batch fermentation yield and kinetics. Biotechnol. Bioeng. 25:991–998.

28. Diekert, G., M. Hansch, and R. Conrad. 1984. Acetate synthesis from 2 CO_2 in acetogenic bacteria: is carbon monoxide an intermediate? Arch. Microbiol. 138:224–228.

29. Diekert, G., and M. Ritter. 1982. Nickel requirement of *Acetobacterium woodii*. J. Bacteriol. 151:1043–1045.

30. Diekert, G., and M. Ritter. 1983. Carbon monoxide fixation into the carboxyl group of acetate during growth of *Acetobacterium woodii* on H_2 and CO_2. FEMS Microbiol. Lett. 17:229–302.

31. Diekert, G. B., and R. K. Thauer. 1978. Carbon monoxide oxidation by *Clostridium thermoaceticum* and *Clostridium formicoaceticum*. J. Bacteriol. 136:597–606.

32. Drake, H. L. 1980. Demonstration of hydrogenase in extracts of the homoacetate-fermenting bacterium *Clostridium theromaceticum*. J. Bacteriol. 150:702–709.

33. Dürre, P., and J. R. Andreesen. 1982. Pathway of carbon dioxide reduction to acetate without a net energy requirement in *Clostridium purinolyticum*. FEMS Microbiol. Lett. 15:51–56.

34. Dürre, P., and J. R. Andreesen. 1983. Purine and glycine metabolism by purinolytic clostridia. J. Bacteriol. 154:192–199.

35. Dürre, P., and R. Spahr, and J. R. Andreesen. 1983. Glycine fermentation via a glycine reductase in *Peptococcus glycinophilus* and *Peptococcus magnus*. Arch. Microbiol. 134:127–135.

36. Eden, G., and G. Fuchs. 1982. Total synthesis of acetyl coenzyme A involved in autotrophic CO_2 fixation in *Acetobacterium woodii*. Arch. Microbiol. 133:66–74.

37. El Ghazzawi, E. 1967. Neuisolierung von *Clostridium aceticum* Wieringa und stoffwechselphysiologische Untersuchhungen. Arch. Mikrobiol. 57:1–19.

38. El Ghazzawi, E., and K. Schmidt. 1967. Hexosen-Umwandlungen in Nährlösungen: Fructose-Verwertung durch *Clostridium aceticum* Wieringa. Zentralbl. Bakteriol. Parasitenkunde, Infektionskrankheiten und Hygiene. Abteilung II. 121:569–575.

39. Ferry, J. G. and R. S. Wolfe. 1976. Anaerobic degradation of benzoate to methane by a microbial consortium. Arch. Microbiol. 107:33–40.

40. Fina, L. R., R. L. Bridges, T. H. Coblentz, and F. F. Roberts. 1978. The anaerobic decomposition of benzoic acid during methane fermentation. III. The fate of carbon four and the identification of propanoic acid. Arch. Microbiol. 118:169–172.

41. Fontaine, F. E., W. H. Peterson, E. McCoy, M. J. Johnson, and G. J. Ritter. 1942. A new type of glucose fermentation by *Clostridium thermoaceticum* n. sp. J. Bacteriol. 43:701–715.

42. Fuchs, G., U. Schnitker, and R. K. Thauer. 1974. Carbon monoxide oxidation by growing cultures of *Clostridium pasteurianum*. Eur. J. Biochem. 49:111–115.

43. Ghambeer, R. K., H. G. Wood, M. Schujman, and L. G. Ljungdahl. 1971. Total synthesis of acetate from CO_2. III. Inhibition by alkylhalides of the synthesis from CO_2, methyltetrahydrofolate, and methyl-B_{12} by *Clostridium thermoaceticum*. Arch. Biochem. Biophys. 143:471–484.

44. Gujer, W., and A. J. B. Zehnder. 1983. Conversion processes in anaerobic digestion. Water Sci. Technol. 15:127–167.

45. Henderson, C. 1980. The influence of extracellular hydrogen on the metabolism of *Bacteroides ruminicola*, *Anaerovibrio lipolytica* and *Selenomonas ruminantium*. J. Gen. Microbiol. 119:485–491.

46. Heyes, R. H., and R. J. Hall. 1983. Kinetics of two subgroups of propionate-using organisms in anaerobic digestion. Appl. Environ. Microbiol. 46:710–715.

47. Hilpert, W., B. Schink, and P. Dimroth. 1984. Life by a new decarboxylation-dependent energy conservation mechanism with Na^+ as coupling ion. EMBO J. 3:1665–1670.

48. Hu, S. I., H. L. Drake, and H. G. Wood. 1982. Synthesis of acetyl coenzyme A from carbon monoxide, methyltetrahydrofolate, and coenzyme A by enzymes from *Clostridium thermoaceticum*. J. Bacteriol. 149:440–448.

49. Hungate, R. E. 1966. The rumen and its microbes. Academic Press, New York.

50. Hungate, R. E. 1967. Hydrogen as an intermediate in the rumen fermentation. Arch. Mikrobiol. 59:158–164.

51. Ianotti, E. L., P. Kafkewitz, M. J. Wolin, and M. P. Bryant. 1973. Glucose fermentation products of *Ruminococcus albus* grown in continuous culture with *Vibrio succinogenes*: changes caused by interspecies transfer of H_2. J. Bacteriol. 114:1231–1240.

52. Johns, A. T., and H. A. Barker. 1960. Methane formation; fermentation of ethanol in the absence of carbon dioxide by *Methanobacillus omelianskii*. J. Bacteriol. 80:837–841.

53. Karlsson, J. L., B. E. Volcani, and H. A. Barker, 1948. The nutritional requirements of *Clostridium aceticum*. J. Bacteriol. 56:781–782.

54. Kaspar, H. F., and K. Wuhrmann. 1978. Kinetic parameters and relative turnovers of some important catabolic reactions in digesting sludge. Appl. Environ. Microbiol. 36:1–7.

55. Keith, C. L., R. L. Bridges, L. R. Fina, K. L. Iverson, and J. A. Cloran. 1978. The anaerobic decomposition of benzoic acid during methane fermentation. IV. Dearomatization of the ring and volatile fatty acids formed on ring rupture. Arch. Microbiol. 118:173–176.

56. Kerby, R., W. Niemczura, and J. G. Zeikus. 1983. Single-carbon catabolism in acetogens: analysis of carbon flow in *Acetobacterium woodii* and *Butyribacterium methylotrophicum* by fermentation and ^{13}C nuclear magnetic resonance measurement. J. Bacteriol. 155:1208–1218.

57. Kerby, R., and J. G. Zeikus. 1983. Growth of *Clostridium thermoaceticum* on H_2/CO_2 or CO as energy source. Curr. Microbiol. 8:27–30.

58. Koch, M., J. Dolfing, K. Wuhrmann, and A. J. B. Zehnder. 1983. Pathways of propionate degradation by enriched methanogenic cultures. Appl. Environ. Microbiol. 45:1411–1414.

58a. Krumholz, L. R., and M. P. Bryant. 1986. *Syntrophococcus sucromutans* sp. nov. gen. nov. uses carbohydrates as electron donors and formate, methoxymonobenzenoids or *Methanobrevibacter* as electron acceptor systems. Arch. Microbiol. 143:313–318.

59. Laanbroek, H. J., T. Abee, and J. L. Voogd. 1982. Alcohol conversions by *Desulfobulbus propionicus* Lindhorst in the presence and absence of sulfate and hydrogen. Arch. Microbiol. 133:178–184.

60. Laanbroek, H. J., H. J. Geerlings, A. A. C. M. Peijnenburg, and J. Siesling. 1983. Competition for L-lactate between *Desulfovibrio*, *Veillonella*, and *Acetobacterium* species isolated from anaerobic intertidal sediments. Microb. Ecol. 9:341–354.

61. Leigh, J. A., F. Mayer, and R. S. Wolfe. 1981. *Acetogenium kivui*, a new thermophilic hydrogen-oxidizing, acetogenic bacterium. Arch. Microbiol. 129:275–280.

62. Lentz, K., and H. G. Wood. 1955. Synthesis of acetate from formate and carbon dioxide by *Clostridium thermoaceticum*. J. Biol. Chem. 215:645–654.

63. Lettinga, G., W. de Zeeuw, and E. Ouborg. 1981. Anaerobic treatment of wastes containing methanol and higher alcohols. Water Res. 15:171–182.

64. Lettinga, G., A. T. van der Geest, S. Hobma, and J. van der Laan. 1979. Anaerobic treatment of methanolic wastes. Water Res. 13:725–737.

65. Ljungdahl, L. G., and H. G. Wood. 1982. Acetate biosynthesis, p. 165–202, in D. Dolphin (ed.), B_{12}, Vol. 2. Wiley, New York.

66. Ljungdahl, L. G. 1979. Formate dehydrogenases: role of molybdenum, tungsten and selenium, p. 463–486, in M. P. Coughlan (ed.), Molybdenum and molybdenum-containing enzymes. Pergamon Press, Elmsford, N.Y.

67. Lovley, D. R., and M. J. Klug. 1983. Methanogenesis from methanol and methylamines and acetogenesis from hydrogen and carbon dioxide in the sediments of a eutrophic lake. Appl. Environ. Microbiol. 45:1310–1315.

68. Lynd, L. H., and J. G. Zeikus. 1983. Metabolism of H_2-CO_2, methanol, and glucose by *Butyribacterium methylotrophicum*, J. Bacteriol. 153:1415–1423.

69. Lynd, L., R. Kerby, and J. G. Zeikus. 1982. Carbon monoxide metabolism of the

methylotrophic acidogen *Butyribacterium methylotrophicum.* J. Bacteriol. 149:255–263.

70. Mackie, R. I., and M. P. Bryant. 1981. Metabolic activity of fatty acid-oxidizing bacteria and the contribution of acetate, propionate, butyrate, and CO_2 to methanogenesis in cattle waste at 40 and 60°C. Appl. Environ. Microbiol. 41:1363–1373.

71. Martin, D. R., L. L. Lundie, R. Kellum, and H. L. Drake. 1983. Carbon monoxide-dependent evolution of hydrogen by the homoacetate-fermenting bacterium *Clostridium thermoaceticum.* Curr. Microbiol. 8:337–340.

72. McInerney, M. J., and M. P. Bryant. 1981. Anaerobic degradation of lactate by syntrophic association of *Methanosarcina barkeri* and *Desulfovibrio* species and effect of H_2 on acetate degradation. Appl. Environ. Microbiol. 41:346–354.

73. McInerney, M. J., M. P. Bryant, and N. Pfennig. 1979. Anaerobic bacterium that degrades fatty acids in syntrophic association with methanogens. Arch. Microbiol. 122:129–135.

74. McInerney, M. J., M. P. Bryant, and D. A. Stafford. 1980. Metabolic stages and energetics of microbial anaerobic digestion, p. 91–98, in D. A. Stafford, B. I. Wheatly, and D. E. Hughes (eds.), Anaerobic digestion. Applied Science Publishers, Barking, Essex, England.

75. McInerney, M. J., M. P. Bryant, R. B. Hespell, and J. W. Costerton. 1981. *Syntrophomonas wolfei* gen. nov. sp. nov., an anaerobic, syntrophic, fatty acid-oxidizing bacterium. Appl. Environ. Microbiol. 41:1029–1039.

76. McInerney, M. J., R. I. Mackie, and M. P. Bryant. 1981. Syntrophic association of a butyrate-degrading bacterium and *Methanosarcina* enriched from bovine rumen fluid. Appl. Environ. Microbiol. 41:826–828.

77. Meyer, O. 1982. Chemical and spectral properties of carbonmonoxide: methylene blue oxidoreductase; the molybdenum-containing iron sulfur flavoprotein from *Pseudomonas carboxydovorans.* J. Biol. Chem. 257:1333–1341.

78. Moore, W., and E. Cato. 1965. Synonomy of *Eubacterium limosum* and *Butyribacterium rettgeri.* Int. Bull. Bacteriol. Nomencl. Taxon. 15:69–80.

79. Mountfort, D. O., W. J. Brulla, L. R. Krumholz, and M. P. Bryant. *Syntrophus buswellii* gen. nov., sp. nov.: a benzoate catabolizer from methanogenic ecosystems. Int. J. Syst. Bacteriol. 34:216–217.

80. Mountfort, D. O., and M. P. Bryant. 1982. Isolation and characterization of an anaerobic syntrophic benzoate-degrading bacterium from sewage sludge. Arch. Microbiol. 133:249–256.

81. Nottingham, P. M., and R. E. Hungate. 1969. Methanogenic fermentation of benzoate. J. Bacteriol. 98:1170–1172.

82. Odelson, D. A., and J. A. Breznak, 1983. Volatile fatty acid production by the hindgut microbiota of xylophagous termites. Appl. Environ. Microbiol. 45:1602–1613.

83. Odom, J. M., and H. D. Peck, Jr. 1981. Hydrogen cycling as a general mechanism for energy coupling in the sulfate-reducing bacteria, *Desulfovibrio* sp. FEMS Microbiol. Lett. 12:47–50.

84. Odom, J. M., and H. D. Peck, Jr. 1984. Hydrogenase, electron-transfer proteins, and energy coupling in the sulfate-reducing bacteria *Desulfovibrio.* Annu. Rev. Microbiol. 38:551–592.

85. Pezacka, E., and H. G. Wood. 1984. The synthesis of acetyl-CoA by *Clostridium*

thermoaceticum from carbon dioxide, hydrogen, coenzyme A and methyltetrahydrofolate. Arch. Microbiol. 137:63–69.

85a. Phelps, T. J., and J. G. Zeikus. 1984. Influence of pH on terminal carbon metabolism in anoxic sediments from a mildly acidic lake. Appl. Environ. Microbiol. 48:1088–1095.

86. Poston, J. M., K. Kuratomi, and E. R. Stadtman. 1964. Methyl-vitamin B_{12} as a source of methylgroups for the synthesis of acetate by cell-free extracts of *Clostridium thermoaceticum*. Ann. N.Y. Acad. Sci. 112:804–806.

87. Ragsdale, S. W., L. G. Ljungdahl, and D. V. DerVartanian. 1982. EPR evidence for nickel-substrate interaction in carbon monoxide dehydrogenase from *Clostridium thermoaceticum*. Biochem. Biophys. Res. Commun. 108:658–663.

88. Ragsdale, S. W., J. E. Clark, L. G. Ljungdahl, L. L. Lundie, and H. L. Drake. 1983. Properties of purified carbon monoxide dehydrogenase from *Clostridium thermoaceticum*, a nickel, iron-sulfur protein. J. Biol. Chem. 258:2364–2369.

89. Ragsdale, S. W., L. G. Ljungdahl, and D. V. DerVartanian. 1983. Isolation of the carbon monoxide dehydrogenase from *Acetobacterium woodii* and comparison of its properties with the *Clostridium aceticum* enzyme. J. Bacteriol. 155:1224–1237.

90. Reddy, C. A., M. P. Bryant, and M. J. Wolin. 1972. Ferredoxin- and nicotinamide adenine dinucleotide-dependent H_2 production from ethanol and formate in extracts of S organism isolated from ''*Methanobacillus omelianskii*''. J. Bacteriol. 110:126–132.

91. Reddy, C. A., M. P. Bryant, and M. J. Wolin. 1972. The ferredoxin-dependent conversion of acetaldehyde to acetate and H_2 in extracts of S organism. J. Bacteriol 110:133–138.

92. Robinson, J. A., and J. M. Tiedje. 1984. Competition between sulfate-reducing and methanogenic bacteria for H_2 under resting and growing conditions. Arch. Microbiol. 137:26–32.

93. Samain, E., G. Albagnac, H. C. Dubourguier, and J. P. Touzel. 1982. Characterization of a new propionic acid bacterium that ferments ethanol and displays a growth factor dependent association with a Gram negative homoacetogen. FEMS Microbiol. Lett. 15:69–74.

94. Sansone, F. J., and C. S. Martens. 1981. Methane production from acetate and associated methane fluxes from coastal sediments. Science 211:707–709.

95. Scheifinger, C. C., B. Linehan, and M. J. Wolin. 1975. H_2 production by *Selenomonas ruminantium* in the absence and presence of methanogenic bacteria. Appl. Microbiol. 29:480–483.

96. Schink, B. 1984. Fermentation of 2,3-butanediol by *Pelobacter carbinolicus* sp. nov. and *Pelobacter propionicus* sp. nov., and evidence for propionate formation from C_2 compounds. Arch. Microbiol. 137:33–41.

97. Schink, B. 1984. *Clostridium magnum* sp. nov., a non-autotrophic homoacetogenic bacterium. Arch. Microbiol. 137:250–255.

98. Schink, B., and N. Pfennig, 1982. *Propionigenium modestum* gen. nov. sp. nov. a new strictly anaerobic, nonsporing bacterium growing on succinate. Arch. Microbiol. 133:209–216.

99. Schink, B., T. E. Thompson, and J. G. Zeikus. 1982. Characterization of *Propion-*

ispira arboris gen. nov. sp. nov., a nitrogen-fixing anaerobe common to wetwoods of living trees. J. Gen. Microbiol. 128:2771–2779.

100. Schulman, M., R. K. Ghambeer, L. G. Ljungdahl, and H. G. Wood. 1973. Total synthesis of acetate from CO_2. VII. Evidence with *Clostridium thermoaceticum* that the carboxyl of acetate is derived from the carboxyl of pyruvate by transcarboxylation and not by fixation of CO_2. J. Biol. Chem. 248:6255–6261.

101. Sharak-Genthner, B. R., and M. P. Bryant. 1982. Growth of *Eubacterium limosum* with carbon monoxide as the energy source. Appl. Environ. Microbiol. 43:70–74.

102. Sharak-Genthner, B. R., C. L. Davis, and M. P. Bryant. 1981. Features of rumen and sewage sludge strains of *Eubacterium limosum*, a methanol- and H_2-CO_2-utilizing species. Appl. Environ. Microbiol. 42:12–19.

102a. Shelton, D. R., and J. M. Tiedje. 1984. Isolation and partial characterization of bacteria in an anaerobic consortium that mineralizes 3-chlorobenzoic acid. Appl. Environ. Microbiol. 48:840–848.

103. Shlomi, E. R., A. Lankhorst, and R. A. Prins. 1978. Methanogenic fermentation of benzoate in an enrichment culture. Microb. Ecol. 4:249–261.

104. Smith, P. H., and R. A. Mah. 1966. Kinetics of acetate metabolism during sludge digestion. Appl. Microbiol. 14:368–371.

105. Stadtman, T. C., and H. A. Barker. 1951. Studies on methane fermentation. III. Tracer experiments on fatty acid oxidation by methane bacteria. J. Bacteriol. 61:67–80.

106. Stams, A. J. M., and T. A. Hansen. 1984. Fermentation of glutamate and other compounds by *Acidaminobacter hydrogenoformans* gen. nov. sp. nov., an obligate anaerobe isolated from black mud. Studies with pure cultures and mixed cultures with sulfate-reducing and methanogenic bacteria. Arch. Microbiol. 137:329–337.

106a. Stieb, M., and B. Schink. 1985. Anaerobic oxidation of fatty acids by *Clostridium bryantii* sp. nov., a sporeforming, obligately syntrophic bacterium. Arch. Microbiol. 140:387–390.

106b. Stieb, M., and B. Schink. 1986. Anaerobic degradation of isovalerate by a defined methanogenic coculture. Arch. Microbiol. 144:291–295.

107. Stumm, C. K., H. J. Gijzen, and G. D. Vogels. 1982. Association of methanogenic bacteria with ovine rumen ciliates. Br. J. Nutr. 47:95–99.

108. Tanner, R. S., E. Stackebrandt, G. E. Fox, R. Gupta, L. J. Magrum, and C. R. Woese. 1982. A phylogenetic analysis of anaerobic eubacteria capable of synthesizing acetate from carbon dioxide. Curr. Microbiol. 7:127–132.

109. Tanner, R. S., R. S. Wolfe, and L. G. Ljungdahl. 1978. Tetrahydrofolate enzyme levels in *Acetobacterium woodii* and their implication in the synthesis of acetate from CO_2. J. Bacteriol. 134:668–670.

110. Tarvin, D., and A. M. Buswell. 1934. The methane fermentation of organic acids and carbohydrates. J. Am. Chem. Soc. 56:1751–1755.

111. Tempest, D. W., and O. M. Neijssel. 1984. The status of Y_{ATP} and maintenance energy as biologically interpretable phenomena. Annu. Rev. Microbiol. 38:459–486.

112. Thauer, R. K., K. Jungermann, and K. Decker. 1977. Energy conservation in chemotrophic anaerobic bacteria. Bacteriol. Rev. 41:100–180.

113. Thauer, R. K., and J. G. Morris. 1984. Metabolism of chemoautotrophic anaerobes. Old views and new aspects, p. 123–168, in D. P. Kelly and N. G. Carr (eds.) The

microbe 1984, Part. 2 (Symp. 100th Meet. Soc. Gen. Microb., Warwick 1984). Cambridge University Press, Cambridge.

114. Thompson, T. E., R. Conrad, and J. G. Zeikus. 1984. Regulation of carbon and electron flow in *Propionispira arboris*: physiological function of hydrogenase and its role in homopropionate formation. FEMS Microbiol. Lett. 22:265–271.

115. Traore, A. S., C. Gaudin, C. E. Hatchikian, J. Le Gall, and J. P. Belaich. 1983. Energetics of growth of a defined mixed culture of *Desulfovibrio vulgaris* and *Methanosarcina barkeri*: maintenance energy coefficient of the sulfate-reducing organism in the absence and presence of its partner. J. Bacteriol. 155:1260–1264.

116. Van Bruggen, J. J. A., C. K. Stumm, and G. D. Vogels. 1983. Symbiosis of methanogenic bacteria and sapropelic protozoa. Arch. Microbiol. 136:89–95.

117. Van Bruggen, J. J. A., K. B. Zwart, R. M. van Assema, C. K. Stumm, and G. D. Vogels. 1984. *Methanobacterium formicicum*, an endosymbiont of the anaerobic ciliate *Metopus striatus* McMurrich. Arch. Microbiol. 139:1–7.

118. Van der Meijden, P., C. van der Drift, and G. D. Vogels. 1984. Methanol conversion in *Eubacterium limosum*. Arch. Microbiol. 138:360–364.

119. Yarlett, N., A. C. Hann, and D. Lloyd. 1983. Hydrogenosomes in a mixed isolate of *Isotricha prostoma* and *Isotricha intestinalis* from ovine rumen contents. Comp. Biochem. Physiol. B74:357–364.

120. Waber, J. L., and H. G. Wood. 1979. Mechanism of acetate synthesis from CO_2 by *Clostridium acidiurici*. J. Bacteriol. 140:468–478.

121. Warford, A. L., D. R. Kosiur, and P. R. Doose. 1979. Methane production in Santa Barbara basin sediments. Geomicrobiol. J. 1:117–137.

122. Widdel, F. 1980. Anaerober Abbau von Fettsäuren und Benzoesäure durch neu isolierte Arten Sulfat-reduzierender Bakterien. Dissertation, University of Göttingen.

123. Widdel, F., and N. Pfennig. 1982. Studies on dissimilatory sulfate-reducing bacteria that decompose fatty acids. II. Incomplete oxidation of propionate by *Desulfobulbus propionicus* gen. nov., sp. nov. Arch. Microbiol. 131:360–365.

124. Wiegel, J., M. Braun, and G. Gottschalk. 1981. *Clostridium thermoautotrophicum* species novum, a thermophile producing acetate from molecular hydrogen and carbon dioxide. Curr. Microbiol. 5:255–260.

125. Wieringa, K. T. 1936. Over het verdwijnen van waterstof en koolzuur onder anaerobe voorwaarden. Antonie van Leeuwenhoek 3:263–273.

126. Wieringa, K. T. 1939/1940. The formation of acetic acid from carbon dioxide and hydrogen by anaerobic spore-forming bacteria. Antonie van Leeuwenhoek 6:251–262.

127. Winter, J., and R. S. Wolfe. 1979. Complete degradation of carbohydrate to carbon dioxide and methane by syntrophic cultures of *Acetobacterium woodii* and *Methanosarcina barkeri*. Arch. Microbiol. 121:97–102.

128. Winter, J. U., and R. S. Wolfe. 1980. Methane formation from fructose by syntrophic associations of *Acetobacterium woodii* and different strains of methanogens. Arch. Microbiol. 124:73–79.

128a. Wofford, N. G., P. S. Beaty, and M. J. McInerney. 1986. Preparation of cell-free extracts and the enzymes involved in fatty acid metabolism in *Syntrophomonas wolfei*. J. Bacteriol. 167:179–185.

129. Wolin, M. J. 1974. Metabolic interactions among intestinal microorganisms. Am. J. Clin. Nutr. 27:1320–1328.

130. Wolin, M. J. 1976. Interactions between H_2-producing and methane-producing species, p. 141–150, in H. G. Schlegel, G. Gottschalk, and N. Pfennig (eds.), Microbial formation and utilization of gases (H_2, CH_4, CO). E. Goltze KG, Göttingen, West Germany.

131. Wolin, M. J., and T. L. Miller. 1982. Interspecies hydrogen transfer: 15 years later. ASM News 48:561–565.

132. Wolin, M. J. 1982. Hydrogen transfer in microbial communities, p. 323–356, in A. T. Bull and J. H. Slater (eds.), Microbial interactions and communities Vol. 1. Academic Press, London.

133. Wood, H. G. 1952. Fermentation of $[3,4 - C^{14}]$- and $[1 - C^{14}]$- labelled glucose by *Clostridium thermoaceticum*. J. Biol. Chem. 199:579–583.

134. Wood, H. G. 1952. A study of carbon dioxide fixation by mass determination of the types of C^{13}-acetate. J. Biol. Chem. 194:905–931.

135. Wood, H. G., H. L. Drake, and S. I. Hu. 1982. Studies with *Clostridium thermoaceticum* and the resolution of the pathway used by acetogenic bacteria that grow on carbon monoxide or carbon dioxide and hydrogen. Proc. Biochem. Symp., Pasadena, Ann. Rev., p. 29–56.

136. Zehnder, A. J. B., B. A. Huser, T. D. Brock, and K. Wuhrmann. 1980. Characterization of an acetate-decarboxylating, non-hydrogen-oxidizing methane bacterium. Arch. Microbiol. 124:1–11.

137. Zehnder, A. J. B., K. Ingvorsen, and T. Marti. 1981. Microbiology of methane bacteria, p. 45–68, in D. E. Hughes, D. A. Stafford, B. I. Wheatley, W. Baader, G. Lettinga, E. J. Nyns, and W. Verstraete (eds.), Anaerobic digestion 1981. Elsevier, Amsterdam.

138. Zeikus, J. G. 1980. Chemical and fuel production by anaerobic bacteria. Annu. Rev. Microbiol. 34:423–464.

139. Zeikus, J. G. 1982. Microbial intermediary metabolism in anaerobic digestion, p. 23–44, in D. E. Hughes, D. A. Stafford, B. I. Wheatley, W. Baader, G. Lettinga, E. J. Nyns, and W. Verstraete (eds.), Anaerobic digestion 1981. Elsevier, Amsterdam.

140. Zeikus, J. G. 1983. Metabolism of one-carbon compounds by chemotrophic anaerobes. Adv. Microb. Physiol. 24:215–299.

141. Zeikus, J. G., L. H. Lynd, T. E. Thompson, J. A. Krzycki, P. J. Weimer, and P. W. Hegge. 1980. Isolation and characterization of a new, methylotrophic, acidogenic anaerobe, the Marburg strain. Curr. Microbiol. 3:381–386.

142. Zinder, S. H. 1984. Microbiology of anaerobic conversion of organic wastes to methane: recent developments. ASM News 50:294–298.

143. Zinder, S. H., and R. A. Mah. 1979. Isolation and characterization of a thermophilic strain of *Methanosarcina* unable to use H_2-CO_2 for methanogenesis. Appl. Environ. Microbiol. 38:996–1008.

144. Zinder, S. H., and M. Koch. 1984. Non-aceticlastic methanogenesis from acetate: acetate oxidation by a thermophilic syntrophic coculture. Arch. Microbiol. 138:263–272.

10

MICROBIOLOGY AND ECOLOGY OF SULFATE- AND SULFUR-REDUCING BACTERIA*

FRIEDRICH WIDDEL

Mikrobiologie, Fachbereich Biologie, Philipps-Universität, D-3550 Marburg, Federal Republic of Germany

10.1 INTRODUCTION
    10.1.1 Sulfur in Living Organisms
    10.1.2 Dissimilatory Metabolism of Sulfur Compounds
10.2 MICROBIOLOGY OF SULFATE-REDUCING BACTERIA
    10.2.1 Classification
    10.2.2 Chemical and Genetic Relationships
        10.2.2.1 Gram-Staining Behavior and Cell Wall
        10.2.2.2 Nucleic Acids
        10.2.2.3 Enzymes and Electron Carriers
    10.2.3 Electron Donors and Degradative Capacities
        10.2.3.1 Lactate
        10.2.3.2 Hydrogen, Formate
        10.2.3.3 Acetate
        10.2.3.4 Propionate
        10.2.3.5 Butyrate, Higher Straight-Chain Fatty Acids
        10.2.3.6 Branched-Chain Fatty Acids
        10.2.3.7 Monovalent Alcohols
        10.2.3.8 Dicarboxylic Acids

*Dedicated to Professor John R. Postgate on the occasion of his sixty-fifth birthday.

10.2.3.9   Aromatic Compounds

10.2.3.10   Saturated Cyclic Organic Acids

10.2.3.11   Choline, Amino Acids, Sugars, Glycerol

10.2.3.12   Hydrocarbons

10.2.4   Fermentative Capacities and Syntrophic Growth with Methanogenic Bacteria

10.2.5   Electron Acceptors

10.2.6   Disproportionation of Inorganic Sulfur Compounds

10.2.7   Assimilatory Capacities and Autotrophic Growth

10.2.8   Nitrogen Sources

10.3   ECOLOGY OF SULFATE-REDUCING BACTERIA

10.3.1   Sulfate Reduction and Alternative Degradation Processes

10.3.1.1   Consumption of Hydrogen and Acetate

10.3.1.2   Consumption of Propionate and Butyrate

10.3.1.3   Consumption of Fermentable Substrates

10.3.1.4   Incomplete or Complete Oxidation by Sulfate Reducers

10.3.1.5   Consumption of C_1-Methyl Compounds

10.3.1.6   Consumption of Sulfate

10.3.1.7   Methanogenesis in Digestors in the Presence of Sulfate

10.3.2   Sulfate Reduction in Various Habitats

10.3.3   Significance of Metabolic Types of Sulfate Reducers in Natural Habitats

10.3.4   Significance of Environmental Factors for Development of Sulfate Reducers

10.3.4.1   Effect of Temperature

10.3.4.2   Salt Requirement and Salt Tolerance

10.3.4.3   Effect of pH

10.3.4.4   Survival in Oxic Environments

10.3.4.5   Morphological Adaptations: Aggregating Cells and Gliding Filaments

10.3.5   Detrimental Effects Caused by Sulfate Reducers

10.3.5.1   Sulfate Reduction in Waters and Flooded Soils

10.3.5.2   Sulfate Reduction in Industrial Water Systems

10.3.5.3   Anaerobic Metal Corrosion

10.3.5.4   Comments on Control of Sulfate Reducers

10.4   NONSULFATE-REDUCING BACTERIA WITH DISSIMILATORY REDUCTION OF INORGANIC SULFUR

10.4.1   Eubacterial Sulfur Reducers

10.4.1.1   *Desulfuromonas* Species

10.4.1.2   Spirilloid Sulfur Reducers

10.4.2   Archaebacterial Sulfur Reducers

10.4.3   Further Organisms Reducing Sulfur Compounds

10.5 PRINCIPLES OF THE BIOLOGICAL SULFUR CYCLE
    10.5.1   Reoxidation of Hydrogen Sulfide
            10.5.1.1   Aerobic Bacteria Oxidizing Sulfur Compounds
            10.5.1.2   Chemical Oxidation of Sulfide
            10.5.1.3   Biological and Chemical Sulfide Oxidation in Natural
                       Habitats
            10.5.1.4   Phototrophic Sulfide-Oxidizing Bacteria
            10.5.1.5   Anaerobic Sulfur Cycle
    10.5.2   Precipitation of Iron Sulfides
    10.5.3   The Sulfur Cycle on a Global Scale and Through Time
            10.5.3.1   General Aspects of Sulfur Isotope Fractionation
            10.5.3.2   Oceanic Sulfates
            10.5.3.3   Deposits of Metal Sulfides
            10.5.3.4   Deposits of Elemental Sulfur
            10.5.3.5   Carbonate Deposits
            10.5.3.6   Age of Dissimilatory Sulfate Reduction
REFERENCES
NOTE ADDED IN PROOF

10.1 INTRODUCTION

Sulfate-reducing bacteria are notable for their end product, hydrogen sulfide (often briefly termed *sulfide*), which by its chemical properties and physiological effects is a far more conspicuous substance than the substrate sulfate. Sulfate is a chemically rather inert, nonvolatile, and nontoxic compound. It is widespread in rocks, soils, and waters. In contrast, hydrogen sulfide is chemically reactive. In aqueous habitats where it is formed, it often blackens the sediment due to the formation of ferrous sulfide from iron-containing materials. As a reductant, hydrogen sulfide traps oxygen and may be converted to sulfur. Hydrogen sulfide, which even at low concentration (≥ 0.2 ppm) is recognized by its distinctive smell, is toxic to plants, animals, and humans. Thus hydrogen sulfide appears as a strange form of sulfur in the part of the biosphere with which we are most familiar. Indeed, sulfate-reducing bacteria thrive outside the aerobic environment, in niches where oxygen has no access. However, organically bound reduced sulfur is an indispensible constituent of every living organism.

10.1.1 Sulfur in Living Organisms

The content of sulfur in living organisms is on the average about 1% (w/w) of their dry mass. Variations depend on the types of plants, animals, or bacteria (10,

35, 92). With respect to the abundance of atoms, sulfur in organisms ranges behind phosphorus, calcium, potassium, nitrogen, oxygen, carbon, and hydrogen (NaCl not considered), the approximate ratios of the atoms being 1 (S) : 1–4 (P) : 0.1–5 (Ca) : 1–5 (K) : 13–40 (N) : 50–180 (O) : 70–270 (C) : 110–540 (H).

Most of the sulfur in organisms occurs in proteins, namely as a component of cysteine and methionine. In these amino acids, sulfur is in its most reduced state (-2, or -1 in disulfide bonds). Further reduced sulfur compounds of general biological importance are the cofactors coenzyme A, thiamine (vitamin B_1), biotin (vitamin H), lipoic acid, and the ferredoxins. Organisms also contain oxidized organically bound sulfur as present in sulfate esters or sulfonates (385). In our oxic biosphere, the energetically stable and most abundant sulfur source for organisms is sulfate at the highest oxidation state ($+6$). All green plants, fungi, and most bacteria are able to reduce sulfate to the level of sulfide which is incorporated into the various organic molecules (119, 271–273, 284, 339). This process is known as *assimilatory sulfate reduction*. Humans, animals, and those bacteria that cannot reduce sulfate obtain reduced organic sulfur compounds as essential nutrients via the food chain.

By the metabolic degradation processes in the higher organisms or decomposing bacteria, the reduced sulfur from the organic molecules is reoxidized and finally enters the large pool of inorganic sulfate (284, 339).

10.1.2 Dissimilatory Metabolism of Sulfur Compounds

The awareness of a potentially complete mineralization of dead biomass led to the concept of cycles of matter. In these cycles, which are maintained by sunlight, the elements change in a permanent flow between organically bound and free, simple inorganic forms. The involvement of carbon, oxygen, nitrogen, and sulfur in biological cycles implies a change between reduced and oxidized states.

Whereas animals, protozoa, fungi, and most decomposing bacteria in their energy metabolism take advantage of the oxidation of the carbon moiety of nutrients, specialized aerobic bacteria have exploited reduced inorganic compounds as energy sources and hydrogen donors for cell synthesis. Ammonia is oxidized by nitrifying bacteria to nitrate. Hydrogen sulfide is oxidized by various aerobic bacteria, comprised as colorless sulfur bacteria, which form sulfate as an end product. Unlike ammonia, hydrogen sulfide is also oxidized as a hydrogen donor under anaerobic conditions in the light, namely by the photosynthetic purple and green sulfur bacteria (Chapter 2).

The oxidation of ammonia or hydrogen sulfide in the energy and hydrogen metabolism of the lithotrophic bacteria mentioned has counterparts in the anaerobic environment. In the absence of oxygen, the oxidized forms of nitrogen and sulfur are used as electron acceptors for the oxidation of organic substrates by nitrate-reducing bacteria (Chapter 4) or sulfate-reducing bacteria (sulfate reducers), respectively. Since this reduction of the inorganic compounds serves for energy conservation, the process is distinguished as dissimilatory reduction from the assimilatory reduction of nitrate or sulfate in plants and bacteria. Whereas many nitrate-reducing bacteria are facultatively anaerobic, the sulfate-reducing bacteria are all obligate anaerobes.

The amount of hydrogen sulfide formed by dissimilatory sulfate reduction far exceeds the amount that is used for cell synthesis of sulfate-reducing bacteria. As anaerobes, these bacteria have relatively low cell yields (cell dry mass formed per amount of energy substrate used). If, for example, *Desulfobacter* forms 100 g of hydrogen sulfide by sulfate reduction with the stoichiometric amount of 174 g of acetate as electron donor, 14 g of cell dry mass is synthesized (368). With a sulfur content of 1.3% measured in sulfate reducers (248), synthesis of the indicated amount of cell mass would require only 0.2 g of hydrogen sulfide. Similarly, the bacterial degradation of biomass in oxygen-free sulfate-containing waters via dissimilatory sulfate reduction as terminal anaerobic process may yield far more hydrogen sulfide than the mere liberation (desulfuration) of reduced sulfur from the organic compounds. As explained later (Section 10.2.3), all reducing equivalents from anaerobically degradable biomass can potentially flow into dissimilatory sulfate reduction and finally appear as hydrogen sulfide. If the average oxidation state of biomass is written approximately as a carbohydrate, the overall degradation reaction with sulfate as terminal electron acceptor would be as follows:

$$2\langle CH_2O \rangle + SO_4^{2-} \longrightarrow 2HCO_3^- + H_2S \qquad (1)$$

Thus 1000 g of biomass would yield 570 g of H_2S, provided that the biopolymers are completely degradable under anaerobic conditions and sulfate is not limiting. Desulfuration of the organically bound reduced sulfur from the same amount of biomass would yield about 10 g of hydrogen sulfide.

Hydrogen sulfide is usually the only product excreted from the dissimilatory sulfur metabolism of sulfate-reducing bacteria (Section 10.2.5). Under anaerobic, reduced conditions, hydrogen sulfide is the energetically stable form of sulfur, as sulfate is the stable form under aerobic conditions. Intermediary oxidation states of sulfur may be formed in natural habitats by incomplete chemical or biological oxidation of sulfide that diffuses into the oxic environment, or during anaerobic sulfide oxidation by purple sulfur bacteria (Section 10.5.1). Furthermore, intermediary oxidation states are introduced into the biosphere by vulcanic activities or human technology. The most important of such forms of sulfur are sulfite (or sulfur dioxide), thiosulfate, and elemental sulfur.

If the intermediary oxidation states of sulfur get to or get back to anaerobic conditions in aqueous environments, these sulfur species may also be used by anaerobic bacteria. Specialized but probably widespread types of sulfate-reducing bacteria can disproportionate sulfite or thiosulfate to sulfide and sulfate (20a). Many species of sulfate-reducing bacteria can reduce sulfite or thiosulfate (18, 69, 240, 371), and a few elemental sulfur also (31). The most important reducers of elemental sulfur in nature are probably special anaerobes that may be designated as true sulfur-reducing bacteria; these carry out the dissimilatory reduction of sulfur as their primary or even obligate metabolic reaction during oxidation of organic substrates. The true sulfur-reducing bacteria do not reduce sulfate, and only some reduce other oxygen-containing sulfur compounds. Whereas all known sulfate reducers are eubacteria, sulfur-reducing bacteria occur in both prokaryotic kingdoms, the eubacteria and the archaebacteria. The majority of the eubacterial sulfur reducers are mesophilic, as are most sulfate reducers. In contrast, all described archaebacterial sulfur reducers are extreme thermophiles the temperature optima of which are the highest known in the living world (313, 314, 317, 318).

In the present chapter we treat the sulfate-reducing bacteria more extensively than the sulfur-reducing bacteria; the latter have been discovered relatively recently (235, 391), and much more microbiological and ecological research has been done with sulfate-reducing bacteria since their discovery by Beijerinck (27). From the voluminous literature on sulfate-reducing bacteria and their activities in ecosystems, only a selection of topics and some principal insights are possible within the scope of this chapter. For more detailed information, the reader is referred to the comprehensive monograph of Postgate (248) and to surveys on nutrition physiology and classification (238, 371), on biochemistry (173, 174, 223, 233, 284, 339; and Chapter 11 of this volume), and on the involvement of sulfate-reducing bacteria in the natural sulfur and carbon cycle (92, 142, 143, 213, 338).

10.2 MICROBIOLOGY OF SULFATE-REDUCING BACTERIA

The production of hydrogen sulfide from sulfate in waters was recognized as a biological process in 1864 by Meyer, who supposed that the reaction was mediated by certain algae (194). A few years later, Cohn attributed sulfide formation from sulfate to filamentous gliding microorganisms classified as Beggiatoae (57). With the isolation of a bacterium reducing sulfate to sulfide, Beijerinck clearly demonstrated the correlation of the observed process to a special kind of bacteria (27). At the end of his report, Beijerinck precisely characterized the problems and experiences of a microbiologist who had to isolate and cultivate an obligate anaerobe without our modern anaerobic techniques (27):

> Hat man *Spirillum desulfuricans* einmal in Reinkultur gebracht, so muß man nicht glauben, daß dann die Schwierigkeiten für die weitere Untersuchung verschwunden sind. Ganz im Gegenteil, dieselben werden dann noch viel größer wie beim Rohmaterial. Dieses resultiert aus der Notwendigkeit der Durchführung von Kautelen, wodurch der Sauerstoff vollständig entfernt wird, und Jeder, welcher sich mit solchen Versuchen beschäftigt hat, welche sehr oft wiederholt werden müssen, weiß, wie aufreibend es ist, in dieser Beziehung radikal zu handeln, wenn man dabei nicht aërobische Mikrobien zu Hilfe nehmen kann. (If one has once isolated *Spirillum desulfuricans* in pure culture, one must not assume that the difficulties have disappeared in further investigation. Quite in contrast, these become even greater than with the crude material. This results from the necessity of taking precautions to remove oxygen completely, and everybody who has dealt with such experiments, which have to be repeated very often, knows how exhausting it is to be strict in this respect if one cannot make use of aerobic microbes.)

The isolated type of sulfate reducer, which had curved motile cells, was known for a while as *Vibrio desulfuricans* (14), and then finally named *Desulfovibrio desulfuricans* (154). The genus designation *Desulfovibrio* was maintained for non-sporing sulfate reducers usually having curved motile cells and growing on a relatively limited range of organic substrates; preferred substrates are lactate or pyruvate, which are incompletely oxidized to acetate (250). *Desulfovibrio* species are still the best studied sulfate reducers. Sporing sulfate-reducing bacteria with a similar metabolism were classified within the genus *Desulfotomaculum* (46). Later, additional types of sulfate-reducing bacteria were described, several of which differ markedly physiologically and morphologically from the known *Desulfovibrio* and *Desulfotomaculum* species (237, 238, 371). Hence, the term "sulfate-reducing bacteria" describes a rather heterogeneous assemblage of microorganisms having in common merely dissimilatory sulfate metabolism and obligate anaerobiosis.

10.2.1 Classification

Classification of the presently known, different types of sulfate-reducing bacteria is based primarily on nutritional and morphological characteristics; this classification is supported by some chemical criteria, such as the guanine plus cytosine content of the DNA and the presence of special pigments. Detailed data from nucleic acid studies such as from DNA/rRNA hybridization or rRNA oligonucleotide cataloging were not available or insufficiently available when the species and genera of sulfate-reducing bacteria were established. The classification is complicated by the fact that morphologically similar types may differ by their nutrition, whereas nutritionally similar types may have different morphology. Also, the presence of pigments does not always agree with nutritional and morphological groups. This complexity of features made compromises necessary in classification of the sulfate-reducing bacteria. The species and genera were established for determinative purposes and do not necessarily indicate phylogenetic relationships.

As in the case of many other bacteria, the genus names of sulfate reducers, apart from the prefix "Desulfo," have been derived from features observed by microscopy. However, cell shape and motility are not always constant features of strains and may vary with growth conditions (248). Moreover, when the range of known sulfate reducers was enlarged more and more by finding new variants and types, it could not be avoided that genus names were used less precisely than before. For example, the designation *Desulfovibrio* was originally given to the mentioned sulfate reducers having curved cells with striking vibrating motility. Later, the rod-shaped *Desulfovibrio* strain "Norway 4" (199, 248), *Desulfovibrio thermophilus* (262), and *Desulfovibrio baculatus* (263) were included in the same genus since these sulfate reducers behave nutritionally as the curved representatives of *Desulfovibrio*. The situation was reversed in the classification of *Desulfovibrio baarsii* (237, 249). This sulfate reducer has the same motile curved cells as many typical *Desulfovibrio* species, but is nutritionally completely different from the latter. Typical *Desulfovibrio* species grow well on lactate that is incompletely oxidized to acetate; fatty acids are not used. For growth with hydrogen or formate, these species require acetate as carbon source for cell synthesis (17, 20, 36, 305–307). In contrast, *D. baarsii* grows on fatty acids up to stearate that are completely oxidized, but does not take lactate. If *D. baarsii* grows on formate, no additional organic carbon source is necessary (131).

Table 10.1 summarizes major characteristics of presently classified sulfate-reducing bacteria. Morphological types are shown in Fig. 10.1, and some submicroscopical structures in Fig. 10.2

Most of the different types of sulfate-reducing bacteria were isolated from batch enrichment cultures; these have often been selective for one type of sulfate reducer, depending on the kind of added electron donor and carbon source, on the sample used as inoculum, or on the temperature and salt concentration. For example, sporeforming species of the genus *Desulfotomaculum* may be obtained from pasteurized samples or from dry, aerobic soil (46, 58, 63a, 153). Thermophilic species enrich at 50°C or at higher temperature, up to about 70°C (262, 386). The selection of species by different substrates is explained in Section 10.2.3.

Not all types of observed sulfate-reducing bacteria have been classified so far. For example, an unnamed sulfate reducer with thin, extremely long cells has been described that physiologically resembled the common *Desulfovibrio* species (134); another isolate of this type (65) is shown in Fig. 10.1*h*. A common sulfate reducer that has not been affiliated to a genus is shown in Fig. 10.1*y*. This type is often enriched from freshwater muds with butyrate or higher fatty acids that are oxidized to acetate; other electron donors are not utilized. One group that may comprise far more than the presently classified species is the genus *Desulfotomaculum*. Chemostat and batch enrichment cultures with acetate (149) or long-chain fatty acids (153), respectively, showed that there are sporing sulfate reducers that morphologically or nutritionally differ from the classified species. Just as little is known about the range of thermophilic sulfate reducers, or about types that grow very slowly and yield low cell densities under laboratory conditions. Future systematic studies on sulfate-reducing bacteria and isolation of further species may necessitate rearrangements in the present classification scheme.

10.2.2 Chemical and Genetic Relationships

Comparative studies on the chemistry of cell components (chemotaxonomy) and on nucleic acids from sulfate-reducing bacteria agree in many cases with the classification based on nutrition and morphology. However, there are also exceptions. With respect to certain molecular properties, some sulfate reducers appear to be less related to members of their own genera than to other sulfate-reducing genera. Such similarities interlace the classification scheme of sulfate reducers. Moreover, comparative studies on the molecular level reveal relationships of sulfate reducers to groups of nonsulfate-reducing bacteria. The relationships are interpreted in terms of evolution.

TABLE 10.1 Properties of classified sulfate-reducing bacteria[a]

Species	Form	Width (μm)	Length (μm)	Motility[b]	G + C of DNA (mol %)	Sulfite reductases Desulfoviridin	Sulfite reductases Other	Major Menaquinone	Cytochromes (Minor Species in Parentheses)
Desulfotomaculum (sporing)									
nigrificans	Rod	0.5–0.7	2–4	+ (pe)	49	–	P582	MK-7	*b* (*c*)
orientis	Slightly curved rod	0.7–1	3–5	+ (pe)	45	–	P582	MK-7	*b* (*c*)
ruminis	Rod	0.5–0.7	2–4	+ (pe)	49	–	P582	MK-7	*b*
antarcticum	Rod	1–1.2	4–6	+ (pe)	nr	–	nr	nr	*b*
acetoxidans	Slightly curved rod	1–1.5	3.5–9	+ (sp)	38	–	P582	MK-7	*b*
guttoideum	Rod	1.0	2.3	+ (pe)	52	–	nr	nr	*c*
sapomandens	Rod	1.2–2	5–7	+	48	–	nr	nr	nr
Desulfovibrio									
desulfuricans	Vibrio	0.5–0.8	1.5–4	+ (sp)	59	+	nr	MK-6	(*b*) c_3
vulgaris	Vibrio	0.5–0.8	1.5–4	+ (sp)	65	+	2 further	MK-6	(*b*) c_3
gigas	Large vibrio	0.8–1	6–11	+ (lo)	65	+	nr	MK-6	(*b*) c_3
africanus	Vibrio	0.5–0.6	2–3	+ (lo)	65	+	nr	MK-6(H_2)[d]	(*b*) c_3
salexigens	Vibrio	0.5–0.8	1.3–2.5	+ (sp)	49	+	nr	MK-6(H_2)[d]	(*b*) c_3
baculatus	Rod	0.6	1.3	+ (sp)	57	–	nr	nr	*b c*
sulfodismutans	Vibrio	0.5–1	3–5	+	64	+	nr	nr	Present
thermophilus	Rod	0.5	1.2–2.5	+ (sp)	nr	–[e]	Desulfofuscidin[f]	MK-7	*c*
sapovorans	Vibrio	1.5	3–5.5	+ (sp)	53	–	P582?[g]	MK-7	*b c*
baarsii	Vibrio	0.5–0.7	1.5–4	+ (sp)	66	–	nr	MK-7(H_2)[d]	*b c*
Desulfomonas									
pigra	Rod	0.8–1.3	1.2–5	–	66	+	nr	MK-6	*c*
Thermodesulfobacterium									
commune	Rod	0.3	0.9	–	34	–	Desulfofuscidin	MK-7	c_3
Desulfobulbus									
propionicus	Oval or lemon shape	1–1.3	1.8–2	–[h]	60	–	Desulforubidin	MK-5(H_2)[d]	*b c*
elongatus	Rod	0.6–0.7	1.5–2.5	+ (sp)	59	–	Desulforubidin	MK-5(H_2)[d]	*b* c_3

TABLE 10.1 (*Continued*)

Species	Form	Width (μm)	Length (μm)	Motility[b]	G + C of DNA (mol %)	Sulfite reductases Desulfoviridin	Sulfite reductases Other	Major Menaquinone	Cytochromes (Minor Species in Parentheses)
Desulfobacter postgatei	Oval rod	1–1.5	1.7–2.5	+/−[i] (sp)	46	−	Desulforubidin	MK-7	b c
hydrogenophilus[e]	Rod	1–1.3	2–3	−	45	−	nr	MK-7(H$_2$)[d]	Present
latus[e]	Large oval rod	1.6–2.4	5–7	+/−[i]	44	−	nr	MK-7	Present
curvatus[e]	Vibrio	0.5–1	1.7–3.5	+	46	−	nr	MK-7(H$_2$)	Present
Desulfococcus multivorans	Sphere	1.5–2.2		−	57	+	nr	MK-7	b c
niacini[j]	Irregular sphere	1.5–3		+ (sp)	46	−	nr	MK-7	b c
Desulfosarcina variabilis	Oval rod, packages	1–1.5	1.5–2.5	+/−[i] (sp)	51	−	nr	MK-7	b c
Desulfobacterium autotrophicum[k]	Oval rod	0.9–1.3	1.5–3	+ (sp)	48	−	nr	MK-7	b c
vacuolatum[e]	Oval rod or sphere	1.5–2	2–2.5	−	45	−	nr	MK-7(H$_2$)[d]	Present
phenolicum	Oval rod	1–1.5	2–3	+ (sp)	41	−	nr	MK-7(H$_2$)[d]	Present
indolicum	Oval rod	0.7–1.5	2–2.5	+ (sp)	47	−	nr	MK-7(H$_2$)[d]	Present
catecholicum	Lemon shape	1.3–1.8	2.2–2.8	−	52	−	nr	nr	Present
Desulfonema limicola	Filament	2.5–3	50–1000, one cell: 2.5–3	Gliding	35	+	nr	MK-7	(b) c
magnum	Filament	6–8	100–2000, one cell: 9–13	Gliding	42	−	P582 ?[g]	MK-9	b c

TABLE 10.1 (Continued)

Species	Temperature Optimum (°C)	Oxidation	Hydrogen	Formate	Lactate	Ethanol	Acetate	Fatty acids: C-atoms	Isobutyrate	2-Methylbutyrate	3-Methylbutyrate	Fumarate	Malate	Benzoate	Other Readily Used	Growth Factor Requirement	NaCl Requirement (g/l)	References
Desulfotomaculum																		
nigrificans	55 (max. 70)	i	+	+	+	+	–	–	nr	nr	nr	–	–	–	Fructose	Unknown	–	46, 153, 248
orientis	37	i	+*	+*	+	+	–	–	nr	nr	nr	–	–	–	Methanol	Unknown	–	46,153, 248
ruminis	37	i	+	+	+	+	–	–	nr	nr	nr	–	–	–	Alanine	pa, bi	–	46, 153, 248
antarcticum	20–30	i	nr	–	–	nr	–	–	nr	nr	nr	nr	nr	nr	Glucose	Unknown	–	116, 248
acetoxidans	35	c	–	–	+	+	+	4–5	+	–	–	–	–	–	Butanol	bi	–	248, 367, 369
guttoideum	31	i	+	–	+	–	–	nr	nr	nr	nr	nr	–	nr	–	Unknown	–	91
sapomandens	38	c	nr	+	–	+	(+)	4–18	+	–	+	(+)	(+)	+	Phenylacetate, phenylpropionate	Unknown	–	63a
Desulfovibrio																		
desulfuricans	30–36	i	+	+	+	+	–	–	nr	nr	nr	+	+	–	Choline	None	–	248, 249, 250
vulgaris	30–36	i	+	+	+	(+)[c]	–	–	nr	nr	nr	+	+	–	1 subsp. oxamate	None	–	248, 249, 250
gigas	30–36	i	+	+	+	(+)[c]	–	–	nr	nr	nr	–	–	–	nr	bi[c]	–	248, 249, 250
africanus	30–36	i	+	+	+	+[c]	–	–	nr	nr	nr	+	+	–	nr	None	–	133, 248, 249, 250
salexigens	30–36	i	+	+	+	+[c]	–	–	nr	nr	nr	+	+	–	nr	None	20	248, 249, 250
baculatus	28–37	i	(+)	+	+	–	–	nr	nr	nr	nr	+	+	nr	nr	Unknown	–	248, 249, 263
sulfodismutans	30–35	i	(+)	–	+	+	–	–	nr	nr	nr	–	–	–	nr	bi, pt	–	20a
thermophilus	65 (max. 85)	i	+	+	+	–	–	–	nr	nr	nr	nr	–	nr	nr	Unknown	–	248, 249, 262
sapovorans	34	i	–	–	+	–	–	4–16	–	+	–	–	–	–	nr	None	–	237, 248, 249
baarsii	35–39	c	–	+*	–	–	(+)	(3)–18	+	(+)	+	–	–	–	nr	None	–	131, 237, 248, 249
Desulfomonas																		
pigra	37	i	+[e]	–	+	+	–	–	nr	nr	nr	–	nr	r		pa[e]	–	201, 202

479

TABLE 10.1 (Continued)

Species	Temperature Optimum (°C)	Oxidation	Hydrogen	Formate	Lactate	Ethanol	Acetate	Fatty acids: C-atoms	Isobutyrate	2-Methylbutyrate	3-Methylbutyrate	Fumarate	Malate	Benzoate	Other Readily Used	Growth Factor Requirement	NaCl Requirement (g/l)	References
Thermodesulfobacterium commune	70 (max. 85)	i	+	nr	+	–	–	–	nr	nr	nr	nr	–	nr	nr	Unknown	–	104, 386
Desulfobulbus propionicus	28–39	i	+	–	+	+	–	3	–	–	nr	–	–	–	nr	pa	–	238, 370, 371
elongatus	35	i	+	–	+	+	–	3	nr	nr	nr	–	–	nr	nr	Unknown	–	266
Desulfobacter postgatei	28–32	c	–	–	–	–	+	–	–	–	–	–	–	–	nr	pa, bi	7^p	238, 368, 371
hydrogenophilus	28–32	c	+*	–	–	(+)	+	–	–	–	–	–	–	–	nr	pa, bi	20^p	
latus	28–32	c	–	–	–	–	+	–	–	–	–	–	–	–	nr	bi, th	20^p	
curvatus	28–30	c	(+)	–	–	+	+	–	–	–	–	–	–	+	nr	bi	10^p	
Desulfococcus multivorans	35	c	–	+*	+	+	(+)	3–16	+	+	+	–	–	+	Phenylacetate, phenylpropionate	pa, bi, th	5^p	238, 371
niacini	29	c	+*	+*	–	+	(+)	(3)–16	–	–	–	+	+	–	Nicotinate, succinate, glutarate, pimelate	bi, th	15^p	118
Desulfosarcina variabilis	33	c	+*	+*	+	+	(+)	3–14	–	(+)	(+)	+	–	+	Phenylacetate, phenylpropionate	None	15^p	238, 371

Species	Temp (°C)	Oxid.[m]							Fatty acids							n-Oxidized substrate[n]	Growth factors[o]		Refs.
Desulfobacterium																			
autotrophicum	20–26	c	+*	+*	+	+	+	(+)	(3)–16	+	+	+	−	+	−	Succinate	bi, ni, th	20[o]	
vacuolatum	25–30	c	+*	+*	+	(+)	(+)	(+)	(3)–16	+	+	+	+	−	−	Succinate	None	20[o]	
phenolicum	28	c	−	(+)	−	−	(+)	(+)	(4)	+	+	(+)	(+)	+	+	Phenol, p-cresol, glutarate	None	20[o]	20c
indolicum	28	c	−	(+)	−	−	(+)	(+)	(3)	+	−	−	(+)	(+)	−	Indole	B₁₂	20[o]	20b
catecholicum	28	c	(+)	(+)	(+)	(+)	nr	nr	(3–20)	nr	+	(+)	(+)	(+)	+	Catechol	Unknown	−	325a
Desulfonema																			
limicola	30	c	+*	+*	+	+	+	(+)	3–14	+	+	+	−	+	−	Succinate	bi	15[p]	238, 366, 371
magnum	32	c	−	+	−	(+)	(+)	(+)	3–10	+	+	+	+	+	+	Succinate	pa, bi, B₁₂	20[q]	238, 366, 371

[a] nr, not reported or not determined.

[b] Flagellation in parentheses: pe, peritrichous; sp, single, polar; lo, lophotrichous.

[c] *Desulfovibrio* "*desulfuricans*" strain Norway 4 contains desulforubidin, but no desulfoviridin.

[d] Terminal saturation in the isoprenoid side chain.

[e] Widdel, unpublished data.

[f] Fauque and LeGall, personal communication.

[g] Speculative (from spectra).

[h] Type strain nonmotile; other strains may be motile (single, polar flagellum).

[i] Motility may disappear after isolation.

[j] Considered to be reclassified as *Desulfobacterium niacini*.

[k] Brysch, Schneider, Fuchs, and Widdel, unpublished data.

[l] Symbols: +, utilized; +*, autotrophic growth; ( ), poorly utilized; —, not utilized.

[m] i, incomplete oxidation to acetate as an end product; c, complete oxidation to carbon dioxide.

[n] Not completely listed; for further substrates, see references and text.

[o] pa, *para*-aminobenzoate; bi, biotin; pt, pantothenate; th, thiamine; ni, nicotinate.

[p] In addition, 1.2 to 3 g of MgCl₂ · 6H₂O per liter required or routinely added to media.

[q] In addition, 5 g of MgCl₂ · 6H₂O and 1.3 g of CaCl₂ · 2H₂O per liter required.

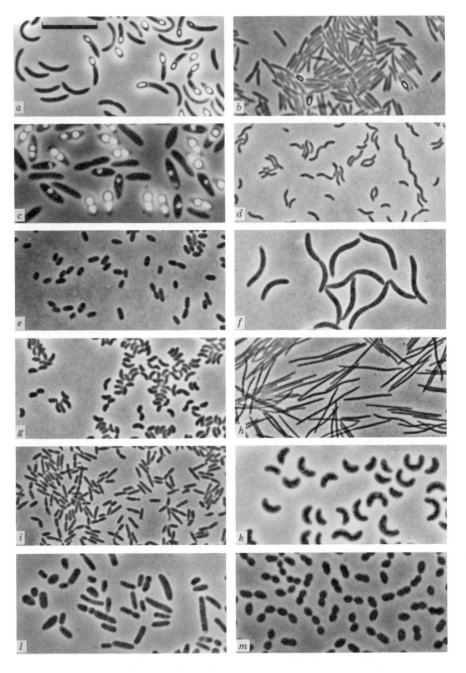

Figure 10.1 (*See p. 484 for legend.*)

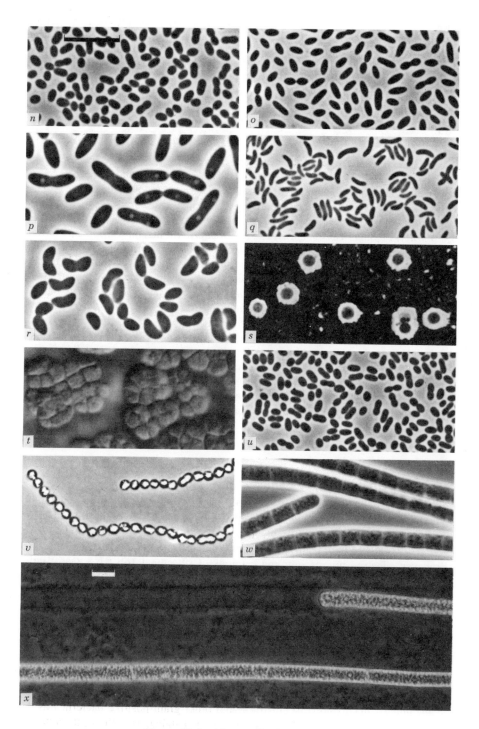

Figure 10.1 (*See p. 484 for legend.*)

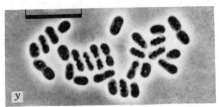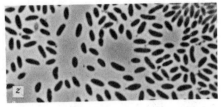

Figure 10.1 Phase-contrast photomicrographs of viable cells of diverse sulfate-reducing bacteria. Bars equal 10 μm; bars in (*a*), (*n*), and (*y*) applicable to all figures except for (*x*). (*a*) *Desulfotomaculum orientis*, with spores formed during sulfate-limited growth. (*b*) *Desulfotomaculum ruminis*, with few spores. (*c*) *Desulfotomaculum acetoxidans*, with spherical spores and cone-shaped gas vesicles, formed during growth in acetate/sulfate agar. (*d*) *Desulfovibrio desulfuricans* strain Essex. (*e*) *Desulfovibrio desulfuricans* strain Norway. (*f*) *Desulfovibrio gigas*. (*g*) *Desulfovibrio salexigens*. (*h*) Unnamed isolate with long, thin cells, physiologically similar to *Desulfovibrio* spp. (*i*) *Desulfovibrio thermophilus*. (*k*) *Desulfovibrio sapovorans*; cells contain granules of poly-β-hydroxybutyric acid. (*l*) *Desulfomonas pigra*. (*m*) *Desulfobulbus propionicus*. (*n*) *Desulfobacter postgatei*. (*o*) *Desulfobacter hydrogenophilus*. (*p*) *Desulfobacter latus*. (*q*) *Desulfobacter curvatus*. (*r*) Unnamed isolate with large curved cells, similar to *Desulfobacter* spp. (*s*) *Desulfococcus multivorans*, negative staining of cells surrounded by slime. (*t*) *Desulfosarcina variabilis*, cell packages. (*u*) *Desulfobacterium autotrophicum*. (*v*) *Desulfobacterium vacuolatum*. (*w*) *Desulfonema limicola*, from an enrichment with isobutyrate. (*x*) *Desulfonema magnum*; one filament leaves a trail behind in artificial sediment of aluminum phosphate. (*y*) Unnamed, long-chain fatty acid–degrading isolate; cells contain granules of poly-β-hydroxybutyric acid. (*z*) Unnamed benzoate-degrading sulfate reducer isolated from an oil tank. Photomicrograph (*a*) by H. Cypionka.

10.2.2.1 *Gram-Staining Behavior and Cell Wall*

Most sulfate-reducing bacteria were described to stain gram-negative. Young cultures of *Desulfotomaculum nigrificans* sometimes showed a proportion of gram-positive cells (248). The cell filaments of *Desulfonema* species exhibited an irregularly distributed Gram stain and thus simulated to be gram positive (238, 366). However, electron microscopy and catalogs of ribosomal 16S RNA (see Section 10.2.2.2) revealed that the Gram-staining behavior of sulfate-reducing bacteria is diagnostically unreliable. Like representatives of other nonsporing genera of sulfate reducers (20, 266, 332), *Desulfonema* species exhibited cell walls characteristic of gram-negative bacteria (366). In contrast, the cell wall structure of *Desulfotomaculum nigrificans* showed that this sulfate reducer is a true gram-positive (211, 292). This finding agrees with the ability of *Desulfotomaculum* species to form spores that have been observed so far only among gram-positive bacteria and the closely related *Sporomusa* species. The affiliation of *Desulfotomaculum nigrificans* and *D. acetoxidans* to clostridia and other gram-positive bacteria was evident from 16S ribosomal RNA obligonucleotide catalogs (see the next section).

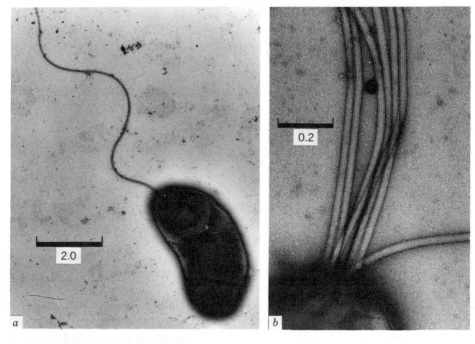

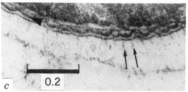

Figure 10.2 Electron micrographs of cell structures connected with motility of sulfate-reducing bacteria. Dimensions are given in micrometers. (*a*) *Desulfovibrio sapovorans* with single polar flagellum; negative staining. (*b*) Polar bundle of flagella of an unnamed fatty acid–degrading *Desulfovibrio*-like isolate (strain 3bu10); negative staining. (*c*) Undulated outer membrane (arrowhead) and radial fibers (thin arrows) in the gliding *Desulfonema magnum*; ultrathin section. Electron micrograph (*a*) by F. Mayer, (*b*) by E. Spiess, and (*c*) by G.-W. Kohring.

10.2.2.2 Nucleic Acids

DNA base ratios (mol % guanine + cytosine, abbreviated G + C) are known from representatives of each genus of sulfate reducers (Table 10.1). The values may vary somewhat with the manner of determination and calculation (289). The G + C values of the "classical" *Desulfovibrio* species isolated with lactate range from 49 to 65 mol %. Postgate (248) pointed at three tight clusters of DNA base ratios within this group, namely around 49, 59, and 65 mol % G + C. Whereas great differences between G + C values of species may indicate a low degree of relationship, a coincidence of values is not necessarily due to a genetic overlap.

In a number of sulfate-reducing bacteria from different genera, oligonucleotides from 16S ribosomal RNA (16S rRNA) were determined and compared to determine genealogical relationships (85). There is a deep gap between the sporing and the nonsporing sulfate reducers (Fig. 10.3). *Desulfotomaculatum nigrificans* and *D. acetoxidans* cluster with the *Clostridium* branch of the gram-positive division of eubacteria, a finding that agrees with the ability of these sulfate reducers to form spores and with the cell wall structure observed in the former species (211, 292). All other, nonsporing sulfate-reducing bacteria examined form one distinctive but not very coherent division of gram-negative bacteria. Within this division, *Desulfovibrio desulfuricans* and *D. gigas* represent one branch with a low degree of relatedness to *Desulfobacter postgatei*, *Desulfobulbus propionicus*, *Desulfococcus niacini*, *Desulfonema limicola*, and *Desulfosarcina variabilis*. Within the latter five species, only the morphologically very different *Desulfonema limicola* and *Desulfosarcina variabilis* show a relatively high degree of relatedness. The sulfur-reducing bacteria of the genus *Desulfuromonas* appear as a coherent group among completely oxidizing sulfate reducers. Furthermore, the division includes the aerobic *Myxococcus* and *Bdellovibrio* species that lack a dissimilatory sulfur metabolism. The oligonucleotide catalogs of 16 rRNA cannot satisfactorily refine the deep branching points (i.e., the exact relationships among the major bacterial branches). These can be resolved by complete sequencing. This has been carried out with 16S rRNA of *Desulfovibrio desulfuricans* as a representative of the gram-negative sulfate reducers (228a). The results not only confirm the relationship to *Myxococcus*, as detected by oligonucleotide catalogs; on a deep level, the sulfate reducer was also related to the purple bacteria that themselves consist of three subdivisions. On the one hand, the existence of a gram-positive and gram-negative group of sulfate reducers and the interweaving with physiologically and biochemically very unsimilar bacteria are not satisfactorily understood from the viewpoint of a metabolic evolution. On the other hand, the relatedness between *Desulfovibrio desulfuricans* and the anoxygenic phototrophs agrees with the assumption that dissimilatory sulfate reduction evolved as an inversion of the oxidative dissimilatory sulfur metabolism in the purple sulfur bacteria (342).

Some genetic relationships of sulfate-reducing bacteria among themselves and to other bacteria were also shown by comparative rRNA/DNA hybridization. The rRNA of different species was allowed to compete with rRNA of *Desulfovibrio vulgaris* for hybridization with DNA from the latter (229). For rRNA of *Desulfovibrio desulfuricans*, nearly 100% homology with rRNA from *D. vulgaris* was calculated. The degrees of homology of the other rRNAs to the rRNA of *D. vulgaris* were as follows: *Desulfovibrio salexigens*, 67%; *Desulfovibrio africanus*, 59%; *Desulfotomaculum nigrificans*, 38%; *Bacillus stearothermophilus*, 49%; *Alcaligenes faecalis*, 41%; *Vibrio marinus*, 37%; *Escherichia coli*, 31%.

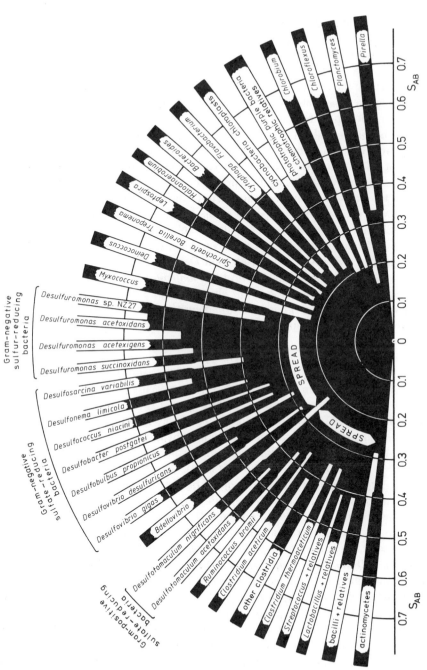

Figure 10.3 Relationships of sulfate-reducing and sulfur-reducing bacteria among themselves and to other eubacteria. Relationships are based on 16S rRNA oligonucleotide catalogues and expressed as similarity coefficients (S_{AB} values). The "*Clostridium*-branch" (bacilli to *Desulfotomaculum nigrificans*) and the branch with the gram-negative sulfate-reducing and sulfur-reducing bacteria (*Bdellovibrio* to *Myxococcus*) are spread to show the lines within these branches in detail.

487

10.2.2.3 Enzymes and Electron Carriers

Redox proteins and other enzymes that are involved in dissimilatory sulfate reduction have been investigated mainly in *Desulfovibrio* and *Desulfotomaculum* species (173, 223, 233, 248; Chapter 11).

ATP sulfurylase, the enzyme that activates sulfate to APS, is not found only in sulfate-reducing bacteria. This enzyme catalyzes also the first step in the assimilatory sulfate reduction in bacteria and plants, or in the biosynthesis of sulfate esters in plants and animals. There are major chemical differences among ATP sulfurylases from bacteria, plants, and animals (7, 71, 339). Electrophoretic characterization of ATP sulfurylases from *Desulfovibrio* and *Desulfotomaculum* strains exhibited considerable variations of this enzyme even within these genera (291).

APS reductase, that in sulfate-reducing bacteria allows the reduction of APS to sulfite and AMP, is also found in species of the sulfide-oxidizing thiobacilli and in phototrophic sulfur bacteria. In the oxidative sulfur metabolism of the latter bacteria, the APS reductase catalyzes the inverse reaction, the AMP-dependent sulfite oxidation to APS. APS reductases from *Desulfovibrio vulgaris*, *Desulfobulbus propionicus*, thiobacilli, and *Chlorobium limicola* resemble each other by containing flavin and nonheme iron-sulfur clusters (320, 339, 343). APS reductases from various *Desulfovibrio* and *Desulfotomaculum* strains behaved electrophoretically rather similar within each genus, but different between the two genera (291).

Sulfate-reducing bacteria contain different types of sulfite reductases (233, 248, 333, 339); if these have pH optima lower than 7, the designation bisulfite reductase is more precise. The bisulfite reductase desulfoviridin occurs in most of the classical lactate- or hydrogen-utilizing *Desulfovibrio* species (248, 249) and in *Desulfomonas pigra* (201, 202). Desulfoviridin can be simply identified by a red fluorescence if cell suspensions are treated with sodium hydroxide solution under UV light (243, 248, 249). Desulfoviridin occurs furthermore in *Desulfococcus multivorans* (238, 371) and *Desulfonema limicola* (366), which nutritionally and morphologically differ significantly from *Desulfovibrio* species. Desulfoviridin is not present in a number of rod-shaped *Desulfovibrio* species, such as *Desulfovibrio* strain Norway (199), *D. baculatus* (263), and *D. thermophilus* (Widdel, unpublished), or so far unclassified facultatively sulfur-reducing *Desulfovibrio* strains (31); nutritionally, these desulfoviridin-negative species behave as typical *Desulfovibrio* species. *Desulfovibrio sapovorans* and *Desulfovibrio baarsii*, which oxidize long-chain fatty acids, also lack desulfoviridin. In *Desulfovibrio* strain Norway (172) and another rod-shaped *Desulfovibrio* species (strain 9974; Fauque, personal communication), a bisulfite reductase termed desulforubidin was detected. This pigment also occurs in *Desulfobulbus* species (266) and *Desulfobacter postgatei* (223). The carbon monoxide–reacting pigment P582 is the bisulfite reductase of *Desulfotomaculum* species (6, 291, 367). A carbon monoxide difference spectrum resembling that of P582 was obtained with the cytoplasm of the nonsporing filamentous *Desulfonema magnum* (366) and *Desulfovibrio sapovorans* (Widdel, unpublished). *Desulfovibrio thermophilus* and *Thermodesulfo-*

bacterium commune contain the same type of bisulfite reductase termed desulfo-fuscidin (104; Fauque and LeGall, personal communication). Another type of bi-sulfite reductase occurs besides desulfoviridin in a *Desulfovibrio vulgaris* strain (75). The same strain also contains a sulfite reductase that by its catalytic properties resembles an assimilatory enzyme (171). Bisulfite reductases from *Desulfovibrio* species (desulfoviridin) and from *Desulfotomaculum* species (P582) were electro-phoretically very similar within each genus, but not between the genera (291). Although the bisulfite reductases from sulfate reducers are different with respect to absorption spectra or chemical properties, these enzymes resemble each other by the presence of a highly carboxylated heme, siroheme, as a prosthetic group which is also present in assimilatory sulfite reductases (284, 285, 339). Further-more, sulfite reductases containing siroheme were found in the aerobic *Thiobacil-lus denitrificans* and the photosynthetic *Chromatium vinosum* (341, 343). In the oxidative sulfur metabolism of these bacteria, sulfite reductase catalyzes sulfide oxidation to sulfite.

Flavodoxins, rubredoxins, and ferredoxins have been studied in *Desulfovibrio* species. Similar spectral and structural properties were observed within the fla-vodoxins or the rubredoxins (101, 209, 355). Among the nonsulfate- or nonsulfur-reducing bacteria studied, *Clostridium* species appear to be the closest relatives of *Desulfovibrio* species with respect to flavodoxin or rubredoxin structure (76, 185, 355). A number of different ferredoxins is known from *Desulfovibrio* species. Some species contain more than one type of ferredoxin (96, 101, 248). Most ferredoxins from *Desulfovibrio* species contain one [4Fe–4S] cluster. *Desulfovibrio* desulfur-icans strain Norway contains an additional ferredoxin with two [4Fe–4S] clusters, as they occur in clostridia. The amino acid sequence of ferredoxin from *Desulfo-vibrio gigas* has been compared to the sequences obtained from nonsulfate-reduc-ing gram-positive and gram-negative bacteria (255). Ferredoxins of *Desulfovibrio* and *Bacillus stearothermophilus* were considered as having evolved from an an-cestral clostridial ferredoxin (255, 278). A certain biochemical relatedness of *De-sulfovibrio* species to clostridia as gram-positives may be also suspected from the occurrence of dissimilatory sulfate reduction among gram-positive bacteria, namely in the sporing genus *Desulfotomaculum*. However, detailed protein analyses from *Desulfotomaculum* species are not available so far.

The independent discovery by Ishimoto et al. (124) and Postgate (242) of a cytochrome in *Desulfovibrio* strains in 1954 was the first finding of this kind of pigments in obligate anaerobes and a fundamental biochemical hint on a respira-tory type of metabolism in sulfate-reducing bacteria. The detected type of cyto-chrome, c_3, turned out to be characteristic of all *Desulfovibrio* species studied (173, 223, 233, 248; Chapter 11); furthermore, it has been detected also in *De-sulfobulbus elongatus* (265a). The cytochromes c_3 are a unique class of hemopro-teins; they contain four or eight hemes with two histidine ligands per iron, exhibit an unusually low standard potential (below −200 mV) and are autooxidizable (173, 223, 233, 248, 252). Similarities and differences between cytochromes c_3 from *Desulfovibrio* species have been studied (73, 173, 223, 286). Besides a few other types of cytochromes (173, 223, 248), *Desulfovibrio* species also contain *b*-type cytochromes (19, 103, 135). The majority of sulfate reducers examined from other nonsporing genera also contained both *b*- and *c*-type cytochromes (Table 10.1). In most *Desulfotomaculum* species, the principal cytochromes are of the *b*-type. *D. ruminis* and *D. acetoxidans* contain even only *b*-type cytochromes (135, 369). *D. guttoideum* is a striking exception since it contains only *c*-type cytochrome (91).

Examination of respiratory quinones in various sulfate-reducing bacteria revealed that all types studied contained menaquinones; ubiquinones were not detected (61). In most cases, species of one genus contained menaquinones with isoprenoid side chains of the same length (Table 10.1). All *Desulfobulbus* strains possessed a menaquinone with five isoprene units and a terminal saturation in the side chain [i.e., MK-5 (V-H$_2$)]. The majority of the classical lactate- and hydrogen-utilizing *Desulfovibrio* species contained menaquinones with a chain of six unsaturated isoprene units (i.e., MK-6); in some of these species, a menaquinone with terminal saturation, MK-6 (VI-H$_2$), was found. The most frequently found menaquinones in the sulfate reducers were MK-7 and MK-7 (VII-H$_2$). Menaquinones with a terminal saturation have been observed so far only in sulfate-reducing bacteria.

Two enzymes the presence of which in sulfate-reducing bacteria appears enigmatic are catalase and superoxide dismutase. These enzymes that otherwise function in oxygen-respiring organisms were observed in *Desulfovibrio vulgaris*, *Desulfovibrio gigas*, and *Desulfovibrio desulfuricans* strain Norway (40, 102). Explanations for the presence or the role of catalase and superoxide dismutase in the strict anaerobes are speculative (69, 102, 248).

10.2.3 Electron Donors and Degradative Capacities

The electron donors oxidized by sulfate-reducing bacteria are always low-molecular-weight compounds; nearly all of these are known to be fermentation products from the anaerobic bacterial degradation of carbohydrates, proteins, and other components of dead biomass. By using an external, inorganic electron acceptor, the sulfate reducers perform a net oxidation of their substrates. Energy is conserved by two major processes: The oxidation of most organic molecules includes substrate-level phosphorylation, whereas the reduction of sulfate, as a respiratory type of metabolism ("sulfate respiration"), is associated with electron transport phosphorylation (18, 19, 233, 248; Chapter 11). The latter is the only ATP-yielding process if species of sulfate-reducing bacteria grow with hydrogen (18, 19, 36) or acetate (37, 88). To our present knowledge, oxidation of these substrates cannot be coupled to a net ATP production by substrate-level phosphorylation. Similarly, species of sulfur-reducing bacteria can grow with hydrogen or acetate by possessing a respiratory type of metabolism (Sections 10.4.1 and 10.4.2). Besides sulfate reducers, sulfur reducers, and denitrifying bacteria, only a few other metabolic types of anaerobes can grow with hydrogen or acetate as sole energy sources: Hydrogen may be used also by methane bacteria (Chapter 13) or homoacetogenic bacteria (Chapter 9), acetate by special methane bacteria or by a thermophilic obligately syntrophic anaerobe (394); the latter as an "inverse homoacetogen" oxidizes acetate to carbon dioxide and liberates hydrogen if this is kept at low partial pressure by a hydrogen-scavenging partner. A further special nutritional feature of species of sulfate-reducing bacteria is the ability to grow with reduced organic compounds that cannot be utilized in pure cultures of fermentative bacteria. Such compounds are propionate, butyrate, higher fatty acids, or phenyl-substituted organic acids; from a bioenergetically point of view, these compounds represent "low-energy substrates" under anaerobic conditions. By using an external electron acceptor, sulfate-reducing bacteria can exploit the reduced products as energy sources. The only other bacteria that anaerobically degrade fatty acids in the dark are some sulfur-reducing bacteria (Section 10.4.1), denitrifying bacteria (Chapter 4), or obligately syntrophic hydrogen-producing acetogenic bacteria in co-cultures with methanogens (Chapter 9). Sulfate-reducing bacteria oxidize also fermentation products which as "high-energy substrates" can be fermented further by other anaerobes; examples are lactate or ethanol. In habitats or in enrichments with these substrates, sulfate reducers have to compete versus fermentative bacteria (162, 169).

Within the nutritionally diverse sulfate-reducing bacteria, two major metabolic groups may be distinguished. Group (a) comprises species oxidizing their substrates incompletely to acetate that cannot be oxidized further. Group (b) embraces the sulfate reducers, which are on principle able to oxidize their organic substrates, including acetate completely to carbon dioxide.

(a) The "classical" lactate-oxidizing species of the genera *Desulfotomaculum* (46) and *Desulfovibrio* (250), which do not oxidize fatty acids, are unable to carry out a terminal oxidation of their organic substrates. Further incomplete oxidizers are *Desulfomonas pigra*, which physiologically resembles the "classical" *Desulfovibrio* species; *Desulfovibrio thermophilus*; *Thermodesulfobacterium commune*; *Desulfobulbus* species; *Desulfovibrio sapovorans* and similar so-far-unnamed fatty acid–oxidizing sulfate reducers with vibrio-shaped cells (237, 238) or oval cells (Fig. 10.1y). Apparently, all these sulfate reducers do not have an operating enzymatic mechanism, like the citric acid cycle, that allows the oxidation of the acetate unit (acetyl-CoA) originating from substrates and which is, therefore, excreted as acetate. Accordingly, acetate added to the medium is not oxidized. Only C_2 compounds such as oxalate and glycine, which are more oxidized than acetate, and C_1 compounds, such as formate and methanol, are completely oxidized. However, acetate may be assimilated as an organic carbon source into cell material if the species grow with hydrogen or formate as electron donors. Some electron donors yield products or fragments other than acetate that may not be metabolized further by the respective sulfate reducer. Examples are the oxidation of *n*-propanol, *n*-butanol, or isobutanol by *Desulfovibrio desulfuricans* to propionate, *n*-butyrate or isobutyrate, respectively (193), or the oxidation of C-odd fatty acids by *Desulfovibrio sapovorans* to acetate plus propionate (237, 238). On the whole, the incompletely oxidizing sulfate reducers are nutritionally less versatile than the completely oxidizing species. The probably most versatile incomplete oxidizer is a *Desulfurovibrio* using not only hydrogen, lactate, pyruvate, malate, fumarate, or ethanol, which are the typical substrates of the "classical" species, but also glycerol, choline, and a number of amino acids (310). The incomplete oxidizers may grow significantly faster than the complete oxidizers (b). Under optimum conditions, *Desulfovibrio* species using hydrogen, lactate, or pyruvate (18, 310), or *Desulfotomaculum* species using pyruvate (Widdel, unpublished) may reach a doubling time of 3 to 4 h. Therefore, batch enrichments with electron donors that can be utilized by both incompletely and completely oxidizing sulfate reducers usually select for the former.

(b) The species of the genera *Desulfobacter, Desulfococcus, Desulfosarcina, Desulfonema*, and *Desulfobacterium* are on principle able to oxidize all their substrates completely to carbon dioxide. *Desulfotomaculum acetoxidans, D. sapomandens*, and *Desulfovibrio baarsii* are also complete oxidizers and differ, therefore, from other representatives of the same genera. There are further so far not classified sporing and nonsporing sulfate reducers that are able to oxidize acetate, fatty acids, or benzoate completely (65, 149, 153). The effectivity of acetate utilization is not the same in all complete oxidizers. Specialized types (*Desulfobacter* species) that use acetate as their only or preferred substrate may grow on this substrate relatively fast (doubling time 20 h or somewhat less) and form high cell yields. Versatile complete oxidizers prefer other organic acids or alcohols and may use acetate from the medium very slowly with low cell yields. Despite their capability of a terminal oxidation, completely oxidizing sulfate reducers may form acetate from organic substrates. Examples are *Desulfotomaculum acetoxidans* growing on ethanol or butyrate (369) or *Desulfobacter* species growing on ethanol (163). Probably, the oxidation of the substrates to the level of acetyl-CoA is faster than its terminal oxidation, so that acetate excretion via acetyl phosphate and a substrate-level phosphorylation is favored. *Desulfobacterium autotrophicum* was shown to form 1 mol of acetate per mole of butyrate used (268a). From this molar ratio it was concluded that acetyl-CoA is used to activate butyrate by a CoA transfer reaction. With limiting amounts of substrates, the acetate excreted by complete oxidizers may be used again. Some species (e.g., *Desulfococcus niacini*) may leave the acetate once formed from a substrate almost untouched, due to a poor and uneffective oxidation of acetate from the medium (118). In such cases, the capacity for a complete oxidation can be recognized only by exact stoichiometric measurements with different electron donors. Formation of propionate has been observed with *Desulfotomaculum acetoxidans* if growing on valerate; unlike other complete oxidizers, this species cannot oxidize proprionate. Among the completely oxidizing sulfate-reducing bacteria, there are nutritionally specialized types, such as *Desulfobacter* species, and nutritionally very versatile species. Nearly all electron donors used by species of incompletely oxidizing sulfate reducers are also degraded by completely oxidizing species. Moreover, the latter may degrate substrates that have not been observed so far to be used by an incompletely oxidizing sulfate reducer. Such compounds (besides acetate) are the branched-chain fatty acids isobutyrate and isovalerate, acetone, phenyl-substituted organic acids, phenolic compounds, and indole, which in enrichments have been selective for complete oxidizers. Most completely oxidizing sulfate reducers grow significantly slower than incompletely oxidizing types. The doubling time reached by the former under optimal conditions is often 20 h or more. Only on a few substrates, some complete oxidizers may reach a doubling time of approximately 10 h, for example, *Desulfotomaculum acetoxidans* growing on butyrate or *Desulfobacterium vacuolatum* growing on fumarate.

The significance of different chemical substances as electron donors for enrichments or pure cultures of sulfate-reducing bacteria is explained in the following.

10.2.3.1 Lactate

For a long time lactate (or pyruvate) has been used as an excellent organic substrate for enrichment, isolation, and cultivation or for determining cell numbers of *Desulfovibrio* and *Desulfotomaculum* species (230, 248, 249). *Desulfovibrio* species were shown to use both L-lactate and D-lactate (233). Lactate is also oxidized by several completely oxidizing sulfate reducers (Table 10.1). The equations of incomplete and complete lactate oxidation are as follows*

$$2CH_3CHOH\ COO^- + SO_4^{2-} \rightarrow 2CH_3COO^- + 2HCO_3^- + HS^- + H^+$$
$$\Delta G^{\circ\prime} = -160.1\ kJ \tag{2}$$
$$2CH_3CHOH\ COO^- + 3SO_4^{2-} \rightarrow 6HCO_3^- + 3HS^- + H^+$$
$$\Delta G^{\circ\prime} = -255.3\ kJ \tag{3}$$

A number of incompletely fatty acid–oxidizing sulfate reducers (exception: *Desulfovibrio sapovorans*) with vibrio-shaped (238) or oval cells (Fig. 10.1 y) are unable to use lactate. Among the complete oxidizers, most *Desulfobacter* species, *Desulfotomaculum acetoxidans*, *Desulfovibrio baarsii*, *Desulfococcus niacini*, some *Desulfobacterium* species, and *Desulfonema magnum* do not grow with lactate.

10.2.3.2 Hydrogen, Formate

Incompletely oxidizing sulfate reducers using lactate are usually able to grow just as well with hydrogen as electron donor (exception: *Desulfovibrio sapovorans*), as demonstrated with different *Desulfovibrio* species (36), *Desulfotomaculum* species (153), *Desulfobulbus* species (370), or *Thermodesulfobacterium commune* (386). Accordingly, *Desulfovibrio* strains isolated from natural sources with hydrogen were all able to grow on lactate (20, 166). *Desulfovibrio* species may grow rather fast on hydrogen, which is, therefore, an almost unfailing electron donor for their enrichment. The utilization of hydrogen by *Desulfovibrio* species was the first hint from nutrition physiology that sulfate reducers can conserve energy solely

---

*All $\Delta G^{\circ\prime}$ values calculated from Ref. 334.

by electron transport phosphorylation. Detailed studies on the bioenergetics of this thermodynamically rather favorable anaerobic reaction [equation (4)] were carried out with *Desulfovibrio vulgaris* strain Marburg (18, 19, 217).

$$4H_2 + SO_4^{2-} + H^+ \rightarrow 4H_2O + HS^- \qquad \Delta G^{\circ\prime} = -152.2 \text{ kJ} \qquad (4)$$

Hydrogen is also used, with relatively slow growth, by several completely oxidizing sulfate-reducing bacteria that may grow autotrophically (see Section 10.2.7). Many sulfate reducers using hydrogen may grow with formate. There are, however, species that only use either hydrogen or formate, for example, *Desulfobulbus propionicus* or *Desulfovibrio baarsii*, respectively.

10.2.3.3 Acetate

Best growth on acetate is observed with *Desulfobacter* species which are nutritionally very specialized sulfate reducers. Some strains use no further substrate besides acetate; other strains may use ethanol in addition (163, 166, 368). Hydrogen was found to be utilized by only two *Desulfobacter* isolates, namely by *D. hydrogenophilus* and, rather slowly, by *D. curvatus* (Widdel, unpublished). *Desulfobacter* species are most easily enriched from brackish or marine sediment in media with sodium and magnesium chloride concentrations that approximately correspond to those in the habitat. However, *Desulfobacter* species, mainly curved cell types, may also develop in enrichments from freshwater sludges; growth of these enrichments is accelerated by using saline media (368). The physiological role of sodium and magnesium chloride for growth of *Desulfobacter* species is not yet understood. An acetate-utilizing sulfate reducer that grows preferentially in freshwater media is *Desulfotomaculum acetoxidans* (369); this sulfate reducer uses acetate more slowly but is nutritionally more versatile than *Desulfobacter* species (Table 10.1). The oxidation of acetate [equation (5)] in *Desulfobacter postgatei* was shown to occur via the citric acid cycle (37, 88). There is also evidence for the same mechanism in *Desulfobacter hydrogenophilus* (268a).

$$CH_3COO^- + SO_4^{2-} \rightarrow 2HCO_3^- + HS^- \qquad \Delta G^{\circ\prime} = -47.6 \text{ kJ} \qquad (5)$$

D. postgatei couples the activation of acetate to acetyl-CoA with the formation of succinate from succinyl-CoA as an intermediate of the citric acid cycle. By this CoA transfer, the oxidative pathway of acetate does not require ATP. Completely oxidizing sulfate reducers other than *Desulfobacter* species, for example *Desulfovibrio baarsii*, *Desulfococcus* species, *Desulfosarcina variabilis*, or *Desulfobacterium* species, oxidize acetate slowly and sometimes even without substantial formation of cell mass. Successful enrichment of the latter types with acetate is unlikely or nearly impossible, since they are overgrown by the faster-growing sulfate reducers or even by methanogens. In the mentioned slow acetate utilizers and in *Desulfotomaculum acetoxidans*, the absence of 2-ketoglutarate dehydrogenase and the incorporation pattern of [^{14}C]carbon sources into amino acids indicated an incomplete citric acid cycle; therefore, an alternative mechanism of acetyl-CoA oxidation was expected (268a). Indeed, these sulfate reducers exhibited carbon monoxide dehydrogenase activity that catalyzed an exchange between free carbon dioxide and the C_1 position of acetyl-CoA or even free acetate. It was concluded that the 2-ketoglutarate dehydrogenase-lacking complete oxidizers cleave activated acetate directly into a carrier-bound methyl group and carrier-bound carbon monoxide that are separately oxidized to carbon dioxide. This mechanism represents an inversion of the acetate synthesis in homoacetogenic bacteria. The appearance of a bound methyl group during acetyl-CoA oxidation would also explain the observed "mini-methane formation" (on the average approximately 1 μmol per liter of grown culture); this was not observed with *Desulfobacter* species.* Acetate activation in the 2-ketoglutarate dehydrogenase-lacking species should require ATP. Among the complete oxidizers that lack a citric acid cycle, utilization of acetate is fastest by *Desulfotomaculum acetoxidans* (doubling time approximately 30 h), although this sulfate reducer grows slower than the *Desulfobacter* species. The relatively slow acetate oxidation by *Desulfonema* species, in which the oxidation mechanism is still unknown, is promoted by additional electron donors (366).

10.2.3.4 Propionate
Incomplete oxidation of propionate to acetate [equation (6)] is a characteristic feature of *Desulfobulbus* species.

$$4CH_3CH_2COO^- + 3SO_4{}^{2-} \rightarrow 4CH_3COO^- + 4HCO_3{}^- + 3HS^- + H^+$$
$$\Delta G^{\circ\prime} = -150.6 \text{ kJ} \quad (6)$$

There is evidence that *Desulfobulbus propionicus* degrades propionate via a randomizing pathway. This includes carboxylation of propionyl-CoA to methylmalonyl-CoA, isomerization to succinyl-CoA, and oxidation of succinate via fumarate and malate finally to acetate (311). *Desulfobulbus* species isolated so far are a physiologically homogeneous group. Growth with lactate may be somewhat faster than with propionate, but only the latter allows selective enrichment. Nearly all *Desulfobulbus* isolates described use ethanol, propanol, or hydrogen (166, 266, 370); a few strains slowly degrade butyrate or 2-methylbutyrate (Widdel, unpublished) to acetate. Among the completely oxidizing sulfate reducers, many species oxidize propionate (Table 10.1).

10.2.3.5 Butyrate and Higher Straight-Chain Fatty Acids
Like other aliphatic substrates, butyrate and higher fatty acids are used by species of both the incompletely and completely oxidizing sulfate-reducing bacteria. En-

---

*However, the earlier reported mini-methane formation from pyruvate by disrupted cells of incompletely oxidizing and nonautotrophic *Desulfovibrio* strains (245a) cannot be explained by the model presented here.

richments with these fatty acids up to palmitate yield in most cases incompletely oxidizing sulfate reducers with small or large vibrio-shaped cells ["sapovorans group" (238)] or with oval cells (Fig. 10.1 y). Whereas fatty acids up to palmitate (C_{16}) allow good growth of these sulfate reducers with a doubling time of 10 to 15 h, stearate (C_{18}) is used distinctively slower. Some species may slowly use arachinate (C_{20}), but no pure culture has been observed so far to oxidize behenate (C_{22}). The C-even fatty acids are degraded to acetate as the entire oxidation product [equation (7)]; C-odd fatty acids, to acetate plus propionate [equation (8)].

$$CH_3(CH_2)_{2n}COO^- + \frac{n}{2}SO_4^{2-} \rightarrow (n+1)CH_3COO^- + \frac{n}{2}HS^- + \frac{n}{2}H^+$$

$$(7)$$

$$CH_3(CH_2)_{2n+1}COO^- + \frac{n}{2}SO_4^{2-} \rightarrow nCH_3COO^- + CH_3CH_2COO^-$$
$$+ \frac{n}{2}HS^- + \frac{n}{2}H^+ \quad (8)$$

These degradation balances can be explained by the presence of a β-oxidation pathway. A few species of sulfate reducers that incompletely oxidize long-chain fatty acids grow on lactate (e.g., *Desulfovibrio sapovorans*); utilization of hydrogen by these types of sulfate reducers was never observed. Stearate (in freshwater media) is a selective substrate for enrichment of *Desulfovibrio baarsii*, which completely oxidizes only monocarboxylic acids from formate to stearate. Other nutritionally more versatile complete oxidizers, such as *Desulfosarcina variabilis*, *Desulfococcus*, *Desulfonema*, and *Desulfobacterium* species, may grow on fatty acids up to chain lengths of 10 to 16 carbon atoms. The general equation for a complete oxidation of monocarboxylic acids is as follows:

$$4H(CH_2)_nCOO^- + (3n+1)SO_4^{2-} + H^+ \rightarrow (4n+4)HCO_3^-$$
$$+ (3n+1)HS^- + nH^+ \quad (9)$$

There are also sporing sulfate reducers, for example, *Desulfotomaculum sapomandens* (63a) and so far unclassified species (153), that oxidize long-chain fatty acids. *Desulfotomaculum acetoxidans* oxidizes butyrate and valerate but not higher fatty acids.

10.2.3.6 Branched-Chain Fatty Acids

Oxidation of isobutyrate and isovalerate (3-methylbutyrate) has only been observed with species of completely oxidizing sulfate reducers. Isobutyrate as an organic substrate in marine enrichments promotes development of the gliding filamentous *Desulfonema limicola* (Fig. 10.1 w); isovalerate may select for *Desulfococcus multivorans* or types with oval cells, probably *Desulfobacterium* species

(Widdel, unpublished). Of the branched-chain fatty acids, isovalerate is by far the most slowly utilized. 2-Methylbutyrate is oxidized not only by completely oxidizing but also by incompletely oxidizing sulfate reducers, namely, by *Desulfovibrio sapovorans* and by special *Desulfobulbus* strains (Widdel, unpublished).

10.2.3.7 Monovalent Alcohols

The ability to grow with ethanol as electron donor is very common among incompletely and completely oxidizing sulfate reducers. The higher homologs such as propanol, *n*-butanol, or isobutanol may be used, too; whereas *Desulfovibrio* strains that cannot oxidize propionate or other fatty acids form the corresponding acids as end products from the higher alcohols (193), fatty acid–oxidizing sulfate reducers may perform a further oxidation. Batch enrichments or pure cultures on ethanol or the higher alcohols sometimes cease to grow after a while and produce intensely smelling organic sulfur compounds that seem to affect the bacteria; these compounds are probably formed from aldehydes and hydrogen sulfide (223a). 2-Propanol and especially the corresponding ketone, acetone, allow rather selective, but slow enrichment of *Desulfococcus multivorans*. In enrichments with 2-propanol, *Desulfovibrio*-like species that form acetone may also appear (Widdel, unpublished). Another alcohol scarcely used by sulfate reducers is methanol; until now, only *Desulfotomaculum orientis*, another strain of the genus (153), two *Desulfovibrio* strains (38, 65b), and *Desulfobacterim catecholicum* (325a) were reported to oxidize this alcohol. Growth of the sulfate reducers on methanol was slow, with a doubling time of around 1 day or much longer. One of the *Desulfovibrio* strains did not even grow when metabolizing the methanol (38). Other, classified species, even if they grew well with ethanol, did not metabolize methanol (Widdel, unpublished).

10.2.3.8 Dicarboxylic Acids

Some *Desulfovibrio* species (248, 249) and a number of completely oxidizing sulfate reducers (e.g., *Desulfococcus niacini* and *Desulfobacterium* species) grow with succinate, fumarate, or malate. Growth with the latter two compounds is often faster than with succinate. Reports on the utilization of lower or higher dicarboxylic acids are rare. *Desulfovibrio vulgaris* subsp. *oxamicus* oxidizes oxalate (C_2) or its monamide, oxamate (245, 248, 249, 250). *Desulfococcus niacini* grows on glutarate (C_5) or pimelate (C_7).

10.2.3.9 Aromatic Compounds

Degradation of aromatic compounds has been observed so far only with completely oxidizing sulfate reducers. Many of these have been directly enriched and isolated with an aromatic substrate, for example, *Desulfococcus multivorans*, *Desulfosarcina variabilis*, and *Desulfonema magnum* with benzoate, *Desulfococcus niacini* with nicotinate, *Desulfobacterium phenolicum* with phenol, *Desulfobacterium indolicum* with indole, or *Desulfobacterium catecholicum* with catechol. There are also sporeforming sulfate reducers that oxidize benzoate (63a, 153; Table 10.1).

The *Desulfococcus* species, *Desulfosarcina variabilis*, *Desulfobacterium catecholicum*, and *Desulfotomaculum sapomandens* may be the nutritionally most versatile types of sulfate reducers (Table 10.1). Enrichment and counting series revealed that anaerobic habitats may harbor further benzoate-utilizing sulfate reducers, mainly species with oval cells, that have not yet been classified (64; Fig. 10.1z). The stoichiometry of benzoate oxidation is as follows:

$$4C_6H_5COO^- + 15SO_4^{2-} + 16H_2O \rightarrow 28HCO_3^- + 15HS^- + 9H^+ \quad (10)$$

Benzoate-degrading species often use phenylacetate, phenylpropionate, or one or two of the three different hydroxybenzoates (371). *Desulfococcus multivorans* and *D. niacini* require an addition of selenite to the medium for growth on phenyl-substituted acids or nicotinate, respectively, but not for growth on aliphatic substrates (118, 371). Selenium, in a functional group of enzymes, appears to play a role in ring cleavage. Similar functions of selenium have been observed in fermentative bacteria growing on xanthine (358) or nicotinate (117). Phenylpropionate may also be used by some sulfate reducers that cannot oxidize benzoate, for example, by *Desulfococcus niacini* (118) or *Desulfobacterium vacuolatum* (Widdel, unpublished). These species oxidize phenylpropionate, probably by one β-oxidation step, to benzoate as an end product.

10.2.3.10   *Saturated Cyclic Organic Acids*

Cyclopentane- and cyclohexane-substituted carboxylic acids may constitute a part of the acidic compounds (naphthenic acids) in crude mineral oil. Cylcohexanecarboxylate has been discussed as an intermediate in anaerobic benzoate degradation (80). Indeed, this compound is used by some (but not all) benzoate-utilizing sulfate reducers and allows enrichment of *Desulfococcus multivorans* (Widdel, unpublished). Cyclohexanecarboxylate may be also used by a few fatty acid–oxidizing *Desulfobacterium* types that do not grow on benzoate (Widdel, unpublished). Cyclopentanecarboxylate, cyclopentylacetate, and cyclohexylacetate were not oxidized in growth tests with different sulfate reducers isolated on fatty acids or benzoate. Direct enrichments with these three cyclic acids yielded so far unknown, very slowly growing sulfate-reducing bacteria (Widdel, unpublished).

10.2.3.11   *Choline, Amino Acids, Sugars, Glycerol*

Choline is oxidized by certain *Desulfovibrio* strains to trimethylamine and acetate (82, 283, 310). Amino acids, sugars, and glycerol, although quantitatively important constituents in biomass, have seldom been reported to be used by sulfate-reducing bacteria. Enrichments with these well-fermentable compounds usually yield fast-growing fermentative bacteria before any sulfate reducer that may use the same substrates can develop. Sulfate reducers that in substrate tests turned out to grow on amino acids were *Desulfotomaculum ruminis* using alanine (58), a sulfate reducer, presumably a *Desulfovibrio*, using cysteine (282), *Desulfococcus niacini* using glutamate (118), and a versatile *Desulfovibrio* type that exhibited reasonable growth on alanine, serine, or glycine, and slow growth on aspartate, cysteine, or threonine (310). The latter sulfate reducer also oxidized glycerol. A *Desulfovibrio* isolate (strain JJ) and *Desulfotomaculum nigrificans* were able to grow on fructose* (65b, 153); *Desulfotomaculum antarcticum* has been reported to use glucose (116).

10.2.3.12 Hydrocarbons

An oxidation of methane or higher, saturated hydrocarbons by sulfate-reducing bacteria or other anaerobes was the subject of controversial observations and discussions (9a, 9b, 127a, 222, 256, 260, 304, 382; Chapter 14). Thermodynamically, an oxidation of these hydrocarbons would be possible, although in the case of an assumed methane oxidation, the free energy change would be rather low:

$$CH_4 + SO_4{}^{2-} \rightarrow HCO_3{}^- + HS^- + H_2O \qquad \Delta G^{\circ\prime} = -16.6 \text{ kJ} \qquad (11)$$

The assumption of an anaerobic methane oxidation is first based on observed methane profiles in anoxic sediments; the profile shape was best explained by postulating anaerobic methane oxidizers (9a, 9b). Secondly, studies with ^{14}C-labeled methane in sediments suggested an anaerobic methane oxidation (9a, 9b, 127a, 382). In anaerobic marine sediments investigated in Denmark, the maximum of the methane oxidation coincided with the maximum of sulfate reduction (127a). It was calculated that methane could account for 10% of the total electron donors used for sulfate reduction. Nevertheless, at the present time, no species of sulfate reducers is known to oxidize methane. From observations of Zehnder and Brock it was concluded that anaerobic methane oxidation could be a reverse reaction catalyzed by methanogenic bacteria themselves to a low extent during methane formation (382). However, inhibition studies in sediments with molybdate and 2-bromoethanesulfonate suggested that sulfate reducers or methanogens, respectively, did not perform the anaerobic methane oxidation, but that other, unknown anaerobes were responsible for the reaction. Biochemically, there is no hypothesis so far that could explain a reaction of the nonpolar methane or higher hydrocarbons in the absence of oxygen. Nevertheless, in recent saline enrichments with hexadecane that were inoculated with oil-field water, a sulfate-reducing culture developed after 4 months of incubation (Bak and Widdel, unpublished results). Subcultures (10% inoculum) grew within 4 weeks and produced more than 10 mmol of sulfide per liter. Pentadecane, heptadecane, and octadecane could also serve as the only organic substrates, whereas with hydrocarbons of lower molecular weight the sulfide production decreased significantly. The enrichments grew only under strictly anaerobic conditions. The dominant cell types were small rods mostly attached to the hexadecane phase. Via dilution series in the presence of hexadecane, this type of bacterium could be separated from acetate- and fatty acid–oxidizing

---

*A formerly observed glucose utilization by *Desulfotomaculum nigrificans* (8) was not reproducible if the substrate was filter-sterilized; only autoclaved glucose allowed growth (153).

sulfate reducers that were also present in the enrichment. In the absence of the latter, the small rod-shaped bacterium continued to produce sulfide and in addition some waxlike film attached to the hexadecane. As final prove of a hexadecane utilization by the observed anaerobe, quantitative measurements with substrates and products have to be carried out. An anaerobic degradation of unsaturated hydrocarbons, namely, of hexadecene-(1) and squalene by methane-forming enrichment cultures, has been described by Schink (275; Chapter 14). The double bonds may facilitate or enable an anaerobic enzymatic attack. Under intermittent aerobic/anaerobic conditions, hydrocarbons can also indirectly serve as substrates for anaerobic bacteria. Aerobic hydrocarbon- and petrol-oxidizing bacteria are known to release monocarboxylic acids that may get under anaerobic conditions and serve as substrates for sulfate reducers or methanogenic populations (13, 29, 161, 264). Furthermore, it has to be considered that the cells of the aerobic hydrocarbon-oxidizing bacteria may be anaerobically degraded and yield electron donors for sulfate reducers (Cord-Ruwisch, personal communication).

10.2.4 Fermentative Capacities and Syntrophic Growth with Methanogenic Bacteria

In the absence of sulfate or other inorganic electron acceptors, several types of sulfate reducers can grow by fermentation of special organic substrates. Some *Desulfovibrio* species (198, 200, 281, 310), *Desulfobacterium* species (Widdel, unpublished), and *Desulfosarcina variabilis* (371) ferment fumarate and, with the exception of the latter species, malate; the fermentation products are succinate, acetate, carbon dioxide, and sometimes propionate (371). Pyruvate is easily fermented by many sulfate reducers belonging to the genera *Desulfovibrio* (241, 248–250, 281), *Desulfotomaculum* (46, 248), *Desulfobulbus* (370), *Desulfococcus* (118, 371), and *Desulfobacterium* (Widdel, unpublished). As products of pyruvate fermentation by *Desulfovibrio desulfuricans*, acetate, carbon dioxide, and hydrogen were found (248, 281, 310). In *Desulfovibrio sapovorans*, which does not possess a hydrogenase (237, 249), lactate, acetate, and carbon dioxide are expected to be the fermentation products. Lactate or ethanol plus carbon dioxide allow good fermentation growth of several *Desulfobulbus* strains that form propionate plus acetate (162, 370). Thus *Desulfobulbus* cannot only oxidize but also form propionate. Ethanol-limited, sulfate-free chemostat enrichments inoculated with anaerobic sediment samples may be selective for fermenting *Desulfobulbus* strains (162). Propionate is formed by *Desulfobulbus* via a randomizing pathway as occurring in propionic acid bacteria (311). *Desulfovibrio desulfuricans* can ferment choline to trimethylamine, ethanol, and acetate (21, 82, 283). A *Desulfovibrio* strain and *Desulfotomaculum nigrificans* fermented fructose in the absence of sulfate (65b, 153). The *Desulfovibrio* formed succinate, acetate, and ethanol as products. Furthermore, a fermentation of cysteine with liberation of sulfide and ammonia has been reported for a sulfate reducer, probably a *Desulfovibrio* strain (282). *Desulfotomaculum orientis* grew slowly by converting formate, methanol, or the methoxyl groups of 3,4,5-trimethoxybenzoate via a homoacetogenic metabolism to acetate (153).

Desulfovibrio species, that cannot ferment ethanol or lactate may grow with these compounds in the absence of sulfate if co-cultured with hydrogen-scavenging methanogenic bacteria (41, 190, 336). In these syntrophic associations, the sulfate reducers serve as hydrogen-producing acetogenic bacteria (Chapter 9). The reactions are as follows:

Desulfovibrio:

$$2CH_3CHOH\ COO^- + 2H_2O \rightarrow 2CH_3COO^- + 2CO_2 + 4H_2 \quad (12)$$

Methanogen:

$$4H_2 + CO_2 \rightarrow CH_4 + 2H_2O \quad (13)$$

Total reaction:

$$2CH_3CHOH\ COO^- \rightarrow 2CH_3COO^- + CH_4 + CO_2 \quad (14)$$

From a methanogenic enrichment culture with choline, a *Desulfovibrio* species was isolated (82). In co-culture with a methanogenic bacterium, the *Desulfovibrio* converted choline to trimethylamine, acetate, and hydrogen; the latter was scavenged by the methanogenic partner. Fatty acids or benzoate were not degraded by co-cultures of sulfate reducers with methanogens, even if the sulfate reducers tested possessed hydrogenases and grew syntrophically with lactate (Schneider and Widdel, unpublished). Apparently, in the case of fatty acids or benzoate, sulfate reducers cannot substitute for the obligately syntrophic acetogenic bacteria that have been obtained in co-cultures with methanogens (32, 191, 192, 207, 208, 319).

10.2.5 Electron Acceptors

Sulfate-reducing bacteria reduce their electron acceptor, sulfate, usually to sulfide as the entire, final product. There is only one report that sulfite and thiosulfate were excreted at low concentrations (thiosulfate up to approximately 0.1 mmol/L) in addition to sulfide by *Desulfovibrio desulfuricans* during the first 4 days of growth (353).

Most sulfate-reducing bacteria can grow with sulfite or thiosulfate as electron acceptors instead of sulfate (18, 69, 240, 371). Sulfite and probably also thiosulfate are intermediates in sulfate reduction (Chapter 11). If sulfite or thiosulfate are added as electron acceptors to the medium, an ATP-consuming activation as occurring with sulfate is not necessary; therefore, the cell yields per amount of electron donor oxidized may be higher than with sulfate (Table 10.4). *Desulfovibrio sapovorans* does not reduce thiosulfate (69); *Desulfotomaculum acetoxidans* (369), *Desulfobacter latus* (Widdel, unpublished), and *Desulfonema magnum* (366) reduce neither sulfite nor thiosulfate, probably because the species lack an appropriate transport system for these sulfur compounds. Trithionate, another assumed intermediate of sulfate reduction, was not reduced when tested (125, 240). In contrast, tetrathionate and dithionite, which are not known to be intermediates of sulfate reduction, were reduced.

Growth of sulfate-reducing bacteria with elemental sulfur (sulfur flower) as electron acceptor has been observed with some rod-shaped isolates (including *Desulfovibrio* strain Norway 4) that by their carbon metabolism resembled common *Desulfovibrio* sp. (31). These facultatively sulfur-reducing strains all lacked desulfoviridin. The only vibrio-shaped, desulfoviridin-positive sulfate reducers that grew with elemental sulfur were an isolate resembling *Desulfovibrio gigas* (31, 81) and a facultatively nitrate-reducing strain considered to be classified as *Desulfovibrio multispirans* (Fauque and LeGall, and personal communication). Colloidal sulfur may allow weak growth of *Desulfovibrio desulfuricans* that does not reduce sulfur flower (240). Sulfur may also inhibit growth of sulfate-reducing bacteria in the presence of sulfate, as shown with *Desulfotomaculum acetoxidans* (367, 369), *Desulfovibrio sapovorans* (Widdel, unpublished), *Desulfobacter postgatei* (368), or *Desulfonema* species (366). Obviously, the presence of elemental sulfur is incompatible with the strongly reducing conditions required by these species in their media.

Desulfovibrio desulfuricans, a few other *Desulfovibrio* strains, including the isolate with the proposed name *Desulfovibrio multispirans* (Fauque and LeGall, personal communication), *Desulfobulbus propionicus*, and *Desulfobacterium catecholicum* were shown to use nitrate as electron acceptor (26, 176, 187, 325a, 370). The end product of this type of dissimilatory nitrate reduction was ammonia, not dinitrogen as in true denitrifiers (Chapter 4). *Desulfobacterium catecholicum* is the only bacterium known so far that couples a dissimilatory nitrate reduction to ammonia with a complete oxidation of acetate and other organic substrates (325a).

Some *Desulfovibrio* species can grow with fumarate as an alternative electron acceptor for oxidation of hydrogen or lactate (25, 95, 200, 248). Fumarate is reduced to succinate. *Desulfovibrio desulfuricans* reduces fumarate even prior to sulfate.

Oxygen is inhibitory to sulfate reducers as strictly anaerobic bacteria. Nevertheless, oxygen may allow sulfate-free growth of sulfate-reducing bacteria in zones of opposed oxygen-sulfide gradients. In these, the sulfate reducers use a chemical oxidation product of sulfide, most probably thiosulfate, as electron acceptor, without getting into direct contact with oxygen (69). The oxidation product is again reduced to sulfide. The resulting sulfur cycle mediates between the sulfate-reducing bacteria and the otherwise harmful oxygen that serves indirectly as final electron acceptor.

10.2.6 Disproportionation of Inorganic Sulfur Compounds

A metabolically unique capacity was recently detected by Bak in counting series and enrichment cultures originally designed to quantify expected thiosulfate-reducing bacteria that oxidize acetate. However, the cultures yielded almost exclusively vibrio-shaped anaerobes that did not oxidize the added acetate, but gained energy for growth solely by dismutation (disproportionation) of thiosulfate to sulfate and sulfide (20a). Acetate served merely as an organic carbon source for cell synthesis. The species isolated in pure culture was also able to dismutate sulfite or dithionite. In the presence of sulfate as electron acceptor, organic substrates such as lactate or ethanol were incompletely oxidized (Table 10.1); by such a metabolism and its curved cell shape the isolate resembled the classical *Desulfovibrio* species and was thus named *Desulfovibrio sulfodismutans*. In the presence of organic electron donors and thiosulfate or sulfite, *D. sulfodismutans* first disproportionated the sulfur compounds and then oxidized the organic substrates with the formed sulfate.

Examination of various sulfate reducers also showed a weak capacity to disproportionate thiosulfate or sulfite in *Desulfobacter curvatus*, whereas *Desulfovibrio vulgaris* (strain Marburg) slowly disproportionated only sulfite (20a). Further enrichment series yielded rod-shaped forms that only disporportionated thiosulfate and grew in the presence of carbon dioxide without an organic carbon source [autotrophic growth (Section 10.2.7)].

The disproportionation of thiosulfate and sulfite occurred according to the following equations (20a):

$$S_2O_3{}^{2-} + H_2O \rightarrow SO_4{}^{2-} + HS^- + H^+ \qquad \Delta G°' = -21.9 \text{ kJ} \qquad (15)$$
$$4SO_3{}^{2-} + H^+ \rightarrow 3SO_4{}^{2-} + HS^- \qquad \Delta G°' = -235.6 \text{ kJ} \qquad (16)$$

The reactions demonstrate that the coexistance of the most reduced and the most oxidized state of sulfur is energetically more stable than one of the intermediate oxidation states alone. However, a disproportionation of elemental sulfur to the same products at pH 7 would be energetically unfavorable:

$$4S + 4H_2O \rightarrow SO_4{}^{2-} + 3HS^- + 5H^+ \qquad \Delta G°' = +40.9 \text{ kJ} \qquad (17)$$

This suggests that the thermodynamically most stable form of sulfur at neutral or low pH in the absence of hydrogen donors is the element itself. Nevertheless, organisms that "comproportionate" sulfate and sulfide to elemental sulfur have not been observed. By the scavenge of H^+ ions at high pH, a disproportionation of elemental sulfur becomes exergonic and occurs as a chemical reaction; the products are sulfide, polysulfide, and diverse, barely identified oxygen–sulfur compounds (see, e.g., Ref. 28). Vice versa, at low pH, bisulfite, sulfur dioxide, or also thiosulfate comproportionate spontaneously with hydrogen sulfide to elemental sulfur and varying proportions of tetrathionate and other polythionates ("Wackenroder's liquid"; 90a).

10.2.7 Assimilatory Capacities and Autotrophic Growth

All sulfate-reducing bacteria can be cultivated in defined media without complex supplements such as yeast extract, even if addition of the latter is sometimes useful for stimulation of growth; only addition of a few vitamins as growth factors may be necessary. Thus sulfate reducers are able to synthesize their cell material from the simple organic compounds that also serve as electron donors, and in addition from carbon dioxide.

The assimilatory capacities of sulfate-reducing bacteria are of special interest since they have been cultivated with hydrogen as an inorganic electron donor (372). The first investigations on pure cultures of *Desulfovibrio* strains in mineral media in the presence of hydrogen and carbon dioxide led to the assumption that these anaerobes were able to synthesize all their cell material from carbon dioxide and thus were facultative autotrophs (42). However, in later detailed analytical measurements with *Desulfovibrio* strains growing on hydrogen, an autotrophic growth could not be confirmed (193, 244). It was assumed that traces of organic carbon sources were present in the media that had been used in the former studies on hydrogen-consuming sulfate reducers. Indeed, further investigations on *Desulfovibrio* species growing on hydrogen revealed that they required at least an organic C_2-compound such as acetate in addition to carbon dioxide for cell synthesis. Only about one-third of the cell material was derived from carbon dioxide; two-thirds was derived from acetate (17, 20, 36, 305–307). The assimilation ratio is explained by a reductive carboxylation of the activated acetate, acetyl-CoA, to pyruvate (pyruvate synthase reaction) as the first step in cell synthesis, a reaction observed in many anaerobic bacteria (17). Thus, the nutrition status of *Desulfovibrio* species growing with hydrogen has to be designated as chemolithoheterotrophic. Further typical chemolithoheterotrophic sulfate reducers are strains of the genus *Desulfobulbus* (370). The carbon source acetate is also required by nonautotrophic *Desulfovibrio* species if they grow with electron donors that do not lead to usable organic carbon sources. Such an electron donor is, for example, isobutanol; since it is oxidized by *Desulfovibrio* species to isobutyrate as an end product, addition of acetate or yeast extract for cell synthesis is necessary (193). The same carbon sources have to be added if the electron donor is converted exclusively to carbon dioxide, as in the case of formate, oxalate, or oxamate (245, 248, 250, 305–307). Interestingly, a nonautotrophic *Desulfovibrio* type did not require an additional organic carbon source when grown on glycine, which is more oxidized than acetate; in contrast, glycine was partly converted to acetate, possibly via serine (310).

The observed simultaneous assimilation of carbon dioxide together with acetate into cell material has implications as to the cultivation of sulfate-reducing bacteria. If these grow with organic electron donors that are exclusively oxidized to acetate, a requirement for carbon dioxide is expected. Therefore, for the growth of incompletely oxidizing sulfate reducers on ethanol or fatty acids, carbon dioxide (or bicarbonate) should be present. But even complete oxidizers that form carbon dioxide during growth may require a certain initial concentration of carbon dioxide for synthetic purposes.

After the classical *Desulfovibrio* species had been shown to be chemolithoheterotrophs, autotrophic growth of hydrogen-using sulfate reducers was reported for *Desulfosarcina variabilis* (238), *Desulfonema limicola* (366), and *Desulfococcus niacini* (118); however, growth was rather slow.

Desulfococcus multivorans and *Desulfovibrio baarsii*, which do not take hydrogen, were shown to grow with formate without an additional organic carbon source (130, 131, 238). Later, *Desulfotomaculum orientis*, which has been cultivated since its isolation in complex media, turned out to be capable of autotrophic growth with hydrogen as electron donor (153). Newly isolated types of sulfate reducers growing under autotrophic conditions with hydrogen are *Desulfobacterium autotrophicum* and a number of related strains, and *Desulfobacter hydrogenophilus* (Table 10.1). The former was isolated by direct dilution of natural samples in agar tubes with hydrogen. Enrichments with hydrogen in liquid mineral media mostly yield mixed cultures of nonautotrophic *Desulfovibrio* sp. and autotrophic homoacetogens of the *Acetobacterium* type (Chapter 9) that feed the sulfate reducers with acetate. Such enrichments may simulate cultures of autotrophic sulfate reducers, especially because of the low cell numbers of *Acetobacterium* compared to the cell density of *Desulfovibrio* (Brysch, Schneider, Fuchs, and Widdel, unpublished).

Desulfobacterium strains and *Desulfotomaculum orientis* were grown for labeling studies with [^{14}C]carbon dioxide under autotrophic conditions. Chemical reoxidation of washed grown cells to carbon dioxide and radioactivity measurements revealed that 91 to 99% of the cell carbon was derived from carbon dioxide (Brysch, Schneider, Fuchs, and Widdel, unpublished).

The pathway of assimilation of formate and carbon dioxide was studied in *Desulfovibrio baarsii* (130, 131). The results suggest a synthesis of an acetate unit by condensation of a carrier-bound methyl group with bound carbon monoxide, both being derived from carbon dioxide. This pathway seems to invert the reactions that during growth on organic substrates are used for the terminal oxidation to carbon dioxide (Section 10.2.3). The carbon dioxide assimilation is analogous to the acetate formation in homoacetogenic bacteria (Chapter 9). Indeed, *Desulfovibrio baarsii*, *Desulfobacterium* strains, and *Desulfotomaculum orientis* may excrete acetate (around 1 mmol/L) into the medium if growing with hydrogen plus carbon dioxide or with formate in the presence of sulfate. Whereas the former species do not grow in the absence of sulfate, *Desulfotomaculum orientis* can grow slowly without sulfate by converting formate, methanol, or the methoxyl groups of 3,4,5-trimethoxybenzoate to acetate; sulfate-free growth with hydrogen and carbon dioxide was very poor (153). Among the facultatively autotrophic sulfate reducers, *Desulfotomaculum orientis* is the only one that is unable to perform a complete oxidation. Acetate is always stoichiometrically formed as an end product (153). There is no explanation why this species cannot use the pathway of acetyl-CoA synthesis in an inversed, oxidative manner.

Desulfobacter hydrogenophilus is an autotrophe that lacks carbon monoxide dehydrogenase but possesses 2-ketoglutarate dehydrogenase, a key enzyme of an operative citric acid cycle (Section 10.2.3). One may, therefore, expect that carbon dioxide is assimilated via a type of reductive citric acid cycle. Most of the postulated enzymes and an ATP citrate lyase have been found in preliminary enzymatic studies (Schauder, personal communication).

Since, like in other anaerobes, the part of substrate being assimilated in sulfate reducers is low in comparison to the catabolized amounts (ratio on the average 1 : 10), a shortage of organic carbon sources for cell synthesis is very unlikely in anaerobic habitats. It appears that the autotrophic growth of sulfate reducers, except that of *Desulfotomaculum orientis*, is a secondary effect of enzymes that primarily function dissimilatory (i.e., in the oxidation of acetyl-CoA). In case of the incompletely oxidizing *Desulfotomaculum orientis*, the autotrophic growth may reflect a relationship to *Acetobacterium woodii* and homoacetogenic clostridia; these are all gram-positives (Section 10.2.2).

10.2.8 Nitrogen Sources

Ammonium ions are commonly used as nitrogen sources by sulfate-reducing bacteria. *Desulfobulbus propionicus* also grew with glutamate as the nitrogen source while using lactate or propionate as electron donors and sulfate as an electron acceptor (Bomar and Widdel, unpublished). Sulfate reducers capable of a dissimilatory nitrate reduction to ammonia (Section 10.2.5) are expected to use nitrate as a nitrogen source. Fixation of molecular nitrogen was observed with species of the genera *Desulfovibrio* (248, 251, 251a, 257), *Desulfotomaculum* (246; Fauque, personal communication), *Desulfobulbus*, and *Desulfobacter* (Bomar and Widdel, unpublished).

10.3 ECOLOGY OF SULFATE-REDUCING BACTERIA

The growth physiology of sulfate-reducing bacteria reflects their distribution and role in nature. They require an aerobic environment in which organic substrates and dissolved sulfate are available. Such conditions are most likely in the depths of aquatic habitats. Most natural waters contain sulfate. Furthermore, the availability of oxygen for biological degradation processes in water is limited by the solubility of this gas. If organic matter accumulates in aquatic environments, oxygen becomes depleted rather quickly; the degradation of the biomass is then taken over by anaerobic bacterial communities, a part of which are the sulfate-reducing bacteria.

The reactions by which sulfate-reducing bacteria are involved in anaerobic degradation are indicated by their metabolic capacities. As far as tested, these bacteria use low-molecular-weight compounds as electron donors (Section 10.2.3) and therefore depend on fermentative bacteria that degrade the original polymers from biomass. Thus sulfate reducers are terminal degraders and their role is analogous to that of methanogenic bacteria that form methane and carbon dioxide as final anaerobic products. The cognition of dissimilatory sulfate reduction and methanogenesis as two alternative terminal processes has contributed a lot to our understanding of anaerobic mineralization.

10.3.1 Sulfate Reduction and Alternative Degradation Processes

In 1886, Hoppe-Seyler reported on degradation balances in anaerobic enrichment cultures with cellulose in the absence or the presence of gypsum (111). Without sulfate, the reducing equivalents from the organic substrate appeared as methane, whereas in the presence of sulfate they were quantitatively recovered as hydrogen sulfide. Later, this effect of sulfate, which obviously changed the anaerobic degradation from methanogenesis to sulfide formation, became the subject of several investigations (e.g., Refs. 44, 47–49, 183, and 375). The apparent inhibition of methanogenesis is not sufficiently explained by either a toxic effect of sulfide on methanogens or by an unfavorable change of redox potentials due to sulfate. Methanogens may tolerate relatively high sulfide concentrations of several millimoles per liter (204, 359, 383), and if limiting sulfate concentrations are added to sediments, methane production starts again after sulfate depletion. Also, sulfate as a chemically inert molecule does not, like an oxidant, affect pure cultures of methanogens. A sulfate-dependent methane oxidation that could account for the almost complete cessation of methane formation in the presence of sulfate appears plausible only if effective methane-scavenging sulfate reducers exist; however, the existence of methane-oxidizing sulfate reducers has not been shown so far by pure culture experiments (Section 10.2.3.12). The effect of sulfate is most plausibly explained if methanogenesis and sulfate reduction are regarded as alternative degradation reactions that compete for common substrates (359a, 375). As long as sulfate is present, electron flow into sulfate reduction would be the favored reaction. This hypothesis of a dissimilatory sulfate reduction versus methanogenesis agrees well with many studies on substrate flow in the presence and absence of sulfate in sediments of lakes and marine environments (4, 178, 180, 205, 206, 359a).

If sulfate reduction is compared to other degradation processes with external electron acceptors, it appears in a "hierarchy" of alternative microbial degradation pathways (303, 326; Chapter 1). As long as oxygen is present as the electron acceptor, denitrification, sulfate reduction, and methanogenesis are inhibited. Denitrification starts after consumption of oxygen, and sulfate reduction starts after consumption of nitrate. Finally, methanogenesis occurs that is partly a dissimilatory carbon dioxide reduction (Chapter 13). The sequence of these reactions agrees with the sequence of their free energy changes, which decrease from respiration to methanogenesis. This is also reflected in the fact that the product from an electron acceptor of a "lower rank" can, at least theoretically (Section 10.2.3.12), be oxidized with any "higher-ranking" electron acceptor. Nevertheless, thermodynamics alone are insufficient to explain the preference of one reaction over another in microbial populations. The physiological effects of electron acceptors and the kinetics of the corresponding reactions both have to be considered. Oxygen and probably also nitrite as an intermediate from denitrification are potent oxidants that affect the strictly anaerobic bacteria. Since similar physiological effects of sulfate on methanogens have not been observed, the halt in methanogenesis in sediments upon sulfate addition should allow an interpretation in terms of substrate consumption kinetics.

In the following simplified derivation, m_1 and m_2 are the cell masses of sulfate reducers or methanogens, respectively, that compete in a sediment or a defined mixed culture for a common substrate of the (constant) concentration S. If the rates of substrate consumption (units substrate per time) by the sulfate reducer and the methanogen, R_1 and R_2, respectively, are calculated from Michaelis–Menten kinetics*, the ratio of the rates is

$$\frac{R_1}{R_2} = \frac{V_{max1}(K_{M2} + S)}{V_{max2}(K_{M1} + S)} \frac{m_1}{m_2} \tag{18}$$

V_{max1} and V_{max2} are the corresponding specific maximum consumption rates (units of substrate per time and cell mass, at saturation), and K_{M1} and K_{M2} are the substrate concentrations at which the consumption rates are half of the maximum. At limiting, very low substrate concentration ($S \ll K_M$) such as that present in sediments, the following approximation is derived:

$$\frac{R_1}{R_2} \approx \frac{V_{max1}/K_{M1}}{V_{max2}/K_{M2}} \frac{m_1}{m_2} \tag{19}$$

According to this equation, the ratio V_{max}/K_M is a parameter that, together with cell mass, appears to govern the partition of a species in the total turnover of a limiting substrate, provided that growth is negligible during the period being examined. As explained elsewhere (105, 259), the competitive advantage of one growing species over another is predicted in terms of Monod kinetics, that is, by the ratio μ_{max}/K_s (μ_{max} = maximum growth rate). This ratio reflects the slope of the Monod equation at the very low concentrations at which the growth rate depends on the substrate concentration by a first-order reaction.

At high substrate concentration ($S \gg K_{M1}, K_{M2}$), the maximum consumption rate of nongrowing cells appears to be the only kinetic parameter that governs competition, as derived from equation (18):

---

*$V = V_{max}S/(K_m + S)$ (V is the specific consumption rate in units of substrate per time and cell mass; for other symbols, see the text).

$$\frac{R_1}{R_2} \approx \frac{V_{max1}}{V_{max2}} \frac{m_1}{m_2} \tag{20}$$

10.3.1.1 Consumption of Hydrogen and Acetate

In the competition between sulfate reducers and methanogens in nature, primarily the consumption kinetics of hydrogen and acetate have to be visualized. These are used by methanogens directly (Chapters 12 and 13) and are the key intermediates through which organic matter in sediments and sewage digesters is channeled into methanogenesis (4, 48, 49, 132, 148, 178, 179, 297, 322, 373, 375, 376). Also, for sulfate reduction in sediments, acetate and hydrogen are major electron donors (4, 22, 23, 178–180, 375). Tables 10.2 and 10.3 summarize kinetic data reported for *Desulfovibrio* strains and *Desulfobacter postgatei* using hydrogen or acetate, respectively, in comparison to data from methanogens. The data show variations and appear to depend greatly on the experimental conditions. Consumption rates estimated from growth parameters of proliferating cells (see the footnotes for Tables 10.2 and 10.3) are significantly higher than consumption rates measured with resuspended cells. Nevertheless, sulfate reducers tend to have (in most cases) higher V_{max}/K_M and μ_{max}/K_S values than methanogens. Hence the results from the kinetic calculations and measurements agree fairly well with the competitive advantage of the former bacteria that is observed in nature. The maximum substrate consumption or growth rates of sulfate reducers and methanogens are on the average rather similar. Judging only from these rates, one would predict a substrate partitioning between nongrowing cells, or a coexistence of growing cells over several generations at high substrate concentrations. However, only the prediction for the nongrowing cells agrees with experimental results, as shown by short-term tests in which substrates were added at nonlimiting concentrations to sediments (375) or mixed cultures of sulfate reducers and methanogens (259, 276). If growth takes place, sulfate reducers completely outcompete methanogens after some generations on hydrogen or acetate, as observed with batch enrichment cultures (20, 367, 368). These observations may be explained if the yield coefficients (grams of cell dry weight per mole of substrate consumed) are taken into consideration (Table 10.4); these correlate, although not very strictly, with the free energy changes of the reactions. Thus the higher free energy change of dissimilatory sulfate reduction in comparison to methanogenesis appears to contribute to the dominance of sulfate reducers over methanogens after a certain time of growth in batch enrichment cultures.

An explanation that comprises both a kinetic and a thermodynamic aspect in competition between sulfate reducers and methanogens has been given elsewhere (177, 178): Because of their more effective consumption, sulfate reducers maintain the substrate concentration (activity) such low that methane formation is energetically unfavorable.

Actually, minimum concentrations below which a substrate is not scavenged

TABLE 10.2 Kinetic parameters (mainly average values) for hydrogen utilization by mesophilic sulfate-reducing and methanogenic bacteria in pure cultures or in natural systems[a,b]

Species (Strain) or System	Temperature (°C)	For H$_2$ consumption[c]			For growth with H$_2$			References
		V_{max} (μmol·g^{-1}·h^{-1})	K_M (μmol·L^{-1})	V_{max}/K_M (L·g^{-1}·h^{-1})	μ_{max} (h^{-1})	K_S (μmol·L^{-1})	μ_{max}/K_S (L·μmol^{-1}·h^{-1})	
Sulfate reducers								
Desulfovibrio								
vulgaris (Marburg)	35	880	1.3	680	nd	nd	nd	157
	35	79,000*	nd	nd	0.23	nd	nd	217
vulgaris[d]	37	1,770	1.9	930	nd	nd	nd	259
desulfuricans[d]	37	5,280	1.8	2930	nd	nd	nd	259
sp. (G11)	37	3,300	1.1	3000	0.057	3.3	0.017	259
sp. (PS1)	37	3,300	0.7	4710	nd	nd	nd	259
Lake sediment + SO$_4{}^{2-}$ (freshwater, eutrophic)	20	nd[e]	1.1[f]	nd	nd	nd	nd	178
Methanogens								
Methanobrevibacter	35	8,510	6.6	1290	nd	nd	nd	157
arboriphilus (AZ)	33	215,000*	nd	nd	0.144	nd	nd	384
	23	nd	nd	nd	nd	5	nd	383

510

	Temp. (°C)							Ref.
Methanospirillum hungatei (JF-1)	37	4,200	5.0	840	0.053	6.6	0.008	259
sp. (PM1)	37	5,400	2.5	2160	nd	nd	nd	259
Methanococcus maripaludis	37	nd	nd	nd	0.18	nd	nd	136
Methanosarcina barkeri (MS)	37	6,600	13	510	0.058	nd	nd	259, 361
Lake sediment (freshwater, eutrophic)	20	nd[e]	4.7[f]	nd	nd	nd	nd	178
	10–14	nd[e]	1.1–4.1[f]	nd	nd	nd	nd	322
Sewage digestor sludge	33	nd[e]	78	nd	nd	nd	nd	383

[a] Values reported before 1977 not considered.

[b] nd, not determined in the study.

[c] Most of the values were obtained from resuspended ("resting") cells. Marked V_{max} values (*) were estimated from μ_{max} and the corresponding (real) growth yield, Y, according to $V = \mu/Y$. Grams refer to cell dry mass; assumption for conversion of protein mass: 1 g protein corresponds to 2 g cell dry mass.

[d] Strain of H. D. Peck.

[e] No data, since cell densities in the muds could not be determined.

[f] Calculated via solubility of H_2 at 101,325 Pa (1 atm): 848 μmol $\cdot$ L^{-1} at 10°C; 804 μmol $\cdot$ L^{-1} at 20°C.

511

TABLE 10.3 Kinetic parameters (mainly average values) for acetate utilization by mesophilic sulfate-reducing and methanogenic bacteria in pure culture or in natural systems[a]

Species (Strain) or System	Temperature (°C)	For acetate consumption[b]			For growth with acetate			References
		V_{max} (μmol·g⁻¹·h⁻¹)	K_M (μmol·L⁻¹)	V_{max}/K_M (L·g⁻¹·h⁻¹)	μ_{max} (h⁻¹)	K_S (μmol·L⁻¹)	μ_{max}/K_S (L·μmol⁻¹·h⁻¹)	
Sulfate reducers								
Desulfobacter postgatei (2ac9)	30	830	230	3.6	nd	nd	nd	276
	30	3100[c]	64[c]	48[c]	nd	nd	nd	122
	30	3200[d]	77[d]	42[d]	nd	nd	nd	122
	30	7200* to 9600*	nd	nd	0.035 to 0.046	nd	nd	368
Lake sediment (freshwater, oligotrophic)	10	nd[e]	5[f]	nd	nd	nd	nd	180
Methanogens								
Methanosarcina barkeri (Fusaro)	30	2240	3000	0.75	nd	nd	nd	276
barkeri (227)	37	6800*	nd	nd	0.023	5000	4·10⁻⁶	295, 296
mazei[g] (S-6)	35	nd	nd	nd	0.042	nd	nd	182
Methanothrix soehngenii (Opfikon)	37	1620	nd	nd	0.0085	700	12·10⁻⁶	115
Lake sediment (freshwater, oligotrophic)	10	nd[e]	33[f]	nd	nd	nd	nd	180
Sewage digestor sludge	37	nd[e]	320	nd	nd	nd	nd	383

[a]nd, not determined in the study.

[b]Marked V_{max} values (*) were estimated from μ_{max} and the corresponding (real) growth yield, Y, according to $V = \mu/Y$. Other values were obtained from resuspended cells. Grams refer to cell dry mass.

[c]Cells from sulfate-limited chemostat.

[d]Cells from acetate-limited chemostat.

[e]No data, since cell densities in the muds could not be determined.

[f]Value for K_M plus *in situ* concentration.

[g]Originally named *Methanococcus mazei*.

512

TABLE 10.4 Yield coefficients[a] (Y) of mesophilic sulfate-reducing, sulfur-reducing, and methanogenic bacteria on hydrogen or acetate

Species (strain)	Energy Substrates	Y(g cell dry mass per mol H_2 or acetate dissimilated)	References
Desulfovibrio	$H_2 + SO_4^{2-}$	2.1; 2.9[b]	18, 217
vulgaris (Marburg)	$H_2 + S_2O_3^{2-}$	4.2	18
Desulfotomaculum	$H_2 + SO_4^{2-}$	1.9; 3.1[b]	68, 153
orientis (Singapore I)	$H_2 + SO_3^{2-}$	4.0[b]	68
	$H_2 + S_2O_3^{2-}$	4.5[b]	68
Methanosarcina	$H_2 + CO_2$	0.7–2.2	295, 361
barkeri strains			
Other H_2-utilizing	$H_2 + CO_2$	0.5–1.0	364
methanogens			
Desulfobacter	Acetate + SO_4^{2-}	4.8	368
postgatei (2ac9)	Acetate + SO_3^{2-}	11.2	368
	Acetate + $S_2O_3^{2-}$	10.4	368
Desulfotomaculum	Acetate + SO_4^{2-}	3.4–4.8	
acetoxidans (5575)			
Desulfovibrio	Acetate + SO_4^{2-}	0.7–1.4	
baarsii (2st14)			
Desulfuromonas	Acetate + S	4.2	235
acetoxidans (11070)	Acetate + malate	16.5	235
Methanosarcina	Acetate	1.6–3.2	295, 296, 362
barkeri strains			
Methanothrix	Acetate	1.4	115
soehngenii (Opfikon)			

[a]Only real Y values are listed, but no extrapolated Y_{max} values ($\mu = \infty$).
[b]From chemostat culture.

have been observed in pure cultures and are known as minimum thresholds. However, the few available threshold data from cultures of sulfate-reducing and methanogenic bacteria and from natural habitats (Table 10.5) are insufficient for confirmation of competition models. Michaelis-Menten or Monod kinetics would not predict minimum thresholds. An explanation of thresholds from the viewpoint of bioenergetics may be possible; however, the concentrations (activities) of all reactants, including products, have to be considered in the thermodynamic calculations.

A problem in quantification of the actually available substrate at very low concentration arises from the fact that the total (i.e., chemically determined) amount of a substrate such as acetate in sediments occurs in distinguishable pools of different biological availability (54, 231). These have been characterized as (a) an absorbed pool not exchanging with excess substrate; (b) an absorbed pool exchang-

TABLE 10.5 Lowest measured concentrations of hydrogen or acetate in pure cultures (= minimum thresholds) and *in situ* under conditions of sulfate reduction or methanogenesis[a]

Substrate Measured	Conditions	Species/System	H₂ or Acetate Concentration (μmol/L)	Reference
H₂	Sulfate reduction[b] (SO_4^{2-} present)	Marine sediment	0.04^c	302
		Cariaco Trench (anoxic water)	≥ 0.0002	279
		Lake sediment, SO_4^{2-} added (freshwater, eutrophic)	0.0013^c	178
	Methanogenesis (SO_4^{2-} absent)	*Methanobacterium formicicum*	$\geq 0.046^c$	177
		Methanospirillum hungatei	$\geq 0.067^c$	177
		Lake sediment (freshwater, eutrophic)	0.0088^c	178
		Lake sediment (freshwater, eutrophic)	0.031	63
		Sewage digestor sludge	0.205	63
Acetate	Sulfate reduction (SO_4^{2-} present)	*Desulfobacter postgatei*	≥ 1	122
		Marine sediment (pore water)	0.1–70	Summarized by 231
	Methanogenesis[b] (SO_4^{2-} absent)	Lake sediment (freshwater, eutrophic)	2.7–4.5	376

[a] For critical consideration of acetate *in situ* concentrations see Section 10.3.1.1.
[b] No data available from pure cultures.
[c] Calculated via solubility of H₂ at 101,325 Pa (1 atm): 848 μmol $\cdot$ L^{-1} at 10°C; 804 μmol $\cdot$ L^{-1} at 20°C; 714 μmol $\cdot$ L^{-1} at 39°C.

ing with excess substrate; (c) a slowly metabolized, rather recalcitrant pore water pool; and (d) an available pore water pool that was rapidly oxidized by *Desulfobacter*. The recalcitrant pore water pool of acetate, which was shown to be not a residual threshold, amounted to 20 to 50% of the total pore water acetate (231). The recalcitrant pool would account for an overestimate in turnover rate calculations based on the total substrate pool. The chemical factors causing recalcitrance of dissolved acetate are unknown.

10.3.1.2 Consumption of Propionate and Butyrate

Next to acetate, propionate and butyrate are probably the most important organic products from the fermentative decomposition of biomass in sediments and other anaerobic sludges (22, 24, 148, 179, 299, 302). Further degradation of propionate and butyrate to methane is possible only by syntrophic associations of methanogens with hydrogen-producing acetogenic bacteria, since the former are specialized bacteria that do not directly use organic acids higher than acetate (Chapter 13). In a sulfate-dependent mineralization of propionate and butyrate, two pathways are theoretically possible:

1. The first possibility is a syntrophy between hydrogen- and acetate-consuming sulfate reducers and the hydrogen-producing acetogens, as already obtained in defined cultures on propionate, butyrate, or benzoate (32, 191, 192, 207, 208, 319). The co-cultures with the hydrogen-scavenging sulfate reducers always grew faster than the co-cultures with methanogens, a further finding indicating the competitive advantage of sulfate reduction versus methanogenesis.

2. The second possibility is that these fatty acids are used directly by sulfate reducers, as observed with pure cultures (Section 10.2.3).

Considering both possibilities, Banat and Nedwell (24) determined that propionate and butyrate turnover in salt marsh sediments was inhibited by molybdate as an antagonist of sulfate, but not by an increased concentration of molecular hydrogen. From this finding the authors concluded that proprionate and butyrate were, in fact, directly oxidized by sulfate reducers. Apparently, proprionate or butyrate oxidation via a sulfate reducer alone is faster and more effective than via a syntrophy. If this is also true for the oxidation of all other fermentation products, a complete sulfate-dependent mineralization of biomass would involve only the fermentative and the sulfate-reducing bacteria; an intervention of syntrophic hydrogen producers, as necessary for a complete degradation of biomass to methane (Chapters 9 and 12), would be superfluous. Nevertheless, a syntrophism between sulfate reducers and hydrogen producing acetogens should on principle be possible in nature; this appears most likely in sediments of low or fluctuating sulfate concentration where methanogenesis takes place in addition to sulfate reduction and where, therefore, an active population of the syntrophic acetogenic bacteria is present.

10.3.1.3 *Consumption of Fermentable Substrates*
In the utilization of degradation products that can be fermented further (e.g, lactate, ethanol, or amino acids), sulfate reducers have to compete versus fermentative bacteria. At growth-limiting concentrations in chemostat enrichments, both lactate and ethanol selected for *Desulfovibrio* species (162, 169). Studies on inhibition of substrate consumption upon molybdate addition to a lake sediment suggested that in natural sediments, sulfate reducers directly use lactate at low concentration (299). There was evidence in the same study that even amino acids were directly used by sulfate reducers. Lactate at high concentration has long been used for selective, fast enrichment of *Desulfovibrio* species (248, 249). However, lactate added to natural estuarine sediment at high concentration (10 mmol/L) was not oxidized with sulfate but fermented, due to an abundant population of lactate-fermenting bacteria (166, 169, 330); in a second phase, the fermentation products proprionate and acetate were oxidized as electron donors for sulfate reduction. Like lactate, ethanol at high concentration in batch enrichments was reported to select for *Desulfovibrio* species (162). Selection of sulfate reducers with amino acids has never been reported (Section 10.2.3.11). The importance of lactate as a

fermentation product in natural anaerobic populations is uncertain. In the carbohydrate-rich animal rumen, lactate is not a significant product (114).

10.3.1.4 Incomplete or Complete Oxidation by Sulfate Reducers

The direct utilization of fermentation products other than acetate or hydrogen by sulfate reducers again includes two possible pathways. First, fermentation products such as propionate, higher fatty acids, lactate, or alcohols may be degraded by incompletely oxidizing sulfate reducers to acetate; this product would then be consumed by acetate-oxidizing sulfate reducers. Second, complete oxidizers may degrade the fermentation products to carbon dioxide as the sole end product, provided that acetate is not excreted as an intermediate (Section 10.2.3). Competition between an incomplete oxidizer, *Desulfovibrio* sp., and a complete oxidizer, *Desulfobacter* sp., was investigated in a chemostat with limitation by ethanol as the common substrate. By its capacity for complete oxidation, *Desulfobacter* should be able to synthetize more cell mass per amount of ethanol than *Desufovibrio*; however, the latter become dominant, although *Desulfobacter* remained present (163). In batch enrichments, propionate, higher fatty acids up to palmitate, and alcohols usually select for incompletely oxidizing sulfate reducers which grow faster than completely oxidizing species (Section 10.2.3). Growth of the former may be followed by acetate-utilizing sulfate reducers (370). Counting of propionate-utilizing cells in an estuarine sediment (166) and of butyrate-utilizing cells in a ditch sediment (Widdel, unpublished) yielded incomplete oxidizers (*Desulfobulbus* and *Desulfovibrio sapovorans*-like sulfate reducers, respectively) as the dominant types. Degradation products that in nature appear to be used only by complete oxidizers are isobutyrate, isovalerate, acetone, or aromatic compounds; no incomplete oxidizer able to use these substances has been isolated so far.

10.3.1.5 Consumption of C_1-Methyl Compounds

A special case of competition between sulfate reducers and methanogens or other bacteria is found in the degradation of C_1–methyl compounds. Naturally occurring methyl compounds are, for example, methanol derived from pectin degradation (274), trimethylamine from decomposition of choline (82) or reduction of trimethylamine oxide from sea animals (323), the methoxyl groups of the aromatic lignin monomers (Chapter 7), and the methyl group of methionine. Very few types of sulfate reducers are known to use methanol or methoxyl groups from aromatic compounds (38, 65b, 153, 325a) whereas a utilization of methylamines or methionine by sulfate reducers has never been observed. Growth of the sulfate reducers on methanol or methoxybenzoates is slower than the growth of special methane bacteria or homoacetogenic bacteria, respectively, on these compounds (16, 108, 182); therefore, enrichments with methanol or methoxybenzoates select for methanogens or homoacetogens, respectively. In sediments, a conversion of methanol, methylamines, or the methyl group of methionine to methane has been observed despite the presence of sulfate (181, 225, 226). The utilization of methanol by nonsulfate-reducing bacteria is an example for the preference of a less exergonic over a more exergonic reaction in natural habitats:

$$4CH_3OH \longrightarrow 3CH_4 + HCO_3^- + H^+ + H_2O$$
$$\Delta G^{\circ\prime} = -314.6 \text{ kJ} \tag{21}$$
$$4CH_3OH + 3SO_4^{2-} \longrightarrow 4HCO_3^- + 3HS^- + 4H_2O + H^+$$
$$\Delta G^{\circ\prime} = -364.4 \text{ kJ} \tag{22}$$

There are, however, reports that methanol, the methyl group of methionine, and to a certain extent methylamines in marine sediments, were oxidized in a sulfate-dependent mineralization to carbon dioxide (152, 374). The oxidation of methanol in these habitats is not necessarily a direct consumption by sulfate reducers (153). Homoacetogens may convert methanol first to acetate (16) or, in syntrophic association with *Desulfovibrio*, to hydrogen plus carbon dioxide (65a). Also, *Methanosarcina* in the presence of *Desulfovibrio* may convert methanol partially to hydrogen plus carbon dioxide (239). Quantitatively, C_1-methyl compounds are less significant anaerobic intermediates than hydrogen, acetate, or fatty acids.

10.3.1.6 Consumption of Sulfate

The extent to which sulfate reducers in competition versus methanogens can take over the terminal degradation depends on the availability of dissolved sulfate. The kinetic data that explain the competitive advantage of sulfate reducers versus methanogens were obtained at nonlimiting sulfate concentration. At limiting concentration of the electon acceptor, sulfate reducers become less effective as competitors. The substrate consumption rate, V, of a sulfate reducer is a function of both the electron donor concentration, S_D, and the electron acceptor concentration, S_A (213, 254):

$$V = V_{max} \frac{S_D}{K_{MD} + S_D} \frac{S_A}{K_{MA} + S_A} \tag{23}$$

V_{max} is the maximum consumption rate at nonlimiting concentrations of the electron donor and the electron acceptor. K_{MD} and K_{MA} are the half-saturation constants for the electron donor and the electron acceptor, determined at nonlimiting concentration of the electron acceptor or electron donor, respectively. This equation is not applicable, of course, at electron donor or sulfate concentrations near thresholds.

Kinetic data for sulfate consumption by pure cultures have been determined with *Desulfovibrio* species (121) and *Desulfobacter postgatei* (122). The freshwater species *Desulfovibrio vulgaris* (Marburg) and *Desulfovibrio sapovorans* exhibited rather low K_M values (Table 10.6) that may indicate an effective sulfate uptake in freshwater habitats. However, the K_S value determined in a freshwater sediment was somewhat higher (Table 10.6). *Desulfobacter postgatei* had a rather high K_M value (216 μmol/L), and it was concluded that this species is not an important sulfate reducer in freshwater sediments, but an effective acetate scavenger in sulfate-rich brackish or marine habitats. *Desulfobacter* stopped to consume sulfate if a threshold concentration of 5 to 20 μmol/L was reached, probably

TABLE 10.6 Kinetic parameters (mainly average values) for sulfate consumption by sulfate-reducing bacteria in pure culture or a natural system

Species (Strain) or System	Temperature (°C)	Electron Donor	For SO_4^{2-} consumption[a]			Reference
			V_{max} ($\mu mol \cdot g^{-1} \cdot h^{-1}$)	K_M ($\mu mol \cdot L^{-1}$)	V_{max}/K_M ($L \cdot g^{-1} \cdot h^{-1}$)	
Desulfovibrio						
vulgaris (Hildenborough)	20	Lactate	1100	32.3	34	121
vulgaris (Marburg)[b]	20	Lactate	1500	4.8	313	121
salexigens (Brit. Guiana)	20	Lactate	2600	76.7	34	121
sapovorans (1pa3)	20	Lactate	800	7.3	110	121
Desulfobacter postgatei (2ac9)	30	Acetate	4200	216	19	122
Lake sediment (freshwater, eutrophic)	10	Natural	—	68	—	298

[a] Grams refer to cell dry mass.
[b] For *D. vulgaris* growing on $H_2 + SO_4^{2-}$ (35°C) with $\mu_{max} = 0.23$ h^{-1}, a K_s for sulfate of 10 $\mu mol \cdot L^{-1}$ was determined (217).

because sulfate uptake at lower concentrations becomes energetically unfavorable (122). In competition experiments in sulfate-limited chemostat cultures, a *Desulfobacter* species was inferior to a *Desulfobulbus* and a *Desulfovibrio* species (163).

10.3.1.7 *Methanogenesis in Digestors in the Presence of Sulfate*

Anaerobic mesophilic enrichments in batch or continuous systems with organic acids, ethanol, higher alcohols, or also complex organic matter in the presence of sulfate usually select for sulfate-reducing bacteria as the terminal degraders. Methanogenesis under such conditions may occur if C_1-methyl compounds such as methanol or methylamines are present or liberated from pectin or choline. However, by special processing of the digestion, substrates other than C_1-methyl compounds may be channeled into methanogenesis and allow a significant methane production despite of the presence of sulfate. First, this possibility exists in thermophilic digestors. At increasing temperature above approximately 55°C, the range of active sulfate-reducing species and their abilities as competitors should become more and more limited (Section 10.3.4). In contrast, many methanogens as archaebacteria are active at higher temperatures. However, in case of complex substrates, a temperature limit may also be given by the fermentative and syntrophic bacteria. Second, methanogenesis in the presence of sulfate can be achieved under mesophilic conditions if the process is carried out in floculated sludge digestors; in these, most of the organic substrates can be converted to methane in the presence of excess sulfate (123a, 123b). Nevertheless, the biogas produced is rich in hydrogen sulfide because sulfate reducers are not eliminated as long as sulfate is present. An important factor allowing the methanogens to compete against the sulfate reducers is probably the preferred attachment of the latter, especially of the acetate-cleaving *Methanothrix* species; sulfate reducers appear to attach less tightly to floculated sludge and to be kept at relatively low density by washout (123a, 123b). Furthermore, the sulfate reducers inside the flocs could be more energy limited than the methanogens and their partners, since the former depend also on the diffusion of the electron acceptor sulfate. Finally, it has to be considered that the relatively few species of acetate-utilizing sulfate reducers that live in or tolerate freshwater environments grow somewhat slower than species in saline habitats and are therefore less effective competitors of methanogens.

10.3.2 *Sulfate Reduction in Various Habitats*

Anaerobic habitats may be characterized by whether sulfate reduction or methanogenesis is the major terminal degradation process. Sulfate reduction dominates in marine or other sulfate-rich saline habitats, whereas the anaerobic biodegradation in sulfate-poor environments such as freshwater habitats ends up mainly as methane.

The sulfate concentration in seawater is on the average 28 mmol or 2.7 g SO_4^{2-} per liter (346). Under anaerobic conditions, this concentration enables the complete mineralization of 1.7 g of a carbohydrate, (CH_2O), per liter, via fermentative and sulfate-reducing bacteria. In comparison, the oxygen concentration in water saturated with air at 20°C is 0.28 mmol of O_2 per liter; in water saturated with oxygen from photosynthesis the concentration is 1.34 mmol of O_2 per liter. These amounts allow a complete mineralization of 0.008 g or 0.040 g, respectively, of a carbohydrate per liter. In marine environments the dissimilatory sulfate reduction normally takes place in the sediment, since the water in nearly all parts of the sea is oxic down to the ground. Also, the upper part of the sediment is usually oxic; the thickness of the oxic layer in different areas ranges from a few millimeters to several meters of depth, depending on the income of organic matter (141, 143). Between the oxic layer and the layer of sulfate reduction, a zone of denitrification may be observed (303). Thus, the sediment may reflect the mentioned sequence of reduction of different electron donors (Section 10.3.1). Denitrification is, however, quantitatively less important in mineralization than respiration or sulfate reduction (303). The layer of sulfate reduction is often visible by the black appearance of the sediment due to ferrous sulfide. Also, a change in the redox potential from about $+200$ mV to -200 mV is observed when passing a platinum electrode from the aerobic to the anaerobic part (303). A substantial formation of methane in marine sediments is observed only if sulfate becomes limiting in some meters of depth or in parts where high amounts of organic matter accumulate (143, 183). The sulfate concentration and the methane production rate in a profile of eutrophic intertidal sediment are shown in Fig. 10.4.

An important contribution of sulfate reduction to mineralization is found in coastal regions where the bulk of the detritus settles before a significant aerobic mineralization can take place, or in salt marshes in which grasses may provide organic material in the sediment (213). Comparison of respiration and sulfate reduction rates in sediment from Limfjord (Denmark) revealed that 53% of the total mineralization of the incoming detrius occurred via dissimilatory sulfate reduction (137). From the coastal regions to the deep-sea basins, the average sulfate reduc-

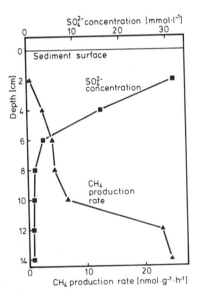

Figure 10.4 Dependence of methane production on sulfate concentration in the vertical profile of an organic-rich intertidal sediment at Delaware Inlet, Nelson, New Zealand. *In situ* temperature 30°C, sediment dry mass 70 to 80%. Redrawn from Ref. 206.

tion rate per unit area decreases significantly, as explained elsewhere (141, 143). The sulfate reduction rate in shallow marine habitats may reach some 100 mmol SO_4^{2-} m^{-2} · day^{-1} (143, 290). The rates in parts of the continental slope and deep sea is estimated to be around 0.1 mmol SO_4^{2-} m^{-2} · day^{-1}. In sediments of coastal areas, most of the total activity of sulfate reduction takes place near the surface within a layer of a few centimeters, which may therefore exhibit very high rates if calculated per unit volume. In some intertidal sediments, the maximum rates are 200 to 6000 nmol SO_4^{2-} cm^{-3} · day^{-1} (213). The activity of sulfate reduction in sediments of the continental shelf and continental slope is distributed over several meters of depth; determined rates per unit volume were not higher than 0.1 to 9 nmol SO_4^{2-} cm^{-3} · day^{-1} (143, 213).

In some habitats, reduced conditions do not occur only in the sediment but also in the overlaying water. This may happen if the transport of oxygen into the deeper water is reduced, mainly due to stratification, and if the concentration of oxidizable substrates exceeds that of oxygen. The Black Sea and the Cariaco Trench have anaerobic, sulfide-containing water from below about 150 and 375 m, respectively, to the maximum depths of 2000 and 1400 m, respectively (308, 350). Furthermore, in seawater lagoons that are rich in biomass, the sulfide-containing zone may temporarily extend upward from the sediment into the whole water column (51).

The sulfate concentration in freshwater habitats is usually low and ranges from about 0.01 to 0.2 mmol/L (123). The sulfate limitation enables a significant methanogenesis in aerobic parts of most fresh waters. In lakes, anaerobiosis may occur not only in the sediment, but also in the water column, namely in the hypolimnia during stratification. The hypolimnia are relatively closed water systems in which the biological oxidation of sedimenting organic matter and reduced degradation products from the sediment, mainly methane and hydrogen sulfide, scavenges oxygen; hydrogen sulfide also reacts chemically with oxygen. Anaerobic, sulfide-containing hypolimnia often occur in small lakes about 5 to 20 m of depth (Fig. 10.5). The contribution of dissimilatory sulfate reduction to the total anaerobic mineralization depends on the ratio of the amount of incoming organic matter to the amount of available sulfate. In an oligotrophic lake, up to 81% of the total terminal metabolism in the sediment proceeded through sulfate reduction (180). In the eutrophic lakes, Wintergreen Lake (Michigan) and Lake Mendota (Wisconsin), only 13% and 20%, respectively, of the terminal anaerobic electron flow occurred via sulfate reduction (120, 178). The sulfate pool in the anaerobic hypolimnion of Lake Mendota was not completely reduced during stratification. In the sediment, the sulfate concentration was below detection level and the availability of sulfate was limited by diffusion from the overlaying water. Nevertheless, sulfate reduction in the sediment accounted for at least 75% of the total sulfate depletion (120). The obvious accumulation of the anaerobic activity in the upper sediment appears to be characteristic of lakes in which the water column is temporarily oxic. In a meromictic (i.e,, permanently stratified) lake, a significant sulfate reduction was also observed in the water column (232). The hypolimnia of deep and great lakes, for example, Lake Constance, are usually aerobic.

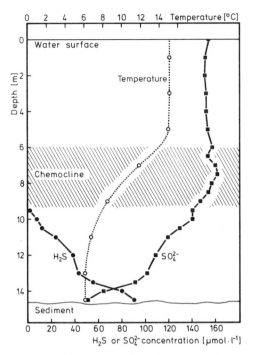

Figure 10.5 Vertical profile of the stratified Lake Rotsee near Luzern, Switzerland, at the end of September. Redrawn from Ref. 155.

10.3.3 Significance of Metabolic Types of Sulfate Reducers in Natural Habitats

Every compound that in natural sediments has been observed so far to serve as electron donor for sulfate reduction (4, 22–24, 179, 180, 267, 299, 302, 375) is also used by sulfate-reducing bacteria isolated and studied in pure culture (Section 10.2.3). It appears desirable to correlate a measured substrate turnover in a habitat to a particular type of bacterium and in this way estimate its role in the overall degradation. A hint as to the contribution of a bacterium to the overall process is given by the cell number of the concerned species in the habitat. It has to be considered, however, that the presence of a bacterium at a cell density that could explain the observed substrate turnover is a necessary, but not a sufficient criterion for regarding the species as responsible for the process. The observed cells could also be mostly dormant or metabolize substrates other than those used for counting.

Cell counts of sulfate-reducing bacteria in different habitats have been determined so far in dilution series mainly with lactate as electron donor (230, 248, 249). Since lactate usually causes rapid development of *Desulfovibrio* species, most determined cell numbers refer to types belonging to this genus. In estuarine and marine sediments, the counts of lactate-utilizing sulfate reducers range at 10^3

to 10^5 colony-forming units per cubic centimeter of sample (e.g., Refs. 3, 97, 107, 140, 166, and 184). In most cases, the cell counts were highest near the sediment surface below the oxic layer, where the highest rates of sulfate reduction were also observed.

Desulfovibrio strains isolated with lactate were able to grow well with hydrogen (36, 166), and strains isolated with hydrogen grew on lactate (20, 155). This metabolic coincidence and the evidence that hydrogen in sediments is a quantitatively more important intermediate than lactate (Section 10.3.1) suggest that the *Desulfovibrio* strains obtained with lactate mainly consume hydrogen under natural conditions. With its low K_s value for hydrogen and the high maximum growth rate, *Desulfovibrio* would be an effective hydrogen scavenger that could account for the low partial pressures observed in sulfate-containing anaerobic habitats (Table 10.5). However, the cell counts determined in marine habitats appear relatively low, since similar cell densities of *Desulfovibrio* were determined in habitats of low sulfate concentration (155, 175); moreover, the sheep rumen, a rather sulfate-poor habitat, contained 10^7 colony-forming units of *Desulfovibrio* (112). In fact, it was concluded from *in situ* reduction rates and the turnover capacity of one cell that the common counting techniques using lactate underestimate the true numbers of sulfate reducers in marine sediment by a factor of at least 1000 (140). Even if the presence of sulfate reducers performing reactions other than *Desulfovibrio* is considered for a habitat, the density of the latter type is expected to be higher than determined, due to the importance of hydrogen as an anaerobic intermediate. One factor that contributes to the underestimation may be the presence of lactate-fermenting bacteria (166, 169, 330); if these occur in counting series at higher densities than sulfate reducers, most lactate may be fermented prior to its utilization as an electron donor for sulfate reduction. Numbers of cells are also underestimated if these are associated in clumps or attached to particles. Homogenization of the sample before counting may separate the bacteria but may also destroy a number of cells.

Counting of acetate-utilizing sulfate reducers in an estuarine sediment revealed up to 6×10^3 colony-forming units of *Desulfobacter* per cubic centimeter (166). In view of the importance of acetate, these counts again appear to underestimate the true number of acetate oxidizers; marine *Desulfobacter* strains, especially, tend to form clumps (368), which would explain an underestimation. The rather fast development of *Desulfobacter* in saline acetate enrichments and the low K_s value of *Desulfobacter postgatei* for acetate (Table 10.3) suggest that this type of sulfate reducer is an important acetate scavenger in marine sediments. Acetate-oxidizing sulfate reducers in freshwater sediments have not yet been quantified. By using enrichment media with sodium and magnesium chloride at brackish-water concentration, the presence of *Desulfobacter* has also been demonstrated in eutrophic freshwater sediments (368) and activated sludge (Widdel, unpublished). Enrichments from freshwater mud in freshwater media sometimes yield acetate-utilizing *Desulfobacter*-like sulfate reducers with curved, vibrioid cells resembling those in Fig. 10.1q and r; but these acetate oxidizers grow best in brackish media (Widdel, unpublished). It remains unclear whether *Desulfobacter* types, which require saline media for optimum growth, are really significant in acetate metabolism in freshwater sediments.

Desulfotomaculum acetoxidans is an acetate oxidizer that prefers freshwater media. Due to the rapid enrichment from animal dung and a temperature optimum around 36°C, *D. acetoxidans* has been regarded as primarily an intestinal sulfate reducer (369). Hence a typical acetate oxidizer in aerobic lake sediments may still have to be discovered. The possibility exists that even hydrogen-utilizing *Desulfovibrio* species participate in acetate oxidation. *Methanosarcina barkeri* converts about one-third of its metabolized acetate to hydrogen plus carbon dioxide when co-cultivated with *Desulfovibrio* (239). A real syntrophic hydrogen-forming acetate-cleaving organism such as that observed in a moderately thermophilic co-culture (394) has not been found in mesophilic environments.

To overcome the difficulties inherent in the counting techniques, the distribution of lipid fatty acids as chemical markers was investigated in strains of *Desulfobacter*, *Desulfobulbus*, and *Desulfovibrio* (74a, 329). In *Desulfobacter* species C-even straight-chain fatty acids dominated, followed by cyclopropyl fatty acids and 10-methylhexadecanoic acid. The 10-methyl fatty acid, especially, is characteristic of *Desulfobacter* and represents a specific biomarker. *Desulfobulbus* species contained mainly C-odd or C-even straight-chain fatty acids, depending on whether propionate or hydrogen and carbon dioxide in the presence of acetate were the growth substrates. High proportions of saturated or monounsaturated branched-chain iso or anteiso fatty acids were characteristic of *Desulfovibrio* species. Volatile fatty acids (e.g., propionate, isobutyrate, 2-methylbutyrate, or 3-methylbutyrate) in the growth medium of sulfate reducers can be used to a certain extent as chain initiators during lipid synthesis and cause shifts in the fatty acid patterns (74a, 329). In marine sediment slurries, the increase in marker fatty acids upon addition of different electron donors was examined (330). The results suggested that the *in situ* consumption of acetate and propionate was due mainly to *Desulfobacter* and *Desulfobulbus*, respectively. The latter also appeared to be important in the removal of hydrogen in the sediment examined. Counting techniques, however, revealed that *Desulfobulbus* in an estuarine sediment occured at lower numbers than *Desulfovibrio* (166).

Direct microscopic identification of sulfate-reducing bacteria in samples is almost impossible; the cell forms of most sulfate reducers (vibrio, rod, oval, coccus) are found in many other bacterial groupings. Furthermore, viable cells or cell clumps between the particles of marine sediments are scarcely visible by microscopy, even after preenrichment by addition of an electron donor (Widdel, unpublished). Only the long cell filaments of *Desulfonema* species may be identified in samples; these filaments exhibit slow gliding movement or jerky swinging. The organic granules in *Desulfonema* filaments are nonrefractile, in contrast to the sulfur globules in *Beggiota*. However, significant densities of *Desulfonema* are found only in special sulfate-containing habitats that are rich in dead algal material (365). A useful technique analogous to the direct identification of methanogens by epifluorescence microscopy (74) is not known for sulfate-reducing bacteria. The applicability of immunofluorescence techniques for the *in situ* quantification of different types of sulfate reducers is limited. Since antisera turned out to be highly specific for strains but not for taxonomic or physiological groups of sulfate reducers (3, 294), a quantification of these by immunofluorescence would imply an "experimental circle": The goal—namely, the recognition of all serotypes in a sample—would be the presupposition for preparation of the antisera.

10.3.4 Significance of Environmental Factors for Development of Sulfate Reducers

In natural habitats, the role of a bacterium may not be determined only by its nutritional and growth kinetical properties as observed in batch or chemostat cultures under defined conditions; in the natural environment, the capacities to adapt to changing physical, chemical, and biological factors may also be decisive for the activity and growth of a bacterium.

10.3.4.1 *Effect of Temperature*

One of the less dramatic changes of physical factors for bacteria in nontropical habitats is the annual drop in temperature. Freshwater bacteria in hypolimnia of stratified lakes live at low temperatures (often $< 10°C$) all year. The sulfate reduction rate in sediments strongly depends on the temperature, the effect of which within a certain range is described by the Arrhenius equation (213). The Q_{10} values reported for sulfate reduction in marine sediments range between 2.0 and 3.9 (2, 137, 213, 214, 287) (i.e., a temperature increase of 10°C below the optimum stimulates the sulfate reduction rate 2.0- to 3.9-fold). The temperature optima of most pure cultures of sulfate reducers are in the same range (238, 371; Table 10.1). The existence of true psychrophilic sulfate reducers has not been shown so far by the isolation of pure cultures. The lowest optima among sulfate reducers were observed with some facultatively autotrophic *Desulfobacterium* strains and a curved *Desulfobacter* strain that grew well at 24 to 28°C and were affected at higher temperature (Brysch, Schneider, Fuchs, and Widdel, unpublished). *Desulfobacter hydrogenophilus* has an optimum of about 32°C, but still grows slowly at 0°C (Widdel, unpublished). On the basis of our present knowledge it appears that sulfate reducers with temperature optima around 30°C are also responsible for sulfate reduction in colder sediments.

None of the presently known sulfate reducers is able to grow at extremely high temperature, near 100°C, where archaebacterial sulfur reducers still thrive (83, 313–315, 318). The temperature optima and maxima of the known thermophilic species *Desulfotomaculum nigirificans* (46), *Desulfovibrio thermophilus* (262), and *Thermodesulfobacterium commune* (386) are significantly lower than 100°C (Table 10.1). A competition of sulfate reducers versus methanogens has been observed at a temperature up to 55°C in a hot spring (359).

10.3.4.2 Salt Requirement and Salt Tolerance

Sulfate reducers from marine environments often need saline media for growth and may be damaged in freshwater media. *Desulfovibrio salexigens* requires sodium chloride; the ionic species required appeared to be the Cl^- ions (248). For *Desulfobacter* strains from brackish or marine sediments, a requirement for both sodium chloride and magnesium chloride was observed (368; Widdel, unpublished). Salt demands of marine sulfate reducers may be satisfied by 20 g of NaCl and 1.5 g of $MgCl_2$ per liter; brackish-water strains may require less. *Desulfonema magnum* has, in addition, a relatively high demand for calcium ions and needs at least 0.5 g of $CaCl_2$ per liter (366). Freshwater species (e.g., *Desulfovibrio vulgaris*, *Desulfotomaculum acetoxidans*, or the type strain of *Desulfobulbus propionicus*) may be inhibited by marine NaCl concentration (248, 369, 370). Growth at NaCl concentrations higher than in seawater (about 27 g/L) has been observed with *Desulfovibrio desulfuricans*, *D. salexigens* (248), and a number of sulfate reducers that were isolated with lactate, hydrogen, acetate, fatty acids, or benzoate from oil-field water containing about 100 g of NaCl per liter (65). *D. desulfuricans* is rather adaptable with respect to salt concentrations and grows without NaCl as well as with 60 g of NaCl per liter (248). *D. salexigens* and the isolates from the oil field (65) required 20 to 50 g of NaCl per liter for optimum growth. Slow growth was still observed with about 100 to 120 g NaCl per liter, whereas 150 g of NaCl per liter inhibited nearly all strains. Only one acetate- and fatty acid–oxidizing type of sulfate reducer from the oil field exhibited slow growth at salt concentrations up to 270 g of NaCl per liter (65). ZoBell reported on sulfate reducers that grew optimally with about 30 g of NaCl per liter, but that were still active near saturation (395). All these observations question whether there are true extremely halophilic sulfate reducers in nature. Sulfate reducers in highly saline habitats appear to thrive far beyond their salt optimum, which is near the concentration in seawater.

10.3.4.3 Effect of pH

Sulfate-reducing bacteria prefer an environment around pH 7 and are usually inhibited at pH values lower than 6 or higher than 9 (62, 238, 371, 395). Changing of the pH with acid or alkali has been suggested as a method of diminishing sulfate reduction in industrial plants (395). Nevertheless, sulfate reduction has been observed in a peat bog (30) and acid mine water (348) that exhibited pH values of about 3 to 4. Sulfate-reducing bacteria isolated from these habitats were inhibited below pH 6. Therefore, it was supposed that the sulfate reducers in the acid environments occurred in microniches of a higher, more favorable pH. The less acid conditions in such microniches may be partly due to the activity of the sulfate reducers themselves. The overall reactions with different electron donors may consume protons [e.g., equation (4)] or lead to the proton-scavenging ions, HCO_3^- and HS^-. Also, high pH values may be compensated if, for example, long-chain fatty acids serve as electron donors [see, e.g., equations (7), (8), and (9)]. If, however, CO_2 and H_2S escape as gases, sulfate reduction always causes an alkalization (1).

10.3.4.4 Survival in Oxic Environments

Although sulfate reducers are obligate anaerobes, they may survive a temporary exposure to oxygen and again become active under anaerobic conditions (69, 100). It was shown with pure cultures that species of sulfate reducers differ significantly with respect to oxygen sensitivity; if sulfide is present at the same time, the cells may be more affected by oxygen than in a nonreduced medium (69). Sulfate reducers of the *Desulfovibrio* type survive many hours in an oxygenated, reductant-free medium (69). If a habitat becomes homogeneously oxic over a long period, vegetative cells of sulfate reducers die off, and only spores of *Desulfotomaculum* species survive. Therefore, *Desulfotomaculum* species may be selectively enriched from dry or oxic soil samples. If an environment turns from aerobic to anaerobic conditions, desulfotomacula are expected to be the first active sulfate reducers. A selection for *Desulfotomaculum* species due to dry periods has been observed in paddy fields (Jacq, Dakar, Sénégal, personal communication).

If aerobic sediments or waters are rich in organic particles, sulfate-reducing bacteria may be active in anaerobic microniches despite the aerobic surroundings. *Desulfovibrio*, *Desulfobacter*, and other species of sulfate reducers were enriched from well-aerated activated sewage sludge (Widdel, unpublished). Jørgensen (138) measured sulfate reduction in the upper oxidized marine sediment layer and also in detritus particles suspended in oxic seawater. In the oxic sediment, 2×10^2 colony-forming units of lactate-utilizing sulfate reducers were determined. The sulfate reduction took place in reduced organic-rich sediment pellets 50 to 200 μm in diameter. Formation and maintenance of the anaerobic microniches is ascribed to two processes. First, the local high concentration of degradable matter leads to scavenge of oxygen by aerobic organisms and thus favors growth conditions for anaerobic bacteria. The fermentation products of the latter may also serve as substrates for the aerobes. Second, the end product of sulfate reducers, hydrogen sulfide, causes consumption of oxygen (138). Sulfide is not only biologically oxidized by colorless sulfur bacteria, but as a reductant also reacts chemically with oxygen. The autoxidation of sulfide yields products that may be reduced again by sulfate reducers. Pure cultures of sulfate reducers were able to grow in zones of oxygen-sulfide gradients without added sulfate, solely by using autoxidation products of sulfide as electron acceptors (69). An important product was probably thiosulfate, which appeared to be continuously regenerated in a cycling process.

10.3.4.5 Morphological Adaptations: Aggregating Cells and Gliding Filaments

Sulfate reducers cultivated in the laboratory sometimes aggregate in clumps and stick to surfaces (238, 248, 365, 366, 368, 371). On the one hand, this behavior may be due to unfavorable growth conditions, such as an extreme pH, temperature above the optimum, too high concentrations of electron donors or mineral salts, or to the presence of toxic compounds. Under such unfavorable conditions, the bacteria often exhibit a "morbid," irregular, swollen cell shape, and cells normally motile may become immotile. On the other hand, clump formation and surface growth may be a normal feature that is significant in natural habitats. In such cases, the cell aggregates may release individual cells of normal shape and motility. Enrichments or pure cultures of sulfate reducers from saline habitats, especially, tend to grow in clumps (e.g., Ref. 368). *Desulfosarcina* (Fig. 10.1*t*) was named for its pronounced growth in cell packets (238, 371). One advantage of clump formation in nature is probably a certain protection against unfavorable changes of the redox potential or a temporary penetration of oxygen. Furthermore, associated and surface-attached cells often scavenge growth-limiting substrates more effectively than free, suspended cells. The gliding filamentous sulfate reducers of the genus *Desulfonema* (366) represent an ideal combination of motility with attachment to surfaces. The gliding motility allows spreading between sediment particles and in viscous organic matter; the motion may be directed by chemical gradients and thus enable the filaments to reach favorable growth conditions. Another advantage of the filamentous bacteria is their resistance against phagocytosis by ciliates and amoebae, species of which also occur in anaerobic habitats (365). A morphological feature of so far unknown advantage for sulfate reducers in their anaerobic sediments is the formation of gas vesicles (Fig. 10.1*c* and *v*).

10.3.5 Detrimental Effects Caused by Sulfate Reducers

As terminal anaerobic degraders, sulfate reducers also colonize aqueous systems that are made or economically exploited by humans. The activity of sulfate reducers in such systems may be detrimental. Nearly all disastrous effects of these bacteria are due to the production of hydrogen sulfide as a toxic and reactive reductant. The subject and the literature have been surveyed (98, 247, 248). Only a few principles of the economic activities of sulfate reducers can be described here.

10.3.5.1 Sulfate Reduction in Waters and Flooded Soils

In eutrophic ponds, lakes, or seawater lagoons, the anaerobic zone in which sulfate reduction takes place may temporarily expand upward from the bottom and fill the whole water column (51). Seasonal factors such as enhanced sedimentation of detritus from primary production may initiate such events. The hydrogen sulfide causes a rapid consumption of oxygen and poisons plants and animals, which are usually abundant in nutrient-rich waters. The killed organisms provide new substrates for the anaerobic process, which, therefore, is self-stimulating. There are observations of sudden dying of fish at high numbers due to biogenic sulfide (248). But even if the anaerobiosis remains restricted to the lower part (e.g., the hypolimnion) of normally oxic waters, the ecosystem may be deranged; the toxicity of sulfide and the oxygen deficiency threatens ground-dwelling water animals (e.g., crayfish) and fish species with larvae hatched on the sediment surface.

Conditions for development of sulfate reducers are also met in waterlogged soils. Reducing conditions are common in paddy fields (360), but have been also observed in tropical maize fields after rain or irrigation (129). If the soil is rich in gypsum, or if seawater has access, the organic material from the roots may stimulate a bacterial sulfate reduction. The sulfide often causes the death of the plants. Iron compounds in the soil may scavenge sulfide and thus protect the plants (129).

10.3.5.2 Sulfate Reduction in Industrial Water Systems

The production of hydrogen sulfide by sulfate-reducing bacteria in industrial water systems, for example, in the papermaking industry (300) or in oil technology (64, 65), is a well-known phenomenon. The detrimental effects caused by sulfate reducers in such systems are numerous. Formed hydrogen sulfide as a rather volatile compound easily escapes from the aqueous phase and may become dangerous to humans under poorly ventilated conditions. Furthermore, hydrogen sulfide precipitates iron and other heavy metals as dark sulfides. In oil technology, ferrous sulfide plugs injection wells and, due to its lipophilic character, stabilizes oil–water emulsions and hence causes troubles during separation of these. Free hydrogen sulfide contaminates fuel gas and oil. Above all, sulfate reducers play a key role in anaerobic corrosion (see Section 10.3.5.3).

The activity of sulfate reducers in the papermaking industry can easily be explained in view of the abundantly present cellulose fibers that via fermentative bacteria lead to electron donors for sulfate reduction. The processes leading to sulfate reduction in oil fields appear to be rather complex and are not yet satisfactorily understood. An anaerobic utilization of aliphatic hydrocarbons, which are the major constituents of petrol, may be considered (Section 10.2.3.12). If air has access in oil fields, for example with the injection water, aerobic hydrocarbon-oxidizing bacteria may release organic acids, as observed in the critical zone of injection wells (29). Such acids may serve as electron donors for sulfate reduction. Furthermore, polar compounds present in petrol may be extracted by intense mixing of oil with water and yield some electron donors for sulfate reducers in the aqueous phase. It is, however, not known how much polar oil constituents or products from an aerobic hydrocarbon oxidation can contribute to sulfate reduction in oil fields. Oil field chemicals are probably another source of electron donors for sulfate reduction. Even if sulfate reducers do not use these chemicals directly, they may be converted to suitable electron donors via fermentative bacteria. Examples for biodegradable oil field chemicals are methanol being commonly used as solvent for other chemicals, the bipolymer xanthan applied in enhanced oil recovery (EOR) projects, and the chelating agent citric acid. Xanthan and citric acid are easily fermented (65); methanol may be used directly by sulfate reducers (65b, 153, 325a) or first be converted to acetate by homoacetogenic bacteria (16; Chapter 9). Microbiological examination of an oil–water separation tank (treater), in which sulfide was formed, revealed many different metabolic types of sulfate reducers (64). With 6.3×10^6 colony-forming units per milliliter of treater water, acetate-oxidizing sulfate reducers of the *Desulfobacter* type (counted at 34°C) were dominant by numbers. Next to these, a so far unknown sulfate reducer (Fig. 10.1z) that grew best on benzoate was the most abundant type with a density of 1.1×10^6 colony-forming units per milliliter. Lactate- and hydrogen-consuming *Desul-*

fovibrio species were present with 1.4×10^5 colony-forming units per milliliter. The cell density of butyrate- and higher fatty acid–oxidizing sulfate reducers was not significant. Since no significant amounts of chemicals had been applied in the oil treatment plant, the origin of acetate and benzoate that apparently were important substrates remained unclear.

10.3.5.3 Anaerobic Metal Corrosion

Microbial corrosion of metals is the result of oxidative reactions initiated or maintained by the metabolic activity of microorganisms. These produce corrosive metabolites, such as acids, or build up electrolytic cells with cathodic and anoxic sites with oxidative destruction of the latter (98, 197). The biological processes usually take place in surface-attached films (biofilms) that may harbor communities of aerobic, microaerobic, and anaerobic microorganisms (98). Hence the destruction of the metal occurs within a highly inhomogeneous system with microgradients of oxygen, organic nutrients, and other reactants involved in corrosion, and the mechanisms are not yet known in detail. Some aerobic organisms have been identified as primary culprits in different types of metal corrosion; however, the economically most important metal-corroding microbes are the anaerobic sulfate-reducing bacteria (98, 197). The various theories and viewpoints that explain the factors and mechanisms in corrosion due to sulfate reducers have been reviewed (67, 98, 197, 248, 335). The most widely accepted explanation for the principal mechanism by which sulfate reducers are corrosive is based on the cathodic depolarization theory of von Wolzogen Kühr and van der Vlugt (357). Metal becomes polarized in water by loosing positive metal ions (anodic reaction). The electrons left in the metal reduce protons from water to atomic hydrogen (cathodic reaction). Atomic or molecular hydrogen remains on the metal surface, where a dynamic equilibrium is established. Sulfate reducers are supposed to remove molecular hydrogen permanently from the metal surface by oxidation with sulfate as electron acceptor (cathodic depolarization) so that a net oxidation of the metal takes place. The theory of a depolarization of iron surfaces by sulfate reducers has been confirmed experimentally (65c, 99). *Desulfovibrio* species may be effective depolarizers, due to their high affinity to hydrogen (Section 10.3.1). The oxidation of 4 mol of iron would yield 1 mol of hydrogen sulfide that precipatates 1 mol of ferrous sulfide. The remaining ferrous ions may form ferrous hydroxide. If organic electron donors are present, sulfide reduction with these would lead to further sulfide, so that all ferrous iron may be precipitated as sulfide (Fig. 10.6). From the viewpoint of thermodynamics, the additional sulfide is expected to favor the overall corrosion process since ferrous ions or hydroxide are effectively removed. The additional sulfide may not only be formed by the hydrogen-consuming, depolarizing sulfate reducers, but also by species that cannot use hydrogen. In quantitative experiments with steel wool in cultures of hydrogen-utilizing *Desulfovibrio* species, a significant consumption of cathodically formed hydrogen was observed only if organic electron donors were present in addition (65c).

In an improved depolarization theory, the mere action of sulfate reducers on the

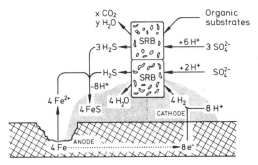

Figure 10.6 Reactions during anaerobic corrosion as suggested by the depolarization theory of von Wolzogen-Kühr and van der Vlugt, with additional sulfate reduction with organic substrates. Dotted area indicates attached ferrous sulfide. The organotrophically growing sulfate-reducing bacteria (SRB) may or may not be the same as the depolarizing (hydrogen consuming) ones.

iron surface has been regarded as insufficient to account for the striking anaerobic destruction of iron (33). There was evidence that solid ferrous sulfide in contact with iron acts as cathode, like a noble metal, and facilitates the depolarization significantly. Hydrogen sulfide would be removed by sulfate reducers in the attached layers of ferrous sulfide. A corrosion scheme that considers this spatial separation of the anodic and the cathodic reaction is shown in Fig. 10.6.

In another corrosion theory, reduced phosphorus compounds were postulated to be involved in the anaerobic destruction of iron (128). The depolarizing action of sulfate reducers has also been questioned (66). Instead, sulfide itself was supposed to corrode iron by receiving electrons and yielding hydrogen.

Another type of corrosion due to the activity of sulfate reducers may occur in environments with intermittent anaerobic/aerobic conditions or sulfide-oxygen gradients. The sulfide produced reacts with oxygen and yields sulfur or polysulfide, which are highly corrosive (98, 268).

Anaerobic corrosion tends to continue at the same spot on the iron surface where the process has started (197, 248). The iron in the anaerobic environment is pitted rather than evenly corroded (Fig. 10.7). Establishment of the anodic, corroding sites may be initiated by irregularities in the iron crystal and surface structure and by electrochemically different microenvironments in the aqueous surroundings.

10.3.5.4 Comments on Control of Sulfate Reducers

Development of sulfate reducers has to be expected whenever organic matter accumulates in sulfate-containing waters, especially if access of oxygen is limited. Attempts to avoid contamination (infection) of waters with sulfate-reducing bacteria are useless in most cases and may only be successful in completely closed water systems after sterilization. Nonclosed water systems probably receive sulfate reducers from entering water, soil particles, dust, or excreta of animals. Even if

Figure 10.7 Scanning electron micrograph of pits in anaerobically corroded iron specimen that has been incubated for 10 months in a *Desulfovibrio* culture. Bar equals 100 μm. Courtesy of W. Kleinitz, Preussag AG, Hannover.

the entering water is aerobic, it may contain dormant sulfate reducers of *Desulfotomaculum* spores that revive under suitable conditions. Spores are also present in soil or dust. In nearly all natural waters, sulfate reducers occur in the sediment, even if their activity is not obvious, and they multiply after increase of organic material.

A causative control of sulfate reducers is achieved if their energy sources can be limited, for example by keeping the access of organic matter or nutrients for algae in ponds or lakes as minute as possible. In flooded soils, sulfate reduction may be limited by exclusion of sulfate-rich waters and by avoiding excessive fertilization with sulfates. Also, in oil fields, the limitation of added chemicals that may serve as substrates for anaerobes, and the injection of sulfate-poor waters would help to diminish the biogenic sulfide production.

Furthermore, sulfate-reducing bacteria may be controlled by maintaining the environmental conditions unfavorable for these bacteria. If practicable in waters, aeration or oxygenation is a simple effective method to prevent sulfate reduction (248). In small technical water systems, hydrogen peroxide may be applied. Changing of the pH to about 5 by addition of hydrochloric acid in an oil-field water system diminished sulfate reduction and precipitation of ferrous sulfide (65). However, further lowering of the pH in technical water systems may cause chemical corrosions. Alkalization also inhibits sulfate reducers, but often leads to unwanted inorganic precipitates.

Biocides are often used for inhibition of sulfate reduction in industrial plants. The secondary effects of an application of biocides should be considered carefully. Recalcitrant biocides in waters lead to environmental problems. Biocides that decompose after a while to more harmless products may yield additional electron donors for sulfate reducers. Inhibitory concentrations of biocides determined in laboratory experiments are sometimes insufficient *in situ* where other than the laboratory strains are present and where bacteria occur in protected niches. Furthermore, biocides may be inactivated by absorption to minerals or by reaction of aldehyde groups with hydrogen sulfide. However it is also possible that the *in situ* conditions, for example, high pressures or temperatures in oil wells, increase the effectivity of biocides (106).

Possibilities for a control of anaerobic corrosion have been summarized (197, 248). The use of inert, noncorroding materials appears to be the best solution but is often not practicable because of the cost or less suitable mechanical properties if compared to iron or steel. Pipes may be protected by coating or by avoiding waterlogged surroundings in the soil (248). Cathodic protection of iron by sacrificial anodes or application of an external electromotive force, first used for inhibiting nonbiological corrosion, has also been successful against corrosion by sulfate-reducing bacteria (197). This method, however, although protecting the iron, does not inhibit the activity of sulfate reducers. In contrast, the increased negative potential imposed on the iron provides further hydrogen as electron donor for these bacteria (Widdel, unpublished).

10.4 NON-SULFATE-REDUCING BACTERIA WITH DISSIMILATORY REDUCTION OF INORGANIC SULFUR

Sulfur compounds other than sulfate may be reduced by many different microorganisms to sulfide. Among these, the significance of the dissimilatory sulfur metabolism ranges, via several transitions, from an incidental reduction in otherwise fermenting or respiring organisms to a main or even obligate catabolic reaction in strictly anaerobic bacteria. Thus organisms that reduce sulfur compounds other than sulfate cannot be classified, like the sulfate-reducing bacteria, as one nutritionally, biochemically, or ecologically defined group. Nevertheless, in this diversity of ''sulfide-forming'' microorganisms, two outstanding physiological assemblages of bacteria can be distinguished: the eubacterial and the archaebacterial reducers of elemental sulfur. The sulfur reduction in these mostly obligately anaerobic bacteria is their principal catabolic feature. Therefore, these sulfur reducers may be differentiated as true sulfur reducers from those bacteria in which the reduction of sulfur is a side reaction of little or no obvious bioenergetic significance.

10.4.1 Eubacterial Sulfur Reducers

Two major groups of eubacteria with a pronounced dissimilatory sulfur metabolism have been described (Table 10.7 Fig. 10.8). The first group is identical with the genus, *Desulfuromonas*, that comprises rod-shaped acetate-oxidizing sulfur reducers (235, 236, 371). The second group is represented by small spirilloid bacteria using hydrogen or formate as electron donors. Sulfur reducers of these two

TABLE 10.7 Properties of mesophilic (eubacterial) sulfur-reducing bacteria[a]

Species/Strain	Form	Width (μm)	Length (μm)	Motility[b]	G + C of DNA (mol %)	Major Menaquinone	Cytochromes
Desulfuromonas							
acetoxidans	Rod	0.4–0.7	1–3	+ (sl)	52	MK-8	b, c_7, and other
acetexigens	Rod	0.8–1.2	1–2	+ (sp)	62	MK-8	b, c
succinoxidans[c]	Rod; some strains curved	0.5	1–2	+ (sl)	45	MK-8	b, c
sp. strains NZ3, NZ27, and other[c]	Oval rod	0.8–1	1–2	+ (sl)	60	nr	b, c
Spirillum 5175 or free-living *Campylobacter* spp.	Vibrio	0.5	1.5–4.5	+ (sp)	38–42	MK-6 + TPQ-6[d]	b, c

Source: Data from Refs. 164, 167, 235, 236, and 379.

[a] nr, not reported.

[b] Flagellation in parentheses: sl, single, lateral to subpolar; sp, single, polar.

[c] Bache and Pfennig, personal communication.

[d] Thermoplasmaquinone-6.

Table 10.7. *(Continued)*

Species/Strain	Temperature Optimum (°C)	Electron donors[e]							Electron acceptors[e]								Growth Factor Requirement	NaCl Requirement (g/L)
		Hydrogen	Formate	Acetate	Propionate	Ethanol	Succinate	Other Readily Used	Sulfur	Sulfite	Thiosulfate	Nitrate	Oxygen	Fumarate	Malate	Aspartate		
Desulfuromonas																		
acetoxidans	30	−	−	+	+	+	−	*Pyruvate*	+	−	−	−	−	+	+	+	Biotin	20[f]
acetexigens	30	−	−	+	+	−	−	nr	+	−	−	−	−	+	+	−	Biotin	−
succinoxidans	30	−	−	+	−	−	+	Glutamate, lactate	+	−	−	+[g]	−	+	+	+	Biotin	Some: 20[f]
sp. strains NZ3, NZ27, and other	30	−	−	+	−	−	−	Pyruvate	+	−	−	−	−	−	−	−	Biotin	Some: 20[f]
Spirillum 5175 or free-living *Campylobacter* spp.	30	+	+	−	−	−	−	Lactate, pyruvate	+	+	+	+	+[h]	+	+	+[g]	None	−

[e] +, utilized; −, not utilized.
[f] In addition 3 g of MgCl$_2$ · 6H$_2$O per liter required or routinely added to media.
[g] Not utilized by all strains.
[h] Only at low O$_2$ partial pressure.

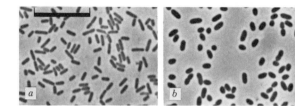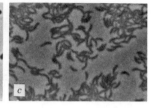

Figure 10.8   Phase-contrast photomicrographs of eubacterial (mesophilic) sulfur-reducing bacteria. Bar equals 10 μm, applicable to all three figures. (*a*) *Desulfuromonas acetoxidans*. (*b*) Obligately sulfur-reducing *Desulfuromonas* sp. strain NZ3. (*c*) Spirillum 5175. Photomicrograph (*b*) by R. Bache and (*c*) by N. Pfennig.

groups are mesophiles. Furthermore, one eubacterial thermophilic isolate has been described (28).

10.4.1.1 Desulfuromonas Species

The genus *Desulfuromonas* is a rather homogeneous group of rod-shaped to slightly curved motile sulfur reducers; most possess a laterally or subpolarly inserted single flagellum, causing a propeller-like movement. All *Desulfuromonas* species oxidize acetate with elemental sulfur to carbon dioxide:

$$CH_3COO^- + 4S + 4H_2O \rightarrow$$

$$2HCO_3^- + 4H_2S + H^+ \qquad \Delta G^{\circ\prime} = -6.9 \text{ kJ} \quad (24)$$

An oxidation of hydrogen or formate, and a reduction of sulfur compounds more oxidized than sulfur, has never been observed. Species may be distinguished by nutritional and morphological features, by the G + C content of the DNA (Table 10.7), and by comparing oligonucleotide catalogs obtained from 16S rRNA (Fig. 10.3). Instead of acetate, only a few organic acids were found to be oxidized. The type species, *D. acetoxidans*, is the only species using ethanol and higher alcohols. For growth on butanol that is oxidized only to the butyrate level, an organic carbon source such as succinate has to be added (for a similar case among sulfate reducers, see Sections 10.2.3 and 10.2.7). A freshwater type for which the species designation *D. acetexigens* has been proposed (236) does not take electron donors other than acetate.

The oxidation of acetate with sulfur in *D. acetoxidans* was shown to proceed via the citric acid cycle (89). In this mechanism the dehydrogenation of succinate to fumarate ($E_0' = +33$ mV) is of special interest since this reaction has to be coupled to the reduction of sulfur to hydrogen sulfide ($E_0' = -270$ mV). Due to the redox potentials, only the coupling of the inverse reactions, as observed in spirilloid sulfur reducers [equation (27)], would be possible. It has therefore been assumed that the succinate oxidation during growth of *Desulfuromonas* on acetate is an energy-driven reaction in the catabolism. The same problem exists in a species growing with succinate plus sulfur (Table 10.7).

Most *Desulfuromonas* species can grow in the absence of sulfur if their electron donor (e.g., acetate) is combined with fumarate or malate:

$$\text{acetate}^- + 4\text{ fumarate}^{2-} + 4\text{H}_2\text{O} \rightarrow$$
$$2\text{HCO}_3^- + 4\text{ succinate}^{2-} + \text{H}^+ \qquad \Delta G^{\circ\prime} = -239.5\text{ kJ} \qquad (25)$$

A possible pathway using sequences of the citric acid cycle (89) for the conversion of acetate and fumarate or malate to succinate and carbon dioxide has been proposed (93). In media with low bicarbonate/carbon dioxide concentration, a fermentative growth on fumarate or malate alone has been observed (Kümmel and Pfennig, personal communication). There are also *Desulfuromonas* strains that obligately depend on elemental sulfur as electron acceptor (Table 10.7 and Fig. 10.8*b*).

Desulfuromonas cells possess large quantities of cytochromes, by which colonies or cell pellets appear pink or brownish (15, 235, 236). A cytochrome from *Desulfuromonas acetoxidans* has been described as c-551.5 or c_7 (252); it is a low-potential cytochrome with three hemes and exhibits structural homologies with cytochrome c_3 of *Desulfovibrio* species (173). Also, rubredoxin and ferredoxin have been characterized from *Desulfuromonas acetoxidans* (210, 253). With respect to amino acid composition, the former protein resembles the rubredoxin from *Desulfovibrio* species (210). The ferredoxin from *Desulfuromonas acetoxidans* has two [4Fe–4S] clusters, like ferredoxins from clostridia and a ferredoxin from the sulfate reducer, *Desulfovibrio* strain Norway, which can also reduce sulfur (31). All *Desulfuromonas* species examined contain menaquinone-8 (61).

If arranged in a dendrogram based on 16S rRNA oligonucleotide sequences, the *Desulfuromonas* species appear as a rather coherent group among the versatile, completely oxidizing sulfate reducers (Fig. 10.3).

In their natural habitats, *Desulfuromonas* species may primarily scavenge acetate and reduce elemental sulfur. This must be assumed, since acetate and sulfur are more abundant in nature than the other organic compounds dissimilated by *Desulfuromonas* species. Moreover, there are species that are obligate acetate oxidizers (*D. acetexigens*) or obligate sulfur reducers (*Desulfuromonas* sp. strains NZ3, NZ27, and others). Species with alternative dissimilation pathways grow fastest with elemental sulfur, despite the rather low free energy yield [equation (24)]: With acetate plus sulfur at 28°C, the doubling time is about 2.5 h, whereas it is about 7 h with acetate plus malate (236). *Desulfuromonas* strains can always be enriched within a few days from anaerobic marine sediments. In contrast, samples from freshwater habitats yield fast-growing enrichments less frequently (236; Bache and Pfennig, personal communication). The apparent abundance of *Desulfuromonas* species in marine sediments may be explained by the special conditions that are met in these habitats. In marine sediments, an intense sulfate reduction provides sulfide that may yield sulfur as an oxidation product (142). Furthermore, the marine environment has the most active anaerobic bacterial communities in the sediments of shallow coastal or intertidal regions; here, light can reach the anaerobic zone and enable development of phototrophic sulfur bacteria. *Desulfuro-monas* species form robust, well-growing syntrophic cultures with phototrophic green sulfur bacteria (235, 236). *Desulfuromonas acetoxidans* has originally been detected as the chemotrophic partner in a marine mixed culture named "Chlorop-seudomonas ethylica" which contained *Prosthecochloris aestuarii* as phototrophic bacterium (235). In this syntrophy, the *Desulfuromonas* converts sulfur and or-ganic substrates to sulfide and carbon dioxide; these serve as electron donor and carbon source for *Chlorobium*, which uses light as energy source and reoxidizes the sulfide. Marine *Chlorobium* species especially tend to end up with the reoxi-dation of sulfide at the level of extracellular elemental sulfur (234) and can there-fore cooperate well with sulfur reducers in a light-driven sulfide–sulfur cycle. *Chlorobium* species grow mainly photolithotrophically with sulfide and carbon dioxide and poorly assimilate acetate.

A reduction of elemental sulfur in a eutrophic freshwater sediment has been reported by Smith and Klug (298). The rate of sulfur reduction was lower than the rate of sulfate reduction.

10.4.1.2 Spirilloid Sulfur Reducers

A spirilloid sulfur-reducing bacterium oxidizing hydrogen or formate was first iso-lated by Wolfe and Pfennig and designated spirillum 5175 (379).

$$H_2 + S \rightarrow H_2S \quad \Delta G^{\circ\prime} = -27.9 \text{ kJ} \tag{26}$$

Independently, an aspartate-fermenting spirilloid bacterium, described as a free-living *Campylobacter* species, was isolated which also turned out to reduce sulfur with hydrogen or formate (164, 167). Later, another spirilloid bacterium, known many years before, *Wolinella* (formerly *Vibrio*) *succinogenes*, was shown to re-duce elemental sulfur with hydrogen or formate (181a). Actually, spirillum 5175, the free-living *Campylobacter* species, and *Wolinella succinogenes* share most morphological and physiological features. They have tiny, curved cells (Fig. 10.8*c*) with a single, polar flagellum (164) and oxidize hydrogen, formate, and often lactate or pyruvate with sulfur, fumarate, malate, or nitrate as electron acceptors. Several isolates can also reduce sulfite, thiosulfate, or aspartate (167, 379). Sulfate is never reduced. In contrast to the strictly anaerobic *Desulfuromonas* species, the spirilloid sulfur reducers were reported to use oxygen as terminal acceptor under microaerophilic conditions (164, 167, 379, 380). At the air, however, growth is inhibited. The inorganic sulfur species are reduced to sulfide, the organic dicarb-oxylic acids to succinate. A unique organic sulfur compound used as electron ac-ceptor by the spirilloid sulfur reducers is dimethylsulfoxide (DMSO), which in enrichments with lactate was selective for these organisms (393). DMSO is re-duced to dimethylsulfide (DMS). Spirillum 5175 also reduced DMSO (Pfennig, personal communication). Nitrate reduction by the spirilloid sulfur reducers led to ammonia and sometimes small quantities of nitrate as a possible intermediate (167, 168). *Wolinella succinogenes* was shown to use nitrous oxide (N_2O) as electron acceptor, which, in contrast to nitrate, was reduced to dinitrogen (381).

If the spirilloid sulfur reducers are transferred into media containing hydrogen sulfide and fumarate, the cells catalyze a rapid oxidation of the former to elemental sulfur, with fumarate being converted to succinate (379):

$$H_2S + \text{fumarate}^{2-} \rightarrow S + \text{succinate}^{2-} \quad \Delta G^{\circ\prime} = -58.2 \text{ kJ} \quad (27)$$

Only *Wolinella succinogenes* has been shown to grow by this reaction (181a).

In the absence of an electron donor, fumarate, malate, or aspartate can be fermented by the spirilloid sulfur reducers; however, the purely fermentative growth is slower than growth with hydrogen or formate as electron donors (167). Aspartate fermentation yielded succinate plus acetate in a molar ratio of 8:1 (165). An aspartate fermentation (disproportionation) in which succinate is formed solely via fumarate reduction would yield a succinate:acetate ratio of 2:1. The ratio of 8:1 is explained by an additional formation of succinate via an oxidative pathway with acetyl-CoA and oxaloacetate as intermediates; these may be condensed to citrate, finally being oxidized to succinate. The acceptor for the reducing equivalents is always fumarate derived from aspartate (165).

Spirilloid sulfur reducers, *Campylobacter* species and *Wolinella succinogenes*, appear to lack an operating citric acid cycle, despite their facultative microaerobic growth. These bacteria were never observed to oxidize acetate, and *Campylobacter sputorum* ssp. *bubulus* oxidized lactate with oxygen as electron acceptor quantitatively to acetate (219).

The spirilloid sulfur reducers cannot grow autotrophically. If hydrogen or formate and inorganic electron acceptors are the energy sources, acetate has to be added as a carbon source in addition to carbon dioxide.

The original classification of the spirilloid sulfur-reducing bacteria as *Campylobacter* species appears justified, although the other species of this genus were described as commensals or pathogens in animals and humans (293). The sulfur reducers exhibit physiological similarities with subspecies of *Campylobacter sputorum* (170, 293) and the species *C. concisus* (293, 327) that can reduce fumarate, malate, or nitrate with formate or hydrogen, or grow microaerobically (72, 219–221). These *Campylobacter* types have not been tested so far for sulfur reduction. For a new or final classification of the spirilloid sulfur reducers, the genus *Wolinella* also has to be considered.

The G + C content (mol %) of the DNA of the bacteria mentioned is as follows: spirillum 5175, 38.4; free-living *Campylobacter* species isolated with aspartate, 41.6; *Campylobacter sputorum* subspecies, 29 to 34; *Campylobacter concisus*, 34 to 38; *Wolinella succinogens*, 46 to 49.

There are also biochemical similarities between the spirilloid sulfur reducers, *Campylobacter sputorum* subspecies and *Wolinella succinogenes*. They possess *b*- and *c*-type cytochromes (15, 72, 79, 170, 220, 380). In *Wolinella* and probably also in other spirilloid sulfur reducers, *b*-type cytochrome is involved in fumarate reduction, whereas *c*-type cytochromes appear to be mainly involved in nitrate reduction or oxygen respiration under microaerobic conditions (72, 158, 165, 220). The sulfur-reducing spirillum 5175, *Campylobacter* species, and *Wolinella suc-*

cinogenes also resemble each other by their quinones; all contain menaquinone-6 as major quinone and, in addition, thermoplasmaquinone-6 (59–61).

In nature, spirilloid sulfur reducers are widespread. They can be enriched with formate or hydrogen plus carbon dioxide and sulfur (plus acetate as carbon source) within a few days from nearly every small or large accumulation of freshwater or saline waters (236). The range of electron acceptors used by these bacteria and their capacity of microaerobic growth suggest that they thrive mainly near the oxic/ anoxic interface in sediments. The spirilloid sulfur reducers cannot ferment carbohydrates and thus are not primary degraders of biomass. Rather, these bacteria appear to be commensals, and find utilizable substrates in many different communities of microorganisms.

10.4.2   Archaebacterial Sulfur Reducers

The archaebacterial sulfur reducers are all extreme thermophiles. They are nutritionally, morphologically, and genetically rather diverse, and for detailed information, the original literature of Stetter and Zillig, who discovered and characterized these unique bacteria, should be consulted (313–315, 317, 318, 387, 388, 390, 391). Species of eight genera have been described that reduce sulfur: *Thermoproteus*, *Thermofilum*, *Desulfurococcus*, *Thermococcus*, *Thermodiscus*, *Pyrodictium*, *Acidothermus* and species classified so far as *Sulfolobus*.

Representatives of the archaebacterial sulfur reducers exhibit the highest temperature optima and maxima that have been ever observed among living organisms. *Pyrodictium occultum* has its optimum at 105°C and a maximum at 110°C and is cultivated in "autoclaves" under pressure (313, 317); the species does not grow below 80°C. The other species grow optimally between 85 and 90°C and do not multiply below 60 to 70°C (315, 318). Some archaebacterial sulfur reducers exhibit a relatively low pH optimum between 5 and 6. *Thermoproteus tenax* grows down to a pH value of 2.5 (391). A recently isolated sulfur reducer, *Acidothermus infernus*, that grows also aerobically by sulfur oxidation was active under anaerobic conditions within a pH range of 1 to 5 with an optimum at pH 2 (280, 315).

The cell forms observed among the thermophilic sulfur reducers are more or less regular spheres, disks, and rods of various lengths (Fig. 10.9). The diameters of the spherical *Desulfurococcus* cells and of the disk-shaped *Thermodiscus* and *Pyrodictium* species may vary between about 0.3 and 2.5 μm (314, 317, 318, 390). The cells of *Pyrodictium* are embedded in a network of ultrathin fibrils (Fig. 10.9g). *Thermofilum pendens* forms thin, long rods with a diameter of 0.17 to 0.35 μm and a length of sometimes more than 100 μm. *Desulfurococcus mobilis* and *Thermococcus celer* are motile by monopolar bundles of flagella (315, 318, 390). The cell envelopes of the archaebacterial sulfur reducers probably contain heat-stable glycoprotein subunits (314, 318). The cell multiplication usually occurs by budding, fragmentation, or constriction, but not by formation of septa.

The archaebacterial sulfur reducers are very diverse with respect to their electron donors and carbon sources utilized. *Thermoproteus tenax* uses a wide range of organic compounds, including glucose, starch, amylose, amylopectin, glyco-

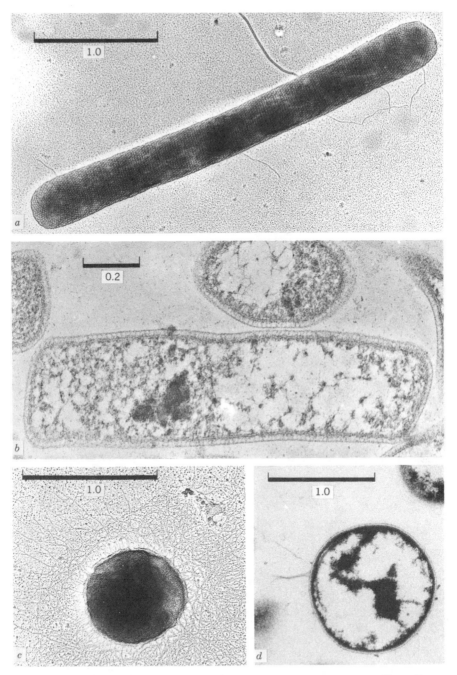

Figure 10.9 Electron micrographs of archaebacterial (extremely thermophilic) sulfur-reducing bacteria. Dimensions are given in micrometers. (*a*) *Thermoproteus tenax*, platinum-shadowed. (*b*) *Thermoproteus neutrophilus*, ultrathin section. (*c*) *Desulfurococcus* species, strain HVV-13, platinum-shadowed. (*d*) *Desulfurococcus mobilis*, ultrathin section with cut through flagellum. (*e*) *Thermococcus celer*, negative staining. (*f*) *Thermococcus celer*, ultrathin section. (*g*) *Pyrodictium occultum*, platinum-shadowed. (*b*) and (*g*) courtesy of K. O. Stetter; (*a*) and (*c*)–(*f*) courtesy of W. Zillig.

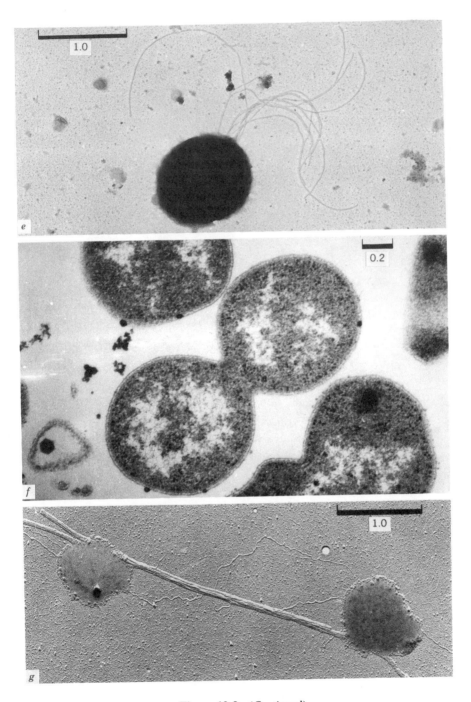

Figure 10.9 (*Continued*)

gen, malate, fumarate, ethanol, methanol, formate, and carbon monoxide (83, 391). *Desulfurococcus* species are mainly degraders of proteins and peptides (314, 318, 390). Facultatively autotrophic growth with hydrogen as electron donor and carbon dioxide as sole carbon source has been demonstrated with *Thermoproteus* species (83) and *Thermodiscus maritimus* (Stetter, personal communication). *Thermoproteus neutrophilus* and *Pyrodictium occultum* are obligately autotrophic (83). Archaebacterial sulfur reducers have not been observed so far to reduce oxygen-containing sulfur compounds such as sulfite, thiosulfate, or sulfate. *Desulfurococcus* and *Thermococcus* species may grow by fermentation of organic compounds without the electron acceptor, however, sulfur promotes growth significantly (318, 388, 390). *Acidothermus infernus*, *Sulfolobus brierleyi*, and *Sulfolobus ambivalens* are facultatively aerobic/anaerobic (amphiaerobic); under anaerobic conditions sulfur is reduced with hydrogen, whereas at the air sulfur is oxidized to sulfuric acid (280, 392). The other archaebacterial sulfur reducers studied are obligate anaerobes.

It has so far not been shown in detail whether the extremely thermophilic sulfur reducers oxidize all their organic electron donors completely to carbon dioxide, or whether organic end products such as acetate are formed. Growth tests with *Thermoproteus* on acetate were negative (391). Hence little is known about the significance of acetate in extremely thermophilic bacterial communities. Under moderately thermophilic conditions around 60°C, complete oxidation of acetate would be possible via an acetate-oxidizing, hydrogen-producing bacterium (394) in syntrophic association with a hydrogen-consuming sulfur or sulfate reducer.

The phylogenetic relationships of the archaebacterial sulfur reducers have been investigated by hybridization of DNA and 16S rRNA (318, 345, 389), comparison of 16S rRNA oligonucleotide sequences (378), and the composition of DNA-dependent RNA polymerases (389). As in the case of other archaebacteria, the molecular data are also the basis of the taxonomy of the sulfur reducers. The genera *Thermoproteus*, *Thermofilum*, *Desulfurococcus*, and *Thermococcus* are rather closely related to each other and are therefore grouped as Thermoproteales (318). The exact phylogenetic position of *Thermodiscus* and *Pyrodictium*, which are clearly related to each other, is still uncertain, but a classification within the Thermoproteales has been suggested (318). The Thermoproteales again exhibit relationships to the aerobic sulfur-oxidizing *Sulfolobus* species and form with these the group of "sulfur-dependent" thermophiles, which is one of the two major branches of the archaebacteria. The other major branch embraces the methanogens and the extreme halophiles. *Thermoplasma* remains rather isolated. A relationship between the anaerobic and the aerobic sulfur-dependent archaebacteria is also indicated by the mentioned thermophiles that can facultatively reduce or oxidize sulfur (280, 392).

The natural habitats of the archaebacterial sulfur reducers are anaerobic aqueous parts of continental or submarine volcanic areas. *Thermoproteus*, *Thermofilum*, *Desulfurococcus*, and *Thermococcus* were isolated from terrestrial hot solfataric springs, *Thermodiscus* and *Pyrodictium* from submarine volcanic areas (314, 318). In the submarine habitats, sulfur reducers such as *Pyrodictium* may thrive, due to elevated boiling point of water, at temperatures above 100°C (313, 317, 318). Microscopical examination of mud samples from solfatares and other hot springs revealed up to 10^8 bacterial cells per milliliter, many of which morphologically resembled types of archaebacterial sulfur reducers (314, 318, 390, 391). In comparison to the eubacterial, mesophilic sulfur reducers, the archaebacterial sulfur reducers utilize a wider range of electron donors, ranging from low-molecular-weight compounds, including carbon monoxide and hydrogen, to proteins and polysaccharides. The archaebacterial sulfur reducers, together with the thermophilic methanogens (314, 315, 364), can, therefore, constitute a purely anaerobic "food chain" maintained by inorganic volcanic energy and carbon sources (hydrogen, sulfur, carbon monoxide, carbon dioxide). The autotrophic sulfur reducers or methanogens would act as primary producers, whereas the heterotrophic sulfur reducers would be the degraders of dead cell material (83, 318). A remarkable physiological dependence in such a food chain has been demonstrated with *Thermofilum pendens*; this peptide-oxidizing species obligately depends on a polar lipid fraction of another sulfur reducer, *Thermoproteus tenax* (387). If oxygen has access, the facultatively lithotrophic *Sulfolobus* and the organotrophic *Thermoplasma* are further participants in the thermophilic food chain (318).

10.4.3 Further Organisms Reducing Sulfur Compounds

A reduction of sulfur to sulfide by organisms that normally respire oxygen or ferment carbohydrates was known as early as the last century; the reaction has been demonstrated with bacteria, yeasts, other fungi, plants, and animal tissues (27, 261, 312). In most of these organisms, the sulfur reduction appears to be without metabolic and bioenergetic significance and is therefore regarded as "incidental." A formation of high sulfide concentrations from sulfur has been observed with a number of mesophilic and thermophilic methanogenic bacteria (316). Some of the thermophilic strains produced higher sulfide concentrations than observed among true sulfur reducers. The methane production from hydrogen and carbon dioxide continued in the presence of sulfur, but was in many cases much less than in sulfur-free controls. It is, however, not known whether the methanogens really conserve energy by sulfur reduction. Nevertheless, by consuming sulfur the methanogens may actively establish a low, favorable redox potential in their environment. There is evidence that the incidental sulfur reduction to sulfide is a reaction of sulfur with sulfhydryl groups of proteins or glutathione that are oxidized to the disulfides (261, 312). In the methanogenic bacteria, one of the hydrogen donors for the reduction of elemental sulfur may be coenzyme M (2-mercaptoethanesulfonate). An enzyme that reduces the disulfide, 2,2'-dithiodiethanesulfonate, back to active coenzyme M with NADPH as electron donor has been investigated in *Methanobacterium* species (Smith and Rouvière, personal communication). The general mechanism of the incidental sulfur reduction is as follows:

$$2RSH + S \rightarrow RSSR + H_2S \tag{28}$$

Some prokaryotes are supposed to profit from the sulfur reduction, although they do not multiply under anaerobic conditions by this reaction. The cyanobacterium *Oscillatoria limnetica* is capable of a facultatively anoxygenic photosynthesis with hydrogen sulfide that is oxidized to extracellular sulfur (227). In the dark under anaerobic conditions, sulfur is reduced to sulfide with polyglucose as electron donor (228). A similar reaction in the dark has been observed in a *Chromatium* species that reduced intracellular sulfur with storage carbohydrate that was converted to poly-β-hydroxybutyric acid (354). The sulfur reduction in the cyanobacterium and *Chromatium* may allow energy conservation for a maintenance metabolism under anaerobic conditions in the dark. Furthermore, colorless sulfur bacteria of the genus *Beggiatoa* may reduce stored sulfur globules under anaerobic conditions; during sulfur reduction, a distinct increase of the cell dry mass of *Beggiotoa* was measured (215). A thermophilic facultatively sulfur-reducing anaerobe (optimum 77°C) that was a eubacterium has been isolted by Belkin and colleagues (28). The bacterium also grew in a complex medium in the absence of sulfur, but cell yields increased if sulfur was added.

Belkin and colleagues (28) also demonstrated that at temperatures higher than 80°C an abiological sulfide formation from sulfur takes place. The reaction, a sulfur disproportionation, occurs in mineral solutions and even in distilled water and is accelerated by increasing temperature. At pH values below 7, the sulfide formation is minute, but it increases drastically with increasing pH above 7.

Also, oxygen-containing sulfur compounds may be reduced by nonsulfate-reducing organisms that otherwise grow fermentatively or aerobically. *Clostridium pasteurianum* produced sulfide form sulfite in sucrose medium (189). Three thermophilic carbohydrate-fermenting clostridia, *C. thermohydrosulfuricum*, *C. thermosaccharolyticum*, and *C. tartarivorum* (also classified as *C. thermosaccharolyticum*), reduced sulfite or thiosulfate to sulfide (109). The purple phototrophic bacterium *Thiocapsa floridana* forms sulfide from thiosulfate with endogenous hydrogen donors in the dark (344). In experiments with aerobic and facultatively anaerobic bacteria, a number of *Salmonella* species, some *Proteus* species, and some *Escherichia coli* strains reduced thiosulfate or sulfite to sulfide (12, 55, 159). In clinical microbiology, the thiosulfate reduction by *Salmonella* is of diagnostic value. An example of a eukaryotic organism reducing sulfite to sulfide is the yeast *Saccharomyces cerevisiae* (188, 321); the sulfite reduction was observed in an aerated culture. There is, so far, no evidence for a bioenergetic advantage from the sulfite or thiosulfate reduction in the organisms mentioned. One may assume that the reduction is catalyzed by enzymes of the assimilatory sulfur metabolism (261, 271, 339). Since sulfite and thiosulfate do not have to be activated, like sulfate, at the expense of ATP, a noncontrolled excess reduction to free sulfide is likely. A yeast was shown to reduce preferentially the sulfane sulfur of thiosulfate to sulfide, so that sulfite was formed in addition (218, 261):

$$2[H] + SSO_3^{2-} \rightarrow H_2S + SO_3^{2-} \tag{29}$$

The reaction resembles that proposed as the ultimate step in dissimilatory sulfate reduction (248, 284). As in case of sulfur reduction, sulfite and thiosulfate reduction may also be advantageous for certain bacteria even if this is a side reaction. Tuttle and Jannasch isolated an aerobic marine pseudomonad that under anaerobic conditions produced sulfide from sulfite or thiosulfate in pyruvate or lactate media (351). The dissimilation of sulfite and thiosulfate yielded relatively low sulfide concentrations, but enabled anaerobic growth on lactate that was not used in the absence of an electron acceptor. In a number of facultatively anaerobic bacteria, an incomplete reduction of tetrathionate to thiosulfate has been observed (261, 351).

10.5 PRINCIPLES OF THE BIOLOGICAL SULFUR CYCLE

The hydrogen sulfide produced in various habitats may undergo further reactions in the environment (Fig. 10.10). Different aerobic microorganisms are able to oxidize sulfide with oxygen. Furthermore, sulfide is one of the few bacterial degradation products that are "autooxidizable" (i.e., that chemically react with oxygen at normal temperature). If light has access to the anaerobic zone, hydrogen sulfide is oxidized as electron donor by purple and green sulfur bacteria that use light as energy source for cell synthesis (Chapter 2). Another characteristic reaction of hydrogen sulfide is its binding to heavy metal ions, yielding almost insoluble sulfides. Depending on the environmental conditions, the heavy metal sulfides may be oxidized after a while or may be preserved, sometimes over geological periods.

The entirety of the transformations between the reservoirs of terrestrial sulfur species is known as the "sulfur cycle," analogous to cycles of other elements. This designation, however, expresses an idealized view. On a global scale and over long periods, the transformations are unidirectional changes rather than cyclic

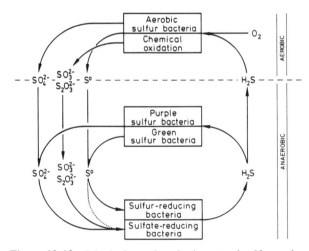

Figure 10.10 Principal reactions in the natural sulfur cycle.

restorations of the sulfur reservoirs. Only over some months or a few years in spatially limited systems, the biological reductions and oxidations of sulfur compounds may approximate the ideal cycling of sulfur. Habitats in which the reductions and oxidations of sulfur compounds are the predominating bacterial processes are termed "sulfureta."

10.5.1 Reoxidation of Hydrogen Sulfide

The overall anaerobic degradation of organic substances is a far less exergonic process than their oxidation with oxygen. Therefore, every degradation product escaping from anaerobic habitats still carries a significant part of the energy that has originally been conserved in the organic matter. The part of the energy carried by sulfide [equation (32)] is estimated by thermodynamic comparison of the aerobic oxidation of glucose [equation (30)] with its anaerobic mineralization via fermentative and sulfate-reducing bacteria [equation (31)]:

$$\tfrac{1}{3} C_6H_{12}O_6 + 2O_2 \rightarrow 2HCO_3^- + 2H^+ \qquad \Delta G^{\circ\prime} = -947.7 \text{ kJ} \qquad (30)$$

$$\tfrac{1}{3} C_6H_{12}O_6 + SO_4^{2-} \rightarrow 2HCO_3^- + H_2S \qquad \Delta G^{\circ\prime} = -151.2 \text{ kJ} \qquad (31)$$

Difference:

$$H_2S + 2O_2 \rightarrow SO_4^{2-} + 2H^+ \qquad \Delta G^{\circ\prime} = -796.5 \text{ kJ} \qquad (32)$$

Under anaerobic conditions in the absence of nitrate, sulfide is the final, energetically stable form of sulfur. Reoxidation of sulfide under these conditions with compounds present in sediments (e.g., carbon dioxide) is possible only with an input of energy [exception: see equation (27)], such as occurring in phototrophic sulfur bacteria.

10.5.1.1 Aerobic Bacteria Oxidizing Sulfur Compounds

The aerobic, colorless sulfur bacteria have exploited the free energy available from sulfide oxidation [equation (32)]. Mesophilic sulfide-oxidizing bacteria whose ecophysiology has been investigated are the *Thiobacillus* species (34, 160), *Thiosphaera pantotropha* (258), the filamentous *Beggiatoa* species (e.g., Refs. 195 and 216), and the veil-forming *Thiovulum* species (145, 377). As an intermediate of sulfide oxidation, *Thiobacillus* species or *Beggiatoa* species may form extracellular or intracellular sulfur, respectively. The final oxidation product is sulfate. Some specialized *Thiobacillus* species and apparently also *Thiovulum* are obligate lithoautotrophs that do not oxidize or assimilate organic carbon. Other *Thiobacillus* species and *Beggiotoa* species are facultatively lithoautrophic sulfide oxidizers and can in addition, or alternatively, use organic substrates, mainly acetate (160, 216). An oxidation of intermediary oxidation states of sulfur (e.g., elemental sulfur or thiosulfate) has been observed in a variety of organisms. These range from obligate autotrophs such as specialized thiobacilli, via facultatively autotrophic bacteria, to heterotrophs in which the oxidation of the sulfur species is more or less incidental. Facultatively autotrophic growth with thiosulfate has not only been observed among thiobacilli, but also among some hydrogen-oxidizing aerobes (Knallgasbakterien), for example, *Paracoccus denitrificans* (87). Also, some phototrophic purple sulfur bacteria can grow under microaerobic conditions in the dark as chemoautotrophs on thiosulfate (146, 156). In special heterotrophic pseudomonads, growth rate and organic carbon incorporation were increased by thiosulfate (347, 349, 352); apparently, thiosulfate served as an additional energy source for the heterotrophs, although they obligately depended on organic substrates. Molds and other fungi may oxidize elemental sulfur or thiosulfate and form sulfate, but have not been shown so far to benefit from the reaction (151, 261).

10.5.1.2 Chemical Oxidation of Sulfide

The autooxidation of sulfide in waters may be nearly as fast as the biologically mediated reaction (69, 144, 308). The mechanism of sulfide autooxidation appears to be rather complex and is influenced by the pH value, the concentration of the reactants, and the presence of catalytic concentrations of heavy metals. The time course of sulfide disappearance in the presence of excess oxygen agrees well with a first-order mechanism, and the half-life of sulfide under various conditions ranges from some minutes to some hours (9, 385). At pH values lower than 7, at which mainly undissociated H_2S is present, elemental sulfur is a major oxidation product. At pH values above 7, at which HS^- ions dominate, thiosulfate and sulfite are major oxidation products (9, 53, 56, 90, 385); furthermore, tetrathionate or even sulfate may be chemically formed. Heavy metals, especially manganese(II), nickel(II), and cobalt(II) compounds, stimulate the autooxidation of sulfide. The sulfur, thiosulfate, or sulfite formed may be oxidized further by colorless sulfur bacteria. If getting back into the anaerobic environment, these sulfur species can be reduced biologically to sulfide; thiosulfate and sulfite are also microbially disproportionated to sulfide and sulfate (Section 10.2.6).

10.5.1.3 Biological and Chemical Sulfide Oxidation in Natural Habitats

It is an amazing fact that just hydrogen sulfide as a simple inorganic and rather toxic compound feeds more peculiar and bizarre forms of bacteria than any organic fermentation product. Many of these microbial forms may be understood as strategies for an effective sulfide utilization and as adaptations to the special conditions under which sulfide becomes available. The aerobic sulfide oxidizers have to live in zones where oxygen and sulfide coexist and must therefore compete with chemical oxidation of sulfide. Many purple sulfur bacteria (Section 10.5.1.4) have to position themselves in the water not only at a tolerable sulfide concentration below the oxic zone, but also at a suitable light intensity.

The development of sulfide-oxidizing species in a habitat depends strongly on the kind of the anoxic/oxic transition zone, known as chemocline. The chemocline may be in the water column, at the sediment surface, or in the sediment (142). In stratified lakes, in the Black Sea, and in the Cariaco Trench, the chemocline is in the water column; in this, H_2S and oxygen may coexist over zones of several centimeters up to many meters, due to the free diffusion and the turbulences in water. In such chemoclines, sulfide appears to undergo preferentially autoxidation. In the Black Sea chemocline, there was no significant contribution to sulfide oxidation from bacteria (142, 308); these may, however, oxidize the chemically formed thiosulfate. *Thiovulum* exhibits a special ''strategy'' for an effective sulfide utilization in the water. The fine veils, in which the motile, rather large, oval *Thiovulum* cells are slacky interlinked, surround decaying plant or animal material, usually at a distance of some millimeters. By this kind of growth, *Thiovulum* creates and maintains a sharp sulfide–oxygen transition zone between the reduced water on the side of decomposition and the oxic water on the other side of the veil. If the sulfide–oxygen interface is at the sediment surface, as occurs in organic-rich sediments of gently flowing waters or of marine habitats, *Beggiatoa* species find suitable growth conditions (142). These filamentous, gliding bacteria are typical gradient organisms that actively position themselves at favorable sulfide and oxygen concentrations; *Beggiatoa* species prefer a low oxygen concentration (216). By converting sulfide into intracellular sulfur which is not autoxidizable, *Beggiatoa* withdraws its energy source from the chemical oxidation process. In most coastal marine sediments with sulfate reduction, the chemocline is below the surface and does not exhibit an overlapping of oxygen and sulfide. In contrast, free oxygen and free sulfide are usually separated by a zone of a few centimeters in which neither of these compounds is present at detectable concentration (142). Nevertheless, oxygen is the final oxidant. It has been suspected that iron–sulfur compounds or bioturbation mediate between free sulfide and oxygen in these sediments, but a real understanding of the apparently complex process is still lacking.

10.5.1.4 *Phototrophic Sulfide-Oxidizing Bacteria*

Green and purple sulfur bacteria use sulfide as electron donor for cell synthesis from carbon dioxide with light as sole energy source for anoxygenic photolithoautotrophic growth (Chapter 2). The final product is sulfate. As intermediate, significant amounts of sulfur may appear; green sulfur bacteria and *Ectothiorhodospira* species form extracellular sulfur, whereas *Chromatium* species store intracellular sulfur globules. Furthermore, the normally oxygenic cyanobacteria were shown to perform an anoxygenic photosynthesis in the presence of sulfide as electron donor that is oxidized to extracellular sulfur (50, 144, 227, 228). In stratified lakes with a chemocline not deeper than about 15 m, the penetrating light may enable a development of purple and green sulfur bacteria at high densities. These often cause a reddish or greenish appearance of the water (155, 232). Very dense layers of phototrophic bacteria are observed in shallow sulfureta containing high amounts of degradable organic matter, for example, from primary production of algal or cyanobacterial mats (142, 309). Under the cyanobacterial and algal mats covering the sediment of some intertidal areas, a pink, dense layer of purple sulfur bacteria and sometimes an adjacent layer of green sulfur bacteria may be observed. These oxidize the sulfide from the underlying black zone of sulfate reduction (309). This phenomenon of successive, distinctive zones of the different organisms was named ''Farbstreifensandwatt.'' Also, the colorless sulfide-oxidizing *Beggiatoa* species may occur in such stratified microbial mats. At the surface of a sulfide-rich sediment at Kalø Vig near Århus (Denmark), the zonations of the cyanobacteria, *Beggiatoa* species, and *Chromatium* species changed periodically by vertical movements of these organisms (142); the movements were governed by the availability of light, oxygen, and sulfide.

10.5.1.5 Anaerobic Sulfur Cycle

Although sulfide in natural sulfureta is partly or even exclusively oxidized with oxygen, the phototrophic sulfur bacteria would on principle allow a purely anaerobic cycling of sulfur. A well-operating anaerobic sulfur cycle has been demonstrated with a defined mixed culture of a marine, green sulfur bacterium and *Desulfuromonas acetoxidans* (Section 10.4.1.1). The sulfur cycle in these syntrophies includes two states, sulfur and sulfide, and thus represents a "short circuit" of the "great" sulfur cycle via sulfate. A cycle via sulfide and sulfate has been demonstrated by mixing pure cultures of *Desulfovibrio* and *Chromatium* (43). A real sulfur cycle occurs in such mixed cultures only if the organic electron donor is not used directly by the phototroph. The cycling of sulfur continues as long as organic substrate is present for sulfur or sulfate reduction. Mixed cultures of *Chromatium* with completely oxidizing sulfate reducers using fatty acids grew only poorly (Biebl, personal communication). Obviously, the purple bacteria, which in addition to sulfate formed intracellular sulfur globules, elevated the redox potential in the medium to levels that were disadvantageous for the fatty acid–oxidizing sulfate reducers. A cooperation of *Chromatium* with these sulfate reducers would, therefore, be most effective if the organisms occur in separate layers, as under natural conditions.

With the natural entire community of phototrophic, sulfur-reducing, sulfate-reducing and fermentative bacteria, a cycling not only of sulfur (Fig. 10.10) but also of carbon appears possible under completely anaerobic conditions. Various fermentative bacteria would convert dead bacterial cell material into new electron donors for reduction of sulfate or sulfur to sulfide; the sulfide, carbon dioxide, and parts of the fermentation products would again be assimilated by the phototrophs. However, due to the absence of oxygenic photosynthesis, the light energy conserved by such a closed sulfuretum cannot increase the totally available reducing equivalents for biomass production. As in every closed system, the sum of the reducing equivalents present as sulfide and sulfur plus the organic compounds would be constant. Only a new input of these substances can increase the potential yield of biomass in the anaerobic sulfuretum.

10.5.2 Precipitation of Iron Sulfides

Ions or compounds of heavy metals (Me) of the oxidation state $+2$ precipitate with hydrogen sulfide to yield sulfides of the general formula MeS.* Some metals or metalloids react in other oxidation states to sulfides (e.g., Ag_2S, As_2S_3, Sb_2S_3).

---

*The sulfides of the alkaline earth metals, MgS, CaS, SrS, and BaS, which are known as solid compounds, are not stable in the presence of water and therefore never precipitate.

Heavy metal sulfides belong to the ionic compounds with the lowest solubility. Although hydrogen sulfide as a very weak acid ($pK_1 = 7.0$, $pK_2 = 12.0$, at 20°C) yields only very small portions of S^{2-} ions in pure water or neutral salt solutions, the concentration of these ions is sufficient for an almost complete precipitation of heavy metal sulfides. Some of these (e.g., CuS) even precipitate in acid solution. Thus free sulfide is an effective scavenger of heavy metals under various environmental conditions.

Iron is by far the most abundant heavy metal in soils, sediments, and waters. Under oxic conditions, it usually occurs as ferric iron as the most stable state. Common species are hydrates of ferric oxide, $Fe_2O_3 \cdot H_2O$, such as goethite, $FeO(OH) = Fe_2O_3 \cdot H_2O$. A defined hydroxide $Fe(OH)_3$ is not known. The reactions leading to iron sulfides in sediments have been investigated (for an overview and literature, see Refs. 92 and 356). In the presence of hydrogen sulfide, the almost insoluble ferric minerals are reduced to the ferrous state (e.g., to ferrous hydroxide or ferrous ions) [equation (33)]. The ferrous state reacts immediately with further hydrogen sulfide to ferrous sulfide [equation (34)].

$$2FeO(OH) + H_2S + 4H^+ \rightarrow 2Fe^{2+} + S + 4H_2O \qquad (33)$$
$$2Fe^{2+} + 2H_2S \rightarrow 2FeS + 4H^+ \qquad (34)$$

The freshly precipitated, black ferrous sulfide is mostly amorphous. In addition, varying proportions of the crystalline form, the tetragonal mackinawite, may be formed (92). Further mackinawite results from aging of the amorphous sulfide. The amorphous sulfide and mackinawite were shown to be sulfur-deficient compounds the composition of which is best approximated by the formula $FeS_{0.9}$ (92). Amorphous ferrous sulfide or crystalline mackinawite are supposed to react slowly with elemental sulfur [equation (35)] and also with sulfur donors such as polysulfide or thiosulfate to the cubic pyrite, FeS_2, the end product of sulfide precipitation in sediments (92, 356):

$$FeS + S \rightarrow FeS_2 \qquad (35)$$

During pyrite formation, the iron keeps its oxidation state ($+2$), since the pyrite crystal lattice contains S_2^{2-} ions. There is evidence that at undersaturation of ferrous sulfide (i.e., at very low Fe^{2+} and S^{2-} concentrations), pyrite is directly precipitated from ferrous ions and hydrogen sulfide plus sulfur of polysulfide (92, 113, 356):

$$Fe^{2+} + H_2S + S \rightarrow FeS_2 + 2H^+ \qquad (36)$$

Such conditions are likely at very low hydrogen sulfide concentration, at low pH values, or in the presence of natural organic compounds chelating ferrous iron. The direct pyrite formation was shown to occur much faster than the formation via ferrous sulfide (113).

If hydrogen sulfide in sediments is formed at low rates, most may be bound to iron in the sediment (143). At high activities of sulfate reduction, as occurs in coastal sediments, the major part of the hydrogen sulfide diffuses into the aerobic zone (143). In a coastal marine sediment of Limfjorden (Denmark), only about 10% of the formed hydrogen sulfide went into the iron sulfide pool (137). Nevertheless, the pyrite concentration in the uppermost part of the reduced sediment zone reached 300 to 400 μmol S cm^{-3}, whereas the sulfate concentration was 18 μmol cm^{-3}. Thus the sediment can act as a sulfur trap (Fig. 10.11). Under anaerobic conditions, pyrite is stable. If the overlying oxic sediment zone expands (e.g., in winter after decrease of the degradation of organic matter), the pyrite is biologically oxidized, probably by lithotrophic bacteria such as *Thiobacillus ferrooxidans* (see, e.g., Ref. 11). The products are ferric hydroxide and sulfate. At the air, pyrite is chemically much more stable than ferrous sulfide. Pyrite is readily dissolved only in oxidizing acids. The insolubility of pyrite in nonoxidizing acids has to be considered if pools and turnover rates of sulfur compounds are determined in sediment samples (139).

Via the iron cycle, the cycle of sulfur is interlinked with that of phosphorus. Ferric hydroxide in waters binds dissolved orthophosphate (HPO_4^{2-}, $H_2PO_4^{-}$). The equilibrium concentration of free phosphate in the presence of iron hydroxide is lowest at about pH 5, but binding is still efficient at pH 7 (324). Under neutral conditions, as in natural waters, the iron–phosphate precipitate does not reach a Fe/P ratio of 1 (i.e., the formula $FePO_4$) but forms ferric hydroxyphosphates, for

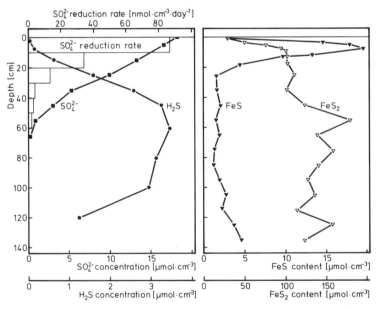

Figure 10.11   Distribution of sulfur compounds in the vertical sediment profile of Limfjorden, Denmark, in July 1975. Redrawn from Ref. 140.

example, $Fe_4(OH)_9PO_4$ (86, 331). The phosphate is removed from the water column and transported into the sediment (77, 78, 224, 363). It is generally observed that a change of sediments or soils from aerobic to anaerobic conditions causes a liberation of phosphate from ferric phosphates (e.g., Ref. 203). Obviously, the ferric hydroxyphosphate is reduced to the ferrous state, which is more soluble. The mechanisms by which the nearly insoluble ferric iron is reduced under anaerobic conditions are not yet fully understood. There is evidence that sediment bacteria reduce the ferric compounds directly (84, 301). If sulfate reduction occurs, the hydrogen sulfide formed appears to be the most powerful agent that liberates phosphate by reduction of ferric hydroxyphosphate and precipitation of ferrous sulfide (77, 224). The liberation of phosphate under anaerobic conditions is of practical importance in sewage plants if the phosphate is precipitated with iron salts and gets into the anaerobic digestor. Phosphates of aluminum and calcium (hydroxylapatite) which are other highly insoluble phosphates are not redox-active and therefore are not influenced by reducing conditions or hydrogen sulfide. Liberation of phosphate from these compounds may be possible only by complexing agents or, in the case of calcium phosphate, by lowering the pH below 7 (324).

10.5.3 The Sulfur Cycle on a Global Scale and through Time

Within the sulfur cycle two main categories of transformations can be distinguished, physical processes such as transports, and chemical processes, mainly oxidations and reductions. Often, the transformation of a sulfur species includes both types of processes, for example, the oxidation of iron sulfides to soluble sulfate and its leaching by water. For calculations or estimations of turnovers in the global sulfur cycle, the fluxes (i.e., transport processes) and the chemical net transformations between the sulfur reservoirs have usually been considered. However, redox transformation between sulfur species may also occur without obvious spatial fluxes or net changes. An important example are the sulfureta, where within a narrow zone the sulfur compounds are reduced approximately at the same rate as they are oxidized again. The quantification of such processes as part of the entire sulfur cycle is more problematic than quantification of fluxes and chemical net transformations. For detailed information about measured or estimated magnitudes of sulfur reservoirs and the natural and anthropogenic transformations between them, the literature should be consulted (e.g., Refs. 127 and 385).

Estimates (since 1952) of the sulfur content in the earth's crust range between 0.030 and 0.113 wt % S (196). One major sulfur reservoir on the earth is the oxidized state (i.e., sulfate); it is present as dissolved sulfate in oceans with 1.3×10^9 Tg of sulfate-S, and as solid sulfate in evaporites (mostly gypsum and anhydrite) with 5×10^9 Tg of sulfate-S. The other main reservoir is the reduced sulfur in metal sulfides; it amounts to 2.7×10^9 Tg in shales and other sediments, and to 7×10^9 Tg in metamorphic and igneous rocks. The estimated amount of reduced sulfur in living organisms plus dead organic matter derived from them is 5.6×10^3 Tg sulfide-S (all estimates summarized in Ref. 338). From global primary production, a cycling of about 5×10^3 Tg S yr^{-1} through living organisms has been calculated (288). Dissimilatory sulfate reduction probably contributes far less to the biological annual cycling of sulfur than the assimilatory reduction and desulfuration; only a small part of the globally synthesized organic matter is degraded in anoxic parts of aquatic habitats where sulfate serves as terminal electron acceptor. However, with respect to deposition of sulfur species, dissimilatory sulfate reduction appears to count more than the assimilatory process. Most of the assimilated reduced sulfur is liberated during degradation under aerobic conditions (284, 339) and thus recycled to the sulfate pool in soils or waters. In contrast, dissimilatory sulfate reduction is a characteristic process of anaerobic aquatic sediments. The conditions in these are generally favorable for a preservation of all kinds of biologically or chemically formed water-insoluble matter. Also, iron sulfides and sulfur formed from hydrogen sulfide are buried in sediments (137, 142, 143) and are thus withdrawn from the rapid turnover in the biological sulfur cycle. The estimated annual accumulation of reduced sulfur, derived mostly from dissimilatory sulfate reduction, in the sediments of the world oceans (shelf, and continental slope and rise) is 97 Tg S yr^{-1} (356). It is not known how much of this reduced sulfur is likely to be preserved over geological periods. In coastal areas, a significant part of precipitated sulfide is reoxidized (143, 340), so that the precipitation and reoxidation of iron sulfide represents a slowly operating "bypass" in the sulfur cycle. Nevertheless, there is evidence from the geological record that dissimilatory sulfate reduction during the history of the earth has led to deposits of sediment-hosted metal sulfides and of elemental sulfur. First, the structures of the deposits can be explained by transformation and sedimentation processes that are observed in recent sulfureta. There are even analogies in the microscopical structures. For example, pyritic framboids, which are raspberry-like spherules (1 to 100 μm in diameter) of many pyritic crystals, have been observed in ancient sulfidic deposits as well as in presently formed sediments; the formation of these framboids appears to be favored in environments of sulfate reduction (for a discussion, see Ref. 337). Second, evidence for geochemically significant activities of sulfate-reducing bacteria is furnished by sulfur isotope data from desposited sulfur species, as explained next.

10.5.3.1 General Aspects of Sulfur Isotope Fractionation

Sulfur found in nature consists of four stable isotopes (^{32}S, ^{33}S, ^{34}S, ^{36}S), ^{32}S and ^{34}S being the most abundant. The isotope ratios in samples of different origin are, however, not constant but reveal differences that are obviously due to an isotope fractionation during formation of the sulfur species. A fractionation may occur in a thermodynamic equilibrium between two sulfur species (equilibrium or thermodynamic isotope fractionation) or in a unidirectional conversion of one sulfur compound into another (kinetic isotope effect). At moderate temperature, only the latter type of fractionation is significant (52, 270, 337). By many biological, but also chemical reactions, compounds of the lighter isotope are converted somewhat faster than those of the heavier one. As a result, the product is enriched in ^{32}S compared to the substrate as long as the reservoir of the latter does not approach depletion. The fractionation is highest in an open system where the sulfur compound serving as substrate occurs at non-limiting concentration and is continuously supplied from a reservoir. If the concentration in an open system limits the reaction (i.e., at a steady-state concentration near zero), fractionation cannot be expected. In a closed system, the isotopic composition of the forming product changes continuously. The latter situation is analogous to the distillation of an amount of two mixed substances differing by their volatility; the change in the isotope ratio of the reactant or the product is described formally by the Rayleigh distillation equation (52, 337, 338).

The isotopic composition of a sample is expressed as δ^{34}S, which is the grade of enrichment of ^{34}S compared to a standard:

$$\delta^{34}S = \left[\frac{(^{34}S/^{32}S)_{sample}}{(^{34}S/^{32}S)_{standard}} - 1 \right] \times 1000‰ \tag{37}$$

Troilite from Cañon Diabolo meteorite with a ^{32}S/^{34}S ratio of 22.21 is assumed to represent the average isotopic composition of terrestrial sulfur (i.e, to represent the compositon of primordial magmatic sulfur). The δ^{34}S of the standard is 0‰ by definition. The isotopic composition of a product is often also compared to the composition of the remaining reactant.

Figure 10.12 summarizes some fractionation data obtained with various mainly biological oxidation or reduction reactions of sulfur compounds. The highest enrichment of ^{32}S (i.e., the most negative δ^{34}S values) has been observed in sulfide formed by dissimilatory sulfate reduction (52, 147, 270, 338). The pure culture measurements were carried out with *Desulfovibrio* (147). The isotope fractionation was greatly influenced by the substrate and the reduction rate. With lactate or ethanol as electron donor, the fractionation decreased with increasing reduction rate; in contrast, with hydrogen the fractionation increased at higher reduction rates (147). The highest fractionation measured at slow sulfate reduction with ethanol was −46‰ relative to the starting sulfate. In sediments even a fractionation of −62‰ has been observed (52, 150).

Evidence from the geological record (for eras, see Table 10.8) for bacterial sulfate reduction is most convincing if δ^{34}S values are available not only from deposited sulfides but also from the remaining coevally buried sulfate. However, such conditions are rarely met since sulfate evaporites from seawater tend to disappear from rocks (270). In many cases, δ^{34}S values can only be compared to the standard sulfur. In reconstruction of the history of bacterial sulfate reduction, it must not be forgotten that the evidence from isotope data is only indirect. Fractionation of sulfur isotopes has also been observed during chemical sulfate reduction (338, 340), and not all experiments with sulfate reducers yielded the high fractionation rates mentioned (150, 186).

10.5.3.2 Oceanic Sulfates

The best known indication from the "sulfate record" to globally significant activities of sulfate-reducing bacteria through time is the isotope ratio in present sea-

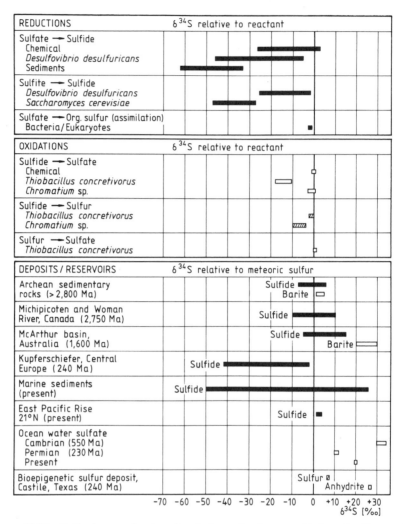

Figure 10.12 Sulfur isotope fractionation during reduction or oxidation of sulfur species, and isotopic composition of some sulfur species from the geological record. Drawn after data from Refs. 70, 147, 150, 188, and 325 and data cited in Refs. 52, 270, 288, and 338.

water sulfate, with a $\delta^{34}S$ value of about $+20‰$ (270, 338). It is assumed that sulfate was not present under the reducing conditions of the primeval earth but has gradually been formed by biological or chemical oxidation of reduced sulfur species (39). By these oxidations, however, the high enrichment of ^{34}S in the ocean sulfate in comparison to primordial sulfur ($\delta^{34}S = 0‰$) cannot be explained (270). Therefore, major net transformations of the formed, dissolved sulfate by sulfate reducers to sulfides in sediments must have occurred. The preferred removal of the lighter isotope in the sediments would have caused a relative enrichment of the

TABLE 10.8 **Geologic eons, eras, and suberas (from 277)**

Eon	Era	Subera	Duration (Ma B.P.)
Phanerozoic	Cenozoic		63–0
	Mesozoic		240–63
	Paleozoic		570–240
Precambrian	Proterozoic	Late Proterozoic	900–570
		Middle Proterozoic	1600–900
		Early Proterozoic	2500–1600
	Archean	Late Archean	2900–2500
		Middle Archean	3300–2900
		Early Archean	3900–3300
		Hadean	4500–3900

heavier one in the pore water, which exchanged with the overlying water. Thus seawater is expected to average the enrichment of ^{34}S-sulfate in all the areas of sulfate reduction (270). Gypsum and anhydrite directly orginating from seawater can be traced back to 1,200 Ma B.P.* (270). The record of these evaporites shows variations of the δ^{34}S values through time, for example, a maximum of +32‰ about 550 Ma ago in the Cambrian, and a minimum of +11‰ about 230 Ma ago in the Permian (270, 338). A possible explanation for the increases in δ^{34}S values is an enhanced primary production and burial of organic matter yielding substrates for sulfate reducers in sediments (270). As another possible reason, catastrophic influxes of brines enriched in ^{34}S-sulfate from mediterranean seas into the ocean have been discussed (110). A decrease of the ^{32}S content in ocean sulfates may reflect periods of intense oxidation or weathering of sulfides from sediments or rocks of magmatic origin. Evaporites older than 1,200 Ma have not been found; they have probably disappeared as a result of their facilitated recycling compared to other sedimentary materials (270). Isotopic data from older sulfates, such as barite, which could have been derived from evaporites, have to be interpreted reservedly.

10.5.3.3 Deposits of Metal Sulfides

The typical anaerobic sediment as known from modern marine habitats is probably represented in the geological record by the black shales, in which the sulfide content is relatively low (340). A sulfur fractionation and the presence of framboidal pyrite indicating activities of sulfate reducers have been traced in black shales to the beginning of the Proterozoic, 2500 Ma B.P. (340). Deposits with a high content of iron and other metal sulfides exhibit δ^{34}S values near 0‰. Therefore, these high-grade sulfide deposits appear to be of magmatic rather than of biologic origin (338, 340). This is also expected in view of the environmental conditions for bi-

*1200 Ma B.P. = 1200 × 10^6 years before present.

ological sulfide precipitation. The simultaneous presence of heavy metal compounds and organic electron donors at high concentrations that could account for the formation of high-grade ore deposits by bacterial sulfate reduction is very unlikely. The content of bound sulfide in marine sediments is, at best, 1 to 2% S (143, 338). Several of the low-grade deposits exhibit δ^{34}S values significantly below 0‰, although there is usually a large spread among them. Well-known examples are the Copperbelt deposit of Zambia and Zaire, and the White Pine copper deposit in Michigan, both from the late Proterozoic (900 to 570 Ma B.P.). An example from the Phanerozoic era (since 570 Ma B.P.) is the about-250-Ma-old Permian Kupferschiefer in Central Europe (270, 340). The sulfide content in these copper deposits is below 10% S. One explanation for their formation is that iron in biologically precipitated pyrite has been replaced by copper (340). Framboidal pyrite is still present in these deposits. Also lead–zinc deposits such as that from the Middle Proterozoic (1600 to 900 Ma B.P.) at the north Australian McArthur River have been discussed to be of potentially biologic origin; however, isotopic patterns consistent with a bacterial sulfate reduction have been observed only in the pyrite found in this deposit, but not in the sphalerite, ZnS, and galena, PbS (338, 340). The oldest sample with a sulfur fractionation of possible biologic origin is from the about-2800-Ma-old iron formations of Michipicoten and Woman River of the Canadian Shield (270).

Compared to modern aquatic environments, it appears that in former eras sites of sulfate reduction were more widespread, and that conditions for long-term preservation of sulfides were more favorable. On the earth today, there are only a few sites, as for example the Black Sea, with extensive sulfate reduction and deposition of lasting sulfidic sediments. Even the high-grade sulfides that are deposited today from abiotic hydrothermal brines in oceanic spreading centers (Atlantis II deep in the Red Sea, and East Pacific Rise near 21°N, 109°W) are unlikely to be preserved in future eras. In contact with the oxic seawater, these sulfides are chemically or biologically oxidized (340). One may assume that in former times, when oxygen was less abundant, the conditions were more favorable for large-scale developments of sulfate-reducing bacteria and for preservation of reduced sulfur species.

10.5.3.4 Deposits of Elemental Sulfur
There is little doubt that deposits of elemental sulfur other than those of volcanic origin were formed from biologically produced hydrogen sulfide (70, 126, 265, 338). The sulfur may have been formed by sulfide oxidation with oxygen, or by phototrophic bacteria as observed in Cyrenaikan lakes in the Libyan desert (43). Among the biogenic deposits, two types were distinguished (265): In deposits classified as biosyngenetic, the sulfur was formed prior to burial in the sediment. Bioepigenetic deposits were formed by postsedimentary reduction of gypsum or anhydrite. The sulfide was aerobically oxidized to sulfur, whereas the carbon dioxide formed from the organic electron donor combined with calcium ions to yield cacite in a replacement process. It was suggested that hydrocarbons from oil, which is still found in association with sulfur deposits, served as electron donors for reduction of sedimentary sulfates (265). The economically important sulfur deposits are

probably of bioepigenetic origin (265). The sulfur in biosyngenetic and bioepigenetic deposits may exhibit $\delta^{34}S$ values relative to the coexisting sulfates of more than about $-20‰$, although with respect to standard sulfide the values may be positive (70, 126, 338).

10.5.3.5 Carbonate Deposits

Another geochemically relevant process attributed to the activity of sulfate-reducing bacteria is the formation of soda deposits, as studied in the Egyptian Wadi Natrûn (1). There is evidence that sulfate from infiltrating water and soil was reduced by sulfate-reducing bacteria. By gas exchange between the water and the atmosphere, the volatile, oxidizable hydrogen sulfide was completely lost, whereas carbon dioxide formed from the organic substrates came into an equilibrium with the atmospheric pool. The result was a replacement of SO_4^{2-} ions by CO_3^{2-} or HCO_3^- ions and thus an increase in alkalinity. Evaporation of water lead to deposition of Na_2CO_3 and $NaHCO_3$. The process resembles the assumed replacement of anhydrite or gypsum by calcite during formation of epigenetic sulfur deposits (265).

10.5.3.6 Age of Dissimilatory Sulfate Reduction

A prerequisite for development of sulfate reducers on earth was the availability of sulfate, which was probably not present under the assumed primeval reducing conditions [39]. The oldest detected sedimentary sulfate, which is barite from the Early Archean, is about 3500 Ma old (270). The first sulfate may have been formed by oxidation of sulfide with oxygen from photolysis of water in the atmosphere, and by phototrophic sulfur bacteria (39, 269). However, the sulfate concentration in the Archaen oceans could not have been significant. Since the crust was probably highly mobile until the end of the Archean (2500 Ma B.P.), not only traces of oxygen but also formed sulfate must have been reduced rather soon by chemical reaction during the intense circulations through the mantle (45). Therefore, if bacterial sulfate reduction occurred in the Archean water, it could neither have been a geochemically significant process nor have led to detectable isotope fractionations. In accordance with this assumption, isotopic evidence for sulfate reduction is lacking in the geological record until the late Archean (269, 270, 340). According to present views, dissimilatory sulfate reduction became widespread sometime between 2800 and 2500 Ma B.P. (45, 269, 270, 339, 340). Thus dissimilatory sulfate reduction would be about as old as oxygenic photosynthesis by cyanobacteria or their ancestors, but older than oxygen respiration (269, 277). Anoxygenic photosynthesis as a sulfate-providing process appears to be older than sulfate reduction. In a view of Skyring and Donnelly (288), the spreading of sulfate reducers did not occur until 2000 Ma B.P. It was assumed that bacteria reducing sulfite were responsible for earlier sulfur fractionations, and that sulfate reducers evolved from these sulfite reducers. Sulfite (or bisulfite) may have been derived from sulfur dioxide vented from the crust into the sea. Sulfur dioxide is generated at deep crustal sites by high-temperature reactions of sulfides with water (94).

A new aspect in the evolution of sulfate-reducing bacteria is given by the relatedness to purple bacteria as shown by 16S rRNA sequencing (Section 10.2.2.2) and by the recent finding of sulfate reducers that disproportionate sulfite or thiosulfate to sulfide and sulfate (Section 10.2.6). The oxidative part of this disproportionation resembles the sulfate formation by phototrophic sulfur bacteria (341–343).

Bacteria performing a dissimilatory reduction of elemental sulfur may be older than sulfate reducers. Reduction of elemental sulfur as a rather reactive compound does not require a highly evolved enzymatic activating system such as in dissimilatory sulfate reduction. Sulfur reduction appears as a simple, primitive reaction that can be associated with energy conservation. Elemental sulfur formed by *Chlorobium*-like phototrophs has been regarded as the earliest biological oxidation product of sulfide (342).

ACKNOWLEDGMENTS

I wish to thank Norbert Pfennig for his support of my work on sulfate-reducing bacteria. Futhermore, I am indebted to Regina Bache, Friedhelm Bak, Katharina Brysch, Ralf Cord-Ruwisch, Guy Fauque, Georg Fuchs, Christian Schneider, and Erko Stackebrandt for providing unpublished data, and to Karl O. Stetter and Wolfram Zillig for photographs. The skillful help of Lori Nappe and Heribert Cypionka in the computerized completion of the manuscript is gratefully acknowledged.

REFERENCES

1. Abd-el-Malek, Y., and S. G. Rizk. 1963. Bacterial sulphate reduction and the development of alkalinity. III. Experiments under natural conditions in the Wadi Natrûn. J. Appl. Bacteriol. 26:20–26.

2. Abdollahi, H., and D. B. Nedwell. 1979. Seasonal temperature as a factor influencing bacterial sulfate reduction in a saltmarsh sediment. Microb. Ecol. 5:73–79.

3. Abdollahi, H., and D. B. Nedwell. 1980. Serological characteristics within the genus *Desulfovibrio*. Antonie Leeuwenhoek J. Microbiol. Serol. 46:73–83.

4. Abram, J. W., and D. B. Nedwell. 1978. Hydrogen as a substrate for methanogenesis and sulphate reduction in anaerobic saltmarsh sediment. Arch. Microbiol. 117:93–97.

5. Akagi, J. M., and V. Adams. 1967. Electron carriers for the phosphoroclastic reaction of *Desulfovibrio desulfuricans*. J. Biol. Chem. 242:2578–2483.

6. Akagi, J. M., and V. Adams. 1973 Isolation of a bisulfite reductase activity from *Desufotomaculum nigrificans* and its identification as the carbon monoxide-binding pigment P582. J. Bacteriol. 116:392–396.

7. Akagi, J. M., and L. L. Campbell. 1962. Studies on thermophilic sulfate-reducing bacteria. III. Adenosine triphosphate-sulfurylase of *Clostridium nigrificans* and *Desulfovibrio desulfuricans*. J. Bacteriol. 84:1194–1201.

8. Akagi, J. M., and G. Jackson. 1967. Degradation of glucose by proliferating cells of *Desulfotomaculum nigrificans*. Appl. Microbiol. 15:1427–1430.

9. Almgren, T., and I. Hagström. 1974. The oxidation rate of sulphide in sea water. Water Res. 8:395–400.

9a. Alperin, M. J., and W. S. Reeburgh. 1984. Geochemical observations supporting anaerobic methane oxidation, p. 282–289, in R. L. Crawford and R. S. Hansen (eds.), Microbial growth on C_1-compounds. American Society of Microbiology, Washington D.C.

9b. Alperin, M. J., and W. S. Reeburgh. 1985. Inhibition experiments on anaerobic methane oxidation. Appl. Environ. Microbiol. 50:940–945.

10. Anderson, J. W. 1978. Sulphur in biology (Studies in biology, No. 101). Arnold, London.

11. Arkesteyn, G. J. M. W. 1979. Pyrite oxidation by *Thiobacillus ferrooxidans* with special reference to the sulphur moiety of the mineral. Antonie Leeuwenhoek J. Microbiol. Serol. 45:423–435.

12. Artman, M. 1956. The production of hydrogen sulphide from thiosulphate by *Escherichia coli*. J. Gen. Microbiol. 14:315–322.

13. Atlas, R. M., and R. Bartha. 1973. Inhibition by fatty acids of the biodegradation of petroleum. Antonie Leeuwenhoek J. Microbiol. Serol. 39:257–271.

14. Baars, J. K. 1930. Over sulfaatreductie door bacteriën. Doctoral thesis, Hoogeschool Delft.

15. Bache, R., P. M. H. Kroneck, H. Merkle, and H. Beinert. 1983. A survey of EPR-detectable components in sulfur-reducing bacteria. Biochim. Biophys. Acta 722:417–426.

16. Bache, R., and N. Pfennig. 1981. Selective isolation of *Acetobacterium woodii* on methoxylated aromatic acids and determination of growth yields. Arch. Microbiol. 130:255–261.

17. Badziong, W., B. Ditter, and R. K. Thauer. 1979. Acetate and carbon dioxide assimilation by *Desulfovibrio vulgaris* (Marburg), growing on hydrogen and sulfate as sole energy source. Arch. Microbiol. 123:301–305.

18. Badziong, W., and R. K. Thauer. 1978. Growth yields and growth rates of *Desulfovibrio vulgaris* (Marburg) growing on hydrogen plus sulfate and hydrogen plus thiosulfate as the sole energy sources. Arch. Microbiol. 117:209–214.

19. Badziong, W., and R. K. Thauer. 1980. Vectorial electron transport in *Desulfovibrio vulgaris* (Marburg) growing on hydrogen plus sulfate as sole energy source. Arch. Microbiol. 125:167–174.

20. Badziong, W., R. K. Thauer, and J. G. Zeikus. 1978. Isolation and characterization of *Desulfovibrio* growing on hydrogen plus sulfate as the sole energy source. Arch. Microbiol. 116:41–49.

20a. Bak, F., and N. Pfennig. 1987. Chemolithotrophic growth of *Desulfovibrio sulfodismutans* sp. nov. by disproportionation of inorganic sulfur compounds. Arch. Microbiol. 147:184–189.

20b. Bak, F., and F. Widdel. 1986. Anaerobic degradation of indolic compounds by sulfate-reducing enrichment cultures, and description of *Desulfobacterium indolicum* gen. nov. sp. nov. Arch. Microbiol. 146:170–176.

20c. Bak, F., and F. Widdel. 1986. Anaerobic degradation of phenol and phenol derivatives by *Desulfobacterium phenolicum* sp. nov. Arch. Microbiol. 146:177–180.

21. Baker, F. D., H. R. Papiska, and L. L. Campbell. 1962. Choline fermentation by *Desulfovibrio desulfuricans*. J. Bacteriol. 84:973–978.

22. Balba, M. T., and D. B. Nedwell. 1982. Microbial metabolism of acetate, propionate and butyrate in anoxic sediment from the Colne Point Saltmarsh, Essex, U.K. J. Gen. Microbiol. 128:1415–1422.

23. Banat, I. M., E. B. Lindstrom, D. B. Nedwell, and M. T. Balba. 1981. Evidence for coexistence of two distinct functional groups of sulfate-reducing bacteria in salt marsh sediment. Appl. Environ. Microbiol. 42:985–992.

24. Banat, I. M., and D. B. Nedwell. 1983. Mechanisms of turnover of C_2-C_4 fatty acids in high-sulphate and low-sulphate anaerobic sediments. FEMS Microbiol. Lett. 17:107–110.

25. Barton, L. L., J. LeGall, and H. D. Peck, Jr. 1970. Phosphorylation coupled to oxidation of hydrogen with fumarate in extracts of the sulfate reducing bacterium, *Desulfovibrio gigas*. Biochem. Biophys. Res. Commun. 41:1036–1042.

26. Baumann, A., and V. Denk. 1950. Zur Physiologie der Sulfatreduktion. Arch. Mikrobiol. 15:283–307.

27. Beijerinck, W. M. 1895. Über *Spirillum desulfuricans* als Ursache von Sulfatreduction. Zentralbl. Bakteriol. 2 Abt. 1:1–9, 49–59, 104–114.

28. Belkin, S., C. O. Wirsen, and H. W. Jannasch. 1985. Biological and abiological sulfur reduction at high temperatures. Appl. Environ. Microbiol. 49:1057–1061.

29. Belyaev, S. S., K. S. Laurinavichus, A. Ya. Obraztova, S. N. Gorlatov, and M. V. Ivanov. 1982. Microbiological processes in the critical zone of injection wells. Microbiology (Engl. Transl. Mikrobiologiya) 51:793–797 (Mikrobiologiya 51:997–1001).

30. Benda, I. 1957. Mikrobiologische Untersuchungen über das Auftreten von Schwefelwasserstoff in den anaeroben Zonen des Hochmoores. Arch. Mikrobiol. 27:337–374.

31. Biebl, H., and N. Pfennig. 1977. Growth of sulfate-reducing bacteria with sulfur as electron acceptor. Arch. Microbiol. 112:115–117.

32. Boone, D. R.,and M. P. Bryant. 1980. Propionate-degrading bacterium, *Syntrophobacter wolinii* sp. nov. gen. nov., from methanogenic ecosystems. Appl. Environ. Microbiol. 40:626–632.

33. Booth, G. H., L. Elford, and D. S. Wakerley. 1968. Corrosion of mild steel by sulphate-reducing bacteria: an alternative mechanism. Br. Corros. J. 3:242–245.

34. Bos, P., and J. G. Kuenen. 1983. Microbiology of sulphur-oxidizing bacteria, p. 18–27, in Microbial corrosion. The Metals Society, London.

35. Bowen, H. J. M. 1979. Environmental chemistry of the elements. Academic Press, New York.

36. Brandis, A., and R. K. Thauer. 1981. Growth of *Desulfovibrio* species on hydrogen and sulphate as sole energy source. J. Gen. Microbiol. 126:249–252.

37. Brandis-Heep, A., N. A. Gebhardt, R. K. Thauer, F. Widdel, and N. Pfennig. 1983. Anaerobic acetate oxidation to CO_2 by *Desulfobacter postgatei*. 1. Demonstation of all enzymes required for the operation of the critic acid cycle. Arch. Microbiol. 136:222–229.

38. Braun, M., and H. Stolp. 1985. Degradation of methanol by a sulfate reducing bacterium. Arch. Microbiol. 142:77–80.

39. Broda, E. 1975. The evolution of the bioenergetic process. Pergamon Press, Oxford.

40. Bruschi, M., E. C. Hatchikian, J. Bonicel, G. Bovier-Lapierre, and P. Couchoud. 1977. The N-terminal sequence of superoxide dismutase from the strict anaerobe *Desulfovibrio desulfuricans*. FEBS Lett. 76:121–124.

41. Bryant, M. P., L. L. Campbell, C. A. Reddy, and M. R. Crabill. 1977. Growth of *Desulfovibrio* in lactate or ethanol media low in sulfate in association with H_2-utilizing methanogenic bacteria. Appl. Environ. Microbiol. 33:1162–1169.

42. Butlin, K. R., and M. E. Adams. 1947. Autotrophic growth of sulphate-reducing bacteria. Nature (Lond.) 160:154–155.

43. Butlin, K. R., and J. R. Postgate. 1954. The microbiological formation of sulphur in Cyrenaican lakes, p. 112–122, in J. L. Cloudsley-Thompson (ed.), Biology of deserts. Institute of Biology, London.

44. Butlin, K. R., S. C. Selwyn, and D. S. Wakerley. 1956. Sulphide production from sulphate-enriched sewage sludges. J. Appl. Bacteriol. 19:3–15.

45. Cameron, E. M. 1982. Sulphate and sulphate reduction in early Precambrian oceans. Nature (Lond.) 296:145–148.

46. Campbell, L. L., and J. R. Postgate. 1965. Classification of the spore-forming sulfate-reducing bacteria. Bacteriol. Rev. 29:359–363.

47. Cappenberg, T. E. 1974. Interrelations between sulfate-reducing and methane-producing bacteria in bottom deposits of a fresh-water lake. I. Field observations. Antonie Leeuwenhoek J. Microbiol. Serol. 40:285–295.

48. Cappenberg, T. E. 1974. Interrelations between sulfate-reducing and methane-producing bacteria in bottom deposits of a fresh-water lake. II. Inhibition experiments. Antonie van Leeuwenhoek J. Microbiol. Serol. 40:297–306.

49. Cappenberg, T. E. and R. A. Prins. 1974. Interrelations between sulfate-reducing and methane-producing bacteria in bottom deposits of a fresh-water lake. III. Experiments with ^{14}C-labeled substrates. Antonie van Leeuwenhoek J. Microbiol. Serol. 40:457–469.

50. Castenholz, R. W. 1977. The effect of sulfide on the blue-green algae of hot springs. II. Yellowstone National Park. Microb. Ecol. 3:79–105.

51. Caumette, P. 1978. Participation des bactéries phototrophes sulfo-oxidantes dans le métabolisme du soufre en milieu lagunaire méditerranéen (Étang du Prévost). Étude des crises dystrophiques (Malaigues). Doctoral thesis, Université Languedoc, Académie Montpellier.

52. Chambers, L. A., and P. A. Trudinger. 1979. Microbiological fractionation of stable sulfur isotopes: a review and critique. Geomicrobiol. J. 1:249–293.

53. Chen, K. Y., A. M. Asce, and J. C. Morris. 1972. Oxidation of sulfide by O_2: catalysis and inhibition. J. Sanit. Eng. Div. Proc. Am. Soc. Civ. Eng. 98:215–227.

54. Christensen, D., and T. H. Blackburn. 1982. Turnover of ^{14}C-labelled acetate in marine sediments. Mar. Biol. 71:113–119.

55. Clark, P. H. 1953. Hydrogen sulphide production by bacteria. J. Gen. Microbiol. 8:397–407.

56. Cline, J. D., and F. A. Richards. 1969. Oxygenation of hydrogen sulfide in seawater at constant salinity, temperature, and pH. Environ. Sci. Technol. 3:838–843.

57. Cohn, F. 1867. Beiträge zur Physiologie der Phycochromaceen und Florideen. Arch. Mikrosk. Anat. 3:1–60.

58. Coleman, G. S. 1960. A sulphate-reducing bacterium from the sheep rumen. J. Gen. Microbiol. 22:423–436.

59. Collins, M. D., M. Costas, and R. J. Owen. 1984. Isoprenoid quinone composition of representatives of the genus *Campylobacter*. Arch. Microbiol. 137:168–170.

60. Collins, M. D., and F. Fernandez. 1984. Menaquinone-6 and thermoplasmaquinone-6 in *Wolinella succinogenes*. FEMS Microbiol. Lett. 22:273–276.

61. Collins, M. D., and F. Widdel. 1986. Respiratory quinones of sulphate-reducing and sulphur-reducing bacteria: a systematic investigation. Syst. Appl. Microbiol. 8:8–18.

62. Connell, W. E., and W. H. Patrick, Jr. 1968. Sulfate reduction in soil: effects of redox potential and pH. Science 159:86–87.

63. Conrad, R., T. J. Phelps, and J. G. Zeikus. 1985. Gas metabolism evidence in support of the juxtaposition of hydrogen-producing and methanogenic bacteria in sewage sludge and lake sediments. Appl. Environ. Microbiol. 50:595–601.

63a. Cord-Ruwisch, R., and J.-L. Garcia. 1985. Isolation and characterization of an anaerobic benzoate-degrading spore-forming sulfate-reducing bacterium, *Desulfotomaculum sapomandens* sp. nov. FEMS Microbiol. Lett. 29:325–330.

64. Cord-Ruwisch, R., W. Kleinitz, and F. Widdel. 1986. Sulfatreduzierende Bakterien in einem Erdölfeld—Arten und Wachstumsbedingungen. Erdöl, Erdgas, Kohle, 102. Jahrgang, Heft 6:281–289.

65. Cord-Ruwisch, R., W. Kleinitz, and F. Widdel. 1987. Sulfate-reducing bacteria and their activities in oil production. J. Petrol. Technol. January:97–106.

65a. Cord-Ruwisch, R., and B. Ollivier. 1986. Interspecies hydrogen transfer during methanol degradation by *Sporomusa acidovorans* and hydrogenophilic anaerobes. Arch. Microbiol. 144:163–165.

65b. Cord-Ruwisch, R., B. Ollivier, and J.-L. Garcia. 1986. Fructose degradation by *Desulfovibrio* sp. in pure culture and in coculture with *Methanospirillum hungatei*. Curr. Microbiol. 13:285–289.

65c. Cord-Ruwisch, R., and F. Widdel. 1986. Corroding iron as a hydrogen source for sulphate reduction in growing cultures of sulphate-reducing bacteria. Appl. Microbiol. Biotechnol. 25:169–174.

66. Costello, J. A. 1974. Cathodic depolarization by sulphate-reducing bacteria. S. Afr. J. Sci. 70:202–204.

67. Crombie, D. J., G. J. Moodie, and J. D. R. Thomas. 1980. Corrosion of iron by sulphate-reducing bacteria. Chem. Ind. (Lond.) 500–504.

68. Cypionka, H., and N. Pfennig. 1986. Growth yields of *Desulfotomaculum orientis* with hydrogen in chemostat culture. Arch. Microbiol. 143:396–399.

69. Cypionka, H., F. Widdel, and N. Pfennig. 1985. Survival of sulfate-reducing bacteria after oxygen stress, and growth in sulfate-free oxygen-sulfide gradients. FEMS Microbiol. Ecol. 31:39–45.

70. Davis, J. B., and D. W. Kirkland. 1979. Bioepigenetic sulfur deposits. Econ. Geol. 74:462–468.

71. De Meio, R. H. 1975. Sulfate activation and transfer, p. 287–358, in D. M. Greenberg (ed.), Metabolic pathways, 3rd ed., Vol. 7, Metabolism of sulfur compounds. Academic Press, New York.

72. De Vries, W., H. G. D. Niekus, H. Berchum, and A. H. Stouthamer. 1982. Electron transport-linked proton translocation at nitrite reduction in *Campylobacter sputorum* subspecies *bubulus*. Arch. Microbiol. 131:132–139.

73. Dobson, C. M., N. J. Hoyle, C. F. Geraldes, P. E. Wright, R. J. P. Williams, M. Bruschi, and J. LeGall. 1974. Outline structure of cytochrome c_3 and consideration of its properties. Nature (Lond.) 249:425–429.

74. Doddema, H. J., and G. D. Vogels. 1978. Improved identification of methanogenic bacteria by fluorescence microscopy. Appl. Environ. Microbiol. 36:752–754.

74a. Dowling, N. J. E., F. Widdel, and D. C. White. 1986. Phospholipid ester-linked fatty acid biomarkers of acetate-oxidizing sulphate-reducers and other sulphide-forming bacteria. J. Gen. Microbiol. 132:1815–1825.

75. Drake, H. L., and J. M. Akagi. 1976. Purification of a unique bisulfite-reducing enzyme from *Desulfovibrio vulgaris*. Biochem. Biophys. Res. Commun. 71:1214–1219.

76. Dubourdieu, M., and J. L. Fox, 1977. Amino acid sequence of *Desulfovibrio vugaris* flavodoxin. J. Biol. Chem. 252:1453–1463.

77. Einsele, W. 1936. Über die Beziehung des Eisenkreislaufs zum Phosphatkreislauf im eutrophen See. Arch. Hydrobiol. 29:664–686.

78. Einsele, W. 1938. Über chemische and kolloidchemische Vorgänge in Eisen-Phosphat-Systemen unter limnischen and limnogeologischen Gesichtpunkten. Arch. Hydrobiol. 33:361–367.

79. ElKurdi, A. B., J. L. Leaver, and G. W. Pettigrew. 1982. The *c*-type cytochromes of *Campylobacter sputorum* ssp. *mucosalis*. FEMS Microbiol. Lett. 14:177–182.

80. Evans, W. C. 1977. Biochemistry of the bacterial catabolism of aromatic compounds in anaerobic environment. Nature (Lond.) 270:17–22.

81. Fauque, G. D., L. L. Barton, and J. LeGall. 1980. Oxidative phosphorylation linked to dissimilatory reduction of elemental sulphur by *Desulfovibrio*, p. 71–86, in Sulphur in biology (Ciba Foundation Symposium No. 72). Excerpta Medica, Amsterdam.

82. Fiebig, K., and G. Gottschalk. 1983. Methanogenesis from choline by a coculture of *Desulfovibrio* sp. and *Methanosarcina barkeri*. Appl. Environ. Microbiol. 45:161–168.

83. Fischer, F., W. Zillig, K. O. Stetter, and G. Schreiber. 1983. Chemolithoautotrophic metabolism of anaerobic extremely thermophilic archaebacteria. Nature (Lond.) 301:511–513.

84. Fleischer, S. 1978. Evidence for the anaerobic release of phosphorus from lake sediments as a biological process. Naturwissenschaften 65:109.

85. Fowler, V. J., F. Widdel, N. Pfennig, C. R. Woese, and E. Stackebrandt. 1986. Phylogenetic relationships of sulfate- and sulfur-reducing eubacteria. Syst. Appl. Microbiol. 8:32–41.

86. Frevert, T. 1979. Phosphorus and iron concentrations in the interstitial water and dry substance of sediments of Lake Constance (Obersee). Arch. Hydrobiol. (Suppl.) 55:298–323.

87. Friedrich, C. G., and G. Mitrenga. 1981. Oxidation of thiosulfate by *Paracoccus denitrificans* and other hydrogen bacteria. FEMS Microbiol. Lett. 10:209–212.

88. Gebhardt, N. A., D. Linder, and R. K. Thauer. 1983. Anaerobic acetate oxidation to CO_2 by *Desulfobacter postgatei*. 2. Evidence from ^{14}C-labelling studies for the operation of the citric acid cycle. Arch. Microbiol. 136:230–233.

89. Gebhardt, N. A., R. K. Thauer, D. Linder, P.-M. Kaulfers, and N. Pfennig. 1985. Mechanism of acetate oxidation to CO_2 with elemental sulfur in *Desulfuromonas acetoxidans*. Arch. Microbiol. 141:392–398.

90. Gmelins Handbuch der anorganischen Chemie. 1953. 8. Aufl., Schwefel, Teil B, Lieferung 1, p. 91–93. Verlag Chemie, Weinheim, West Germany.

90a. Gmelins Handbuch der anorganischen Chemie. 1960. 8. Aufl., Schwefel, Teil B. Lieferung 2, p. 523–534. Verlag Chemie, Weinheim, West Germany.

91. Gogotova, G. I., and M. B. Vainshtein. 1983. Spore-forming, sulfate-reducing bacterium *Desulfotomaculum guttoideum* sp. nov. Microbiology (Engl. Transl. Mikrobiologiya) 52:618–622 (Mikrobiologiya 52: 789–793).

92. Goldhaber, M. B., and I. R. Kaplan. 1974. The sulfur cycle, p. 569–655, in E. D. Goldberg (ed.), The sea, Vol. 5, Marine chemistry. Wiley, New York.

93. Gottschalk, G., and J. R. Andreesen. 1979. Energy metabolism in anaerobes, p. 85–115, in J. R. Quayle (ed.), International review of biochemistry, vol. 21, Microbial biochemistry. University Park Press, Baltimore.

94. Grinenko, V. A., and M. V. Ivanov. 1983. Principal reactions of the global biogeochemical cycle of sulphur, p. 1–23, in M. V. Ivanov and J. R. Freney (eds.), The global biogeochemical sulphur cycle. Wiley, New York.

95. Grossman, J. P., and J. R. Postgate. 1955. The metabolism of malate and certain other compounds by *Desulphovibrio desulphuricans*. J. Gen. Microbiol. 12:429–445.

96. Guerlesquin, F., J. J. G. Moura, and R. Cammack. 1982. Iron-sulphur cluster composition and redox properties of two ferredoxins from *Desulfovibrio desulfuricans* Norway strain. Biochim. Biophys. Acta. 679:422–427.

97. Gunnarsson, L. Å. H., and P. H. Rönnow. 1982. Interrelationships between sulfate-reducing and methane-producing bacteria in costal sediments with intense sulfide production. Mar. Biol. 69:121–128.

98. Hamilton, W. A. 1985. Sulphate-reducing bacteria and anaerobic corrosion. Annu. Rev. Microbiol. 39:195–217.

99. Hardy, J. A. 1983. Utilization of cathodic hydrogen by sulphate-reducing bacteria. Br. Corros. J. 18:190–193.

100. Hardy, J. A., and W. A. Hamilton, 1981. The oxygen tolerance of sulfate-reducing bacteria isolated from North Sea waters. Curr. Microbiol. 6:259–262.

101. Hatchikian, E. C., H. E. Jones, and M. Bruschi. 1979. Isolation and characterization of a rubredoxin and two ferredoxins from *Desulfovibrio africanus*. Biochim. Biophys. Acta 548:471–483.

102. Hatchikian, E. C., J. LeGall, and G. R. Bell. 1977. Significance of superoxide dismutase and catalase activities in the strict anaerobes, sulfate reducing bacteria, p. 159–172, in A. M. Michelson, J. M. McCord and I. Fridovich (eds.), Superoxide and superoxide dismutases. Academic Press, New York.

103. Hatchikian, E. C., J. LeGall, and N. Forget. 1972. Evidence for the presence of a *b*-type cytochrome in the sulfate-reducing bacterium *Desulfovibrio gigas*, and its role

in the reduction of fumarate by molecular hydrogen. Biochim. Biophys. Acta 267:479–484.

104. Hatchikian, E. C., and J. G. Zeikus. 1983. Characterization of a new type of dissimilatory sulfite reductase present in *Thermodesulfobacterium commune*. J. Bacteriol. 153:1211–1220.

105. Healey, F. P. 1980. Slope of the Monod equation as an indicator of advantage in nutrients competition. Microb. Ecol. 5:281–286.

106. Herbert, B. N., and F. D. J. Stott. 1983. The effects of pressure and temperature on bacteria in oilfield water injection systems, p. 7–17, in Microbial corrosion. The Metals Society, London.

107. Hines, M. E., and J. D. Buck, 1982. Distribution of methanogenic and sulfate-reducing bacteria in near-shore marine sediments. Appl. Environ. Microbiol. 43:447–453.

108. Hippe, H., D. Caspari, K. Fiebig, and G. Gottschalk. 1979. Utilization of trimethylamine and other N-methyl compounds for growth and methane formation in *Methanosarcina barkeri*. Proc. Natl. Acad. Sci. USA 76:494–498.

109. Hollaus, F., and A. Sleytr. 1972. On the taxonomy and fine structure of some hyperthermophilic saccharolytic clostridia. Arch. Microbiol. 86:129–146.

110. Holser, W. T. 1977. Catastrophic chemical events in the history of the ocean. Nature (Lond.) 267:403–408.

111. Hoppe-Seyler, F. 1886. Über die Gährung der Cellulose mit Bildung von Methan und Kohlensäure. II. Der Zerfall der Cellulose durch Gährung unter Bildung von Methan und Kohlensäure und die Erscheinungen, welche dieser Process veranlasst. Z. Physiol. Chem. 10:401–440.

112. Howard, B. H., and R. E. Hungate. 1976. *Desulfovibrio* of the sheep rumen. Appl. Environ. Microbiol. 32:598–602.

113. Howarth, R. W. 1979. Pyrite: its rapid formation in a salt marsh and its importance in ecosystem metabolism. Science 203:49–51.

114. Hungate, R. E. 1966. The rumen and its microbes. Academic Press, New York.

115. Huser, B., K. Wuhrmann, and A. J. B. Zehnder. 1982. *Methanothrix soehngenii* gen. nov. sp. nov., a new acetotrophic non-hydrogen-oxidizing methane bacterium. Arch. Microbiol. 132:1–9.

116. Iizuka, H., H. Okahazi, and N. Seto. 1969. A new sulfate-reducing bacterium isolated from Antarctica. J. Gen. Appl. Microbiol. 15:11–18.

117. Imhoff, D., and J. R. Andreesen. 1979. Nicotinic acid hydroxylase from *Clostridium barkeri*: selenium-dependent formation of active enzyme. FEMS Microbiol. Lett. 5:155–158.

118. Imhoff-Stuckle, D., and N. Pfennig, 1983. Isolation and characterization of a nicotinic acid-degrading sulfate-reducing bacterium, *Desulfococcus niacini*, sp. nov. Arch. Microbiol. 136:194–198.

119. Imhoff, J. F. 1982. Occurrence and evolutionary significance of two sulfate assimilation pathways in the Rhodospirillaceae. Arch. Microbiol. 132:197–203.

120. Ingvorsen, K., and T. D. Brock. 1982. Electron flow via sulfate reduction and methanogenesis in the anaerobic hypolimnion of Lake Mendota. Limnol. Oceanogr. 27:559–564.

121. Ingvorsen, K., and B. B. Jørgensen. 1984. Kinetics of sulfate uptake by freshwater and marine species of *Desulfovibrio*. Arch. Microbiol. 139:61–66.

122. Ingvorsen, K., A. J. B. Zehnder, and B. B. Jørgensen. 1984. Kinetics of sulfate and acetate uptake by *Desulfobacter postgatei*. Appl. Environ. Microbiol. 47:403–408.

123. Ingvorsen, K., J. G. Zeikus, and T. D. Brock. 1981. Dynamics of bacterial sulfate reduction in a eutrophic lake. Appl. Environ. Microbiol. 42:1029–1036.

123a. Isa, Z., S. Grusenmeyer, and W. Verstraete. 1986. Sulfate reduction relative to methane production in high-rate anaerobic digestion: technical aspects. Appl. Environ. Microbiol. 51:572–579.

123b. Isa, Z., S. Grusenmeyer, and W. Verstraete. 1986. Sulfate reduction relative to methane production in high-rate anaerobic digestion: microbiological aspects. Appl. Environ. Microbiol. 51:580–587.

124. Ishimoto, M., J. Koyama, and Y. Nagai. 1954. Biochemical studies on sulfate-reducing bacteria. IV. The cytochrome system of sulfate-reducing bacteria. J. Biochem. 41:763–770.

125. Ishimoto, M., J. Koyama, T. Omura, and Y. Nagai. 1954. Biochemical studies on sulfate-reducing bacteria. III. Sulfate reduction by cell suspensions. J. Biochem. 41:537–546.

126. Ivanov, M. V. 1968. Microbiological processes in the formation of sulfur deposits. Israel Program for Scientific Translations, Jerusalem.

127. Ivanov, M. V., and J. R. Freney. 1983. The global biogeochemical sulphur cycle. Wiley, New York.

127a. Iversen, N., and B. B. Jørgensen. 1985. Anaerobic methane oxidation rates at the sulfate-methane transition in marine sediments from Kattegat and Skagerrak (Denmark). Limnol. Oceanogr. 30:944–955.

128. Iverson, W. P., and G. J. Olson. 1984. Anaerobic corrosion of iron and steel: a novel mechanism, p. 623–627, in M. J. Klug and C. A. Reddy (eds), Current perspectives in microbial ecology. American Society for Microbiology, Washington, D.C.

129. Jacq, V., and Y. Dommergues. 1971. Sulfato-réduction spermatosphérique. Ann. Inst. Pasteur (Paris) 121:199–206.

130. Jansen, K., G. Fuchs, and R. K. Thauer. 1985. Autotrophic CO_2 fixation by *Desulfovibrio baarsii*: demonstration of enzyme activities characteristic for the acetyl-CoA pathway. FEMS Microbiol. Lett. 28:311–315.

131. Jansen, K., R. K. Thauer, F. Widdel, and G. Fuchs. 1984. Carbon assimilation pathways in sulfate reducing bacteria. Formate, carbon dioxide, carbon monoxide, and acetate assimilation by *Desulfovibrio baarsii*. Arch. Microbiol. 138:257–262.

132. Jeris, J. S., and P. L. McCarty. 1965. The biochemistry of methane formation using C^{14} tracers. J. Water Pollut. Control Fed. 37:178–192.

133. Jones, H. E., 1971. A re-examination of *Desulfovibrio africanus*. Arch. Mikrobiol. 80:78–86.

134. Jones, H. E. 1971. Sulfate-reducing bacterium with unusual morphology and pigment content. J. Bacteriol. 106:339–346.

135. Jones, H. E. 1972. Cytochromes and other pigments of dissimilatory sulphate-reducing bacteria. Arch. Mikrobiol. 84:207–224.

136. Jones, W. J., M. J. B. Paynter, and R. Gupta. 1983. Characterization of *Methano-*

coccus maripaludis sp. nov., a new methanogen isolated from salt marsh sediment. Arch. Microbiol. 135:91–97.

137. Jørgensen, B. B. 1977. The sulfur cycle of a coastal marine sediment (Limfjorden, Denmark). Limnol. Oceanogr. 22:814–832.

138. Jørgensen, B. B. 1977. Bacterial sulfate reduction within reduced microniches of oxidized marine sediments. Mar. Biol. 41:7–17.

139. Jørgensen, B. B. 1978. A comparison of methods for the quantification of bacterial sulfate reduction in coastal marine sediments. I. Measurement with radiotracer techniques. Geomicrobiol. J. 1:11–28.

140. Jørgensen, B. B. 1978. A comparison of methods for the quantification of bacterial sulfate reduction in coastal marine sediments. III. Estimation from chemical and bacteriological field data. Geomicrobiol. J. 1:49–64.

141. Jørgensen, B. B. 1982. Mineralization of organic matter in the sea bed: the role of sulfate reduction. Nature (Lond.) 296:643–645.

142. Jørgensen, B. B. 1982. Ecology of the bacteria of the sulphur cycle with special reference to anoxic-oxic interface environments. Philos. Trans. R. Soc. Lond. B298:543–561.

143. Jørgensen, B. B. 1983. The microbial sulphur cycle, p. 91–124, in W. E. Krumbein (ed.), Microbial geochemistry. Blackwell Scientific, Oxford.

144. Jørgensen, B. B., J. G. Kuenen, and Y. Cohen. 1979. Microbial transformations of sulfur compounds in a stratified lake (Solar Lake, Sinai). Limnol. Oceanogr. 24:799–822.

145. Jørgensen, B. B., and N. P. Revsbech. 1983. Colorless sulfur bacteria, *Beggiatoa* spp. and *Thiovulum* spp., in O_2 and H_2S microgradients. Appl. Environ. Microbiol. 45:1261–1270.

146. Kämpf, C., and N. Pfennig, 1980. Capacity of Chromatiaceae for chemotrophic growth. Specific respiration rates of *Thiocystis violacea* and *Chromatium vinosum*. Arch. Microbiol. 127:125–135.

147. Kaplan, I. R., and S. C. Rittenberg. 1964. Microbiological fractionation of sulphur isotopes. J. Gen. Microbiol. 34:195–212.

148. Kaspar, H. F., and K. Wuhrmann. 1978. Kinetic parameters and relative turnovers of some important catabolic reactions in digesting sludge. Appl. Environ. Microbiol. 36:1–7.

149. Keith, S. M., R. A. Herbert, and C. G. Harfoot. 1982. Isolation of new types of sulphate-reducing bacteria from estuarine and marine sediments using chemostat enrichments. J. Appl. Bacteriol. 53:29–33.

150. Kemp, A. L. W., and H. G. Thode. 1968. The mechanism of the bacterial reduction of sulphate and of sulphite from isotope fractionation studies. Geochim. Cosmochim. Acta 32:71–91.

151. Killham, K., N. D. Lindley, and M. Wainwright. 1981. Inorganic sulfur oxidation by *Aureobasidium pullulans*. Appl. Environ. Microbiol. 42:629–631.

152. King, G. M., M. J. Klug, and D. R. Loveley. 1983. Metabolism of acetate, methanol, and methylated amines in intertidal sediments of Lowes Cove, Maine. Appl. Environ. Microbiol. 45:1848–1853.

153. Klemps, R., H. Cypionka, F. Widdel, and N. Pfennig. 1985. Growth with hydrogen,

and further physiological characteristics of *Desulfotomaculum* species. Arch. Microbiol. 143:203–208.

154. Kluyver, A. J., and C. B. van Niel. 1936. Prospects for a natural system of classification of bacteria. Zentralbl. Bakteriol. 2 Abt. 94:369–403.

155. Kohler, H.-P., B. Åhring, C. Abella, K. Ingvorsen, H. Keweloh, E. Laczkó, E. Stupperich, and F. Tomei. 1984. Bacteriological studies on the sulfur cycle in the anaerobic part of the hypolimnion and in the surface sediments of Rotsee in Switzerland. FEMS Microbiol. Lett. 21:279–286.

156. Kondratieva, E. N., V. G. Zhukov, R. N. Ivanovsky, and Yu. P. Petushkova. 1976. The capacity of phototrophic sulfur bacterium *Thiocapsa reseopersicina* for chemosynthesis. Arch. Microbiol. 108:287–292.

157. Kristjansson, J. K., P. Schönheit, and R. K. Thauer. 1982. Different K_s values for hydrogen of methanogenic bacteria and sulfate-reducing bacteria: an explanation for the apparent inhibition of methanogenesis by sulfate. Arch. Microbiol. 131:278–282.

158. Kröger, A. 1978. Fumarate as terminal acceptor of phosphorylative electron transport. Biochim. Biophys. Acta 505:129–145.

159. Krouse, H. R., R. G. L. McCready, S. A. Husain, and J. N. Campbell. 1967. Sulfur isotope fractionation by *Salmonella* species. Can. J. Microbiol. 13:21–25.

160. Kuenen, J. G., and R. F. Beudeker. 1982. Microbiology of thiobacilli and other sulphur-oxidizing autotrophs, mixotrophs and heterotrophs. Philos. Trans. R. Soc. Lond. B298:473–497.

161. Kuznetsova, V. A., and V. M. Gorlenko. 1965. Effect of temperature on the development of microorganisms from flooded strata of the Romashkino oil field. Microbiology (Engl. Transl. Mikrobiologiya) 34:274–278 (Mikrobiologiya 34:329–334).

162. Laanbroek, H. J., T. Abee, and I. L. Voogd. 1982. Alcohol conversions by *Desulfobulbus propionicus* Lindhorst in the presence and absence of sulfate and hydrogen. Arch. Microbiol. 133:178–184.

163. Laanbroek, H. J., H. J. Geerlings, L. Sijtsma, and H. Veldkamp. 1984. Competition for sulfate and ethanol among *Desulfobacter*, *Desulfobulbus*, and *Desulfovibrio* species isolated from intertidal sediments. Appl. Environ. Microbiol. 47:329–334.

164. Laanbroek, H. J., W. Kingma, and H. Veldkamp. 1977. Isolation of an aspartate-fermenting, free-living *Campylobacter* species. FEMS Microbiol. Lett. 1:99–102.

165. Laanbroek, H. J., J. T. Lambers, W. M. De Vos, and H. Veldkamp. 1978. L-Aspartate fermentation by a free-living *Campylobacter* species. Arch. Microbiol. 117:109–114.

166. Laanbroek, H. J., and N. Pfennig, 1981. Oxidation of short-chain fatty acids by sulfate-reducing bacteria in freshwater and in marine sediments. Arch. Microbiol. 128:330–335.

167. Laanbroek, H. J., L. J. Stal, and H. Veldkamp. 1978. Utilization of hydrogen and formate by *Campylobacter* spec. under aerobic and anaerobic conditions. Arch. Microbiol. 119:99–102.

168. Laanbroek, H. J., and H. Veldkamp. 1979. Growth yield and energy generation in anaerobically-grown *Campylobacter* spec. Arch. Microbiol. 120:47–51.

169. Laanbroek, H. J., and H. Veldkamp. 1982. Microbial interactions in sediment communities. Philos. Trans. R. Soc. Lond. B297:533–550.

170. Lawson, G. H. K., J. L. Leaver, G. W. Pettigrew, and A. C. Rowland. 1981. Some features of *Campylobacter sputorum* subsp. *mucosalis* subsp. nov., nom. rev. and their taxonomic significance. Int. J. Syst. Bacteriol. 31:385–391.

171. Lee, J.-P., J. LeGall, and H. D. Peck, Jr. 1973. Isolation of assimilatory- and dissimilatory-type sulfite reductase from *Desulfovibrio vulgaris*. J. Bacteriol. 115:529–542.

172. Lee, J.-P., C.-S. Yi, J. LeGall, and H. D. Peck, Jr. 1973. Isolation of a new pigment, desulforubidin, from *Desulfovibrio desufuricans* (Norway strain) and its role in sulfite reduction. J. Bacteriol. 115:453–455.

173. LeGall, J., D. V. DerVartanian, and H. D. Peck, Jr. 1979. Flavoproteins, iron proteins, and hemoproteins as electron-transfer components of the sulfate-reducing bacteria, p. 237–265, in D. R. Sanadi (ed.), Current topics in bioenergetics, Vol. 9. Academic Press, New York.

174. LeGall, J., and J. R. Postgate. 1973. The physiology of sulfate-reducing bacteria, p. 81–133, in A. H. Rose and D. W. Tempest (eds.), Advances in microbial physiology, Vol. 10. Academic Press, New York.

175. Lighthart, B. 1963. Sulfate-reducing bacteria in San Vincente Reservoir, San Diego County, California. Limmol. Oceanogr. 8:349–351.

176. Liu, M. C., and H. D. Peck, Jr. 1981. The isolation of a hexaheme cytochrome from *Desulfovibrio desulfuricans* and its identification as a new type of nitrite reductase. J. Biol. Chem. 256:13159–13164.

177. Loveley, D. R. 1985. Minimum threshold for hydrogen metabolism in methanogenic bacteria. Appl. Environ. Microbiol. 49:1530–1531.

178. Loveley, D. R., D. F. Dwyer, and M. J. Klug. 1982. Kinetic analysis of competition between sulfate reducers and methanogens for hydrogen in sediment. Appl. Environ. Microbiol. 43:1373–1379.

179. Loveley, D. R., and M. J. Klug. 1982. Intermediary metabolism of organic matter in the sediments of a eutrophic lake. Appl. Environ. Microbiol. 43:552–560.

180. Loveley, D. R., and M. J. Klug. 1983. Sulfate reducers can outcompete methanogens at freshwater sulfate concentrations. Appl. Environ. Microbiol. 45:187–192.

181. Loveley, D. R., and M. J. Klug. 1983. Methanogenesis from methanol and methylamines and acetogenesis from hydrogen and carbon dioxide in the sediments of a eutrophic lake. Appl. Environ. Microbiol. 45:1310–1315.

181a. Macy, J. M., I. Schröder, R. K. Thauer, and A. Kröger. 1986. Growth of *Wolinella succinogenes* on H_2S plus fumarate and on formate plus sulfur as energy sources. Arch. Microbiol. 144:147–150.

182. Mah, R. A. 1980. Isolation and characterization of *Methanococcus mazei*. Curr. Microbiol. 3:321–326.

183. Martens, C. S., and R. A. Berner. 1974. Methane production in the interstitial waters of sulfate-depleted marine sediments. Science 185:1167–1169.

184. Marty, D. 1981. Distribution of different anaerobic bacteria in Arabian Sea sediments. Mar. Biol. 63:277–281.

185. Mayhew, S. G., and M. L. Ludwig. 1975. Flavodoxins and electron-transferring flavoproteins, p. 57–118, in P. D. Boyer (ed.), The enzymes, 3rd ed., Vol. 12, Pt. B. Academic Press, New York.

186. McCready, R. G. L. 1975. Sulphur isotope fractionation by *Desulfovibrio* and *Desulfotomaculum* species. Geochim. Cosmochim. Acta 39:1395–1401.

187. McCready, R. G. L., W. D. Gould, and F. D. Cook. 1983. Respiratory nitrate reduction by *Desufovibrio* sp. Arch. Microbiol. 135:182–185.

188. McCready, R. G. L., and I. R. Kaplan. 1974. Fractionation of sulfur isotopes by the yeast *Saccharomyces cerevisiae*. Geochim. Cosmochim. Acta 38:1239–1253.

189. McCready, R. G. L., E. J. Laishley, and H. R. Krouse. 1975. Stable isotope fractionation by *Clostridium pasteurianum*. 1. ^{34}S/^{32}S:inverse isotope effects during SO_4^{2-} and SO_3^{2-} reduction. Can. J. Microbiol. 21:235–244.

190. McInerney, M. J., and M. P. Bryant. 1981. Anaerobic degradation of lactate by syntrophic associations of *Methanosarcina barkeri* and *Desulfovibrio* species and effect of H_2 on acetate degradation. Appl. Environ. Microbiol. 41:346–354.

191. McInerney, M. J., M. P. Bryant, R. B. Hespell, and J. W. Costerton. 1981. *Syntrophomonas wolfei* gen. nov. sp. nov., an anaerobic, syntrophic, fatty acid-oxidizing bacterium. Appl. Environ. Microbiol. 41:1029–1039.

192. McInerney, M. J., M. P. Bryant, and N. Pfennig. 1979. Anaerobic bacterium that degrades fatty acids in syntrophic association with methanogens. Arch. Microbiol. 122:129–135.

193. Mechalas, B. J., and S. C. Rittenberg. 1960. Energy coupling in *Desulfovibrio desulfuricans*. J. Bacteriol. 80:501–507.

194. Meyer, L. 1864. Chemische Untersuchungen der Thermen zu Landeck in der Grafschaft Glatz. J. Prakt. Chem. 91:1–15.

195. Mezzino, M. J., W. R. Strohl, and J. M. Larkin. 1984. Characterization of *Beggiatoa alba*. Arch. Microbiol. 137:139–144.

196. Migdisov, A. A., A. B. Ronov, and V. A. Grinenko. 1983. The sulphur cycle in the lithosphere, p. 25–127, in M. V. Ivanov and J. R. Freney (eds.), The global biogeochemical sulphur cycle. Wiley, New York.

197. Miller, J. D. A. 1981. Metals, p. 149–202, in A. H. Rose (ed.), Economic microbiology, Vol. 6, Microbial biodeterioration. Academic Press, London.

198. Miller, J. D. A., P. M. Neumann, L. Elford, and D. S. Wakerley. 1970. Malate dismutation by *Desulfovibrio*. Arch. Mikrobiol. 71:214–219.

199. Miller, J. D. A., and A. M. Saleh. 1964. A sulphate-reducing bacterium containing cytochrome c_3 but lacking desulfoviridin. J. Gen. Microbiol. 37:419–423.

200. Miller, J. D. A., and D. S. Wakerley. 1966. Growth of sulphate-reducing bacteria by fumarate dismutation. J. Gen. Microbiol. 43:101–107.

201. Moore, W. E. C., and L. V. Holdeman. 1984. Genus *Desulfomonas*, p. 672–673, in N. R. Krieg and J. G. Holt (eds.), Bergey's manual of systematic bacteriology, Vol. 1. Williams & Wilkins, Baltimore.

202. Moore, W. E. C., J. L. Johnson, and L. V. Holdeman. 1976. Emendation of Bacteroidaceae and *Butyrivibrio* and descriptions of *Desulfomonas* gen. nov. and ten new species in the genera *Desulfomonas*, *Butyrivibrio*, *Eubacterium*, *Clostridium*, and *Ruminococcus*. Int. J. Syst. Bacteriol. 26:238–252.

203. Mortimer, C. H. 1941. The exchange of dissolved substances between mud and water in lakes. J. Ecol. 29:280–329.

204. Mountfort, D. O., and R. A. Asher. 1979. Effect of inorganic sulfide on the growth

and metabolism of *Methanosarcina barkeri* strain DM. Appl. Environ. Microbiol. 37:670–675.

205. Mountfort, D. O., and R. A. Asher, 1981. Role of sulfate reduction versus methanogenesis in terminal carbon flow in polluted intertidal sediment of Maimea inlet, Nelson, New Zealand. Appl. Environ. Microbiol. 42:252–258.

206. Mountfort, D. O., R. A. Asher, E. L. Mays, and J. M. Tiedje. 1980. Carbon and electron flow in mud and sandflat intertidal sediments at Delaware inlet, Nelson, New Zealand. Appl. Environ. Microbiol. 39:686–694.

207. Mountfort, D. O., W. J. Brulla, L. R. Krumholz, and M. P. Bryant. 1984. *Syntrophus buswellii* gen. nov., sp. nov.: a benzoate catabolizer from methanogenic ecosystems. Int. J. Syst. Bacteriol. 34:216–217.

208. Mountfort, D. O., and M. P. Bryant. 1982. Isolation and characterization of an anaerobic syntrophic banzoate-degrading bacterium from sewage sludge. Arch. Microbiol. 133:249–256.

209. Moura, I., J. J. G. Moura, M. Bruschi, and J. LeGall. 1980. Flavodoxin and rubredoxin from *Desulfovibrio salexigens*. Biochim. Biophys. Acta 591:1–8.

210. Moura, I., J. J. G. Moura, M. H. Santos, A. V. Xavier, and J. LeGall. 1979. Redox studies on rubredoxins from sulphate and sulphur reducing bacteria. FEBS Lett. 107:419–421.

211. Nazina, T. N., and T. A. Pivovarova. 1979. Submicroscopic organization and sporulation in *Desulfotomaculum nigrificans*. Microbiology (Engl. Transl. Mikrobiologiya) 48:241–245 (Mikrobiologiya 48:302–306).

212. Nazina, T.N., E. P. Rozanova, and T. A. Kalininskaya. 1979. Fixation of molecular nitrogen by sulfate-reducing bacteria from oil strata. Microbiology (Engl. Transl. Mikrobiologiya) Microbiology 48:102–104 (Mikrobiologiya 48:133–136).

213. Nedwell, D. B. 1982. The cycling of sulphur in marine and freshwater sediments, p. 73–106, in D. B. Nedwell and C. M. Brown (eds.), Sediment microbiology. Society for General Microbiology, Academic Press, London.

214. Nedwell, D. B., and J. W. Abram. 1979. Relative influence of temperature and electron donor and electron acceptor concentrations on bacterial sulfate reduction in saltmarsh sediment. Microb. Ecol. 5:67–72.

215. Nelson, D. C., and R. W. Castenholz. 1981. Use of reduced sulfur compounds by *Beggiatoa* sp. J. Bacteriol. 147:140–154.

216. Nelson, D. C., and H. W. Jannasch. 1983. Chemoautotrophic growth of a marine *Beggiatoa* in sulfide-gradient cultures. Arch. Microbiol. 136:262–269.

217. Nethe-Jaenchen, R., and R. K. Thauer. 1984. Growth yields and saturation constant of *Desulfovibrio vugaris* in chemostat culture. Arch. Microbiol. 137:236–240.

218. Neuberg, C., and E. Welde. 1914. Phytochemische Reduktionen. IX. Die Umwandlung von Thiosulfat in Schwefelwasserstoff und Sulfit durch Hefen. Biochem. Z. 67:111–118.

219. Niekus, H. G. D., W. De Vries, and A. H. Stouthamer. 1977. The effect of different dissolved oxygen tensions on growth and enzyme activities of *Campylobacter sputorum* subspecies *bubulus*. J. Gen. Microbiol. 103:215–222.

220. Niekus, H. G. D., E. van Doorn, W. De Vries, and A. H. Stouthamer. 1980. Aerobic growth of *Campylobacter sputorum* subspecies *bubulus* with formate. J. Gen. Microbiol. 118:419–428.

221. Niekus, H. G. D., E. Van Doorn, and A. H. Stouthamer. 1980. Oxygen consumption by *Campylobacter sputorum* subspecies *bubulus* with formate as substrate. Arch. Microbiol. 127:137–143.

222. Novelli, G. D., and C. E. ZoBell. 1944. Assimilation of petroleum hydrocarbons by sulfate-reducing bacteria. J. Bacteriol. 47:447–448.

223. Odom, J. M., and H. D. Peck, Jr. 1984. Hydrogenase, electron-transfer proteins, and energy coupling in the sulfate-reducing bacteria *Desulfovibrio* Annu. Rev. Microbiol. 38:551–592.

223a. Ogata, Y. and A. Kawasaki, 1970. Equilibrium additions to carbonyl compounds, p. 1–69, in J. Zablicky (ed.), The chemistry of the carbonyl group, Vol. 2. Wiley Interscience, New York.

224. Ohle, W. 1954. Sulfat als "Katalysator" des limnischen Stoffkreislaufs. Vom Wasser 21:13–32.

225. Oremland, R. S., L. M. Marsh, and S. Polcin, 1982. Methane production and simultaneous sulfate reduction in anoxic, salt marsh sediments. Nature (Lond.) 296:143–145.

226. Oremland, R. S., and S. Polcin. 1982. Methanogenesis and sulfate reduction: competitive and noncompetitive substrates in estuarine sediments. Appl. Environ. Microbiol. 44:1270–1276.

227. Oren, A., and E. Padan. 1978. Induction of anaerobic, photoautotrophic growth in the cyanobacterium *Oscillatoria limnetica*. J. Bacteriol. 133:558–563.

228. Oren, A., and M. Shilo. 1979. Anaerobic heterotrophic dark metabolism in the cyanobacterium *Oscillatoria limnetica*: sulfur respiration and lactate fermentation. Arch. Microbiol. 122:77–84.

228a. Oyaizu, H. and C. R. Woese. 1985. Phylogenetic relationships among the sulfate respiring bacteria, myxobacteria and purple bacteria. Syst. Appl. Microbiol. 6:257–263.

229. Pace, B., and L. L. Campbell. 1971. Homology of ribosomal ribonucleic acid of *Desulfovibrio* species with *Desulfovibrio vulgaris*. J. Bacteriol. 106:717–719.

230. Pankhurst, E. S. 1971. The isolation and enumeration of sulphate-reducing bacteria, p. 223–240, in D. A. Shapton and R. G. Board (eds.), Isolation of anaerobes. Academic Press, New York.

231. Parkes, R. J., J. Taylor, and D. Jørck-Ramberg. 1985. Demonstration, using *Desulfobacter* sp., of two pools of acetate with different biological availabilities in marine pore water. Mar. Biol. 83:271–276.

232. Parkin, T. B., and T. D. Brock. 1981. The role of phototrophic bacteria in the sulfur cycle of a meromictic lake. Limnol. Oceanogr. 26:880–890.

233. Peck, H. D., Jr., and J. LeGall. 1982. Biochemistry of dissimilatory sulphate reduction. Philos. Trans. R. Soc. Lond. B298:443–466.

234. Pfennig, N. 1975. The photolithotrophic bacteria and their role in the sulfur cycle. Plant Soil 43:1–16.

235. Pfennig, N., and H. Biebl. 1976. *Desulfuromonas acetoxidans* gen. nov. and sp. nov., a new anaerobic, sulfur-reducing, acetate-oxidizing bacterium. Arch. Microbiol. 110:3–12.

236. Pfennig, N., and H. Biebl. 1981. The dissimilatory sulfur-reducing bacteria, p. 941–

947, in M. P. Starr, H. Stolp, H. G. Trüper, A. Balows, and H. G. Schlegel (eds.), The prokaryotes, Vol. 1. Springer-Verlag, Heidelberg.

237. Pfennig, N., and F. Widdel. 1981. Ecology and physiology of some anaerobic bacteria from the microbial sulfur cycle, p. 169–177, in H. Bothe and A. Trebst (eds.), Biology of inorganic nitrogen and sulfur. Springer-Verlag, Heidelberg.

238. Pfennig, N., F. Widdel, and H. G. Trüper. 1981. The dissimilatory sulfate-reducing bacteria, p. 926–940, in M. P. Starr, H. Stolp, H. G. Trüper, A. Balows, and H. G. Schlegel (eds.), The prokaryotes, Vol. 1. Springer-Verlag, Heidelberg.

239. Phelps, T. J., R. Conrad, and J. G. Zeikus. 1985. Sulfate-dependent interspecies H_2 transfer between *Methanosarcina barkeri* and *Desulfovibrio vugaris* during coculture metabolism of acetate or methanol. Appl. Environ. Microbiol. 50:589–594.

240. Postgate, J. R. 1951. The reduction of sulphur compounds by *Desulphovibrio desulphuricans*. J. Gen. Microbiol. 5:725–738.

241. Postgate, J. R. 1952. Growth of sulphate-reducing bacteria in sulphate-free media. Research 5:189–190.

242. Postgate, J. R. 1954. Presence of cytochrome in an obligate anaerobe. Biochem. J. 56:xi–xii.

243. Postgate, J. R. 1959. A diagnostic reaction of *Desulphovibrio desulphuricans*. Nature (Lond.) 183:481–482.

244. Postgate, J. R. 1960. On the autotrophy of *Desulphovibrio desulphuricans*. Z. Allg. Mikrobiol. 1:53–56.

245. Postgate, J. R. 1963. A strain of *Desulfovibrio* able to use oxamate. Arch. Mikrobiol. 46:287–295.

245a. Postgate, J. R. 1969. Methane as a minor product of pyruvate metabolism by sulphate-reducing and other bacteria. J. Gen. Microbiol. 57:293–352.

246. Postgate, J. R. 1970. Nitrogen fixation by sporulating sulphate-reducing bacteria including rumen strains. J. Gen. Microbiol. 63:137–139.

247. Postgate, J. R. 1982. Economic importance of sulfur bacteria. Philos. Trans. R. Soc. Lond. B298:583–600.

248. Postgate, J. R. 1984. The sulphate-reducing bacteria, 2nd ed. Cambridge University Press, Cambridge.

249. Postgate, J. R. 1984. Genus *Desulfovibrio*, p. 666–672, in N. R. Krieg and J. G. Holt (eds.), Bergey's manual of systematic bacteriology, Vol. 1, Williams & Wilkins, Baltimore.

250. Postgate, J. R., and L. L. Campbell. 1966. Classification of *Desulfovibrio* species, the nonsporulating sulfate-reducing bacteria. Bacteriol. Rev. 30:732–738.

251. Postgate, J. R., and H. M. Kent. 1985. Diazotrophy within *Desulfovibrio*. J. Gen. Microbiol. 131:2119–2122.

251a. Postgate, J. R., H. M. Kent, and R. L. Robson. 1986. DNA from diazotrophic *Desulfovibrio* strains is homologous to *Klebsiella pneumoniae* structural *nif* DNA and can be chromosomal or plasmid born. FEMS Microbiol. Lett. 33:159–163.

252. Probst, I., M. Bruschi, N. Pfennig, and J. LeGall. 1977. Cytochrome *c*-551.5 (c_7) from *Desulfuromonas acetoxidans*. Biochim. Biophys. Acta 460:58–64.

253. Probst, I., J. J. G. Moura, I. Moura, M. Bruschi, and J. LeGall. 1978. Isolation and

characterization of a rubredoxin and an (8Fe–8S) ferredoxin from *Desulfuromonas acetoxidans*. Biochim. Biophys. Acta 502:38–44.

254. Ramm, A. E., and D. A. Bella. 1974. Sulfide production in anaerobic microcosms. Limnol. Oceanogr. 19:110–118.

255. Rao, K. K., and R. Cammack. 1981. The evolution of ferredoxin and superoxide dismutase in microorganisms, p. 175–213, in M. J. Carlile, J. F. Collins, and B. E. B. Moseley (eds.), Molecular and cellular aspects of microbial evolution (32nd Symp. Soc. Gen. Microbiol. Edinburgh). Cambridge University Press, Cambridge.

256. Reeburgh, W. S. 1980. Anaerobic methane oxidation: rate depth distributions in Skan Bay sediments. Earth Planet. Sci. Lett. 47:345–352.

257. Riederer-Henderson, M.-A., and P. W. Wilson. 1970. Nitrogen fixation by sulphate-reducing bacteria. J. Gen. Microbiol. 61:27–31.

258. Robertson, L. A., and J. G. Kuenen. 1983. *Thiosphaera pantotropha* gen. nov. sp. nov., a facultatively anaerobic, facultatively autotrophic sulphur bacterium. J. Gen. Microbiol. 129:2847–2855.

259. Robinson, J. A., and J. M. Tiedje. 1984. Competition between sulfate-reducing and methanogenic bacteria for H_2 under resting and growing conditions. Arch. Microbiol. 137:26–32.

260. Rosenfeld, W. D. 1947. Anaerobic oxidation of hydrocarbons by sulfate-reducing bacteria. J. Bacteriol. 54:664–665.

261. Roy, A. B., and P. A. Trudinger. 1970. The biochemistry of inorganic compounds of sulfur. Cambridge University Press, Cambridge.

262. Rozanova, E. P., and A. I. Khudyakova. 1974. A new nonspore-forming thermophilic sulfate-reducing organism, *Desulfovibrio thermophilus* nov. spec. Microbiology (Engl. Transl. Mikrobiologiya) 43:908–912 (Mikrobiologiya 43:1069–1075).

263. Rozanova, E. P., and T. N. Nazina. 1976. A mesophilic, sulfate-reducing, rod-shaped, nonsporeforming bacterium. Microbiology (Engl. Transl. Mikrobiologiya) 45:711–716 (Mikrobiologiya 45:825–830).

264. Rozanova, E. P., and T. N. Nazina. 1982. Hydrocarbon-oxidizing bacteria and their activity in oil pools. Microbiology (Engl. Transl. Mikrobiologiya) 51:287–293 (Mikrobiologiya 51:342–348).

265. Ruckmick, J. C., B. H. Wimberly, and A. F. Edwards. 1979. Classification and genesis of biogenic sulfur deposits. Econ. Geol. 74:469–474.

265a. Samain, E., G. Albagnac, and J. LeGall. 1986. Redox studies of the tetraheme cytochrome c_3 isolated from the propionate-oxidizing, sulfate-reducing bacterium *Desulfobulbus elongatus*. FEBS 204:247–250.

266. Samain, E., H. C. Dubourguier, and G. Albagnac. 1984. Isolation and characterization of *Desulfobulbus elongatus* sp. nov. from a mesophilic industrial digester. Syst. Appl. Microbiol. 5:391–401.

267. Sansone, F. J., and C. S. Martens. 1982. Volatile fatty acid cycling in organic-rich marine sediment. Geochim. Cosmochim. Acta 46:1575–1589.

268. Schaschl, E. 1980. Elemental sulphur as a corrodent in deaerated neutral aqueous environments. Mater. Perform. 19:9–12.

268a. Schauder, R., B. Eikmanns, R. K. Thauer, F. Widdel, and G. Fuchs. 1986. Acetate oxidation to CO_2 in anaerobic bacteria via a novel pathway not involving reactions of the citric acid cycle. Arch. Microbiol. 145:162–172.

269. Schidlowski, M. 1979. Antiquity and evolutionary status of bacterial sulfate reduction: sulfur isotope evidence. Origins Life 9:299–311.

270. Schidlowski, M., J. M. Hayes, and I. R. Kaplan. 1983. Isotopic inferences of ancient biochemistries: carbon, sulfur, hydrogen, and nitrogen, p. 149–186, in J. W. Schopf (ed.), Earth's earliest biosphere. Princeton University Press, Princeton, N.J.

271. Schiff, J. A. 1980. Pathways of assimilatory sulphate reduction in plants and microorganisms, p. 49–69, in Sulphur in biology (Ciba Foundation Symposium No. 72). Excerpta Medica, Amsterdam.

272. Schiff, J. A., and H. Frankhauser. 1981. Assimilatory sulfate reduction, p. 153–168, in H. Bothe and A. Trebst (eds.), Biology of inorganic nitrogen and sulfur. Springer-Verlag, Heidelberg.

273. Schiff, J. A., and R. C. Hodson. 1973. The metabolism of sulfate. Annu. Rev. Plant Physiol. 24:381–414.

274. Schink, B. 1984. Microbial degradation of pectin in plants and aquatic environments, p. 580–587, in M. J. Klug and C. A. Reddy (eds.) Current perspectives in microbial ecology. American Society for Microbiology, Washington, D.C.

275. Schink, B. 1985. Degradation of unsaturated hydrocarbons by methanogenic enrichment cultures. FEMS Microbiol. Ecol. 31:69–77.

276. Schönheit, P., J. K. Kristjansson, and R. K. Thauer. 1982. Kinetic mechanism for the ability of sulfate reducers to out-compete methanogens for acetate. Arch. Microbiol. 132:285–288.

277. Schopf, J. W., J. M. Hayes, and M. R. Walter. 1983. Evolution of Earth's earliest ecosystems: recent progress and unsolved problems, p. 361–384, in J. W. Schopf (ed.), Earth's earliest biosphere. Princeton University Press, Princeton, N.J.

278. Schwartz, R. M., and O. Dayhoff. 1978. Origins of prokaryotes, eukaryotes, mitochondria, and chloroplasts. Science 199:395–403.

279. Scranton, M. I., P. C. Novelli, and P. A. Loud. 1984. The distribution and cycling of hydrogen gas in the waters of two anoxic marine environments. Limnol. Oceanogr. 29:993–1003.

280. Segerer, A., K. O. Stetter, and F. Klink. 1985. Two contrary modes of chemolithotrophy in the same archaebacterium. Nature (Lond.) 313:787–789.

281. Senez, J. C. 1954. Fermentation d l'acide pyruvique et des acides dicarboxyliques par les bactéries anaérobies sulfato-réductrices. Bull. Soc. Chim. Biol. Paris 36:541–552.

282. Senez, J. C., and J. Leroux-Gilleron. 1954. Note préliminaire sur la dégradation anaérobie de la cysteine et de la cystine par les bactéries sulfato-réductrices. Bull. Soc. Chim. Biol. Paris 36:553–559.

283. Senez, J. C., and M.-C. Pascal. 1961. Dégradation de la choline par les bactéries sulfatoréductrices. Identification de *Desulfovibrio desulfuricans* et de *Vibrio cholinicus*. Z. Allg. Mikrobiol. 1:142–149.

284. Siegel, L. M. 1975. Biochemistry of the sulfur cycle, p. 217–286, in D. M. Greenberg (ed.), Metabolic pathways, Vol. 7, Metabolism of sulfur compounds. Academic Press, New York.

285. Siegel, L. M. 1978. Structure and function of siroheme and the siroheme enzymes, p. 201–214, in T. P. Singer (ed.), Developments in biochemistry, Vol. 1, Mechanisms of oxidizing enzymes. Elsevier, Amsterdam.

286. Singleton, R., L. L. Campbell, and F. M. Hawkridge. 1979. Cytochrome c_3 from the sulfate-reducing anaerobe *Desulfovibrio africanus* Benghazi: purification and properties. J. Bacteriol. 140:893–901.

287. Skyring, G. W., L. A. Chambers, and J. Bauld. 1983. Sulfate reduction in sediments colonized by cyanobacteria, Spencer Gulf, South Australia. Aust. J. Mar. Freshwater Res. 34:359–374.

288. Skyring, G. W., and T. H. Donnelly. 1982. Precambrian sulfur isotopes and a possible role for sulfite in the evolution of biological sulfate reduction. Precambrian Res. 17:41–61.

289. Skyring, G. W., and H. E. Jones. 1972. Guanine plus cytosine contents of the deoxyribonucleic acids of some sulfate-reducing bacteria: a reassessment. J. Bacteriol. 109:1298–1300.

290. Skyring, G. W., R. L. Oshrain, and W. J. Wiebe. 1979. Sulfate reduction rates in Georgia marshland soils. Geomicrobiol. J. 1:389–400.

291. Skyring, G. W., and P. A. Trudinger. 1973. A comparison of the electrophoretic properties of the ATP-sulfurylases, APS-reductases, and sulfite reductases from cultures of dissimilatory sulfate-reducing bacteria. Can. J. Microbiol. 19:375–380.

292. Sleytr, U., H. Adam, and H. Klaushofer. 1969. Die Feinstruktur der Zellwand und Cytoplasmamembran von *Clostridium nigrificans*, dargestellt mit Hilfe der Gefrierätz- und Ultradünnschnittechnik. Arch. Mikrobiol. 66:40–58.

293. Smibert, R. M. 1984. Genus *Campylobacter*, p. 111–118, in N. R. Krieg and J. G. Holt (eds.), Bergey's manual of systematic bacteriology, Vol. 1. Williams & Wilkins, Baltimore.

294. Smith, A. D. 1982. Immunofluorescence of sulphate-reducing bacteria. Arch. Microbiol. 133:118–121.

295. Smith, M. R., and R. A. Mah. 1978. Growth and methanogenesis by *Methanosarcina* strain 227 on acetate and methanol. Appl. Environ. Microbiol. 36:870–879.

296. Smith, M. R., and R. A. Mah. 1980. Acetate as sole carbon and energy source for growth of *Methanosarcina* strain 227. Appl. Environ. Microbiol. 39:993–999.

297. Smith, P. H., and R. A. Mah. 1966. Kinetics of acetate metabolism during sludge digestion. Appl. Microbiol. 14:368–371.

298. Smith, R. L., and M. J. Klug. 1981. Reduction of sulfur compounds in the sediments of a eutrophic lake basin. Appl. Environ. Microbiol. 41:1230–1237.

299. Smith, R. L., and M. J. Klug, 1981. Electron donors utilized by sulfate-reducing bacteria in eutrophic lake sediments. Appl. Environ. Microbiol. 42:116–121.

300. Soimajärvi, J., M. Pursiainen, and J. Korhonen. 1978. Sulfate-reducing bacteria in paper machine waters and in suction roll perforations. Eur. J. Appl. Microbiol. Biotechnol. 5:87–93.

301. Sørensen, J. 1982. Reduction of ferric iron in anaerobic, marine sediment and interaction with reduction of nitrate and sulfate. Appl. Environ. Microbiol. 43:319–324.

302. Sørensen, J., D. Christensen, and B. B. Jørensen. 1981. Volatile fatty acids and hydrogen as substrates for sulfate-reducing bacteria in anaerobic marine sediment. Appl. Environ. Microbiol. 42:5–11.

303. Sørensen, J., and B. B. Jørgensen, and N. P. Revsbech. 1979. A comparison of oxygen, nitrate, and sulfate respiration in coastal marine sediments. Microb. Ecol. 5:105–115.

304. Sorokin, Yu. I. 1957. On the ability of sulfate-reducing bacteria to utilize methane for the reduction of sulfates to hydrogen sulfide. Dokl. Biol. Sci. 115:713–715. (Dokl. Akad. Nauk SSSR 115:816–818).

305. Sorokin, Yu. I. 1966. Sources of energy and carbon for biosynthesis in sulfate-reducing bacteria. Microbiology (Engl. Transl. Mikrobiologiya) 35:643–647 (Mikrobiologiya 35:761–766).

306. Sorokin, Yu. I. 1966. Investigation of the structural metabolism of sulfate-reducing bacteria with C^{14}. Microbiology (Engl. Transl. Mikrobiologiya) 35:806–814 (Mikrobiologiya 35:967–977).

307. Sorokin, Yu. I. 1966. Role of carbon dioxide and acetate in the biosynthesis by sulfate reducing bacteria. Nature (Lond.) 210:551–552.

308. Sorokin, Yu. I. 1972. The bacterial population and the process of hydrogen sulphide oxidation in the Black Sea. J. Conseil Int. Explor. Mer 34:423–455; through reference (142).

309. Stal, L. J., H. van Gemerden, and W. Krumbein. 1985. Structure and development of a benthic marine microbial mat. FEMS Microbiol. Ecol. 31:111–125.

310. Stams, A. J. M., T. A. Hansen, and G. W. Skyring. 1985. Utilization of amino acids as energy substrates by two marine *Desulfovibrio* strains. FEMS Microbiol. Ecol. 31:11–15.

311. Stams, A. J. M., D. R. Kremer, K. Nicolay, G. H. Weenk, and T. A. Hansen. 1984. Pathway of propionate formation in *Desulfobulbus propionicus*. Arch. Microbiol. 139:167–173.

312. Starkey, R. L. 1937. Formation of sulfide by some sulfur bacteria. J. Bacteriol. 33:545–571.

313. Stetter, K. O. 1982. Ultrathin mycelia-forming organisms from submarine volcanic areas having an optimum growth temperature of 105°C. Nature (Lond.) 300:258–260.

314. Stetter, K. O. 1984. Anaerobic life at extremely high temperatures. Origins Life 14:809–815.

315. Stetter, K. O. 1985. Extrem thermophile Bakterien. Naturwissenschaften 72:291–301.

316. Stetter, K. O., and G. Gaag. 1983. Reduction of molecular sulphur by methanogenic bacteria. Nature (Lond.) 305:309–311.

317. Stetter, K. O., H. König, and E. Stackebrandt. 1983. *Pyrodictium* gen. nov., a new genus of submarine disc-shaped sulphur-reducing archaebacteria growing optimally at 105°C. Syst. Appl. Microbiol. 4:535–551.

318. Stetter, K. O., and W. Zillig. 1985. *Thermoplasma* and the thermophilic sulfur-dependent archaebacteria, p. 85–170, in C. R. Woese and R. S. Wolfe (eds.), The bacteria, Vol. 8, Archaebacteria. Academic Press, Orlando, FL.

319. Stieb, M., and B. Schink. 1985. Anaerobic oxidation of fatty acids by *Clostridium bryantii* sp. nov., a sporeforming, obligately syntrophic bacterium. Arch. Microbiol. 140:387–390.

320. Stille, W., and H. G. Trüper. 1984. Adenylylsulfate reductase in some new sulfate-reducing bacteria. Arch. Microbiol. 137:145–150.

321. Stratford, M., and A. H. Rose. 1985. Hydrogen sulphide production from sulphite by *Saccharomyces cerevisiae*. J. Gen. Microbiol. 131:1417–1424.

322. Strayer, R. F., and J. M. Tiedje. 1978. Kinetic parameters of the conversion of methane precursors to methane in a hypereutrophic lake sediment. Appl. Environ. Microbiol. 36:330–340.

323. Strøm, A. R., J. A. Olafsen, and H. Larsen. 1979. Trimethylamine oxide: a terminal electron acceptor in anaerobic respiration of bacteria. J. Gen. Microbiol. 112:315–320.

324. Stumm, W., and J. J. Morgan. 1981. Aquatic chemistry, 2nd ed. Wiley, New York.

325. Styrt, M. M., A. J. Brackmann, H. D. Holland, B. C. Clark, V. Pisutha-Arnond, C. S. Eldridge, and H. Ohmoto. 1981. The mineralogy and the isotopic composition of sulfur in hydrothermal sulfide/sulfate deposits on the East Pacific Rise, 21° N latitude. Earth Planet. Sci. Lett. 53:382–390.

325a. Szewzyk, R., and N. Pfennig. 1987. Complete oxidation of catechol by the strictly anaerobic sulfate-reducing *Desulfobacterium catecholicum* sp. nov. Arch. Microbiol. 147:163–168.

326. Takai, Y., and T. Kamura. 1966. The mechanism of reduction in waterlogged paddy soil. Folia Microbiol. 11:304–313.

327. Tanner, A. C. R., S. Badger, C.-H. Lai, M. A. Listgarten, R. A. Visconti, and S. S. Socransky. 1981. *Wolinella* gen. nov., *Wolinella succinogenes* (*Vibrio succinogenes* Wolin et al.) comb. nov., and description of *Bacteroides gracilis* sp. nov., *Wolinella recta* sp. nov., *Campylobacter concisus* sp. nov., and *Eikenella corrodens* from humans with periodontal disease. Int. J. Syst. Bacteriol. 31:432–445.

328. Tanner, A. C. R., and S. S. Socransky. 1984. Genus VIII. *Wolinella*, p. 646–650, in N. R. Krieg and J. G. Holt (eds.), Bergey's manual of systematic bacteriology, Vol. 1. Williams & Wilkins, Baltimore.

329. Taylor, J., and R. J. Parkes. 1983. The cellular fatty acids of the sulphate-reducing bacteria, *Desulfobacter* sp., *Desulfobulbus* sp. and *Desulfovibrio desulfuricans*. J. Gen. Microbiol. 129:3303–3309.

330. Taylor, J., and R. J. Parkes. 1985. Identifying different populations of sulphate-reducing bacteria within marine sediment systems, using fatty acid biomarkers. J. Gen. Microbiol. 131:631–642.

331. Tessenow, U. 1974. Lösungs-, Diffusions-, und Sorptionsprozesse in der Oberschicht von Seesedimenten. IV. Reaktionsmechanismen und Gleichgewichte im System Eisen-Mangan-Phosphat im Hinblick auf die Vivianitakkumulation im Ursee. Arch. Hydrobiol. (Suppl.) 47:1–79.

332. Thauer, R. K. 1982. Dissimilatory sulphate reduction with acetate as electron donor. Philos. Trans. R. Soc. Lond. B298:467–471.

333. Thauer, R. K., and W. Badziong. 1980. Respiration with sulfate as electron acceptor, p. 65–85, in C. J. Knowles (ed.), Diversity of bacterial respiratory systems, Vol. 2. CRC Press, Boca Raton, FL.

334. Thauer, R. K., K. Jungermann, and K. Decker. 1977. Energy conservation in chemotrophic anaerobic bacteria. Bacteriol. Rev. 41:100–180.

335. Tiller, A. K. 1983. Electrochemical aspects of microbial corrosion: an overview, p. 54–65, in Microbial corrosion. The Metals Society, London.

336. Traore, A. S., M.-L. Fardeau, E. C. Hatchikan, J. LeGall, and J.-P. Belaich. 1983. Energetics of growth of a defined mixed culture of *Desulfovibrio vulgaris* and *Meth-*

anosarcina barkeri: interspecies hydrogen transfer in batch and continuous culture. Appl. Environ. Microbiol. 46:1152–1156.

337. Trudinger, P. A. 1976. Microbiological processes in relation to ore genesis, p. 135–190, in K. H. Wolf (ed.), Handbook of strata-bound and stratiform ore deposits, Vol. 2. Elsevier, Amsterdam.

338. Trudinger, P. A. 1982. Geological significance of sulphur oxidoreduction by bacteria. Philos. Trans. R. Soc. Lond. B 298:563–581.

339. Trudinger, P. A., and R. E. Loughlin. 1981. Metabolism of simple sulphur compounds, p. 165–256, in A. Neuberger and L. L. M. van Deenen (eds.), Comprehensive biochemistry, Vol. 19A. Elsevier, Amsterdam.

340. Trudinger, P. A., and N. Williams. 1982. Stratified sulfide deposition in modern and ancient environments, p. 177–198, in H. D. Holland and M. Schidlowski (eds.), Mineral deposits and the evolution of the biosphere. Springer-Verlag, Heidelberg.

341. Trüper, H. G. 1981. Photolithotrophic sulfur oxidation, p. 199–211, in H. Bothe and A. Trebst (eds.), Biology of inorganic nitrogen and sulfur. Springer-Verlag, Heidelberg.

342. Trüper, H. G. 1982. Microbial processes in the sulfur cycle through time, p. 5–30, in H. D. Holland and M. Schidlowski (eds.), Mineral deposits and the evolution of the biosphere. Springer-Verlag, Heidelberg.

343. Trüper, H. G., and U. Fischer. 1982. Anaerobic oxidation of sulphur compounds as electron donors for bacterial photosynthesis. Philos. Trans. R. Soc. Lond. B298:529–542.

344. Trüper, H. G., and N. Pfennig. 1966. Sulphur metabolism in Thiorhodaceae. III. Storage and turnover of thiosulphate sulphur in *Thiocapsa floridana* and *Chromatium* species. Antonie Leeuwenhoek J. Microbiol. Serol. 32:261–276.

345. Tu, J., D. Prangishvilli, H. Huber, G. Wildgruber, W. Zillig, and K. O. Stetter. 1982. Taxonomic relations between archaebacteria including 6 novel genera examined by cross hybridization of DNAs and 16S rRNAs. J. Mol. Evol. 18:109–114.

346. Turekian, K. K. 1969. The oceans, streams, and atmosphere, p. 295–323, in K. H. Wedepohl (ed.), Handbook of geochemistry. Springer-Verlag, Heidelberg.

347. Tuttle, J. H. 1980. Organic carbon utilization by the resting cells of thiosulfate-oxidizing marine heterotrophs. Appl. Environ. Microbiol. 40:516–521.

348. Tuttle, J. H., P. R. Dugan, C. B. Macmillan, and C. I. Randles. 1969. Microbial dissimilatory sulfur cycle in acid mine water. J. Bacteriol. 97:594–602.

349. Tuttle, J. H., P. E. Holmes, and H. W. Jannasch. 1974. Growth rate stimulation of marine pseudomonads by thiosulfate. Arch. Microbiol. 99:1–14.

350. Tuttle, J. H., and H. W. Jannasch. 1973. Sulfide and thiosulfate-oxidizing bacteria in anoxic marine basins. Mar. Biol. 20:64–70.

351. Tuttle, J. H., and H. W. Jannasch. 1973. Dissimilatory reduction of inorganic sulfur by facultatively anaerobic marine bacteria. J. Bacteriol. 115:732–737.

352. Tuttle, J. H., and H. W. Jannasch. 1977. Thiosulfate stimulation of microbial dark assimilation of carbon dioxide in shallow marine waters. Microb. Ecol. 4:9–25.

353. Vainshtein, M. B., A. G. Matrosov, V. P. Baskunov, A. M. Zyakun, and M. V. Ivanov. 1980. Thiosulfate as an intermediate product of bacterial sulfate reduction. Microbiology (Engl. Transl. Mikrobiologiya) 49:672–675 (Mikrobiologiya 49:855–858).

354. van Gemerden, H. 1968. On the ATP generation by *Chromatium* in darkness. Arch. Microbiol. 64:118–124.

355. Vogel, H., M. Bruschi, and J. LeGall. 1977. Phylogenetic studies of two rubredoxins from sulfate reducing bacteria. J. Mol. Evol. 9:111–119.

356. Volkov, I. I., and A. G. Rozanov. 1983. The sulphur cycle in oceans, p. 357–448, in M. V. Ivanov and J. R. Freney (eds.), The global biochemical sulphur cycle. Wiley, New York.

357. von Wolzogen Kühr, C. A. H., and I. S. van der Vlugt. 1934. The graphitization of cast iron as an electrobiochemical process in anaerobic soils. Water 18:147–165.

358. Wagner, R., and J. R. Andreesen. 1979. Selenium requirement for active xanthine dehydrogenase from *Clostridium acidiurici* and *Clostridium cylindrosporum*. Arch. Microbiol. 121:255–260.

359. Ward, D. M., and G. J. Olson. 1980. Terminal processes in the anaerobic degradation of an algal-bacterial mat in a high-sulfate hot spring. Appl. Environ. Microbiol. 40:67–74.

359a. Ward, D. M., and M. R. Winfrey. 1985. Interaction between methanogenic and sulfate-reducing bacteria in sediments, p. 141–179, in H. W. Jannasch and P. J. LEB. Williams (eds.), Advances in aquatic microbiology, vol. 3. Academic Press, Orlando, FL.

360. Watanabe, I., and C. Furusaka. 1980. Microbial ecology of flooded rice soils. Adv. Microb. Ecol. 4:125–168.

361. Weimer, P. J., and J. G. Zeikus. 1978. One carbon metabolism in methanogenic bacteria. Arch. Microbiol. 119:49–57.

362. Weimer, P. J., and J. G. Zeikus. 1978. Acetate metabolism in *Methanosarcina barkeri*. Arch. Microbiol. 119:175–182.

363. Wetzel, R. G. 1983. Limnology, 2nd ed. Saunders College Publishing, Philadelphia.

364. Whitman, W. B. 1985. Methanogenic bacteria, p. 3–84, in C. R. Woese and R. S. Wolfe (eds.), The bacteria, Vol. 8, Archaebacteria. Academic Press, Orlando, Fla.

365. Widdel, F. 1983. Methods for enrichment and pure culture isolation of filamentous gliding sulfate-reducing bacteria. Arch. Microbiol. 134:282–285.

366. Widdel, F., G.-W. Kohring, and F. Mayer. 1983. Studies on dissimilatory sulfate-reducing bacteria that decompose fatty acids. III. Characterization of the filamentous gliding *Desulfonema limicola* gen. nov. sp. nov., and *Desulfonema magnum* sp. nov. Arch. Microbiol. 134:286–294.

367. Widdel, F., and N. Pfennig. 1977. A new anaerobic, sporing, acetate-oxidizing, sulfate-reducing bacterium, *Desulfotomaculum* (emend.) *acetoxidans*. Arch. Microbiol. 112:119–122.

368. Widdel, F., and N. Pfennig. 1981. Studies on dissimilatory sulfate-reducing bacteria that decompose fatty acids. I. Isolation of new sulfate-reducing bacteria enriched with acetate from saline environments. Description of *Desulfobacter postgatei* gen. nov., sp. nov. Arch. Microbiol. 129:395–400.

369. Widdel, F., and N. Pfennig. 1981. Sporulation and further nutritional characteristics of *Desulfotomaculum acetoxidans*. Arch. Microbiol. 129:401–402.

370. Widdel, F., and Pfennig. 1982. Studies on dissimilatory sulfate-reducing bacteria that decompose fatty acids. II. Incomplete oxidation of propionate by *Desulfobulbus propionicus* gen. nov., sp. nov. Arch. Microbiol. 131:360–365.

371. Widdel, F., and N. Pfennig. 1984. Dissimilatory sulfate- or sulfur-reducing bacteria, p. 663–679, in N. R. Krieg and J. G. Holt (eds.), Bergey's manual of systematic bacteriology, Vol. 1. Williams & Wilkins, Baltimore.

372. Wight, K. M., and R. L. Starkey. 1945. Utilization of hydrogen by sulfate-reducing bacteria and its significance in anaerobic corrosion. J. Bacteriol. 50:238.

373. Winfrey, M. R., D. R. Nelson, S. C. Klevickis, and J. G. Zeikus. 1977. Association of hydrogen metabolism with methanogenesis in Lake Mendota sediments. Appl. Environ. Microbiol. 33:312–318.

374. Winfrey, M. R., and D. M. Ward. 1983. Substrates for sulfate reduction and methane production in intertidal sediments. Appl. Environ. Microbiol. 45:193–199.

375. Winfrey, M. R., and J. G. Zeikus. 1977. Effect of sulfate on carbon and electron flow during microbial methanogenesis in freshwater sediments. Appl. Environ. Microbiol. 33:275–281.

376. Winfrey, M. R., and J. G. Zeikus. 1979. Anaerobic metabolism of immediate methane precursors in Lake Mendota. Appl. Environ. Microbiol. 37:244–253.

377. Wirsen, C. O., and H. W. Jannasch. 1978. Physiological and morphological observations on *Thiovulum* sp. J. Bacteriol. 136:765–774.

378. Woese, C. R., R. Gupta, C. M. Hahn, W. Zillig, and J. Tu. 1984. The phylogenetic relationships of three sulfur dependent archaebacteria. Syst. Appl. Microbiol. 5:97–105.

379. Wolfe, R. S., and N. Pfennig. 1977. Reduction of sulfur by spirillum 5175 and syntrophism with *Chlorobium*. Appl. Environ. Microbiol. 33:427–433.

380. Wolin, M. J., E. A. Wolin, and N. J. Jacobs. 1961. Cytochrome-producing anaerobic vibrio, *Vibrio succinogenes*, sp. n. J. Bacteriol. 81:911–917.

381. Yoshinari, T. 1980. N₂O reduction by *Vibrio succinogenes*. Appl. Environ. Microbiol. 39:81–84.

382. Zehnder, A. J. B., and T. D. Brock. 1980. Anaerobic methane oxidation: occurrence and ecology. Appl. Environ. Microbiol. 39:194–204.

383. Zehnder, A. J. B., K. Ingvorsen, and T. Marti. 1982. Microbiology of methane bacteria, p. 45–68, in D. E. Hughes, D. A. Stafford, B. I. Wheatley, W. Baader, G. Lettinga, E. J. Nyns, and W. Verstraete (eds.), Anaerobic digestion 1981. Elsevier, Amsterdam.

384. Zehnder, A. J. B., and K. Wuhrmann. 1977. Physiology of a *Methanobacterium* strain AZ. Arch. Microbiol. 111:199–205.

385. Zehnder, A. J. B., and S. H. Zinder. 1980. The sulfur cycle, p. 105–145, in O. Hutzinger (ed.), The handbook of environmental chemistry, Vol. 1, Pt. A. Springer-Verlag, Heidelberg.

386. Zeikus, J. G., M. A. Dawson, T. E. Thompson, K. Ingvorsen, and E. C. Hatchikian. 1983. Microbial ecology of volcanic sulphidogenesis: isolation and characterization of *Thermodesulfobacterium commune* gen. nov. and sp. nov. J. Gen. Microbiol. 129:1159–1169.

387. Zillig, W., A. Gierl, G. Schreiber, S. Wunderl, D. Janekovic, K. O. Stetter, and H. P. Klenk. 1983. The archaebacterium *Thermofilum pendens* represents a novel genus of the thermophilic, anaerobic sulfur respiring Thermoproteales. Syst. Appl. Microbiol. 4:79–87.

388. Zillig, W., I. Holz, D. Janekovic, W. Schäfer, and W. D. Reiter. 1983. The archae-

bacterium *Thermococcus celer* represents a novel genus within the thermophilic branch of the archaebacteria. Syst. Appl. Microbiol. 4:88–94.

389. Zillig, W., R. Schnabel, J. Tu., and K. O. Stetter. 1982. The phylogeny of archaebacteria, including novel anaerobic thermoacidophiles in the light of RNA polymerase structure. Naturwissenschaften 69:197–204.

390. Zillig, W., K. O. Stetter, D. Prangishvilli, W. Schäfer, S. Wunderl, D. Janekovic, I. Holz, and P. Palm. 1982. Desulfurococcaceae, the second family of the extremely thermophilic, anaerobic, sulfur-respiring Thermoproteales. Zentralbl. Bakteriol. Mikrobiol. Hyg. 1. Abt. Orig. C3:304–317.

391. Zillig, W., K. O. Stetter, W. Schäfer, D. Janekovic, S. Wunderl, I. Holz, and P. Palm. 1981. Thermoproteales: a novel type of extremely thermoacidophilic anaerobic archaebacteria isolated from Icelandic solfataras. Zentralbl. Bakteriol. Mikrobiol. Hyg. 1. Abt. Orig. C2:205–227.

392. Zillig, W., S. Yeats, I. Holz, A. Böck, F. Gropp, M. Rettenberger, and S. Lutz. 1985. Plasmid-related anaerobic autotrophy of the novel archaebacterium *Sulfolobus ambivalens*. Nature (Lond.) 313:789–791.

393. Zinder, S. H., and T. D. Brock. 1978. Dimethyl sulfoxide as an electron acceptor for anaerobic growth. Arch. Microbiol. 116:35–40.

394. Zinder, S. H., and M. Koch. 1984. Non-aceticlastic methanogenesis from acetate: acetate oxidation by a thermophilic syntrophic coculture. Arch. Microbiol. 138:263–272.

395. ZoBell, C. E. 1958. Ecology of sulfate reducing bacteria. Prod. Mon. 22:12–29.

Note Added in Proof

After completion of the manuscript, publications appeared on the following topics:

Description of the First Extremely Thermophilic Sulfate-Reducing Archaebacterium

Achenbach-Richter, L., K. O. Stetter, and C. R. Woese. 1987. A possible biochemical missing link among archaebacteria. Nature (Lond.) 327:348–349.
Stetter, K. O., G. Lauerer, M. Thomm, and A. Neuner. 1987. Isolation of extremely thermophilic sulfate reducers: evidence for a novel branch of archaebacteria. Science 236:822–824.

Phylogenetic Relationships Among Bacteria (Including Representatives of Sulfate- and Sulfur-Reducing Eubacteria)

Woese, C. R. 1987. Bacterial evolution. Microbiol. Rev. 51:221–271.

Growth Physiology and Substrate Utilization by Sulfate-Reducing Eubacteria

Brysch, K., C. Schneider, G. Fuchs, and F. Widdel. 1987. Lithoautotrophic growth of sulfate-reducing bacteria, and description of *Desulfobacterium autotrophicum* gen. nov. sp. nov. Arch. Microbiol. 148:264–274.
Nanninga, H. J., and J. C. Gottschal. 1986. Isolation of a sulfate-reducing bacterium growing with methanol. FEMS Microbiol. Ecol. 38:125–130.
Nanninga, H. J., and J. C. Gottschal. 1987. Properties of *Desulfovibrio carbinolicus* sp. nov. and other sulfate-reducing bacteria isolated from an anaerobic-purification plant. Appl. Environ. Microbiol. 53:802–809.
Pankhania, I. P., A. N. Moosavi, and W. A. Hamilton. 1986. Utilization of cathodic hydrogen by *Desulfovibrio vulgaris* (Hildenborough). J. Gen. Microbiol. 132:3357–3365.

Samain, E., H. C. Dubourguier, J. LeGall, and G. Albagnac. 1986. Regulation of hydrogenase activity in the propionate oxidizing sulfate-reducing bacterium *Desulfobulbus elongatus*, p. 23–27, in H. C. Dubourguier, G. Albagnac, J. Montreuil, C. Romond, P. Sautière, and J. Guillaume (eds.), Progress in biotechnology, Vol. 2, Biology of anaerobic bacteria. Elsevier, Amsterdam.

Seitz, J., and H. Cypionka. 1986. Chemolithotrophic growth of *Desulfovibrio desulfuricans* with hydrogen coupled to ammonification of nitrate or nitrite. Arch. Microbiol. 146:63–67.

Widdel, F. 1987. New types of acetate-oxidizing. sulfate-reducing *Desulfobacter* species, *D. hydrogenophilus* sp. nov., *D. latus* sp. nov., and *D. curvatus* sp. nov. Arch. Microbiol. 148:286–291.

Description of an Extremely Thermophilic Sulfur-Reducing Eubacterium

Huber, R., T. A. Langworthy, H. König, M. Thomm, C. R. Woese, U. B. Sleytr, and K. O. Stetter. 1986. *Thermotoga maritima* sp. nov. represents a new genus of unique extremely thermophilic eubacteria growing up to 90° C. Arch. Microbiol. 144:324–333.

Description of Extremely Thermophilic Sulfur-Reducing Archaebacteria

Fiala, G., and K. O. Stetter. 1986. *Pyrococcus furiosus* sp. nov. represents a novel genus of marine heterotrophic archaebacteria growing optimally at 100°C. Arch. Microbiol. 145:56–61.

Fiala, G., K. O. Stetter, H. W. Jannasch, T. A. Langworthy, and J. Madon. 1986. *Staphylothermus marinus* sp. nov. represents a novel genus of extremely thermophilic submarine heterotrophic archaebacteria growing up to 98°C. Syst. Appl. Microbiol. 8:106–113.

Segerer, A., A. Neuner, J. K. Kristjansson, and K. O. Stetter. 1986. *Acidianus infernus* gen. nov., sp. nov., and *Acidianus brierleyi* comb. nov.: facultatively aerobic, extremely acidophilic thermophilic sulfur-metabolizing archaebacteria. Int. J. Syst. Bacteriol. 36:559–564.

Zillig, W., I. Holz, H.-P. Klenk, J. Trent, S. Wunderl, D. Janekovic, E. Imsel, and B. Haas. 1987. *Pyrococcus woesei*, sp. nov., an ultra-thermophilic marine archaebacterium, representing a novel order, Thermococcales. Syst. Appl. Microbiol. 9:62–70.

Zillig, W., S. Yeats, I. Holz, A. Böck, M. Rettenberger, F. Gropp, and G. Simon. 1986. *Desulfurolobus ambivalens*, gen. nov., sp. nov., an autotrophic archaebacterium facultatively oxidizing or reducing sulfur. Syst. Appl. Microbiol. 8:197–203.

Biochemistry/Bioenergetics of Sulfate- and Sulfur-Reducing Bacteria

Cypionka, H. 1987. Uptake of sulfate-, sulfite and thiosulfate by proton-anion symport in *Desulfovibrio desulfuricans*. Arch. Microbiol. 148:144–149.

Cypionka, H., and W. Dilling. 1986. Intracellular localization of the hydrogenase in *Desulfotomaculum orientis*. FEMS Microbiol. Lett. 36:257–260.

Kremer, D. R., and T. A. Hansen. 1987. Glycerol and dihydroxyacetone dissimilation in *Desulfovibrio* strains. Arch. Microbiol. 147:249–256.

Möller, D., R. Schauder, G. Fuchs, and R. K. Thauer. 1987. Acetate oxidation to CO_2 via a citric acid cycle involving an ATP-citrate lyase: a mechanism for the synthesis of ATP via substrate level phosphorylation in *Desulfobacter postgatei* growing on acetate and sulfate. Arch. Microbiol. 148:202–207.

Paulsen, J., A. Kröger, and R. K. Thauer. 1986. ATP-driven succinate oxidation in the catabolism of *Desulfuromonas acetoxidans*. Arch. Microbiol. 144:78–83.

Schauder, R., F. Widdel, and G. Fuchs. 1987. Carbon assimilation pathways in sulfate-reducing bacteria. II. Enzymes of a reductive citric acid cycle in the autotrophic *Desulfobacter hydrogenophilus*. Arch. Microbiol. 148:218–225.

Stams, A. J. M., and T. A. Hansen. 1986. Metabolism of L-alanine in *Desulfotomaculum ruminis* and two marine *Desulfovibrio* strains. Arch. Microbiol. 145:277–279.

Bacterial Reduction of Tetrathionate and Thiosulfate

Barrett, E. L., and M. A. Clark. 1987. Tetrathionate reduction and production of hydrogen sulfide from thiosulfate. Microbiol. Rev. 51:192–205.

11

DISSIMILATORY REDUCTION OF SULFUR COMPOUNDS

JEAN LEGALL

Biochemistry Department, The University of Georgia, Athens, GA 30602, and ARBS-Equipe Commune d'Enzymologie CNRS-CEA, CEN-Cadarache, Association pour la Recherche en Bioénergie Solaire, 13108-Saint-Paul-Lez-Durance Cédex, France

GUY FAUQUE

ARBS-Equipe Commune d'Enzymologie CNRS-CEA, CEN-Cadarache, Association pour la Recherche en Bioénergie Solaire, F-13108-Saint-Paul-Lez-Durance Cédex, France

11.1 INTRODUCTION
11.2 PERIPLASMIC ENZYMES AND ELECTRON TRANSFER PROTEINS
    11.2.1 Hydrogenases
    11.2.2 Formate Dehydrogenase
    11.2.3 Electron Carriers
            11.2.3.1 Tetraheme Cytochrome c_3
            11.2.3.2 Cytochrome c_{553}
11.3 MEMBRANE ORGANIZATION
    11.3.1 *Desulfovibrio* Nitrite Reductase: A Transmembrane, Hexaheme Cytochrome *c*
11.4 CYTOPLASMIC OR INTERNAL ENZYMES AND ELECTRON TRANSFER PROTEINS
    11.4.1 Internal or Cytoplasmic Hydrogenases
    11.4.2 Reductases
            11.4.2.I Activation of Sulfate
            11.4.2.2 APS Reductase

11.4.2.3 Bisulfite Reductases

11.4.2.4 Thiosulfate Reductase

11.4.3 Dehydrogenases

11.4.3.1 D-Lactate and L-Lactate Dehydrogenases

11.4.3.2 Pyruvate Dehydrogenase

11.4.4 Electron Carriers

11.4.4.1 Ferredoxins

11.4.4.2 Flavodoxin

11.4.4.3 Octaheme Cytochrome c_3

11.5 ELECTRON TRANSFER PROTEINS OF UNCERTAIN LOCALIZATION AND FUNCTION

11.5.1 Molybdenum-Containing Iron-Sulfur Protein from *Desulfovibrio gigas*

11.5.2 Rubredoxin

11.5.3 Desulforedoxin

11.5.4 Cobalt-Containing Protein

11.5.5 Triheme Cytochrome c_3

11.6 ENERGETICS

11.7 CONCLUSIONS

REFERENCES

11.1 INTRODUCTION

The capability of reducing sulfur compounds is not an exceptional feature among living organisms, although it seems to be limited to the plant and bacterial kingdoms. Therefore, the terms "sulfate-reducing bacteria" and "sulfur-reducing bacteria" which are utilized to designate the microorganisms that are capable of dissimilatory sulfate reduction are somewhat misleading. However, they will be retained in this chapter. The term "sulfidogenic bacteria," which has recently been coined (125), is not satisfactory either since many bacteria will produce hydrogen sulfide from organic sulfur compounds (102). The terms "sulfate/sulfur respiration" will be used throughout this chapter to describe processes leading to energy conservation through oxidation of energy substrates concomitant with the reduction of sulfur compounds, inasmuch as respiration has been defined as "the set of functions ensuring the processes of oxidation of a living organism" (164).

The two processes, assimilation and dissimilation, use similar enzymes, cofactors, and prosthetic groups to activate or reduce sulfur and its oxgenated compounds. However, the first process, which is limited to the building of sulfur-containing molecules such as amino acids and cofactors, is regulated in such a way that the final product, hydrogen sulfide, will not accumulate under normal conditions. In contrast, the second process, which is a true respiration, is performed by specialized organisms, the sulfate and sulfur-reducing bacteria, and under the proper conditions will lead to massive and spectacular accumulations of sulfide. As a consequence, the differences between the two processes are to be found in the way electrons are transferred from the carbon and energy substrates and in the spatial organization of the redox proteins and the terminal reductases, since the main result of the respiratory process is the net production of ATP.

The particular metabolism of the sulfate-reducing bacteria makes them a very important agent in industrial and environmental processes. These bacteria have been found to be the principal agent for the corrosion of steel, through a number of mechanisms which are still largely unknown. The economic consequences of bacterial corrosion are immense. It has been calculated that for the offshore oil industry alone, a replacement job due mainly to metal corrosion can cost as much as $3 million dollars, with lost production of $9 million dollars (70). On the other hand, controlled reduction of sulfates by the bacteria could lead to interesting industrial processes, since both organic and gypsum wastes could be utilized as energy and sulfur sources. Sulfate-reducing bacteria also participate in interspecies hydrogen transfer in the complex fermentations leading to methanogenesis (37). Finally, they are engaged in sulfur interspecies transfer with photosynthetic bacteria (28, 29). Recent evidence indicates that the sulfate-reducing bacteria encompass a much larger variety of organisms than suspected, previously, at least eight genera (157), and have a number of growth modes other than sulfate reduction. Thus some but not all species of these bacteria can grow fermentatively (159) and utilize fumarate (65) or nitrate (175) as terminal electron acceptors. A better knowledge of how these bacteria obtain their energy and how they utilize the interspecific energetic vectors H_2 and S^{2-} is needed to control the detrimental effect of their metabolism and to utilize the bacteria for useful purposes.

The reader is probably convinced after reading Chapter 10 that it is very difficult to define these bacteria as a homogeneous group of organisms. As will be seen in the following pages, study of their biochemical pathways, although limited almost entirely to two genuses, *Desulfovibrio* (*D.*) and *Desulfotomaculum* (*Dt.*), also demonstrates their extreme diversity. These observations point to one possible conclusion: Dissimilatory sulfate and sulfur reductions are performed by a large variety of microorganisms, which, as a result of convergent evolution, are unified by these common properties.

As a special bonus resulting in part from that situation, the sulfate-reducing bacteria belonging to the genus *Desulfovibrio* contain a splendid group of electron transfer proteins of low molecular mass as well as a number of unique reductases and dehydrogenases of higher molecular mass, which are a delight to bio-inorganic chemists and physicists. A large amount of information which has recently been reviewed (152) is available concerning their structures and oxidation-reduction mechanisms; however, information concerning their function, particularly as far as electron carrier proteins are concerned, has been contradictory and inconclusive.

TABLE 11.1 Enzyme localization in *Desulfovibrio*[a,b]

	D. gigas	D. vulgaris Hildenborough	D. desulfuricans Norway 4
Periplasm			
Hydrogenase	1Ni, 1[3Fe-xS], 2[4Fe-4S]	No Ni, 3[4Fe-4S]	None
Formic dehydrogenase	Present	Present	Not reported
Membrane			
Hydrogenase	None	Present	1Ni, 6Fe
Fumarate reductase	Present	None	Not reported
Nitrite reductase	Present	Present	Present
Cytoplasm			
Hydrogenase	—	—	1Ni, 3[4Fe-4S]
Pyruvate dehydrogenase	Present	Present	—
Thiosulfate reductase	Present	Present	Present
Sulfite reductase	Present, desulfoviridin	Present, desulfoviridin, also low-spin sulfite reductase	Present, desulforubidin
APS reductase	Present	Present	Present
Unknown localization or questionable	Malic enzyme, fumarase		Formic dehydrogenase, pyruvic dehydrogenase
Lactate dehydrogenase	Present	Reported both soluble and membrane bound	Present

Source: Data from Refs. 145 and 152 and from our own unpublished results.

[a] —, questionable.

[b] Although no precise localization have been reported, it is most probable that the enzymes involved in the activation of sulfate, ATP sulfurylase, pyrophosphatase, and acetate kinase are located in the cytoplasm and will be treated as such.

TABLE 11.2 Electron carrier localization in *Desulfovibrio*[a]

	D. gigas	*D. vulgaris* Hildenborough	*D. desulfuricans* Norway 4
Periplasm			
Cytochromes			
Tetraheme cytochrome c_3	Present	Present	—
Cytochrome c_{553}	Not detected	Present	Not detected
Membrane			
Cytochrome c^b	Present	Present	Present
Cytochrome b	Present	Not detected	Not reported
Menaquinone (MK6)	Present	Present	Not reported
Cytoplasm			
Octaheme cytochrome c_3	Present	Present	—
Ferredoxin I[c]	Present	Very small amounts only when grown in high-Fe media	Present
Ferredoxin II[c]	Present		Present
Flavodoxin	Present	Present	Not detected
Rubredoxin	Present	Present	Present
Unknown localization			
Mo-Fe protein	Present	Not detected	Present
Cytochrome $c_{553(550)}$	Not detected	Not detected	Present
Cytochrome c_3			
Tetraheme	Not 100% in periplasm	—	Present[d]
Octaheme	—	—	Present

[a] — questionable.

[b] Could be a nitrite reductase.

[c] In *D. gigas* ferredoxin I and II are different oligomers of the same protein; in *D. desulfuricans* Norway 4 they are two different proteins.

[d] Function is the sulfur reductase of *D. desulfuricans* Norway 4.

591

This lack of information is due in part to the fact that not all the electron carrier proteins are uniformly distributed among *Desulfovibrio* and to the contradictory results coming from cell-free reconstitution experiments. It has been proposed (112) that the difficulties in the latter case have arisen from the disruption of cellular structures that localized or compartmentalized many of the electron carriers. Thus the classical idea that ferredoxin and flavodoxin function as universal or nonspecific electron carriers in anaerobic bacteria appear to be a simplification no longer warranted by existing data. Since enzymes and electron carrier localization are very important for the energetics of the cells, we have, in the following sections, used all the present data allowing their localization in either the periplasm, membrane, or cytoplasmic portion of the cell. However, the reader must be aware of the fact that the technology involved in these localization experiments is still very unsatisfactory. In our laboratory we have found that these experiments are not exactly reproducible and that the results may vary according to growth conditions, physiological medium, and so on. A tentative localization of the main enzymes and electron carrier proteins of three species of *Desulfovibrio* is presented in Tables 11.1 and 11.2: the results must be viewed cautiously and are open to future research.

11.2 PERIPLASMIC ENZYMES AND ELECTRON TRANSFER PROTEINS

11.2.1 Hydrogenases

Periplasmic or external hydrogenases have been demonstrated conclusively in *D. desulfuricans* (ATCC 27774) (175), *D. vulgaris* strains Marburg (14) and Hildenborough (192), *D. gigas* (22), and *D. salexigens* (48). Evidence for the external location of the hydrogenase of *D. desulfuricans* (ATCC 27774) consists of whole-cell and extract assays coupled with the demonstration of a half-life for the change in pH accompanying H_2 oxidation with benzyl viologen, which is 10 times faster than the corresponding half-life of the proton gradient generated by proton translocation with nitrite (175). In the other four cases, the external location of hydrogenase was demonstrated by selective removal of hydrogenase from spheroplasts or intact cells. Three types of periplasmic hydrogenases have been characterized from *Desulfovibrio* species. The physicochemical properties of iron, nickel–iron, and nickel–iron–selenium periplasmic hydrogenases from *Desulfovibrio* are represented in Table 11.3. The purified hydrogenase from *D. vulgaris* strain Hildenborough (127) is composed of two polypeptide chains of molecular mass 46,000 and 13,5000 Da (162c, 198) and contains 12 iron and 12 sulfide atoms per molecule, arranged in three [4Fe–4S] nonheme clusters. This enzyme does not contain nickel (92), has a high specific activity (4800 μmol H_2 evolved $\cdot$ min^{-1} $\cdot$ mg^{-1}), and is bidirectional.

The hydrogenase from *D. gigas* has a molecular mass of 89,000 Da and contains two subunits of molecular mass 62,000 and 26,000 Da (82). It is not known

TABLE 11.3 Physicochemical properties of iron, nickel–iron, and nickel–iron–selenium periplasmic hydrogenase from *Desulfovibrio*

Property	D. vulgaris Hildenborough	D. gigas	D. salexigens British Guiana
Molecular weight (kDa)	56	89.5	98
Subunits	2	2	2
Metal content			
Nonheme iron	12	11	12–15
Nickel	0	1	1
Selenium	NR[a]	0	1
Specific activity[b]			
Uptake[c]	50,000 (8.4)	1,500 (8.0)	1,300 (7.75)
Production[c]	4,800 (5.8)	440 (4.25)	1,830 (4.75)
D_2–H^+ exchange reaction[d]	2,700 (4.75)	267 (7.75)	900 (4.0)
References	92, 118b, 162c, 192, 198	82, 118b, 181	48, 48a, 118b, 182a

[a] Not reported.
[b] pH optimum for the specific activity in parentheses.
[c] In μmol H_2/min/mg protein.
[d] In μmol HD + H_2 produced/min/mg protein.

whether these two subunits are necessary for activity. The protein also contains 12 iron and 12 sulfide atoms per molecule, but they are arranged in two [4Fe–4S] clusters and one [3Fe–xS] cluster (181). In the oxidized state at 77K, the hydrogenase exhibits EPR signals with g values at 2.31, 2.20, and 2.0 (38, 117). Under hydrogen, these EPR signals disappear and are replaced by a new set of EPR signals with g values at 2.19, 2.08, and 2.04. Based on the EPR observations of Lancaster (105), the presence of 1 gram atom of nickel per mole of enzyme (117), and isotope substitution studies with ^{61}Ni (141), these unique EPR signals have been attributed to the presence of a redox active nickel in this hydrogenase. It has also been proposed that nickel is an important functional unit of this enzyme (181). How the atom of nickel is bound to the protein is still unknown. However, recent X-ray absorption spectroscopic (XAS) studies of the enzyme indicate that the nickel is reduced from Ni(III) to Ni(II) upon hydrogen reduction of the oxidized enzyme and that the ligands are bound to the nickel through sulfur atoms (171). The *D. gigas* enzyme is also bidirectional but has a lower specific activity (440 μmol H_2 evolved $\cdot$ min^{-1} $\cdot$ mg^{-1}) than the hydrogenase of *D. vulgaris* Hildenborough.

Current data on *D. gigas* hydrogenase have recently been reviewed and completed (182). Available data do not allow a clear distinction to be made between the two hypotheses that have been put forward to account for the complicated set of EPR spectra corresponding to the different redox states of the enzyme. As shown on Table 11.4, several steps are clearly distinguishable during the reduction of the protein from the "as prepared" oxygenated form to the fully reduced state (which is inactive in the sense that it cannot evolve molecular hydrogen). The nickel atom can either undergo redox transitions from Ni(III) to Ni(II), Ni(I), and Ni(0), or,

TABLE 11.4 Possible oxidation states of nickel in *D. gigas* periplasmic hydrogenase (adapted from Ref. 182)

Redox States of Hydrogenase	First Hypothesis	Second Hypothesis
Oxidized (EPR active)	Ni(III)	Ni(III)
Active state (EPR silent)	Ni(III) (coupled)	Ni(II)
Hydride intermediate (EPR active)	Ni(III)–hydride	Ni(I)
Reduced state (EPR silent)	Ni(II)	Ni(O)

more simply, from Ni(III) to Ni(II). In the latter hypothesis, the intermediary silent state is interpreted as resulting from a strong magnetic coupling between Ni(III) and one of the [4Fe–4S] clusters rather than a nickel redox transition. Basing their conclusions on spectrometric and chemical evidences, the authors favor the second hypothesis and propose a working model as a frame-work for future research (182).

The hydrogenase purified from the halophilic *D. salexigens* strain British Guiana has a molecular mass of 98,000 Da and is composed of two nonidentical subunits. This protein contains approximately 1 nickel, 12 to 15 iron, and 1 selenium atom(s) per molecule (48a, 182a). This enzyme is bidirectional and the specific activity in the H_2 evolution is 1830 μmol $\cdot$ min^{-1} $\cdot$ mg^{-1} and is 1300 μmol $\cdot$ min^{-1} $\cdot$ mg^{-1} in H_2 consumption (Table 11.3). In the native state (as isolated., under aerobic conditions), this hydrogenase is almost EPR silent at 100 K and below. However, upon reduction under H_2 atmosphere, a rhombic EPR signal assigned to nickel develops at *g* values of 2.22, 2.16, and around 2.0. During the course of a redox titration under H_2., this EPR signal attains a maximal intensity around -380 mV (182a).

In spite of these differences in molecular weight, subunit structure, metal content, and specific activity, their cellular localization and requirement of cytochrome c_3 (M_r 13,000) for activity with other electron transfer proteins suggest that these periplasmic hydrogenases serve analogous functions in these three species of *Desulfovibrio*. Both types of enzyme are oxygen stable and can be purified and stored in the presence of air. The remarkable stability of *D. gigas* hydrogenase has made it a favorite tool for the production of molecular hydrogen by coupling with chloroplastic membranes (4, 43).

Using the deuterium-proton exchange reaction (25, 25a, 62a, 62b, 118b), it has been shown that the Fe-Ni-hydrogenase of *D. gigas* differentiates from the Fe-hydrogenase of *D. vulgaris* Hildenborough by the presence of a slow activation phase in the "as prepared" enzyme. Being linked to a redox reaction (124), this activation is probably due to a modification of the electronic state of the redox center(s).

Evidence for the existence of periplasmic hydrogenase in other species of *Desulfovibrio* in less persuasive (147); however, a periplasmic hydrogenase containing nickel and selenium has been eluted from intact cells of *D. baculatus* strain 9974, which is similar to the Norway 4 strain of *D. desulfuricans* (180, 182b).

11.2.2 Formate Dehydrogenase

Formate dehydrogenase activity has been demonstrated to be periplasmic in *D. gigas* and *D. desulfuricans* (ATCC 27774) (145). This conclusion is also supported by a rapid rate of proton production by intact cells in the presence of formate plus benzyl viologen (175). The formate dehydrogenase has been only partly purified (163a), and no definitive evidence is available about its physical properties. This localization of formate dehydrogenase is similar to that reported for the formate dehydrogenase of *Wolinella (W.) succinogenes* (formerly *Vibrio succinogenes*) (100), where it is postulated to be involved in the generation of a proton gradient by vectorial electron transfer during growth on formate and fumarate. The existing evidence on the localization of formate dehydrogenase in *Desulfovibrio* suggests a similar function in these organisms.

11.2.3 Electron Carriers

The nomenclature of multiple-heme cytochromes such as cytochrome c_3 poses many problems. We propose here that the term "cytochrome c_3" should be reserved for any multiple-heme cytochromes c having exclusively histidinyl residues as heme external ligands. The former use of having the name of the cytochrome followed by its molecular weight should be dropped because important variations can occur in the same family of cytochromes. When the number of amino acid residues varies from 118 to 106, in six tetraheme cytochromes c_3, the situation is even worse in the so-called cytochrome c_3 (M_r 26,000) since it contains 250 residues in *D. desulfuricans* Norway 4 and 214 residues in *D. gigas* (67). To overcome these difficulties we shall state the number of hemes per mole of cytochrome: The former cytochrome c_3 (M_r 13,000) becomes "tetraheme cytochrome c_3," the former cytochrome c_3 (M_r 26,000) becomes "octaheme cytochrome c_3," and so on.

11.2.3.1 *Tetraheme Cytochrome* c_3

This cytochrome was first discovered in 1954 by Postgate (160) and by Ishimoto et al. (94) in *D. vulgaris* strains Hildenborough and Miyazaki. It is present in all species of *Desulfovibrio* and is characteristic of the genus. The presence of a homologous cytochrome in the thermophilic sulfate reducer *Thermodesulfobacterium (T.) commune* (84) shows that this name was premature and that the bacterium probably should be classified as another strain of the previously described *D. thermophilus* (166), especially as the latter strain also contains desulfofuscidin as well as tetraheme cytochrome c_3 (G. D. Fauque, M. H. Czechowski, L. Kang, D. V. DerVartanian and J. LeGall, Abstr. Annu. Meet. Am. Soc. Microbiol. 1986, K50, p. 201).

It is the first cytochrome discovered in a strict anaerobic, nonphotosynthetic bacterium. It differs from mitochondrial cytochrome c by its low redox potential, its reactivity with oxygen, and the presence of four hemes per molecule. Both external heme ligands of the four hemes are provided by histidinyl residues (49).

EPR measurements coupled with potentiometric titrations were performed to determine the midpoint redox potentials of the individual hemes of *D. gigas* tetraheme cytochrome c_3 (203): -235, -235, -306, -315 mV; *D. vulgaris* Hildenborough (50): -284, -310, -319, -324 mV; *D. desulfuricans* Norway 4 (39): -125, -125, -305, -325 mV, and *T. commune* (84): -140, -280, -280, -280 mV.

The comparison of six amino acid sequences of tetraheme cytochromes c_3 shows that although the direct homologies are poor, they all retain sufficiently characteristic to be considered as a distinct family of homologous proteins. The amino acid residues that are common in the six sequences determined so far are shown in Fig. 11.1. It can be seen that the residues which are directly involved in the binding of the four heme irons (cysteinyl and histidinyl residues) represent 64% of the residues found in the same positions in the six sequences.

The three dimensional structures of tetraheme cytochrome c_3 from *D. desulfuricans* Norway 4 (73) and one of the numerous Miyazaki strains of *D. vulgaris* (87) are known (Fig. 11.2). The most striking feature is that the four hemes are arranged in a cluster, their edges being no more than 0.2 nm apart. Although their amino acid sequences are not identical, the cluster of hemes is the same in both cytochromes. The differences of heme exposure to solvent could provide an explanation for the individual redox midpoint potential of the hemes.

The remarkable stability of four paramagnetic centers in a relatively small protein has prompted several proton NMR studies. A detailed analysis of *D. gigas* tetraheme cytochrome c_3 reoxidation pattern was based on a model taking into account heme-heme redox interaction (168). The establishment of such a model is particularly difficult since it necessitates the use of 32 Nernst equations for the 32 redox couples involved among the four hemes. It was found that the intramolecular electron exchange is fast on the NMR scale (greater than 10^5 s^{-1}). The data on the pH dependence of the chemical shifts for heme resonances in different oxidation steps and resonance intensities are compatible with an interacting model with redox interactions between the hemes. Two analyses at pH 7.2 and 9.6 show that the heme-heme interacting potentials cover a range from -50 to $+60$ mV. The midpoint redox potentials of some of the hemes, as well as some of their interacting potentials, are pH dependent. Thus protons produced from dehydrogenases could temporarily increase the midpoint redox potential of the hemes, facilitating the acceptance of electrons; also, the release of protons from the molecule could facilitate its subsequent oxidation. This proposed mechanism is extremely important in view of the reactivity of *Desulfovibrio* periplasmic hydrogenase with tetraheme cytochrome c_3; proton transfer would help in electron transfer between the two proteins. Hydrogenase itself contains redox centers which are pH dependent (38).

The first evidence for the periplasmic location of this cytochrome was a report (114) indicating that a substantial amount of tetraheme cytochrome c_3 of *D. gigas* could be removed by washing intact cells. More recently, these initial observations have been confirmed for *D. vulgaris* strains Marburg (14) and Hildenborough (127).

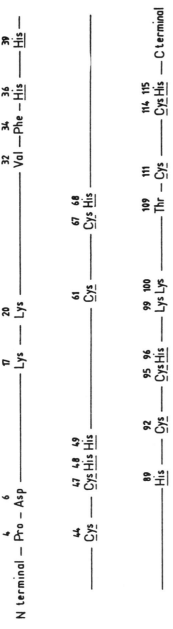

Figure 11.1 Amino acid residues common in *Desulfovibrio* tetraheme cytochromes c_3 of known sequences. The amino acid sequence of the tetraheme cytochrome c_3 from *Desulfovibrio desulfuricans* Norway 4 has been used as reference for residue numeration (73, 158). The residues that are directly involved in heme binding are underlined.

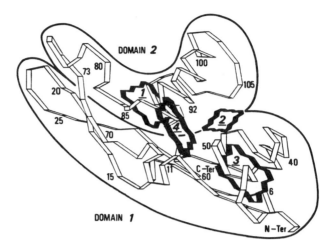

Figure 11.2 Polypeptide chain organization of tetraheme cytochrome c_3 from *Desulfovibrio desulfuricans* Norway 4 and the heme cluster. Reproduced with permission from Ref. 158.

In the former strain, 92% of the tetraheme cytochrome c_3 was found in the periplasm, and in the latter strain, the authors clearly indicated that the tetraheme cytochrome c_3 is periplasmic and present in a 4.5:1 ratio with hydrogenase. Other workers (145) have found substantial amounts of c-type cytochromes in the cytoplasm and on the membrane; however, this situation may well result from the incomplete removal of cytochromes from the external aspect of the cytoplasmic membrane, or the presence of octaheme cytochrome c_3 in the cytoplasm.

Tetraheme cytochrome c_3 is a cofactor for the hydrogenase of *Desulfovibrio* because it is required for the reduction of ferredoxin, flavodoxin and rubredoxin by hydrogenase plus hydrogen (23, 205). NMR and EPR data suggest a specific interaction between tetraheme cytochrome c_3 and the high-potential ferredoxin II purified from *D. gigas* (138, 203), and a similar observation has been made concerning ferredoxin I and tetraheme cytochrome c_3 from *D. desulfuricans* Norway 4 (39).

It has been pointed out (202) that such a reaction between tetraheme cytochrome c_3 which is located in the periplasm, and ferredoxin which is found in the cytoplasm, is not physiologically significant. The interaction could indeed be due to unspecific complementarity between the numerous charged lysine residues which are found at the surface of the cytochrome (73, 87) and the acidic residues of ferredoxins which are known to have very low isolelectric points. It has in fact been noted that the interaction between *D. gigas* tetraheme cytochrome c_3 and ferredoxin II is of an electrostatic character (203). The real physiological reaction could then be between a transmembrane cytochrome (see below the description of the membrane-bound nitrite reductase) which has retained some analogies with the tetraheme cytochrome c_3 and the cytoplasmic ferredoxins. Another possibility would be that tetraheme cytochrome c_3 itself is a transmembrane protein. This can be checked by calculating the distribution of hydrophobic residues ("hydropathy" function) along the aminoacid sequence of the the protein. This has been done with success with chloroplast cytochrome b_6 and other membrane-bound cytochromes b (201). It was shown by this method that the membrane-folding pattern involves eight to nine membrane-spanning domains in this family of cytochromes. The same method was used for the tetraheme cytochrome c_3 of D. *desulfuricans* Norway 4 (A. Trebst, personal communication) and showed too little hydrophobicity to get a hydrophobic helical span across the membrane. Other tetraheme cytochromes c_3 gave identical results (J. E. Wampler, personal communication). Then earlier models for electron transfer in *Desulfovibrio* in which tetraheme cytochrome c_3 plays a central role in the lactate/hydrogen/sulfate metabolism must no longer be considered as valid (68, 132). Even if the interaction between ferredoxins and the tetraheme cytochrome c_3 is not physiologically significant, it constitutes an excellent theoretical model for the study of electron transfer mechanisms between hemes and nonheme iron clusters.

A third function of tetraheme cytochrome c_3 is that of a sulfur reductase in species of *Desulfovibrio* that can grow on organic substrates plus colloidal sulfur (60). With the use of inverted membrane vesicle preparations from D. *gigas*, it has been possible to demonstrate electron transfer–coupled phosphorylation with this simple system (62). The tetraheme cytochrome c_3 from D. *vulgaris* Hildenborough, a strain that is incapable of growing in the presence of sulfur as a respiratory substrate, is inhibited by the product of the reaction, hydrogen sulfide (60). This sensibility of D. *vulgaris* Hildenborough tetraheme cytochrome c_3 to sulfide has been used as a possible explanation for the fact that this organism diverts an important amount of its reducing power to the reduction of protons instead of utilizing it exclusively for the reduction of sulfate (185; see Section 11.6 on energetics).

The triheme cytochrome c_3 from *Desulfuromonas (Drm.) acetoxidans* is incapable of reducing colloidal sulfur (39). Significantly, this cytochrome is lacking the low-potential hemes characteristic of tetraheme cytochromes c_3. Since *Drm. acetoxidans* is an obligate anaerobic sulfur reducer (156), a sulfur reductase has to be present in this organism and will most probably be found in one of the numerous cytochromes that are found in this very specialized bacterium.

An EPR study of tetraheme cytochrome c_3 from D. *desulfuricans* Norway 4 failed to provide any clue to the identity of the sulfur binding site (39). The following mechanism for the reduction of colloidal sulfur was proposed: The polysulfide chains of colloidal sulfur are rapidly attacked by ferrous cytochromes c_3, leading to a collapse of the micelles and the precipitation of S_8 molecules. This explains the appearance of the abundant precipitate at the beginning of the reaction (60); the sulfide that has been formed by reduction of the polysulfides opens up the S_8 rings by nucleophilic attack (165), again leading to the appearance of new molecules of polysulfides which are rapidly reduced to S^{2-} by cytochrome c_3. This scheme is based on the assumption that colloidal sulfur micelles are formed by a stacking of S_8 and polysulfides and on the observation that tetraheme cytochrome c_3 can reduce the latter molecules (59).

Tetraheme cytochrome c_3 has also been observed to stimulate a number of reactions in extracts of *Desulfovibrio* involving either the production or utilization of H_2; however, this cytochrome has never been shown to be the immediate electron acceptor or donor in these reactions. The data now available indicate that these stimulations reflect the effect of tetraheme cytochrome c_3 on hydrogenase activity rather than specific dehydrogenases or reductases. Finally, during the purification of hydrogenases from *Desulfovibrio*, there is always a close association with tetraheme cytochrome c_3, so that the cytochrome may be required to bind hydrogenase to the cytoplasmic membrane as well as to transfer electrons across the membrane.

11.2.3.2 Cytochrome c_{553}

Cytochrome c_{553} was first detected in *D. vulgaris* Hildenborough (111): it is an autooxidizable monoheme cytochrome related to the cytochrome c family (11). It is found in most species of *Desulfovibrio* (58, 61, 204) except *D. gigas*. In *D. vulgaris* Hildenborough it has been reported to be present in a ratio of 3 cytochrome molecules per molecule of hydrogenase among the periplasmic proteins (127). Although there have been a number of suggestions (148, 204), the function of cytochrome c_{553} remains unknown.

The redox potentials of cytochrome c_{553} from *D. vulgaris* Hildenborough and the related cytochrome $c_{553(550)}$ from *D. desulfuricans* Norway 4 (61) have been determined by voltametric and spectrophotometric methods (27). Their midpoint redox potentials are 20 and 40 mV, respectively. This may be compared with the value of 250 mV found for *Pseudomonas aeruginosa* cytochrome c_{551}. The value found for *D. vulgaris* Hildenborough cytochrome c_{553} was also in agreement with an independent EPR measurement (26).

The correlation between the primary structures of cytochrome c_{553} from one of the *D. vulgaris* Miyazaki strains and *Pseudomonas* cytochrome c_{551} (144) confirms the observation that was made 12 years earlier with *D. vulgaris* Hildenborough cytochrome c_{553} (31). Since a homology exists between the primary structures of the oxygen-unreactive, high-potential cytochrome c_{551} and the autooxidizable, low-potential cytochrome c_{553}, the general distinctive properties of the latter protein cannot be explained adequately in terms of amino acid sequence, especially since histidinyl and methionyl residues are the external heme ligands in both hemoproteins. The tertiary structure of cytochrome c_{553} is awaited impatiently.

11.3 MEMBRANE ORGANIZATION

Present ideas about the organization of proteins around the membrane of *Desulfovibrio* have recently been reviewed (152). Both electron transfer components—cytochrome b, c-type cytochromes, menaquinone-6, flavin, and nonheme iron (14, 79, 145)—and specific enzymes—nitrite reductase (122), fumarate reductase (77), nitrate reductase (122), *L*-lactate dehydrogenase (175), and *D*-lactate dehydrogenase (148)—appear to be localized in the membrane. Evidence has been presented that the latter two enzymes are located on the internal side of the cytoplasmic membrane (147). It should be emphasized that the terminal reductases (i.e., APS reductase and the bisulfite reductase) are found exclusively among the soluble cytoplasmic proteins; however, an effect of the membranes on product formation by bisulfite reductase has been described (53).

The presence of *b*-type cytochromes and menaquinone allow the existence for some type of Q cycle (189); however, the presence of cytochrome *b* in all species of *Desulfovibrio* is uncertain (79). Its presence has been detected in *Desulfotomaculum* and it constitutes the one cytochrome in common with the two genera since no cytochrome *c* is present in the latter genus. Cytochrome *b* has also been detected in cells of *Drm. acetoxidans*, but only when the bacterium was grown on fumarate or malate (156). A correlation between the increase of both fumaric reductase activity and cytochrome *b* has been found in *D. gigas* when the cells are grown in the presence of fumarate (79). Menaquinone-6 seems to act as a redox carrier together with cytochrome *b* in particles with hydrogen-fumarate reductase activity (74). The same particles were found to be able to perform oxidative phosphorylations linked to fumarate reduction from hydrogen (16). However, *D. vulgaris* strain Marburg lacks a fumarate reductase but does synthetize cytochrome *b* (14).

11.3.1 *Desulfovibrio* Nitrite Reductase: A Transmembrane, Hexaheme Cytochrome *c*

A new type of nitrite reductase has been found in *D. desulfuricans* (ATCC 27774) grown on lactate with nitrate as terminal electron acceptor. It was found to have a molecular weight of 65,000 Da and to contain six *c*-type hemes per molecule (122). In contrast to the reduction of nitrite by nitrite reductase cytochrome *cd* found in the true denitrifiers, the product is not nitric nor nitrous oxides but rather ammonia, thus corresponding to a six electrons reduction. This reductase from *D. desulfuricans* (ATCC 27774) is membrane-bound and seems to be closely associated with hydrogenase (175), although it has been reported that reduction of nitrite requires the addition of FAD to couple the two enzymes when molecular hydrogen is to be used as the reductant (122). It is to be pointed out that these experiments were done using the Fe hydrogenase from *D. vulgaris* Hildenborough when it has been shown that *D. desulfuricans* (ATCC 27774) contains a Ni-Fe hydrogenase (103): thus direct coupling between the two *D. desulfuricans* proteins cannot be ruled out.

Proton translocation studies with hydrogen and nitrite suggest that the proton-binding and nitrite-binding sites are located on the external side of the membrane, while the binding site for benzyl viologen appears to be internal, thus suggesting the transmembrane location of the protein (175).

Although *D. gigas* cannot grow on nitrate since the enzyme nitrate reductase is absent from this organism, it does contain a high nitrite reductase activity. This activity is membrane-bound, and synthesis of ATP has been demonstrated to occur, coupled to the oxidation of molecular hydrogen and reduction of either nitrite or hydroxylamine (18). Although a membrane-bound cytochrome c is present in *D. gigas*, its identity as a hexaheme nitrite reductase remains to be demonstrated.

EPR studies of the *D. desulfuricans* enzyme (123) have revealed that this cytochrome is not a typical cytochrome c_3 since it contains both high- and low-spin hemes. The nature of the heme external ligands is still unknown. However, because of its transmembrane location, it may have sufficient specificity to be able to react with hydrogenase on the external side of the membrane and with cytoplasmic electron carriers such as octaheme cytochrome c_3, ferredoxins, and flavodoxin. It would be interesting to look for this protein in several other species of *Desulfovibrio*. Nitrite reductase activity in species that do not utilize nitrate could be residual and comparable to the sulfur reductase activity of some tetraheme cytochromes c_3 found in species of sulfate-reducing bacteria unable to use sulfur as a respiratory substrate (60). However, the presence of nitrite reductase in *D. gigas* (175) could serve to detoxify the environment of the nitrite which is produced by some anaerobic nitrate reducers (149).

11.4 CYTOPLASMIC OR INTERNAL ENZYMES AND ELECTRON TRANSFER PROTEINS

11.4.1 Internal or Cytoplasmic Hydrogenases

The first evidence for the existence of multiple forms of hydrogenase in a large number of microorganisms, including *D. desulfuricans*, was provided by Ackrell et al. (1), and these preliminary data have been extended in a number of laboratories (2, 96, 169, 183, 193). Recently, three different genes for hydrogenase have been identified in *Escherichia (E.) coli* and provided the most convincing evidence for multiple forms of hydrogenase (167). Preliminary enzymological data also exist for multiple forms of hydrogenase in *Desulfovibrio* (103, 126, 206). The existence of more than one hydrogenase in the same species of *Desulfovibrio* has very recently been reported in *D. vulgaris* strains Hildenborough (64a, 124a, 148a) and Miyazaki F (206a). There are some indications that *D. vulgaris* strain Hildenborough contains the three types of hydrogenases (iron, nickel–iron, and nickel–iron–selenium enzymes) that could be differentially expressed during the growth cycle or depending upon culture conditions (124a). The periplasmic enzyme from *D. vulgaris* strain Hildenborough appears to be three orders of magnitude more active than the membrane-bound hydrogenases (124a). Differences in the specific activity of hydrogenase have been observed between intact cells or spheroplasts and extracts, depending on growth conditions (146, 151). By using four different experimental variations (sulfate limitation in a chemostat, growth phase, different substrates, and growth by interspecies H_2 transfer), consistent ratios of 2.0 for extract to intact cell hydrogenase activities with benzyl viologen have been observed with *D. vulgaris* and *D. gigas*.

Employing the newly isolated species *D. multispirans*, a nickel-containing cytoplasmic hydrogenase has been purified (47). Although its structure (two subunits of molecular weight of 58,000 and 24,500 Da, respectively), metal content (1 nickel and 11 ± 1 atoms of iron), and EPR characteristics are quite similar to the periplasmic *D. gigas* hydrogenase, its enzymatic properties are somewhat different. With a value of 790 μmol $\cdot$ min^{-1} $\cdot$ mg^{-1}, the hydrogenase evolution activity is about twice that of *D. gigas* hydrogenase. In contrast to the *D. vulgaris* Hildenborough and *D. gigas* periplasmic enzymes, which have a hydrogen consumption rate much higher than the evolution rate, the rate of benzyl viologen reduction in the presence of hydrogen was lower than that of hydrogen formation (586 μmol $\cdot$ min^{-1} $\cdot$ mg^{-1}). These differences in activity between the periplasmic enzymes and this cytoplasmic hydrogenase could be related to the different physiological function and localization of the two types of enzymes. The fact that anaerobically lysed cells of *D. gigas* and *D. multispirans* gave results similar to those of the purified enzyme for the ratios of hydrogen evolution versus hydrogen consumption does not favor the presence of two different hydrogenases with different properties in these two species.

However, several existing data suggest that the *Desulfovibrio* species have at least one external and one internal hydrogenase; definitive evidence for the existence of two hydrogenases must await purification and characterization of the specific enzymes from a single species of *Desulfovibrio* or identification of specific genes encoding for the different hydrogenases. Employing multiclonal antibodies to the purified periplasmic hydrogenase of *D. vulgaris* Hildenborough, a gene encoding for a hydrogenase of *D. vulgaris* has been isolated using a hydrogenase-deficient mutant of *E. coli* and plasmid pUC[9] (198) and sequenced (197). Since there was no evidence for a leader sequence in the large subunit, it was suggested that either a special mechanism for translocation of the hydrogenase across the cytoplasmic membrane exists or that the gene encodes for a cytoplasmic hydrogenase. However, convincing evidence for the presence in the small subunit of this hydrogenase of a leader sequence bearing similarities to those of known procaryotic signal peptides has recently been published (162c). This suggests that the mechanism of translocation across the cytoplasmic membrane of this complex enzyme is actually analogous to the translocation of other proteins and enzymes.

11.4.2 Reductases

11.4.2.1 Activation of Sulfate

A scheme representing the current view on the respiratory reduction of sulfate is given on Fig. 11.3. Owing to the chemical inertia of the molecule of sulfate, it needs first to be activated. Thus the bacteria are faced with a paradoxical situation of needing to spend energy to be able to utilize their respiratory substrate. This activation is accomplished by two enzymes. The first, ATP sulfurylase (8, 15), forms adenosine phosphosulfate (APS) according to the reaction

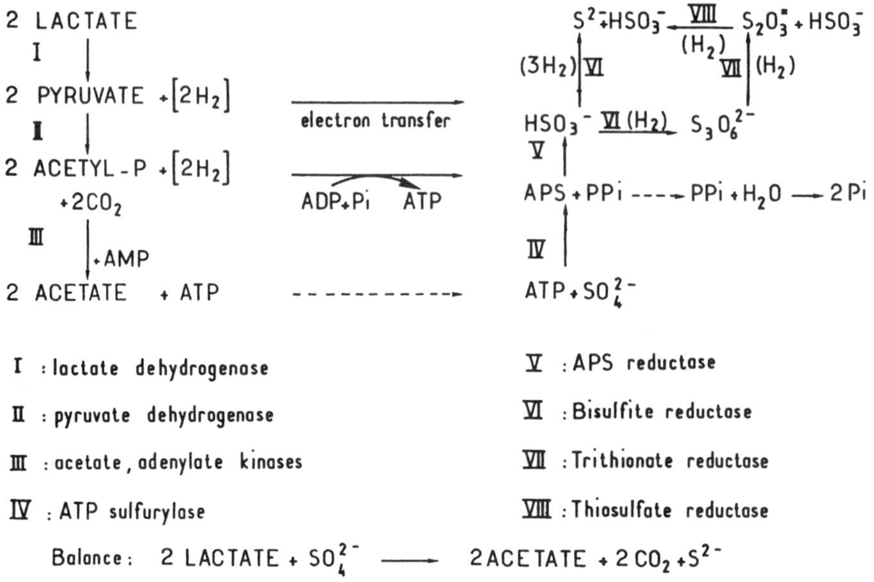

2 LACTATE

I |

2 PYRUVATE $+[2H_2]$

I | electron transfer

2 ACETYL-P $+[2H_2]$

    $+2CO_2$ ADP+Pi   ATP

III |$+$AMP

2 ACETATE $+$ ATP   - - - - - - - - - - ▸

$S^{2-}+HSO_3^- \xrightarrow[(H_2)]{VIII} S_2O_3^{=}+HSO_3^-$

$(3H_2)|$VI   VII$|(H_2)$

$HSO_3^- \xrightarrow{VII(H_2)} S_3O_6^{2-}$

V |

$APS+PPi ---\!\!\!▸ PPi+H_2O \longrightarrow 2Pi$

IV |

$ATP+SO_4^{2-}$

I : lactate dehydrogenase

II : pyruvate dehydrogenase

III : acetate, adenylate kinases

IV : ATP sulfurylase

V : APS reductase

VI : Bisulfite reductase

VII : Trithionate reductase

VIII : Thiosulfate reductase

Balance: $2\ LACTATE + SO_4^{2-} \longrightarrow 2\ ACETATE + 2\ CO_2 + S^{2-}$

Figure 11.3 Lactate–sulfate metabolism in sulfate-reducing bacteria. The balance corresponds to experimental data for *Desulfovibrio gigas* and *Desulfovibrio africanus* (186).

$$SO_4{}^{2-} + ATP \rightarrow APS + PP_i$$

Since the formation of pyrophosphate (PP_i) is thermodynamically unfavorable, the reaction needs to be pulled to the right by a second enzyme, a pyrophosphatase (150), which hydrolyzes PP_i according to the reaction

$$H_2O + PP_i \rightarrow 2P_i$$

11.4.2.2 APS Reductase

APS reductase is a cytoplasmic enzyme constituting 2 to 3% of soluble proteins, and no evidence exists to indicate any association with the cytoplasmic membrane. It is a nonheme iron flavoprotein (30) which has also been found in the new genera of sulfate-reducing bacteria: *Desulfobulbus*, *Desulfobacter*, *Desulfococcus*, and *Desulfosarcina* (176). Tetraheme cytochrome c_3, ferredoxin, and flavodoxin stimulate the reduction of sulfate in crude extracts but have not been demonstrated to function as immediate electron donors for the reductase. A scheme has been proposed for the mechanism of action of APS reductase (153) utilizing reduced methyl viologen, oxygen, ferricyanide, and cytochrome c for the assay of enzymatic activity. The function of the nonheme iron is of interest, as it can largely be removed without affecting enzymatic activities with oxygen, cytochrome c, or reduced methyl viologen. Enzymatic activity with ferricyanide is destroyed; however, it is highly unlikely that the physiological function of the nonheme iron centers is the reduction of ferricyanide. In all probability the nonheme iron clusters represent the site at which the natural electron donor(s) interact with APS reductase.

It has been demonstrated (128) that the reaction of sulfite with APS reductase results in the formation of an FAD-sulfite adduct causing the bleaching of the FAD and the appearance in the electronic spectrum of the enzyme of a maximum at 320 nm. This reflects a reaction of sulfite at the N-5 position of the isoalloxazine ring of FAD (Fig. 11.4, reaction 1). The subsequent addition of AMP (Fig. 11.4, reaction 2) results in a decrease in absorbance at 320 nm, the partial reduction of nonheme iron, and the formation of APS. The exact nature and role of the iron–sulfur clusters in this protein are still unknown.

11.4.2.3 Bisulfite Reductase

It has been difficult to demonstrate that sulfite is an intermediate in the reduction of sulfate (129). The reason is that, in contrast to nitrite, which is sometimes accumulated in nitrate reduction (149), sulfite is never found as a free form in growing cultures of sulfate-reducing bacteria. Although no precise determination has been made, APS reductase and bisulfite reductase, which represent several percent of the total cell proteins, appear to be present at a ratio close to 1. It can be speculated that they actually form *in situ* an enzymatic complex allowing a fast transfer and reduction of the molecule of sulfite. This idea is consistent with the fact that, as was pointed out by Dickerson and Timkovich (51), the flavin-bound sulfite of APS reductase is energetically primed for subsequent reactions in a manner that a solitary sulfite ion is not.

The common prosthetic group of all sulfite reductases is an iron-octacarboxylic tetrahydroporphyrin of the isobacteriochlorin type (142, 143) which has been termed siroheme. The structure proposed earlier has recently been revised and the new structure is shown on Fig. 11.5 (20, 170). Siroheme is also the cofactor for

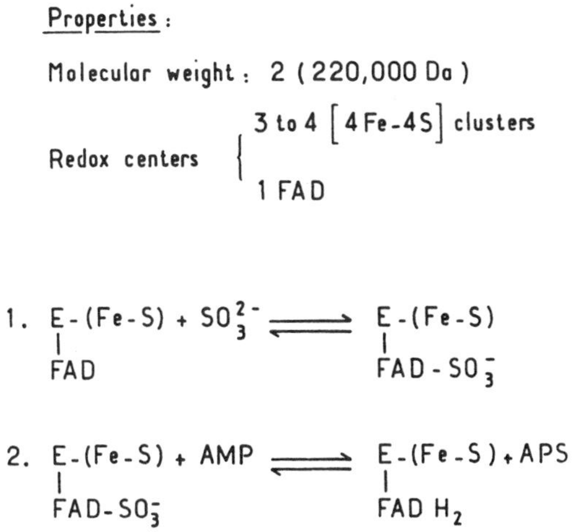

Properties :

Molecular weight : 2 (220,000 Da)

Redox centers $\begin{cases} 3 \text{ to } 4 \left[4\,Fe\text{-}4S \right] \text{ clusters} \\ 1\ FAD \end{cases}$

1. $E\text{-}(Fe\text{-}S) + SO_3^{2-} \rightleftharpoons E\text{-}(Fe\text{-}S)$
   $\quad\ \ |$
   $\quad\ FAD \qquad\qquad\qquad\quad |$
   $\qquad\qquad\qquad\qquad\qquad FAD\text{-}SO_3^-$

2. $E\text{-}(Fe\text{-}S) + AMP \rightleftharpoons E\text{-}(Fe\text{-}S) + APS$
   $\quad\ \ |$
   $\quad FAD\text{-}SO_3^- \qquad\qquad\qquad |$
   $\qquad\qquad\qquad\qquad\qquad FAD\ H_2$

Figure 11.4 APS reductase: properties and formation of APS from sulfite.

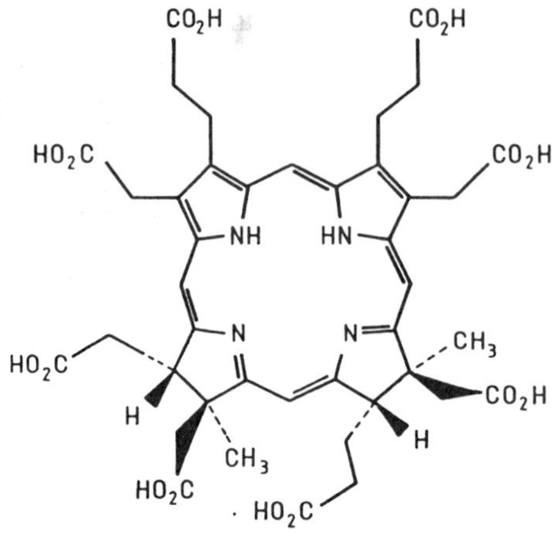

Figure 11.5 Structure of syrohydrochlorin, the metal-free prosthetic group of siroheme (20, 177).

several nitrite reductases (88, 194, 195), and sirohydrochlorin has been shown to be on the biosynthetic pathway to vitamin B_{12} (21, 170). The similarities between the two reactions, nitrite and sulfite reductions,

$$NO_2^- + 8H^+ + 6e^- \rightarrow NH_4^+ + 2H_2O$$
$$SO_3^{2-} + 8H^+ + 6e^- \rightarrow H_2S + 3H_2O$$

have prompted extensive research on chemical models of the sirohydrochlorin ring able to perform the same reaction (177).

It has previously been seen that the nitrite reductase of *Desulfovibrio* species, although performing the same reaction with nitrite, is significantly unreactive toward sulfite and contains mesohemes instead of sirohemes (122).

There are two types of bisulfite reductases in sulfate and sulfur-reducing bacteria (the optimum pH of activity for these enzymes being acidic, their substrate is actually the bisulfite ion). They are readily distinguished by the spin state of their heme iron. The first class is constituted by the high-spin sulfite reductases, which have a high molecular weight and a complex structure. The low-spin sulfite reductases, which have a low molecular weight and have only one polypeptide chain, comprise the second class. A comparison of the properties of the two types of enzymes is given in Table 11.5.

i. High-Spin Bisulfite Reductases. Four different enzymes of this class have been described so far; two, desulfoviridin and desulforubidin (106, 107), belong to the genus *Desulfovibrio*. Another, termed P582, has been found in *Dt. nigrificans* (188) and *Desulfonema magnum* (200), while a fourth, desulfofuscidin, is the bisulfite reductase of *T. commune* (81) and *D. thermophilus* (G. D. Fauque,

TABLE 11.5   Physical and chemical properties of high-spin and low-spin bisulfite reductases

	High-spin			Low-spin
	Desulfoviridin *D. gigas*	Desulforubidin *D. desulfuricans* Norway 4	Desulfofuscidin *T. commune*	Assimilatory S.R.[a] *D. vulgaris* Hildenborough
Molecular weight	200,000	225,000	167,000	27,200
Structure	$\alpha_2\beta_2$	$\alpha_2\beta_2$	$\alpha_2\beta_2$	α
Major absorption peak (nm)	628	545	576	580
[4Fe-4S] cluster	4	4	4	1
Siroheme	—	2	4	1
Siroporphyrin	2	—	—	—
Labile sulfide	14	14.7	16	5
Reaction with CO	—	+	+	+
Reference	152	107	81	108

[a] S.R.; sulfite reductase.

M. H. Czechowski, L. Kang, D. V. Der Vartanian, and J. LeGall, Abstr. Annu. Meet. Am. Soc. Microbiol. 1986, K50, p. 201).

The principal characteristics of desulfoviridin, desulforubidin, and desulfofuscidin are given in Table 11.5. The reductases differ with regard to their major absorption peaks, EPR spectra (121), and behavior of their siroheme moieties. Although EPR spectroscopy indicates the presence of a heme group in each reductase, desulfoviridin yields a siroporphyrin (142) which is responsible for the characteristic red fluorescence observed in alkali, a test that has been often used as diagnostic of *Desulfovibrio* sp. (161). All other bisulfite reductases, including the low-spin enzymes, yield a siroheme upon denaturation. Comparative Mössbauer studies of desulfoviridin from *D. gigas* and *D. desulfuricans* (ATCC 27774) and desulforubidin from *D. desulfuricans* Norway 4 (90) show that the former contain much less Mössbauer absorption of siroheme than the latter. This, taken together with the fact that desulforividin reacts with neither CO nor with pyridine (152), could be an indication that the iron is loosely bound to the siroporphyrin. In this view it is interesting to note that it has been shown that a reduction of sulfite can be obtained by the siroporphyrin alone in the presence of ferrous iron and that this iron does not bind to the porphyrin (172). A similar result has been obtained with an isolated siroheme (173). It has been pointed out (42) that the isolation of a lactone as the main product of sirohydroporphyrin extracted from *D. gigas* desulfoviridin (20) could be of special significance for the mechanism of this enzyme. As seen on Fig. 11.6, the lactonization corresponds to a redox reaction involving two protons and two electrons. Since the absorption spectra of such lactones do not deviate significantly from those of the parent compounds, the reaction, should it happen in the enzyme, would not induce significant spectral changes (42).

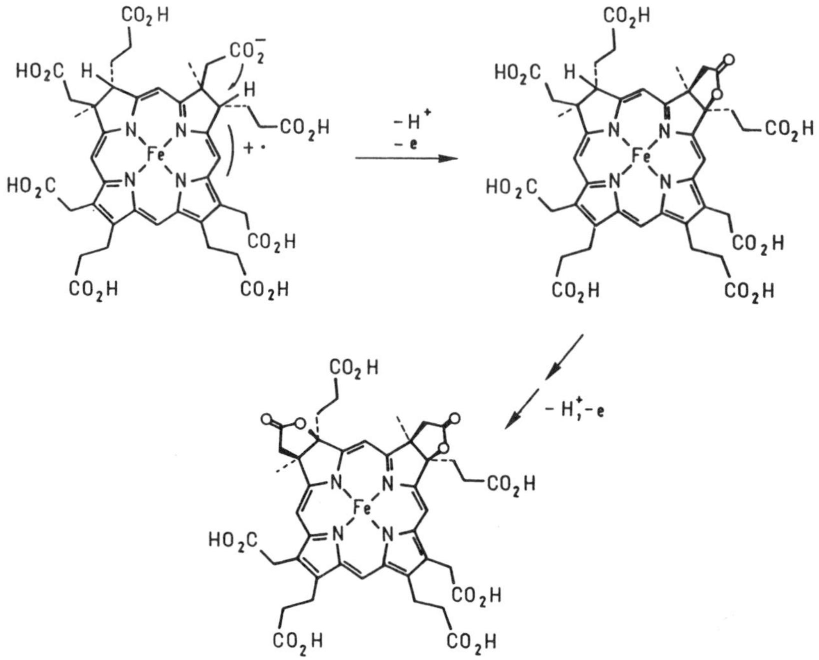

Figure 11.6   Lactonization of siroheme. Reproduced with permission from C. K. Chang, Hemes of hydroporphyrins, p. 313–334, in H. B. Dunford, D. Dolphin, K. N. Raymond, and L. Sieker (eds.), The biological chemistry of iron. Copyright © 1982 by D. Reidel Publishing Company, Dordrecht, Holland.

There is no physical evidence to indicate that the bisulfite reductases are membrane associated; however, it has been proposed that the reductases are membrane bound to account for a shift in products observed in the presence of membranes (53). Ferredoxin and flavodoxin have been shown to stimulate the reduction of sulfite with hydrogen (83). More recently, it has been shown that ferredoxin I is not very efficient in coupling hydrogenase and bisulfite reductase in extracts of *D. gigas*, but ferredoxin II is fully active in the same system (132).

It is firmly established that desulfoviridin, desulforubidin, and P582 from *Dt. nigrificans* (9), irreversibly produce trithionate and thiosulfate in addition to sulfide, the proportions depending on assay conditions. Because of the irreversible formation of trithionate and thiosulfate, the pathway requires the presence of specific reductases for trithionate and thiosulfate. A specific thiosulfate reductase has been found (95) and purified from extracts of several sulfate-reducing bacteria (71, 75). The demonstration of a specific trithionate reductase has been more elusive, but an enzyme has been described (52) that forms thiosulfate from trithionate and sulfite by a mechanism analogous to the cyanolysis of trithionate, with the resulting thiosulfate molecule derived from added sulfite and the sulfane atom from the trithionate. Evidence for the trithionate pathway as opposed to a direct six electrons reduction of sulfite to sulfide has recently been discussed (6, 152). Although the formation of trithionate and thiosulfate (83, 99) has important implications for the mechanism of sulfite reduction, the available evidence suggests that they are not free intermediates in respiratory sulfate reduction. However, this observation could be limited to *Desulfovibrio* species since it has been shown (7) that trithionate inhibits thiosulfate reductase activity in crude extracts of *Dt. nigrificans*, whereas it has no effect on the reduction of either bisulfite or thiosulfate by extracts of *D. vulgaris*. These results, together with the fact that trithionate has no effect on purified P582, are in favor of an active trithionate pathway in *Desulfotomaculum*. The main arguments for and against the trithionate pathway are listed in Table 11.6.

ii Low-Spin Bisulfite Reductases. A low-molecular-weight "assimilatory-type" sulfite reductase has been isolated from *D. vulgaris* Hildenborough (108). The recent discovery of a similar enzyme in the methanogenic bacterium *Methanosarcina (M.) barkeri* strain DSM 800 (136) and the obligate sulfur reducer *Drm. acetoxidans* (119) shows that these proteins form a new class of enzyme and are better termed low-spin bisulfite reductases. The reductase from *D. vulgaris* Hildenborough has a molecular weight of 27,200 Da. EPR and Mössbauer spectroscopy have shown that it contains a single [4Fe–4S] cluster coupled to a siroheme (89). Since the protein contains a total of five labile sulfides, it was postulated that the extra sulfur atom provides for the covalent binding between the cluster and the heme iron. The physiological significance of this enzyme is still not understood.

11.4.2.4 Thiosulfate Reductase

Thiosulfate reductase from *D. gigas* (75) and *D. vulgaris* strains Miyazaki and Hildenborough (10, 71) differs significantly from APS reductase and the bisulfite reductase. It has a similar molecular mass of about 200,000 Da, but does not contain flavin, heme, or nonheme iron and uses octaheme cytochrome c_3 as its electron donor (83). These considerations, plus the fact that it has not been possible to observe the formation of a proton gradient during the reduction of thiosulfate with hydrogen, suggest that thiosulfate reduction per se is not coupled to the generation of energy and that electron transfer between hydrogen and thiosulfate is cytoplasmic (152).

11.4.3 Dehydrogenases

11.4.3.1 D-Lactate and L-Lactate Dehydrogenases

Lactate dehydrogenase activities have been shown to be localized on the cytoplasmic side of the membrane of *D. desulfuricans* (ATCC 27774) (175). The presence of a pyridine nucleotide-independent D-lactate dehydrogenase has been reported in the soluble and membrane fractions of *D. vulgaris* strains Miyazaki and Hildenborough (148) and *D. desulfuricans* (ATCC 7757) (46). Intact cells of *D. vulgaris* Miyazaki oxidize D, L-lactate to completion (148) and an L(+)-lactate dehydrogenase has been reported in *D. desulfuricans* strain HL 21 (174). These

TABLE 11.6 Arguments for and against a trithionate pathway

Direct reduction: $SO_3^{2-} \xrightarrow{6e^-} S^{2-}$ (I)

Trithionate pathway: $SO_3^{2-} \xrightarrow[(1)]{2e^-} S_3O_6^{2-} \xrightarrow[(2)]{2e^-} SO_3^{2-} + S_2O_3^{2-} \xrightarrow[(3)]{2e^-} S^{2-} + SO_3^{2-}$ (II)

For

$S_3O_6^{2-}$ and $S_2O_3^{2-}$ are irreversibly formed by bisulfite reductases (9).
A specific $S_2O_3^{2-}$ reductase has been isolated (71, 75).
Step (2), equation (II) having a high redox potential (+225 mV) is energetically favorable for ATP formation (83).
A thiosulfate-forming enzyme has been found, but in *D. vulgaris* only (52).
$S_3O_6^{2-}$ inhibition of thiosulfate reductase, but in *Desulfotomaculum* only (7).

Against

$S_3O_6^{2-} \rightarrow S^{2-}$ depend on assay conditions.
$S_3O_6^{2-}$, $S_2O_3^{2-}$ are not free intermediates.
$^{35}SO_4^{2-}$ labeling experiments are not consistent with the trithionate pathway (41).
Formation of a proton gradient with H_2–SO_3^{2-}, but not with $S_3O_6^{2-}$, $S_2O_3^{2-}$ (152).
No true $S_3O_6^{2-}$ reductase has been characterized so far.

observations suggest that the utilization of D, L-lactate involves two specific dehydrogenases. Hydrogen is produced from the oxidation of lactate to pyruvate; however, the redox potential of the lactate–pyruvate couple is -190 mV, while that of the H^+/H_2 couple is -420 mV. Although sulfate reducers of the genus *Desulfovibrio* do not exhibit significant growth on lactate in the absence of sulfate, in the presence of a H_2-utilizing methanogenic bacterium, lactate is converted to H_2 and used for the formation of methane in interspecies H_2 transfer (37). The electron transfer proteins involved in the formation of H_2 from lactate have not yet been investigated.

11.4.3.2 *Pyuvate Dehydrogenase*

All species of *Desulfovibrio* contain pyruvate phosphoroclastic activity which converts pyruvate to acetyl phosphate, CO_2, and H_2. The system is cytoplasmic (77), requires thiamine pyrophosphate, coenzyme A, and Mg^{2+} for activity, and is stimulated by the presence of ATP (207). This pyruvate dehydrogenase has not been purified from any of the sulfate-reducing bacteria and consequently has not been extensively studied. However, it appears to be similar to the clostridial pyruvate dehydrogenase, which is a nonheme iron protein and uses ferredoxin as its immediate electron acceptor (191). Both the pyruvate–CO_2 exchange (178) and the evolution of H_2 from pyruvate (5) have been shown to require tetaheme cytochrome c_3 and ferredoxin for maximum activity. An explanation for this result would be that the stimulation of the reaction by tetaheme cytochrome c_3 is due to the mixing of the external hydrogenase and tetaheme cytochrome c_3 with the internal hydrogenase and ferredoxin. The crude extract would therefore not reflect the physiological situation. In *D. gigas* the stimulation by ferredoxin is complicated by the fact that the ferredoxin used in these and other studies may have been a mixture of ferredoxins containing both four-iron and three-iron centers (93). It has been demonstrated that only the ferredoxin containing the four-iron center is active in coupling the pyruvate dehydrogenase with hydrogenase in extracts devoid of ferredoxin and flavodoxin (132). Flavodoxin also stimulates the formation of hydrogen from pyruvate in extracts of *D. gigas* (78).

11.4.4 Electron Carriers

11.4.4.1 *Ferredoxins*

The ferredoxin from *D. gigas* (109) was the first such protein to be shown to contain four iron atoms and four sulfide residues per molecule (104), and this observation has been confirmed by determination of the amino acid sequence (33, 187). The ferredoxin is active in both the phosphoroclastic reaction (78) and the reduction of sulfite (109) and thiosulfate (83) to sulfide.

Ferredoxin from *D. gigas* has been shown to exist in two forms, ferredoxin I and ferredoxin II, which can be separated by chromatographic procedures. Both forms of ferredoxin contain the same basic monomeric unit, but ferredoxin I was shown to be a trimer and ferredoxin II a tetramer (36). The ferredoxin I is a typical ferredoxin in that it is diamagnetic in its oxidized state and has an E_0' of -440 mV; on the other hand, the ferredoxin II is paramagnetic in its oxidized form and has an E_0' of -130 mV, approximately 300 mV more positive than the trimeric form (40). Ferredoxin II has now been demonstrated to contain the three-iron-sulfur cluster (184) and is not active in catalyzing the production of H_2 from pyruvate in the phosphoroclastic reaction; however, it is active in stimulating the oxidation of H_2 in the presence of sulfite (132). Conversely, ferredoxin I containing mainly the four-iron centers is less efficient in catalyzing the oxidation of H_2 with sulfite but is fully active in the phosphoroclastic reaction. It should be cautioned that the stimulation of activity was observed with crude preparations and it is possible that additional electron transfer proteins are required for the H_2–sulfite couple.

From the aspect of regulation, it is most interesting that ferredoxin II can enzymologically be converted to ferredoxin I (116) and that the three- and four-nonheme iron centers can be chemically interconverted (97, 140). In our laboratory (M. Czechowski, L. Kang, and J. LeGall, unpublished results) we have demonstrated that pure *D. gigas* hydrogenase can reduce ferredoxin II, under hydrogen, without any other intermediate redox carrier. During the reaction, the [3Fe-4S] cluster of ferredoxin II is converted into a [4Fe-4S] cluster such as that found in ferredoxin I [these results have been confirmed by EPR (J. J. G. Moura and I. Moura, personal communication)]. The interesting fact is that the [4Fe-4S] cluster of ferredoxin I is not directly reducible by hydrogenase and requires the presence of tetraheme cytochrome c_3. Whatever the real physiological significance of these reactions, they demonstrate that a fine tuning of electron flow between different pathways can be obtained by regulation of the [3Fe-4S]/[4Fe-4S] interconversions.

The biophysical studies required for the characterization of ferredoxins have only been completed with the ferredoxins from *D. gigas*. However, the presence of multiple forms of ferredoxin has been reported in *D. desulfuricans* stains Norway 4 (35, 69) and Berre-Eau (M. Teixeira, G. Fauque, I. Moura, B. H. Huynh, A. V. Xavier, J. LeGall, and J. J. G. Moura, unpublished results), *D. africanus* (80), and *D. baculatus* strain 9974 (M. Teixeira, G. Fauque, I. Moura, B. H. Huynh, A. V. Xavier, J. LeGall and J. J. G. Moura, unpublished results). This suggests that different mechanisms apply to the regulation of electron transfer in members of the genus *Desulfovibrio*.

11.4.4.2 *Flavodoxin*

The flavodoxin of *D. vulgaris* Hildenborough has been studied extensively and both its primary (54) and its tertiary structures (199) have been established (Fig. 11.7). Its spectrum of biological activities is very similar to that of ferredoxin, in that it replaces ferredoxin in both H_2-producing and H_2-utilizing reactions (78, 110) and it is assumed to be cytoplasmic (152). A similarity in function is also indicated by the reciprocal relation that exists between iron and the biosynthesis

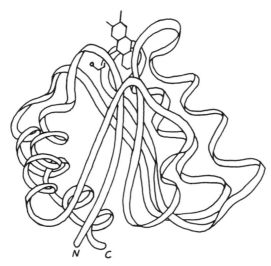

Figure 11.7   Three-dimensional structure of *Desulfovibrio vulgaris* Hildenborough flavo-doxin. Special thanks are due to Dr. L.C. Sieker for an original drawing of the molecule.

of flavodoxin and ferredoxin (98); however, this relation has not been established for the sulfate-reducing bacteria (115). There also exists an important biochemical similarity between flavodoxin and ferredoxin of *D. gigas* in that flavodoxin has two stable oxidation-reduction states; the semiquinone-hydroquinone form ($E_0' = -440$ mV) and the semiquinone–quinone form ($E_0' = -150$ mV) (55). This should be compared with the potentials of ferredoxin I ($E_0' = -440$ mV) and ferredoxin II ($E_0' = -130$ mV) (132), and this oxidation-reduction behavior of the two elec-tron-transfer proteins is entirely consistent with the observed biological inter-changeability. The possible importance of the two oxidation-reduction states of flavodoxin in biological systems is indicated by the report of Petitdemange et al. (155) concerning the substitution in *Clostridium (C.) tyrobutiricum* of flavodoxin for ferredoxin in the oxidation of NADH by NADH-ferredoxin oxidoreductase. Barton et al. (17) reported that electron transfer between hydrogenase and sulfite reductase of *D. gigas* could be restored by the addition of either ferredoxin or flavodoxin; however, electron transfer phosphorylation was only reconstituted by the addition of ferredoxin. The existence of phosphorylated forms of flavodoxin (56) suggests that there may be more specificity in the reactions of flavodoxin than suspected previously.

It is also noteworthy that *D. desulfuricans* Norway 4, a strain that does not contain flavodoxin (35), possesses both a one [4Fe-4S]-cluster ferredoxin analo-gous to *D. gigas* ferredoxin, and a two [4Fe-4S]-cluster ferredoxin (69) similar to clostridial ferredoxins. It is possible that the one-cluster ferredoxin cannot substi-tute for flavodoxin in some metabolic pathway and that a two-cluster protein is required.

11.4.4.3 Octaheme Cytochrome c₃

A cytochrome c containing eight hemes per molecule was first isolated from *D. gigas* (34) and has been found in all species of *Desulfovibrio* investigated. It has a low redox potential and an amino acid composition that distinguishes it from tetraheme cytochrome c_3. After removing the hemes, prior to SDS electrophoresis, it has been shown that octaheme cytochrome c_3 from *D. desulfuricans* Norway 4 is a dimer consisting of two identical subunits of 13,500 Da (67). It reacts with hydrogenase and has a K_m value of 10^{-6} M^{-1}, identical to that of tetraheme cytochrome c_3 (23). In crude extracts of *D. gigas* it specifically stimulates the reduction of thiosulfate and is not found among the periplasmic proteins. Tetraheme cytochrome c_3, which is found in the periplasm of *D. gigas*, is not active in the reduction of thiosulfate, a result that is consistent with the previously noted observation that the reduction of thiosulfate by molecular hydrogen is not coupled to the generation of energy (152). Consequently, the role of octaheme cytochrome c_3 could be limited to cytoplasmic electron transfer reactions, for example, as a cofactor for the putative cytoplasmic hydrogenase.

11.5 ELECTRON TRANSFER PROTEINS OF UNCERTAIN LOCALIZATION AND FUNCTION

11.5.1 Molybdenum-Containing Iron–Sulfur Protein from *Desulfovibrio gigas*

The molybdenum-containing iron–sulfur protein purified from *D. gigas* has a molecular weight of 120,000 Da and contains 12 atoms of iron and labile sulfides and one atom of molybdenum per molecule (134). The absorption spectrum of the protein revealed maxima at 425 and 467 nm and a shoulder at 545 nm. It was most similar to the [2Fe-2S]-type iron–sulfur cluster and, in fact, the protein has been shown to contain a total of six [2Fe-2S] clusters. The properties of the redox centers have been studied extensively (45, 135). Fluorescence detection and synchrotron radiation showed that in the oxidized molecule the molybdenum is surrounded by two terminal oxo groups and two long Mo-S bonds (2.47 Å). There was also evidence for an oxygen or nitrogen ligand at 1.90 Å. Reduction of the protein by dithionite resulted in the loss of a terminal oxo group and a 0.1-Å contraction of the Mo-S bond. These results are similar to those obtained with "desulfo" xanthine dehydrogenase (44). Thus the protein from *D. gigas* shows similarities not only to xanthine oxidase but also to sulfite oxidase and certain forms of nitrate reductase. None of these enzymatic activities are relevant for the metabolism of *D. gigas*. Recently, it has been found that this molybdenum-containing iron-sulfur protein possesses aldehyde oxidase activity (190a). This finding is relevant to the fact that *D. gigas* can grow with ethanol as carbon and energy sources and sulfate as terminal electron acceptor (see Chapter 10). There is a close similarity in amino acid composition between this protein and plant-type [2Fe-2S]-containing ferredoxins.

A molybdenum-containing iron–sulfur protein has also been isolated and characterized in several other species of sulfate reducers from the genus *Desulfovibrio*: *D. africanus* strain Benghazi 1 (79a), *D. desulfuricans* strain Berre-Eau (B. A. S. Barata, I. Moura, A. V. Xavier, G. Fauque, J. LeGall, and J. J. G. Moura, Abstr. Setimo Encontro Anu. Soc. Port. Quim., 1984, PC₃), *D. salexigens* strain British Guiana (48a), and *D. desulfuricans* strains Berre-Sol and ATCC 27774 (B. A. S. Barata, I. Moura, A. V. Xavier, G. Fauque, J. LeGall, and J. J. G. Moura, unpublished data).

11.5.2   Rubredoxin

Rubredoxin ($M_r = 6,000$ Da) can be routinely isolated from most strictly anaerobic bacteria; however, a specific function, if any, has not been identified. The structure of the rubredoxin from *D. vulgaris* Hildenborough has been established to a 2 Å resolution (3) and compared with the well-known structure of rubredoxin from *C. pasteurianum* (86). In particular, it shows that the four Fe-S bonds range in length from 2.15 to 2.35 Å and do not differ significantly from the mean value of 2.29 Å (Fig. 11.8). Despite this consistency, there are remarkable specificities shown by the rubredoxins from *D. gigas*, *D. vulgaris* Hildenborough, and *C. pasteurianum* with the NADH : rubredoxin oxidoreductase from *D. gigas* (J. M. Odom, M. Bruschi, H. D. Peck, Jr., and J. LeGall, 1976, Fed. Proc. Fed. Am. Soc. Exp. Biol. 35:1360). A large number of charged residues have been changed from one rubredoxin to another (196) and may explain the differences in the affinity of the oxidoreductase for the various rubredoxins.

Considering the efforts that have been made to resolve the structure of rubredoxins from strict anaerobes and the success that has followed them, it is paradoxical that their general physiological function is still unknown. The relatively high redox potential of rubredoxins (between 0 and −50 mV) (131) makes it dif-

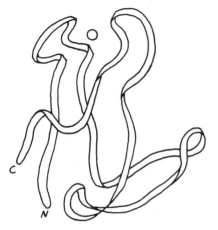

Figure 11.8 Three-dimensional structure of *Desulfovibrio vulgaris* Hildenborough rubredoxin. Special thanks are due to Dr. L. C. Sieker for an original drawing of the molecule.

ficult to fit them into the physiological reactions which are commonly utilized to study these activities in sulfate-reducing bacteria.

11.5.3 Desulforedoxin

A new rubredoxin-like protein has been found in *D. gigas* (130) which contains two iron atoms and eight cysteine residues per molecule which can be redox cycled. Although its optical spectrum is reminiscent of rubredoxin, it presents important differences. The polypeptide chain is composed of 73 amino acid residues and lacks six amino acids: histidine, arginine, proline, isoleucine, phenylalanine, and tryptophan. The N-terminal sequence shows no homology with that of other non-heme iron proteins. No physiological function is known for desulforedoxin. The data indicate that this protein represents a new class of nonheme iron protein.

11.5.4 Cobalt-Containing Protein

A protein-containing cobalt bound to a porphyrin has been found in *D. gigas* (133) and *D. desulfuricans* Norway 4 (76). The claim that the cobalt is reducible in the latter protein remains to be verified since the reversibility of the reaction was not demonstrated. The porphyrin binding the cobalt ^{III} has been identified as sirohy-drochlorin. It is to be noted that the presence of corrinoids has never been reported in sulfate-reducing bacteria, although sirohydroporphyrin is an intermediate on the pathway leading to the synthesis of B_{12} (19).

A protein containing cobalt and molybdenum has recently been isolated and characterized from the thermophilic strain of the sulfate reducer *D. thermophilus* DSM 1276 (G. Fauque, M. Czechowski, I, Moura, J. J. G. Moura, and J. LeGall, unpublished data).

11.5.5 Triheme Cytochrome c₃

A *c*-type cytochrome containing three hemes per molecule and for which the name cytochrome c_7 was given (12) has been found in the obligate anaerobic sulfur reducer *Drm. acetoxidans* (163). This protein has an interesting history because it was discovered, purified, and sequenced before the organism that synthetized it was identified. In fact, *Drm. acetoxidans* was later found (156) to be part of a mixed culture of the so-called *Chloropseudomonas ethylica*.

Its amino acid sequence (12) shows that it is homologous to the tetraheme cy-tochromes c_3. The midpoint redox potentials of the hemes were determined by optical spectroscopy (63), showing that two heme groups have a similar midpoint redox potential (-112 mV and -117 mV, respectively) when the third has a redox potential 60 mV more positive. Thus there are no low-potential hemes such as those present in tetraheme cytochromes c_3. Since it has been shown that triheme cytochrome c_3 from *Drm. acetoxidans* is not active in the reduction of colloidal sulfur (39) and that extensive NMR studies indicate that the heme ligands are histidinyl groups as in tetraheme cytochromes c_3, it can be postulated that it is the low redox potential of the hemes rather that the heme ligand arrangement that governs the reactivity toward S^0.

Only preliminary X-ray crystallographic data exist for the triheme cytochrome c_3 of *Drm. acetoxidans* (72). However, the analysis of its amino acid sequence (12) shows that a peptide corresponding to the peptide containing the binding site of heme 4 of the *D. desulfuricans* Norway 4 tetraheme cytochrome c_3 is missing (see Fig. 11.2).

NMR studies of this triheme cytochrome c_3 (137) have shown that the hemes undergo a low-spin/high-spin transition below pH 4.5. The relatively fast electron exchange that is observed contrasts sharply with tetraheme cytochromes c_3 from *D. gigas* (168) or *D. vulgaris* Hildenborough (139). It is not the number of hemes that can explain this difference in electron exchange rate: The tetraheme cytochrome c_3 of *D. desulfuricans* Norway 4 behaves in a manner similar to that of triheme cytochrome c_3 from *Drm. acetoxidans*. There are then particular and yet unknown features of the protein structure that control the electron exchange rates.

11.6 ENERGETICS

Hydrogenase had no place in the classical scheme for the growth of sulfate-reducing bacteria growing on lactate/sulfate media, as shown in Fig. 11.3. The now classical work of Bryant et al. (37) showing that *Desulfovibrio* could grow without sulfate in the presence of methanogenic bacteria through molecular interspecies hydrogen transfer radically changed our ideas concerning the role of hydrogenase in the metabolism of sulfate-reducing bacteria. As shown on Fig. 11.9, hydrogen-

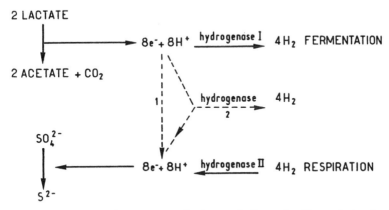

Figure 11.9 Importance of hydrogenase in the metabolism of sulfate-reducing bacteria. The questions which are still debated are: 1. Is the presence of two hydrogenases (solid lines) with different localizations necessary for energy conversion? 2. Is one hydrogenase sufficient (dashed line 2)? 3. Are the electrons coming from lactate and pyruvate passed directly to the sulfate reductases system without the participation of any hydrogenase (dashed line 1)? 4. Is it possible that the three hypotheses are operative in different sulfate-reducing bacteria?

ase plays a central role in the metabolism of these microorganisms. When growing on lactate by hydrogen interspecies transfer, the bacteria can be described as purely fermentative organisms, deriving their energy from ATP produced through substrate phosphorylation. They can also behave as purely respiratory organisms when growing on lactate/sulfate media.

The question now posed is: Is molecular hydrogen a residual product during growth on lactate/sulfate, or is it an essential metabolite? This question, which has not yet been answered correctly, has for several years been a major problem in our understanding of sulfate-reducing bacteria energetics.

Studies of the bioenergetics of *W. succinogenes* (100, 101) have provided evidence that a proton gradient is generated across the cytoplasmic membrane by the periplasmic oxidation of formate. This is coupled to the transfer of electrons only across the membrane and the cytoplasmic reduction of fumarate to succinate with the consumption of two protons. A proton gradient is generated with a $H^+/2e$ ratio of 1.0. This process has been proposed (13) to account for the growth yields of *D. vulgaris* Marburg on hydrogen/sulfate or thiosulfate: The external oxidation of hydrogen is responsible for the generation of a proton gradient and ATP synthesis via a reversible ATPase (66). As noted previously (154), neither of these mechanisms require a direct coupling of proton translocation to electron transfer for the dissimilatory sulfate reduction (202).

Following these earlier observations, two mechanisms have been described for the formation of a proton gradient in *Desulfovibrio*: vectorial electron transfer coupled to the oxidation of hydrogen by hydrogenase (14) and proton translocation coupled to the reduction of specific substrates (175). From observations of H_2 production during growth (185, 190), enzyme localization (14, 145), and vectorial electron transfer (14, 100, 202), a chemiosmotic hydrogen cycle has been proposed as a mechanism by which *Desulfovibrio* produces the ATP required for growth on lactate plus sulfate (146). The scheme involves the formation of molecular hydrogen, a permeant molecule, from lactate and pyruvate in the cytoplasm or on the cytoplasmic surface of the cell membrane and the rapid diffusion of H_2 across the cytoplasmic membrane (Fig. 11.10). On the external surface of the membrane, H_2 is oxidized by the periplasmic hydrogenase (23), which requires tetraheme cytochrome c_3, and the electrons produced from this oxidation are transferred back across the membrane, leaving the protons at the external surface of the membrane. The electrons are used in the cytoplasm for the reduction of sulfate to sulfide, which results ideally in the consumption of eight protons. The net effect is the transfer of eight protons across the cytoplasmic membrane without direct coupling of proton translocation to electron transfer. The proton gradient would then drive the synthesis of ATP in the conventional fashion via a reversible ATPase (66). Alternatively, (Fig. 11.10) a trace hydrogen transformation scheme has been proposed (125) in which hydrogen production is visualized as controlling the redox state of internal electron carriers linked to energy conservation and hydrogen utilization by the periplasmic hydrogenase is proposed as preventing the loss of energy in the form of hydrogen (190). The biochemical and enzymological requirements of the two schemes are qualitatively similar, but their physiological

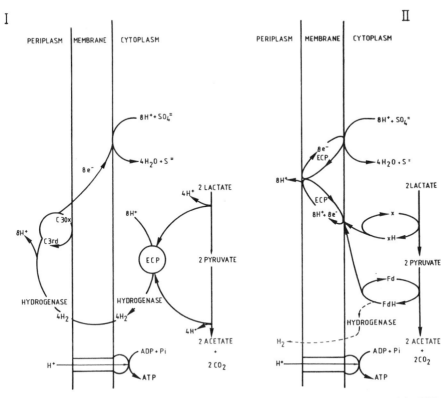

Figure 11.10 The obligate H$_2$ cycling and the trace hydrogen transformation models (125, 146). Model I: hydrogen is produced by an internal hydrogenase that diffuses freely across the membrane and is recycled by a periplasmic hydrogenase that assures charges separation, electrons only being allowed to cross the membrane. Model II: a cytoplasmic hydrogenase controls the redox level of ferredoxin or flavodoxin, thus regulating electron and proton transfer which generates a membrane-bound proton motive force. Fd, ferredoxin; $C3$, tetraheme cytochrome c_3; ECP, electron carrier protein.

significance is completely different. The first requires only vectorial electron transfer for charge separation, and in the second energy conservation is attained through a typical Mitchell loop.

Neither scheme accounts for the failure of molecular hydrogen to inhibit growth on lactate plus sulfate, for the accumulation of hydrogen by *D. vulgaris* Hildenborough (185), and does not take into consideration the complex array of electron transfer proteins which are present in sulfate-reducing bacteria. However, a direct demonstration of hydrogen cycling has recently been made: Employing membrane-inlet mass spectrometry, direct evidence is presented that substrate amounts of hydrogen are produced and consumed simultaneously during the metabolism of pyruvate plus sulfate by washed intact cells of *D. vulgaris* strain Hildenborough (154a).

A specific requirement for the H_2-cycling mechanism is that reducing equivalents from organic substrates which lead to H_2 production in the cytoplasm must not be diverted to the sulfate- or sulfur-reducing enzymes. This separation of electron transfer might be achieved by compartmentation, electron carrier specificity or regulation or a combination of these factors. It should always be kept in mind that microorganisms in the genus *Desulfovibrio* may not be closely related, and that generalizations concerning the function of electron transfer proteins may not be valid for every species of *Desulfovibrio*. This is illustrated by a detailed study of the balance of different species of *Desulfovibrio* grown on lactate/sulfate media (185, 186). The products of lactate oxidation are variable (Table 11.7): Only *D. gigas* and *D. africanus* will utilize all the reducing power from lactate to reduce sulfate according to the reaction that has been proposed so far for all dissimilatory sulfate reducers. In contrast, *D. desulfuricans* Norway 4 utilizes some of the electrons for the production of butanol, while *D. vulgaris* Hildenborough produces 1 mol of molecular hydrogen from 2 mol of lactate. The latter result is in accord with the trace hydrogen transformation model (125) and has been interpreted in terms of the regulation of electron flow by tetraheme cytochrome c_3 during growth. It could be that the production of molecular hydrogen prevents the accumulation of hydrogen sulfide and would keep its concentration below a level that would be irreversibly toxic for tetraheme cytochrome c_3. We have already noted (60) that hydrogen sulfide inhibits the reduction of colloidal sulfur by *D. vulgaris* Hildenborough tetraheme cytochrome c_3 when the homologous proteins from *D. desulfuricans* Norway 4, *D. gigas*, or *D. baculatus* strain 9974 show no inhibition in the same experimental conditions. The observation that *D. vulgaris* Hildenborough derives 25% of its reductive power for the reduction of protons at the very early stage of its growth on lactate/sulfate (185) is not the same phenomenon that was described in a study using *D. vulgaris* Miyazaki strain (190). In this case, only trace amounts of molecular hydrogen were formed (0.014 mol of H_2 per mol of lactate) and only at the early stage of the growth. A significant comparison between the two *D. vulgaris* strains would require the utilization of identical growth conditions (185).

Desulfovibrio and *Desulfotomaculum* species are clearly unrelated and the former classification of *Desulfotomaculum* as a specialized species of *Clostridia* should

---

TABLE 11.7   Energetics balances of *D. vulgaris* Hildenborough (reaction I), *D. gigas* and *D. africanus* (reaction II), and *D. desulfuricans* Norway 4 (reaction III) (185, 186)

$CH_3CHOHCOO^- + 0.37SO_4^{2-} + 0.56H^+ \rightarrow CO_2 + 0.98CH_3COO^-$
$+ 0.02CH_3CH_2OH + 0.16H_2S + 0.215HS^- + 0.5H_2O + 0.48H_2$ [reaction (I)]

$CH_3CHOHCOO^- + 0.5SO_4^{2-} + 0.71H^+ \rightarrow CO_2 + CH_3COO^- + 0.218H_2S$
$+ 0.282HS^- + H_2O$ [reaction (II)]

$CH_3CHOHCOO^- + 0.375SO_4^{2-} \rightarrow CO_2 + 0.8CH_3COO^- + 0.375S^{2-}$
$+0.1CH_3(CH_2)_2CH_2OH + 0.8H_2O$ [reaction (III)]

always be kept in mind. Since *Desulfotomaculum* does not contain tetraheme cytochrome c_3, the central role of this hemoprotein in the energetics of sulfate-reducing bacteria has been questioned and a scheme was proposed in which it was limited to some ill-defined "periplasmic oxidoreduction" (202). An interesting hypothesis is that *Desulfotomaculum* does not perform oxidative phosphorylations but derives all its energy from substrate phosphorylations through the formation of acetyl phosphate from pyruvate and also from acetate and pyrophosphate; this would account for the absence of tetraheme cytochrome c_3 in these sporulated organisms (120). However, a definitive proof of the existence of significant levels of acetate pyrophosphate kinase, the enzyme that forms acetyl phosphate from acetate and pyrophosphate, remains to be given. Also, the presence of cytochrome *b* in *Desulfotomaculum* shows that a typical Mitchell loop could be operative in these organisms, and recent evidence shows that they are capable of conserving energy by electron transport phosphorylations (F. Widdel, personal communication).

11.7 CONCLUSIONS

In this chapter we have limited most of our discussions to the well-described members of the genus *Desulfovibrio* and to the "classical" lactate/sulfate metabolism. This is because most of the information on electron carriers and bioenergetics has been obtained with bacteria grown under these conditions. Needless to say, the exploration of the newly isolated species of *Desulfovibrio* and of new genera of sulfate-reducing bacteria (28, 157, 200) which are able to grow on a large variety of substrates, including fatty acids up to 18 carbon atoms long and aromatic compounds, will provide new and exciting information as to the energetics and the significance of sulfate-reducing bacteria in anaerobic environments.

The schemes that have been put forward to explain the energetics of these organisms are still very speculative and are completely open to future research. It appears that there is no model bacterium that can be used the way *E. coli* K_{12} has been utilized (and still is) as far as dissimilatory reduction of sulfate is concerned. This is indeed very fortunate since every research group interested in this very active field should be able to pick up enough original strains to prevent unfortunate and sterile duplicative research.

More data have accumulated on *D. gigas* electron carrier proteins and energetics than with any other strain of sulfate-reducing bacteria. This is why, in Fig. 11.11, we have schematically put together all the existing data concerning the localization of the significant enzymes and electron carrier proteins that have been reviewed in this chapter. This is not a working model as those presented in Fig. 11.10. Readers are left with the most recent data and are welcome to decide for themselves which hypothesis fits best and eventually to come up with their own.

A schematic representation of *D. gigas* periplasmic hydrogenase and its possible relationships with the membrane and cytoplasmic enzymes and electron carriers is shown in Fig. 11.12. In this scheme, tetraheme cytochrome c_3 accepts both

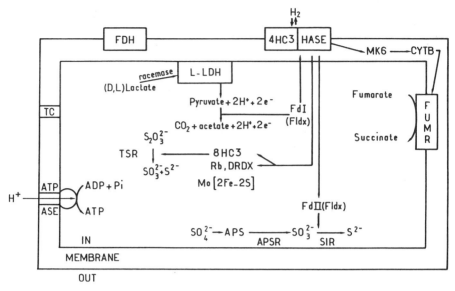

Figure 11.11 A schematic representation of the localization of enzymes and electron carrier proteins in *Desulfovibrio gigas*. The arrows indicate reactions in which a specific electron carrier has been demonstrated to react. FDH, formic dehydrogenase; 4HC3, tetraheme cytochrome c_3; FdI, FdII; ferredoxins I and II; TC, transmembrane cytochrome c; FUMR, fumaric reductase; TSR, thiosulfate reductase; 8HC3, octaheme cytochrome c_3; Fldx, flavodoxin; APS, adenosine phosphosulfate; APSR, APS reductase; SIR, sulfite reductase; MK6, menaquinone 6; CYTB, cytochrome b; HASE, hydrogenase; ATPASE, adenosine triphosphatase; L-LDH, L-lactate dehydrogenase; Rb, rubredoxin; DRDX, desulforedoxin; Mo[2Fe-2S]; Molybdenum-containing iron–sulfur protein.

protons and electrons from a transmembrane electron carrier (MK6?) and transfers them to the low-potential [4Fe-4S] cluster which finally gives them to the "active center" of hydrogenase, the [4Fe-4S]---Ni complex. Molecular hydrogen can be formed or, alternatively, the [3Fe-xS] center, which has been reduced independently of the other centers (182), can transfer electrons, leaving protons out, to a transmembrane electron carrier (nitrite reductase ?) that is itself connected with the electron transfer chains for the reduction of sulfate and sulfite. This scheme is highly tentative and, if valid, would apply only to *D. gigas*. Recently data show that a [3Fe-xS] center is not present either in the periplasmic hydrogenase of *D. baculatus* strain 9974 (180, 182b) or in the soluble hydrogenase of *D. desulfuricans* Norway 4 (24).

Confusing conclusions have been reached in the search for phylogenetical relationships between sulfate-reducing bacteria and other microorganisms using sequence determinations of homologous proteins. By comparing rubredoxins primary structures, it was concluded that *Desulfovibrio* species are related to *Clostridia* and other strict anaerobes (32); flavodoxins indicated a possible relation with *Azotobacter* (64); the amino-acid sequence of cytochrome c_{553} was noted to

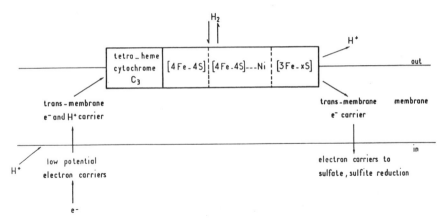

Figure 11.12 A hypothetic mechanism for the relations between *Desulfovibrio gigas* hydrogenase, periplasmic, membrane-bound, and cytoplasmic electron carriers. Low-potential electrons are accepted by a transmembrane proton and electron carrier (MK6?) and are transferred to tetraheme cytochrome c_3, which acts as a proton and electron acceptor (168); electrons are transferred to the low-potential [4Fe-4S] cluster of hydrogenase; molecular hydrogen is synthetized, in the absence of sulfate, at the level of the active center of hydrogenase formed by the [4Fe-4S]---Ni couple. In the presence of sulfate, electrons coming from H_2 or from the oxidation of lactate, are accepted by the high-potential [3Fe-xS] cluster. The protons are left out and electrons are passed through the membrane by a transmembrane cytochrome (nitrite reductase ?) and transported to the sulfate/sulfite reducing enzymes via cytoplasmic redox carriers. This scheme implies that the redox potentials of the tetraheme cytochrome c_3 are lowered by a specific interaction with hydrogenase.

resemble cytochrome c_{551} from some *Pseudomonas* species (31) and it was further concluded that *Desulfovibrio* spp. are closely related to nitrate-respiring and aerobic bacteria (144). Finally, although *D. gigas* ferredoxin (33, 187) is clearly related to ferredoxin I from *C. thermoaceticum* (57), it bears little homology with ferredoxin II from *D. desulfuricans* strain Noway 4 (68). In contrast, the latter protein shows a large homology with the so-called "archaebacterium" *M. barkeri* DSM 800 ferredoxin (85), a fact that has escaped the attention of microbiologists specialized in the classification of methane-producing bacteria (Fig. 11.13). Of course, *Desulfovibrio* cannot be closely related to aerobic as well as archaebacteria; however, this does not mean that valuable information cannot be extracted from these studies, but not the kind that was expected, that is, establishing relationships between whole organisms from an extrapolation of the linear structure of one family of redox proteins. What should be looked for are similar pathways in the bacteria for which a particular protein is specifically required. This would give interesting insights into the evolution of physiological functions if not into the evolution of bacterial species. Unfortunately, although redox-carrying proteins are very often small molecules and thus relatively easy to sequence, their functions are not well established or in many cases even completely unknown as in the case

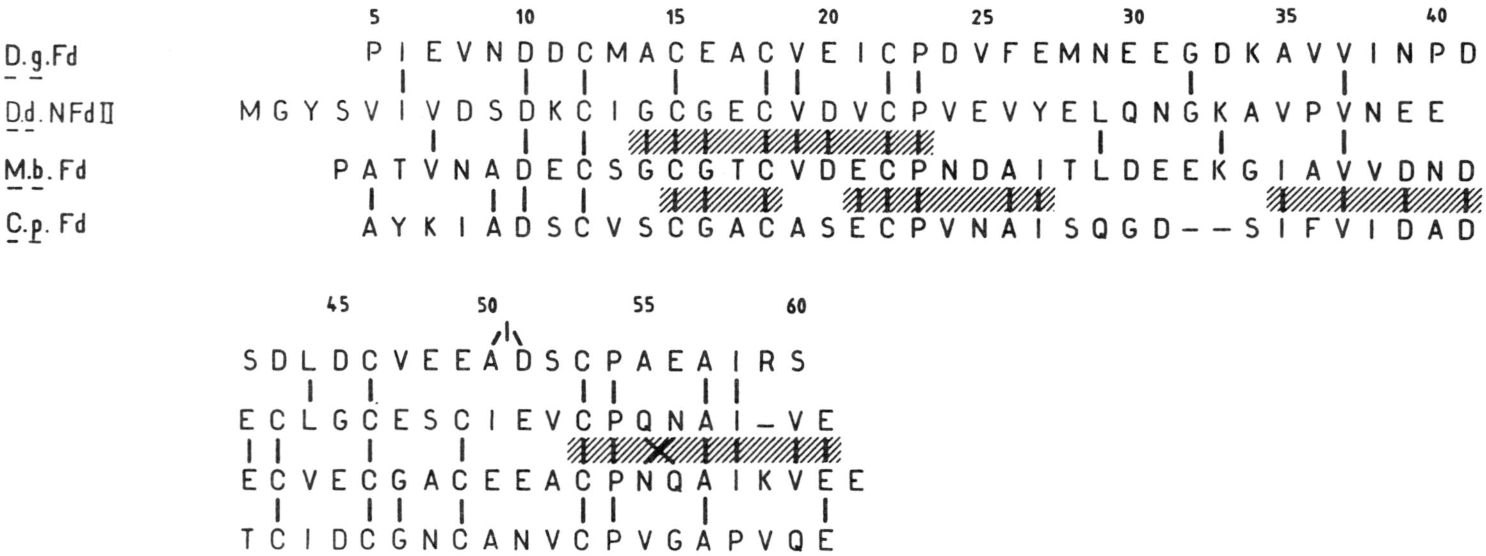

```
              5         10        15        20        25        30        35        40
D.g.Fd    P I E V N D D C M A C E A C V E I C P D V F E M N E E G D K A V V I N P D
Dd.NFdII  M G Y S V I V D S D K C I G C G E C V D V C P V E V Y E L Q N G K A V P V N E E
M.b.Fd    P A T V N A D E C S G C G T C V D E C P N D A I T L D E E K G I A V V D N D
C.p.Fd    A Y K I A D S C V S C G A C A S E C P V N A I S Q G D - - S I F V I D A D

              45        50        55        60
          S D L D C V E E A D S C P A E A I R S
          E C L G C E I C I E V C P Q N A I - V E
          E C V E C G A C E E A C P N Q A I K V E E
          T C I D C G N C A N V C P V G A P V Q E
```

Figure 11.13 Comparison of four amino acid sequences of two 2[4Fe-4S] ferredoxins, one 1[4Fe-4S] ferredoxin, and one 1[3Fe-3S] ferredoxin. The archaebacterium ferredoxin of *Methanosarcina barkeri* DSM 800 (85) appears to be equally close to *Clostridium pasteurianum* ferredoxin (179) and *Desulfovibrio desulfuricans* Norway 4 ferredoxin II (68). Since *Desulfovibrio gigas* ferredoxin contains only one [4Fe-4S] cluster, it appears less related to any of these proteins. However, it is more related to the one [4Fe-4S] ferredoxin from *Clostridium thermoaceticum* (57) and ferredoxin I from *Desulfovibrio desulfuricans* Norway 4 (68) (not shown). Comparing structures is not a sterile game, however; note, for example, that the residues which are homologous are clustered in different peptides when *Desulfovibrio desulfuricans* Norway 4 ferredoxin II is compared with *Methanosarcina barkeri* ferredoxin and *Clostridium pasteurianum* ferredoxin with the latter protein. These "not random" homologies will find their real significance when the relations between ferredoxins and their physiological electron carrier partners are known in detail. Fd, ferredoxin; *D.g; D. gigas; D. d N; D. desulfuricans* Norway 4; *M. b; M. barkeri; C. p; C. pasteurianum*

of rubredoxins of anaerobic bacteria. Undoubtedly, future research in the field of comparative biochemistry will provide sufficient answers so that the information that is still hidden within the primary structures of redox proteins will be unveiled.

From a biochemical point of view, sulfate-reducing bacteria are a gold mine. For example, as we have seen, they contain all known nonheme iron sites: 1Fe as in rubredoxin, [2Fe-2S] clusters as in the *D. gigas* molybdenum protein, [3Fe-4S] clusters ferredoxins, and [4Fe-4S] clusters ferredoxins. They also contain unique iron–sulfur centers such as the one contained in *D. vulgaris* Hildenborough hydrogenase that resembles the "P" cluster in nitrogenase (91, 92). Since pure cultures of *Desulfovibrio* have been shown to be able to fix nitrogen (113, 118, 118a, 118c, 162, 162a, 162b) they should be able to synthetize the "P" cluster of nitrogenase. Some other sulfate-reducing bacteria redox proteins contain original and yet poorly understood iron clusters such as the two rubredoxin-like irons of desulforedoxin and the iron–sulfur clusters of APS reductase.

One of us used to say that he would die happily if the structure of tetraheme cytochrome c_3 were to be known before this highly probable event. Since this structure *is* known (73, 87), he now wishes to see the structure of all sulfate-reducing bacteria electron-carrying proteins. This should guarantee him a very long life span.

ACKNOWLEDGMENTS

This work was made possible in part by Grant NSF DMB-8415632 and SERI subcontract D-1-1155-1 (J. LeGall) and in part thanks to a special collaborative agreement between the University of Georgia (USA), the Centre National de la Recherche Scientifique and the Commissariat à l'Energie Atomique (France). We wish to thank our colleagues Drs. I. Moura, J. J. G. Moura, A. Trebst, J. E. Wampler, and F. Widdel for their personal communications.

REFERENCES

1. Ackrell, B. A. C., R. N. Asato, and H. F. Mower. 1966. Multiple form of bacterial hydrogenases. J. Bacteriol. 92:828–839.

2. Adams, M. W. W., and L. E. Mortenson. 1984. The physical and catalytic properties of hydrogenase II of *Clostridium pasteurianum*. J. Biol. Chem. 259:7045–7055.

3. Adman, E. T., L. C. Sieker, L. H. Jensen, M. Bruschi, and J. LeGall. 1977. A structural model of rubredoxin from *D. vulgaris* at 2Å resolution. J. Mol. Biol. 112:113–120.

4. Aguirre, R., M. F. Cocquempot, E. C. Hatchikian, P. Monsan, and T. Lissolo. 1982. Utilization of free and immobilized *Desulfovibrio* hydrogenase in H_2 production; coupling efficiency of cytochrome c_3, ferredoxin and flavodoxin. Biotechnol. Lett. 4:297–302.

5. Akagi, J. M. 1967. Electron carriers for the phosphoroclastic reaction of *Desulfovibrio desulfuricans*. J. Biol. Chem. 242:2478–2483.

6. Akagi, J. M. 1981. Dissimilatory sulfate reduction: mechanistic aspects, p. 169–177, in H. Bothe and A. Trebst (eds.), Biology of inorganic nitrogen and sulfur. Springer-Verlag, Heidelberg.

7. Akagi, J. M. 1983. Reduction of bisulfite by the trithionate pathway by cell extracts from *Desulfotomaculum nigrificans*. Biochem. Biophys. Res. Commun. 117:530–535.

8. Akagi, J. M., and L. L. Campbell. 1963. Studies on thermophilic sulfate-reducing bacteria. III. Adenosine triphosphate sulfurylase of *Clostridium nigrificans* and *Desulfovibrio desulfuricans*. J. Bacteriol. 84:1194–1201.

9. Akagi, J. M., and V. Adams. 1973. Isolation of a bisulfite reductase activity from *Desulfotomaculum nigrificans* and its identification as the carbon monoxide-binding pigment, P 582. J. Bacteriol. 116:392–398.

10. Aketagawa, J., K. Kobayaski, and M. Ishimoto. 1985. Purification and properties of thiosulfate reductase from *Desulfovibrio vulgaris*, Miyazaki. J. Biochem. (Tokyo) 97:1025–1032.

11. Almassy, R. S., and R. E. Dickerson. 1978. *Pseudomonas* cytochrome c_{551} at 2.0 Å resolution: enlargement of the cytochrome *c* family. Proc. Natl. Acad. Sci. USA 75:2674–2678.

12. Ambler, R. P. 1971. The amino-acid sequence of cytochrome $c_{551.5}$ (cytochrome c_7) from the green photosynthetic bacterium *Chloropseudomonas ethylica*. FEBS Lett. 18:351–353.

13. Badziong, W., and R. K. Thauer. 1978. Growth yields and growth rates of *Desulfovibrio vulgaris* (Marburg) growing on hydrogen plus sulfate and hydrogen plus thiosulfate as the sole energy source. Arch. Microbiol. 117:209–214.

14. Badziong, W., and R. K. Thauer. 1980. Vectorial electron transport in *Desulfovibrio vulgaris* (Marburg) growing on hydrogen plus sulfate as sole energy source. Arch. Microbiol. 125:167–184.

15. Baliga, B. S., H. G. Vartak, and V. Jagan-Nathan. 1961. Purification and properties of sulfurylase from *Desulfovibrio desulfuricans*. J. Sci. Ind. Res. (India) C20:33–40.

16. Barton, L. L., J. LeGall, and H. D. Peck, Jr. 1970. Phosphorylation coupled to the oxidation of hydrogen with fumarate in extracts of the sulfate reducing bacterium *Desulfovibrio gigas*. Biochem. Biophys. Res. Commun. 41:1036–1042.

17. Barton, L. L., J. LeGall, and H. D. Peck, Jr. 1972. Oxidative phosphorylation in the obligate anaerobe, *Desulfovibrio gigas*, p. 33–51, in A. San Pietro and H. Gest (eds.), Horizons of bioenergetics. Academic Press, New York.

18. Barton, L. L., J. LeGall, J. M. Odom, and H. D. Peck, Jr. 1983. Energy coupling to nitrite respiration in the sulfate-reducing bacterium *Desulfovibrio gigas*. J. Bacteriol. 154:867–871.

19. Battersby, A. R., and Z. C. Sheng. 1982. Preparation and spectroscopic properties of Co^{III}-isobacteriochlorins: relationship to the cobalt-containing proteins from *Desulphovibrio gigas* and *D. desulphuricans*. J. Chem. Soc. Chem. Commun. 1393–1394.

20. Battersby, A. R., K. Jones, E. McDonald, J. A. Robinson, and H. R. Morris. 1977. The structures and chemistry of isobacteriochlorins from *Desulfovibrio gigas*. Tetrahedron Lett. 25:2213–2216.

21. Battersby, A. R., E. McDonald, H. R. Morris, M. Thompson, and D. C. Williams. 1977. Biosynthesis of vitamin B_{12}: structural study on the corriphyrins from *Proprionibacterium shermanii* and the link with sirohydrochlorin. Tetrahedron Lett. 25:2217–2220.

22. Bell, G. R., J. LeGall, and H. D. Peck, Jr. 1974. Evidence for the periplasmic location of hydrogenase in *Desulfovibrio gigas*. J. Bacteriol. 120:994–997.

23. Bell, G. R., J. P. Lee, H. D. Peck, Jr., and J. LeGall. 1978. Reactivity of *Desulfovibrio gigas* hydrogenase towards natural and artificial electron donors. Biochimie (Paris) 60:315–320.

24. Bell, S. H., D. P. E. Dickson, R. Rieder, R. Cammack, D. S. Patil, D. O. Hall, and K. K. Rao. 1984. Spectroscopic studies of the nature of the iron clusters in the soluble hydrogenase form *Desulfovibrio desulfuricans* (strain Norway 4). Eur. J. Biochem. 145:645–651.

25. Berlier, Y. M., G. Fauque, P. A. Lespinat, and J. LeGall. 1982. Activation, reduction and proton-deuterium exchange reaction of the periplasmic hydrogenase from *Desulfovibrio gigas* in relation with the role of cytochrome c_3. FEBS Lett. 140:185–188.

25a. Berlier, Y. M., B. Dimon, G. Fauque, and P. A. Lespinat. 1985. Direct mass-spectrometric monitoring of the metabolism and isotope exchange in enzymic and microbiological investigations, p. 17–35, in H. Degn, R. P. Cox, and H. Toftlund (eds.), Gas enzymology. D. Reidel, The Hague.

25b. Berlier Y., G. D. Fauque, J. LeGall, E. S. Choi, H. D. Peck, Jr., and P. A. Lespinat. 1987. Inhibition studies of three classes of *Desulfovibrio* hydrogenase: application to the further characterization of the mutliple hydrogenases found in *Desulfovibrio vulgaris* Hildenborough. Biochem. Biophys. Res. Commun. 146:147–153.

26. Bertrand, P., M. Bruschi, M. Denis, J. P. Gayda, and F. Manca. 1982. Cytochrome c_{553} from *Desulfovibrio vulgaris*: potentiometric characterization by optical and EPR studies. Biochem. Biophys. Res. Commun. 106:756–760.

27. Bianco, P., J. Haladjian, M. Loufti, and M. Bruschi. 1983. Comparative studies of monohemic bacterial *c*-type cytochromes. Redox and optical properties of *Desulfovibrio desulfuricans* Norway cytochrome $c_{553(550)}$ and *Pseudomonas aeruginosa* cytochrome c_{551}. Biochem. Biophys. Res. Commun. 113:526–530.

28. Biebl, H., and N. Pfennig. 1977. Growth of sulfate reducing bacteria with sulfur as electron acceptor. Arch. Microbiol. 112:115–117.

29. Biebl, H., and N. Pfennig. 1978. Growth yields of green sulfur bacteria in mixed cultures with sulfur and sulfate reducing bacteria. Arch. Microbiol. 117:9–16.

30. Bramlett, R. N., and H. D. Peck, Jr. 1975. Some physical and kinetic properties of adenylyl sulfate reductase from *Desulfovibrio vulgaris*. J. Biol. Chem. 250:2979–2986.

31. Bruschi, M., and J. LeGall. 1972. C-type cytochromes of *D. vulgaris*: the primary structure of cytochrome c_{553}. Biochim. Biophys. Acta 271:48–60.

32. Bruschi, M. and J. LeGall. 1978. Phylogenetic studies of three families of electron transfer proteins from sulfate reducing bacteria: cytochrome c_3, ferredoxins and rubredoxins, p. 221–232, in H. Matsubara and T. Yamanaka (eds.), Evolution of protein molecules, Japan Scientific Society Press, Tokyo.

33. Bruschi, M., and P. Couchoud. 1979. Amino acid sequence of *Desulfovibrio gigas* ferredoxin: revisions. Biochem. Biophys. Res. Commun. 91:623–628.

34. Bruschi, M., J. LeGall, E. C. Hatchikian, and M. Dubourdieu. 1969. Cristallisation et propriétés d'un cytochrome intervenant dans la réduction du thiosulfate par *D. gigas*. Bull. Soc. Fr. Physiol. Veg. 15:381–390.

35. Bruschi, M., E. C. Hatchikian, L. Golovleva, and J. LeGall. 1977. Purification and characterization of cytochrome c_3, ferredoxin, and rubredoxin isolated from *D. desulfuricans* Norway. J. Bacteriol. 129:30–38.

36. Bruschi, M., E. C. Hatchikian, J. LeGall, J. J. G. Moura, and A. V. Xavier. 1976. Purification, characterization and biological activities of three forms of ferredoxin from the sulfate reducing bacterium, *Desulfovibrio gigas*. Biochim. Biophys. Acta 449:275–284.

37. Bryant, M. P., L. L. Campbell, C. A. Reddy, and M. R. Crabill. 1977. Growth of *Desulfovibrio* in lactate or ethanol media low in sulfate in association with H_2-utilizing methanogenic bacteria. Appl. Environ. Microbiol. 33:1162–1169.

38. Cammack, R., D. Patil, R. Aguirre, and E. C. Hatchikian. 1982. Redox properties of the ESR detectable nickel in hydrogenase from *Desulfovibrio gigas*. FEBS Lett. 142:289–292.

39. Cammack, R., G. Fauque, J. J. G. Moura, and J. LeGall. 1984. ESR studies of cytochrome c_3 from *Desulfovibrio desulfuricans* Norway 4. Midpoint potentials of the four haems and interactions with ferredoxin and colloidal sulphur. Biochim. Biophys. Acta. 784:68–74.

40. Cammack, R., K. K. Rao, D. O. Hall, J. J. G. Moura, A. V. Xavier, M. Bruschi, J. LeGall, A. Deville, and J. P. Gayda. 1977. Spectroscopic studies of the oxidation-reduction properties of three forms of ferredoxin from *D. gigas*. Biochim. Biophys. Acta. 784:68–74.

41. Chambers, L. A., and P. A. Trudinger. 1975. Are thiosulfate and trithionate intermediates in dissimilatory sulfate reduction? J. Bacteriol. 123:36–40.

42. Chang, C. K. 1981. Hemes of hydroporphyrins, p. 313–334, in H. B. Dunford, D. Dolphin, K. N. Raymond, and L. Sieker (eds.), The biological chemistry of iron. D. Reidel, Dordrecht, The Netherlands.

43. Cocquempot, M. F., T. Lissolo, R. Aguirre, E. C. Hatchikian, and P. Monsan. 1982. Co-immobilization effect on H_2 production by a chloroplast membrane-hydrogenase system. Biotechnol. Lett. 4:313–318.

44. Cramer, S. P., R. Wahl, and K. V. Rajagopalan. 1981. Molybdenum sites of sulfite oxidase and xanthine dehydrogenase. A comparison by EXAFS. J. Am. Chem. Soc. 103:7721–7727.

45. Cramer, S. P., J. J. G. Moura, A. V. Xavier, and J. LeGall. 1984. Molybdenum EXAFS of the *Desulfovibrio gigas* Mo(2Fe-2S) protein; structural similarity to "desulfo" xanthine dehydrogenase. J. Inorg. Biochem. 20:275–280.

46. Czechowski, M. H., and H. W. Rossmore. 1980. Factors affecting *Desulfovibrio desulfuricans* lactate dehydrogenase. Dev. Ind. Microbiol. 21:349–356.

47. Czechowski, M. H., S. H. He, M. Nacro, D. V. DerVartanian, H. D. Peck, Jr. and J. LeGall. 1984. A cytoplasmic nickel-iron hydrogenase with high specific activity from *Desulfovibrio multispirans* sp. n., a new species of sulfate-reducing bacterium. Biochem. Biophys. Res. Commun. 125:1024–1032.

48. Czechowski, M., G. Fauque, Y. Berlier, P. A. Lespinat, and J. LeGall. 1985. Purification of an hydrogenase from an halophilic sulfate-reducing bacterium: *Desulfovibrio salexigens* strain British Guiana. Rev. Port. Quim. 27:196–197.

48a. Czechowski, M., G. Fauque, N. Galliano, B. Dimon, I. Moura, J. J. G. Moura, A. V. Xavier, B. A. S. Barato, A. R. Lino, and J. Le Gall. 1986. Purification and characterization of three proteins from a halophilic sulfate-reducing bacterium, *Desulfovibrio salexigens*. J. Ind. Microbiol. 1:139–147.

49. DerVartanian, D. V., and J. LeGall. 1974. A monomolecular electron transfer chain: structure and function of cytochrome c_3. Biochim. Biophys. Acta. 346:79–99.

50. DerVartanian, D. V., A. V. Xavier, and J. LeGall. 1978. EPR determination of the oxidation-reduction potentials of the hemes in cytochrome c_3 from *D. vulgaris*. Biochime (Paris) 60:321–325.

51. Dickerson, R. E., and R. Timkovich. 1975. Cytochromes *c*, p. 397–547, in P. D. Boyer (ed.), The enzymes, 3rd ed., Vol. 11. Academic Press, New York.

52. Drake, H. L., and J. M. Akagi. 1977. Characterization of a novel thiosulfate forming enzyme isolated from *Desulfovibrio vulgaris*. J. Bacteriol. 132:132–138.

53. Drake, H. L., and J. M. Akagi. 1977. Bisulfite reductase of *Desulfovibrio vulgaris*: explanation for product formation. J. Bacteriol. 132:139–143.

54. Dubourdieu, M., J. LeGall, and J. L. Fox. 1973. The amino-acid sequence of *Desulfovibrio vulgaris* flavodoxin. Biochem. Biophys. Res. Commun. 52:1418–1425.

55. Dubourdieu, M., J. LeGall, and V. Favaudon. 1975. Physico-chemical properties of flavodoxin from *D. vulgaris*. Biochim. Biophys. Acta. 376:519–532.

56. Edmonson, D. E., and T. L. James. 1979. Covalently bound non-coenzyme phosphorus residues in flavoproteins. Proc. Natl. Acad. Sci. USA 76:3786–3789.

57. Elliot, J. I., S. S. Yang, L. G. Ljungdhal, J. Travis, and C. F. Reilly. 1982. Complete amino acid sequence of the 4Fe-4S, thermostable ferredoxin from *Clostridium thermoaceticum*. Biochemistry 21:3294–3298.

58. Fauque, G. 1979. Réduction physiologique du soufre colloidal chez les bactéries sulfato-réductrices. Doctoral thesis, University of Aix-Marseille I.

59. Fauque, G. 1985. Relations structure-fonction de différentes protéines d'oxydo-réduction isolées de bactéries sulfato-réductrices et méthanigènes. Doctorat d'Etat thesis, University of Technology of Compiègne.

60. Fauque, G., D. Herve, and J. LeGall. 1979. Structure-function relationships in hemoproteins: the role of cytochrome c_3 in the reduction of colloidal sulfur by sulfate reducing bacteria. Arch. Microbiol. 121:261–264.

61. Fauque, G., M. Bruschi, and J. LeGall. 1979. Purification and some properties of cytochrome $c_{553(550)}$ isolated from *Desulfovibrio desulfuricans* Norway 4. Biochem. Biophys. Res. Commun. 86:1020–1029.

62. Fauque, G. D., L. L. Barton, and J. LeGall. 1980. Oxidative phosphorylation linked to the dissimilatory reduction of elemental sulfur by *Desulfovibrio*, p. 71–86, in Sulphur in biology (CIBA Foundation Symposium No. 72). Excerpta Medica, Amsterdam.

62a. Fauque, G. D., Y. M. Berlier, M. H. Czechowski, B. Dimon, P. A. Lespinat, and J. LeGall. 1987. A proton-deuterium exchange study of three types of *Desulfovibrio* hydrogenases. J. Ind. Microbiol. 2:15–23.

62b. Fauque, G., Y. Berlier, E. S. Choi, H. D. Peck, Jr., J. LeGall, and P. A. Lespinat. 1987. The carbon monoxide inhibition of the proton–deuterium exchange activity of iron, nickel–iron, and nickel–iron–selenium hydrogenases from *Desulfovibrio vulgaris* Hildenborough. Biochem. Soc. Trans. 15:1050–1051.

63. Fiechtner, M. D., and R. J. Kassner. 1979. The redox properties and heme environment of cytochrome $c_{551.5}$ from *Desulfuromonas acetoxidans*. Biochim. Biophys. Acta. 579:269–278.

64. Fox, J. L. 1976. Structure function and evolution of flavodoxin, p. 261–272, in J. L. Fox, Z. Deyl, and A. Blazej (eds.), Protein structure and evolution. Marcel Dekker, New York.

64a. Gow, L. A., I. P. Pankhania, S. P. Ballantine, D. H. Boxer, and W. A. Hamilton. 1986. Identification of a membrane-bound hydrogenase of *Desulfovibrio vulgaris* (Hildenborough). Biochim. Biophys. Acta. 851:57–64.

65. Grossman, J. P., and J. R. Postgate. 1955. The metbolism of malate and certain other compounds by *D. desulfuricans*. J. Gen. Microbiol. 12:429–445.

66. Guarraia, L. J., and H. D. Peck, Jr. 1971. Dinitrophenol-stimulated adenosine triphosphatase activity in extracts of *Desulfovibrio gigas*. J. Bacteriol. 106:890–895.

67. Guerlesquin, F., G. Bovier-Lapierre, and M. Bruschi. 1982. Purification and characterization of cytochrome c_3 (Mr 26.000) isolated from *Desulfovibrio desulfuricans* Norway strain. Biochem. Biophys. Res. Commun. 105:530–538.

68. Guerlesquin, F., M. Bruschi, and G. Bovier-Lapierre. 1984. Electron transfer mechanism and interaction studies between cytochrome c_3 and ferredoxin. Biochimie (Paris) 66:93–99.

69. Guerlesquin, F., M. Bruschi, G. Bovier-Lapierre, and G. Fauque. 1980. Comparative study of two ferredoxins from *Desulfovibrio desulfuricans* Norway. Biochim. Biophys. Acta. 626:127–135.

70. Hamilton, W. A. 1983. Sulphate-reducing bacteria and the offshore oil industry. Trends Biotechnol. 1:36–40.

71. Haschke, R. H., and L. L. Campbell. 1971. Thiosulfate reductase from *Desulfovibrio vulgaris*. J. Bacteriol. 106:603–607.

72. Haser, R., F. Payan, R. Bache, M. Bruschi, and J. LeGall. 1979. Crystallization and preliminary crystallographic data for cytochrome $c_{551.5}(c_7)$ from *Desulfuromonas acetoxidans*. J. Mol. Biol. 130:97–98.

73. Haser, R., M. Pierrot, M. Frey, F. Payan, J. P. Astier, M. Bruschi, and J. LeGall. 1979. Structure and sequence of cytochrome c_3, a multiheme cytochrome. Nature (Lond.) 282:806–810.

74. Hatchikian, E. C. 1974. On the role of menaquinone-6 in the electron transport of the hydrogen: fumarate reductase system in the strict anaerobe *Desulfovibrio gigas* J. Gen. Microbiol. 81:261–266.

75. Hatchikian, E. C. 1975. Purification and properties of thiosulfate reductase from *Desulfovibrio gigas*. Arch. Microbiol. 105:249–256.

76. Hatchikian, E. C. 1981. A cobalt porphyrin containing protein reducible by hydrogenase isolated from *Desulfovibrio desulfuricans* (Norway). Biochem. Biophys. Res. Commun. 103:521–530.

77. Hatchikian, E. C., and J. LeGall. 1970. Etude du métabolisme des acides dicarboxyliques et du pyruvate chez les bactéries sulfato-réductrices. I. Etude enzymatique de l'oxydation du fumarate en acétate. Ann. Inst. Pasteur (Paris) 118:125–136.

78. Hatchikian, E. C., and J. LeGall. 1970. Etude du métabolisme des acides dicarboxyliques et du pyruvate chez les bactéries sulfato-réductrices. II. Transport des électrons-accepteurs finaux. Ann. Inst. Pasteur (Paris) 118:288–301.

79. Hatchikian, E. C., and J. LeGall. 1972. Evidence for the presence of a b-type cytochrome in the sulfate-reducing bacterium *Desulfovibrio gigas* and its role in the reduction of fumarate by molecular hydrogen. Biochim. Biophys. Acta. 267:479–484.

79a. Hatchikian, E. C., and M. Bruschi. 1979. Isolation and characterization of a molybdenum iron-sulfur protein from *Desulfovibrio africanus*. Biochem. Biophys. Res. Commun. 86:725–734.

80. Hatchikian, E. C., and M. Bruschi. 1981. Characterization of a new type of ferredoxin from *Desulfovibrio africanus*. Biochim. Biophys. Acta. 634:41–51.

81. Hatchikian, E. C., and J. G. Zeikus. 1983. Characterization of a new type of dissimilatory sulfite reductase present in *Thermodesulfobacterium commune*. J. Bacteriol. 153:1211–1220.

82. Hatchikian, E. C., M. Bruschi, and J. LeGall. 1978. Characterization of the periplasmic hydrogenase from *Desulfovibrio gigas*, p. 93–106, in H. G. Schlegel and K. Schneider (eds.), Hydrogenases: their catalytic activity, structure and function. E. Göltze KG, Göttingen, West Germany.

83. Hatchikian, E. C., J. LeGall, M. Bruschi, and M. Dubourdieu. 1972. Regulation of the reduction of sulfite and thiosulfate by ferredoxin, flavodoxin and cytochrome c_3 in extracts of the sulfate reducer *D. gigas*. Biochim. Biophys. Acta. 263:272–282.

84. Hatchikian, E. C., P. Papavassiliou, P. Bianco, and J. Haladjian. 1984. Characterization of cytochrome c_3 from the thermophilic sulfate reducer *Thermodesulfobacterium commune*. J. Bacteriol. 159:1040–1046.

85. Hausinger, R. P., I. Moura, J. J. G. Moura, A. V. Xavier, H. Santos, J. LeGall, and J. B. Howard. 1982. Amino-acid sequence of a 3Fe–3S ferredoxin from the ''archaebacterium'' *Methanosarcina barkeri*. J. Biol. Chem. 257:14192–14194.

86. Herriot, J. R., L. C. Sieker, L. H. Jensen, and W. Lovenberg. 1970. Structure of rubredoxin: an X-ray study to 2.5 Å resolution. J. Mol. Biol. 50:391–406.

87. Higushi, Y., S. Bando, M. Kusunoki, Y. Matsuura, N. Yasuoka, M. Kadudo, T. Yamanaka, T. Yagi, and M. Inokushi. 1981. The structure of cytochrome c_3 from *Desulfovibrio vulgaris* Miyazaki at 2.5 Å resolution. J. Biochem. (Tokyo) 89:1659–1662.

88. Hucklesby, D. P., D. M. James, M. J. Banwell, and E. J. Hewitt. 1976. Properties of nitrite reductase from *Cucurbita pepo*. Phytochemistry 15:599.

89. Huynh, B. H., L. Kang, D. V. DerVartanian, H. D. Peck, Jr., and J. LeGall. 1984. Characterization of a sulfite reductase from *Desulfovibrio vulgaris*. Evidence for the presence of a low-spin siroheme and an exchange coupled [4Fe–4S] unit. J. Biol. Chem. 259:15373–15376.

90. Huynh, B. H., L. Kang, D. V. DerVartanian, H. D. Peck. Jr., and J. LeGall. 1985. A comparison study of sulfite reductases from sulfate-reducing bacteria: a Mössbauer investigation. Rev. Port. Quim. 27:23–24.

91. Huynh, B. H., M. T. Henzl, J. A. Christner, R. Zimmermann, W. H. Orme-Johnson, and E. Münck. 1980. Nitrogenase. XII. Mössbauer studies of the Mo–Fe protein from *Clostridium pasteurianum*. Biochim. Biophys. Acta. 623:124–138.

92. Huynh, B. H., M. H. Czechowski, H. J. Krüger, D. V. DerVartanian, H. D. Peck, Jr., and J. LeGall. 1984. *Desulfovibrio vulgaris* hydrogenase a non-heme iron enzyme lacking nickel that exhibits anomalous EPR and Mössbauer spectra. Proc. Natl. Acad. Sci. USA 81:3728–3732.

93. Huynh, B. H., J. J. G. Moura, I. Moura, T. A. Kent, J. LeGall, A. V. Xavier, and E. Münck. 1980. Evidence for a three-iron centre in a ferredoxin from *Desulfovibrio gigas*. J. Biol. Chem. 255:3242–3244.

94. Ishimoto, M., J. Koyama, and Y. Nagai. 1954. Role of a cytochrome in thiosulfate reduction by a sulfate-reducing bacterium. Seikagaku Zasshi 26:303.

95. Ishimoto, M., J. Koyama, and Y. Nagai. 1955. Biochemical studies on sulfate reducing bacteria. IV. Reduction of thiosulfate by cell free extract. J. Biochem. (Tokyo) 42:41–53.

96. Jacobson, F. S., L. Daniels, J. A. Fox, C. T. Walsh, and W. Orme-Johnson. 1982. Purification and properties of an 8-hydroxy-5-deazaflavin-reducing hydrogenase from *Methanobacterium thermoautotrophicum*. J. Biol. Chem. 257:3385–3388.

97. Kent, T. A., I. Moura, J. J. G. Moura, J. D. Lipscomb, B. H. Huynh, J. LeGall, A. V. Xavier, and E. Münck. 1982. Conversion of 3Fe clusters into 4Fe–4S clusters in a *D. gigas* ferredoxin and isotopic labeling of cluster subsites. FEBS Lett. 138:55–58.

98. Knight, E., and R. W. F. Hardy. 1966. Isolation and characteristics of flavodoxin from nitrogen-fixing *Clostridium pasteurianum*. J. Biol. Chem. 241:2752–2755.

99. Kobayaski, K., Y. Seki, and M. Ishimoto. 1974. Biochemical studies on sulfate reducing bacteria. XIII. Sulfite reductase from *Desulfovibrio vulgaris*: mechanism of trithionate, thiosulfate and sulfite formation and enzymatic properties. J. Biochem. (Tokyo) 74:519–529.

100. Kröger, A. 1978. Fumarate as terminal electron acceptor of phosphorylative electron transport. Biochim. Biophys. Acta 505:129–145.

101. Kröger, A. 1980. Bacterial electron transport to fumarate, p. 1–17, in C. J. Knowles (ed.), Diversity of bacterial respiration systems, Vol. 2. CRC Press, Boca Raton, Fla.

102. Krouse, H. R., and R. G. L. McCready. 1979. Biogeochemical cycling of sulfur, p. 401–430, in P. A. Trudinger and D. J. Swaine (eds.), Biogeochemical cycling of mineral-forming elements. Elsevier, Amsterdam.

103. Krüger, H.-J., B. H. Huynh, P. O. Ljungdahl, A. V. Xavier, D. V. DerVartanian, I. Moura, H. D. Peck, Jr., M. Teixeira, J. J. G. Moura, and J. LeGall. 1982. Evidence for nickel and a three-iron center in the hydrogenase of *Desulfovibrio desulfuricans*. J. Biol. Chem. 257:14620–14623.

104. Laishley, E. J., J. Travis, and H. D. Peck, Jr. 1969. Amino acid composition of ferredoxin and rubredoxin isolated from *Desulfovibrio gigas*. J. Bacteriol. 98:302.

105. Lancaster, R. J. 1980. Soluble and membrane bound paramagnetic centers in *Methanobacterium bryantii*. FEBS Lett. 115:285–288.

106. Lee, J. P., and H. D. Peck, Jr. 1973. Purification of the enzyme reducing bisulfite to trithionate from *Desulfovibrio gigas* and its identification as desulfoviridin. Biochem. Biophys. Res. Commun. 45:583–589.

107. Lee, J. P., C. S. Yi, J. LeGall, and H. D. Peck, Jr. 1973. Isolation of a new pigment from *Desulfovibrio desulfuricans* (Norway 4) and its role in sulfite reduction. J. Bacteriol. 115:453–455.

108. Lee, J. P., C. S. Yi, J. LeGall, and H. D. Peck, Jr. 1973. Isolation of assimilatory and dissimilatory type sulfite reductases from *Desulfovibrio vulgaris*. J. Bacteriol. 115:529–542.

109. LeGall, J., and N. Dragoni. 1966. Dependence of sulfite-reduction on a crystallized ferredoxin from *D. gigas*. Biochem. Biophys. Res. Commun. 23:145–149.

110. LeGall, J., and E. C. Hatchikian. 1967. Purification et propriétés d'une flavoprotéine intervenant dans la réduction du sulfite par *D. gigas*. C. R. Acad. Sci. Paris. 264:2580–2583.

111. LeGall, J., and M. Bruschi. 1968. Purification and some properties of a new cytochrome from *D. desulfuricans*, p. 467–470, in K. Okuniki, M. D. Kamen, and I. Sekusu (eds.), Structure and function of cytochromes. University Park Press, College Park, Md.

112. LeGall, J., and J. R. Postgate. 1973. The physiology of the sulfate reducing bacteria. Adv. Microb. Physiol. 10:81–133.

113. LeGall, J., J. C. Senez, and F. Pichinoty. 1959. Fixation de l'azote par les bactéries sulfato-réductrices. Isolement et caractérisation de souches actives. Ann. Inst. Pasteur (Paris) 96:223–230.

114. LeGall, J., G. Mazza, and N. Dragoni. 1965. Le cytochrome c_3 de *Desulfovibrio gigas*. Biochim. Biophys. Acta 99:385–387.

115. LeGall, J., D. V. DerVartanian, and H. D. Peck, Jr. 1979. Flavoproteins, iron proteins, and hemoproteins as electron-transfer components of the sulfate-reducing bacteria, p. 237–265, in D. R. Sanadi, (ed.), Current topics in bioenergetics, Vol. 9. Academic Press, New York.

116. LeGall, J., J. J. G. Moura, H. D. Peck, Jr., and A. V. Xavier. 1982. Hydrogenase and other iron–sulfur proteins from sulfate-reducing and methane-forming bacteria, p. 177–248, in T. G. Spiro (ed.), Iron–sulfur proteins. Wiley, New York.

117. LeGall, J., P. O. Ljungdahl, I. Moura, H. D. Peck, Jr., A. V. Xavier, J. J. G. Moura, M. Teixeira, B. H. Huynh, and D. V. DerVartanian. 1982. The presence of redox sensitive nickel in the periplasmic hydrogenase from *Desulfovibrio gigas*. Biochem. Biophys. Res. Commun. 106:610–616.

118. Lespinat, P. A., Y. Berlier, G. Fauque, R. Toci, G. Denariaz, and J. LeGall. 1984. Nitrogenase and hydrogenase activities in sulfate-reducing bacteria, p. 223, in C. Veeger and W. E. Newton (eds)., Advances in nitrogen fixation research. Martinus Nijhoff, The Hague.

118a. Lespinat, P. A., G. Denariaz, G. Fauque, R. Toci, Y. Berlier, and J. LeGall. 1985. Fixation de l'azote atmosphérique et métabolisme de l'hydrogène chez une bactérie sulfato-réductrice sporulante. *Desulfotomaculum orientis*. C. R. Acad. Sci. Paris 301:707–710.

118b. Lespinat, P. A., Y. Berlier, G. Fauque, M. Czechowski, B. Dimon, and J. LeGall. 1986. The pH dependence of proton–deuterium exchange, hydrogen production and uptake catalyzed by hydrogenases from sulfate-reducing bacteria. Biochimie (Paris) 68:55–61.

118c. Lespinat, P. A., Y. M. Berlier, G. D. Fauque, R. Toci, G. Denariaz, and J. LeGall. 1987. The relationship between hydrogen metabolism, sulfate reduction and nitrogen fixation in sulfate reducers. J. Ind. Microbiol. 1:383–388.

119. Lino, A. R., J. J. G. Moura, A. V. Xavier, I. Moura, G. Fauque, and J. LeGall. 1985. Isolation of two low-spin bacterial siroheme proteins. Rev Port. Quim. 27:215.

120. Liu, C. L., and H. D. Peck, Jr. 1981. Comparative bioenergetics of sulfate reduction in *Desulfovibrio* and *Desulfotomaculum* sp. J. Bacteriol. 145:966–973.

121. Liu, C. L., D. V. DerVartanian, and H. D. Peck, Jr. 1979. On the redox properties of three bisulfite reductases from the sulfate-reducing bacteria. Biochem. Biophys. Res. Commun. 91:962–970.

122. Liu, M. C., and H. D. Peck, Jr. 1981. Isolation of a hexaheme cytochrome from *Desulfovibrio desulfuricans* and its identification as a new type of nitrite reductase. J. Biol. Chem. 256:13159–13164.

123. Liu, M. C., D. V. DerVartanian, and H. D. Peck, Jr. 1980. On the nature of the oxidation–reduction properties of nitrite reductase from *Desulfovibrio desulfuricans*. Biochem. Biophys. Res. Commun. 96:278–285.

124. Lissolo, T., S. Pulvin, and D. Thomas. 1984. Reactivation of the hydrogenase from *Desulfovibrio gigas* by hydrogen. Influence of redox potential. J. Biol. Chem. 259:11725–11729.

124a. Lissolo, T., E. S. Choi, J. LeGall, and H. D. Peck, Jr. 1986. The presence of multiple intrinsic membrane nickel-containing hydrogenases in *Desulfovibrio vulgaris* (Hildenborough). Biochem. Biophys. Res. Commun. 139:701–708.

125. Lupton, F. S., R. Conrad, and J. G. Zeikus. 1984. Physiological function of hydrogen metabolism during growth of sulfidogenic bacteria on organic substrates. J. Bacteriol. 159:843–849.

126. Martin, S. M., B. R. Glick, and W. G. Martin. 1980. Factors affecting the production of hydrogenase by *Desulfovibrio desulfuricans*. Can. J. Microbiol. 26:1209–1213.

127. Mayhew, S. G., C. Van Dijk, and H. M. Van der Westen. 1978. Properties of hydrogenases from the anaerobic bacteria *Megasphaera elsdenii* and *Desulfovibrio vulgaris* (Hildenborough), p. 125–140, in H. G. Schegel and K. Schneider (eds.), Hydrogenases: their catalytic activity, structure, and function. Erich Göltze KG, Göttingen, West Germany.

128. Michaels, G. B., J. T. Davidson, and H. D. Peck, Jr. 1971. Studies on the mechanism of adenylyl sulfate reductase from the sulfate reducing bacterium *Desulfovibrio vulgaris*, p. 555–580, in H. Kamin (ed.), Flavins and flavoproteins. University Park Press, Baltimore.

129. Millet, J. 1956. Contribution à l'étude du métabolisme des bactéries réductrices de sulfate. Thesis, University of Paris.

130. Moura, I., M. Bruschi, J. LeGall, J. J. G. Moura, and A. V. Xavier. 1977. Isolation and characterization of desulforedoxin, a new type of non-heme iron protein from *D. gigas*. Biochem. Biophys. Res. Commun. 75:1037–1044.

131. Moura, I., J. J. G. Moura, M. H. Santos, A. V. Xavier, and J. LeGall. 1979. Redox studies on rubredoxins from sulphate and sulphur reducing bacteria. FEBS Lett. 107:419–421.

132. Moura, J. J. G., A. V. Xavier, E. C. Hatchikian, and J. LeGall. 1978. Structural control of the redox potentials and of the physiological activity by oligomerization of ferredoxin. FEBS Lett. 89:177–179.

133. Moura, J. J. G., I. Moura, M. Bruschi, J. LeGall, and A. V. Xavier. 1980. A cobalt-containing protein isolated from *D. gigas*, a sulfate reducer. Biochem. Biophys. Res. Commun. 92:962–970.

134. Moura, J. J. G., A. V. Xavier, M. Bruschi, J. LeGall, D. O. Hall, and R. Cammack.

1976. A molybdenum-containing iron–sulfur protein from *D. gigas*. Biochem. Biophys. Res. Commun. 72:782–789.

135. Moura, J. J. G., A. V. Xavier, R. Cammack, D. O. Hall, M. Bruschi, and J. LeGall. 1978. Oxidation reduction studies of the Mo-[2Fe–2S] protein from *D. gigas*. Biochem. J. 173:419–425.

136. Moura, J. J. G., I. Moura, H. Santos, A. V. Xavier, M. Scandellari, and J. LeGall. 1982. Isolation of P_{590} from *Methanosarcina barkeri*, evidence for the presence of sulfite reductase activity. Biochem. Biophys. Res. Commun. 108:1002–1009.

137. Moura, J. J. G., G. K. Moore, R. J. P. Williams, I. Probst, J. LeGall, and A. V. Xavier. 1984. Nuclear-magnetic-resonance studies of *Desulfuromonas acetoxidans* cytochrome $c_{551.5}(c_7)$. Eur. J. Biochem. 144:433–440.

138. Moura, J. J. G., A. V. Xavier, D. J. Cookson, G. R. Moore, R. J. P. Williams, M. Bruschi, and J. LeGall. 1977. Redox states of cytochrome c_3 in the absence and presence of ferredoxin. FEBS Lett. 81:275–280.

139. Moura, J. J. G., H. Santos, I. Moura, J. LeGall, G. R. Moore, R. J. P. Williams, and A. V. Xavier. 1982. NMR studies of *Desulfovibrio vulgaris* cytochrome c_3. Electron transfer mechanisms. Eur. J. Biochem. 127:151–155.

140. Moura, J. J. G., I. Moura, T. A. Kent, J. D. Lipscomb, B. H. Huynh, J. LeGall, A. V. Xavier, and E. Münck. 1982. Interconversion of [3Fe–3S] and [4Fe–4S] clusters: Mössbauer and electron paramagnetic resonance studies of *Desulfovibrio gigas* ferredoxin II. J. Biol. Chem. 257:6259–6267.

141. Moura, J. J. G., I. Moura, B. H. Huynh, H. J. Krüger, M. Teixeira, R. C. Duvarney, D. V. DerVartanian, A. V. Xavier, H. D. Peck, Jr., and J. LeGall. 1982. Unambiguous identification of the nickel EPR signal in [61]Ni-enriched *Desulfovibrio gigas* hydrogenase. Biochem. Biophys. Res. Commun. 108:1381–1383.

142. Murphy, M. J., and L. M. Siegel. 1973. Siroheme and sirohydrochlorin. The basis of a new type of porphyrin-related prosthetic group common to both assimilatory and dissimilatory sulfite reductases. J. Biol. Chem. 248:6911–6919.

143. Murphy, M. J., L. M. Siegel, H. Kamin, D. V. DerVartanian, J. P. Lee, J. LeGall, and H. D. Peck, Jr. 1973. An iron tetrahydroporphyrin prosthetic group common to both assimilatory and dissimilatory sulfite reductases. Biochem. Biophys. Res. Commun. 54:82–88.

144. Nakano, K., Y. Kikumoto, and T. Yagi. 1983. Amino-acid sequence of cytochrome c_{553} from *Desulfovibrio vulgaris* Miyazaki. J. Biol. Chem. 258:12409–12412.

145. Odom, J. M., and H. D. Peck, Jr. 1981. Localization of dehydrogenases, reductases, and electron transfer components in the sulfate-reducing bacterium *Desulfovibrio gigas*. J. Bacteriol. 147:161–169.

146. Odom, J. M., and H. D. Peck, Jr. 1981. Hydrogen cycling as a general mechanism for energy coupling in the sulfate reducing bacteria *Desulfovibrio* sp. FEMS Lett. 12:47–50.

147. Odom, J. M., and H. D. Peck, Jr. 1984. Hydrogenase, electron-transfer proteins, and energy coupling in the sulfate-reducing bacteria *Desulfovibrio*. Annu. Rev. Microbiol. 38:551–592.

148. Ogata, M., K. Arihara, and T. Yagi. 1981. D-Lactate dehydrogenase of *Desulfovibrio vulgaris*. J. Biochem. (Tokyo) 89:1423–1431.

148a. Pankhania, I. P., L. A. Gow, and W. A. Hamilton. 1986. Extraction of periplasmic hydrogenase from *Desulfovibrio vulgaris* (Hildenborough). FEMS Lett. 35:1–4.

149. Payne, W. J. 1981. The denitrifiers, p. 48–49, in Denitrification. Wiley, New York.

150. Peck, H. D., Jr. 1959. The ATP-dependent reduction of sulfate with hydrogen in extracts of *Desulfovibrio desulfuricans*. Proc. Natl. Acad. Sci. USA 45:701–708.

151. Peck, H. D., Jr., and J. M. Odom. 1981. Anaerobic fermentations of cellulose to methane, p. 375–395, in A. Hollaender (ed.), Trends in the biology of fermentations for fuels and chemicals. Plenum Press, New York.

152. Peck, H. D., Jr., and J. LeGall. 1982. Biochemistry of dissimilatory sulphate reduction. Philos. Trans. R. Soc. Lond. B298:443–466.

153. Peck, H. D., Jr., and R. N. Bramlett. 1982. Flavoproteins in sulfur metabolism, p. 851–858, in V. Massey and G. Williams (eds.), 7th international symposium on flavins and flavoproteins. University Park Press, Tokyo.

154. Peck, H. D., Jr., and J. M. Odom. 1984. Hydrogen cycling in *Desulfovibrio*: a new mechanism for energy coupling in anaerobic microorganisms, p. 215–243, in H. O. Halvorsen and Y. Cohen (eds.), Microbial mats: stomatolites. Alan R. Liss, New York.

154a. Peck, H. D., Jr., J. LeGall, P. A. Lespinat, Y. Berlier, and G. Fauque. 1987. A direct demonstration of hydrogen cycling by *Desulfovibrio vulgaris* Hildenborough employing membrane-inlet mass spectrometry. FEMS Lett. 40:295–299.

155. Petitdemange, H., R. Marczak, and R. Gay. 1979. NADH-flavodoxin oxidoreductase activity in *Clostridium tyrobutiricum*. FEMS Lett. 5:291–294.

156. Pfennig, N., and H. Biebl. 1976. *Desulfuromonas acetoxidans* gen. nov. and sp. nov., a new anaerobic, sulfur reducing, acetate oxidizing bacterium. Arch. Microbiol. 110:3–12.

157. Pfennig, N., and F. Widdel. 1981. Ecology and physiology of some anaerobic bacteria from the microbiological sulfur cycle, p. 169–177, in H. Bothe and A. Trebst (eds.), Biology of inorganic nitrogen and sulfur. Springer-Verlag, Heidelberg.

158. Pierrot, M., R. Haser, M. Frey, F. Payan, and J. P. Astier. 1982. Crystal structure and electron transfer properties of cytochrome c_3. J. Biol. Chem. 257:14341–14348.

159. Postgate, J. R. 1952. Growth of sulfate reducing bacteria in sulfate free media. Res. Lond. 5:189–190.

160. Postgate, J. R. 1954. Presence of cytochrome in an obligate anaerobe. Biochem. J. 56:xi.

161. Postgate, J. R. 1959. A diagnostic reaction of *Desulphovibrio desulphuricans*. Nature (Lond.) 183:481–482.

162. Postgate, J. R., and H. M. Kent. 1984. Derepression of nitrogen fixation in *Desulfovibrio gigas* and its stability to ammonia or oxygen stress in vivo. J. Gen. Microbiol. 130:2825–2831.

162a. Postgate, J. R., and H. M. Kent. 1985. Diazotrophy within *Desulfovibrio*. J. Gen. Microbiol. 131:2119–2122.

162b. Postgate, J. R., H. M. Kent, and R. L. Robson. 1986. DNA from diazotrophic *Desulfovibrio* strains is homologous to *Klebsiella pneumoniae* structural *nif* DNA and can be chromosomal or plasmid-borne. FEMS Lett. 33:159–163.

162c. Prickril, B. C., M. H. Czechowski, A. E. Przybyla, H. D. Peck, Jr., and J. LeGall.

1986. Putative signal peptide on the small subunit of the periplasmic hydrogenase from *Desulfovibrio vulgaris*. J. Bacteriol. 167:722–725.

163. Probst, I., M. Bruschi, N. Pfennig, and J. LeGall. 1977. Cytochrome $c_{551.5}$ (c_7) from *D. acetoxidans*. Biochim. Biophys. Acta 460:58–64.

163a. Riederer-Henderson, M. A., and H. D. Peck, Jr. 1986. Properties of formate dehydrogenase from *Desulfovibrio gigas*. Can. J. Microbiol. 32:430–435.

164. Robert, P. 1984. Le Petit Robert. 1. Dictionnaire alphabétique et analogique de la langue française, p. 1688. Société du Nouveau Littré, Paris.

165. Roy, A. B., and P. A. Trudinger. 1970. The biochemistry of inorganic compounds of sulfur. Cambridge University Press, Cambridge.

166. Rosanova, E. P., and A. I. Khudyakova. 1974. A new nonsporeforming thermophilic sulfate-reducing organism, *Desulfovibrio thermophilus* nov. sp. Microbiol. (Engl. Trans. Microbiologiya) 43:908–912.

167. Sankar, P., J. H. Lee, and K. T. Shanmugam. 1985. Cloning of hydrogenase genes and fine structure analysis of an operon essential for H_2 metabolism in *Escherichia coli*. J. Bacteriol. 162:353–360.

168. Santos, H., J. J. G. Moura, I. Moura, J. LeGall, and A. V. Xavier. 1984. NMR studies of electron transfer mechanisms in a protein with interacting redox centers: *Desulfovibrio gigas* cytochrome c_3. Eur. J. Biochem. 141:283–296.

169. Schinck, B., and H. G. Schlegel. 1980. The membrane-bound hydrogenase of *Alcaligenes eutrophus*. II. Localization and immunological comparison with other hydrogenase systems. Antonie Leeuwenhoek J. Microbiol. 46:1–14.

170. Scott, A. I., A. J. Irwin, L. M. Siegel, and J. N. Shoolery. 1978. Sirohydrochlorin, prosthetic group of a sulfite reductase enzyme and its role in the biosynthesis of vitamin B_{12}. J. Am. Chem. Soc. 100:316–318.

171. Scott, R. A., S. A. Wallin, M. Czechowski, D. V. DerVartanian, J. LeGall, H. D. Peck, Jr., and I. Moura. 1984. X-ray absorption spectroscopy of the nickel in the hydrogenase from *Desulfovibrio gigas*. J. Am. Chem. Soc. 106:6864–6865.

172. Seki, Y., and M. Ishimoto. 1979. Catalytic activity of the chromophore of desulfoviridin, sirohydrochlorin, in sulfite reduction in the presence of iron. J. Biochem (Tokyo) 86:273–276.

173. Seki, Y., N. Sogawa, and M. Ishimoto. 1981. Siroheme as an active catalyst in sulfite reduction. J. Biochem. (Tokyo) 90:1478–1492.

174. Stams, A. J. M., and T. A. Hansen. 1982. Oxygen-labile L(+) lactate dehydrogenase activity in *Desulfovibrio desulfuricans*. FEMS Lett. 13:389–394.

175. Steenkamp, D. J., and H. D. Peck, Jr. 1981. Proton translocation associated with nitrite respiration in *Desulfovibrio desulfuricans*. J. Biol. Chem. 256:5450–5458.

176. Stille, W., and H. G. Trüper. 1984. Adenylsulfate reductase in some new sulfate-reducing bacteria. Arch. Microbiol. 137:145–150.

177. Stolzenberg, A. M., L. O. Spreer, and R. H. Holm. 1980. Octaethylisobacteriochlorin, a model for the sirohydrocholorin ring system of siroheme enzymes: preparation and spectroscopic properties of the free base and its Zn(II) complex. J. Am. Chem. Soc. 102:364–370.

178. Suh, B., and J. M. Akagi. 1966. Pyruvate carbon dioxide exchange reaction of *Desulfovibrio desulfuricans*. J. Bacteriol. 91:2281–2285.

179. Tanaka, M., T. Nakashima, A. Benson, H. Mower, and K. Yasunobu. 1966. The amino acid sequence of *Clostridium pasteurianum* ferredoxin. Biochemistry 5:1666–1681.

180. Teixeira, M., I. Moura, A. V. Xavier, J. J. G. Moura, G. Fauque, B. Prickril, and J. LeGall. 1985. Nickel–iron–sulfur–selenium containing hydrogenases isolated from *Desulfovibrio baculatus* strain 9974. Rev. Port. Quim. 27:194–195.

181. Teixeira, M., I. Moura, A. V. Xavier, D. V. DerVartanian, J. LeGall, H. D. Peck, Jr., B. H. Huynh, and J. J. G. Moura. 1983. *Desulfovibrio gigas* hydrogenase-redox properties of the nickel and iron–sulfur centers. Eur. J. Biochem. 130:481–484.

182. Teixeira, M., I. Moura, A. V. Xavier, B. H. Huynh, D. V. DerVartanian, H. D. Peck, Jr., J. LeGall, and J. J. G. Moura. 1985. Electron paramagnetic resonance study on the mechanism of activation and the catalytic cycle of the nickel-containing hydrogenase from *Desulfovibrio gigas*. J. Biol. Chem. 260:8942–8950.

182a. Teixeira, M., I. Moura, G. Fauque, M. Czechowski, Y. Berlier, P. A. Lespinat, J. LeGall, A. V. Xavier, and J. J. G. Moura. 1986. Redox properties and activity studies on a nickel-containing hydrogenase isolated from a halophilic sulfate reducer *Desulfovibrio salexigens*. Biochimie (Paris) 68:75–84.

182b. Teixeira, M., G. Fauque, I. Moura, P. A. Lespinat, Y. Berlier, B. Prickril, H. D. Peck, Jr., A. V. Xavier, J. LeGall, and J. J. G. Moura. 1987. Nickel–iron–sulfur-selenium-containing hydrogenases from *Desulfovibrio baculatus* (DSM 1743). Redox centers and catalytic properties. Eur. J. Biochem. 167:47–58.

183. Tel-Or, E., L. W. Luijk, and L. Packer. 1978. Hydrogenase in N_2-fixing cyanobacteria. Arch. Biochem. Biophys. 185:185–194.

184. Thomson, A. J., A. E. Robinson, M. K. Johnson, J. J. G. Moura, I. Moura, A. V. Xavier, and J. LeGall. 1981. The three-iron cluster in a ferredoxin from *Desulfovibrio gigas*. A low temperature magnetic circular dichroism study. Biochim. Biophys. Acta 670:93–100.

185. Traore, A. S., C. E. Hatchikian, J. P. Belaich, and J. LeGall. 1981. Microcalorimetric studies of the growth of sulfate-reducing bacteria: energetics of *Desulfovibrio vulgaris* growth. J. Bacteriol. 145:191–199.

186. Traore, A. S., E. C. Hatchikian, J. LeGall, and J. P. Belaich. 1982. Microcalorimetric studies of the growth of sulfate-reducing bacteria: comparison of the growth parameters of some *Desulfovibrio* species. J. Bacteriol. 149:606–611.

187. Travis, J., D. J. Newman, J. LeGall, and H. D. Peck, Jr. 1971. The amino acid sequence of ferredoxin from the sulfate reducing bacterium *Desulfovibrio gigas*. Biochem. Biophys. Res. Commun. 45:452–458.

188. Trudinger, P. A. 1970. Carbon monoxide-reacting pigment from *Desulfotomaculum nigrificans* and its possible relevance to sulfite reduction. J. Bacteriol. 104:158–170.

189. Trumpower, B. L. 1981. New concepts on the role of ubiquinone in the mitochondrial respiratory chain. J. Bioenerg. Biomembr. 13:1–24.

190. Tsuji, K., and T. Yagi. 1980. Significance of hydrogen burst from growing cultures of *Desulfovibrio vulgaris* Miyazaki and the role of hydrogenase and cytochrome c_3 in energy producing system. Arch. Microbiol. 125:35–42.

190a. Turner, N., B. Barata, R. C. Bray, J. Deistung, J. LeGall, and J. J. G. Moura. 1987. The molybdenum iron-sulphur protein from *Desulfovibrio gigas* as a form of aldehyde oxidase. Biochem. J. 243:755–761.

191. Uyeda, K., and J. C. Rabinowitz. 1971. Pyruvate ferredoxin oxidoreductase. III. Purification and properties of the enzyme. J. Biol. Chem. 246:3111–3119.

192. Van der Westen, H., S. G. Mayhew, and C. Veeger. 1978. Separation of hydrogenase from intact cells of *Desulfovibrio vulgaris*. Purification and properties. FEBS Lett. 86:122–126.

193. Van Dijk, C., H. J. Grande, S. G. Mayhew, and C. Veeger. 1979. Purification and properties of hydrogenase form *Megasphaera elsdenii*. Eur. J. Biochem. 102:317–330.

194. Vega, J. M., and H. Kamin. 1977. Spinach nitrite reductase. Purification and properties of a siroheme-containing iron–sulfur enzyme. J. Biol. Chem. 252:896–909.

195. Vega, J. M., R. H. Garrett, and L. M. Siegel. 1975. Siroheme: a prosthetic group of the *Neurospora crassa* assimilatory nitrite reductase. J. Biol. Chem. 250:7980–7989.

196. Vogel, H., M. Bruschi, and J. LeGall. 1977. Phylogenetic studies of two rubredoxins from sulfate reducing bacteria. J. Mol. Evol. 9:111–119.

197. Voordouw, G., and S. Brenner. 1985. Nucleotide sequence of the gene encoding the hydrogenase from *Desulfovibrio vulgaris* (Hildenborough). Eur. J. Biochem. 148:515–520.

198. Voordouw, G., J. E. Walker, and S. Brenner. 1985. Cloning of the gene encoding of the hydrogenase from *Desulfovibrio vulgaris* (Hildenborough) and determination of the NH$_2$-terminal sequence. Eur. J. Biochem. 148:509–514.

199. Watenpaugh, K. D., L. C. Sieker, L. H. Jensen, J. LeGall, and M. Dubourdieu. 1972. Structure of the oxidized form of a flavodoxin at 2.5 Å resolution. Proc. Natl. Acad. Sci. USA 69:3185–3188.

200. Widdel, F. 1980. Anaerober Abbau von Fettsäuren und Benzoesäure durch neu isolierte Arten Sulfat-reduzierender Bakterien. Doctoral thesis, University of Göttingen.

201. Widger, W. R., W. A. Cramer, R. G. Herrmann, and A. Trebst. 1984. Sequence homology and structural similarity between cytochrome *b* of mitochondrial complex III and the chloroplast b_6-f complex: position of the cytochrome *b* hemes in the membrane. Proc. Natl. Acad. Sci. USA 81:674–678.

202. Wood, P. M. 1978. A chemiosmotic model for sulfate respiration. FEBS Lett. 95:12–18.

203. Xavier, A. V., J. J. G. Moura, J. LeGall, and D. V. DerVartanian. 1979. Oxidation reduction potentials of the hemes in cytochrome c_3 from *D. gigas* in the presence and absence of ferredoxin by EPR spectroscopy. Biochimie (Paris) 61:689–695.

204. Yagi, T. 1979. Purification and properties of cytochrome c_{553}, an electron acceptor for formate dehydrogenase of *Desulfovibrio vulgaris* Miyazaki. Biochim. Biophys. Acta 548:96–105.

205. Yagi, T., M. Honya, and N. Tamiya. 1968. Purification and properties of hydrogenases of different origins. Biochim. Biophys. Acta 153:699–705.

206. Yagi, T., A. Endo, and K. Tusji. 1978. Properties of hydrogenase from particulate fraction of *Desufovibrio vulgaris*, p. 107–124, in H. G. Schlegel and K. Schneider, (eds.), Hydrogenases: their catalytic activity, structure, and function. Erich Göltze KG, Göttingen.

206a. Yagi, T., K. Kimura, and H. Inokuchi. 1985. Analysis of the active center of hydrogenase from *Desulfovibrio vulgaris* Miyazaki by magnetic measurements. J. Biochem. 97:181–187.

207. Yates, M. G. 1976. Stimulation of the phosphoroclastic system of *Desulfovibrio* by nucleotide triphosphates. Biochem. J. 103:326–340.

12

BIOGEOCHEMISTRY OF METHANOGENIC BACTERIA

RONALD S. OREMLAND

U.S. Geological Survey, 345 Middlefield Road, Menlo Park, CA 94025

12.1 INTRODUCTION

12.2 MICROBIOLOGICAL ASPECTS

    12.2.1 Historical Perspective

    12.2.2 Taxonomy and Characteristics

    12.2.3 Substrates

    12.2.4 Nutrients and Growth Requirements

    12.2.5 Inhibitors

12.3 ECOLOGICAL ASPECTS

    12.3.1 Methanogenic Interactions

    12.3.2 Competitive and Noncompetitive Substrates

    12.3.3 Natural Habitats

        12.3.3.1 Within Living Organisms

        12.3.3.2 Aquatic Environments

        12.3.3.3 Extreme Enviroments

12.4 GEOCHEMICAL ASPECTS

    12.4.1 Geolipids of Methanogens

    12.4.2 Classification of Natural Gases

    12.4.3 Production of Higher Molecular Weight (C_2^+) Gaseous Alkanes

    12.4.4 Stable Carbon Isotope Fractionation

    12.4.5 Anaerobic Methane Oxidation

12.4.6   Depth of Sedimentary Penetration by Active Methanogens
12.4.7   Atmospheric Methane
12.4   SUMMARY
REFERENCES

12.1   INTRODUCTION

Over the past 15 years, methanogenic bacteria have received increased scrutiny from the scientific community. The discoveries of their unique biochemical and genetic properties ultimately led to both the concept of "Archaebacteria" and to their prominent classification within that kingdom. However, important economic factors have also placed these organisms in the anaerobic "limelight." The cattle industry, for example, has an interest in rumen fermentation because the methane produced therein represents a net loss of energy which might otherwise be channeled into higher yields. Oil companies have a stake in being able to distinguish between natural gas formed by methanogenic bacteria as opposed to that formed by the thermocatalytic reactions associated with petroleum generation. In addition, both the unreliability of the supply of petroleum and its quadrupled price since 1973 has led to serious interest in alternative forms of energy, including recovery of methane via anaerobic biomass digestion of waste materials. Improved digestor designs, operations, and their utilization to supplement energy supplies have been made possible, in part, by advances in understanding the physiology and ecology of methanogens. Finally, concerns about the complex sequence of chemical reactions involving methane in the earth's atmosphere and its possible role(s) in reversing ozone depletion, and enhancement of the "greenhouse" effect, have prompted studies on the global distribution of this gas. Subsequent studies became concerned with the biogeochemistry of methane formation and the global estimates of its release rates from the biosphere to the atmosphere.

Several major reviews have been published on the methanogenic bacteria. These include the early works by Barker (1956), Stadtman (1967), McCarty (1964), and Wolfe (1971). More recent contributions have been made by Mah et al. (1977), Zeikus (1977), Balch et al. (1979), Daniels et al. (1984), Kirsop (1984), Winfrey (1984), and Whitman (1985). In addition, there are several articles concerned with aspects of methanogenesis or with the geochemistry of methane. These publications are cited in the text. Therefore, much has already been written concerning the biology, physiology, and ecology of methanogenic bacteria, and a constant stream of publications steadily updates these topics. To avoid redundancy, I have decided to write this chapter from a different perspective—that of biogeochemistry. In taking this approach I have attempted to cover general aspects of microbiology and ecology which are also relevant to the topic of methane geochemistry. It is my goal, therefore, to give both microbiologists and geochemists an appreciation for the continuity between these disciplines with respect to the topic of methanogenesis.

12.2 MICROBIOLOGICAL ASPECTS

12.2.1 Historical Perspective

Summaries of the early research work on methanogenesis have been written by Barker (1956) and Zehnder et al. (1982). The first description of methane emanating from aquatic muds is credited to Volta in 1776. However, it was not until almost a century later that Béchamp demonstrated a microbial origin for the gas, using an ethanol-based media inoculated with rabbit feces. Several microbiologists, including Popoff, Hoppe-Seyler, and Omelianskii pursued studies on methanogenesis in the latter nineteenth and early twentieth centuries. These workers were able to establish that methanogenesis was apparently linked to the anaerobic breakdown of polymeric materials, such as cellulose. However, it was not particularly clear whether the bacteria responsible for the evolution of methane were capable of directly attacking cellulosic compounds, or were involved in the consumption of simpler molecules (such as ethanol and acetate) that were liberated during the degradation of polymeric compounds by other microorganisms. Söhngen (1906) was able to expand on Béchamp's earlier observations. He demonstrated that a variety of simple organic molecules (formate, acetate, butyrate, ethanol) as well as the combination of hydrogen and carbon dioxide could be converted into methane and that these substances arose from the anaerobic decomposition of cellulose. Although these early researchers made microscopic observations of what are today recognized as various species of methanogens, no cultures were obtained. The necessary techniques simply did not exist at that time.

The story picks up again in the 1930s with the pioneering work of Barker. In 1936 he reported isolating the first ''pure''culture of a methanogen, namely ''*Methanobacillus omelianskii*.'' The anerobic techniques that he employed were crude by today's standards, yet he and his students were able to perform the first biochemical studies on methanogenic bacteria. Working with ''*M. omelianskii*,'' Stadtman and Barker (1949) reported that the organism oxidized ethanol to acetate. The electrons generated by this oxidation were utilized to reduce the bicarbonate ions present in the media to methane. ''*M. omelianskii*'' became the most studied organism but was subsequently shown to be impure (see below). The first pure cultures of methanogens were isolated by Schnellen (1947). Experiments by workers in the late 1940s and throughout the 1950s were able to show that the methyl groups of acetate and methanol were converted to methane by the available cultures.

The development of laboratory procedures for culturing strict anaerobes (Hungate, 1969) was an important technical advance and provided a great impetus for further studies on methanogenesis. Use of Hungate's ''roll-tube'' technique allowed workers to isolate several new species of methanogens in pure culture. Although various improvements and modifications have been made to the basic procedure (e.g., Bryant, 1972; Balch and Wolfe, 1976), including use of anaerobic glove boxes, it is still in widespread use. By employing this technique, Bryant et al. (1967) revealed that "*M. omelianskii*" really consisted of an association of two different microorganisms: the nonmethanogenic "S" organism, which oxidized ethanol to acetate plus hydrogen, while the other member of the association, *Methanobacterium bryantii* (formerly called *Methanobacterium* strain MoH), reduced the bicarbonate present in the medium with the hydrogen evolved by the S organism. Thermodynamically, the reactions may be described as follows:

Microbe	Reaction	$\Delta G^{\circ\prime}$ (kJ)
S organism	$2C_2H_5OH + 2H_2O \longrightarrow 2CH_3COO^- + 4H_2 + 2H^+$	$+19.2$
M. bryantii	$4H_2 + HCO_3^- + H^+ \longrightarrow CH_4 + 3H_2O$	-135.6
Sum	$2C_2H_5OH + HCO_3^- \longrightarrow 2CH_3COO^- + H^+ + CH_4 + H_2O$	-116.4

Removal of hydrogen by *M. bryantii* kept the partial pressure of the gas below 10^{-3} atm. This made the first reaction thermodynamically favorable and allowed for growth of the S organism. The resolution of the culture of "*M. omelianskii*," together with the demonstration of the importance of hydrogen as an intermediate in the rumen (Hungate, 1967), helped give rise to the concept of "interspecies hydrogen transfer" (Ianotti et al., 1973). This concept has provided the foundation for much of the progress on the ecology of methanogenic interactions made over the past decade.

The subsequent biochemical characterization of the unique pathways and cofactors involved in methanogenesis (e.g., see Wolfe, 1985) raised the question of their phylogenetic relations to other bacteria, as well as their role in the evolution of life on earth. By examining patterns of 16S ribosomal RNA oligonucleotide sequences, Fox et al. (1977) found that methanogenic bacteria were phylogenetically distinct from other prokaryotes (Eubacteria) and eukaryotic cells. There was also great internal genetic diversity within the methanogens themselves. These results suggested that an ancient divergence occurred in the primitive biotic earth which ultimately gave rise to ancestral lines of the kingdoms of Eukaryota, Eubacteria, and Archaebacteria (Balch et al., 1977a).

12.2.2 Taxonomy and Characteristics

Bryant (1974) made the first formal taxonomic treatment of methanogens, based on Barker's 1956 outline. Bryant used morphological characteristics and substrate affinites to arrange the family "Methanobacteriaceae" into three genera (*Methanobacterium, Methanosarcina,* and *Methanococcus*). Nine species of methanogens were described in this work, of which there were only seven cultures available. Subsequently, Woese and Fox (1977) proposed that methanogens (along with thermoacidophiles and extreme halophiles) be included as representative members of the separate, novel kingdom of "Archaebacteria." On the basis of their 16S ribosomal RNA oligonucleotide sequence catalogs, archaebacteria were shown to be of sufficent phylogenetic difference from other prokaryotes ("eubacteria") to warrant the creation of a new kingdom. By using these oligonucleotide sequence patterns, as well as biochemical and morphological characteristics, Balch et al. (1979) proposed a major revision of the taxonomy of methanogens. This work was based on the characteristics of the 13 species (17 strains) available at the time. Methanogens were grouped into three orders (Methanobacteriales, Methanomicrobiales, and Methanococcales) comprised of four families and seven genera. This proposed revision was substantiated by independent immunological investigations (Conway de Macario et al., 1981). At the time of this writing (1984), the number of formal literature reports of methanogenic species in pure culture has risen to 29 and there are preliminary reports of several new isolates. Two new families (Methanoplanaceae and Methanothermaceae) and five new genera have been added since the 1979 revision (Table 12.1).

Morphologically, the methanogens exhibit a wide variety of shapes and sizes, including various rods, cocci (regular and irregular), long-chained spirilla, sarcina, and even an unusual flattened plate seen in *Methanoplanus limicola* (Fig. 12.1). Both motile and nonmotile species are evident. *Methanosarcina barkeri* and *Methanococcus mazei* grow in clusters of "packets" and "aggregates," respectively. Species of *Methanosarcina* can contain gas vacuoles (Zhilina, 1971; Archer and King, 1983). In addition, *Methanoplasma elizabethii* (Rose and Pirt, 1981) is an unusual species that lacks a cell wall. This organism's taxonomic location has not yet been established. For more information on classification and properties of methanogens, see Chapter 13.

Various genera of methanogens exhibit both gram-positive (e.g., *Methanobrevibacter* and *Methanobacterium*) and gram-negative reactions (e.g., *Methanogenium* and *Methanococcus*). The chemical makeup of the cell walls of methanogens differs greatly from those of the eubacteria, and there is considerable diversity among the methanogens themselves in the makeup of their cell envelopes. Methanogens do not contain murein, the peptidoglycan of eubacteria which contains muramic acid (Kandler and Hippe, 1977). Instead, the order Methanobacteriales contains "pseudomurein," a substance in which muramic acid is replaced by N-acetyl L-talosaminouronic acid. Different amino acids, which exhibit only the L-form, are present in the peptide side chains of pseudomurein as opposed to murein (Kandler and Konig, 1978; Konig and Kandler, 1979; Konig et al., 1982). Cell walls of the Methanococcales and several genera of the Methanomicrobiales (*Methanogenium, Methanospirillum*, and *Methanomicrobium*) consits of protein subunits (Kandler and Konig, 1978; Jones et al., 1977), while the *Methnosarcina* contain a nonsulfonated heteropolysaccharide (Kandler and Hippe, 1977). Methanogens are therefore insensitive to the antibiotics that block murein or peptidoglycan synthesis in eubacteria, such as penicillin, cycloserine, and valinomycin (Hilpert et al., 1981). Isolation of new species of methanogens has been

TABLE 12.1 Classification scheme of current isolates of methanogic bacteria[a]

Order: Methanobacteriales
  Family: Methanobacteriaceae
    Genus: *Methanobacterium*
      Species: *M. formicicum*
             M. bryantii
             M. thermoautotrophicum
  Family: Methanothermacea
    Genus: *Methanothermus*
      Species: *M. fervidus* (Stetter et al., 1981)

Order: Methanococcales
  Family: Methanococcaceae
    Genus: Methanococcus
      Species: *Mcs. vannielii*
          Mcs. voltae
          Mcs. thermolithotrophicus (Huber et al., 1982)
          Mcs. maripaludis (Jones et al., 1983b)
          Mcs. jannaschii (Jones et al., 1984)
          Mcs. halophilus (Zhilina, 1983)
          Mcs. deltae (Corder et al., 1983)
          Mcs. mazei (Mah, 1980)

Order: Methanomicrobiales
  Family: Methanomicrobiaceae
    Genus: *Methanomicrobium*
      Species: *M. mobile*
              M. paynteri (Rivard et al., 1983)
    Genus: *Methanogenium*
      Species: *M. cariaci*
              M. marisnigri
              M. olentangyi (Corder et al., 1983)
              M. tatii (Zabel et al., 1984)
              M. thermophilum (Rivard and Smith, 1982)
    Genus: *Methanococcoides*
      Species: *M. methylmutens* (Sowers and Ferry, 1983)

  Family: Methanoplanaceae
    Genus: *Methanoplanus*
      Species: *M. limicola* (Wildgruber et al., 1982)

  Family: Methanosarcinaceae
    Genus: *Methanosarcina*
      Species: *Ms. barkeri*
          Ms. TM-1 (Zinder and Mah, 1979)
          Ms. acetivorans (Sowers et al., 1984)
    Genus: *Methanothrix*
      Species: *M. soehngenii* (Huser et al., 1982)
    Genus: *Methanolobus*
      Species: *M. tindarius* (Konig and Stetter, 1982)

Source: Based on Balch et al. (1979).

[a]Recent isolates are listed with appropriate reference.

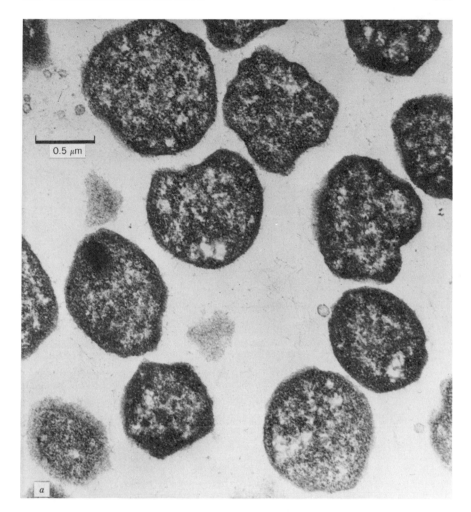

Figure 12.1 Thin-section electron micrographs and phase-contrast photomicrographs of some novel species of methanogens. (*A*) *Methanococcoides methylmutens* (courtesy of J. G. Ferry). (*B*) *Methanoplanus limicola*; magnification 75,000× (courtesy of K. O. Stetter). (*C*) *Methanothermus fervidus*; magnification 60,000 (courtesy of K. O. Stetter). (*D*) Phase-contrast micrograph of filaments and bundles of *Methanothrix soehngenii*. Bar indicates 45 μm (courtesy of A. J. B. Zehnder).

greatly facilitated by using these anti-cell-wall antibiotics to cleanse cultures of contaminating heterotrophic eubacteria. In addition, methanogens can be readily identified under the microscope (or on agar plates) by exposing them to ultraviolet light. Individual cells autofluoresce a blue-green color under these conditions, owing to their content of $Factor_{420}$, a low-potential electron carrier found only in methanogens (Cheeseman et al., 1972; Edwards and McBride, 1975; Mink and Dugan, 1977; Doddema and Vogels, 1978).

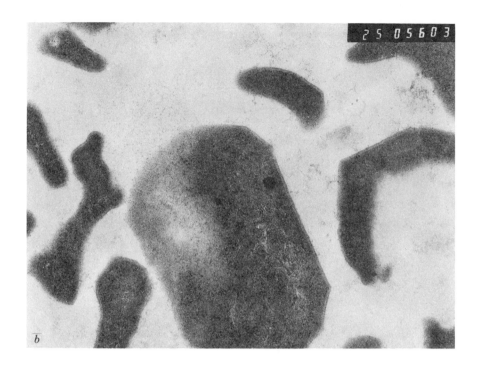

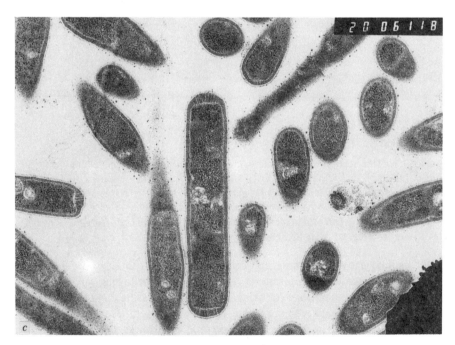

Figure 12.1 (*Continued*)

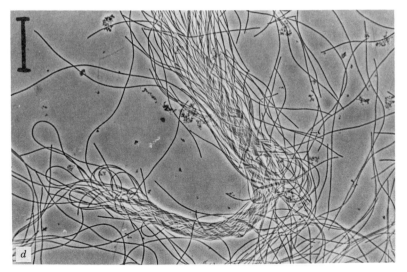

Figure 12.1 (*Continued*)

The cell envelope lipids of methanogens and other archaebacteria are comprised of an unusual group of both polar and neutral forms which account for 2 to 5% of cellular dry weight (see review by Langworthy et al., 1982). The membrane glyco- and phospholipids of eubacteria and eukaryotes consist primarily of fatty acid esters of glycerol. However, archaebacterial polar lipids contain isoprenoid (branched) hydrocarbon chains which are attached to the glycerol molecules by ether linkages (DeRosa et al., 1977). In methanogens, this isoprenoid chain consists of different arrays of the hydrocarbon phytane. Phytane is incorporated into the glycerol moiety as either diphytanylglycerol diethers or in "head-to-head" bondings (separated by two unbranched carbons) of two phytyl groups to form dibiphytanyldiglycerol tetraethers (Fig. 12.2). Recently, a novel macrocylcic diphytanyl glycerol diether was found to account for 95% of the polar lipids of *Methanococcus jannaschii* (Comita and Gagosian, 1983). However, unlike certain thermoacidophilic archaebacteria (e.g., Sulfolobus), methanogens do not contain cyclopentane rings in their isoprenoid side chains. The relative content of di- and tetraethers varies with respect to species. Organisms of the genus *Methanococcus* contain almost exclusively the diether, while other genera are characterized by roughly equivalent amounts of the two forms. *Methanobacterium* strain AZ has the highest tetraether (62.4%) and lowest diether (37.5%) content. Polar lipids make up the dominant fraction (80 to 90%) of the cell-membrane lipid bilayers of methanogens.

Acyclic isoprenoid hydrocarbons also characterize the neutral lipids of methanogens (Tornabene et al., 1978, 1979.) These principally consist of C_{20} and C_{30} isoprenoids (Fig. 12.2), although C_{15}, C_{18}, and C_{19} are also present. The isopren-

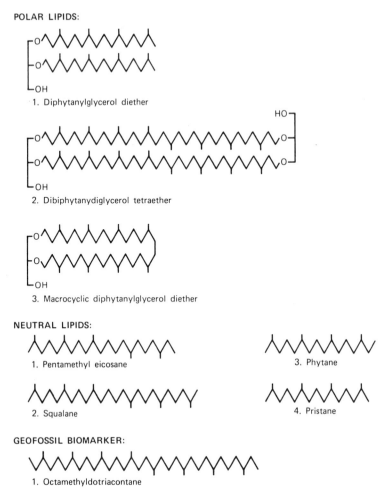

POLAR LIPIDS:

1. Diphytanylglycerol diether

2. Dibiphytanydiglycerol tetraether

3. Macrocyclic diphytanylglycerol diether

NEUTRAL LIPIDS:

1. Pentamethyl eicosane

2. Squalane

3. Phytane

4. Pristane

GEOFOSSIL BIOMARKER:

1. Octamethyldotriacontane

Figure 12.2 Some examples of polar and neutral cell wall lipids and geofossil biomarkers of methanogenic bacteria.

oids are either in the form of alkanes (e.g., pentamethyleicosane) or as alkenes with varying numbers of unsaturated bonds. Squalenes (C_{30}) are the most abundant neutral lipids of methanogens and are made up of "tail-to-tail" (separated by four unbranched carbons) linkages of farnesyl (C_{15}) derivatives. The function of squalene is as yet uncertain; however, it may represent the evolutionary precursor of the oxygenated sterols present in eubacteria and eukaryotes. The alkane squalane ($C_{30}H_{62}$) has been found only in *Methanobacterium thermoautotrophicum* (Holzer and Oró, 1979). Isoprenoids containing 25 or less carbon atoms consist of head-to-tail linkages (methylated carbon separated by three unbranched carbons). The alkanes pristane ($C_{19}H_{40}$) and phytane ($C_{20}H_{42}$) occur in some methanogens, the former found in *Methanococcus vannielii* and *Methanobrevibacter* strain PS, while the latter was detected in *Methanobacterium bryantii* and in *Methanobacterium* strain AZ (Holzer and Oró, 1979). Because the lipids of methanogenic bacteria have important geochemical ramifications, they will also be discussed in Section 12.4.

12.2.3 Substrates

Compounds providing sources of energy and carbon for the growth of methanogenic bacteria are simple and of limited scope. Currently, recognized substrates include H_2 plus CO_2, formate, acetate, methanol, and methylated amines (Table 12.2). Carbon monoxide can also be used by some species as a substrate (Daniels et al., 1977); however, it supports only poor growth and is not thought to be important in nature. Methylmercaptan has been reported as a methane precursor in aquatic sediments (Zinder and Brock, 1978), but pure cultures examined in that study were incapable of forming methane from that compound. Recently, conversion of methylated reduced sulfur compounds (dimethylsulfide, methylmercaptan, dimethyldisulfide) to methane was shown to be carried out by methanogenic bacteria present in sediments and a pure culture was isolated which grew on dimethylsulfide (Kiene et al., 1986).

Until quite recently it was believed that all methanogens shared the common ability to form methane via reduction of carbon dioxide with hydrogen (Bryant, 1979). This concept must now be modified to state the most species of methanogens held in culture are capable of hydrogen utilization. Two types of nonhydrogen utilizers have been described thus far. One type is "acetotrophic" and includes *Methanothrix soehngenii*, Methanosarcina TM-1, and the recently isolated *Methanosarcina acetivorans*. The second group (obligate methylotrophs) metabolizes only methanol, methylated amines, or dimethylsulfide and includes the species *Methanolobus tindarus*, *Methanococcoides methylmutens*, and *Methanococcus halophilus* (see Table 12.1 for the appropriate references). In addition, there is a preliminary report of an alkalophilic methylotroph that appears incapable of utilizing H_2 (Oremland et al., 1982a). Apparently, ecological niches exist in nature, such as in sulfate-rich environments, where methanogens may have lost their ability to consume hydrogen.

TABLE 12.2 Energy yields by important methanogenic substrates

Reaction	ΔG (kJ/mol)
1. $4H_2 + CO_2 \rightarrow CH_4 + 2H_2O$	-139
2. $4HCOO^- + 2H^+ \rightarrow CH_4 + CO_2 + 2HCO_3^-$	-127
3. $4CH_3OH \rightarrow 3CH_4 + CO_2 + 2H_2O$	-103
4. $4CH_3NH_2 + 2H_2O + 4H^+ \rightarrow 3CH_4 + CO_2 + NH_4^+$	-102
5. $CH_3COO^- + H_2O \rightarrow CH_4 + HCO_3^-$	-28
6. $(CH_3)_2S + H_2O \rightarrow 1.5CH_4 + 0.5CO_2 + H_2S$	-74

Source: Modified from Zehnder et al., 1982 and Kiene et al., 1986.

Several species of methanogens, such as *Methanobacterium thermoautotrophicum*, *M. bryantii*, and *Methanobrevibacter arboriphilus*, grow only on hydrogen plus carbon dioxide, thus giving them the characteristics of autotrophs. *M. thermoautotrophicum* is an obligate autotroph because its growth is not stimulated by inclusion of growth factors or nutrient supplements (e.g., vitamins, fatty acids, coenzyme M, yeast extract, etc.) in the media. Other species may be regarded as facultative autotrophs either because they respond to such nutrient supplements, or because they metabolize other methanogenic substrates such as formate (e.g., *Methanococcus thermolithotrophicus*). Hydrogen provides only a source of electrons for the reduction of carbon dioxide to methane, while protons are derived from water molecules (Daniels, et al., 1980). Membrane-associated hydrogenase activity creates a proton gradient by which ATP can be formed by chemiosmotic mechanisms (Doddema et al., 1979). The reported K_m values for H_2 by whole cells of methanogens are usually low (about 10^{-6} M) but range from 1 to 80 × 10^{-6} M. Hydrogen metabolism is associated with F_{420}, which all methanogens appear to contain.

Formate supports growth of about half the known species of methanogens. The compound provides about the same amount of energy when converted to methane as does carbon dioxide reduction by hydrogen (Table 12.2) and will provide for excellent growth in culture provided that the pH is regulated. The formate hydrogenlyase system of methanogens contains F_{420} and converts formate to hydrogen plus carbon dioxide. However, hydrogen does not accumulate to significant concentrations (under physiological conditions) during growth of *Methanobacterium formicicum* on formate (Schauer and Ferry, 1980). Cells can metabolize formate and hydrogen simultaneously (provided that they are at high enough concentrations) and therefore do not display diauxic growth patterns. However, when ambient formate levels are in excess of the apparent K_m for formate (e.g., 0.58 mM for *M. formicicum*), metabolism of low concentrations of hydrogen (below 6 μM) is strongly repressed (Schauer et al., 1982). The lowest concentrations of formate able to be utilized by *M. formicicum* and *Methanospirillum hungateii* were reported by these authors to be 26 and 15 μM, respectively.

Acetate is an important precursor of methane in nature and it has been estimated to give rise to about 70% of the methane formed in digested sludge and freshwater lake sediments (Jeris and McCarty 1965; Cappenberg, 1974b). However, the energetics of acetate decarboxylation to methane and carbon dioxide are quite poor (Table 12.2). Thus it is not clear how acetoclastic methanogens can generate sufficient ATP to sustain growth upon acetate. Zehnder and Brock (1979a) hypothesized that a chemiosmotic mechanism might be involved. At least five species of methanogens are capable of growth upon acetate as the sole carbon and energy source: *Methanosarcina barkeri*, Methanosarcina TM-1, *Methanothrix soehngenii*, *Methanosarcina acetivorans*, and *Methanococcus mazei* (Stadtman and Barker, 1951; Zinder and Mah, 1979; Zehnder et al., 1980; Touzel and Albagnac, 1983; Sowers et al., 1984). Doubling times for growth on acetate are usually long and range from 12 h for the thermophilic *Methanosarcina TM-1* (55°C) to as much

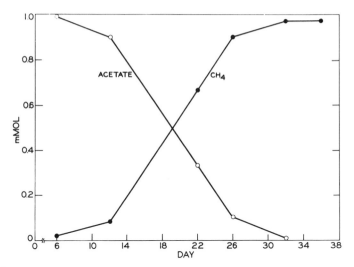

Figure 12.3 Growth of *Methanothrix soehngenii* on acetate. From Zehnder et al. (1980), reproduced with permission from the Archives of Microbiology.

as 9 days for *Methanothrix soehngenii* (Fig. 12.3). Apparent K_m values for acetate are also high, with reported values ranging from 0.46 m*M* (Mts. soehngenii) to 5 m*M* for *Methanosarcina* strain 227 (Smith and Mah, 1980).

Methanosarcina barkeri and *Methanococcus mazei* are metabolically the most versatile of the methanogens and, with the exception of formate, are capable of growth on any of the recognized methanogenic substrates. *M. barkeri* will preferentially metabolize hydrogen plus carbon dioxide, methanol, or trimethylamine over acetate, and therefore can exhibit diauxic growth patterns when given such a choice (Smith and Mah, 1978; Mah et al., 1978; Mah, 1979; Blaut and Gottshalk, 1982; Ferguson and Mah, 1983). However, interpretation of diauxic growth can be complicated by pH changes of the medium or by inhibition of dissolved substrate metabolism by hydrogen when carbon dioxide is not present (Hutten et al., 1980). Both methanol and hydrogen plus carbon dioxide can therefore be metabolized simultaneously. However, methanol is consumed more rapidly.

Recently, methylated amines have been identified as substrates for methanogenesis (Hippe et al., 1979). Metabolism of methylated amines to methane also results in the production of ammonia, which is then available as a nitrogen source to methanogens and other bacteria (Patterson and Herspell, 1979). Initially, only species of *Methanosarcina* were identified as having the ability to metabolize methylated amines. However, as mentioned previously, several other newly isolated species have been described which grow only upon methylated amines or methanol and may be considered as obligately methylotrophic methanogens. The significance of these compounds as precursors of methane in natural environments is a current area of investigation (see Section 12.3).

12.2.4 Nutrients and Growth Requirements

Methanogens require a very low reducing potential (E_h below -400 mV) to achieve growth. In culturing methanogens, strong reducing agents (e.g., sulfide, cysteine, titanium III, etc.) must be added to poise the medium at the proper E_h prior to inoculation. Utilization of reduced sulfur compounds as reducing agents will also satisfy the sulfur requirements of methanogens (most lack assimilatory sulfate reductases; Daniels et al., 1986). Although there is variability with respect to species, usually a combination of inorganic (sulfide) with an organic (cysteine) S source is chosen as reductant. Cysteine is an essential amino acid for the growth of *Methanobacterium* strain AZ (Zehnder and Wurhmann, 1977); however, the presence of sulfide ions, glutathione, or coenzyme M ($HSCH_2CH_2SO_3^-$) will replace cysteine (Wellinger and Wurhmann, 1977). Coenzyme M is found in all methanogens, but not all methanogens can synthesize the compound, and some must therefore obtain it from their surroundings. *Methanobacterium ruminantium* is permeable to coenzyme M (Balch and Wolfe, 1979b) which it obtains from rumen fluid (Taylor et al., 1974). Thus coenzyme M must be included in the media for certain methanogens (usually, 10^{-6} M). Other organic growth factors that stimulate some methanogens include fatty acids (e.g., acetate, 2-methylbutyrate, etc.), vitamins, and complex nutrient supplements, such as yeast extract or trypticase (Bryant et al., 1971). As in the case of sulfur, there is considerable variability among methanogenic species with respect to requirements (if any) for these compounds.

Methanogens need ammonium ions to satisfy the bulk of their N requirements, but some species (*M. ruminantium*, *Mcs. voltae*) have additional requirements for certain amino acids. Until quite recently, no known species of methanogen appeared capable of fixing atmospheric nitrogen. Pine and Barker (1954) reported that *M. omelianskii* incorporated $^{15}N_2$ into cellular nitrogen and that incorporation was eliminated by high levels of ammonium ions. The question as to which member of the association fixed nitrogen has never been resolved. Acetylene-reduction assays for nitrogen fixation cannot be employed because the gas inhibits methanogenic bacteria (see below). Recently, the ability to fix nitrogen has been observed in *Methanosarcina* st. 227 and *Methanococcus thermolithotrophicus* (Murray and Zinder, 1984; Belay et al., 1984).

Trace elements needed by methanogens include cobalt, iron, molybdenum, nickel, magnesium, and potassium (Schönheit et al., 1979; Sprott and Jarrell, 1981). In addition, growth of *Methanococcus vannielii* is stimulated by tungsten and selenium (Jones and Stadtman, 1977). Sodium chloride (5 to 15 mM) is required for growth (Patel and Roth, 1977; Perski et al., 1982), and higher levels are needed for marine isolates (0.52 to 0.70 M; Corder et al., 1983) and the moderate halophile, *Methanococcus halophilus* (1.2 M; Zhilina, 1982). These organisms also have elevated requirements for magnesium (20 mM).

Most methanogens grow over a relatively narrow pH range (about 6 to 8). Methane production occurs in acidic environments such as peat bogs (Williams and Crawford, 1984), although there are not reports as yet of acidophilic isolates.

However, an enrichment culture of a methanogen was recovered from an alkaline lake. The enrichment had a pH optimum of 9.7 for methane formation from methanol (Oremland et al., 1982a). Recently, Worakit et al. (1986) isolated four strains of alkaliphilic (pH optima 8.1 to 9.1) methanogens.

The temperature range for methanogens is quite broad. Zeikus and Winfrey (1976) found that lake sediments actively evolved methane over a temperature range of 4 to 55°C, with an optimum between 35 and 42°C. The majority of known methanogens are mesophilic and have temperature optima of about 35°C. *Methanobacterium thermoautotrophicum* was the first thermophilic methanogen isolated (Zeikus and Wolfe, 1972), and the organism had an optimum temperature of 65 to 70°C. Subsequently, there have been several reports of thermophilic methanogens isolated with temperature optima either below (e.g., 55°C for *Methanosarcina TM-1* and *Methanogenium thermophilicum*). at (e.g., *Methanococcus thermolithotrophicus*), or well above the range of *M. thermoautotrophicum*. In the latter category of extremely thermophilic methanogens are *Methanothermus fervidus* and *Methanococcus jannaschii* (Stetter et al., 1981; Jones et al., 1984). These organisms have temperature optima at or exceeding 83°C and *M. fervidus* exhibits growth at 97°C. *Mcs. jannaschii* has an extremely rapid doubling time (26 min) at 83°C. Despite the fact that methanogenesis is operative in nature at temperatures near 4°C, there are no reports as yet of psychrophilic methanogens.

12.2.5 Inhibitors

Metabolic inhibitors can be particularly useful tools when performing biochemical or ecological experiments. A variety of chemical substances will physiologically block the formation of methane. Some of these substances are toxic to methanogens either without or with varying degrees of toxicity to other bacteria. Others are either less toxic or cause a decrease of methanogenic activity by altering patterns of electron flow. This topic has recently been reviewed by Oremland and Capone (1988).

Methanogens are very sensitive to low levels of oxygen (e.g., a few ppm), exposure to which lowers their adenylate charge (Roberton and Wolfe, 1970) and causes death. Oxygen causes an irreversible disassociation of the F_{420}-hydrogenase enzyme complex. In cells grown under an Fe^{2+} limitation, F_{420} degrades rapidly upon exposure to oxygen. This may be due to the lack of protective superoxide dismutase (Schönheit et al., 1981).

Alternate electron acceptors (e.g., nitrate and sulfate) inhibit methanogenesis in mixed microbial ecosystems (such as sediments), by channeling electron flow to microorganisms that are thermodynamically more efficient than methanogens [e.g., denitrifiers or sulfate reducers (see Section 12.3]. However, although sulfate ions themselves have no affect on methanogenic bacteria in pure culture, both sulfite (Prins et al., 1972; Balderston and Payne, 1976) and elemental sulfur (Stetter and Gaag, 1983) strongly inhibit methane formation by pure cultures of methanogens growing on hydrogen. In the case of sulfur, the methanogens tested pro-

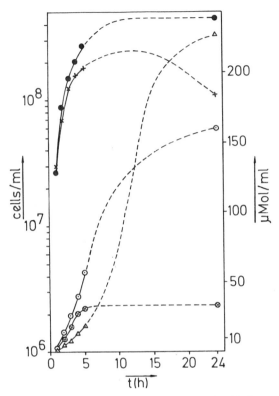

Figure 12.4 Production of methane and hydrogen sulfide by *Methanococcus thermolith-otrophicus*. Cells were grown under H_2 at 65°C with and without sulfur. ●, growth without sulfur; X, growth with sulfur; ○, methane formation without sulfur; ⊗, methane formation with sulfur; △, H_2S formation with sulfur. From Stetter and Gaag (1983), reproduced with permission from Nature, 305:309–311, copyright © 1983 Macmillan Journals Limited.

duced hydrogen sulfide in lieu of methane (Fig. 12.4). Oxides of nitrogen (nitrate, nitrite, nitrous oxide, nitric oxide) inhibit methanogenesis by whole-cell suspensions of *M. thermoautotrophicum* and *M. formicicum* (Balderston and Payne, 1976). Although the mechanism(s) of inhibition by nitrogen oxides is not known, it is apparently more complicated than an increase in the oxidation/reduction potential of the medium.

Use of 2-bromoethanesulfonic acid (BES; $BrCH_2CH_2SO_3^-$) as an inhibitor of methanogenic bacteria has recently gained popularity. The compound is a structural analog of coenzyme M, and coenzyme M is found in all methanogens but in no other forms of life tested (Balch and Wolfe, 1979a). Thus BES may be employed to inhibit only the methanogenic components of mixed bacterial assemblages. It is therefore considered a ''selective'' or ''specific'' inhibitor when used to dissect ecosystems. The methylcoenzyme M reductase enzyme complex is inhibited by 10^{-5} M BES (Gunsalus et al., 1978), however, somewhat higher levels may be a needed for whole cells (10^{-4} M). Sediments require considerably more BES (at least 10^{-3} to 10^{-2}) to achieve inhibition (Oremland, 1981). Mutants of *Methanosarcina* strain 227 and of *M. formicicum* were isolated which were resistant to 2×10^{-4} M BES (Smith and Mah, 1981). Resistance to BES in these mutants appears to be caused by impermeability to the compound rather than any alteration in biochemical pathways (Smith, 1983). There appears to be a differential susceptibility to BES by the various components of the methanogenic population present in digestors. Thus acetoclastic methane production was totally inhibited by 1×10^{-3} M BES, but complete blockage of CO_2 reduction by H_2 required 50×10^{-3} M BES (Zinder et al., 1984).

Chlorinated methanes (e.g., chloroform, methylene chloride, etc.) are competitive inhibitors of methane formation from methycobalamin in cell extracts (Wood et al., 1968). The compounds block methanogenesis in mixed bacterial systems and are effective at low concentrations (below 10 μM; Bauchop, 1967; Prins et al., 1972). However, unlike BES, which is highly water soluble, chlorinated methanes have extremely low solubilities and are usually dissolved in ethanol before usage to enhance dispersion and effectiveness. This practice is not advisable if one needs to examine the intermediates of methane production by mixed populations because ethanol may be preferentially metabolized by these microbes. Surprisingly, both chloroform and carbon tetrachloride can be degraded to methane by enrichment cultures exposed to low levels of the compounds (100 $\mu g/L$) for long periods (Bouwer and McCarty, 1983). Micromolar levels of the agricultural nitrification inhibitor "nitrapyrin" (2-chloro 6-trichloromethyl pyridine) also inhibits methanogenesis in sediments, apparently by action of the trichloromethyl group (Salvas and Taylor, 1980). Low-potential viologen dyes blocked methane formation by "*M. omelianskii*" (Wolin et al., 1964) but not by rumen enrichments (Bauchop, 1967). High levels of methylviologen (2.5 g/L) were needed to inhibit methane formation by freshwater sediments (Pedersen and Sayler, 1981). Thus use of this compound to study natural systems appears to have limited scope.

Methanogenic bacteria are also sensitive to unsaturated carbon–carbon bonds. Both acetylene (Fig. 12.5) and ethylene (Fig. 12.6) inhibited methane formation in sediment slurries; however, the inhibition by ethylene was reversible (Oremland and Taylor, 1975). Raimbault (1975) reported that acetylene inhibited growth and methane formation by a *Methanosarcina*. Exposure of *M. hungatei* or *M. bryantii* to acetylene caused a rapid drop of cellular ATP and the destruction of their transmembrane pH gradients (Sprott et al., 1982; Jarrell and Sprott, 1983), implying that the compound blocks the chemiosmotic functioning of methanogenic membranes rather than a part of the biochemical pathway of methane formation. Although acetylene has been employed to inhibit methanogenesis in mixed systems, it is not a selective inhibitor and will disrupt a variety of other bacterial processes, including growth of sulfate reducers and clostridia as well as preventing nitrification, methane oxidation, and the hydrogenase activity of nonmethanogens. This subject has recently been reviewed by Payne (1984). Unsaturated fatty acids (e.g.,

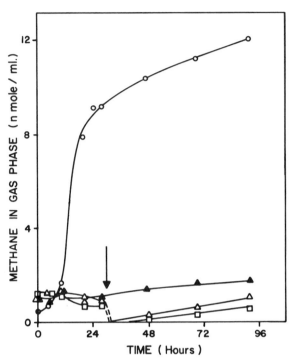

Figure 12.5 Inhibition of methanogenesis in marine sediment slurries by C_2H_2. The flasks contained 20 mL of sediment homogenate and a 30-cm headspace of H_2. $\bigcirc$, uninhibited control; $\blacktriangle$, 1.25% C_2H_2; $\triangle$, 1.25% C_2H_2 gassed at 30 h (arrow) with H_2; $\square$, 1.25% C_2H_2 and 0.05% $CHCl_3$ gassed as 30 h (arrow) with H_2. From Oremland and Taylor (1975), reproduced with permission from Applied Microbiology.

linolenate), as well as levulinic acid, have inhibitory effects upon methanogenesis in pure cultures, the latter compound by blocking synthesis of F_{430} (Prins et al., 1972; Jaenchen et al., 1981).

Other inhibitors of methanogenic bacteria include corrinoid antagonists, such as iodopropane, which prevent the functioning of B_{12} enzymes as methyl group carriers (Wood and Wolfe, 1966; Kenealy and Zeikus, 1981). Monensin, an ionophore antibiotic used to increase cattle feed efficiency, appears to inhibit some methanogens as well as other rumen bacteria (Chen and Wolin, 1979; Hilpert et al., 1981). However, studies with mixed rumen populations indicate that the drug functions by interfering with the components of the bacterial population responsible for providing substrates to the methanogens (e.g., hydrogen) rather than the methanogens themselves (van Nevel and Demeyer, 1977). Heavy metals such as copper can block methane formation by *M. formicicum* (Hobson and Shaw, 1976). However, addition of heavy metals to marine sediments may actually stimulate methane production by eliminating metal-sensitive substrate competitors such as sulfate reducers (Capone et al., 1983). Methanogens are also capable of alkylating toxic metals (e.g., mercury, selenium, and arsenic) and may therefore have a role

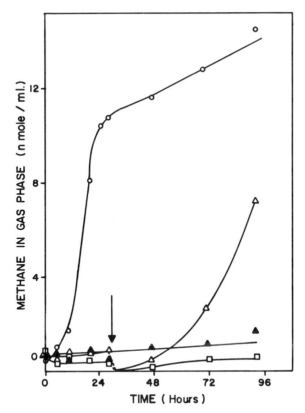

Figure 12.6 Reversible inhibition of methanogenesis by C_2H_4. The flasks contained 20 mL of sediment homogenate and a 30-cm^3 headspace of H_2. $\bigcirc$, uninhibited control; $\blacktriangle$, 5% C_2H_4; $\triangle$, 5% C_2H_4 gassed at 30 h (arrow) with H_2; $\square$, 5% C_2H_4 and 0.05% $CHCl_3$ gassed at 30 h with H_2. From Oremland and Taylor (1975), reproduced with permission from Applied Microbiology.

in the metal contamination of food chains (McBride and Wolfe, 1971; Ridley et al., 1977). However, Oremland and Zehr (1986) recently demonstrated that methanogens in sediments and pure culture could demethylate dimethylselenide. These observations, as well as the observed accumulation of CH_3-Hg^+ in sediments in the presence of BES (Compeau and Bartha, 1985), led Oremland and Zehr (1986) to theorize the methanogens were involved primarily in demethylation of metals in nature.

12.3 ECOLOGICAL ASPECTS

12.3.1 Methanogenic Interactions

Methanogens are entirely dependent on the metabolic activities of other anaerobes for providing their growth substrates. The breakdown of organic matter in anaer-

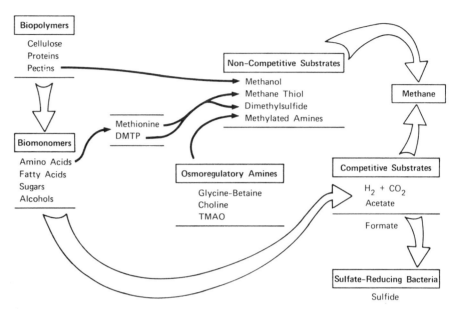

Figure 12.7 Anaerobic food web for aquatic microbial ecosystems. Biopolymers are degraded to simpler compounds by fermentative bacteria. Osmoregulatory amines, including trimethylamine oxide (TMAO), are converted to methylated amines. Similarly, sulfur-containing osmoregulatory compounds such as dimethylpropiothetin (DMTP) are converted to dimethylsulfide. These osmoregulatory solutes assume greater importance as precursors of methanogenic substrates in saline and hypersaline environments. Substrates such as trimethylamine and dimethylsulfide are considered "noncompetitive" in that sulfate-reducing bacteria do not have as strong an affinity for them as they do for acetate or hydrogen.

obic ecosystems proceeds sequentially from the complex to the simple (Fig. 12.7). Thus biopolymers such as cellulose undergo initial attack to yield biomonomers (e.g., sugars) which are eventually degraded to the level of methanogenic substrates. In saline or hypersaline environments bacterial degradation of osmoregulatory compounds (e.g., glycinebetaine, dimethylsulfoniopropionate) yields the substrates TMA or DMS (Fig. 12.7). Methanogens therefore occupy the terminal position in a complex anaerobic food web, although they can be displaced from this position by bacteria that use other electron acceptors (e.g., nitrate, sulfate). The nature of methanogenic interactions varies with the type of organisms involved, the chemical milieu of the environment, and the particular organic compound(s) degraded. These interactions or "partnerships" of methanogens with other anaerobes can range from obligate syntrophies, if there is an absolute mutual dependency for survival, to more "casual" groupings in which the growth of one or more members does not require the others. Only when methanogenic substrates are channeled into anoxic environments from nonbiological sources are methanogens released from their substrate dependence on anaerobic food webs. Examples of such situations are thought to occur when geothermally produced hydrogen enters hot springs (Zeikus et al., 1980), geological rift zones such as Lake Kivu (Deuser et al., 1973), and perhaps seafloor hydrothermal vents (Baross et al., 1982; Jones et al., 1984).

The discovery that *M. omelianskii* really consisted of a syntrophic association helped to reveal the subtle nature of H_2 as an intermediate in some methanogenic partnerships. Because the methanogen efficiently consumed the H_2 produced by the S-organism, the gas was never evident during normal growth of the association. In this case, the S-organism had an obligate dependence on *M. bryantii* to keep the H_2 partial pressure low enough to allow for growth. This phenomenon was subsequently termed "interspecies transfer of H_2" and the process is not restricted solely to methanogenic syntrophies (Ianotti et al., 1973). An important aspect of interspecies H_2 transfer is that a shift from reduced to more oxidized soluble metabolic products occurs when a fermentative organism is co-cultured with a H_2-consuming methanogen. Thus, provided that the fermenter possesses hydrogenase, it will produce reduced compounds (e.g., ethanol) when cultured axenically, but will generate compounds such as acetate and H_2 when co-cultured with a H_2 consumer. Hungate (1966) first theorized that this mechanism was the basis for the oxidized soluble intermediates he detected during rumen methanogenesis. The subject has been an important focal point of research on methanogenesis for the past 17 years and has been reviewed in detail (Wolin, 1982; Wolin and Miller, 1982; Mah, 1982).

The basis for the shift from reduced to oxidized products lies in the fermenter's mechanism of NADH oxidation. For example, in pure culture, *Ruminococcus albus* oxidizes NADH generated during glycolysis by reducing acetyl-CoA to form ethanol. However, when *R. albus* was co-cultured with a H_2-consuming bacterium (*Vibrio succinogenes*), *R. albus* oxidized NADH by forming H_2 (Ianotti et al., 1973). Because *V. succinogenes* maintained a low H_2 partial pressure, H_2 did not inhibit the hydrogenase activity of *R. albus*. This allowed for acetyl-CoA to be channeled into acetate (rather than ethanol) and resulted in the production of an additional molecule of ATP. Thus, in addition to forming different products, growth can be enhanced by interspecies H_2 transfer. It should be borne in mind, however, that there are mechanisms other than NADH oxidation by which fermentative bacteria can evolve H_2. In these cases, H_2 accumulation does not inhibit the growth of the fermenter and the organism is therefore not reliant on a H_2 consumer (e.g., methanogen) to remove the gas. However, shifts from reduced to more oxidized soluble products, as well as enhanced H_2 production, can occur when these types of bacteria are co-cultured with H_2-oxidizing methanogens (Fig. 12.8). Interspecies H_2 transfer has now been demonstrated with defined co-cultures of various methanogens with a wide assortment of anaerobic bacteria. Some examples of methanogenic interactions during carbohydrate degradation were achieved with *Acetobacterium woodii* (Winter and Wolfe, 1980), *Clostridium thermocellum* (Weimer and Zeikus, 1977), *Ruminococcus flavefaciens* (Latham and Wolin, 1977), *Selenomonas ruminantium* (Schiefinger et al., 1975), and *Anaeroplasma* sp. (Rose and Pirt, 1981). There is also a suggestion that anaerobic protozoa may have a H_2-based interaction with methanogens (Krumholz et al., 1983) and en-

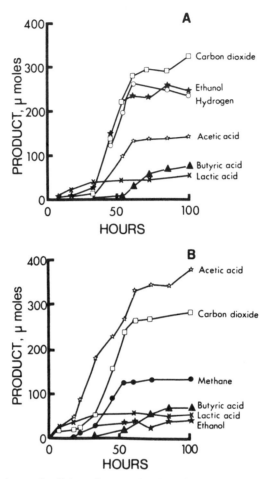

Figure 12.8 Products of cellulose fermentation by *Clostridium thermocellum* grown in (A) the absence and (B) the presence of *Methanobacterium thermoautotrophicum*. Results are expressed as μmol of product formed per 12 mL of culture volume. From Weimer and Zeikus (1977), reproduced with permission from Applied and Environmental Microbiology.

dosymbiotic methanogens are present within anaerobic protozoans (van Bruggen et al., 1983).

Degradation of fatty acids (e.g., butyrate, caproate, valerate, etc.) to methane and CO_2 is first mediated by a special group of anaerobes referred to as "H_2-producing acetogens" (Bryant, 1979). These organisms degrade long-chain fatty acids to H_2, acetate, and in the case of odd-length chains, propionate. However, because of their inability to either respire anaerobically or to ferment, these organisms are obligate proton reducers. Because H_2 inhibits the hydrogenase of these bacteria, they must remove the gas by interspecies H_2 transfer with a methanogen or other H_2-consuming anaerobe. Therefore, these organisms occupy restricted ecological niches in which syntrophic removal of their waste H_2 by other bacteria is critical for their survival (McInerney et al., 1979). Recently, Zinder and Koch (1984) reported methanogenesis from acetate oxidation by a syntrophic co-culture. Thus the acetoclastic reaction carried out by pure cultures of methanogens (e.g., *Methanosarcina*) is not the only mechanism by which acetate degrades to methane in nature.

Acetate has long been recognized as an important precursor of methane formed in various anaerobic environments. However, studies demonstrating defined syntrophic associations by which methane is formed from acetate (as opposed to H_2) have been relatively few. In part, this was due to the difficulties in working with acetate-utilizing methanogens (e.g., *Methanosarcina barkeri*) which must first be adapted to the substrate. Methanogenic interactions based on acetate were demonstrated in a co-culture of *Acetobacterium woodii* and *Methanosarcina barkeri* which completely degraded glucose to CH_4 and CO_2 (Winter and Wolfe, 1979). Subsequently, other workers have demonstrated a nearly complete degradation of cellulose by employing tricultures composed of acetate-utilizing *Ms. barkeri* with either *Acetovibrio cellulolyticus* plus *Desulfovibrio* sp. (Laube and Martin, 1981) or an anaerobic fungus plus *Methanobrevibacter* sp. (Mountfort et al., 1982).

Much progress has been made by using defined, artificial syntrophic associations, like the examples cited, to study methanogenic partnerships. However, this approach is of only limited utility when a different type of question is posed; namely what types of compounds are subject to methanogenic routes of biodegradation? This question is of particular environmental importance because of the introduction of hazardous organic waste products into anaerobic ecosystems such as sediments, agricultural soils, and groundwaters. The approach taken is to determine whether sustained enrichment cultures can degrade the compound in question to CH_4 and CO_2. Once this is achieved, the chemical intermediates involved as well as some (if not all) of the participant microbes are usually identified. The term "consortium" is used to refer to the group of organisms involved in the degradation of a particular compound. Ferry and Wolfe (1976) first employed this term to describe the bacterial flora which degraded benzoate to CH_4 and CO_2. Intermediates produced by this consortium included H_2, CO_2, formate, and acetate. Both *Methanobacterium formicicum* and *Methanospirillum hungatei* were identified in and isolated from the consortium. Subsequently, methanogenic degradation of benzoate has been studied in more detail (Fina et al., 1978; Keith et al., 1978; Shlomi et al., 1978) and extended to other types of substituted aromatic compounds (Healy and Young, 1979; Boyd et al., 1983; Grbić-Galić, 1983; Kaiser and Hanselmann, 1982; etc.). Furthermore, halogenated substituted aromatics (i.e., 2-bromobenzoate, 3-iodobenzoate, and 3,5-dichlorobenzoate) can be reductively dehalogenated and degraded to CH_4 and CO_2 via the benzoate pathway (Sulflita et al., 1982; Horowitz et al., 1983).

Other types of pollutant organic compounds are also susceptible to methanogenic degradation. For example, polyethylene glycol can be fermentatively attacked by *Pelobacter venetianus* to yield ethanol and acetate (Schink and Stieb,

1983). Methanogenic consortia completely degraded both ethylene glycol and diethylene glycol to acetate, ethanol, and methane (Dweyer and Tiedje, 1983). Thus the consortia approach to determining biodegradability of pollutant organics has proven to be a useful tool in defining pathways leading to methane, and will probably assume even more importance with time. Economic aspects of methane recovery from fermentation reactions were discussed by Klass (1984).

12.3.2 Competitive and Noncompetitive Substrates

Other anaerobes besides methanogens are capable of metabolizing hydrogen and/or acetate. Hydrogen-consuming acetogens such as *Acetobacterium woodii* have been described only relatively recently (Balch et al., 1977b). Because they have the potential to remove H_2 and thereby shift methanogenesis to acetoclastic routes, these organisms may be of considerable ecological importance. However, they are capable of autotrophic, mixotrophic, or heterotrophic growth (Braun and Gottshalk, 1981), and can form methanogenic partnerships based on either H_2 or acetate (Winter and Wolfe, 1979, 1980). Thus their ecological role is as yet unclear. Studies with lake sediments indicate that they do not compete significantly with methanogens for H_2 (Lovley and Klug, 1983b).

In contrast, the complex array of interactions between sulfate reducers and methanogens has been well studied and encompasses competition for substrates (e.g., see Ward and Winfrey, 1985). Although sulfate-reducing bacteria can provide substrates to methanogens in the absence of sulfate (Bryant et al., 1979; McInerney and Bryant, 1981), some species are capable of growing on H_2 plus CO_2 and/or acetate (e.g., Badzoing et al., 1978; Widdel and Pfennig, 1977, 1981). Inhibition studies with sediments have clearly demonstrated that sulfate reducers (in the presence of sulfate) can outcompete methanogens for these substances (e.g., Abram and Nedwell, 1978; Oremland and Taylor, 1978; Sørensen et al., 1981; Oremland and Polcin, 1982; Lovley and Klug, 1983a; King et al., 1983). Furthermore, kinetic investigations with pure cultures revealed that sulfate reducers have higher affinities (i.e., lower K_m values) for either H_2 (Kristjansson et al., 1982; Robinson and Tiedje, 1984) or acetate (Schönheit et al., 1982) than do methanogens. These types of kinetic results were also obtained for H_2 in freshwater sediments (Lovley et al., 1982). Thus competitive exclusion of methanogenesis by the gastronomically greedy sulfate-reducing bacteria has been well established and supports the models of thermodynamic preclusion of methane by sulfate in sediment porewaters (Claypool and Kaplan, 1974; Martens and Berner, 1974; Figs. 12.7 and 12.9).

There have been several reports of sediment incubation experiments in which the two processes were not mutually exclusive and occurred simultaneously. The basis for the phenomenon was shown to be methane production from methanol and/or methylated amines, substrates for which sulfate reducers show relatively little affinity (Oremland et al., 1982b; Oremland and Polcin, 1982). It has been shown subsequently that sulfate reducers can outcompete methanogens for methanol when the compound is at low (but not high) ambient concentrations (King,

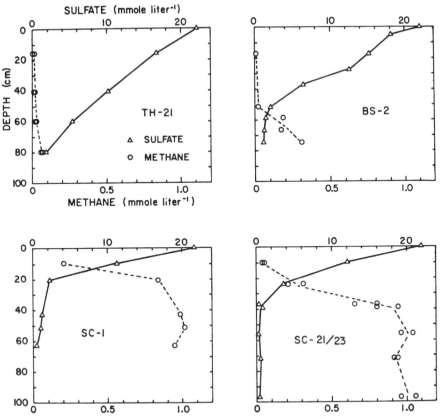

Figure 12.9 Profiles of porewater sulfate and methane in four cores recovered from Long Island Sound. From C. S. Martens and R. A. Berner, Methane production in the interstitial waters of sulfate-depleted marine sediments. Science 185:1167–1169, copyright © 1974 by AAAS. Reproduced with permission.

1984a). Trimethylamine, however, has been confirmed as a noncompetitive methanogenic substrate at *in situ* concentrations (King, 1984b). Methylated amines can be derived via bacterial attack of osmoregulatory quaternary amines, such as choline, betaine, and trimethylamine oxide. Methanol is formed during anaerobic decomposition of plant pectins (Schink and Zeikus, 1980, 1982). Nearly complete degradation of choline to ammonia and methane occurred in a co-culture of *Ms. barkeri* with a desulfovibrio (Fiebig and Gottshalk, 1983). However, in sediments choline and betaine may undergo initial attack by fermentative bacteria rather than sulfate reducers. This results in the production of both trimethylamine and acetate which are then channeled to methanogenesis and sulfate reduction, respectively (King, 1984b). The topic of competitive and noncompetitive methanogenic substrates was reviewed by King (1984a). Recently, dimethylsulfide has been shown to be a "noncompetitive"-type substrate (Kiene et al., 1986).

12.3.3 Natural Habitats

12.3.3.1 Within Living Organisms

Although the involvement of methanogenic bacteria in the digestive processes of ruminants and other animals is well established (see below), methanogens have also been encountered in other anoxic environments present in living organisms. For example, the heartwood tissues of a variety of trees can become infected with soil bacteria, especially if the trees are located in poorly drained areas. Eventually, the tissues become structurally weak (''wetwood'') and devoid of oxygen. Methanogenesis occurs with a concommitant buildup of pressurized, flammable gases (Zeikus and Ward, 1974). This occurs at the expense of the degradation of cellulose and pectins (Schink et al., 1981) and methanogens (e.g., *Methanobrevibacter arboriphilus*) can be isolated from these tissues (Zeikus and Henning, 1975). Another anoxic environment in which methanogens are present is the dental plaque of primates (Kemp et al., 1983). It has not as yet been established, however, whether methanogens are involved in tooth decay.

Considerably more is known about the activities of methanogenic bacteria within the digestive tracts of animals (Hungate, 1966; Bryant, 1977; Wolin, 1981). To profit nutritionally from a diet rich in cellulose, herbivorous mammals (because they lack cellulytic enzymes) have evolved specialized structures such as the rumen (present in cows, goats, sheep, etc.) and the cecum (present in horses, rabbits, etc.) in which anaerobic microbial degradation of the polymer takes place. An important difference between the methane fermentation occurring in gastrointestinal tracts as opposed to other environments (e.g., sediments, digestors, etc.) is that methane production from acetate is not important in the former. This is because the fatty acids produced during gastrointestinal fermentations (e.g., acetate, propionate, butyrate, etc.) are absorbed into the bloodstream and used for the animal's nutrition. Thus about 82% of the methane formed in the rumen is thought to be formed from H_2 reduction of CO_2, while about 18% is derived from formate (Hungate et al., 1970). Acetoclastic organisms (*Methanosarcina*) are present in significant numbers (10^5 to 10^6 mL^{-1}) in rumen fluid (Patterson and Hespell, 1979). However, these organisms are apparently linked to methane formation from methylated amines (or methanol) rather than acetate. Methylated amines are derived from compounds such as choline (Neill et al., 1978) and may be an important source of ammonia for rumen bacteria. The contribution of methylated amines to methane produced in the rumen has not been estimated.

The large bowel of humans is a site of postgastric fermentation and methanogenic bacteria are present in feces taken from that location (Nottingham and Hungate, 1968). The most prominent species of methanogen is *Methanobrevibacter smithii* (Miller et al., 1982), which can be present at high levels (10^{10} cells per gram of feces) in some individuals (Miller and Wolin, 1983). Surprisingly, only about one-third of the adult population actually harbors methanogens. When methane is produced in the large bowel, about 20% of the gas is transported via the bloodstream to the lungs, where it is respired. The remainder is expelled as flatus.

As most undergraduates are well aware, human flatus is composed of flammable gases. However, the "Volta-type" experiments conducted in fraternity houses do not reveal the major components of flatus. Actually, hydrogen is an important fermentative gas produced in the large bowel and can cause considerable intestinal discomfort when produced in excess. An active methanogenic population in the bowel of individuals suffering this affliction would prove a blessing, considering the volume drop by H_2 reduction of CO_2. *In vitro* models of the large intestine indicate that the bacterial flora may change in relation to diet (Miller and Wolin, 1981), but this was not borne out for *M. smithii* populations studied *in vivo* (Miller and Wolin, 1983). Certain foods, such as beans, contain the carbohydrate stachyose. This compound is not digested, but it is subjected to postgastric bacterial degradation to CO_2. In a now classical study, Steggerda and Dimmick (1966) were able to show a tenfold increase in flatus volume (from 17.6 to 171.5 mL/h) when individuals were placed on a 51% bean diet. The fact that these results have not been repeated may be due to misgivings by the volunteers tested concerning the method of gas collection.

Production of methane from animal gut environments has geochemical ramifications, especially with respect to atmospheric chemistry. The hindgut region of xylophagus insects, for example, harbors an active and diverse microbial flora (Breznak, 1982). Methanogenic bacteria have been identified in the guts in various insects, including the cockroach (Bracke et al., 1979) and termites (Odelson and Breznak, 1983). It has been proposed that termites can potentially supply about 150×10^{12} g CH_4 per year to the earth's atmosphere, or roughly half the estimated global flux from the biosphere (Zimmerman et al., 1982). Because these experiments were conducted with only one species of termite and under laboratory conditions, these results have been challenged (Rasmussen and Khalil, 1983). Actual field experiments with termite nests suggests that their contribution to atmospheric methane is much smaller (2 to 5×10^{12} g per year) than estimated in lab studies (Seiler et al., 1984a).

Concentrations versus depth profiles of methane in the open ocean's aerobic surface layers frequently exhibit anomalous maxima (Fig. 12.10) which are well above equilibrium values with the atmosphere (Scranton and Brewer, 1977; Brooks et al., 1981; Traganza et al., 1979). These observations suggest an *in situ* biological source in an aerobic environment. Oremland (1979) postulated that the methane was derived from the activities of methanogens located within the intestinal tracts of marine animals, and was able to show methanogenic activity in plankton and within gut portions of two fishes. In addition, preliminary examination of the preserved contents of the forestomachs of baleen whales indicates the presence of bacterial cells (10^8 to 10^{10} cells per milliliter) and volatile fatty acids (Herwig et al., 1984). The authors hypothesized that fermentation of chitin occurs in this pregastric location, a situation that is analogous to the rumen. Therefore, baleen whales might have a methanogenic bacterial component in the flora of their forestomachs, and expulsion of methane by cetaceans could certainly elevate local concentrations and provide a spectacle worth witnessing.

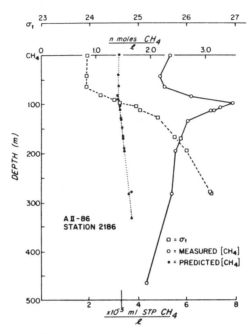

Figure 12.10 Vertical profile of dissolved methane concentrations in the oxygenated waters of the subtropical North Atlantic. Predicted methane values were calculated by assuming equilibrium with atmospheric methane. "σ_t" indicates temperature-associated water density. From Scranton and Brewer (1977), reproduced with permission from Deep-Sea Research, Vol. 24, Occurrence of methane in the near-surface waters of the western subtropical North Atlantic, copyright © 1977, Pergamon Press, Ltd.

12.3.3.2 Aquatic Environments

Aspects of the topic of methanogenesis in aquatic environments were reviewed by Rudd and Taylor (1980) and Ward and Winfrey (1985). Considerable attention has been devoted to the question of the interactions between sulfate-reducing and methanogenic bacteria, as well as the distribution of both sulfate and methane in sediments. A detailed discussion of this subject is beyond the scope of this chapter and the reader is referred to the above-mentioned reviews as well as Chapter 10 by Widdel. In addition, Lovley and Klug (1986) recently devised a model for freshwater sediments.

The first quantitative ecological fieldwork on methanogenesis in aquatic environments was done in Lake Vechten. By employing ^{14}C-labeled substrates and various chemical inhibitors, Cappenberg and his associates estimated that 70 to 75% of the methane formed in these sediments was derived from the methyl group of acetate (Cappenberg, 1974a, b; Cappenberg and Prins, 1974). Other workers have subsequently illustrated the importance of acetate as a methane precursor in both freshwater and sulfate-depleted marine sediments (e.g., Winfrey and Zeikus, 1977; Sansone and Martens, 1981). However, the amount of methane derived from acetate is variable from location to location. More attention has been paid to demonstrating that H_2 is a rate-limiting factor in the process of methanogenesis in sediments (Winfrey et al., 1977; Abram and Nedwell, 1978; Oremland and Taylor, 1978; Strayer and Tiedje, 1978). Conversion of acetate to methane by eutrophic lake sediments was estimated to be operating close to its V_{max}, while H_2 oxidation was operating well below its V_{max} (Strayer and Tiedje, 1978). This allowed for a low partial pressure of H_2 to be maintained. Interspecies H_2 transfer occurred in these sediments because the addition of H_2 inhibited the metabolism of C_3 to C_5 fatty acids (Lovley and Klug, 1982) and shifted the end products of glucose metabolism from acetate to lactate (King and Klug, 1982). Formate was rapidly metabolized to methane by these sediments, and formate additions also enhanced methanogenesis in salt marsh slurries (Strayer and Tiedje, 1978; Jones and Paynter, 1980). However, formate additions to sediment slurries from either estuarine muds (Oremland and Polcin, 1982) or from the bottom deposits of Big Soda Lake (Oremland et al., 1982a) caused only minor enhancement of methanogenesis. Thus hydrogen and acetate are important methane precursors in aquatic sediments, although their percent relative contributions appears variable and is still a matter of debate. Formate may be of importance in some environments, but its contribution to methane produced has not been assessed.

Both trimethylamine and methanol accumulated when salt marsh sediment slurries were inhibited with BES (Fig. 12.11; Oremland et al., 1982b). Accumulations of trimethylamine could account for all the methane produced in these experiments, with about 10% coming from methanol. In contrast to these results, meth-

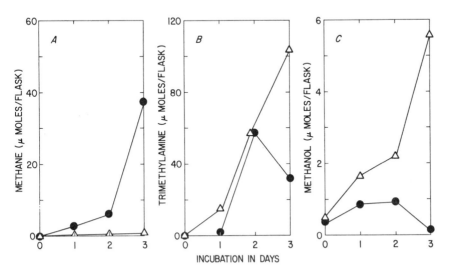

Figure 12.11 Accumulation of (*A*) methane, (*B*) trimethylamine, and (*C*) methanol in uninhibited salt marsh sediment slurries (●), and in slurries in which methanogenesis was inhibited by BES (△). Slurry vol, 80 ml. From Oremland et al. (1982b), reproduced with permission from Nature, 296:143–145, copyright © 1982 Macmillan Journals Limited.

anol and methylated amines comprised only about 5% and 1%, respectively, of the methane formed in freshwater Wintergreen Lake (Lovley and Klug, 1983b). However, trimethylamine was an important methane precursor (31.5 to 61.1%) in the estuarine, intertidal sediments of Lowes Cove, Maine (King et al., 1983). The variation in the importance of these compounds as methane precursors is probably due to the abundance of decomposing plant materials in the sediment systems. The salt marsh had abundant stands of the seagrass *Spartina foliosa*. This could provide the sediments with an ample supply of pectins, choline, dimethylsulfoniopropionate, and betaine from which to produce large pools of methylated amines, methanol, and dimethylsulfide.

Methanogenesis also occurs in the anoxic water columns of meromictic lakes. Winfrey and Zeikus (1979) reported production of $^{14}CH_4$ when the anoxic waters of Knaack Lake were supplied with ^{14}C-labeled formate, bicarbonate, methanol, or acetate. Maximum rates of methane formation (from HCO_3^-) in this low sulfate lake were 1.4 μmol CH_4 L^{-1} day^{-1}. Oremland and DesMarais (1983) were able to recover methanogens near the deep chemocline of alkaline Big Soda Lake (SO_4^{2-} = 58 to 68 mM). *In situ* incubations of anoxic Big Soda Lake waters revealed that production rates in the monimolimnion (2 to 10.5 nmol CH_4 L^{-1} day^{-1}) were higher than that in the mixolimnion (0.1 to 0.8 nmol CH_4 L^{-1} day^{-1}) (Iversen et al., 1987).

12.3.3.3 Extreme Environments

Methane of apparently biogenic origin has been detected in several anoxic, hypersaline environments. These include the Great Salt Lake (Schink et al., 1983), Big Soda Lake (Oremland and DesMarias, 1983), the Orca Basin (Sackett et al., 1979), the Red Sea brines, (Burke et al., 1981), Solar Lake (Giani et al., 1984), Mono Lake (Oremland, et al., 1987), and Soap Lake (L. Miller, unpublished data). Methane is also present in the Dead Sea, but its origin is not clear (A. Nissenbaum, personal communication). Despite these reports, there have been very few studies on the biochemistry of methane formation in these environments. Zeikus (1983) reported that Great Salt Lake sediments produced methane from methylmercaptan and the methyl group of methionine. Zeikus et al. (1983) isolated a fermentative anaerobe, *Haloanaerobium praevalens*, which could grow at 25% NaCl and produce methylmercaptan from methionine. Big Soda Lake is a moderately hypersaline (87 g salts per liter) alkaline lake. Methanogenesis by the lake's bottom sediments was stimulated by methanol, trimethylamine, and methionine, but not by acetate, hydrogen, or formate (Oremland et al., 1982a). However, this substrate restriction was apparently due to the elevated sulfate content (58 to 68 mM) of the lake water and its active populations of sulfate reducers (Smith and Oremland, 1987) rather than by the lake's high pH (9.7). Mono Lake methanogenesis was stimulated by dimethylsulfide and methane thiol (Kiene et al., 1986).

Thermophilic environments, such as hot springs, are sites of active methanogenesis. Species isolated from hot springs, such as *Methanothermus fervidus* (Stetter et al., 1981), have high-temperature optima (83°C) and can grow up to 97°C. However, field studies in both Yellowstone National Park and in Iceland indicate that much lower temperature maxima occur *in situ*. Algal-bacterial mats in Yellowstone produced methane over a temperature range of 30 to 70°C (Ward and Olson, 1980). Optimum *in situ* temperatures were between 50 and 60°C, although artificial incubation of mat material at higher temperatures increased the optimum to 70°C (Sandbeck and Ward, 1982). Competition between thermophilic methanogens and sulfate reducers for acetate and H_2 also occurs within hot-spring mats that contain 6 to 8 mM sulfate (Ward and Olson, 1980).

Methane has also been detected emanating from the high-temperature (300 to 400°C) "black smokers" located at seafloor spreading centers, such as 21°N (Welhan and Craig, 1979). It has been suggested that the gas is of bacterial origin (Baross et al., 1982), and to support this, a thermophilic methanogen (*Methanococcus jannaschii*) was isolated from the 21°N site (Jones et al., 1983a). Baross and Deming (1983) claimed that bacterial growth and methane production occurred in a vent enrichment culture incubated under high hydrostatic pressure at 250°C. They hypothesized that growth at these high temperatures was sustained because of the counteracting effects of high hydrostatic pressures. However, Trent et al. (1984) challenged these results and were able to reproduce "growth" data using sterile controls. Thus the claims of growth by Baross and Deming (1983) may have been due to experimental artifacts and should therefore be viewed with caution.

12.4 GEOCHEMICAL ASPECTS

Organic compounds formed in the biosphere are stored in living organisms, decomposed and transferred to the oceans and atmosphere, or buried in sediments. Sedimentary organics are attacked by microorganisms (including methanogens), and if environmental conditions are correct (e.g., high sedimentation rates), methane may become entrapped within the sediment matrix. Therefore, methanogenic activity occurring during the diagenesis of recent sediments can ultimately result in commercially exploitable deposits of natural gas (Rice and Claypool, 1981). The specific nature of the complex organic matter buried in sediments reflects both the types of organisms and paleoenvironmental conditions present during initial deposition (Didyk et al., 1978). Some organic compounds will be microbially attacked during diagenesis and converted into new molecules (e.g., bacterial cell wall lipids), thereby adding a microbial "signature" upon the sedimentary organic matter. Eventually, as the sediments are buried deeper and lithified over geologic time periods, significant bacterial activities cease. This is due in part to the increased refractory nature of the buried organics as they polymerize into "kerogen" (organic matter that is insoluble in organic solvents). As burial increases to greater depths (> 1 km), higher temperatures and pressures bring about the final demise of biological reactions. It is at these higher temperatures (80 to 120°C) and pressures (>> 200 atm) where the thermogenic and catagenic reactions occur whereby the kerogens in rock are converted into petroleum and "wet" (methane plus higher alkanes) natural gases (see Tissot and Welte, 1978; Hunt, 1979). At very high

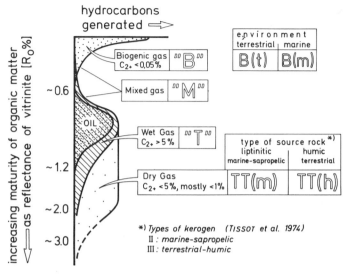

Figure 12.12 Classification scheme for the formation of natural gases from bacterial and/ or thermogenic sources. Vitrinite reflectance is a measure of the thermal maturity of sedimentary organic matter as it becomes more "coal-like" at higher temperatures. From Schoell (1980), reproduced with permission from Geochimica et Cosmochimica Acta.

temperatures ($> 150°C$), petroleum itself will be converted into graphite and the "dry" gaseous hydrocarbons (mainly methane) of thermogenically formed natural gas (Fig. 12.12). An important economic question, therefore, is how do gases formed by thermocatalytic processes differ from those produced by methanogenic activity associated with early diagenesis? Because both types of gases eventually migrate from depth (along microfractures, etc.) to the earth's surface, they can easily be collected and analyzed. Thus the ability to discriminate between thermocatalytic and bacterially formed natural gases improves the chance of discovering petroleum deposits by employing inexpensive surface geochemical techniques. To address this and related questions, we must consider the extent to which methanogens are involved in the process of petroleum formation, their ability to form gaseous hydrocarbons other than methane, their depth distribution in the earth's crust, and the microbial reactions that influence the stable isotopic composition of methane.

12.4.1 Geolipids of Methanogens

The basic molecular structure and stereochemistry of many lipids are remarkably well preserved over geologic time, thereby retaining the initial biochemical imprint of their original source (Mackenzie et al., 1982). Prior to the discovery of isoprenoids in the lipids of archaebacteria, various types of isoprenoid hydrocarbons had been detected in both recent and ancient sediments, sedimentary rocks, and petroleum. Compounds like pristane and phytane, for example, were known to be present in both sedimentary samples and petroleum; however, they were thought to be derived solely from the phytol side chain of chlorophyll (Didyk et al., 1978) and were considered as "biomarkers" (molecular fossils) of photosynthetic organisms. A nonarchaebacterial source may also be attributed to the squalenes and squalane found in sediments and petroleum (Samman et al., 1981; Brassell et al., 1981). The discovery of the phytyl side chains in the lipids of archaebacteria (DeRosa et al, 1977; Holzer and Oró, 1979; Tornabene et al., 1978, 1979) questioned whether C_{18} to C_{20} isoprenoid biomarkers were derived solely from photosynthetic sources. The fact that pristane and phytane can be found in rocks dating from the Archaen ($> 2.7 \times 10^9$ years old), which is thought to precede the "invention" of photosynthesis, and from depositional environments from which all but archaebacteria should be excluded (e.g., hypersaline, thermoacidic) gave circumstantial evidence that certain isoprenoid biomarkers originate from archaebacteria (Hahn, 1982).

Recently, additional evidence has confirmed that the cell wall lipids of methanogens contribute to the acyclic isoprenoids found in geologic samples. In these cases, molecular structures that are presently found only in either methanogenic or archaebacterial lipids have been clearly identified. For example 2, 6, 10, 15, 19-pentamethyleicosane ($C_{25}H_{52}$) is a neutral lipid found only in methanogens (Langworthy et al., 1982) and the compound was detected (0.1 to 250 ppb) in marine sediment samples that varied in age from 10^4 to 10^8 years old (Brassell et al., 1981). Furthermore, 3, 7, 11, 15, 18, 22, 26, 30-octamethyldotriacontane ($C_{40}H_{82}$) is an isoprenoid which can only be derived from the unique "head-to-head" biphytanyl groups of archaebacterial diglycerol tetraethers (Fig. 12.2). This compound was detected in samples of petroleum (Moldowan and Seifert, 1979; Albaigés, 1979; Chappe et al., 1982), sediments (Chappe et al., 1982), and oil shale kerogens treated with boron tribromide to cleave ether linkages (Chappe et al., 1979; Michaelis and Albrecht, 1979). Apparently, the ether bonds are a conservative feature of archaebacterial polar lipids. In many of these samples it was clear that the depositional environment favored an archaebacterial input from methanogenic rather than thermoacidophilic bacteria. Thus, although certain sedimentary isoprenoids having "head-to-tail" (e.g., phytane) or "tail-to-tail" (e.g., squalenes and squalane) linkages may be derived from either eubacterial, eucaryotic, or archaebacterial sources (or all), there is now clear evidence that methanogenic isoprenoids (e.g., pentamethyleicosane, octamethyldotriacontane) comprise some of the hydrocarbons found in geologic samples. This fact, taken in light of the known presence in oil of hopanoid compounds (derived from the cell wall steroids of eubacteria), implicates an important contribution of microbially derived cell materials to petroleum (Brassell et al., 1981; Ourisson et al., 1984).

12.4.2 Classification of Natural Gases

For a geochemical perspective on this subject the reader is referred to the reviews by Claypool and Kvenvolden (1983), Bernard (1979), and Claypool and Kaplan

(1974). Several parameters are used to classify natural gases with respect to their mechanism(s) of formation. The two most frequently employed criteria to distinguish between thermogenically formed gaseous hydrocarbons and those produced by microorganisms are the relative abundance of the stable carbon isotopes of methane (e.g., ^{12}C versus ^{13}C or "$\delta^{13}CH_4$") and the abundance of methane relative to higher gaseous alkanes (e.g., $CH_4/[C_2H_6 + C_3H_8]$). The assumption made in the former case is that because biochemical reactions favor the light isotope over the heavy one, methane evolved by methanogenic bacteria will be enriched in ^{12}C relative to ^{13}C. Thus isotopically "light" $\delta^{13}CH_4$ (more negative than about $-50‰$) is thought to characterize a bacterial origin while isotopically "heavy" methane (less negative than $-50‰$) indicates a thermogenic source (Fuex, 1977). In the case of the abundance of methane relative to higher alkanes, it is believed that although methane can be formed by either bacterial or thermogenic reactions, significant concentrations ($>1.0\%$) of C_2^+ alkanes (e.g., ethane, propane, etc.) can only be produced by thermogenic mechanisms. Thus, natural gases having $CH_4/(C_2H_6 + C_3H_8)$ ratios greater than 1000 indicate deposits of bacterially produced gases while lower values (<100) are typical of a thermogenic origin (Bernard, 1978). Values in between 100 and 1000 may reflect mixtures of biogenic and thermogenic gases. Use of both these parameters ($\delta^{13}CH_4$ and $CH_4/[C_2H_6 + C_3H_8]$) is frequently employed to determine the mechanism of origin for natural hydrocarbon gases collected from sedimentary systems (Fig. 12.13). Other classification parameters used in conjunction with $\delta^{13}CH_4$ values include examination of the deuterium content of the methane (Schoell, 1980). Here it is assumed that biogenic CH_4 is depleted in deuterium relative to the surrounding waters. The importance of methanogenic bacterial activities with regard to the assumptions made in these classification schemes will now be discussed.

12.4.3 Production of Higher-Molecular-Weight (C_2^+) Gaseous Alkanes

Although C_2^+ alkanes are known to be formed during thermogenic attack of kerogens (Durand-Souron et al., 1982), there is mounting evidence that these gases may also originate from bacterial sources. Significant concentrations of dissolved ethane, propane, and butanes have been detected in a variety of anaerobic aquatic environments which appear to have an active methanogenic flora. These include both marine regions and saline lakes with permanently anoxic water columns, such as the Black Sea (Hunt, 1974), the Cariaco Trench (Hammond, 1974), the Orca Basin (Sackett et al., 1979), Lake Kivu (Tietze, 1978), Big Soda Lake (Oremland and DesMarais, 1983), Mono Lake (Oremland et al., 1987), and Soap Lake (Miller, unpublished data). In addition, appreciable levels of C_2^+ alkanes are commonly found in more typical anoxic environments, such as marine and estuarine sediments (Bernard et al., 1978; Kvenvolden and Redden, 1980; Whelan et al., 1980; Vogel et al., 1982). Because some of these systems are located in regions having geothermal activities (e.g., Lake Kivu, Big Soda Lake, and Mono Lake), thermogenic inputs of C_2^+ alkanes into these lakes cannot be unequivocally ruled

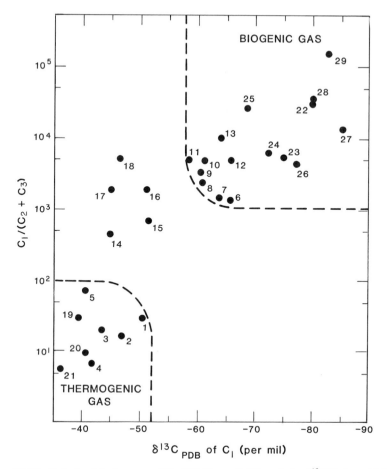

Figure 12.13 Relationship between $CH_4/(C_2H_6 + C_3H_8)$ and the $\delta^{13}CH_4$ by which hydrocarbon gases from various marine seeps, vents, and sediments are assigned to either biogenic or thermogenic origins. From Claypool and Kvenvolden (1983), reproduced with permission from the Annual Review of Earth and Planetary Sciences, Vol. 11. © 1983 by Annual Reviews Inc.

out. However, most of the sites mentioned occur where there are clearly no thermogenic sources present. Therefore, there is good circumstantial evidence that bacterial production of C_2^+ alkanes accompanies methanogenic activity in anoxic aquatic ecosystems. There is considerable debate, however, whether the levels of the gases produced by microbes are sufficiently great enough to invalidate the classification schemes [e.g., $CH_4/(C_2H_6 + C_3H_8)$] used to designate the origin of the gas.

Experimental evidence also exists which indicates that anaerobic bacteria (including methanogens) can produce gaseous alkanes higher than methane. Davis and Squires (1954) first reported that bacterial enrichment cultures produced traces of ethane and propane during methanogenic degradation of cellulose. More recently, Hunt et al. (1980) found that low levels of C_4 to C_7 hydrocarbons were evolved during growth of anaerobic enrichment cultures on terpenoid compounds. Vogel et al. (1982) were able to show that apparent microbial production of ethane, propane, n-butane, and i-butane occurred during prolonged incubation (7 months) of anoxic estuarine sediments at 27°C. Production was greatly inhibited in controls incubated at 4°C (Fig. 12.14). However, evidence for the direct involvement of methanogens in C_2^+ alkane formation in these experiments is only circumstantial.

A biochemical basis by which methanogens could evolve higher gaseous al-

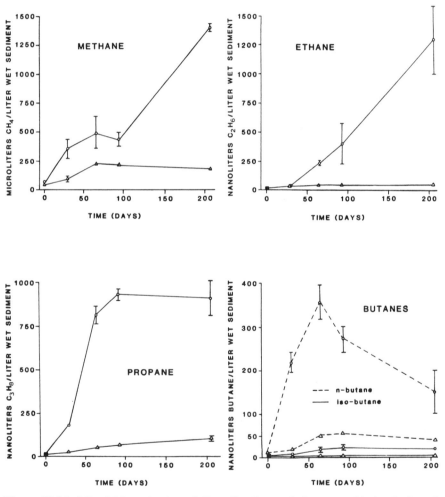

Figure 12.14 Microbial production of C_1 to C_4 alkanes during anaerobic incubation of sediments at 27°C ($\bigcirc$) and at 4°C ($\triangle$). From Vogel et al. (1982), reproduced with permission from Chemical Geology.

kanes was established by Gunsalus et al. (1978). These workers found that cell-free extracts of *Methanobacterium thermoautotrophicum* could evolve ethane when ethyl-S-CoM was provided as a substrate for the methyl-S-CoM reductase enzyme complex. However, the enzyme complex of *M. thermoautotrophicum* could not produce propane from propyl-S-CoM. Involvement of methanogenic bacteria in the formation of ethane in estuarine sediments was demonstrated by Oremland (1981). Formation of both methane and ethane by sediment slurries was stimulated by H_2 and inhibited by BES. Addition of ethyl-S-CoM to H_2-incubated slurries enhanced ethane production sixfold (Fig. 12.15). In addition, a highly purified enrichment culture of a *Methanococcus* sp. was isolated from the sediments that could produce traces of ethane from ethyl-S-CoM during growth on H_2. Some of the results obtained with estuarine sediments were reproduced with anoxic sediments from a freshwater lake (Fig. 12.16). Thus an ecological basis for ethano-genesis by methanogenic bacteria was established. It is possible that ethyl-S-CoM, or a structurally similar substance, may occur naturally in recent anoxic sediments. Gollakota and Jayalakshmi (1983) speculated that an alkylated coenzyme M derivatives may have been responsible for the production of ethane, propane, and butanes they reported during anaerobic digestion of oil cake. Recently, ethane thiol and diethylsulfide have been implicated as the likely compounds (Oremland et al., submitted).

It remains to be answered if methanogens can form more than just trace quan-

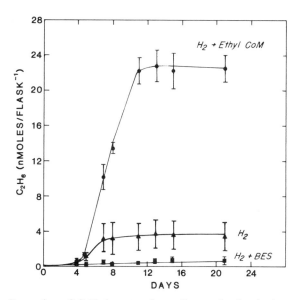

Figure 12.15 Formation of C_2H_6 by estuarine sediment slurries (volume, 75 ml) incubated under H_2. Levels of ethyl-S-CoM and BES were 2.1×10^{-3} M and 2.8×10^{-3} M, respectively. No ethane production occurred in flasks that contain both BES and ethyl-S-CoM (not shown). Adapted from Oremland (1981).

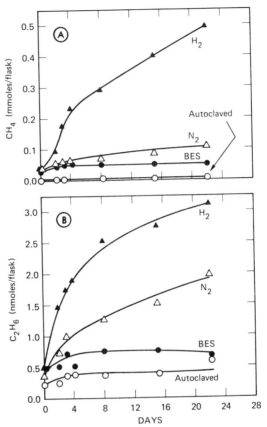

Figure 12.16 Production of (*A*) CH$_4$ and (*B*) C$_2$H$_6$ during anaerobic incubation of sediment slurries from freshwater Searsville Lake (slurry volume, 75 ml).

tities of ethane. Possibly this question can be tackled by examining pure cultures obtained from anoxic environments in which the gas is present at high concentrations. In addition, it is still not clear whether methanogens have the capacity to form propane and butanes. Although the methyl reductase enzyme complex of *M. thermoautotrophicum* is incapable of this feat, this does not rule out the methyl reductases of other methanogens. This is more than just a question of academic interest. Both ethane and propane can easily be converted in ethylene and propylene, which are important industrial feedstocks (Pruett, 1981). As supplies of petroleum dwindle or are cut off because of political reasons, an economically feasible method to produce ethane and propane could help maintain supplies of these chemical feedstocks.

12.4.4 Stable Carbon Isotope Fractionation

When grown under the proper experimental conditions, methanogenic bacteria will express a kinetic isotope effect by favoring the light ^{12}C over the heavier ^{13}C. Thus methane produced will be enriched in ^{12}C relative to the $\delta^{13}C$-value of the starting substrate. Fractionation factors express the maximum theoretical enrichment of product in ^{12}C when the organism is grown under ideal conditions. Thus a fractionation factor of 1.050 means that an organism is capable of a 50 per mil (‰) enrichment in ^{12}C in the methane relative to the substrate. In this case, for example, if the substrate $\delta^{13}C$ was -20‰, the $\delta^{13}CH_4$ would be -70‰. Table 12.3 lists some of the fractionation factors reported for methanogens grown on either H_2 plus CO_2 or methanol. As can be seen, fractionation factors decrease at higher temperatures. Thus the extreme thermophile *Methanococcus jannaschii* has a fractionation factor of only 1.032 at 83°C (J. Hayes, personal communication), which is low when compared with some of the values listed for mesophilic methanogens (Table 12.3). There are as yet no published reports concerning fractionation factors achieved during growth of methanogens on methylated amines, formate, or acetate.

It should be borne in mind that fractionation factors are derived by extrapolating experimental results to conditions of an infinite dilution of the growth substrate. This is because at conditions less than infinite dilution, the residual unreacted substrate will be somewhat isotopically heavier than the starting values, due to preferential removal of ^{12}C during growth of the methanogen. Thus as experimental gassing rates of $H_2 + CO_2$ are decreased, the enrichment of methane in ^{12}C relative to the $\delta^{13}C - CO_2$ values progressively diminishes until the effect is nearly abolished at low flow rates (Fuchs et al., 1979). This phenomenon is known as a "closed system" effect and is characterized by progressively heavier values of $\delta^{13}CH_4$ as the substrate is depleted and converted to methane. For example, Rosenfeld and Silverman (1959) reported that when methanol ($\delta^{13}C = -16.3$) was converted by an enrichment culture to methane, initial $\delta^{13}CH_4$ values were light (-84‰). However, as methanol was depleted as bacterial growth progressed, the $\delta^{13}CH_4$ values became increasingly enriched in $\delta^{13}C$. Final $\delta^{13}CH_4$ values were actually positive (e.g., $+49.9$‰) and therefore isotopically very heavy. This is

TABLE 12.3 Stable carbon isotopic fractionation factors reported for some pure and enrichment cultures of methanogens

Organism	Substrate	T(°C)	Fractionation Factor	Reference
1. Enrichment culture (marine)	Methanol	30	1.081 ⎤	Rosenfeld and Silverman
	Methanol	23	1.094 ⎦	(1959)
2. *M. barkerii*	$H_2 + CO_2$	40	1.045	Games et al. (1978)
3. *M. bryantii*	$H_2 + CO_2$	40	1.061	Games et al. (1978)
4. *M. thermoautotrophicum*	$H_2 + CO_2$	65	1.025	Games et al. (1978)
5. *M. thermoautotrophicum*	$H_2 + CO_2$	65	1.040	Fuchs et al. (1979)
6. Enrichment culture (alkalophilic)	Methanol	20	>1.077	Oremland et al. (1982a)
7. *Methanobacterium* sp. strain ivanov	$H_2 + CO_2$	37	1.033	Belyaev et al. (1983)

an important fact that should be considered when attempting to apply fractionation factors for methanogens to a natural system. In nature, bacterial reactions are frequently substrate limited, and therefore methanogens will not discriminate between ^{12}C and ^{13}C to the extent that their fractionation factors predict. It is possible, therefore, that closed system effects can occur in nature if rates of methanogenic reactions quickly deplete the pool sizes of methanogenic precursors. Unfortunately, there have been very few studies which have monitored the $\delta^{13}CH_4$ values displayed during methanogenesis in either defined mixed cultures or in sediment/soil systems. Silverman and Oyama (1968) reported that $\delta^{13}CH_4$ values were actually heavier than the substrate $\delta^{13}CO_2$ during methanogenesis in soil samples. They attributed this phenomenon to a competitive effect caused by other soil anaerobes which preferentially consumed most of the $H_2 + CO_2$.

The range of $\delta^{13}CH_4$ values found in the biological environment is quite broad, sometimes falling well into the category of thermogenic gas ($\delta^{13}CH_4$ less negative than -50‰). Table 12.4 lists some examples of $\delta^{13}CH_4$ values reported in natural systems where the methane had an obvious bacterial origin. Although classification of gases by using $\delta^{13}CH_4$ values frequently is correct, there are numerous examples of cases in which the scheme does not work. As mentioned above, closed-system effects and microbial competitors may prevent methanogenic bacteria from performing any significant ^{12}C enrichment. Other factors, however, can also play a key role in the determination of $\delta^{13}CH_4$. The isotopic composition of the starting organic matter and the methanogenic substrates must be considered, although they rarely are in natural systems. Thus Rust (1981) reported that corn-fed cows emitted methane that was about 13 per mil heavier than that emitted by alfalfa-fed cows (Table 12.4). This was because alfalfa is a C_3 plant and had a lighter $\delta^{13}C$ value (-13‰) than corn (-27‰), which is a C_4 plant.

Finally, consumption of methane by aerobic methane-oxidizing bacteria can also result in a fractionation that is opposite to that of methanogenesis (Silverman and Oyama, 1968; Barker and Fritz, 1981; Coleman et al., 1981). Thus, when methane encounters oxic/anoxic interfaces (as it frequently does in soils, sediments, rhizomes and in stratified water columns) it will be oxidized by aerobic methanotrophs. This will result in the production of isotopically light CO_2, and the residual, unoxidized methane will become enriched in ^{13}C, sometimes to the extent of having the appearance of thermogenic gas.

Isotopically heavy values of methane have also been reported in anoxic environments from which aerobic methane oxidizers are excluded. In these environments, progressive shifts from isotopically light methane to heavier values is displayed with transit from the sulfate-depleted layers to the sulfate-containing horizons of cores (Bernard, 1978). Similar shifts were seen in that anoxic sediments and waters of Big Soda Lake (Fig. 12.17; Oremland and DesMarais, 1983). One possible explanation for this phenomenon is that methane is consumed by anaerobic bacteria (possibly sulfate reducers), with the result that unoxidized, residual methane becomes isotopically heavy. Evidence for this controversial subject is discussed next.

TABLE 12.4 Stable carbon isotope ratios ($\delta^{13}CH_4$)[a] for methane of biogenic origin

Light values		
Orca Basin sediments	−88 to −105	Sackett et al. (1979)
Orca Basin brinewaters	−72 to −74.5	Sackett et al. (1979)
Marine sediments (Norway)	−71.2 to −87.3	Morris (1976)
Black Sea sediments	−68	Hunt (1974)
Steer rumen	−76	Rust (1981)
Intermediate values		
Big Soda Lake sediments	−74 to −55	Oremland and DesMarais (1983)
Big Soda Lake anoxic deep water	−55 to −59	Oremland and DesMarais (1983)
Alfalfa-fed cows	−63 ± 7	Rust (1981)
Corn-fed cows	−50 ± 3	Rust (1981)
Marine sediments (British Columbia)	−55.6	Nissenbaum et al. (1972)
Heavy values		
Big Soda Lake anoxic midwater	−20.3 to −48	Oremland and DesMarais (1983)
Lake sediments (Utah)	−45.8	Rice and Claypool (1981)
Lake Kivu sediments	−45	Deuser et al. (1973)
Dairy cow rumen	−47.4	Rust (1981)
Sludge gas	−47.1	Nissenbaum et al. (1972)

$$^a\delta^{13}CH_4 = \left[\frac{(^{13}C/^{12}C)_{unknown}}{(^{13}C/^{12}C)_{standard}} - 1 \right] \times 1000.$$

12.4.5 Anaerobic Methane Oxidation

Concentration profiles of methane in anoxic marine waters and sediments frequently have ''concave upward'' appearances (Barnes and Goldberg, 1976; Reeburgh, 1976; Reeburgh and Heggie, 1977; Martens and Berner, 1977; Bernard, 1979). Kinetic models of such profile data indicate that there is less methane in these regions than would be predicted by simple diffusion. The models achieve a better ''fit'' to the data when a term for methane consumption is included (Fig. 12.18). Thus there is circumstantial evidence that microbial consumption of methane can take place in the absence of oxygen. Because sulfate is the only apparent oxidant present in sufficient quantity to cause a significant removal of methane, this process is theorized to be mediated by sulfate-reducing bacteria. Thermodynamic calculations indicate that methane oxidation by sulfate-reducing bacteria is favorable but inadequate for bacterial growth (Martens and Berner, 1977; Wake et al., 1977).

Experimental evidence with bacterial cultures, however, is conflicting. Sorokin (1957) was unable to demonstrate any such activity in either pure or mixed cultures of sulfate reducers. However, Davis and Yarbrough (1966) reported that a small amount of $^{14}CH_4$ could be converted to $^{14}CO_2$ by *Desulfovibrio desulfuricans*, and that this activity was enhanced by co-metabolism of lactate. More recently, Panganiban et al. (1979) were able to demonstrate a time-dependent conversion of

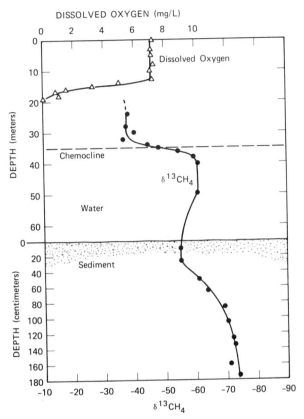

Figure 12.17 Values of $\delta^{13}CH_4$ detected in the anaerobic water column and sediments of Big Soda Lake during October 1982 [for details see Oremland and DesMarais (1983)].

$^{14}CH_4$ to $^{14}CO_4$ by anaerobic lake waters and crude enrichment cultures. Oxidation of CH_4 by the anaerobic enrichment cultures required the presence of both acetate and sulfate. Subsequently, there have been several reports of oxidation of $^{14}CH_4$ to $^{14}CO_4$ by anaerobic marine sediments obtained from diverse locations (Reeburgh, 1980; Iversen and Blackburn, 1981; Devol, 1983; Alperin and Reeburgh, 1985). However, there has been disagreement on the potential contribution of methane as a possible electron donor for sulfate reduction. Iversen and Blackburn (1981) estimated that anaerobic methane oxidation could account for less than 0.06% of the observed rates of sulfate reduction, while the data presented by Devol (1983) and Reeburgh (1980) imply a contribution of about 30% at the depths of maximum oxidation.

A further complicating feature of this debate was the demonstration that methanogenic bacteria were capable of oxidizing $^{14}CH_4$ to $^{14}CO_2$ (Zehnder and Brock, 1979b). Of the nine strains of methanogens tested, all were capable of methane

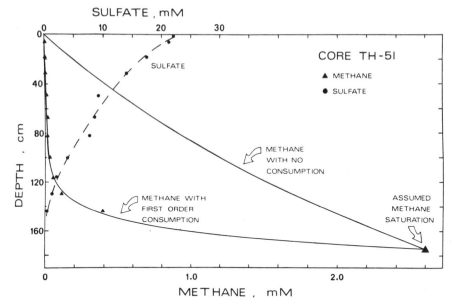

Figure 12.18 Methane and sulfate porewater concentrations in a core recovered from Long Island Sound. Solid lines represent plots of theoretical curves with and without anaerobic methane oxidation by first-order kinetics. Reproduced with permission from C. S. Martens and R. A. Berner, 1977, Interstitial water chemistry of anoxic Long Island Sound sediments. I. Dissolved gases. Limnology and Oceanography, 22:10–25.

oxidation, which they achieved via a biochemical pathway that was somewhat different from the route of methanogenesis. However, oxidation rates were only a small percentage (0.001 to 0.3%) of the rates of production, and therefore a net production (rather than consumption) of methane was always observed. In subsequent experiments with lake sediments and sewage sludge, $^{14}CH_4$ oxidation rates were only 2% and 8%, respectively, of the rates of methane formation (Zehnder and Brock, 1980). Inhibition of both methane production and oxidation by BES indicated that methanogenic bacteria were the causative agents of both the reactions. Although oxidation rates could be increased by addition of ferrous sulfate and by boosting the partial pressure of methane, a net consumption of methane was never observed. These results had several important implications to the question of anaerobic methane oxidation. First, sulfate-reducing bacteria need not be directly involved in the phenomenon. Second, and most important, is that mere demonstration of $^{14}CO_2$ production from $^{14}CH_4$ is insufficient evidence for a net consumption of methane. It is essential that a concurrent assessment be made of the *in situ* rate of methanogenesis to determine whether a true net consumption of methane takes place or whether production exceeds the measured oxidation. In addition, the purity of the $^{14}CH_4$ must be carefully checked and cleansed to remove contaminant $^{14}CO_2$.

Results with anoxic Big Soda Lake waters indicate that methane oxidation rates exceed production by several-fold (Iversen, et al., 1987). In addition, unlabeled methane disappeared during incubation of some anoxic Framvaren Fjord water samples, whereas no loss was observed in controls containing formalin (Lidstrom, 1983). These recent results indicate that reports of anaerobic methane oxidation cannot be dismissed as experimental artifacts and bear further scrutiny. It is significant that anaerobic oxidation has been conclusively demonstrated for the gaseous hydrocarbon acetylene (Culbertson et al., 1981). This compound may serve as a model for future investigations into the problem of anaerobic methane oxidation.

12.4.6 Depth of Sedimentary Penetration by Active Methanogens

Most studies concerning the microbial formation of methane in sediments scrutinize only the very recent portion (upper few meters) of the sediment column. Relatively little work has been done with materials recovered from greater sedimentary depths (e.g., tens to hundreds of meters). Circumstantial geochemical evidence points to the active presence of methanogens in deep sediments obtained during the Deep Sea Drilling Project. For example, Hathaway et al. (1979) published profiles of interstitial sulfate and methane that were reminiscent of those encountered in nearshore sediments. The only difference was that sulfate depletion and subsequent methane appearance extended over hundreds of meters instead of centimeters. In addition, isotopically "light" methane can be found at depths of hundreds of meters in these DSDP materials (Fig. 12.19), sometimes penetrating the zones of methane gas-hydrate (clathrate) formation (Claypool and Kvenvolden, 1983). Methanogenic bacteria have been recovered from deep aquifers in Montana (Olson et al., 1981) and Florida (Godsy, 1980), suggesting that the organisms may be active within the aquifers. No methanogenic activity or viable cells were found in DSDP materials taken during Leg 50 (Ivanov et al., 1980); however, methanogenic activity was found in samples down to 12 m in DSDP Leg 64 (Oremland et al., 1982c) as well as in samples from 42 m depth in Leg 95 (Tarafa et al., 1986) and down to 167 m in Leg 96 (Whelan et al., 1986). Belyaev et al. (1980) reported that sediments taken from a subbottom depth of 90 m in the Caspian Sea had low, but detectable methanogenic activity. Subsequently, these workers have reported both the presence of viable cells of methanogens (~ 60 mL^{-1}) as well as detectable methanogenic activity in deep (720 to 750 m) groundwaters of the Caspian depression (Belyaev and Ivanov, 1983).

The active presence of an anaerobic flora in subsurface environments should prove to be of importance with respect to degradation of organic pollutants. In addition, there are important implications for the field of petroleum geochemistry. Volatile components of petroleum can be biodegraded by anaerobes in oil reservoirs (Hunt, 1979), and it is possible that methanogens may be involved (Muller, 1957). It is significant, therefore, that methanogenic activity was recently reported in an oil field (Ivanov et al., 1983) and the methane produced greatly altered the

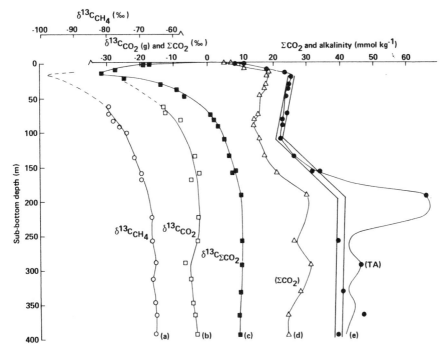

Figure 12.19 Stable carbon isotope values for CH_4 and CO_2 in cores taken from the Blake Outer Ridge by the Deep Sea Drilling Project, Leg 76. Data can be interpreted to indicate bacterial methanogenesis occurring at great sedimentary depths. From Claypool and Threlkeld (1983), reproduced with permission from the Initial Reports of the Deep Sea Drilling Project.

$\delta^{13}CH_4$ signal of the methane found in the deposits, thereby giving the methane a "biogenic" isotopic signature.

12.4.7 Atmospheric Methane

Methane is one of several important trace gases that participate in the complex chemical reactions that occur in the earth's atmosphere. Global tropospheric methane concentrations average about 1.6 ppm; however, short-term temporal and spatial variability is common and is influenced by geographic location, seasons, mixing patterns, and diurnal cycles (Ehhalt and Heidt, 1973; Khalil and Rasmussen, 1983; Blake and Rowland, 1986). Average methane concentrations are lower in the southern hemisphere (1.5 ppm), possibly due to less anthropogenic input (Lamontagne et al., 1974). Analyses of air bubbles entrapped in polar ice indicate that there has been a two- to threefold increase in the abundance of methane in the atmosphere over the past 100–200 years (Robbins et al., 1973; Khalil and Rasmussen, 1982), presumably due to human activities and the onset of the industrial revolution. Currently, the rate of increase has risen to 1 to 2% per year (Khalil and Rasmussen, 1983). These results have generated much concern because of possible global climatic changes (e.g., warming trend or a runaway "greenhouse" effect) induced by altered abundances of important atmospheric trace gases. Because methane absorbs in the infrared, it plays a role that is analogous to carbon dioxide and may therefore be an additive factor to global warming phenomena. In addition, methane is destroyed in the atmosphere by reaction with hydroxyl radicals, with the production of carbon monoxide and hydrogen. Methane may therefore be the major source of atmospheric carbon monoxide (Levy, 1973). On the positive side, elevated levels of methane may help protect the stratospheric ozone layer from destruction by chlorine or nitrous oxide. Methane may react with these gases, which are derived from either propellents (e.g., chlorofluorocarbons) or from increased fertilizer application, respectively (Owens et al., 1982). These findings have heightened the need for better data concerning the release of methane from the geosphere in order to obtain accurate estimates of its residence time in the atmosphere. Currently, the residence time is believed to be about 8 years (Khalil and Rasmussen, 1983).

Atmospheric methane subjected to ^{14}C dating indicates that about 80% of the gas is derived from "recent" carbon (mainly via bacterial decay), while only 20% originates from older sources, such as fossil fuel or possibly volcanic emissions (Ehhalt, 1976). An estimate of total geosphere methane release is 544 to 861 $\times$ 10^{12} g year^{-1}, of which 528 to $8^{12} \times 10^{12}$ g is derived from bacterial decay of organic matter (Ehhalt, 1976). The surface oceans are slightly supersaturated with respect to methane, and therefore this vast area represents a minor source as opposed to a major sink for atmospheric methane (Lamontagne et al., 1973). Major inputs from biogenic sources include animal gastrointestinal fermentations (101 to 220 $\times 10^{12}$ g year^{-1}), rice paddy soils (280 $\times 10^{12}$ g year^{-1}) and marshes (130 to 260 $\times 10^{12}$ g year^{-1}). It should be borne in mind, however, that these flux estimates are based on a relatively few, narrowly focused studies. Extrapolation to a global basis of laboratory or field methane flux measurements quite naturally involves several tenuous assumptions concerning weather, seasons, soil types, agricultural practices, and so on. Recent research efforts have concentrated on improving these estimates by making field measurements over a variety of locations, seasons, and biological cycles.

Koyama (1962) estimated methane flux from Japanese rice paddies to be about 9.2 mg m^{-2} h^{-1}. This estimate was derived from laboratory incubations of anaerobic soil columns held in the dark at constant temperature. More recently, Seiler et al. (1984b) measured methane flux from a Spanish rice paddy. Not surprisingly, their field data were more variable than Koyama's work and they reported that rates ranged between 2 and 14 mg m^{-2} h^{-1}. Low rates were observed during the tillering stage, while high rates were evident during the plant's flowering stage. Therefore, emission rates were influenced by the biological status of the system rather than by physical factors like temperature. In addition, 95% of the methane emanating from the paddies was transported and emitted to the atmosphere via the plant rather than by bubble ebullition or diffusion from the waterlogged soils. Similarly, water lilies serve as conduits to the atmosphere for methane formed in shallow, anoxic lake sediments (Dacey and Klug, 1979). This phenomenon effectively bypasses removal of methane by aerobic methane-oxidizing bacteria which occupy the water column of lakes and thus increases net flux to the atmosphere from the regions. Not all aquatic plants, however, serve as methane conduits. Although methane in the rhizosphere gases of the tropical seagrass *Thalassia testudinum* exhibited large diurnal fluctuations, very little methane was present in the gases of the rhizomes themselves (Oremland and Taylor, 1977). Because the plant "pumped" oxygen into the rhizosphere during daytime illumination, bacterial methane oxidation may have occurred on the rhizome surface, thereby lowering the quantity of the gas entering the rhizome. Finally, photosynthetic oxygen production by epibenthic diatoms in shallow, tropical sediments can limit methane efflux measurements by maintaining aerobic conditions in flux chambers (Oremland, 1975). In this study, methane flux rates were ten-fold greater in dark-incubated chambers (0.34 mg m^{-2} h^{-1}) compared with chambers exposed to normal diurnal light patterns (0.033 mg m^{-2} h^{-1}).

12.5 SUMMARY

Methanogenic bacteria are of significant biological and evolutionary interest because of their unique biochemical properties as well as their classification as Archaebacteria. In addition, these organisms play an important role in the degradation of organic matter in nature, and an understanding of their ecological niche provides an economic incentive for the recovery of methane from society's waste products. Furthermore, methanogens play an important role in the geochemistry of carbon in the earth's crust and atmosphere. They are involved in the formation of natural gas and petroleum deposits as well as the increase of methane abundance in the atmosphere. More research investigations concerned with all aspects of methanogenesis should therefore be vigorously pursued.

ACKNOWLEDGMENTS

I wish to express my thanks to J. G. Ferry, A. J. B. Zehnder, K. O. Stetter, K. A. Kvenvolden, and M. Schoell for providing certain graphic materials used in the chapter. I am grateful to A. Catena, L. Miller, J. Whelan, B. F. Taylor, R. L. Smith, and D. Capone for helpful discussions and manuscript reviews. Finally, a special thanks to F. Strohmaier for typing the manuscript.

REFERENCES

Abram, J. W., and D. B. Nedwell. 1978. Hydrogen as a substrate for methanogenesis and sulfate reduction in anaerobic saltmarsh sediments. Arch. Microbiol. 117:93–97.

Albaigés, J. 1979. Identification and geochemical significance of long chain acyclic isoprenoid hydrocarbons in crude oils, p. 19–28, in A. G. Douglas and J. R. Maxwell (eds.), Advances in organic geochemistry 1979. Pergamon Press, Oxford.

Alperin, M. J., and W. S. Reeburgh. 1985. Inhibition experiments on anaerobic methane oxidation. Appl. Environ. Microbiol. 50:940–945.

Archer, D. B., and N. R. King. 1983. A novel ultrastructural feature of a gas-vacuolated Methanosarcina. FEMS Microbiol. Lett. 16:217–223.

Badzoing, W., R. K. Thauer, and J. G. Zeikus. 1978. Isolation and characterization of Desulfovibrio growing on hydrogen plus sulfate as the sole energy sources. Arch. Microbiol. 116:41–50.

Balch, W. E., and R. S. Wolfe. 1976. New approach to the cultivation of methanogenic bacteria: 2-mercaptoethane sulfonic acid (HS-CoM)-dependent growth of Methanobacterium ruminantium in a pressurized atmosphere. Appl. Environ. Microbiol. 32:781–791.

Balch, W. E., and R. S. Wolfe. 1979a. Specificity and biological distribution of coenzyme M (2-mercaptoethane sulfonic acid). J. Bacteriol. 137:256–263.

Balch, W. E., and R. S. Wolfe. 1979b. Transport of coenzyme M (2-mercapto-ethanesulfonic acid) in Methanobacterium ruminantium. J. Bacteriol. 137:264–273.

Balch, W. E., L. J. Magrum, G. E. Fox, R. S. Wolfe, and C. R. Woese. 1977a. An ancient divergence among the bacteria. J. Mol. Evol. 9:305–311.

Balch, W. E., S. Schoberth, R. S. Tanner, and R. S. Wolfe. 1977b. Acetobacterium, a new genus of hydrogen-oxidizing, carbon dioxide-reducing anaerobic bacteria. Int. J. Syst. Bacteriol. 27:355–361.

Balch, W. E., G. E. Fox, L. J. Magrum, C. R. Woese, and R. S. Wolfe. 1979. Methanogens: reevaluation of a unique biological group. Microbiol. Rev. 43:260–296.

Balderston, W. L., and W. J. Payne. 1976. Inhibition of methanogenesis in salt marsh sediments and whole-cell suspensions of methanogenic bacteria by nitrogen oxides. Appl. Environ. Microbiol. 32:264–269.

Barker, H. A. 1936. Studies on methane producing bacteria. Arch. Microbiol. 7:420–438.

Barker, H. A. 1956. Bacterial fermentations. Wiley, New York, p. 1–95.

Barker, J. F., and P. Fritz. 1981. Carbon isotope fractionation during microbial methane oxidation. Nature (Lond.) 293:389–391.

Barnes, R. O., and E. D. Goldberg. 1976. Methane production and consumption in anoxic marine sediments. Geology (Boulder) 4:297–300.

Baross, J. A., and J. W. Deming. 1983. Growth of "black smoker" bacteria at temperatures of at least 250°C. Nature (Lond.) 303:423–426.

Baross, J. A., M. D. Lilley, and L. I. Gordon. 1982. Is the CH_4, H_2 and CO venting from submarine hydrothermal systems produced by thermophilic bacteria? Nature (Lond.) 298:366–368.

Bauchop, T. 1967. Inhibition of rumen methanogenesis by methane analogues. J. Bacteriol. 94:171–175.

Belay, N., R. Sparling, and L. Daniels. 1984. Dinitrogen fixation by a thermophilic methanogenic bacterium. Nature (Lond.) 312:286–288.

Belyaev, S. S., and M. V. Ivanov. 1983. Bacterial methanogenesis in underground waters, p. 273–280, in R. Hallberg (ed.), Environmental biogeochemistry, Proc. 5th Int. Symp.

Env. Biogeochemistry (ISEB). Ecol. Bull. (Stockholm) No. 35. Swedish Natural Scientific Research Council, Stockholm.

Belyaev, S. S., A. Y. Lein, and M. V. Ivanov. 1980. Role of methane-producing and sulfate-reducing bacteria in the process of organic matter destruction, in P. A. Trudinger and M. R. Walter (eds), Biogeochemistry of ancient and modern environments. Proc. 4th Int. Symp. Environ. Biogeochem., pp. 235–242.

Belyaev, S. S., R. Wolkin, W. R. Kenealy, M. J. DeNiro, S. Epstein, and J. G. Zeikus. 1983. Methanogenic bacteria from the Bondyuzhskoe oil field: general characterization and analysis of stable-carbon isotopic fractionation. Appl. Environ. Microbiol. 45:691–697.

Bernard, B. B. 1978. Light hydrocarbons in marine sediments. Dissertation, Texas A & M University.

Bernard, B. B. 1979. Methane in marine sediments. Deep-Sea Res. 26:429–443.

Bernard, B. B., M. M. Brooks, and W. M. Sackett. 1978. Light hydrocarbons in recent Texas continental shelf and slope sediments. J. Geophys. Res. 83:4053–4061.

Blake, D. R. and F. S. Rowland. 1986. Worldwide increase in tropospheric methane, 1978–1983. J. Atmos. Chem. 4:43–62.

Blaut, M., and G. Gottschalk. 1982. Effect of trimethylamine on acetate utilization by *Methanosarcina barkeri*. Arch. Microbiol. 133:230–235.

Bouwer, E. J., and P. L. McCarty. 1983. Transformations of 1- and 2-carbon halogenated aliphatic organic compounds under methanogenic conditions. Appl. Environ. Microbiol. 45:1295–1299.

Boyd, S. A., D. R. Shelton, D. Berry, and J. M. Tiedje. 1983. Anaerobic biodegradation of phenolic compounds in digested sludge. Appl. Environ. Microbiol. 46:50–54.

Bracke, J. W., D. L. Cruden, and A. J. Markovetz. 1979. Intestinal microbial flora of the American cockroach, *Periplaneta americana*. Appl. Environ. Microbiol. 38:945–955.

Brassell, S. C., A. M. K. Wardroper, I. D. Thomson, J. R. Maxwell, and G. Eglington. 1981. Specific acyclic isoprenoids as biological markers of methanogenic bacteria in marine sediments. Nature (Lond.) 290:693–696.

Braun, K., and G. Gottschalk. 1981. Effect of molecular hydrogen and carbon dioxide on chemo-organotrophic growth of *Acetobacterium woodii* and *Clostridium aceticum*. Arch. Microbiol. 128:294–298.

Breznak, J. A. 1982. Intestinal microbiota of termites and other xylophageous insects. Annu. Rev. Microbiol. 32:323–343.

Brooks, J. M., D. F. Reid, and B. B. Bernard. 1981. Methane in the upper water column of the Northwestern Gulf of Mexico. J. Geophys. Res. 86:11,029–11,040.

Bryant, M. P. 1972. Commentary on the Hungate technique for culture of anaerobic bacteria. Am. J. Clin. Nutr. 25:1324–1328.

Bryant, M. P. 1974. Part 13. Methane-producing bacteria, p. 472–477, in R. E. Buchanan and N. E. Gibbons (eds.) Bergey's manual of determinative bacteriology, 8th ed. Williams & Wilkins, Baltimore.

Bryant, M. P. 1977. Microbiology of the rumen, p. 287–304, in M. J. Swenson (ed.), Physiology of domestic animals, 9th rev. ed. Cornell University Press, Ithaca, N.Y.

Bryant, M. P. 1979. Microbial methane production—theoretical aspects. J. Anim. Sci. 48:193–201.

Bryant, M. P., E. A. Wolin, M. J. Wolin, and R. S. Wolfe. 1967. *Methanobacterium omelianskii*, a symbiotic association of two species of bacteria. Arch. Microbiol. 59:20–31.

Bryant, M. P., S. F. Tzeng, I. M. Robinson, and A. E. Joyner. 1971. Nutrient requirements of methanogenic bacteria. Adv. Chem. Ser. 105:23–40.

Bryant, M. P., L. L. Campbell, C. A. Reddy, and M. R. Crabill. 1977. Growth of *Desulfovibrio* on lactate or ethanol media low in sulfate in association with H_2-utilizing methanogenic bacteria. Appl. Environ. Microbiol. 33:1162–1169.

Burke, R. A., Jr., J. M. Brooks, and W. M. Sackett. 1981. Light hydrocarbons in Red Sea brines and sediments. Geochim. Cosmochim. Acta 45:627–634.

Capone, D. G., D. D. Reese, and R. P. Kiene. 1983. Effects of metals on methanogenesis, sulfate reduction, carbon dioxide evolution, and microbial biomass in anoxic salt marsh sediments. Appl. Environ. Microbiol. 45:1586–1591.

Cappenberg, T. E. 1974a. Interrelations between sulfate-reducing and methane-producing bacteria in bottom deposits of a fresh-water lake. I. Field observation. Antonie Leeuwenhoek J. Microbiol. Serol. 40:285–295.

Cappenberg, T. E. 1974b. Interrelations between sulfate-reducing and methane-producing bacteria in bottom deposits of a fresh-water lake. II. Inhibition experiments. Antonie Leeuwenhoek J. Microbiol. Serol. 40:297–306.

Cappenberg, T. E., and R. A. Prins. 1974. Interrelations between sulfate-reducing and methane-producing bacteria in bottom deposits of a fresh-water lake. III. Experiments with ^{14}C-labelled substrates. Antonie Leeuwenhoek J. Microbiol. Serol. 40:457–469.

Chappe, B., W. Michaelis, and P. Albrecht. 1979. Molecular fossils of archaebacteria as selective degradation products of kerogen, p. 265–274, in A. G. Douglas and J. R. Maxwell (eds.), Advances in organic geochemistry 1979. Pergamon Press, Oxford, 1980.

Chappe, B., P. Albrecht, and W. Michaelis. 1982. Polar lipids of archaebacteria in sediments and petroleums. Science 217:65–66.

Cheeseman, P., A. Toms-Wood, and R. S. Wolfe. 1972. Isolation and properties of a fluorescent compound, factor$_{420}$ from *Methanobacterium* strain M.O.H. J. Bacteriol. 112:527–531.

Chen, M., and M. J. Wolin. 1979. Effect of monensin and lasalocid-sodium on the growth of methanogenic and rumen saccharolytic bacteria. Appl. Environ. Microbiol. 38:72–77.

Claypool, G. E., and I. R. Kaplan. 1974. The origin and distribution of methane in marine sediments, p. 99–140, in I. R. Kaplan (ed.), Natural gases in marine sediments. Plenum Press, New York.

Claypool, G. E., and K. A. Kvenvolden. 1983. Methane and other hydrocarbon gases in marine sediments. Annu. Rev. Earth Planet. Sci. 11:299–327.

Claypool, G. E., and C. N. Threlkeld. 1983. Anoxic diagenesis and methane generation in sediments of the Blake Outer Ridge, Deep Sea Drilling Project site 533, Leg 76, p. 391–402, in initial reports of the deep sea drilling project, Vol. 76. U.S. Government Printing Office, Washington, D.C.

Coleman, D. D., J. B. Risatti, and M. Schoell. 1981. Fractionation of carbon and hydrogen isotopes by methane-oxidizing bacteria. Geochim. Cosmochim. Acta 45:1033–1037.

Comita, P. B. and R. B. Gagosian. 1983. Membrane lipid from deep sea hydrothermal vent methanogen: a new macrocyclic glycerol diether. Science 222:1329–1331.

Compeau, G. C., and R. Bartha. 1985. Sulfate-reducing bacteria: Principal methylators of mercury in anoxic estuarine sediments. Appl. Environ. Microbiol. 50:498–502.

Conway de Macario, E., M. J. Wolin, and A. J. L. Macario. 1981. Immunology of archaebacteria that produce methane gas. Science 214:74–75.

Corder, R. E., L. A. Hook, J. M. Larkin, and J. I. Frea. 1983. Isolation and characterization of two new methane-producing cocci: *Methanogenium olentangyi* sp. nov., and *Methanococcus deltae*, sp. nov. Arch. Microbiol. 134:28–32.

Culbertson, C. W., A. J. B. Zehnder, and R. S. Oremland. 1981. Anaerobic oxidation of acetylene by estuarine sediments and enrichment cultures. Appl. Environ. Microbiol. 41:396–403.

Dacey, J. W. H., and M. Klug. 1979. Methane efflux from lake sediments through water lilies. Science 203:1253–1255.

Daniels, L., G. Fuchs, R. K. Thauer, and J. G. Zeikus. 1977. Carbon monoxide oxidation by methanogenic bacteria. J. Bacteriol. 132:118–126.

Daniels, L., G. Fulton, R. W. Spencer, and W. H. Orme-Johnson. 1980. Origin of hydrogen in methane-produced by *Methanobacterium thermoautotrophicum*. J. Bacteriol. 141:694–698.

Daniels, L., R. Sparling, and G. D. Sprott. 1984. The bioenergetics of methanogenesis. Biochim. Biophys. Acta 768:113–163.

Daniels, L., N. Belay, and B. S. Rajagopal, 1986. Assimilatory reduction of sulfate and sulfite by methanogenic bacteria. Appl. Environ. Microbiol. 51:703–709.

Davis, J. B., and R. M. Squires. 1954. Detection of microbially produced gaseous hydrocarbons other than methane. Science 119:381–382.

Davis, J. B., and H. F. Yarbrough. 1966. Anaerobic oxidations of hydrocarbons by *Desulfovibrio desulfuricans*. Chem. Geol. 1:137–144.

DeRosa, M., S. DeRosa, A. Gambocorta, L. Minale, and J. D. Bu'lock. 1977. Chemical structure of the ether lipids of thermophilic acidophilic bacteria of the Caldariella groups. Phytochemistry (Oxf.) 16:1961–1965.

Deuser, W. G., E. T. Degans, and G. R. Harvey. 1973. Methane in Lake Kivu: new data bearing on its origin. Science 181:51–54.

Devol, A. H. 1983. Methane oxidation rates in the anaerobic sediments of Saanich Inlet. Limnol. Oceanog. 28:738–742.

Didyk, B. M., B. R. T. Simoneit, S. C. Brassel, and G. Eglington. 1978. Organic geochemical indicators of paleoenvironmental conditions of sedimentation. Nature (Lond.) 272:216–222.

Doddema, H. J., and G. D. Vogels. 1978. Improved identification of methanogenic bacteria by fluorescence microscopy. Appl. Environ. Microbiol. 36:752–754.

Doddema, H. J., C. van der Drift, G. D. Vogels, and M. Veenhuis. 1979. Chemiosmotic coupling in *Methanobacterium thermoautotrophicum*: hydrogen-dependent adenosine 5'-triphosphate synthesis by subcellular particles. J. Bacteriol. 140:1081–1089.

Durand-Souron, C., R. Boulet, and B. Durand. 1982. Formation of methane and hydrocarbons by pyrolysis of immature kerogens. Geochim. Cosmochim. Acta 46:1193–1202.

Dwyer, D. F., and J. M. Tiedje. 1983. Degradation of ethylene glycol and polyethylene glycols by methanogenic consortia. Appl. Environ. Microbiol. 46:185–190.

Edwards, T., and B. C. McBride. 1975. A new method for the isolation and identification of methanogenic bacteria. Appl. Microbiol. 29:540–545.

Ehhalt, D. H. 1976. The atmospheric cycle of methane, p. 13–22, in H. G. Schlegel, G. Gottschalk, and N. Pfennig (eds.), Microbial formation and utilization of gases (H_2, CH_4, CO). E. Geltze KG, Göttingen, West Germany.

Ehhalt, D. H., and L. E. Heidt. 1973. Vertical profiles of CH_4 in the troposphere and stratosphere. J. Geophys. Res. 78:5265–5272.

Ferguson, T. J., and R. A. Mah. 1983. Effect of H_2-CO_2 on methanogenesis from acetate or methanol in *Methanosarcina* spp. Appl. Environ. Microbiol. 46:348–355.

Ferry, J. G., and R. S. Wolfe. 1976. Anaerobic degradation of benzoate to methane by a microbial consortium. Arch. Microbiol. 107:33–40.

Fiebig, K., and G. Gottschalk. 1983. Methanogenesis from choline by a co-culture of *Desulfovibrio* sp. and *Methanosarcina barkeri*. Appl. Environ. Microbiol. 45:161–168.

Fina, L. R., R. L. Bridges, T. H. Coblentz, and F. F. Roberts. 1978. The anaerobic decomposition of benzoic acid during methane fermentation. III. The fate of carbon four and the identification of propanoic acid. Arch. Microbiol. 118:169–172.

Fox, G. E., L. J. Magrum, W. E. Balch, R. S. Wolfe, and C. R. Woese. 1977. Classification of methanogenic bacteria by 16S ribosomal RNA classification. Proc. Natl. Acad. Sci. USA 74:4537–4541.

Fuchs, G., R. Thauer, H. Ziegler, and W. Stichler. 1979. Carbon isotope fractionation by *Methanobacterium thermoautotrophicum*. Arch. Microbiol. 120:135–139.

Fuex, A. N. 1977. The use of stable carbon isotopes in hydrocarbon exploration. Geochem. Explor. 7:155–188.

Games, L. M., J. M. Hayes, and R. P. Gunsalus. 1978. Methane producing bacteria: natural fractionations of the stable carbon isotopes. Geochim. Cosmochim. Acta 42:1295–1297.

Giani, D., L. Giani, Y. Cohen, and W. E. Krumbein. 1984. Methanogenesis in the hypersaline Solar Lake (Sinai). FEMS Microbiol. Lett. 25:219–224.

Godsy, E. M. 1980. Isolation of *Methanobacterium bryantii* from a deep aquifer by using a novel broth-antibiotic disk method. Appl. Environ. Microbiol. 39:1074–1075.

Gollakota, K. G., and B. Jayalakshmi. 1983. Biogas (natural gas?) production by anaerobic digestion of oil cake by a mixed culture isolated from cow dung. Biochem. Biophys. Res. Commun. 110:32–35.

Grbić-Galić, D. 1983. Anaerobic degradation of coniferyl alcohol by methanogenic consortia. Appl. Environ. Microbiol. 46:1442–1446.

Gunsalus, R. P., J. A. Romesser, and R. S. Wolfe. 1978. Preparation of coenzyme M analogues and their activity in the methyl coenzyme M reductase system of *Methanobacterium thermoautotrophicum*. Biochemistry 17:2374–2377.

Hahn, J. 1983. Geochemical fossils of a possibly archaebacterial origin in ancient sediments. Zentralbl. Bakteriol. Mikrobiol. Hyg. 1 Abt. Orig. C3:40–52.

Hammond, D. E. 1974. Dissolved gases in Cariaco Trench sediments: anaerobic diagenesis, p. 71–90, in I. R. Kaplan (ed.), Natural gases in marine sediments. Plenum Press, New York.

Hathaway, J. C., C. W. Poag, P. C. Valentine, R. E. Miller, D. M. Schultz, F. T. Manheim, F. A. Kohout, M. H. Bothner, and D. W. Sangrey. 1979. U.S. Geological Survey core drilling on the Atlantic shelf. Science 206:515–527.

Healy, J. B., Jr., and L. Y. Young. 1979. Anaerobic biodegradation of eleven aromatic compounds to methane. Appl. Environ. Microbiol. 38:84–89.

Herwig, R. P., J. T. Staley, M. K. Nerini, and H. W. Braham. 1984. Baleen whales: preliminary evidence for forestomach microbial fermentation. Appl. Environ. Microbiol. 47:421–423.

Hilpert, R., J. Winter, W. Hammes, and O. Kandler. 1981. The sensitivity of archaebacteria to antibiotics. Zentralbl. Bakteriol. Mikrobiol. Hyg. 1. Abt. Orig. C2:11–20.

Hippe, H., D. Caspari, K. Fiebig, and G. Gottschalk. 1979. Utilization of trimethylene and other n-methyl compounds for growth and methane formation by *Methanosarcina barkeri*. Proc. Natl. Acad. Sci. USA 76:494–498.

Hobson, P. N., and B. G. Shaw. 1976. Inhibition of methane production by *Methanobacterium formicicum*. Water Res. 10:849–852.

Holzer, G., and J. Oró. 1979. Gas chromatographic-mass spectrometric analysis of neutral lipids from methanogenic and thermoacidophilic bacteria. J. Chromatogr. 186:795–809.

Horowitz, A., J. M. Suflita, and J. M. Tiedje. 1983. Reductive dehalogenations of halobenzoates by anaerobic lake sediment microorganisms. Appl. Environ. Microbiol. 45:1459–1469.

Huber, H., M. Thomm, H. König, G. Thies, and K. O. Stetter. 1982. *Methanococcus thermolithotrophicus*, a novel thermophilic lithotrophic methanogen. Arch. Microbiol. 132:47–50.

Hungate, R. E. 1966. The rumen and its microbes. Academic Press, New York.

Hungate, R. E. 1967. Hydrogen as an intermediate in the rumen fermentation. Arch. Microbiol. 59:158–164.

Hungate, R. E. 1969. A roll tube method for the cultivation of strict anaerobes, p. 117–132, in J. R. Norris and D. W. Ribbons (eds.), Methods in microbiology, Vol. 3B. Academic Press, New York.

Hungate, R. E., W. Smith, T. Bauchop, I. Yu, and J. C. Rabinowitz. 1970. Formate as an intermediate in the bovine rumen fermentation. J. Bacteriol. 102:389–397.

Hunt, J. M. 1974. Hydrocarbon geochemistry of the Black Sea, p. 449–504, in E. T. Degens and D. A. Ross (eds.), The Black Sea—geology, chemistry and biology. American Association Petroleum Geologists, Tulsa, Okla.

Hunt, J. M. 1979. Petroleum geochemistry and geology. W. H. Freeman, San Francisco.

Hunt, J. M., R. J. Miller, and J. K. Whelan. 1980. Formation of C_4-C_7 hydrocarbons from bacterial degradation of naturally occurring terpenoids. Nature (Lond.) 288:577–578.

Huser, B. A., K. Wuhrmann, and A. J. B. Zehnder. 1982. *Methanotrix soehngenii* gen. nov. sp. nov., a new acetotrophic nonhydrogen-oxidizing methane bacterium. Arch. Microbiol. 132:1–9.

Hutten, T. J., H. C. M. Bongaerts, C. van der Drift, and G. D. Vogels. 1980. Acetate, methanol and carbon dioxide as substrates for growth of *Methanosarcina barkeri*. Antonie Leeuwenhoek J. Microbiol. 46:601–610.

Ianotti, E. L., P. Kafkewitz, M. J. Wolin, and M. P. Bryant. 1973. Glucose fermentation products of *Ruminococcus albus* grown in continuous culture with *Vibrio succinogenes*: changes caused by interspecies transfer of H_2. J. Bacteriol. 114:1231–1240.

Ivanov, M. V., S. S. Belyaev, and K. S. Laurinavichus. 1980. A short report on microbiology of sediments from Deep Sea Drilling Project holes 415, 415A, and 416A, p.

665–666, in Initial reports of the deep sea drilling project, Vol. 50. U.S. Government Printing Office, Washington, D.C.

Ivanov, M. V., S. S. Belyaev, A. M. Zyakun, V. A. Bondar, and K. S. Laurinavichus. 1983. Microbiological formation of methane in the oil-field development. Geokhimiya 11:1647–1654.

Iversen, N., and T. H. Blackburn. 1981. Seasonal rates of methane oxidation in anoxic marine sediments. Appl. Environ. Microbiol. 41:1295–1300.

Iversen, N., R. S. Oremland, and M. J. Klug. 1987. Big Soda Lake (Nevada). 3. Pelagic methanogenesis and anaerobic methane oxidation. Limnol. Oceanogr., 32:804–814.

Jaenchen, R., H. H. Gilles, and R. K. Thauer. 1981. Inhibition of factor F_{430} synthesis by levulinic acid in *Methanobacterium thermoautotrophicum*. FEMS Microbiol. Lett. 12:167–170.

Jarrell, K. F., and G. D. Sprott. 1983. The effects of ionophores and metabolic inhibitors on methanogenesis and energy-related properties of *Methanobacterium bryantii*. Arch. Biochem. Biophys. 225:33–41.

Jeris, J. S., and P. L. McCarty. 1965. The biochemistry of methane fermentation using ^{14}C tracers. J. Water Pollut. Control. Fed. 37:178–192.

Jones, W. J., and M. J. B. Paynter. 1980. Populations of methane-producing bacteria and in vitro methanogenesis in saltmarsh and estuarine sediments. Appl. Environ. Microbiol. 39:864–871.

Jones, J. B., and T. C. Stadtman. 1977. *Methanococcus vannielii*: culture and effects of selenium and tungsten on growth. J. Bacteriol. 130:1404–1406.

Jones, J. B., B Bowers, and T. C. Stadtman. 1977. *Methanococcus vannielii*: ultrastructure and sensitivity to detergents and antibiotics. J. Bacteriol. 130:1357–1363.

Jones, W. J., J. A. Leigh, F. Mayer, C. R. Woese, and R. S. Wolfe. 1983a. *Methanococcus jannaschii* sp. nov., an extremely thermophilic methanogen from a submarine hydrothermal vent. Arch. Microbiol. 136:254–261.

Jones, W. J., M. J. B. Paynter, and R. Gupta. 1983b. Characterization of *Methanococcus maripaludis* sp. nov., a new methanogen isolated from saltmarsh sediment. Arch. Microbiol. 135:91–97.

Kaiser, J. P., and K. W. Hanselmann. 1982. Fermentative metabolism of substituted monoaromatic compounds by a bacterial community from sediments. Arch. Microbiol. 133:185–194.

Kandler, O., and H. Hippe. 1977. Lack of peptidoglycan in the cell walls of *Methanosarcina barkeri*. Arch Microbiol. 113:57–60.

Kandler, O., and H. König. 1978. Chemical composition of the peptido-glycan-free cell walls of methanogenic bacteria. Arch. Microbiol. 118:141–152.

Keith, C. L., R. L. Bridges, L. R. Fina, K. L. Iverson, and J. A. Cloran. 1978. The anaerobic decomposition of benzoic acid during methane fermentation. IV. Dearomatization of the ring and volatile fatty acids formed on ring rupture. Arch. Microbiol. 118:173–176.

Kemp, C. W., M. A. Curtis, S. A. Robrish, and W. H. Bowen. 1983. Biogenesis of methane in primate dental plaque. FEBS Lett. 155:61–64.

Kenealy, W., and J. G. Zeikus. 1981. Influence of corrinoid antagonists on methanogen metabolism. J. Bacteriol. 146:133–140.

Khalil, M. A. K., and R. A. Rasmussen. 1982. Secular trends of atmospheric methane (CH_4). Chemosphere 11:877–883.

Khalil, M. A. K., and R. A. Rasmussen. 1983. Sources, sinks and cycles of atmospheric methane. J. Geophys. Res. 88:5131–5144.

Kiene, R. P., R. S. Oremland, A. Catena, L. G. Miller, and D. G. Capone. 1986. Metabolism of reduced methylated sulfur compounds by anaerobic sediments and a pure culture of an estuarine methanogen. Appl. Environ. Microbiol. 52, 1037–1045.

King, G. M., 1984a. Utilization of hydrogen, acetate and "non-competitive" substrates by methanogenic bacteria in marine sediments. Geomicrobiol. J. 3:275–306.

King, G. M. 1984b. On the metabolism of trimethylamine, choline, and glycine betaine by sulfate-reducing and methanogenic bacteria in marine sediments. Appl. Environ. Microbiol. 48:719–725.

King, G. M., and M. J. Klug. 1982. Glucose metabolism in sediments of a eutrophic lake: tracer analysis of uptake and product formation. Appl. Environ. Microbiol. 44:1308–1317.

King, G. M., M. J. Klug, and D. R. Lovley. 1983. Metabolism of acetate, methanol, and methylated amines in intertidal sediments of Lowes Cove, Maine. Appl. Environ. Microbiol. 45:1848–1853.

Kirsop, B. H. 1984. Methanogenesis. CRC Crit. Rev. Biotechnol. 1:109–159.

Klass, D. L. 1984. Methane from anaerobic fermentation. Science. 223:1021–1028.

König, H., and O. Kandler. 1979. The amino acid sequence of the peptide moiety of the pseudomurein from *Methanobacterium thermoautotrophicum*. Arch. Microbiol. 121:271–275.

König, H., and K. O. Stetter. 1982. Isolation and characterization of *Methanolobus tindarius* sp. nov., a coccoid methanogen growing only on methanol and methylamines. Zentralbl. Bakteriol. Hyg. Mikrobiol. 1 Abt. Orig. C3:478–490.

König, H., R. Knalik, and O. Kandler. 1982. Structure and modification of pseudomurein in *Methanobacteriales*. Zentralbl. Bakteriol. Mikrobiol. Hyg. 1 Abt. Orig. C3:179–191.

Koyama, T. 1962. Gaseous metabolism in lake sediments and paddy soils, p. 363–375, in U. Columbo and G. D. Hobson (eds.), Advances in organic geochemistry. Macmillan, New York.

Kristjansson, J. K., P. Schönheit, and R. K. Thauer. 1982. Different K_s values for hydrogen of methanogen bacteria and sulfate reducing bacteria: an explanation for the apparent inhibition of methanogenesis by sulfate. Arch. Microbiol. 131:278–282.

Krumholz, L. R., C. W. Forsberg, and D. M. Veira. 1983. Association of methanogenic bacteria with rumen protozoa. Can. J. Microbiol. 29:676–680.

Kvenvolden, K. A., and G. D. Redden. 1980. Hydrocarbon gas in sediment from the shelf, slope and basin of the Bering Sea. Geochim. Cosmochim. Acta 44:1145–1150.

Lamontagne, R. A., J. W. Swinnerton, V. J. Linnenbom, and W. D. Smith. 1973. Methane concentrations in various marine environments. J. Geophys. Res. 78:5316–5324.

Lamontagne, R. A., J. W. Swinnerton, and V. J. Linnenbom. 1974. C_1-C_4 hydrocarbons in the North and South Pacific. Tellus 26:71–77.

Langworthy, T. A., T. G. Tornabene, and G. Holzer. 1982. Lipids of archaebacteria. Zentralbl. Bakteriol. Mikrobiol. Hyg. 1 Abt. Orig. C3:228–244.

Latham, M. J., and M. J. Wolin. 1977. Fermentation of cellulose by *Ruminococcus flavefaciens* in the presence and absence of *Methanobacterium ruminantium*. Appl. Environ. Microbiol. 34:297–301.

Laube, U. M., and S. M. Martin. 1981. Conversion of cellulose to methane and carbon dioxide by triculture of *Acetovibrio cellulolyticus*, *Desulfovibrio sp.* and *Methanosarcina barkeri*. Appl. Environ. Microbiol. 42:413–420.

Levy, H., II. 1973. Tropospheric budgets for methane, carbon monoxide, and related species. J. Geophys. Res. 78:5325–5332.

Lidstrom, M. E. 1983. Methane consumption in Framvaren, an anoxic marine fjord. Limnol. Oceanogr. 28:1247–1251.

Lovley, D. R., and M. J. Klug. 1983a. Sulfate reducers can outcompete methanogens at freshwater sulfate concentration. Appl. Environ. Microbiol. 45:187–192.

Lovley, D. R., and M. J. Klug. 1983b. Methanogenesis from methanol and from hydrogen and carbon dioxide in the sediments of a eutrophic lake. Appl. Environ. Microbiol. 45:1310–1315.

Lovley, D. R., and M. J. Klug. 1986. Model for the distribution of sulfate reduction and methanogenesis in freshwater sediments. Geochim. Cosmochim. Acta 50:11–18.

Lovley, D. R., D. F. Dwyer, and M. J. Klug. 1982. Kinetic analysis of competition between sulfate reducers and methanogens for hydrogen in sediments. Appl. Environ. Microbiol. 43:1373–1379.

McBride, B. C., and R. S. Wolfe. 1971. Biosynthesis of dimethylarsine by *Methanobacterium*. Biochemistry 10:4312–4317.

McCarty, P. L. 1964. The methane fermentation, p. 314–343, in H. Heukelekian and N. C. Dondero (eds.), Principles and applications in aquatic microbiology. Wiley, New York.

McInerney, M. J., and M. P. Bryant. 1981. Anaerobic degradation of lactate by syntrophic associations of *Methanosarcina barkeri* and *Desulfovibrio* species and effect of H_2 on acetate degradation. Appl. Environ. Microbiol. 41:346–354.

McInerney, M. J., M. P. Bryant, and N. Pfennig. 1979. Anaerobic bacterium that degrades fatty acids in syntrophic association with methanogens. Arch. Microbiol. 122:129–135.

Mackenzie, A. S., S. C. Brassell, G. Eglington, and J. R. Maxwell. 1982. Chemical fossils: the geological fate of steroids. Science 217:491–504.

Mah, R. A. 1980. Isolation and characterization of *Methanococcus mazei*. Curr. Microbiol. 3:321–326.

Mah, R. A. 1982. Methanogenesis and methanogenic partnerships. Philos. Trans. R. Soc. Lond. 297:599–616.

Mah, R. A., D. M. Ward, L. Baresi, and T. Glass. 1977. Biogenesis of methane. Annu. Rev. Microbiol. 31:309–341.

Martens, C. S., and R. A. Berner. 1974. Methane production in the interstitial waters of sulfate-depleted marine sediments. Science 185:1167–1169.

Martens, C. S., and R. A. Berner. 1977. Interstitial water chemistry of anoxic Long Island Sound sediments. I. Dissolved gases. Limnol. Oceanogr. 22:10–25.

Michaelis, W., and P. Albrecht. 1979. Molecular fossils of archaebacteria in kerogen. Naturwissenschaften 66:420–422.

Miller, T. L., and M. J. Wolin. 1981. Fermentation by the human large intestine microbial community in an in vitro semicontinuous culture system. Appl. Environ. Microbiol. 42:400–407.

Miller, T. L., and M. J. Wolin. 1983. Stability of *Methanobrevibacter smithii* populations in the microbial flora excreted from the human large bowel. Appl. Environ. Microbiol. 45:317–318.

Miller, T. L., M. J. Wolin, E. Conway de Macario, and A. J. L. Macario. 1982. Isolation of *Methanobrevibacter smithii* from human feces. Appl. Environ. Microbiol. 43:227–232.

Mink, R. W., and P. R. Dugan. 1977. Tentative identification of methanogenic bacteria by fluorescence microscopy. Appl. Environ. Microbiol. 33:713–717.

Moldowan, J. M., and W. K. Seifert. 1979. Head-to-head linked isoprenoid hydrocarbons in petroleum. Science 204:169–171.

Morris, D. A., 1976. Organic diagenesis of Miocene sediments from site 341, Voring Planeu, Norway, p. 809–814, in S. M. White (ed.), Initial reports of deep sea drilling project, Vol. 38. U.S. Government Printing Office, Washington, D.C.

Mountfort, D. O., R. A. Asher, and T. Bauchop. 1982. Fermentation of cellulose to methane and carbon dioxide by a rumen anaerobic fungus in a triculture with *Methanobrevibacter* sp. strain RA1 and *Methanosarcina barkeri*. Appl. Environ. Microbiol. 44:128–134.

Muller, F. M. 1957. On methane fermentation of higher alkanes. Antonie Leeuwenhoek J. Microbiol. Serol. 23:369–384.

Murray, P. A., and S. H. Zinder. 1984. Nitrogen fixation by a methanogenic bacterium. Nature (Lond.) 312:284–286.

Neill, A. R., D. W. Grime, and R. M. C. Dawson. 1978. Conversion of choline methyl groups through trimethylamine into methane in the rumen. Biochem. J. 170:529–535.

Nissenbaum, A., B. J. Presley, and I. R. Kaplan. 1972. Early diagenesis in a reducing fjord, Saanich Inlet, British Columbia. I. Chemical and isotopic changes in major components of interstitial water. Geochim. Cosmochim. Acta 36:1007–1027.

Nottingham, P. M., and R. E. Hungate. 1968. Isolation of methanogenic bacteria from feces of man. J. Bacteriol. 96:2178–2179.

Odelson, D. A., and J. A. Breznak. 1983. Volatile fatty acid production by the hindgut microbiota of xylophagus termites. Appl. Environ. Microbiol. 45:1602–1613.

Olson, G. J., W. S. Dockins, G. A. McFeters, and W. P. Iverson. 1981. Sulfate-reducing and methanogenic bacteria from deep aquifers in Montana. Geomicrobiol. J. 2:327–340.

Oremland, R. S. 1975. Methane production in shallow-water, tropical marine sediments. Appl. Microbiol. 30:602–608.

Oremland, R. S. 1979. Methanogenic activity in plankton samples and fish intestines: a mechanism for *in situ* methanogenesis in oceanic surface waters. Limnol. Oceanogr. 24:1136–1141.

Oremland, R. S. 1981. Microbial formation of ethane in anoxic estuarine sediments. Appl. Environ. Microbiol. 42:122–129.

Oremland, R. S., and D. G. Capone. 1988. Use of ''specific'' inhibitors in biogeochemistry and microbial ecology. Adv. Microb. Ecol. 10, 285–383.

Oremland, R. S., and D. J. DesMarais. 1983. Distribution, abundance and carbon isotopic composition of gaseous hydrocarbons in Big Soda Lake, Nevada: an alkaline, meromictic lake. Geochim. Cosmochim. Acta 47:2107–2114.

Oremland, R. S., and S. Polcin. 1982. Methanogenesis and sulfate-reduction: competitive and noncompetitive substrate in estuarine sediments. Appl. Environ. Microbiol. 44:1270–1276.

Oremland, R. S., and B. F. Taylor. 1975. Inhibition of methanogenesis in marine sediments by acetylene and ethylene: validity of the acetylene reduction assay for anaerobic microcosms. Appl. Microbiol. 30:707–709.

Oremland, R. S., and B. F. Taylor. 1977. Diurnal fluctuations of O_2 N_2, and CH_4 in the rhizosphere of *Thalassia testudinum*. Limnol. Oceanogr. 22:566–570.

Oremland, R. S., and B. F. Taylor. 1978. Sulfate reduction and methanogenesis in marine sediments. Geochim. Cosmochim. Acta 42:209–214.

Oremland, R. S., and J. R. Zehr. 1986. Formation of methane and carbon dioxide from dimethylselenide and in anoxic sediments and by a methanogenic bacterium. Appl. Environ. Microbiol. 52, 1031–1036.

Oremland, R. S., L. Marsh, and D. J. DesMarais. 1982a. Methanogenesis in Big Soda Lake, Nevada: an alkaline, moderately hypersaline desert lake. Appl. Environ. Microbiol. 43:462–468.

Oremland, R. S., L. M. Marsh, and S. Polcin. 1982b. Methane production and simultaneous sulfate reduction in anoxic saltmarsh sediments. Nature (Lond.) 296:143–145.

Oremland, R. S., L. G. Miller, and M. J. Whitiear. 1987. Sources and flux of natural gases from Mono Lake, California. Geochim. Cosmochim. Acta 51:2915–2929.

Oremland, R. S., C. W. Culbertson, and B. R. T. Simoneit. 1982c. Methanogenic activity in sediment from leg 64, Gulf of California, p. 759–762, in J. R. Curray, D. G. Moore, et al. (eds), Initial reports of the deep sea drilling project, U.S. Government Printing Office, Washington, D.C. Vol. 64.

Ourisson, G., P. Albrecht, and M. Rohmer. 1984. The microbial origin of fossil fuels. Sci. Am. 251:44–51.

Owens, A. J., J. M. Steed, D. L. Filkin, C. Miller, and J. P. Jesson. 1982. The potential effects of increased methane on atmospheric ozone. Geophys. Res. Lett. 9:1105–1108.

Panganiban, A. T., Jr., T. E. Patt, W. Hart, and R. S. Hanson. 1979. Oxidation of methane in the absence of oxygen in lake water samples. Appl. Environ. Microbiol. 37:303–309.

Patel, G. B., and L. A. Roth. 1977. Effect of sodium chloride on growth and methane production of methanogens. Can. J. Microbiol. 23:893–897.

Patterson, J. A., and R. B. Hespell. 1979. Trimethylamine and methylamine as growth substrates for rumen bacteria and *Methanosarcina barkeri*. Curr. Microbiol. 3:79–83.

Payne, W. J. (1984). Influence of acetylene on microbial and enzymatic assays. J. Microbiol. Methods 2:117–133.

Pedersen, D., and G. S. Sayler. 1981. Methanogenesis in freshwater sediments: inherent variability and effects of environmental contaminants. Can. J. Microbiol. 27:198–205.

Perski, H. J., P. Schönheit, and R. K. Thauer. 1982. Sodium dependence of methane formation in methanogenic bacteria. FEBS Lett. 143:323–326.

Pine, M. J., and H. A. Barker. 1954. Studies on the methane bacteria. XI. Fixation of atmospheric nitrogen by *Methanobacterium omelianskii*. J. Bacteriol. 68:589–591.

Prins, R. A., C. J. van Nevel, and D. I. Demeyer. 1972. Pure culture studies of inhibitors for methanogenic bacteria. Antonie Leeuwenhoek J. Microbiol. Serol. 38:281–287.

Pruett, R. L. 1981. Synthesis gas: a raw material for industrial chemicals. Science 211:11–16.

Raimbault, M. 1975. Étude de l'influence inhibitrice de l'acetylene sur la formation biologique du méthane dans un dol de rizière. Ann. Inst. Pasteur (Paris) A126:247–258.

Rasmussen, R. A., and M. Khalil. 1983. Global production of methane by termites. Nature (Lond.) 301:700–702.

Reeburgh, W. S. 1976. Methane consumption in Cariaco Trench waters and sediments. Earth Planet. Sci. Lett. 28:337–344.

Reeburgh, W. S. 1980. Anaerobic methane oxidation: rate depth distributions in Skan Bay sediments. Earth Planet. Sci. Lett. 47:345–352.

Reeburg, W. S., and D. T. Heggie. 1977. Microbial methane consumption reactions and their effect on methane distributions in freshwater and marine environments. Limnol. Oceanogr. 22:1–9.

Rice, D. D., and G. E. Claypool. 1981. Generation, accumulation and resource potential of biogenic gas. Am Assoc. Pet. Geol. Bull. 65:5–25.

Ridley, W. P., L. J. Dizikes, and J. M. Wood. 1977. Biomethylation of toxic elements in the environment. Science 197:329–332.

Rivard, C. J., and P. H. Smith, 1982. Isolation and characterization of a thermophilic marine methanogenic bacterium, *Methanogenium thermophilicum* sp. nov. Int. J. Syst. Bacteriol. 32:430–436.

Rivard, C. J., J. M. Henson, M. V. Thomas, and P. H. Smith. 1983. Isolation and characterization of *Methanomicrobium paynteri* sp. nov., a mesophilic methanogen isolated from marine sediments. Appl. Environ. Microbiol. 46:484–490.

Robbins, R. C., L. A. Cavanagh, and L. J. Salas. 1973. Analysis of ancient atmospheres. J. Geophys. Res. 78:5341–5344.

Roberton, A. M., and R. S. Wolfe. 1970. ATP pools in *Methanobacterium*. J. Bacteriol. 102:43–51.

Robinson, J. A., and J. M. Tiedje. 1984. Competition between sulfate-reducing and methanogenic bacteria for H_2 under resting and growing conditions. Arch. Microbiol. 137:26–32.

Rose, C. S., and S. J. Pirt. 1981. Conversion of glucose and fatty acids to methane: roles of two mycoplasmal agents. J. Bacteriol. 147:248–254.

Rosenfeld, W. D., and S. R. Silverman. 1959. Carbon isotopic fractionation in bacterial production of methane. Science 130:1658–1659.

Rudd, J. W. M., and C. D. Taylor. 1980. Methane cycling in aquatic environments. Adv. Aquat. Microbiol. 2:77–150.

Rust, F. 1981. Ruminant methane ($^{13}C/^{12}C$) values: relation to atmospheric methane. Science 211:1044–1046.

Sackett, W. M., J. M. Brooks, B. B. Bernard, C. R. S. Schwab, and H. Chung. 1979. A carbon inventory for Orca Basin brines and sediment. Earth Planet. Sci. Lett. 44:73–81.

Salvas, P. L., and B. F. Taylor. 1980. Blockage of methanogenesis in marine sediments by the nitrification inhibitor 2-chloro-6-(trichloromethyl) pyridine (nitrapyrin or n-serve). Curr. Microbiol 4:305–308.

Samman, N., T. Ignasiak, C. J. Chen, O. P. Strauz, and D. S. Montgomery. 1981. Squalene in petroleum asphaltene. Science 213: 1381–1382.

Sandbeck, K. A., and D. M. Ward. 1982. Temperature adaptions in the terminal processes of anaerobic decomposition of Yellowstone National Park and Icelandic hot spring microbial mats. Appl. Environ. Microbiol. 44:844–851.

Sansone, F. J., and C. S. Martens. 1981. Methane production from acetate and associated methane fluxes from anoxic coastal sediments. Science 211:707–709.

Schauer, N. L., and J. G. Ferry. 1980. Metabolism of formate in *Methanobacterium formicicum*. J. Bacteriol. 142:800–807.

Schauer, N. L., D. P. Brown, and J. G. Ferry. 1982. Kinetics of formate metabolism in *Methanobacterium formicicum* and *Methanospirillum hungatei*. Appl. Environ. Microbiol. 44:549–554.

Scheifinger, C. C., B. Linehan, and M. J. Wolin. 1975. H_2 production by *Selenomonas ruminantium* in the absence and presence of methanogenic bacteria. Appl. Microbiol. 29:480–483.

Schinck, B., and M. Stieb. 1983. Fermentative degradation of polyethylene glycol by a strictly anaerobic, gram-negative, nonspore forming bacterium, *Pelobacter venetianus* sp. nov. Appl. Environ. Microbiol. 45:1905–1913.

Schink, B., and J. G. Zeikus. 1980. Microbial methanol formation: a major endproduct of pectin metabolism. Curr. Microbiol. 4:387–389.

Schink, B., and J. G. Zeikus. 1982. Microbial ecology of pectin decomposition in anoxic lake sediments. J. Gen. Microbiol. 128:393–404.

Schink, B., J. C. Ward, and J. G. Zeikus. 1981. Microbiology of wetwood: importance of pectin degradation and Clostridium species in living trees. Appl. Environ. Microbiol. 42:526–532.

Schink, B., F. S. Lupton, and J. G. Zeikus. 1983. Radioassay for hydrogenase activity in viable cells and documentation of aerobic hydrogen-consuming bacteria living in extreme environments. Appl. Environ. Microbiol. 45:1491–1500.

Schnellen, C. G. T. P. 1947. Onderzoekingen over de methanungisting. Dissertation, Delft University of Technology.

Schoell, M. 1980. The hydrogen and carbon isotopic composition of methane from natural gas of various origins. Geochim. Cosmochim. Acta 44:649–661.

Schönheit, P., J. Moll, and R. K. Thauer. 1979. Nickel, cobalt and molybdenum requirement for growth of *Methanobacterium thermoautotrophicum*. Arch. Microbiol. 123:105–107.

Schönheit, P., J. Moll, and R. K. Thauer. 1980. Growth parameters (K_s, μmax, Y_s) of *Methanobacterium thermoautotrophicum*. Arch. Microbiol. 127:59–65.

Schönheit, P., H. Keweloh, and R. K. Thauer. 1981. Factor F_{420} degradation in *Methanobacterium thermoautotrophicum* during exposure to oxygen. FEMS Microbiol. Lett. 12:347–349.

Schönheit, P., J. K. Kristjansson, and R. K. Thauer. 1982. Kinetic mechanism for the ability of sulfate reducers to out-compete methanogens for acetate. Arch. Microbiol. 132:285–288.

Scranton, M. I., and P. G. Brewer. 1977. Occurrence of methane in the near-surface waters of the western subtropical North-Atlantic. Deep Sea Res. 24:127–138.

Seiler, W., R. Conrad, and D. Scharffe. 1984a. Field studies of methane emission from termite nests into the atmosphere and measurements of methane uptake by tropical soils. J. Atmos. Chem. 1:171–186.

Seiler, W., A. Holzapfel-Pschorn, R. Conrad, and D. Scharffe. 1984b. Methane emission from rice paddies. J. Atmos. Chem. 1:241–268.

Shlomi, E. R., A. Lankhorst, and R. A. Prins. 1978. Methanogenic fermentation of benzoate in an enrichment culture. Microb. Ecol. 4:249–261.

Silverman, M. P., and V. I. Oyama. 1968. Automatic apparatus for sampling and preparing gases for mass spectral studies of carbon isotope fractionation during methane metabolism. Anal. Chem. 40:1883–1877.

Smith, M. R. 1983. Reversal of 2-bromoethanesulfonate inhibition of methanogenesis in *Methanosarcina* sp. J. Bacteriol. 156:516–523.

Smith, M. R., and R. A. Mah. 1978. Growth and methanogenesis by *Methanosarcina* strain 227 on acetate and methanol. Appl. Environ. Microbiol. 36:870–879.

Smith, M. R., and R. A. Mah. 1980. Acetate as a sole carbon and energy source for growth of *Methanosarcina* strain 227. Appl. Environ. Microbiol. 39:993–999.

Smith, M. R., and R. A. Mah. 1981. 2-bromoethanesulfonate: a selective agent for isolating resistant *Methanosarcina* mutants. Curr. Microbiol. 6:321–326.

Smith, R. L., and R. S. Oremland. 1987. Big Soda Lake (Nevada). 2. Pelagic sulfate reduction. Limnol. Oceanogr. 32:794–803.

Söhngen, N. L. 1906. Het onstaan en verdwijnen van waterstof en methaan onder den invloed van het organische leven. Dissertation, Technical University Delft, J. Vis Jr.

Sørensen, J., D. Christensen, and B. B. Jørgensen. 1981. Volatile fatty acids and hydrogen as substrates for sulfate-reducing bacteria in anaerobic marine sediment. Appl. Environ. Microbiol. 42:5–11.

Sorokin, Y. I. 1957. Ability of sulfate reducing bacteria to utilize methane for reduction of sulfate to hydrogen sulfide. Dokl. Akad. Nauk. SSSR 115:816–818.

Sowers, K. R., and J. G. Ferry. 1983. Isolation and characterization of a methylotrophic marine methanogen, *Methanococcoides methylmutens* gen. nov., sp. nov. Appl. Environ. Microbiol. 45:684–690.

Sowers, K., S. F. Baron, and J. G. Ferry. 1984. *Methanosarcina acetivorans* sp. nov., an acetotrophic methane producing bacterium isolated from marine sediments. Appl. Environ. Microbiol. 47:971–978.

Sprott, G. D., and K. F. Jarrell. 1981. K^+, Na^+, and Mg^{2+} content and permeability of *Methanospirillum hungatei and Methanobacterium thermoautotrophicum*. Can. J. Microbiol. 27:444–451.

Sprott, G. D., K. F. Jarrell, K. M. Shaw, and R. Knowles. 1982. Acetylene as an inhibitor of methanogenic bacteria. J. Gen. Microbiol. 128:2453–2462.

Stadtman, T. C., 1967. Methane fermentation. Annu. Rev. Microbiol. 21:121–142.

Stadtman, T. C., and H. A. Barker. 1949. Studies on the methane fermentation. VII. Tracer studies. Arch. Biochem. 21:256–264.

Steggerda, F. R., and J. F. Dimmick. 1966. Effects of bean diets on concentration of carbon dioxide in flatus. Am. J. Clin. Nutr. 19:120–124.

Stetter, K. O., and G. Gaag. 1983. Reduction of molecular sulfur by methanogenic bacteria. Nature (Lond.) 305:309–311.

Stetter, K. O., M. Thomm, J. Winter, A. Wildgruber, H. Huber, W. Zillig, D. Jane-Covic, H. König, P. Palm, and S. Wunderl. 1981. *Methanothermus fervidus* sp. nov., a novel extremely thermophilic methanogen isolated from an Icelandic hot spring. Zentralbl. Bakteriol. Mikrobiol. Hyg. 1 Orig. C2:166–178.

Strayer, R. F., and J. M. Tiedje. 1978. Kinetic parameters of the conversion of methane precursors to methane in a hypereutrophic lake sediment. Appl. Environ. Microbiol. 36:330–340.

Suflita, J. M., A. Horowitz, D. R. Shelton, and J. M. Tiedje. 1982. Dehalogenation: a novel pathway for the anaerobic biodegradation of haloaromatic compounds. Science 218:1115–1117.

Tarafa, M. E., J. K. Whelan, R. S. Oremland, and R. L. Smith. 1986. Evidence of microbiological activity in leg 95 (New Jersey transect) sediments, in C. W. Poag and A. B. Watts (eds.), Initial reports of the deep sea drilling project, Vol. 95. U.S. Government Printing Office, Washington, D.C., 635–640.

Taylor, C. D., B. C. McBridge, R. S. Wolfe, and M. P. Bryant. 1974. Coenzyme M, essential for growth of a rumen strain of *Methanobacterium ruminantium*. J. Bacteriol. 120:974–975.

Tietze, K. 1978. Geophysikalische Untersuhung des Kivusees und seiner ungewöhnlichen Methanogaslagerstätte—Schichtung, Dynamik und Gasgehalt des Seewassers. Dissertation, University of Kiel.

Tissot, B. P., and D. H. Welte. 1978. Petroleum formation and occurrence. Springer-Verlag, New York.

Tornabene, T. G., R. S. Wolfe, W. E. Balch, G. Holzer, G. E. Fox, and J. Oró. 1978. Phytanyl-glycerol ethers and squalenes in the archaebacterium *Methanobacterium thermoautotrophicum*. J. Mol. Evol. 11:259–266.

Tornabene, T. G., T. A. Langworthy, G. Holzer, and J. Oró. 1979. Squalenes, phytanes and other isoprenoids as major neutral lipids of methanogenic and thermoacidophilic "archaebacteria." J. Mol. Evol. 13:73–83.

Touzel, J. P., and G. Albagnac. 1983. Isolation and characterization of *Methanococcus mazei* strain MC3. FEMS Microbiol. Lett. 16:241–245.

Traganza, E. D., J. W. Swinnerton, and C. H. Cheek. 1979. Methane supersaturation and ATP-zooplankton blooms in near-surface waters of the Western Mediterranean and the Subtropical North Atlantic Ocean. Deep Sea Res. A26:1237–1245.

Trent, J. D., R. A. Chastain, and A. A. Yayanos. 1984. Possible artefactual basis for apparent bacterial growth at 250°C. Nature (Lond.) 307:737–740.

van Bruggen, J. J. A., C. K. Stumm, and G. D. Vogels. 1983. Symbiosis of methanogenic bacteria and sapropelic protozoa. Arch. Microbiol. 136:89–95.

van Nevel, C. J., and D. I. Demeyer. 1977. Effect of monensin on rumen metabolism in vitro. Appl. Environ. Microbiol. 34:251–257.

Vogel, T. M., R. S. Oremland, and K. A. Kvenvolden. 1982. Low-temperature formation of hydrocarbon gases in San Francisco Bay sediment (California, USA). Chem. Geol. 37:289–298.

Wake, L. V., R. K. Christopher, P. A. D. Rickard, J. E. Andersen, and B. J. Ralph. 1977. A thermodynamic assessment of possible substrates for sulfate-reducing bacteria. Aust. J. Biol. Sci. 30:155–172.

Ward, D. M., and G. J. Olson. 1980. Terminal processes in the anaerobic degradation of an algal-bacterial mat in a high-sulfate hot spring. Appl. Environ. Microbiol. 40:67–74.

Ward, D. M., and M. R. Winfrey. 1985. Interactions between methanogenic and sulfate-reducing bacteria in sediments. Adv. Aquat. Microbiol. 3:141–179.

Weimer, P. J., and J. G. Zeikus. 1977. Fermentation of cellulose and cellobiose by *Clostridium thermocellum* in the absence and presence of *Methanobacterium thermoautotrophicum*. Appl. Environ. Microbiol. 33:289–297.

Welhan, J. A., and H. Craig. 1979. Methane and hydrogen in East Pacific Rise hydrothermal fluids. Geophys. Res. Lett. 6:829–831.

Wellinger, A., and K. Wuhrmann. 1977. Influence of sulfide compounds on the metabolism of *Methanobacterium* strain AZ. Arch. Microbiol. 115:13–17.

Whelan, J. K., J. M. Hunt, and J. Berman. 1980. Volatile C_1-C_7 organic compounds in surface sediments from Walvis Bay. Geochim. Cosmochim. Acta 44:1767–1785.

Whelan, J. K., R. Oremland, M. Tarafa, R. Smith, R. Howarth, and C. Lee. 1986. Evidence for sulfate-reducing and methane-producing microorganisms in sediments from sites 618, 619, and 622, p. 767–775, in A. Bouma and J. Coleman (eds.), Initial reports of the deep sea drilling project, Vol. 96, U.S. Government Printing Office, Washington, D.C.

Whitman, W. B. 1985. Methanogenic bacteria, p. 3–84, in C. R. Woese and R. S. Wolfe (eds.), The bacteria, Vol. 8. Archaebacteria. Academic Press, New York.

Widdel, F., and N. Pfennig. 1977. A new anaerobic sporing, acetate-oxidizing, sulfate-reducing bacterium, *Desulfotomaculum* (emend.) *acetoxidans*. Arch. Microbiol. 112:119–122.

Widdel, F., and N. Pfennig. 1981. Studies on dissimilatory sulfate-reducing bacteria that decompose fatty acids. I. Isolation of a new sulfate-reducing bacteria enriched with acetate from saline environments. Description of *Desulfobacter postgatei* gen. nov., sp. nov. Arch. Microbiol. 129:275–281.

Wildguber, G., M. Thomm, H. König, K. Ober, T. Ricchiuto, and K. O. Stetter. 1982. *Methanoplanus limicola*, a plate-shaped methanogen representing a novel family, the Methanoplanacea. Arch. Microbiol. 132:31–36.

Williams, R. T., and R. L. Crawford. 1984. Methane production in Minnesota peatlands. Appl. Environ. Microbiol. 47:1266–1271.

Winfrey, M. R. 1984. Microbial production of methane, p. 153–219, in R. M. Atlas (ed.), Petroleum microbiology. Macmillan, New York.

Winfrey, M. R., and J. G. Zeikus. 1977. Effect of sulfate on carbon and electron flow during microbial methanogenesis in fresh water sediments. Appl. Environ. Microbiol. 33:275–281.

Winfrey, M. R., and J. G. Zeikus. 1979. Microbial methanogenesis and acetate metabolism in a meromictic lake. Appl. Environ. Microbiol. 37:213–221.

Winfrey, M. R., D. R. Nelson, S. C. Klevickis, and J. G. Zeikus. 1977. Association of hydrogen metabolism with methanogenesis in Lake Mendota sediments. Appl. Environ. Microbiol. 33:312–318.

Winter, J., and R. S. Wolfe. 1979. Complete degradation of carbohydrate to carbon dioxide and methane by syntrophic cultures of *Acetobacterium woodii* and *Methanosarcina barkeri*. Arch. Microbiol. 121:97–102.

Winter, J. V., and R. S. Wolfe. 1980. Methane formation from fructose by syntrophic associations of *Acetobacterium woodii* and different strains of methanogens. Arch. Microbiol. 124:73–79.

Woese, C. R., and G. E. Fox. 1977. Phylogenetic structure of the procaryotic domain: the primary kingdoms. Proc. Natl. Acad. Sci. USA 74:5088–5090.

Wolfe, R. S. 1971. Microbial formation of methane, p. 107–146, in A. H. Rose and J. F. Wilkinson (eds.), Advances in microbial physiology, Vol. 6. Academic Press, New York.

Wolfe, R. S. 1985. Unusual coenzymes of methanogenesis. Trends Biochem. Sci. 10:396–399.

Wolin, M. J. 1981. Fermentation in the rumen and human large intestine. Science 213:163–168.

Wolin, M. J. 1982. Hydrogen transfer in microbial communities, p. 323–356, in A. T. Bull and J. H. Slater (eds.), Microbial interactions and communities, Vol. 1. Academic Press, New York.

Wolin, M. J., and T. L. Miller. 1982. Interspecies hydrogen transfer: 15 years later. ASM News 48:561–565.

Wolin, E. A., R. S. Wolfe, and M. J. Wolin. 1964. Viologen dye inhibition of methane formation of *Methanobacillus omelianskii*. J. Bacteriol. 87:993–998.

Wood. J. M., and R. S. Wolfe. 1966. Alkylation of an enzyme in the methane-forming system of Methanobucillus omelianskii. Biochem. Biophys. Res. Commun. 22:119–123.

Wood, J. M., F. S. Kennedy, and R. S. Wolfe. 1968. The reaction of multi halogenated hydrocarbons with free and bound reduced vitamin B12 Biochemistry 7:1707–1713.

Worakit, S., D. R. Boone, R. A. Mah, M. E. Abdel-Samie, and M. M. El Halwagi. 1986. *Methanobacterium alcaliphilum* sp. nov., an H_2-utilizing methanogen that grows at high pH values. Int. J. Syst. Bacteriol. 36:380–382.

Zabel, H. P., H. König, and J. Winter. 1984. Isolation and characterization of a new coccoid methanogen, *Methanogenium tatii* spec. nov. from a solfataric field on Mount Tatio. Arch. Microbiol. 137:308–315.

Zehnder, A. J. B., and T. D. Brock. 1979a. Biological energy production in the apparent absence of electron transport and substrate level phosphorylation. FEMS Lett. 107:1–3.

Zehnder, A. J. B., and T. D. Brock. 1979b. Methane formation and methane oxidation by methanogenic bacteria. J. Bacteriol. 137:420–432.

Zehnder, A. J. B., and T. D. Brock. 1980. Anaerobic methane oxidation: occurrence and ecology. Appl. Environ. Microbiol. 39:194–204.

Zehnder, A. J. B., and K. Wuhrmann. 1977. Physiology of a *Methanobacterium* strain AZ. Arch. Microbiol. 111:199–205.

Zehnder, A. J. B., B. A. Huser, T. D. Brock, and K. Whuhrmann. 1980. Characterization of an acetate-decarboxylating, non-hydrogen-oxidizing methane bacterium. Arch. Microbiol. 124:1–11.

Zehnder, A. J. B., K. Ingvorsen, and T. Marti. 1982. Microbiology of methane bacteria, p. 45–68, in D. E. Hughes, D. A. Stafford, B. I. Wheatley, W. Baader, G. Lettinga, E. J. Nyns, and W. Verstraeten (eds.), Anaerobic Digestion 1981. Elsevier, Amsterdam.

Zeikus, J. G. 1977. The biology of methanogenic bacteria. Bacteriol. Rev. 41:514–541.

Zeikus, J. G. 1983. Metabolic communication between biodegradative populations in nature, p. 423–462, in H. Slater, E. Whittenbury, and J. Wimpenny (eds.), Microbes in their natural environments. Soc. Gen. Microbiol. Sym. 34. Cambridge University Press, Cambridge.

Zeikus, J. G., and D. L. Henning. 1975. *Methanobacterium arborophilicum* sp. n., an obligate anaerobe isolated from wetwood in trees. Antonie Leeuwenhoek J. Microbiol. Serol. 41:171–180.

Zeikus, J. G., and J. C. Ward. 1974. Methane formation in living trees: a microbial origin. Science 184:1181–1183.

Zeikus, J. G., and M. Winfrey. 1976. Temperature limitation of methanogenesis in aquatic sediments. Appl. Environ. Microbiol. 31:99–107.

Zeikus, J. G., and R. S. Wolfe. 1972. *Methanobacterium thermoautotrophicus* sp. n., an anaerobic, autotrophic, extreme thermophile. J. Bacteriol. 109:707–713.

Zeikus, J. G., A. Ben-Bassat, and P. W. Hegge. 1980. Microbiology of methanogenesis in thermal, volcanic environments. J. Bacteriol. 143:432–440.

Zeikus, J. G., P. W. Hegge, T. E. Thompson, T. J. Phelps, and T. A. Langworthy. 1983. Isolation and description of *Haloanaerobium praeualens* gen. nov. and sp. nov., an obligately anaerobic halophile common to Great Salt Lake sediments. Curr. Microbiol. 9:225–234.

Zhilina, T. 1971. The fine structure of *Methanosarcina*. Mikrobiologiya 40:674–680.

Zhilina, T. N. 1983. New oligate halophilic methane-producing bacterium. Mikrobiologiya 52:375–382.

Zimmerman, P. R., J. P. Greenberg, S. O. Wandiga, and P. J. Crutzen. 1982. Termite: a potentially large source of atmosphere methane, carbon dioxide, and molecular hydrogen. Science 218:563–565.

Zinder, S. H., and T. D. Brock. 1978. Production of methane from methane thiol and dimethyl sulfide by anaerobic lake sediments. Nature (Lond.) 273:226–228.

Zinder, S. H., and M. Koch. 1984. Non-aceticlastic methanogenesis from acetate by a thermophilic syntrophic coculture. Arch. Microbiol. 138:263–272.

Zinder, S. H., and R. A. Mah. 1979. Isolation and characterization of a thermophilic strain of *Methanosarcina* unable to use H_2-CO_2 for methanogenesis. Appl. Environ. Microbiol. 38:996–1008.

Zinder, S. H., T. Anguish, and S. C. Cardwell. 1984. Selective inhibition by 2-bromoethanesulfonic acid of methanogenesis from acetate in a thermophilic anaerobic digestor. Appl. Environ. Microbiol. 47:1343–1345.

13

BIOCHEMISTRY OF METHANE PRODUCTION

GODFRIED D. VOGELS, JAN T. KELTJENS, AND
CHRIS VAN DER DRIFT

Department of Microbiology, Faculty of Science, University of Nijmegen, Toernooiveld, NL-6525 ED Nijmegen, The Netherlands

13.1 INTRODUCTION
13.2 ONE-CARBON METABOLISM
    13.2.1 Introduction
    13.2.2 Conversion of Hydrogen and Carbon Dioxide
        13.2.2.1 Initial Steps of Reduction
        13.2.2.2 Terminal Step of Methanogenesis
        13.2.2.3 Activation and Reduction of Carbon Dioxide
    13.2.3 Conversion of Formate
    13.2.4 Carbon Monoxide, a Physiological Substrate of Methanogenesis?
    13.2.5 Conversion of Methanol
        13.2.5.1 Methyltransferases Involved
        13.2.5.2 Disproportionate Reactions
    13.2.6 Conversion of Methylamines
    13.2.7 Acetate, the Most Abundant Substrate of Methanogenesis in Nature
13.3 ELECTRON TRANSFERRING PROCESSES
    13.3.1 Introduction. Membraneous Constituents of Methanogenic Bacteria
    13.3.2 Iron–Sulfur Proteins: Hydrogenase and Ferredoxin
    13.3.3 Cytochromes

13.3.4   Coenzyme F_{420}

13.3.5   Oxidoreductases

13.4   ENERGY CONSERVATION AND ATP SYNTHESIS

13.4.1   Energy Conservation

13.4.2   Energy Consumption for Growth

13.4.3   Electron Transport and ATP Synthesis

13.4.4   Transport Systems

13.5   SYNTHESIS OF CELL CARBON COMPOUNDS

13.5.1   Introduction

13.5.2   Synthesis of Acetate

13.5.3   Intermediary Cell Carbon Synthesis

13.5.4   Nitrogen Assimilation

13.5.5   Sulfur Assimilation

13.5.6   Biosynthesis of Coenzymes

13.5.7   Nucleic Acids and Genetics

REFERENCES

13.1   INTRODUCTION

Methanogenic bacteria belong to the primary kingdom of the archaebacteria (4, 83, 322, 323) and share this property with some other organisms living in extreme environments such as the halobacteria, the thermoacidophiles and the thermophilic anaerobes *Thermoproteus* and *Desulfurococcus*. The separate classification of the methanogenic bacteria occurred on the basis of a number of aberrant characteristics as compared to the primary kingdoms of the eubacteria and the eukaryotes, that is, the nucleotide pattern of 16S rRNA, the lack of thymine and dihydrouracil in tRNA, the composition of the cell walls lacking murein, the composition of the cell membranes, and the length of the genome (for a review, see Ref. 4). In addition a number of unique coenzymes are involved in their metabolism.

Taxonomically methanogenic bacteria, which are all obligate anaerobes, comprise a rather diverse group. As proposed by Balch et al. (4), three orders may be distinguished—the methanobacteriales, the methanococcales, and the methano-

---

Abbreviations used—*Coenzyme of methanogenesis*: coenzyme F_{420}, 7,8-didemethyl-8-hydroxy-5-deazaflavin derivative; coenzyme M, HS-CoM, 2 mercaptoethanesulfonic acid; methylcoenzyme M, CH_3S-CoM, 2-(methylthio)ethanesulfonic acid; component B, 7-mercaptoheptanoylthreoninephosphate, compound involved in the methylcoenzyme M reductase system; factor III, B_{12}-HBI, B_{12} compound containing 5-hydroxybenzimidazole as the α-ligand; factor F_{430}, nickel(II) containing hexahydrouroporphinogen III corphin; methanopterin, 7-methyl-6-(N-ethyl)pterin derivative; H_4MPT (methenyl-H_4MPT, methylene-H_4MPT, methyl-H_4MPT), 5,6,7,8-tetrahydromethanopterin (the three latter compounds contain C_1-units at the formyl, formaldehyde, and methyl state of reduction; methanofuran, furan-containing one-carbon carrier involved in the primary steps of CO_2 reduction.

microbiales—subdivided into six families (Table 13.1). The 49 species described in Table 13.1 belong to no fewer than 10 genera.

The common property is the production of methane from a limited number of one-carbon compounds and the two-carbon compound acetate according to the following equations:

$$4H_2 + CO_2 \rightarrow CH_4 + 2H_2O$$
$$\Delta G°' = -130.4 \text{ kJ/mol } CH_4 \tag{1}$$

$$4HCOO^- + 4H^+ \rightarrow CH_4 + 3CO_2 + 2H_2O$$
$$\Delta G°' = -119.5 \text{ kJ/mol } CH_4 \tag{2}$$

$$4CO + 2H_2O \rightarrow CH_4 + 3CO_2$$
$$\Delta G°' = -185.5 \text{ kJ/mol } CH_4 \tag{3}$$

$$4CH_3OH \rightarrow 3CH_4 + CO_2 + 2H_2O$$
$$\Delta G°' = -103 \text{ kJ/mol } CH_4 \tag{4}$$

$$4CH_3NH_3^+ + 2H_2O \rightarrow 3CH_4 + CO_2 + 4NH_4^+$$
$$\Delta G°' = -74 \text{ kJ/mol } CH_4 \tag{5}$$

$$2(CH_3)_2NH_2^+ + 2H_2O \rightarrow 3CH_4 + CO_2 + 2NH_4^+$$
$$\Delta G°' = -74 \text{ kJ/mol } CH_4 \tag{6}$$

$$4(CH_3)_3NH^+ + 6H_2O \rightarrow 9CH_4 + 3CO_2 + 4NH_4^+$$
$$\Delta G°' = -74 \text{ kJ/mol } CH_4 \tag{7}$$

$$CH_3COO^- + H^+ \rightarrow CH_4 + CO_2$$
$$\Delta G°' = -32.5 \text{ kJ/mol } CH_4 \tag{8}$$

The different species are more or less specialized with respect to the number of substrates for conversion into methane. For some species the substrate serves as both the energy and the sole carbon source. Other organisms grow only when media are supplemented with additional carbon compounds. These characteristics and other requirements for optimal growth are enlisted in Table 13.1.

In Section 13.2 the process of methanogenesis from the different one-carbon compounds is described, with emphasis on the unique coenzymes involved. Oxidation and reductive processes, which are important in creating a proton motive force and for the generation of reducing equivalents for cell carbon synthesis, are outlined in Section 13.3. Section 13.4 deals with energetical aspects of methanogenic biochemistry, that is, energy-yielding and energy-consuming steps in methanogenesis, energy requirement for growth, ATP synthesis, and active transport. A discussion of cell carbon synthesis, as far as relevant with respect to the preceding sections, is given in Section 13.5.

The results of the investigations on the biochemistry of methanogenesis during the past 80 years have been reviewed thoroughly. The reader may find reviews of the early work in Refs. 9, 265, and 324. The major breakthrough in understanding the process took place during the past decade (for reviews, see Refs. 3, 4, 38, 72,

TABLE 13.1 Classification and properties of methanogens[a]

Species	Morphology	Cell Wall	Substrate	G + C (mol %)	Optimal Temperature (°C)	Essential Organic Growth Factor	References
Family I. Methanobacteriaceae							
Genus I. *Methanobacterium*							
M. alcaliphilum	Long rods to filaments	NA	H_2/CO_2	57	37	Yeast extract or Trypticase peptone	333
M. bryantii	Long rods to filaments	Pseudomurein	H_2/CO_2	32.7–38	37–39	—	4, 181
M. formicicum	Long rods to filaments	Pseudomurein	H_2/CO_2, formate	40.7–42	37–45	—	4, 181
M. ivanovii	Rods to filaments	NA	H_2/CO_2	NA	45	—	10
M. thermoaggregans	Long rods to filaments forming aggregates	NA	H_2/CO_2	42	65	—	20
M. thermoalcaliphilum	Long rods to filaments	NA	H_2/CO_2	38.8	58–62	Yeast extract	21
M. thermoautotrophicum	Long rods to filaments	Pseudomurein	H_2/CO_2	49.7–52	65–70	—	4, 181
M. thermoformicicum	Rods	NA	H_2/CO_2, formate	NA	55	—	350
M. uliginosum	Long rods	Pseudomurein	H_2/CO_2	29.4–33.8	40	—	153
M. wolfei	Rods	Pseudomurein	H_2/CO_2	61	55–65	Yeast extract	321
Genus II. *Methanobrevibacter*[b]							
M. arboriphilus[b]	Short rods, short chains	Pseudomurein	H_2/CO_2	27.5–31.6	37–39	Cysteine	4, 181
M. ruminantium	Short rods, short chains	Pseudomurein	H_2/CO_2, formate	30.6	37–39	Coenzyme M, acetate, α-methylbutyrate, amino acids	4, 181

ORDER I. METHANOBACTERIALES

M. smithii	Short rods, short chains	Pseudomurein	H₂/CO₂, formate	31–32	37–39	Acetate	4, 181

Family II. Methanothermaceae
Genus I. *Methanothermus*

M. fervidus	Short rods	Pseudomurein, protein	H₂/CO₂	33	83	Yeast extract	268

ORDER II. METHANOCOCCALES

Family I. Methanococcaceae
Genus I. *Methanococcus*

M. aeolicus	NA	Protein	Formate	NA	NA	NA	245
M. deltae	Irregular cocci	NA	H₂/CO₂, formate	40.5	37	—	33
M. frisius	Regular cocci	NA	H₂/CO₂, methanol, methylamines	38.2	36	—	22
M. halophilus	Irregular cocci forming aggregates	NA	Methanol, methylamines	NA	26–36	—	349
M. jannaschii	Irregular cocci	NA	H₂/CO₂	31	85	—	132
M. maripaludis	Irregular cocci	NA	H₂/CO₂, formate	33	35–39	—	133
M. thermolithotrophicus	Regular to irregular cocci	Protein	H₂/CO₂, formate	31.3	65	—	107
M. vannielii	Regular to irregular cocci	Protein	H₂/CO₂, formate	28.3–31.1	36–40	—	4, 181, 324
M. voltae	Regular to irregular cocci	NA	H₂/CO₂, formate	30.7	35–40	Acetate, leucine, isoleucine	4, 181, 316

TABLE 13.1 (*Continued*)

Species	Morphology	Cell Wall	Substrate	G + C (mol %)	Optimal Temperature (°C)	Essential Organic Growth Factor	References
ORDER III. METHANOMICROBIALES							
Family I. Methanomicrobiaceae							
Genus I. Methanomicrobium							
M. mobile	Short rods	Protein	H_2/CO_2, formate	48.8	40	Rumen Fluid	4, 181
M. paynteri	Short irregular rods	NA	H_2/CO_2	44.9	40	Acetate	220
Genus II. Methanogenium							
M. aggregans	Irregular cocci forming aggregates	NA	H_2/CO_2, formate	52	35	Acetate, yeast extract or Trypticase peptone	207
M. bourgense	Irregular cocci	NA	H_2/CO_2, formate	59	37	Acetate	206
M. cariaci	Irregular cocci	Protein	H_2/CO_2, formate	50.0–51.6	20–25	Acetate, yeast extract	4, 181, 342
M. fritonii	Irregular cocci	Protein	H_2/CO_2, formate	49.2	57	—	99
M. marisnigri	Irregular cocci	(Glyco)protein	H_2/CO_2, formate	58.8–61.2	20–25	Trypticase	4, 181, 342
M. olentangyi	Irregular cocci	NA	H_2/CO_2	54.4	37	Acetate	33
M. tatii	Regular to irregular cocci	(Glyco)protein	H_2/CO_2, formate	54	37–40	Acetate	342
M. thermophilicum	Irregular cocci	(Glyco)protein	H_2/CO_2, formate	56–60	55–58	Acetate, Trypticase	221, 343
M. wolfei[c]	Irregular cocci	NA	H_2/CO_2, formate	61.1	45	Acetate, arginine	

	Morphology	Cell wall	Substrates	mol% G+C	Temp (°C)	Growth factor	References
Genus III. Methanospirillum							
M. hungatei	Regular curved rods to long spiral filaments	Protein	H_2/CO_2, formate	45–46.5	30–40	—	4, 181
Family II. Methanosarcinaceae							
Genus I. Methanosarcina							
M. acetivorans	Irregular cocci, forming cysts	Protein	Methanol, methylamines, acetate	41	35–40	—	255
M. barkeri	Irregular cocci, forming packets	Heteropoly saccharide	H_2/CO_2, methanol, methylamines, acetate	38.8–51	35–40	—	4, 181
M. mazei	Irregular cocci, forming cysts	Heteropoly saccharide	Methanol, methylamines, acetate, H_2/CO_2 very slowly or not	NA	30–40	—	180, 181
M. thermophila	Irregular cocci, forming aggregates	Heteropoly saccharide	H_2/CO_2, methanol, methylamines acetate	42	50	p-Aminobenzoic acid	354
M. vacuolata[d]	Irregular cocci, forming cysts	NA	H_2/CO_2, methanol, methylamines, acetate	51	40	—	181, 351
Genus II. Methanococcoides							
M. methylutens	Irregular cocci	Protein	Methanol, methylamines	42	30–35	—	256

TABLE 13.1 *(Continued)*

Species	Morphology	Cell Wall	Substrate	G + C (mol %)	Optimal Temperature (°C)	Essential Organic Growth Factor	References
ORDER III. METHANOMICROBIALES							
Family III. Methanoplanaceae							
Genus I. Methanoplanus							
M. endosymbiosus	Plate-shaped	(Glyco)protein	H_2/CO_2, formate	38.7	32	—	298
M. limicola	Plate-shaped	(Glyco)protein	H_2/CO_2, formate	47.5	40	Acetate	320
Family not assigned							
Genus Methanothrix							
M. concilii	Rods to filaments	NA	Acetate	61.2	35–40	—	209
M. soehngenii	Rods to filaments	Protein	Acetate	51.9	37	—	110
M. thermoacetophila	Rods to filaments	NA	Acetate	NA	65	—	204
Order and family not assigned							
Genus Methanolobus							
M. tindarius	Irregular cocci	(Glyco)protein	Methanol, methylamines	45.9	25	—	155
Genus Halomethanococcus							
H. mahi	Irregular cocci	NA	Methanol, methylamines	48.5	35	—	210
Genus Methanosphaera							
M. stadmanae	Spherical	Pseudomurein	Methanol	26	37	Acetate, thiamin, isoleucine	190

[a] NA, data not available.

[b] Recently named as *M. arboriphilicus* (98).

[c] T. B. Moore and L. D. Eirich, Characterization of a new freshwater *Methanogenium* isolate, *Methanogenium wolfei*, sp. nov. Abstr. Annu. Meet. Am. Soc. Microbiol., 1985, p. 160.

[d] This strain is also classified as a vacuolated *Methanosarcina* strain (181).

142, 181, 182, 284, 315, 325, 327, 346, 347, and 348). Most of the papers cited here date from this period.

13.2 ONE-CARBON METABOLISM

13.2.1 Introduction

In the reduction to methane, methanogenic bacteria use a limited number of one-carbon compounds as the electron acceptor (Table 13.1). The conversion of CO_2 to methane can formally be considered to proceed through the stages formate, formaldehyde, and methanol. In the early phase of the investigation it was noted that these compounds did not occur as free intermediates (for a review, see Ref. 9). Hence it was proposed by Barker in 1956 (9) that the one-carbon units might be bound to one or more carriers. Moreover, results of labeling experiments led Barker to propose a common pathway for the conversion of the various substrates (Fig. 13.1). Although knowledge concerning the nature of the one-carbon carriers is emerging rapidly, Barker's concept is still valid.

The role of tetrahydrofolic acid derivatives gained some attention in the 1960s by analogy to their role in CO_2 reduction by certain clostridia. Methanogenesis from methylene- and methyltetrahydrofolate proceeds readily in cell-free extracts of *M. bryantii* and *M. barkeri* (16, 26, 332). However, the results conflict with the low activities found for formyltetrahydrofolate synthetase and methylenetetrahydrofolate dehydrogenase (77). A possible role of folic acid derivatives was eroded by the discovery of a novel pterin, called methanopterin (see Section 13.2.2.1). Alternatively, it was thought that methylcorrinoids were the general methyl donor in the terminal step of reduction (for a review, see Ref. 265). At present, the function of these compounds in methanogenic biochemistry seems to be restricted to the transfer of the methyl group of methanol to coenzyme M (see Section 13.2.5), and possibly also to the aceticlastic reaction in methanogenesis from acetate (see Section 13.2.7) and the C_1-C_1 condensation reaction in cell carbon synthesis (see Section 13.5.2).

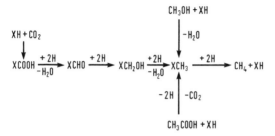

Figure 13.1 Unifying scheme of the process of methanogenesis according to Barker (9).

13.2.2 Conversion of Hydrogen and Carbon Dioxide

13.2.2.1 Initial Steps of Reduction

The pathway from the formyl to the methyl level is best understood in *M. thermoautotrophicum*. For the methylcoenzyme M (CH_3S-CoM)–stimulated reduction of CO_2, the so-called RPG effect (see Section 13.2.2.3), an enzyme mixture, component B (see Section 13.2.2.2), and a fraction containing oxygen-stable low-molecular-weight cofactors were required (176, 227). The fraction was resolved into two components, a compound designated CDR ("carbon dioxide reduction") factor and a compound previously described as methanopterin. CDR factor, renamed methanofuran, was purified and its structure was elucidated by Leigh et al. (174). Methanofuran is composed of a long side chain attached to a para-substituted phenol and a furan moiety (Fig. 13.2), and it does not have a known biochemical analogon. Methanofuran occurs in two forms: as a free amine (Fig. 13.2*A*) and as a formylamide (Fig. 13.2*B*). By labeling experiments it was shown that the formyl group of the latter is derived from CO_2 (175).

Methanopterin was first identified (141, 143) as the metabolic precursor of YFC ("yellow fluorescent compound"), now called 5,10-methenyl-5,6,7,8-tetrahydromethanopterin (methenyl-H_4MPT), which becomes labeled within 2 to 20 s during short-term labeling experiments performed with $^{14}CO_2$ under H_2 atmosphere or with $^{14}CH_3OH$ (40). Methanopterin could be identified as a pterin derivative (141, 143) and its structure was elucidated by use of high-resolution ^{1}H-NMR and ^{13}C-NMR techniques supported by degradational studies (292, 293); the structure is shown in Fig. 13.3*A*. Methanopterin is a complex structural analogon of folic acid with respect to the presence of the pterin moiety and the para-substituted aniline. Apart from the ribityl group and the structural elements attached to it (instead of a para-substituted carboxyl glutamyl moiety), methanopterin deviates from folic acid by the presence of a methyl group at C-7 and an ethyl group at C-6 of the pterin ring. Methanopterin, or more properly tetrahydromethanopterin (H_4MPT), is a one-carbon carrier. This was proven by van Beelen et al. (290, 295) by the structural elucidation of methenyl-H_4MPT, which was at first erroneously identified as carboxy-H_4MPT (140). The analogy of the structures of methenyl-H_4MPT (Fig. 13.3*C*) and methenyltetrahydrofolic acid is straightforward. After short-

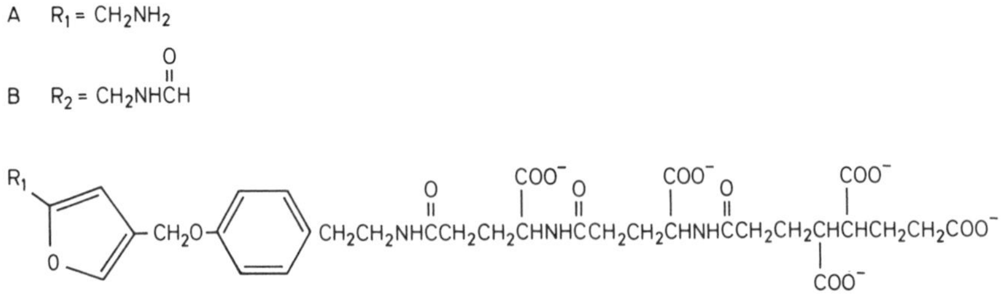

Figure 13.2 Structure of methanofuran. (*A*) Amine form. (*B*) formylamide form (174).

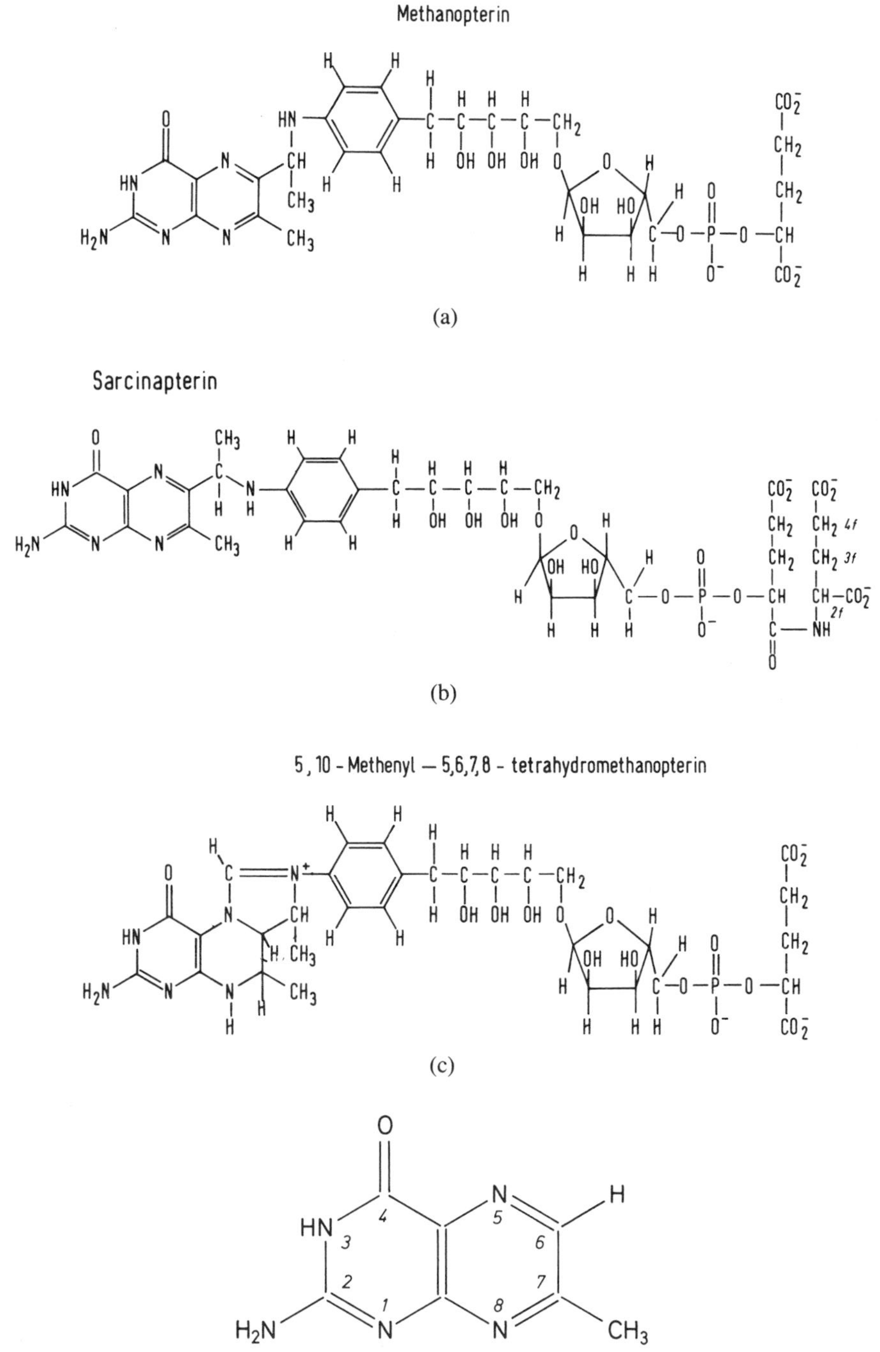

Figure 13.3. Structures of 7-methyl-substituted pterins of methanogenic bacteria. (*A*) Methanopterin (292, 293). (*B*) Sarcinapterin (292). (*C*) 5,10-Methenyl-5,6,7,8-tetrahydromethanopterin (methenyl-H$_4$MPT) (290, 295). (*D*) 7-Methylpterin (137).

term incubation of whole cells of *M. thermoautotrophicum* with $H^{13}CO_3^-$ under 80% H_2/20% CO_2 atmosphere and subsequent purification of labeled methenyl-H_4MPT, an increased intensity of the 5,10-methenyl carbon atom was observed in the ^{13}C-NMR spectrum (295).

H_4MPT functions in methanogenic biochemistry in a way analogous to tetrahydrofolate in other organisms (108; Fig. 13.4). Evidence herefore was presented by Escalante-Semerena et al. (70, 71), who isolated H_4MPT from *M. thermoautotrophicum* and demonstrated that the compound could operate as a carrier of the C_1-unit in three successive reduction levels: the formyl, formaldehyde, and methyl state (73). H_4MPT binds nonenzymatically to formaldehyde, giving methylene-H_4MPT (71, 138). The latter compound is oxidized to methenyl-H_4MPT, and coenzyme F_{420} (see Section 13.3.4) serves as the electron acceptor in this reaction (100). In *M. thermoautotrophicum*, methenyl-H_4MPT is converted to 5-formyl-H_4MPT by a cyclohydrolase (54). The latter two reactions are reversible. 5-Formyl-H_4MPT is also formed in *M. thermoautotrophicum* by the formyl transfer of formylmethanofuran to H_4MPT; the reaction is independent of ATP (138).

Under hydrogen atmosphere, methylene-H_4MPT is reduced to methyl-H_4MPT (73, 138). The methyl group is subsequently transferred to HS-CoM (215, 232) by a 5-hydroxybenzimidazolylcobamide (B_{12}-HBI)–containing protein (215). This ATP-dependent reaction shows some striking similarities to the methanol: HS-CoM methyltransferase system in *M. barkeri* (see Section 13.2.5.1).

M. barkeri grown on methanol contains a methanopterin derivative, called sarcinapterin (Fig. 13.3*B*), in considerable amounts (291, 292) indicating that the compound is also involved in the oxidative pathway from methanol to CO_2 (see Section 13.2.5.2). This is corroborated by the finding that labeled methenyl-H_4MPT is readily formed from ^{14}CH$_3$OH (40). Remarkably, cyclohydrolase of *M. barkeri* produces a formyl-H_4MPT derivative from methenyl-H_4MPT, which is different from the one observed in *M. thermoautotrophicum* (138), the more energy-rich 10-formyl-H_4MPT.

Since all methanogenic bacteria investigated contain methanofuran and methanopterin derivatives, the intermediary reactions presumably proceed along a common line (131; Fig. 13.4).

Pterins in methanogenic bacteria are characterized by the presence of a methyl

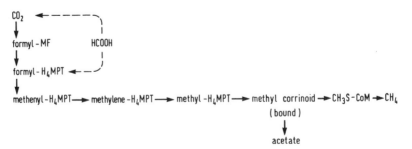

Figure 13.4 Conversion of carbon dioxide and formate into methane. MF, methanofuran.

group at C-7. This not only holds for the methanopterin derivatives: The most simple representative of the family, 7-methylpterin (Fig. 13.3D), could also be isolated (137). The compound lends the organism the typical blue fluorescence when studied by epifluorescence microscopy (53, 57).

13.2.2.2 Terminal Step of Methanogenesis

In all methanogenic bacteria tested, the terminal step of methanogenesis involves the reductive demethylation of methylcoenzyme M (CH_3S-CoM) to coenzyme M and methane [equation (9)]. Coenzyme M derivatives were discovered by McBride and Wolfe (186) and their structures were established by Taylor and Wolfe (277). The compounds are quite simple molecules: HS-CoM is 2-mercaptoethanesulfonic acid (Fig. 13.5A) and CH_3S-CoM is 2-(methylthio)ethanesulfonic acid (Fig. 13.5B). Coenzyme M derivatives occur only in methanogenic bacteria, and in amounts ranging from 0.2 to 16 μmol per gram of dry weight (6). The compounds function as a vitamin for growth of M. ruminantium strain M_1 (5, 276).

In the methylcoenzyme M reductase reaction

$$CH_3S\text{-CoM} + H_2 \xrightarrow[\text{methylcoenzyme M reductase}]{\text{comp. A, FAD, ATP, Mg}^{2+}} CH_4 + HS\text{-CoM} \qquad (9)$$

CH_3S-CoM is the only substrate to yield methane except for CH_3CH_2S-CoM, which produces ethane, though at a considerably lower rate (94). Coenzyme M derivatives modified at the sulfide moiety, the sulfonate group, or at the ethylene bridge were inactive (94).

The reductive demethylation of CH_3S-CoM is catalyzed by a complex system of proteins and cofactors (96, 97, 198). Component A represents a hydrogenase system (197), which was further resolved into three protein fractions (198): an oxygen-stable coenzyme F_{420}-dependent hydrogenase (component A_1) that uses FAD as a coenzyme (see Section 13.3.2) and two enzyme fractions A_2 and A_3 of unknown function; the latter is oxygen sensitive and contains a number of proteins. Component A_2 is oxygen stable and binds firmly to ATP-agarose, which provides a simple means for its purification (228). The hydrogenase system could be substituted by an electron-generating system composed of NADPH: coenzyme F_{420} oxidoreductase; either NADPH or reduced coenzyme F_{420} could serve as the electron donor (67, 68).

Component B could be identified as 7-mercaptoheptanoylthreonine phosphate (203). The role in the methylcoenzyme M reductase reaction is unknown.

Methylcoenzyme M reductase is an acidic, oxygen-stable protein showing an absorption maximum around 425 nm (66, 68). The enzymes of M. thermoautotro-

A $HS-CH_2CH_2SO_3^-$

B $CH_3-S-CH_2CH_2SO_3^-$

Figure 13.5 Structures of (A) coenzyme M and (B) methylcoenzyme M according to Taylor and Wolfe (277).

phicum (68, 69) and *M. voltae* (P. Hartzell and R. S. Wolfe, Abstr. Annu. Meet. Am. Soc. Microbiol., 1983, p. 141) are composed of three subunits in an $\alpha_2\beta_2\gamma_2$ arrangement showing molecular weights of 68,000, 45,000, and 38,500 Da for the former organism and 70,000, 48,000, and 45,000 Da for *M. voltae*. The reductase of *M. barkeri* contains four subunits (72,000, 57,000, 42,000, and 28,000 Da; Hartzell and Wolfe). The absorbance in the visible light is due to the presence of a novel nickel-containing tetrapyrrole, called factor F_{430}: methylcoenzyme M reductase of *M. thermoautotrophicum* contains 2 mol of factor F_{430} per mole of enzyme (66). Supported by the results of incorporation studies (see Section 13.5.6) the structure of factor F_{430} was elucidated by Pfaltz et al. (213). The structure of a derivative (F_{430} M) modified by methylation of the carboxymethyl and carboxyethyl groups is shown in Fig. 13.6. Although a number of isomers and oxidation products were described (43, 139, 317), the structure of the compound as shown in Fig. 13.6 represents the actual prosthetic group of methylcoenzyme M reductase (178). Apart from the presence of nickel, factor F_{430} is a remarkable molecule in several aspects: (a) the highly reduced character of the uroporphinoid ligand skeleton; (b) the biologically unique corphin-type of conjugation, which is a structural hybrid between corrins and porphyrins; and (c) the presence of an additional carbocyclic ring (213). The function of the nickel tetrapyrrole is unclear. However, the finding that coenzyme M derivatives are associated with factor F_{430} in a 1:1 stoichiometry (143, 144) together with some interesting oxidation and reduction properties (139, 156) and the presence of coenzyme M derivatives associated (144) make it an attractive candidate to function in the active center of methylcoenzyme M reductase.

Although the final step of methanogenesis is energetically most favorable (see Section 13.4.1), ATP is required for the terminal reaction in cell-free extracts,

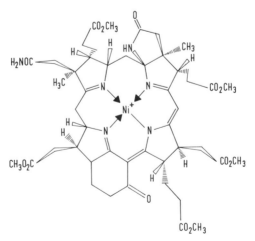

Figure 13.6 Structure of the methylated factor F_{430} derivative (F_{430}M) proposed by Pfaltz et al. (213).

although only in catalytic amounts: about 15 mol of methane is produced at the expense of 1 mol of ATP (96, 186). Other nucelotide triphosphates, except TTP, could replace ATP to some extent; AMP, cyclic AMP, and α,β- or β,γ-methylene ATP were inactive (96, 194). ATP is also required in the resolved methylcoenzyme M reductase reaction and in the system at which NADPH acts as the electron donor (67, 68, 97, 198). From detailed experiments it was concluded by Whitman and Wolfe (318) that ATP presumably performs an activation of an enzyme or enzyme-bound cofactor by adenylation or phosphorylation, rather than acting as an allosteric effector as suggested by Mountfort (194). An inherent problem in studying the effect of ATP is the labile nature of the enzyme in the activated state (318). Optimal activity of the methylcoenzyme M reductase system requires the presence of a relatively high concentration of Mg^{2+}: the ion presumably acts as a physiological effector (96, 318).

The reduction of CH_3S-CoM by hydrogen is inhibited by a whole series of compounds, such as inhibitors of the electron transfer (NaN_3, 2,4-dinitrophenol, methylviologen, benzylviologen, 1,10-phenantroline), the sulfhydryl reagent 5,5'-dithiobis(2-nitrobenzoic acid), oxygen, nitromethane, nitrite, and halogenated hydrocarbons (chloral hydrate, chloroform, methylene chloride, dichlorodifluoromethane) (96). In addition, 2-bromoethanesulfonic acid and 2-chloroethanesulfonic acid proved to be potent inhibitors showing 50% inhibition at $7.9 \times 10^{-6}\ M$ and $7.5 \times 10^{-5}\ M$, respectively (94).

13.2.2.3 Activation and Reduction of Carbon Dioxide

By resolving the participation of methanofuran, H_4MPT derivatives, and coenzyme M discussed in the preceding section, a major part of the reduction pathway of CO_2 to methane appears to be elucidated (Fig. 13.4). However, the first step, the activation of CO_2 and the reduction to the formate level, remains obscure. Isotope fractionation experiments showed that CO_2 rather than bicarbonate is involved in the initial step (88).

In connection with the activation of CO_2 a little understood phenomenon, the RPG effect named after its discoverer R. P. Gunsalus (95), should be mentioned. The rate of methanogenesis from hydrogen and CO_2 was stimulated 30-fold by the addition of CH_3S-CoM. The reaction stopped when 2 to 20 mol of methane were produced for each mole of CH_3S-CoM added and the effect could be reproduced by repetitive additions of CH_3S-CoM (95). Not only CH_3S-CoM stimulates the reduction of CO_2, but also a variety of compounds that seem to have in common the property to donate one-carbon intermediates at the methylene (formaldehyde, serine, hydroxymethylcoenzyme M) (226) and methyl (CH_3S-CoM) level of reduction. A second class of stimulants, that include pyruvate, α-ketoglutarate, malate, isocitrate, and carbon monoxide (185, 294, 299) stimulate by providing electrons for the activation of the methyltransferase and methylreductase systems (299). Apart from CO_2 reduction the rate of conversion of methanopterin derivatives (294) and the production of acetyl-CoA (271) were stimulated by CH_3S-CoM.

The RPG effect was interpreted as a direct coupling between the first step, the activation of CO_2, and the final step of methanogenesis (72, 225, 226, 325, 328). Recent investigations indicate that the effect is connected to the activation of the methyltransferase and methylreductase systems (299). The RPG effect was observed only for *M. thermoautotrophicum* and *M. bryantii* (185), organisms that can use only CO_2 as the electron acceptor. Formate-utilizing bacteria such as *M. formicicum* and *M. ruminantium* and the most versatile organism, *M. barkeri*, did not show the RPG effect (111, 185). The presence or the absence of the effect may reflect differentiations in the way methanogenic bacteria activate CO_2.

13.2.3   Conversion of Formate

About half the species of methanogenic bacteria are able to grow on formate. Formate can serve as an electron donor by the action of formate dehydrogenase. It is still unclear whether formate is directly channeled into the reduction pathway or indirectly via CO_2. Low activities of formyltetrahydrofolate synthetase (formylase) and only very low activities of methylenetetrahydrofolate dehydrogenase were found in a number of organisms (77, 238). However, these studies were performed with tetrahydrofolate as the coenzyme and should be repeated with the methanogenic coenzyme methanopterin to establish the possible involvement of a formylase in the conversion of formate. If incorporation of formate proceeds via CO_2, the latter compound presumably does not equilibrate with the medium (78, 80, 127, 237, 238).

Reducing equivalents from formate are generated by formate dehydrogenase. By the concerted action of this enzyme and hydrogenase formate is split into hydrogen and CO_2. Hydrogen could be detected only during growth on formate when cultures of *M. vannielii* or *M. hungatei* were exposed to high pH values or high temperatures (78, 128, 238).

The presence of an oxygen-sensitive formate dehydrogenase that uses coenzyme F_{420} as the electron acceptor was demonstrated in *M. smithii* (288), *M. hungatei* (78), *M. vannielii* (129), and *M. formicicum* (239). The enzymes of *M. vannielii* and *M. formicicum* were purified and studied in some detail. These, presumably soluble, cytoplasmatic proteins (238, 239) contain molybdenum and iron–sulfur clusters in the active centers. *M. vannielii* may synthesize two forms of formate dehydrogenase: one is a protein of Mr 105,000, which contains about 1 mol of Mo and 6 mol of Fe-S per mole of protein and is the predominant form when the organism is cultured in the absence of selenite and wolframate. When grown in the presence of the latter two compounds, a labile high-molecular-weight enzyme containing selenocysteine (127, 130) is the major form; molybdenum is then partially replaced by wolfram (130). Formate dehydrogenase of *M. formicicum* is a nonheme FeS flavoprotein of M_r 177,000 Da and is composed of two subunits showing molecular weights of 85,000 and 53,000 Da, respectively (239, 241). The enzyme contains 1 mol of FAD, 2 mol of Zn, 21 to 24 mol of Fe, and 25 to 29 mol of acid-labile S (241). FAD is noncovalently bound and the coenzyme is readily dissociated from the reduced enzyme (240, 241). In addition, a molybdopterin cofactor is present (183, 239). The C-6-substituted pterin, which contains covalently bound phosphate, is structurally different from the pterin found in other molybdoenzymes, (i.e., sulfite oxidase, nitrate reductase, and xanthine oxidase) (183).

It is not clear whether formate-utilizing methanogenic bacteria growing on hydrogen and CO_2 bind CO_2 by the reversed action of formate dehydrogenase ("CO_2 reductase") as was found for a number of clostridia (330). For CO_2 reductase CO_2, rather than HCO_3^-, is the substrate (283). In this respect it is noteworthy that *M. vannielii* shows optimal growth on hydrogen and CO_2 at pH values (6.5 to 7.5) distinctly below the values of pH 8 to 8.5 for growth on formate (128).

13.2.4 Carbon Monoxide, a Physiological Substrate of Methanogenesis?

Kluyver and Schnellen (150) demonstrated the conversion of carbon monoxide into methane by cell suspensions of *M. barkeri* and *M. formicicum*. Daniels et al. (37) demonstrated that CO is oxidized during growth of methanogenic bacteria on gas mixtures containing CO together with CO_2 and hydrogen and they reported on the presence of CO dehydrogenase (coupled to coenzyme F_{420}) in cell extracts of *M. thermoautotrophicum*. Later, Zeikus (347) and O'Brien et al. (205) were able to grow *M. barkeri* either with CO as the sole carbon and energy source or mixotrophically with CO and methanol. Growth was rather slow; the doubling time was 65 h in the most rapid growth phase. Remarkably, large amounts of hydrogen, together with CO_2, accumulated during growth and methanol metabolism was inhibited by CO. These results indicate that CO is a poor substrate and inhibits the use of the common methanogenic substrates. This is the more surprising, as high activities of CO dehydrogenase are reported to be commonly found in *M. barkeri* (158, 160). The physiological meaning of CO consumption and the role of CO dehydrogenase need further studies (see Sections 13.2.5.1 and 13.5.2).

CO dehydrogenases from *M. arboriphilicus* (98) and *M. barkeri* (160) were purified and studied in detail. The highly labile enzyme from *M. arboriphilicus* is a nickel protein; its synthesis during growth required nickel, and the 21-fold purified enzyme from ^{63}Ni-labeled cells comigrated with the label and contained 0.20 to 0.25 nickel per 140,000-Da protein molecule. About 5% of the soluble cell protein of acetate-grown cells of *M. barkeri* was CO dehydrogenase (160). The enzyme is extremely sensitive to oxygen, and when CO dehydrogenase was isolated from cells labeled with ^{63}Ni, the radioactivity copurified with the enzymic activity, indicating the presence of tightly bound Ni. The 20-fold purified enzyme showed a specific activity of 216 μmol of CO oxidized per minute and milligram of protein; it displayed a molecular weight of 232,000 Da and a composition of two subunits of each 92,000 and 18,000 Da. The broad shoulder in the absorption spectrum extends from 340 to 420 nm and decreased when CO was added. The presence of Ni and the rather large K_m values (about or above 1 mM) for CO are striking common properties of CO dehydrogenase from anaerobic acetogens (44, 48, 55, 56, 218) and methanogens.

In anaerobic acetogenic bacteria CO dehydrogenase is the enzyme at which the

C—C condensation reaction occurs (for reviews see Refs. 85 and 348). Analogously, in methanogenic bacteria the enzyme may play this central role both in acetate catabolism (see Section 13.2.7) and in acetyl-CoA synthesis (see Section 13.5.2).

13.2.5 Conversion of Methanol

In 1938, Schnellen (246) obtained a pure culture of a sarcina-like methanogen, which he named *Methanosarcina barkeri* and which was able to grow at the expense of methanol. From that time on, many attempts were made to elucidate the pathway by which methanol was reduced to methane. After the discovery that methylcobalamin could serve as a precursor for methanogenesis by extracts of *M. barkeri*, Blaylock and Stadtman (17) suggested that methanol fermentation may involve an enzymic transfer of the methyl group from methanol to cobalamin, followed by a reductive cleavage to yield methane. Later they reported the enzymic formation of methylcobalamin from cob(I)alamin and methanol (18), and the responsible enzyme system was resolved into four components (16, 19). A high-molecular-weight corrinoid-containing enzyme, a ferredoxin-like compound, a protein with unknown function, and a heat- and acid-stable cofactor were needed in the formation of methylcobalamin from methanol and electrochemically prepared cob(I)alamin in the presence of ATP and H_2. Another indication of the possible involvement of corrinoids was the presence of rather copious amounts of an aberrant corrinoid (factor III) in methanogens (Fig. 13.7). In *M. barkeri* grown on methanol 4.1 nmol of corrinoid per milligram of cell dry weight was found (159); the corrinoids present differ from commonly found cobalamins as to the structure of the α-ligand, which is 5-hydroxybenzimidazole (HBI) in *M. barkeri* instead of 5,6-dimethylbenzimidazole (DMBI) (217). After the discovery of coenzyme M and its central role as methyl carrier in the process of methanogenesis from both CO_2 and methanol the role of corrinoids in methanogenesis became uncertain. Present knowledge, based primarily on the studies of van der Meijden et al. (300–305), shows a definite role of corrinoids in the conversion of methanol; the possible role of these compounds in the synthesis of acetyl groups is discussed in Section 13.5.2.

At present 11 species of methanogens are known to grow on methanol (Table 13.1). All studies on the process of methanogenesis from methanol were performed with *M. barkeri*.

13.2.5.1 *Methyltransferases Involved*

In aerobic organisms methanol is converted to formaldehyde by bacterial methanol dehydrogenases involving pyrroloquinoline quinone as a prosthetic group or by a peroxisomal alcohol oxidase in yeasts. In methanogens and the anaerobic acetogen *Eubacterium limosum* (304), methanol is converted to an enzyme-bound methyl-corrinoid by a methyltransferase.

A scheme of the reactions involved in the conversion of methanol in *M. barkeri*

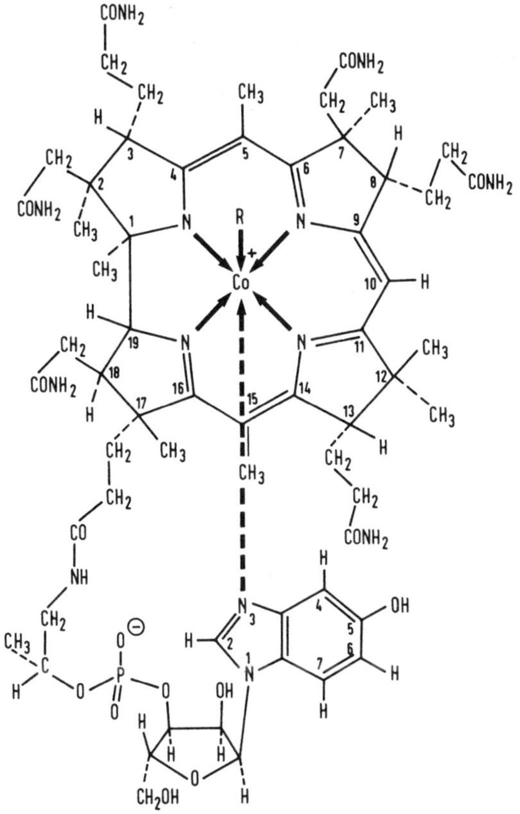

Figure 13.7 Structure of the corrinoid of methanogenic bacteria (217).

is presented in Fig. 13.8. Two methyltransferases are involved and CH_3S-CoM is produced; this product is used as a substrate of methanogenesis (see Section 13.2.2.2). The first methyltransferase MT_1 (methanol: 5-hydroxybenzimidazolyl-cobamide methyltransferase) contains a firmly bound corrinoid which is methylated by methanol when it is present in its most reduced state (Co^I). The methyl group is then transferred to HS-CoM by Co-methylcobamide: HS-CoM methyltransferase (MT_2). Both enzymes are insensitive to 2-bromoethanesulfonic acid, the typical inhibitor of the CH_3S-CoM reductase system.

About 15% of the protein in the crude extract of methanol-grown cells of *M. barkeri* consists of MT_1, which has a molecular weight of 122,000 Da and an $\alpha_2\beta$ composition with subunits of 34,000 and 53,000 Da, respectively (303). Approximately three to four molecules of B_{12}-HBI (the corrinoid with 5-hydroxybenzimidazole as α-ligand) were found per molecule of protein. The corrinoid is partly lost upon treatment of MT_1 with 2-mercaptoethanol and completely upon dissociation of the enzyme into its subunits by SDS plus 2-mercaptoethanol (303).

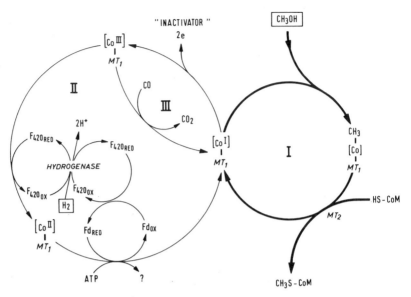

Figure 13.8 Conversion of methanol into CH_3S-CoM in *M. barkeri*. The carbon cycle (I) involves the action of methanol: 5-hydroxybenzimidazolylcobamide methyltransferase (MT_1), which is methylated when it is present in its most reduced state ($[Co^I]$). The methyl group is then transferred to HS-CoM by Co-methylcobamide: HS-CoM methyltransferase (MT_2). Cycle II represents the inactivation and activation reactions of MT_1. The presence of an oxidator causes the conversion of the enzyme-bound corrinoid to the Co^{III} level and inactivation of MT_1. Activation is brought about by reduction. First, the corrinoid is reduced to the Co^{II} level and a subsequent reduction to the active Co^I level is mediated by H_2, hydrogenase, coenzyme F_{420}, and ferredoxin. Cycle III shows the chemical reduction and activation of the enzyme by CO. If both CO and a suitable electron acceptor are present, MT_1 (or solely the MT_1-bound corrinoid) catalyzes the oxidation of CO to CO_2. In this way CO dehydrogenase-like activity is displayed by MT_1 and/or its bound corrinoid.

The enzyme is active only if the corrinoid is present in its most reduced state (Co^I). Oxidation of the reduced corrinoid occurs readily, for example, in the presence of viologen dyes, flavins, or oxygen and the inactivated enzyme obtained in this way may be activated by reduction (Fig. 13.8). ATP is required in the activation process, but how it is involved is yet unknown (303). Ferredoxin reduced by either hydrogen or pyruvate (301) is needed to obtain reduction of the corrinoid beyond the Co^{II} state. This role of ferredoxin may explain the rather high concentrations of this compound found in *M. barkeri* (see Section 13.3.2).

The occurrence of MT_1 is not restricted to methanogens, and an enzyme with similar properties and function is found in *E. limosum* (304). Two other methyl-transferring enzymes with firmly bound corrinoids are known: the "corrinoid" enzyme involved in the synthesis of acetate by *Clostridium thermoaceticum* (314) and 5-methyltetrahydrofolate: homocysteine methyltransferase, which is involved in methionine biosynthesis of many microorganisms and animal liver. Reductive preactivation of the latter enzyme was reported before (93, 229, 309, 313) and a similar process was suggested for the former enzyme (314). Therefore, prereduction of bound corrinoids and participation of ATP or S-adenosylmethionine in this process appear to be a common prerequisite in the functioning of methyltransfer (307).

Inactive MT_1 can be activated by the presence of CO (303). This activation is inhibited by cyanide and probably involves a reduction of the corrinoid to the Co^I level that might be brought about by CO in a chemical way. The demonstration of CO dehydrogenase-like activity with purified samples of MT_1 (303) is in agreement with a reaction sequence in which the enzyme-bound corrinoid is reduced by CO and subsequently oxidized by electron acceptors such as methylviologen (Fig. 13.8).

As a result of the MT_1 reaction, an enzyme containing a firmly bound methyl-corrinoid is obtained (302, 304, 305). The enzyme-bound methylcorrinoid from *M. barkeri* (300, 302) and *E. limosum* (304), and free Co-methyl-5,6-dimethyl-benzimidazolylcobamide (300), are substrates of the second methyltransferase MT_2 present in *M. barkeri*. As a result of the presence of this enzyme, Wood et al. (331) could demonstrate the production of methane from a chemically methylated B_{12} protein from *M. barkeri*.

MT_2 is insensitive to oxygen, has a molecular weight of 43,000 Da, and was purified 86-fold (300). The enzyme is also present in *M. bryantii* (278), *M. formicium* (304), and *M. thermoautotrophicum* (van der Meijden, unpublished results), and in these organisms MT_2 may be involved in the methyl-H_4MPT:HS-CoM methyltransferase reaction (215). Interestingly, methane is produced from methanol by the mixed cell-free extracts of *E. limosum* and *M. formicium* (304). The former organism contains MT_1 and is able to activate methanol, whereas *M. formicium* is unable to use methanol but contains MT_2 and CH_3S-CoM reductase system.

13.2.5.2 Disproportionate Reactions

As a result of the consecutive reactions catalyzed by MT_1 and MT_2, CH_3S-CoM is formed from methanol. In the presence of hydrogen CH_3S-CoM is completely converted to methane (111). If methanol is used as a sole substrate, part of it must be oxidized to CO_2 to allow reduction in the CH_3S-CoM reductase system. When $^{14}CH_3OH$ was added to cell-free extract of *M. barkeri*, methenyl-H_4MPT was formed as a short-term labeling product (40; see Section 13.2.2.1). The oxidation route is not yet clear (Fig. 13.9). Methanol may be oxidized after transfer of the methyl group to HS-CoM (306, 307) or to H_4MPT, giving CH_3S-CoM or methyl-H_4MPT, respectively. Alternatively, methanol is first oxidized to the formaldehyde level by an energy- and Na^+-dependent reaction (14, 15). Conversion of methenyl-H_4MPT into formate and H_4MPT could involve substrate-level phosphorylation analogous to results obtained with methenyltetrahydrofolate (103) in *Clostridium* species (35, 184). A compound such as formyl phosphate could also

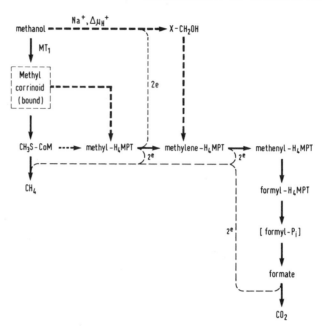

Figure 13.9 Conversion of methanol into methane and CO_2. The initial steps of methanol oxidation remain to be established. The methyl group may either be transferred to HS-CoM or to H_4MPT prior to oxidation. Alternatively, methanol is first oxidized to the formaldehyde state of reduction; at this oxidation the C_1 unit becomes bound to an unknown C_1 carrier.

be involved in the conversion of methenyl-H_4MPT. The oxidation of formate by formate dehydrogenase might be coupled to cytochromes as in *Desulfovibrio vulgaris* (244). Cytochromes of the *b*- and *c*-type are found in methanogens which use methyl-containing substrates (methanol, methylamines, and acetate) and produce CO_2 (110, 165, see Section 13.3.3). In this way the disproportional conversion of methanol, and of CH_3S-CoM formed of it, would result in both substrate-level phosphorylation and the establishment of a proton motive force (see Section 13.4.3).

13.2.6 Conversion of Methylamines

Methylated amines are used as growth substrates by the same species that use methanol (Table 13.1). Tri-, di-, and monomethylamine are converted to methane, CO_2, and ammonia by *M. barkeri* (104), while dimethylethylamine is converted to methane, CO_2, and ethylamine (74). The methyl groups of trimethylamine and monomethylamine are reduced directly to methane by *M. barkeri* without an exchange with the protons of water (308), and free methanol could not be demonstrated as an intermediate (104). The enzymes for methylamine utilization in *M. barkeri* are inducible (200).

It was demonstrated (200) that a trimethylamine : HS-CoM methyltransferase system is present in trimethylamine-grown cells of *M. barkeri*. The enzyme system has many properties in common with that involved in methanol conversion (see Section 13.2.5; 301).

13.2.7 Acetate, the Most Abundant Substrate of Methanogenesis in Nature

Acetate is the first recognized (106) substrate of methanogenesis, and about two-thirds of methane is produced from this substrate in nature (125, 254). Nevertheless, only eight species (*Methanosarcina* and *Methanothrix* species) are known to use acetate as a substrate of methanogenesis (Table 13.1).

Studies on the biochemistry of methanogensis from acetate were limited mainly to growing cells or cell suspensions of *M. barkeri* and demonstrated that the methyl group of acetate ends up almost completely in methane at which CH_3S-CoM could be identified as an intermediate (157), whereas the carboxylic group is oxidized to CO_2 (252, 266, 311). However, Krzycki et al. (158) reported that the reverse is true for 15% of the substrate converted by an acetate-adapted strain of *M. barkeri*. High concentrations of CO dehydrogenase (158), acetate kinase, and phosphate acetyltransferase (312) are present in acetate-adapted cells. ATP is needed for acetate activation (151, 312) and the compound will be produced in the final step of methanogenesis (see Section 13.4.3). To allow the net production of ATP, an additional energy-generating reaction must be present. In this respect Bott et al. (24) made the important observation that ATP is synthesized in acetate-grown *M. barkeri* through the establishment of a proton motive force at the oxidation of CO to hydrogen and CO_2.

Progress in the study of acetate conversion has been hampered by the lack of suitable cell-free systems. At present only two of such systems were described using different strains of *M. barkeri* (8, 157, 161). One, containing a particulate fraction, was able to convert acetate to methane and CO_2 under nitrogen atmosphere in the absence of added ATP (8). In contrast, a soluble enzyme fraction was strictly dependent on hydrogen and ATP to perform methanogenesis from acetate (157, 161). In the latter system particulate enzymes needed to couple CO oxidation to CH_3S-CoM reduction and to synthesize ATP presumably are lacking. Further studies in this area should be focused on the peculiar properties of CO dehydrogenase and of corrinoid-enzymes as observed in studies on methanol conversion (see Section 13.2.5.1) and on the results obtained in studies on the process of acetate formation (see Section 13.5.2).

As to the discongruity of numbers of acetophilic methanogens known and the relative importance of natural methanogensis from acetate, the specific attributions of symbiotic bacteria and the (hydrogenosomes of) anaerobic protozoa (296, 297) in the oxidation process of acetate have to be considered, since these symbiotic interactions allow the conversion of acetate into methane by hydrogenotrophic methanogens (296, 352).

13.3 ELECTRON TRANSFERRING PROCESSES

13.3.1 Introduction. Membraneous Constituents of Methanogenic Bacteria

Electron-transferring processes are of special importance for the organisms in the various methanogenic redox reactions (142) and as a result in creating a proton motive force across the cell membrane that may be used for ATP synthesis by a proton-translocating ATP synthetase (see Section 13.4). In addition, redox reactions take place for cell carbon synthesis (see Section 13.5).

The finding that growth of *M. thermoautotrophicum* on hydrogen and CO_2 in the presence of ^2H_2O resulted in ^2H-enriched methane (39, 257) implicates the involvement of one or more carriers that exclusively transfer electrons and that are situated in a proton-impermeable cell membrane. Little information exists on the total composition of cell membranes in methanogenic bacteria, since spheroplasts are difficult to prepare by conventional methods due to the presence of unique cell walls (for a review see Ref. 4). Recently, it was found (154) that *Methanobacterium wolfei* is able to synthesize pseudomurein-degrading proteins; these enzymes may provide a valuable tool for the preparation of spheroplasts. Cell membranes isolated from spheroplasts of dithiothreitol-treated *M. hungatei* contained 45 to 50% protein, 36% glycolipid and lipid, and 5.7% nonlipid carbohydrates, among them glucose, rhamnose, galactose, and mannose (259, 263). The composition of the lipids in methanogens has been investigated extensively (for reviews see Refs. 4 and 286). Polar lipids account for about 80% of the total amount of lipids. These polar lipids are C_{20} diphytanyl glycerol diethers and C_{40} dibiphytanyl diglycerol tetraethers. The presence of glycerol ethers, rather than esters, is a common property of archaebacteria (4). Neutral lipids, mainly squalenes, compose about 20% of the total cell lipid (4, 287).

Hydrogenase was localized on the internal membrane of *M. thermoautotrophicum* by use of cytochemical stains (51). The bound position of coenzyme F_{420}-dependent hydrogenase was also evident from the fact that the enzyme comigrates with a particulate fraction containing membrane-bound ATPase (91). EPR spectroscopy revealed the presence of at least five components exhibiting paramagnetic properties in membranes of *M. bryantii* (166). These include iron–sulfur centers (either HiPIP-type Fe_3-S_3 or Fe_4-S_4), a low-spin Ni(III) in an octahedral geometry (166, 168, 224), which is also present in hydrogenase (see Section 13.3.2), and a $g = 2.03$ signal attributable to flavin semiquinone. The presence of noncovalently bound FAD could be demonstrated in membranes of *M. bryantii* (167). Hydrogenase of *M. formicicum* also appears to be associated with the internal membranes, whereas formate dehydrogenase of the organism is located within invaginations of these membranes (S. F. Baron, N. L. Schauer, and J. G. Ferry, Abstr.

Annu. Meet. Am. Soc. Microbiol., 1985, p. 149). In addition, CH₃S-CoM reductase of *M. vannielii* and of *M. thermoautotrophicum* was shown to be membrane bound (208).

Remarkable is the absence in methanogenic bacteria of conventional electron carriers, such as menaquinones and ubiquinones (162, 284, 326) and cytochromes, which are however present in methylotrophic methanogens (see Section 13.3.3). A quinone derivative, α-tocopherolquinone (Fig. 13.10) of unknown function and only present in low amounts, was isolated from *M. thermoautotrophicum* and *M. bryantii* (109). Its role may be restricted to some specific reaction(s) in cell carbon synthesis.

13.3.2 Iron–Sulfur Proteins: Hydrogenase and Ferredoxin

Initial studies to characterize and purify hydrogenase from various methanogenic bacteria showed a rather diverse picture as to molecular weight, oxygen sensitivity, sensitivity toward CO and CN^-, localization (membrane bound versus "soluble"), and specificity of electron acceptors. At present, compelling evidence indicates that the organisms contain two types of hydrogenase: one that uses coenzyme F_{420} as the electron acceptor and a second one that depends on the (artificial) electron acceptor methylviologen (Table 13.2). The enzymes tend to be stable toward aerobic purification, but they are active only after treatment with dithiothreitol and hydrogen (335) or after treatment with high-salt concentrations (112). Both types of hydrogenase are iron-sulfur proteins. In the coenzyme F_{420}–dependent hydrogenase of *M. barkeri* (76) and in the methylviologen-dependent enzyme of *M. formicicum* (1), $[Fe_4-S_4]$ clusters are present; the nature of the Fe-S clusters in the other hydrogenases is yet unclear. Coenzyme F_{420}–dependent hydrogenase of *M. vanniellii* contains selenocysteine bound to the 42,000-Da subunit (334, 335, 338). Hydrogenases of methanogenic bacteria are nickel proteins (Table 13.2), a property they share with hydrogenases of sulfate-reducing bacteria (28, 173), knallgas bacteria (84), and nitrogen-fixing organisms. The presence of nickel(III) showing EPR signals comparable to Ni(III)-tetraglycine complexes (171) was demonstrated in oxidized inactive preparations (2, 152, 224). In the oxidized state nickel is bound to three sulfur atoms (177). In addition, the presence of a nitrogen atom could be demonstrated in the neighborhood of the paramagnetic Ni(III) in

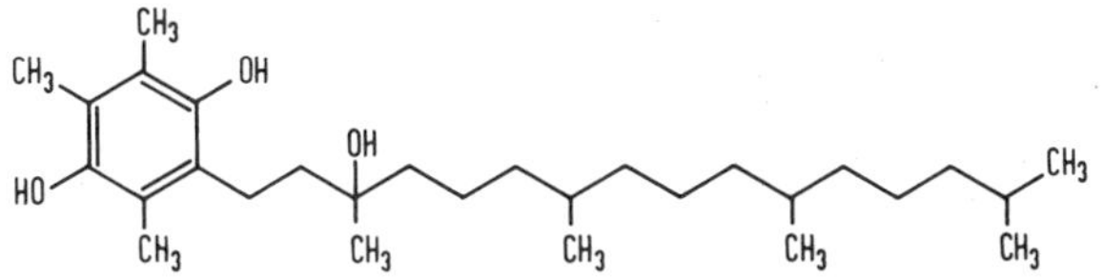

Figure 13.10 Structure of a quinone derivative, α-tocopherolquinone, present in methanogenic bacteria (109).

the coenzyme F_{420}-dependent hydrogenase of *M. thermoautotrophicum* (275). The nitrogen atom was suggested to be either a part of the protein or of FAD (275). However, the nitrogen atom was absent in the methylviologen-dependent hydrogenase of the same organism (275). The paramagnetic Ni(III) signals disappeared upon reductive activation though catalytic activity was not kinetically correlated with the disappearance of the signals (152). The precise role of nickel, its change in valence state and its coordination chemistry during catalysis are not yet clear. Nickel might be the site of interaction of hydrogen with the enzyme (2).

The molecular weights of coenzyme F_{420}-dependent hydrogenase vary over a fairly wide range (Table 13.2). It is not clear if and how the enzymes and their subunits from the different organisms are inter-related. Remarkable is the presence of a flavin, either FAD or FMN, in this type of hydrogenase (Table 13.2). The coenzyme, which is both a one-electron and a two-electron carrier, was suggested (112, 113) to intermediate between the one-electron-carrying iron–sulfur moieties and the two-electron acceptor coenzyme F_{420}.

The properties of coenzyme F_{420}-dependent hydrogenase of *M. thermoautotrophicum* described by Jacobson et al. (112) correspond (198) with those of component A_1 of the methylcoenzyme M reductase complex (see Section 13.2.2.2), suggesting that the hydrogenase is, among others, an electron-donating system of the terminal step of methanogenesis. In the resolved methylcoenzyme M reductase system FAD was required as a coenzyme (198). Since coenzyme F_{420} is an obligate cofactor of methylene-H_4MPT dehydrogenase (100), coenzyme F_{420}-dependent hydrogenase may provide reducing equivalents for the reduction of methenyl-H_4MPT.

The function of the methylviologen-dependent hydrogenases remains to be established. This type of hydrogenase occurs in various organisms in lower amounts compared to the enzyme that uses coenzyme F_{420} as the electron acceptor. The properties of another class of iron–sulfur proteins, formate dehydrogenase, were discussed in Section 13.2.3.

The presence of ferredoxin is documented only for *M. barkeri* (16, 101, 102, 305) and *M. formicicum* (304). One type with a hitherto unknown function contains [Fe_3-S_3] clusters (101, 196). The protein has a molecular weight of 20,000 to 22,000 Da and is composed of 59 amino acid residues, among them eight molecules of cysteine; aromatic amino acids, methionine, arginine, and histidine are absent. Ferredoxin isolated from *M. barkeri* strain Fusaro is the electron carrier of pyruvate dehydrogenase (102). The enzyme is composed of two identical subunits of 6000 Da and shows an N-terminal sequence homologous to ferredoxin present in *Clostridium pasteurianum*, *Desulfovibrio gigas*, and *Desulfovibrio africanus*. The complete molecule contains about seven atoms of iron, and seven to eight atoms of sulfur, in addition to eight cysteine residues. A third ferredoxin of M_r 12,000 having a [Fe_4-S_4] cluster was found (305) to participate in *M. barkeri* at the reductive activation of methanol: 5-hydroxybenzimidazolylcobamide methyltransferase (see Section 13.2.5.1).

TABLE 13.2 Properties of hydrogenases of methanogenic bacteria

Type	Organism	Molecular Weight (Da)	Subunits (Da)	Functional Groups[a]	References
F_{420}-dependent hydrogenase	M. thermoautotrophicum strain ΔH	170,000	40,000 (2); 31,000 (1); 26,000 (1)	Fe (33); S (24); Ni (2.1–2.3); FAD (2.3)	112, 152
	M. formicicum strain MF, JF	600,000	42,600; 34,000; 23,500	FAD	126, 201
	M. bryantii strain M.o.H.	600,000–680,000		Ni?; FAD?	126
	M. vannielii	340,000	56,000; (2 × 27,000); 42,000; 35,000	Ni (2); selenocysteine (3.8)	334, 335, 336, 338
	M. barkeri	800,000	60,000	Fe (8–10 subunit); S; Ni (0.6–0.8/subunit); FMN (1/subunit)	76
Methylviologen-dependent hydrogenase	M. thermoautotrophicum strain ΔH		52,000; 42,000	Fe; S; Ni	25
	M. thermoautotrophicum strain Marburg	60,000		Ni (0.8)	2, 86, 92
	M. formicicum strain MF, JF	70,000	43,000; 38,000	Fe (10); S (8); Ni (1); Zn (1); Cu (2)	1, 126
	M. bryantii strain M.o.H.	70,000			126

[a]Values in parentheses show the number of mol/mol enzyme.

13.3.3 Cytochromes

Representatives of the various families of methanogenic bacteria were carefully checked for the presence of cytochromes by Kühn et al. (162–165). Interestingly, cytochromes could be detected only in organisms that are able to convert methanol, methylamines, or acetate.

Particulate fractions of different species of *Methanosarcina* investigated contain two types of cytochrome b, occurring in amounts of 0.3 to 0.5 μmol/g membrane protein, when cells were grown on methanol, methylamines, acetate, or H_2/CO_2 (163–165). The cytochromes show extremely low midpoint potentials of about -325 mV and -180 mV, respectively. In cells grown on acetate a third cytochrome b ($E_0' = -250$ mV) may be observed (165). In addition, a cytochrome c is present that occurs only in 5 to 20% concentrations as compared to the cytochromes b. In contrast, cytochromes c are the predominant forms in the marine organisms *Methanococcoides methylutens* and *Methanolobus tindarius* (134). *Methanothrix soehngenii* contains cytochromes b and c in about equal amounts, 0.14 and 0.12 μmol per gram of membrane protein, respectively (163). The content of cytochromes is comparable to that found in aerobic methylotrophic bacteria (172, 319).

13.3.4 Coenzyme F_{420}

A central electron carrier in methanogenic biochemistry is coenzyme F_{420}, first described by Cheeseman et al. (30). The compound is the electron acceptor of hydrogenase (see Section 13.3.2) and formate dehydrogenase (see Section 13.2.3) and it is involved in NADP:coenzyme F_{420} oxidoreductase (see Section 13.3.5). In crude systems it functions as the electron acceptor of the CO dehydrogenase reaction (see Section 13.2.4). Coenzyme F_{420} is the electron donor in the methenyl-H_4MPT reduction (see Section 13.2.2.1) and in the terminal step of methanogenesis (see Section 13.2.2.2). Moreover, the compound plays a role in two important steps in cell carbon synthesis: α-keto acid reductive carboxylations (see Section 13.5.3).

The structure of coenzyme F_{420} was elucidated by Eirich et al. (60) and is shown in Fig. 13.11. The compound is not unique for methanogenic bacteria, but it is found in a rapidly growing number of organisms: *Streptomyces aureofaciens*, in

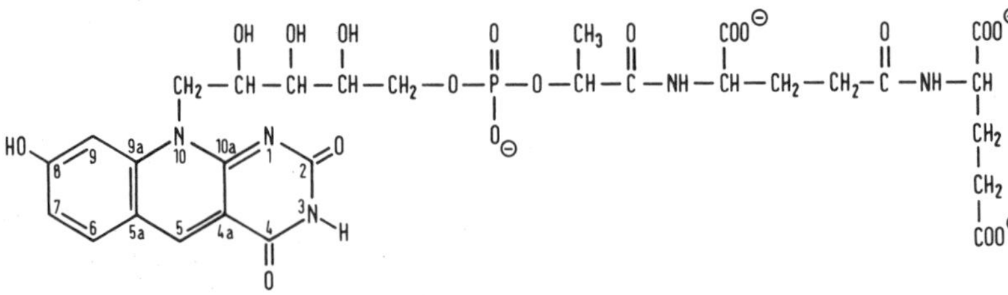

Figure 13.11 Structure of coenzyme F_{420} proposed by Eirich et al. (60).

which it functions in chlortetracycline synthesis (187); and in *Streptomyces griseus* (64, 65), *Anacystis nidulans*, and *Agmenella quadruplicata* (62, 65), where coenzyme F_{420} is involved in the photoreactivating system. In *M. thermoautotrophicum* (135) and possibly *Halobacterium cutirubrum* (63), the compound is the chromophore of the photoreactivating enzyme. Its presence was also established in *Mycobacterium* species (199) and actinomycetes (36), but the function is yet unclear.

Coenzyme F_{420} is a 7,8-didemethyl-8-hydroxy-5-deazariboflavin derivative (60). The modification of the central pyrazine ring to a pyridine ring lends the compound chemical properties that are more consistent to nicotinamide- than to riboflavin-based coenzymes (112, 113). The compound is only a two-electron carrier (112, 113), showing a midpoint potential $E_0' = -340$ mV (113, 216). Reduction takes place in a stereospecific way (113, 337) by hydride transfer to C-5 (113, 337). Hydrogenases from *M. vannielii* and *M. thermoautotrophicum* reduce the compound at the *Si* face of the molecule, resulting in the introduction of *pro*-S hydrogen at C-5 (282, 341). For the enzymatic activity in the NADP:coenzyme F_{420} oxidoreductase reaction the presence of a 2,4-dioxopyrimid [4, 5-b] quinoline moiety is a minimal requirement (339). No activity was found when 8-OCH$_3$ was substituted for 8-OH, and activity was lowered when a side chain was introduced at C-7, suggesting that the phenol group is in direct contact with the active center of the enzyme (339). In addition, 5-CH$_3$-deazariboflavin was not reduced. Substitution at N-10 has a limited impact on the catalytic activity: FO, a coenzyme F_{420} derivative lacking the *N*-(*N*-L-lactyl-γ-L-glutamyl)-L-glutamic acid phosphate moiety, shows an apparent K_m in the hydrogenase assay about fourfold higher than the complete coenzyme, whereas the K_m values for FO and coenzyme F_{420} in the NADP:coenzyme F_{420} oxidoreductase differed only slightly (61).

13.3.5 Oxidoreductases

The presence of NADP:coenzyme F_{420} oxidoreductase was demonstrated in *M. thermoautotrophicum* (59, 67, 68, 113), *M. smithii* (289), *M. barkeri* (311), and *M. vannielii* (337, 339, 340). NAD, FMN, and FAD could not function as a substrate (129, 311). The oxidoreductase of *M. vanniellii* was purified to homogeneity (337). The protein ($M_r = 85,000$) is composed of two identical subunits and contains at least one sulfhydryl moiety, which is essential for catalysis (337). NADP:coenzyme F_{420} oxidoreductase of *M. thermoautotrophicum* is a 95,000-Da protein (59).

NADPH is involved in a number of reactions in cell carbon synthesis, it may function as an electron donor in the methylcoenzyme M reductase reaction (see Section 13.2.2.2), and the compound is a substrate in the enzymatic reduction of 2,2'-di (mercaptoethanesulfonic acid) (186, 278), which is an oxidation product of HS-CoM. At present it is not clear how NADH is generated by methanogenic bacteria. In none of the electron transfer reactions described here was NAD(H) converted. Nor could the presence of an NAD reductase or of an NADP-dependent transhydrogenase be demonstrated (188). This is remarkable since NAD is a cofactor of malate dehydrogenase (see Section 13.5.3) and of diaphorase of *M. hungatei*.

FAD is a coenzyme of formate dehydrogenase of *M. formicicum* (see Section 13.2.3) and of coenzyme F_{420}–dependent hydrogenases (see Section 13.3.2). In all coenzyme F_{420}–dependent hydrogenases and formate dehydrogenases, FAD as well as FMN could substitute for the deazaflavin. FAD is a coenzyme of NAD-dependent diaphorase purified from cells of *M. hungatei* (188). The enzyme has a molecular weight of 117,500 Da and sulfhydryl groups are required for catalysis. *M. bryantii* contains a superoxide dismutase ($M_r = 91,000$) composed of four identical subunits in which iron is the functional group (148). The latter enzyme may protect methanogenic bacteria from oxygen toxicity.

13.4 ENERGY CONSERVATION AND ATP SYNTHESIS

13.4.1 Energy Conservation

As described above, the reduction of CO_2 to methane proceeds formally via formate, formaldehyde, and methanol. Because the nature of the one-carbon carrier was unknown at the time, these formal redox couples were used by Kell et al. (136) and by Blaut and Gottschalk (14) to make an estimation of energy-yielding and energy-consuming steps during methanogenesis. Now we reevaluate these data, taking into account the one-carbon carriers described in the preceding section. The free energy data were calculated from published data of methanogenic and analogous nonmethanogenic reactions (Table 13.3; for a more extensive list of free energy changes, see Ref. 142).

Assuming a value of $\Delta G^{\circ\prime} = 31.8$ to 43.9 kJ/mol for ATP formed from ADP and inorganic phosphate (284), the conversions of CO_2 ($\Delta G^{\circ\prime} = -130.4$ kJ/mol) and of formate ($\Delta G^{\circ\prime} = -119.5$ kJ/mol) to methane are exergonic enough to allow the synthesis of maximally about 3 mol of ATP each. As can be seen from Table 13.3, only the reduction of CH_3S-CoM is exergonic enough to allow the synthesis of (more than) 1 mol of ATP. The energy yield of the methylene-H_4MPT reduction ($\Delta G^{\circ\prime} = -19.8$ kJ/reaction) is only limited. However, it was shown by Bott et al. (24) that a reaction giving about the same free energy change, that is, the oxidation of CO to CO_2 and hydrogen ($\Delta G^{\circ\prime} = -20$ kJ/mol), is coupled to ATP synthesis in *M. barkeri*. On the contrary the formation of formyl methanofuran from hydrogen and CO_2 is endergonic (Table 13.3), but in *M. thermoautotrophicum* no appreciable amount of ATP is used in this reaction (299).

The formation of methane from methanol:

$$4CH_3OH \rightarrow 3CH_4 + CO_2 + 2H_2O \qquad (\Delta G^{\circ\prime} = -103 \text{ kJ/mol methane}) \quad (10)$$

is exergonic enough to yield 3 mol of ATP. If the oxidation of methanol proceeds via the reversed reactions (3) to (6), such reactions may be coupled directly to the

TABLE 13.3 Heats of reaction and redox potentials at standard conditions for the steps of methanogenesis[a,b]

Equation Number	Substrate	Products	$\Delta G^{\circ\prime}$ (kJ/reaction)	E_0' (mV)
(1)	$H_2 + CO_2$ + methanofuran	Formylmethanofuran + H_2O	$+16^c$	-497
(2)	Formylmethanofuran + H_4MPT	5-Formyl-H_4MPT + methanofuran	-5.0^d	
(3)	5-Formyl-H_4MPT	Methenyl-H_4MPT + H_2O	-2.1^e	
(4)	Methenyl-H_4MPT + H_2	Methylene-H_4MPT	-4.8^f	-390
(5)	Methylene-H_4MPT + H_2	Methyl-H_4MPT	-19.8^g	-311
(6)	Methyl-H_4MPT + HS-CoM	CH_3S-CoM + H_4MPT	-29.7^h	
(7)	CH_3S-CoM + H_2	CH_4 + HS-CoM	-85^i	$+26$
Overall (1)–(7)	$4H_2 + CO_2$	$CH_4 + 2H_2O$	-130.4	-245
(8)	CH_3OH + HS-CoM	CH_3S-CoM + H_2O	-27.5^i	
Overall (7) and (8)	$CH_3OH + H_2$	$CH_4 + H_2O$	-112.5	$+169$
(9)	$CH_3COO^- + H^+$ + HS-CoM	CH_3S-CoM + CO + H_2O	$+72.5$	
(10)	$CO + H_2O$	$CO_2 + H_2$	-20	-518
Overall (7), (9) and (10)	$CH_3COO^- + H^+$	$CH_4 + CO_2$	-32.5	

[a] Gibbs free energies of formation of hydrogen, CO_2, H_2O, formate, acetate, methanol, and CO were taken from Thauer et al. (284).

[b] Redox potentials were calculated from the relationship $\Delta G^{\circ\prime} = -nF\,\Delta E_0'$, where n is the number of electrons transferred in the reaction, F [Faraday constant] $= 96.55$ kJ/mV equivalent, and E_0' (H^+/H_2) $= -414$ mV (Chapter 1).

[c] Assuming that $\Delta G^{\circ\prime} = -12.5$ kJ/mol for the hydrolysis of formyl methanofuran, analogous to the hydrolysis of formyl glutamate (223) and $\Delta G^{\circ\prime}$ ($H_2 + CO_2 \rightarrow$ formate) $= +3.5$ kJ/mol.

[d] This value follows from the overall $\Delta G^{\circ\prime}$ ($2H_2 + CO_2 \rightarrow$ formaldehyde + H_2O) $= +26.9$ kJ/mol, the heats of reaction equations (1), (3), and (4), and the heat of reaction of methylene-H_4MPT + $H_2O \rightleftharpoons$ formaldehyde + H_4MPT, for which a $K_{diss} = 11 \times 10^{-5}$ M was measured (138).

[e] Calculated from Ref. 54.

[f] Calculated from Refs. 71 and 73.

[g] This value follows from the overall $\Delta G^{\circ\prime}$ (H_2 + formaldehyde $\rightarrow$ methanol) $= -44.8$ kJ/mol, the heats of reaction equations (6) and (8), and the heat of reaction $\Delta G^{\circ\prime}$ (methylene-H_4MPT + $H_2O \rightleftharpoons$ formaldehyde + H_4MPT) (see d).

[h] Taken from Ref. 230 for CH_3-H_4folate + homocysteine $\rightleftharpoons H_4$folate + methionine.

[i] Calculated from the bond energies at 25°C D (H—OH), D (H—H), D (CH_3—SR), D (H—SR), and D (H—CH_3) as given by Refs. 147 and 179; the values may differ somewhat for coenzyme M derivatives.

final reduction step. Moreover, the oxidation of formyl-H$_4$MPT [reaction (9)] might yield an additional ATP by substrate phosphorylation via formyl phosphate (see Section 13.2.5). The conversion of acetate ($\Delta G^{\circ\prime} = -32.5$ kJ/mol) might yield maximally 1 mol of ATP.

ATP is required for the activation of methylcoenzyme M reductase (see Section 13.2.2.2), methyl-H$_4$MPT: HS-CoM methyltransferase (215) and methanol: 5-hydroxybenzimidazolylcobamide methyltransferase (see Section 13.2.5), but only in catalytic amounts. This ATP requirement is negligible for the overall yields.

From these considerations it cannot be predicted how many ATP molecules are produced per mole of methane formed. In addition, the values may differ among various types of methanogenic bacteria, depending on the way ATP synthesis is organized (see Section 13.4.3) and the presence and complexity of (membrane-bound) electron transport chains.

13.4.2 Energy Consumption for Growth

The amount of ATP used for biosynthesis depends on the routes of carbon fixation and on the cellular composition. The calculated yield of cell mass per mole of ATP used for biosynthesis only ($Y_{\text{ATP}}^{\text{max}}$) varies between about 29 g of dry cells per mole of ATP for heterotrophic growth on glucose (269) and 4.75 g/mol ATP for autotrophs using the Calvin cycle (81). The $Y_{\text{ATP}}^{\text{max}}$ for heterotrophic growth on acetate (269) and on acetate supplemented with CO$_2$ (41) was calculated to be 10 and 15.4 g dry cells per mole of ATP, respectively. Methanogenic bacteria synthesize their cell components starting from activated acetate (acetyl-CoA) and CO$_2$ (see Section 13.5). The formation of acetate from the substrates is an exergonic process:

$$4H_2 + 2CO_2 \rightarrow \quad CH_3COO^- + H^+ + 2H_2O \quad (11)$$

$$\Delta G^{\circ\prime} = -95 \text{ kJ/mol}$$

$$4HCOO^- + 4H^+ \rightarrow CH_3COO^- + H^+ + 2CO_2 + 2H_2O \quad (12)$$

$$\Delta G^{\circ\prime} = -99 \text{ kJ/mol}$$

$$2CH_3OH \rightarrow \quad CH_3COO^- + H^+ + 2H_2 \quad (13)$$

$$\Delta G^{\circ\prime} = -59 \text{ kJ/mol}$$

The energy produced is conserved by the formation of acetyl-CoA rather than acetate (see Section 13.5.2). Since no energy is required for the transport of the substrates hydrogen, CO$_2$, and methanol, the organisms produce their cellular matter in a rather efficient way. If one assumes a similar pattern of components on which the calculation by Stouthamer (269) and Forrest and Walker (81) were based, and if one assumes that synthesis starting from the intermediates described in Sec-

TABLE 13.4 ATP requirement for the formation of cells during heterotrophic growth on acetate and methanogenic growth on H_2/CO_2 [a]

	ATP required (mol $\times$ 10^{-4}/g cells)	
	Acetate (Heterotrophic Organisms)	H_2/CO_2 (Methanogenic Bacteria)
Polysaccharide		
Glucose-6-phosphate formation	82	41
Polymerization	10	10
Protein		
Amino acid formation	236	110
Polymerization	191	191
Lipid	50	28[b]
RNA		
Nucleoside monophosphate formation	78	53
Polymerization	9	9
DNA		
Deoxynucleoside monophosphate formation	14	12.5
Polymerization	2	2
Turnover mRNA	14	14
ATP required for transport of:		
Carbon source	254	—
Ammonium ions	42	42[c]
Potassium ions	2	2
Phosphate	8	8
Total ATP requirement	995	522.5
Y_{ATP}^{max}	10	19.1

Source: Table and data for growth on acetate from Stouthamer (269).

[a] For methanogenic bacteria the same pattern of cell carbon is assumed.
[b] Calculated as C_{20} fatty acids.
[c] It is not known if methanogenic bacteria perform active transport of these ions.

tion 13.5.3 proceeds via similar biochemical pathways, $Y_{ATP}^{max} = 19.1$ g/mol ATP can be calculated (Table 13.4). Although unique cellular components are present and the composition of the cells may vary among the different species, such deviations may exert only marginal effects. When grown on acetate Y_{ATP}^{max} will be about 10 g/mol ATP, but the value may be larger if no energy is required for active transport of the substrate.

In view of these considerations the low amounts of cellular mass formed per mole of methane (Y_{CH_4}) are striking (Table 13.5). However, only a part of the energy generated is used for biosynthetic purposes, as shown by the equation according to Pirt (214, 270)

TABLE 13.5 Growth yields and specific growth rates of methanogenic bacteria

Substrate	Organism	Y_{CH_4} (g dry weight/mol CH_4)	μ (h^{-1})	References
H_2/CO_2	M. thermoautotrophicum strain Marburg	1.6	0.312	250
	M. thermoautotrophicum strain ΔH	0.6–1.6	0.173–0.231	280
	M. arboriphilicus strain AZ	2.68	0.144	345
	M. bryantii strain M.o.H.	2.48	0.098	222
	M. formicicum	3.5	0.082	238
	M. barkeri	6.37	0.058	310, 311
	Methanosarcina strain 227	8.7	—	252
Formate	M. formicicum	4.8	0.055	238
Methanol	M. barkeri	5.5	0.0693	104
		7.2	0.07	310, 311
	Methanosarcina strain 227	5.1	—	252
	Methanosarcina thermophila	4.5	0.099	353, 354
Methylamine	M. barkeri	5.8	0.058	104
Acetate	M. soehngenii	1.4	0.0032	110, 344
	Methanosarcina strain 227	2.1	0.057	225
		3.17	0.0183	253
		3.41	0.0231	253
	M. thermophila	1.8	0.0578	353, 354

$$\frac{1}{Y_{CH_4}} = \frac{1}{Y_{CH_4}^{max}} + \frac{m_s}{\mu} \tag{14}$$

In the equation μ is the specific growth rate and m_s is the so-called maintenance coefficient (mole of CH_4 used per gram of dry weight per hour), a function that takes into account the part of energy consumed by processes other than biosynthesis (e.g., electrochemical processes). Methanogenic bacteria tend to grow slowly (Table 13.5). This implicates that the term m_s/μ will be large even if the maintenance coefficient does not deviate much from values commonly found in other organisms. The maximal growth yield $Y_{CH_4}^{max}$ and m_s may be obtained from experiments, generally performed as chemostat cultures. Presumably due to experimental difficulties, few of those experiments were done with methanogenic bacteria (31, 75, 250) and the results are in accordance with the Pirt equation; Reciprocal plots of μ versus Y_{CH_4} yielded straight curves.

It can be seen from Table 13.5 that Y_{CH_4} values of different organisms even when grown on the same substrate (e.g., hydrogen and CO_2) vary considerably. Part of this variation is, of course, due to the m_s/μ term. However, the differences in Y_{CH_4} may also reflect differences in $Y_{CH_4}^{max}$ values. Indeed, from steady-state experiments $Y_{CH_4}^{max}$ values of 3 and 3.33 dry weight per mole of CH_4 at low gassing rates were found for *M. thermoautotrophicum* (250) and *M. thermolithotrophicus* (75) growing on hydrogen and CO_2. A twofold higher value, $Y_{CH_4}^{max} = 5.8$, was obtained for formate-utilizing *M. formicicum* (31). An even higher theoretical maximal cell yield will be reached by methylotrophic organisms, since Y_{CH_4} values of about 6 g dry weight per mole of CH_4 and higher were measured at low growth rates (Table 13.5). Taking into account the low energy gain for growth on acetate and the lower $Y_{CH_4}^{max}$ (Table 13.4), aceticlastic methanogens conserve their energy in a relatively efficient way. The amount of ATP used for biosynthetic purposes will not vary much among various types, except for growth on acetate. Hence it is tempting to speculate that the differences in $Y_{CH_4}^{max}$ reflect differences in energy conservation (38). More growth experiments have to be done to support this hypothesis, and even then the results have to be interpreted with utmost care (281). In this respect it is important to note that the calculated $Y_{CH_4}^{max}$ rapidly decreased when the hydrogen and CO_2 supply was increased (75, 250). This implicates that catabolic and anabolic processes become uncoupled. The rationale of this uncoupling, which is also found for nonmethanogenic bacteria (281), is unknown.

13.4.3 Electron Transport and ATP Synthesis

Based on theoretical considerations (284) and experimental evidence outlined below, the current opinion is that methanogenic bacteria conserve energy by electron transport phosphorylation (ETP; for an extensive review, see Ref. 38). Through electron transport a proton motive force ($\Delta\bar{\mu}_{H^+}$) would be generated, which is composed of a membrane potential ($\Delta\psi$), inside negative, and an osmotic component (ΔpH), inside alkaline, according to the well-known equation derived by Mitchell (191):

$$\Delta\bar{\mu}_{H^+} = \Delta\psi - 2.3\frac{RT}{F}\Delta pH \qquad (15)$$

Evidence obtained from experiments performed with *M. barkeri* is compelling that the organism synthesizes ATP by use of ETP (14). However, from studies with *M. thermoautotrophicum* and *M. voltae*, results were found that oppose such a mechanism (see below) and an alternative mechanism involving substrate-level phosphorylation was proposed (169). In general, one could state that methanogenic bacteria have an architecture to perform chemiosmotic processes.

The organisms contain membrane-bound redox carriers (see Section 13.3.1). In fact, methanogenesis and ATP synthesis can be brought about with a membraneous fraction of *M. thermoautotrophicum* supplemented with a soluble cofactor(s) (233, 236). Many aspects of the orientation of (electron transfer) proteins in the membrane are unknown. The membrane-associated nature of methanogenic enzymes does not necessarily fulfill a function in energy conservation. It could also serve a more organizational purpose: to accommodate the individual steps to one and another in an optimal way. It is remarkable (38) that cell breakage is accompanied with a dramatic loss in methanogenic activity. *M. thermoautotrophicum* strain ΔH contains a membrane-bound ATPases that shows properties resembling those of other bacterial ATPases (52). ATPase activity was also demonstrated in *M. barkeri* (146, 193), *M. thermoautotrophicum* strain Marburg (25), *M. bryantii* (318), and *M. hungatei* (262). The ATPase of the marine organism *M. voltae* is deviant since it translocates sodium instead of protons (29). The activity was inhibited by *N, N'*-dicyclohexylcarbodiimide (DCCD) in cells of *M. barkeri* (12, 13, 193) and *M. hungatei* (259) but not in a number of other organisms (261). Individual differences in the composition of cell walls and cell membranes may be of influence on the effectivity of the inhibitor: this factor may apply to other commonly employed ionophores as well.

Methanogenic bacteria can maintain a ΔpH across the cell membrane (121, 258, 260, 263). An internal pH_i of about 6.8 was determined for *M. thermoautotrophicum* (118, 234), *M. bryantii* (121), and *M. hungatei* (118). ATP synthesis could be induced by an acid pH shift in *M. thermoautotrophicum* (52) and *M. barkeri* (193). In *M. bryantii* an imposed pH gradient (interior alkaline) was dissipated by the protonophore carbonylcyanide-*m*-chlorophenylhydrazone (CCCP), gramicidin (a general ionophore), nigericin (a K^+/H^+-antiporter), monensin (a Na^+/H^+ antiporter), and valinomycin + KCl (121). In general, addition of ionophores (CCCP, 2,4-dinitrophenol, pentachlorophenol, tetrachlorosalicylanilide) to cells or cell-free preparations caused a drop in the ATP pool, which is expected, and a concomitant drop in the rate of methane synthesis (193, 222, 234, 235).

The near-neutral pH_i implicates that at physiological conditions the membrane potential is the major contributor to the proton motive force. $\Delta\psi$ values in respiring *M. thermoautotrophicum* (118), *M. bryantii* (121), and *M. barkeri* (13) were determined to be about -120 to -150 mV; *M. hungatei* showed a membrane potential of only -89 mV (118). These data may seem low with respect to the commonly reported values of -170 to -230 mV (79). However, as pointed out by Jarrell and Sprott (121), the values may be an underestimate due to the method employed: the distribution of [^{3}H] triphenylmethylphosphonium (TPMP$^+$). By use of tetraphenylphosphonium (TPP$^+$) Butsch and Bachofen (27) obtained $\Delta\psi$ values ranging from -155 to -173 mV for *M. thermoautotrophicum*. Strikingly, ATP synthesis coupled to methanogenesis also occurred in *M. thermoautotrophicum* (248) and in *M. voltae* (34) when the proton motive force was dissipated by the addition of uncoupler, which is apparent evidence against ETP.

Methanogenic bacteria are capable of maintaining high internal potassium concentrations, varying between 150 and 1225 m*M* for organisms representing diverse phylogenetic groupings (121, 123, 260). The compound can be accumulated in the cytoplasm, resulting in concentrations up to 50,000-fold (249). It is proposed that K^+ is involved in osmoregulation (264). Addition of the K^+ ionophore valinomycin to cells of *M. thermoautotrophicum* induced ATP synthesis and extrusion of potassium into the medium (52). When used in high concentration ($17\ \mu M$), valinomycin acted as an H^+/K^+ exchanger (235). Addition of sodium drastically stimulated the valinomycin-induced ATP synthesis in *M. thermoautotrophicum* strain Marburg (251). The organism (247) and presumably also *M. barkeri* (13, 14) contain a Na^+/H^+ antiporter, which plays a role in pH regulation. Since growth and methanogenesis of cell suspensions of all organisms tested by Perski et al. (211, 212) was strictly dependent on low concentrations of sodium, Na^+/H^+ antiporter activity may be a common phenomenon in methanogenic bacteria. Na^+ is a central component of electrochemical processes in *M. voltae* (122). Apart from the presence of a Na^+-translocating ATPase, active transport in the organism is dependent on the ion (see Section 13.4.4). The precise way in which ATP synthesis is coupled to the formation of methane remains to be established. Roberton and Wolfe (222) observed a close correlation between the rate of methanogenesis in cultures of *M. bryantii* and the level of ATP. Conversely, a decline in the ATP pool induced by protonophores also resulted in a drop of the rate of methanogenesis. In this respect, an interesting observation was made by Blaut and Gottschalk (12, 13): DCCD inhibited the formation of methane in cells of *M. barkeri* from CO_2, formaldehyde, and methanol under H_2 atmosphere. The inhibition of methanogenesis from formaldehyde and methanol, but not from CO_2 and from methanol under nitrogen atmosphere, was relieved by addition of the uncoupler tetrachlorosalicylanilide. As pointed out by the authors (13, 14), this is consistent with the view that oxidation of methanol to the formaldehyde level is endergonic and may be driven by the terminal energy-conserving reaction. In acetate-grown *M. barkeri*, CO oxidation to hydrogen and CO_2 is also coupled to phosphorylation of ADP (24). ATP synthesis was inhibited by DCCD and tetrachlorosalicylanilide. Remarkably, methanol conversion under hydrogen (12) and CO oxidation (24) was stimulated by the addition of uncoupler in a way that bears an analogy with respiratory control in mitochondria.

A number of authors (50, 136, 234, 235) have suggested that ATP synthesis would take place on specialized intracellular vesicles ("methanochondrions"). Evidence for the existence of such organelles was derived from electron microscopy, cytochemical staining, the anomalous behavior of uncouplers, and the presence of an ADP/ATP translocase in vesicles of *M. thermoautotrophicum* (50); for a review see Ref. 136).

13.4.4 Transport Systems

Only a few transport systems for methanogens have been described up until now. The first was that of HS-CoM uptake by a *M. ruminantium* strain which was dependent on exogenous HS-CoM for growth (7). The transport system requires energy, shows saturation kinetics, accumulates HS-CoM against a concentration gradient, and is dependent on both hydrogen and CO_2. Uptake is inhibited by oxygen and 2-bromoethanesulfonic acid. CH_3S-CoM is also transported very efficiently, but the specificity of the carrier is restricted to a limited range of HS-CoM analogs. *M. hungatei, M. mobile,* and an HS-CoM-independent strain of *M. ruminantium* show rates of uptake which are about 30%, less than 10%, and less than 10% of the rate with the HS-CoM-dependent strain, respectively (7).

Nickel is required for growth of methanogens (47) and thus it is to be expected that Ni^{2+} transport systems are present. In *M. bryantii* a high-affinity uptake system is present (119) which transports Ni^{2+} against a concentration gradient and is dependent on both hydrogen and CO_2. Uptake is inhibited by oxygen. The system is specific for Ni^{2+}, and of all the cations tested, only Co^{2+} was inhibitory. From the effects of metabolic inhibitors, ionophores, and artificially imposed pH gradients it was suggested that Ni^{2+} transport is not dependent on the membrane potential or intracellular ATP level, but is coupled to proton movement (120).

The presence of an adenine nucleotide translocase in membrane vesicles of *M. thermoautotrophicum* (50) indicates that a nucleotide transport might play a role in methanogens. *M. voltae,* which requires exogenous leucine and isoleucine for growth (316), possesses a high-affinity isoleucine transport system (117). Transport is dependent on both hydrogen and CO_2 and is inhibited by oxygen. On the basis of the energetics of transport, a Na^+-isoleucine transport model was suggested (117).

13.5 SYNTHESIS OF CELL CARBON COMPOUNDS

13.5.1 Introduction

Present knowledge on the pathways of the synthesis of cellular components indicates that methanogenic bacteria use well-known routes but with many surprises as to the coenzymes and enzymes involved and with aberrant cellular consitutents as products (e.g., in membranes and cell walls).

The synthesis of the two-carbon building block out of one-carbon units (see Section 13.5.2) is still a most intriguing aspect of these anabolic reactions. Moreover, peculiar coenzymes, which are known to be involved in methanogenesis, also participate in other cellular processes. A specific role for coenzyme F_{420} in the synthesis of pyruvate and α-ketoglutarate has already been demonstrated, and methanopterin and 5-hydroxybenzimidazolyl corrinoids function in reactions normally performed in the presence of folic acid and 5,6-dimethylbenzimidazolyl corrinoids, respectively. This section deals with those aspects of anabolic processes that are relevant to the biochemistry of the process of methanogenesis.

13.5.2 Synthesis of Acetate

The first condensation product in cell carbon synthesis is acetyl-CoA. This was concluded from studies with growing cells, Triton-X-100-treated cells, and cell extracts of *M. thermoautotrophicum* (105, 170, 231, 271–274) and *M. barkeri*

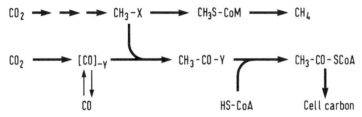

Figure 13.12 Tentative scheme of autrotrophic acetyl-CoA synthesis from CO_2 in *Methanobacterium* adapted from (105, 271). In this scheme X and Y represent a corrinoid-containing methyltransferase and carbon monoxide dehydrogenase, respectively.

(146). The overall pathway for acetyl-CoA formation (Fig. 13.12) shows a striking analogy to the one established for acetogenic bacteria (for reviews see Refs. 85 and 348).

In *M. thermoautotrophicum* the methyl group of acetate is produced by the reduction of CO_2 following the methanogenic pathway described in Section 13.2.2.1, [^{14}C]Methenyl-H$_4$MPT, [^{14}C]formaldehyde, and [3-^{14}C]serine are exclusively incorporated into C-2 of acetate (105, 170). Methanogenesis and acetate formation are tightly coupled processes. Stimulation of methane formation by the addition of RPG effectors (see Section 13.2.2.3) stimulates acetogenesis (105, 271–273), although the latter process proceeds at a 1000-fold lower rate than methane formation (105, 271–273). Conversely, inhibition of methane production by 2-bromoethanesulfonic acid also inhibits the production of acetyl-CoA (105, 272, 273).

The carboxyl group of acetate results from the reduction of CO_2 by the action of CO dehydrogenase (see Section 13.2.4). When the CO_2-reducing activity of the latter enzyme was inhibited by cyanide, CO could substitute for CO_2, and labeled CO was specifically incorporated into C-1 of acetyl-CoA (42, 272, 273). CO binds reversibly to CO dehydrogenase, and during growth on hydrogen and CO_2, CO is evolved in trace amounts (32).

In acetogenic bacteria the condensation of the methyl group, the carbonyl moiety, and coenzyme A occurs at the CO dehydrogenase (85, 348). Although this seems most probable, the role of the enzyme in this condensation reaction is yet to be proven in methanogens. A corrinoid protein is involved in acetogens in the methyltransferase reaction of methyltetrahydrofolate to CO dehydrogenase. Since propyliodide inhibited growth (58, 105) and acetate synthesis but not methanogenesis, it was proposed that a corrinoid protein plays an analogous role (Fig. 13.12) in methanogenic bacteria. However, this conclusion can only be drawn with care, since the action of propyliodide may also be directed to some other enzyme (e.g., CO dehydrogenase). Moreover, the corrinoid-depending methyl transferase reactions of methanol (see Section 13.2.5.1) and of methyl-H$_4$MPT (S. Kengen, unpublished results) to HS-CoM are not inhibited by the compound.

13.5.3 Intermediary Cell Carbon Synthesis

The usual pathways for the assimilation of C_1 compounds, such as the Calvin cycle, the ribulose monophosphate pathway, the reductive tricarboxylic acid cycle, and the serine pathway, are absent in methanogens (87, 274, 279). Most studies concerning the possible presence of these pathways are confined to *M. thermoautotrophicum* and *M. barkeri* (Fig. 13.13).

Acetyl-CoA, the earliest detectable intermediate in autotrophic CO_2 assimilation by *M. thermoautotrophicum* (87), is formed by a mechanism described in Section 13.5.2.

The synthesis of oxaloacetate from pyruvate in *M. barkeri* has not been elucidated. Phosphoenolpyruvate is not an intermediate since PEP carboxylase is absent (347). Moreover, citrate lyase is absent, which rules out the presence of a glyoxylate cycle (312). Also, the pathway of succinate synthesis in this methanogen is unknown (312).

In *M. thermoautotrophicum* the reductive reactions of the tricarboxylic acid cycle are used in the synthesis of glutamate, whereas in *M. barkeri* the oxidative reactions are used. The electron carriers used in the pyruvate synthase and α-ketoglutarate synthase reactions differ from those used by eubacteria; that is, F_{420} is used in both methanogens (87, 347). However, the involvement of ferredoxin, which recently was demonstrated in *M. barkeri* (101, 196), cannot yet be ruled out completely. For fumarate reduction the electron carrier has not yet been identified (87, 347).

13.5.4 Nitrogen Assimilation

In bacteria, glutamate and glutamine are the main precursors for the synthesis of other nitrogen-containing compounds. In *M. thermoautotrophicum* and *M. barkeri*, glutamate accounts for about 52% and 65%, respectively, of the total amino acids in the soluble pool (145). Moreover, glutamate is one of the first readily labeled intermediates in short-term $^{14}CO_2$ fixation studies (40). Synthesis of glutamate in *M. thermoautotrophicum* and *M. barkeri* proceeds via the glutamine synthetase (GS)/glutamate synthase (GOGAT) pathway. Glutamate dehydrogenase is absent (145). Formation of other amino acids is secured by the presence of glutamate-pyruvate transaminase and glutamate-oxaloacetate transaminase (145).

Following glutamate, alanine is the most abundant amino acid present in *M. thermoautotrophicum* and *M. barkeri*. Alanine also belongs to the first labeled products after short-term $^{14}CO_2$ fixation (40). Alanine dehydrogenase (ADH) activity could be demonstrated in *M. thermoautotrophicum* (145).

When *M. thermoautotrophicum* was grown under ammonia-limiting conditions, the GS activity increased while the ADH activity decreased compared to an ammonia-excess condition (145). Labeling studies showed that in *M. thermoautotrophicum* ammonia was assimilated via the GS/GOGAT pathway, not via ADH (145).

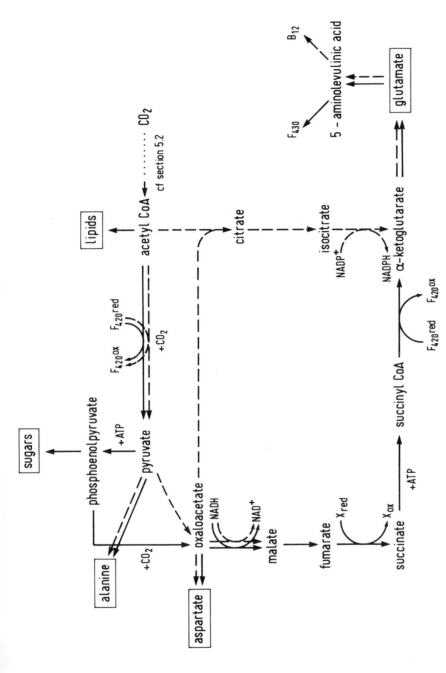

Figure 13.13 CO_2 assimilation pathways in *M. thermoautotrophicum* (solid line) and *M. barkeri* (dashed line) (87, 347). Evidence for the biosynthesis of factor F_{430} (89, 90) and B_{12} (242) from glutamate was obtained recently.

13.5.5 Sulfur Assimilation

Methanogens require reduced sulfur (e.g., sulfide or cysteine) for growth (346). The metabolism of sulfur is still obscure, notwithstanding its importance in the biosynthesis of HS-CoM, which contains both a reduced and an oxidized sulfur atom. Sulfide is important not only as a reducing agent, but is essential for growth and methanogenesis (195, 243). Cysteine stimulates growth in methanogens (4, 243, 346) but is an essential amino acid for growth of *M. arboriphilus* strain AZ (345).

Moura et al. (197) isolated a protein, called P_{590}, from *M. barkeri* which possessed sulfite reductase activity. The spectral characteristics of this protein are similar to those of siroheme-containing sulfite reductases. Sulfite reductases have been associated with a dissimilatory type of metabolism. However, a respiratory utilization of sulfur compounds by *M. barkeri* does not seem very likely. It was suggested (197) that sulfite reductase is utilized to gain energy from reduced sulfur compounds or, as an alternative, is involved in assimilation.

The addition of elemental sulfur to exponentially growing cultures of methanogens results in the formation of H_2S (267). At the same time methane production is greatly lowered in Methanobacteriales and Methanococcales, but is hardly affected in Methanomicrobiales.

13.5.6 Biosynthesis of Coenzymes

The results of incorporation studies performed with ^{63}Ni (46, 49, 317) and labeled δ-aminolevulinic acid and succinate (43, 45, 114, 115) laid the foundation of the structural elucidation of factor F_{430} (see Section 13.2.2.2). The biosynthesis of the compound starting from α-ketoglutaric acid and resulting in uroporphinogen III proceeds along known biochemical pathways (89, 90); subsequently, two methyl groups derived from methione are incorporated (114). The sequence of reactions from urophorphinogen III to the complete molecule is unknown. The tetrapyrrole of corrinoids (see Section 13.2.5) is synthesized from glutamate; glycine is no intermediate (242). This suggests the presence of the same biochemical pathway resulting in uroporphinogen III as observed for factor F_{430}. The nitrogenous base of corrinoids, 5-hydroxybenzimidazole, is synthesized from glycine; guanosine triphosphate is not an intermediate (242). However, the latter compound is involved in the biosynthesis of the deazaflavin moiety of coenzyme F_{420} (116).

13.5.7 Nucleic Acids and Genetics

The genome sizes of methanogens range from about 1.0×10^9 to 1.8×10^9 Da, somewhat less (40 to 65%) as compared to *Escherichia coli* (149, 192). It was suggested that repetitive sequences are present (192), although not in a higher amount as compared to *E. coli* (149). The mol % G + C of methanogens, ranging from 27.5 to 61, spans a range about as broad as that reported for eubacteria (4).

Furthermore, in some species fractions of DNA were present with a higher A-T content than the average DNA of the species. (149). Recently, the presence of a cryptic plasmid in *M. thermoautotrophicum* (189) and a coccoid methanogen (285) was reported.

The structure of the mRNA is still unknown. The rRNAs of methanogens differ in primary sequence and in the post-transcriptionally modification pattern compared to eubacteria (82). The tRNAs lack the universal common arm sequence GTψCG; instead, they contain either G$\psi\psi\dot{C}$G or G$\dot{U}\psi$CG (4) (the superscript dot denotes a modification of the base). The modified bases ribothymidine and 7-methylguanosine, typical of *E. coli*, are absent in the tRNAs of methanogens (4, 11).

In *M. thermoautotrophicum* (202) and *M. vannielii* (124) two and four rRNA gene clusters are present in the genome. Moreover, an extra 5S rRNA gene unlinked to the rRNA gene cluster was found in *M. vannielii* (124). The rRNA genes are arranged in the order 5'-16S-23S-5S-3', which is also typical for eubacteria (124, 202).

The most important breakthrough in the genetics of methanogens has been the demonstration that DNA from methanogens can be functionally expressed in *E. coli* (23, 219, 329). These results led to the conclusion that the genetic code of methanogens is the same as that of *E. coli*.

Expression of eubacterial genes in methanogens has not yet been possible due mainly to a lack of a transformation system and suitable mutants of methanogens.

REFERENCES

1. Adams, M. W. W., S.-L. C. Jin, J.-C. Chen, and L. E. Mortenson. 1986. The redox properties and activation of the F_{420}-non-reactive hydrogenase of *Methanobacterium formicicum*. Biochim. Biophys. Acta 869:37–47.

2. Albracht, S. P. J., E.-G. Graf, and R. K. Thauer. 1982. The EPR properties of nickel in hydrogenase from *Methanobacterium thermoautotrophicum*. FEBS Lett. 140:311–313.

3. Anthony, C. 1982. Biochemistry of methylotrophs. Academic Press, New York, p. 296–327.

4. Balch, W. E., G. E. Fox, L. J. Magrum, C. R. Woese, and R. S. Wolfe. 1979. Methanogens: reevaluation of a unique biological group. Microbiol. Rev. 43:260–296.

5. Balch, W. E., and R. S. Wolfe. 1976. New approach to the cultivation of methanogenic bacteria: 2-mercaptoethanesulfonic acid (HS-CoM)-dependent growth of *Methanobacterium ruminantium* in a pressurized atmosphere. Appl. Environ. Microbiol. 32:781–791.

6. Balch, W. E., and R. S. Wolfe, 1979. Specificity and biological distribution of coenzyme M (2-mercaptoethanesulfonic acid). J. Bacteriol. 137:256–263.

7. Balch, W. E., and R. S. Wolfe. 1979. Transport of coenzyme M (2-mercaptoethanesulfonic acid) in *Methanobacterium ruminantium*. J. Bacteriol. 137:264–273.

8. Baresi, L. 1984. Methanogenic cleavage of acetate by lysates of *Methanosarcina barkeri*. J. Bacteriol. 160:365–370.

9. Barker, H. A. 1956. Bacterial fermentations. Wiley, New York, p. 1–27.

10. Belyaev, S. S., R. Wolkin, W. R. Kenealy, M. J. DeNiro, S. Epstein, and J. G. Zeikus. 1983. Methanogenic bacteria from the Bondyuzhskoe oil field: general characterization and analysis of stable-carbon isotopic fractionation. Appl. Environ. Microbiol. 45:691–697.

11. Best, A. N. 1978. Composition and characterization of tRNA from *Methanococcus vannielii*. J. Bacteriol. 133:240–250.

12. Blaut, M., and G. Gottschalk, 1984. Coupling of ATP synthesis and methane formation from methanol and molecular hydrogen in *Methanosarcina barkeri*. Eur. J. Biochem. 141:217–222.

13. Blaut, M., and G. Gottschalk. 1984. Protonmotive force driven synthesis of ATP during methane formation from molecular hydrogen and formaldehyde or carbon dioxide in *Methanosarcina barkeri*. FEMS Microbiol. Lett. 24:103–107.

14. Blaut, M., and G. Gottschalk. 1985. Evidence for a chemiosmotic mechanism of ATP synthesis in methanogenic bacteria. Trends Biochem. Sci. 10:486–489.

15. Blaut, M., V. Müller, K. Fiebig, and G. Gottschalk. 1985. Sodium ions and an energized membrane required by *Methanosarcina barkeri* for the oxidation of methanol to the level of formaldehyde. J. Bacteriol. 164:95–101.

16. Blaylock, B. A. 1968. Cobamide-dependent methanol-cyanocob(I)alamin methyltransferase of *Methanosarcina barkeri*. Arch. Biochem. Biophys. 124:314–324.

17. Blaylock, B. A., and T. C. Stadtman. 1963. Biosynthesis of methane from the methyl moiety of methylcobalamin. Biochem. Biophys. Res. Commun. 11:34–38.

18. Blaylock, B. A., and T. C. Stadtman. 1964. Enzymic formation of methylcobalamin in *Methanosarcina barkeri* extracts. Biochem. Biophys. Res. Commun. 17:475–480.

19. Blaylock, B. A., and T. C. Stadtman. 1966. Methane biosynthesis by *Methanosarcina barkeri*. Properties of the soluble enzyme system. Arch. Biochem. Biophys. 116:138–152.

20. Blotevogel, K.-H., and U. Fischer. 1985. Isolation and characterization of a new thermophilic and autotrophic methane producing bacterium: *Methanobacterium thermoaggregans* spec. nov. Arch. Microbiol. 142:218–222.

21. Blotevogel, K.-H., U. Fischer, M. Mocha, and S. Jannsen. 1985. *Methanobacterium thermoalcaliphilum* spec. nov., a new moderately alkaliphilic and thermophilic autotrophic methanogen. Arch. Microbiol. 142:211–217.

22. Blotevogel, K. H., U. Fischer, and K. H. Lüpkes. 1986. *Methanococcus frisius* sp. nov., a new methylotrophic marine methanogen. Can. J. Microbiol. 32:127–131.

23. Bollschweiler, C., and A. Klein. 1982. Polypeptide synthesis in *Escherichia coli* directed by cloned *Methanobrevibacter arboriphilicus* DNA. Zentralbl. Bakteriol. Mikrobiol. Hyg. 1 Abt. Orig. C3:101–109.

24. Bott, M., B. Eikmanns, and R. K. Thauer. 1986. Coupling of carbon monoxide oxidation to CO_2 and H_2 with phosphorylation of ADP in acetate-grown *Methanosarcina barkeri*. Eur. J. Biochem. 159:393–398.

25. Brandis, A., R. K. Thauer, and K. O. Stetter. 1981. Relatedness of strains ΔH and Marburg of *Methanobacterium thermoautotrophicum*. Zentralbl. Bakteriol. Mikrobiol. Hyg. 1 Abt. Orig. C2:311–317.

26. Bryant, M. P., B. C. McBride, and R. S. Wolfe. 1968. Hydrogen-oxidizing methane bacteria. I. Cultivation and methanogenesis. J. Bacteriol. 95:1118–1123.

27. Butsch, B. M., and R. Bachofen. 1982. Measurement of the membrane potential in *Methanobacterium thermoautotrophicum*. Experientia (Basel) 38:1377.

28. Cammack, R., D. Patil, R. Aguirre, and E. C. Hatchikian. 1982. Redox properties of the ESR-detectable nickel in hydrogenase from *Desulfovibio gigas*. FEBS Lett. 142:289–292.

29. Carper, S. W., and J. R. Lancaster, Jr. 1986. An electrogenic sodium-translocating ATPase in *Methanococcus voltae*. FEBS Lett. 200:177–180.

30. Cheeseman, P., A. Toms-Wood, and R. S. Wolfe. 1972. Isolation and properties of a fluorescent compound, Factor F_{420}, from *Methanobacterium* strain M.o.H. J. Bacteriol 112:527–531.

31. Chua, H. B., and J. P. Robinson. 1983. Formate-limited growth of *Methanobacterium formicicum* in steady-state cultures. Arch. Microbiol. 135:158–160.

32. Conrad, R., and R. K. Thauer. 1983. Carbon monoxide production by *Methanobacterium thermoautotrophicum*. FEBS Lett. 20:229–232.

33. Corder, R. E., L. A. Hook, J. M. Larkin, and J. I. Frea. 1983. Isolation and characterization of two new methane-producing cocci: *Methanogenium olentangyi*, sp. nov., and *Methanococcus deltae*, sp. nov. Arch. Microbiol. 134:28–32.

34. Crider, B. P., S. W. Carper, and J. R. Lancaster, Jr. 1985. Electron transfer-driven ATP synthesis in *Methanococcus voltae* is not dependent on a proton electrochemical gradient. Proc. Natl. Acad. Sci. USA 82:6793–6796.

35. Curthoys, N. P., and J. C. Rabinowitz. 1972. Formyltetrahydrofolate synthetase. Binding of folate substrates and kinetics of the reverse reaction. J. Biol. Chem. 247:1965–1971.

36. Daniels, L., N. Bakhiet, and K. Harmon. 1985. Widespread distribution of a 5-deazaflavin cofactor in *Actinomyces* and related bacteria. Syst. Appl. Microbiol. 6:12–17.

37. Daniels, L., G. Fuchs, R. K. Thauer, and J. G. Zeikus. 1977. Carbon monoxide oxidation by methanogenic bacteria. J. Bacteriol. 132:118–126.

38. Daniels, L., R. Sparling, and G. D. Sprott. 1984. The bioenergetics of methanogenesis. Biochim. Biophys. Acta 768:113–163.

39. Daniels, L., G. Fulton, R. W. Spencer, and W. H. Orme-Johnson. 1980. Origin of hydrogen in methane produced by *Methanobacterium thermoautotrophicum*. J. Bacteriol. 141:694–698.

40. Daniels, L., and J. G. Zeikus. 1978. One-carbon metabolism in methanogenic bacteria: analysis of short-term fixation products of $^{14}CO_2$ and $^{14}CH_3OH$ incorporated into whole cells. J. Bacteriol. 136:75–84.

41. Dekker, K., K. Jungermann, and R. K. Thauer. 1970. Energy production in anaerobic organisms. Angew. Chem. Int. Ed. Engl. 9:138–158.

42. Diekert, G., G. Fuchs, and R. K. Thauer. 1985. Properties and function of carbon monoxide dehydrogenase from anaerobic bacteria, p. 115–130, in R. K. Poole and C. S. Dow (eds.), Microbiol. gas metabolism. Mechanistic, metabolic and biotechnological aspects. Academic Press, London.

43. Diekert, G., H.-H. Gilles, R. Jaenchen, and R. K. Thauer. 1980. Incorporation of 8

succinate per mol nickel into factors F_{430} by *Methanobacterium thermoautotrophicum*. Arch. Microbiol. 128:256–262.

44. Diekert, G. B., E. G. Graf, and R. K. Thauer. 1979. Nickel requirement for carbon monoxide dehydrogenase formation in *Clostridium pasteurianum*. Arch. Microbiol. 122:117–120.

45. Diekert, G., R. Jaenchen, and R. K. Thauer. 1980. Biosynthetic evidence for a nickel tetrapyrrole structure of factor F_{430} from *Methanobacterium thermoautotrophicum*. FEBS Lett. 119:118–120.

46. Diekert, G., B. Klee, and R. K. Thauer. 1980. Nickel, a component of factor F_{430} from *Methanobacterium thermoautotrophicum*. Arch. Microbiol. 124:103–106.

47. Diekert, G., U. Kohnheiser, K. Piechulla, and R. K. Thauer, 1981. Nickel requirement and factor F_{430} content of methanogenic bacteria. J. Bacteriol. 148:459–464.

48. Diekert, G., and M. Ritter. 1983. Purification of the nickel protein carbon monoxide dehydrogenase of *Clostridium thermoaceticum*. FEBS Lett. 151:41–44.

49. Diekert, G., B. Weber, and R. K. Thauer. 1980. Nickel dependence of factor F_{430} content in *Methanobacterium thermoautotrophicum*. Arch. Microbiol. 127:273–278.

50. Doddema, H. J., C. A. Claesen, D. B. Kell, C. van der Drift, and G. D. Vogels. 1980. An adenine nucleotide translocase in the procaryote *Methanobacterium thermoautotrophicum*. Biochem. Biophys. Res. Commun. 95:1288–1293.

51. Doddema, H. J., C. van der Drift, G. D. Vogels, and M. Veenhuis. 1979. Chemiosmotic coupling in *Methanobacterium thermoautotrophicum*: hydrogen-dependent adenosine 5'-triphosphate synthesis by subcellular particles. J. Bacteriol. 140:1081–1089.

52. Doddema, H. J., T. J. Hutten, C. van der Drift, and G. D. Vogels. 1978. ATP hydrolysis and synthesis by the membrane-bound ATP synthetase complex of *Methanobacterium thermoautotrophicum*. J. Bacteriol. 136:19–23.

53. Doddema, H. J., and G. D. Vogels. 1978. Improved identification of methanogenic bacteria by fluorescence microscopy. Appl. Environ. Microbiol. 36:752–754.

54. Donnelly, M. I., J. C. Escalante-Semerena, K. L. Rinehart, Jr., and R. S. Wolfe. 1985. Methenyl-tetrahydromethanopterin cyclohydrolase in cell extracts of *Methanobacterium*. Arch. Biochem. Biophys. 242:430–439.

55. Drake, H. L. 1982. Occurrence of nickel in carbon monoxide dehydrogenase from *Clostridium pasteurianum* and *Clostridium thermoaceticum*. J. Bacteriol. 149:561–566.

56. Drake, H. L., S.-I. Hu, and H. G. Wood. 1980. Purification of carbon monoxide dehydrogenase, a nickel enzyme from *Clostridium thermoaceticum*. J. Biol. Chem. 255:7174–7180.

57. Edwards, T., and B. C. McBride. 1975. New method for the isolation and identification of methanogenic bacteria. Appl. Microbiol. 29:540–545.

58. Eikmanns, B., and R. K. Thauer. 1985. Evidence for the involvement and role of a corrinoid enzyme in methane formation from acetate in *Methanosarcina barkeri*. Arch. Microbiol. 142:175–179.

59. Eirich, L. D., and R. S. Dugger. 1984. Purification and properties of an F_{420}-dependent NADP reductase from *Methanobacterium thermoautotrophicum*. Biochim. Biophys. Acta 802:454–458.

60. Eirich, L. D., G. D. Vogels, and R. S. Wolfe. 1978. Proposed structure for coenzyme F_{420} from *Methanobacterium*. Biochemistry 17:4583–4593.

61. Eirich, L. D., G. D. Vogels, and R. S. Wolfe. 1979. Distribution of coenzyme F_{420} and properties of its hydrolytic fragments. J. Bacteriol. 140:20–27.

62. Eker, A. P. M. 1980. Photoreactivating enzyme from *Streptomyces griseus*. III. Evidence for the presence of an intrinsic chromophore. Photochem. Photobiol. 32:593–600.

63. Eker, A. P. M. 1983. Photorepair processes, p. 109–132, in G. Montagnoli and B. F. Erlanger, (eds.), Molecular models of photoresponsiveness. Plenum Press, New York.

64. Eker, A. P. M., R. H. Dekker, and W. Berends. 1981. Photoreactivating enzyme from *Streptomyces griseus*. IV. On the nature of the chromophoric cofactor in *Streptomyces griseus* photoreactivating enzyme. Photochem. Photobiol. 33:65–72.

65. Eker, A. P. M., A. Pol, P. van der Meijden, and G. D. Vogels. 1980. Purification and properties of 8-hydroxy-5-deazaflavin derivatives from *Streptomyces griseus*. FEMS Microbiol. Lett. 8:161–165.

66. Ellefson, W. L., W. B. Whitman, and R. S. Wolfe. 1982. Nickel-containing factor F_{430}: chromophore of the methylreductase of *Methanobacterium*. Proc. Natl. Acad. Sci. USA 79:3707–3710.

67. Ellefson, W. L., and R. S. Wolfe. 1980. Role of component C in the methylreductase system of *Methanobacterium*. J. Biol. Chem. 255:8388–8389.

68. Ellefson, W. L., and R. S. Wolfe. 1981. Biochemistry of methylreductase and evolution of methanogens, p. 171–180, in H. Dalton (ed.), Microbial growth on C_1 compounds. Heyden, London.

69. Ellefson, W. L., and R. S. Wolfe. 1981. Component C of the methylreductase system of *Methanobacterium*. J. Biol. Chem. 256:4259–4262.

70. Escalante-Semerena, J. C. 1983. In vitro methanogenesis from formaldehyde. Identification of three C_1 intermediates of the methanogenic pathway. Ph.D. thesis, University of Illinois.

71. Escalante-Semerena, J. C., J. A. Leigh, K. L. Rinehart, Jr., and R. S. Wolfe. 1984. Formaldehyde activation factor, tetrahydromethanopterin, a coenzyme of methanogenesis. Proc. Natl. Acad. Sci. USA 81:1976–1980.

72. Escalante-Semerena, J. C., J. A. Leigh, and R. S. Wolfe. 1984. New insights into the biochemistry of methanogenesis from H_2 and CO_2, p. 191–198, R. L. Crawford and R. S. Hanson (eds.), Microbial growth on C_1 compounds. American Society for Microbiology, Washington, D.C.

73. Escalante-Semerena, J. C., K. L. Rinehart, Jr., and R. S. Wolfe. 1984. Tetrahydromethanopterin, a carbon carrier in methanogenesis. J. Biol. Chem. 259:9447–9455.

74. Fahlbusch, K., H. Hippe, and G. Gottschalk. 1983. Formation of ethylamine and methane from dimethylethylamine by *Methanosarcina barkeri*. FEMS Microbiol. Lett. 19:103–104.

75. Fardeau, M.-L., and J.-P. Belaich. 1986. Energetics of growth of *Methanococcus thermolithotrophicus*. Arch. Microbiol. 144:381–385.

76. Fauque, G., M. Teixeira, I. Moura, P. A. Lespinat, A. V. Xavier, D. V. DerVartanian, H. D. Peck, Jr., J. LeGall, and J. G. Moura. 1984. Purification, charac-

terization and redox properties of hydrogenase from *Methanosarcina barkeri* (DSM 800). Eur. J. Biochem. 142:21–28.

77. Ferry, J. G., D. W. Sherod, H. D. Peck, Jr., and L. G. Ljungdahl. 1976. Levels of formyltetrahydrofolate synthetase and methylenetetrahydrofolate dehydrogenase in methanogenic bacteria, p. 151–155, in H. G. Schlegel, G. Gottschalk, and N. Pfennig (eds.), Microbiol. production and utilization of gases (H_2, CH_4, CO). E. Goltze KG, Göttingen, West Germany.

78. Ferry, J. G., and R. S. Wolfe. 1977. Nutritional and biochemical characterization of *Methanospirillum hungatii*. Appl. Environ. Microbiol. 34:371–376.

79. Fillingame, R. H. 1980. The proton-translocating pumps of oxidative phosphorylation. Annu. Rev. Biochem. 49:1079–1113.

80. Fina, L. R., H. J. Sincher, and D. F. De Cou. 1960. Evidence for production of methane from formic acid by direct reduction. Arch. Biochem. Biophys. 91:159–162.

81. Forrest, W. W., and D. J. Walker. 1971. The generation and utilization of energy during growth. Adv. Microbiol. Physiol. 5:213–274.

82. Fox, G. E., L. J. Magrum, W. E. Balch, R. S. Wolfe, and C. R. Woese. 1977. Classification of methanogenic bacteria by 16S ribosomal RNA characterization. Proc. Natl. Acad. Sci. USA 74:4537–4541.

83. Fox, G. E., E. Stackebrandt, R. B. Hespell, J. Gibson, J. Maniloff, T. A. Dyer, R. S. Wolfe, W. E. Balch, R. S. Tanner, L. J. Magrum, L. B. Zablen, R. Blakemore, R. Gupta, L. Bonen, B. J. Lewis, D. A. Stahl, K. R. Luehrsen, K. N. Chen, and C. R. Woese. 1980. The phylogeny of prokaryotes. Science 209:457–463.

84. Friedrich, C. G., K. Schneider, and B. Friedrich. 1982. Nickel in the catalytically active hydrogenase of *Alcaligenes eutrophus*. J. Bacteriol. 152:42–48.

85. Fuchs, G. 1986. CO_2 fixation in acetogenic bacteria: variations on a theme. FEMS Microbiol. Rev. 39:181–213.

86. Fuchs, G., J. Moll, P. Scherer, and R. K. Thauer. 1978. Activity, acceptor specificity and function of hydrogenase in *Methanobacterium thermoautotrophicum*, p. 83–92, in H. G. Schlegel and K. Schneider (eds.), Hydrogenases: their catalytic activity, structure and function. E. Goltze KG, Göttingen, West Germany.

87. Fuchs, G., and E. Stupperich. 1982. Autotrophic CO_2 fixation pathway in *Methanobacterium thermoautotrophicum*. Zentralbl. Bakteriol. Mikrobiol. Hyg. 1 Abt. Orig. C3:277–282.

88. Fuchs, G., R. Thauer, H. Ziegler, and W. Stichler. 1979. Carbon isotope fractionation by *Methanobacterium thermoautotrophicum*. Arch. Microbiol. 120:135–139.

89. Gilles, H., R. Jaenchen, and R. K. Thauer. 1983. Biosynthesis of 5-aminolevulinic acid in *Methanobacterium thermoautotrophicum*. Arch. Microbiol. 135:237–240.

90. Gilles, H., and R. K. Thauer. 1983. Uroporphyrinogen III, an intermediate in the biosynthesis of the nickel-containing factor F_{430} in *Methanobacterium thermoautotrophicum*. Eur. J. Biochem. 135:109–112.

91. Gillies, K., and J. R. Lancaster. 1983. Association of H_2-dependent factor-420 reductase activity with ATPase-rich particulate fractions in *Methanobacterium thermoautotrophicum*. Fed. Proc. 42:2143.

92. Graf, E.-G., and R. K. Thauer. 1981. Hydrogenase from *Methanobacterium thermoautotrophicum*, a nickel-containing enzyme. FEBS Lett. 136:165–169.

93. Guest, J. R., S. Friedman, M. A. Foster, G. Tejerina, and D. D. Woods. 1964. Transfer of the methyl group from N^5-methyltetrahydrofolates to homocysteine in *Escherichia coli*. Biochem. J. 92:497–504.

94. Gunsalus, R. P., J. A. Romesser, and R. S. Wolfe. 1978. Preparation of coenzyme M analogues and their activity in the methyl coenzyme M reductase system of *Methanobacterium thermoautotrophicum*. Biochemistry 17:2374–2377.

95. Gunsalus, R. P., and R. S. Wolfe. 1977. Stimulation of CO_2 reduction to methane by methylcoenzyme M in extracts of *Methanobacterium*. Biochem. Biophys. Res. Commun. 76:790–795.

96. Gunsalus, R. P., and R. S. Wolfe. 1978. ATP activation and properties of the methyl coenzyme M reductase system in *Methanobacterium thermoautotrophicum*. J. Bacteriol. 135:851–857.

97. Gunsalus, R. P., and R. S. Wolfe. 1980. Methyl coenzyme M reductase from *Methanobacterium thermoautotrophicum*. Resolution and properties of the components. J. Biol. Chem. 255:1891–1895.

98. Hammel, K. E., K. L. Cornwell, G. B. Diekert, and R. K. Thauer. 1984. Evidence for a nickel-containing carbon monoxide dehydrogenase in *Methanobrevibacter arboriphilicus*. J. Bacteriol. 157:975–978.

99. Harris, J. E., P. A. Pinn, and R. P. Davis. 1984. Isolation and characterization of a novel thermophilic, freshwater methanogen. Appl. Environ. Microbiol. 48:1123–1128.

100. Hartzell, P. L., G. Zvilius, J. C. Escalante-Semerena, and M. I. Donnelly. 1985. Coenzyme F_{420} dependence of the methylenetetrahydromethanopterin dehydrogenase of *Methanobacterium thermoautotrophicum*. Biochem. Biophys. Res. Commun. 133:884–890.

101. Hatchikian, E. C., M. Bruschi, N. Forget, and M. Scandellari. 1982. Electron tranport components from methanogenic bacteria: the ferredoxin from *Methanosarcina barkeri* (strain Fusaro). Biochem. Biophys. Res. Commun. 109:1316–1323.

102. Hausinger, R. P., I. Moura, J. J. G. Moura, A. V. Xavier, M. H. Santos, J. LeGall, and J. B. Howard. 1982. Amino acid sequence of a 3Fe:3S ferredoxin from the "archaebacterium" *Methanosarcina barkeri* (DSM 800). J. Biol. Chem. 257:14192–14197.

103. Himes, R. H., and J. C. Rabinowitz. 1962. Formyltetrahydrofolate synthetase. II. Characteristics of the enzyme and the enzymic reaction. J. Biol. Chem. 237:2903–2914.

104. Hippe, H., D. Caspari, K. Fiebig, and G. Gottschalk. 1979. Utilization of trimethylamine and other N-methyl compounds for growth and methane formation by *Methanosarcina barkeri*. Proc. Natl. Acad. Sci. USA 76:494–498.

105. Holder, U., D.-E. Schmidt, E. Stupperich, and G. Fuchs. 1985. Autotrophic synthesis of activated acetic acid from two CO_2 in *Methanobacterium thermoautotrophicum*. III. Evidence for common one-carbon precursor pool and the role of corrinoid. Arch. Microbiol. 141:229–238.

106. Hoppe-Seyler, F. 1887. Die Methangährung der Essigsäure. Z. Physiol. Chem. 9:561–568.

107. Huber, H., M. Thomm, H. König, G. Thies, and K. O. Stetter. 1982. *Methanococcus*

thermolithotrophicus, a novel thermophilic lithotrophic methanogen. Arch. Microbiol. 132:47–50.

108. Huennekens, F. M., and K. G. Scrimgeour. 1964. Tetrahydrofolic acid—the coenzyme of one-carbon metabolism, p. 355–376, in W. Pfleiderer and E. C. Taylor (eds.), Pteridine chemistry. Pergamon Press, Oxford.

109. Hughes, P. E., and S. B. Tove. 1982. Occurrence of α-tocopherolquinone and α-tocopherolquinol in microorganisms. J. Bacteriol. 151:1397–1402.

110. Huser, B. A., K. Wuhrmann, and A. J. B. Zehnder. 1982. *Methanothrix soehngenii* gen. nov. sp. nov., a new acetotrophic non-hydrogen-oxidizing methane bacterium Arch. Microbiol. 132:1–9.

111. Hutten, T. J., M. H. de Jong, B. P. H. Peeters, C. van der Drift, and G. D. Vogels. 1981. Coenzyme M derivatives and their effects on methane formation from carbon dioxide and methanol by cell extracts of *Methanosarcina barkeri*. J. Bacteriol. 145:27–34.

112. Jacobson, F. S., L. Daniels, J. A. Fox, C. T. Walsh, and W. H. Orme-Johnson. 1982. Purification and properties of an 8-hydroxy-5-deazaflavin-reducing hydrogenase from *Methanobacterium thermoautotrophicum*. J. Biol. Chem. 257:3385–3388.

113. Jacobson, F., and C. Walsh. 1984. Properties of 7,8-didemethyl-8-hydroxy-5-deazaflavins relevant to redox coenzyme function in methanogen metabolism. Biochemistry 23:979–988.

114. Jaenchen, R., G. Diekert, and R. K. Thauer. 1981. Incorporation of methionine-derived methyl groups into factor F_{430} by *Methanobacterium thermoautotrophicum*. FEBS Lett. 130:133–136.

115. Jaenchen, R., H.-H. Gilles, and R. K. Thauer. 1981. Inhibition of factor F_{430} synthesis by levulinic acid in *Methanobacterium thermoautotrophicum*. FEMS Microbiol. Lett. 12:167–170.

116. Jaenchen, R., P. Schönheit, and R. K. Thauer. 1984. Studies on the biosynthesis of coenzyme F_{420} in methanogenic bacteria. Arch. Microbiol. 137:362–365.

117. Jarrell, K. F., S. E. Bird, and G. D. Sprott. 1984. Sodium-dependent isoleucine transport in the methanogenic archaebacterium *Methanococcus voltae*. FEBS Lett. 166:357–361.

118. Jarrell, K. F., and G. D. Sprott. 1981. The transmembrane electrical potential and intracellular pH in methanogenic bacteria. Can. J. Microbiol. 27:720–728.

119. Jarrell, K. F., and G. D. Sprott. 1982. Nickel transport in *Methanobacterium bryantii*. J. Bacteriol. 151:1195–1203.

120. Jarrell, K. F., and G. D. Sprott. 1983. Measurement and significance of the membrane potential in *Methanobacterium bryantii*. Biochim. Biophys. Acta 725:280–288.

121. Jarrell, K. F., and G. D. Sprott. 1983. The effects of ionophores and metabolic inhibitors on methanogenesis and energy-related properties of *Methanobacterium bryantii*. Arch. Biochem. Biophys. 225:33–41.

122. Jarrell, K. F., and G. D. Sprott. 1985. Importance of sodium to the bioenergetic properties of *Methanococcus voltae*. Can. J. Microbiol. 31:851–855.

123. Jarrell, K. F., G. D. Sprott, and A. T. Matheson. 1984. Intracellular potassium concentration and relative activity of the ribosomal proteins of methanogenic bacteria. Can. J. Microbiol. 30:663–668.

124. Jarsch. M., J. Altenbuchner, and A. Böck. 1983. Physical organization of the genes for ribosomal RNA in *Methanococcus vannielii*. Mol. & Gen. Genet. 189:41–47.

125. Jeris, J. S., and P. L. McCarthy. 1965. The biochemistry of methane fermentation using C^{14} tracers. J. Water Pollut. Control Fed. 37:178–192.

126. Jin, S.-L. C., D. K. Blanchard, and J.-S. Chen. 1983. Two hydrogenases with distinct electron-carrier specificity and subunit composition in *Methanobacterium formicicum*. Biochim. Biophys. Acta 748:8–20.

127. Jones, J. B., G. L. Dilworth, and T. C. Stadtman. 1979. Occurrence of selenocysteine in the selenium-dependent formate dehydrogenase of *Methanococcus vannielii*. Arch. Biochem. Biophys. 195:255–260.

128. Jones, J. B., and T. C. Stadtman. 1977. *Methanococcus vannielii*: culture and effects of selenium and tungsten on growth. J. Bacteriol. 130:1404–1406.

129. Jones, J. B., and T. C. Stadtman. 1980. Reconstitution of a formate-NADP$^+$ oxidoreductase from formate dehydrogenase and a 5-deazaflavin-linked NADP$^+$ reductase isolated from *Methanococcus vannielii*. J. Biol. Chem. 255:1049–1053.

130. Jones, J. B., and T. C. Stadtman. 1981. Selenium-dependent and selenium-independent formate dehydrogenases of *Methanococcus vannielii*. Separation of the two forms and characterization of the purified selenium-independent form. J. Biol. Chem. 256:656–663.

131. Jones, W. J., M. I. Donnelly, and R. S. Wolfe. 1985. Evidence of a common pathway of carbon dioxide reduction to methane in methanogens. J. Bacteriol. 163:126–131.

132. Jones, W. J., J. A. Leigh, F. Mayer, C. R. Woese, and R. S. Wolfe. 1983. *Methanococcus jannaschii* sp. nov., an extremely thermophilic methanogen from a submarine hydrothemal vent. Arch. Microbiol. 136:254–261.

133. Jones, W. J., M. J. B. Paynter, and R. Gupta. 1983. Characterization of *Methanococcus maripaludis* sp. nov., a new methanogen isolated from salt marsh sediment. Arch. Microbiol. 135:91–97.

134. Jussofie, A. 1984. Cytochromuntersuchungen aus methanogenen und acetogenen Bakterien. Thesis, University of Göttingen.

135. Kiener, A., R. Gall, T. Rechsteiner, and T. Leisinger. 1985. Photoreactivation in *Methanobacterium thermoautotrophicum*. Arch. Microbiol. 143:147–150.

136. Kell, D. B., H. J. Doddema, J. G. Morris, and G. D. Vogels. 1981. Energy coupling in methanogens, p. 159–170, in H. Dalton (ed.), Microbial growth on C_1 compounds. Heyden, London.

137. Keltjens, J. T., P. van Beelen, A. M. Stassen, and G. D. Vogels. 1983. 7-Methylpterin in methanogenic bacteria. FEMS Microbiol. Lett. 20:259–262.

138. Keltjens, J. T., G. C. Caerteling, C. van der Drift, and G. D. Vogels. 1986. Methanopterin and the intermediary steps of methanogenesis. Syst. Appl. Microbiol. 7:370–375.

139. Keltjens, J. T., C. G. Caerteling, A. M. van Kooten, H. F. van Dijk, and G. D. Vogels. 1983. Chromophoric derivatives of coenzyme MF_{430}, a proposed coenzyme of methanogenesis in *Methanobacterium thermoautotrophicum*. Arch. Biochem. Biophys. 223:235–253.

140. Keltjens, J. T., L. Daniels, H. G. Janssen, P. J. Borm, and G. D. Vogels. 1983. A

novel one-carbon carrier (carboxy-5, 6, 7, 8-tetrahydromethanopterin) isolated from *Methanobacterium thermoautotrophicum* and derived from methanopterin. Eur. J. Biochem. 130:545–552.

141. Keltjens, J. T., M. J. Huberts, W. H. Laarhoven, and G. D. Vogels. 1983. Structural elements of methanopterin, a novel pterin present in *Methanobacterium thermoauto-trophicum*. Eur. J. Biochem. 130:537–544.

142. Keltjens, J. T., and C. van der Drift. 1986. Electron transfer reactions in methanogens. FEMS Microbiol. Rev. 39:259–302.

143. Keltjens, J. T., and G. D. Vogels. 1981. Novel coenzymes of methanogens, p. 152–158, in H. Dalton (ed.), Microbial growth on C_1 compounds. Heyden, London.

144. Keltjens, J. T., W. B. Whitman, C. G. Caerteling, A. M. van Kooten, R. S. Wolfe, and G. D. Vogels. 1982. Presence of coenzyme M derivatives in the prosthetic group (coenzyme MF_{430}) of the methylcoenzyme M reductase from *Methanobacterium thermoautotrophicum*. Biochem. Biophys. Res. Commun. 108:495–503.

145. Kenealy, W. R., T. E. Thompson, K. R. Schubert, and J. G. Zeikus, 1982. Ammonia assimilation and synthesis of alanine, aspartate, and glutamate in *Methanosarcina barkeri* and *Methanobacterium thermoautotrophicum*, J. Bacteriol. 150:1357–1365.

146. Kenealy, W. R., and J. G. Zeikus. 1982. One-carbon metabolism in methanogens: evidence for synthesis of a two-carbon cellular intermediate and unification of catabolism and anabolism in *Methanosarcina barkeri*. J. Bacteriol. 151:932–941.

147. Kerr, J. A., and A. F. Trotman-Dickenson. 1972. Bond strengths of some organic molecules, p. F-186, in R. C. Weast (ed.), Handbook of chemistry and physics, 52nd ed. Chemical Rubber Company, Cleveland, Ohio.

148. Kirby, T. W., J. R. Lancaster, Jr., and I. Fridovich. 1981. Isolation and characterization of the iron-containing superoxide dismutase of *Methanobacterium bryantii*. Arch. Biochem. Biophys. 210:140–148.

149. Klein, A., and M. Schnorr, 1984. Genome complexity of methanogenic bacteria. J. Bacteriol. 158:628–631.

150. Kluyver, A. J., and C. G. T. P. Schnellen. 1947. On the fermentation of carbon monoxide by pure cultures of methane bacteria. Arch. Biochem. 14:57–70.

151. Kohler, H.-P. E., and A. J. B. Zehnder. 1984. Carbon monoxide dehydrogenase and acetate thiokinase in *Methanothrix soehngenii*. FEMS Microbiol. Lett. 21:287–292.

152. Kojima, N., J. A. Fox, R. P. Hausinger, L. Daniels, W. H. Orme-Johnson, and C. Walsh. 1983. Paramagnetic centers in the nickel-containing, deazaflavin-reducing hydrogenase from *Methanobacterium thermoautotrophicum*. Proc. Natl. Acad. Sci. USA 80:378–382.

153. König, H. 1984. Isolation and characterization of *Methanobacterium uliginosum* sp. nov. from a marshy soil. Can. J. Microbiol. 30:1477–1481.

154. König, H., R. Semmler, C. Lerp, and J. Winter. 1985. Evidence for the occurrence of autolytic enzymes in *Methanobacterium wolfei*. Arch. Microbiol. 141:177–180.

155. König, H., and K. O. Stetter. 1982. Isolation and characterization of *Methanolobus tindarius*, sp. nov., a coccoid methanogen growing only on methanol and methylamines. Zentralbl. Bakteriol. Mikrobiol. Hyg. 1 Abt. Orig. C3:478–490.

156. Kratky, C., A. Fässler, A. Pfaltz, B. Kräutler, B. Jaun, and A. Eschenmoser. 1984. Chemistry of corphinoids: structural properties of corphinoid nickel (II) complexes related to coenzyme F_{430}. J. Chem. Soc. Chem. Commun. 1368–1371.

157. Krzycki, J. A., L. J. Lehman, and J. G. Zeikus. 1985. Acetate catabolism by *Methanosarcina barkeri*: evidence for involvement of carbon monoxide dehydrogenase, methyl coenzyme M, and methylreductase. J. Bacteriol. 163:1000–1006.

158. Krzycki, J. A., R. H. Wolkin, and J. G. Zeikus. 1982. Comparison of unitrophic and mixotrophic substrate metabolism by an acetate-adapted strain of *Methanosarcina barkeri*. J. Bacteriol. 149:247–254.

159. Krzycki, J. A., and J. G. Zeikus. 1980. Quantification of corrinoids in methanogenic bacteria. Curr. Microbiol. 3:243–245.

160. Krzycki, J. A., and J. G. Zeikus. 1984. Characterization and purification of carbon monoxide dehydrogenase from *Methanosarcina barkeri*. J. Bacteriol. 158:231–237.

161. Krzycki, J. A., and J. G. Zeikus. 1984. Acetate catabolism by *Methanosarcina barkeri*: hydrogen-dependent methane production from acetate by a soluble cell protein fraction. FEMS Microbiol. Lett. 25:27–32.

162. Kühn, W. 1982. Erstbeschreibung und Charakterisierung von Cytochromen in methanogenen Bakterien. Ph.D. thesis. University of Göttingen.

163. Kühn, W., K. Fiebig, H. Hippe, R. A. Mah, B. A. Huser, and G. Gottschalk. 1983. Distribution of cytochromes in methanogenic bacteria. FEMS Microbiol. Lett. 20:407–410.

164. Kühn, W., K. Fiebig, R. Walther, and G. Gottschalk. 1979. Presence of a cytochrome b_{559} in *Methanosarcina barkeri*. FEBS Lett. 105:271–274.

165. Kühn, W., and G. Gottschalk. 1983. Characterization of the cytochromes occurring in *Methanosarcina* species. Eur. J. Biochem. 135:89–94.

166. Lancaster, J. R., Jr. 1980. Soluble and membrane-bound paramagnetic centers in *Methanobacterium bryantii*. FEBS Lett. 115:285–288.

167. Lancaster, J. R., Jr. 1981. Membrane-bound flavin adenine dinucleotide in *Methanobacterium bryantii*. Biochem. Biophys. Res. Commun. 100:240–246.

168. Lancaster, J. R., Jr. 1982. New biological paramagnetic center: octahedrally coordinated nickel (III) in the methanogenic bacteria. Science 216:1324–1325.

169. Lancaster, J. R., Jr. 1986. A unified scheme for carbon and electron flow coupled to ATP synthesis by substrate-level phosphorylation in the methanogenic bacteria. FEBS Lett. 199:12–18.

170. Länge, S., and G. Fuchs. 1985. Tetrahydromethanopterin, a coenzyme involved in autotrophic acetyl coenzyme A synthesis from 2 CO_2 in *Methanobacterium*. FEBS Lett. 181:303–307.

171. Lappin, A. G., C. K. Murray, and D. W. Margerum. 1978. Electron paramagnetic resonance studies of nickel (III)-oligopeptide complexes. Inorg. Chem. 17:1630–1634.

172. Large, P. J., J. B. M. Meiberg, and W. Harder. 1979. Cytochrome C_{co} is not a primary electron acceptor for the amine dehydrogenases of *Hyphomicrobium* X. FEMS Microbiol. Lett. 5:281–286.

173. LeGall, J., P. O. Ljungdahl, I. Moura, H. D. Peck, Jr., A. V. Xavier, J. J. G. Moura, M. Teixera, B. H. Huynh, and D. V. DerVartanian. 1982. The presence of redox-sensitive nickel in the periplasmic hydrogenase from *Desulfovibrio gigas*. Biochem. Biophys. Res. Commun. 106:610–616.

174. Leigh, J. A., K. L. Rinehart, Jr., and R. S. Wolfe. 1984. Structure of methanofuran, the carbon dioxide reduction factor of *Methanobacterium thermoautotrophicum*. J. Am. Chem. Soc. 106:3636–3640.

175. Leigh, J. A., K. L. Rinehart, Jr., and R. S. Wolfe. 1985. Methanofuran (carbon dioxide reduction factor), a formyl carrier in methane production from carbon dioxide in *Methanobacterium*. Biochemistry 24:995–999.

176. Leigh, J. A., and R. S. Wolfe. 1983. Carbon dioxide reduction factor and methanopterin, two coenzymes required for CO_2 reduction to methane by extracts of *Methanobacterium*. J. Biol. Chem. 258:7536–7540.

177. Lindahl, P. A., N. Kojima, R. P. Hausinger, J. A. Fox, B. K. Teo, C. T. Walsh, and W. H. Orme-Johnson. 1984. Nickel and iron EXAFS of F_{420}-reducing hydrogenase from *Methanobacterium thermoautotrophicum*. J. Am. Chem. Soc. 106:3062–3064.

178. Livingston, D. A., A. Pfaltz, J. Schreiber, A. Eschenmoser, D. Ankel-Fuchs, J. Moll, R. Jaenchen, and R. K. Thauer. 1984. Zur Kenntnis des Faktors F_{430} aus methanogenen Bakterien: Struktur des proteinfreien Faktors. Helv. Chim. Acta. 67:334–351.

179. Mackle, H. 1963. The thermochemistry of sulphur-containing molecules and radicals. II. The dissociation energies of bonds involving sulphur: the heats of formation of sulphur-containing radicals. Tetrahedron 19:1159–1170.

180. Mah, R. A. 1980. Isolation and characterization of *Methanococcus mazei*. Curr. Microbiol. 3:321–326.

181. Mah, R. A., and M. R. Smith. 1981. The methanogenic bacteria, p. 948–977, in M. P. Starr, H. Stolp, H. G. Trüper, A. Balows, and H. G. Schlegel (eds.), The prokaryotes. Springer-Verlag, Heidelberg.

182. Mah, R. A., D. M. Ward, L. Baresi, and T. L. Glass. 1977. Biogenesis of methane. Annu. Rev. Microbiol. 31:309–341.

183. May, H. D., N. L. Schauer, and J. G. Ferry. 1986. Molybdopterin cofactor from *Methanobacterium formicicum* formate dehydrogenase. J. Bacteriol. 166:500–504.

184. Mayer, F., J. I. Elliott, D. Sherod, and L. G. Ljungdahl. 1982. Formyltetrahydrofolate synthetase from *Clostridium thermoaceticum*. An electron microscopic study and specific interaction of the enzyme with ATP and ADP. Eur. J. Biochem. 124:397–404.

185. McBride, B. C. 1970. Transmethylation reactions in *Methanobacterium* strain M.o.H. Ph.D. thesis. University of Illinois.

186. McBride, B. C., and R. S. Wolfe. 1971. A new coenzyme of methyl transfer, coenzyme M. Biochemistry 10:2317–2324.

187. McCormick, J. R. D., and G. O. Morton. 1982. Identity of cosynthetic factor I of *Streptomyces aureofaciens* and fragment FO from coenzyme F_{420} of *Methanobacterium* species. J. Am. Chem. Soc. 104:4014–4015.

188. McKellar, R. C., K. M. Shaw, and G. D. Sprott. 1981. Isolation and characterization of a FAD-dependent NADH diaphorase from *Methanospirillum hungatei*. Can. J. Biochem. 59: 83–91.

189. Meile, L., A. Kiener, and T. Leisinger. 1983. A plasmid in the archaebacterium *Methanobacterium thermoautotrophicum*. Mol. & Gen. Genet. 191:480–484.

190. Miller, T. L., and M. J. Wolin. 1985. *Methanosphaera stadtmaniae* gen. nov., sp. nov.: a species that forms methane by reducing methanol with hydrogen. Arch. Microbiol. 141:116–122.

191. Mitchell, P. 1966. Chemiosmotic coupling in oxidative and photosynthetic phosphorylation. Biol. Rev. 41:445–502.

192. Mitchell, R. M., L. A. Loeblich, L. C. Klotz, and A. R. Loeblich III. 1979. DNA organization of *Methanobacterium thermoautotrophicum*. Science 204:1082–1084.

193. Mountfort, D. O. 1978. Evidence for ATP synthesis driven by a proton gradient in *Methanosarcina barkeri*. Biochem. Biophys. Res. Commun. 85:1346–1351.

194. Mountfort, D. O. 1980. Effect of adenosine 5'-monophosphate on adenosine 5'-triphosphate activation of methyl coenzyme M methylreductase in cell extracts of *Methanosarcina barkeri*. J. Bacteriol. 143:1039–1041.

195. Mountfort, D. O., and R. A. Asher. 1979. Effect of inorganic sulfide on the growth and metabolism of *Methanosarcina barkeri* strain DM. Appl. Environ. Microbiol. 37:670–675.

196. Moura, I., J. J. G. Moura, B.-H. Huynh, H. Santos, J. LeGall, and A. V. Xavier. 1982. Ferredoxin from *Methanosarcina barkeri*: evidence for the presence of a three-iron center. Eur. J. Biochem. 126:95–98.

197. Moura, J. J. G., I. Moura, H. Santos, A. V. Xavier, M. Scandellari, and J. LeGall. 1982. Isolation of P_{590} from *Methanosarcina barkeri*: evidence for the presence of sulfite reductase activity. Biochem. Biophys. Res. Commun. 108:1002–1009.

198. Nagle, D. P., Jr., and R. S. Wolfe. 1983. Component A of the methyl coenzyme M methylreductase system of *Methanobacterium*: resolution into four components. Proc. Natl. Acad. Sci. USA 80:2151–2155.

199. Naraoka, T., K. Momoi, K. Fukasawa, and M. Goto. 1984. Isolation and identification of a naturally occurring 7,8-didemethyl-8-hydroxy-5-deazariboflavin derivative from *Mycobacterium avium*. Biochim. Biophys. Acta 797:377–380.

200. Naumann, E., K. Fahlbusch, and G. Gottschalk. 1984. Presence of a trimethylamine: HS-coenzyme M methyltransferase in *Methanosarcina barkeri*. Arch. Microbiol. 138:79–83.

201. Nelson, M. J. K., D. P. Brown, and J. G. Ferry. 1984. FAD requirement for the reduction of coenzyme F_{420} by hydrogenase from *Methanobacterium formicicum*. Biochem. Biophys. Res. Commun. 120:775–781.

202. Neumann, H., A. Gierl, J. Tu, J. Leibrock, D. Staiger, and W. Zillig. 1983. Organization of the genes for ribosomal RNA in archaebacteria. Mol. & Gen. Genet. 192:66–72.

203. Noll, K. M., K. L. Rinehart, Jr., R. S. Tanner, and R. S. Wolfe. 1986. Structure of component B (7-mercaptoheptanoylthreonine phosphate) of the methylcoenzyme M methylreductase system of *Methanobacterium thermoautotrophicum*. Proc. Natl. Acad. Sci. USA 83:4238–4242.

204. Nozhevnikova, A. N., and V. I. Chudina. 1984. Morphology of the thermophilic acetate methane bacterium *Methanothrix thermoacetophila* sp. nov. Microbiology (Eng. Transl. Mikrobiologiya) 53:618–623.

205. O'Brien, J. M., R. H. Wolkin, T. T. Moench, J. B. Morgan, and J. G. Zeikus. 1984. Association of hydrogen metabolism with unitrophic or mixotrophic growth of *Methanosarcina barkeri* on carbon monoxide. J. Bacteriol. 158:373–375.

206. Ollivier, B. M., R. A. Mah, J. L. Garcia, and D. R. Boone. 1986. Isolation and characterization of *Methanogenium bourgense* sp. nov. Int. J. Syst. Bacteriol. 36:297–301.

207. Ollivier, B. M., R. A. Mah, J. L. Garcia, and R. Robinson. 1985. Isolation and characterization of *Methanogenium aggregans* sp. nov. Int. J. Syst. Bacteriol. 35:127–130.

208. Ossmer, R., T. Mund, P. L. Hartzell, U. Konheiser, G. W. Kohring, A. Klein, R. S. Wolfe, G. Gottschalk, and F. Mayer. 1986. Immunochemical localization of component C of the methylreductase system in *Methanococcus voltae* and *Methanobacterium thermoautotrophicum*. Proc. Natl. Acad. Sci. USA 83:5789–5792.

209. Patel, G. B. 1984. Characterization and nutritional properties of *Methanothrix concilii* sp. nov., a mesophilic, aceticlastic methanogen. Can. J. Microbiol. 30:1383–1396.

210. Paterek. J. R., and P. H. Smith. 1985. Isolation and characterization of a halophilic methanogen from Great Salt Lake. Appl. Environ. Microbiol. 50:877–881.

211. Perski, H. J., J. Moll, and R. K. Thauer. 1981. Sodium dependence of growth and methane formation in *Methanobacterium thermoautotrophicum*. Arch. Microbiol. 130:319–321.

212. Perski, H. J., P. Schönheit, and R. K. Thauer. 1982. Sodium dependence of methane formation in methanogenic bacteria. FEBS Lett. 143:323–326.

213. Pfaltz, A., B. Jaun, A. Fässler, A. Eschenmoser, R. Jaenchen, H.-H. Gilles, G. Diekert, and R. K. Thauer. 1982. Zur Kenntnis des Faktors F$_{430}$ aus methanogenen Bakterien: Struktur des porphinoiden Ligandsystems. Helv. Chim. Acta 65:828–865.

214. Pirt, S. J. 1965. The maintenance energy of bacteria in growing cultures. Proc. Soc. Lond. B163:224–231.

215. Poirot, C. M., S. W. M. Kengen, E. Valk, J. T. Keltjens, C. van der Drift, and G. D. Vogels. 1987. Formation of methaylcoenzyme M from formaldehyde by cell-free extracts of *methanobacterium thermoautotrophicum*. Evidence for the involvement of a corrinoid-containing methyltransferase. FEMS Microbiol. Lett. 40:7–13.

216. Pol, A., C. vander Drift, G. D. Vogels, T. J. H. M. Cuppen, and W. H. Laarhoven. 1980. Comparison of coenzyme F$_{420}$ from *Methanobacterium bryantii* with 7- and 8-hydroxyl-10-methyl-5-deazaisoalloxazine. Biochem. Biophys. Res. Commun. 92:255–260.

217. Pol, A., C. van der Drift, and G. D. Vogels. 1982. Corrinoids from *Methanosarcina barkeri*: structure of the α-ligand. Biochem. Biophys. Res. Commun. 108:731–737.

218. Ragsdale, S. W., J. E. Clark, L. G. Ljungdahl, L. L. Lundie, and H. L. Drake. 1983. Properties of purified carbon monoxide dehydrogenase from *Clostridium thermoaceticum*, a nickel, iron-sulfur protein. J. Biol. Chem. 258:2364–2369.

219. Reeve, J. N., N. J. Trun, and P. T. Hamilton. 1982. Beginning genetics with methanogens, p. 233–244 in A. Hollaender, R. D. DeMoss, S. Kaplan, J. Konisky, D. Savage, and R. S. Wolfe (eds.), Genetic engineering of microorganisms for chemicals. Plenum Press, New York.

220. Rivard, C. J., J. M. Henson, M. V. Thomas, and P. M. Smith. 1983. Isolation and characterization of *Methanomicrobium paynteri* sp. nov., a mesophilic methanogen isolated from marine sediments. Appl. Environ. Microbiol. 46:484–490.

221. Rivard, C. J., and P. H. Smith, 1982. Isolation and characterization of a thermophilic marine methanogenic bacterium, *Methanogenium thermophilicum* sp. nov. Int. J. Syst. Bacteriol. 32:430–436.

222. Roberton, A. M., and R. S. Wolfe, 1970. Adenosine triphosphate pools in *Methanobacterium*. J. Bacteriol. 102:43–51.

223. Robinson, D. R. 1971. The nonenzymatic hydrolysis of N^5,N^{10}-methenyltetrahydrofolic acid and related reactions, p. 716–725, in D. B. McCormick and L. D. Wright (eds.), Methods in enzymology, Vol. 18B. Academic Press, New York.

224. Rogers, K., K. Gillies, and J. R. Lancaster, Jr. 1983. EPR studies of electron transfer reactions in the methanogenic bacteria. Fed. Proc. 42:2267.

225. Romesser, J. A. 1978. The activation and reduction of carbon dioxide to methane in *Methanobacterium thermoautotrophicum*. Ph.D. thesis, University of Illinois.

226. Romesser, J. A., and R. S. Wolfe. 1982. Coupling of methyl coenzyme M reduction with carbon dioxide activation in extracts of *Methanobacterium thermoautotrophicum*. J. Bacteriol. 152:840–847.

227. Romesser, J. A., and R. S. Wolfe. 1982. CDR factor, a new coenzyme required for carbon dioxide reduction to methane by extracts of *Methanobacterium thermoautotrophicum*. Zentralbl. Bakteriol. Mikrobiol. Hyg. 1 Abt. Orig. C3:271–276.

228. Rouvière, P. E., J. C. Escalante-Semerena, and R. S. Wolfe. 1985. Component A_2 of the methylcoenzyme M methylreductase system from *Methanobacterium thermoautotrophicum*. J. Bacteriol. 162:61–66.

229. Rüdiger, H. 1971. The vitamin B_{12}-dependent methionine synthetase. The cycle of transmethylation. Eur. J. Biochem. 21:264–268.

230. Rüdiger, H., and L. Jaenicke. 1969. Methionine synthesis: demonstration of the reversibility of the reaction. FEBS Lett. 4:316–318.

231. Rühlemann, M., K. Zeigler, E. Stupperich, and G. Fuchs. 1985. Detection of acetyl coenzyme A as an early assimilation intermediate in *Methanobacterium*. Arch. Microbiol 141:399–406.

232. Sauer, F. D. 1986. Tetrahydromethanopterin methyltransferase, a component of the methane synthetisizing complex of *Methanobacterium thermoautotrophicum*. Biochem. Biophys. Res. Commun. 136:542–547.

233. Sauer, F. D., J. D. Erfle, and S. Mahadevan. 1980. Methane production by the membraneous fraction of *Methanobacterium thermoautotrophicum*. Biochem. J. 190:177–182.

234. Sauer, F. D., J. D. Erfle, and S. Mahadevan. 1981. Evidence for an internal electrochemical proton gradient in *Methanobacterium thermoautotrophicum*, J. Biol. Chem. 256:9843–9848.

235. Sauer, F. D., S. Mahadevan, and J. D. Erfle. 1980. Valinomycin inhibited methane synthesis in *Methanobacterium thermoautotrophicum*. Biochem. Biophys. Res. Commun. 95:715–721.

236. Sauer, F. D., S. Mahadevan, and J. D. Erfle. 1984. Methane synthesis by membrane vesicles and a cytoplasmic cofactor isolated from *Methanobacterium thermoautotrophicum*. Biochem. J. 221:61–69.

237. Schauer, N. L., D. P. Brown, and J. G. Ferry. 1982. Kinetics of formate metabolism in *Methanobacterium formicicum* and *Methanospirillum hungatei*. Appl. Environ. Microbiol. 44:549–554.

238. Schauer, N. L., and J. G. Ferry. 1980. Metabolism of formate in *Methanobacterium formicicum*. J. Bacteriol. 142:800–807.

239. Schauer, N. L., and J. G. Ferry. 1982. Properties of formate dehydrogenase in *Methanobacterium formicicum*. J. Bacteriol. 150:1–7.

240. Schauer, N. L., and J. G. Ferry. 1983. FAD requirement for the reduction of coen-

zyme F_{420} by formate dehydrogenase from *Methanobacterium formicicum*. J. Bacteriol. 155:467–472.

241. Schauer, N. L., and J. G. Ferry. 1986. Composition of the coenzyme F_{420}-dependent formate dehydrogenase from *Methanobacterium formicicum*. J. Bacteriol. 165:405–411.

242. Scherer, P., V. Höllriegl, C. Krug, M. Bokel, and P. Renz. 1984. On the biosynthesis of 5-hydroxybenzimidazolylcobamide (vitamin B_{12}-Factor III) in *Methanosarcina barkeri*. Arch. Microbiol. 138:354–359.

243. Scherer, P., and H. Sahm. 1981. Influence of sulphur-containing compounds on the growth of *Methanosarcina barkeri* in a defined medium. Eur. J. Appl. Microbiol. Biotechnol. 12:28–35.

244. Scherer, P. A., and R. K. Thauer. 1978. Purification and properties of reduced ferredoxin: CO_2 oxidoreductase from *Clostridium pasteurianum*, a molybdenum nonsulfur-protein. Eur. J. Biochem. 85:125–135.

245. Schmid, K., M. Thomm, A. Laminet, F. G. Laue, C. Kessler, K. O. Stetter, and R. Schmitt. 1984. Three new restriction endonucleases *Mae I*, *Mae II* and *Mae III* from *Methanococcus aeolicus*. Nucl. Acids Res. 12:2619–2628.

246. Schnellen, C. G. T. P. 1947. Onderzoekingen over de methaangisting. Ph.D. thesis, Delft University of Technology.

247. Schönheit, P., and D. P. Beimborn. 1985. Presence of a Na^+/H^+ antiporter in *Methanobacterium thermoautotrophicum* and its role in Na^+-dependent methanogenesis. Arch. Microbiol 142:354–361.

248. Schönheit, P., and D. B. Beimborn. 1985. ATP synthesis in *Methanobacterium thermoautotrophicum* coupled to CH_4 formation from H_2 and CO_2 in the apparent absence of an electrochemical proton potential across the cytoplasmic membrane. Eur. J. Biochem. 148:545–550.

249. Schönheit, P., D. B. Beimborn, and H. Perski. 1984. Potassium accumulation in growing *Methanobacterium thermoautotrophicum* and its relation to the electrochemical proton gradient. Arch. Microbiol. 140:247–251.

250. Schönheit, P., J. Moll, and R. K. Thauer. 1980. Growth parameters (K_s, μ_{max}, Y_s) of *Methanobacterium thermoautotrophicum*. Arch. Microbiol. 127:59–65.

251. Schönheit, P., and H. J. Perski. 1983. ATP synthesis driven by a potassium diffusion potential in *Methanobacterium thermoautotrophicum* is stimulated by sodium. FEMS Microbiol. Lett. 20:263–267.

252. Smith, M. R., and R. A. Mah. 1978. Growth and methanogenesis by *Methanosarcina* strain 227 on acetate and methanol. Appl. Environ. Microbiol. 36:870–879.

253. Smith, M. R., and R. A. Mah. 1980. Acetate as sole carbon and energy source for growth of *Methanosarcina* strain 227. Appl. Environ. Microbiol. 39:993–999.

254. Smith, P. H., and R. A. Mah. 1966. Kinetics of acetate metabolism during sludge digestion. Appl. Microbiol. 14:368–371.

255. Sowers, K. R., S. F. Baron, and J. G. Ferry. 1984. *Methanosarcina acetivorans* sp. nov., an acetotrophic methane-producing bacterium isolated from marine sediments. Appl. Environ. Microbiol. 47:971–978.

256. Sowers, K. R., and J. G. Ferry. 1983. Isolation and characterization of a methylotrophic marine methanogen, *Methanococcoides methylutens* gen. nov., sp. nov. Appl. Environ. Microbiol. 45:684–690.

257. Spencer, R. W., L. Daniels, G. Fulton, and W. H. Orme-Johnson. 1980. Product isotope effects on *in vivo* methanogenesis by *Methanobacterium thermoautotrophicum*. Biochemistry 19:3678–3683.

258. Sprott, G. D., S. E. Bird, and I. J. McDonald. 1985. Proton motive force as a function of the pH at which *Methanobacterium bryantii* is grown. Can. J. Microbiol. 31:1031–1034.

259. Sprott, G. D., J. R. Colvin, and R. C. McKellar. 1979. Spheroplasts of *Methanospirillum hungatii* formed upon treatment with dithiothreitol. Can. J. Microbiol. 25:730–738.

260. Sprott, G. D., and K. F. Jarrell. 1981. K^+, Na^+, and Mg^{2+} content and permeability of *Methanospirillum hungatei* and *Methanobacterium thermoautotrophicum*. Can. J. Microbiol. 27:444–451.

261. Sprott, G. D. and K. F. Jarrell. 1982. Sensitivity of methanogenic bacteria to dicyclohexylcarbodiimide. Can. J. Microbiol. 28:982–986.

262. Sprott, G. D., K. F. Jarrell, K. M. Shaw, and R. Knowles. 1982. Acetylene as an inhibitor of methanogenic bacteria. J. Gen. Microbiol. 128:2453–2462.

263. Sprott, G. D., K. M. Shaw, and K. F. Jarrell. 1983. Isolation and chemical composition of the cytoplasmic membrane of the archaebacterium *Methanospirillum hungatei*. J. Biol. Chem. 258:4026–4031.

264. Sprott, G. D., K. M. Shaw, and K. F. Jarrell. 1985. Methanogenesis and the K^+ transport system are activated by divalent cations in ammonia-treated cells of *Methanospirillum hungatei*. J. Biol. Chem. 260:9244–9250.

265. Stadtman, T. C. 1967. Methane fermentation. Annu. Rev. Microbiol. 21:121–142.

266. Stadtman, T. C., and H. A. Barker. 1951. Studies on the methane fermentation. IX. The origin of methane in the acetate and methanol fermentations by *Methanosarcina*. J. Bacteriol. 61:81–86.

267. Stetter, K. O., and G. Gaag. 1983. Reduction of molecular sulphur by methanogenic bacteria. Nature (Lond.) 305:309–311.

268. Stetter, K. O., M. Thomm, J. Winter, G. Wildgruber, H. Huber, W. Zillig, D. Janékovic, H. König, P. Palm, and S. Wunderl. 1981. *Methanothermus fervidus*, sp. nov., a novel extremely thermophilic methanogen isolated from an icelandic hot spring. Zentralbl. Bakteriol. Mikrobiol. Hyg. 1 Abt. Orig. C2:166–178.

269. Stouthamer, A. H. 1973. A theoretical study on the amount of ATP required for synthesis of microbial cell material. Antonie Leeuwenhoek J. Microbiol. Serol. 39::545–565.

270. Stouthamer, A. H., and C. Bettenhausen. 1973. Utilization of energy for growth and maintenance in continuous and batch cultures of microorganisms. A reevaluation of the method for the determination of ATP production by measuring molar growth yields. Biochim. Biophys. Acta. 301:53–70.

271. Stupperich, E., and G. Fuchs. 1983. Autotrophic acetyl coenzyme A synthesis *in vitro* from two CO_2 in *Methanobacterium*. FEBS Lett. 156:345–348.

272. Stupperich, E., and G. Fuchs. 1984. Autotrophic synthesis of activated acetic acid from two CO_2 in *Methanobacterium thermoautotrophicum*. I. Properties on in vitro system. Arch. Microbiol. 139:8–13.

273. Stupperich, E., and G. Fuchs. 1984. Autotrophic synthesis of activated acetic acid

from two CO_2 in *Methanobacterium thermoautotrophicum*. II. Evidence for different origins of acetate carbon atoms. Arch. Microbiol. 139:14–20.

274. Stupperich, E., K. E. Hammel, G. Fuchs, and R. K. Thauer. 1983. Carbon monoxide fixation into the carboxyl group of acetyl coenzyme A during autotrophic growth of *Methanobacterium*. FEBS Lett. 152:21–23.

275. Tan, S. L., J. A. Fox, N. Kojima, C. T. Walsh, and W. H. Orme-Johnson. 1984. Nickel coordination in deazaflavin and viologen-reducing hydrogenases from *Methanobacterium thermoautotrophicum* investigation by electron spin echo spectroscopy. J. Am. Chem. Soc. 106:3064–3066.

276. Taylor, C. D., B. C. McBride, R. S. Wolfe, and M. P. Bryant. 1974. Coenzyme M, essential for growth of a rumen strain of *Methanobacterium ruminantium*. J. Bacteriol. 120:974–975.

277. Taylor, C. D., and R. S. Wolfe. 1974. Structure and methylation of coenzyme M ($HSCH_2CH_2SO_3$). J. Biol. Chem. 249:4879–4885.

278. Taylor, C. D., and R. S. Wolfe. 1974. A simplified assay for coenzyme M. Resolution of methylcobalamin-coenzyme M methyltransferase and use of sodium borohydride. J. Biol. Chem. 249:4886–4890.

279. Taylor, G. T. 1982. The methanogenic bacteria, p. 231–239 in M. J. Bull (ed.), Progress in industrial microbiology. Elsevier, Amsterdam.

280. Taylor, G. T., and S. J. Pirt. 1977. Nutrition and factors limiting the growth of a methanogenic bacterium (*Methanobacterium thermoautotrophicum*). Arch. Microbiol. 113:17–22.

281. Tempest, D. W., and O. M. Neijssel. 1984. The status of Y_{ATP} and maintenance energy as biologically interpretable phenomena. Annu. Rev. Microbiol. 38:459–486.

282. Teshima, T., A. Nakaji, T. Shiba, L. Tsai, and S. Yamazaki. 1985. Elucidation of stereospecificity of a selenium-containing hydrogenase from *Methanococcus vannielii*: synthesis of (R)- and (S)-4 [H]-3,4-dihydro-7-hydroxy-1-hydroxyethylquinoline. Tetrahedron Lett. 26:351–354.

283. Thauer, R. K., G. Fuchs, and P. Scherer. 1976. The active species of "CO_2" utilized by enzymes involved in "CO_2" reduction to formate, p. 157–162, in H. Schlegel, G. Gottschalk, and N. Pfennig (eds.), Microbial production and utilization of gases (H_2, CH_4, CO). E. Goltze KG, Göttingen, West Germany.

284. Thauer, R. K., K. Jungermann, and K. Decker. 1977. Energy conservation in chemotrophic anerobic bacteria. Bacteriol. Rev. 41:100–180.

285. Thomm, M., J. Altenbuchner, and K. O. Stetter. 1983. Evidence for a plasmid in a methanogenic bacterium. J. Bacteriol. 153:1060–1062.

286. Tornabene, T. G., and T. A. Langworthy. 1978. Diphytanyl and dibiphytanyl glycerol ether lipids of methanogenic archaebacteria. Science 203:51–53.

287. Tornabene, T. G., T. A. Langworthy, G. Hölzer, and J. Oró. 1979. Squalenes, phytanes, and other isoprenoids as major neutral lipids of methanogenic and thermoacidophilic "archaebacteria". J. Mol. Evol. 13:1–8.

288. Tzeng, S. F., M. P. Bryant, and R. S. Wolfe. 1975. Factor 420-dependent pyridine nucleotide-linked formate metabolism of *Methanobacterium ruminantium*. J. Bacteriol. 121:192–196.

289. Tzeng, S. F., R. S. Wolfe, and M. P. Bryant. 1975. Factor 420-dependent pyridine

nuleotide-linked hydrogenase system of *Methanobacterium ruminantium*. J. Bacteriol. 121:184–191.

290. Van Beelen, P., R. M. de Cock, W. Guijt, C. A. G. Haasnoot, and G. D. Vogels. 1984. Isolation and identification of 5,10-methenyl-5,6,7,8,-tetrahydromethanopterin, a coenzyme involved in methanogenesis. FEMS Microbiol. Lett. 21:159–163.

291. Van Beelen, P., W. J. Geerts, A. Pol, and G. D. Vogels, 1983. Quantification of coenzymes and related compounds from methanogenic bacteria by high-performance liquid chromatography. Anal. Biochem. 131:285–290.

292. Van Beelen, P., J. F. A. Labro, J. T. Keltjens, W. J. Geerts, G. D. Vogels, W. H. Laarhoven, W. Guijt, and C. A. G. Haasnoot. 1984. Derivatives of methanopterin, a coenzyme involved in methanogenesis. Eur. J. Biochem. 139:359–365.

293. Van Beelen, P., A. P. M. Stassen, J. W. G. Bosch, G. D. Vogels, W. Guijt, and C. A. G. Haasnoot. 1984. Elucidation of the structure of methanopterin, a coenzyme from *Methanobacterium thermoautotrophicum*, using two-dimensional nuclear-magnetic-resonance techniques. Eur. J. Biochem. 138:563–571.

294. Van Beelen, P., H. L. Thiemessen, R. M. de Cock, and G. D. Vogels. 1983. Methanogenesis and methanopterin conversion by cell-free extracts of *Methanobacterium thermoautotrophicum*. FEMS Microbiol. Lett. 18:135–138.

295. Van Beelen, P., J. W. van Neck, R. M. de Cock, G. D. Vogels, W. Guijt, and C. A. G. Haasnoot. 1984. 5,10-Methenyl-5,6,7,8-tetrahydromethanopterin, a one-carbon carrier in the process of methanogenesis. Biochemistry 23:4448–4454.

296. Van Bruggen, J. J. A., C. K. Stumm, and G. D. Vogels. 1983. Symbiosis of methanogenic bacteria and sapropelic protozoa. Arch. Microbiol. 136:89–95.

297. Van Bruggen, J. J. A., K. B. Zwart, R. M. van Assema, C. K. Stumm, and G. D. Vogels. 1984. *Methanobacterium formicicum*, an endosymbiont of the anaerobic ciliate *Metopus striatus* McMurrich. Arch. Microbiol. 139:1–7.

298. Van Bruggen, J. J. A., K. B. Zwart, J. G. F., Hermans, E. M. van Hove, C. K. Stumm, and G. D. Vogels. 1986. Isolation and characterization of *Methanoplanus endosymbiosus* sp. nov., an endosymbiont of the marine sapropelic ciliate *Metopus contortus* Quennerstedt. Arch. Microbiol. 144:367–374.

299. Van der Drift, C., and J. T. Keltjens. 1986. Intermediary steps of methanogenesis, p. 62–69, in J. A. Duine and H. M. van Verseveld (eds.), Microbial growth on C_1 compounds, Martinus Nijhoff, Dordrecht.

300. Van der Meijden, P., H. J. Heythuysen, A. Pouwels, F. Houwen, C. van der Drift, and G. D. Vogels. 1983. Methyltransferases involved in methanol conversion by *Methanosarcina barkeri*. Arch. Microbiol. 134:238–242.

301. Van der Meijden, P., H. J. Heythuysen, H. T. Sliepenbeek, F. P. Houwen, C. van der Drift, and G. D. Vogels. 1983. Activation and inactivation of methanol: 2-mercaptoethanesulfonic acid methyltransferase from *Methanosarcina barkeri*. J. Bacteriol. 153:6–11.

302. Van der Meijden, P., L. P. J. M. Jansen, C. van der Drift, and G. D. Vogels. 1983. Involvement of corrinoids in the methylation of coenzyme M (2-mercaptoethanesulfonic acid) by methanol and enzymes from *Methanosarcina barkeri*. FEMS Microbiol. Lett. 19:247–251.

303. Van der Meijden, P., B. W. te Brömmelstroet, C. M. Poirot, C. van der Drift, and

G. D. Vogels. 1984. Purification and properties of methanol: 5-hydroxybenzimidazolylcobamide methyltransferase from *Methanosarcina barkeri*. J. Bacteriol. 160:629–639.

304. Van der Meijden, P., C. van der Drift, and G. D. Vogels. 1984. Methanol conversion in *Eubacterium limosum*. Arch. Microbiol. 138:360–364.

305. Van der Meijden, P., C. van der Lest, C. van der Drift, and G. D. Vogels. 1984. Reductive activation of methanol: 5-hydroxybenzimidazolylcobamide methyltransferase of *Methanosarcina barkeri*. Biochem. Biophys. Res. Commun. 118:760–766.

306. Vogels, G. D., P. van Beelen, J. T. Keltjens, and C. van der Drift. 1984. Structure and function of methanopterin and other 7-methylpterins of methanogenic bacteria, p. 182–187, in R. L. Crawford and R. S. Hanson (eds.), Microbial growth on C_1 compounds. American Society for Microbiology, Washington, D.C.

307. Vogels, G. D., and C. M. Visser. 1983. Interconnection of methanogenic and acetogenic pathways. FEMS Microbiol. Lett. 20:291–297.

308. Walther, R., K. Fahlbusch, R. Sievert, and G. Gottschalk. 1981. Formation of trideuteromethane from deuterated trimethylamine or methylamine by *Methanosarcina barkeri*. J. Bacteriol. 148:371–373.

309. Walton, N. J., and M. A. Foster, 1979. Cobalamin and methionine formation in *Paracoccus denitrificans*. FEMS Microbiol. Lett. 51:305–307.

310. Weimer, P. J., and J. G. Zeikus. 1978. One carbon metabolism in methanogenic bacteria. Cellular characterization and growth of *Methanosarcina barkeri*. Arch. Microbiol. 119:49–57.

311. Weimer, P. J., and J. G. Zeikus, 1978. Acetate metabolism in *Methanosarcina barkeri*. Arch. Microbiol. 119:175–182.

312. Weimer, P. J., and J. G. Zeikus. 1979. Acetate assimilation pathway of *Methanosarcina barkeri*. J. Bacteriol. 137:332–339.

313. Weisbach, H., A. Peterkofsky, B. G. Redfield, and H. Dickerman. 1963. Studies on the terminal reaction in the biosynthesis of methionine. J. Biol. Chem. 238:3318–3324.

314. Welty, F. K., and H. G. Wood. 1978. Purification of the "corrinoid" enzyme involved in the synthesis of acetate by *Clostridium thermoaceticum*. J. Biol. Chem. 253:5832–5838.

315. Whitman, W. B. 1985. Methanogenic bacteria p. 3–84, in C. R. Woese and R. S. Wolfe (eds.), The bacteria, Vol. 8. Academic Press, New York.

316. Whitman, W. B., E. Ankwanda, and R. S. Wolfe. 1982. Nutrition and carbon metabolism of *Methanococcus voltae*. J. Bacteriol. 149:852–863.

317. Whitman, W. B., and R. S. Wolfe. 1980. Presence of nickel in factor F_{430} from *Methanobacterium bryantii*. Biochem. Biophys. Res. Commun. 92:1196–1201.

318. Whitman, W. B., and R. S. Wolfe. 1983. Activation of the methylreductase system from *Methanobacterium bryantii* by ATP. J. Bacteriol. 154:640–649.

319. Widdowson, D., and C. Anthony. 1975. The microbial metabolism of C_1 compounds. The electron-transport chain of *Pseudomonas* AM1. Biochem. J. 152:349–356.

320. Wildgruber, G., M. Thomm, H. König, K. Ober, T. Ricchiuto, and K. O. Stetter. 1982. *Methanoplanus limicola*, a plate-shaped methanogen representing a novel family, the Methanoplanaceae. Arch. Microbiol. 132:31–36.

321. Winter, J., C. Lerp, H.-P. Zabel, F. X. Wildenauer, H. König, and F. Schindler. 1984. *Methanobacterium wolfei*, sp. nov., a new tungsten-requiring, thermophilic, autotrophic methanogen. Syst. Appl. Microbiol. 5:457–466.

322. Woese, C. R. and G. E. Fox. 1977. Phylogenetic structure of the prokaryotic domain: the primary kingdoms. Proc. Natl. Acad. Sci. USA 74:5088–5090.

323. Woese, C. R., L. J. Magrum, and G. E. Fox. 1978. Archaebacteria. J. Mol. Evol. 11:245–252.

324. Wolfe, R. S. 1971. Microbial formation of methane, p. 107–146, in A. H. Rose and J. F. Wilkinson (eds.), Advances in microbial physiology, Vol. 6. Academic Press, New York.

325. Wolfe, R. S. 1979. Methanogens: a surprising microbial group. Antonie Leeuwenhoek, J. Microbiol. 45:353–364.

326. Wolfe, R. S. 1980. Respiration in methanogenic bacteria, p. 161–186, in J. C. Knowles (ed.), Diversity of bacterial respiratory systems, Vol. 1C. CRC Press, Boca Raton, Fl.

327. Wolfe, R. S. 1985. Unusual coenzymes of methanogenesis. Trends Biochem. Sci. 10:396–399.

328. Wolfe, R. S., and I. J. Higgins. 1979. Microbial biochemistry of methane—a study in contrasts, p. 267–352, in J. R. Quayle (ed.) International review of biochemistry, Vol. 21, Microbial biochemistry. University Park Press, Baltimore.

329. Wood, A. G., A. H. Redborg, D. R. Cue, W. B. Whitman, and J. Konisky. 1983. Complementation of *arg* G and *his* A mutations of *Escherichia coli* by DNA cloned from the archaebacterium *Methanococcus voltae*. J. Bacteriol. 156:19–29.

330. Wood, H. G., H. L. Drake, and S.-I. Hu. 1982. Studies with *Clostridium thermoaceticum* and the resolution of the pathway used by acetogenic bacteria that grow on carbon monoxide or carbon dioxide and hydrogen. Proc. Biochem. Symp., Pasadena, Annu. Rev., p. 29–56.

331. Wood, J. M., I. Moura, J. J. G. Moura, M. H. Santos, A. V. Xavier, J. LeGall, and M. Scandellari. 1982. Role of vitamin B_{12} in methyl transfer for methane biosynthesis of *Methanosarcina barkeri*. Science 216:303–305.

332. Wood, J. M., and R. S. Wolfe. 1965. The formation of CH_4 from N^5-methyltetrahydrofolate monoglutamate by cell-free extracts of *Methanobacillus omelianskii*. Biochem. Biophys. Res. Commun. 19:306–311.

333. Worakit, S., D. R. Boone, R. A. Mah, M.-E. Abdel-Samie, and M. M. El-Halwagi. 1986. *Methanobacterium alcaliphilum* sp. nov., an H_2-utilizing methanogen that grows at high pH values. Int. J. Syst. Bacteriol. 36:380–382.

334. Yamazaki, S. 1982. Selenium-containing hydrogenase from *Methanococcus vannielii*. Fed. Proc. 41:888.

335. Yamazaki, S. 1982. A selenium-containing hydrogenase from *Methanococcus vannielii*. Identification of the selenium moiety as a selenocysteine residue. J. Biol. Chem. 257:7926–7929.

336. Yamazaki, S. 1983. Properties of selenium-containing hydrogenase from *Methanococcus vannielii*. Fed. Proc. 42:2267.

337. Yamazaki, S., and L. Tsai. 1980. Purification and properties of 8-hydroxy-5-deazaflavin-dependent $NADP^+$ reductase from *Methanococcus vannielii*. J. Biol. Chem. 255:6462—6465.

338. Yamazaki, S., and L. Tsai. 1981. Purification of a hydrogenase from *Methanococcus vannielii*. Fed. Proc. 40:1546.

339. Yamazaki, S., L. Tsai, and T. C. Stadtman. 1982. Analogues of 8-hydroxy-5-dea-zaflavin cofactor: relative activity as substrates for 8-hydroxy-5-deazaflavin-dependent NADP$^+$ reductase from *Methanococcus vannielii*. Biochemistry 21: 934–939.

340. Yamazaki, S., L. Tsai, T. C. Stadtman, F. S. Jacobson, and C. Walsh. 1980. Stereochemical studies of 8-hydroxy-5-deazaflavin-dependent NADP$^+$ reductase from *Methanococcus vannielii*. J. Biol. Chem. 255:9025–9027.

341. Yamazaki, S., L. Tsai, T. C. Stadtman, T. Teshima, A. Nakaji, and T. Shiba. 1985. Stereochemical studies of a selenium-containing hydrogenase from *Methanococcus vannielii*: determination of the absolute configuration of C-5 chirally labeled dihydro-8-hydroxy-5-deazaflavin cofactor. Proc. Natl. Acad. Sci. USA 82:1364–1366.

342. Zabel, H. P., H. König, and J. Winter. 1984. Isolation and characterization of a new coccoid methanogen, *Methanogenium tatii* spec. nov. from a solfataric field on Mount Tatio. Arch. Microbiol. 137:308–315.

343. Zabel, H.-P., H. König, and J. Winter. 1985. Emended description of *Methanogenium thermophilicum*, Rivard and Smith, and assignment of new isolates to this species. Syst. Appl. Microbiol. 6:72–78.

344. Zehnder, A. J. B., B. A. Huser, T. D. Brock, and K. Wuhrmann. 1980. Characterization of an acetate-decarboxylating, non-hydrogen-oxidizing methane bacterium. Arch. Microbiol. 124:1–11.

345. Zehnder, A. J. B., and K. Wuhrmann. 1977. Physiology of a *Methanobacterium* strain AZ. Arch. Microbiol. 111:199–205.

346. Zeikus, J. G. 1977. The biology of methanogenic bacteria. Bacteriol. Rev. 41:514–541.

347. Zeikus, J. G. 1983. Metabolism of one-carbon compounds by chemotrophic anaerobes. Adv. Microb. Physiol. 25:215–299.

348. Zeikus, J. G., R. Kerby, and J. A. Krzycki. 1985. Single-carbon chemistry of acetogenic and methanogenic bacteria. Science 227:1167–1173.

349. Zhilina, T. N. 1983. A new obligate halophilic methane-producing bacterium. Mikrobiologiya 52:375–382.

350. Zhilina, T. N., and S. A. Ilarionov. 1984. Isolation and comparative characteristics of methanogenic bacteria assimilating formate with the description of *Methanobacterium thermoformicicum* sp. nov. Mikrobiologiya 53:785–790.

351. Zhilina, T. N., and G. A. Zavarzin. 1979. Comparative cytology of Methanosarcinae and description of *Methanosarcina vacuolata* n. sp. Mikrobiologiya 48:279–285.

352. Zinder, S. H., and M. Koch. 1984. Non-aceticlastic methanogenesis from acetate: acetate oxidation by a thermophilic syntrophic coculture. Arch. Microbiol. 138:263–272.

353. Zinder, S. H., and R. A. Mah. 1979. Isolation and characterization of a thermophilic strain of *Methanosarcina* unable to use H_2-CO_2 for methanogenesis. Appl. Environ. Microbiol. 38:996–1008.

354. Zinder, S. H., K. R. Sowers, and J. G. Ferry. 1985. *Methanosarcina thermophila* sp. nov., a thermophilic, acetotrophic, methane-producing bacterium. Int. J. Syst. Bacteriol. 35:522–523.

14

PRINCIPLES AND LIMITS OF ANAEROBIC DEGRADATION: ENVIRONMENTAL AND TECHNOLOGICAL ASPECTS

BERNHARD SCHINK

Lehrstuhl Mikrobiologie I, Eberhard-Karls-Universität, D-7400 Tübingen, Federal Republic of Germany

14.1 INTRODUCTION

14.2 MICROBIAL DEGRADATION REACTIONS IN THE PRESENCE AND ABSENCE OF OXYGEN

  14.2.1 Chemistry of Oxygen and Oxygenase Reactions

  14.2.2 Degradative Reactions in the Absence of Oxygen

14.3 ANAEROBIC DEGRADATION OF UNCOMMON SUBSTRATES AND THE ANAEROBIC DEGRADATION POTENTIAL

  14.3.1 Degradation of Aliphatic Hydrocarbons

  14.3.2 Degradation of Alcohols and Ketones

  14.3.3 Degradation of Aromatic Compounds and the Lignin Problem

  14.3.4 Cleavage of Ether Linkages

  14.3.5 Dehalogenations

  14.3.6 Elimination of Nitrogen Substituents and N-Dealkylations

  14.3.7 Anaerobic Cometabolism

14.4 THE ENVIRONMENT'S NATURE: CHANCE AND CHALLENGE

  14.4.1 Surfaces

  14.4.2 Spatial and Temporal Heterogeneity

  14.4.3 Chemical and Physiological Interactions

14.5 CARBON AND ELECTRON FLOW IN METHANOGENIC
       DEGRADATION
14.6 METHANOGENIC WASTE TREATMENT
       14.6.1   Aerobic versus Anaerobic Waste Treatment
       14.6.2   Reactor Design
       14.6.3   Microbiological Aspects: Films and Flocs
       14.6.4   Examples of Application
       14.6.5   Process Manipulation for Efficiency Enhancement
       14.6.6   Further Processing of Anaerobically Treated Wastewater
       14.6.7   Digestion of Solid Wastes
       14.6.8   Feasibility of Methanogenic Waste Treatment
14.7 NONMETHANOGENIC FERMENTATIONS
       14.7.1   Mass Production of Alcohols and Other Solvents
       14.7.2   Anaerobic Production of Organic Acids
       14.7.3   Further Products of Anaerobic Metabolism
14.8 ENVIRONMENTAL IMPLICATIONS OF ANAEROBIC DEGRADATION
14.9 CONCLUSIONS
REFERENCES

14.1 INTRODUCTION

Green plants, algae, and cyanobacteria are the primary producers on earth which reduce annually about 1.3×10^{17} g of carbon as carbon dioxide to form organic matter. This enormous photosynthetic activity is counteracted by degradative processes, including consumption by animals and final mineralization by lower fungi and bacteria, and to some degree, by the plants themselves. The degradation of organic matter ends primarily with CO_2, H_2S, and NH_3 as products, and thus recycles the mineral substrates necessary for photosynthesis. However, photosynthesis and degradation do not operate at exactly the same annual rates. Vast amounts of biogenic reduced carbon compounds (about 6×10^{21} g; figures after Refs. 27 and 270) are buried in sediments, rocks, coal, peat, oil, and natural gas, whereas the standing crop of living plants (about 99% of total biomass) amounts to only about 7.4×10^{17} g of bound organic carbon. The resources of biogenically reduced fossil carbon in the earth's crust correlate to a total reserve of photosynthetically produced molecular oxygen of 2.8×10^{22} g in the earth's atmosphere and the ocean waters.

This amount of molecular oxygen ensures aerobic degradation for the overwhelming proportion of the organic matter produced annually. Anaerobic degradation processes are restricted to a few niches, such as sediments and isolated water bodies of lakes and oceans, and the gastrointestinal tracts of animals and man. Nevertheless, anaerobic degradation processes are of basic interest for an understanding of the cycling of organic matter in nature and the various processes and organisms involved.

Because of its low solubility and diffusibility in water, oxygen transport through a water body depends nearly exclusively on convection due to physical mixing processes (e.g., eddy diffusion). As soon as mixing is impeded due to thermal stratification or prevented by particle accumulation in a sediment layer, oxygen will rapidly become limiting. If organic matter is present in excess, anaerobic degradation takes over in a specific sequence of events which is dealt with in more detail in Chapter 1 of this book. Sediments are the main metabolizing organs of our freshwater bodies. With increasing pollution of rivers and lakes and human-enhanced eutrophication of surface waters, oxygen deprivation in lower water layers will expand and render anaerobic degradation more and more important, especially in densely populated areas of the world.

Moreover, a deeper insight into microbial ecology in the recent past has revealed that most apparently oxygen-saturated environments, such as oxygenated water bodies, soil particles, and even the human skin are inhabited by strictly anaerobic bacteria which apparently thrive in these habitats in close connection with aerobes and profit from their respiratory activity as a means of protecting them from toxic reduced-oxygen species (191). Thus an apparently aerobic environment may contain innumerable anoxic microniches which contribute significantly to its overall metabolism. The gastrointestinal tract of higher animals represents another anaerobic environment which is maintained in an oxygenic atmosphere by a well-organized cooperation of aerobic and anaerobic partners. The latter case also exemplifies how little is really known about the relationship and mutual benefit of aerobes and anaerobes in such a heterogeneous system.

Although the importance of understanding anaerobic processes in our environment increases, our knowledge of this matter is scarce compared to that which accumulated on aerobic metabolism during the past century. Reliable information exists only on the physiology and metabolic activities of numerous strains of anaerobic bacteria that could be isolated in pure culture. However, nearly nothing is known about the interrelationship of various metabolic types of anaerobes in a natural ecosystem except that anaerobes depend on symbiotic cooperations to a by far larger degree than do their aerobic counterparts. Studies on the ecology of anaerobes in nature continually provide new and unexpected insights into the complex network of substrate and energy transfer in an anoxic environment.

The lack of information on anaerobes as compared to aerobes is basically due to two reasons. First, the cultivation of strictly anaerobic bacteria needs some special laboratory techniques which are not difficult to use, but depend on some basic equipment and the availability of oxygen-free gases. Reliable methods for cultivation of fastidious anaerobes were developed only about 20 years ago by R. S. Hungate and his school (117). Second, most anaerobic bacteria grow only slowly compared to their aerobic counterparts, and studies on their metabolic activity, nutrition, physiology, and genetics require a lot more time and patience than comparable studies with aerobes. An experiment that requires 2 days to complete with *E. coli* can easily take several weeks or months for a methanogenic bacterium.

The degradative potential of microbes is one of the most intriguing phenomena in nature. The diversity of known naturally synthesized compounds is still increasing, and the secondary metabolism of plants and fungi seems especially to be a never-ending source of bizarre chemical structures. Since none of these numerous creations of nature's retort seems to have accumulated to significant amounts during the earth's history, we have to assume that a suitable microbe and a biochemical pathway exist for the degradation of every natural compound, no matter how unusual. The perfection with which nature operates in this respect has been termed the "principle of biological infallibility." However, strictly speaking, this principle is true only for the microbial degradative potential in the presence of molecular oxygen. Large accumulations of coal, oil, and oil shales testify that biogenic organic matter could sometimes be preserved in earth's history over comparably long periods. It is generally accepted that lack of oxygen prevented degradation of these fossil remnants of biomass; however, this may not be the only reason. Nevertheless, the real limits of the degradative capacity of anaerobes are not easy to define, and the literature is full of contradictory reports on the "ifs" and "hows" of, for example, anaerobic hydrocarbon degradation. The lack of suitable techniques for the cultivation of strictly anaerobic bacteria before the 1970s caused considerable confusion about their metabolic capacities. The elder literature often reported on mixed cultures which probably used traces of the oxygen present in experimental flasks for primary attack on substrates which, in our opinion today, are recalcitrant under strictly anaerobic conditions. The use of reliable cultivation methods together with highly purified substrate preparations narrowed the range of substrates open to anaerobic attack in the 1960s and 1970s. This trend together with the well-known problems of cultivating anaerobes and their apparent slow growth, have helped to establish the widespread opinion that anaerobic degradation is always inferior to aerobic degradation and not suited for biotechnological processes such as waste treatment.

A change in this view was brought about by a political event: As a result of the so-called oil crisis in 1973, it became obvious to everyone, including politicians, that energy in the form of combustible fossils is limited on earth. The rising oil prices caused scientists and engineers to look for other sources of energy and chemical feedstocks. The fascinating economy of anaerobic energy metabolism and the possible application of fermentation products (methane, ethanol, solvents, acids) from waste materials revived the interest of science and science politics in anaerobic microbiology and biotechnology. As a consequence, the last decade has advanced to a considerable extent our knowledge of the microbiology and biochemistry of anaerobic metabolism. The discovery of interspecies hydrogen transfer in methanogenic ethanol degradation (45) opened the door to an understanding of the complex cooperation of bacteria in an anoxic environment and its bioenergetic implications. The consequences of this approach are even now, many years later, not yet fully realizable, with respect to fermentation intermediates other than hydrogen.

In this chapter I want to summarize our present understanding of anaerobic metabolism from an environmentalist's point of view. I also want to outline how such an understanding can contribute to improvements in the biotechnological application of anaerobic metabolic capacities. Consequently, we proceed from basic microbiology via ecology to biotechnology with the focus on the biological rather than the technological point of view. It is not intended to provide a complete review of every aspect of the applied microbiology of anaerobes, and the picture drawn will have blanks in those areas remote from my experience and research interest as a biologist. Nevertheless, I hope to provide some insight into the underestimated metabolic capacity of anaerobes and their promising potential in the welfare of nature and humankind.

14.2 MICROBIAL DEGRADATION REACTIONS IN THE PRESENCE AND ABSENCE OF OXYGEN

For an understanding of many problems of anaerobic degradation processes it appears necessary to summarize those reactions in which molecular oxygen is directly or indirectly involved and which cannot, therefore, take place in its absence.

Oxygen serves two different functions in the degradation of organic matter: that of a terminal acceptor of electrons which are released during oxidation of organic carbon, and that of a reactant in a primary attack on the substrate molecules themselves. Whereas the first function may be transferred in the absence of oxygen to other oxidized compounds, such as nitrate, metal ions, sulfate, or carbon dioxide [although with smaller energy gains (see Chapter 1)], there is no equivalent to oxygen which could fulfill its functions as a reactant in the primary transformation of some important kinds of substrates. For a comprehension of this unique role of oxygen, a short characterization of its chemistry and of the biochemical reactions involving molecular oxygen follows.

14.2.1 Chemistry of Oxygen and Oxygenase Reactions

The metabolic importance of oxygen is due basically to its high oxidation potential ($E'_0 = +0.81$ V) and its biradical character combined with relative kinetic inertness (168). Therefore, its oxidation potential, unlike that of fluorine or chlorine, can, at physiological temperatures, be brought into play only by the presence of a suitable catalyst. Introduction of external electrons into the π-orbital of oxygen to form O_2^- and O_2^{2-} weakens the $O—O$ bond considerably. Electron transfer from, for example, a transition metal catalyst activates the oxygen molecule and produces a complex-bound superoxide radical of high reactivity. This reaction can be a spontaneous process as, for example, with flavoproteins, which produces a highly reactive and therefore toxic superoxide molecule. This is the molecular basis of oxygen toxicity in anaerobes which needs to be overcome by protection devices such as superoxide dismutase and catalase enzymes (84). On the other hand, the high reactivity of superoxides, peroxides, and hydroxyl radicals appears to be a prerequisite for efficient degradation of inert complex organic materials, such as lignin (see Section 14.3.3).

The enzymes attacking comparably inert substrates by insertion of oxygen atoms into the substrate molecule itself are termed oxygenases. Depending on whether only one or both atoms of an oxygen molecule are inserted, the enzyme is classified as a monooxygenase (hydroxylase, mixed function oxygenase) or a dioxygenase. Monooxygenases react with aliphatic or aromatic substrates and introduce a hydroxyl substituent into the substrate molecule while the other oxygen atom is reduced to water (Fig. 14.1a and b). Dioxygenases react with aromatic substrates lacking a hydroxyl substituent in a suitable position and form orthodiols (Fig. 14.1c). Cleavage of aromatic rings is catalyzed by a further type of dioxygenases and leads to an unsaturated dioic acid (Fig. 14.1d). The reaction mechanisms and the chemistry of oxygen activation by these enzymes have been reviewed recently (168, 281). Monooxygenases of the cytochrome P450 type have rather broad substrate spectra and catalyze hydroxylations of carbon residues as well as N-oxidations, sulfoxidations, epoxidations, dealkylations, peroxidations, desulfurations, dehalogenations, and reduction reactions such as reduction of azo groups, nitro groups, N-oxides, or epoxides. The substrate is first bound to a ferric

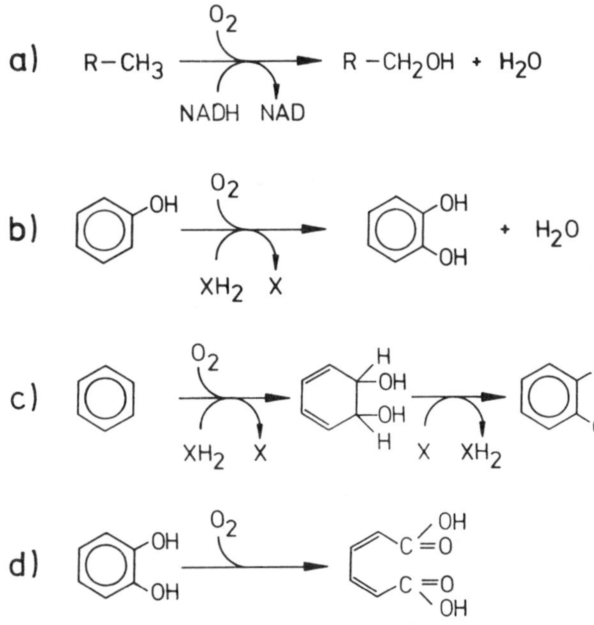

Figure 14.1 Mode of action of oxygenases. (*a*) Hydroxylation of a methyl group by a monoxygenase. (*b*) Insertion of a second hydroxyl substituent into a phenol molecule by a monooxygenase. (*c*) Transformation of an aromatic hydrocarbon into an ortho-diol by a dioxygenase. (*d*) Cleavage of an aromatic ortho-diol by a dioxygenase. After Ref. 224.

iron atom which is reduced to ferrous iron by pyridin nucleotide electrons. Oxygen is bound by the same center, and both water and the hydroxylated substrate molecule are released in one step. Much less is known about the reaction mechanism of other monooxygenases, such as tyrosinase, which contains copper instead of heme-bound iron. In a further type of monooxygenases, oxygen is bound to a flavin prosthetic group as, for example, in p-hydroxybenzoate hydroxylase. Dioxygenases always contain iron in the active center. This may be heme-bound and undergo valency change, as in tryptophane dioxygenase or indolamine deoxygenase, and oxygen binds to the metal itself. In the ring-cleaving protocatechuate 3,4-dioxygenase, a nonheme ferric iron appears to react with a semiquinone form of the substrate which is formed by direct interaction with the oxygen molecule.

Thus there is no general mechanism of oxygenase reaction, and nearly every type of oxygenase enzyme profits from the biradicalic character of oxygen and its high oxydation potential by a different way of activation. It is evident from this short survey that neither sulfate, nitrate, nor carbonate, none of which contain unpaired electrons, could substitute for oxygen in the activation of inert molecules in reactions similar to those described above.

14.2.2 Degradative Reactions in the Absence of Oxygen

The versatility of oxygenase enzymes, especially of the monooxygenases of the cytochrome P450 type, allows a broad range of reactions to be catalyzed in the presence of oxygen. The range of reactions possible in the absence of oxygen is rather restricted. Of the long list of reaction types observed in natural environments [compiled by Menzie (177)], only a few can be catalyzed in the absence of oxygen: namely, hydrogenations, dehydrogenations, hydrations, dehydrations, hydrolyses, condensations, and, in some cases, photoreactions. I should also mention carboxylations, decarboxylations, and lyase reactions and coenzyme B_{12}–dependent rearrangements of a carbon skeleton. In the following section we see how much these reactions can accomplish in anaerobic degradation of some important classes of substrates, and where the limits of anaerobic degradation may be expected.

14.3 ANAEROBIC DEGRADATION OF UNCOMMON SUBSTRATES AND THE ANAEROBIC DEGRADATION POTENTIAL

In this section I compare aerobic and anaerobic degradation paths of some classes of substrates which have long been considered to be degradable only in the presence of oxygen. It turns out that many of these substrates are also degradable without the participation of molecular oxygen, but that the reaction mechanisms are different. The necessary question concerning the limits of anaerobic degradation will be asked again, and answers will follow on the basis of experimental results as well as geochemical and environmental analyses.

14.3.1 Degradation of Aliphatic Hydrocarbons

Anaerobic degradation of hydrocarbons has been a matter of dispute among biochemists, microbiologists, and environmental chemists for several decades. Conversion of saturated hydrocarbons to methane and carbon dioxide is thermodynamically possible, as the following calculation for ethane demonstrates (calculation after Ref. 258):

$$8C_2H_6 + 6H_2O \rightarrow 14CH_4 + 2HCO_3^- + 2H^+$$
$$\Delta G°' = -34.4 \text{ kJ} + \text{mol ethane}^{-1}$$

This reaction becomes even more exergonic with increasing chain length (86). Nonetheless, no conclusive evidence exists for net degradation of a saturated hydrocarbon in the absence of molecular oxygen.

Anaerobic oxidation of methane with sulfate as electron acceptor has been postulated on the basis of geochemical data, but except for a small backward oxidation of radiolabeled methane by methanogenic bacteria concomitant with methane formation (293), no reliable data exist to substantiate this postulate. The problem is dealt with in detail in Chapter 12 and does not need further evaluation here.

Anaerobic degradation of other aliphatic hydrocarbons was published repeatedly in the literature before 1960 (179, 183, 206, 304). However, the experiments reported either dealt with not completely anoxic cultures or with degradation of added emulsifiers (47) or substrate impurities (305). That aerobic and anaerobic bacteria can cooperate in petroleum degradation in the presence of traces of oxygen was shown in a recent publication (131). An often quoted paper (64) reports anaerobic degradation of methane, ethane, and n-octadecane by a *Desulfovibrio desulfuricans* culture which, however, also utilizes glucose and appears to be an impure culture. The amounts of substrate metabolized in these experiments are extremely small and can easily be attributed to tracer impurities and oxygen traces in the cultures. The general opinion about microbial degradation of saturated hydrocarbons is that it depends on the presence of molecular oxygen (7, 83, 86, 92, 188). This opinion agrees with the assumption that the primary attack on the saturated hydrocarbon molecule has to be catalyzed by an oxygenase enzyme. Reports of desaturative attack to form a terminal double bond have been published (56, 121, 227); however, they were probably misinterpretations of the monooxygenase reaction mechanism (9).

The situation becomes different if there is at least one double bond in the hydrocarbon molecule. Hydration of this double bond would form an alcohol which, depending on its position, could be oxidized to a ketone or, via an aldehyde, to the corresponding fatty acids. Thermodynamically, methanogenic degradation becomes more exergonic as, for example, with ethene:

$$2C_2H_4 + 3H_2O \rightarrow 3CH_4 + HCO_3^- + H^+$$
$$\Delta G°' = -102.1 \text{ kJ} \cdot \text{mol ethene}^{-1}$$

and the free energy change is again higher with an alkine, such as acetylene:

$$4C_2H_2 + 9H_2O \rightarrow 5CH_4 + 3HCO_3^- + 3H^+$$
$$\Delta G^{\circ\prime} = -208.7 \text{ kJ} \cdot \text{mol ethine}^{-1}$$

Experiments performed in the author's lab (217) demonstrated that 1-hexadecene was completely converted to methane and carbon dioxide by enrichment cultures in the absence of oxygen. The hydrocarbon skeleton was apparently hydrated to 1-hexadecanol and further degraded by β-oxidation via acetyl residues (Fig. 14.2). Thus a single double bond appears to be sufficient for complete anaerobic degradation of a linear hydrocarbon. The degradation pathway is different from that used by aerobic microorganisms which would either epoxidize the double bond to form a 1,2-diol via the epoxide (43, 119), or start from the saturated end with a monooxygenase reaction (123, 244).

Shorter hydrocarbons were less effectively degraded in anaerobic enrichment cultures, and ethene itself was not found to be degraded. Incomplete methanogenic degradation was found with squalene, a highly unsaturated branched hydrocarbon which is a constituent of bacterial membrane lipids. A degradation pathway that involves hydrations, dehydrogenations, and carboxylations analogous to reactions reported for squalene and branched-chain fatty acid metabolism by some aerobic bacteria (228, 229) was suggested (217). Since about half of the maximum amount of methane calculated was recovered, branching of the hydrocarbon skeleton was probably not the only reason for the incomplete degradation observed. It appears more probable that at least some of the double bonds were reduced by hydrogen

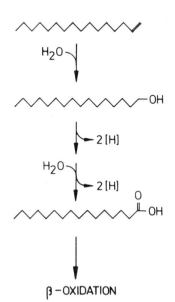

β – OXIDATION

Figure 14.2 Hypothetical pathway of anaerobic hexadecene degradation.

equivalents from substrate oxidation. The resulting saturated branched hydrocarbon probably resisted any further degradation. Unfortunately, the residual degradation intermediates can not yet be identified, and further speculation on the degradation mechanism is premature. Nonetheless, anaerobic degradation of branched hydrocarbons with subterminal double bonds appears to present much more difficulty for anaerobic degradation than, for example, 1-hexadecene.

Anaerobic sulfate-dependent oxidation of an alkine, acetylene, was observed in estuarine sediment samples (60). The authors were able to maintain an enrichment culture with acetylene as the sole source of carbon and energy. Pure cultures of fast-growing acetylene-fermenting anaerobes were isolated in the author's lab (217a). Ethanol and acetate were the only products'detected and were probably formed by disproportionation of acetaldehyde after an initial hydration of acetylene. Hydration was also reported to be the primary reaction in aerobic acetylene degradation (65).

The results discussed so far allow the conclusion that some unsaturated hydrocarbons, unlike saturated ones, have been found to be degraded to a varying extent in enrichment and defined cultures in the complete absence of molecular oxygen. Terminally unsaturated substrates were degraded best, whereas subterminal double bonds and branching of the substrate molecule restricted anaerobic biodegradability. The reaction mechanisms involved were different from those of aerobic degradation of the same substrates: Instead of oxygenase reactions, only hydrations, dehydrogenations, and probably carboxylations were implied. It should be noted that the anaerobic degradation processes described are slow compared to common microbiological growth experiments: whereas the acetylene-fermenting bacteria have doubling times of a few hours, degradation of hexadecene and squalene requires several weeks to be completed, and the primary enrichment cultures had to be incubated for 3 to 4 months until methane formation was observed.

Enrichments with other unsaturated hydrocarbons, such as β-carotene and styrene, were unsuccessful in our lab, but this should not be taken as a proof that these substrates are not degradable in the absence of oxygen. Different cultivation conditions, including pH, temperature, sulfide content, trace elements provided, and so on, could allow successful degradation, no matter to what extent. Our enrichment media did not contain sulfate or nitrate as electron acceptors, and at least intermediately formed nitrite might be helpful in anaerobic attack on hydrocarbon substrates.

The relative instability of unsaturated hydrocarbons compared to saturated ones in the absence of oxygen is reflected in the results of chemical analyses of anoxic sediment contents and sedimentary rock strata. Saturated hydrocarbons are usually found in by far greater amounts than unsaturated ones, and the ratio increases with depth and age. Although this could also be due to reduction of double bonds by microbial activities as observed with fatty acids (202), this distribution is in agreement with the results of the laboratory experiments mentioned above. Saturated hydrocarbons also prevail in kerogen and the hydrocarbon fraction of Precambrian rocks (176). Aside from linear hydrocarbons, these fractions are rich in branched-chain and cyclic aliphatic hydrocarbons, which probably originated from the isoprenoid components of chlorophylls (172) and the isoprenoid lipids of archaebacteria (265). Also, carotenoids were found to be very stable: Only slight modifications due to partial reductions were observed in 340,000-year-old sediment samples (276), and the same seems to be true for steroids and hopanoids (93, 172). The stability of carotenoids in sediments allows their use in tracing back the history of surface waters (306). Sediment studies recently gave hints of anaerobic degradation of a saturated hydrocarbon, *n*-heptadecane (94); however, this result awaits further microbiological examination. For the present it can be concluded that saturated hydrocarbons are stable in the absence of oxygen, and that this is also true for many branched-chain unsaturated hydrocarbons. These conclusions agree basically with the results of enrichment experiments.

14.3.2 Degradation of Alcohols and Ketones

Anaerobic degradation of primary aliphatic alcohols is a well-known process. First studies on ethanol degradation go back to Omelianski (184), who enriched a methanogenic mixed culture from rabbit intestinal contents. Barker (17) studied methanogenic ethanol degradation with a similar culture which was later identified as an obligately syntrophic association of an alcohol-oxidizing and a hydrogen-scavenging methanogenic bacterium (45). Oxidation to acetate with concomitant reduction of either sulfate (see Chapter 10) or protons and interspecies hydrogen transfer to methanogens (see Chapter 9) are the most important paths of ethanol degradation in anoxic habitants. Recently, methanogenic bacteria were isolated which can oxidize some primary alcohols with concomitant methane formation in pure culture (281a). Other pathways, such as homoacetogenic acetate formation (39, 73), butyrate formation (18), and propionate formation (153, 208, 214), are probably only of minor environmental importance (see Section 14.5).

Degradation of 1,2-diols by pure cultures of anaerobes proceeds via dehydration to the corresponding aldehydes and further fermentation to fatty acids and primary alcohols (1, 87, 234) or syntrophic oxidation to the fatty acids. The diol dehydrase enzyme contains a B_{12}-corrinoid prosthetic group (264). Several diol-fermenting bacteria were recently isolated which could also operate as partners of methanogens in the syntrophic oxidation of primary alcohols to fatty acids (214, 220), an activity which is possibly their main role in their natural habitats. 2,3-Butanediol is cleaved to two C_2-units which are subsequently fermented to acetate, ethanol, or propionate (214, 215). Anaerobic degradation of aliphatic 1,3- and 1,4-diols has not yet been studied.

The known fermentations of glycerol lead to either 1,3-propanediol or propionate as reduced end products. Recently also, sulfate-reducing bacteria were isolated which can combine the oxidation of glycerol to acetate with the reduction of sulfate or with syntrophic hydrogen transfer to methanogens (241a). Also, some homoacetogenic bacteria convert glycerol to acetate as the sole product (73).

Oxidation of secondary alcohols leads to the corresponding ketones. This reaction has so far not been demonstrated with pure cultures. Two species of methanogenic bacteria were recently isolated which combine isopropanol oxidation with methane formation (281a).

The further fate of ketones is largely unknown. Acetone is aerobically converted to acetol (2-oxopropanol-1), probably by the action of a monooxygenase (254). Anaerobic methanization of acetone was observed in undefined enrichment cultures (173, 283), but the organisms involved have never been identified. Methanogenic enrichment cultures in the author's laboratory convert acetone to methane and carbon dioxide via an intermediate carboxylation of acetone to form two acetate residues (197a). This culture represents a new type of syntrophic bacterial association in which acetate is the only metabolite transferred.

The anaerobic degradation of ketones also needs further study with respect to their possible involvement in anaerobic degradation of unsaturated branched-chain hydrocarbons. It should be mentioned that the carboxylation reaction described here as a way of acetone activation depends on the presence of a hydroxyl- or oxo function at the β-carbon atom. Such a reaction cannot occur with saturated hydrocarbons.

14.3.3 Degradation of Aromatic Compounds and the Lignin Problem

In the minds of microbiologists, degradation of aromatic compounds is so intimately associated with the action of dioxygenases that anaerobic degradation of these compounds was long believed to be impossible. Today, this assumption holds true only for aromatic hydrocarbons such as benzene and its higher homologs: naphthalene, anthracene, naphthacene, phenanthrene, and so on. Once the aromatic ring carries any substituent function, anaerobic degradation appears to be possible in many cases.

Methyl substituents do not significantly influence the mesomeric π-electron system of the benzene nucleus. Nonetheless, anaerobic degradation of toluene with nitrate as electron acceptor was recently achieved in enrichment cultures (Zehnder, personal communication). The degradative mechanism is unknown, but participation of nitrite formed as an intermediate of nitrate reductions seems likely. Toluene, xylene, and styrene (ethenylbenzene) were not degraded in methanogenic enrichment cultures in the author's laboratory, but evidence was recently reported that these compounds can be transformed anaerobically (97a, 151a).

Benzoate is comparably rapidly converted to methane and carbon dioxide in suitable enrichment cultures, and our present knowledge of the anaerobic degradation of aromatics is based mostly on studies with this compound. The subject has been reviewed recently (74, 288) and is also discussed in detail in Chapter 7. Therefore, few key reactions will be emphasized here, which appear to be basically similar in nitrate-dependent, phototrophic and syntrophic methanogenic cultures. Benzoate is first activated to the coenzyme A-bound derivative. The ring is assumed to be saturated in a one-step reduction reaction which transfers six electrons to form cyclohexane carboxylate. The subsequent dehydrogenation and hydration reactions are basically analogous to the well-known β-oxidation of fatty acids and end up with either the pimelyl or adipyl residue, depending on whether or not the *o*-oxycyclohexanecarboxylate intermediate is decarboxylated. This is the now generally accepted pathway of anaerobic benzoate degradation, which, however, has never been substantiated by enzymological experiments.

The pathways of anaerobic degradation of the higher benzoate homologs (phenylacetate, phenylpropionate) have not yet been studied in detail. Phenylpropionate is probably cleaved to acetate and benzoate, whereas β-oxidation of the saturated phenylacetate derivative would form cyclohexanone as an intermediate (15).

The phthalic acids (benzenedicarboxylic acids) are not naturally occurring compounds but important industrial products. The phthalic acid alkyl esters are used as plasticizing constituents of synthetics. Aerobic degradation of phthalates and their alkyl esters was repeatedly reported in the recent past (21, 249). Whereas *o*-phthalate is degraded via benzoate by decarboxylation which involves a still obscure redox reaction (253), isophthalate (*m*-phthalate) was shown to be oxidatively decarboxylated by an oxygenase reaction yielding protocatechuate (10). Anaerobic degradation of phthalates has so far been observed only with nitrate-reducing bacteria and proceeds via decarboxylation to benzoate (2, 3).

Phenols and their carboxylated derivatives (hydroxybenzoates) take rather different paths for anaerobic degradation depending on the kind, number, and position of the respective substituents. Degradation experiments have so far been carried out with methanogenic and nitrate-reducing cultures, and the results do not yet provide a coherent picture of anaerobic degradation of this important group of natural and synthetic compounds. A tentative scheme of the degradation paths and their interrelationships is presented in Fig. 14.3, which is based on results from the author's laboratory and from other authors. It should be noted that the hydroxyl substituent has a strong inductive effect on the π-electron system and thus affects the symmetry and stability of the aromatic ring structure much more than, for example, a methyl or carboxyl substituent. This inductive effect stabilizes the quinoid tautomers of the respective phenols to a considerable extent, and reduction to the hypothesized saturated oxocyclohexanes may take shorter paths than with benzoate, as Fig. 14.3 indicates. It is also interesting to note that the preference for anaerobic degradation among the mono-and bivalent phenols is contrary to that for aerobic degradation: Whereas aerobically catechol and hydroquinone are degraded much more rapidly than resorcinol, resorcinol and phenol were the preferred substrates for degradation in anaerobic sludge enrichments. Catechol and hydroquinone required three to five times more time for complete degradation in enrichment cultures than the former two. The slow anaerobic degradation of catechol and its carboxyl derivatives has supported the assumption that these compounds are anaerobically recalcitrant (133, 134).

These observations can to some extent be explained by the difference of degradation pathways involved. Phenol is supposed to be degraded via cyclohexanol and cyclohexanone in a reaction sequence analogous to benzoate degradation (14,

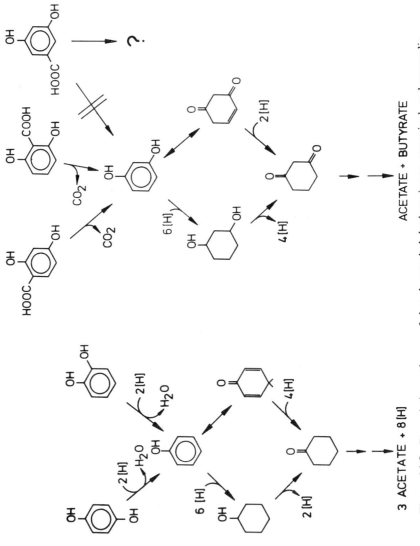

Figure 14.3 Degradation pathways of phenol, catechol, hydroquinone, resorcinol, and resorcylic acids. Explanations are given in the text.

16), although the initial reduction reaction is still an enigma, as is the activation for the assumed hydrolytic cleavage. In the absence of light, nitrate, or sulfate, phenol degradation depends on the presence of hydrogen-scavenging methanogenic bacteria in a manner similar to benzoate degradation (79). This is also true for catechol and hydroquinone degradation, and neither one has so far been studied in defined methanogenic cultures. Both catechol and hydroquinone are converted to phenol by reductive dehydroxylation (16, 251). This is a very interesting and so far not sufficiently investigated reaction which was observed to occur also with other hydroxylated aromatics (28, 62, 187, 212, 226). It is tempting to speculate that the dehydroxylating bacterium may be different from the ring-cleaving one and may thus profit from the overall degradation reaction only marginally. This would explain why hydroquinone- and catechol-degrading bacteria are much more difficult to enrich than are phenol-degrading bacteria. In both catechol and hydroquinone the position of the hydroxyl substituents relative to each other is not suited for β-oxidation of the carbon skeleton, and that is why the dehydroxylations above may have to occur.

With resorcinol the situation is different because the second hydroxyl function is in a suitable position for β-oxidation. Defined resorcinol-degrading cultures were studied in the author's laboratory and were found to use 2,4-dihydroxybenzoate and 2,6-dihydroxybenzoate also but no other hydroxy- or carboxybenzenes. Resorcinol and its derivatives were fermented to acetate and butyrate without intermediary phenol formation and without interspecies hydrogen transfer or reduction of external electron acceptors. The two carboxy resorcinols mentioned are probably transformed to resorcinol by decarboxylation (Fig. 14.3). This reaction does not appear to be possible with 3,5-dihydroxybenzoate, which takes a different, so far unknown method of degradation (268a).

The phenol carboxylates (hydroxybenzoates) can either be decarboxylated to the corresponding phenols and further metabolized like those, or dehydroxylated and degraded via benzoate. With monohydroxybenzoates, both pathways exist, and the relative position of the two substituents determines which pathway is used (268b).

Degradation of the trihydroxybenzenes and their carboxylated derivatives is comparably easy to understand. Gallic acid and 2,4,6-trihydroxybenzoate are first decarboxylated to the corresponding trivalent phenol, pyrogallol, or phloroglucinol. Pure cultures of strictly anaerobic bacteria have been isolated which ferment the carboxylated and decarboxylated trivalent phenols completely to acetate and carbon dioxide as sole products (218). Pyrogallol degradation involves a transhydroxylation to phloroglucinol (208a), which should be easily accessible to hydrolysis in its tautomeric trioxocyclohexane form (Fig. 14.4). The postulated 1,4-transhydroxylation is another reaction of basic interest which deserves further investigation. In this context it is interestng to note that the third trihydroxybenzene, hydroxyhydroquinone, is not degraded by these cultures: This compound cannot be transformed to phloroglucinol by 1,4-transhydroxylation. Other anaerobes degrading phloroglucinol were isolated (266); however, the fermentation balances

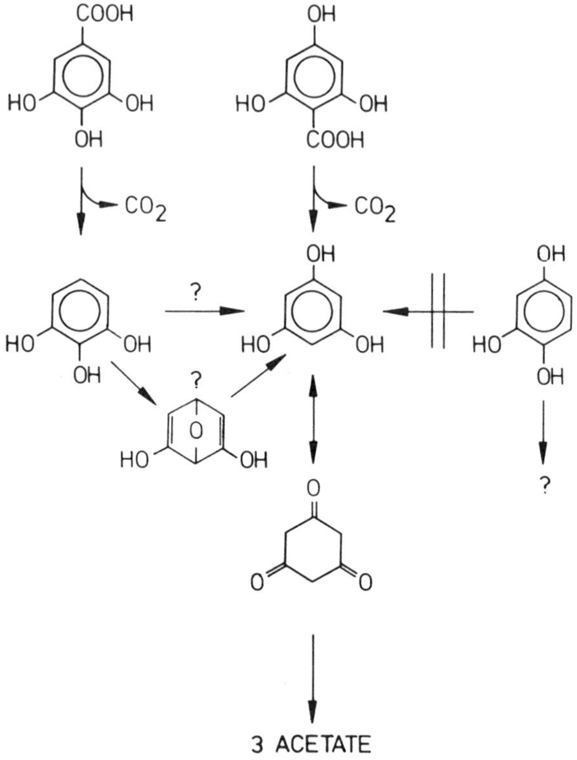

Figure 14.4 Hypothetical degradation pathways of trihydroxybenzenes and their carboxyl derivatives. Explanations in the text.

reported do not allow conclusions as to whether degradation was purely fermentative or depended on external electron acceptors (267).

Discussion of the environmental importance of anaerobic phenol and hydroxybenzoate degradation has long been based only on the question of anaerobic lignin degradation. In my opinion, this argument has needlessly narrowed the importance of phenol degradation considerably. Plant matter is full of mono- and oligonuclear phenolic substances, including flavonoids, tannins, and precursors of lignin biosynthesis, as well as phytoalexins and their precursors (199). The browning of an apple bite surface makes this evident. Moreover, all lignins are not alike. The lignin contents of hardwoods are much higher than those of weeds, and due to its polymerization mechanism, the texture density of the former is much higher than that of the latter. Anaerobic degradation of lignin di- and oligomers (58, 302, Chapter 7) and very slow anaerobic degradation of polymeric lignin and other phenolic derivatives from weeds and herbaceous plants (22) should not be taken as proof that "lignin" is anaerobically degradable. On the other hand, lack of significant degradation of a synthetic lignin analog during anaerobic incubation experiments (99) is not proof that "lignin" is "biologically inert in the absence of molecular oxygen'' (297, 299). The densely intricate lignin in hardwoods has preserved the lignocellulosic wood substance of medieval British and classical Roman warships in anoxic marine mud for several centuries, but surface erosion of these wood samples demonstrates that it was not completely ''inert'' during these involuntary incubation experiments. There is no doubt that aerobic lignin degradation which involves highly reactive reduced-oxygen species (82, 261) is a far more efficient method of lignin degradation which may also produce small amounts of monomeric degradation products accessible to anaerobic degradation (50).

It is worth mentioning in this context that other polyanellated phenol derivatives, such as melanins, are insufficiently degraded in the presence or absence of molecular oxygen and persist in variant quantities in soils, waters, and sediments (100, 274). Obviously, the extent and kinetics of anaerobic polyphenol degradation are critically determined by the complexity of internuclear ring linkages.

Heteroatoms in aromatic rings apparently enhance anaerobic biodegradability, and degradation becomes easier with increasing numbers of heteroatoms. Thus naphthalene is apparently not attacked under anaerobic conditions, indol (1 N-atom) is degraded by sulfate-reducing and fermentative anaerobes (13a), and pyrimidine and purine derivatives are even fermentable in pure cultures (71, 112). Sulfate-dependent degradation was reported for furfural (2-furaldehyde) (42), and pyridin was fermentatively degraded in methanogenic mixed cultures in the author's laboratory. Thus heteroatoms within the ring structure have an effect on its anaerobic biodegradability similar to those of polarizing peripheral substituents.

14.3.4 Cleavage of Ether Linkages

Oxygen is the active reactant via monooxygenase action in aerobic cleavage of ether linkages. Phenylmethyl ethers in, for example, lignin monomers are split by oxidation of the methylcarbon group to form the hydroxymethyl derivative (Fig. 14.5). Thus the ether linkage is changed to a half-acetal linkage with very low stability, and the methyl carbon is released as formaldehyde. A variation of this principle is the transformation of the phenylmethylether into an acylmethyl ester by action of a dioxygenase (69). Transformation of a phenolic into a quinoid structure by partial oxidation has a similar effect on phenylmethyl ether stability (6).

Stability of ether linkages in the absence of oxygen was often taken as an ar-

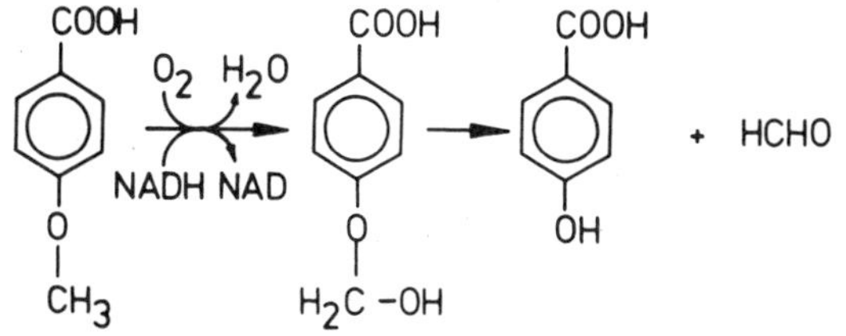

Figure 14.5 Cleavage of a phenylmethyl ether linkage by a monooxygenase. Taken from Ref. 24.

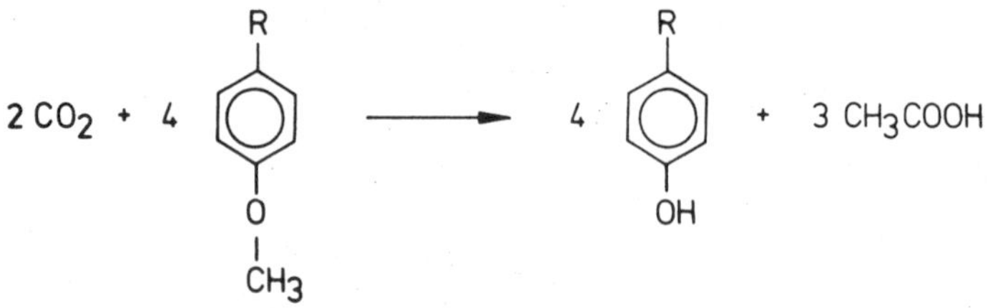

$$2\,CO_2 \;+\; 4\;\; \text{(ring with R, O–CH}_3\text{)} \;\longrightarrow\; 4\;\; \text{(ring with R, OH)} \;+\; 3\,CH_3COOH$$

Figure 14.6 Anaerobic demethoxylation of methoxylated phenyl compounds by *Aceto-bacterium woodii*.

gument for lignin recalcitrance under anaerobic conditions, although radiotracer studies with gastrointestinal contents of rodents long ago demonstrated anaerobic phenol demethoxylations (211, 212). Bache and Pfennig (11) enriched *Acetobacterium woodii* in anaerobic cultures with methoxylated phenol derivatives and demonstrated that the methoxyl groups are removed by this bacterium and fermented to acetate in a manner comparable to methanol fermentation (Fig. 14.6). Alkyl methyl ethers were not cleaved. Studies on growth kinetics and yields with the same organism suggest that the ether bond is not hydrolyzed, but the methyl residue is transferred directly to a carrier system (e.g., a coenzyme B_{12} derivative) (268).

The obvious susceptibility of phenylmethyl ethers to anaerobic cleavage raised the question of whether dialkylethers are also anaerobically accessible. Enrichment cultures with polyethylene glycol led to the isolation of new anaerobes which obviously degrade this artificial polymer by shifting the ultimate hydroxyl group to the penultimate one, thus transforming the ether linkage into a halfacetal bond (Fig. 14.7; 220, 244a). This reaction bears some resemblance to the aerobic demethoxylation reactions mentioned above (24), and transformation of the ether

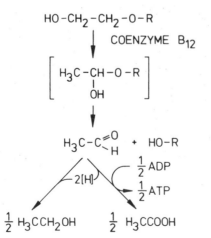

Figure 14.7 Anaerobic polyethylene glycol degradation by *Pelobacter venetianus*. Modified after Ref. 220.

linkage into a halfacetal linkage by action of a monooxygenase enzyme is also likely to underlie aerobic polyethylene glycol degradation (135).

Ether linkages therefore appear to be accessible to anaerobic cleavage if one of the linkage partners is a phenol. Due to the higher acidity of phenols compared to aliphatic alcohols, these ethers have some properties in common with esters and may thus facilitate anaerobic degradation. The presence of hydroxyl functions vicinal to the ether linkage and their possible shift to form a halfacetal bond is at least one condition that allows anaerobic cleavage of dialiphatic ethers, and reactions of this kind may be important, for example, in the mineralization of archaebacterial glycerol ether lipids.

14.3.5 Dehalogenations

Halogenated aliphatic and aromatic compounds are produced in huge amounts as solvents and precursors of herbicides, insecticides, and so on. They are also formed from organic contaminants during chlorination of fresh water and wastewater (205). Their unphysiological structure often makes us forget that halogenated aliphatics and aromatics also occur as natural secondary products of plants and microbes (148).

Degradation of halogenated aromatic compounds has long been studied only with aerobic mixed and pure cultures. In most cases examined so far, the aromatic ring is first cleaved by dioxygenases, and the halogen substituent is eliminated at a late stage of ring degradation (70, 75, 107, 108, 201). These reactions are obviously possible only in the presence of molecular oxygen, and the principle of primary polarization of the halogenated carbon atom to facilitate dehalogenation resembles in some respect the mechanism of aerobic methanol release from partially degraded phenylmethyl ethers mentioned above (69). Different degradation pathways for halogenated aromatics were reported which convert the halogen into a hydroxyl substituent prior to ring degradation (132, 142, 143, 230), but the biochemistry of these dehalogenation reactions has not yet been studied in detail. Thus it is not clear whether molecular oxygen participates in dehalogenation (e.g., by means of a monooxygenase reaction).

Removal of halogen substituents from aromatic compounds in the absence of oxygen was a revolutionizing discovery in 1982 (247). Unlike the aerobic processes mentioned above, anaerobic dehalogenation is a reductive process which forms hydrochloric acid and the corresponding hydrogen-substituted derivative (Fig. 14.8). A mixed methanogenic culture could be maintained which transformed a variety of halogenated benzoates either partially or completely to methane, carbon dioxide, and hydrochloric acid. It was shown in a recent study that the dehalogenating bacterium is different from the ring-degrading ones, and thus benefits only marginally from substrate degradation (231). The reductive dehalogenation bears some resemblance to the reductive dehydroxylation reactions mentioned in Section 14.3.3. Further studies on anaerobic dehalogenation of various halogenated benzoate and phenol derivatives (33, 34, 113, 248) indicate that de-

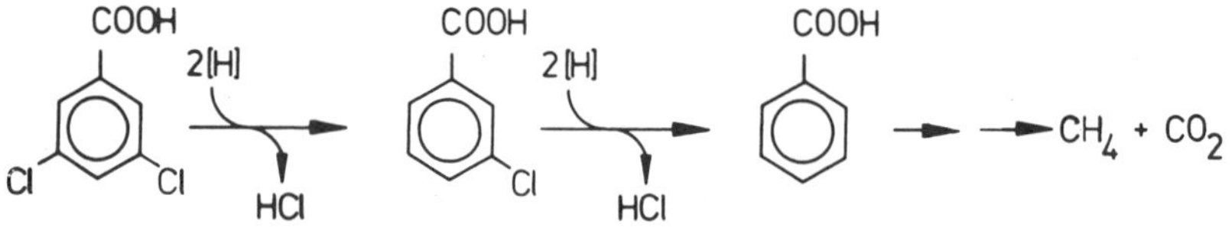

Figure 14.8 Reductive dehalogenation of 3,5-dichlorobenzoate by an anaerobic enrichment culture. After Ref. 247.

halogenation is an oxygen-sensitive process that prefers certain halogen positions at the aromatic ring over others. Two or more substituents are removed sequentially, and unlike aerobic degradation of certain chloroaromatics (20), no intermediates more toxic than the original substrate are formed during anaerobic dehalogenation. Even pentachlorophenol was completely degraded in an anaerobic reactor that was clearing paper industry wastes (101). Thus anaerobic dehalogenation of haloaromatic compounds looks like a promising new way of efficient detoxification of aromatic halogen derivatives in the future.

Degradation of halogenated aliphatic compounds was recently reported. A *Pseudomonas* strain (246) hydrolyzes 2-chloroacetic acid to glycolate prior to terminal oxidation. Anaerobic degradation of chloroform and tetrachloromethane in sediment samples leads finally to carbon dioxide, but the primary attack is probably a reductive step (29, 30, 32). Reductive dehalogenation was also observed with tetrachloroethylene and 1,1,2,2-tetrachloroethane, and led to trichloroethylene and 1,1,2-trichloroethane, respectively (30). Debromination of 2-bromoethanesulfonic acid and 2-chloroethanesulfonic acid (31) is probably the reason for the apparent inactivation in anaerobic digestors of these inhibitors of methanogenic metabolism. Nothing is known so far about the biochemistry of these reductive dehalogenation reactions.

14.3.6 Elimination of Nitrogen Substituents and N-dealkylations

Nitroaromatic compounds are readily reduced by microorganisms to the corresponding amino derivatives, probably via unstable nitroso compounds, and these reactions are far more efficient in the absence of oxygen than in its presence (103, 164). Stoichiometric reduction of aromatic nitro substituents to amino groups was demonstrated with trinitrotoluene in crude cell extracts of *Veillonella alcalescens* with molecular hydrogen as reductant (175). Anaerobic microorganisms are thus even able to deactivate explosives! Reductive degradation of an aliphatic nitro compound, 3-nitropropionic acid, was demonstrated with pure cultures of rumen bacteria, but the mechanism of degradation is unknown (167).

Aerobic degradation of aniline proceeds via dioxygenase reactions (165). The degradation of amino aromatics in the absence of oxygen is less clear. Aniline was not degraded in sediments and digested sludge (14), and efforts in the author's laboratory to enrich for aniline-degrading anaerobes failed. A recent publication

(38) reports degradation of anthranilic acid (2-aminobenzoate) by denitrifying bacteria. Nothing is known so far about the mechanism of deamination: whether it is a reductive or hydrolytic reaction or whether it occurs before or after ring cleavage. Also, an interaction of nitrite with the amino group to yield a diazo compound appears to be possible with denitrifying bacteria. Anthranilate was also degraded in methanogenic enrichment cultures in the author's laboratory. In this case, benzoate was identified as a degradation intermediate, suggesting a reductive elimination of the amino group (Tschech and Schink, unpublished).

The best-known aliphatic amines are the amino acids, which are deaminated by dehydrogenation to the corresponding imines and subsequent hydrolysis. A similar degradation mechanism is involved in the degradation of ethanolamine. This compound is transformed by a coenzyme B_{12}–dependent shift of the hydroxyl function and dehydration into an imine and further to acetaldehyde (35). Aerobic deamination reactions are again a domain of cytochrome P450 enzymes (281).

Reactions similar to deaminations are the N-demethylations. Trimethylamine-N-oxide is an antifreeze agent in the tissue of fish which is reduced during early oxygen-limited decomposition of fish to trimethylamine, a substance of very characteristic smell, the typical fish odor (140, 245). Aerobic degradation of this compound is again catalyzed by cytochrome P450 enzymes (281). Among the anaerobes, methanogenic bacteria able to grow with methanol were found to demethylate these compounds (see Chapter 12). It was recently reported that the methyl groups are transferred to coenzyme M by a methyltransferase in *Methanosarcina barkeri* (180). Betain is demethylated by a new *Acetobacterium* species (73). The mechanism of this reaction could be a methyl group transfer to coenzyme B_{12} similar to the O-demethylation of aromatic compounds discussed in Section 14.3.4.

14.3.7 Anaerobic Cometabolism

The list of anaerobic degradation reactions discussed in this chapter is far from complete. Only those reactions were mentioned which are catalyzed under aerobic conditions by oxygenase enzymes and thus need to take different degradation paths in the absence of oxygen. It appears from this short survey that a number of these reactions are managed under anaerobic conditions by a series of hydrations, dehydrations, hydrogenations, dehydrogenations, carboxylations, decarboxylations, and coenzyme B_{12}–dependent shift reactions. Some new reaction types, such as reductive dehydroxylations, dehalogenations, deaminations, transhydroxylations, and demethylations, were discovered in the recent past and deserve thorough investigations in the future, also as promising means of xenobiotic degradation.

The degradation processes mentioned were based either on investigations with radiolabeled tracer compounds in natural microbial populations or on studies with pure or mixed cultures enriched with the respective compound as carbon or energy source. Whereas the former method can detect nearly every reaction type possible in an ecosystem, the latter approach concentrates on those microbial transformations which are of use for the carbon or energy metabolism of the organisms involved, and "cometabolic" activities are neglected. A question arises as to whether cometabolic processes occur in anaerobic environments, and how important they may be.

The terms "cooxidation" and later, "cometabolism" were coined to characterize microbial metabolism of a compound which the organism involved is unable to use as a source of energy or as an essential nutrient (4, 5, 115). This phenomenon was evident in the case of oxidation of long-chain hydrocarbons by a methylotrophic *Pseudomonas* sp. (159): the bacterium oxidized ethane, propane, and butane to the corresponding ketones and carboxylic acids, but was unable to assimilate these substrates or to obtain energy from these oxidations. Later, the term "cometabolism" was used to describe microbial transformation of, for example, labeled xenobiotics added at trace concentrations to ill-defined natural samples such as soil or sewage sludge, and more or less complete recovery of the substrate carbon and lack of substrate incorporation into cell material were taken as proof for a cometabolic process (124, 275). This view, however, implies that microbial metabolism has always to be associated with growth and thus "is an expression of the bias of the experimenter rather than of the microbes involved" (116). Nearly every substrate supplied at a sufficiently low concentration will cover only the maintenance energy needs of a microbial population, and growth is possible only beyond a certain minimum substrate concentration. Thus even glucose supplied at a very low concentration will appear to be only cometabolically degraded by the criteria mentioned above. Therefore, the term "cometabolism" should *sensu stricto* be applied only to a defined degradation process for which the mode of microbial transformation and its implications on energy and carbon metabolism are known.

Such a clear definition is necessary before the question of cometabolic activities under anaerobic conditions is discussed. First, growth yields of anaerobic bacteria are comparably low as a consequence of low energy yields, and usually less than 10% of the substrate carbon can be incorporated into cell matter, as opposed to around 50% with aerobic bacteria. It may be unnecessary to mention that these ratios are by far lower under environmental conditions of low substrate supply. Second, many ecologically important groups of anaerobes use different substrates for energy and for carbon metabolism: *Desulfovibrio vulgaris* or *Methanospirillum hungatei* oxidize hydrogen to sulfide or methane with either sulfate or carbon dioxide, but need acetate and carbon dioxide as substrates for cell carbon synthesis. Even a substrate as attractive as glycerol is often only dissimilated, not assimilated into cell carbon (214, 220). Third, the role of the oxidant as a "co-substrate" in the metabolic process under study needs to be defined. In the case of oxygen, nitrate, or sulfate, this is not a difficult task. Things become more complicated if the oxidant is an organic substrate or carbon dioxide, both of which may contribute to cell carbon synthesis. *Wolinella succinogenes* uses fumarate only as an electron acceptor (151). A *Campylobacter* sp. can ferment fumarate to acetate and succinate and also assimilates fumarate carbon into cell material (271).

The recently described reductive dehydroxylations, dehalogenations, and so on, mentioned above are important steps in the degradation of the respective compounds; however, they may be of only limited advantage for the bacterium involved. If the reducing bacterium does not also cleave and degrade the ring system—and this appears to be the case in the only process studied so far in defined cultures (231)—the substrate carbon skeleton is not changed by this bacterium, and no substrate carbon is assimilated. Although such a process will clearly be defined as cometabolic by the criteria mentioned above, the respective bacterium will to some extent benefit from it as a means of releasing reducing equivalents and thus changing its fermentation balance to an energetically more favorable pattern, whether or not electron transport phosphorylation is involved. The numerous activities of anaerobes involved in the modification of unsaturated fatty acids (77), steroids (77, 171), and aromatic compounds (212) can from this point of view be classified as cometabolic or "coreductive." They may be advantageous especially for those anaerobes which lack hydrogenases and thus cannot participate in interspecies hydrogen transfer.

The ambivalence of the question of whether or not an apparently cometabolic activity is of advantage for the bacterium involved was demonstrated recently in studies on anaerobic succinate degradation. Conversion of succinate to propionate by *Selenomonas ruminantium*, propionibacteria, or *Veillonella alcalescens* (210, 290) has so far been considered to be a cometabolic activity without implications for energy metabolism. Hilpert and Dimroth (109) succeeded recently in showing that the decarboxylation of methylmalonyl-CoA during succinate metabolism by *V. alcalescens* is coupled to the establishment of a sodium-ion gradient which may help to cover energy needs for transport activities. The same decarboxylation reaction allows growth in the case of *Propionigenium modestum* by using the sodium-ion gradient for ATP synthesis (110, 219). Thus *Veillonella* and *Propionigenium* differ only by the absence or presence of a sodium-dependent ATPase which allows growth and substrate assimilation in the latter but not in the former case.

It appears from these few instances that the limits between cometabolic and "ordinary" metabolic activities are extremely difficult to assess among anaerobic degradative processes. The energy gain of an individual bacterium in a specific reaction step during the complex ballgame of interspecies intermediate transfer is often very small, and cannot be judged simply from substrate assimilation ratios obtained under defined culture conditions. Moreover, the extent to which an organism benefits from an individual degradation reaction will to a greater or lesser extent depend also on the presence or absence of other bacteria which modify the environmental conditions, for example, by production or utilization of protons and other metabolic intermediates (147). Differentiation between advantageous and nonadvantageous reactions thus comes down to splitting hairs without any practical or theoretical significance. Thus, if use of the term "cometabolism" is questionable even when applied to aerobic degradation processes in several instances (116), this is far more true where anaerobic degradative activities in complex reaction chains in a natural ecosystem are concerned. We have to learn that microbial degradation activities and microbial growth in a natural environment are only rarely linked to each other in a way assessible by simple laboratory methods.

14.4 THE ENVIRONMENT'S NATURE: CHANCE AND CHALLENGE

The degradative potential of an anaerobic microbial community has so far been discussed in relation to biotic activities alone. However, in a natural environment as well as in a sewage fermenter, abiotic factors also influence considerably the kind and extent of microbial activities, which may be advantageous or detrimental to the bacteria involved. Some of these abiotic conditions of an anaerobic environment are dealt with in this section.

14.4.1 Surfaces

Since, unlike oxic water bodies, anoxic habitats such as sediments or sludge digestors are characterized by the accumulation of digestible and nondigestible organic and inorganic solids, which both contribute to the prevention of oxygen diffusion, the typical anoxic environment will contain an enormous amount of surface which can serve as a support for adhesion of anaerobic microbial communities. These surfaces can be predominantly inorganic or organic, depending on substrate supply, community structure and so on. A remarkable amount of literature concerning microbial interference with any kind of surfaces has accumulated in the last 10 years, and surveys are available in recent books and review articles (23, 54, 59, 80, 169, 170) which cover mechanistic, physiological, ecological, and technological aspects of microbial adhesion to surfaces. So far, nearly all investigations on microbial interactions with surfaces have centered on aerobic bacteria, and our knowledge of anaerobic bacterial surface interference is scarce. I therefore restrict myself to aspects of microbe–surface interactions which are of specific interest for anaerobic bacteria. I also concentrate on microbial interference with surfaces considered to be inert to microbial degradation since the ecological importance of degradable surfaces on microbial activities is largely self-evident.

Accumulation of nutrients close to surfaces is of general advantage for microbes adsorbed to surfaces under conditions of nutrient limitation (126). For anaerobes, this enhanced metabolic activity of an adsorbed microbial population also has the advantage of an efficient protection from oxygen. Oxygen can diffuse basically only from one direction and will be used up very efficiently by aerobic microbes. Precipitates of metal sulfides which commonly cover nearly every surface in an anoxic habitat serve as an additional oxygen protectant and redox buffer. Ferrous sulfide has been used successfully as a reductant in the cultivation of fastidious anaerobes (40). Metal sulfides also may act as ion exchangers which maintain a well-equilibrated balance of dissolved trace elements, toxic heavy metals, and free hydrogen sulfide. Clay and other insoluble minerals may function in a similar manner and have been found to enhance the metabolic activities of some sulfate-reducing bacteria (152). Similar results were obtained by Söhngen (236), who studied the influence of silica and humic compounds on aerobic and anaerobic microbial activities.

Surfaces may also contribute significantly to the establishment and maintenance of stable anaerobic microbial communities which depend on or benefit from any kind of interspecies metabolite transfer. Evidence for close juxtapositioning of anaerobes cooperating in interspecies hydrogen transfer has been obtained by measurements of hydrogen fluxes in anoxic sediment and sludge samples (58a). It should be unneccessary to stress that processes like this are of far greater importance among anaerobic than among aerobic bacteria. Many anaerobic bacteria form clumps or filaments which lend themselves to adsorption to surfaces. Others form intergeneric aggregates which are of metabolic interest for both partners: The highly organized consortia of *Chlorochromatium*, *Pelochromatium*, or *Chloroplana* (192) are only some of the examples of bacterial aggregates which establish metabolic cooperation by close juxtaposition.

The bacterial tendency to stick to surfaces has probably been largely underestimated, because conditions in most liquid cultures select for a loss of adsorption abilities during a few generations of growth (59). In a surface-rich habitat such as a lake sediment, the majority of the bacterial community is probably adsorbed to surfaces rather than floating free in the pore water. Interest in the conditions and modes of microbial attachment to surfaces was revived recently because of its exploitation in fixed-film bioreactors (see Section 14.6).

14.4.2 Spatial and Temporal Heterogeneity

Solid surfaces are only one kind of discontinuity in an aqueous solution, although perhaps the most conspicuous. The natural environment is nearly nowhere completely homogeneous with respect to solutes, suspended material, and so on, and the profundals of large water bodies such as lakes and oceans are no exception. The natural environment is usually characterized by uneven, gradually changing distribution in space of nutrients, oxidants, proton activity, salinity, metals, complexing agents, temperature, light, or others. These various gradients are the ecological challenge for the development of positioning and orientation mechanisms (53, 193). It is evident that the classical enrichment in liquid medium which usually provides very homogeneous conditions will only rarely mimic the situation in the natural environment. It is not astonishing that typical gradient bacteria such as *Beggiatoa*, the methane-oxidizing or the magnetotactic bacteria, are so difficult to cultivate and to study, and indications exist that those strains cultivated in the laboratory are not typical of the respective metabolic groups in the natural environment. Some new experimental approaches to the investigation of spatially organized environments have been developed in the recent past (51, 259, 284, 285) which so far provide only a limited insight into the extreme complexity of such structures. A deeper insight into microbial population structures in the undisturbed natural habitat by new and very sensitive analytical techniques (279, 280) will perhaps help to base our assumptions on the spatial organization of anoxic habitats on a more reliable ground and to improve experimental approaches to study the organisms involved.

Heterogeneity of the natural environment applies not only to its spatial but also to its temporal dimension. Except perhaps for deep ocean sediments, nearly every natural environment is subject to fluctuations of temperature, light, nutrient supply, acidity, and so on, in time. Anaerobic environments are usually rich in water, which helps to buffer changes of the foregoing parameters better than air in a comparably dry and extremely heterogeneous environment such as soil. Changes of the parameters mentioned are common and force the microbial community either to adapt to changes sufficiently fast, or to tolerate them. Fluctuations in the environmental conditions can also be advantageous for degradative processes: Sequential changes from primarily anoxic to oxic conditions enhanced degradation of the insecticide methoxychlor in soil samples significantly compared to continuously aerobically or anaerobically incubated samples (81). This result can be interpreted as being due to the difference between aerobic and anaerobic attack on many recalcitrant compounds as discussed above: Extremely slow reaction steps which represent bottlenecks in the respective compound's degradation under one condition can perhaps easily be catalyzed when conditions change (e.g., from anaerobic to aerobic, or vice versa).

14.4.3 Chemical and Physiological Interactions

There is perhaps no environment in nature in which a single substrate is metabolized by a single organism. It has been mentioned repeatedly in this chapter that several groups of bacteria usually cooperate in the anaerobic degradation of one organic compound. Microorganisms, on the other hand, do not feed on one single substrate, at least not for a longer time. Even if an organism is specialized to one single substrate to maintain its energy metabolism, it is usually able to assimilate other organic compounds into cell material, or it is impeded by the presence of toxic compounds or supported by trace elements or vitamins. Different compounds will therefore influence their respective metabolism not so much by direct chemical interactions, although the presence of highly reactive agents such as oxygen on one side or hydrogen sulfide on the other side will, to some extent, modify susceptible functional groups of substrate molecules. Beyond such purely chemical interactions, one compound can modify the metabolism of a different one via a metabolically active microbe. Some instances of such interactions on microbial attachments to surfaces which often also act as adsorbents, ion exchangers and so on, to promote or impede microbial activities were given in Section 14.4.1. In sediments, suspended or dissolved humic colloids may have a similar function (236).

Monomeric phenols were found to inhibit methanogenic degradation of other anaerobic sludge contents (78). Polymeric phenols such as lignin residues, tannins, fulvic acids, and so on, may bind toxic pollutants in anaerobic habitats in a manner similar to that of humic substances in soil (165, 243) and may release them at low rates. They could thus ensure slow but constant degradation of these compounds rather than destruction of the microbial community. Degradation of isopropyl-*N*-phenylcarbamate in aerobic water samples was enhanced by inorganic, but not by organic nutrients (275). The simultaneous utilization of xenobiotics and easily degradable organic substrates by one organism was one of the reasons for creating the concept of cometabolism mentioned in Section 14.3.7. Today it appears to be more useful to talk about primary and secondary substrate utilization and to define the role of either one (155). There is no doubt that one substrate in such a manner can enhance utilization of another one. The various effects of chemical and physical environmental factors on microbial metabolic activity have been discussed excellently by Bull (47). Many of the points raised in this treatise apply to aerobic and anaerobic environments as well.

14.5 CARBON AND ELECTRON FLOW IN METHANOGENIC DEGRADATION

Methanogenesis is the final process in terminal anaerobic degradation once inorganic electron acceptors such as nitrate and sulfate are exhausted. It is therefore the most important process in anoxic freshwater lake sediments, sewage sludge, or the rumen, in which the supply of nitrate or sulfate is very small compared with the input of organic substrates. Nonetheless, sulfate reduction is of basic importance for the establishment and maintenance of a sufficiently low electron potential which allows proliferation of the complex methanogenic microbial community. It also provides sufficient hydrogen sulfide for the synthesis of cell material.

Provided that anaerobic degradation is possible at all (see Section 14.3), complex organic matter can be transformed into methane and carbon dioxide according to the following equation (250, 252):

$$C_nH_aO_b + \left(n - \frac{a}{2} - \frac{b}{4}\right) H_2O \longrightarrow \left(\frac{n}{2} - \frac{a}{8} + \frac{b}{4}\right) CO_2 + \left(\frac{n}{2} + \frac{a}{8} - \frac{b}{4}\right) CH_4$$

It appears that the percentage of methane in the gas mixture formed depends on the oxidation state of the substrate used. Chemically speaking, the entire rather complex process is a disproportionation of organic carbon into its most oxidized and its most reduced form. Carbohydrates are converted to equal amounts of methane and carbon dioxide, methanol and lipids to more methane than carbon dioxide, formic acid and oxalic acid to more carbon dioxide than methane, and no methane can be obtained from urea hydrolysis (Fig. 14.9). Due to the better solubility of carbon dioxide in water and its partial conversion into bicarbonate as the main buffering agent in anoxic environments, the mean portion of methane in digestor gas varies between 60 and 75%.

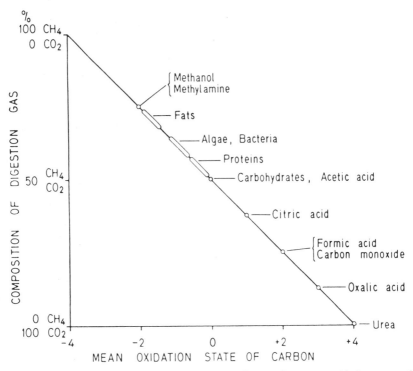

Figure 14.9 Composition of digestion gas depending on the mean oxidation state of the substrate carbon, under conditions of complete mineralization. Reprinted with permission from Ref. 98.

The paths of carbon and electron flow in methanogenic degradation have been reviewed recently in detailed articles to which the interested reader is referred (44, 287, 292, 295, 296). A flow scheme is represented in Fig. 14.10. Three major metabolic groups of bacteria are involved: I. the hydrolytic and fermentative bacteria which convert complex biopolymers, sugars, amino acids, fatty acids, and so on, to acetate, hydrogen and carbon dioxide, on the one hand, and a mixture of alcohols, fatty acids, succinate, and lactate, on the other hand. The latter substrates have to be converted to acetate, hydrogen, and carbon dioxide by the organisms of group II, the proton-reducing acetogenic bacteria, which depend for thermodynamic reasons on efficient removal of their fermentation products. Group III, the methanogenic bacteria, convert acetate, hydrogen, and carbon dioxide as well as the less important one carbon compounds methanol, methylamines, and formate to methane and carbon dioxide. The importance of the organisms of group IV, the homoacetogenic bacteria, is less obvious in a methanogenic ecosystem. The population sizes of the four groups mentioned so far were estimated in an anaerobic digestor by most-probable number and other enumeration techniques (Table 14.1). It appears that the first group is by far the largest, although all these enumeration

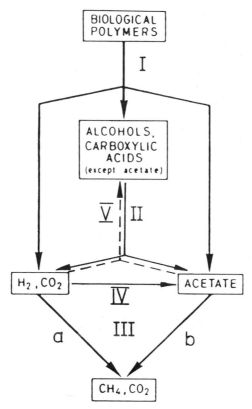

Figure 14.10 Carbon and electron flow in methanogenic environments. Metabolic groups involved: I, hydrolytic and fermentative bacteria; II, proton-reducing acetogenic bacteria; III, methanogenic bacteria (a) hydrogenophilic, (b) acetophilic; IV. homoacetogenic bacteria; V. fatty acid–synthetizing bacteria. Scheme modified after Ref. 295; explanations in the text.

TABLE 14.1 Numbers of anaerobic bacterial populations in sewage sludge digestors

Group	Numbers (mL^{-1})
I. Hydrolytic and fermentative bacteria	10^8–10^9
Proteolytic	10^7
Cellulolytic	10^5
II. Proton-reducing acetogenic bacteria (butyrate-oxidizing)	10^6
III. Methanogenic bacteria (hydrogen-oxidizing)	10^6–10^8
IV. Homoacetogenic bacteria (hydrogen-oxidizing)	10^5–10^6
Sulfate-reducing bacteria (lactate-oxidizing)	10^4

Source: After Ref. 296.

experiments may underestimate the real population sizes by one or two orders of magnitude, due to the well-known difficulties of cultivating environmentally important bacteria.

Whereas many of the bacteria of groups I, III, and IV have been cultivated and studied in detail in the past, not much is known about the bacteria of group II. Since they catalyze reactions which are thermodynamically endergonic under standard conditions, they can be cultivated only in the presence of hydrogen-scavenging bacteria such as methanogens, which maintain a sufficiently low hydrogen partial pressure to allow substrate degradation. These problems are treated in detail in Chapters 8, 9, and 12; a scheme that visualizes the thermodynamically delicate situation of ethanol and propionate degrading bacteria is given in Fig. 14.11. The thermodynamic implications of hydrogen and acetate concentrations on the turnover of propionate, acetate, and hydrogen in an anaerobic digestor are discussed in detail elsewhere (98, 294).

Although the scheme in Fig. 14.10 is now basically accepted, controversy remains concerning the relative importance of the formation of reduced intermediary products. Whereas the older literature in this field attributes considerable importance to propionate, butyrate, and alcohols as fermentation intermediates (263), more recent studies with radiolabeled tracers revealed that only a comparably small proportion of carbon is converted via these compounds (98, 166). Laboratory experiments demonstrated that the hydrolytic and fermentative bacteria (group I, Fig. 14.10) change their fermentation pattern dramatically mainly toward hydrogen, carbon dioxide, and acetate as products if the hydrogen partial pressure in the

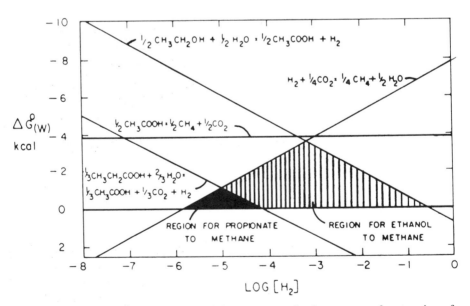

Figure 14.11 Effect of hydrogen partial pressure on the free energy of conversion of ethanol, propionate, acetate, and hydrogen during methane formation. After Ref. 174.

environment is kept low by, for example, methanogenic bacteria (55, 120, 256). Under such conditions, the carbon and electron flux through the reduced intermediates would be small, and this situation is probably typical of a substrate-limited, well-balanced microbial community in an anoxic lake sediment. At times of rather high substrate inputs (e.g., by dying algal blooms), the metabolic capacity of the methanogenic bacterial population does not suffice to balance the increasing metabolic activity of the primary fermenters and reduced metabolites accumulate which have to be degraded by proton-reducing acetogenic bacteria (group II). Thus the importance of the group II organisms depends basically on the substrate input on the one hand and the activity of the methanogenic population on the other, which both regulate the flux of carbon and electrons through the reduced intermediates. This relationship is illustrated by a simple model in Fig. 14.12.

The rain barrel represents the community of fermentative bacteria by which the complex substrate flowing in at A is transformed to the products flowing out at B, C, or D. B represents the low-energy products of fermentation (acetate, hydrogen, and carbon dioxide), C the reduced intermediates propionate, butyrate, and so on. Under normal conditions (water level E), nearly all substrate carbon goes exclusively through outlet B; outlet C is used only if the capacity of B (as a function of the methanogenic bacterial activity) is not sufficient for balancing the inflow. The same would happen if at normal substrate inflow outlet B is partially or completely plugged as a consequence of inhibition of methanogenic activities (see below). A complete blocking of this outlet could lead to a high accumulation of diverse fatty acids, which would lower the pH and, as a further consequence, would even cause

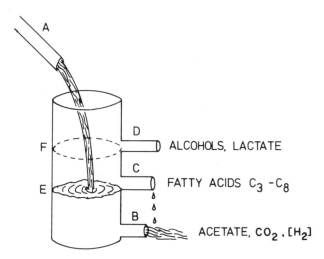

Figure 14.12 Rain barrel model of carbon and electron flow in methanogenic degradation. *A*, inflow; *B*, low outlet representing low-energy products; *C*, medium outlet, representing fatty acids; *D*, high outlet, representing high-energy products; *E*, normal water level; *F*, high water level. Hydrogen at the lower outlet is written in brackets to indicate that its partial pressure has to be kept very low. After an idea by Fritz Widdel.

ethanol and lactate to be formed (outlet D). It should be kept in mind, however, that a small amount of fatty acids always forms during lipid and protein degradation, and that the flux through fatty acids will never be zero if a complex substrate mixture is degraded.

Substrate supply and its consequences for carbon and electron flow as outlined above represent the one important difference between a sediment ecosystem and an anaerobic sludge digestor. Contrary to the former, the latter always receives high amounts of comparably easy-to-digest substrates, and the flux of carbon through reduced intermediates will be higher. Another basic difference relates to the residence times of the microbial biomass. Whereas microbes in a sediment can establish and persist for nearly unlimited times, the mean residence time in a sludge digestor is, for economic reasons, usually limited to 2 to 3 weeks. Thus substrate conversion in a digestor depends to a great extent on the growth rates of metabolically important microbial groups, and minimum residence times are dictated by the doubling times of the acetate-, propionate-, and butyrate-degrading communities (158). Rumen contents of cattle have a mean residence time of only 15 to 50 h. Establishment of acetate-, propionate-, and butyrate-degrading bacterial populations is thus prevented in the rumen, and methane is formed only from hydrogen and carbon dioxide. The short-chain fatty acids are absorbed by the animal host (286).

In sediments as well as in digestors, about two-thirds of the methane formed is derived from acetate cleavage, and one-third from hydrogen oxidation (235). This balance agrees with the assumption that carbohydrates are degraded mainly via these two intermediates.

A further metabolic group of bacteria (group V) mentioned in Fig. 14.10 has not yet been discussed. At least some of the energetically very delicate reactions that oxidize fatty acids to acetate and carbon dioxide can be reverted in the presence of excess ethanol or hydrogen. The formation of butyrate from acetate and ethanol by *Clostridium kluyveri* is a well-known process of this kind (18). Recently, propionate-forming bacteria were isolated which can convert acetate or ethanol to propionate and act as a hydrogen sink by means of hydrogenases (153, 214, 222, 260). The ecological importance of these organisms has not yet been studied in detail; they are probably involved to some extent in ethanol degradation (223) and may also cooperate with homoacetogenic bacteria in propionate formation from C_1-compounds (96).

Another interesting observation concerns the importance of homoacetogenic bacteria (group IV). Obviously, these bacteria can take over the role of hydrogen-oxidizing methanogens if those are inhibited by high proton activity. Thus more than 95% of all methane was derived from acetate in a mildly acidic lake sediment at pH 6.1 (194), indicating that carbon and electron flow went nearly exclusively through homoacetogenesis and acetate cleavage.

Comparison of aerobic and methanogenic degradation with respect to thermodynamics demonstrates the enormous energetical efficiency of the anaerobic process:

$$C_6H_{12}O_6 + 6O_2 \longrightarrow 6CO_2 + 6H_2O \qquad \Delta G°' = -2826 \text{ kJ}$$
$$C_6H_{12}O_6 \longrightarrow 3CH_4 + 3CO_2 \qquad \Delta G°' = -394 \text{ kJ}$$

Whereas aerobic sugar degradation yields about 38 ATP, methanogenic degradation only allows synthesis of 5 to 6 ATP, and this energy has to be shared by two to three different metabolic groups of bacteria. This comparison illustrates how promising methanogenic processes are with respect to application in waste treatment processes: More than 85% of the potential oxidation energy of organic substrates is conserved in methane. Therefore, the following section deals with technological applications of methanogenic degradation.

14.6 METHANOGENIC WASTE TREATMENT

14.6.1 Aerobic versus Anaerobic Waste Treatment

Although anaerobic systems for wastewater treatment have been used since the beginning of the twentieth century ("Emscherbrunnen" in Germany), they were long considered to be inefficient and too slow to serve the needs of a quickly expanding wastewater volume, especially in industrialized and densely populated areas. Aerobic techniques such as trickling filters and oxidation ponds with more or less intense mixing devices were installed for wastewater treatment in small communities. Larger treatment plants today depend nearly exclusively on the activated sludge process, in which a mixed and so far only marginally identified population of bacteria and protozoa degrades the organic material in intensely aerated basins. The biomass produced (activated sludge) is recycled to a varying proportion within the system to enhance its operational efficiency. The excess activated sludge is often stabilized in an anaerobic fermentation step. The digestor gas obtained in this process was mostly flared off; in recent years, it has been applied to cover some part of the enormous energy needs invested in stirring or aeration. The activated sludge can usually adapt sufficiently well to qualitative and quantitative changes in wastewater composition, although management problems such as bulking are still an ill-understood phenomenon which is counteracted on an empirical rather than an analytical basis (156, 195). The function of the protozoa is mainly the maintenance of a stable and well-settling bacterial community, and the reduction of pathogens (61).

The disadvantages of this technology can be summarized as follows:

1. Stirring or aeration of activated sludge basins requires vast amounts of energy, and the energy needs increase with the load of easy-to-degrade organic contents.
2. Due to the high energy yield of aerobic degradation, large amounts of sludge biomass are produced, which create increasing disposal problems.

3. Many substrates, such as phenolic and other aromatic compounds, are converted chemically into polymeric adducts which resist further degradation.

4. Maintenance of a constant oxygen tension in the aeration basin requires costly continuous regulation and control.

Anaerobic waste treatment could avoid some of these disadvantages. No aeration is necessary, and instead of investing energy, some energy could be recovered in the form of methane gas for heating and pumping purposes. Sludge production would be low since energy yields of anaerobic bacteria are small. Many substrates could be degraded successfully instead of being polymerized into barely degradable humic substances. Last, but not least, the technical expenditure for regulation and maintenance devices could be kept minimal. Nonetheless, low growth rates and yields of anaerobes would require very long residence times for wastewaters unless care were taken to retain the catalytic cell mass within the degradation system. Several methods have been developed in recent years to maintain high catalytic activities of anaerobic bacteria in anaerobic sludge digestors. They are described in the following section.

14.6.2 Reactor Design

The simplest anaerobic digestor design is the septic tank as it is still used as a cheap device for disposal of domestic sewage in rural areas (Fig. 14.13a). A mixed freight of solid and dissolved organic material enters the reactor and separates during degradation into a sedimenting layer of sludge which also contains most of the bacterial biomass, and a scum layer which floats on the surface by means of trapped gas bubbles. No mixing occurs; the quality of the effluent water changes with the influent freight, and the sludge produced has to be removed from time to time. Improvements on this simple system were the Travis tank and the Imhoff tank, which both tried to destroy the scum layer and to separate inflowing and outflowing material by screens and baffles (174). In the conventional anaerobic digestor (Fig. 14.13b) as it is used today for stabilization of activated sludge residues, the entire reactor content is mixed by internal stirring or cyclic external pumping. Fresh and digested materials are thus kept in close contact, but separation of biomass and treated water is not possible. Such separation can be achieved by combination with a settling device (Fig. 14.13c): The treated effluent water leaves the system, and the undigested material, together with the bacterial biomass, is retained and recycled to enhance digestor efficiency and allow further degradation of particulate waste material. This system thus fulfills the mean requirement for successful anaerobic waste treatment: Recycling of biomass overcomes the general drawbacks of anaerobic versus aerobic treatment, namely, the low growth rates and yields of anaerobes. Retaining the active catalyst within the reactor thus makes the hydraulic retention time to some extent independent of the bacterial growth rate; a high standing population of sedimenting anaerobes will

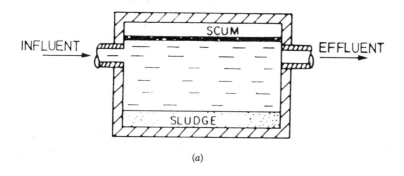

(a)

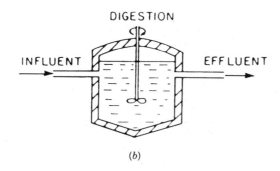

(b)

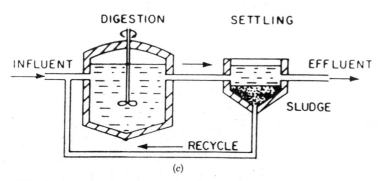

(c)

Figure 14.13 Basic types of anaerobic digestors. (*a*) Septic tank. (*b*) Conventional anaer-obic digestor. (*c*) Anaerobic contact process. (*d*) Upflow anaerobic sludge blanket reactor. (*e*) Anaerobic filter. (*f*) Anaerobic fluidized bed reactor. (*a*)–(*c*) and (*e*)–(*f*) from Ref. 174; (*d*) from Ref. 190.

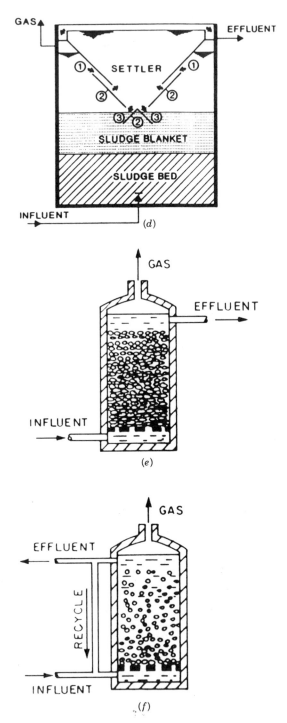

Figure 14.13 (*Continued*)

also be able to adapt to environmental fluctuations in sufficiently short periods of time. The combination of a conventional anaerobic digestor with a settling device and sludge recycling mimics the aerobic activated sludge process; it has also been applied successfully for direct treatment of high-strength wastewaters. Unfortunately, the bacterial biomass does not sediment as easily as a well-operating activated sludge does: Anaerobic protozoa are not as efficient as their aerobic counterparts in reducing the numbers of free-floating single bacterial cells. Chemical flocculants therefore often have to be added in the settling device.

The upflow anaerobic sludge blanket (UASB) reactor (Fig. 14.13*d*) combines digestion and separation in a single vessel. This reactor type was developed by Lettinga (160) for treatment of high-strength wastewaters from the food industry. The substrate solution enters the reactor from the bottom through a layer of sedimented bacterial sludge. Gas bubbles produced during substrate degradation carry sludge flocs up into the liquid phase, in which they form a floating blanket of highly active bacterial flocs. Settler screens in the upper part of the reactor separate gas bubbles, flocs, and particulate material, and the latter two sediment back to the bottom. Suspended bacterial cells leave the reactor together with the treated water at the top. This reactor type thus selects for floc-forming bacteria already in the digestor vessel, and usually does not require the addition of flocculating agents. The gradient of suspended solids coincides with the gradient of substrate distribution if the system is well equilibrated.

Other devices to maintain a high anaerobic microbial population in the reactor are the anaerobic filter and anaerobic fluidized bed reactor (Fig. 14.13*e* and *f*). Analogous with the aerobic trickling filter, both take advantage of the microbial tendency to stick to surfaces. A solid or floating bed material of rock, gravel, or sand (in laboratory experiments, also glass or plastic materials with large internal surfaces) serves as support for the microbial biomass. It is evident that such a system is suited mainly for treatment of dissolved organic matter; particles and fibers would soon plug the fine channels in the bed material and thus decrease its efficiency. The stationary-bed anaerobic filter (Fig. 14.13*e*) also traps gas bubbles to varying degrees, depending on the bed material. As a consequence, the microbially colonized surface involved in substrate degradation cannot work to its full extent, and the degradation efficiency decreases. This disadvantage is avoided in the fluidized bed reactor: a strong beam of recirculating water keeps the bed material floating and allows sufficient release of gas bubbles. Recycling circles have been applied to stationary bed filters to expel gas bubbles from the bed material, but with limited success. Recirculation of the effluent water is advisable if the influent wastewater is not suited for direct undiluted treatment because of high organic loads, low or high pH, toxic compounds, or other factors. Without recirculation, both types of bed reactors are mainly suited for the treatment of low-strength wastewaters, analogous to the aerobic trickling filter.

14.6.3 Microbiological Aspects: Films and Flocs

A short microbiological intermezzo appears desirable concerning the formation and properties of films and flocs. The tendency of many bacteria to stick to surfaces has already been mentioned in Section 14.4.1. In biotechnology, the formation of biofilms is often an undesired process which promotes corrosion of metallic surfaces, impedes heat transfer through heat exchanger walls, or restricts the transport capacities of tubes (54). In fixed-film bioreactors, the growth of biofilms is essential; however, the microbiology and physiology of this process are little understood as yet. A scheme of biofilm growth, maturation, and decay is presented in Fig. 14.14. This scheme, although developed for an aerobic biofilm, exemplifies the various stages of film development. With increasing thickness, a multilayered biofilm becomes a quite inhomogeneous, structured microbial community in which the lower, older layers are faced with an environment very different from that of the surface cells. Substrates change qualitatively and quantitatively from the outer to the innermost layers, and cells in the latter may starve and die while the outer layers are metabolically still highly active. Autolysis of lower cell layers may be one of the reasons for sudden and uncontrolled sloughing off of biofilms, but our knowledge of these processes is very limited (46, 54). Bivalent cations apparently play a pivoting role in the maintenance of biofilm stability (269).

It is evident that organization into a complex biofilm structure protects the individual cell from many adverse environmental effects. Recent studies have demonstrated that biofilms in fixed-bed reactors are highly resistant to substrate shock loads (48, 136) or to pH fluctuations (19). Enhancement of colonizable surfaces by the addition of powdered activated carbon to stressed digestor populations enhanced methane formation and stabilized the microbial activity (240). Protection from pH stresses can also be improved by using a buffering support material such as limestone chips (19).

In laboratory experiments it is not always easy to find a suitable carrier matrix for surface cultivation of microbial communities adapted to the degradation of a certain compound. One also has to realize that the pore size of the support material has to be suited for microbial colonization. Thus most of the enormous internal surface of activated carbon (500 to 1600 $m^2 \cdot g^{-1}$) cannot be colonized by bacteria

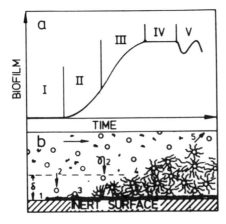

Figure 14.14 (*a*) Scheme of biofilm development with time. I, lag phase; II, exponential growth; III, decreasing growth rate phase; IV, plateau; V, sloughing. (*b*) Events in biofilm development. 1, adsorption of dissolved organics; 2, transport of cells to the surface; 3, microbial adhesion to the surface; 4, biofilm growth; 5, biofilm detachment. Arrow indicates flow of aqueous medium. δ represents the viscous sublayer thickness. Black spots counterfeit organic detritus particles. From Ref. 46, reproduced by permission.

because most pores are smaller than 50 nm in diameter (25). In our own laboratory studies, many so far unsuited surfaces were quickly colonized by anaerobic bacteria if they were first covered by a thin layer of ferrous sulfide which may not only represent the most prominent adsorbent in nature but may also act as a simple redox buffer system (40).

Flocs differ from films by the lack of a nonbiological surface. However, differentiation is often difficult because many flocs develop from biofilms which originated at very small inert particles. Nonetheless, films and flocs have many properties in common, especially their differentiation into outer and inner layers and the resulting consequences for the various microbial populations involved. Figure 14.15 shows in increasing magnification steps the structure and microbial heterogeneity of a bacterial floc from a mesophilic UASB-reactor degrading a mixture of acetate and propionate. *Methanothrix soehngenii* is the predominant acetate degrader which forms a dense network with short rods and bigger lemon-shaped cells which may cooperate in syntrophic propionate degradation.

As mentioned above for biofilms, very little is known about the forces that hold the cells of a floc together (52). Extracellular polymers are often visible in scanning electron micrographs of flocs and probably act as gluing material between the cells (Fig. 14.16). However, it is not clear which factors regulate exopolymer formation and degradation, and to which extent exopolysaccharides, for example, are simply residues of substrate degradation. Cations such as manganese may again play an important role as bridging ions or regulatory effectors.

14.6.4. Examples of Application

Organic loads in wastewaters are usually measured as the chemical oxygen demand (COD) or the biological oxygen demand (BOD) of a complete (COD) or nearly complete (BOD_5, after 5 days of incubation) conversion of the organic carbon into carbon dioxide and water. These measures, therefore, apply to the total amount of reducing equivalents present and also involve, strictly speaking, the reduced forms of nitrogen and sulfur, ammonia and sulfide. The COD differs from the BOD by the amount of reducing equivalents that cannot be oxidized by aerobic microorganisms during the incubation period. Thus the BOD value is linked intrinsically to the aerobic biodegradation potential, and is always lower than the COD value. Large differences between COD and BOD point to a high proportion of nonbiodegradable compounds. As demonstrated in Section 14.3, the aerobic and anaerobic degradation potential are not identical. Nevertheless, the BOD values are usually also applicable for the calculation of anaerobic degradation processes, except for some special cases of industrial wastewaters. It is important to note that the BOD value of a given organic waste substrate does not significantly change during the initial fermentation steps in anaerobic degradation. Only the final release of methane changes the BOD balance significantly. The maximum amount of methane that can be formed during complete anaerobic degradation can be calculated directly from the BOD value by the relationship

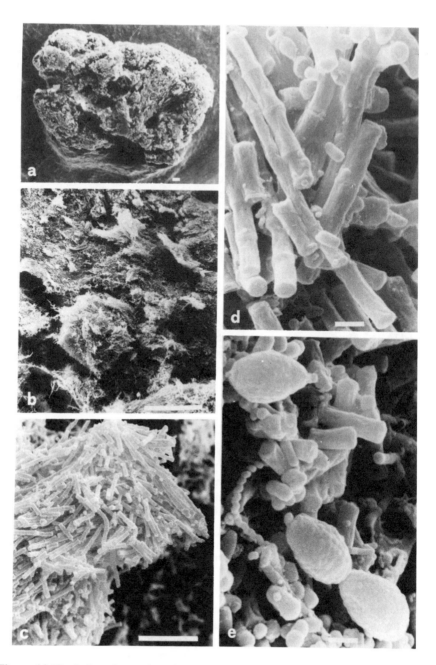

Figure 14.15 Series of scanning electron micrographs of a bacterial floc taken from a methanogenic enrichment degrading a mixture of acetate and propionate. Bar represents (*a*) and (*b*), 100 μm; (*c*) 10 μm; (*d*) and (*e*), 1 μm. From A. D. Adamse, M. H. Deinema, and A. J. B. Zehnder, Studies on bacterial activities in aerobic and anaerobic waste water purification. Antonie Leeuwenhoek, J. Microbiol. 50:665–682, © Martinus Nijhoff Publishers, Dordrecht, The Netherlands, reproduced by permission.

Figure 14.16 Scanning electron micrographs of a bacterial floc grown anaerobically in a sugar factory wastewater. Note extensive formation of extracellular material in (*b*). Bar represents 100 μm in (*a*) and 1 μm in (*b*). Reproduced from Ref. 68a, reproduced with permission from the Canadian Journal of Microbiology.

$$1 \text{ kg BOD} \triangleq 15.625 \text{ mol CH}_4 \text{ or } 350 \text{ L CH}_4$$

at standard conditions. The total amount of biogas formed (including CO_2 and N_2) depends on the redox balance of the substrate and the amounts of nitrate and sulfate provided as alternative electron acceptors. *In praxi*, the theoretically possible methane yields are only rarely reached.

Bioreactors as simple as the septic tank are still in use and represent for many people the state of the art of anaerobic biotechnology. However, they are easy to construct, and advanced versions of this reactor type function in rural areas of India, China, and Taiwan for the disposal of domestic and agricultural wastes. Their simple architecture allows their application on a small scale with a 5- to 30-m^3 working volume in small villages and thus avoids transport expenses for collection purposes. A construction scheme for a very widespread type of rural biogas reactor is presented in Fig. 14.17.

The conventional homogeneously mixed sludge digestor is the most generally accepted system today for anaerobic treatment of sewage sludge residues in wastewater treatment plants. It is also widely applied for the treatment of wastes from cattle and swine farming, slaughterhouse offals, agricultural wastes, or generally speaking, any kind of semisolid waste material with a comparably high solids content.

Since separation of bacterial biomass and undergraded solids is not possible, the efficiency of methane production depends largely on the detention time of the material in the reactor. Nonetheless, conversion efficiencies of 70 to 90% were achieved in these reactors, depending on the biodegradability of the influent substrate. It is important to mention that simultaneous treatment of different kinds of substrates often enhances methane yields. This was found with straw and cattle manure mixtures (204) as well as with mixtures of manure from different sources (125).

The anaerobic contact process has been applied successfully in the recent past for treatment of highly concentrated sewage. Especially the very rich wastewaters produced in the sugar, alcohol, potato, and yeast industry, with BOD loads in the

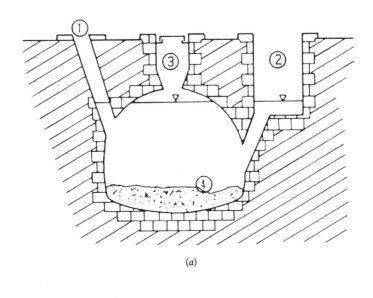

(a)

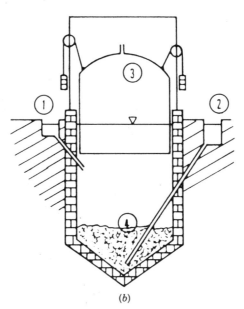

(b)

Figure 14.17 Construction scheme of simple rural biogas reactors used for disposal of domestic and agricultural wastes in third-world countries. (a) Chinese type. (b) Indian type. 1, inflow; 2, outflow; 3, gas reservoir; 4, solid sludge residues. From Ref. 36, reproduced by permission.

range 2 to 20 g/L, were treated in anaerobic contact reactors, and the BOD load decreased by 80 to 95%. In combination with a subsequent aerobic polishing step, the system was patented as the ANAMET process, which today is sold by industrial manufacturers in the form of complete plants (118). An operation scheme of such a combined system is shown in Fig. 14.18. Similar systems were applied in the pulp and paper industry for the treatment of sulfite evaporator condensates (85), liquors from heat-treated sewage sludge (225), or sanitary landfill leachates (49).

The upflow anaerobic sludge blanket (UASB) process has found broad application in the food industry (161, 162). Conversion efficiencies up to 95% at hydraulic retention times of several hours to a few days were achieved, and application of this process for the primary treatment of municipal wastewaters was considered (190). With distillery wastewaters as organic substrate, the UASB system was also successfully applied for wastewater denitrification (144).

With organic loads lower than about 1 g BOD per liter, the anaerobic treatment systems above become inefficient because the comparably high water volumes transported wash too much of the biomass out of the reactor. For treatment of those lower-strength wastewaters, fixed-film reactors proved effective and can handle high-strength dissolved wastes as well (289). Fixed-film anaerobic filters filled with quartzite or limestone gravel, ceramics, porous polyvinyl chloride, sinter glass, activated carbon, wood chips, or even shrimp chitin coatings were applied for treatment of pharmaceutical wastewater (127), distillery sewage (37, 209), and silage effluents (19). The results were usually convincing, with high-strength influents yielding BOD decreases of 90 to 98%. Treatment of low-strength domestic wastewater in an anaerobic filter with a hydraulic retention time of 24 h decreased the COD from 288 mg $\cdot$ L^{-1} to 78 mg $\cdot$ L^{-1}. The gas composition obtained indicated that nitrate reduction played an important role in the degradation process, and the methane yield was comparably low (146). Nonetheless, the system was very insensitive to daily fluctuations in the influent wastewater quality, and its operation was far cheaper than that of a comparable aerobic treatment system.

Even better results were obtained for domestic wastewater treatment in a fluidized bed reactor. Around 75% of COD decrease was achieved at a hydraulic retention time of less than 30 min (128). Combined with an aerobic polishing step, this system yielded sufficient wastewater cleaning up to the nitrification step, with comparably low operation costs and minimal sludge production.

Anaerobic filters with fixed or suspended beds also hold promise for treatment of industrial wastewaters or sanitary landfill leachates which contain compounds that are either toxic or refractory during aerobic degradation. The discovery of anaerobic biodegradability of an entire series of such compounds as halogenated aliphatic and aromatic hydrocarbons, furfurals, pesticide residues, and so on (29, 31, 34, 41, 101) opens the door to a whole new world for anaerobic degradation in special industrial applications (102).

14.6.5 Process Manipulation for Efficiency Enhancement

What can be done to enchance the efficiency of a methanogenic digestor? This question has been asked many times, and answers have been sought by various

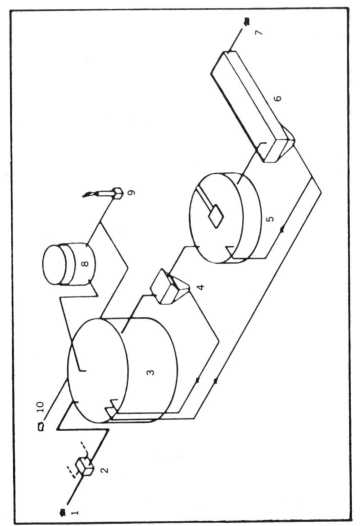

Figure 14.18 Schematic diagram of the ANAMET sewage treatment system. 1, Influent wastewater; 2, heat exchanger; 3, anaerobic tank; 4, anaerobic sludge separator; 5, aerobic tank; 6, aerobic sedimentation; 7, treated effluent; 8, gas holder; 9, torch for excess gas; 10, gas to factory for thermal energy production. From Ref. 118, reproduced by permission.

approaches. It should be evident from the considerations above that retaining of the microbial biomass within the digestor is the most important prerequisite of reactor efficiency.

Our discussion of the carbon and electron flow during methanogenic degradation (Section 14.5) should also have illustrated that an intrinsic connection between fermentative and methanogenic bacteria is of extreme importance for an efficient substrate methanization. Nonetheless, separation of a primary substrate fermentation step (liquefaction, acidification) from the terminal methanization process (gasification) is often postulated by sanitary engineers to be helpful in increasing conversion efficiency. Such a two-step process has several disadvantages: first, the accumulation of long-chain fatty acids and acetate in the fermentation step limits substrate conversion at an early stage unless this process is really acidophilic. Second, degradation of the intermediately formed fermentation products is kinetically far less favorable than an immediate conversion of complex organic matter directly to acetate and hydrogen. Third, the intermediate accumulation of acids necessitates a considerable effort of pH measurement and control. Fourth, some of the reducing equivalents are lost as hydrogen gas in the first step, which could otherwise be transformed into methane. In the author's opinion, two-stage methanization is useful only if (a) the substrate solution contains high amounts of easily fermentable compounds (e.g., monomeric sugars), which are readily converted to easily degradable acids such as lactic acid without hydrogen formation, or (b) the substrate is supplied only seasonally for a short period of time, as in the sugar beet industry. In both cases, the acidification step serves the function of intermediate substrate preservation with subsequent continuous methanogenic treatment.

Raising the reactor temperature is another way to increase the methane formation rate. Methanogenic populations exhibit two temperature optima: one around 35°C and another around 60°C. It should be noted that these two optima are due to completely different populations, including not only the methanogenic but also the various types of fermentative bacteria involved. Most anaerobic digestors are operated at 30 to 35°C because methane formation at 20°C or lower is extremely slow. Operation at 55 to 60°C increases the methane formation rates about twofold compared to 35°C (273); the resulting halving of residence times or reactor volume, however, has to be paid for by high heating expenses to maintain the reactor temperature. Therefore, thermophilic biomethanation appears useful only if the digestor influent is hot on its own (e.g., in the distillery industry) or if inexpensive heat is available (e.g., close to power plants). Nonetheless, thermophilic methane formation is still a matter of basic interest because at least some metabolic processes appear to take different paths at elevated temperatures than at ambient temperature (e.g., acetate degradation) (303). Thermophilic operation is also an interesting alternative in ethanol fermentation (see Section 14.7.1).

Chemical additions were also tried for improvement of methane formation efficiency. Although the cell yields during anaerobic degradation are low and the nutrient requirements for the substrate mixture composition are low as well, equilibration of a too-one-sided diet can greatly increase the efficiency of methanogenic degradation. Sugar factory wastewaters contain 1 to 4 g dissolved organic carbon but only 20 to 50 mg of nitrogen and 0 to 5 mg of phosphorus per liter (190). It is evident that addition of some municipal wastewater can easily cover the nitrogen and phosphate needs of the bacterial community and can thus enhance process efficiency. Addition of nickel as a trace element required by methanogenic bacteria was recently found to stimulate anaerobic digestion significantly (239). Iron additions had a similar enhancing effect (111), and addition of sulfur compounds could also be helpful (137, 213). It should be noted, however, that metal sulfides are soluble only to a very limited extent, and metal or sulfide additions are often precipitated and thus are lost to the bacterial community. Whether sodium can also be limiting in methanogenic degradation has still to be examined. In pure cultures, a sodium chloride concentration of 10 mmol/L (0.6 g/L) was found to be optimal for the growth of some methanogens (189).

The pH should be at 6.8 to 7.5. At higher proton activities (pH 6.5), homoacetogenic anaerobes outcompete the methanogens for hydrogen, and short-chain fatty acids accumulate (106, 194).

It was recently reported that traces of oxygen can also stimulate methanogenesis (196). It is not yet clear whether oxygen helps to activate otherwise undegradable substrates or allows aerobic bacteria to produce growth features for the anaerobic community. The stimulating effect of activated carbon on methane production (240) has already been mentioned.

The triumphant march of genetical engineering through biotechnology does not bypass methanogenic degradation, and projects are in progress in several laboratories to "improve" methane production by genetical manipulations of the microbes involved. In the author's opinion, such efforts do not promise success. The shortcomings of anaerobic bacteria, low growth rates and yields, cannot be overcome by genetical tricks but are determined by the thermodynamics of the conversion processes concerned. Methanogenic degradation of organic matter belongs to the oldest process in evolution; if any "improvements" were possible, they would already be in operation. Moreover, it is doubtful whether a genetically manipulated bacterial strain could win through in competition with the natural populations. So far, manipulated bacteria have had a chance only in strictly controlled artificial laboratory environments, and have failed to survive in a natural system. The only promising field of genetical engineering therefore appears to be the adaptation of existing anaerobes to the degradation of substrates which so far fail to be degraded under anaerobic conditions (e.g., xenobiotics). All other efforts of improvement have to deal with improvement of the conditions for selection in the degradative system, not with artificial changes of the existing populations.

14.6.6 Further Processing of Anaerobically Treated Wastewater

Wastewaters that have passed a primary anaerobic treatment by any of the processes mentioned in the foregoing section differ considerably from aerobically treated wastewater. They contain all inorganic nitrogen and sulfur in their reduced forms, and due to the high solubility of phosphates in the presence of excess sulfide (see Chapter 10), they are comparably rich in phosphates. Unless direct application for irrigation purposes for example, is possible, the treated water needs a further "polishing" treatment before release into a surface water. Mostly, this polishing step is an aerobic oxidation analogous to the classical activated sludge process. Energy expenditure in this treatment will be comparably low because the bulk of the organic freight has been removed during anaerobic treatment. A combination of primary anaerobic with secondary aerobic treatment has been introduced as the ANAMET process (Fig. 14.18). Because of the low residual organic freight of the entering water, the aerobic treatment can easily reach a complete oxidation of ammonia to nitrate which can be converted into nitrogen gas in a subsequent de-nitrification step. Chemical precipitation of phosphate appears desirable as a final treatment; the phosphate-rich precipitates should not be mixed with the bacterial sludges in order to avoid resolubilization of phosphates.

The various techniques of secondary wastewater treatment cannot be reviewed here in detail. It should be mentioned, however, that fixed-bed reactors have also been applied successfully for wastewater nitrification and denitrification (72, 150). Oxidation of reduced sulfur compounds does not necessarily require molecular oxygen as reactant. Provided that enough space and enough sunlight are available, as in tropical or subtropical regions, sulfide oxidation can also be carried out by purple sulfur bacteria in open oxidation ponds (145). Unfortunately, these two prerequisites usually do not apply to densely populated industrialized areas.

14.6.7 Digestion of Solid Wastes

Our discussion of anaerobic waste degradation has so far centered on the treatment of wastewaters of industrial or municipal origin. However, huge amounts of solid wastes are produced all over the world, especially as by-products of agriculture, and the disposal technology in these cases is often unsatisfactory. Specialization of agriculture into separate branches has often increased the disposal problem. In the classical mid-European farm system, straw was produced in corn production, used within the same farm unit for cattle and swine feeding and bedding, and the mixed straw dung was dispersed on the field for fertilization, where it was suffi-ciently decomposed within one vegetation period. Specialized farms today produce huge amounts of plant litter on the one hand and manure on the other hand, and they both pose considerable disposal problems. Straw is very slowly degraded in field soil since the nitrogen/carbon balance is very low; it is often burned in the field or in special stoves. Manure from cattle, swine, and hen farming can be applied to the field in only limited amounts because of possible groundwater con-tamination.

Disposal of agricultural wastes by methanogenic fermentation does not require dilution in water prior to fermentation. Surprisingly, a high solids content in the decomposing matter is advantageous for methanogenic fermentation, as experi-ments with cow manure have demonstrated: Methane formation was optimal at 10 to 30% moisture content, and declined at a moisture content above 35% (129). Methane production from other agricultural wastes, such as grass, cornstalks (57), or straw (104), appears feasible as long as they cannot be used as animal food. The lignin content of straw (10 to 15% w/w) is probably the main obstacle in anaerobic degradation. Delignification of straw is nearly as difficult as delignification of wood, and the various techniques applied included solvent treatment (8, 91), steam explosion (200), and milling with subsequent acid extraction (90). All these techniques are costly and probably too expensive to operate on a farm-scale basis. Han (104) finds feeding to animals to still be the best way to utilize straw. Unfortunately, most of the cellulose synthesized in nature is present as lignocellulose (i.e., in intrinsic association with lignin, which prevents significant cellulose degradation under anaerobic conditions). Once lignin is removed by any of the treatments mentioned above, the resulting pure cellulose polymer is easily accessible to anaerobic microbial attack.

The principles of cellulose depolymerization are treated in Chapter 7. Technical cellulose depolymerization proceeds either via hydrolysis by cell-free enzyme preparations (68, 237) or through direct attack by mesophilic or thermophilic mixed microbial populations. It has to be noted that natural cellulose exists to a high proportion in a crystalline arrangement which has been made by the producer for the purpose of structural rigidity and resistance to microbial attack. The various purified cellulose preparations used in laboratory experiments (paper, Solka Floc, carboxymethyl cellulose, cellulose powder) are all already partially denatured cellulose derivatives, and experimental results obtained with these substrates cannot be transferred directly to natural cellulosic biomass. Nonetheless, several mesophilic organisms suited for conversion of natural cellulose to methane in mixed defined cultures have been described in the recent past, and include gram-negative anaerobes (138, 157), *Clostridium* sp. (233), and lower fungi (178, 185). A two-stage combination fermenter for methanogenic cellulose degradation was recently described in which cellulose is depolymerized in the suspended liquid phase, while methanogenic degradation of the soluble intermediates occurs in a fixed bed within the same reactor vessel. This combined system was far superior to a fully mixed reactor system with or without solids recycling, and reduced the hydraulic retention time from 28 days to 6 days at a feeding rate of 4.8 g of cellulose per liter per day (139). These results agree well with the considerations in Section 14.6.5 that retainment in the reactor vessel of those bacteria involved in the slowest and least energy-yielding process greatly enhances the conversion efficiency. The applicability of such a mixed reactor system for the treatment of mixed cellulosic waste materials remains to be examined. Thermophilic cellulose degradation was another field of intense research in recent years. The cellulolytic enzyme system of *Clostridium thermocellum* has been characterized in detail (181). Cellobiose is released in small amounts as the only soluble sugar intermediate, and is fermented in pure culture to ethanol, lactate, acetate, hydrogen, and carbon dioxide. Mixed methanogenic cultures with this bacterium and a methanogenic component were composed (277), but their technical application has still to be proven.

Hemicelluloses are an ill-defined group of natural noncellulosic polysaccharides which have only recently gained the attention of biotechnologists due to their high contribution to the biomass of plant litter (16 to 27% of dry weight; 299). Enzymic and physicochemical treatments have been used for its degradation (66–68). Recently, thermophilic anaerobic bacteria able to ferment xylane have been isolated (278). Apparently, hemicelluloses as well as pectin (216) do not represent major obstacles in methanogenic degradation of complex plant biomass in mixed anaerobic fermenters, but our knowledge of this matter is poor.

14.6.8 Feasibility of Methanogenic Waste Treatment

The various aspects of methanogenic waste treatment discussed in this section can be summarized as follows:

1. Anaerobic treatment with methane as the main reduced end product can compete successfully with aerobic treatment provided that the reactor system allows sufficient retention of the active biomass.
2. Anaerobic treatment is by far the most economical since energy can be recovered from biomass instead of being invested in costly aeration devices.
3. Sludge production and the consequent disposal problems are by far smaller during anaerobic than during aerobic treatment.
4. Anaerobic digestors can often be constructed from very simple building material and usually do not need expensive control devices.

Despite these obvious advantages of anaerobic over aerobic waste treatment, anaerobic treatment of wastes is still of minor importance in industrialized countries. This is due basically to the fact that suitable anaerobic reactors have only recently been constructed and require more improvement before they can compete successfully with the widely used aerobic treatment devices. In Fig. 14.19 a scheme is presented which recommends anaerobic waste treatment in the range 1 to 300 g of dry matter content per liter of liquid volume. Advanced techniques will soon be at hand which allow wastewaters with solids contents down to 100 mg/L to be treated anaerobically, and devices for fermentation of solid material with water contents lower than 30% are being developed. Some special wastes from chemical industry rich in phenols, halogenated and aminated aromatics, or surface-active substances lend themselves to anaerobic degradation because of technical difficulties in aerobic treatment. In general, any reducing equivalent transferred to nitrate or carbon dioxide to form nitrogen gas or methane is a success for the treatment plant since it does not consume oxygen in expensive aeration devices.

Nonetheless, anaerobic treatment also has some drawbacks. Hydrogen sulfide is a necessary by-product of anaerobic treatment, depending on the sulfate content of the influent water. Its toxicity, chemical reactivity, and unpleasant odor require suitable precautions to prevent poisoning or corrosion of building material, including pipes and control devices. The main reduced end product, methane, has to be handled with caution to avoid explosions and fires. Moreover, all substrates containing lignin as a major component will be only incompletely digested in the absence of molecular oxygen.

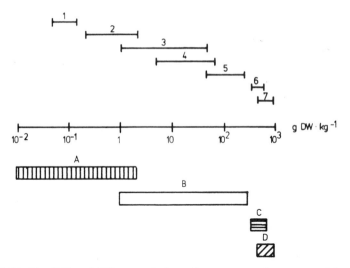

Figure 14.19 Feasibility of different techniques for treatment of waste materials with various solid contents. 1, municipal wastewaters; 2, wastewaters from paper and pulp industry; 3, wastewaters from food industry; 4, sewage sludge; 5, manure; 6, municipal refuse; 7, plant litter; *A*, aerobic treatment; *B*, anaerobic treatment; *C*, composting; *D*, incineration. From *P. Präve et al. (eds.), Handbuch der Biotechnologie*, 2. Aufl. München, Wien: R. Oldenbourg Verlag, 1984, p. 525, reproduced by permission.

Also, hygienic problems remain to be solved. Our knowledge of anaerobic protozoa and other eukaryotes that feed on bacteria is very limited, but so far they seem to be less efficient in the reduction of numbers of suspended bacteria in anaerobic sewage sludge than do their aerobic counterparts in activated sludge. Even less is known about the fate of viruses during anaerobic treatment. Perhaps the mixed population of bacteria and protozoa established in the secondary aerobic polishing step after a primary anaerobic treatment will take sufficient care of those pathogens that survive the ill-smelling anaerobic mix. Research in this field is needed before anaerobic processes can find broader application for the treatment of municipal wastewaters.

14.7 NONMETHANOGENIC FERMENTATIONS

In the foregoing sections, complete methanogenic degradation of complex organic matter was described as a desirable process for inexpensive waste disposal. However, complete degradation is not always desired: From ancient times, humankind has used various fermentation processes just for protection of food from microbial degradation. Various sour milk products, sauerkraut, silage, cheese, and alcohol beverages are conserved foods of this kind which are preserved by anaerobic fermentation products (e.g., lactic acid, propionic acid, or ethanol). The microorganisms involved, mainly lactic and propionic acid bacteria and yeasts, modify the respective media in a very short time in such a way that other bacteria competing for the same substrates are strictly inhibited by high acid or alcohol concentrations. Natural or artificially stabilized enrichment cultures have been maintained over centuries and still do their work at a sufficient level of quality. Recent technological improvements in applying pure and defined cultures in the milk and cheese industry improved the reproducibility of the fermentation process but not product quality.

New developments have recently been achieved in the field of mass production of raw chemicals from biomass to substitute for petrochemicals as sources of chemical production. Old processes such as butanol–acetone fermentation were revived after a long time, which was characterized by the complete takeover of petrochemicals. Rising oil prices made biomass an economically feasible alternative as a renewable source of raw chemicals, and this development is progressing.

A simple scheme for production of some important raw materials from biomass is presented in Fig. 14.20. Space will allow to discuss only a few recent advances in this field; the interested reader is referred to the literature mentioned.

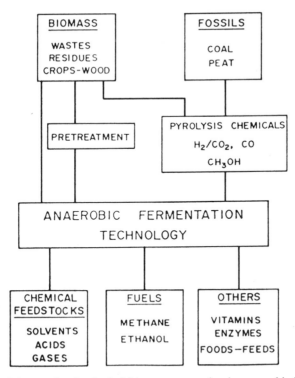

Figure 14.20 Scheme of technological biomass conversion by anaerobic bacteria. From Ref. 298, reproduced with permission from the Annual Revew of Microbiology, Vol. 34, © 1980 by Annual Reviews Inc.

14.7.1 Mass Production of Alcohols and Other Solvents

Ethanol production from sugars by yeasts is an old and well-known process. This and other techniques of ethanol fermentation have recently been reviewed (149). The main limitation of ethanol fermentation by yeasts is their limited spectrum of fermentable substrates: Most yeasts are only able to use monomeric or dimeric hexoses. Addition of fungal or bacterial amylases to yeast cultures fermenting starch-containing plant residues increased ethanol yields considerably. Another approach was genetic manipulation to yield starch-digesting yeasts with high metabolic activity (154). In continuous fermentation systems, substrate conversion efficiency was considerably increased by immobilizing the yeast cells in tubular fermentors (88) or by cell recycling (255), processes that are of great importance with yeasts since their growth in the absence of oxygen is very slow.

Although yeasts are rather tolerant to their product and are usually inhibited by ethanol only in the range of 13% (v/v; $\triangleq$ 2.5 mol/L), this content is still too low for direct ethanol recovery by distillation. Other means of product recovery are the selective absorption and desorption of ethanol by polystyrene beads (197) or the use of hollow-fiber reactors in which the ethanol produced is concentrated by selective extraction through the fiber walls (122).

Still, bacterial ethanol production has several drawbacks. Their substrate range is greater than that of yeasts and allows utilization of pentoses, hemicelluloses, and other substrates such as whey or spent sulfite liquor from the wood pulping industry, but the fermentation does not lead exclusively to ethanol, and the tolerance to ethanol is far lower than with yeasts (149, 163). Mixed culture fermentations with bacteria and yeasts promise to combine the metabolic advantages of either partner organism, but the successful application of such a system has still to be proven. Thermophilic bacterial ethanol fermentation looks more promising. Enhanced process temperatures (55 to 70°C) help to avoid contaminants, allow faster substrate turnovers, and direct recovery of ethanol from the hot fermentation broth by continuous vacuum extraction (282). Some thermophilic bacteria exhibit fermentation patterns of 1.8 to 1.9 mol of ethanol per mole of glucose fermented, and thus reach nearly the efficiency of yeasts (76). The combination of cellulose-degrading thermophiles with sugar-utilizing bacteria with high ethanol production gave good results (182, 301); the application of new techniques to new bacterial mixed culture combinations looks promising. Ethanol fermentation may also be a suitable technique for the utilization of waste products such as whey and sewage from the fruit canning industry.

Acetone–butanol fermentation experienced a revival in recent years. Bagasse, cornstarch, or molasses are converted by enzymatic hydrolysis into sugar solutions which are fermented by *Clostridium acetobutylicum* to a n-butanol/acetone/ethanol mixture in a 6:3:1 weight ratio (241). The process consists of two steps: During the first, the bacteria grow and produce mainly acetate, butyrate, hydrogen, and carbon dioxide, with concomitant lowering of the pH from 7.0 to 4.5 to 5.0; in the second phase, more sugar and part of the already excreted acetate and butyrate are converted to acetone and butanol. It has been shown in recent experiments that the drop in pH is a necessary but not sufficient prerequisite for solvent production; accumulation (or addition) of acetate and butyrate trigger the production of acetone and butyrate, respectively (13, 89, 97). Solvent production was enhanced significantly by adding the respective acids from the beginning (291); continuous operation of the process was tried but without really convincing success so far (12, 130). Poisoning of the hydrogenase system by carbon monoxide shifted the fermentation balance to mainly butanol (141). On the other hand, one could think of a fermentation of sugars to nearly exclusively acetone by combination of *C. acetobutylicum* with a hydrogen-oxidizing homoacetogenic bacterium unable to ferment sugars: This would perhaps change the fermentation pattern during the first stage exclusively to acetate, which is condensated later to acetone.

Production of butanol and 2,3-butanediol from steam-exploded wood preparations by a mixture of *C. acetobutylicum* and *Klebsiella pneumoniae* was recently reported (207). Other fermentative solvent productions include 2,3-butanediol formation from whey by *Bacillus polymyxa* or *Streptococcus faecalis* (238), and glycerol production from K-carageenan by immobilized *S. cerevisiae* cells in the presence of some sulfite to change the fermentation pattern to high proportions of glycerol (26).

Methanol is not a major product of anaerobic microbial fermentations; it is formed from the methoxyl groups of pectin (221). Technical production proceeds via catalytic oxidation of methane; a biological alternative to this reaction has not yet been experienced technically.

14.7.2 Anaerobic Production of Organic Acids

The production of acetate from ethanol by *Acetobacter* species is an old but not really an anaerobic process. Homoacetogenic bacteria oxidizing ethanol could achieve the same transformation with higher acetate yields, but their tolerance to acetate and high proton activities is limited. Homoacetogens could also convert pyrolysis gases, sugar-containing waste materials, or methanol to acetate (300).

Propionate is produced from lactose, other sugars, and starch by propionic acid bacteria (203). The propionate yields can be enhanced considerably if hydrogenase-containing species are used and hydrogen is provided as an additional reductant (260). Butyrate production can start from the same substrates as acetate production if *Butyribacterium methylotrophicum* is used instead of a homoacetogenic bacterium (63, 300). Production of lactate from glucose has been improved by trapping the biocatalyst *Lactobacillus delbrueckii* in a calcium alginate gel (242). With a different *Lactobacillus* strain, xylose could be converted to lactate (262).

14.7.3 Further Products of Anaerobic Metabolism

The lignin component of wood has so far usually been disregarded as a substrate for biotechnological conversions. Fermentation of the lignin-containing extract from the steam-exploded wood leads to catechol as the main aromatic product,

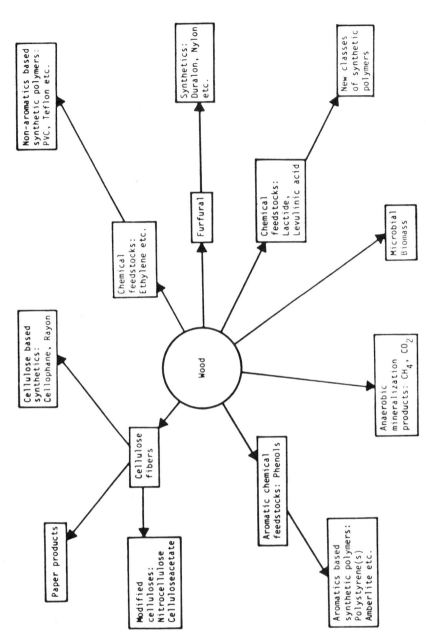

Figure 14.21 Possible products from wood. From Ref. 105, reproduced by permission.

which is only very slowly degraded (134). An entire chemical industry could be based on wood as the sole source, as Fig. 14.21 illustrates. However, neither industrialized nor developing countries could provide the amount of wood required to base their entire chemical industry on wood (105). Nonetheless, the possibilities for utilization of waste materials from wood processing should be explored to save the world's dwindling wood resources and to minimize pollution of surface water with, for example, sulfite liquors. It was calculated that many other waste materials could also be fermented to useful raw products for chemical synthesis and could thus substitute for oil and petrochemicals (186).

Production of hydrogen from biomass has been explored in the last decade. Fermentative hydrogen evolution is for thermodynamic reasons limited to a very small fraction of the reducing equivalents provided (257); the various efforts of photocatalyzed hydrogen evolution did not fulfill the hopes people had for it 15 years ago. Due to the small energy gains of anaerobic metabolism, anaerobes are not suited for single-cell protein production. The only exception could be the photosynthetic bacteria, which could convert treated wastewater containing organic compounds or sulfides to protein materials rich in essential amino acids or vitamins provided that sufficient light is available (232).

14.8 ENVIRONMENTAL IMPLICATIONS OF ANAEROBIC DEGRADATION

The foregoing chapters illustrate that the degradation potential of anaerobes is not identical with that of aerobes. Nonetheless, all procedures employed in testing the biodegradability of new products of the chemical industry so far cover only their degradation under aerobic conditions, and very often, legislation demands only partial degradability. As a consequence, chemical products are currently sold which are only incompletely degraded during aerobic wastewater treatment, and the remnants accumulate in anoxic sediments, sludges and so on. This is the case with, for example, the nonionic detergent builder alkylphenolethoxylate (Fig. 14.22). The usual chemical detection reaction detects only the polyethylene glycol moiety, which is nearly completely degraded in, for example, the activated sludge treatment. The residual alkylphenol mono- or diethoxylate is rather lipophilic and adsorbs to sludge components, with which it reaches anaerobic sewage fermenters or anoxic sediments. The anaerobic microbial community cleaves the last one or two ethoxyl residues off and releases alkylphenol, a highly toxic substance which is neither aerobically nor anaerobically degradable and accumulates in the sediments of lakes and rivers in alarming concentrations (95). Although the production of this surfactant type has been reduced in the last two years, there is still a considerable increase in alkylphenols in natural habitats.

This example should demonstrate that not only are well-known environmental pollutants such as halogenated hydrocarbons threatening our environment, but substances which are harmless in the form in which they are sold can be converted to

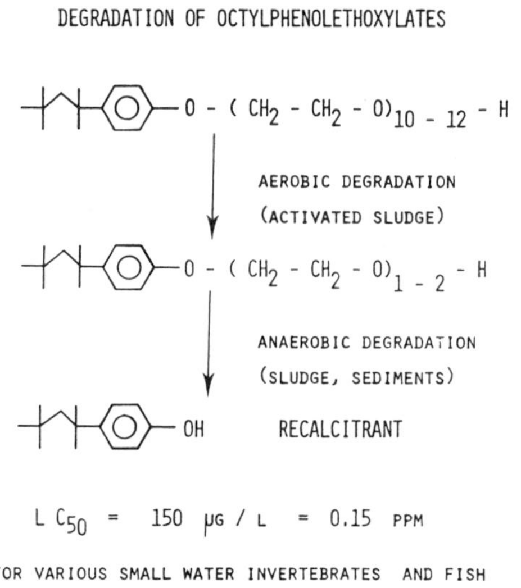

DEGRADATION OF OCTYLPHENOLETHOXYLATES

$$L\ C_{50}\ =\ 150\ \mu G\ /\ L\ =\ 0.15\ \text{PPM}$$

FOR VARIOUS SMALL WATER INVERTEBRATES AND FISH

Figure 14.22 Degradation of alkylphenolethoxylates under aerobic and anaerobic conditions.

toxic or otherwise harmful derivatives by microbial action. It has to be postulated that the environmental behavior of any new sold product is checked under both aerobic *and* anaerobic conditions, and that *complete* degradation is possible in both cases. We can no longer afford to convert our sediments and sewage sludges into huge, highly toxic cemeteries for the products of a too inventive chemical industry.

These considerations also apply to the planned replacement of phosphates as a detergent builder by nitrilotriacetate (NTA). There is no convincing evidence yet for complete degradation of this compound in anoxic sediments. The possible consequences of NTA accumulation in anoxic sediments (e.g., remobilization of heavy metals) are too troubling to allow widespread use of this compound without sufficient checks on its anaerobic biodegradability.

The widespread use of antibiotics in raising cattle and hogs is another matter of concern for the environmental biologist. The advantages of these feed additions are doubtful; the disadvantages (accumulation of chemically modified antibiotics in food and the environment) are obvious. Monensin addition to cattle feed to prevent methane belching and to transform the lost reducing equivalents into utilizable acetate is one of the most dubious misuses of scientific achievements in this respect. The resulting problems in methanogenic degradation of cattle waste (272) is a minor but, on a large-scale, disastrous consequence.

14.9 CONCLUSIONS

The biodegradative potential of anaerobic bacteria has seen a tremendous expansion in the last 10 to 15 years. The old view that compounds not degradable aerobically will never be degraded anaerobically is no longer true. Many aerobically problematic substances proved to be degradable in the absence of oxygen, and in some cases anaerobic biodegradation was even more efficient than aerobic breakdown. Most substrates which in the presence of oxygen are attacked by oxygenases undergo basically different degradation reactions in the absence of oxygen. Nonetheless, the principles of some cleavage reactions (e.g., the cleavage of ethers) are comparable in the absence and the presence of oxygen.

Among the natural substrates, only lignin and ligninaceous plant matter resist anaerobic biodegradation. The conversion of wood or straw into useful fermentation products therefore requires costly physical or chemical pretreatment.

The main drawbacks of anaerobic bacteria—low growth rates and yields—which hampered their widespread use for wastewater treatment in the past have been overcome by the construction of various types of anaerobic reactors which either retain or recycle the active biomass to maintain substrate conversion rates which are competitive with aerobic processes. The production of methane as a useful end product, especially from high-strength wastewaters, makes methanogenic treatment an attractive alternative to the widespread energy-intensive aerobic wastewater treatment, with its increasing waste sludge disposal problems in industrialized countries.

Anaerobic biodegradation of toxic or otherwise problematic environmental pollutants which in the past were considered to resist anaerobic attack can be applied to the treatment of special industrial wastewaters or landfill leachates. Nonetheless, anaerobic biotechnology cannot perform miracles. The success of all efforts to recycle products from waste materials, especially to recycle drinking water, will depend to a high extent on the composition of waste materials. Accumulation of heavy metals and other toxic compounds in sediments and sewage sludges precludes recycling of otherwise useful organic substrates and energy carriers, and endangers our groundwater resources.

The fascinating economy of anaerobic energy metabolism allows us to produce useful chemical raw products from waste materials which until now have usually been dumped, wasted, or burned. The production of fuels and chemicals from biomass, however, finds its limitations in biomass availability and should be restricted primarily to the economic utilization of waste materials. Growing wood for fuel production at times which are characterized by alarming worldwide forest losses is a technological concept as dubious as ethanol production from corn starch in a world in which millions of people are starving.

The relative ease of constructing and maintaining anaerobic sewage digestors on a small scale renders this technology well suited for waste treatment and energy recovery in rural areas of third-world countries which lack the money to buy oil on the international market. Anaerobic biotechnology could thus help to save the world's forest resources and could be a helpful developmental aid from the industrial countries.

We microbiologists and biotechnologists are encouraged to spread our knowledge of economic methods of waste treatment and resource recycling in those countries which so far learned from us only how to destroy nature and waste its gifts.

ACKNOWLEDGMENTS

I am indebted to Prof. Dr. Norbert Pfennig for generous support of my experimental work and for useful hints for improvement of the manuscript. I also gratefully acknowledge the help of Dr. Marion Stieb and Dr. Andreas Tschech, who provided unpublished results and suggestions. Dr. Jan Dolfing generously made pictures of microbial flocs available prior to publication. Skilful and careful typing by Renate Gimmi is gratefully acknowledged. Experimental work in the author's laboratory was supported by the Fakultät für Biologie der Universität Konstanz and the Deutsche Forschungsgemeinschaft.

REFERENCES

1. Abeles, R. H., and H. A. Lee, Jr. 1961. An intramolecular oxidation reduction requiring a cobamide coenzyme. J. Biol. Chem. 236:2347–2350.

1a. Adamse, A. D., M. H. Deinema, and A. J. B. Zehnder. 1984. Studies on bacterial activities in aerobic and anaerobic waste water purification. Antonie Leeuwenhoek, J. Microbiol. 50:665–682.

2. Aftring, R. P., and B. F. Taylor. 1981. Aerobic and anaerobic catabolism of phthalic acid by a nitrate-respiring bacterium. Arch. Microbiol. 130:101–104.

3. Aftring, R. P., B. E. Chalker, and B. F. Taylor. 1981. Degradation of phthalic acids by denitrifying, mixed cultures of bacteria. Appl. Environ. Microbiol. 41:1171–1183.

4. Alexander, M. 1965. Biodegradation: problems of molecular recalcitrance and microbial fallibility, p. 35–80, in W. W. Umbreit (ed.), Advances in Applied Microbiology, Vol. 7. Academic Press, New York.

5. Alexander, M. 1977. Introduction to soil microbiology, 2nd ed. Wiley, New York.

6. Ander, P., K.-E. Eriksson, and H.-S. Yu. 1983. Vanillic acid metabolism by *Sporotrichum pulverulentum*: evidence for demethoxylation before ring cleavage. Arch. Microbiol. 136:1–6.

7. Atlas, R. M. 1981. Microbial degradation of petroleum hydrocarbons: an environmental perspective. Microbiol. Rev. 45:180–209.

8. Avgerinos, G. C., and D. I. C. Wang. 1983. Selective solvent delignification for fermentation enhancement. Biotechnol. Bioeng. 25:67–83.

9. Azoulay, E., J. Chouteau, and G. Davidovics. 1963. Isolement et caractérisation des enzymes responsables de l'oxydation des hydrocarbures. Biochim. Biophys. Acta 77:554–567.

10. Babu, B. H., and C. S. Vaidyanathan. 1982. Catabolism of isophthalic acid by a soil bacterium. FEMS Microbiol. Lett. 14:101–106.

11. Bache, R., and N. Pfennig. 1981. Selective Isolation of *Acetobacterium woodii* on

methoxylated aromatic acids and determination of growth yields. Arch. Microbiol. 130:255–261.

12. Bahl, H., W. Andersch, and G. Gottschalk. 1982. Continuous production of acetone and butanol by *Clostridium acetobutylicum* in a two stage phosphate limited chemostat. Eur. J. Appl. Microbiol. Biotechnol. 15:201–205.

13. Bahl, H., W. Andersch, K. Braun, and G. Gottschalk. 1982. Effect of pH and butyrate concentration on production of acetone and butanol by *Clostridium acetobutylicum* grown in continuous culture. Eur. J. Appl. Microbiol. Biotechnol. 14:17–20.

13a. Bak, F., and F. Widdel. 1986. Anaerobic degradation of indolic compounds by sulfate-reducing enrichment cultures, and description of *Desulfobacterium indolicum* gen. nov. sp. nov. Arch. Microbiol. in press.

14. Bakker, G. 1977. Anaerobic degradation of aromatic compounds in the presence of nitrate. FEMS Microbiol. Lett. 1:103–108.

15. Balba, M. T., and W. C. Evans. 1979. The methanogenic fermentation of ω-phenyl alkane carboxylic acids. Biochem. Soc. Trans. 7:403–405.

16. Balba, M. T., and W. C. Evans. 1980. The methanogenic biodegradation of catechol by a microbial consortium: evidence for the production of phenol through *cis*-benzenediol. Biochem. Soc. Trans. 8:452–454.

17. Barker, H. A. 1936. On the biochemistry of the methane fermentation. Arch. Microbiol. 7:404–419.

18. Barker, H. A. 1937. The production of caproic and butyric acids by the methane fermentation of ethyl alcohol. Arch. Microbiol. 8:416–421.

19. Barry, M., and E. Colleran. 1984. Silage effluent digestion by an upflow anaerobic filter. Water Res. 18:827–832.

20. Bartels, I., H.-J. Knackmuss, and W. Reinecke. 1984. Suicide inactivation of catechol 2,3-dioxygenase from *Pseudomonas putida* mt-2 by halocatechols. Appl. Environ. Microbiol. 47:500–505.

21. Benckiser, G., and J. C. G. Ottow. 1982. Metabolism of the plasticizer di-*n*-butylphthalate by *Pseudomonas pseudoalcaligenes* under anaerobic conditions, with nitrate as the only electron acceptor. Appl. Environ. Microbiol. 44:576–578.

22. Benner, R., A. E. Maccubin, and R. E. Hodson. 1984. Anaerobic biodegradation of the lignin and polysaccharide components of lignocellulose and synthetic lignin by sediment microflora. Appl. Environ. Microbiol. 47:998–1004.

23. Berkeley, R. C. W., J. M. Lynch, J. Melling, P. R. Rutter, and B. Vincent (eds.). 1980. Microbial adhesion to surfaces. Ellis Horwood, Chichester, West Sussex, England.

24. Bernhardt, F.-H., H. Staudinger, and V. Ullrich. 1970. Eigenschaften einer *p*-Anisat-O-Demethylase im zellfreien Extrakt von *Pseudomonas* species. Hoppe-Seyler's Z. Physiol. Chem. 351:467–478.

25. Beverloo, W. A., and S. Bruin. 1978. Kinetics of activated carbon absorption, p. 307–323, in New processes of waste water treatment and recovery. Ellis Horwood, Chichester, West Sussex, England.

26. Bisping, B., and H. J. Rehm. 1982. Glycerol production by immobilized cells of *Saccharomyces cerevisiae*. Eur. J. Appl. Microbiol. Biotechnol. 14:136–139.

27. Böger, P. 1975. Photosynthese in globaler Sicht. Naturwiss. Rundsch. 28:429–435.

28. Booth, A. N., and R. T. Williams. 1963. Dehydroxylation of caffeic acid by rat and rabbit caecal contents and sheep rumen liquor. Nature (Lond.) 198:684–685.

29. Bouwer, E. J., and P. L. McCarty. 1983. Transformations of 1- and 2-carbon halogenated aliphatic organic compounds under methanogenic conditions. Appl. Environ. Microbiol. 45:1286–1294.

30. Bouwer, E. J., and P. L. McCarty. 1983. Transformation of halogenated organic compounds under denitrification conditions. Appl. Environ. Microbiol. 45: 1295–1299.

31. Bouwer, E. J., and P. L. McCarty. 1983. Effects of 2-bromoethanesulfonic acid and 2-chloroethanesulfonic acid and acetate utilization in a continuous-flow methanogenic fixed-film column. Appl. Environ. Microbiol. 45:1408–1410.

32. Bouwer, E. J., B. E. Rittmann, and P. L. McCarty. 1981. Anaerobic degradation of halogenated 1- and 2-carbon organic compounds. Environ. Sci. Technol. 15:596–599.

33. Boyd, S. A., and D. R. Shelton. 1984. Anaerobic biodegradation of chlorophenols in fresh and acclimated sludge. Appl. Environ. Microbiol. 47:272–277.

34. Boyd, S. A., D. R. Shelton, D. Berry, and J. M. Tiedje. 1983. Anaerobic biodegradation of phenolic compounds in digested sludge. Appl. Environ. Microbiol. 46:50–54.

35. Bradbeer, C. 1965. The clostridial fermentations of choline and ethanolamine. J. Biol. Chem. 240:4669–4681.

36. Braun, R. 1982. Biogas—Methangärung organischer Abfallstoffe. Grundlagen und Anwendungsbeispiele. Springer-Verlag, New York.

37. Braun, R., and S. Huss. 1982. Anaerobic filter treatment of molasses distillery slops. Water Res. 16:1167–1171.

38. Braun, K., and D. T. Gibson. 1984. Anaerobic degradation of 2-aminobenzoate (anthranilic acid) by denitrifying bacteria. Appl. Environ. Microbiol. 48:102–107.

39. Braun, M., F. Mayer, and G. Gottschalk. 1981. *Clostridium aceticum* (Wieringa), a microorganism producing acetic acid from molecular hydrogen and carbon dioxide. Arch. Microbiol. 128:288–293.

40. Brock, T. D., and K. O'Dea. 1977. Amorphous ferrous sulfide as a reducing agent for culture of anaerobes. Appl. Environ. Microbiol. 33:254–256.

41. Brune, G., S. M. Schoberth, and H. Sahm. 1982. Anaerobic treatment of an industrial wastewater containing acetic acid, furfural and sulphite. Process Biochem. 17(3):20–35.

42. Brune, G., S. M. Schoberth, and H. Sahm. 1983. Growth of a strictly anaerobic bacterium on furfural (2-furaldehyde). Appl. Environ. Microbiol. 46:1187–1892.

43. Bruyn, J. 1954. An intermediate product in the oxidation of hexadecene-1 by *Candida lipolytica*. Koninkl. Ned. Akad. Wetenschap. Proc. C57:41–45.

44. Bryant, M. P. 1979. Microbial methane production—theoretical aspects. J. Anim. Sci. 48:193–201.

45. Bryant, M. P., E. A. Wolin, M. J. Wolin, and R. S. Wolfe. 1967. *Methanobacillus omelianskii*, a symbiotic association of two species of bacteria. Arch. Microbiol. 59:20–31.

46. Bryers, J. D., and G. W. Characklis. 1982. Processes governing primary biofilm formation. Biotechnol. Bioeng. 24:2452–2476.

47. Bull, A. T. 1980. Biodegradation: some attitudes and strategies of microorganisms and microbiologists, p. 107–136, D. C. Ellwood, J. N. Hedger, M. J. Latham, J. H. Slater, and J. M. Lynch (eds.), in Contemporary microbial ecology. Academic Press, London.

48. Bull, M. A., R. M. Sterritt, and J. N. Lester. 1983. Response of the anaerobic fluidized bed reactor to transient changes in process parameters. Water Res. 17:1563–1568.

49. Bull, P. S., J. V. Evans, R. M. Wechsler, and K. J. Cleland. 1983. Biological technology of the treatment of leachate from sanitary landfills. Water Res. 17:1473–1481.

50. Cain, R. B. 1980. The uptake and catabolism of lignin-related aromatic compounds and their regulation in microorganisms, p. 21–60, in T. K. Kirk, T. Higushi, and H. Chang (eds.), Lignin biodegradation: microbiology, chemistry and potential applications, Vol. 1. CRC Press, Boca Raton, Fla.

51. Caldwell, D. E., and P. Hirsch. 1973. Growth of microorganisms in two-dimensional steady-state diffusion gradients. Can. J. Microbiol. 19:53–58.

52. Calleja, G. B., B. Atkinson, D. R. Garrod, P. Hirsch, D. Jenkins, B. F. Johnson, H. Reichenbach, A. H. Rose, B. Schink, B. Vincent, P. A. Wilderer. 1984. Microbial aggregation. Group report, p. 303–321, in K. C. Marshall (ed.), Microbial adhesion and aggregation (Dahlem Konferenzen, Berlin). Springer-Verlag, Heidelberg.

53. Carlile, M. J. 1980. Positioning mechanisms—the role of motility, taxis and tropism in the life of microorganisms. p. 55–74, in D. C. Ellwood, J. N. Hedger, M. J. Latham, J. H. Slater, and J. M. Lynch (eds.), Contemporary microbial ecology. Academic Press, London.

54. Characklis, W. G., and K. E. Cooksey. 1983. Biofilms and microbial fouling, p. 93–138, in A. J. Laskin (ed.), Advances in applied microbiology, Vol. 29. Academic Press, New York.

55. Chen, M., and M. J. Wolin. 1977. Influence of CH_4 production by *methanobacterium ruminantium* on the fermentation of glucose and lactate by *Selenomonas ruminantium*. Appl. Environ. Microbiol. 34:756–759.

56. Chouteau, J., E. Azoulay, and J. C. Senez. 1962. Anaerobic formation of *n*-Hept-1-ene from n-heptane by resting cells of *Pseudomonas aeruginosa*. Nature (Lond.) 194:576–578.

57. Clausen, E. C., O. C. Sitton, and J. L. Gaddy. 1979. Biological production of methane from energy crops. Biotechnol. Bioeng. 21:1209–1219.

58. Colberg, P. J., and L. Y. Young. 1982. Biodegradation of lignin-derived molecules under anaerobic conditions. Can. J. Microbiol. 28:886–889.

58a. Conrad, R., T. J. Phelps, and J. G. Zeikus. 1985. Gas metabolism evidence in support of the juxtaposition of hydrogen-producing and methanogenic bacteria in sewage sludge and lake sediments. Appl. Environ. Microbiol. 50:595–601.

59. Costerton, J. W., R. T. Irvin, and K. J. Cheng. 1981. The bacterial glycocalyx in nature and disease. Annu. Rev. Microbiol. 35:299–324.

60. Culbertson, C. W., A. J. B. Zehnder, and R. S. Oremland. 1981. Anaerobic oxidation of acetylene by estuarine sediments and enrichment cultures. Appl. Environ. Microbiol. 41:396–403.

61. Curds, C. R. 1982. The ecology and role of protozoa in aerobic sewage treatment processes. Annu. Rev. Microbiol. 36:27–46.

62. Dacre, J. C., and R. T. Williams. 1962. Dehydroxylation of (^{14}C) protocatechuic acid in the rat. Biochem. J. 84:81.

63. Datta, R., and J. Ogeltree. 1983. Methanol bioconversion by *Butyribacterium methylotrophicum*—batch fermentation yield and kinetics. Biotechnol. Bioeng. 25:991–998.

64. Davis, J. B., and H. F. Yarbrough. 1966. Anaerobic oxidation of hydrocarbons by *Desulfovibrio desulfuricans*. Chem. Geol. 1:137–144.

65. De Bont, J. A. M., and M. W. Peck. 1980. Metabolism of acetylene by *Rhodococcus* A 1. Arch. Microbiol. 127:99–104.

66. Dekker, R. F. H. 1983. Bioconversion of hemicellulose: aspects of hemicellulose production by *Trichoderma reesei* QM 9414 and enzymic saccharification of hemicellulose. Biotechnol. Bioeng. 25:1127–1146.

67. Dekker, R. F. H., and A. F. A. Wallis. 1983. Enzymic saccharification of sugarcane bagasse pretreated by autohydrolysis-steam explosion. Biotechnol. Bioeng. 25:3027–3048.

68. De Menezes, T. J. B., C. L. M. Dos Santos, and A. Azzini. 1983. Saccharification of bamboo carbohydrates for the production of ethanol. Biotechnol. Bioeng. 25:1071–1082.

68a. Dolfing, J., A. Griffioen, A. R. W. van Neerven, and L. P. T. M. Zevenhuizen. 1985. Chemical and bacteriological composition of granular methanogenic sludge. Can. J. Microbiol. 31:744–750.

69. Donnelly, M. I., and S. Dagley. 1980. Production of methanol from aromatic acids by *Pseudomonas putida*. J. Bacteriol. 142:916–924.

70. Dorn, E., M. Hellwig, W. Reineke, and H.-J. Knackmuss. 1974. Isolation and characterization of a 3-chlorobenzoate degrading pseudomonad. Arch. Microbiol. 99:61–70.

71. Dürre, P., and J. R. Andreesen. 1982. Anaerobic degradation of uric acid via pyrimidine derivatives by selenium-starved cells of *Clostridium purinolyticum*. Arch. Microbiol. 131:255–260.

72. Du Toit, P. J., and T. R. Davies. 1973. Denitrification. Studies with laboratoryscale continuous-flow units. Water Res. 7:489–500.

73. Eichler, B., and B. Schink. 1984. Oxidation of primary aliphatic alcohols by *Acetobacterium carbinolicum* sp. nov., a homoacetogenic anaerobe. Arch. Microbiol. 140:147–152.

74. Evans, W. C. 1977. Biochemistry of the bacterial catabolism of aromatic compounds in anaerobic environments. Nature (Lond.) 270:17–22.

75. Evans, W. C., S. W. Smith, P. Moss, and H. N. Fernley. 1971. Bacterial metabolism of 4-chlorophenoxyacetate. Biochem. J. 122:509–517.

76. Eveleigh, D. E. 1984. Biofuels and oxychemicals from natural polymers: a perspective, p. 553–557, in M. J. Klug, and C. A. Reddy (eds.), Current perspectives in microbial ecology. American Society for Microbiology, Washington, D.C.

77. Eyssen, H., and A. Verhulst. 1984. Biotransformation of linoleic acid and bile acids by *Eubacterium lentum*. Appl. Environ. Microbiol. 47:39–43.

78. Fedorak, P. M., and S. E. Hrudey. 1984. The effects of phenol and some alkyl phenolics on batch anaerobic methanogenesis. Water Res. 18:361–367.

79. Ferry, J. G., and R. S. Wolfe. 1976. Anaerobic degradation of benzoate to methane by a microbial consortium. Arch. Microbiol. 107:33–40.

80. Fletcher, M., and K. C. Marshall. 1982. Are solid surfaces of ecological significance to aquatic bacteria, P. 199–236, in K. C. Marshall (ed.), Advances in microbial ecology, Vol. 6. Plenum Press, New York.

81. Fogel, S., R. L. Lancione, and A. E. Sewall. 1982. Enhanced biodegradation of methoxychlor in soil under sequential environmental conditions. Appl. Environ. Microbiol. 44: 113–120.

82. Forney, L. J., C. A. Reddy, M. Tien, and S. D. Aust. 1982. The involvement of hydroxyl radical derived from hydrogen peroxide in lignin degradation by the white rot fungus *Phanerochaete chrysosporium*. J. Biol. Chem. 257:11455–11462.

83. Foster, J. W. 1962. Hydrocarbons as substrates for microorganisms. Antonie Leeuwenhoek J. Microbiol. Serol. 28:241–274.

84. Fridovich, I. 1974. Superoxide and evolution. Horiz. Biophys. 1:1–37.

85. Frostell, B. 1984. Anaerobic-aerobic pilot-scale treatment of a sulfite evaporator condensate. Pulp Pap. Mag. Can. 85(3):T57–62.

86. Fuhs, G. W. 1961. Der mikrobielle Abbau von Kohlenwasserstoffen. Arch. Mikrobiol. 39:374–422.

87. Gaston, L. W., and E. R. Stadtman. 1963. Fermentation of ethylene glycol by *Clostridium glycolicum* sp. n. J. Bacteriol. 85:356–362.

88. Gencer, M. A., and R. Mutharasan. 1983. Ethanol fermentation in a yeast immobilized tubular fermenter. Biotechnol. Bioeng. 25:2243–2262.

89. George, H. A., and J.-S. Chen. 1983. Acidic conditions are not obligatory for onset of butanol formation by *Clostridium beijerinckii* (Synonym, *C. butylicum*). Appl. Environ. Microbiol. 46:321–327.

90. Gharpuray, M. M., Y-H. Lee, and L. T. Fan. 1983. Structural modification of lignocellulosics by pretreatments to enhance enzymatic hydrolysis. Biotechnol. Bioeng. 25:157–172.

91. Ghose, T. K., P. V. Pannir Selvam, and P. Ghosh. 1983. Catalytic solvent delignification of agricultural residues: organic catalysts. Biotechnol. Bioeng. 25:2577–2590.

92. Gibson, D. T. 1975. Microbial degradation of hydrocarbons, p. 667–696, in E. D. Goldberg (ed.), The nature of seawater (Dahlem Konferenzen, Berlin). Springer-Verlag, Heidelberg.

93. Giger, W., and C. Schaffner. 1981. Unsaturated steroid hydrocarbons as indicators of diagenesis in immature Monterey shales. Naturwissenschaften 68:37–39.

94. Giger, W., C. Schaffner, and S. G. Wakeham. 1980. Aliphatic and olefinic hydrocarbons in recent sediments of Greifensee, Switzerland. Geochim. Cosmochim. Acta 44:119–129.

95. Giger, W., P. H. Brunner, and C. Shaffner. 1984. 4-nonylphenol in sewage sludge: accumulation of toxic metabolites from anionic surfactants. Science 225:623–625.

96. Goldberg, I., and C. L. Cooney. 1981. Formation of short-chain fatty acids from H_2 and CO_2 by a mixed culture of bacteria. Appl. Environ. Microbiol. 41:148–154.

97. Gottschal, J. C., and J. G. Morris. 1981. The induction of acetone and butanol production in cultures of *Clostridum acetobutylicum* by elevated concentrations of acetate and butyrate. FEMS Microbiol. Lett. 12:385–389.

97a. Grbić-Galić, D. 1986. Anaerobic production and transformation of aromatic hydrocarbons and substituted phenols by ferulic acid-degrading BESA-inhibited methanogenic consortia. FEMS Microbiol. Ecol. 38:161–169.

98. Gujer, W., and A. J. B. Zehnder. 1983. Conversion processes in anaerobic digestion. Water Sci. Technol. 15:127–167.

99. Hackett, W. F., W. J. Connors, J. K. Kirk, and J. G. Zeikus. 1977. Microbial decomposition of synthetic ^{14}C-labeled lignins in nature: lignin biodegradation in a variety of natural materials. Appl. Environ. Microbiol. 33:43–51.

100. Haider, K., and J. P. Martin. 1979. Abbau und Umwandlung von Pflanzenrückständen und ihren Inhaltsstoffen durch Mikroflora des Bodens. Z. Pflanzenernaehr. 142:456–475.

101. Hakulinen, R., and M. Salkinoja-Salonen. 1982. Treatment of pulp and paper industry wastewaters in an anaerobic fluidized bed reactor. Proc. Biochem. 2:18–22.

102. Hall, E. R., and H. Melcer. 1983. Biotechnology developments for the treatment of toxic and inhibitory wastewaters. Biotechnol. Adv. 1:59–71.

103. Hallas, L. E., and M. Alexander. 1983. Microbial transformation of nitroaromatic compounds in sewage effluent. Appl. Environ. Microbiol. 45:1234–1241.

104. Han. Y. W. 1978. Microbial utilization of straw (a reveiw), p. 119–153, in D. Perlman (ed.), Advances in applied microbiology, Vol. 23. Academic Press, New York.

105. Hanselmann, K. W. 1982. Lignochemicals. Experientia (Basel) 38:176–189.

106. Hansson, G. 1979. Effects of carbon dioxide and methane on methanogenesis. Eur. J. Appl. Microbiol. Biotechnol. 6:351–359.

107. Harper, D. B., and E. R. Blakley. 1971. The metabolism of p-fluorophenylacetic acid by a *Pseudomonas* sp. II. The degradative pathway. Can. J. Microbiol. 17:645–650.

108. Hartmann, J., W. Reineke, and H.-J. Knackmuss. 1979. Metabolism of 3-chloro-, 4-chloro-, and 3,5-dichlorobenzoate by a pseudomonad. Appl. Environ. Microbiol. 37:421–428.

109. Hilpert, W., and P. Dimroth. 1982. Conversion of the chemical energy of methylmalonyl-CoA decarboxylation into a Na$^+$ gradient. Nature (Lond.) 296:584–585.

110. Hilpert, W., B. Schink, and P. Dimroth. 1984. Life by a new decarboxylation-dependent energy conservation mechanism with Na$^+$ as coupling ion. EMBO J. 3:1665–1670.

111. Hoban, D. J., and L. van den Berg. 1979. Effect of iron on conversion of acetic acid to methane during methanogenic fermentations. J. Appl. Bacteriol. 47:153–160.

112. Holcenberg, J. S., and E. R. Stadtman. 1969. Nicotinic acid metabolism. III. Purification and properties of a nicotinic acid hydroxylase. J. Biol. Chem. 244:1194–1203.

113. Horowitz, A., J. M. Suflita, and J. M. Tiedje. 1983. Reductive dehalogenations of halobenzoates by anaerobic lake sediment microorganisms. Appl. Environ. Microbiol. 45:1459–1465.

114. Horowitz, A., D. R. Shelton, C. P. Cornell, and J. M. Tiedje. 1982. Anaerobic degradation of aromatic compounds in sediments and digested sludge, p. 435–444, in Development in industrial microbiology. Dev. Ind. Microbiol. 23:435–444.

115. Horvath, R. S. 1972. Microbial cometabolism and the degradation of organic compounds in nature. Bacteriol. Rev. 36:146–155.

116. Hulbert, M. H., and S. Krawiec. 1977. Co-metabolism: a critique. J. Theor. Biol. 69:287–291.

117. Hungate, R. E. 1969. A roll tube method for cultivation of strict anaerobes, p. 117–132, in J. R. Norris, and D. W. Ribbons (ed.), Methods in microbiology, Vol. 3B. Academic Press, New York.

118. Huss, L. 1981. Application of the Anamet Process for waste water treatment, p. 137–150, in D. E. Hughes, D. A. Stafford, B. J. Wheatley, W. Baader, G. Lettinga, E. J. Nyns, and W. Verstraeten (eds.), Anaerobic digestion 1981. Elsevier, Amsterdam.

119. Huybregtse, R., and A. C. van der Linden. 1964. The oxidation of α-olefins by a *Pseudomonas*. Reactions involving the double bond. Antonie Leeuwenhoek J. Microbiol. Serol. 30:185–196.

120. Iannotti, E. T., D. Kafkewitz, M. J. Wolin, and M. P. Bryant. 1973. Glucose fermentation products of *Ruminococcus albus* grown in continuous culture with *Vibrio succinogenes*: changes caused by interspecies transfer of H_2. J. Bacteriol. 114:1231–1240.

121. Iizuka, H., M. Ilida, and S. Jujita. 1969. Formation of n-decene by resting cells of *Candida rugosa*. Z. Allg. Microbiol. 9:223–226.

122. Inloes, D. S., D. P. Taylor, S. N. Cohen, A. S. Michaels, and C. R. Robertson. Ethanol production by *Saccharomyces cerevisiae* immobilized in hollow-fiber membrane bioreactors. Appl. Environ. Microbiol. 46:264–278.

123. Ishikura, T., and J. W. Foster. 1961. Incorporation of molecular oxygen during microbial utilization of olefins. Nature (Lond.) 192:892–893.

124. Jacobson, S. N., N. L. O'Mara, and M. Alexander. 1980. Evidence for cometabolism in sewage. Appl. Environ. Microbiol. 40:917–921.

125. Jain, M. K., R. Singh, and P. Tauro. 1982. Biochemical changes during anaerobic digestion of animal wastes. Water Res. 16:411–415.

126. Jannasch, H. W. 1978. Microorganisms and their aquatic environment, p. 17–24, in W. E. Krumbein (ed.), Environmental biogeochemistry and geomicrobiology. Ann Arbor Science Publishers, Ann Arbor, Mich.

127. Jennett, J. C., and N. D. Dennis, Jr. 1975. Anaerobic filter treatment of pharmaceutical waste. J. Water Pollut. Control. Fed. 47:104–121.

128. Jewell, W. J., M. S. Switzenbaum, and J. W. Morris. 1981. Municipal waste water treatment with the anaerobic attached microbial film expanded bed process. J. Water Pollut. Control. Fed. 53:482–489.

129. Jewell, W. J., S. Dell'Orto, K. J. Fanfoni, S. J. Fast, E. J. Gotting, D. A. Jackson, and R. M. Kabrick. 1981. Agricultural and high strength wastes, p. 151–168, in D. E. Hughes, D. A. Stafford, B. J. Wheatley, W. Baader, G. Lettinga, E. J. Nyns, and W. Verstraeten (eds.), Anaerobic digestion 1981. Elsevier, Amsterdam.

130. Jöbses, J. M. L., and J. A. Roels. 1983. Experience with solvent production by *Clostridium beijerinckii* in continuous culture. Biotechnol. Bioeng. 25:1187–1194.

131. Jobson, A. M., F. D. Cook, and D. W. S. Westlake. 1979. Interaction of aerobic and anaerobic bacteria in petroleum biodegradation. Chem. Geol. 24:355–365.

132. Johnston, H. W., G. G. Briggs, and M. Alexander. 1972. Metabolism of 3-chlorobenzoic acid by a pseudomonad. Soil Biol. Biochem. 4:187–190.

133. Kaiser, J. P., and K. W. Hanselmann. 1982. Fermentative metabolism of substituted

monoaromatic compounds by a bacterial community from anaerobic sediments. Arch. Microbiol. 133:185–194.

134. Kaiser, J. P., and K. W. Hanselmann. 1982. Aromatic chemicals through anaerobic microbial conversion of lignin monomers. Experientia (Basel) 38:167–176.

135. Kawai, F., T. Kimura, Y. Tani, and H. Yamada. 1984. Involvement of a polyethylene glycol (PEG)-oxidizing enzyme in the bacterial metabolism of PEG. Agric. Biol. Chem. 48:1349–1351.

136. Kennedy, K. J., and L. van den Berg. 1982. Stability and performance of anaerobic fixed film reactors during hydraulic overloading at 10–35°C. Water Res. 16:1391–1398.

137. Khan, A. W., and T. M. Trottier. 1978. Effect of sulfur-containing compounds on anaerobic degradation of cellulose to methane by mixed cultures obtained from sewage sludge. Appl. Environ. Microbiol. 35:1027–1034.

138. Khan, A. W., T. M. Trottier, G. B. Patel, and S. M. Martin. 1979. Nutrient requirement for the degradation of cellulose to methane by a mixed population of anaerobes. J. Gen. Microbiol. 112:365–372.

139. Khan, A. W., S. S. Miller, and W. D. Murray. 1983. Development of a two-phase combination fermenter for the conversion of cellulose to methane. Biotechnol. Bioeng. 25:1571–1579.

140. Kim, K. E., and G. W. Chang. 1974. Trimethylamine oxide reduction by *Salmonella*. Can. J. Microbiol. 20:1745–1748.

141. Kim, B. H., P. Bellows, R. Datta, and J. G. Zeikus. 1984. Control of carbon and electron flow in *Clostridium acetobutylicum* fermentations: utilization of carbon monoxide to inhibit hydrogen production and to enhance butanol yields. Appl. Environ. Microbiol. 48:764–770.

142. Klages, U., and F. Lingens. 1980. Degradation of 4-chlorobenzoic acid by a *Pseudomonas* species. Zentralbl. Bacteriol. Parasitenkd. Infektionskr. Hyg. 1 Abt. Orig. C1:215–223.

143. Klages, U., A. Markus, and F. Lingens. 1981. Degradation of 4-chlorophenylacetic acid by a *Pseudomonas* species. J. Bacteriol. 146:64–68.

144. Klapwijk, A., J. C. M. van den Hoeven, and G. Lettinga. 1981. Biological denitrification in an upflow sludge blanket reactor. Water Res. 15:1–6.

145. Kobayaski, H. A., M. Stenstrom, and R. A. Mah. 1983. Use of photosynthetic bacteria for hydrogen sulfide removal from anaerobic waste treatment effluent. Water Res. 17:579–587.

146. Kobayashi, H. A., M. K. Stenstrom, and R. A. Mah. 1983. Treatment of low strength domestic wastewater using the anaerobic filter. Water Res. 17:903–909.

147. Konigs, W. N., and H. Veldkamp. 1980. Phenotypic responses to environmental change, p. 161–191, in D. C. Ellwood, J. N. Hedger, M. J. Latham, J. H. Slater, and J. M. Lynch (eds.), Contemporary microbial ecology. Academic Press, London.

148. Korte, F. 1980. Ökologische Chemie. Georg Thieme, Stuttgart, West Germany.

149. Kosaric, N., D. C. M. Ng, I. Russel, and G. C. Stewart. 1980. Ethanol production by fermentation: an alternative liquid fuel, p. 147–227, in D. Perlman (ed.), Advances in applied microbiology, Vol. 26. Academic Press, New York.

150. Kowalski, E., and Z. Lewandowski. 1983. Nitrification process in a packed bed reactor with a chemically active bed. Water Res. 17:157–160.

151. Kröger, A., and E. Winkler. 1981. Phosphorylative fumarate reduction in *Vibrio succinogenes*: stoichiometry of ATP synthesis. Arch. Microbiol. 129:100–104.

151a. Kuhn, E. P., P. J. Colberg, J. L. Schnoor, O. Wanner, A. J. B. Zehnder, and R. P. Schwarzenbach. 1985. Microbial transformations of substituted benzenes during infiltration of river water to groundwater: laboratory column studies. Environ. Sci. Technol. 19:961–968.

152. Laanbroek, H. J., and H. J. Geerligs. 1983. Influence of clay particles (Illite) on substrate utilization by sulfate-reducing bacteria. Arch. Microbiol. 134:161–163.

153. Laanbroek, H. J., T. Abee, and I. L. Voogd. 1982. Alcohol conversions by *Desulfobulbus propionicus* Lindhorst in the presence and absence of sulfate and hydrogen. Arch. Microbiol. 133:178–184.

154. Laluce, C., and J. R. Mattoon. 1984. Development of rapidly fermenting strains of *Saccharomyces diastaticus* for direct conversion of starch and dextrins to ethanol. Appl. Environ. Microbiol. 48:17–25.

155. La Pat-Polasko, L. T., P. L. McCarty, and A. J. B. Zehnder. 1984. Secondary substrate utilization of methylene chloride by an isolated strain of *Pseudomonas* sp. Appl. Environ. Microbiol. 47:825–830.

156. La Rivière, J. W. M. 1977. Microbial ecology of liquid waste treatment, p. 215–259, in M. Alexander (ed.), Advances in microbial ecology, Vol. 1. Plenum Press, New York.

157. Laube, V. M., and S. M. Martin. 1983. Effect of some physical and chemical parameters on the fermentation of cellulose to methane by a coculture system. Can. J. Microbiol. 29:1475–1480.

158. Lawrence, A. W., and P. L. McCarty. 1969. Kinetics of methane fermentation in anaerobic treatment. J. Water Pollut. Control Fed. 41:R1–R17.

159. Leadbetter, E. R., and J. W. Foster. 1959. Oxidation products formed from gaseous alkanes by the bacterium *Pseudomonas methanica*. Arch. Biochem. Biophys. 82:491–492.

160. Lettinga, G., and A. F. M. van Velsen. 1974. Toepassing van methaangisting voor de behandeling van minder geconcentreerd afvalwater. H₂O 7:281–289.

161. Lettinga, G., W. de Zeeuw, and E. Ouborg. 1981. Anaerobic treatment of wastes containing methanol and higher alcohols. Water Res. 15:171–182.

162. Lettinga, G., A. F. M. van Velsen, S. W. Hobma, W. de Zeeuw, and A. Klapwijk. 1979. The use of the upflow sludge blanket (USB) reactor for biological wastewater treatment, especially for anaerobic treatment. Biotechnol. Bioeng. 22:699–734.

163. Linko, P. 1983. Fuels and industrial chemicals. Biotechnol. Adv. 1:47–58.

164. Lin, D., K. Thomson, and A. C. Anderson. 1984. Identification of nitroso compounds from biotransformation of 2,4-dinitrotoluene. Appl. Environ. Microbiol. 47:1295–1298.

165. Lyons, C. D., S. Katz, and R. Bartha. 1984. Mechanisms and pathways of aniline elimination from aquatic environments. Appl. Environ. Microbiol. 48:491–496.

166. Mackie, R. J., and M. P. Bryant. 1981. Metabolic activity of fatty acid-oxidizing bacteria and the contribution of acetate, propionate, butyrate, and CO₂ to methanogenesis in cattle waste at 40 and 60°C. Appl. Environ. Microbiol. 41:1363–1373.

167. Majak, W., and K.-J. Cheng. 1981. Identification of rumen bacteria that anaerobically degrade aliphatic nitrotoxins. Can. J. Microbiol. 27:646–650.

168. Malmström, B. G. 1982. Enzymology of oxygen. Annu. Rev. Biochem. 51:21-59.

169. Marshall, K. C., 1979. Growth at interfaces, p. 281-290, in M. Shilo (ed.), Strategies of microbial life in extreme environments (Dahlem Konferenzen, Berlin). Springer-Verlag, Heidelberg.

170. Marshall, K. C. (ed.). 1984. Microbial adhesion and aggregation (Dahlem Konferenzen, Berlin). Springer-Verlag, Heidelberg.

171. Masuda, N., H. Oda, S. Hirano, M. Masuda, and H. Tanaka. 1984. 7 α-dehydroxylation of bile acids by resting cells of a *Eubacterium lentum*-like intestinal anaerobe, strain c-25. App. Environ. Microbiol. 47:735-739.

172. Maxwell, J. R., and A. M. K. Wardroper. 1982. Early transformations of isoprenoid compounds in surface sedimentary environments, p. 203-229, in D. B. Nedwell and C. M. Brown (eds.), Sediment microbiology. Academic Press, London.

173. Mazé, P. 1915. Ferment forménique. Fermentation forménique de l'acétone. Procédé de culture simple du ferment forménique. C. R. Soc. Biol. 78:398-405.

174. McCarty, P. L. 1981. One hundred years of anaerobic treatment, p. 3-22, in D. E. Hughes, D. A. Stafford, B. J. Wheatley, W. Baader, G. Lettinga, E. J. Nyns, and W. Verstraeten (eds.), Anaerobic digestion 1981. Elsevier, Amsterdam.

175. McCormick, N. G., F. E. Feeherry, and H. S. Levinson. 1976. Microbial transformation of 2,4,6-trinitrotoluene and other nitroaromatic compounds. Appl. Environ. Microbiol. 31:949-958.

176. McKirdy, D. M., and J. H. Hahn. 1982. The composition of kerogen and hydrocarbons in precambrian rocks, p. 123-154, in H. D. Holland and M. Schidlowski (eds.), Mineral deposits and the evolution of the biosphere (Dahlem Konferenzen). Springer-Verlag, Heidelberg.

177. Menzie, C. M. 1980. Reaction types in the environment, 2 A, p. 247-302, in O. Hutzinger (ed.), The handbook of environmental chemistry. Springer-Verlag, Heidelberg.

178. Mountfort, D. O., R. A. Asher, and T. Bauchop. 1982. Fermentation of cellulose to methane and carbon dioxide by a rumen anaerobic fungus in a triculture with *Methanobrevibacter* sp. strain RA 1 and *Methanosarcina barkeri*. Appl. Environ. Microbiol. 44:128-134.

179. Muller, F. M. 1957. On methane formation of higher alkanes. Antonie Leeuwenhoek. J. Microbiol. Serol. 23:369-384.

180. Naumann, E., K. Fahlbusch, and G. Gottschalk. 1984. Presence of a trimethylamine: HS-coenzyme M methyltransferase in *Methanosarcina barkeri*. Arch. Microbiol. 138:79-83.

181. Ng, T. K., and J. G. Zeikus. 1981. Comparison of extracellular cellulase activities of *Clostridium thermocellum* LQRI and *Trichoderma reesei*. Appl. Environ. Microbiol. 42:231-240.

182. Ng, T. K., A. Ben-Bassat, and J. G. Zeikus. 1981. Ethanol production by thermophilic bacteria: fermentation of cellulosic substrates by cocultures of *Clostridium thermocellum* and *Clostridium thermohydrosulfuricum*. Appl. Environ. Microbiol. 41:1337-1343.

183. Novelli, G. D., and C. E. ZoBell. 1944. Assimilation of petroleum hydrocarbons by sulfate-reducing bacteria. J. Bacteriol 47:447-448.

184. Omelianski, W. 1916. Fermentation méthanique de l'alcool éthylique. Ann. Inst. Pasteur (Paris) 30:56–60.

185. Orpin, C. G. 1981. Isolation of cellulolytic phycomycete fungi from the caecum of the horse. J. Gen. Microbiol. 123:287–296.

186. Palsson, B. O., S. Fathi-Afshar, D. F. Rudd, and E. N. Lightfoot. 1981. Biomass as a source of chemical feedstocks: an economic evaluation. Science 213:513–517.

187. Peppercorn, M., and P. Goldman. 1971. Caffeic acid metabolism by bacteria of the human gastrointestinal tract. J. Bacteriol. 108:996–1000.

188. Perry, J. J. 1979. Microbial cooxidations involving hydrocarbons. Microbiol. Rev. 43:59–79.

189. Perski, H.-J., J. Moll, and R. K. Thauer. 1981. Sodium dependence of growth and methane formation in *Methanobacterium thermoautotrophicum*. Arch. Microbiol. 130:319–321.

190. Pette, K. C., and A. I. Versprille. 1981. Application of the U.A.S.B.-concept for wastewater treatment, p. 121–133, in D. E. Hughes, D. A. Stafford, B. I. Wheatley, W. Baader, G. Lettinga, E. J. Nyns, and W. Verstraeten (eds.), Anaerobic digestion 1981. Elsevier, Amsterdam.

191. Pfennig, N. 1979. Formation of oxygen and microbial processes establishing and maintaining anaerobic environments, p. 137–148 in M. Shilo (ed.), Strategies of microbial life in extreme environments (Dahlem Konferenzen, Berlin). Springer-Verlag, Heidelberg.

192. Pfennig, N. 1980. Syntrophic mixed cultures and symbiotic consortia with phototrophic bacteria: a review, p. 127–131, in G. Gottschalk, N. Pfennig, and H. Werner (eds.), Anaerobes and anaerobic infections. Gustav Fischer, Stuttgart, West Germany.

193. Pfennig N. 1984. Microbial behaviour in natural environments, p. 23–50, in D. P. Kelly and N. G. Starr (eds.), The Microbe 1984: Part II. Prokaryotes and eukaryotes (Society for General Microbiology Symposia No. 36). Cambridge University Press, Cambridge.

194. Phelps, T. J., and J. G. Zeikus. 1984. Influence of pH on terminal carbon metabolism in anoxic sediments from a mildly acidic lake. Appl. Environ. Microbiol. 48:1088–1095.

195. Pipes, W. O. 1978. Microbiology of activated sludge bulking, p. 85–127, in D. Perlman (ed.), Advances in applied microbiology, Vol. 24. Academic Press, New York.

196. Pirt, S. J., and Y. K. Lee. 1983. Enhancement of methanogenesis by traces of oxygen in bacterial digestion of biomass. FEMS Microbiol. Lett. 18:61–63.

197. Pitt, W. W., Jr., G. L. Haag, and D. D. Lee. 1983. Recovery of ethanol fermentation broths using selective sorption-desorption. Biotechnol. Bioeng. 25:123–131.

197a. Platen, H., and B. Schink. 1987. Methanogenic degradation of acetone by an enrichment culture. Arch. Microbiol. 149:136–141.

198. Präve, P., U. Faust, W. Sittig, and D. A. Sukatsch. 1984. Handbuch der Biotechnologie. 2. Aufl., p. 616. R. Oldenbourg, Munich, West Germany.

199. Pridham, J. B. 1965. Low molecular weight phenols in higher plants. Annu. Rev. Plant Physiol. 16:13–36.

200. Puri, V. P., and H. Mamers. 1983. Explosive pretreatment of lignocellulosic residues

with high-pressure carbon dioxide for the production of fermentation substrates. Biotechnol. Bioeng. 25:3149–3161.

201. Reineke, W., and H.-J. Knackmuss. 1984. Microbial metabolism of haloaromatics: isolation and properties of a chlorobenzene-degrading bacterium. Appl. Environ. Microbiol. 47:395–402.

202. Rhead, M. M., G. Eglinton, G. H. Draffan, and P. J. England. 1971. Conversion of oleic acid to saturated fatty acids in Severn Estuary sediment. Nature (Lond.) 232:327–330.

203. Rivière, J. 1977. Industrial applications of microbiology. Surrey University Press, Glasgow.

204. Robbins, J. E., M. T. Arnold, and S. L. Lacher. 1979. Methane production from cattle waste and delignified straw. Appl. Environ. Microbiol. 38:175–177.

205. Rook, J. J. 1974. Formation of haloforms during chlorination of natural waters. Water Treat. Examin. 23:234–243.

206. Rosenfeld, W. D. 1947. Anaerobic oxidation of hydrocarbons by sulfate reducing bacteria. J. Bacteriol. 54:664–665.

207. Saddler, J. N., E. K. C. Yu, M. Mes-Hartree, N. Levitin, and H. H. Brownell. 1983. Utilization of enzymatically hydrolyzed wood hemicelluloses by microorganisms for production of liquid fuels. Appl. Environ. Microbiol. 45:153–160.

208. Samain, E., G. Albagnac, H. C. Dubourgier, and J. P. Touzel. 1982. Characterization of a new propionic acid bacterium that ferments ethanol and displays a growth factor dependent association with a gram negative homoacetogen. FEMS Microbiol. Lett. 15:69–74.

208a. Samain, E., G. Albagnac, and H.-C. Dubourgier. 1986. Initial steps of catabolism of trihydroxybenzenes in *Pelobacter acidigallici*. Arch. Microbiol. 144:242–245.

209. Sánchez Riera, F., S. Valz-Gianinet, D. Callieri, and F. Siñeriz. 1982. Use of a packed-bed reactor for anaerobic treatment of stillage of sugar cane molasses. Biotechnol. Lett. 4:127–132.

210. Scheifinger, C. C., and M. J. Wolin. 1973. Propionate formation from cellulose and soluble sugars by combined cultures of *Bacteroides succinogenes* and *Selenomonas ruminantium*. Appl. Microbiol. 26:789–795.

211. Scheline, R. R. 1966. Decarboxylation and demethylation of some phenolic benzoic acid derivatives by rat cecal contents. J. Pharm. Pharmacol. 18:664–669.

212. Scheline, R. S. 1973. Metabolism of foreign compounds by gastrointestinal microorganisms. Pharmacol. Rev. 25:451–523.

213. Scherer, P., and H. Sahm. 1981. Influence of sulphur-containing compounds on the growth of *Methanosarcina barkeri* in a defined medium. Eur. J. Appl. Microbiol. 12:28–35.

214. Schink, B. 1984. Fermentation of 2,3-butanediol by *Pelobacter carbinolicus* sp. nov. and *Pelobacter propionicus*, sp. nov., and evidence for propionate formation from C_2 compounds. Arch. Microbiol. 137:33–41.

215. Schink, B. 1984. *Clostridium magnum* sp. nov. a non-autotrophic homoacetogenic bacterium. Arch. Microbiol. 137:250–255.

216. Schink, B. 1984. Microbial degradation of pectin in plants and aquatic environments, p. 580–587 in M. J. Klug and C. A. Reddy (eds.), Current perspectives in microbial ecology. American Society for Microbiology, Washington, D.C.

217. Schink, B. 1985. Degradation of unsaturated hydrocarbons by methanogenic enrichment cultures. FEMS Microbiol. Ecol. 31:69–77.

217a. Schink, B. 1985. Fermentation of acetylene by an obligate anaerobe, *Pelobacter acetylenicus* sp. nov. Arch. Microbiol. 142:295–301.

218. Schink, B., and N. Pfennig. 1982. Fermentation of trihydroxybenzenes by *Pelobacter acidigallici* gen. nov. sp. nov., a new strictly anaerobic, non-sporeforming bacterium. Arch. Microbiol. 133:195–201.

219. Schink, B., and N. Pfennig. 1982. *Propionigenium modestum* gen. nov. sp. nov., a new strictly anaerobic, nonsporing bacterium growing on succinate. Arch. Microbiol. 133:209–216.

220. Schink, B., and M. Stieb. 1983. Fermentative degradation of polyethylene glycol by a new, strictly anaerobic, Gram negative, non-sporeforming bacterium, *Pelobacter venetianus* sp. nov. Appl. Environ. Microbiol. 45:1905–1913.

221. Schink, B., and J. G. Zeikus. 1980. Microbial methanol formation: a major end product of pectin metabolism. Curr. Microbiol. 4:387–390.

222. Schink, B., T. E. Thompson, and J. G. Zeikus. 1982. Characterization of *Propionispira arboris* gen. nov. sp. nov., a nitrogen-fixing anaerobe common to wetwoods of living trees. J. Gen. Microbiol. 128:2771–2779.

223. Schink, B., T. J. Phelps, B. Eichler, and Z. G. Zeikus. 1985. Comparison of ethanol degradation pathways in anoxic freshwater environments. J. Gen. Microbiol. 131:651–660.

224. Schlegel, H. G. 1981. Allgemeine Mikrobiologie. 5. Aufl. Georg Thieme, Stuttgart, West Germany.

225. Schlegel, S., and K.-H. Kalbskopf. 1981. Treatment of liquors from heat-treated sludge using the anaerobic contact process, p. 169–184, in D. E. Hughes, D. A. Stafford, B. J. Wheatley, W. Baader, G. Lettinga, E. J. Nyns, and W. Verstraeten (eds.), Anaerobic digestion 1981. Elsevier, Amsterdam.

226. Scott, T. W., P. F. V. Ward, and R. M. C. Dawson. 1964. The formation and metabolism of phenyl-substituted fatty acids in the ruminant. Biochem. J. 90:12–23.

227. Senez, J. C., and E. Azoulay. 1961. Déshydrogénation d'hydrocarbures parafiniques par les suspensions non-proliférantes et les extraits de *Pseudomonas aeruginosa*. Biochim. Biophys. Acta 47:307–316.

228. Seo, C., Y. Yamada, N. Takada, and H. Okada. 1981. Hydration of squalene and oleic acid by *Corynebacterium* sp. Agric. Biol. Chem. 45:2025–2030.

229. Seubert, W., and U. Remberger. 1963. Untersuchungen über den bakteriellen Abbau von Isoprenoiden. II. Die Rolle der Kohlen-säure. Biochem. Z. 338:245–264.

230. Shailubhai, K., S. R. Sahasrabudhe, K. A. Vora, and V. V. Modi. 1984. Degradation of chlorobenzoates by *Aspergillus niger*. Experientia (Basel) 40:406–407.

231. Shelton, D. R., and J. M. Tiedje. 1984. Isolation and partial characterization of bacteria in an anaerobic consortium that mineralizes 3-chlorobenzoic acid. Appl. Environ. Microbiol. 48:840–848.

232. Shipman, R. H., L. T. Fan, and I. C. Kao. 1977. Single-cell protein production by photosynthetic bacteria, p. 161–183, in D. Perlman (ed.), Advances in Applied Microbiology, Vol. 21. Academic Press, New York.

233. Sleat, R., R. A. Mah, and R. Robinson. 1984. Isolation and characterization of an

anaerobic cellulolytic bacterium, *Clostridium cellulovorans* sp. nov. Appl. Environ. Microbiol. 48:88–93.

234. Slininger, P. J., R. J. Bothast, and K. L. Smiley. 1983. Production of 3-hydroxypropionaldehyde from glycerol. Appl. Environ. Microbiol. 46:62–67.

235. Smith, P. H., and R. A. Mah. 1966. Kinetics of acetate metabolism during sludge digestion. Appl. Microbiol. 14:368–371.

236. Söhngen, N. S. 1913. Einfluß von Kolloiden auf mikrobiologische Prozesse. Zentralbl. Bakteriol. 2, 38:621–647.

237. Spano, L. A. 1976. Enzymatic hydrolysis of cellulosic materials, p. 157–177, in H. G. Schlegel and J. Barnea (eds.), Microbial energy conversion. E. Goltze KG, Göttingen, West Germany.

238. Speckman, R. A., and E. B. Collins. 1982. Microbial production of 2,3-butylene glycol from cheese whey. Appl. Environ. Microbiol. 43:1216–1218.

239. Speece, R. E., G. F. Parkin, and D. Gallagher. 1983. Nickel stimulation of anaerobic digestion. Water Res. 17:677–683.

240. Spencer, R. R. 1978. Enhancement of methane production in the anaerobic digestion of sewage sludges. Biotechnol. Bioeng. Symp. 8:257–268.

241. Spivey, M. J. 1978. The acetone/butanol/ethanol fermentation. Process Biochem. 13(11):2–4, 25.

241a. Stams, A. J. M., T. A. Hansen, and G. W. Skyring. 1985. Utilizations of amino acids as energy substrates by two marine *Desulfovibrio* strains. FEMS Microbiol. Ecol. 31:11–15.

242. Stenroos, S.-L., Y.-Y. Linko, and P. Linko. 1982. Production of L-lactic acid with immobilized *Lactobacillus delbrueckii*. Biotechnol. Lett. 4:159–164.

243. Stevenson, F. J. 1972. Role and function of humus in soil with emphasis on adsorption of herbicides and chelation of micronutrients. Bioscience 22:643–650.

244. Stewart, J. E., W. R. Finnerty, R. E. Kallio, and D. P. Stevenson. 1960. Esters from bacterial oxidation of olefins. Science 132:1254–1255.

244a. Straß, A., and B. Schink. 1986. Fermentation of polyethylene glycol via acetaldehyde in *Pelobacter venetianus*. Appl. Microbiol. Biotechnol., 25:37–42.

245. Strøm, A. R., J. A. Olafsen, and H. Larsen. 1979. Trimethylamine oxide: a terminal electron acceptor in anaerobic respiration of bacteria. J. Gen. Microbiol. 112:315–320.

246. Stucki, G., and T. Leisinger. 1983. Bacterial degradation of 2-chloroethanol proceeds via 2-chloroacetic acid. FEMS Microbiol. Lett. 16:123–126.

247. Suflita, J. M., A. Horowitz, D. R. Shelton, and J. M. Tiedje. 1982. Dehaologenation: a novel pathway for the anaerobic bio-degradation of haloaromatic compounds. Science (Washington, D.C.) 218:1115–1117.

248. Suflita, J. M., J. A. Robinson, and J. M. Tiedje. 1983. Kinetics of microbial dehalogenation of haloaromatic substrates in methanogenic environments. Appl. Environ. Microbiol. 45:1466–1473.

249. Sugatt, R. H., D. P. O'Grady, S. Banerjee, P. H. Howard, and W. E. Gledhill. 1984. Shake flask biodegradation of 14 commercial phthalate esters. Appl. Environ. Microbiol. 47:601–606.

250. Symons, G. E., and A. M. Buswell. 1933. The methane fermentation of carbohydrates. J. Am. Chem. Soc. 55:2028–2036.

251. Szewzyk, U., R. Szewzyk, and B. Schink. 1985. Methanogenic degradation of hydroquinone and catechol via reductive dehydroxylation to phenol. FEMS Microbiol. 31:79–87.

252. Tarvin, D., and A. M. Buswell. 1934. The methane formation of organic acids and carbohydrates. J. Am. Chem. Soc. 56:1751–1755.

253. Taylor, B. F., and D. W. Ribbons. 1983. Bacterial decarboxylation of o-phthalic acids. Appl. Environ. Microbiol. 46:1276–1281.

254. Taylor, D. G., P. W. Trudgill, R. E. Cripps, and P. R. Harris. 1980. The microbial metabolism of acetone. J. Gen. Microbiol. 118:159–170.

255. Terrell, S. L., A. Bernard, and R. B. Bailey. 1984. Ethanol from whey: continuous fermentation with a catabolite repression-resistant *Saccharomyces cerevisiae* mutant. Appl. Environ. Microbiol. 48:577–580.

256. Tewes, F. J., and R. K. Thauer. 1980. Regulation of ATP-synthesis in glucose fermenting bacteria involved in interspecies hydrogen transfer, p. 97–104, in G. Gottschalk, N. Pfennig, and H. Werner (eds.), Anaerobes and anaerobic infections. Gustav Fischer, Stuttgart, West Germany.

257. Thauer, R. 1976. Limitation of microbial H_2-formation via fermentation, p. 201–204, in H. G. Schlegel and J. Barnea (eds.), Microbial energy conversion. E. Goltze KG, Göttingen, West Germany.

258. Thauer, R. K., K. Jungermann, and K. Decker. 1977. Energy conservation in chemotrophic anaerobic bacteria. Bacteriol. Rev. 41:100–180.

259. Thompson, L. A., D. B. Nedwell, M. T. Balba, I. M. Banat, and E. Senior. 1983. The use of multiple-vessel, open flow systems to investigate carbon flow in anaerobic microbial communities. Microb. Ecol. 9:189–199.

260. Thompson, T. E., R. Conrad, and J. G. Zeikus. 1984. Regulation of carbon and electron flow in *Propionispira arboris*: physiological function of hydrogenase and its role in homopropionate formation. Fems Microbiol. Lett. 22:265–271.

261. Tien, M., and T. K. Kirk. 1984. Lignin-degrading enzyme from *Phanerochaete chrysosporium*: purification, characterization, and catalytic properties of a unique H_2O_2-requiring oxygenase. Proc. Natl. Acad. Sci. USA. 81:2280–2286.

262. Tipayang, P., and M. Kozaki. 1982. Lactic acid production by a new *Lactobacillus* sp., *Lactobacillus vaccinosterus* Kozaki and Okada sp. nov., immobilized in calcium alginate. J. Ferment. Technol. 60:595–598.

263. Toerien, D. F., and W. H. J. Hattingh. 1969. Anaerobic digestion. I. The microbiology of anaerobic digestion. Water Res. 3:385–416.

264. Toraya, T., and S. Fukui. 1982. Diol dehydrase, p. 233–262, in D. Dolphin (ed.), B_{12}, Vol. 2. Wiley, New York.

265. Tornabene, T. G., T. A. Langworthy, G. Holzen, and T. Oro. 1979. Squalenes, phytanes and other isoprenoids as major neutral lipids of methanogenic and thermoacidophilic archaebacteria. J. Mol. Evol. 13:73–83.

266. Tsai, C.-G., and A. Jones. 1975. Isolation and identification of rumen bacteria capable of anaerobic phloroglucinol degradation. Can. J. Microbiol. 21:794–801.

267. Tsai, C.-G., D. M. Gates, W. M. Ingledew, and G. A. Jones. 1976. Products of anaerobic phloroglucinol degradation by *Coprococcus* sp. Pe₁5. Can. J. Microbiol. 22:159–164.

268. Tschech, A., and N. Pfennig. 1984. Growth yield increase linked to caffeate reduction in *Acetobacterium woodii*. Arch. Microbiol. 137:163–167.

268a. Tschech, A., and B. Schink. 1985. Fermentative degradation of resorcinol and resorcylic acids. Arch. Microbiol. 143:52–59.

268b. Tschech, A., and B. Schink. 1986. Fermentative degradation of monohydroxybenzoates by defined syntrophic cocultures. Arch. Microbiol. 145:396–402.

269. Turakhia, M. H., K. E. Cooksey, and W. G. Characklis. 1983. Influence of a calcium-specific chelant on biofilm removal. Appl. Environ. Microbiol. 46:1236–1238.

270. Vallentyne, J. R. 1960. Fossil pigments: the fate of carotenoids, p. 83–105, in M. B. Allen (ed.), Comparative biochemistry of photoreactive systems (Symposia on Comparative Biology, Vol. 1). Academic Press, New York.

271. Van Palenstein Heldermann, W. H., and J. Rosman. 1976. Hydrogen-dependent organisms from the human gingival crevice resembling *Vibrio succinogenes*. Antonie Leeuwenhoek, J. Microbiol. Serol. 42:107–118.

272. Varel, V. H., and A. G. Hashimoto. 1981. Effect of dietary monensin or chlortetracycline on methane production from cattle waste. Appl. Environ. Microbiol. 41:29–34.

273. Varel, V. H., H. R. Isaacson, and M. P. Bryant. 1977. Thermophilic methane production from cattle waste. Appl. Environ. Microbiol. 33:298–307.

274. Wakeham, S. G., C. Schaffner, and W. Giger. 1980. Polycyclic aromatic hydrocarbons in recent lake sediments. II. Compounds derived from biogenic precursors during early diagenesis. Geochim. Cosmochim. Acta 44:415–429.

275. Wang, Y.-S., R. V. Subba-Rao, and M. Alexander. 1984. Effect of substrate concentration and organic and inorganic compounds on the occurrence and rate of mineralization and cometabolism. Appl. Environ. Microbiol. 47:1195–1200.

276. Watts, D. C. and J. R. Maxwell. 1977. Carotenoid diagenesis in a marine sediment. Geochim. Cosmochim. Acta 41:493–497.

277. Weimer, P. J., and J. G. Zeikus. 1977. Fermentation of cellulose and cellobiose by *Clostridium thermocellum* in the absence and presence of *Methanobacterium thermoautotrophicum*. Appl. Environ. Microbiol. 33:289–297.

278. Weimer, P. J., W. Wagner, S. Knowlton, and T. K. Ng. 1984. Thermophilic anaerobic bacteria which ferment hemicellulose: characterization of organisms and identification of plasmids. Arch. Microbiol. 138:31–36.

279. White, D. C. 1983. Analysis of microorganisms in terms of quantity and activity in natural environments. Soc. Gen. Microbiol. Symp. 34:37–66.

280. White, D. C. 1984. Chemical characterization of films, p. 159–176, K. C. Marshall (ed.), Microbial adhesion and aggregation (Dahlem Konferenzen, Berlin). Springer-Verlag, Heidelberg.

281. White, R. E., and M. J. Coon. 1980. Oxygen activation by cytochrome P-450. Annu. Rev. Biochem. 49:315–356.

281a. Widdel, F. 1986. Growth of methanogenic bacteria in pure culture with 2-propanol and other alcohols as hydrogen donors. Appl. Environ. Microbiol. 51:1056–1062.

282. Wiegel, J. 1980. Formation of ethanol by bacteria. A pledge for the use of extreme thermophilic anaerobic bacteria in industrial ethanol fermentation processes. Experientia (Basel) 36:1434–1446.

283. Wikén, T. 1940. Untersuchungen über Methangärung und die dabei wirksamen Bakterien. Arch. Mikrobiol. 11:312–317.

284. Wimpenny, J. W. T., J. P. Coombs, R. W. Lovitt, and S. G. Whittaker. 1981. A gel-stabilized model ecosystem for investigating microbial growth in spatially ordered solute gradients. J. Gen. Microbiol. 127:277–287.

285. Wimpenny, J. W. T., R. W. Lovitt, and J. P. Coombs. 1983. Laboratory model systems for the investigation of spatially and temporally organised microbial ecosystems. Soc. Gen. Microbiol. Symp. 34:67–117.

286. Wolin, M. J. 1979. The rumen fermentation: a model for microbial interactions in anaerobic systems, p. 49–77, in M. Alexander (ed.), Advances in microbial ecology, Vol. 3. Plenum Press, New York.

287. Wuhrmann, K. 1982. Ecology of methanogenic systems in nature. Experientia (Basel) 38:193–198.

288. Young, L. Y. 1984. Anaerobic degradation of aromatic compounds, p. 349–376, in D. T. Gibson (ed.), Microbial degradation of organic compounds. Marcel Dekker, New York.

289. Young, J. C., and P. L. McCarty. 1969. The anaerobic filter for waste treatment. J. Water Pollut. Control Fed. 41(5):R160–R173.

290. Yousten, A. A., and E. A. Delwiche. 1961. Biotin and vitamin B_{12} coenzymes in succinate decarboxylation by *Propionibacterium pentosaceum* and *Veillonella alcalescens*. Bacteriol. Proc. 61:175.

291. Yu, E. K. C., and J. N. Saddler. 1983. Enhanced acetone-butanol fermentation by *Clostridium acetobutylicum* grown on D-xylose in the presence of acetic or butyric acid. FEMS Microbiol. Lett. 18:103–107.

292. Zehnder, A. J. B. 1978. Ecology of methane formation, p. 349–376, in R. Mitchell (ed.), Water pollution microbiology, Vol. 2. Wiley, New York.

293. Zehnder, A. J. B., and T. D. Brock. 1979. Methane formation and methane oxidation by methanogenic bacteria. J. Bacteriol. 137:420–432.

294. Zehnder, A. J. B., and M. E. Koch. 1983. Thermodynamic and kinetic interactions of the final steps in anaerobic digestion, p. 86–96. In Proceedings of the European Symposium on anaerobic waste water treatment, TNO Corporate Communication Department, The Hague.

295. Zehnder, A. J. B., K. Ingvorsen, and T. Marti. 1982. Microbiology of methane bacteria, p. 45–68, in D. E. Hughes, D. A. Stafford, B. I. Wheatley, W. Baader, G. Lettinga, E. J. Nyns, and W. Verstraeten (eds.), Anaerobic digestion 1981. Elsevier, Amsterdam.

296. Zeikus, J. G. 1979. Microbial populations in digestors, p. 61–85, in D. A. Stafford, B. J. Wheatley, and D. E. Hughes (eds.), Anaerobic digestion. Applied Science Publishers, Barking, Essex, England.

297. Zeikus, J. G. 1980. Fate of lignin and related aromatics in anaerobic environments, p. 101–110, in T. Kirk, T. Higushi, and H. M. Chung (eds.), Lignin biodegradation: microbiology, chemistry and potential applications. CRC Press, Boca Raton, Fla.

298. Zeikus, J. G. 1980. Chemical and fuel production by anaerobic bacteria. Annu. Rev. Microbiol. 34:423–464.

299. Zeikus, J. G. 1981. Lignin metabolism and the carbon cycle. Polymer biosynthesis,

biodegradation, and environmental recalcitrance, p. 211–243, in M. Alexander (ed.), Advances in microbial ecology, Vol. 5. Plenum Press, New York.

300. Zeikus, J. G. 1983. Metabolism of one carbon compounds by chemotrophic anaerobes. Adv. Microb. Physiol. 24:215–299.

301. Zeikus, J. G., A. Ben-Bassat, T. K. Ng, and R. Lamed. 1981. Thermophilic ethanol fermentations, p. 441–461, in A. Hollaender, R. Rabson, P. Rogers, A. San Pietro, R. Valentine, and R. Wolfe (eds.), Trends in the biology of fermentations for fuels and chemicals. Plenum Press, New York.

302. Zeikus, J. G., A. L. Wellstein, and T. K. Kirk. 1982. Molecular basis for the biodegradative recalcitrance of lignin in anaerobic environments. FEMS Microbiol. Lett. 15:193–197.

303. Zinder, S. H., and M. Koch. 1984. Non-aceticlastic methanogenesis from acetate: acetate oxidation by a thermophilic syntrophic coculture. Arch. Microbiol. 138:263–272.

304. ZoBell, C. E. 1946. Action of microorganisms on hydrocarbons. Bacteriol. Rev. 10:1–49.

305. ZoBell, C. E., and J. F. Prokop. 1966. Microbial oxidation of mineral oils in Barataria Bay bottom deposits. Z. Allg. Mikrobiol. 6:143–162.

306. Züllig, H. 1981. On the use of carotenoid stratigraphy in lake sediments for detecting past developments of phytoplankton. Limnol. Oceanogr. 26:970–976.

INDEX

Acetaldehyde, disproportionation, 780
Acetate:
  competition, 664
  conversion, 729
  decarboxylation, 32
  from ethanol, 643
  formation from pyruvate, 394
  formation in sulfate reducing bacteria, 491,
    494, 504–509, 523
  from glycine, 450
  homoacetate fermentation, 442
  from lysine, 397
  from one-carbon compound, 709
  oxidation, 428, 438, 729
  production from H_2–CO_2, 445
  reductive condensation, 454
  in sediments, 667
  stable isotope fractionation, 679
  as substrate for green sulfur bacteria, 55
  in sulfate reduction, 473
  synthesis, 744
  from two-carbon compound, 709
Acetobacterium, 791
  wieringae, 446
  woodii, 25, 445, 446, 450, 661, 664, 788
Acetogenesis, 418, 446
  acetogenic bacteria, 446
  inhibition, 745
  stimulation, 745
Acetogenic bacteria, 446, 450, 745, 798
  biotechnological applications, 456
  ecological implications, 456
  as hydrogen scavengers, 446
  methanol as substrate, 446
  phylogenetic analysis, 450
Acetogenic dehydrogenation, 419, 432, 436
  ecology, 436

Acetogenic hydrogenation, 441, 450
  ecology, 450
Acetogens, 30, 383, 661, 663, 723, 745
  hydrogen consuming, 664
Acetovibrio cellulolyticus, 663
Acetylene:
  as inhibitor to methanogens, 657
  inhibition method to measure
    denitrification, 230
Achromobacter:
  cycloclastes, 252
  fischeri, 262
Acidaminobacter, 458
  hydrogenoformans, 401, 458
Acidaminococcus fermentans, 396
Acidothermus, 540
  infernus, 540
Actinomycetes, 735
Adenosine phosphosulfate, 603
Adenylsulfate, 153
ADP sulfurylase, 153
Aerobic conversion, of acetone to acetol,
  782
Aerobic degradation:
  of benzoate homologs, 783
  of cellulose, 339
  of lignin, 334
Aerobic respiration, 18
Agmenella quadruplicata, 735
Agrobacterium, 190
Alanine:
  enzymes from degradation of, 399
  enzymes from fermentation of, 399
  uptake in *Chromatium vinosum,* 138
Alcaligenes:
  faecalis, 252, 486
  sp., 253

847

Alcohol:
    fermentation, 16, 19
    mass production, 822
    in sulfate reducing bacteria, 497
Aldehyde oxidase, 250
Alpha-ketoglutarate, 390
Amino acids:
    anaerobic degradation, 386–399
    degradation in rumen, 387, 390
    fermentation, 386
    oxidative deamination, 390
    sequence in complex B800-850, 121
    in sulfate-reducing bacteria, 498
    transamination, 390
    utilization by *Clostridia,* 387
Amino aromatics, degradation, 790
Ammonium, dissimilation from nitrate, 197
Amoebobacter, 43, 53
Amytal, 149
Anacalochloris, 43
Anacystic nidulans, 735
Anaerobic bacteria:
    in human skin, 773
    in soil particles, 773
Anaerobic cometabolism, 792
Anaerobic degradation:
    of aromatic compounds, 341–348, 782
    of benzoate homologs, 783
    of cellulose:
        by anaerobic fungi, 337
        in anoxic sediments, 340
        in aquatic ecosystems, 339
        enzymes, 337
        fermentation products, 337
        by fungal zoospores, 339
        by protozoa, 337
        in rumen, 337
        in termite hindgut, 339
    of 1,2-diols, 781
    environmental implications, 825
    of hydrocarbons, 782
    legislation, 825
    of lignin:
        in aquatic ecosystems, 334
        cleavage of intermonomeric bonds, 334
        rates of mineralization, 334
    of lipids, 374
    of oligolignols, radiolabeled preparations,
        350
    potential, 777, 825
    of protein, 374
        in digestor sludge, 381
        proteolytic activity, 382

    in rumen, 381
    of uncommon substrates, 777
Anaerobic fermentation, 25
    ethanol, 820
    lactic acid, 820
    products, 820
    propionic acid, 820
    yeasts, 821
Anaerobic filters:
    activated carbon, 813
    ceramics, 813
    with limestone gravel, 813
    polyvinyl chloride, 813
    schrimp chitin coating, 813
    sinter glass, 813
    wood chips, 813
Anaerobic habitat, formation, 3
Anaerobic microbial degradation,
    environmental aspects, 794
Anaerobic respiration process, 25
Anaerobic wastewater treatment:
    corrosion problems, 819
    drawbacks, 819
    further processing, 816
    hygienic problems, 819
    toxicity problems, 819
Anaeroplasma sp., 661
Anaerovibrio, 454
    lipolytica, 376
Ancalochloris, perfilievii, 50
Anhydrite, 553, 557
Anoxic basins and fjords, 21
Antimycin A, 140–147, 261–264, 283, 314,
    317
    cytochrome *c,* 262
Aphanocapsa, 393
APS, 604, 609
    reductase, 488, 601, 604, 609
Aquaspirillum:
    itersonii, 316
    magnetotacticum, 189, 191, 316
        strain MS-1, 189
Archaebacteria, 14, 34, 383, 474, 540, 585,
    642, 730
    thermophilic, 383
Archean water, 559
Arginine, degradation, 393
Arginine deiminase, 393
Aromatic compounds, 334, 341
    aerobic catabolism, 341
    aerobic metabolism, 341
    anaerobic degradation, 192, 341, 782, 819
    to bind toxic pollutants, 797

degradation, 392, 497, 796
formation of soil humus, 341
fulvic acids, 797
isopropyl-*N*-phenylcarbamate, 797
lignin, 786
lignin fermentation, 825
lignin residues, 797
tannins, 797
Arthrobacter siderocapsulatus, 318
Aspartate:
    fermentation, 399
    spirilloid bacterium fermenting, 538
    in sulfur reducing bacteria, 538
Aspergillus nidulans, 249
Assembly of nitrate reductase in membrane, 248, 281
Assimilation:
    of nitrate, 274
    of sulfur, 587, 748
Assimilatory sulfate reduction, 472
Association of nitrate reductase with cell membrane, 248
Atlantis II, 558
Atmospheric methane:
    abundance and reactions, 686
    flux rates, 686
    in plant conduits, 687
ATP sulfurylase, 488, 603
Autogenous regulation, of gene expression, 283
Azospirillum, 189

Bacillaceae, denitrification, 191
Bacillus, 29, 189, 308, 313, 314, 319
    licheniformis, 25, 247
    polymyxa, 324, 823
    stearothermophilus, 486, 489
    subtilus, 316
Bacterial photosynthesis, 115
Bacteriochlorophyll, 41, 42
    binding site, 121
    esterifying alcohol, 44
    structure, 116
Bacteriochlorophyll *a*, 118, 119, 130, 131
    adsorption bands, 120
    fluorescence, 120
    reduction, 128
Bacteriochlorophyll *a* protein, X-ray diffraction, 122
Bacteriochlorophyll *b*, 116, 120, 123, 125, 130
Bacteriochlorophyll *c:*
    arrangement, 119

chlorosomes, 118
d or *e*, 117
electron acceptor, 131
Bacteriochlorophyll *g*, 117, 132
Bacteriopheophytin, 124, 125
Bacteriopheophytin *a:*
    Chloroflexus aurantiacus, 132
    reduction, 132
Bacteroides, 454
    amylophilus, 382, 385
    fragilis, 392
    gingivalis, 383
    melaninogenicus, 383, 385, 399
    ruminicola, 382, 385, 392
    sp., 382
    succinogenes, 381
B800-850 complex, 121
B$_{12}$-corrinoids, 781
b-cycle model, 268
Bdellovibrio, 486
Beggiatoa, 28, 88, 474, 547, 549, 795
Benzenes, degradation, 782
Benzoate:
    anaerobic metabolism, 192
    conversion, 782
    defluorination, 192
    degradation pathway, 429
    dehalogenation, 192
Benzylviologen, 262, 263
Beta- and alpha-carotenes, 46
Betaine, in Stickland reaction, 390
Bifidobacterium sp., 383
Bioenergetics, of nitrate reduction, 258
Biomass:
    economical aspects, 822
    production of raw chemicals, 821
    recycling, 804
Biomethanation, thermophilic, 815
Bioturbation, 28
Bisulfite, reductase, 601, 605, 609
Bisulfite reductase, 488, 606
    high-spin, 606
    low-spin, 609
Black Sea, 558
Borrelia, 378
    sp., 380
Bradyrhizobium:
    japonicum, 190
    lupini, 190
    meliloti, 190
Bromoethanesulfonic acid:
    in ethane formation, 678
    as inhibitor in methane oxidation, 682

Bromoethanesulfonic acid (*Continued*)
  mercury metabolism, 659
  trimethylamine metabolism, 669
b-type cytochromes:
  associated with formate reductase, 260
  associated with nitrate reductase, 260
  characterization, 260
  midpoint potential, 260
Burke Lake, 68
Butanediol, 454
Butyrate:
  degradation pathway, 429, 435
  formation, 781
    from hexoses, 452
    from lysine, 397
    from methanol, 452
  in sulfate-reducing bacteria, 495
Butyribacterium, 454
  methylotrophicum, 25, 447, 452, 456, 823
  rettgeri, 452
Butyrivibrio:
  fibrisolvens, 377, 380, 385
  sp., 382
  strain S2, 377, 378, 381
Butyrogenic bacteria, 418

Calvin cycle mechanisms, in green bacteria, 57
Campylobacter, 538
  as denitrifier, 199
  sp., 26, 399, 792
  sputorum, 199, 258, 262
    scheme for electron transport, 275
    scheme for proton translocation, 275
Carbamate kinase, 393
Carbonate, deposits, 559
Carbon compound:
  assimilation pathways, 746
  cell synthesis in *Methanogenium,* 744
  influence on growth of denitrifiers, 200
  one-carbon metabolism, 715
  as regulator in denitrification, 203, 217, 224
  role:
    in DNRA activity, 209
    in population distribution of denitrifiers, 203
    in soil denitrification, 217
  transformation of, 25
Carbon cycle, 24, 334, 341, 361
  carbon budget in aquatic ecosystems, 349
Carbon dioxide:
  activation, 721

reduction, 721
  to methane, 736
Carbon monoxide, 444, 448, 495, 448, 721
  dehydrogenase, 444, 448, 495, 505, 507
  homoacetogenesis, 444
  in methanogenesis, 723
  as poison, 823
Carboxylation, 777
Carotenoid, 46, 121
  chlorosomes, 118
  electrochromic band shifts, 147
  in green bacteria, 74
Casein, 382
Cellulose, 335
  aerobic degradation studies, 339
  anaerobic degradation in rumen, 337
  carboxymethyl, 818
  cellulolytic enzyme, 818
  depolymerization, 818
  digestion in termites, 339
  hydrolysis, 336
  preparation, 818
  radiolabeled preparations, 339, 340
  structure, 335
  thermophilic degradation, 818
Charge separation, 274
  efficiency of oxidative phosphorylation, 269
  influence of sites on, 269
  and location of reductases, 269
Chemical markers, 524
Chemical wastes:
  aminated aromatics, 819
  halogenated, 819
  phenols, 819
  surface-active substances, 819
Chemodenitrification, 182
Chemostat experiment, enrichment technique, 437
Chemotaxonomy, of sulfate reducers, 476
chl C gene, structural gene for nitrate reductase subunits, 248
chl C-*lac,* 282
chl C operon, 281
chl operon, information, 248
Chlorate-resistant mutant, 282
Chlorobactene, 46
Chlorobium, 43, 46, 49, 53, 549
  autotrophic CO_2 fixation, 56
  brown-colored species, 46, 74
  citrate lyase activity in cells, 56
  isorenieratene, 47
  limicola, 119, 131, 141, 149, 151, 152, 488

limicola f. thiosulfatophilum, 118, 139
membrane arrangement, 47
reductive citric acid cycle, 57
Chlorochromatium, 50, 70, 795
aggregatum, 72
photo and electron micrographs, 71
Chloroflexus, 42, 52, 53
aurantiacus, 44, 66, 81, 82, 117, 119, 122, 132, 139, 141, 145
autotrophy, 66
bacteriochlorophyll c_s, 44
chlorosome, 47, 118
cytochrome oxidase, 146
heterotrophy, 66
light adaptation, 82
physiology, 66
temperature strains, 81
Chloroherpeton, 43, 50, 53
Chloronema, 43, 50, 66
Chlorophyll, isoprenoid components of, 781
Chloroplana, 795
Chloropseudomonas ethylica, 50, 617
Chlorosome, 74, 117, 118, 123
chemical composition, 118
energy transfer, 119
transfer of absorbed light energy, 119
Chlortetracycline, 735
Choline, 378
decomposed to trimethylamine, 516
sulfate-reducing bacteria, 498
Chromatium, 28, 43, 53, 69, 545
okenii, 55
sp., 47
growth requirements, 55
tepidum, 55
vinosum, 48, 120–145, 149–153, 489
weissei, 55, 89
Chromatophore:
membrane, 147
purple bacteria, 133
Chromobacterium violaceum, 189
Ciliates:
endosymbiontic interactions with methanogens, 439
episymbiontic interactions with methanogens, 439
Ciso lake, 72
dial metabolism of phototrophs, 75
Citric acid cycle, 494
Citrobacter, 258
Clathrochloris, 70
Clostridia, 25, 484
fermenting amino acid, 386

proteolytic, 380
Clostridium, 25, 53, 378
aceticum, 25, 445, 448, 450
acetobutylicum, 823
acidiurici, 450
bifermentans, 383
botulinum, 377, 383, 387
type G, 387
types A, B, F, 387
butyricum, 306
cellobioparum, 432
CV–AAI, 448
cylindrosporum, 450
difficile, 392
histolyticum, 383, 385, 394
kluyveri, 436, 453, 802
magnum, 454
pasteurianum, 63, 393, 615, 732
perfringens, 262, 378, 383
propionicum, 393, 399
proteolyticus, 383
purinolyticum, 394, 450
putrefaciens, 386
sp., 400
sporogenes, 383, 387, 389
sputorum, 264
sticklandii, 397
subterminale, 387, 397
tertium, 262
tetanomorphum, 386, 392
thermoaceticum, 442, 443, 445, 448, 450, 453, 726
thermoautotrophicum, 444, 447
thermocellum, 661, 818
tyrobutiricum, 613
Cobalt, 616
protein containing, 616
Coenzyme B_{12}, 777, 788
Coenzyme M, 654, 715
inhibitors, 656
role in ethanogenesis, 677
Coenzymes:
of methanogenesis, 708
biosynthesis, 748
coenzyme F_{420}, 734
coenzyme M, 715
component B, 716, 719
cytochromes, 734
factor F_{430}, 720
ferredoxin, 732
methanofuran, 716
methanopterin, 715
sarcinapterin, 718

Coenzymes (*Continued*)
 tetrahydromethanopterin (H₄MPT), 716,
   718
 regeneration, 18
Complex B800-850:
 amino acid sequence, 121
 circular dichroism, 121
 fluorescence, 121
 linear dichroism, 121
 polypeptides, 121
Complex B875:
 circular dichroism, 122
 peptides, 122
 Rhodopseudomonas sphaeroides, 122
 Rhodospirillum rubrum, 122
Condensation, 777
Control of electron flow to nitrogenous
   oxides and oxygen, 283
Copper deposit, 558
Corphin, 720
Corrin, 720
Corrinoids, 724, 748
Corrosion, anaerobic, 530, 533
Corynebacterium nephridii, 252
c-Cytochromes, 137
c-type cytochromes, cytochrome *c552*, 260
Cyanide, 448, 727
Cyanobacteria, 40
Cyclic organic acid, 498
Cyclohexanecarboxylate, 498
Cyclohexanecarboxylate, 498
Cyclopentanecarboxylate, 498
Cysteine, fermentation, 399
Cytochrome *aa₃*, 146, 260
 as a proton pump, 268
Cytochrome *b*, 247, 282
 role in electron transport to nitrate, 261
Cytochrome *bc₁*, 145
 complex, 139, 141, 143, 145, 147
  amino acid sequences, 142
  electron transport, 142
  Rhodopseudomonas sphaeroides, 142
  stoichiometry, 142
Cytochrome *b₆*, 599
Cytochrome *b* cycle, 260, 265
Cytochrome *b₅₀, Rhodopseudomonas
   sphaeroides,* 140
Cytochrome *b*-90, *Rhodopseudomonas
   sphaeroides,* 140
Cytochrome *b*-type, 599, 601, 728, 734
Cytochrome *b₅₆₀*, 140, 144
Cytochrome *b₅₆₆*, 140, 143, 144
 alpha-band spectrum, 140

Cytochrome *c*, 128, 283
 mutants of *Paracoccus denitrificans* lacking
   cyt *c*, 261
 role in electron transport:
  to nitrite, 261
  to nitrous oxide, 261
Cytochrome *c*-type, 595, 596, 604, 614, 616,
   728, 734
 high-potential, 600
 low-potential, 600
 mitochondrial, 596
 monoheme, 600
Cytochrome *c₁*, 142, 143, 144
 Rhodospirillum rubrum, 141
 Rhodospirillum sphaeroides, 140
Cytochrome *c₂*, 136, 138, 144, 145, 611
 absorbance spectra, 133
 binding, 137
 cyclic electron flow, 140
 electron donor, 136
 Eₘ value, 136, 140
 lysine residues, 137
 mutants, 137
 oxidation, 137
 structures, 133
Cytochrome *c₃*, 489, 594, 595, 598, 604, 609,
   611
 octaheme, 595, 614
 tetraheme, 595, 617, 618
 three dimensional structure, 596
 triheme, 616
Cytochrome *c₅₅₀*, 152, 153
 Chromatium vinosum, 138
Cytochrome *c₅₅₁*, 152, 600
Cytochrome *c₅₅₂*, 137, 138
 Chromatium vinosum, 138
 oxidation, 137
Cytochrome *c₅₅₃*, 600
 Chromatium vinosum, 138
 green sulfur bacteria, 139
Cytochrome *c₅₅₄, Chloroflexus aurantiacus,*
   139
Cytochrome *c₅₅₅*, 139, 152
 cyclic electron flow, 138
 oxidation, 137
Cytochrome *c₅₅₈*, 138
Cytochrome *cd*, 252, 601
 role in denitrification, 182
Cytochrome *o*, 260, 277
Cytochrome oxidase, 252, 260, 277
 Chloroflexus aurantiacus, 146
 inhibition by hydroxylamine, 285
 inhibition by nitrite, 285

Rhodopseudomonas sphaeroides, 146
Cytochrome P450, 777, 791
Cytochromes, 260, 595, 728
  multiple-heme, 595
Cytophagaceae, denitrification, 191
Cytoplasmic membrane, 123
  invaginations, 115
Cytoplasmic side of membrane, in
  Clostridium sputorum, 264

DBMIB, 145
Dealkylation, 776
Deamination, 791, 792
Deazaflavin, 735
Decarboxylation, 777
Degradation:
  of alcohols, 781
  of aliphatic hydrocarbons, 778
  of aromatic compounds:
    by anaerobic photometabolism, 341
    by fermentative bacteria, 342
    methanogenic conditions, 343
    mixed culture studies, 341
    nitrate-reducing conditions, 341
    reductive ring cleavage, 341, 342
    sulfate-reducing conditions, 341
  of hydrocarbon, 774
  of ketones, 781
  of lignin:
    in anaerobic environments, 358
    in aquatic ecosystems, 358
    by organisms, 355
    rates of mineralization, 359
    in rumen, 354
    in termite hindgut, 356
  of oligolignols, 350, 351
  of xenobiotics, 792
Degradative potential, of anaerobic microbial
  communities, 794
Dehalogenation, 27, 776, 789, 792
  anaerobic, 789
  reductive, 789
Dehydration, 777
Dehydrogenase, 589, 595, 596, 600, 609
  "desulfo" xanthine, 614
Dehydrogenation, 348, 419, 432, 436, 777
Dehydroxylation, reductive, 785, 789, 792
Delta-aminovalerate, degradation, 400
Demethylation, 792
De-molybdoenzyme, 249
Denitrification, 14, 19, 29, 180, 246, 280, 817
  acetylene inhibition method, 230
  aerobic, 14, 279

concentration effects of oxygen on, 215
ecological and economical importance, 185
effect of oxygen on, 216
enzyme activity in air-dried soils, 205
growth of denitrifying bacteria, 185
influence of carbon compounds, 203
inhibition by oxygen, 191
intermediates accumulation, 194, 195
macrofeatures affecting regulators, 225
methods used to measure, 226
model of environment regulation, 217
nonrespiratory, 183
organisms responsible of, 185
pathway between nitrite and nitrous oxide,
  252
proximate regulators of, 217
regulation:
  of intermediates production, 194
  by nitrate, 209, 222
  by oxygen, 191, 210, 215, 219
regulatory factors, 217, 219
respiratory, 182, 185
by *Rhizobiaceae,* 190
role of NO, 252
scheme for, 254
in soil, 258
synthesis of enzymes, 215
Denitrifiers, 185, 190, 191
  adaption to environmental stress, 204
  in bioconversion reactions, 194
  competition for carbon, 203
  degradation:
    of chlorinated solvents, 193
    of nonionic detergents, 193
  energy sources, 187
  growth requirements, 193, 203
  habitats, 201
  lithotrophs, 187
  natural habitats, 189
  organotrophs, 187
  oxygen as regulator, 210
  phototrophs, 187
  physiological features, 188, 199
  in pollutant treatment, 192, 194
  population ecology, 201
  population selection, 203
  in sediments, 203
  in soils, 203
  under stress conditions, 205
Depolarization, cathodic, 530
Desulfobacter, 473, 604
  curvatus, 503
  hydrogenophilus, 494

Desulfobacter (*Continued*)
  postgatei, 486
  sp., 26
Desulfobacterium:
  autotrophicum, 492
  catecholicum, 497
  indolicum, 497
  phenolicum, 497
Desulfobulbus, 604
  elongatus, 486, 489
  propionicus, 453, 486, 488
Desulfococcus, 604
  multivorans, 496
  niacini, 486, 493, 497
Desulfofuscidin, 489, 596, 607
Desulfomonas:
  acetoxidans, 26
  pigra, 491
Desulfonema:
  limicola, 486, 496
  magnum, 493, 497, 606
  sp., 26, 484
Desulforedoxin, 616
Desulforubidin, 488, 606
Desulfosarcina, 604
  variabilis, 486, 497
Desulfotomaculum, 476
  acetoxidans, 490
  antarcticum, 499
  guttoideum, 490
  nigrificans, 484, 606
  ruminis, 490
  sapomandens, 496, 497
  sp., 475
Desulfovibrio, 589
  africanus, 486, 612, 615, 732
    africanus strain Benghazi, 615
  baarsii, 476, 493
  baculatus, 475, 594, 612
    baculatus strain 9974, 594, 612
  desulfuricans, 262, 378, 437, 475, 490, 592–615
    strain ATCC 27774, 592, 595, 607, 609, 615
    strain ATCC 7757, 609
    strain Berre-Eau, 612, 615
    strain Berre-Sol, 615
    strain HL 21, 609
    strain Norway 4, 594, 595, 607, 612
  gigas, 199, 449, 486, 489, 592, 593, 603, 607, 732
  growth requirements of, 476
  multispirans, 603

  salexigens, 486, 592, 594, 615
    strain British Guiana, 615
  sapovorans, 488, 491
  strain Norway 4, 475
  sulfodismutans, 503
  thermophilus, 475, 491, 596, 607, 616
    thermophilus strain DSM 1276, 616
  vulgaris, 432, 437, 486, 490–495, 602, 609, 792
    strain Hildenborough, 594, 595, 602, 609
    strain Hiyazaki, 602
    strain Marburg, 596
    strain Miyazaki, 595
Desulfovibrio desulfuricans:
  formate dehydrogenase, 263
  location of hydrogenase, 263
  nitrite reductase, 264
Desulfovibrio sp., 26
  measurement of oxidative phosphorylation, 279
  nitrate reduction by, 279
  proton translocation during nitrite reduction in, 279
Desulfoviridin, 488, 606
Desulfuration, 473, 554, 776
Desulfurococcus, 540, 708
  sp., 383
Desulfuromonas, 486
  acetoxidans, 50, 484, 600, 609, 617
Deuterium-proton exchange reaction, 594
Diabolic acids, 377
Diagenesis, early, 4
Diagenic processes, 24
Dicarboxylic acids, 497
Dicumarol, 314
Digestor:
  bacterial activity in sludge, 383
  high-strength acid whey, 383
Dimethylsulfide, 651
  as noncompetitive substrate, 664
Dimethylsulfoxide (DMSO):
  utilization by purple bacteria, 61
  utilization by sulfur-reducing bacteria, 538
Diol dehydrase, 781
Dioxygenase, 789
Disporportionation:
  of acetaldehyde, 780
  of sulfite, 503
  of sulfur, 503, 545
  of thiosulfate, 503
Dissimilatory nitrate reduction, 197
  population ecology, 201
  population location of organisms, 208

role of cytochrome c in, 261
to ammonium (DNRA), 197
to ammonium (DNRA), 197, 208, 261
  organisms responsible, 198
  role of cytochrome c_{552} in, 261
Dissimilatory sulfate reduction, 472, 588,
  589, 618, 620
  electron carriers, 488
  enzymes, 488
Diurnal cycle, 28
Diurnal light-dark cycle, 28
DNRA, 198
  dissimilatory nitrate reduction to
    ammonium, 19, 183, 197, 208, 261
  population selection of organisms, 209
Domestic sewage sludge, 376, 813

East Pacific Rise, 558
Ectothiorhodospira, 43, 48, 52, 549
  halochloris, 120
  mobilis, 139
Efficiency of oxidative phosphorylation, 273
Electron acceptor, 18, 24, 28, 32, 131, 427,
  501
  alternative, 306
  influence in DNRA activity, 209
  sequence, 24, 323
Electron activity, 1, 5, 8
Electron carrier, 18, 592
  localization, 592, 596, 598, 602, 613, 614
  protein, 589, 595, 611
Electron donation, 277
  at cytoplasmic side of membrane, 264
  at periplasmic side of membrane, 264
Electron donor, 19, 28, 256, 277, 490
Electron-free energy diagram, 1, 9, 11, 19
Electron levels, 10
Electron paramagnetic resonance, 247, 249
Electron pressure, 8
Electron sink, 18
Electron-titration diagrams, 10
Electron transfer, 5, 618
  coupled to phosphorylation, 199, 599, 613,
    621
  localization, 731
  from P*, 125
  processes in methanogenesis, 730
  protein, 589, 592, 594, 602, 611–614, 619,
    730, 731, 741, 775
  vectorial, 595, 618, 619
Electron transport phosphorylation, 182, 199
Electron tunneling, reaction center, 138
Embden–Meyerhof pathway, 442

Emulsifiers, 778
Endergonic process, 30
Endosymbiosis, 439
Energy conservation:
  in methanogenesis, 736
  via sodium gradient, 436
Energy consumption, for growth of
  methanogenic bacteria, 738
Energy transfer:
  chlorosome, 123
  to BChl a, 120
Enrichment technique, 437
Enterobacteria, 25
Enterobacteriaceae, 183
  in denitrification processes, 199
Enzymes:
  cytochrome P450, 776, 777
  denitrifying activity, 207
  diol dehydrase, 781
  dioxygenase, 776
  p-hydroxybenzoate hydroxylase, 777
  hydroxylase, 776
  indolamine deoxygenase, 777
  of methanogenesis, 716
    CO dehydrogenase, 723
    CO-methylcobamide: HS-CoM
      methyltransferase, 725
    component A, 719
    cyclohydrolase, 718
    formate dehydrogenase, 722
    formyltetrahydrofolate synthetase, 715
    hydrogenase, 719
    methanol: 5-hydroxybenzimidazolyl-
      cobamidemethyltransferase, 725
    methylcoenzyme M reductase, 719
    methylene-H_4MPT dehydrogenase, 732
    methylene-H_4MPT reductase, 718
    methylene tetrahydrofolate
      dehydrogenase, 715
    methyl-H_4MPT: HSCoM
      methyltransferase, 727
    NADH: coenzymes F_{420} oxidoreductase,
      719
    trimethylamine: HS-CoM
      methyltransferase, 729
  monoxygenase, 776
  oxygenase, 776
  triptophane dioxygenase, 777
  tyrosinase, 777
Episymbiosis, 439
Epoxidation, 776
Equilibrium:
  concentrations, 13

Equilibrium (*Continued*)
  constant, 7, 8, 10
  hypothetical, 21
  internal, 23
  local, 20
  partial, 20
  redox potential, 8
  reversible, 16
Escherichia:
  coli, 189, 200, 247, 258, 316, 486, 602
  coli respiratory chain of, 260
Ethanol, 19
  fermentation, 815, 822
  production drawbacks, 822
  from sugars by yeasts, 822
Ether linkages, cleavage, 787
Ethylene as inhibitor to methanogens, 657
Eubacteria, 14, 474, 533, 585
Eubacterium, 379
  lentum, 379
  limosum, 447, 452, 724
  ruminantium, 382
Eukaryotes, 14, 34
Eutrophic lake sediment, 383
Evaporites, 553, 555
Evolution, 14, 34
  of anaerobes, 34
  of life, 34
Exciton annihilation, 121
Exciton interaction, 126

Facultative green bacteria, 51
  physiology, 66
Facultative purple bacteria, 51, 57
  in aquatic habitats, 79
  autotrophic growth, 57
  culture media, 93
  enrichment, 94
  fermentative growth, 61
  genetics, 64
  heterotrophic (aerobic) growth, 60
  lithotrophic growth, 58
  photoheterotrophic growth, 58
  physiology, 57
  in sewage plant, 80
FAD (flavin adenine dinucleotide), 18, 601
  sulfite, 605
Faraday constant, 9
Fatty acids, 374
  from C_2-compounds, 451
  composition of proteolytic clostridia,
    380
  degradation, 662
  exogenous utilization, 380

Fermentation:
  industrial use, 822
  alcohol, 820
  beverages, 820
  butanol-acetone, 821
  cheese, 820
  of glycerol, 781
  nonmethanogenic, 820
  petrochemicals, 821
  processes, 19, 32
  sauerkraut, 820
  silage, 820
  sour milk products, 820
  substrate conversion, 822
  of thermophilic bacterial ethanol, 822
Fermenting bacteria, 29, 500
Ferredoxin, 380, 489, 592, 597, 598, 604,
    608, 611, 613, 724, 726, 731
  I and II *D. desulfuricans,* 597
  I and II *D. gigas,* 597, 608, 611, 613
  NADH-oxidoreductase, 613
  NAD(P)$^+$-oxidoreductase, 151
  structures, 133
Ferric iron, reduction, 24
Ferric oxide, amorphous oxyhydroxides, 324
Ferricyanide, effect on nitrate reduction, 283
Ferromanganese, Baltic Sea, 308
Fish Lake, 76
Fixed-film bioreactors, 795
Flavin, 601, 605, 609, 726
Flavobacterium, 195, 280
Flavocytochrome c_{552}, 152, 153
Flavocytochrome c_{553}, 152
Flavodoxin, 489, 592, 597, 604, 608, 611, 612
  three dimensional structure, 612
Flavoprotein, 604, 722, 775
Floc-forming bacteria, 807
fnr gene, and nitrate reductase formation,
    281
Food chain, anaerobic, 544
Formate:
  bacteria utilizing, 722
  conversion, 722
  dehydrogenase, 264, 443, 595
  as intermediate in homoacetogenesis
    dehydrogenase, 443
  methanogenic bacteria utilizing, 723
  oxidation, 728
  in rumen, 665
  in sediments, 669
  in sulfate-reducing bacteria, 493
  in sulfur-reducing bacteria, 538
Formate dehydrogenase:
  in *C. fetus* subsp. *je/juni,* 277

in *D. desulfuricans,* 277
in *D. gigas,* 277
electron-donating site, 277
from *E. coli,* 247
in *V. succinogenes,* 277
Formate dehydrogenase and hydrogenase, electron donation, 264
Formate hydrogenlyase, 281
Formation of nitrate reductase, by autogenous regulation, 282
Framboids, 554
Free energy:
  changes, 7
  of formation, 8
  standard, 7
Fumarate:
  as electron acceptor, 427
  reductase, 601
  in sulfur-reducing bacteria, 538
Fungi, 818
Fusobacterium sp., 400
Fusocillus, 379

Gamma-aminobutyrate, degradation, 400
Gastrointestinal system, 33
Geochemistry of methane, 671
  anaerobic oxidation, 681
  atmospheric methane, 685
  ethane formation by methanogens, 677
  natural gas, 673
  sedimentary lipids, 672
  sedimentary penetration, 684
  stable istopes, 679
Gliding bacteria, 527, 528
Glucose, complete oxidation, 432
Glutamate:
  degradation, 394–397
  dehydrogenase, 390
  synthesis, 746
Glycerol, sulfate-reducing bacteria, 498, 499
Glycerol esters:
  galactolipids, 376
  hydrolysis, 376
  phospholipids, 376
  sulpholipids, 377
  triglycerides, 376
Glycerol ether lipids, 788
Glycine:
  decarboxylase pathway, 450
  degradation, 394
  as electron acceptor, 389
  fermentation, 394

hydrogen production, 394
reductase, 389
reduction, 389
Glycogen, 56
Graphite, 16
Green sulfur bacteria, 40, 43, 50, 117, 122, 472, 538
  absorption spectra, 117
  autotrophic CO_2 fixation, 56
  chlorosomes, 47
  consortia, 50
  electron transport, 130
  enrichment and isolation, 91
  growth conditions, 46
  isotope discrimination, 56
  membrane arrangement, 47
  photoheterotrophic metabolism, 56
  physiology, 55
  syntrophy, 50
  taxonomy, 47
    of facultative, 51
Growth requirement:
  of *Chromatium* sp., 55
  of denitrifiers, 186, 193, 203
  of *Desulfovibrio,* 476
  of methanogens, 645, 654
  of Rhodospirillaceae, 59
Gypsum, 553, 557

Habitats:
  of anaerobic bacteria:
    sediments, 794
    sludge digestors, 794
    surfaces, 794
  of denitrifiers:
    aquatic environments, 201
    sediments, 201
    soils, 201
      sandy-desert, 204
    waters, 204
  extremely heterogeneous, 796
  of methanogens:
    aquatic sediments, 668
    baleen whales, 667
    bowel, 666
    cecum, 666
    gastrointestinal tract, 666
    hindgut of insects, 667
    hot springs, 670
    hypersaline lakes, 670
    living organisms, 666
    meromictic lakes, 670
    plankton, 667
    rumen, 666

Habitats, of methanogens (*Continued*)
trees, 666
of phototrophic bacteria, 53, 93
hot springs, 81
marine microbial mats, 86
sewage and waste water, 79
soil, 796
spatial and temporal heterogeneity, 795
of sulfate reducers, 519, 558
Black sea, 521
Cariaco Trench, 521
in lakes, 521
Limfjord, Denmark, 521
in marine sediments, 520
in salt marshes, 520
significance of metabolic types, 522
Half-reaction, 7
Halobacteriaceae, denitrification, 191
Halobacterium, 189
cutirubrum, 735
Heliobacillus, 44
sp., 44
Heliobacterium, 44, 53
chlorum, 44, 45, 65, 117
electron acceptor, 132
nitrogen fixation, 65
physiology, 65
Heliospirillum, 44, 45, 65
sp., 44
Heliothrix, 44
oregonensis, 87
Heme, 595, 596, 599, 601
high-spin, 602
transition, 617
iron, 606
low-potential, 600, 617
low-spin, 602
transition, 617
Hemicelluloses, 818
Hemoprotein, 600, 621
Histidine, fermentation, 400
Homoacetogenesis, 442, 445, 446
carbon monoxide:
as intermediate, 445, 448
as substrate, 448
formate as intermediate, 443
methanol as substrate, 446
pyruvate as intermediate, 442, 448
of sulfate-reducing bacteria, 505, 507
Homoacetogenic bacteria, 418, 442, 516, 798
oxidizing ethanol, 823
Homopropionate fermentation, 454
HQNO, 260, 262, 267

Human gut, bacterial activity, 382
Hydration, 777
Hydrocarbon:
aliphatic, 778
beta-carotene, 780
degradation, 774, 782
halogenated, 72
kerogen, 780
styrene, 780
in sulfate reducing bacteria, 499, 529
Hydrogen:
cathodic reaction, 530
consumption, 594, 603
cycle, 618
cycling, 449, 619
evolution, 594
interspecies transfer, 30, 401, 419, 430,
439, 453, 589, 602, 611, 618, 643, 644,
661
metabolism, 599
oxidation, 592, 612, 618
partial pressure, 30–32, 800
production, 401, 594, 612, 613, 618, 619,
620
reductase, 601
in sulfate-reducing bacteria, 493, 509
in sulfur-reducing bacteria, 538
sulfide, 589, 620
transformation model, 618, 620
utilization, 611, 613, 618
Hydrogenase, 264, 449, 454, 592, 593, 600–
618, 731
bidirectional, 593
coenzyme F_{420}-dependent, 731
cytoplasmic, 602
EPR signal, 593, 594
evolution, 603
in *C. fetus* subsp. *je/juni,* 277
in *D. desulfuricans,* 277
in *D. gigas,* 277
in *V. succinogenes,* 277
iron, 592, 602
localization, 594
membrane-bound, 602
nickel-containing cytoplasmic, 603
nickel-iron, 592, 602
nickel-iron-selenium, 592, 602
periplasmic, 592
at periplasmic side of membrane, 277
redox active nickel, 593
specific activity, 593, 602
Hydrogenation, 441, 450, 777
of polyunsaturated fatty acids, 380

by ruminal protozoa, 379
of unsaturated fatty acids, 378, 381
Hydrogen plus carbon dioxide:
  competition, 670
  in hot springs, 670
  in sediments, 667
  stable isotope fractionation, 679
Hydrogenosomes, 441, 729
  protozoa: *Metopus striatus*, 441
Hydrogenotrophic methanogen, syntrophic
  association, 422
Hydrogen-producing reactions, 31
Hydrogen sulfide:
  purple and green sulfur bacteria, 47
  reoxidation, 547
  from sulfate, 474
  from sulfate dissimilation, 473
  toxicity, 41
Hydrolysis, 777
  enzymatic, 822
Hydroxylamine, 252
Hyphomicrobium, 189
  sp., 279
Hypolimnion, 75
Hyponitrite, intermediate in denitrification,
  254

Inhibitor of methanogens:
  alternate electron acceptors, 655, 659
  2-bromoethansulfonic acid, 656
  chlorinated methane, 657
  heavy metals, 658
  monensin, 658
  unsaturated carbon–carbon bonds, 657
  unsaturated fatty acids, 657
Interaction, microbe-surface, 794
Interactions of methanogens:
  competition, 664, 668
    with sulfate reducers, 664
  consortia, 663
  demethylation of Hg, Se, 658
  experimental syntrophies, 661
  fatty acids degradation, 662
  noncompetitive substrates, 664
  osmoregulatory compounds, 660
Interspecies hydrogen transfer, 30, 33, 401,
  419, 439, 453, 643, 644, 661, 774, 781
  effect on growth yield, 430
Interspecies reductant transfer, 436
Intracellular redox potential, 280
Intracytoplasmic membrane, 115
  chromatophores, 47
  vesicles, 47

Invaginations, purple bacteria, 115
Iron, 28, 594, 596, 601, 603, 604
  assimilation, 316
  nonheme, 611
  reaction center, 130
  reductase, 316, 324
  reduction, 29, 306
  respiration, 317
  system, 15
Iron compounds, reduction, 26
Iron oxide, 306
  cellular binding, 319
  mineral type, 319
Iron reduction, 306
  aerobic, 317
  anaerobic, 317
  Aquaspirillum itersonii, 316
  Aquaspirillum magnetotacticum, 317
  Bacillus polymyxa, 324
  Bacillus subtilis, 317
  by bacteria, 308
  in cell free systems, 316
  detection, 312
  by end products of metabolism, 317
  Escherichia coli, 317
  by fungi, 308
  by hydrogen oxidizing bacteria, 324
  inhibition, 310, 317
  isolation of microorganisms, 311
  lake sediments, 323
  marine sediments, 322
  mechanisms, 312
  by mixed populations, 312
  in nature, 307, 322
  nitrate reduction, 323
  Pseudomonas aeruginosa, 316
  rate, 320
  redox potential, 311
  redox sequence, 323
  respiratory chain, 317
  significance, 322
  Staphylococcus aureus, 316
  Sulfolobus acidocaldarius, 313
  by syntrophic consortia, 312
  Thiobacillus denitrificans, 313
  Thiobacillus ferrooxidans, 313
  Thiobacillus thiooxidans, 313
Iron–sulfur center, 131, 149, 151, 614
  in reduction of nitrate, 268
  in reduction of nitric oxide, 268
  in reduction of nitrite, 268
  in reduction of nitrous oxide, 268
  role in proton translocation at site I, 268

Iron–sulfur centers/clusters, 247, 488, 592, 594, 599, 605, 609, 611, 722, 730
Iron–sulfur protein, 131, 149, 731
  Chlorobium limicola f. *thiosulfatophilum,* 141
  Chromatium vinosum, 141
  containing molybdenum, 614
  Rieske, 141
Isoprenoid ether, 34
Isoprenoid hydrocarbon, 34
Isorenieratene, 46
Isotope fractionation, 554

K-carageenan, glycerol production, 823
Kingella, 190
Kinnert Lake, 75
Klebsiella:
  aerogenes, 247, 266
  pneumoniae, 258, 266, 281, 823
Knaack Lake, 76
Knallgas bacteria, 731
Kupferschiefer, 558

Lachnospira multiparus, 382
Lactate:
  dehydrogenase, 601, 609
  in sulfate-reducing bacteria, 493
  metabolism, 599
Lactate dehydrogenase, electron donation, 264
Lactobacillus, 25, 823
  delbrueckii, 823
  sp., 392
Laguna Figueroa, 87
Lake Mendota, 521
Lakes:
  bacteriochlorophyll, 68
  eutrophic, 67
  holomictic, 67
  hypolimnion, 67
  meromictic, 67
  phototrophic bacteria, 67
  sediments, 28
Lamprobacter, 43
Lamprocystis, 43
  roseopersicinia, 72
Leptothrix pseudoochraceae, 318
Leucine:
  de novo synthesis from glucose, 392
  in Stickland reaction, 389
Leuconostoc, 25
Lignin, 334, 350, 353, 360
  aerobic degradation, 334, 349

anaerobic degradation, 334, 354, 786
arylglycerol-beta-aryl ether bond, 336, 350
bound to hemicellulose, 336
bound to holocellulose, 336
model structure, 336
radiolabeled preparations, 354, 357
synthetic, 334, 354, 358
Lignocellulose, 334, 335
  defined, 335
  hemicellulose, 335
  pectin, 336
Lipase, 377
Lipids:
  alpha-oxidation, 381
  anaerobic degradation, 374
  de novo synthesis, 380
  sediments, 649
  synthesis by ruminal bacteria, 381
Lipolytic activity, 377
Lipolytic bacteria, 374
Location of nitrate reductase, 277
Lyase reaction, 777
Lycopenal, 46
Lycopene, 46
Lyngbya, 87
Lysine:
  cytochrome, 152
  cytochrome *c,* 137
  enzymes from fermentation of, 399
  fermentation, 397

Mackinawite, 551
Macrofeatures, affecting denitrifying regulators, 225
Magnesium tetrapyrrole, 42
Magnetotactic bacteria, 795
Maintenance energy, 271
Manganese compounds, reduction, 26, 306
Manganese oxide, 306
  cellular binding, 319
  mineral type, 319
Manganese reduction, 28, 306
  aerobic, 311, 317, 325
  anaerobic, 317
  Arthrobacter siderocapsulatus, 318
  ATP formation, 316
  Bacillus 29, 313
  Bacillus polymyxa, 324
  by bacteria, 308
  Baltic Sea, 308
  catalase activity, 315
  in cell-free systems, 313
  deposition, 318

by electron transport system, 314
by end products of metabolism, 317
enumeration of microorganisms, 307
by fungi, 308
induced cells, 314
inhibition, 310, 314
isolation of microorganisms, 308
Leptothrix pseudoochraceae, 318
marine sediments, 324
mechanisms, 312
Metallogenium, 318
mixed populations, 312, 325
in nature, 307, 322
nonenzymatic reaction, 313
peroxide formation, 315
by peroxides, 318
Pichia guillermondii, 313
rate, 320
redox potential, 311
reducing microorganisms, 310
respiration, 316
role of superoxide dismutase, 319
significance, 322
stimulation, 314
sulfate reducers, 325
by syntrophic consortia, 312
system, 15
terminal reductase, 314
uninduced cells, 314
Veillonella alcalescens, 313
Measurement of oxidative phosphorylation in
    cell-free preparations, 279
Mellum, 88
Membrane organization, 600
    of *Desulfovibrio* sp., 600
Menaquinone, 125, 126
    Chloroflexus aurantiacus, 132
    electron acceptor, 130
    pool, 135
    reaction center, 125
Menaquinone-G, 601
Mesoheme, 606
Metabolic flexibility, 459
Metallogenium, 318
Metal sulfides, deposits, 557
Metanogenic substrates, 651
Methane:
    bacteria, 29, 34, 437, 450, 609, 618, 642,
        708, 723
    bacteria oxidizing, 499, 795
    fermentation, 16
    formation, 32
        formation in reactors, 815

oxidation, 499
producing bacteria, 30
synthesis, 742
Methanobacillus omelianskii, 422, 453, 456,
    643
Methanobacterium, 644
    bryantii, 644, 715, 722
    formicicum, 439, 652, 656, 663, 722, 723
    propionicum, 438
    ruminantium, 654
    strain AZ, 649
    suboxydans, 429
    thermoautotrophicum, 428, 650, 716, 718,
        723
    wolfei, 730
Methanobrevibacter:
    arboriphilus, 666, 723
    arboriphilus strain AZ, 748
    ruminantium, 437, 719, 744
    ruminantium strain M_1, 719
    smithii, 666, 722
    strain PS, 650
Methanococcoides methylutens, 651, 734
Methanococcus, 644
    halophilus, 651, 654
    jannaschii, 649, 655
    mazei, 645, 652
    thermolithotrophicus, 654, 741
    vannielii, 650, 654, 722
    voltae, 654, 720
Methanofuran, 718
Methanogenesis, 29, 734
    from acetate, 729
    in aquatic environment, 668
    cell synthesis, 744, 748
    consumption for growth, 738
    in digestors, 519
    electron-transferring processes, 731
    energy conservation, 736
    inhibition, 745
    in presence of sulfate, 519
    stimulation, 745
    substrates, 709
Methanogenic bacteria, 29, 34, 437, 450, 609,
    618, 642, 708, 723
    acetophilic, 729
    activity inhibition, 742
    cell envelope lipids of, 649
    cell wall characteristics, 645
    chemiosmotic processes, 741
    composition of lipids in, 730
    demethylation of Hg, Se, 659
    depth distribution, 684

Methanogenic bacteria (*Continued*)
  eubacterial genes, 749
  genetics, 749
  growth, 740
    growth requirements, 651, 654
    growth with sulfur, 748
  hydrogenase, 731
  hydrogenotrophic electron transferring
    processes, 730
  as hydrogen scavengers, 422, 424, 437, 450
  inhibitors, 655
  membraneous constituents, 730
  methylotrophic, 731
  natural habitats, 666–670
  potassium concentrations in, 742
  taxonomy, 644
Methanogenic degradation:
  biotechnology, 816
  carbon and electron flow, 797
  of ethanol, 774
  genetic engineering, 816
Methanogenic ecosystems, population size,
  798
Methanogenic substrates, 30, 651–653
  acetate, 651, 652
  dimethylsulfide, 651
  formate, 651
  hydrogen plus carbon dioxide, 651
  methylated amines, 653
Methanogenic waste treatment, 803
  aerobic versus anaerobic, 803
  efficiency enhancement, 815
  feasibility, 819
  process manipulation, 815
  reactor design, 804
  reactor efficiency, 815
  technological advantages, 803
Methanogenium thermophilicum, 655
Methanol:
  cobalamin methyltransferase, 448
  conversion, 724
  fermentation, 724
  reductive activation, 732
  in sediments, 665
  stable isotope fractionation, 679
  as substrate for acetogenesis, 446
  as substrate for methanogenesis, 446
  in sulfate-reducing bacteria, 516
  utilization by purple bacteria, 59
Methanolobus tindarius, 651, 734
Methanomicrobium mobile, 744
Methanoplanus limicola, 645
Methanoplasma elizabethii, 645

Methanopterin, 715
Methanosarcina, 437, 448, 644, 734
  acetivorans, 651, 652
  barkeri, 25, 378, 432, 452, 609, 645, 652,
    715, 718, 791
  mazei, 25
  strain DSM 800, 609
  TM-1, 652
Methanospirillum, 437
  hungatei, 424, 429, 437, 663, 722, 792
Methanothermus fervidus, 655
Methanothrix, 437
  soehngenii, 25, 441, 651, 652, 734, 782,
    809
Methionine, 727, 748
Methylamine:
  conversion, 728
Methylated amines:
  in hypersaline environment, 670
  as noncompetitive substrate, 664
  in rumen, 664
  in sediments, 670
Methyltetrahydrofolate, 443, 448, 715
Methyltransferase, 791
Metopus striatus, 441
Microbial biomass, residence times, 802
Micrococcus, 26
Microcoleus, 87
Microelectrode, 28, 29
Microenvironment, 3, 13
Microsite, 29
Mini-methane formation, 495
Mirror Lake, 76
Mitchell loop, 619
Molar growth yield:
  of *Pa. denitrificans,* 271
  of *Ps. denitrificans,* 271
  of *T. denitrificans,* 271
  of *Thiobacillus* A2, 271
  of *Vibrio succinoges,* 278
Molybdate, formate dehydrogenase, 443
Molybdenum, 614, 722
  in nitrate reduction, 248
  protein containing, 616
Molybdenum cofactor, 282
  extraction procedure, 250
  in nitrate reduction, 248
  regulation of formation of, 280
  sensitivity toward oxygen, 250
Molybdoenzymes, 249
Molybdopterin, 250, 251, 285
Monooxygenase, 788
Moraxella, 190

Mycobacterium
sp., 735
Mycoplasma arthritidis, 393
Myxococcus, 486
Myxothiazol, 140, 142, 143, 144

NAD (nicotinamide adenine dinucleotide),
18
NAD$^+$, *Rhodospirillum rubrum,* 149
NADH:
dehydrogenase, 149
electron sink, 199
N-demethylation, 791
Neisseria, 190
role in nitrate dissimilation to ammonium,
199
subflava, 199
Neisseriaceae, denitrification, 191
Nernst equation, 9
Neurospora:
crassa, 249
crassa nit mutants, 250
Neurosporene, 46
Nickel, 593, 603, 732
low-spin, 730
redox transition, 594
tetrapyrrole, 720
Nitrate:
ammonification, 19
as regulator in denitrification, 197
assimilating bacteria, 183
assimilation, 265, 274, 472
assimilatory reduction, 180
conversion, 182
dissiumilating bacteria, 183, 208, 602
dissimilatory reduction, 14, 180, 183, 246
to ammonium, 19, 183, 197, 209, 261,
502
in sulfate-reducing bacteria, 278, 501
in sulfur-reducing bacteria, 538
reducing-bacteria, 208, 602
reductase, 14, 183, 601, 614, 723
reduction, 180, 208, 258, 323, 502, 602
to ammonium, 197, 199, 209, 502
to nitrite, 199
respiration, 182, 208, 246
Nitrate reductase, 247, 280
control of activity by redox state of
ubiquinone, 284
effect of oxygen on activity, 280
formation:
influence of oxygen, 279
regulatory aspects, 279

inhibition of activity by nitrite, 284
inhibition of activity by nitrous oxide, 284
Nitrate reduction:
to ammonia, 502
bioenergetics, 258
at cytoplasmic side of membrane, 262
energy production during, 265
in *Escherichia coli,* 258
by facultative anaerobes, 323
at periplasmic side of membrane, 264
stoichiometry of proton translocation,
265
by sulfate-reducing bacteria, 278
Nitrate transport, 264, 274
active uptake process, 264
nitrate/nitrite antiport, 264, 274
Nitric oxide:
intermediate in denitrification, 252
proton translocation during reduction, 270
reductase, 252, 255
role of nitric oxide, 252
Nitrification, 817
Nitrite:
conversion to ammonium, 199
detoxification, 199
effect on nitrate reductase activity, 284
as inhibitor of cytochrome oxidase, 275
reductase, 182, 601, 605
toxic effect on bacteria, 274
Nitrite reductase, 251, 252, 256
involved in dissimilatory nitrate reduction,
261
regulation of formation, 280
Nitrite reduction:
influence of inhibition of cytochrome
oxidase, 284
Nitrite/nitrous oxide reduction, at
periplasmic side, 262
Nitro-aromatic compounds, 790
Nitrobacteraceae, denitrification, 191
Nitrogen:
compounds containing, 746
cycle, 14
dealkylation, 790
fixing organism, 14, 62, 65, 88, 281, 506,
732
system, 12
Nitrogen compounds, reduction, 25
Nitrogen fixation, 14, 62, 281, 506, 732
cyanobacteria, 88
in sulfate-reducing bacteria, 506
Nitrogen oxide, as electron acceptor, 182
Nitrogen substituents, elimination of, 790

Nitrogenous compounds, anaerobic
    fermentation, 374
Nitromethane, 721
Nitrosomonas europeae, 191
Nitrous oxide, 252, 538
    effect on nitrate reductase activity, 284
    production:
        by denitrifiers, 184
        by nondenitrifiers, 258
    reductase, 256, 257, 258, 280
Nitroxyl, as intermediate in denitrification,
    253
Noncyclic electron flow, 148
Nonsulfur purple bacteria,
    photoheterotrophic growth, 58
Nonyl, 2-*n*-nonyl-4-hydroxy-quinoline-*N*-
    oxide (NOQNO), 314
North Sea, 23
N-oxidation, 776

Obligate proton reducer, 27–33, 401, 419–
    439, 453, 643, 644, 661, 774
Ocean, sediments, 28
O-demethylation, 791
Okenone, 46
Oligocarbophiles, 189
Oligolignols, 335, 349
    anaerobic degradation of, 349
    biodegradation studies, 350
    Phanerochaete chrysosporium, 354
    produced by aerobic lignolytic
        microorganisms, 349, 351
    Streptomyces spp., 354
    water-soluble intermediates, 349, 354
Oligomers on oligolignols, 349
Organic acids, anaerobic production, 823
Organic matter:
    decomposed, 21
    input, 21
    oxidation, 16
Organisms:
    chemoautotrophic, 18
    heterotrophic, 18
Organotrophic activity, 22
Ornithine transcarbonylase, 393
Oscillatoria, 87
    limnetica, 78, 545
Oscillochloris, 43, 50, 66
Oxidant-dependent fermentation, 61
Oxidase:
    aldehyde, 614
    inhibition, 275
    sulfite, 250, 614
    xanthine, 614

Oxidation:
    ponds, 817
    of secondary alcohols, 781
    state of element, 5
Oxidative phosphorylation, 265, 601, 621
    efficiency of and charge separation, 269
    measurement in cell-free preparations,
        269, 278, 279
    proton translocation, 279
Oxide surface, enzyme contact, 307
Oxidoreductase, 735
Oxygen:
    in atmosphere, 4
    chemistry, 775
    as dominant regulator in denitrification,
        210
    effect on denitrification, 210, 216
    effect on nitrate reductase formation and
        activity, 280
    effect on nitrate reduction, 283
    function in degradation, 775
    inhibition:
        of denitrifying enzyme activity, 215
        of reduction of nitrous oxide, 285
    of organic matter, 775
    regulation of denitrification, 210
    as regulator of denitrification, 210, 216
    repression of denitrifying enzyme
        synthesis, 215
    in sulfate-reducing bacteria, 527
    threshold concentrations for
        denitrification, 210
Oxygenase, 775
Oxyhyponitrite, as intermediate in
    denitrification, 254

P/2e-ratios, 273
P582, 488, 609
P798, 132
P840, 131, 139, 152
    midpoint potential, 131
    optical difference spectrum, 131
P865, 132
P870, 123, 124, 130, 145
    absorption band, 126
    difference spectrum, 126
    electron donor to P$^+$, 136
    triplet state, 128
P985, 126, 139
    triplet state, 128
Pacific, 24
Paracoccus:
    denitrificans, 25, 247, 252, 260, 262, 268,
        280, 282, 548

electron transport, 268
molar growth yield, 271
mutants lacking cyt.*c*, 261
nitrate reduction, 262
nitrite and nitrous oxide reduction, 262
proton-consuming site, 268
proton translocation, 268
respiratory chain, 261
halodenitrificans, 189, 247
Y_{ATP} for, 274
Partial pressure of hydrogen, 30
Pectin, 819
degradation, 516
Pedoscope, 29
Pelobacter:
acidigallii, 392
carbinolicus, 454
propionicus, 453
venetianus, 663
Pelochromatium, 50, 795
roseum, 72
Pelodictyon, 43, 50
clathratiforme, 50
Peloscope, 29
Peptococcus:
aerogenes, 396
anaerobicus, 383
glycinosphilus, 450
magnus, 450
Peptococcus (Diplococcus), glycinophilus,
394
Peptococcus (Micrococcus):
anaerobicus, 394
magnus, 394
Periplasmic side of membrane, in
Clostridium sputorum, 264
Peroxidation, 776
Peroxide production, *Bacillus* 29, 315
Phanerochaete, chrysosporium, 354
Phanerozoic, 558
Phenolic substances, 786
Phenols, anaerobic degradation, 783
Phospholipases, 377
A; B; C activities, 378
alpha-toxin, 378
glycerophosphorylcholine disterase
activities, 378
lysophospholipase (phospholipase B), 378
Phosphoroclastic reaction, 611
Phosphorylation potential, 272
Photoautotrophic:
oxygen producers, 28
sulfide oxidizers, 28
Photobacterium, 199

Photophosphorylation, 146
Photoreaction, 777
Photosynthesis, 2, 3
Photosynthetic bacteria, 114
primary charge separation, 123
Phototrophic bacteria:
aerobic, 41
anaerobic dark metabolism, 55
anaerobic respiratory metabolism, 61
anoxygenic, 40
bacteriochlorophyll, 41, 42, 44
bacteriochlorophyll *c,d,e*, 42
bacteriochlorophyll *c,d,e,c$_s$*, 47
in Burke Lake, 68
carbon monoxide utilization, 61
carotenoids, 46
cellulose degradation, 64
chemostat studies, 88
Chlorobiaceae, 42
Chloroflexaceae, 42
Chromatiaceae, 42
in Cisó Lake, 72
citric acid cycle, 60
in colored Wisconsin lakes, 76
culture media, 90
ecology, 67
Ectothiorhodospiraceae, 42
enrichment and isolation, 90
evolutionary relationship, 52
genetics, 64
green bacteria, 42
growth conditions, 41
habitats, 53
heterotrophic growth, 54
in hot springs, 81, 86
intracytoplasmic membranes, 47
in Lagune Figueroa (Baja, California), 87
lithotrophic growth, 54
in marine microbial mats, 86
nitrogen fixation, 62
photoautotroph, 41
photoheterotroph, 41
photosynthesis studies on natural
populations, 76
physiology, 53
poly-beta-hydroxybutyric acid, 55
preservation, 95
purple bacteria, 42
purple sulfur bacteria, 486
Rhodospirillaceae, 42
ribosomal RNA sequences, 52
in Rotsee, 72
in sewage and waste water, 79
in solar lake, Sinai, 78

Phototrophic bacteria (*Continued*)
  in stratified lakes, 67
  sulfide oxidation, 48, 89
  sulfur cycle, 51
  taxonomy, 42, 43
  types of habitats, 92
  in Wintergreen Lake, 63
  Yellowstone National Park, 52
Phototrophic sulfur bacteria, 559
Phylogenetic relationships, 34
  in sulfate-reducing bacteria, 486, 584
Phylogenetic tree, 34
Pichia, guillermondii, 313
Piericidin A, 149
Pigment–protein complexes, 119
Pollutants metabolism, 192
Polysaccharide, 340, 360
  noncellulosic, 818
Polysulfide, 600
Porphyrin, 605, 607, 616, 720
Postulated cleavage, of intermonomeric
    bonds, 350
Potassium, in methanogenesis, 742
Principle of biological infallibility, 774
Prochlorophytes, 40
Prokaryotes, 14
Proline reductase, 389
Propionate:
  degradation pathway, 30, 428, 437
  formation from acetate, 453
  formation from succinate, 436, 781
  with hydrogen as reductant, 823
  from lactose, 823
  in sulfate-reducing bacteria, 495, 514, 516
Propionibacteria, 25, 258
Propionibacterium:
  acidi-propionici, 189
  sp., 383
Propionigenium, modestum, 436, 459, 793
Propionispira, arboris, 454
Prosthecochloris, 43, 70
  aestuarii, 50, 131
  aestuarii primary electron acceptor, 131
  sp., 50
Proteases, 383
  aspartic acid, 385
  cell-associated, 385
  clostripain (clostridio-peptidase B), 385
  cysteine, 385
  inhibition characteristics, 385
  leaf fraction 1 (ribulosebisphosphate
    carboxylase), 385
  metalloproteinase, 385

  in presence of metal ions, 385
  serine, 385
  substrate specificity, 385
  trypsin, 385
Proteins:
  anaerobic degradation, 374, 381, 386, 802
  factors affecting degradation, 385
  modification, 386
  solubility, 385
Proteolytic activity:
  amino acid arylamidase, 382
  collagenolytic, 383
Proteolytic bacteria, 374, 382
  activity in human gut, 382
  activity in the rumen, 382
  in digestor sludge, 383
Proteolytic clostridia, 380
Proterozoic, 557
Proteus, mirabilis, 247, 280, 282
Proton:
  facultative reducing-bacteria, 430
  obligate reducing-bacteria, 27–33, 401,
    419–439, 453, 643, 774
  translocation, 601, 618
Proton motive Q cycle, 260, 265
Proton-consuming site:
  of nitrite and nitrous oxide, 268
  in reduction of nitrate, 262
  in reduction of nitrite, 262
  in reduction of nitrous oxide, 262
Protons, reduction, 27, 419–439
Protons translocation, stoichiometry, 146,
    260–279
Protozoa, 661
  anaerobic, 807, 820
Pseudomonacea, denitrification, 191
Pseudomonas, 25
  aeruginosa, 247, 250, 252, 253, 316, 600
  aeruginosa mutant defective in heme
    biosynthesis, 283
  aureofaciens, 253, 257
  chlororaphis, 257
  denitrificans, 253, 271
  fluorescens, 189, 257, 265
  fluorescens biotype II, 189
  perfectomarinus, 189, 252
  putida, 393
  role in nitrate dissimilation to ammonium,
    199
  solanacearum, 190
  sp. strain 200, 317
  stutzeri, 189, 253
Pteridin, 250

Pterin, 715, 722
  6-carboxylic acid, 250
Purple bacteria, 40, 43, 559
  absorption spectra, 117
  growth conditions, 46
  physiology of facultative, 57
  physiology of non sulfur, 57
  pigment-protein complexes, 119
  primary charge separation, 123
  reaction center complexes, 119, 124
  taxonomy, 47
  taxonomy of facultative, 51
Purple nonsulfur bacteria, 43, 145
  respiration, 145
Purple sulfur bacteria, 43, 145, 472, 486, 817
  carbon assimilation, 55
  enrichment and isolation, 91
  habitats, 53
  photoheterotrophic growth, 55
  physiology, 53
  respiration, 145
  thermophilic, 86
Pyrazine, 735
Pyrite, 554, 558
Pyrobaculum, islandicum, 26
Pyrodictium, 540
  occultum, 540
  sp., 26
Pyrophosphatase, 604
Pyrophosphate, 604, 611, 621
Pyruvate:
  dehydrogenase, 611
  ferredoxin oxidoreductase, 391
  as intermediate in homoacetogenesis, 442,
    448
  phosphoroclastic reaction, 611
  synthesis of oxalvacetate from, 746

Q-cycle, 143, 145, 147
  Chromatium vinosum, 145
  Rhodopseudomonas sphaeroides, 143
  Rhodospirillum rubrum, 145
Qinone:
  dicumarol-sensitive, 314
  pool, 133, 134

Reaction center:
  aborption spectrum, 132
  adsorption spectrum, 124
  crystals, 125
  cytochrome c_2 binding, 137
  electron transfer, 125
  EPR spectroscopy, 124

gliding green bacteria, 123
green sulfurbacteria, 131
isolated, 128
midpoint potential, 124
organization, 125
purple bacteria, 124
Rhodopseudomonas sphaeroides, 128
Rhodopseudomonas viridis, 125
subunits, 125
ultrafast spectroscopy, 128
X-ray analysis, 125
Reactors for wastes treatment:
  anaerobic contact process, 811
  anaerobic filter, 807
  ANAMET process, 813
  application, 809
  biodegradation potential, 809
  biofilm formation, 808
  biotechnological aspects, 808
  degradation:
    of alcohol, 811
    of sugar, 811
  distillery wastewaters, 813
  exopolymer formation, 809
  fluidized bed, 813
  fludized bed reactor, 807
  hydraulic retention time, 813
  Imhoff tank, 804
  mixed sludge diagestor, 811
  pulp and paper industry, 813
  rural biogas reactor, 811
  sanitary landfill leachates, 813
  septic tanks, 804
  sewage-sludge, 813
  sulfate evaporator condensates, 813
  technological study, 808
  Travis tank, 804
  trickling filter, 807
  UASB-reactor, 807
  of yeast industry, 811
Redox:
  couples, 10
  environmental sequences, 24
  equilibrium, 1, 5
  equilibrium potential, 8
  gradients, 20
  indicator, 13
  loops in *E. coli,* 266
  loops in *Pa. denitrificans,* 267
  natural processes, 9
  potentials, 8
  processes, 5, 10
  reaction, 1, 18, 24

Redox (*Continued*)
  reaction progress models, 20
  sequence, 9, 19, 28
  species, 21, 23
  state of ubiquinone, 284
Reductase, 153, 252, 488, 599, 600, 605, 609,
  620
  membrane-bound, 608
Reduction:
  of CO_2, 16, 422, 651, 742
  of ferromanganese, 317
  of iron, 305
  of manganese, 305
  of nitrate, 180, 258, 322, 502
  of sulfate, 29, 35, 471
  of sulfur, 152, 384, 533, 566, 589
Reductive carboxylation reaction, 391
Reductive dechlorination, 27, 776–792
Reductive dehydroxylation, 785
Regulation of nitrate reductase formation,
  282
  induction by nitrate, 280
Regulators of denitrification, 219
  carbon as electron donor, 224
  nitrate, 222
  oxygen, 219
Resistance against chlorate, 250
Respiration:
  of nitrate, 342
  purple nonsulfur bacteria, 145
  purple sulfur bacteria, 145
  of sulfate, 342
Respiratory chain:
  diversity in the composition of, 265
  of *Escherichia coli,* 260
  influence of place of reduction, 261
  as linear array of electron carriers, 261
  of *Paracoccus denitrificans,* 260
  in *Thiobacillus denitrificans,* 261
Retention time, hydraulic, 804
Reversed electron flow, 149, 459
Rhizobiaceae, denitrification, 191
Rhizobium:
  fredii, 190
  leguminosarum, 190
  meliloti, 250
  phaseoli, 190
  trifiollii, 190
Rhodobacter, 43, 51
  capsulans, 64
  capsulatus, 58, 59, 64, 120
  sphaeroides, 47, 51, 59, 64, 120
  sulfidophila, 58

Rhodobacter capsulatus:
  gene transfer agent, 64
  nitrogen fixation, 64
Rhodobacter sphaeroides, nitrogen fixation,
  64
Rhodocyclus, 43, 51, 52
  gelatinosus, 47, 59, 62
  purpureus, 58, 60
  tenuis, 47, 58
Rhodomicrobium, 43, 46, 51
  vannielii, 47, 52
Rhodophila
  globiformis, 43, 51, 52
Rhodopin, 46
Rhodopinal, 46
Rhodopseudomonas, 43, 51
  acidophila, 52, 58, 120
  capsulata, 121, 124, 125, 136, 139, 145,
    149, 151
  capsulata, nitrate transport in, 265
  gelatinosa, 124, 136, 137
  globiformis, 52
  palustris, 58, 115
  rubrum, 136
  sp., 47
  sphaeroides, 121, 124, 134–147, 151, 191,
    252
  sphaeroides nitrate reduction, 264
  sulfoviridis, 58
  viridis, 62, 123–142
Rhodospirillaceae, 191
  in aquatic habitats, 51
  autotrophic growth, 57
  culture media, 93
  denitrification, 191
  enrichment, 94
  fermentative growth, 61
  glycerol catabolism, 60
  growth inhibition by sulfide, 57
  heterotrophic (aerobic) growth, 60
  isolation on membrane filters, 94
  lithotrophic growth, 58
  nitrogen fixation, 62
  photoheterotrophic growth, 57, 58
  sulfide oxidation by, 58
  utilization of aromatic compounds, 59
  vitamin requirements, 58
Rhodospirillum, 43, 51
  fulvum, 59
  molischianum, 47
  rubrum, 47, 57, 61, 65, 120–124, 136, 139–
    142, 146, 149
Ribosomal RNA sequences, 34, 57

methanogenic bacteria, 644
phototrophic bacteria, 52
Rieske iron–sulfur protein, 142, 143
Rotenone, 149
Rotsee, 23, 72
rRNA, 16S, 34, 52, 57, 644
Rubredoxin, 489, 597, 615
  (NADH) oxidoreductase, 615
Rumen, 33
  bacterial activity, 337, 381
Ruminant diets, 376
Ruminococcus:
  albus, 379, 430, 661
  flavefaciens, 661

S-organism, 422, 436, 437, 644
Saccharomyces, cerevisiae, 545, 823
Salmonella, 545
Sapropel, 439
Sarcina, ventriculi, 25
Scheldt estuary, 21, 22
Sedimentary environments, 4, 28
Sedimentary penetration
  by active methanogens, 684
Selenite, 722
  formate dehydrogenase, 443
Selenium, 594
Selenocysteine, 722
Selenomonas, 454
  ruminantium, 381, 661, 793
  sp., 382
Sequence of redox reactions, 19
Serine, fermentation, 400
Siderocapsa, 274
Siderochromes, 274
Siroheme, 153, 489, 605, 607
Sirohydrochlorin, 606, 616
Sirohydroporphyrin, 616
Siroporphyrin, 607
Skatole, 392
Sodium:
  dodecyl sulfate detergent, 250
  in methanogenesis, 743
Sodium-ion-gradient, 793
Soil, 29
  aggregates, 29
  environments, 28
Solar lake, 78
Solid wastes:
  acid extraction, 818
  cellulose, 818
  delignification, 818
  digestion, 817

lignin, 818
moisture content, 817
solid content, 817
solvent treatment, 818
Solvents, mass production, 822, 823
Spartina, foliosa, 670
Spheroidene, 46
  Rhodopseudomonas sphaeroides, 121
Spheroidenone, 46
Spheroplasts, *Chromatium vinosum,* 138
Spirillaceae, dentrification, 191
Spirilloxanthin, 46, 122
Spirillum, desulfuricans, 474
Spirochaeta, isovalerica, 390
Sporomusa, sp. 484
Staphylococcus:
  aureus, 316
  sp., 382
Steroids, biotransformation, 376
Stickland reaction, 388
  in *C. sporogenes,* 390
Stoichiometry of ATP generation, influence
  of place of reduction, 261
Stoichiometry of proton translocation, 263,
  264
Streptococcus:
  bovis, 382
  faecalis, 393, 823
  lactis, 393
Streptomyces:
  aureofaciens, 734
  griseus, 735
  spp., 354
Substrate:
  competition for, 508
  phosphorylation, 618, 621
Succinate oxidation, 151
Succinic dehydrogenase, 151
Sulfate:
  activation, 603
  assimilation, 472
  cell wall structure, 484
  dissimilation, 472, 589
  as electron acceptor, 427
  metabolism, 599
  in oceans, 553, 555
  reducing bacteria, 28, 35, 41, 374, 471,
    588–616, 731
  reductase, 620
  reduction, 28, 35, 519, 559, 588, 589, 599,
    603, 609, 618
  reduction and methanogenesis, 507, 508
  sedimentary, 559

Sulfate reducing bacteria, 28, 35, 471
  acetate, 504, 509, 523
  acetate formation, 491, 505, 507
  acetone, 497
  aggregation, 527
  alcohols, 497
  amino acids, 498
  ammonia, 501
  archaebacterial, 584
  assimilatory capacities, 504
  autotrophic growth, 504
  butyrate, 495, 514, 516
  catalase, 490
  cell synthesis, 473
  cell wall structure, 484
  choline, 498
  classification, 475
  complete oxidizers, 492, 516
  control of, 531
  counting, 522, 523, 527
  cyclic organic acids, 498
  cytochromes, 489
  degradation of aromatic compounds, 497
  degradative capacities, 490
  dicarboxylic acids, 497
  ecology, 506, 508
  economic activities, 528
  electron acceptors, 501
  electron donor, 490
  ethanol, 515
  factors for development, 525
  fatty acids, 495, 496
  fermentative capacities, 500
  ferredoxin, 489
  fixation, 507
  flavodoxin, 489
  formate, 493
  gliding motility, 528
  glycerol, 498
  gram-staining behavior, 484
  growth on acetate, 494
  guanine + cytosine content, 485
  habitats, 519–521, 558
  homoacetogenesis, 505, 507
  hydrocarbons, 499, 529
  hydrogen, 523, 493, 509
  as hydrogen scavengers, 424
  incomplete oxidizers, 491, 516
  isolation, 476
  lactate, 493, 515, 522
  menaquinones, 490
  methane oxidation, 499
  methanol, 516
  morphological adaptations, 528
  nitrate, 501
  nitrate reduction by, 278
  nitrogen fixation, 506
  nitrogen sources, 506, 507
  in oil technology, 529
  oxygen effect, 527
  in paddy fields, 528
  in papermaking industry, 529
  pH effect, 526
  phylogenetic relationships, 486, 584
  propionate, 514, 516
  propionate oxidation, 495
  IGS ribosomal RNA, 14, 34, 486
  rRNA/DNA hybridization, 486
  rubredoxin, 489
  salt requirement, 526
  salt tolerance, 526
  substrate utilization, 490, 584
  sugars, 498
  sulfate consumption kinetics, 517
  sulfite, 501
  sulfur, 501
  superoxide dismutase, 490
  syntrophic growth, 500
  temperature, 525
  thiosulfate, 501
Sulfate reduction, 28, 35, 41, 374, 471, 559,
  588–618, 731
  age, 559
Sulfide:
  chemical oxidation of, 548
  as growth inhibitor, 57
  growth requirements by *Chromatium,* 55
  oxidation, 152
    by aerobic bacteria, 547
    by *Chromatium vinosum,* 48
    by phototrophic bacteria, 549
  preparation for culture media, 92
Sulfite:
  high-spin, 606
  low-spin, 606
  oxidase, 250, 723
  oxidation, 153
  reductase, 153, 488, 605, 609
  reduction, 153, 473
  c-type heme groups, 256
  electron donor, 256
  oxidase, 250
  reductase, 153
  in sulfur-reducing bacteria, 538
Sulfolobus, 540
  acidocaldarius, 313
  ambivalens, 543
  brierleyi, 543

Sulfoxidation, 776
Sulfur:
  archaebacterial reducers, 474
  assimilation, 589, 748
  in the biosphere, 473
  colloidal, 599, 617, 620
  compounds reduction, 26, 588
  cycle, 502, 527, 538, 546, 550, 553
  cycle in phototrophic, 51
  deposits of elemental, 558
  disproportionation, 545
  dissimilation, 472, 589
  elemental, 15
  eubacterial reducers, 474
  intermediary oxidation states, 473
  interspecies transfer, 589
  isotope fractionation, 554
  in living organisms, 472
  natural formation, 15
  organisms reducing, 544
  reducing bacteria, 41, 471, 533, 588, 599,
    606, 609, 616
  reductase, 599, 620
  reduction, 152, 383, 533, 544, 589
  reservoirs, 553
  syntrophic cycling bacteria, 51
  system, 14
Sulfur reducing bacteria, 41, 471, 533, 588,
    599, 606, 609, 616
  archaebacterial, 540, 585
  aspartate, 538
  dimethylsulfoxide, 538
  eubacterial, 533, 585
  formate, 538
  fumarate, 538
  hyrdogen, 538
  microaerobic growth, 540
  nitrate, 538
  nitrous oxide, 538
  oxygen, 538
  spirilloid, 538
  succinate, 538
  sulfite, 538
  thiosulfate, 538
Sulfuretum, 29, 547, 549
Symbiotic association, 422
Synechococcus, lividus, 81
Syntrophic associations, 31–33, 207, 422,
    501, 538
  thermophilic, 30
Syntrophic consortia, 29, 392–437
  cycling sulfur, 51
  iron reduction, 312
  manganese reduction, 312

Syntrophic relationships, 29
Syntrophobacter:
  sp., 27
  wolinii, 434
Syntrophomonas:
  sp., 27
  wolfei, 422, 427, 429, 435
  wolinii, 437
Syntrophus:
  buswellii, 392, 422, 427, 429, 434
  sp., 27

Taxonomy of methanogens:
  cell wall characteristics, 645
  lipids, 649
  16S rRNA, 34, 644
Terminal reductase, 601
Termites, 339, 356, 451
Tetrathionte, 153, 585
Thalassia, testudinum, 687
Thermoanaerobium, brockii, 25, 437
Thermococcus, 540
  celer, 383
Thermodesulfobacterium, commune, 493
Thermodiscus, 540
  sp., 26
Thermofilum, 540
  pendens, 383
Thermophilic:
  archaebacteria, 383
  extremely, 540
Thermoproteales, proteolyticus, 402
Thermoproteus, 540, 708
  sp., 26
  tenax, 26, 540
Thermosdesulfobacterium, commune, 596,
    607
Thiobacillus, 25, 271, 547
  denitrificans, 11, 191, 216, 252, 272, 313
    respiratory chain, 261
  ferrooxidans, 313, 552
  thiooxidans, 313
Thiocapsa, 43, 53
  pfennigii, 47, 124, 136, 137
  roseopersicina, 54, 152
Thiocynate, 258, 268
Thiocystis, 43, 53, 69
  violacea, 54
Thiodictyon, 43
Thiomicrospira, 191
Thiopedia, 43, 69
  rosea, 72
Thiosphaera, pantotropha, 14, 25, 191, 192,
    215, 279, 547

Thiospirillum, 43, 69
Thiosulfate, 585
  oxidation, 152, 153
  reductase, 608, 609
  reduction, 473
Thiovulum, 547, 549
Threonine:
  aldolase, 393
  degradation, 393
Thresholds, 513
Thrithionate:
  pathway, 608
  reductase, 608
Transaminase, 392
Transhydrogenation, 274
Transhydroxylation, 785, 792
Transmembrane cytochrome hexaheme, 601
Treponema, 378
  denticola, 378
  minutum, 378
  phagedenis, 378
  refringins, 378
  vincentii, 378
Tricarboxylic acid cycle, 18
Trihydroxybenzenes, degradation, 785
Trimethoxybenzoate, 500
Trimethylamine, 378
  from decomposition of choline, 516
  metabolism in sediments, 669
Trimethylamine-N-oxide (TMAO), utilization
    by purple bacteria, 61
Triphenylmethylphosphonium, 264, 268
Troilite, 555
Trypsin, 385
TTFA, 151
Tungstate, formate dehydrogenase, 443
Tyrosine-phenol lyase, 392

Ubiquinol:cytochrome c oxidoreductase, 142
Ubiquinone, 126, 128, 130, 136, 142, 267
  effect on regulation of electron flow to
    nitrate, 284
  electron acceptors, 130
  E_m value, 136
  pool, 135
  reaction center, 125, 128
  role in proton translocation, 267
  species in proton motive Q cycle, 267

UHDBT, 142, 143, 145
Unsaturated fatty acids
  alpha-linolenic acid, 378
  gamma-linolenic acid, 378
  linoleic acid, 378
Uptake of nitrate, 274
Urkingdom, 34
Uroporphinogens, 748
Uroporphinoids, 720
Urothione, 250

Veillonella, alcalescens, 313, 790, 793
Vibrio:
  desulfuricans, 475
  marinus, 486
  succinogenes, 258, 278, 595, 661
  succinogenes molar growth yields of, 278
Vibrionaceae, role in denitrification
    processes, 199
Viologen, 726
Vitamin B_{12}, 93, 616
Volcanic activity, 28

Waste treatment, 804
  of chemical industry, 819
  digestion of solid wastes, 817
  domestic, 813
  methanogenic, 803, 815, 819
Waste water, domestic, 813
Water-sediment interface, 28
Wintergreen Lake, 68, 521
Wisconsin lakes, 76
Wolfram, 722
Wolinella, 187
  succinogenes, 189, 199, 258, 427, 430, 539,
    595, 792

Xanthine:
  dehydrogenase, 250
  oxidase, 250, 723
Xylane, 819

Y_{ATP} for *Pa. denitrificans,* 274
Yeast, 25, 545
Yellowstone National Park, 82
  phototrophic bacteria in, 81

Zymonas, 25